	I B	II B	III A	IV A	V A	VI A	VII A	
								2 **He** Helium 4.00260[y]
			5 **B** Boron 10.81[y]	6 **C** Carbon 12.011	7 **N** Nitrogen 14.0067	8 **O** Oxygen 15.9994	9 **F** Fluorine 18.998403	10 **Ne** Neon 20.179[y]
			13 **Al** Aluminum 26.98154	14 **Si** Silicon 28.0855	15 **P** Phosphorus 30.97376	16 **S** Sulfur 32.06	17 **Cl** Chlorine 35.453	18 **Ar** Argon 39.948[y]
28 **Ni** Nickel 58.70	29 **Cu** Copper 63.546	30 **Zn** Zinc 65.38	31 **Ga** Gallium 69.72	32 **Ge** Germanium 72.59	33 **As** Arsenic 74.9216	34 **Se** Selenium 78.96	35 **Br** Bromine 79.904	36 **Kr** Krypton 83.80[y]
46 **Pd** Palladium 106.4[y]	47 **Ag** Silver 107.868[y]	48 **Cd** Cadmium 112.41[y]	49 **In** Indium 114.82[y]	50 **Sn** Tin 118.69	51 **Sb** Antimony 121.75	52 **Te** Tellurium 127.60[y]	53 **I** Iodine 126.9045	54 **Xe** Xenon 131.30[y]
78 **Pt** Platinum 195.09	79 **Au** Gold 196.9665	80 **Hg** Mercury 200.59	81 **Tl** Thallium 204.37	82 **Pb** Lead 207.2[y]	83 **Bi** Bismuth 208.9804	84 **Po** Polonium (209)	85 **At** Astatine (210)	86 **Rn** Radon (222)

metals ← → nonmetals

64 **Gd** Gadolinium 157.25[y]	65 **Tb** Terbium 158.9254	66 **Dy** Dysprosium 162.50	67 **Ho** Holmium 164.9304	68 **Er** Erbium 167.26	69 **Tm** Thulium 168.9342	70 **Yb** Ytterbium 173.04	71 **Lu** Lutetium 174.97
96 **Cm** Curium (247)	97 **Bk** Berkelium (247)	98 **Cf** Californium (251)	99 **Es** Einsteinium (254)	100 **Fm** Fermium (257)	101 **Md** Mendelevium (258)	102 **No** Nobelium (259)	103 **Lr** Lawrencium (260)

GENERAL CHEMISTRY

General Chemistry

Jerry March AND *Stanley Windwer*

ADELPHI UNIVERSITY

Macmillan Publishing Co., Inc.

NEW YORK

Collier Macmillan Publishers

LONDON

Printed in the United States of America

Macmillan Publishing Co., Inc.
866 Third Avenue, New York, New York 10022

Collier Macmillan Canada, Ltd.

Library of Congress Cataloging in Publication Data

March, Jerry, (date)
General chemistry.

Includes index.
1. Chemistry. I. Windwer, Stanley, joint author. II. Title.
QD31.2.M365 540 77-28622
ISBN 0-02-375860-0

Printing: 1 2 3 4 5 6 7 8 Year: 9 0 1 2 3 4 5

This book is dedicated to our children and grandchildren: Charlie, David, Esther, Gale, Jennifer, June, Justin, Nancy, Sharon, and Steven.

Preface

Chemistry can be defined as what chemists do. Today chemists are involved with and are studying a much greater variety of things than in the past. In writing this book we have tried to take account of this, so that the reader will find topics as diverse as the mechanism of heredity, eutrophication, entropy and the action of semiconductors.

We have written this book for students and in doing so have tried to make every explanation as clear as we possibly could. No previous knowledge of chemistry has been assumed. Because of this the first chapter contains an extensive detailed discussion of background material in mathematics and physics. Of course, where students already have studied some of this material, instructors will be able to skip this chapter or some of its sections.

In this book we have tried to be rigorous, but without burdening the student with too much advanced material more suitable for later courses. We use no calculus. Due attention is given both to theoretical and to descriptive chemistry, and in accord with recent trends we have included a considerable amount of material on the environment, mostly in Chapter 17 and the Interlude between Chapters 8 and 9.

More historical material is given than is common in most general chemistry books. The historical material is used primarily so that the student will have a better understanding of the chemical principals involved and an appreciation of how the scientific method works. We believe that the scientific method is best taught through actual examples rather than by stating abstract rules. In addition, the historical material enables students to see that chemical principles did not suddenly spring into being, but are part of a historical process that has been going on for hundreds of years. Even more important, they will come to see that today's principles are not immutable, but will be modified and improved in the future, by continued application of the scientific method.

We have organized this book in a somewhat different manner from most recent general chemistry books. We cover equilibrium in the first half of the book and do not introduce atomic structure until Chapter 12. We have found that this arrangement allows easier matching of laboratory experiments with the material taught in the lectures. Historically, most of the principles in Chapters 1 to 11 were known before the structure of atoms and molecules. However, the chapters do not have to be taught in the order given. Instructors will find it perfectly feasible to cover Chapters 12, 13 and 14 immediately after Chapter 3 or 4 and to postpone the material on equilibrium and kinetics until much later in the course.

The book certainly contains more material than can be conveniently covered in most one-year general chemistry courses. Instructors will therefore have a choice of chapters or sections to cover in class, assign for reading, or omit entirely. In particular, Chapters 17 to 22 lend themselves well to selective assignment. Several chapters have been divided into two parts and either or both can be covered. For example, some teachers will want to cover only Part I of Chapter 8 on equilibrium; others will also wish to cover all or a portion of Part II.

It has been our experience and the experience of many other chemistry teachers that the best way for students to learn chemistry is to do problems. We have included a substantial number of problems at the end of each chapter (about 1100 altogether), covering virtually every subject discussed in the book. Each problem set at the end of a chapter is in two parts. The problems in the first part of each set are arranged approximately in the order that the topics are covered in the chapter.

Because this sometimes gives students a gratuitous clue as to how to solve the problem, we have added, at the end of each set, additional problems in no particular order. Some of the most difficult problems in the book are in these "Additional Problem" sections but the sections also contain many that are fairly simple. Altogether the problems range from very easy to very hard, and the instructor will be able to choose from a wide range of levels of difficulty. Answers to selected problems are given at the end of the book.

We wish to acknowledge the assistance of many people who helped us in various ways during the several years we have been working on this book and earlier. Most important are the generations of students we have had over the years, from whom we have learned a great deal. A number of our colleagues at Adelphi have proved especially helpful, and we wish to thank them. These are Professors Frederick Bettelheim, Donald Davis, Stephen Goldberg, Robert Halliday, Donald Opalecky, Reuben Rudman, Madelyn Todd and Frank Vogel. In addition, we thank the following colleagues at other universities, who read sections of the manuscript and whose comments were exceedingly helpful: Professors Stanley Manahan, Peter Yankwich, Floyd W. Kelly, Linda M. Sweeting, Edward K. Mellon, Jr, Harold H. Pokras, Otto Theodore Benfey, Verne A. Simon, Richard S. Treptow, Quentin R. Petersen, Harold M. Kolenbrander, Arthur S. Miller, Wayne Moxley, James O. Schreck and Gary W. Valentine. A number of secretaries have typed portions of the final manuscript and of earlier drafts as well as rendered assistance in numerous other ways. We wish to express our particular thanks to Lee D'Angelo, Cathleen Forte, Sue Goddard, Joanne Palumbo, Jean Phinney and Mrs. Regan. Thanks are also due to Beverly March for help with the photographic work. In addition we wish to thank some friends who have contributed in a personal way: Robert G. DelGadio, Dennis Whyte and Howard Pashenz.

J. M.
S. W.

Brief Contents

Detailed Contents

4 Gases 82

5 Energy Changes and Heats of Reaction 116

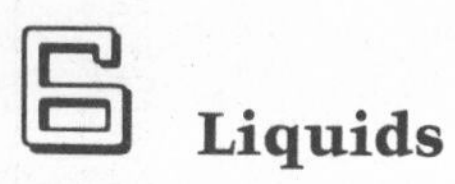

Liquids 147

7 Solutions 170

8 Equilibrium and Thermodynamics 199

Interlude Energy Resources 241

9 Chemical Kinetics 251

Acids and Bases 283

11 Equilibria Involving Solubility 330

12 Atomic Structure and the Quantum Theory: Experimental and Historical Foundations 346

13 The Electronic Structure of Atoms 378

14 The Chemical Bond 412

Organic Chemistry *659*

21 Biochemistry *703*

22 Polymers *736*

23 Nuclear Chemistry 751

Appendix 784

1 Introduction/Physical and Mathematical Background

Antoine Laurent Lavoisier, 1743–1794

The Bettmann Archive

As an introduction to our study of chemistry, it will be useful to discuss what scientists do and how science works. Scientists are, first and foremost, observers. They observe the universe and everything in the universe, especially our own planet, earth. What science is all about is trying to understand how things operate in the universe. Let us see if we can make this clearer by putting it another way. Suppose that you have never seen a baseball game and are watching one for the first time. Just from observing what is happening on the playing field, you are to figure out the rules of the game. We shall also suppose that you do not just have to sit and watch the game, but can play in the game yourself and find out what happens when you play. This is what scientists do. They observe how stars give off light; how birds fly; how bees produce honey; how rivers, lakes, and oceans are polluted; how the human body transforms food into energy; and many, many other things that happen on our earth and in the rest of the universe. As in the baseball game analogy, scientists also participate by doing all kinds of experiments and then observing what happens. While a scientist is doing the experiments and recording the data, he or she is constantly looking for trends in the data and for connections between these results and the results of other scientists, so that some insights can be gained and perhaps some understanding of how nature works. When an insight is gained, it is called a *theory,* and this theory is then used to predict the outcome of other experiments. If the theory is successful for some long period of time it is called a *law,* or a *fundamental law.* The history of science shows that no specific period of time must elapse before a theory becomes a law. After a time, people just call it a law. Knowing these fundamental laws allows us to predict the outcome of various experiments that we do and of natural events. For example, if we knew all the laws of meteorology we could predict the weather far in advance. We want to learn these laws and make these predictions not only to satisfy our curiosity about the working of the universe, but to use our knowledge to benefit mankind in various ways.

At the present time we think we know some of the fundamental laws of the universe—but certainly not all of them. These laws have changed over the years as scientists have produced situations (experiments) that the old laws could not predict. Actually, the laws themselves have not changed—fundamental laws of the universe never change—it is our knowledge of them that changes. It has occasionally happened that the majority of scientists (or even all scientists living at a particular time) have believed that a certain statement was a law when later evidence showed it to be untrue, or partially untrue. We shall see examples of that in this book. It is even possible that some of the laws given in this book will later turn out to be untrue. It is fair to say that we know enough of the fundamental laws to use our knowledge to benefit mankind, but unfortunately some of this knowledge is also used (knowingly or unknowingly) to the detriment of mankind. We shall see examples of this, too.

Other questions that some scientists ask are: Why are the laws of the universe the way they are? What is the fundamental nature of the makeup of our universe which gives rise to, for example, the law of gravitation (Section 1-8)? What sort of world would we have if there were no positive and negative charges (Section 1-8)—if all particles were neutral? This knowledge is in its infancy. We may speculate about many things, and such speculation can be a great deal of fun, but really to understand these things, we must first discover additional fundamental laws so that eventually we might be able to understand why the universe operates the way it does.

Science is a study of the universe whose goal is to understand the fundamental laws that operate within the universe. Scientists are people who spend their time trying to find these laws. As part of their work, they construct experiments (situations) to help them in their search.

Among the sciences, chemistry occupies a central position. **Chemistry is the science that deals with matter: the structure of matter, the properties of matter, and the transformations of one form of matter to another.** Since everything we see and touch is made of matter, chemistry has a very wide scope. By the time you complete this course in general chemistry, you will have some idea of this scope. You will see that chemistry is involved with the food we eat, the way our bodies digest that food, the operation of automobile engines, air and water pollution, the evaporation of water from the streets and roads after it rains; with flashlight batteries, antifreeze, natural and synthetic rubber, pain-killing drugs, artificial fibers, the transistors used in pocket calculators; the light coming from the sun; and even with the very nature of life itself.

Because of its central position, chemistry overlaps most other branches of science to a considerable extent—especially physics, biology, geology, and astronomy. Like the rest of these sciences, chemistry depends a great deal on measurements and on mathematics. Before beginning the study of chemistry, it is necessary to be familiar with certain material from mathematics and physics. In this chapter we shall discuss those aspects of mathematics and physics that we consider prerequisites for studying general chemistry, and also the system of units that we shall use in this book.

1-1 Measurements in Chemistry

Chemistry is a quantitative subject. Chemists must make measurements, record them, and communicate them to others. In this section we discuss the system of units that is used to describe these measurements.

THE METRIC SYSTEM. Most numbers measured in a laboratory have *units* or *dimensions*. Science is an international activity and scientists in all countries must be able to communicate their work to each other. For this reason, the scientists of the world more than a century ago agreed to report all their measurements in the same system of units: the *metric system*. This system is far more logical and easier to work with than the common-unit systems that had evolved in various countries over many centuries. Everyone in the United States knows that converting from one common unit to another can involve long and difficult calculations. There are 12 inches in a foot, 3 feet in a yard, 1760 yards in a mile, not to mention such units as leagues, rods, and furlongs. If we want to know how many inches in 384 yards or how many feet in 33.4 miles, a great deal of arithmetic is involved. Furthermore, it is easy to make mistakes. In the metric system everything is much simpler and the chances of mistakes much less. Instead of conversion factors such as 12, 3, or 1760, which must be memorized or looked up, the only conversion factors in the metric system are 10 and multiples of 10. To make the system even more convenient, the same prefixes are used throughout the system. Some of these prefixes are

1,000,000	mega
1000	kilo
0.1	deci
0.01	centi
0.001	milli
0.000001	micro
0.000000001	nano

For other prefixes, see Appendix B.

For example, we shall soon see that the basic units of mass and length are, respectively, the gram (g) and the meter (m). We then have

1000 g = 1 kilogram (kg)	1000 m = 1 kilometer (km)
0.01 g = 1 centigram (cg)	0.01 m = 1 centimeter (cm)
0.001 g = 1 milligram (mg)	0.001 m = 1 millimeter (mm) etc.

If we learn the meaning of a few prefixes we can easily calculate that there are 1000 millimeters in 1 meter; 1,000,000 millimeters in 1 kilometer, and problems such as how many centimeters in 384 meters, or how many meters in 33.4 kilometers become so simple that we can do them by inspection. The simplicity of this system has caused it to be accepted as the basic system of units in almost all countries. Only in the United States and a few other countries are common-unit measurements still widely used, and it is likely that even these countries will convert to the metric system in the not-too-distant future. Of course, even in the United States, the metric system is used in *scientific* work.

However, even within the metric system the same quantity can often be expressed by several different kinds of units. (We shall see that energy can be measured in calories, ergs, and joules, as well as in other units.) For this reason, international scientific organizations have adopted an even more simplified system, though still based on the metric system, called the *SI system*. While the SI system has advantages over the metric system, it also has disadvantages: for example, many quantities have to be expressed by very large or very small numbers; and, so far, the system has not been widely used by scientists in the United States. In this book we shall use the metric system. The SI system is given in Appendix B.

MASS. *Mass* is a measure of the amount of matter in a substance. Usually mass and *weight* are used interchangeably but they are really different quantities. The weight of an object depends on the gravitational force exerted on the object, whereas mass is independent of the gravitational force. This is easily seen by looking at Figure 1-1. In part (a), we see a spring balance, which determines weight and not mass. If the object were weighed on the moon, the spring balance would read approximately one-sixth of its measured weight on earth, because the gravitational force of the moon is approximately one-sixth that of the earth. In Figure 1-1b the mass of the object is determined. Here the object is measured against a known weight. The pull of gravity on both objects is the same so that it would make no difference if the mass were measured on the moon or the earth. In a space satellite an object has no weight at all, since there is no gravity, but its mass is still the same as it was on earth. The *mass* of an object does not change no matter where it is. The *weight* does change, and decreases as the gravitational force on the object decreases. In the rest of this book, we shall treat mass and weight interchangeably, even though they are fundamentally different things. We are able to do so because everything we talk about will take place on the surface of the earth, where the gravitational force is relatively constant, so that the amount of mass possessed by any object is exactly proportional to its weight. If we were comparing objects on the earth with those on other planets or in space satellites, we would not be able to use these terms interchangeably.

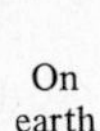

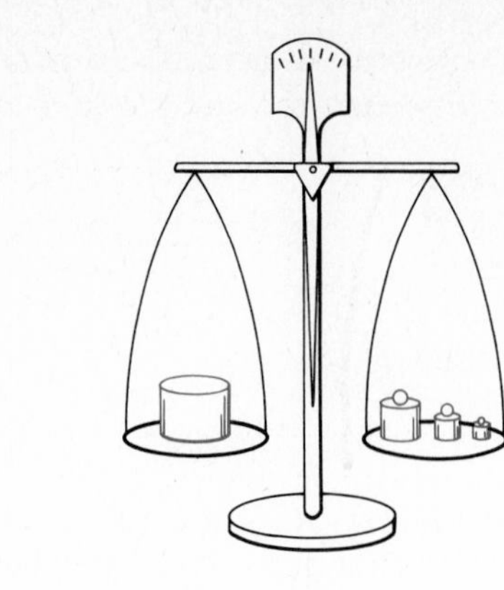

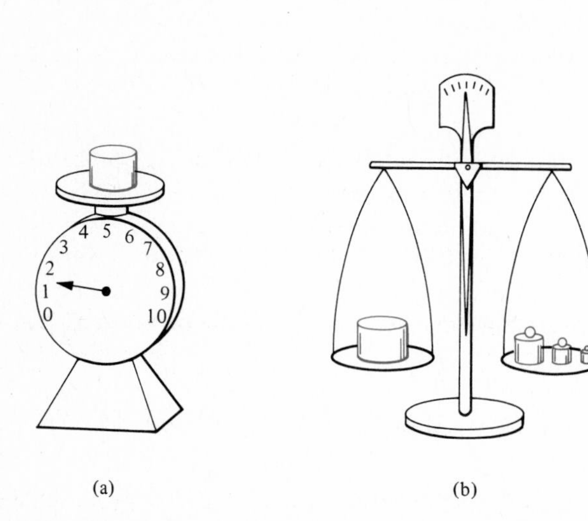

Figure 1-1 Methods of measuring weight and mass. In (a), the weight of the object is determined by a spring balance. In (b), the mass is determined by an equal-arm balance. In an equal-arm balance the force of gravity acts equally on both objects.

In the metric system the basic unit of mass is the *gram* (g), which is a small quantity (there are 453.59 g in an avoirdupois pound). The standard of mass is the standard kilogram, which is a cylinder of an alloy of platinum–iridium which is housed at the International Bureau of Weights and Measures at Sèvres, France.

LENGTH. The basic unit of length, the *meter* (m), is somewhat longer than the U.S. yard (1 m = 1.0936 yd). Kilometers, centimeters, and millimeters are all frequently used as well. In chemistry we need units to measure the actual sizes of atoms and molecules. These are so small that even a millimeter would be far too large. A common unit currently in use for this purpose is the angstrom unit (Å), which is 0.0000001 mm:

$$1\ \text{Å} = 0.0000001\ \text{mm} = 10^{-7}\ \text{mm}$$
$$1\ \text{Å} = 0.00000001\ \text{cm} = 10^{-8}\ \text{cm}$$

(See Section 1-4 for an explanation of exponential numbers.) Another unit sometimes employed is the nanometer:

$$1\ \text{nm} = 0.000001\ \text{mm} = 10^{-6}\ \text{mm}$$
$$1\ \text{nm} = 0.0000001\ \text{cm} = 10^{-7}\ \text{cm}$$
$$1\ \text{nm} = 10\ \text{Å}$$

VOLUME. The volume of an object is the amount of space that it occupies. When we come to volume we meet our first **compound unit.** How do we measure the volume of a rectangular space? We measure the three distances *a*, *b*, and *c*, and multiply them together. If the three distances are measured in centimeters, then the product has the unit centimeters cubed, or cubic centimeters:

$$\text{cm} \times \text{cm} \times \text{cm} = \text{cm}^3 \quad \text{(also abbreviated cc)}$$

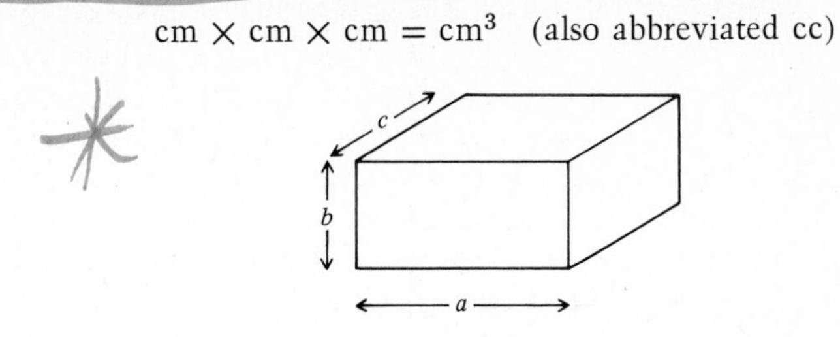

This is a compound unit: it is the product of three simple units. Compound units are frequently used in science, and we shall be dealing with several others later. (Similarly, *area* can be measured in cm^2.)

Another important volume unit is *liters*. A liter is defined as 1000 cm^3, which means that 1 cm^3 equals 1/1000 of a liter, or 1 ml:

$$1 \text{ liter} = 1000 \text{ cm}^3$$

$$1 \text{ ml} = 1 \text{ cm}^3$$

A liter is slightly larger than a U.S. quart (1 liter = 1.0567 qt).

DENSITY. The **density** (d) of any object is defined as its mass divided by its volume:

$$d = \frac{m}{V}$$

If the mass is measured in grams and the volume in milliliters, the unit of density is g/ml (grams per milliliter). Thus, density is another property that has a compound unit, in this case the quotient of two units. We do not have to measure the mass in grams and the volume in milliliters; we could use any other units of mass and volume, but then the numerical value and units of the density would be different.

Example: A sample of pure copper with a volume of 2.00 ml has a mass of 17.9 g at 20°C. What is its density in g/ml and in g/liter?

Answer:

$$d = \frac{m}{V}$$

$$= \frac{17.9 \text{ g}}{2.00 \text{ ml}} = 8.95 \text{ g/ml}$$

2.00 ml is equal to 0.00200 liter, so density in g/liter is

$$d = \frac{17.9 \text{ g}}{0.00200 \text{ liter}} = 0.00895 \text{ g/liter}$$

The density of the sample of copper in the above example is 8.95 g/ml and 0.00895 g/liter. Both numbers are correct—the units are different.

The density of any pure solid or liquid is a property that is a constant at a given temperature. If the density of one sample of pure copper is 8.95 g/ml at 20°C, the density of *any* sample of pure copper is 8.95 g/ml at 20°C. We can simply say that the substance copper has this density at 20°C. As a property, density is quite different from mass or volume. Both mass and volume change when the amount of substance changes. If you add more copper to your sample, the mass and volume will increase. The density does not change when the amount changes. Because density is mass divided by volume, any quantity of pure copper will always have a density of 8.95 g/ml at 20°C. The density of pure water at 25°C is 1.00 g/ml. The reason it is exactly 1.00 is that the gram was originally defined as the mass of 1.00 ml of water.

TIME. Time is ? How do we define time? Let us recognize that time is one of those things that we probably cannot define in a scientific sense. However, we know what we mean by it and how to measure it. The basic unit of time is the same in all systems, including the common U.S. system and the SI system. It is, of course, the *second* (s). The standard second is derived from some naturally occurring oscillations taking place in atoms. These "atomic clocks" give us a more constant time reference than anything we have used in the past. Today's atomic clock is based on the oscillations occurring in the cesium-133 atom.

TEMPERATURE. We will discuss temperature in Section 5-2, but here we can say that the most common temperature scale for scientific work is the *Celsius* (centigrade) scale, named for Anders Celsius (1701–1744). In this scale 0°C is taken as the freezing point of water and 100°C as the boiling point of water. The *Fahrenheit* temperature scale is named after Gabriel Daniel Fahrenheit (1686–1736). On this scale the freezing and boiling points of water are called 32°F and 212°F, respectively. The Fahrenheit scale is not used in scientific work. The Celsius and Fahrenheit temperature scales are compared in Figure 1-2. To convert from one temperature scale to the other, we use the following formulas:

$$F = \tfrac{9}{5}C + 32$$
$$C = \tfrac{5}{9}(F - 32)$$

Example: Convert 212°F to degrees Celsius.

Answer:

$$C = \tfrac{5}{9}(212 - 32) = \tfrac{5}{9}(180) = 100°\text{C}$$

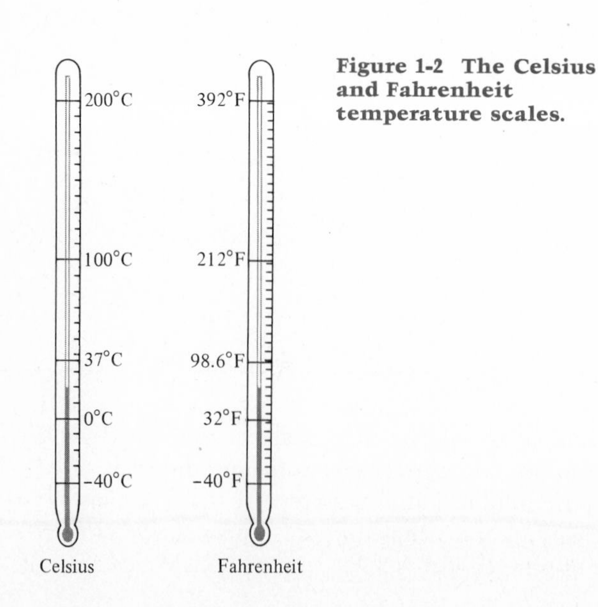

Figure 1-2 The Celsius and Fahrenheit temperature scales.

Example: Convert 20°C to degrees Fahrenheit.

Answer:

$$F = \frac{9}{5}(20) + 32 = 36 + 32 = 68°\text{F}$$

Another temperature scale used in scientific work is the *Kelvin* or *absolute* scale. The Celsius and Kelvin scales will be discussed further in Section 5-2.

1-2 Significant Figures

In Section 1-1, we discussed units of measurement. In this section, we will talk about how to specify how exact our measurements are. In science we use two kinds of numbers. One kind of number, obtained by measuring something, we shall call a *measured* number. The other type we get by counting or by defining. For example, how many sides are there in a triangle? The answer is three. To get this number, we did not have to measure anything; we just had to count the sides. An example of a *defined* number is the number of centimeters in a meter. By definition, there are *exactly* 100 cm in a meter. Once again we do not need to measure anything; we have *defined* this number to be exactly 100.

Most numbers in science are measured, not counted or defined. The difference between counted or defined numbers and measured numbers is that we know the exact values of the former but can never know the *exact* values of the latter. In chemistry, we weigh chemicals, measure volumes of solutions, keep track of time intervals, and do many, many other kinds of measurements. In all measurements there are errors in the measurement no matter who does the measuring. The errors are there because no instrument is exact in its measurement. Compounding the error in the instrument is the human error in making the measurement. Let us see this by use of an example.

Weigh a penny. Figure 1-3 shows this using two different balances. In balance (a), the weight we read is 3.11 g; in balance (b) we get 3.1134 g. Balance (b) gives us more

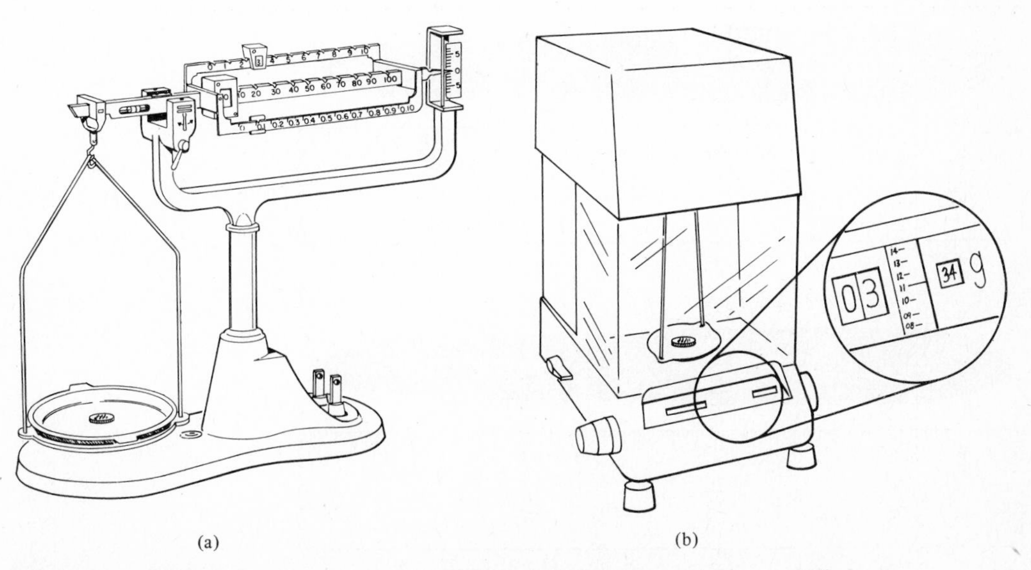

Figure 1-3 Weighing a penny on two different balances may give a different number of significant figures for each measurement.

information. Balance (a) says 3.11 g, but balance (b) says 3.1134 g. If we had only balance (a), we would not know that the third decimal place is 3 or the fourth decimal place 4. Balance (b) tells us this but does not give us the fifth decimal place. A better balance might do that, but no matter how good a balance we used, we could never get an *exact* weight since no balance can give us an infinite number of places. The situation is characteristic of all measurements and therefore of all measured numbers. There are *uncertainties* in our measurements and they depend on the instrument on which we perform the measurement.

Scientists have found it useful to establish a method of identifying the uncertainties in their measurements so that measured numbers are reported as exactly as they are measured and what scientists communicate to each other is completely understood. We define the number of **significant figures** in a measured number as the number of digits that are written down, assuming that we write down everything we know. In Figure 1-3a the number 3.11 has three significant figures, and in Figure 1-3b the number 3.1134 has five significant figures.

In Figure 1-4 we use two different rulers to measure the length of a fountain pen. The results are tabulated below.

Measurement of a fountain pen	*Number of significant figures*
(a) 14 cm	2
(b) 14.2 cm	3

It is important to be able to count correctly the number of significant figures in a measured number. It is not difficult, but we must observe the rules, most of which are concerned with how to treat zeros. A zero in a reported measurement may or may not be significant.

1. All digits on both sides of any decimal point are significant, if zeros are absent.

23.742	5 significant figures
332	3 significant figures
1.4	2 significant figures

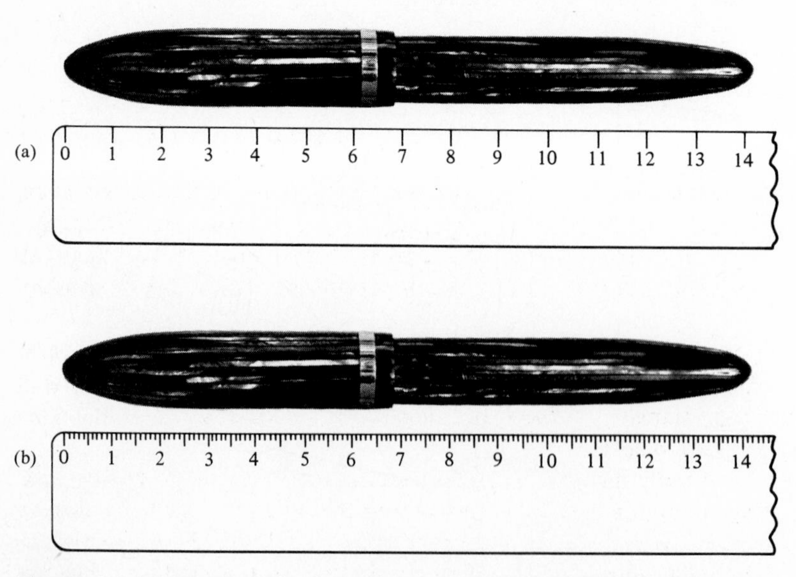

Figure 1-4 Measuring the length of a fountain pen with different rulers.

2. Zeros used to locate a decimal point are not significant.

0.023	2 significant figures
0.23	2 significant figures
0.0000023	2 significant figures

Another way of saying this is that all zeros before the first nonzero are *not* significant.

3. Zeros between numbers *are* significant.

2.003	4 significant figures
1.0008	5 significant figures
0.002034	4 significant figures

4. Zeros to right of the last nonzero digit and to the right of the decimal point *are* significant.

0.00000230	3 significant figures
0.043000	5 significant figures
1.00	3 significant figures
10.0	3 significant figures

5. When an *integer* ends in one or more zeros (that is, when there is nothing written after the decimal point), the zeros that end the integer may or may not be significant. This is the only flaw in the system. In order to tell whether these zeros are significant, one needs additional information. For example, in the last example in rule 4 the measured number is given as 10.0. Since there is a zero after the decimal point, one infers that this figure is significant and so are all of them; that is, there are three significant figures. For the number 2000, we know that the 2 is significant but without additional information about how the number was measured, we do not know whether one, two, or all three of the zeros are significant. If it was given as 2000.0, 2000.3, etc., we would know that all five digits are significant. One way to avoid confusion in this case is to report the number in exponential form (see Section 1–4), reporting only the number of significant figures. For example, if there were only two significant figures in 2000, it would be reported as

2.0×10^3	2 significant figures

If there were three significant figures, we would have 2.00×10^3, etc.

Note that use of this system means that to scientists there is a difference between 21.5, 21.50, and 21.500. Although mathematically these numbers are the same, scientifically they are not. The number 21.5 implies that we do not know the next place. The number 21.50 says that we do know it: it is 0, and not 7, 8, 3, or any other digit.

For numbers that are counted or defined, none of the preceding remarks apply. These numbers have an infinite number of significant figures. When we say that there are 3 sides to a triangle, we know all the places after the 3. They are *all* zero, out to infinity (that is, 3.0000...).

In science we frequently have occasion to use numbers in calculations: we add, subtract, multiply, and divide them. It is important that the answer have the correct number of significant figures; otherwise, we will think we know the answer more accurately (or less accurately) than we actually do. The rules for doing this are different for multiplication and division than they are for addition and subtraction.

For multiplication and division, the number of significant figures in the final result cannot be greater than the *least precise measurement*. When we multiply and/or divide a group of numbers, one of the numbers will have the fewest number of significant figures when compared to the other numbers. *The number of significant figures in the final answer should be the same as in this least precise number.*

Example: Calculate the kinetic energy of a ball of mass 5.0 g traveling with a velocity of 1.15 cm/s. Kinetic energy is obtained from the formula

$$\text{K.E.} = \tfrac{1}{2}mv^2 \qquad m = \text{mass of object}$$
$$v = \text{velocity of object}$$
(see Section 1-7)

Answer:

$$\text{K.E.} = \tfrac{1}{2}(5.0\ \text{g})\left(1.15\ \frac{\text{cm}}{\text{s}}\right)^2 = 3.3\frac{\text{g}\cdot\text{cm}^2}{\text{s}^2}$$

Let us examine this result. If we carry out the multiplication all the way, either by hand or on a calculator, we get

$$\tfrac{1}{2}(5.0)(1.15)^2 = 3.306250 \quad \leftarrow \text{7 significant figures}$$

This number, 3.306250, has seven significant figures. But the rule says that in multiplication and/or division, the answer should have the same number of significant figures as the least precise number. Which number is least precise?

$\tfrac{1}{2}$ (or 0.5)	is not a measured number; it is part of the formula and so has an infinite number of significant figures
5.0	has 2 significant figures
1.15	has 3 significant figures

The least precise number has two significant figures, so our answer must also have two. Since 3.306250 has seven, we must drop the last five digits and round off to two significant figures:

$$3.306250 = 3.3\frac{\text{g}\cdot\text{cm}^2}{\text{s}^2} \qquad \textit{answer}$$

The reason we must drop the five digits 06250 is that they are not true! Our multiplication gives us a false impression that we know these numbers, but we actually do not. For example, we do not know the next digit in 5.0 g. Therefore, if we were able to measure this more precisely, we might find that it was 5.02 g. But in that case, $\tfrac{1}{2}(5.02)(1.15)^2$ would come out to 3.319475. Since we do not know whether the mass was 5.00, 5.02, 5.04, or 4.97 g, etc., how can we be sure of any digit in the answer after 3.3? Since we cannot be sure, we cannot use them and they are dropped in the final answer.

Example: Two students are attempting to determine the density of an unknown liquid in their general chemistry laboratory. The first student measures out 40 ml of the material in a graduated cylinder and the second student determines its weight on an analytical balance and reports the weight as 40.0023 g. What is the density of the unknown liquid in g/ml?

Answer:

$$d = \frac{m}{V}$$

$$= \frac{\overset{\text{6 significant figures}}{40.0023}\ \text{g}}{\underset{\text{2 significant figures}}{40}\ \text{ml}} = 1.0\ \frac{\text{g}}{\text{ml}}$$

The student who weighed out the liquid did not have to measure so precisely because the first student could only manage to get two significant figures in his measurement. The number of significant figures in the result is determined by the least precisely known measurement, which in this case was the volume. (We assume that the 0 in 40 is significant, since with a graduated cylinder it is easy to distinguish 40 from 41 or 39.)

When discarding digits it is important to round off properly. In this book we shall adopt the rule that if the *first digit dropped* is 5 to 9, we raise the *last digit kept* to the next higher number; otherwise we do not.

Example: In each case, drop the last two digits: 33.679, 2.4715, 1.145, 0.001309, 3.52.

Answer:

33.679 = 33.7
2.4715 = 2.47
1.145 = 1.1
0.001309 = 0.0013
3.52 = 4

When we deal with addition and subtraction, the rule is different. It is not the fewest number of significant figures that determines the number of the significant figures in the final result. The rule is that *the last digit kept should correspond to the first uncertainty in the decimal place.*

Example: Suppose that we wish to measure the total volume we will be using in making up a quantity of household paintbrush cleaner. We use 320.04 ml of kerosene, 80.2 ml of oleic acid, 20.020 ml of ammonia, and 20.0 ml of methyl alcohol. What is the total volume of paintbrush cleaner?

Answer:

a b c d	*e f*	
320.0	4	kerosene
80.2		oleic acid
20.0	20	ammonia
20.0		methyl alcohol
440.2	60	

The volume is 440.3 ml.

What counts here is that all the numbers have significant digits in column *d*, but not in columns *e* and *f*. Therefore, the answer cannot have a number in column *e* or *f* either. We add the numbers, including those in columns *e* and *f*, but then we must discard these digits in the answer, rounding off as before. Even though kerosene and ammonia are more precisely measured than the other two substances, their greater precision has no affect on the answer.

1-3 Dimensional Analysis

In chemistry it is necessary to solve a great many numerical problems. Many students have little trouble with the arithmetic or elementary algebra involved but often have difficulty in converting the words given in the problem to the algebraic equations necessary to solve the problem—that is, in setting up the problem. Dimensional analysis is an important tool which not only helps in setting up problems but also in solving them. The method is so powerful that we can sometimes solve complicated problems without having to analyze them fully! We can sum up the rules as follows:

1. When you write a number, always include the *units* of that number. A few numbers have no units (they are dimensionless), but most numbers that we encounter in chemistry do have units.
2. *Addition and substraction*. Numbers that are to be added or subtracted *must have the same units*. We can add 4 g + 3 g, or 4 books + 3 books, but we cannot add 4 g + 3 liters or 4 books + 3 cows. If the units involved can be interconverted (for example, if they are both units of length, or both of volume), they can be added if one unit is changed to the other. Thus, we cannot directly add 40 g + 2.555 kg, but we can add these numbers if we change kg to g (or vice versa): 40 g + 2555 g = 2595 g.
3. *Multiplication and division*. We can multiply and divide numbers no matter whether the units are the same or different, but *when multiplying or dividing numbers, we multiply and divide the units also*. This rule is the heart of the method of dimensional analysis.

Examples:

$$3.2\text{ g} \times 4.1\text{ g} = 13\text{ g}^2$$

$$0.45\text{ liter} \times 1.21\text{ g} = 0.54\text{ liter}\cdot\text{g}$$

$$\frac{13.6\text{ g}}{14.1\text{ liter}} = 0.965\,\frac{\text{g}}{\text{liter}}$$

units cancel

$$\frac{21.1\,\cancel{\text{liter}}}{6.3\,\cancel{\text{liter}}} = 3.3 \leftarrow \text{dimensionless number}$$

$$6.12\,\cancel{\text{liter}} \times 4.41\,\frac{\text{g}}{\cancel{\text{liter}}} = 27.0\text{ g}$$

units cancel

Note that when we divide two numbers with identical units, the units cancel.

4. When dividing compound units, invert the denominator and multiply.

Example:

$$\frac{\dfrac{\text{g}}{\text{liter}}}{\dfrac{\text{mg}}{\text{liter}}} = \frac{\text{g}}{\cancel{\text{liter}}} \times \frac{\cancel{\text{liter}}}{\text{mg}} = \frac{\text{g}}{\text{mg}}$$

5. When we have carried out the arithmetic or algebra necessary to solve the problem, the units remaining may tell us if the method was wrong.

Example: The density of platinum at 20°C is 21.45 g/ml. What is the weight in grams of 3.5 ml of platinum?

Answer: The correct way to set up this problem is to multiply the density by the volume (because $d = m/V$ and $m = dV$):

$$21.45\,\frac{\text{g}}{\cancel{\text{ml}}} \times 3.5\,\cancel{\text{ml}} = 75\text{ g} \qquad \textit{correct answer}$$

Note that when we do this, milliliters cancel out, and we are left with grams, which is what the problem asked for. But suppose we had made a mistake and thought that we should divide 21.45 by 3.5. In that case, if we had written the units out, we would have found

$$\frac{21.45\,\dfrac{\text{g}}{\text{ml}}}{3.5\text{ ml}} = 21.45\,\frac{\text{g}}{\text{ml}} \times \frac{1}{3.5\text{ ml}} = 6.1\,\frac{\text{g}}{\text{ml}^2} \qquad \textit{wrong answer}$$

The units of this answer are g/ml^2, which is obviously absurd in this problem since the question asked for grams. The method of dimensional analysis has told us that our method is incorrect. Similarly, if we had thought that we had to divide 3.5 by 21.45, we would have found

$$\frac{3.5\text{ ml}}{21.45\,\dfrac{\text{g}}{\text{ml}}} = 3.5\text{ ml} \times \frac{1\text{ ml}}{21.45\text{ g}} = 0.16\,\frac{\text{ml}^2}{\text{g}} \qquad \textit{wrong answer}$$

which is equally wrong.

If we solve a problem using dimensional analysis, and the units come out right, it does not necessarily mean that our setup is correct (although it usually is), but if the units come out wrong, the setup *must* be wrong. It is obvious that students who take the trouble to use dimensional analysis in all their problems will do better than those who do not.

Dimensional analysis will be helpful no matter what method the student chooses to solve problems. Many problems in general chemistry can be solved by either of two methods, one involving proportions and one not using proportions. The methods are the same mathematically, but they use different ways of thinking. We shall illustrate with a very simple example.

Example: How many centimeters are in 18.3 in? The conversion factor is 1.00 in = 2.54 cm.

Answer: We can set up the proportion

$$\frac{2.54\text{ cm}}{1.00\text{ in}} = \frac{x}{18.3\text{ in}}$$

Solving, we get

$$x = \frac{(2.54\text{ cm})(18.3\text{ }\cancel{\text{in}})}{1.00\text{ }\cancel{\text{in}}}$$
$$= 46.5\text{ cm}$$

In the other method, we make use of the fact that multiplying any number by 1 does not change the value of the number.

$$(18.3\,\cancel{\text{in}})\left(2.54\,\frac{\text{cm}}{\cancel{\text{in}}}\right) = 46.5\text{ cm}$$

Since 2.54 cm = 1.00 in, then

$$\frac{2.54\text{ cm}}{1.00\text{ in}} = 1$$

It is obvious that the numerical value of 18.3 in cannot be changed by multiplying by 1. When we do this, inches cancel, and the answer comes out in centimeters. The same situation holds for any other conversion factor. All of them must equal 1, and because this is true, we are always allowed to multiply anything by them. Naturally, we will choose to do so in such a way that the units come out right. For example, it is mathematically correct to multiply 18.3 in by 1.00 in/2.54 cm:

$$18.3\text{ in} \times \frac{1.00\text{ in}}{2.54\text{ cm}} = 7.20\,\frac{\text{in}^2}{\text{cm}}$$

but if we do so, we get the obviously incorrect units in^2/cm.

Not all problems in general chemistry can be handled by either of these methods, but many of them can be. Experience has shown that, for most people, the second method is more convenient than the first. If you are used to thinking out problems by means of proportions, then by all means go ahead and do so, but most students will find that the other method is a more useful tool and that is the method that we shall generally use in this book. Some examples follow.

Example. How many inches are in 2.31 miles?
Conversion Factors

$$\left.\begin{array}{r}12\text{ inches} = 1\text{ foot}\\ 5280\text{ feet} = 1\text{ mile}\end{array}\right\}\text{ all of these are defined numbers}$$

Answer:

$$2.31\,\cancel{\text{miles}} \times \frac{5280\,\cancel{\text{ft}}}{1\,\cancel{\text{mile}}} \times \frac{12\text{ in}}{1\,\cancel{\text{ft}}} = 146{,}000\text{ in}$$

Example: The speed of light is 186,000 miles/s (this number has three significant figures). Calculate the speed of light in cm/s.

Answer:

$$186{,}000\,\frac{\cancel{\text{mile}}}{\text{s}} \times \frac{5280\,\cancel{\text{ft}}}{\cancel{\text{mile}}} \times 12\,\frac{\cancel{\text{in}}}{\cancel{\text{ft}}} \times 2.54\,\frac{\text{cm}}{\cancel{\text{in}}} = 29{,}900{,}000{,}000\text{ cm/s}$$

Example: Boyle's law states that the pressure (P) times the volume (V) of a given quantity of a gas at constant temperature is equal to a constant. Set up an equation for this relationship and calculate the constant if 2 g of hydrogen at 0°C occupies 22.4 liters at a pressure of 1.00 atm.

Answer: At a constant T, PV = constant.

$$(1.00 \text{ atm})(22.4 \text{ liter}) = \text{constant of } 22.4 \text{ liter} \cdot \text{atm}$$

Example: If the Boyle's law constant is 22.4 liter · atm, and we define a pressure of 760 torr as equal to 1.00 atm, convert the Boyle's law constant to liter · torr.

Answer: 1.00 atm = 760 torr.

$$22.4 \text{ liter} \cdot \cancel{\text{atm}} \times \frac{760 \text{ torr}}{\cancel{\text{atm}}} = 17{,}000 \text{ liter} \cdot \text{torr}$$

rounded off to three significant figures

Example: If the Boyle's law constant is 22.4 liter · atm and the volume in the first example is changed to 150 liters, what is the final pressure in atmospheres and torr?

Answer: PV = constant.

$$P(150 \text{ liter}) = 2.24 \text{ liter} \cdot \text{atm}$$

$$P = \frac{22.4 \cancel{\text{liter}} \cdot \text{atm}}{150 \cancel{\text{liter}}} = 0.149 \text{ atm}$$

$$0.149 \cancel{\text{atm}} \times \frac{760 \text{ torr}}{1.00 \cancel{\text{atm}}} = 113 \text{ torr}$$

Example: We defined density as mass/volume. What is the volume of 7.25 g of mercury if the density of mercury is 13.6 g/cm³ at 25°C?

Answer: Density = mass/volume.

$$13.6 \frac{\text{g}}{\text{cm}^3} = \frac{7.25 \text{ g}}{V}$$

$$V = \frac{7.25 \cancel{\text{g}}}{13.6 \cancel{\text{g}}/\text{cm}^3} = 0.533 \text{ cm}^3$$

Example: The volume of an unknown mass of mercury is 157 cm³. What is its mass?

Answer: Density = mass/volume.

$$13.6 \frac{\text{g}}{\text{cm}^3} = \frac{m}{157 \text{ cm}^3}$$

$$m = 13.6 \frac{\text{g}}{\cancel{\text{cm}^3}} \times 157 \cancel{\text{cm}^3} = 2140 \text{ g}$$

rounded off to three significant figures

1-4 Exponential Numbers

In chemistry and in other sciences we often must deal with very large numbers and very small numbers. For example, the weight of a single hydrogen atom is 0.00000000000000000000000167 g, and we saw in an example in Section 1-3 that the velocity of light is 29900000000 cm/s. Now it certainly would be easier to express these values as 1.67×10^{-24} g and 2.99×10^{10} cm/s. Since we frequently must manipulate numbers of this kind, this shorthand way of writing them is very handy and convenient. Another advantage of these numbers is that we can use them to express significant figures when an integer ends in one or more zeros.

In writing numbers in exponential form (also called **scientific notation**), the idea is to write the number in terms of integral powers of 10. The exponential number consists of two parts, a coefficient and a power of 10. The coefficient is usually given as a number between 1 and 10, although this is not mandatory. As an example, let us return to the value of the velocity of light.

$$2.99 \times 10^{10}$$

coefficient between 1 and 10 ↗ (2.99); power of 10 ↖ (10^{10})

When a large number (actually any number greater than 10) is written in exponential form, the exponent is always *positive*. The decimal point of the number is moved to the left the same number of places as the power of 10:

$$4780. = 4.78 \times 10^3$$

three places to the left

$$174600000. = 1.746 \times 10^8$$

eight places to the left

$$113.50 = 1.1350 \times 10^2$$

two places to the left

When a small number (actually any number less than 1) is written in exponential form, the exponent is always *negative*. The decimal point of the number is moved to the right the same number of places as the negative power of 10.

$$0.0000793 = 7.93 \times 10^{-5}$$

five places to the right

$$0.00000000044 = 4.4 \times 10^{-10}$$

ten places to the right

Using these rules, we can see that the following are true:

$$1.68 \times 10^4 = 16.8 \times 10^3 = 168 \times 10^2 = 1680 \times 10^1 = 16800$$
$$79.9 \times 10^{-4} = 7.99 \times 10^{-3} = 0.799 \times 10^{-2} = 0.0799 \times 10^{-1} = 0.00799 \quad \text{etc.}$$

Adding and Subtracting Exponential Numbers. The key to adding or subtracting numbers in exponential form is to rewrite all the numbers so that they have the same powers of 10. If we do not do this, it would be like adding numbers in a tens column to numbers in the hundreds column instead of keeping them in their own column (this rule is similar to the rule which says that we can only add or subtract numbers with the same units).

Example: Add

$$6.020 \times 10^{23}$$
$$7.400 \times 10^{24}$$
$$2.700 \times 10^{21}$$

Answer: First, we rewrite the numbers using the rules established earlier in this section so that the exponents are all the same.

$$
\begin{aligned}
6.020 \times 10^{23} &= 6.02|0 \times 10^{23} \\
7.400 \times 10^{24} &= 74.00| \times 10^{23} \\
2.700 \times 10^{21} &= 0.02|7 \times 10^{23}
\end{aligned}
$$

Now we can add them: $80.05| \times 10^{23}$ or 8.005×10^{24}

Example:

$$
\begin{array}{r}
9.02 \times 10^{-8} \\
- \; 8.43 \times 10^{-9} \\
\hline
\end{array}
$$

Answer: Rewriting the numbers so that all powers of 10 are the same, we get

$$
\begin{array}{rr}
 & 9.02| \times 10^{-8} \\
-8.43 \times 10^{-9} = & -\;0.84|3 \times 10^{-8} \\
\hline
 & 8.12| \times 10^{-8}
\end{array}
$$

MULTIPLYING AND DIVIDING EXPONENTIAL NUMBERS. To multiply numbers in exponential form, we algebraically *add* the exponents: for example,

$$10^2 \times 10^3 = 10^{(2+3)} = 10^5 = 1 \times 10^5$$

It is easy to prove that this is correct:

$$
\begin{aligned}
10^2 &= 100 \\
10^3 &= 1000 \\
(100)(1000) &= 100{,}000 = 10^5
\end{aligned}
$$

Examples:

$$
\begin{aligned}
10^5 \times 10^{-2} &= 10^3 = 1 \times 10^3 \\
10^{-5} \times 10^{-6} &= 10^{-11} = 1 \times 10^{-11} \\
10^2 \times 10^{-5} &= 10^{-3} = 1 \times 10^{-3}
\end{aligned}
$$

To divide numbers in exponential form, we algebraically *subtract* the exponents.

$$\frac{10^4}{10^3} = 10^1 = 1 \times 10^1$$

$$\frac{10^6}{10^{-2}} = 10^8 = 1 \times 10^8$$

$$\frac{10^{-3}}{10^{-5}} = 10^2 = 1 \times 10^2$$

In all the preceding examples, the coefficient was 1. When we deal with coefficients other than 1, the coefficients are handled according to the general rules of multiplication and division.

Example: Multiply the following numbers:

$$
\begin{array}{r}
7.47 \times 10^3 \\
\times \; 8.90 \times 10^4 \\
\hline
\end{array}
$$

Answer: $(7.47 \times 8.90) \times 10^{(4+3)} = 66.5 \times 10^7 = 6.65 \times 10^8$.

Example: Divide 9.46×10^{-6} by 3.22×10^{-10}.

Answer:

$$\frac{9.46 \times 10^{-6}}{3.22 \times 10^{-10}} = \frac{9.46}{3.22} \times 10^{-6-(-10)} = 2.94 \times 10^{4}$$

Work out the following examples to test your understanding of this material.

Example: To calculate the weight of one uranium atom, perform the indicated operation.

$$\frac{238.029 \text{ g/mole}}{6.022 \times 10^{23} \text{ atoms/mole}}$$

Answer:

$$\frac{2.38029 \times 10^{2} \text{ g/}\cancel{\text{mole}}}{6.022 \times 10^{23} \text{ atoms/}\cancel{\text{mole}}} = 3.953 \times 10^{-22} \text{ g/atom}$$

The weight of one atom of uranium is 3.953×10^{-22} g (the units will be explained in Chapter 2).

Example: Cube the number 1.08×10^{-8}.

Answer: $(1.08)^3 \times (10^{-8})^3 = (1.08)^3 \times 10^{-24} = 1.26 \times 10^{-24}$.

Example: To calculate approximately the wavelength associated with an electron traveling at a velocity of 3.33×10^{8} cm/s, perform the indicated operation (the unit "erg" will be explained in Section 1-7).

$$x = \frac{6.626 \times 10^{-27} \text{ erg} \cdot \text{s}}{(9.1096 \times 10^{-28} \text{ g})(3.33 \times 10^{8} \text{ cm/s})}$$

Answer: $x = 2.18 \times 10^{-8}$ cm.

1-5 Logarithms

Logarithms were invented by John Napier (1550–1617) as a tool to simplify the multiplication and division of multi-digit numbers (that is, numbers with three or more significant figures). Today this use is declining because we now have pocket calculators that do the job even more quickly. However, it is still necessary to learn something about logarithms in order to study chemistry, because logarithms are part of certain mathematical expressions that are used in chemistry. Logarithms and exponential numbers are related and the proper study of logarithms comes after the study of exponential numbers.

The logarithm of a number to the base 10 is the exponent of 10 that gives the number.

$$\text{If } a = 10^{b} \qquad \text{then } \log a = b$$

This definition will become clearer after studying the following list:

$$1 = 10^{0} \qquad \log 1 = 0$$
$$2 = 10^{0.301} \qquad \log 2 = 0.301$$
$$3 = 10^{0.477} \qquad \log 3 = 0.477$$

$4 = 10^{0.602}$	$\log 4 = 0.602$
$5 = 10^{0.699}$	$\log 5 = 0.699$
$6 = 10^{0.788}$	$\log 6 = 0.778$
$7 = 10^{0.845}$	$\log 7 = 0.845$
$8 = 10^{0.903}$	$\log 8 = 0.903$
$9 = 10^{0.954}$	$\log 9 = 0.954$
$10 = 10^{1.000}$	$\log 10 = 1.000$

From this list we see that any whole number between 1 and 10 can be written as an exponential number. For example, the number 3 can be written as $10^{0.477}$; and 8 is $10^{0.903}$. Although this list shows this only for whole numbers from 1 to 10, it is also true for any number between 1 and 10. For example, the number $3.2 = 10^{0.480}$ and $9.33 = 10^{0.970}$. Under the second column in the list we have the logarithms of the numbers given in the first column. The logarithms of the numbers are the exponents of 10 that give the numbers. So now we have the relationship between logarithms and exponents. **Logarithms are exponents.**

Example: Calculate the logarithms of the following numbers.

$$10^0 \quad 10^1 \quad 10^2 \quad 10^{14} \quad 10^{-1} \quad 10^{-2} \quad 10^{-14}$$

Answer:

$$\log 10^0 = \log 1 = 0$$
$$\log 10^1 = 1$$
$$\log 10^2 = 2$$
$$\log 10^{14} = 14$$
$$\log 10^{-1} = -1$$
$$\log 10^{-2} = -2$$
$$\log 10^{-14} = -14$$

In Section 1-4, we learned the rules for multiplying and dividing exponents; these rules are, of course, the same for logarithms. That is, to multiply numbers, we *add* logarithms, whereas to divide numbers, we *subtract* logarithms.

$$\log(A \cdot B) = \log A + \log B$$
$$\log\left(\frac{A}{B}\right) = \log A - \log B$$
$$\log\left(\frac{ABC}{DE}\right) = \log A + \log B + \log C - \log D - \log E$$

To calculate the logarithm of a product, we first find the logarithm of each number and then add the logarithms.

Example: What is the logarithm of $(1 \times 10^2)(1 \times 10^7)$?

Answer:

$$\log(1 \times 10^2)(1 \times 10^7) = \log(1 \times 10^2) + \log(1 \times 10^7)$$
$$\log(1 \times 10^2) = \log 1 + \log 10^2 = 0 + 2$$
$$\log(1 \times 10^7) = \log 1 + \log 10^7 = 0 + 7$$

so

$$\log(1 \times 10^2)(1 \times 10^7) = (0 + 2) + (0 + 7) = 9$$

This can be solved another way:

$$(1 \times 10^2)(1 \times 10^7) = 1 \times 10^9$$
$$\log(1 \times 10^9) = \log 1 + \log 10^9 = 0 + 9 = 9$$

To calculate the logarithm of a quotient, we find the logarithm of each number and then subtract the logarithms.

Example: What is the logarithm of $\dfrac{1 \times 10^7}{1 \times 10^{-10}}$?

Answer:

$$\log\left(\frac{1 \times 10^7}{1 \times 10^{-10}}\right) = \log(1 \times 10^7) - \log(1 \times 10^{-10})$$
$$\log(1 \times 10^7) = \log 1 + \log 10^7 = 0 + 7$$
$$-\log(1 \times 10^{-10}) = -(\log 1 + \log 10^{-10}) = -(0 - 10)$$

so

$$\log\left(\frac{1 \times 10^7}{1 \times 10^{-10}}\right) = (0 + 7) - (0 - 10) = 17$$

or

$$\frac{1 \times 10^7}{1 \times 10^{-10}} = 1 \times 10^{17}$$
$$\log(1 \times 10^{17}) = \log 1 + \log 10^{17} = 0 + 17 = 17$$

The next example combines multiplication and division.

Example: What is the logarithm of $\dfrac{(1 \times 10^{-3})(1 \times 10^{12})}{1 \times 10^{-7}}$?

Answer:

$$\log\left[\frac{(1 \times 10^{-3})(1 \times 10^{12})}{1 \times 10^{-7}}\right] = \log(1 \times 10^{-3}) + \log(1 \times 10^{12}) - \log(1 \times 10^{-7})$$
$$= (0 - 3) + (0 + 12) - (0 - 7) = 16$$

In our previous examples, the coefficient was always unity. Digits other than unity will, of course, occur in problems, and to see how this works we will consider the following two examples, using logarithms from the list given above.

Example: Calculate the logarithm of 7×10^4.

Answer:

$$\log(7 \times 10^4) = \log 7 + \log 10^4$$
$$\log 7 = 0.845$$
$$\log 10^4 = 4$$
$$\log 7 \times 10^4 = 0.845 + 4 = 4.845$$

Example: What is the logarithm of 4×10^{-4}?

Answer:

$$\log(4 \times 10^{-4}) = \log 4 + \log 10^{-4}$$
$$\log 4 = 0.602$$
$$\log 10^{-4} = -4$$
$$\log(4 \times 10^{-4}) = 0.602 - 4.000 = -3.398$$

This is an important example since it shows that the logarithm of a positive number may turn out to be negative. Logarithms of numbers greater than 1 are always positive. Logarithms of numbers less than 1 are always negative. The logarithm of 1 is, of course, 0.

To find the logarithms of numbers that lie between 1 and 10, we must consult tables of logarithms. Table 1-1 is a four-place logarithm table (handbooks containing five- and seven-place logarithm tables are found in many libraries). We will now show how to use this table to find the logarithms of numbers between 1 and 10. A four-place logarithm table gives the logarithms of numbers to four significant figures. The table is constructed as follows. Along the left-hand side of the table is a row of numbers starting with 10 and ending with 54 on page 22 and starting with 55 on page 23 and ending with 99. It would be to our advantage to look at these numbers as 1.0 to 5.4 and then 5.5 to 9.9. Essentially, then, these are the numbers whose logarithms we wish to find, from 1.0 to almost 10 (9.9). Across the table is a series of numbers from 0 to 9. They constitute the hundredths place when we look up a number like 1.00 or 1.01. The black digits are on the left-hand side and the colored digits are the ones across the table.

We are now ready to look up the logarithms of numbers. We shall choose 1.00 as our first example. Along the left-hand row we find 1.0. It is the first number. We then go across the row to find 1.00 as

N	0	1	2
10	<u>0000</u>	0043	0086

$$\log 1.00 = 0.0000$$

For log 1.01

N	0	1
10	0000	<u>0043</u>

$$\log 1.01 = 0.0043$$

Now let us choose three numbers at random, 7.3, 1.48, and 3.63. For log 7.3

N	0	1	2
—			
—			
—			
73	<u>8633</u>	8639	8645

$$\log 7.3 = 0.8633$$

which, of course, is the same as log 7.30.

Table 1-1 Four-Place Table of Logarithms

N	0	1	2	3	4	5	6	7	8	9
10	0000	0043	0086	0128	0170	0212	0253	0294	0334	0374
11	0414	0453	0492	0531	0569	0607	0645	0682	0719	0755
12	0792	0828	0864	0899	0934	0969	1004	1038	1072	1106
13	1139	1173	1206	1239	1271	1303	1335	1367	1399	1430
14	1461	1492	1523	1553	1584	1614	1644	1673	1703	1732
15	1761	1790	1818	1847	1875	1903	1931	1959	1987	2014
16	2041	2068	2095	2122	2148	2175	2201	2227	2253	2279
17	2304	2330	2355	2380	2405	2430	2455	2480	2504	2529
18	2553	2577	2601	2625	2648	2672	2695	2718	2742	2765
19	2788	2810	2833	2856	2878	2900	2923	2945	2967	2989
20	3010	3032	3054	3075	3096	3118	3139	3160	3181	3201
21	3222	3243	3263	3284	3304	3324	3345	3365	3385	3404
22	3424	3444	3464	3483	3502	3522	3541	3560	3579	3598
23	3617	3636	3655	3674	3692	3711	3729	3747	3766	3784
24	3802	3820	3838	3856	3874	3892	3909	3927	3945	3962
25	3979	3997	4014	4031	4048	4065	4082	4099	4116	4133
26	4150	4166	4183	4200	4216	4232	4249	4265	4281	4298
27	4314	4330	4346	4362	4378	4393	4409	4425	4440	4456
28	4472	4487	4502	4518	4533	4548	4564	4579	4594	4609
29	4624	4639	4654	4669	4683	4698	4713	4728	4742	4757
30	4771	4786	4800	4814	4829	4843	4857	4871	4886	4900
31	4914	4928	4942	4955	4969	4983	4997	5011	5024	5038
32	5051	5065	5079	5092	5105	5119	5132	5145	5159	5172
33	5185	5198	5211	5224	5237	5250	5263	5276	5289	5302
34	5315	5328	5340	5353	5366	5378	5391	5403	5416	5428
35	5441	5453	5465	5478	5490	5502	5514	5527	5539	5551
36	5563	5575	5587	5599	5611	5623	5635	5647	5658	5670
37	5682	5694	5705	5717	5729	5740	5752	5763	5775	5786
38	5798	5809	5821	5832	5843	5855	5866	5877	5888	5899
39	5911	5922	5933	5944	5955	5966	5977	5988	5999	6010
40	6021	6031	6042	6053	6064	6075	6085	6096	6107	6117
41	6128	6138	6149	6160	6170	6180	6191	6201	6212	6222
42	6232	6243	6253	6263	6274	6284	6294	6304	6314	6325
43	6335	6345	6355	6365	6375	6385	6395	6405	6415	6425
44	6435	6444	6454	6464	6474	6484	6493	6503	6513	6522
45	6532	6542	6551	6561	6571	6580	6590	6599	6609	6618
46	6628	6637	6646	6656	6665	6675	6684	6693	6702	6712
47	6721	6730	6739	6749	6758	6767	6776	6785	6794	6803
48	6812	6821	6830	6839	6848	6857	6866	6875	6884	6893
49	6902	6911	6920	6928	6937	6946	6955	6964	6972	6981
50	6990	6998	7007	7016	7024	7033	7042	7050	7059	7067
51	7076	7084	7093	7101	7110	7118	7126	7135	7143	7152
52	7160	7168	7177	7185	7193	7202	7210	7218	7226	7235
53	7243	7251	7259	7267	7275	7284	7292	7300	7308	7316
54	7324	7332	7340	7348	7356	7364	7372	7380	7388	7396
N	0	1	2	3	4	5	6	7	8	9

N	0	1	2	3	4	5	6	7	8	9
55	7404	7412	7419	7427	7435	7443	7451	7459	7466	7474
56	7482	7490	7497	7505	7513	7520	7528	7536	7543	7551
57	7559	7566	7574	7582	7589	7597	7604	7612	7619	7627
58	7634	7642	7649	7657	7664	7672	7679	7686	7694	7701
59	7709	7716	7723	7731	7738	7745	7752	7760	7767	7774
60	7782	7789	7796	7803	7810	7818	7825	7832	7839	7846
61	7853	7860	7868	7875	7882	7889	7896	7903	7910	7917
62	7924	7931	7938	7945	7952	7959	7966	7973	7980	7987
63	7993	8000	8007	8014	8021	8028	8035	8041	8048	8055
64	8062	8069	8075	8082	8089	8096	8102	8109	8116	8122
65	8129	8136	8142	8149	8156	8162	8169	8176	8182	8189
66	8195	8202	8209	8215	8222	8228	8235	8241	8248	8254
67	8261	8267	8274	8280	8287	8293	8299	8306	8312	8319
68	8325	8331	8338	8344	8351	8357	8363	8370	8376	8382
69	8388	8395	8401	8407	8414	8420	8426	8432	8439	8445
70	8451	8457	8463	8470	8476	8482	8488	8494	8500	8506
71	8513	8519	8525	8531	8537	8543	8549	8555	8561	8567
72	8573	8579	8585	8591	8597	8603	8609	8615	8621	8627
73	8633	8639	8645	8651	8657	8663	8669	8675	8681	8686
74	8692	8698	8704	8710	8716	8722	8727	8733	8739	8745
75	8751	8756	8762	8768	8774	8779	8785	8791	8797	8802
76	8808	8814	8820	8825	8831	8837	8842	8848	8854	8859
77	8865	8871	8876	8882	8887	8893	8899	8904	8910	8915
78	8921	8927	8932	8938	8943	8949	8954	8960	8965	8971
79	8976	8982	8987	8993	8998	9004	9009	9015	9020	9025
80	9031	9036	9042	9047	9053	9058	9063	9069	9074	9079
81	9085	9090	9096	9101	9106	9112	9117	9122	9128	9133
82	9138	9143	9149	9154	9159	9165	9170	9175	9180	9186
83	9191	9196	9201	9206	9212	9217	9222	9227	9232	9238
84	9243	9248	9253	9258	9263	9269	9274	9279	9284	9289
85	9294	9299	9304	9309	9315	9320	9325	9330	9335	9340
86	9345	9350	9355	9360	9365	9370	9375	9380	9385	9390
87	9395	9400	9405	9410	9415	9420	9425	9430	9435	9440
88	9445	9450	9455	9460	9465	9469	9474	9479	9484	9489
89	9494	9499	9504	9509	9513	9518	9523	9528	9533	9538
90	9542	9547	9552	9557	9562	9566	9571	9576	9581	9586
91	9590	9595	9600	9605	9609	9614	9619	9624	9628	9633
92	9638	9643	9647	9652	9657	9661	9666	9671	9675	9680
93	9685	9689	9694	9699	9703	9708	9713	9717	9722	9727
94	9731	9736	9741	9745	9750	9754	9759	9763	9768	9773
95	9777	9782	9786	9791	9795	9800	9805	9809	9814	9818
96	9823	9827	9832	9836	9841	9845	9850	9854	9859	9863
97	9868	9872	9877	9881	9886	9890	9894	9899	9903	9908
98	9912	9917	9921	9926	9930	9934	9939	9943	9948	9952
99	9956	9961	9965	9969	9974	9978	9983	9987	9991	9996
N	0	1	2	3	4	5	6	7	8	9

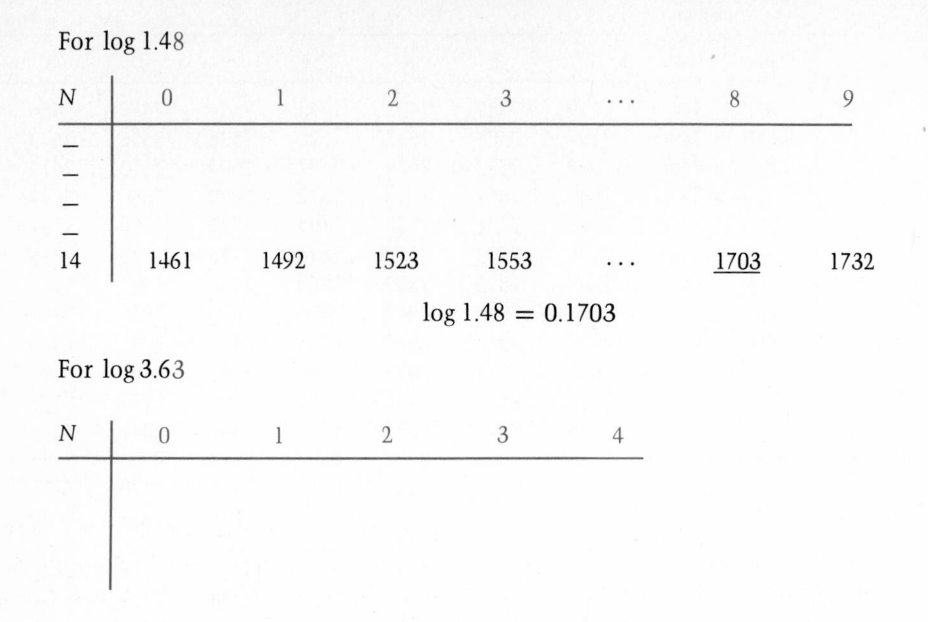

For log 1.48

N	0	1	2	3	...	8	9
–							
–							
–							
–							
14	1461	1492	1523	1553	...	<u>1703</u>	1732

$$\log 1.48 = 0.1703$$

For log 3.63

N	0	1	2	3	4

Now that we know how to find a logarithm, we next wish to reverse the procedure. That is, given the logarithm of a number, we want to be able to find the number itself, or as it is called, the *antilog* of the logarithm. We shall illustrate this technique through a series of examples.

Example: The logarithm of a number is 7. Calculate the number. (Another way to say this is: calculate the antilog of 7.)

Answer:

$$\log x = 7$$
$$x = 10^7 = 1 \times 10^7$$

Example: The logarithm of a number is -3. Calculate the number.

Answer:

$$\log x = -3$$
$$x = 10^{-3} = 1 \times 10^{-3}$$

Example: Calculate the antilog of 7.5465.

Answer:

$$\log x = 7.5465$$
$$x = 10^{7.5465} = (10^7)(10^{0.5465})$$

From Table 1-1, we see that the antilog of 0.5465 is 3.52, so

$$x = 3.52 \times 10^7 \qquad (\log 3.52 = 0.5465)$$

Example: Calculate the antilog of -7.1290.

Answer: In this example, we are required to find the antilog of a negative number. If we follow the previous procedure,

$$\log x = -7.1290$$
$$x = 10^{-7.1290} = (10^{-7})(10^{-0.1290})$$

we will not be able to proceed. We cannot find the antilog of -0.1290 in a table, since log tables do not give antilogs of negative numbers. To solve this problem we use a simple mathematical device. We say that

$$10^{-0.1290} = 10^{(-1+0.8710)} = (10^{-1})(10^{0.8710})$$

If we combine this with 10^{-7}, we have

$$x = (10^{-7})(10^{-0.1290}) = (10^{-7})(10^{-1})(10^{0.8710}) = (10^{-8})(10^{0.8710})$$

Since antilog $0.8710 = 7.43$, the answer is

$$x = 7.43 \times 10^{-8}$$

The logarithms we have discussed above are all to the base 10. A set of logarithms can be computed for any base. One that is frequently used as the base for logarithms is 2.71828, a number that is called *e*. Logarithms to the base *e* are called **natural logarithms** and are given the symbol **ln.** Logarithms to the base 10, called **common logarithms,** are given the symbol **log.**

$$\text{If } a = 2.71828^b \qquad \text{then } \ln a = b$$

Tables of natural logarithms can be found in handbooks, but they can be calculated very simply from the following relationship between log and ln for any number.

$$\log x = 2.303 \ln x$$

Thus, the natural logarithm (ln) for any number can be found by multiplying the common logarithm (log) for that number by 2.303.

1-6 Quadratic Equations

In general chemistry we shall sometimes find it necessary to solve problems involving quadratic equations. A **quadratic equation** in algebra is one that can be put into the form

$$ax^2 + bx + c = 0$$

Because of the x^2 term, quadratic equations generally have two solutions. In chemical work, the most convenient way to find these solutions is by use of the quadratic formula

$$x = \frac{-b \pm \sqrt{b^2 - 4ac}}{2a}$$

or

$$x = \frac{-b + \sqrt{b^2 - 4ac}}{2a} \quad \text{and} \quad x = \frac{-b - \sqrt{b^2 - 4ac}}{2a}$$

It might be supposed that having two solutions to an equation would pose a problem, but actually this seldom happens, because one of the two solutions usually turns out to be physically impossible, so that the other one must be correct. When $b^2 = 4ac$, the quadratic equation has only one solution, given by $x = -b/2a$.

Example: Solve this quadratic equation, in which x is the mass of a given compound:

$$2x^2 + 3x - 11 = 0$$

Answer:

$$x = \frac{-3 \pm \sqrt{9 - 4(2)(-11)}}{2(2)} = \frac{-3 \pm \sqrt{9 + 88}}{4} = \frac{-3 \pm \sqrt{97}}{4}$$

$$x = \frac{-3 \pm 9.8}{4}$$

$$x = \frac{-3 - 9.8}{4} = -3.2 \qquad x = \frac{-3 + 9.8}{4} = 1.7$$

Our two answers are -3.2 and $+1.7$. Since we know that x is the mass of a given compound, it cannot be a negative number. Hence, -3.2 is impossible, and the answer is 1.7 g.

Example: Water has its maximum density a few degrees from its freezing point, 0°C. The temperature for the maximum density of water is given by the quadratic equation

$$2.037 \times 10^{-2}T^2 - 1701T + 6.426 = 0$$

Solve this quadratic equation to obtain the temperature at which water has its maximum density.

Answer:

$$T = \frac{1.701 \pm [(1.701)^2 - 4(2.037 \times 10^{-2})(6.426)]^{1/2}}{2(2.037 \times 10^{-2})}$$

$$= \frac{1.701 \pm (2.89 - 0.524)^{1/2}}{0.0407} = \frac{1.701 \pm 1.54}{0.0407}$$

$$T = 3.96°\text{C} \qquad \text{and} \qquad T = 79.6°\text{C}$$

From the conditions of the problem as given, we conclude that the temperature at which water has its maximum density is 3.96°C; not 79.6°C.

1-7 Energy

Although chemistry is the science that deals with matter, chemists long ago learned that it is impossible to study matter without also studying energy. To gain an understanding of the units of energy, we shall begin with velocity and work our way up to energy. **Velocity** is measured as a distance traveled divided by a time interval. Typical units in the metric system are

$$v = \frac{\text{cm}}{\text{s}}$$

If we take an object moving at a constant velocity and make it move faster, we are subjecting the object to an acceleration. At the end of the acceleration, the object is moving at a higher velocity. **Acceleration** is a measure of how fast the velocity has increased. The units are

$$a = \frac{\text{cm/s}}{\text{s}} \quad \text{or} \quad \frac{\text{cm}}{\text{s}^2}$$

Force is defined as the mass of an object times its acceleration.

$$F = ma$$

The units of force are therefore

$$F = \frac{\text{g} \cdot \text{cm}}{\text{s}^2}$$

The unit $\text{g} \cdot \text{cm/s}^2$ is also called a *dyne:*

$$1 \frac{\text{g/cm}}{\text{s}^2} = 1 \text{ dyn}$$

We now introduce the concept of **work.** In chemistry and physics, work does not have the same meaning as it has in daily life. We define work as the product of a force moving over a distance

$$w = F \cdot x$$

The units of work are

$$w = \frac{\text{g} \cdot \text{cm}}{\text{s}^2}(\text{cm}) = \frac{\text{g} \cdot \text{cm}^2}{\text{s}^2}$$

We are now ready to define **energy** as **the capacity to do work.** If energy is the capacity to do work, it must have the same units as work. The units of energy are

$$E = \frac{\text{g} \cdot \text{cm}^2}{\text{s}^2}$$

This is a rather cumbersome compound unit, and scientists have defined a simpler unit, the *erg,* which means the same thing:

$$1 \text{ erg} = 1 \frac{\text{g} \cdot \text{cm}^2}{\text{s}^2}$$

An erg is a very small quantity of energy, and a larger quantity is the *joule:*

$$1 \text{ J} = 10^7 \text{ erg}$$

Later we shall meet still other energy units, including the one most important to us, the calorie (Section 5-1).

The reason we need different kinds of units for energy (although they are all interconvertible) is that there are so many different kinds of energy, and it is convenient to measure them in different ways. The following are some different forms of energy:

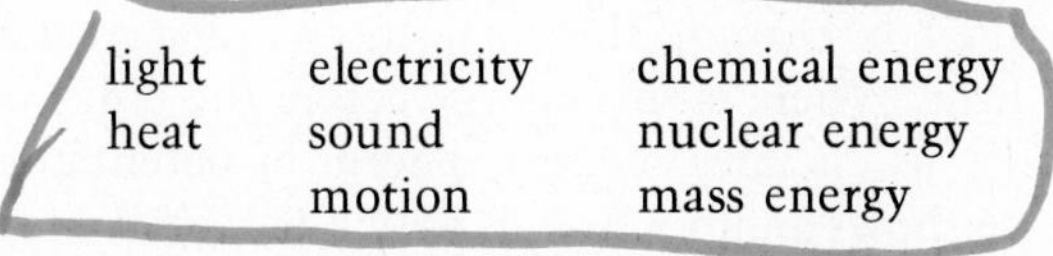

The different forms of energy dealing with matter can be divided into two types: **kinetic energy** and **potential energy.** *Kinetic energy is the energy due to the motion of an object.* It can be calculated by the formula

$$\text{K.E.} = \tfrac{1}{2}mv^2 \qquad m = \text{mass of the object}$$
$$v = \text{velocity of the object}$$

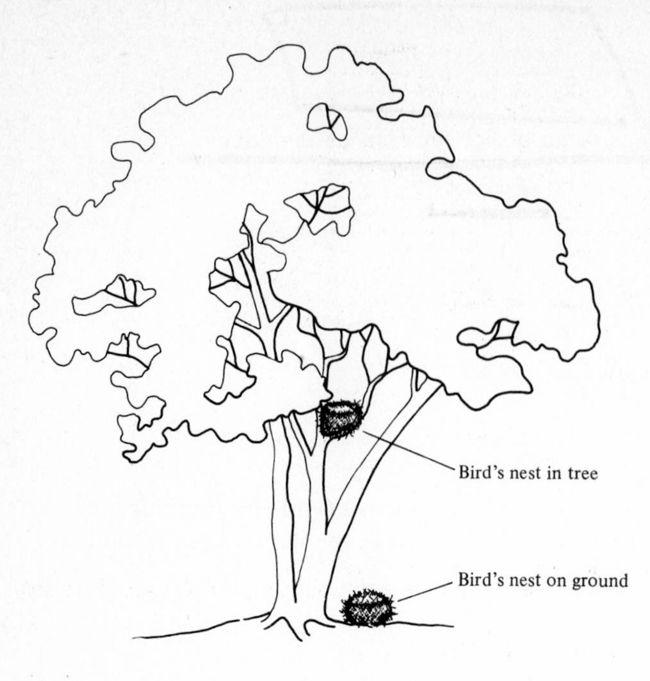

Figure 1-5 The bird's nest in the tree has a potential energy equal to its mass times the height above the ground times the acceleration of gravity. The bird's nest on the ground does not have this form of potential energy.

Note that the unit comes out to be the one we already saw, if m is in grams and v in cm/s.

Example: An object weighing 13.5 g is moving at a velocity of 0.55 cm/s. What is its kinetic energy in ergs and in joules?

Answer:

$$\text{K.E.} = \tfrac{1}{2}(13.5\ \text{g})(0.55\ \text{cm/s})^2$$

$$= 2.0\,\frac{\text{g}\cdot\text{cm}^2}{\text{s}^2} = 2.0\ \text{erg}$$

$$2.0\ \cancel{\text{erg}} \times \frac{1\ \text{J}}{10^7\ \cancel{\text{erg}}} = 2.0 \times 10^{-7}\ \text{J}$$

The other kind of energy related to matter is potential energy. The potential energy possessed by a body arises from its capacity to move or to cause motion. An example is shown in Figure 1-5. The bird's nest in the tree possesses potential energy that comes from its position. Given a slight push, it can fall to the ground. The act of falling is, of course, motion and hence kinetic energy. In falling, the potential energy is changed into kinetic energy. In chemistry we shall be concerned primarily with chemical energy, which is another form of potential energy. For example, both the bird's nest in the tree and the one on the ground, as well as the tree itself, will burn if ignited. In burning, these objects give off a great deal of energy, in the form of heat and light. Where is this energy before the bird's nests are burned? It is stored within them as chemical energy, a form of potential energy.

Figure 1-6 shows another way in which potential energy is converted into kinetic energy, which performs work.

1-8 Gravitational and Electrical Forces

A property possessed by all matter is that of attraction to all other matter by a force, called the **gravitational force.** If m_1 is the mass of object 1 and m_2 the mass of

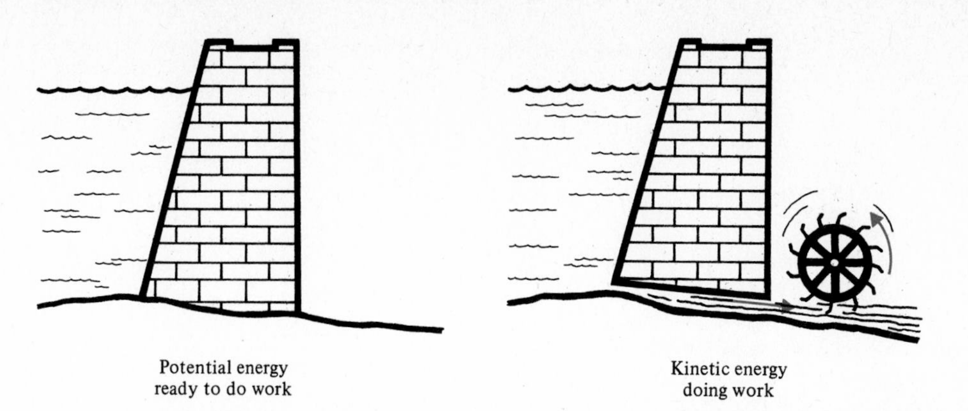

Figure 1-6 The water held back by the dam possesses potential energy, which is converted to kinetic energy when it is released. [Modified from *Energy*, copyright 1976 by American Chemical Society, California Section.]

object 2 and the distance between them is r, then the gravitational force f is given by

$$f \propto \frac{m_1 m_2}{r^2}$$

m_1 = mass of object 1
m_2 = mass of object 2
r = distance between object 1 and object 2
$\propto$ means "is proportional to"

The gravitational force is directly proportional to the product of m_1 and m_2 and inversely proportional to the square of the distance between them. The proportionality constant, called the gravitational constant G, is equal to 6.67×10^{-8} dyn · cm²/g².

There is another force that operates within the universe, called the **electrical force.** This force is like the gravitational force in that it varies inversely as the square of the distance. However, it is enormously more powerful. In fact, it is billions of times more powerful. In the gravitational force everything *attracts* everything else. This is not true in electrical forces. In electrical forces we have two kinds of *charges, positive* and *negative:* like charges repel, whereas unlike charges attract. All matter is made up of positively and negatively charged particles. In most cases, every individual piece of matter contains an equal number of positive and negative charges, so they balance.

When there is an imbalance of charges, the force of attraction or repulsion can be tremendous. This force can be calculated by the use of **Coulomb's law:**

$$F \propto \frac{q_1 q_2}{r^2}$$

where q_1 and q_2 are magnitudes of the charges and r is the distance between the two charges. If q_1 and q_2 are both positive or both negative, then F is positive (a repulsive force). If q_1 and q_2 have opposite signs, then F is negative (an attractive force). The fundamental unit of charge is the *coulomb* (C). The force F is proportional to the two charges q_1 and q_2 and inversely proportional to the square of the distance between them:

$$F = K\frac{q_1 q_2}{r^2}$$

The constant K can be experimentally determined and is

$$9.00 \times 10^{18} \frac{\text{dyn} \cdot \text{cm}^2}{\text{C}^2}$$

Example: Calculate the force of repulsion when two electrons, each of charge -1.60×10^{-19} C, are separated by a distance of 1.00×10^{-8} cm.

Answer:

$$F = K\frac{q_1 q_2}{r^2} \quad \text{and} \quad \begin{aligned} q_1 &= 1.60 \times 10^{-19}\ \text{C} \\ q_2 &= 1.60 \times 10^{-19}\ \text{C} \\ r &= 1.00 \times 10^{-8}\ \text{cm} \\ K &= 9.00 \times 10^{18}\ \frac{\text{dyn} \cdot \text{cm}^2}{\text{C}^2} \end{aligned}$$

Putting these data into the above equation gives

$$\begin{aligned} F &= 9.00 \times 10^{18}\ \frac{\text{dyn} \cdot \text{cm}^2}{\text{C}^2} \times \frac{(-1.60 \times 10^{-19}\ \text{C})(-1.60 \times 10^{-19}\ \text{C})}{(1.00 \times 10^{-8}\ \text{cm})^2} \\ &= \frac{(9.00 \times 10^{18})(2.56 \times 10^{-38})}{1.00 \times 10^{-16}}\ \text{dyn} \\ &= 2.30 \times 10^{-3}\ \text{dyn} \end{aligned}$$

Example: Calculate the ratio of the coulombic force to the gravitational force for the preceding example. The mass of an electron is 9.10×10^{-28} g. The gravitational constant G is equal to 6.67×10^{-8} dyn · cm²/g².

Answer: The force of gravitation is

$$\begin{aligned} F &= \frac{Gm_1m_2}{r^2} = \frac{6.67 \times 10^{-8}\ \text{dyn} \cdot \text{cm}^2/\text{g}^2(9.10 \times 10^{-28}\ \text{g})(9.10 \times 10^{-28}\ \text{g})}{(1.00 \times 10^{-8}\ \text{cm})^2} \\ &= \frac{(6.67 \times 10^{-8})(9.10 \times 10^{-28})(9.10 \times 10^{-28})}{1.00 \times 10^{-16}} = 5.52 \times 10^{-46}\ \text{dyn} \end{aligned}$$

$$\frac{F_{\text{coulombic}}}{F_{\text{gravitational}}} = \frac{2.30 \times 10^{-3}\ \text{dyn}}{5.52 \times 10^{-46}\ \text{dyn}} = 4.17 \times 10^{42}$$

Two electrons repel each other with a force that is approximately 10^{42} times as much as their gravitational attraction.

We do not know the *nature* of these charges. All we know is that there are two opposite kinds, which we call "positive" and "negative," and that like charges attract and unlike charges repel. The phenomenon was originally discovered many centuries ago when it was learned that rubbing a piece of amber with a cloth would make it attractive to small pieces of matter, such as hairs or bits of wool. We still encounter the same type of thing when we walk on a carpet and then touch a piece of metal. Rubbing the amber with the cloth (or the leather of our shoes on the carpet) causes the surface of the rubbed object to become electrically unbalanced, so that it has a surplus of either positive or negative charges, and can attract or repel other objects. We can discharge it by touching it or by touching ourselves to a piece of metal. As we shall see in Chapter 12, the study of electrical charges played an important part in the history of our understanding of the nature of matter. In Section 2-5, we shall discuss the electrical charges present in the elementary constituents of matter.

Problems

For conversion factors, see the inside back cover.

1-1 Define each of the following terms. **(a)** chemistry **(b)** dimensionless number **(c)** compound unit **(d)** density **(e)** significant figure **(f)** dimensional analysis **(g)** logarithm **(h)** natural logarithm **(i)** velocity **(j)** acceleration **(k)** force **(l)** work **(m)** energy **(n)** kinetic energy **(o)** potential energy

1-2 What does it take to get scientists to accept a theory? If they do accept it, does that make it correct?

1-3 What is the difference between mass and weight?

1-4 How many ounces are there in 84.7 gal? How many milliliters in 84.7 liters? Which calculation is easier to do? (1 pint = 16 oz.)

1-5 Change:
(a) 21.5 cm to decimeters and nanometers
(b) 6.44 mg to megagrams and micrograms
(c) 13.5 ml to kiloliters and centiliters
(d) 2.3×10^7 micrometers to meters and Å
(e) 4.1×10^{-5} kg to grams and milligrams
(f) 8.2×10^4 cm^3 to milliliters and liters
(g) 2.54 Å to nanometers, millimeters, and centimeters

1-6 A sample of glycerol, a liquid used as a lubricant, weighs 13.456 g and has a volume of 10.66 ml at 20°C. What is its density at 20°C in g/ml? In g/cm^3? In g/liter?

1-7 The density of the metal bismuth is 9.75 g/ml. What is the volume in milliliters of a piece of bismuth that weighs 251.8 g?

1-8 The antifreeze in your car is probably ethylene glycol, which has a density of 1.11 g/cm^3 at 20°C. **(a)** How many grams of ethylene glycol are there in 100 ml of the substance? **(b)** What volume of ethylene glycol is occupied by 0.230 kg of the substance?

1-9 The density of mercury is 13.6 g/ml. **(a)** What is the density in kg/liter? **(b)** What is the density in g/liter? **(c)** What is the density in kg/ml? **(d)** What volume of mercury contains 144.4 g? **(e)** 100.0 ml of mercury weighs how many grams?

1-10 The density of gold is 19.3 g/ml. Calculate the weight in grams of a pure gold brick 12.0 in long, 5.20 in wide, and 3.10 in high.

1-11 Iodine is a solid with a density of 4.93 g/cm^3 at 20°C. What is the weight in kilograms of a piece of iodine that occupies a volume of 43.2 ml?

1-12 If 1.00 g of butter, density 0.860 g/ml, is thoroughly mixed with 1.00 g of sand, density 2.28 g/ml, what is the density of the mixture?

1-13 What is the density of water in lb/ft^3?

1-14 Convert each of the following to degrees Fahrenheit. **(a)** 5.5°C **(b)** −273°C **(c)** 27°C **(d)** 100°C **(e)** 5000°C

1-15 Convert each of the following to degrees Celsius. **(a)** 98.6°F **(b)** 100°F **(c)** 0°F **(d)** −100°F **(e)** −400°F **(f)** 1000°F **(g)** 32°F

1-16 At what temperature does the number of degrees Celsius equal the number of degrees Fahrenheit?

1-17 How many significant figures are there in each of the following measured numbers? **(a)** 2333 **(b)** 0.023 **(c)** 0.01110 **(d)** 4.3609 **(e)** 40,000 **(f)** 73.001 **(g)** 1001 **(h)** 0.001 **(i)** 49.89099 **(j)** 0.000400 **(k)** 30,000.0 **(l)** 18,930

1-18 How many significant figures are there in each of the following measured numbers? **(a)** 3.2×10^{11} **(b)** 2.405×10^{6} **(c)** 4×10^{-3} **(d)** 7.000×10^{4} **(e)** 8.040×10^{-7} **(f)** 6.02×10^{23}

1-19 Round off each of the following numbers to two significant figures. **(a)** 26.4 **(b)** 3.271 **(c)** 0.037498 **(d)** 46507 **(e)** 0.793926 **(f)** 3.77

1-20 To a scientist, what is the difference between the numbers 6.87 and 6.870?

1-21 What do you think of the following statements? **(a)** A large university library has 5,000,000 volumes. Somebody donates a book. Now they have 5,000,001. **(b)** A radioactive dating experiment carried out in 1970 showed a certain rock to be 27,000,000 years old. In 1979 this rock is 27,000,009 years old. **(c)** *The New York Times* of January 2, 1977, reported that $1,113,623,739 is lost in income each year in the United States because of the murder of males between 25 and 34 years of age (that is, if they had not been murdered, that is how much money they would have earned).

1-22 **(a)** A surveyor calculated the circumference of a square piece of land by measuring one side, which came out to be 324.2 m, and multiplying by 4. What is the circumference and how many significant figures should it have? **(b)** What is the area of the land in m^2?

1-23 Do the indicated mathematical operations on measured numbers, keeping in mind significant figures.

(a) $(73.45/10.0)(7.09)(0.010)$ **(b)** $(7333.3/21.0)(43.02)$ **(c)** $\left[\frac{24.44}{2.3}\right]\left[\frac{6.02}{100.0}\right]$ **(d)** $(4.00)(100)(4.3)$

1-24 Do the indicated calculations on measured numbers, keeping in mind significant figures.

(a) 364.7×8.200 **(b)** $\frac{28.64}{6.0}$ **(c)** $\frac{5.00 \times 1.32}{40\ 652}$

(d) $\frac{2 \times 18.755}{43.617 \times 22.4159}$ **(e)** $\frac{100.00 \times 61 \times 49.717}{7.0}$

1-25 Add each of the following groups of measured numbers.

(a)	(b)	(c)
44.3031	100	96.6
4.202	4.2	100.73
100012.2	0.01	10.0396
1.43	100.034	190
0.00001		7

1-26 **(a)** Add the numbers 31,264 and 0.141, keeping in mind significant figures. **(b)** Subtract the smaller number from the larger. **(c)** What happens when a number is added to or subtracted from a number very much larger than itself?

1-27 Convert each of the following as indicated.
(a) 14 tons to kilograms and megagrams (a megagram is more often called a metric ton)
(b) 37 Å to inches **(c)** 13.5 km to miles and feet **(d)** 95 miles to kilometers and meters
(e) 385 ft to meters **(f)** 16 cubic feet to liters and quarts **(g)** 2.5×10^{-2} liter to gallons and cubic feet **(h)** 3.6×10^{9} erg to calories and joules **(i)** 23.5 atm pressure to torr and pascals **(j)** 83 kilocalories to ergs and Btu

1-28 Make the following conversions. **(a)** 26.0 miles to centimeters **(b)** 100.0 miles/h to meters/minute and to centimeters/second **(c)** 1.34 lb to ounces and grams **(d)** Your weight in kilograms. **(e)** Your height in centimeters. **(f)** The speed of sound in dry air at 0°C is 1087 ft/s. What is it in cm/s? **(g)** A supersonic airplane can travel 1200 miles/h or more. What is this in cm/s?

1-29 Change each of the following as indicated. **(a)** 73.47 kg to grams and milligrams. **(b)** 1.07 cm to meters and millimeters. **(c)** 155.3 g to milligrams and kilograms **(d)** 2500.0 mm to meters and centimeters **(e)** 123.7 cm^3 to milliliters and liters **(f)** 100.0 ml to liters and cm^3

1-30 In April 1977 coffee was selling for $4.00 per pound. What was the price of coffee per gram?

1-31 If the price of gasoline in Germany is 1.08 marks/liter, what would this be in U.S. dollars per liter? In dollars per gallon? (Assume that exactly 2.3 marks = 1 U.S. dollar.)

1-32 Pressure is equal to force divided by area:

$$P = \frac{F}{A}$$

A 115-lb woman exerts a force of 5.12×10^7 dyn. If she stands on the heels of her shoes, what pressure, in dyn/cm^2, does she exert on the ground? (Assume that each heel is a circle with a 1.60-in diameter; the area of a circle is πr^2.)

1-33 From each of the following pairs of numbers, select the one that is smaller.
(a) 0.000011, 1.23×10^{-4}
(b) 100000, 3×10^5
(c) 0.0001, 0.00011
(d) 7.9×10^4, 79×10^4
(e) 8.4×10^7, 8.4×10^{-7}
(f) 1024, 10^4
(g) 10^{-23}, 10^{-24}
(h) 99×10^4, 9.9×10^{-4}
(i) 6.02×10^{23}, $60{,}235 \times 10^{19}$

1-34 Write each of the following numbers in exponential form. **(a)** 100 **(b)** 0.000134 **(c)** 7430 **(d)** 110 **(e)** 0.0000000349 **(f)** 639.5 **(g)** 4.732 **(h)** 7 **(i)** 400000000000000 **(j)** 2222.2

1-35 Rewrite each of the following exponential numbers so that the coefficient is between 1 and 10. **(a)** 37.55×10^8 **(b)** 0.00562×10^7 **(c)** 0.00034×10^2 **(d)** 83.6×10^{-4} **(e)** 7721.3×10^{-2}

1-36 Perform the indicated calculations on measured numbers and write your results with the proper number of significant figures.
(a) $(2.775 \times 10^8) + (2.14 \times 10^6) + (1.32844 \times 10^{10}) + (6.5 \times 10^7)$
(b) $(1.6 \times 10^{-16}) - (1.214 \times 10^{-14})$
(c) $(1.106 \times 10^{-6}) + (3.75 \times 10^{-5}) + (2.8732 \times 10^{-4})$
(d) $(3.7 \times 10^5)(21.0 \times 10^2)$
(e) $(2.41 \times 10^6)(8.3 \times 10^{-4})$
(f) $(21.9 \times 10^{-7})(1.44 \times 10^3)$
(g) $(6.6 \times 10^{-4})(9.7 \times 10^{-6})$
(h) $(13.5)(2.1 \times 10^{-8})$
(i) $\dfrac{3.24 \times 10^{11}}{6.10 \times 10^4}$
(j) $\dfrac{9.33 \times 10^6}{3.7 \times 10^9}$
(k) $\dfrac{2.14}{3.7 \times 10^6}$
(l) $\dfrac{8.40 \times 10^8}{9.20 \times 10^{-4}}$
(m) $\dfrac{3.665 \times 10^{-7}}{2.0 \times 10^5}$
(n) $\dfrac{1.4}{3.207 \times 10^{-11}}$
(o) $\dfrac{1.00 \times 10^{-6}}{3.00 \times 10^{-21}}$
(p) $\dfrac{(3.6 \times 10^5)(8.3761 \times 10}{(2.14 \times 10^{-11})(5.072 \times 10^{10})}$

1-37 Perform the indicated calculations on measured numbers and write your results with the proper number of significant figures.
(a) 732×238.36 **(b)** $0.0127 + 14.32 - 100$
(c) $71.6 - 0.2721$ **(d)** $(4.991 \times 10^3) + (1.20 \times 10^2) - (8.08 \times 10^4)$
(e) $\dfrac{7.63 \times 10^{15}}{6.66 \times 10^{-14}}$ **(f)** $\dfrac{2.497 \times 10^{10}}{1.12 \times 10^{20}}$
(g) $\dfrac{(2.74 \times 10^7)(9.11 \times 10^{28})}{(4.93 \times 10^{-3})(5.753 \times 10^{17})}$

1-38 What is the logarithm of each of the following numbers?
(a) 10^7 **(b)** 10^{-11}
(c) 70000000 **(d)** 0.0000000111
(e) 1.46×7.44 **(f)** $10^3 \times 10^5$
(g) $10^8 \times 10^{-4}$ **(h)** $10^{-3}/10^{-7}$
(i) $(4.77 \times 10^8)(7.01 \times 10^{-9})$ **(j)** $\dfrac{3.60 \times 10^{-4}}{9.61 \times 10^7}$
(k) $\dfrac{(2.01 \times 10^3)(6.02 \times 10^{23})}{6.71 \times 10^{12}}$ **(l)** $\dfrac{(1.73 \times 10^4)(0.021)}{2.91 \times 10^{-12}}$

1-39 Take the antilog of each of the following numbers. **(a)** 14 **(b)** 100 **(c)** 1.0 **(d)** -3.0 **(e)** 0.0 **(f)** 5.3201 **(g)** 2.37701×10^1 **(h)** 4.7404 **(i)** -3.3391 **(j)** -2.1373

1-40 Find the natural logarithm (ln) of each of the following numbers. **(a)** 3.7 **(b)** 2.4×10^3 **(c)** 7.1×10^{-6} **(d)** 0.0 **(e)** 1.00

1-41 Solve each of the following calculations by use of the quadratic formula.
(a) $3.75x^2 + 21.6x + 1.77 = 0$ **(b)** $7.4x^2 - 13x - 8.55 = 0$
(c) $x^2 - 6x - 4.33 = 0$ **(d)** $x^2 + 7x - 15 = 0$
(e) $2x^2 + 0.62x - 0.033 = 0$ **(f)** $-x^2 + 7x + 53 = \frac{11}{3}$

1-42 One kind of potential energy is shown in Figure 1-6. Name some other kinds.

1-43 How much force in dynes does a 2500-lb automobile have when it accelerates from 10.0 miles/h to 30.0 miles/h in the course of 20.0 s?

1-44 How much kinetic energy, in joules and in ergs, is possessed by the 2500-lb automobile in Problem 1-43 when it is traveling 30.0 miles/h?

1-45 An electron has a mass of 9.11×10^{-28} g and a negative charge of 1.60×10^{-19} coulombs. A proton has a mass of 1.67×10^{-24} g and a positive charge of 1.60×10^{-19} coulombs. They are attracted to each other by the gravitational force of their masses as well as by the electrical attraction of their charges. Calculate the ratio of the coulombic force to the gravitational force.

1-46 A point charge of one coulomb is separated from a similar point charge of one coulomb, exerting a force of repulsions of 3.00×10^{12} dyn. Calculate the distance between the two point charges.

Additional Problems

1-47 The density of pure copper is 8.95 g/ml and that of mercury is 13.6 g/ml. If 100 ml of copper is placed on the left-hand pan of an equal-arm balance (Figure 1-1), how many milliliters of mercury must be placed on the right-hand pan so that the weights will be equal?

1-48 The energy available from a normal diet may be taken as 2500 kilocalories per 24-h period. Assuming that a person expends all this energy in activity during 24 h (and therefore gains no weight), how many electron volts does he utilize per second on the average?

1-49 In 1684, Isaac Newton announced his three laws of motion. Why do you suppose these laws were soon accepted by almost all scientists? In 1905, Albert Einstein published his theory of relativity, which stated that there were conditions under which Newton's laws of motion were not valid. Why do you suppose his theory was soon accepted by almost all scientists?

1-50 A candy thermometer has the following markings:

240°F	soft ball
260°F	hard ball
300°F	hard crack
350°F	deep fry

Convert each of these temperatures to degrees Celsius.

1-51 A nugget consisting of gold mixed with quartz, weighing 75.5 g, has a density of 8.20 g/ml. What is the percentage by weight of gold? Density of gold = 19.3 g/ml. Density of quartz = 2.65 g/ml.

1-52 The energy (E) of light is related to its frequency (ν) by the equation $E = h\nu$, where h, called Planck's constant, has the value 6.63×10^{-27} erg · s. Violet light has an energy of 5×10^{-19} joules. What is the frequency of this light, in s^{-1}?

1-53 A container that can hold 350 g of water can hold only 276 g of ethanol. What is the density of ethanol?

1-54 A student had to solve the following equation, where all the quantities were measured numbers:

$$x = \frac{13.774 \times 1.600}{0.000620 \times 180.00}$$

Her calculator gave the answer x = 197.4767. What answer should she report to her chemistry teacher?

1-55 The surface tension of a liquid is determined by how high the liquid rises above its surface when it is placed in a thin tube. The formula is

$$\gamma = \frac{rhdg}{2}$$

where γ = surface tension, r = the radius of the tube, h = the height of the rise, d = the density of the liquid, and g is a constant equal to 9.807 m/s². What is the surface tension of glycerol (density 1.11 g/ml) in ergs/cm² if the liquid rises 5.82 mm in a tube whose diameter is 2.0 mm?

1-56 The great mathematician Archimedes (287–212 B.C.) discovered the concept of density. There is a legend that the king of Syracuse (a Greek city in ancient Sicily) asked Archimedes to determine whether or not a certain crown was pure gold. He did it by measuring the density of the crown and comparing it with that of genuine gold. **(a)** If the crown weighed 4.250 kg and had a volume of 325 ml, was it gold? (density of pure gold = 19.3 g/ml.) **(b)** If you were given this problem, how would you measure the volume of the crown? **(c)** The legend says that the crown in Syracuse was not pure gold. What do you think happened to the man who sold it to the king?

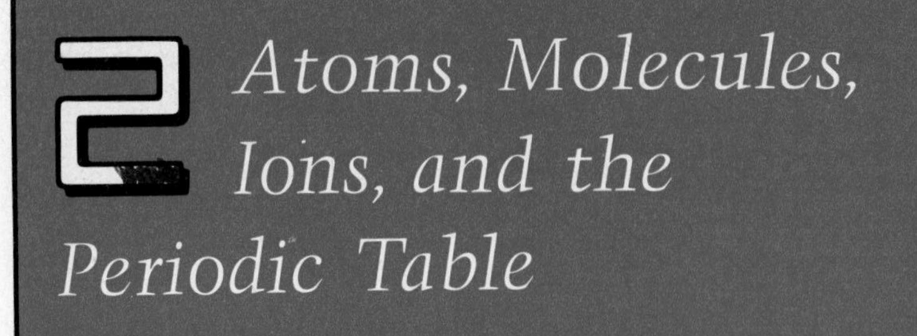

2 Atoms, Molecules, Ions, and the Periodic Table

Dmitri Ivanovich Mendeleev, 1834–1907

2-1 Early Thinking About Atoms

If we take a piece of solid matter, say, a sheet of paper or a rock, or a sample of a liquid such as water, it is obvious that it is possible to divide it into two parts and then to divide each part into two parts again, and so on. How long can this process be continued? Assuming that there existed fine-enough cutting tools, could it go on forever or would there finally come a time when the process would have to stop because we would obtain particles that could no longer be divided? In other words, is matter infinitely divisible or is it made of very tiny particles that are indivisible? It is likely that for many thousands of years some human beings, with more curiosity than their fellows, have considered this question, but the earliest record we have is of the thinking of certain Greeks of the fifth century B.C. Two schools of thought arose. One, under the leadership of Zeno of Elea (born about 450 B.C.), believed and taught that matter was infinitely divisible; the other, whose leading figure was Democritus (~470–380 B.C.), held the view that small indivisible particles, which they called **atoms,** made up all matter; that besides atoms nothing existed except the spaces between atoms. Democritus further taught that atoms of different substances had different properties and that these explained the properties of bulk matter (for example, the atoms of water were smooth and round, while the atoms of fire were thorny).

Today we can look at these ideas and know that Zeno was wrong and Democritus right about the existence of atoms. But before we credit Democritus with the invention of the atomic theory we must realize that neither he and his followers nor Zeno had any evidence whatsoever for their theories. Both schools of thought were based on pure intuition, the notion that a theory is correct because it seems to be correct. Furthermore, neither school had any method of *testing* its theory, nor were they interested in doing so. The difference between the theories of Zeno and Democritus, and that of Dalton, which we will consider shortly, is that the latter was based on evidence that could be tested by anyone who wanted to.

For those Greek and Roman thinkers who believed in the existence of atoms, it was a logical step to the idea that there are only a few different kinds of atoms, and that the many forms of matter we see around us are composed of different proportions or arrangements of these. From this, there grew the idea that there were four elements; earth, air, fire, and water, to which Aristotle (384–322 B.C.) added a fifth: quintessence, which he suggested made up the matter of the heavens. Over the next two thousand years or so, the concept of elements gradually changed, but it was not until 1789 that the first scientific definition of an element was put forth, by Antoine Laurent Lavoisier (1743–1794), who defined an element as *a substance that has not yet been broken down into simpler substances.* Lavoisier also supplied the first list of elements. In this list were 33 elements, 31 of which are still recognized as elements today (the other two were light and heat, which we now know to be forms of energy, not elements at all). Lavoisier's definition was very important because it led to Dalton's atomic theory (to be discussed in the next section) and because it supplied goals for a whole century of chemists to work on—to discover all the elements and their properties.

2-2 Dalton's Atomic Theory

The atomic theory which John Dalton (1766–1844) proposed in 1803 was based chiefly on two laws that had been demonstrated a few years earlier, and on Lavoisier's definition of an element. The two laws were:

1. *The law of conservation of matter*, first demonstrated by Lavoisier,* states that **matter can neither be created nor destroyed.** This had been implicitly believed by many earlier scientists, but Lavoisier made many experiments, consisting chiefly of carrying out chemical and physical changes in sealed vessels and demonstrating that no weight gain or loss occurred.
2. *The law of constant composition*, first put forth by another French chemist, Joseph Proust (1754–1826), states that the elements which make up a given compound are always present in the same proportions. For example, the compound mercuric sulfide is found in nature as a mineral called cinnabar. It can also be made in the laboratory by passing the gas hydrogen sulfide through a solution of mercuric nitrate. According to the law of constant composition, the composition by weight of mercuric sulfide is always the same whether it is manufactured in a laboratory or found in nature; and no matter where found in nature. Analysis of a sample of mercuric sulfide, no matter how or where obtained, always shows 86.2% (by weight) mercury and 13.8% sulfur. A consequence of this law is that one can predict the *relative weights* of elements that combine directly to form compounds.

   Despite the fact that the law of constant composition seems fairly obvious and even natural to us, it was not immediately accepted by all the chemists of the time, some of whom, notably Claude Louis Berthollet (1748–1822), argued that matter has a variable composition. However, in a few years the evidence for the law became overwhelming and it was accepted by all chemists.

We now have two laws that have been tested in the laboratory and shown to be valid under all conditions. John Dalton proposed a model that would explain why these two laws operate the way they do. The model becomes stronger if it also can predict other phenomena. Dalton's theory contains the following statements:

1. All matter is composed of tiny indivisible particles which Dalton called (by the name Democritus had given them) *atoms*. Atoms can neither be created nor destroyed.
2. All atoms of any element are *identical* and have identical properties. Most important, they have the *same weight*.
3. Atoms of different elements are *different* and have different properties. Most important, they have *different weights*.
4. Atoms combine to form molecules.†

Dalton showed that both the law of conservation of matter and the law of constant composition could be readily explained by his atomic theory. If a sample of matter is

*Actually, it had been demonstrated earlier, by a Russian scientist named Mikhail Vasilievich Lomonosov (1711–1765), but his work was not published in Western Europe and, unfortunately, had no effect on the progress of science.

†This is the modern term. Dalton called them "compound atoms."

made of molecules, which in turn are made of atoms, then in a chemical change, molecules may break apart into their constituent atoms, and these may recombine to form different molecules, but the total weights of the products must be the same as that of the starting materials. If atoms are indivisible and indestructible, matter can neither be created nor destroyed. This clearly satisfies the law of conservation of matter. As for the law of constant composition, if all the molecules of a given substance are made up of the same atoms in the same ratios, and if each of these atoms has a definite weight, then the elements of that compound must always be present in the same proportions by weight. For example, take the compound potassium bromide. If the molecules of this compound are made up of one atom of potassium and one atom of bromine (Dalton did not know this, but the argument would hold even if the atoms were present in any other ratio, as long as all the molecules had the same ratio), and if one atom of bromine has about twice the weight of one atom of potassium (this is the case), then potassium bromide must consist, by weight, of 67% bromine and 33% potassium, and no other weight ratio is possible for this compound.

As noted above, Dalton based his atomic theory on the two laws mentioned. Once a theory is advanced, it must be tested. The normal way a theory is tested is by using it to predict the outcome of experiments that have not yet been tried. If the outcome of these experiments is as predicted by the theory, then our belief in the theory is strengthened. If this happens enough times, our confidence in the theory becomes great indeed. However, if the results of any of the experiments are not as predicted by the theory (assuming that the results can be reproduced in several laboratories, because it is always possible for one investigator to make a mistake), the theory must be modified or discarded, even if, prior to these experiments, our confidence in the theory was great indeed. Many times in the history of science it became necessary to discard theories that had been widely accepted because the results of new experiments were in conflict with them.

Dalton tested his theory by using it to predict a third law, the *law of multiple proportions*. At that time, a number of cases were known in which two or more different compounds were made up of exactly the same elements, although in different proportions by weight. Dalton reasoned that if his theory were correct, these compounds contained the same atoms but in different ratios. For example, two gases, methane and ethylene, were known to consist entirely of carbon and hydrogen. Dalton suggested that one compound might be CH, and the other CH_2, C_2H, C_2H_3, or some other simple ratio of small whole numbers. If this were so, the percentages by weight of carbon and hydrogen in methane should be very simply related to the percentages of these elements in ethylene, as the following analysis will show.

To demonstrate the argument, it is not necessary to know the actual atomic weights (or even the weight ratios) or the formulas of the compounds. So we shall invent a hypothetical situation. Suppose that a hydrogen atom has exactly three times the weight of a carbon atom.* Further suppose that the two compounds might have the formulas CH, C_2H, CH_2, and C_2H_3. Figure 2-1 shows that in CH, 1 g of carbon is combined with 3 g of hydrogen, while in the other cases the carbon/hydrogen weight ratios are 1:1.5, 1:6, and 1:4.5, respectively. Now the thing to notice is that the number of grams of hydrogen combined with 1 g of carbon in the four hypothetical cases (Table 2-1) is 3, 1.5, 6, and 4.5, respectively, and that *all these numbers are simple multiples of 1.5*. A similar result would have been obtained if we had used any other weight ratio for carbon and hydrogen or any other set of simple formulas. Whatever the actual formulas of methane and ethylene, if we find out how much hydrogen, by weight, is present for each gram of carbon, then *the*

* Actually, a carbon atom weighs about 12 times as much as a hydrogen atom, but the argument is not altered by the use of any fictitious ratio.

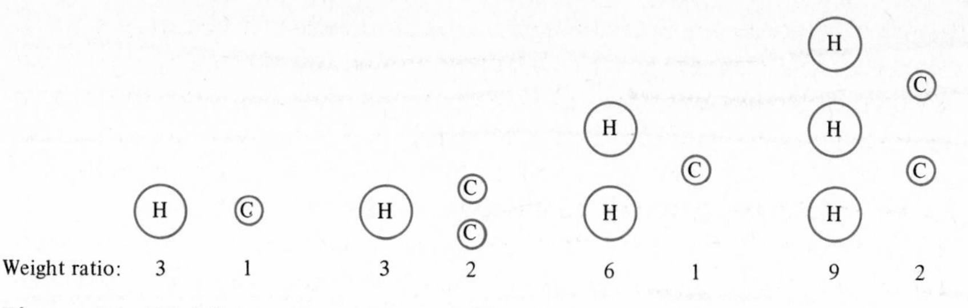

Figure 2-1 Weight ratios corresponding to hypothetical molecules containing carbon and hydrogen.

weight of hydrogen per gram of carbon in one compound should be a simple multiple of the weight of hydrogen per gram of carbon in the other. We are now ready to inspect the actual weight ratios of the real methane and ethylene to see if this statement holds true. In the actual methane the carbon/hydrogen weight ratio is 3:1 while in ethylene it is 6:1. Note that the second number is twice the first, and the relationship holds.

At that time, there were two known oxides of carbon, called fixed air (now known as carbon dioxide) and carbonic oxide (now carbon monoxide). Dalton's argument for the existence of the law of multiple proportions was greatly strengthened by the fact that, although analytical chemistry was yet in its infancy, these two compounds had already been analyzed in other laboratories and he could apply his theory to them. Lavoisier had shown that fixed air contained 28% carbon and 72% oxygen by weight, while other chemists had analyzed carbonic oxide and found 44% carbon and 56% oxygen. A simple calculation will show that fixed air must then contain 2.57 g of oxygen for each gram of carbon, whereas in carbonic oxide there is 1.27 g of oxygen per gram of carbon (Table 2-2). The ratio 2:57:1.27 is 2, within experimental error, so that once again a small whole number ratio is obtained. The law of multiple proportions may now be stated: When two elements unite to form more than one compound, the weights of one element that combine with a *fixed weight of the other* are in the ratio of small whole numbers.

In the history of science, many theories later accepted as correct were given hostile receptions when they were first proposed.* Dalton's atomic theory won quick acceptance by virtually all contemporary chemists, for the simple reason that it predicted a new law; and that law was found by many independent scientists to hold true. Today (as we shall see in this chapter) we know, from evidence which was not

*An example is Avogadro's law (Section 4-6), which was vehemently attacked soon after it was proposed in 1811 by many contemporary chemists, including Dalton.

Table 2-1 Hydrogen/Carbon Weight Ratios in Four Hypothetical Cases

	Weight ratios: hydrogen/carbon	
CH	3:1	3:1
C_2H	3:2	1.5:1
CH_2	6:1	6:1
C_2H_3	9:2	4.5:1

Table 2-2 Carbon/Oxygen Ratios in Fixed Air and in Carbonic Oxide

	Weight ratios: carbon/oxygen	
Fixed air	28:72	1:2.57
Carbonic oxide	44:56	1:1.27

available in the early part of the nineteenth century, that the first three statements of Dalton's theory were incorrect. No matter; the theory was extremely useful in its time and the progress of chemistry would have been most difficult without it.

2-3 Atomic and Molecular Weights

The most fruitful part of Dalton's atomic theory was the concept of **atomic weight.** If the weight of every atom could be known, it would be possible to predict the weights of substances undergoing chemical change and of new substances formed (in Chapter 3, we will see how this is done). Dalton realized that there was no way, at that time, that the actual weight of any atom could be determined, but he also realized that the *actual* weight of atoms was much less important than their weights *relative* to other atoms. If the atomic theory is correct, relative weights can be obtained from the analysis of binary compounds (a **binary compound** consists of only two elements). For example, we saw in Section 2-2 that mercuric sulfide contains 86.2% mercury and 13.8% sulfur. If mercuric sulfide contains one atom of mercury for each atom of sulfur (that is, if the formula is HgS), the ratio by weight of an atom of mercury to an atom of sulfur is 86.2:13.8, or 6.25:1, and each atom of mercury weighs 6.25 times as much as each sulfur atom.

It follows that by the analysis of a large number of compounds one could determine the relative weights of a large number of elements in a similar manner.* The simplest and most convenient way to express the results of such determinations is to choose one element as a standard, give it an arbitrary atomic weight, and relate all the others to it. For example, we might choose sulfur as the standard and assign it an atomic weight of 1. The atomic weight of mercury would then be 6.25, and the atomic weight of a third element would be determined by analyzing a binary compound of it with either mercury or sulfur. At this point, we would know the relative atomic weights of three elements, and we could determine the atomic weight of a fourth by analyzing a binary compound of it with any of the first three. The process could be continued until all atomic weights were known.

During the nineteenth century some chemists chose hydrogen as the standard, while others chose oxygen, but today all these disagreements have been ended. A table of modern atomic weights is given on the inside back cover. The modern table is based on neither hydrogen nor oxygen, but on something else, as we shall discuss in Section 2-6.

It is important to realize that analytical procedures of this kind give us the actual weight ratios of any two *atoms*, even though atoms are so tiny that we cannot weigh an individual atom. Consider a circle ○ and a square □. Assume that the circle weighs 1 g and the square weighs 16 g. The relative weight of the square compared to the circle is 16.

$$\frac{\square}{\bigcirc} = \frac{16\text{ g}}{1\text{ g}} = 16$$

If we have two of each, their relative weights are

$$\frac{\square\square}{\bigcirc\bigcirc} = \frac{32\text{ g}}{2\text{ g}} = 16$$

*Note that the ratio 6.25:1 would not be correct if the actual ratio of atoms in mercuric sulfide were HgS_2, Hg_2S, or some atomic ratio other than 1:1. Since Dalton and the other chemists of his time did not know what the actual ratio was, for this or other compounds, any table of atomic weights produced in their time could only be the result of a considerable amount of guess work. In fact, different chemists made different guesses, and a number of competing tables were used until fairly late in the nineteenth century, when as a result of various advances in chemical knowledge, it became possible to be sure of atomic ratios. From that time on, only one table of atomic weights has been in use, although the numbers are slightly changed from time to time to reflect more accurate analyses.

If we have 100 of each, we get

$$\frac{\square\cdots\square}{\bigcirc\cdots\bigcirc} = \frac{1600\text{ g}}{100\text{ g}} = 16$$

The point should be obvious that as long as we have equal numbers of squares and circles, their *relative* weights are 16:1. Of course, objects other than squares and circles behave the same way—most important, the ones that concern us here, atoms. Knowing the *relative* weights of atoms does not give us any information about *actual* weights of the atoms. Because atomic weights are relative numbers, the unit used to express them is not a real unit (such as grams or seconds) but is simply called *amu*, or *atomic mass unit*.

The fourth statement in Dalton's atomic theory is that atoms combine to form molecules. Dalton did not know what it is that holds atoms together in molecules; he simply assumed that some kind of "glue" was present. Today we know that atoms are held together by forces called *chemical bonds*. We shall discuss the nature of these bonds in Chapter 14. We may define a **molecule** as an uncharged entity in which atoms are held together by chemical bonds. The sum of the atomic weights of all the atoms in a molecule is called the **molecular weight** of that compound.

required definition →

Some molecules consist of only one atom (**monatomic molecules**). Examples are helium, neon, and mercury. In such cases there are no chemical bonds. Other molecules consist of two or more atoms, in which case the atoms are attached to each other by chemical bonds.

The **molecular formula** of a chemical compound shows which atoms are present and how many. Each molecule of water contains one atom of oxygen and two of hydrogen. The formula of water is H_2O. Acetic acid, a compound found in vinegar, has the formula $C_2H_4O_2$. In this case each molecule has two carbon, four hydrogen, and two oxygen atoms.

Example: What is the molecular weight of osmium tetroxide, OsO_4?

Answer: From the modern atomic weight table

Os:	190.2	190.2
O:	15.9994 × 4 =	63.9976
		254.2

mol wt of OsO_4 = 254.2 amu

Example: What is the molecular weight of ammonium phosphate, $(NH_4)_3PO_4$?

Answer: From the table

N:	14.0067 × 3 =	42.0201
H:	1.0079 × 12 =	12.0948
P:	30.97376	30.97376
O:	15.9994 × 4 =	63.9976
		149.0863

mol wt of $(NH_4)_3PO_4$ = 149.0863 amu

2-4 Elements, Compounds, Mixtures

We can divide all matter into three categories, depending on the nature of the atoms and molecules making it up.

* In an **element,** *all the atoms are the same.* This does not necessarily mean that

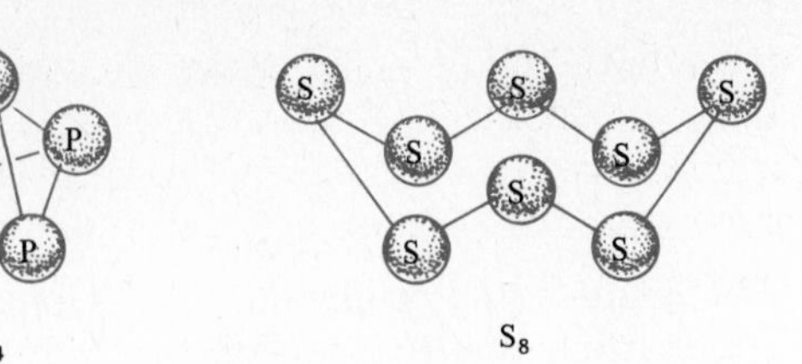

Figure 2-2 Structures of some elements.

only monatomic molecules are present. The element neon does consist of monatomic molecules (formula Ne), but oxygen consists of diatomic molecules; that is, each oxygen molecule is made up of two oxygen atoms (formula O_2). Other elements have even more atoms per molecule (for example, P_4, S_8). But no matter how many atoms per molecule, an element contains only one kind of atom (Figure 2-2).

In a **compound,** *each molecule contains two or more different atoms,* but all molecules are the same. For example, water consists entirely of H_2O molecules—each molecule contains two hydrogen atoms and one oxygen atom, held together by chemical bonds. Similarly, cane sugar consists of $C_{12}H_{22}O_{11}$ molecules (Figure 2-3).

In a **mixture,** *the molecules are not the same.* For example, the two elements iron and sulfur are both solids. Black iron powder and yellow sulfur powder can be mixed in any proportions. In all cases we will have a mixture of the two elements, and can actually see tiny yellow and black particles. This is in contrast to the compound iron sulfide FeS, in which we can see no yellow particles, and in which the iron and sulfur are chemically combined in only one weight ratio.

Both elements and compounds are **homogeneous,** meaning that all the molecules present are the same. The term **substance** is often applied to any pure element or compound. A mixture is not a substance; in fact, it is a mixture of substances. Mixtures may be homogeneous, which means that the *molecules* are thoroughly mixed, or **heterogeneous,** in which case the molecules are not thoroughly mixed. For example, a mixture of sand and sugar is heterogeneous. Although both sand and sugar are present in the mixture, most of the molecules of sugar are next to other molecules of sugar, and there is no thorough mixing of molecules. Homogeneous mixtures are called **solutions,** and we will discuss them further in Chapter 7.

Note that our modern definition of element (a substance all of whose atoms are the same) is not the same as Lavoisier's (a substance that has not yet been broken down into simpler substances). The reason is that we know much more about it then Lavoisier did. He could not be sure that any of his elements would not someday be broken down into simpler substances. Today we can be reasonably certain that the substances we call elements cannot be broken down further.

The definitions of element, compound, and mixture give rise to another set of definitions: physical and chemical change.

A **chemical change** is one in which molecules change their structures. Take, for example, the burning of bottled gas (propane). Each propane molecule has the

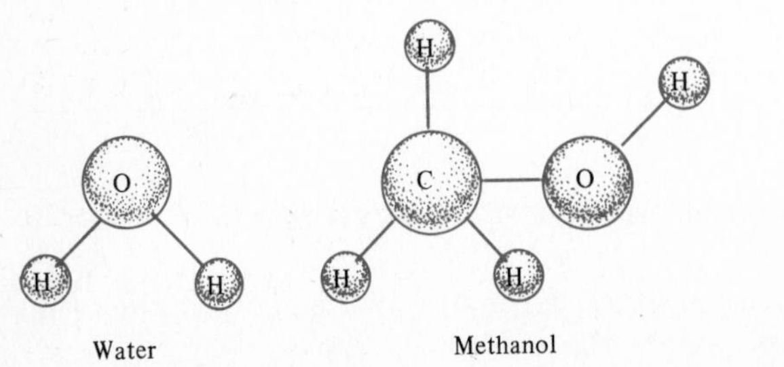

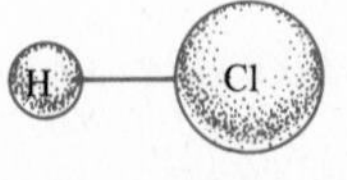

Figure 2-3 Structures of some compounds.

formula C_3H_8. When it burns, it combines with oxygen in the air (O_2). In the process, both the C_3H_8 molecules and the O_2 molecules change their structures (the chemical bonds are broken). The atoms rearrange themselves into new molecules of carbon dioxide (CO_2) and water (H_2O). A chemical change is also called a **chemical reaction.**

A **physical change** is one in which the molecules remain intact. Chemical bonds are not broken. Many physical changes are **changes of state:** a liquid freezes to become a solid; a gas condenses to become a liquid, and so on. Some other physical changes are the magnetizing of a piece of iron, the grinding of coffee beans, and the mixing of two pure liquids.

All substances can be made to undergo chemical and physical changes, each in its own way. Those particular chemical changes which a substance can undergo are called its **chemical properties.** For example, one of the chemical properties of propane is that it burns (undergoes a chemical reaction with oxygen). Each compound and element has its own set of chemical properties. Properties not involving chemical changes are called **physical properties.** Typical physical properties include color, odor, density, and melting point. In this book we shall be introduced to the chemical and physical properties of many substances.

2-5 The Composition of Atoms

When the first tables of atomic weights appeared in the early part of the nineteenth century, it was noted that most atoms had weights which were integers or very close to integers, at least by the comparatively crude analyses of that time.* Some chemists of the time felt that this could not be a coincidence, and one of them, William Prout (1785–1850), suggested in 1815 that atoms of all elements were multiples of hydrogen atoms (this is known as *Prout's hypothesis*). For example, the atomic weight of nitrogen is 14, and Prout postulated that a nitrogen atom is therefore made up of 14 hydrogen atoms, somehow held together.

Prout's hypothesis was not accepted by most chemists of his time, largely because it did not explain why some elements (most notably chlorine, with an atomic weight of 35.5) had atomic weights that were definitely *not* integers. Although nobody had a better explanation of why so many atomic weights were integral, Prout's hypothesis could not be the answer. The fact that many atomic weights are integral or nearly so simply could not be explained with the knowledge then held, and so, as has happened many times during the development of science (and is still happening today), an inexplicable fact was simply noted and filed away until more evidence would come to light, when perhaps it could be explained.

Today we know (the evidence will be discussed in Chapter 12) that atoms are not indivisible [therefore statement 1 of Dalton's theory (Section 2-2) is incorrect]. Atoms consist of an inner **nucleus,** made up of particles called **protons** and **neutrons,** and an outer cloud, made up of other particles called **electrons.** These three, called **elementary particles,** make up all matter.† Protons and neutrons, which are found only in the nucleus, have approximately the same mass, but electrons are much lighter, with a mass about 1/1835 that of a proton or neutron. Almost all the mass of an atom is therefore concentrated in the nucleus. Because they are found in the nucleus, protons and neutrons are collectively referred to as **nucleons.**

*Even with today's much more accurate values, a glance at the atomic weight table will show that many values are very close to integers.

†There are other elementary particles (for example, positrons, mesons, quarks, neutrinos), but they are not found in ordinary matter under ordinary conditions. They are produced, for example, when matter is bombarded with high-energy particles, as in a cyclotron (see Chapter 23).

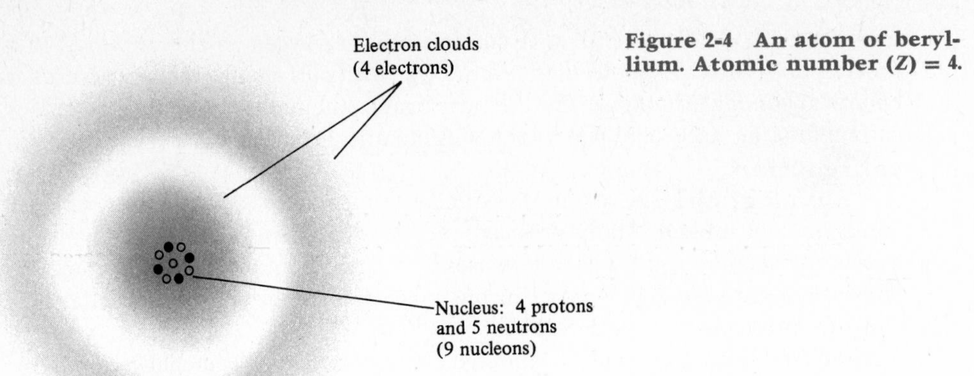

Figure 2-4 An atom of beryllium. Atomic number (*Z*) = 4.

The other important property of elementary particles is their electrical charge. Neutrons have no charge, but protons have a positive charge and electrons a negative charge. Despite the great difference in their masses, the charge on an electron is equal in magnitude to the charge on the proton, though opposite in sign. Figure 2-4 shows a typical atom. Table 2-3 summarizes the masses and charges of the three elementary particles.

The number of protons in an atom is called the **atomic number** and *defines* the element. Thus, any atom with 6 protons is a carbon atom; if it has 92 protons, it is a uranium atom; if it has one proton, it is an atom of hydrogen. The symbol Z is used for atomic number. Atomic numbers for all the elements are given in the table of atomic weights on the inside back cover. Since electrons and protons have equal and opposite charges, then for an atom to be electrically neutral it must have an equal number of protons and electrons. Therefore, a carbon atom has 6 electrons, a uranium atom 92, and a hydrogen atom 1.

2-6 Isotopes

Although we can always tell the number of protons and electons in an atom from the atomic number, things are not so simple for neutrons. In fact, two atoms with the same number of protons (and therefore of the same element) can and often do have different numbers of neutrons. Thus, three kinds of oxygen atom are found in nature. All must have 8 protons and 8 electrons (since the atomic number (Z) of oxygen is 8), but one kind has 8 neutrons, another 9, and a third 10. Two or more atoms of the same element which have different numbers of neutrons are called **isotopes.** The total number of protons and neutrons in an atom is called the **mass number** (symbol = A). The three isotopes of oxygen have mass numbers (A) of 16, 17, and 18, and are referred to as ^{16}O, ^{17}O, and ^{18}O or $^{16}_{8}O$, $^{17}_{8}O$, and $^{18}_{8}O$. In general, isotopes are not given special names, but an exception is made for ^{2}H and ^{3}H, which are called, respectively, **deuterium** and **tritium.** Many isotopes occur in nature, but many others do not and have been produced only in laboratories (for example, ^{13}O, ^{14}O, ^{15}O, ^{19}O, and ^{20}O; we will discuss the artificial production of

Table 2-3 Properties of Elementary Particles

	Mass	*Charge*
Proton	1	+1
Neutron	1	0
Electron	$\frac{1}{1835}$	−1

required

isotopes in Section 23-6). Table 2-4 shows all the naturally occurring isotopes of the elements of atomic numbers from 1 to 10. Note that some elements (for example, fluorine) are **monoisotopic;** they are found in nature only as a single isotope. For most elements, however, two or more isotopes are present in nature.* Only about 20 elements are monoisotopic.

Isotopes were discovered by Frederick Soddy (1877–1956) in 1913, and it subsequently became apparent that a single isotope would serve much better as a standard for atomic weight tables than would a mixture of isotopes. The isotope selected was ^{12}C, which is now the standard upon which the atomic weight table is based. This isotope is assumed to have a weight of 12.00000 amu, and all atomic weights are compared to it.

The existence of isotopes, totally unsuspected by nineteenth-century chemists, makes the second and third of Dalton's statements (Section 2-2) incorrect. All atoms of lithium, for example, are not identical; there are, in fact, two kinds, with different mass numbers. Also, atoms of different elements do not necessarily have different weights. As we can see in Table 2-4, both carbon and nitrogen have an isotope with $A = 14$, and this situation is not uncommon with elements of higher atomic number. It should be noted that the mass number, which is the total number of

*Tin is the leader, with 10 isotopes in nature.

Table 2-4 The Naturally Occurring Isotopes of Elements with Atomic Numbers from 1 to 10

Isotope	*Number of protons (atomic number, Z)*	*Number of neutrons*	*Total number of nucleons (mass number, A)*	*Natural abundance* (%)
$^{1}_{1}H$	1	0	1	99.985
$^{2}_{1}H$	1	1	2	0.015
$^{3}_{1}H$	1	2	3	Trace
$^{3}_{2}He$	2	1	3	0.00013
$^{4}_{2}He$	2	2	4	100
$^{6}_{3}Li$	3	3	6	7.42
$^{7}_{3}Li$	3	4	7	92.58
$^{9}_{4}Be$	4	5	9	100
$^{10}_{5}B$	5	5	10	19.78
$^{11}_{5}B$	5	6	11	80.22
$^{12}_{6}C$	6	6	12	98.89
$^{13}_{6}C$	6	7	13	1.11
$^{14}_{6}C$	6	8	14	Trace
$^{13}_{7}N$	7	6	13	Trace
$^{14}_{7}N$	7	7	14	99.63
$^{15}_{7}N$	7	8	15	0.37
$^{16}_{8}O$	8	8	16	99.759
$^{17}_{8}O$	8	9	17	0.037
$^{18}_{8}O$	8	10	18	0.204
$^{19}_{9}F$	9	10	19	100
$^{20}_{10}Ne$	10	10	20	90.92
$^{21}_{10}Ne$	10	11	21	0.26
$^{22}_{10}Ne$	10	12	22	8.82

protons and neutrons, is not *exactly* the same as the weight, but it is very close to it (the distinction will be discussed in Section 23-7).

Since the mass number must be integral, the relative weight of a single atom (compared to ^{12}C) is always close to integral. It was this fact, and the fact that many elements are monoisotopic (or if they consist of two or more, one greatly predominates, as can be seen in Table 2-4), that led to Prout's hypothesis. Unfortunately, this hypothesis, so near the truth, was destroyed by the existence of some elements in which one isotope does not predominate. Naturally occurring chlorine, for example, consists of 75.53% ^{35}Cl and 24.47% ^{37}Cl. The isotopes are well mixed: samples of chlorine in any compound anywhere on earth contain the two isotopes in these percentages. Thus, the atomic weight of chlorine is 35.453, which is a weighted average of the weights of the two isotopes.

For each monoisotopic element, the atomic weight given in the table is simply a ratio of the mass of that isotope to that of ^{12}C arbitrarily taken as 12.00000. For elements with two or more isotopes, the atomic weight is a weighted average of all the naturally occurring isotopes in their natural abundances.

Example: Table 2-4 shows that the three isotopes of neon are found on Earth in the following percentages:

^{20}Ne	90.92
^{21}Ne	0.26
^{22}Ne	8.82

Calculate the atomic weight of neon.

Answer:

$$\begin{aligned} 20 \times 0.9092 &= 18.18 \\ 21 \times 0.0026 &= 0.055 \\ 22 \times 0.0882 &= 1.94 \\ \hline & \quad 20.18 \end{aligned}$$

The atomic weight of neon, to four significant figures, is 20.18 amu.

This procedure usually leads to a definite value for the atomic weight because for most elements, as for chlorine, *the isotopic ratios are constant throughout nature,* whether the elements are found free, or in compounds or mixtures. However, this is not true for all elements, the most notable example being lead. Because the radioactive decay of uranium and other radioactive elements (Chapter 23) produces only certain isotopes of lead and not others, the isotopic composition of lead does vary, depending on where on earth it is found. As a result, the atomic weight of lead is not always the same, but varies slightly. However, this slight variation does not usually cause much difficulty.

Note that the atomic weight table includes some elements whose atomic weights are given in parentheses. These elements do not occur in nature or they occur in such tiny quantities as to make a weighted average meaningless. For these elements the table gives the mass number of one isotope only.*

Today, atomic weights are no longer determined by analyses of compounds as was the case in the nineteenth century. Today such determinations are performed on *mass spectrometers,* invented by Francis William Aston (1877–1945) in 1919. We shall describe the operation of a mass spectrometer in the next section, after we talk about ions.

*The most stable isotope. The stability of isotopes will be discussed in Section 23-3.

2-7 Ions

We have seen that a neutral atom must contain an equal number of protons and electrons. But it is quite possible (and often happens) that an atom loses or gains one or more electrons.* When that happens, the particle is no longer a neutral atom but is now called an **ion.** All ions therefore carry an electrical charge, equal to the excess of protons over electrons, or vice versa. For example, an atom of copper possesses 29 electrons and 29 protons and is neutral because the charge of +29 balances the charge of −29. If this atom loses an electron, it still has 29 protons but only 28 electrons, and so must have a charge of +1. We write it Cu^{+}. If it lost two electrons, it would have a charge of +2 and be written Cu^{2+}. Similarly, if a fluorine atom gains one electron, it has a charge of −1 and is written F^{-}. Negative ions are called **anions;** positive ions are **cations.**

Not all ions consist of just one atom plus or minus electrons. Many ions are known which contain several atoms. An example is the hydroxide ion, OH^{-}, which contains an oxygen atom and a hydrogen atom held together by a covalent bond (Chapter 14). Since an oxygen atom must have 8 protons, and a hydrogen atom 1, there must be a total of 10 electrons in a hydroxide ion or else it would not have a charge of −1. Other commonly encountered ions that have more than one atom are the sulfate ion, SO_4^{2-}; the nitrate ion, NO_3^{-}; the ammonium ion, NH_4^{+}; the carbonate ion, CO_3^{2-}; and the phosphate ion, PO_4^{3-}. Note that an ion such as PO_4^{2-} or PO_4^{+} is not necessarily impossible (although it would be uncommon), but if it did exist, it could not be called the phosphate ion, since that name belongs to PO_4^{3-}. Some commonly found ions are listed in Table 2-5.

We now see that several kinds of particles can be present in matter. We can have molecules containing just one atom, molecules containing two or more atoms, ions

*Note that atoms do not normally lose or gain protons or neutrons, only electrons.

Table 2-5 Some Commonly Found Ions

Cations

1+		2+		2+		3+	
ammonium	NH_4^{+}	barium	Ba^{2+}	magnesium	Mg^{2+}	aluminum	Al^{3+}
cuprous	Cu^{+}	cadmium	Cd^{2+}	manganese	Mn^{2+}	chromic	Cr^{3+}
hydrogen	H^{+}	calcium	Ca^{2+}	mercuric	Hg^{2+}	ferric	Fe^{3+}
lithium	Li^{+}	chromous	Cr^{2+}	mercurous	Hg_2^{2+}		
potassium	K^{+}	cupric	Cu^{2+}	stannous	Sn^{2+}		
silver	Ag^{+}	ferrous	Fe^{2+}	strontium	Sr^{2+}		
sodium	Na^{+}	lead	Pb^{2+}	zinc	Zn^{2+}		

Anions

3−		2−		1−		1−	
phosphate	PO_4^{3-}	carbonate	CO_3^{2-}	acetate	$C_2H_3O_2^{-}$	hydride	H^{-}
		chromate	CrO_4^{2-}	bromide	Br^{-}	hydroxide	OH^{-}
		dichromate	$Cr_2O_7^{2-}$	chlorate	ClO_3^{-}	hypochlorite	ClO^{-}
		oxalate	$C_2O_4^{2-}$	chloride	Cl^{-}	iodate	IO_3^{-}
		sulfide	S^{2-}	cyanide	CN^{-}	iodide	I^{-}
		sulfate	SO_4^{2-}	fluoride	F^{-}	nitrate	NO_3^{-}
		sulfite	SO_3^{2-}	hydrogen carbonate (bicarbonate)	HCO_3^{-}	nitrite	NO_2^{-}
				hydrogen sulfite (bisulfite)	HSO_3^{-}	perchlorate	ClO_4^{-}
						permanganate	MnO_4^{-}

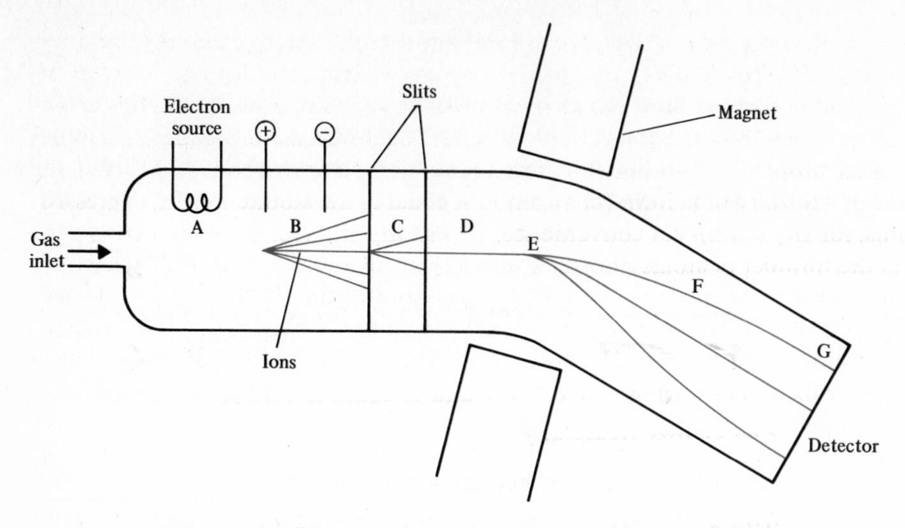

Figure 2-5 Schematic diagram of a mass spectrometer.

consisting of a single atom plus or minus electrons, and ions consisting of two or more atoms plus or minus electrons. We need a word that could stand for any of these things. Such a word is **species.** Thus, the mercury atom is a species; the H_2O molecule another; the Li^+ ion another, and the SO_4^{2-} ion still another. A species is an independant entity, unconnected to any other species, whether it be an atom, a molecule, or an ion.

In a **mass spectrometer** we take advantage of the fact that positive ions are attracted to a negative field. In this device (Figure 2-5) the sample to be analyzed is first converted to a gas if it is not already in the gaseous state. The gas enters the instrument, where it is bombarded by a stream of electrons coming from a heated filament (A). The bombardment causes the atoms to lose one or more electrons (usually just one electron from each atom), turning them into positive ions. The positive ions are attracted by a negative electric field (B) and start to accelerate. Two slits (C) are placed in such a way that only a thin pencil of ions can get through. At this point (D) there is a thin stream of positive ions traveling in a straight line. The stream next enters a magnetic field which is at right angles to the electric field. The attraction of the magnetic field causes the ions to turn in one direction (E). For ions of the same charge the amount of turning is proportional to the mass of the ions.* The lighter ions turn the most; the heavier ions, the least (F). If the original sample consists of three isotopes, then the application of the magnetic field results in three streams of positive ions (G), which strike a detector. The method is very sensitive, and the detector can measure the relative amounts of the three isotopes with high precision. From these data it is easy to calculate the atomic weight, as we did in the previous section.

2-8 The Mole

In Sections 2-3 and 2-6, we discussed the atomic weight scale and how the weights of the atoms on this scale are relative values, compared to the standard ^{12}C. We also showed that if we have the same number of different atoms, their relative weight is always the same regardless of the number.

The table of atomic weights tells us that the atomic weight of fluorine is 18.998403 amu and that of helium is 4.00260 amu. It is clear from our previous

*The amount of turning is proportional to the ratio of charge to mass (e/m). Most of the ions have a charge of +1, and these are the only ones we need to consider.

discussions that the ratio of weight between one atom of fluorine and one atom of helium is 18.998403/4.00260, and this is also the weight ratio for any number of atoms provided that we have the same number of atoms of each. Therefore, if we measure out 18.998403 *grams* of fluorine and 4.00260 *grams* of helium, we must have equal numbers of atoms. But how many atoms are there in 18.998403 g of fluorine or 4.00260 g of helium (or an amount equal to the atomic weight, expressed in grams, for any atom)? For convenience, we call this number of atoms a **mole.** A mole is the number of atoms necessary to make up an amount of a substance equal to the atomic weight in grams. Since ^{12}C is the basis of the atomic weight scale, we may define a mole as being equal to the number of atoms in 12.00000 g of ^{12}C. It may seem surprising, but we actually know how many atoms make up a mole. The number is called **Avogadro's number** and is equal to 6.02×10^{23} atoms.

6.02×10^{23} atoms of fluorine weighs 18.998403 g
6.02×10^{23} atoms of helium weighs 4.00260 g
6.02×10^{23} atoms of carbon weighs 12.011 g
6.02×10^{23} atoms of oxygen weighs 15.9994 g
6.02×10^{23} atoms of uranium weighs 238.029 g

We now have a most interesting situation. Although we cannot measure the weight of a single atom, we can calculate it by dividing the atomic weight in grams by Avogadro's number. This gives us the actual weight of the atom.

Example: Calculate the weight of a carbon atom.

Answer:

$$6.02 \times 10^{23} \text{ atoms of carbon} = 12.011 \text{ g of carbon}$$

$$1 \text{ atom of carbon} = 1 \text{ atom of carbon} \times \frac{12.011 \text{ g of carbon}}{6.02 \times 10^{23} \text{ atoms of carbon}}$$

$$= \frac{12.011}{6.02 \times 10^{23}} \text{ g of carbon}$$

$$= 1.99 \times 10^{-23} \text{ g}$$

We have defined a mole as being equal to the number of atoms in 12.00000 g of ^{12}C. We now know that this number is 6.02×10^{23} and we can now think of a mole as 6.02×10^{23} of anything, just as a dozen is 12 of anything and a score is 20 of anything. Thus:

a mole of cars = 6.02×10^{23} cars
a mole of pencils = 6.02×10^{23} pencils
a mole of H_2O molecules = 6.02×10^{23} H_2O molecules = 18 g of H_2O
a mole of Na^+ ions = 6.02×10^{23} Na^+ ions = 23 g of Na^+ ions etc.

Of course, we do not normally use moles for such large objects as cars or pencils. We use it for tiny particles such as atoms, molecules, and ions, where 6.02×10^{23} particles (1 mole) will weigh the atomic or molecular weight in grams. The purpose of using moles is convenience. We cannot work with one atom or molecule, but we can work with 1 mole of atoms or molecules, and we know that the weight relationships are the same for moles as they are for atoms, molecules, or ions.

Since the weight of a mole of anything is the atomic or molecular weight in grams, we now have another unit for atomic or molecular weight. Besides using amu, we may also express an atomic or molecular weight as *grams/mole.* Furthermore, although Avogadro's number is actually dimensionless (like a dozen), we may give it the unit *molecules/mole*, with the understanding that the word molecules, when used in this way, also means atoms, ions, or any other kind of species.

Example: How many moles of carbon dioxide are present in 85.0 g of carbon dioxide? How many molecules?

Answer: The molecular weight of CO_2 (from the table) is 44.0 g/mole.

$$85.0\,\text{g}\,CO_2 \times \frac{1\ \text{mole}\ CO_2}{44.0\,\text{g}\,CO_2} = 1.93\ \text{moles}\ CO_2$$

$$85.0\,\text{g}\,CO_2 \times \frac{1\ \text{mole}\ CO_2}{44.0\,\text{g}\,CO_2} \times \frac{6.02 \times 10^{23}\ \text{molecules}\ CO_2}{1\ \text{mole}\ CO_2} = 1.16 \times 10^{24}\ \text{molecules}\ CO_2$$

Thus, 85.0 g of CO_2 is 1.93 moles of CO_2 and contains 1.16×10^{24} molecules of CO_2.

Example: How many moles of hydrogen atoms are present in 3.5 moles of ammonium chloride NH_4Cl?

Answer: Each molecule of NH_4Cl contains four hydrogen atoms, so each mole of NH_4Cl contains 4 moles of hydrogen atoms.

$$3.5\ \text{moles of}\ NH_4Cl \times \frac{4\ \text{moles H}}{\text{mole}\ NH_4Cl} = 14\ \text{moles H atoms}$$

2-9 Periodicity of the Elements

The years between 1737 and 1869 were fruitful for the discovery of new elements. Approximately 49 new elements were discovered during this period, an average of more than one every three years, to add to the 13 elements that had previously been known.* In fact, so thick and fast did those new elements come that a mid-nineteenth-century chemist might reasonably wonder where it would all end. Was there a limit to the number of elements in nature, and if so how would he be able to know when that limit was reached? Elements were discovered more or less at random: for example, by analyzing a mineral for all known elements and then looking to see if there was anything left over. Another troubling question was that of relationships among the properties of elements. Certain elements seemed to have similar properties. Was this coincidence, or were there reasons for it?

Although many chemists had noticed the existence of families of elements (groups of elements with similar properties), the most important early formulation of this idea was that of Johann Döbereiner (1780–1849), who in 1829 suggested that certain elements occurred in families of three (which he called triads). Among these were the following five sets:

chlorine	calcium	sulfur	platinum	silver
bromine	strontium	selenium	iridium	lead
iodine	barium	tellurium	osmium	mercury

For each triad, not only were the properties of the three elements similar, but the atomic weight of the middle element was the mean of the other two. The discovery

*The elements Au, Ag, Cu, Fe, Pb, Sn, Hg, S, and C were known in ancient times (before A.D. 500) though of course it was not known that they were elements. As, Sb, Bi, and P were discovered by alchemists. The first element to be discovered by a modern chemist was cobalt, Co, first isolated by a Swedish chemist named Georg Brandt (1694–1768) about 1737.

of Döbereiner was regarded as intriguing but did not immediately lead to anything further, for an interesting reason. For further progress to be made (we now know) it was necessary to arrange the known elements in order of their atomic weights, and as we have previously seen (footnote on page 40), the chemists of 1830–1850 did not agree on what these atomic weights were! Therefore, it was not until the 1860s, when an agreed-upon table of atomic weights was finally arrived at, that a number of chemists began to arrange the elements in order of atomic weights, and when they did, observed that the elements fill into rows, which, when placed under each other, created columns of elements whose properties showed many similarities. Among these chemists were John A. R. Newlands (1838–1898), Alexander Beguyer de Chancourtois (1820–1886), and Julius Lothar Meyer (1830–1895), but the one who is generally given the most credit for the discovery of the periodic table is Dmitri Ivanovich Mendeleev (1834–1907).

Mendeleev published his first table in April 1869 and continued to modify it for several years thereafter. A version of the table which he published in 1871 (after reading the work of Meyer) is shown in Figure 2-6. Column I consists of H, Li, Na, K, Cu, Rb, Ag, Cs, and Au. All of these (except H) are metals, all of them form chlorides with a formula of ECl, and all share a number of other properties. (E is a general symbol for any element.) The elements in column II are also all metals, but in this case all form chlorides whose formula is ECl_2. Similarly, the elements in each of the eight columns form a group with many properties in common. Furthermore, Mendeleev pointed out that many of the properties of the elements formed an orderly progression along a column or a row. Thus, melting points of the elements of column IIA are

Be	1278°C
Ca	842°C
Sr	769°C
Ba	725°C

There are three main reasons why Mendeleev is given the principal credit for the discovery of the periodic table, although other chemists had published similar tables earlier.

1. When Mendeleev published his table, he included with it a lengthy analysis of

	Group I		Group II		Group III		Group IV		Group V		Group VI		Group VII		Group VIII
	A	B	A	B	A	B	A	B	A	B	A	B	A	B	
1		H													
2	Li		Be		B		C		N		O		F		
3		Na		Mg		Al		Si		P		S		Cl	
4	K		Ca		■		Ti		V		Cr		Mn		Fe, Co, Ni, Cu
5		(Cu)		Zn		■		■		As		Se		Br	
6	Rb		Sr		Y?		Zr		Nb		Mo		■		Ru, Rh, Pd, Ag
7		(Ag)		Cd		In		Sn		Sb		Te		I	
8	Cs		Ba		Di?		Ce?		■		■		■		■, ■, ■, ■
9		(■)		■		■		■		■		■		■	
10	■		■		Er?		La?		Ta		W		■		Os, Ir, Pt, Au
11		(Au)		Hg		Tl		Pb		Bi		■		■	
12	■		■		■		Th		■		U				

Figure 2-6 Mendeleev's Table of 1871. The question marks represent elements whose position was uncertain, because their atomic weights and other properties were uncertain. The colored squares are spaces that Mendeleev left for elements not yet discovered. He was unsure of the positions of Cu, Ag, and Au, and so put them in two places. Didymium (Di) was believed to be an element at that time but was later (1885) shown to be a mixture of two elements: praseodymium (Pr) and neodymium (Nd).

the properties of each element and showed in considerable detail how the table correlated a broad range of physical and chemical properties.

2. Unlike some other chemists, Mendeleev was not afraid to reverse the order of certain elements. Thus, tellurium has a higher atomic weight than iodine, but Mendeleev saw that the properties of iodine unquestionably placed it in column VII, because its properties are similar to those of chlorine and bromine (recall that this is one of Döbereiner's triads), while tellurium similarly belonged in column VI. Mendeleev boldly asserted that one or both of the atomic weights must be wrong. We now know that they were not wrong, but the reason for the inversion of position in the table was not discovered until 1913 (see the beginning of Section 2-10).

3. The most important reason is that Mendeleev left spaces in his table for elements that had not yet been discovered (see Figure 2-6), and *predicted the*

Table 2-6 Some of the Predictions Made by Mendeleev, and Properties of the Elements as Reported by Their Discoverers

	Mendeleev's predictions	*Properties reported by discoverer*
	eka-aluminum	*gallium (discovered by Boisbaudran in 1875)*
Atomic weight	68	69.9
Specific gravity	6.0	5.96
Atomic volume[a]	11.5	11.7
	eka-boron	*scandium (discovered by Nilson in 1879)*
Atomic weight	44	43.79
Formula of oxide	Eb_2O_3	Sc_2O_3
Specific gravity of oxide	3.5	3.864
Formula of sulfate	$Eb_2(SO_4)_3$	$Sc_2(SO_4)_3$
	eka-silicon	*germanium (discovered by Winkler in 1886)*
Atomic weight	72	72.3
Specific gravity	5.5	5.469
Atomic volume[a]	13	13.2
Formula of oxide	EsO_2	GeO_2
Specific gravity of oxide	4.7	4.703
Formula of chloride	$EsCl_4$	$GeCl_4$
Boiling point of chloride	$<100°C$	86°C
Density of chloride	1.9	1.887
Formula of ethyl compound	$EsEt_4$	$GeEt_4$
Boiling point of ethyl compound	160°C	160°C
Specific gravity of ethyl compound	0.96	A little less than 1

[a] Atomic volume is the volume, in milliliters, occupied by 1 mole of atoms.

properties of some of these new elements. A scientific theory is most readily acceptable when we can use it to make specific predictions, and these predictions are confirmed. Mendeleev predicted that a new element would be discovered in group IIIA under boron, another in group IIIB under aluminum, and a third in group IVB under silicon. He called these eka-boron, eka-aluminum, and eka-silicon, respectively, and for each predicted a broad range of properties (for example, see Table 2-6). To chemists who until this time had no possible way of predicting the properties of still-undiscovered elements, such predictions must have seemed to approach pure fantasy. Consequently, when all three of these elements were discovered (gallium in 1875, scandium in 1879, germanium in 1886) and found to have properties identical with or very close to those predicted by Mendeleev (Table 2-6), the periodic table was accepted by everyone and Mendeleev was given the most credit for it. In 1955, element 101 was named mendelevium in recognition of his great work.

Mendeleev's table was a major contribution to chemistry because it answered, though in an imperfect manner, both of the questions we discussed at the beginning of this section. It gave a rational basis for understanding the similarities among the properties of the elements, and, for the first time, it showed where there might be a limit to the number of possible elements not yet discovered. After 1869, any undiscovered element could only fall into one of five possible categories.

1. It could fit an existing space in the table. As we have seen, scandium, gallium, and germanium were found later. Still later, other elements (e.g., hafnium and rhenium) were found to fill other spaces.
2. There might exist elements heavier than uranium, the heaviest element known in 1869. We know today that uranium is the heaviest element found in nature, but many heavier elements have been artificially manufactured (see Section 23-6).
3. There might exist elements lighter than hydrogen. None were ever found, and we know now that none ever will be, since there cannot be an atom with less than one proton.
4. An entire new column might be discovered. Actually, this did happen. None of the noble gases were known to Mendeleev in 1869. When they were discovered (five of them were discovered by Rayleigh and Ramsay between 1894 and 1898), it was realized that they had to occupy an entire new column (called column 0).
5. An entire new row might be discovered. This never happened.

end of defi for test

2-10 The Modern Periodic Table

We now know that Mendeleev's table, although in accord with the facts known at that time, was imperfect in a number of ways. Since 1869, there has been a great deal of progress in our understanding of the periodic table. The most important advance was made in 1913, when Henry Moseley (1887–1915) discovered the concept of **atomic number.***

It was immediately understood that the elements in the periodic table should be arranged in order of atomic number rather than atomic weight, and when this is done, the positions of such pairs as iodine and tellurium are no longer anomalous. X-ray experiments show that the atomic numbers of tellurium and iodine are 52 and 53, respectively, so that they clearly belong exactly where Mendeleev put them. A further advantage of atomic number is that it puts even more of a limit on the

*He found that the wavelength of x rays produced when a stream of electrons is allowed to fall on a metallic target depends on the magnitude of the positive charge of the nucleus of the target atoms.

IA	IIA	IIIB														IVB	VB	VIB	VIIB	VIIIB			IB	IIB	IIIA	IVA	VA	VIA	VIIA	0
																1 H															2 He
3 Li	4 Be																									5 B	6 C	7 N	8 O	9 F	10 Ne
11 Na	12 Mg																									13 Al	14 Si	15 P	16 S	17 Cl	18 Ar
19 K	20 Ca	21 Sc															22 Ti	23 V	24 Cr	25 Mn	26 Fe	27 Co	28 Ni	29 Cu	30 Zn	31 Ga	32 Ge	33 As	34 Se	35 Br	36 Kr
37 Rb	38 Sr	39 Y															40 Zr	41 Nb	42 Mo	43 Tc	44 Ru	45 Rh	46 Pd	47 Ag	48 Cd	49 In	50 Sn	51 Sb	52 Te	53 I	54 Xe
55 Cs	56 Ba	57 La	58 Ce	59 Pr	60 Nd	61 Pm	62 Sm	63 Eu	64 Gd	65 Tb	66 Dy	67 Ho	68 Er	69 Tm	70 Yb	71 Lu	72 Hf	73 Ta	74 W	75 Re	76 Os	77 Ir	78 Pt	79 Au	80 Hg	81 Tl	82 Pb	83 Bi	84 Po	85 As	86 Rn
87 Fr	88 Ra	89 Ac	90 Th	91 Pa	92 U	93 Np	94 Pu	95 Am	96 Cm	97 Bk	98 Cf	99 Es	100 Fm	101 Md	102 No	103 Lr	104	105													

Figure 2-7 A Modern Periodic Table (long-long form). The numbers shown are atomic numbers.

discovery of new elements. Thus, all elements with atomic numbers 1 to 105 are now known (either as naturally occurring or as artificially produced), and any new elements discovered or produced from this point on must have atomic numbers of 106 or higher. Other improvements have also been made, and Mendeleev's table of 1871 (Figure 2-6) is now only of historical interest.

A table that gives the modern periodic arrangement is shown in Figure 2-7. However, this style, which may be called the "long-long" form of the periodic table is too long for convenient display, and a much more common form of the modern periodic table, called the *long form*, is shown in Figure 2-8 and on the inside front cover (the original Mendeleev style is called the short form). Figure 2-8 is identical to Figure 2-7, except that the group of elements 58 to 71 and 90 to 103 has been removed from the body of the table and placed in a separate array below. This is done solely to present the table in a more convenient manner.

Period	IA	IIA	IIIB	IVB	VB	VIB	VIIB	VIIIB			IB	IIB	IIIA	IVA	VA	VIA	VIIA	0
1									1 H									2 He
2	3 Li	4 Be											5 B	6 C	7 N	8 O	9 F	10 Ne
3	11 Na	12 Mg											13 Al	14 Si	15 P	16 S	17 Cl	18 Ar
4	19 K	20 Ca	21 Sc	22 Ti	23 V	24 Cr	25 Mn	26 Fe	27 Co	28 Ni	29 Cu	30 Zn	31 Ga	32 Ge	33 As	34 Se	35 Br	36 Kr
5	37 Rb	38 Sr	39 Y	40 Zr	41 Nb	42 Mo	43 Tc	44 Ru	45 Rh	46 Pd	47 Ag	48 Cd	49 In	50 Sn	51 Sb	52 Te	53 I	54 Xe
6	55 Cs	56 Ba	57 La *	72 Hf	73 Ta	74 W	75 Re	76 Os	77 Ir	78 Pt	79 Au	80 Hg	81 Tl	82 Pb	83 Bi	84 Po	85 At	86 Rn
7	87 Fr	88 Ra	89 Ac †	104	105													

Representative Elements — Transition Elements — Representative Elements

*58 Ce	59 Pr	60 Nd	61 Pm	62 Sm	63 Eu	64 Gd	65 Tb	66 Dy	67 Ho	68 Er	69 Tm	70 Yb	71 Lu
†90 Th	91 Pa	92 U	93 Np	94 Pu	95 Am	96 Cm	97 Bk	98 Cf	99 Es	100 Fm	101 Md	102 No	103 Lr

Inner Transition Elements

Figure 2-8 A modern periodic table (long form). The numbers shown are atomic numbers.

As we can see from Figure 2-7, the elements are arranged in columns and rows. The columns (also called **groups**) are given headings IA, IIA, and so on, which are retained from Mendeleev's time, although not all these headings are strictly logical (for example, column VIIIB is actually three columns). There are seven rows, also called **periods.** The elements in the periodic table can be divided into two classes. Elements in the A columns IA to VIIA and column 0 are called **representative elements.** Elements in the B columns IB to VIIIB are called **transition elements.** Elements 58 to 71 and 90 to 103 are also transition elements, but these are given the special designation **inner transition elements.** In Section 13-9, we shall give a more precise definition of transition and of inner transition element.*

In the rest of this chapter we will take a brief look at some of the properties of the representative elements and their compounds. We will put off discussion of the transition elements until Chapter 19.

GROUP IA—ALKALI METALS. The elements of group IA are called the **alkali metals** and are listed in Table 2-7. The name comes from the Arabic word *al-qili,* meaning plant ash. The ashes of plants contain large amounts of sodium and potassium carbonate (Na_2CO_3 and K_2CO_3).

The alkali metals occur in nature only as singly positively charged ions—Li^+, Na^+, and so on. The formulas of some typical binary compounds of the alkali metals are given in Table 2-7.

The table shows compounds of the alkali metals with fluorine (these are called **fluorides**), chlorine (**chlorides**), hydrogen (**hydrides**), and oxygen (**oxides**). We note that the ratio of components within each set of compounds is constant. For example, all the alkali metal fluorides have the formula EF (E is a general symbol for any element), in which the ratio of metal ions to fluoride ions is 1 : 1. A similar ratio holds for the chlorides and hydrides, while for the oxides, the ratio is 2 : 1. The fact that these ratios are constant within a column was of course one of the things that led Mendeleev and his predecessors to the idea of the periodic table. We shall see that similar constant ratios hold for the other columns of representative elements.

GROUP IIA—ALKALINE EARTH METALS. The elements of group IIA are called the **alkaline earth metals** and are listed in Table 2-8. These elements form positive ions with a charge of +2 (for example, Mg^{2+}, Ca^{2+}). We observe from Tables 2-7 and 2-8 that in going from column IA to IIA, the combining capacity of each element has increased by one. Each alkaline earth ion combines with *two* chloride or fluoride ions. Thus, the alkali metal chlorides are ECl, but the alkaline earth chlorides are ECl_2.

*As we shall see in Chapter 13, the elements in columns IB and IIB are not, strictly speaking, transition elements, nor can elements 71 and 103 be strictly classified as inner transition elements.

Table 2-7 Group IA Elements (all metals) and Some Typical Compounds

Element	*Symbol*	*Atomic number*	*Alkali fluoride*	*Alkali chloride*	*Alkali hydride*	*Alkali oxide*
Lithium	Li	3	LiF	LiCl	LiH	Li_2O
Sodium	Na	11	NaF	NaCl	NaH	Na_2O
Potassium	K	19	KF	KCl	KH	K_2O
Rubidium	Rb	37	RbF	RbCl	RbH	Rb_2O
Cesium	Cs	55	CsF	CsCl	CsH	Cs_2O
Francium	Fr	87	FrF	FrCl	FrH[a]	Fr_2O

[a]Francium is available in only trace amounts, and its compounds are not very well known. However, since the other alkali metals form hydrides, we would expect FrH to exist.

Table 2-8 Alkaline Earth Metals and Some Typical Compounds

Element	Symbol	Atomic number	Alkaline earth fluoride	Alkaline earth chloride	Alkaline earth oxide
Beryllium	Be	4	BeF_2	$BeCl_2$	BeO
Magnesium	Mg	12	MgF_2	$MgCl_2$	MgO
Calcium	Ca	20	CaF_2	$CaCl_2$	CaO
Strontium	Sr	38	SrF_2	$SrCl_2$	SrO
Barium	Ba	56	BaF_2	$BaCl_2$	BaO
Radium	Ra	88	RaF_2	$RaCl_2$	RaO

GROUPS IIIA AND IVA. There are no general names for the elements of these groups, comparable to the ones for groups IA and IIA ("alkali metals," "alkaline earth metals") or the ones we shall see for groups VIA, VIIA, and 0. Group IIIA is sometimes called the *boron group;* group IVA, the *carbon group.* Table 2-9 shows these elements and the formulas of some typical binary compounds. Note that once again, the combining power of each element has increased by one in going from group IIA to IIIA, and by one more in going from group IIIA to IVA. Thus, group IIIA and group IVA chlorides are ECl_3 and ECl_4, respectively. In contrast to groups IA and IIA, not all of the elements in these groups are metals. Lead and indium are metals, but carbon and boron are not.

GROUPS VA AND VIA. The elements of group VA have no general name, but the elements of group VIA are sometimes called *chalcogens.* Table 2-10 shows these elements and some of their compounds. Note that the combining capacity now *diminishes* by one for each group. Thus, group VA and VIA chlorides are ECl_3 and ECl_2, respectively. The number of metals also decreases. Only bismuth is a pure metal, although some of the other elements (for example polonium, antimony) also have some metallic properties. It should be evident from what we have seen so far that among the representative elements, the metallic character of the elements *decreases* in going from left to right in the periodic table, and *increases* in going down the table.

GROUP VIIA—THE HALOGENS. The elements of group VIIA are called **halogens,** a word that comes from a Greek root meaning "sea." The most common

Table 2-9 Groups IIIA and IVA Elements and Some Typical Compounds (The metals are marked *m*.)

	Element	Symbol	Atomic number	Compound with F	Compound with Cl	Compound with O
	Group IIIA					
	Boron	B	5	BF_3	BCl_3	B_2O_3
m	Aluminum	Al	13	AlF_3	$AlCl_3$	Al_2O_3
m	Gallium	Ga	31	GaF_3	$GaCl_3$	Ga_2O_3
m	Indium	In	49	InF_3	$InCl_3$	In_2O_3
m	Thallium	Tl	81	TlF_3	$TlCl_3$	Tl_2O_3
	Group IVA					
	Carbon	C	6	CF_4	CCl_4	CO_2
	Silicon	Si	14	SiF_4	$SiCl_4$	SiO_2
	Germanium	Ge	32	GeF_4	$GeCl_4$	GeO_2
m	Tin	Sn	50	SnF_4	$SnCl_4$	SnO_2
m	Lead	Pb	82	PbF_4[a]	$PbCl_4$	PbO_2

[a] PbF_4 has not been made. However, such a compound should be possible, since it is analogous to others in the table.

Table 2-10 Groups VA and VIA Elements and Some Typical Compounds (The metals are marked *m*.)

	Element	Symbol	Atomic number	Compound with F	Compound with Cl	Compound with O
Group VA						
	Nitrogen	N	7	NF_3	NCl_3	N_2O_3
	Phosphorus	P	15	PF_3	PCl_3	P_2O_3
	Arsenic	As	33	AsF_3	$AsCl_3$	As_2O_3
	Antimony	Sb	51	SbF_3	$SbCl_3$	Sb_2O_3
m	Bismuth	Bi	83	BiF_3	$BiCl_3$	Bi_2O_3
Group VIA (chalogens)						
	Oxygen	O	8	OF_2	OCl_2	
	Sulfur	S	16	SF_2	SCl_2	
	Selenium	Se	34	SeF_2[a]	$SeCl_2$	
	Tellurium	Te	52	TeF_2[a]	$TeCl_2$	
	Polonium	Po	84	PoF_2[a]	$PoCl_2$	

[a] These compounds have not been prepared. If they could be, these formulas are those predicted from the periodic table.

Table 2-11 Group VIIA Elements and Some Typical Compounds

Element	Symbol	Atomic number	Compound with F	Compound with Cl	Compound with O
Fluorine	F	9	—	ClF	F_2O
Chlorine	Cl	17	ClF	—	Cl_2O
Bromine	Br	35	BrF	BrCl	Br_2O
Iodine	I	53	IF[a]	ICl	I_2O
Astatine	At	85	AtF[a]	AtCl	At_2O

[a] These compounds have not been prepared. If they are ever prepared, these are the predicted formulas.

halogen, chlorine, occurs in seawater, in the form of the salt sodium chloride. Table 2-11 shows that the combining capacity has diminished once again and that halogen chlorides have the formulas ECl. As we might have expected from the preceding paragraph, none of the halogens is a metal.

Group 0—The Noble Gases. The elements of group 0 (Table 2-12) are called **noble gases.** We have seen the combining capacity decrease by one each time in proceeding along the table in the order IVA → VA → VIA → VIIA. A further decrease by one would result in no combining capacity at all for the elements of group 0. In fact, this is exactly what was observed for many years. The elements of this group were found to form no compounds at all and for that reason were called the *inert gases*. However, this is no longer true. As we shall see in Section 18-7, compounds are now known which contain the elements krypton, xenon, and radon,

Table 2-12 Group 0 Elements: The Noble Gases

Element	Symbol	Atomic number
Helium	He	2
Neon	Ne	10
Argon	Ar	18
Krypton	Kr	36
Xenon	Xe	54
Radon	Rn	86

Table 2-13 Typical Compound Formation of Rows 2 and 3 Elements with Fluorine. Since Fluorine Has A Valence of 1, the Valences of Other Elements May Be Obtained from the Formulas of Their Fluorides.

	IA	*IIA*	*IIIA*	*IVA*	*VA*	*VIA*	*VIIA*	*0*
Row 2	LiF	BeF	BF_3	CF_4	NF_3	OF_2	—	None
Row 3	NaF	MgF_2	AlF_3	SiF_4	PF_3	SF_2	ClF	None
Valence	1	2	3	4	3	2	1	0

and the term "noble gases" is therefore more appropriate. Nevertheless, there is no typical combining capacity here, and no compounds have yet have been prepared which contain helium, neon, or argon.

2-11 Valence

We have seen that the periodic table can give us information about the combining capacity of the elements. The combining capacity of an element is called its **valence** and may be defined as **the number of bonds formed by an atom or ion of that element.** Every element has a characteristic valence or set of valences (many elements have more than one valence). Fluorine (and the other halogens) has a valence of 1, so each fluorine atom or ion forms one bond. This fact and an inspection of the formulas for fluorides enables us to learn the valences of other elements. Table 2-13 (as well as Tables 2-7 to 2-12) shows that the representative elements have valences equal to the column number, or to eight minus the column number. For example, the elements in group IIIA have a valence of 3, while the elements in group VIA have a valence of 2.

The formula of a binary compound can be predicted from a knowledge of the valences. For example, sodium and chlorine each has a valence of 1, so the formula for sodium chloride is NaCl. However, oxygen has a valence of 2, so it forms two bonds. When it bonds with sodium, the formula is Na_2O (since each sodium can only form one bond, two sodiums are needed to bond with one oxygen). In contrast to sodium, the valence of magnesium is 2. Thus, the formula for magnesium chloride must be $MgCl_2$, while magnesium oxide is MgO.

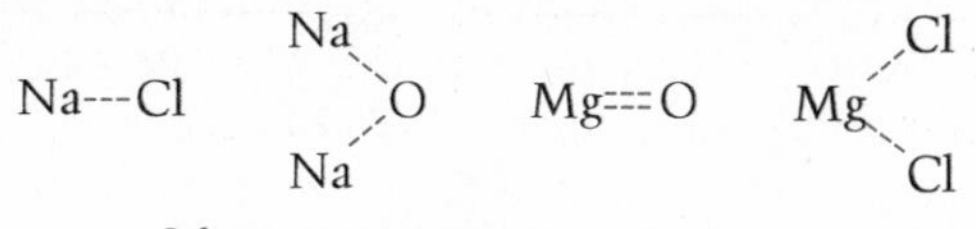

Schematic representation of valences

Example: The valence of aluminum is 3 and the valence of nitrogen is also 3. What are the formulas for aluminum chloride, aluminum oxide, and aluminum nitride?

Answer:

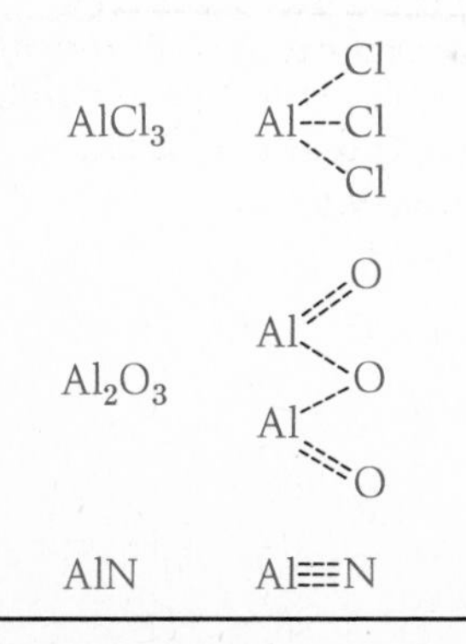

The picture we have just presented, in which the valence of elements and the formulas of binary compounds may be predicted from the periodic table, is oversimplified for the following reasons.

1. As already mentioned, many elements have more than one valence, and these are not always readily predictable from the periodic table. For example, Table 2-9 shows us that lead has a valence of 4, but it also forms compounds in which it has a valence of 2.
2. In Chapter 14, we shall see that there are two kinds of bonds: ionic and covalent. At this point we have not yet learned to distinguish between them, but a complete picture of valence must include this distinction.
3. We have not discussed valence at all for the transition and inner transition elements. Almost all of these show more than one valence, and the valences do not regularly change with position in the periodic table in the same way as do the valences of the representative elements.
4. Besides all of this, there are other exceptions to the simple rules we have presented.

Nevertheless, the valences shown in Tables 2-7 to 2-13 are valid for many compounds, and the rules we have given for combining them are highly useful for predicting the formulas of thousands of binary compounds. A more complete discussion of valence will be presented in Chapter 14.

Mendeleev and the other chemists of his time did not know why the periodic table was able to correlate dozens of properties, including valences. All they knew is that it did. Today we know that the periodic table works because most properties of an element are determined by the configuration of its electrons, and elements in any column of the periodic table have similar electron configurations. In Chapter 13, we shall see just how electron configurations determine the position of each element in the periodic table.

2-12 Abundances of the Elements

At this time, there are 105 known elements, 17 of which do not occur naturally on the surface of the earth and have been made only artificially. The artificial elements are technetium (atomic number 43), promethium (61), astatine (85), francium (87), and all elements with atomic numbers of 93 or more. The other 88 elements, which include all the first 92 except numbers 43, 61, 85 and 87, are found on the earth and are considered naturally occurring elements.

The 88 naturally occurring elements are found in greatly varying amounts, as shown in Table 2-14, which lists the natural abundance by weight of the most common elements in the earth's crust, including the oceans and the atmosphere. Note that 18 elements make up more than 99.9% by weight of the entire crust,

Table 2-14. Relative Abundance of Elements in the Earth's Crust, Including the Oceans and the Atmosphere

Element	*Percentage*	*Element*	*Percentage*
O	49.3	Cl	0.2
Si	25.8	P	0.12
Al	7.6	Mn	0.09
Fe	4.7	C	0.08
Ca	3.4	S	0.06
Na	2.7	Ba	0.05
K	2.4	Cr	0.03
Mg	1.9	N	0.03
H	0.9	All others	<0.1
Ti	0.6		

leaving less than 0.1% for the other 70 elements combined. Also note that oxygen is by far the most abundant element, making up almost half the weight, while oxygen and silicon together make up more than three-fourths. These are percentages by *weight*. A table of percentages by *atoms* would show a somewhat different distribution. Oxygen would still be first, by an even larger percentage, but hydrogen would now be third, behind oxygen and silicon.

Problems

2-1 Define each of the following terms. **(a)** atom **(b)** element (two definitions) **(c)** atomic weight **(d)** binary compound **(e)** molecule **(f)** atomic mass unit **(g)** molecular weight **(h)** molecular formula **(i)** compound **(j)** mixture **(k)** substance **(l)** homogeneous **(m)** heterogeneous **(n)** physical change **(o)** chemical change **(p)** chemical reaction **(q)** nucleus **(r)** proton **(s)** electron **(t)** neutron **(u)** nucleon **(v)** atomic number **(w)** isotopes **(x)** mass number **(y)** deuterium **(z)** species **(aa)** ion **(bb)** anion **(cc)** cation **(dd)** mole **(ee)** Avogadro's number **(ff)** eka-boron **(gg)** representative element **(hh)** group (in the periodic table) **(ii)** alkali metal **(jj)** alkaline earth metal **(kk)** halogen **(ll)** noble gas **(mm)** valence

2-2 **(a)** Ferric oxide is a compound of iron and oxygen containing 69.94% (by weight) iron and 30.06% oxygen. If you had 17.00 g of ferric oxide, calculate the weight in grams of iron and of oxygen you would have. **(b)** Calculate the relative weight in grams of iron to oxygen in this compound.

2-3 The relative weight of oxygen to hydrogen in water is 8:1. Calculate the percentages by weight of hydrogen and oxygen in water.

2-4 A compound of phosphorus and chlorine contains 22.27% phosphorus and 77.73% chlorine. Another contains 14.67% P and 85.33% Cl. Show how these two compounds obey the law of multiple proportions.

2-5 Did Lavoisier's experiments with sealed vessels absolutely *prove* the law of conservation of matter? Could 10,000 such experiments, carried out today, absolutely prove it?

2-6 Calculate the molecular weight of each of the following elements or compounds. **(a)** ozone, O_3 **(b)** potassium chlorate, $KClO_3$ **(c)** nitric acid, HNO_3 **(d)** hydrochloric acid, HCl **(e)** barium nitrate, $Ba(NO_3)_2$ **(f)** sodium aluminum sulfate, $NaAl(SO_4)_2$ **(g)** glucose, $C_6H_{12}O_6$ **(h)** perchloric acid, $HClO_4$ **(i)** testosterone, $C_{19}H_{28}O_2$ **(j)** chlorophyll a, $C_{55}H_{72}MgN_4O_5$ **(k)** $Ni(NH_3)_4(H_2O)_2(NO_3)_2$ **(l)** calcium ferricyanide, $Ca_3[Fe(CN)_6]_2$

2-7 See if you can classify each of the following as an element, a compound, or a mixture. **(a)** air **(b)** mercury **(c)** table salt (noniodized) **(d)** whiskey **(e)** aspirin **(f)** steel **(g)** oxygen **(h)** lead **(i)** seawater **(j)** cane sugar **(k)** sulfur **(l)** blood **(m)** propane

2-8 Classify each of the following as a chemical or a physical change. **(a)** burning a piece of wood **(b)** boiling oil **(c)** melting lead **(d)** digesting food **(e)** ionization of sodium ($Na \rightarrow Na^+$) **(f)** mixing a martini **(g)** rusting of iron **(h)** refining of copper (copper sulfide is converted to pure copper) **(i)** grinding a piece of wood to sawdust

2-9 Fill in the table.

Species	Atomic number (Z)	Number of neutrons	Mass number (A)	Number of electrons
$^{11}_{5}B$	____	____	____	____
$^{A}_{9}X$	____	10	____	____
$^{A}_{Z}X$	____	____	195	78
$^{A}_{Z}Cd$	____	64	____	____
$^{A}_{Z}Cm$	____	____	246	____
$^{A}_{Z}X$	94	145	____	____
$^{18}_{Z}X$	8	____	____	____
$^{232}_{Z}Th$	____	____	____	____

2-10 In each of the following cases you are given the number of protons, neutrons, and electrons. Write the formula for each species (in each case there is only one nucleus).
Example $9p$, $9e$, $10n$ *Ans.* $^{19}_{9}F$
(a) $32p$, $32e$, $42n$ **(b)** $25p$, $23e$, $30n$ **(c)** $1p$, $1n$ **(d)** $96p$, $96e$, $147n$ **(e)** $15p$, $18e$, $16n$
(f) $65p$, $62e$, $94n$

2-11 In Problem 2-10, how many nucleons are present in each species?

2-12 How many protons, electrons, and neutrons are there in each of the following species?
(a) ^{63}Ga **(b)** $^{35}Cl^{-}$ **(c)** $^{27}Al^{3+}$ **(d)** ^{235}U **(e)** $^{202}Po^{2-}$ **(f)** $^{14}N^{1}H_3$ **(g)** $^{32}S^{16}O_4{}^{2-}$

2-13 State all the differences you can among protons, neutrons, and electrons.

2-14 For each species, give the number of protons and of electrons **(a)** F **(b)** F^-
(c) F^+ **(d)** Ce **(e)** Ce^{2+} **(f)** Ce^{4+} **(g)** Ca^{2+} **(h)** Al^{3+} **(i)** S^{2-} **(j)** N^{3-}
(k) $PO_4{}^{3-}$ **(l)** $CO_3{}^{2-}$ **(m)** $C_2O_4{}^{2-}$

2-15 The element boron occurs naturally as two isotopic species, $^{10}_{5}B$ and $^{11}_{5}B$. Its atomic weight is 10.8022. Calculate the percentage natural abundance of the two isotopes.

2-16 Three isotopes of silicon occur in nature, with the following percent abundances: $^{28}_{14}Si$, 92.21%; $^{29}_{14}Si$, 4.70%; and $^{30}_{14}Si$, 3.09%. Calculate the atomic weight of silicon.

2-17 **(a)** Fluorine is a monoisotopic element; the only isotope found in nature is $^{19}_{9}F$. What is the atomic weight of fluorine to four significant figures? **(b)** Cobalt is also a monoisotopic element; the isotope in nature is $^{59}_{27}Co$. What is the atomic weight of cobalt? Compare it with the value in the table on the inside back cover. The reason for the discrepancy may be found in Section 23-7.

2-18 Write formulas for each of the following ions. (Do not forget to write the charge.)
(a) hydroxide **(b)** stannous **(c)** acetate **(d)** nitrate **(e)** nitrite **(f)** calcium ion
(g) potassium ion **(h)** chloride **(i)** dichromate **(j)** sulfite

2-19 Name each of the following ions. **(a)** Cu^+ **(b)** Cu^{2+} **(c)** CN^- **(d)** $NO_2{}^-$
(e) $NO_3{}^-$ **(f)** Hg^{2+} **(g)** $Hg_2{}^{2+}$ **(h)** $CrO_4{}^{2-}$ **(i)** Fe^{2+} **(j)** Fe^{3+}

2-20 Describe the operation of the mass spectrometer.

2-21 How many
(a) moles of NH_3 are there in 57.0 g of NH_3?
(b) moles of CaC_2O_4 are there in 0.55 g of CaC_2O_4?
(c) moles of $Bi_2(SO_4)_3$ are there in 18.5 g of $Bi_2(SO_4)_3$?
(d) moles of quinine ($C_{20}H_{24}N_2O_2$) are there in 83.6 g of quinine?

2-22 How many
(a) moles of C atoms are there in 11.6 moles of CO_2?
(b) moles of Si atoms are there in 0.36 mole of Si_3H_8?
(c) moles of O atoms are there in 2.4×10^4 moles of $Al_2(SO_4)_3$?
(d) moles of Na^+ ions are there in 11.64 moles of Na_2SO_4?
(e) moles of Hg atoms are there in 18.5 g of $Hg(NO_3)_2$?
(f) moles of N atoms are there in 0.043 g of $HAu(CN)_4$?
(g) moles of I^- ions are there in 21.7 g of ScI_3?

2-23 How many
(a) molecules of NO are there in 1.4 moles of NO?
(b) molecules of PCl_5 are there in 3.4×10^{-4} mole of PCl_5?
(c) molecules of C_6H_6 are there in 25.7 moles of C_6H_6?
(d) molecules of Al_2Cl_6 are there in 18.5 g of Al_2Cl_6?
(e) molecules of N_2O_4 are there in 6.5×10^{-3} g of N_2O_4?
(f) molecules of $H_2C_2O_4$ are there in 6.432 g of $H_2C_2O_4$?

2-24 How many
(a) Fe atoms are there in 23 molecules of Fe_2O_3?
(b) O atoms are there in 14.4 moles of $HSeO_4$?
(c) F atoms are there in 2.4×10^{-6} mole of SF_6?
(d) NH_4^+ ions are there in 32.4 moles of $(NH_4)_2MoO_4$?
(e) Os atoms are there in 2.018 g of $OsCl_4$?
(f) Br atoms are there in 2.2×10^{-4} g of $AsBr_3$?
(g) Cl^- ions are there in 24.6 g of $MgCl_2$?

2-25 At one time the lead chamber process was the chief method for producing sulfuric acid, H_2SO_4. The final balanced equation may be written

$$3SO_2 + 2HNO_3 + 2H_2O \longrightarrow 3H_2SO_4 + 2NO$$

This equation says: 3 moles of sulfur dioxide plus 2 moles of nitric acid plus 2 moles of water yields 3 moles of sulfuric acid and 2 moles of nitric oxide.
(a) Change the moles to grams and "read" the equation in grams.
(b) How many moles are contained in 1.000 g of each substance?
(c) How many molecules are there in 2.00 moles of SO_2, 6.00 moles of H_2SO_4, 7.00 moles of H_2O, and 1.76 moles of HNO_3?
(d) How many molecules are there in 1.50 moles of each of the five substances in the reaction above?

2-26 Of the following quantities—17.5 g of $SbCl_3$, 0.0932 mole of Br_2, 4.80 g of C_3H_7I—**(a)** Which weighs the most? **(b)** Which contains the greatest number of molecules? **(c)** Which contains the greatest number of atoms?

2-27 Compare Döbereiner's triads given on page 50 with the periodic table. Do the elements of each triad belong to the same group? Can you suggest a reason why Döbereiner included Ag–Pb–Hg as a triad, even though the atomic weight of lead is not the mean of the other two?

2-28 Discuss the meaning of each of the following terms as applied to the periodic table. Give specific examples where possible. **(a)** group VIA **(b)** group **(c)** period **(d)** inner transition series **(e)** transition series **(f)** noble gases **(g)** alkali metals

2-29 Look in the periodic table and find the pairs of elements whose atomic numbers are not in order of their atomic weights.

2-30 Predict some properties of the elements with atomic numbers 117, 118, and 119. Which elements would each of these most closely resemble?

2-31 Write formulas for binary compounds of the following elements. In each case the element on the left is written first. (If no compound exists, say so.)
(a) K and Br **(b)** Mg and I **(c)** K and Ar **(d)** H and Se
(e) Li and N **(f)** Sr and P **(g)** N and Cl **(h)** Si and O
(i) Be and Te **(j)** Rb and O **(k)** Ge and I **(l)** P and I
(m) Be and Ne **(n)** Al and F **(o)** B and N **(p)** Br and Cl
(q) B and S **(r)** P and S

2-32 For each of the following binary compounds, tell whether each element has a valence equal to the column number, eight minus the column number, or neither. **(a)** $CaCl_2$ **(b)** PCl_3 **(c)** PCl_5 **(d)** SF_6 **(e)** MgS **(f)** $SeBr_2$ **(g)** IF_7 **(h)** $BrCl_3$ **(i)** $SnCl_2$ **(j)** CS_2 **(k)** AlP **(l)** Ba_3N_2 **(m)** SO_2

2-33 If the ratio by weight of Si to H in the earth's crust is 25.8:0.90 (Table 2-14), calculate the ratio by atoms. Which is more abundant by atoms?

Additional Problems

2-34 Hydrogen occurs in nature as ^1_1H with 99.985% natural abundance and ^2_1H with 0.015% natural abundance. Calculate its atomic weight.

2-35 Give the missing atomic number Z, or identify the element. **(a)** $^{65}_{Z}Cu$ **(b)** $^{33}_{16}X$ **(c)** $^{27}_{Z}Al$ **(d)** $^{233}_{90}X$ **(e)** $^{232}_{90}X$ **(f)** $^{212}_{82}X$ **(g)** $^{212}_{83}X$

2-36 How many moles, molecules, and atoms are there in 100.0 g of each of the following species. **(a)** H_2O **(b)** H_3PO_4 (phosphoric acid) **(c)** $C_2H_6O_2$ (ethylene glycol) **(d)** IF_7 (iodine heptafluoride) **(e)** $(NH_4)_2SO_4$ (ammonium sulfate) **(f)** $Al(OH)_3$ (aluminum hydroxide)

2-37 Shortly before Mendeleev, John Newlands proposed a periodic table in which the properties of elements repeat each other after every eighth element (like the white notes on a piano keyboard). Compare this prediction with the modern table (ignoring the noble gases, which were unknown in Newlands' time). Where does it go wrong?

2-38 Potassium iodide is a compound of potassium and iodine consisting of 23.55% potassium and 76.45% iodine. Calculate the relative weights of potassium and iodine in this compound.

2-39 What is the weight in grams of one molecule of aspirin ($C_9H_8O_4$)? (Assume that each atom has the weight shown in the table of atomic weights.)

2-40 **(a)** When Aristotle said that the heavens were made of quintessence, a material not found on earth, it was not just a guess. Can you suggest what evidence he had for his statement? **(b)** It was not possible, in Aristotle's time, to test his theory. Did this make it useless as a scientific theory?

2-41 Using the atomic weight table given on the inside back cover, calculate the molecular weight of the following compounds. **(a)** ferrous oxide, FeO **(b)** ferric oxide, Fe_2O_3
(c) nitroglycerine, $C_3H_5(NO_3)_3$ **(d)** aluminum sulfate, $Al_2(SO_4)_3$
(e) potassium hydrogen tartrate (cream of tartar), $KHC_4H_4O_6$
(f) ethane as a gas, C_2H_6, and as a liquid, C_2H_6

2-42 Why is it better to choose the weight of a single isotope as a standard for atomic weight tables rather than the atomic weight of an element such as oxygen or hydrogen, as had previously been done?

3 Chemical Equations and Stoichiometry

John Dalton, 1766–1844

Engraving by Worthington

3-1 Chemical Equations

A **chemical equation** is just a shorthand notation for expressing what happens when substances react to form other substances. The first set of substances we call **reactants;** the second, **products.** The reactants disappear and are changed into the products. So we have

reactants ⟶ react to form ⟶ products

Let us consider a specific example. A laboratory preparation of hydrogen gas involves the reaction of zinc metal and hydrochloric acid. The products formed are zinc chloride and hydrogen gas. Writing this in terms of their chemical symbols:

$$Zn + HCl \longrightarrow ZnCl_2 + H_2$$

zinc and hydrochloric acid react to form zinc chloride and hydrogen

All chemical equations can be *read* in the same fashion. However, this equation, as it stands, is not complete. It is true that when zinc is added to hydrochloric acid, hydrogen and zinc chloride are formed, but in chemistry we require more from an equation than merely the formulas of the reactants and the products. We require that the equation also tell us the relative *quantities* of the substances that react and are formed. Since the above expression does not tell us that, it is only a skeleton equation. A true equation must obey the law of conservation of matter (matter may neither be created nor destroyed). That is, whatever quantities of material disappear as reactants must appear as products. This is true whether we express quantities in grams or in numbers of atoms or moles of atoms. For example, if 100 g of reactants is used up, then 100 g of products must appear. If there are 10^{20} atoms in those molecules that have disappeared, then there must be 10^{20} atoms in the new molecules which appear.*

With this in mind, let us return to the skeleton equation for the reaction between zinc and hydrochloric acid. (Of course, in order to write chemical equations, we must know exactly what the products are):

$$Zn + HCl \longrightarrow ZnCl_2 + H_2 \qquad \text{(unbalanced)}$$

As we saw previously, the chemical formula for hydrochloric acid is HCl; zinc chloride is $ZnCl_2$, and hydrogen gas is H_2. These formulas tell us that there are two atoms of chlorine in every molecule of $ZnCl_2$ and two atoms of hydrogen in every

*This is not necessarily true with respect to the number of *molecules*. We shall see many examples where the total number of molecules increases or decreases during the course of a reaction, as well as examples in which the number stays the same.

molecule of hydrogen. The formula HCl means that in a molecule of HCl, there is one atom of hydrogen and one of chlorine. The main thrust of our argument is that we cannot produce two atoms of chlorine and two atoms of hydrogen from one molecule of HCl, which contains one atom of hydrogen and one atom of chlorine. Therefore (in order for the atoms to balance), we must use two molecules of HCl (2HCl) to produce one molecule each of $ZnCl_2$ and H_2:

$$Zn + 2HCl \longrightarrow ZnCl_2 + H_2$$

When the number of atoms of every element on the left-hand side of an equation equals the number of atoms of the same element on the right-hand side of the equation, the equation is said to be *balanced* according to mass, and is a true equation. Balancing of equations is generally not difficult, but there is no unique way of doing it. One simply manipulates the coefficients, using small (1, 2, 3, etc.) whole numbers, until the equalities obtain. The only way to learn this is by practice.

Example 1. A laboratory preparation of oxygen is the decomposition (by heating) of mercuric oxide, HgO. The products are mercury, Hg, and oxygen, O_2. The skeleton (unbalanced) equation is

$$HgO \xrightarrow{\Delta} Hg + O_2 \qquad \Delta \text{ means "heat"}$$

There is one oxygen atom in a molecule of HgO. Hence, for each molecule of HgO decomposed, we obtain one oxygen atom. But the product, oxygen, contains two atoms of oxygen per molecule. We overcome this difficulty as soon as we realize that decomposition of *two* molecules of HgO will give us two atoms of oxygen, and these two will make up one molecule of O_2. So an intermediate step is

$$2HgO \longrightarrow Hg + O_2$$

However, the equation is still not balanced, because we note that the left side contains two atoms of mercury and the right side only one, so we must put a 2 in front of the Hg on the right side. Thus, the final, balanced equation is

$$2HgO \longrightarrow 2Hg + O_2 \qquad \text{(Figure 3-1)}$$

Example 2. $Na(s) + Cl_2(g) \longrightarrow NaCl(s)$.

This is a reaction between solid (s) sodium metal and gaseous (g) diatomic chlorine to yield solid sodium chloride. If a liquid had been involved, we would have used the symbol (l).

Intermediate step:

$$Na + Cl_2 \longrightarrow 2NaCl \qquad \text{(balancing the chlorine)}$$

We then balance the sodium.

Balanced equation:

$$2Na(s) + Cl_2(g) \longrightarrow 2NaCl(s)$$

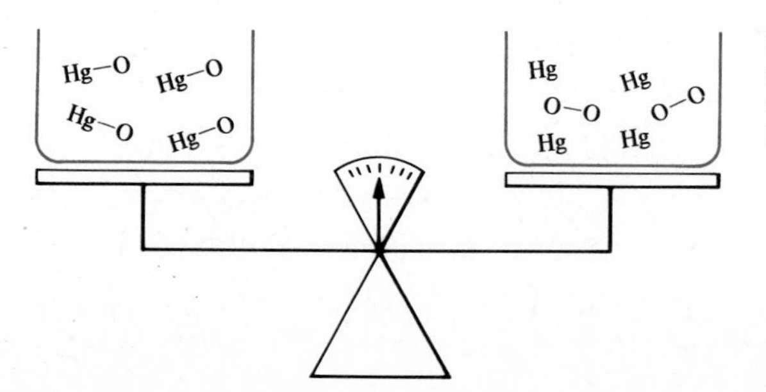

Figure 3-1 The equation is balanced because the number of atoms of every element is the same on the right and on the left.

Example 3. $NO(g) + O_2(g) \longrightarrow NO_2(g)$.

Balanced equation:

$$2NO(g) + O_2(g) \longrightarrow 2NO_2(g)$$

Example 4. $N_2(g) + O_2(g) \longrightarrow NO_2(g)$.

Balanced equation:

$$N_2(g) + 2O_2(g) \longrightarrow 2NO_2(g)$$

Example 5. $KClO_3(s) \longrightarrow KCl(s) + O_2(g)$.

Balanced equation:

$$2KClO_3(s) \longrightarrow 2KCl(s) + 3O_2(g)$$

The balanced equation, $2KClO_3(s) \longrightarrow 2KCl(s) + 3O_2(s)$, is read: "two molecules of potassium chlorate yield two molecules of potassium chloride and three molecules of oxygen." It would also be mathematically and chemically correct to use any multiple of these numbers, although the chemical convention in writing equations is to use *only the lowest integral values*. That is, the equation also could be read as

4 molecules of $KClO_3$ yield 4 molecules of KCl plus 6 molecules of O_2

or

20 molecules of $KClO_3$ yield 20 molecules of KCl plus 30 molecules of O_2

or, since a mole of atoms or molecules is 6.02×10^{23} units of that atom or molecule,

2 moles of $KClO_3$	yield	2 moles of KCl	plus	3 moles of O_2
($2 \times 6.02 \times 10^{23}$ molecules)		($2 \times 6.02 \times 10^{23}$ molecules)	+	($3 \times 6.02 \times 10^{23}$ molecules)

We may also convert moles to weights. That is, since 1 mole of $KClO_3$ weighs 122.55 g, 1 mole of KCl weighs 74.55 g, and 1 mole of oxygen weighs 32.00 g:

245.10 g of $KClO_3$ yields 149.10 g of KCl plus 96.00 g of O_2

or

2.55 g of $KClO_3$ yields 1.55 g of KCl plus 1.00 g of O_2

The preceding examples should clarify the idea that chemical equations can be read *quantitatively* in terms of molecules, moles, or grams. As a final example, we show how these may be interrelated in the same chemical equation.

2 moles of $KClO_3$ yield 149.10 g of KCl plus $3 \times 6.02 \times 10^{23}$ molecules of O_2

The chemical reactions we have been studying proceed to completion. By this, we mean that the reactants are completely used up in forming products, so that at the end of the reaction we have obtained the largest quantity of products possible, given the amount of reactants we started with. However, there are many chemical reactions that do not proceed to completion. Even in these cases, the law of conservation of matter holds, and the amount (whether grams or atoms) of reactants that disappears is exactly equal to the amount of products formed, but in such cases not all of each reactant is used up. Some portion of each of them remains and is not converted to products, even though for all practical purposes the reaction has stopped. Reactions of this kind constitute the subject of chemical equilibrium and will be discussed in Chapter 8.

3-2 Net Ionic Reactions

Many reactions are carried out simply by mixing the reactants together. When the reactants are liquids or gases, this is often a satisfactory procedure, but when two or more solids are mixed, the rate of the reaction is usually too slow to be of value. The reason for this is very simple. Most molecules of a solid are inside and not on the surface, even if the solid is in the form of a very fine powder, and molecules can only react when they touch or nearly touch each other. In order to get a reaction to proceed at a satisfactory rate, chemists often employ a **solvent,** which is a liquid in which all the reactants dissolve. Even for reactions between two liquids a solvent is frequently used, this time for precisely the opposite reason. In many cases, the reaction would be too rapid (and hence difficult to control) if the liquids were not diluted by the solvent. The most common solvent in chemistry is water, not only for reactions carried out by chemists, but also for reactions that take place in the human body and in other biological organisms.

As a solvent, water has a property that many other solvents lack. It was discovered by Svante Arrhenius (1859–1927) that *many species when dissolved in water, either break apart (dissociate) into ions or react with the water molecules to give ions as products.* Thus, reactions that take place in water may involve not only neutral compounds but also ions. As an illustration, consider a reaction between a typical acid, hydrochloric acid (HCl), and a typical base, sodium hydroxide (NaOH), in which the products are a salt (in this case, NaCl) and water. The equation is

$$HCl + NaOH \longrightarrow NaCl + H_2O$$

But hydrochloric acid reacts with water to give ions,

$$HCl + H_2O \longrightarrow H_3O^+ + Cl^-$$

and sodium hydroxide and sodium chloride are both fully ionized in water solution,

$$NaOH \longrightarrow Na^+ + OH^-$$
$$NaCl \longrightarrow Na^+ + Cl^-$$

so in terms of ions we rewrite the above equation as

$$H_3O^+ + Cl^- + Na^+ + OH^- \longrightarrow Na^+ + Cl^- + 2H_2O$$

If we examine the right-hand side (products) and the left-hand side (reactants), we see that the only species actually changing are H_3O^+, OH^-, and H_2O. (The water molecule, H_2O, is a nonionized species.) Nothing happens to the Na^+ or the Cl^-. They were there before the reaction started and are still there after it is ended. Therefore, they are neither reactants nor products and do not enter into the chemistry of the reaction. We call them **spectator ions** and consider them superfluous. It is unnecessary to include them in the reaction. The **net ionic reaction** is therefore

$$H_3O^+ + OH^- \longrightarrow 2H_2O$$

All net ionic reactions follow this principle. **We include in the reaction *only* those species that chemically change.** In order to write a net ionic reaction, we must know:

1. The reactants and products.
2. Those species (reactants and products) that are ionized in solution.

We balance an ionic equation in the same way that we balanced our molecular equations, but this time there is one extra thing we must look for: **An ionic equation must have the same total charge on both sides of the**

equation. The sum of the charges on the left must be equal to the sum of the charges on the right. (Why?)

Example: Write balanced net ionic equations (the arrow ↓ signifies that a **precipitate** (insoluble solid) is formed).

(a) $AgNO_3 + HCl \rightarrow AgCl\downarrow + HNO_3$.

(b) $Mg(ClO_4)_2 + KOH \rightarrow Mg(OH)_2\downarrow + KClO_4$.

Answer:

(a) $$Ag^+ + \cancel{NO_3^-} + \cancel{H_3O^+} + Cl^- \longrightarrow AgCl\downarrow + \cancel{H_3O^+} + \cancel{NO_3^-}$$

$$Ag^+ + Cl^- \longrightarrow AgCl\downarrow$$

The silver chloride (AgCl) forms an insoluble precipitate. (This fact, of course, must be known in order to balance the equation.)

(b) $$Mg^{2+} + \cancel{2ClO_4^-} + \cancel{2K^+} + 2OH^- \longrightarrow Mg(OH)_2\downarrow + \cancel{2K^+} + \cancel{2ClO_4^-}$$

$$Mg^{2+} + 2OH^- \longrightarrow Mg(OH)_2\downarrow$$

When, then, do reactions between ions in water solution (net ionic reactions) take place? They take place when:

1. *Insoluble compounds* are formed, which precipitate out of the solution: for example, $AgCl\downarrow$, $Mg(OH)_2\downarrow$, $Ag_2O\downarrow$, $HgS\downarrow$.
2. *Gases* are formed that leave the solution: for example, CO_2, SO_2, O_2.
3. Species are formed *that do not dissociate in the solvent* (water) used: for example, H_2O, $Ag(NH_3)_2^+$.
4. Charges on the ions change: for example, (Fe^{3+}, Fe^{2+}), (Ce^{3+}, Ce^{4+}), (Cu^+, Cu^{2+}).

If none of these things happens, no reaction takes place. For example, if we mix water solutions of sodium chloride (NaCl) and potassium bromide (KBr), all we have is a water solution of Na^+, K^+, Cl^-, and Br^- (this topic is discussed further in Section 11-1).

3-3 Percentage Composition

In Section 2-3, we learned how to calculate the molecular weight of a compound from its formula. There is another piece of information we can get from the formula of a compound. It is the *percentage composition* of the elements in the compound. The calculation is simple.

Example: What is the percentage of carbon in methane, CH_4?

Answer: In calculating a percentage, we can assume any weight of methane we wish. It is simplest to assume exactly 1 mole. In order to find the percentage of carbon in 1 mole of methane, we need to know (1) the weight of 1 mole of methane and (2) the weight of the carbon in 1 mole of methane.

(1) The weight of 1 mole of methane is just the molecular weight in grams, which is

$$\begin{array}{lll} C: & 12.011 & 12.011 \\ H: & 1.0079 \times 4 = & \underline{4.0316} \\ & & 16.043\ g \end{array}$$

(2) One mole of methane contains 1 mole of carbon atoms, the weight of which is

$$C: \quad 12.011\ g$$

The percentage of carbon in methane is

$$\frac{12.011\ g}{16.043\ g} \times 100 = 74.868\%$$

Example: What is the percentage of hydrogen in methane?

Answer: The molecular weight is the same, but this time we need the total weight of hydrogen in 1 mole of methane. Since there are four atoms of H in every methane molecule, there are 4 moles of hydrogen atoms in each mole of methane. The weight of 4 moles of hydrogen atoms is

$$4H: \quad 1.0079 \times 4 = 4.0316\ g$$

The percentage of hydrogen in methane is

$$\frac{4.0316\ g}{16.043\ g} \times 100 = 25.130\%$$

The procedure in calculating percentage composition is always the same: calculate the *total* weight of the element that is found in 1 mole of the compound and divide by the molecular weight of the compound. The total percentage should equal 100%, except for errors caused by rounding off:

%C in methane	74.868%
%H in methane +	25.130%
	99.998%

Example: What is the percentage composition (percent of each element) in ammonium phosphate, $(NH_4)_3PO_4$?

Answer: In Section 2-3, we calculated the molecular weight of $(NH_4)_3PO_4$ as 149.0863 g/mole. We get the weight of 1 mole of each element and percentage composition as we did before (note that 1 mole of $(NH_4)_3PO_4$ contains 3 moles of nitrogen atoms and 12 moles of hydrogen atoms).

$$3N: \quad 14.0067 \times 3 = 42.0201\ g$$

$$\frac{42.0201\ g}{149.0863\ g} \times 100 = 28.1851\%$$

$$12H: \quad 1.0079 \times 12 = 12.0948\ g$$

$$\frac{12.0948\ g}{149.0863\ g} \times 100 = 8.11262\%$$

$$P: \quad 30.97376\ g$$

$$\frac{30.97376\ g}{149.0863\ g} = 20.77573\%$$

$$O: \quad 15.9994 \times 4 = 63.9976\ g$$

$$\frac{63.9976\ g}{149.0863\ g} = 42.9265\%$$

%N:	28.1851%
%H:	8.11262%
%P:	20.77573%
%O:	42.9265%
	100.0000%

It is easy to calculate the percentage composition from the formula. But a more typical situation in a laboratory requires the reverse. Chemists frequently find the percentage composition of an unknown compound by analyzing it. Their task is then to find the formula from the percentage composition. The method is less simple than the foregoing but still not very difficult.

Example: A compound contains 53.3% C, 11.1% H, and 35.6% O by weight. What is its formula?

Answer: Since we are given the percentages, we can assume any weight we wish. It is simplest to assume 100 g. This weight contains 53.3 g C, 11.1 g H, and 35.6 g O. What we have here is a ratio of *grams*. In a formula we have a ratio of *moles* (for example, the formula Na_2SO_4 means that there are 2 moles of Na atoms and 4 moles of O atoms for every S atom). Our next step is convert the ratio of grams to a ratio of moles, which we can do by calculating how many moles of each element are in the 100 g of compound. To get the number of moles, we simply divide the number of grams by the atomic weight of that element.

$$\text{C:} \quad \frac{53.3 \text{ g}}{12.011 \text{ g/mole}} = 4.44 \text{ moles C}$$

$$\text{H:} \quad \frac{11.1 \text{ g}}{1.0079 \text{ g/mole}} = 11.0 \text{ moles H}$$

$$\text{O:} \quad \frac{35.6 \text{ g}}{15.9994 \text{ g/mole}} = 2.23 \text{ moles O}$$

We might write the formula as $C_{4.44}H_{11.0}O_{2.23}$, but of course it is customary to use integers in formulas and not decimals. We must convert to integers, but *without changing the ratio of moles.* The simplest way to begin this task is to divide each number by the smallest.

$$\text{C:} \quad \frac{4.44}{2.23} = 1.99 \approx 2$$

$$\text{H:} \quad \frac{11.0}{2.23} = 4.93 \approx 5$$

$$\text{O:} \quad \frac{2.23}{2.23} = 1.00 \approx 1$$

$$\text{Formula} = C_2H_5O$$

In the above case, and in many others, when we do this we find that all the numbers come out to be integers (not exactly integers, because the analyses can never give exact results, but close enough), and the formula is C_2H_5O. In other cases, the numbers might not all be integers, in which case we multiply by a common factor until they are all integers. For example, suppose that we had

C: 3.49
H: 7.01
O: 1.00

We would then multiply by 2, to get

C: $6.98 \approx 7$
H: $14.02 \approx 14$
O: $2.00 \approx 2$

and the formula would be $C_7H_{14}O_2$.

Having performed the calculation, do we now have the molecular formula? If we think about it a little, we will see that this calculation alone cannot give us the molecular formula. All it gives us is the *ratio* of atoms in the formula. If the answer comes out C_2H_5O, how do we know that the molecular formula is not $C_4H_{10}O_2$, or $C_6H_{15}O_3$, or some higher formula? After all, in all of these, the atoms have the same *ratio*, and the ratio was all we got from the calculation. We cannot know the molecular formula from this type of information alone. All we get is the *simplest ratio*, which we call the **empirical formula.**

Example: The molecular formula of benzene is C_6H_6. What is the empirical formula?

Answer: The empirical formula is CH. The compound acetylene, C_2H_2, has the same empirical formula, as do other compounds—C_4H_4, C_8H_8, and so on.

In order to get the molecular formula from the percentage composition we need an extra piece of information, namely, the molecular weight. We do not need the exact molecular weight; an approximation will do, because what we really need to know is by what integer to multiply the empirical formula.

Example: The empirical formula of a compound is BH_3. An inexact laboratory investigation shows that the molecular weight is about 27 g/mole. What is the molecular formula?

Answer: Add up the weights of the atoms in the empirical formula.

B:	10.81	10.81 g
H:	1.0079 × 3 =	3.0237 g
BH_3:		13.83 g

Since the molecular weight is about 27 g/mole, the molecular formula cannot be BH_3, but if it were B_2H_6, then the molecular weight would be $2 \times 13.83 = 27.66$ g/mole, which is close to the approximate value obtained in the laboratory. The formula must therefore be B_2H_6. (B_3H_9, B_4H_{12}, and so on, obviously would have molecular weights that are too high.)

Incidentally, now that we know the molecular formula from the empirical formula and the approximate molecular weight, we can calculate the molecular weight much more precisely:

B:	2 × 10.81 =	21.62 g
H:	6 × 1.0079 =	6.0474 g
B_2H_6:		27.67 g/mole

If we have only the percentage composition and not the molecular weight, all we can calculate is the empirical formula, not the molecular formula. If we have both, we could calculate the molecular formula by the procedure shown above, but there is another, simpler procedure.

Example: A compound consisting of 62.02% C, 13.88% H, and 24.11% N has a molecular weight of 116.21. What is the molecular formula?

Answer: Since we know the weight of 1 mole (116.21 g), we can find out how many grams of each element are in 1 mole of the compound by simply using the percentages.

$$\text{C:}\quad 116.21 \times 0.6202 = 72.07 \text{ g C}$$
$$\text{H:}\quad 116.21 \times 0.1388 = 16.16 \text{ g H}$$
$$\text{N:}\quad 116.21 \times 0.2411 = 28.02 \text{ g N}$$

The number of moles of each element is found by dividing by the atomic weights.

$$\text{C:}\quad \frac{72.07 \text{ g}}{12.011 \text{ g/mole}} = 6.00 \text{ moles C/mole of compound}$$
$$\text{H:}\quad \frac{16.13 \text{ g}}{1.0079 \text{ g/mole}} = 16.00 \text{ moles H/mole of compound}$$
$$\text{N:}\quad \frac{28.02 \text{ g}}{14.0067 \text{ g/mole}} = 2.00 \text{ moles N/mole of compound}$$

The molecular formula is $C_6H_{16}N_2$.

This method always gives us the molecular formula directly. We do not divide by the lowest number, or look for a factor by which to multiply. However, remember that in order to use it, we must have the molecular weight as well as the percentage composition. Note that the method does not give us the empirical formula, but if we want that, it is easily obtained by dividing the molecular formula by the greatest common divisor (in the case above, for example, the empirical formula is C_3H_8N).

3-4 Stoichiometry

Given a wide interpretation, **stoichiometry** means weight relationships in chemical equations. We have already discussed balancing chemical equations, and now we shall see how to answer such questions as: Given a certain number of grams of a particular starting material, how many grams of a particular product can be formed? Or: If we want to prepare a certain weight of a given product, what weight of starting materials will we need?

To answer such questions, we *must* begin with a balanced equation. Consider again the decomposition of $KClO_3$ to form KCl and O_2. The balanced equation is

$$2KClO_3 \longrightarrow 2KCl + 3O_2$$

We will use the technique of dimensional analysis (Section 1–3) and so will need conversion factors. As we have already seen, the conversion factors for this reaction are as follows (the symbol $\equiv$ means *equivalent to*):

In terms of moles:

$$2 \text{ moles of } KClO_3 \equiv 2 \text{ moles of KCl} \equiv 3 \text{ moles of } O_2$$

In terms of grams:

$$245.10 \text{ g of } KClO_3 \equiv 149.10 \text{ g of KCl} \equiv 96.00 \text{ g of } O_2$$

In terms of molecules:

$$2(6.02 \times 10^{23}) \text{ molecules } KClO_3 \equiv 2(6.02 \times 10^{23}) \text{ molecules of KCl} \equiv 3(6.02 \times 10^{23}) \text{ molecules of } O_2$$

We can also mix them; that is, it is also true that

$$2 \text{ moles of } KClO_3 \equiv 149.10 \text{ g of KCl} \equiv 3(6.02 \times 10^{23}) \text{ molecules of } O_2$$

Important points to remember are:

1. You must have the *balanced* equation; otherwise, your conversion factors will be incorrect.
2. The conversion factors above apply *only* to the chemical equation $2KClO_3 \rightarrow 2KCl + 3O_2$. Each chemical equation has its own conversion factors.
3. When we write 245.10 g of $KClO_3 \equiv$ 96.00 g of O_2, we understand that this is not a true equality in the mathematical sense. What is meant is that it takes 245.10 g of $KClO_3$ to produce 96.00 g of oxygen by the equation given above.

Example: How many grams of O_2 will be produced by the decomposition of 4.0 g of $KClO_3$?

Answer: The balanced equation is

$$2KClO_3 \longrightarrow 2KCl + 3O_2$$

The conversion factor is 245.10 g $KClO_3 \equiv$ 96.00 g O_2. Therefore,

$$4.0\text{ g }\cancel{KClO_3} \times \frac{96.00\text{ g }O_2}{245.10\text{ g }\cancel{KClO_3}} = \frac{(4.0)(96.00)}{245.10}\text{ g }O_2$$

$$= 1.6\text{ g of }O_2$$

Example: How many grams of $KClO_3$ are needed to produce 10^6 atoms of oxygen? ($2KClO_3 \rightarrow 2KCl + 3O_2$)

Answer: The conversion factors are:

$$2\text{ atoms oxygen} \equiv 1\text{ molecule oxygen}$$
$$3 \times (6.02 \times 10^{23})\text{ molecules }O_2 \equiv 2\text{ moles }KClO_3$$
$$2\text{ moles }KClO_3 \equiv 245.10\text{ g }KClO_3$$

$$10^6\,\cancel{\text{atoms oxygen}} \times \frac{1\,\cancel{\text{molecule }O_2}}{2\,\cancel{\text{atoms oxygen}}} \times \frac{2\,\cancel{\text{moles }KClO_3}}{3 \times 6.02 \times 10^{23}\,\cancel{\text{molecules }O_2}} \times \frac{245.10\text{ g }KClO_3}{2\,\cancel{\text{moles }KClO_3}}$$

$$= \frac{(10^6)(2)(245.10)}{(2)(3 \times 6.02 \times 10^{23})(2)}\text{ g }KClO_3$$

$$= 6.79 \times 10^{-17}\text{ g }KClO_3$$

Example: The total combustion (burning in oxygen) of methane (CH_4) yields CO_2 and H_2O. If 5.0 g of CH_4 is used, how much CO_2 and H_2O are produced?

Answer: The balanced equation is

$$CH_4 + 2O_2 \longrightarrow CO_2 + 2H_2O$$

The conversion factors are (in this example we are taking atomic weights to the nearest integer):

$$16\text{ g }CH_4 \equiv 44\text{ g }CO_2 \equiv 36\text{ g of }H_2O$$

$$5.0\,\cancel{\text{g }CH_4} \times \frac{44\text{ g }CO_2}{16\,\cancel{\text{g }CH_4}} = \frac{5.0 \times 44}{16}\text{ g }CO_2$$

$$= 14\text{ g }CO_2$$

$$5.0\,\cancel{\text{g }CH_4} \times \frac{36\text{ g }H_2O}{16\,\cancel{\text{g }CH_4}} = \frac{5.0 \times 36}{16}\text{ g }H_2O$$

$$= 11\text{ g }H_2O$$

Note that if we know the weight of any compound in the reaction, whether it is a reactant or a product, we can find the weight of any other compound, whether it is written on the same side or the opposite side. Such a calculation, called a **weight-weight calculation,** may be carried out as we shall now describe. For example, suppose that we wish to prepare carbon dioxide by the treatment of calcium carbonate with hydrochloric acid.

$$CaCO_3 + 2HCl \longrightarrow CO_2 + CaCl_2 + H_2O$$

If we begin with 85.00 g of $CaCO_3$, what weight of CO_2 will be produced? The first step is to find out how many *moles* of $CaCO_3$ correspond to 85.00 g. As we have seen, we do this by dividing by the molecular weight of $CaCO_3$ (100.09 g/mole).

$$\frac{85.00\text{ g }\cancel{CaCO_3}}{100.09\text{ g }\cancel{CaCO_3}/\text{mole }CaCO_3} = \text{moles }CaCO_3$$

Second, the equation tells us that 1 mole of $CaCO_3$ is equivalent to 1 mole of CO_2. This is the conversion factor, which we now use to multiply by.

$$\frac{85.00\ \cancel{\text{g }CaCO_3}}{100.09\ \cancel{\text{g }CaCO_3}/\cancel{\text{mole }CaCO_3}} \times \frac{1\text{ mole }CO_2}{1\ \cancel{\text{mole }CaCO_3}} = \text{moles }CO_2$$

This gives us the number of *moles* of CO_2 that will be produced (if this is what we wanted, we could stop right here). Finally, we find the number of *grams* of CO_2 by multiplying the number of moles of CO_2 by the molecular weight of CO_2 (44.010 g/mole).

$$\frac{85.00\ \cancel{\text{g }CaCO_3}}{100.09\ \cancel{\text{g }CaCO_3}/\cancel{\text{mole }CaCO_3}} \times \frac{1\ \cancel{\text{mole }CO_2}}{1\ \cancel{\text{mole }CaCO_3}} \times \frac{44.01\text{ g }CO_2}{\cancel{\text{mole }CO_2}} = 37.37\text{ g of }CO_2$$

So the reaction of 85.00 g of $CaCO_3$ will produce 37.37 g of CO_2. The method consists of the three steps shown:

1. Divide the weight of the given substance by its molecular weight.
2. Multiply by a conversion factor obtained directly from the coefficients of the reaction (the substance whose weight is to be determined is placed in the *numerator* of this factor).
3. Multiply by the molecular weight of the substance whose weight is to be determined.

Note that the balanced equation is consulted only in the second step. The coefficients of the equation play no part in steps 1 or 3. The following are two additional examples.

Example:

$$C_2H_2 + 2Br_2 \longrightarrow C_2H_2Br_4$$

How many grams of bromine will it take to produce 47.3 g of $C_2H_2Br_4$, assuming enough C_2H_2 is present?

Answer:

$$\frac{47.3\ \cancel{\text{g }C_2H_2Br_4}}{345.654\ \cancel{\text{g }C_2H_2Br_4}/\cancel{\text{mole }C_2H_2Br_4}} \times \frac{2\ \cancel{\text{moles }Br_2}}{1\ \cancel{\text{mole }C_2H_2Br_4}} \times \frac{159.808\text{ g }Br_2}{\cancel{\text{mole }Br_2}} = 43.7\text{ g }Br_2$$

Example:

$$N_2 + 3H_2 \longrightarrow 2NH_3$$

How many grams of nitrogen are required to react completely with 15.0 g of hydrogen? How many grams of NH_3 will be produced?

Answer:

$$\frac{15.0\ \cancel{g\ H_2}}{2.0158\ \cancel{g\ H_2}/\cancel{mole\ H_2}} \times \frac{1\ \cancel{mole\ N_2}}{3\ \cancel{moles\ H_2}} \times \frac{28.0134\ g\ N_2}{\cancel{mole\ N_2}} = 69.5\ g\ N_2$$

$$\frac{15.0\ \cancel{g\ H_2}}{2.0158\ \cancel{g\ H_2}/\cancel{mole\ H_2}} \times \frac{2\ \cancel{moles\ NH_3}}{3\ \cancel{moles\ H_2}} \times \frac{17.0304\ g\ NH_3}{\cancel{mole\ NH_3}} = 84.5\ g\ NH_3$$

There are times when one actually performs experiments in the laboratory and, for one reason or another, one chooses to use quantities of chemicals that are not in the exact proportions demanded by the balanced chemical equation. For example, in the reaction

$$\underset{\text{lead}}{Pb} + \underset{\text{chlorine}}{2Cl_2} \longrightarrow \underset{\text{lead tetrachloride}}{PbCl_4}$$

207 g of Pb (1 mole) is equivalent to 142 g of Cl_2 (2 moles). But if only 100 g of Cl_2 were added to 207 g of Pb, then all the Cl_2 would be used up, but not all the Pb. Part of it would remain. And if we wish to calculate the amount of $PbCl_4$ formed, we must use the weight of the Cl_2 and not that of the Pb.

In such cases, one or more of the reactants are in excess, while one or more others are completely converted to products. In these instances, one has to determine which reactant (or reactants) is completely used up, since this determines the weight of the products formed and the weight of the excess reactants remaining.

Example: 10.00 g of Zn and 10.00 g of HCl react to form H_2 and $ZnCl_2$. How much H_2 is produced, which reactant is in excess, and how much of it remains after the reaction goes to completion?

Answer: The balanced equation is

$$\underset{\text{zinc}}{Zn} + \underset{\text{hydrogen chloride}}{2HCl} \longrightarrow \underset{\text{zinc chloride}}{ZnCl_2} + \underset{\text{hydrogen}}{H_2}$$

The conversion factor is

$$65.38\ g\ of\ Zn \equiv 72.922\ g\ of\ HCl \equiv 2.0158\ g\ of\ H_2$$

Since we started with 10.00 g of Zn and 10.00 g of HCl, we must first determine which reactant is in excess. How much HCl will react with 10.00 g of Zn?

$$10.00\ \cancel{g\ Zn} \times \frac{72.922\ g\ HCl}{65.38\ \cancel{g\ Zn}} = \frac{(10.00)(72.922)}{65.38}\ g\ HCl = 11.15\ g\ HCl$$

This tells us that for 10.00 g of Zn to react completely, 11.15 g of HCl is needed. However, we have only 10.00 g of HCl and therefore cannot completely use up the 10.00 g of Zn. We now know which reactant is in excess: it is the Zn. Therefore, we can base the problem on the 10.00 g of HCl, since we now know that this is the reactant that will be completely used up. First, we determine how much zinc will be used up by 10.00 g of HCl.

$$10.00\ \cancel{g\ HCl} \times \frac{65.38\ g\ Zn}{72.922\ \cancel{g\ HCl}} = 8.966\ g\ Zn$$

Hence, 8.966 g of Zn will completely react with 10.00 g of HCl and we will have $(10.00 - 8.97) = 1.03$ g of Zn left over. How much hydrogen is produced?

$$10.00\ \cancel{g\ HCl} \times \frac{2.0158\ g\ H_2}{72.922\ \cancel{g\ HCl}} = 0.2764\ g\ H_2 \text{ produced from 10.00 g of HCl and 8.966 g of Zn}$$

Problems

3-1 Define each of the following terms. **(a)** chemical equation **(b)** reactant **(c)** reaction product **(d)** solvent **(e)** spectator ion **(f)** ionic equation **(g)** precipitate **(h)** empirical formula **(i)** stoichiometry

3-2 What is meant by each of the following symbols? **(a)** $\rightarrow$ **(b)** Δ **(c)** (s) **(d)** (l) **(e)** (g) **(f)** $\downarrow$ **(g)** $\equiv$

3-3 Balance each of the following skeleton equations.
(a) $KClO_3 \rightarrow KCl + O_2$
(b) $H_2 + O_2 \rightarrow H_2O$
(c) $MnCl_2 + KOH \rightarrow Mn(OH)_2 + KCl$
(d) $Fe + H_2O \rightarrow Fe_3O_4 + H_2$
(e) $CO + H_2O \rightarrow H_2 + CO_2$
(f) $Ca(OH)_2 + H_3PO_4 \rightarrow Ca_3(PO_4)_2 + H_2O$
(g) $C_5H_{12} + O_2 \rightarrow CO_2 + H_2O$
(h) $C_6H_6 + O_2 \rightarrow CO_2 + H_2O$
(i) $Cu(NO_3)_2 \rightarrow CuO + NO_2 + O_2$
(j) $C_{24}H_{50} + O_2 \rightarrow CO_2 + H_2O$

3-4 Balance each of the following skeleton equations.
(a) $CH_4 + O_2 \rightarrow CO_2 + H_2O$
(b) $C_2H_2 + O_2 \rightarrow CO_2 + H_2O$
(c) $AlCl_3 + H_2O \rightarrow Al(OH)_3 + HCl$
(d) $P_4 + O_2 + H_2O \rightarrow H_3PO_4$
(e) $Ca_3(PO_4)_2 + C \rightarrow Ca_3P_2 + CO$
(f) $MnO_2 + HCl \rightarrow MnCl_2 + Cl_2 + H_2O$
(g) $CO_2 + H_2O \rightarrow C_6H_{12}O_6 + O_2$
(h) $CoCO_3 + O_2 \rightarrow Co_3O_4 + CO_2$
(i) $FeS + O_2 + H_2O \rightarrow Fe_2O_3 + H_2SO_4$
(j) $Fe_2O_3 + CO \rightarrow Fe + CO_2$

3-5 Write net ionic equations for the following and balance. (All take place in water solution. $\uparrow$ means that a gas is formed.)
(a) $NH_4NO_3 + NaOH \rightarrow NH_3(g) + H_2O + NaNO_3$
(b) $Zn + HCl \rightarrow ZnCl_2 + H_2\uparrow$
(c) $NaOH + H_2SO_4 \rightarrow Na_2SO_4 + H_2O$
(d) $AgNO_3 + KCl \rightarrow AgCl\downarrow + KNO_3$
(e) $Ba(NO_3)_2 + Na_2SO_4 \rightarrow BaSO_4\downarrow + NaNO_3$
(f) $Na_2SO_3 + HCl \rightarrow SO_2\uparrow + NaCl + H_2O$
(g) $H_2SO_4 + Al(OH)_3 \rightarrow H_2O + Al_2(SO_4)_3$

3-6 The following are molecular formulas of some compounds. What are the empirical formulas? **(a)** $C_{14}H_{20}N_4O_2$ **(b)** $C_6H_{12}O_6$ **(c)** $C_6H_8O_2$ **(d)** $C_{12}H_{22}O_{11}$ **(e)** N_2H_4 **(f)** N_2O_5 **(g)** $Na_2S_2O_3$

3-7 In each case you are given the empirical formula and the approximate molecular weight. What are the molecular formulas and more exact molecular weights (to as many places as the atomic weight table allows)?
(a) HO, approx. mol wt = 35
(b) HgCl, approx. mol wt = 468
(c) $C_4H_3N_3O_4$, approx. mol wt = 152
(d) $C_{20}H_{27}$, approx. mol wt = 545
(e) $C_6H_{10}O$, approx. mol wt = 400

3-8 Give the percentage composition of each element in the following compounds. **(a)** Rb_2O **(b)** Al_2S_3 **(c)** $Ca(OH)_2$ **(d)** Li_3AsO_4 **(e)** $Mg_3(PO_4)_2$ **(f)** $(NH_4)_2IrCl_6$ **(g)** $Cu(C_4H_7O_2)_2$ **(h)** $KCo(NH_3)_2(NO_2)_4$ **(i)** $Co[C_2H_4(NH_2)_2]_3Cl_3$

3-9 Consider the compound ammonium sulfate, $(NH_4)_2SO_4$. **(a)** How many oxygen atoms are there in 0.165 mole? **(b)** How many N atoms, H atoms, and O atoms will combine with 800 sulfur atoms to produce $(NH_4)_2SO_4$? **(c)** How many grams of $(NH_4)_2SO_4$ contain 8.255 moles of hydrogen atoms?

3-10 Calculate the empirical formulas, given the following percentage compositions.
(a) 15.34% Na, 84.66% I **(b)** 56.63% La, 43.37% Cl **(c)** 48.96% Pb, 24.57% Cr, 26.47% O
(d) 35.59% C, 5.226% H, 59.18% Br **(e)** 29.08% Na, 40.56% S, 30.36% O.

3-11 Calculate the empirical formulas in each case, given the following weight ratios.
(a) 2.296 g Mg, 3.028 g S, 6.045 g O
(b) 0.5402 g C, 0.05666 g H, 0.1575 g N
(c) 1.781 g C, 0.1708 g H, 0.5934 g N, 1.017 g O
(d) A 2.573-g sample of a manganese bromide contained 0.6582 g of manganese (the rest was bromine).

3-12 Morphine, one of the most powerful analgesics known, has the molecular formula $C_{17}H_{19}NO_3$. **(a)** Calculate its molecular weight. **(b)** Calculate the %C, %H, %N, and %O in morphine. **(c)** A very good grade of opium contains 14% morphine. How many grams of opium will yield 107 billion molecules of morphine?

3-13 Aspirin is a compound of carbon, hydrogen, and oxygen with the molecular formula $C_9H_8O_4$. **(a)** Calculate the %C, %H, and %O in aspirin. **(b)** If one could build aspirin molecules from atoms of carbon, hydrogen, and oxygen, how many molecules of aspirin can be made from 100.0 g C, 200.0 g H, and 86.0 g O?

3-14 Nitrous oxide, N_2O, called laughing gas, is prepared by the decomposition of ammonium nitrate, NH_4NO_3. The balanced equation is

$$NH_4NO_3 \longrightarrow N_2O + 2H_2O$$

How many moles of NH_4NO_3 must be decomposed to yield 111 g of N_2O?

3-15 A certain iron ore contains 23.5% Fe_2O_3. How much ore must be refined to produce 1000 kg of Fe metal, assuming that 100% of the iron can be recovered?

3-16 In the analysis of a compound for carbon, the compound is completely oxidized by burning in oxygen, so that all the carbon in the compound is converted to CO_2. If a sample of a compound weighing 2.737 g is burned in this way, and 3.014 g of CO_2 is obtained, what was the percentage of carbon in the original compound?

3-17 The percentage composition of a certain compound is 35.66% Ba, 47.73% W, and 16.61% O. The molecular weight is 385.19. What is the molecular formula?

3-18 A sample of a certain compound was analyzed and found to contain the following elements, by weight: 1.640 g of C, 0.1032 g of H, 0.4780 g of N, and 1.365 g of O. Its molecular weight was found to be 420.30. What is the molecular formula?

3-19 Lithium chlorate occurs as a hydrate $LiClO_4 \cdot 3H_2O$; that is, there are 3 H_2O molecules in the crystal for every molecule of $LiClO_4$. Give the percent composition of $LiClO_4 \cdot 3H_2O$ in two ways. **(a)** Treat H_2O as a separate molecule; that is, give the percentages of Li, Cl, O, and H_2O. **(b)** Give the total percentages of Li, Cl, O, and H.

3-20 A certain clay was found to contain 38% silica and 15% water. What is the percentage of silica on a dry basis (that is, if all the water were driven off)?

3-21 In the reaction

$$2H_3PO_4 + 3Ba(OH)_2 \longrightarrow Ba_3(PO_4)_2 + 6H_2O$$

how much $Ba(OH)_2$ is required to produce 85.7 g $Ba_3(PO_4)_2$, assuming that excess H_3PO_4 is present?

3-22 Cr_2O_3 can be reduced by treatment with Al, to produce Cr, with Al_2O_3 as a by-product:

$$2Al + Cr_2O_3 \longrightarrow Al_2O_3 + 2Cr$$

If we wish to produce 8.3×10^3 kg of Cr by this method, how much Al_2O_3 will be produced also?

3-23 A 27.0-g sample of impure iron was analyzed and found to contain 21.6 g of iron. How many grams of H_2 can be produced from this sample by the reaction

$$2Fe + 6H_3O^+ \longrightarrow 3H_2 + 2Fe^{3+} + 6H_2O?$$

3-24 When ferrous sulfide reacts with oxygen, ferric oxide and sulfur dioxide are produced. The balanced equation is

$$4FeS + 7O_2 \longrightarrow 2Fe_2O_3 + 4SO_2$$

Fill in the table.

	$4FeS$	+	$7O_2$	$\longrightarrow$	$2Fe_2O_3$	+	$4SO_2$
Moles	100 moles		______		______		______
Grams	______		7.00 g		______		______
Molecules	______		______		______		4 molecules
Molecules	______		______		9.1×10^{20} molecules		______

3-25 Given the balanced equation

$$3CaCl_2 + 2K_3PO_4 \longrightarrow Ca_3(PO_4)_2 + 6KCl$$

fill in the table.

	$3CaCl_2$	+	$2K_3PO_4$	$\longrightarrow$	$Ca_3(PO_4)_2$	+	$6KCl$
Moles	______		71.00 moles		______		______
Grams	62.00 g		______		______		______
Molecules	______		______		9 molecules		______

3-26 Nickel tetracarbonyl, $Ni(CO)_4$, is a poisonous yellow liquid produced by the reaction of nickel and carbon monoxide according to the equation

$$Ni + 4CO \longrightarrow Ni(CO)_4$$

(a) How many grams of carbon monoxide, CO, will react with 15.00 g of nickel?
(b) How many moles of $Ni(CO)_4$ will be produced in the process given in part (a)?

3-27 The reaction

$$CH_3OH(l) + CH_3CO_2H(l) \longrightarrow CH_3CO_2CH_3(l) + H_2O(l)$$

does not go to completion. If 43.5 g of methanol (CH_3OH), treated with excess acetic acid (CH_3CO_2H), produces 47.7 g of methyl acetate ($CH_3CO_2CH_3$), what percentage of the methanol has reacted?

3-28 Under certain conditions of heat and pressure, the reaction

$$N_2(g) + 3H_2(g) \longrightarrow 2NH_3(g)$$

goes 44.5% to completion. Under these conditions, how much H_2 must be used if we want to produce 79.4 kg of NH_3?

3-29 Aluminum reacts with Fe_3O_4 as follows:

$$8Al + 3Fe_3O_4 \longrightarrow 4Al_2O_3 + 9Fe$$

(a) If 8.55 g of Al is treated with 32.5 g of Fe_3O_4, which of the two compounds will be used up, and how many grams of the other compound will remain? **(b)** How many grams of Al_2O_3 will be produced?

3-30 Chlorine can be prepared by the reaction

$$MnO_2 + 4HCl \longrightarrow MnCl_2 + Cl_2 + 2H_2O$$

If 800.0 g of MnO_2 is mixed with 13.3 moles of HCl, how many grams of Cl_2 gas are formed?

3-31 Given the reaction

$$3Cu + 8HNO_3 \longrightarrow 3Cu(NO_3)_2 + 2NO + 4H_2O$$

(a) How many grams of HNO_3 are needed to produce 0.123 mole of NO?
(b) How many moles of Cu will react with 1.00 mole of HNO_3?
(c) For every 1.95×10^{22} molecules of H_2O produced, how many moles of NO and $Cu(NO_3)_2$ will be produced?

3-32 $PbCl_2$ reacts with H_2S as follows:

$$PbCl_2 + H_2S \longrightarrow PbS + 2HCl$$

If 100.0 g of $PbCl_2$ is treated with 75.5 g of H_2S, which of the two compounds will be used up, and how many grams of the other compound will remain?

3-33 Silver can be refined from a silver sulfide ore by the reaction

$$Ag_2S + Zn + 2H_2O \longrightarrow 2Ag + H_2S + Zn(OH)_2$$

If the cost of obtaining the ore is $57.60 per pound, would it be profitable to convert it to Ag if pure Ag sells for $4.50 per troy ounce? (1 troy oz = 31.1 g.)

3-34 Iron ore is converted to iron by a reaction that may be represented as

$$Fe_2O_3 + 3C + 3O_2 \longrightarrow 2Fe + 3CO_2$$

If iron can be sold for $181.50 per ton, how many tons of Fe_2O_3 must be refined to produce $1,000,000 worth of iron?

3-35 H_2SO_4 is produced by the following series of reactions:

$$S + O_2 \longrightarrow SO_2$$
$$2SO_2 + O_2 \longrightarrow 2SO_3$$
$$SO_3 + H_2O \longrightarrow H_2SO_4$$

(a) How many grams of SO_3 can be obtained starting with 145 g of S? **(b)** How many grams of O_2 are required to produce 25.5 g of H_2SO_4? **(c)** How many grams of H_2SO_4 can be obtained starting with 17.5 g of S?

3-36 Carbon reacts with SO_2 as follows:

$$5C + 2SO_2 \longrightarrow CS_2 + 4CO$$

(a) If 33.7 g of C reacts completely with 72.0 g of SO_2, what is the total weight of CS_2 and CO produced?
(b) The equation tells us that 5 carbon atoms react completely with 2 SO_2 molecules. If 500 C atoms are added to 200 SO_2 molecules (total molecules 700), what is the total number of molecules (CS_2 + CO) produced?
(c) In the case given in part (b), 1100 atoms undergo the reaction (500 carbon atoms, 200 S atoms, and 400 O atoms). How many atoms of each kind are found in the products? How many all together?

Additional Problems

3-37 The reaction

$$CO_2(g) + H_2(g) \longrightarrow CO(g) + H_2O(g)$$

does not go the completion. If 27.6 g of CO_2 reacts with 15.0 g H_2 to produce 11.5 g of CO, what percentage of the CO_2 has reacted?

3-38 The percentage composition of a certain compound is 58.63% C, 6.813% H, 10.52% N, and 24.03% O. The molecular weight is 266.30. What is the molecular formula?

3-39 Under certain conditions of heat and pressure, the reaction

$$H_2(g) + I_2(g) \longrightarrow 2HI(g)$$

goes 86.4% to completion. If we begin with 45.6 g of I_2, how much HI will be produced?

3-40 1.00 g of a certain species of alga absorbs 3.50×10^{-3} mole of CO_2 per hour by photosynthesis, and converts it to starch (which may be represented as $C_6H_{10}O_5$):

$$6CO_2 + 5H_2O \longrightarrow C_6H_{10}O_5 + 6O_2$$

How long would it take for the algae to double its own weight? (Assume that the rate of CO_2 intake does not change even though more algae are being produced.)

3-41 In the oxidation of NH_3 to N_2, how many grams of NH_3 must we start with to produce 536.7 g of N_2? Assume that excess O_2 is present.

$$4NH_3 + 3O_2 \longrightarrow 2N_2 + 6H_2O$$

3-42 Lavoisier discovered that mercury (Hg) will react with oxygen in the air to form an orange powder that is mercuric oxide, HgO. The equation is

$$2Hg + O_2 \longrightarrow 2HgO$$

(a) How many grams of mercuric oxide will be formed when 17.00 grams of mercury is completely converted to HgO? How many grams of oxygen must be used to accomplish this? **(b)** If 700.0 g of Hg reacts with 550 g of O_2, how many grams of HgO will be formed?

3-43 When urea, CON_2H_4, is acted upon by the enzyme urease in the presence of water, ammonia and carbon dioxide are formed:

$$CON_2H_4 + H_2O \xrightarrow{\text{urease}} 2NH_3 + CO_2$$

How many moles of ammonia will form when 17.0 g of urea is decomposed?

3-44 **(a)** Calculate the percentages of carbon, hydrogen, nitrogen, and oxygen in trinitrotoluene (TNT), $C_6H_2(CH_3)(NO_2)_3$. **(b)** How many grams of carbon are there in 125 g of TNT?

3-45 Beryl ore contains beryllium in the compound $Be_3Al_2(SiO_3)_6$. The ore contains 65% of the compound plus other inert material. Calculate the number of grams of beryllium that can be obtained from 1.00 ton of beryl ore.

3-46 Hemoglobin is a large protein molecule in the blood that reacts with O_2 to form a complex called oxyhemoglobin, which contains four molecules of O_2 for every molecule of hemoglobin.
(a) Calculate the number of hemoglobin molecules that can react with 25.32 g of oxygen.
(b) Calculate the number of oxygen molecules that are released when 0.00100 mole of oxyhemoglobin forms oxygen and hemoglobin.

3-47 The molecular weight of large molecules can sometimes be obtained by analysis of an element present in small amounts. For example, the protein hemoglobin was found to contain 0.335% iron. What is the molecular weight of hemoglobin, assuming that the protein contains one atom of iron per molecule of hemoglobin? How much would it be if there were four atoms of iron per molecule of hemoglobin?

3-48 How many grams of Fe_2O_3 can be prepared by the complete oxidation of 125 g of FeO?

$$4FeO + O_2 \longrightarrow 2Fe_2O_3$$

3-49 Given the following sequence of reactions used in the preparation of hydrazoic acid (HN_3), how many grams of each of the following are necessary to produce 855 g of HN_3? **(a)** N_2H_4 **(b)** NH_3 **(c)** O_2

$$4NH_3 + 5O_2 \longrightarrow 4NO + 6H_2O$$
$$2NO + O_2 \longrightarrow 2NO_2$$
$$2NO_2 + 2KOH \longrightarrow KNO_2 + KNO_3 + H_2O$$
$$2KNO_2 + H_2SO_4 \longrightarrow K_2SO_4 + 2HNO_2$$
$$HNO_2 + N_2H_4 \longrightarrow HN_3 + 2H_2O$$

3-50 In Section 20-4, we shall see that all the members of a class of compounds called alkenes have the formula C_nH_{2n}, where n is any integer. Do all the members of this class have the same percentage composition? If so, what are the percentages of carbon and hydrogen?

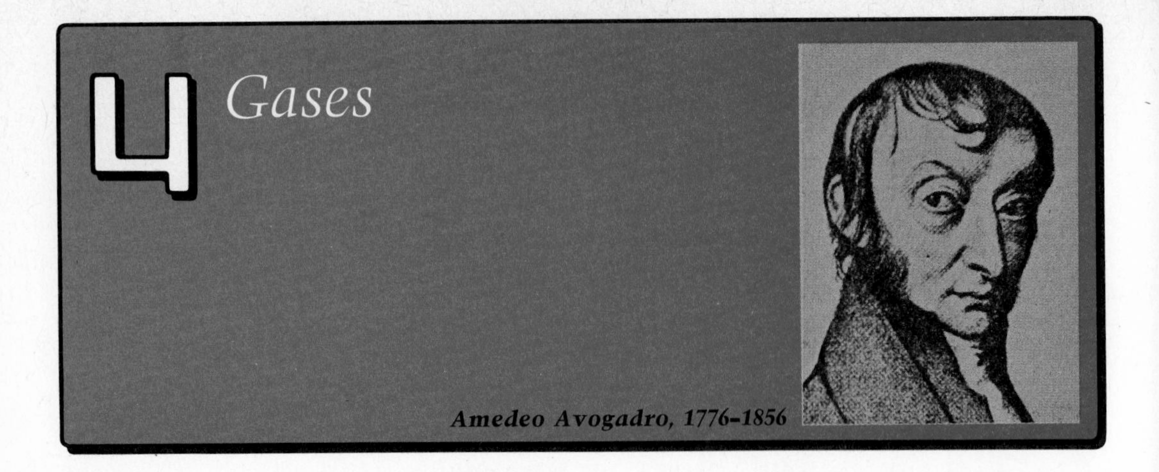

Amedeo Avogadro, 1776–1856

In this chapter, and in Chapters 6 and 15, we shall discuss the three states of matter: gases, liquids, and solids.* There is available a great amount of experimental information concerning all three states of matter. Our present state of knowledge in terms of theoretically interpreting these data is greatest for the gaseous state. The primary reason for this is that gas molecules are relatively far apart and thus influence each other to a lesser extent than they do in the other two states.

VOLUME, PRESSURE AND TEMPERATURE

4-1 Volume and Pressure

We begin our study of gases with the three fundamental measurements we can perform on any sample of a gas: the volume, the pressure, and the temperature.

VOLUME. The **volume** of any sample of gas is considered to be the space of the container that it occupies. This is so because a gas will fill any container in which it is confined, and generally the space taken up by the volumes of the molecules themselves is negligible compared to the volume of the container. The volume of a gas is most often expressed in liters.

PRESSURE. A gas is composed of molecules in rapid random motion (Figure 4-1). They frequently collide with each other and with the walls of the container. It is the collisions of the molecules with the walls of the container that give rise to what is called the **pressure** of the gas. Because the masses of the molecules are so small, they are for all practical purposes unaffected by gravity when confined to a container of normal dimensions. They are equally likely to move in any direction, including up or down, and spread themselves uniformly throughout the container, so that the pressure is the same on all sides of the container.

The gas that constitutes our atmosphere is of course not enclosed in any container. Calculations show that a gas molecule must attain a velocity of the order of 10^6 cm/s in order to escape from the Earth's gravity. Few molecules in the atmosphere can aquire this velocity, so that for discussion purposes, we may assume that the atmosphere also acts like a contained gas. Atmospheric pressure results from collisions of the gas molecules with the surface of the Earth. The weight of the

*There is also a fourth state of matter, **plasma,** but this is found only at extremely high temperatures such as are found in the center of the sun or of a hydrogen-bomb explosion. In the plasma state, all electrons are torn from the nuclei, and atoms as we know them do not exist.

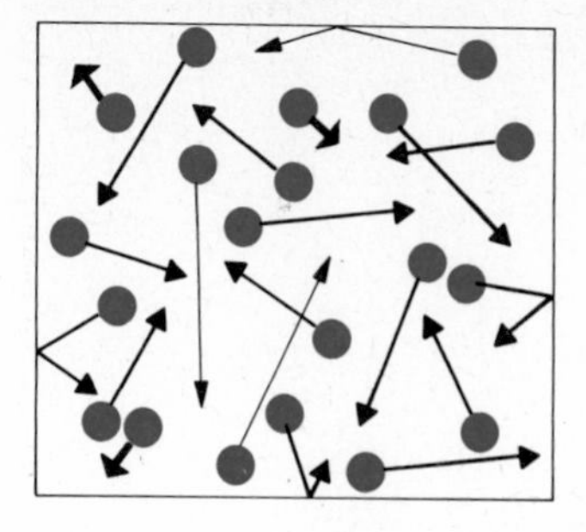

Figure 4-1 Molecules in a gas.

atmosphere is greatest at the surface of the earth and decreases with increasing altitude because of decreasing gravitational attraction (Figure 4-2). It is obvious that the higher we go, the less the amount of air above us. At 5 miles above the earth's surface, the pressure is reduced to approximately 0.3 atm based solely on gravitational forces. Those who climbed Mt. Everest needed oxygen masks, because that height (29,000 ft) is above about three-fourths of the atmosphere, and the oxygen remaining is not enough to keep a human being conscious. Modern commercial airplanes, which frequently fly at even greater heights, must have pressurized cabins. Airplane pilots use air-pressure measurements to tell them their altitudes.

Atmospheric pressure also varies with the weather, being higher in fair weather, but decreasing when a storm arrives. It is safe to say that at sea level the **atmospheric pressure** is generally about 15 lb/in^2 or 1.0×10^6 dyn/cm^2. Atmospheric pressure may be measured with an instrument called a **barometer,** invented by Evangelista Torricelli (1608–1647) in 1643.

A barometer may be constructed by filling a long (at least 80 cm) glass tube with mercury all the way to the top and then inverting the tube into a dish of mercury in such a way that no air is allowed to enter the tube. When this is done, it will be discovered that the mercury does not fill the entire tube. There is a space at the top (Figure 4-3). Since no air was allowed to enter, the space must be a vacuum (not quite a vacuum, since some molecules of mercury vapor are present, but for practical purposes a vacuum). Some mercury therefore has fallen out of the tube and increased the amount of mercury in the dish. The mercury does not rise farther up the tube because the pressure of the atmosphere on the mercury in the dish is not sufficient for that. On days when the atmospheric pressure is higher, the level in the dish will be pushed down, forcing some additional mercury into the tube so that the level in the tube rises. Therefore, the difference in height between the two mercury levels varies directly with the atmospheric pressure. Note that the *width* of the tube

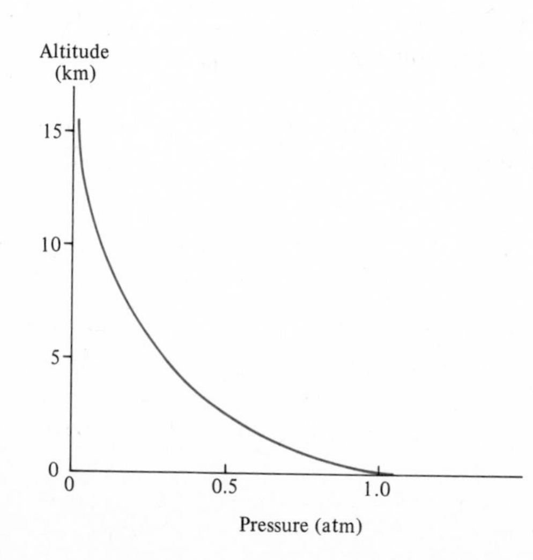

Figure 4-2 The variation of atmospheric pressure with altitude.

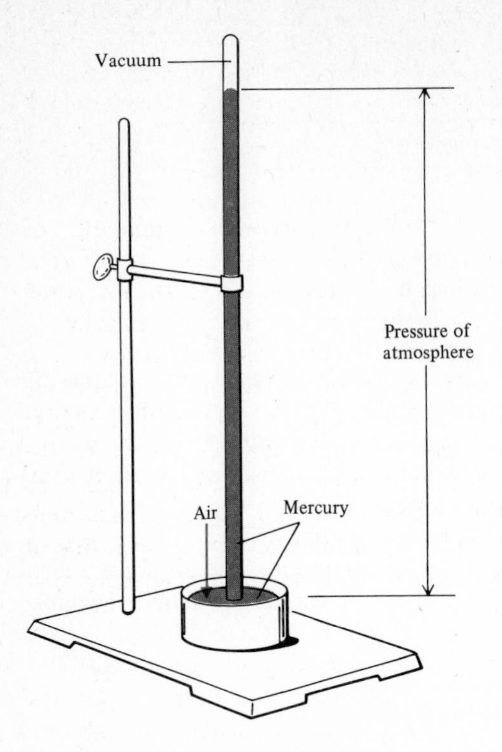

Figure 4-3 A barometer.

is immaterial because pressure is dynes *per* square centimeter, so the number of square centimeters does not affect the *pressure* (weight is a *force*, but pressure is a *force per unit area*). Because this is a direct variation, chemists and meteorologists usually find it most convenient to express atmospheric pressure directly in units of length of the column of mercury. Meteorologists in the United States use inches of mercury while those in the metric countries, and chemists all over the world, use centimeters or, more often, millimeters of mercury. In honor of Torricelli, 1 mm of mercury is also called 1 *torr*. The average pressure at sea level, 760 torr, is defined as 1 *atmosphere*. The relationship among the various pressure units is shown in Table 4-1. In this book, we will express pressure in torr or atm.

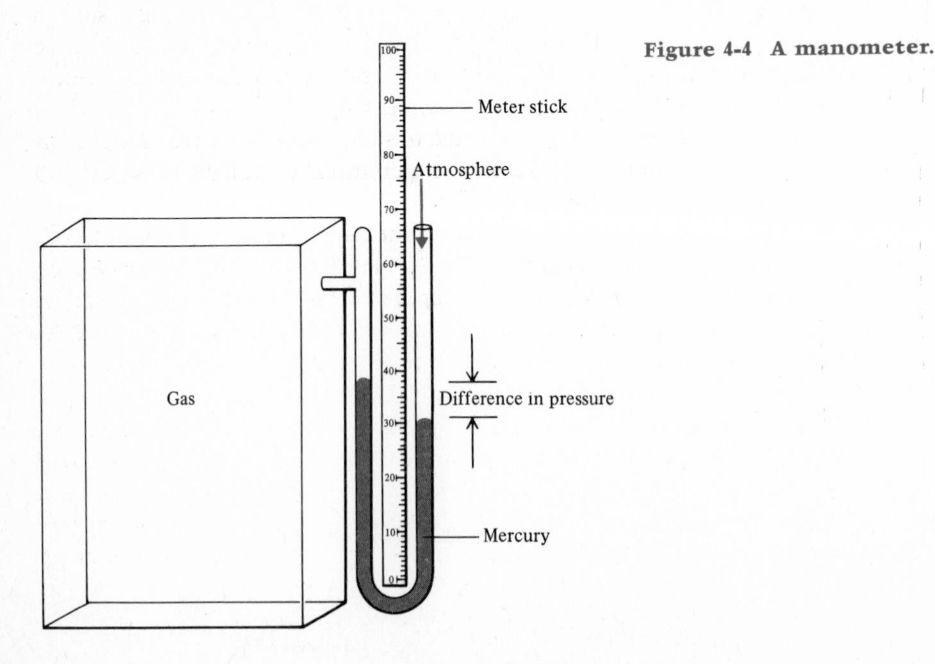

Figure 4-4 A manometer.

Table 4-1 Conversion Factors for Pressure of a Gas

$1 \text{ atm} = 760 \text{ torr} = 760 \text{ mm Hg} = 1.01 \times 10^6 \text{ dyn/cm}^2 = 14.7 \text{ lb/in}^2$

Chemists find it convenient to compare measurements with a standard. For pressure, the standard that has been universally adopted is 760 torr, or 1 atmosphere, which is taken as the *standard pressure*. Atmospheric pressure can also be measured by an *aneroid barometer*, which does not use mercury but measures pressure by the displacement of the flexible lid of a metal box nearly exhausted of air.

In order to measure the pressure of a gas confined in a container, we use a **manometer** (Figure 4-4), which is simply a U tube partially filled with mercury. One end of the tube is connected to the gas, and the other end is open to the atmosphere. The difference in height between the two levels gives directly, in torr, the difference between the pressure of the gas and that of the atmosphere (which must be independently measured by a barometer). If the level on the open side of the tube is lower than that of the closed side, the pressure of the gas is less than that of the atmosphere; otherwise, it is higher.

Example: The level of mercury on the open side of a manometer is 82 mm higher than the level of the side connected to a confined gas. Atmospheric pressure on this particular day in this particular laboratory is 737 torr. What is the pressure of the gas in the container, in torr and in atm?

Answer:

$$737 \text{ mm} + 82 \text{ mm} = 819 \text{ mm} = 819 \text{ torr}$$

$$819 \cancel{\text{torr}} \times \frac{1 \text{ atm}}{760 \cancel{\text{torr}}} = 1.08 \text{ atm}$$

4-2 Temperature

Whereas pressure and volume can readily be defined in physical terms, such a definition of temperature will have to be postponed until a discussion of kinetic molecular theory (Section 4-11). The present discussion is based on what is intuitively known; that is, temperature can be discussed in terms of "hotness" or "coldness." By touching two objects, one is able to state which body is at a higher or lower temperature. Two bodies at different temperatures, if brought together, will attain a temperature intermediate to both.

The measurement of temperature is based on the expansion of certain materials (most often mercury) with increasing temperature. The Celsius scale (the scale used in scientific work) is arbitrary. There is no special reason why water should be chosen as a standard and its freezing point be selected as 0° and its boiling point as 100°. It was done because water is a very common substance, and these are convenient numbers. When the Celsius scale was established in the eighteenth century, it was not known whether there was a highest possible or lowest possible temperature. Today we have still not found any highest possible temperature, but there definitely is a lowest possible temperature. By studying the expansion and compression of gases such as nitrogen, oxygen, or hydrogen, it was found that the volume varies *linearly* with temperature (see Section 4-4). If V_0 and V_T are defined for a specific sample of a given gas as

$$V_0 = \text{volume of gas at } 0°\text{C}$$
$$V_T = \text{volume of gas at any temperature } T$$

where T is the temperature in degrees Celsius, then it is found experimentally that

$$V_T = V_0\left(1 + \frac{T}{273}\right)$$

This means that when T is greater than 0°C, the gas will expand and V_T will be greater than V_0. On the other hand, if T is less than 0°C, the gas will take up less space and V_T will be less than V_0. Since the volume is decreasing with decreasing temperature, we may ask the question: If one continues to compress the gas, at what temperature will the volume of the gas finally reach zero and the gas disappear? In reality, this is physically a meaningless question, since any real gas will turn into a liquid if the temperature is sufficiently lowered.

But let's put the question in the following way: At what T will $V_T = 0$? The equation becomes

$$V_T = 0 = V_0\left(1 + \frac{T}{273}\right)$$

For any sample of a given gas, V_0 is a constant and therefore $1 + T/273 = 0$ is the only way this equation may be satisfied. Solving it, we get $T = -273°C$, which says that if the gas did not liquify, its volume would reach zero and the gas would completely disappear when this temperature was reached. To five significant figures this temperature is −273.15°C, called **absolute zero.** Absolute zero has never been reached by man,* but scientists have been able to come within much less than 1 degree above it.

When we deal with gases we need to use a temperature scale that goes all the way down to absolute zero. To construct a temperature scale, all we need is a fixed point and the size of a degree. The simplest way to construct an absolute temperature scale is to use absolute zero as our fixed point (we call this zero) and the Celsius degree as the size of our degree. This scale is called *Kelvin* or *absolute,* and

$$0°K = -273.15°C$$
$$273.15°K = 0°C$$
$$373.15°K = 100°C \quad \text{etc.}$$

Figure 4-5 shows the relationship between degrees C, K, and F. For most purposes we can ignore the last two places and say that 0°C = 273°K. This temperature, 273°K, or 0°C, is taken as the *standard temperature.*

*One form of the third law of thermodynamics states that it is impossible to reach absolute zero. See Section 8-11.

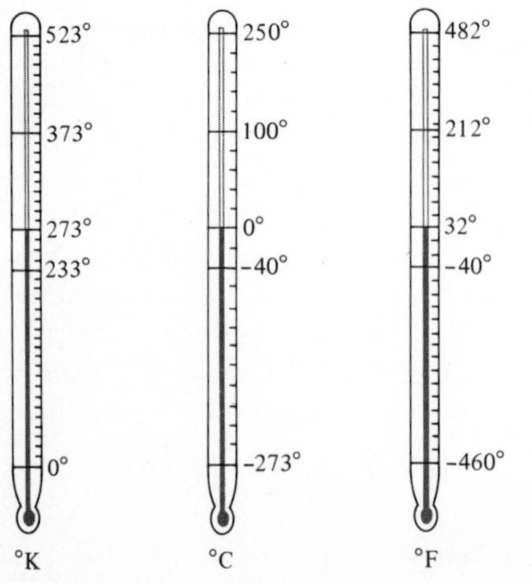

Figure 4-5 Relationships among three temperature scales. Actually, mercury thermometers cannot be used below about −10°C, or above about 360°C, and other types of thermometers must be used above and below these limits.

Standard conditions, for gases, therefore, are 273°K and 1 atm pressure. A gas at this temperature and pressure is said to be at **STP** (standard temperature and pressure).

IDEAL GASES

4-3 Boyle's Law

After Torricelli invented the barometer in 1643, it was possible to measure pressures of gases. It was not long before another scientist, Robert Boyle (1627–1691), discovered that there was a relationship *between the volume and the pressure* of a sample of gas. In 1662, Boyle performed experiments using a simple manometer (Figure 4-6). He allowed some air to be confined above the mercury on one side of the manometer (the other was open to the atmosphere) and measured the pressure and the volume of this sample of air. He obtained the pressure from the difference in height levels between the two sides, as we saw in Figure 4-4. The volume was easily ascertained by calibrating the tube with measured amounts of water before the mercury was added. Boyle then increased the volume, not by adding more air to the sample, but by lowering the right-hand tube while the left-hand tube remained fixed. He then remeasured the volume, which, of course had now changed, and the pressure, which he found had also changed. He repeated this several times, sometimes increasing the volume, sometimes decreasing it, but never adding or removing any air. He found that in all his experiments the volume of the sample multiplied by the pressure always remained the same. In later years, Boyle and other scientists found that this relationship holds not only for air, but for any gas or mixture of gases.

Today, we can express **Boyle's law** as: **At a constant temperature, the pressure of a given quantity of any gas varies inversely with the volume** or, mathematically,

$$PV = k$$

With our better measurements today, we know that Boyle's law is only approximately true, not *exactly*. Because Boyle's law and the other gas laws which we shall discuss subsequently, are so useful (and they are approximately true), chemists have invented the concept of an **ideal gas.** An ideal gas is one that perfectly obeys Boyle's and the other gas laws. No real gas is ideal, but many of them come

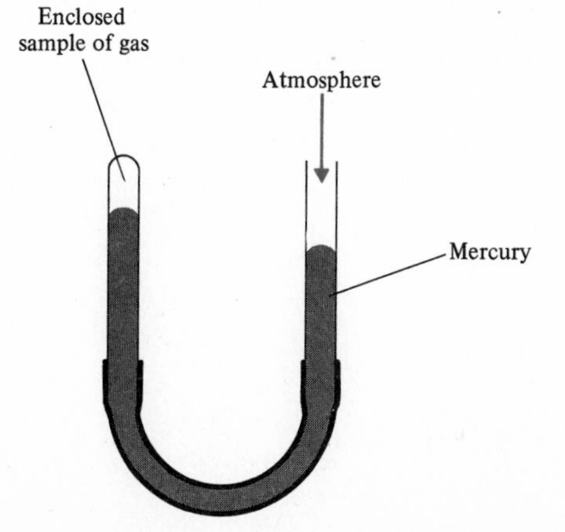

Figure 4-6 Apparatus for demonstrating Boyle's law.

remarkably close to it. Later in this chapter, we shall see why real gases are not ideal, under what conditions they approach ideal behavior, and how to deal with them mathematically.

4-4 Charles's Law

Boyle's law describes the relationship between the volume and pressure of an ideal gas when the temperature and quantity of material are constant. Similar experiments show that there is also a relationship between the volume and the temperature if the pressure and quantity of gas are held constant. Around 1802, the French chemist Joseph Gay-Lussac (1778–1850) showed that when the pressure is kept constant (Figure 4-7), an increase in the temperature of any gas results in an increased volume. No matter which gas one uses, if the experiment is performed at 0°C, each degree of temperature increase or decrease changes the volume by $\frac{1}{273}$ of its original volume.

Today we can state **Charles's law** as: **At a constant pressure the volume of a given quantity of any gas varies directly with the absolute temperature** or, mathematically,

$$V = k'T$$

Note that we must use absolute or Kelvin degrees, not Celsius. We use k' instead of k to emphasize that the constants in Boyle's and Charles's laws are not the same. (In fact, the units of k and k' are obviously different.) The reader may be wondering, if Gay-Lussac discovered the law, where does Charles come in? Jacques Charles (1746–1823) had experimented earlier with gas temperatures and volumes but did not reach conclusions that were as clear as those of Gay-Lussac. More important, he did not publish his work, so only his close acquaintances knew about it. Nevertheless, when Gay-Lussac published his results, he gave credit to Charles and called it Charles's law, and this practice has been followed every since. Furthermore, Gay-Lussac has another gas law named after him (see Section 4-6) and the use of different names helps to avoid confusion. The reader may also be wondering, if Boyle was working with volumes of gases more than a hundred years before Charles

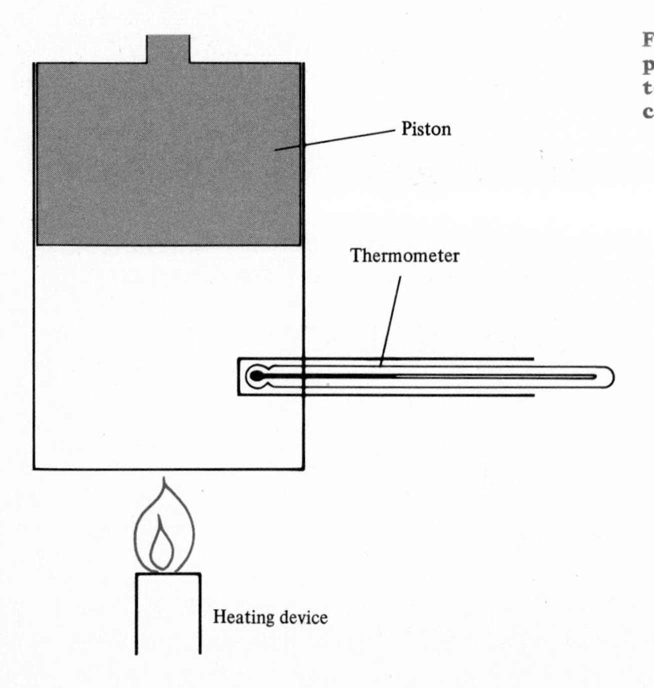

Figure 4-7 Device for keeping the pressure of a gas constant while the temperature and the volume can change.

or Gay-Lussac, why did he not discover Charles's law? The answer is that the first practical thermometers (as well as the Fahrenheit and Celsius temperature scales) were not in use until the eighteenth century, and so Boyle had no way of measuring temperature!

4-5 The Combined Gas Law

Since Boyle's and Charles's laws both hold for any sample of an ideal gas, it is possible to combine them into one law and to say: For any sample of an ideal gas, **the pressure times the volume divided by the absolute temperature is a constant,** or mathematically,

$$\frac{PV}{T} = k''$$

This law (the **combined gas law**) holds for any changes in pressure, volume, or temperature, provided only that no gas is added or taken away. The combined law contains both Boyle's and Charles's laws. It follows that if we have a sample of an ideal gas at a certain pressure, volume, and temperature, and then change any two of these quantities, we can calculate the third. For the first set of conditions, we have

$$\frac{P_1V_1}{T_1} = k''$$

When the conditions are changed, there is a new pressure (P_2), temperature (T_2), and volume (V_2), but the law says that k'' is unchanged, so that

$$\frac{P_2V_2}{T_2} = k''$$

Since both expressions are equal to k'', they are equal to each other,

$$\frac{P_1V_1}{T_1} = \frac{P_2V_2}{T_2}$$

where both sides of the equation refer to the same sample of gas, though at different conditions. If we know any five of these quantities, we can solve for the sixth. In practice, it is usually the temperature and pressure of a gas that are changed, so most often the unknown quantity is V_2. For this type of calculation, we rewrite the equation as

$$V_2 = V_1\left(\frac{P_1}{P_2}\right)\left(\frac{T_2}{T_1}\right)$$

However, we are by no means restricted to calculating a new volume. We can calculate an unknown pressure P_2 or temperature T_2 as follows:

$$P_2 = P_1\left(\frac{V_1}{V_2}\right)\left(\frac{T_2}{T_1}\right)$$

and

$$T_2 = T_1\left(\frac{P_2}{P_1}\right)\left(\frac{V_2}{V_1}\right)$$

if the other five quantities are known.

Example: A gas occupies 13.42 liters at 150°C and 520 torr pressure. How much volume will it occupy if the temperature is changed to 100°C and the pressure to 650 torr?

Answer:

$$V_2 = \frac{13.42 \text{ liters} \times 520 \cancel{\text{torr}} \times 373 \cancel{^\circ\text{K}}}{650 \cancel{\text{torr}} \times 423 \cancel{^\circ\text{K}}} = 9.47 \text{ liters}$$

Example: A gas occupies 144 ml at 0°C and 835 torr. How much volume will it occupy if the temperature is changed to −50°C and the pressure to 1.50 atm?

Answer:

$$V_2 = \frac{144 \text{ ml} \times 835 \cancel{\text{torr}} \times (1 \cancel{\text{atm}}/760 \cancel{\text{torr}}) \times 223 \cancel{^\circ\text{K}}}{1.50 \cancel{\text{atm}} \times 273 \cancel{^\circ\text{K}}} = 86.2 \text{ ml}$$

Example: A gas occupies 18.0 liters at 25°C and 720 torr pressure. How much volume will it occupy at STP?

Answer:

$$V_2 = \frac{18.0 \text{ liters} \times 720 \cancel{\text{torr}} \times 273 \cancel{^\circ\text{K}}}{760 \cancel{\text{torr}} \times 298 \cancel{^\circ\text{K}}} = 15.6 \text{ liters}$$

4-6 Avogadro's Law

In 1809, Gay-Lussac decided to repeat a certain experiment that had been carried out by Henry Cavendish in the early 1780s. Cavendish (1731–1810) had previously discovered hydrogen, and also showed that hydrogen burned in oxygen to produce water. By 1809 the fact that water is a compound made up of the elements hydrogen and oxygen was well known, and the experiment that interested Gay-Lussac was one in which Cavendish reported that when oxygen gas and hydrogen gas at the same temperature and pressure reacted to form water, the *volume* ratio of hydrogen and oxygen gases which completely reacted with each other was almost exactly 2:1. Gay-Lussac was surprised that nature would do anything in exact integers and suspected that Cavendish's measurements were not accurate. So he repeated the experiments a number of times and found that, within experimental error, the ratio *was* exactly 2:1. Furthermore, when the reaction was carried out above 100°C, so that the water would not become a liquid but remain a gas (steam), then exactly 2 volumes of water vapor formed for every 2 volumes of hydrogen used up, when measured at the same temperature and pressure.

The fact that both of these ratios were small whole numbers might have been a coincidence, so Gay-Lussac went on to combine other gaseous elements (for example, nitrogen and oxygen, nitrogen and hydrogen) and found that in each case, **when measured at the same temperature and pressure, the ratios of the volumes of the gases that combined, and of gases that were products, were always small whole numbers.** This is called **Gay-Lussac's law of combining volumes.** Figure 4-8 shows some of these results.

It is important to remember that it is *volumes* which are being discussed here, not weights. Weight relationships hold for all states of matter, but the law of combining volumes holds *only* for gases and *only* when the volumes are measured at the same temperature and pressure.

At the time, Gay-Lussac's results seemed quite astonishing. Why should nature make these volume ratios small whole numbers? The answer was supplied in 1811 by Amedeo Avogadro (1776–1856), who proposed that **equal volumes of gases at the same temperature and pressure contain equal numbers of molecules.** This statement, which is true for ideal gases, is called **Avogadro's**

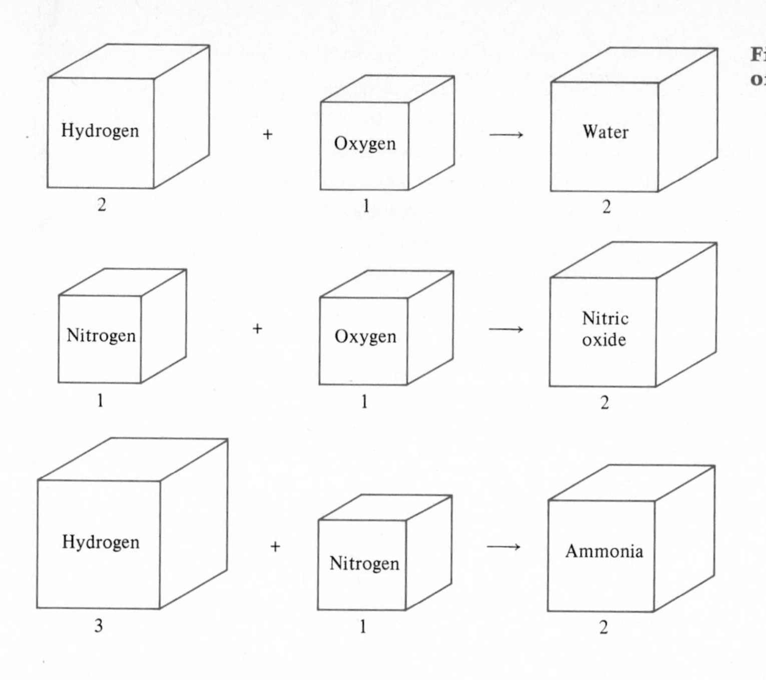

Figure 4-8 Gay-Lussac's law of combining volumes.

law.* It is easy to see that if, for example, 1 liter of nitrogen gas contains the same number of molecules as 1 liter of oxygen gas, *and* if one molecule of nitrogen reacts with one molecule of oxygen, then equal volumes of the two gases must be used up (Figure 4-9). It is important to understand that Avogadro's law holds for *any* ideal gases, no matter what they are. If we compare two equal-volume containers of ideal gas at the same temperature and pressure, then we know that they contain the same number of molecules. This is true whether the two containers hold samples of the same pure gas, or of different pure gases, or if one container holds a pure gas and the other a mixture of gases. The *identity* of the molecules is irrelevant—the *number* is always the same. Real gases do not obey Avogadro's law exactly, but in most cases the approximation is close enough so that Avogadro's law is often used for real gases as well.

*Interestingly enough, this law was not accepted by Dalton or by Gay-Lussac, each for reasons of his own. In fact, it was not accepted by most chemists until about 1860. For the full story of the reasons for this, see "*The Atomic-Molecular Theory*," by L. K. Nash in "*Harvard Case Histories in Experimental Science*", edited by J. B. Conant (Cambridge Mass.: Harvard University Press, 1957), Vol. 1, pp. 215–321.

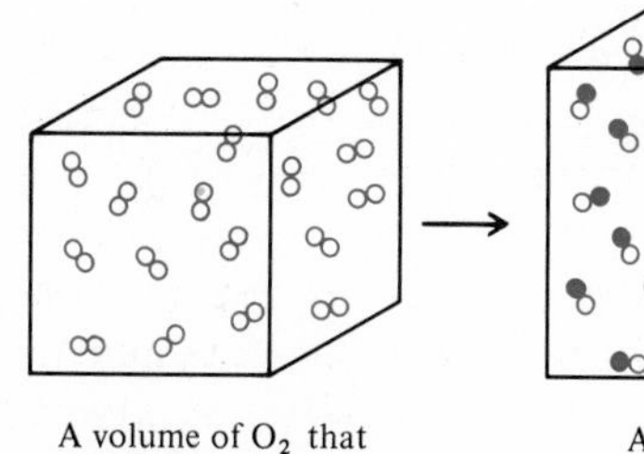
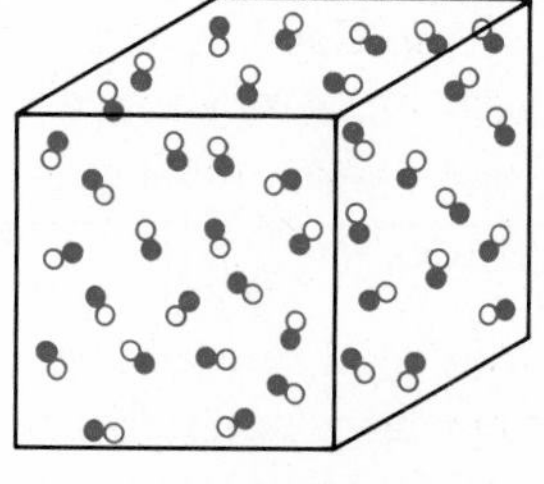

$$N_2 + O_2 \rightarrow 2NO$$

Figure 4-9 Illustrating Avogadro's and Gay-Lussac's Laws. If 20 N_2 molecules react with 20 O_2 molecules, then 40 NO molecules are produced, and because equal volumes of gases at the same temperature and pressure contain equal numbers of molecules, the volume ratio for $N_2/O_2/NO$ is 1:1:2.

As we have discussed previously, molecules are too small to deal with individually and so chemists have invented the mole, which is a unit containing 6.02×10^{23} molecules. It is plain that if two volumes contain the same number of molecules, they must also contain the same number of moles. It is also evident that at a constant temperature and pressure, the volume of a gas is directly proportional to the number of moles (or molecules). *At any given temperature and pressure there must be some volume that will contain exactly 1 mole of any ideal gas.* At STP (0°C and 760 torr) this volume (called the **molar volume**) is 22.4 liters. The molar volume was determined in laboratories by taking samples of 1 mole of known gases at STP and measuring the volume. The units of 22.4 are liters per mole, but it must be remembered that this applies only to liters of gases, and only at STP.

Example: How many liters will 55.7 g of O_2 gas occupy at STP?

Answer: The molecular weight of O_2 is 32.00 g/mole.

$$55.7\ \text{g } O_2 \times \frac{1\ \text{mole } O_2}{32.00\ \text{g } O_2} \times \frac{22.4\ \text{liters } O_2}{\text{mole } O_2} = 39.0\ \text{liters}$$

Example: How many liters will 0.3462 g of NO gas occupy at 50°C and 120 torr?

Answer: The molecular weight of NO is 30.01 g/mole. First, we determine what volume the gas will occupy at STP, and then convert to the conditions required.

$$0.3462\ \text{g NO} \times \frac{1\ \text{mole NO}}{30.01\ \text{g NO}} \times \frac{22.4\ \text{liters } O_2}{\text{mole NO}} = 0.258\ \text{liter at STP}$$

$$V_2 = \frac{0.258\ \text{liter} \times 760\ \text{torr} \times 323°\text{K}}{120\ \text{torr} \times 273°\text{K}} = 1.93\ \text{liters at 50°C and 120 torr}$$

An important use of Avogadro's law is to determine the molecular weight of an unknown gas. Since we know that 1 mole of any gas at STP occupies 22.4 liters, we may calculate the molecular weight of any gas if we know the weight of the gas, the volume, and the temperature and the pressure. If the temperature and pressure are not STP, we can easily convert them to STP.

Example: A liter of diborane gas weighs 1.24 g at STP. Find the molecular weight.

Answer:

$$\frac{1.24\ \text{g diborane}}{\text{liter diborane}} \times \frac{22.4\ \text{liters diborane}}{\text{mole diborane}} = 27.8 \frac{\text{g diborane}}{\text{mole diborane}}$$

Example: A liter of phosgene gas weighs 0.379 g at 75°C and 85.5 torr. Find the molecular weight.

Answer:

$$V_2 = \frac{(1\ \text{liter})(85.5\ \text{torr})(273°\text{K})}{(760\ \text{torr})(358°\text{K})} = 0.0858\ \text{liter at STP}$$

$$\frac{0.379\ \text{g phosgene}}{0.0858\ \text{liter phosgene}} \times \frac{22.4\ \text{liters phosgene}}{\text{mole phosgene}} = 99.0 \frac{\text{g phosgene}}{\text{mole phosgene}}$$

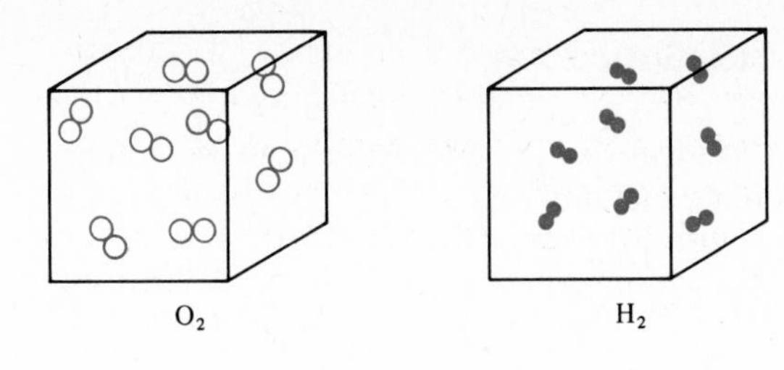

Figure 4-10 Equal volumes of O_2 and H_2 at the same temperature and pressure. If each O_2 molecule weighs 16 times as much as each H_2 molecule, then the entire sample of O_2 weighs 16 times as much as the entire sample of H_2.

If we know the weight and volume of a gas, we know the density, since

$$D = \frac{W}{V}$$

An important result about densities of gases follows from Avogadro's law. Imagine two containers of the same size, one filled with hydrogen and the other with oxygen at the same temperature and pressure (Figure 4-10). According to Avogadro's law, they contain the same number of molecules. We know that each O_2 molecule (molecular weight 32) weighs 16 times as much as each H_2 molecule (molecular weight 2). Therefore, the entire sample of oxygen gas weighs 16 times as much as the entire sample of hydrogen gas. In this case, the volumes of the two containers are the same, so that if the weight ratio of the gases is 16:1, the density ratio is also 16:1. We can demonstrate this by a simple algebraic proof. For the first container,

$$\frac{W_1}{V_1} = D_1 \quad \text{or} \quad V_1 = \frac{W_1}{D_1}$$

Similarly, for the second container:

$$V_2 = \frac{W_2}{D_2}$$

The volumes are the same, so $V_1 = V_2$ and

$$\frac{W_1}{W_2} = \frac{D_1}{D_2}$$

Thus, the ratio of the densities of the two gases is the same as that of the weights of the gases, which is the same as that of the molecular weights of the gases, and we can make the statement that **the density of an ideal gas is proportional to its molecular weight.** Thus, we can also calculate the molecular weight of a gas if we know the density, the temperature, and the pressure, *even if we do not know the weight or the volume.*

Example: The density of hydrazine gas at 800°C and 3.50 atm is 1.274 g/liter. Find the molecular weight.

Answer: We are given the density at 800°C and 3.50 atm, but we need the density at STP. This may be obtained by converting 1 liter at these conditions to STP.

$$V_2 = \frac{(1 \text{ liter})(3.50 \text{ atm})(273°\text{K})}{(1 \text{ atm})(1073°\text{K})} = 0.890 \text{ liter at STP}$$

So the density at STP is 1.274 g/0.890 liter, and

$$\frac{1.274 \text{ g hydrazine}}{0.890 \text{ liter hydrazine}} \times \frac{22.4 \text{ liters hydrazine}}{\text{mole hydrazine}} = 32.0 \frac{\text{g hydrazine}}{\text{mole hydrazine}}$$

4-7 The Ideal Gas Law

We have seen that for any sample of an ideal gas, the pressure times the volume divided by the temperature is equal to a constant:

$$\frac{PV}{T} = k''$$

Let us imagine a 1-liter container filled with a gas at 27°C and 1 atm pressure. For this case the constant is equal to

$$k'' = \frac{(1 \text{ liter})(1 \text{ atm})}{300°\text{K}} = \frac{1}{300} \frac{\text{liter} \cdot \text{atm}}{°\text{K}}$$

Note that it makes no difference which gas we have: k'' will be $\frac{1}{300}$ liter · atm/°K for 1 liter of *any* ideal gas or mixture of gases at 27°C and 1 atm, and Avogadro's law tells us that all ideal gases or mixtures of gases under these conditions have the same number of molecules or moles.

We may change any two of the three variables P, V, or T, and the other will so adjust itself so that PV/T will still be equal to $\frac{1}{300}$ liter · atm/°K *as long as we do not add any more gas to, or remove any gas from, the sample.* If we do add more gas, then, of course, k'' will increase. For example, if we add enough gas so that the number of moles is doubled, then k'' will equal $\frac{2}{300}$ liter · atm/°K. Since the constant k'' is directly proportional to the number of moles of gas present, we may replace it by another constant times the number of moles present. By universal custom this new constant is given the symbol R and is referred to as the **ideal gas constant.** We now have

$$\frac{PV}{T} = nR \qquad \text{or} \qquad PV = nRT$$

where n is the number of moles. The expression we have now derived, $PV = nRT$, is called the **ideal gas law.** It takes into account not only the pressure, volume, and temperature of a gas but also *the amount.* In words, it says that for any ideal gas or mixture of gases, the pressure times the volume divided by the temperature and the number of moles is equal to a constant, and the constant is still the same, even if some of the gas is removed or additional gas (which may be the same or a different gas) is added, since n will change accordingly.

The value of R, which is obtained by measuring P, V, and T for a gas containing a known number of moles, is

$$R = 0.08206 \text{ liter} \cdot \text{atm/mole} \cdot °\text{K}$$

Note that this is the value of R only when P, V, and T are measured in the units given. If we choose to measure P in torr or V in milliliters (or any other units), then the numerical value of R will be different. For example, if P is measured in atmospheres, T in degrees Kelvin, and V in milliliters, then $R = 82.06$ ml · atm/mole · °K. Consequently, students must be careful to match the value of R to the proper units.

Example: We wish to make a container for 2.05 moles of an ideal gas at 300 torr at 50°C. How large shall the container be?

Answer:

$$PV = nRT$$

$$V = \frac{nRT}{P}$$

$$= \frac{(2.05 \text{ mole})(0.08206 \text{ liter} \cdot \text{atm/mole} \cdot °\text{K})(323°\text{K})}{300 \text{ torr}/(760 \text{ torr/atm})}$$

$$= 138 \text{ liters}$$

As we saw in Section 2-8, the number of moles of any substance, gas, liquid, or solid, is equal to the number of grams of the substance divided by its molecular weight.

$$n = \frac{\text{g}}{\text{mol wt}}$$

We may therefore substitute this into the ideal gas law and obtain

$$PV = \frac{\text{g}}{\text{mol wt}} RT$$

It is plain that we may use this expression to find the molecular weight of an unknown gas if we know the weight of a sample of gas and its volume, temperature, and pressure. This provides an alternative method for solving molecular-weight problems of gases. For example, let us look again at two problems we did earlier (pages 92 and 93).

Example: A liter of phosgene gas weighs 0.379 g at 75°C and 85.5 torr. Find the molecular weight.

Answer:

$$PV = \frac{\text{g}}{\text{mol wt}} RT$$

$$\text{mol wt} = \frac{\text{g} \cdot RT}{PV} = \frac{(0.379\ \text{g})(0.08206\ \text{liter} \cdot \text{atm/mole} \cdot {}^\circ\text{K})(358^\circ\text{K})}{(85.5/760\ \text{atm})(1\ \text{liter})}$$

$$= 99.0 \frac{\text{g}}{\text{mole}}$$

Example: The density of hydrazine gas at 800°C and 3.50 atm is 1.274 g/liter. Find the molecular weight.

Answer:

$$PV = \frac{\text{g}}{\text{mol wt}} RT$$

$$\text{mol wt} = \left(\frac{\text{g}}{V}\right)\frac{RT}{P} = \left(1.274 \frac{\text{g}}{\text{liter}}\right) \frac{(0.08206\ \text{liter} \cdot \text{atm/mole} \cdot {}^\circ\text{K})(1073^\circ\text{K})}{3.50\ \text{atm}}$$

$$= 32.1 \frac{\text{g}}{\text{mole}}$$

Note that in this method it is not necessary to change the given conditions to STP.

4-8 Stoichiometrical Relationships Involving Gases

In Section 3-4, we learned how to calculate the weight of any component (reactant or product) of a reaction given the balanced equation and the weight of any other component (a *weight–weight* calculation). The method we used is perfectly valid for gases as well as liquids or solids, but if we are actually carrying out reactions in a laboratory, it is usually inconvenient (though not impossible) to weigh a gas.

Fortunately, the existence of Avogadro's law makes it possible to do stoichiometrical calculations on volumes of gases instead of weights. Note that this is true for *gases only*, not for liquids or solids. Recall, from page 75, that in the equation

$$Pb + 2Cl_2 \longrightarrow PbCl_4$$

1 mole of Pb is equivalent to 2 moles of Cl_2. Pb is a solid at ordinary temperatures, but Cl_2 is a gas. As we have seen, for gases at STP, 1 mole occupies 22.4 liters. Two moles of Cl_2 at STP therefore occupies 44.8 liters, and

207.2 g of Pb is equivalent to 44.8 liters of Cl_2 at STP.

We can now solve any problem of the form: Given the number of liters of any gaseous component of a reaction, how many grams of any other component (gaseous or otherwise) will be formed or used up (a **volume-weight calculation**), or: Given the number of grams of any component (gaseous or otherwise), what volume of any other component (but it must be a gas) will be formed or used up (a **weight-volume calculation**)? We use the same method we used in Chapter 3, but instead of the molecular weight of the gas, we use the constant 22.4 liters/mole. If the gas is not at STP, it can be converted to or from STP.

Example: (a weight-volume calculation)

$$Pb + 2Cl_2 \longrightarrow PbCl_4$$

What volume of Cl_2 at 800 torr and 25°C will completely react with 87.5 g of Pb?

Answer:

$$87.5\ \text{g Pb} \times \frac{1\ \text{mole Pb}}{207.2\ \text{g Pb}} \times \frac{2\ \text{moles Cl}_2}{1\ \text{mole Pb}} \times \frac{22.4\ \text{liters Cl}_2}{\text{mole Cl}_2} = 18.9\ \text{liters Cl}_2\ \text{at STP}$$

Since 2 moles of Cl_2 occupy 44.8 liters only at STP, the answer we have so far obtained, 18.9 liters of Cl_2, must correspond to STP. Since the problem requires us to use the gas at 800 torr and 25°C, we must change from STP to these conditions.

$$18.9\ \text{liters Cl}_2 \times \frac{760\ \text{torr}}{800\ \text{torr}} \times \frac{298°\text{K}}{273°\text{K}} = 19.6\ \text{liters Cl}_2\ \text{at 800 torr and 25°C}$$

Example: (a volume-weight calculation)

$$2B(s) + 3F_2(g) \longrightarrow 2BF_3(g)$$

How many grams of boron are required if we wish to produce 2.05 liters of BF_3 at 0°C and 100 torr?

Answer: First we must change to STP, because only under these conditions does 1 mole occupy 22.4 liters.

$$2.05\ \text{liters} \times \frac{100\ \text{torr}}{760\ \text{torr}} \times \frac{273°\text{K}}{273°\text{K}} = 0.270\ \text{liter at STP}$$

$$0.270\ \text{liter BF}_3 \times \frac{1\ \text{mole BF}_3}{22.4\ \text{liters BF}_3} \times \frac{2\ \text{moles B}}{2\ \text{moles BF}_3} \times \frac{10.81\ \text{g B}}{\text{mole B}} = 0.130\ \text{g B required}$$

If two of the components of a reaction are gases, we can, of course, do a weight-weight calculation, a weight-volume calculation, or a volume-weight calculation; but in this case we can also do a **volume-volume calculation:** Given the volume of one component, we can find the volume of the other. *In this case, it is* *not necessary for the gases to be at STP, as long as they are both at the same temperature and pressure,* which is usually the case. It is not even necessary to know the temperature or the pressure.

Example:

$$2C_3H_6(g) + 9O_2(g) \longrightarrow 6CO_2(g) + 6H_2O(l)$$

How many liters of CO_2 gas will be obtained from the combustion of 17.5 liters of C_3H_6 gas? How many liters of O_2 will be required? (The answer will be the same no matter what the temperature and pressure, as long as the three gases are all at the same temperature and pressure.)

Answer: Two moles of C_3H_6 are equivalent to 9 moles of O_2, which are equivalent to 6 moles of CO_2. Since equal volumes of gases at the same temperature and pressure have equal numbers of moles,

$$2 \text{ liters } C_3H_6 \equiv 9 \text{ liters } O_2 \equiv 6 \text{ liters } CO_2$$

and

$$17.5 \text{ liters } C_3H_6 \times \frac{6 \text{ liters } CO_2}{2 \text{ liters } C_3H_6} = 52.5 \text{ liters } CO_2$$

$$17.5 \text{ liters } C_3H_6 \times \frac{9 \text{ liters } O_2}{2 \text{ liters } C_3H_6} = 78.8 \text{ liters } O_2$$

To make clearer why volume–volume problems can be solved so simply, let us do the CO_2 part of the problem by the former method.

$$17.5 \text{ liters } C_3H_6 \times \frac{1 \text{ mole } C_3H_6}{22.4 \text{ liters } C_3H_6} \times \frac{6 \text{ moles } CO_2}{2 \text{ moles } C_3H_6} \times \frac{22.4 \text{ liters } CO_2}{\text{mole } CO_2} = 52.5 \text{ liters } CO_2$$

Note that in volume–volume calculations, the figure 22.4 always cancels out (provided that the gases are both at the same temperature and pressure) and we can use mole ratios alone.

4-9 Dalton's Law of Partial Pressures

In a mixture of gases, the molecules are so far apart (Figure 4-11) that they have very little influence on each other. Provided that they do not react with each other, *each gas behaves pretty much as if the others were not there;* that is, every gas is a vacuum to every other gas. This behavior is very different from that of liquids and solids, in which the behavior of each molecule is greatly influenced by its neighbors. In particular, for ideal gases, **the total pressure of a mixture of gases is equal to sum of the partial pressures of each gas.**

$$P_T = P_1 + P_2 + P_3 + \cdots$$

where P_T is the total pressure, and $P_1, P_2, P_3, \ldots$ are the partial pressures of each gas. This is called **Dalton's law of partial pressures.**

The **partial pressure** of a gas is defined as **the pressure it would exert if it were alone in the container.** For a mixture of ideal gases, $P_1V = nRT$ holds for *each* gas, independent of the others, where P_1 represents the partial pressure for that gas. This fact enables us to do calculations such as the following.

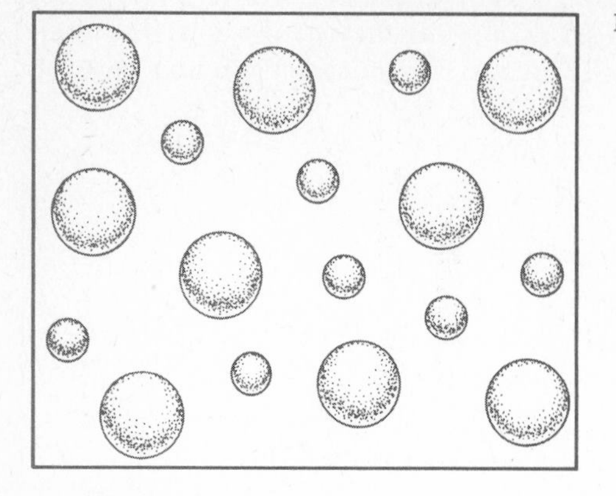

Figure 4-11 A mixture of two different gases.

Example: If 5.40 g CO_2, 3.20 g NO_2, and 15.3 g SO_3 are placed in a 3.50-liter container at 0°C, what is the total pressure in the container and the partial pressure of each gas?

Answer:

$$\frac{5.40\ \cancel{\text{g CO}_2}}{44.010\ \cancel{\text{g CO}_2}/\text{mole CO}_2} = 0.123 \text{ mole CO}_2$$

$$P_{CO_2}V = nRT$$

$$P_{CO_2} = \frac{(0.123\ \cancel{\text{mole}})(0.08206\ \cancel{\text{liter}} \cdot \text{atm}/\cancel{\text{mole}} \cdot \cancel{°\text{K}})(273\cancel{°\text{K}})}{3.50\ \cancel{\text{liters}}}$$

$$= 0.787 \text{ atm}$$

The partial pressure of CO_2 is therefore 0.787 atm. By a similar calculation, P_{NO_2} is 0.445 atm and P_{SO_3} is 1.22 atm. The total pressure is, therefore, 2.45 atm.

A corollary of Dalton's law is that, in a mixture of ideal gases, the ratio of the partial pressure of a single gas to the total pressure of the entire mixture is the same as the ratio of the number of moles of that gas to the total number of moles

$$\frac{P_i}{P} = \frac{n_i}{n} = \chi_i$$

(this number, χ_i, is called the *mole fraction*). Similarly, the ratio of the partial pressures of any two components is equal to the ratio of the number of moles of the same components:

$$\frac{P_i}{P_j} = \frac{n_i}{n_j}$$

One common application of Dalton's law is found when in a laboratory a gas is collected over water (of course, the gas must be essentially insoluble in water). For example, Figure 4-12 shows the generation and collection of oxygen. In such a case the total pressure of the gas collected is measured either by an attached manometer, or by allowing the water level in the flask to equal the water level in the trough. In either case, the pressure measured is the total pressure of the gas in the flask. However, this is not the same as the pressure of oxygen, because the flask also contains water vapor, which is present because some of the water in the flask evaporates into the gas phase above it. To obtain the partial pressure of O_2, it is necessary to determine the partial pressure of the water and subtract it from the total pressure. Fortunately, this is easy to do, because the vapor pressure of water (which is the same as the partial pressure of water vapor in the mixture of O_2 and water

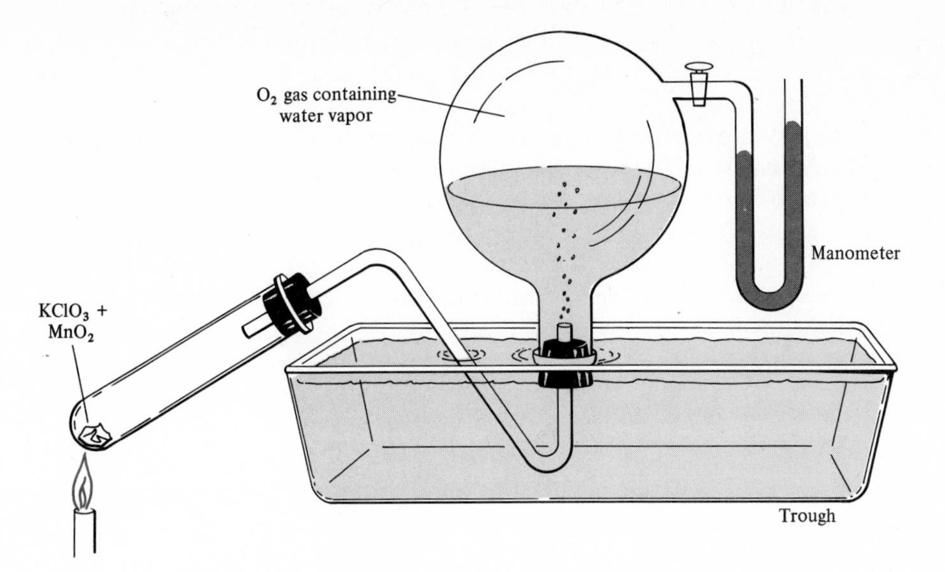

Figure 4-12 Collection of oxygen gas over water. The equation is $2KClO_3 \xrightarrow{MnO_2} 2KCl + 3O_2$.

vapor) is independent of the water level, of the size of the flask, and indeed, of everything *except the temperature* (see Section 6-2 for a discussion of vapor pressure). It is therefore only necessary to measure the temperature and look up the vapor pressure in a table (see Appendix C). Once this is obtained, we subtract to get the partial pressure of oxygen. If we also know the volume and temperature, we can calculate the number of moles and grams.

Example: A sample of 5.45 liters of oxygen is collected over water at a total pressure of 735.5 torr at a temperature of 25°C. How many moles and grams of oxygen have been collected?

Answer: At 25°C, the vapor pressure of water is 23.8 torr.

Total pressure	735.5 torr
Partial pressure, H_2O	23.8 torr
Partial pressure, O_2	711.7 torr

$$PV = nRT$$

$$n = \frac{PV}{RT} = \frac{[711.7 \text{ torr}/(760 \text{ torr/atm})](5.45 \text{ liters})}{(0.08206 \text{ liter} \cdot \text{atm/mole} \cdot {}^\circ\text{K})(298^\circ\text{K})} = 0.209 \text{ mole } O_2$$

$$0.209 \text{ mole } O_2 \times \frac{32.0 \text{ g } O_2}{\text{mole } O_2} = 6.52 \text{ g } O_2$$

4-10 Graham's Law of Diffusion

A gas expands to fill the entire container it is in, even if another gas is already present. For example, if a bit of H_2S (a poisonous gas that smells like rotten eggs) is allowed to escape in one corner of a room, it is soon possible to smell it in every part of the room, even when the room contains no air currents. This process of spreading of a gas is called **diffusion.** Similarly, a gas confined in a container at a pressure higher than the surrounding atmosphere will escape from a small hole which is opened in the container until the pressures outside and inside the container have been equalized. This process is called **effusion.** Neither diffusion nor effusion is instantaneous; both require time to operate. However, this time is not the same for all molecules, but depends on the molecular weight. **Graham's law of diffusion** states: **When compared at the same temperature and pressure, the rates of diffusion (or effusion) of any two gases are inversely proportional to the square roots of their densities.** Since the densities of gases at a given temperature and pressure are proportional to their molecular weights (Section 4-6), the rates of diffusion are also inversely proportional to the square roots of their molecular weights. These relationships may be mathematically stated as follows:

$$\frac{\text{rate}_A}{\text{rate}_B} = \sqrt{\frac{d_B}{d_A}} \quad \text{or} \quad \frac{\text{rate}_A}{\text{rate}_B} = \sqrt{\frac{M_B}{M_A}}$$

Example: What are the comparative rates of diffusion of H_2 and Cl_2 at the same temperature and pressure?

Answer:

$$\frac{\text{rate}_{H_2}}{\text{rate}_{Cl_2}} = \sqrt{\frac{M_{Cl_2}}{M_{H_2}}} = \sqrt{\frac{70.9}{2.02}} = \sqrt{35.1} = 5.92$$

Thus, H_2 diffuses 5.92 times as rapidly as Cl_2.

The fact that diffusion of gases is inversely proportional to molecular weight was used in the Manhattan Project during World War II. The uranium isotope ^{235}U can be used to make atomic bombs, but it is found in nature mixed with a much larger amount of ^{238}U, which cannot be used for this purpose (see Section 23-8). It was therefore necessary to separate these isotopes at a time when not much was known about the separation of isotopes. One of the methods developed for this purpose, which was successfully applied on a very large scale, was to pass the gas UF_6 through a porous solid at low pressure. The $^{235}UF_6$ passes through a little bit faster than $^{238}UF_6$ (the rate ratio is 1.004:1). Although the gas that first passes through has only been slightly enriched in $^{235}UF_6$, many repeated recyclings eventually result in virtually pure $^{235}UF_6$. This method is still is use today.

The Kinetic-Molecular Theory

We have already noted that real gases do not exactly obey the ideal gas laws that we have discussed in this chapter. Nevertheless, the ideal gas laws are highly useful because there are wide ranges of pressure and temperature over which, for all practical purposes, real gases behave as ideal gases and obey the ideal gas laws. In order to try to understand why gases behave the way they do, a model of an ideal gas was developed. This model, called the **kinetic-molecular theory,** assumes certain properties of gases, and from these properties theoreticians attempt to derive and understand ideal gas behavior. As we shall presently see, the assumptions are quite simple and general and yet they are able to explain most of the experimental gas laws. The next step is to develop a theory of real gases. The first approach practitioners use is to modify the existing theory by relaxing some of the original assumptions, which avoids the difficulty of developing a completely new theory.

This approach has been only partially successful for gases. Some modifications of the ideal gas law have been proposed, but they cannot completely explain or predict the behavior of dense or moderately dense gases.

4-11 The Kinetic-Molecular Theory

SKIP

It is not our purpose to present a detailed derivation of this theory but to discuss the assumptions and show qualitatively how the theory explains the ideal gas laws.

The kinetic molecular theory has four fundamental assumptions:

1. Gases are composed of molecules that are in continuous random motion. They frequently collide with each other and with the walls of the container.
2. The molecules of a gas occupy a negligible volume compared to the distance between the molecules. These volumes are so small that we neglect them and call them zero. The volume used in all calculations is the volume of the container.
3. Collisions between molecules and between the molecules and the walls of the container are perfectly elastic. The first two assumptions are easy to understand. We must, however, say something about the third assumption. An **elastic collision** between two molecules is one in which the **sum of the kinetic energies of the two molecules before collision is the same as it is after collision.** The kinetic energy of the molecules is the energy associated with the motion of the molecules. The kinetic energy we are talking about here is the energy ($\frac{1}{2}mv^2$) the molecule has as it is traveling through space. This kind of energy is also called **translational energy.** In an elastic collision, molecule A may be traveling fast and molecule B slowly before the collision; after the collision it may be B that is traveling fast and A slowly, or vice versa; but the sum of the kinetic energies remains unchanged. Elastic collisions occur when there are *no attractive or repulsive forces between molecules.* In fact, an equivalent statement of assumption 3 is: There are no attractive or repulsive forces between the molecules of a gas. Let us see what happens when there *are* attractive or repulsive forces between molecules. If molecules A and B collide and there is an attractive interaction between them, then the molecules may *stick together* for a finite (though probably short) period of time before separating. For that period of time, the number of molecules in the container and their properties are changed. If they spring apart due to repulsive interactions, their kinetic energies have been changed. In elastic collisions no energy is lost to the container or surroundings as heat, nor do the molecules gain energy by absorbing heat from the surroundings. Saying this another way, there are no losses of energy due to *friction.*
4. The average kinetic energy of the molecules in a gas, $\frac{1}{2}m\overline{v^2}$, is constant at constant temperature T. This is really not an assumption but a result of a detailed mathematical analysis of the preceding three assumptions. We shall use it as an assumption since we are not presenting its derivation.

We assume that the average kinetic energy $\frac{1}{2}m\overline{v^2}$ is proportional to the absolute temperature, T.

$$\frac{1}{2}m\overline{v^2} = \text{constant} \times T \qquad (1)$$

where $\overline{v^2}$ is the average of the square of the velocity of the molecules in the container (called the **mean-square velocity**). We use the average since all the molecules do not have the same velocities. The detailed derivation of equation (1) shows that the constant is $3R/2N$, so

$$\tfrac{1}{2}m\overline{v^2} = \frac{3RT}{2N} \qquad \text{where } N = \text{Avogadro's number}, \quad R = \text{gas constant}, \quad m = \text{mass of the molecule} \tag{2}$$

Rearranging, we have

$$\tfrac{1}{2}(Nm)\overline{v^2} = \frac{3RT}{2} \tag{3}$$

Now, Nm is just the molecular weight, M, so that

$$\tfrac{1}{2}M\overline{v^2} = \tfrac{3}{2}RT \tag{4}$$

To use equation (4) properly, the units of R must be correct. Kinetic energy, $\frac{1}{2}m\overline{v^2}$, has the units of ergs (grams · cm²/s²), so the units of R must be erg/mole · °K. The value of R in these units is 8.31×10^7 erg/mole · °K. If we know the absolute temperature and the molecular weight of the gas, we may calculate $\overline{v^2}$, the mean-square velocity, or $(\overline{v^2})^{1/2}$, the root-mean-square velocity, of the molecules in the gas. The root-mean-square velocity provides a good measure of the average speed of the molecules in a gas.

Example: Calculate $\overline{v^2}$ and $(\overline{v^2})^{1/2}$ for gaseous nitrogen at 25°C.

Answer: From equation (4), we have

$$\overline{v^2} = \frac{3RT}{M} = \frac{3(8.31 \times 10^7 \text{ erg/}\cancel{\text{mole}} \cdot \cancel{°\text{K}})(298\cancel{°\text{K}})}{28.01 \text{ g/}\cancel{\text{mole}}}$$

$$= 2.65 \times 10^9 \frac{\text{erg}}{\text{g}}$$

$$(\overline{v^2})^{1/2} = \sqrt{\frac{3RT}{M}} = 5.15 \times 10^4 \frac{\text{cm}}{\text{s}}$$

Thus, the average velocity of a molecule of N_2 at 25°C is approximately 5.15×10^4 cm/s (or 1200 miles/h).

According to the kinetic-molecular theory, the *pressure* of a gas results from the collision of the molecules with the walls of the container. The absolute *temperature* is proportional to the average kinetic (translational) energy of the molecules. Thus, Boyle's law can be explained as follows. When the volume is decreased, the molecules in the container have less space in which to move, but the average kinetic energy is unchanged (because the temperature is unchanged). Therefore, the molecules must hit the walls more often, and the pressure is greater (Figure 4-13). To explain Charles's law, we know that when the temperature of a gas is raised, the

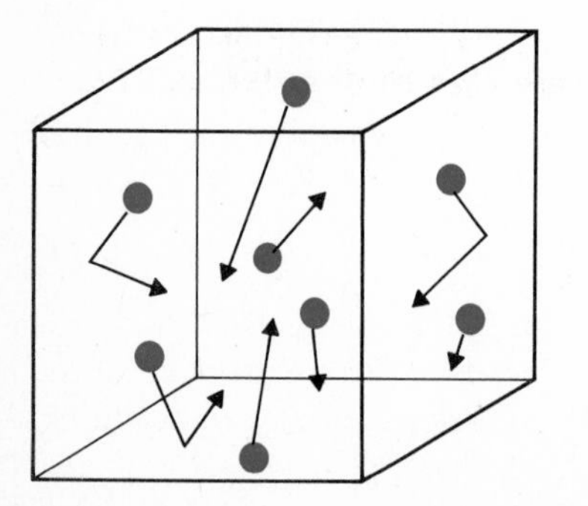

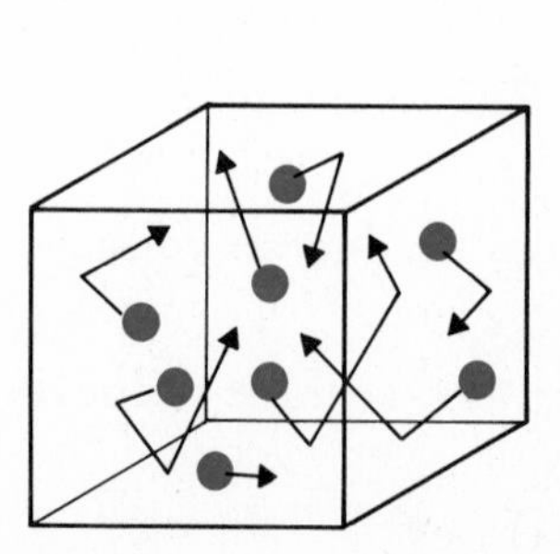

Figure 4-13 Pressure is caused by molecules striking the walls of a container. When the volume is decreased at constant temperature, the molecules hit the walls more often.

molecules are, on the average, moving faster. Thus, if the volume of the container is constant, the pressure must rise, because the faster molecules will hit the walls with greater impact. However, Charles's law is based on the pressure being constant. (For example, the gas might be confined by a piston: Figure 4-7.) In such a case the increased impact of the collisions must raise the volume. Dalton's law of partial pressures is easily explained by the assumptions that molecules occupy essentially no space and exert no forces of attraction or repulsion on each other. In a mixture of gases, the molecules of a given gas will strike the walls of the container just as hard and just as often as if the other gases were not there. Thus, the partial pressure of one gas is independent of the partial pressures of any other gases that may be present. These assumptions also explain why the gas laws are equally valid for any pure gas or any mixture of gases (assuming ideal behavior). If the molecules occupy essentially no space, and do not attract or repel each other, there is nothing to make one gas or mixture of gases behave (as far as the gas laws are concerned) differently from any other.

We have stated that the absolute temperature of a gas is proportional to the average kinetic energy of the molecules. Thus, the average kinetic energy for any two gases at the same temperature is the same. Since kinetic energy is the product of two terms, the mass and the square of the velocity,

$$\text{K.E.} = \tfrac{1}{2}m\overline{v^2}$$

it follows that the molecules of a gas with a large value of m will have a small average value of v^2, and vice versa. In other words, heavy molecules move more slowly, on the average, than do light molecules at the same temperature. Table 4-2 shows average speeds for the molecules of some common gases at 298°K (25°C). Note that even heavy molecules are moving very rapidly at 25°C, and light molecules such as H_2 or He are moving faster than the speed of sound (about 3.4×10^4 cm/s at sea level).

However, it is important to stress that these speeds are *average* speeds. The molecules are by no means all moving at the same velocity. In any sample of a gas, some molecules are moving much faster and others much slower than the average. There is a distribution of velocities. Furthermore, the velocity of a given molecule changes every time it undergoes a collision with another molecule or with the walls of the container, and these collisions are very frequent, typically taking place millions of times per second. Despite this chaotic behavior, it is possible to calculate the distribution of energies or speeds among the molecules of a gas at any temperature. The methods for doing so, which were developed by James Clerk Maxwell (1831–1879) and Ludwig Boltzmann (1844–1906), are beyond the scope of this book, but Figure 4-14 shows the results of a typical calculation: the distribution of energies and speeds of N_2 molecules at two temperatures: 298°K and 1500°K. From this graph we may make a number of statements: (1) The most probable speed at 298°K is about 4×10^4 cm/s, while at 1500°K, as we might expect, the most probable speed

Table 4-2 Average Speed of Some Gas Molecules at 25°C

Gas	*Molecular weight*	*cm/s*	*miles/h*
H_2	2	1.77×10^5	3960
He	4	1.26×10^5	2820
H_2O	18	0.59×10^5	1320
N_2	28	0.47×10^5	1060
O_2	32	0.44×10^5	990
CO_2	44	0.38×10^5	845
Cl_2	71	0.30×10^5	671
Hg	201	0.18×10^5	400

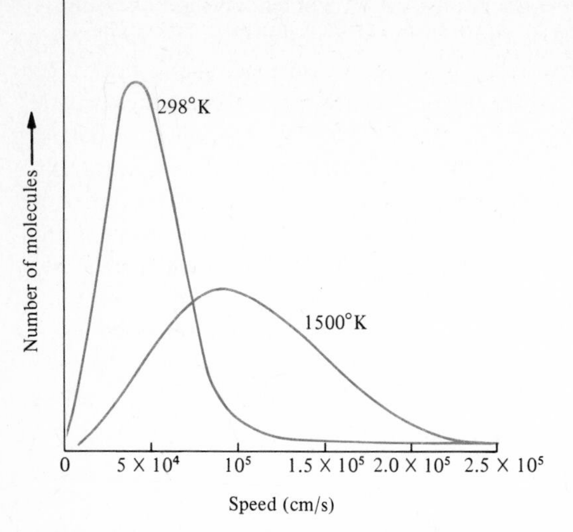

Figure 4-14 The distribution of speeds and kinetic energies of N_2 molecules at 298°K and 1500°K. [Redrawn from *Physical Chemistry*, 3rd ed., by Gordon M. Barrow. Copyright 1973 by McGraw-Hill, Inc. Used with permission of McGraw-Hill Book Company.]

has increased (to about 9×10^4 cm/s). (2) The range of speeds is narrower at the lower temperature, and considerably broadens as the temperature increases. (3) Consequently, at the lower temperature, a larger fraction of the molecules possesses the most probable speed than at the higher temperature. (4) Very few molecules are moving at speeds close to zero at either temperature. (5) At 298°K, very few molecules are moving at very high speeds (for example 2×10^5 cm/s), but at 1500°K there are quite a few molecules at this speed.

REAL GASES

4-12 Deviations from Ideality

That real gases deviate from ideal behavior may be seen in Figures 4-15 and 4-16. Boyle's law states that the pressure times the volume for a given quantity of any gas at a constant temperature should equal a constant. For 1 mole of an ideal gas at standard temperature (273°K), the constant is 22.4 liter · atm. Figures 4-15 and 4-16 show that this ideal behavior does not hold for real gases and that, in fact, the

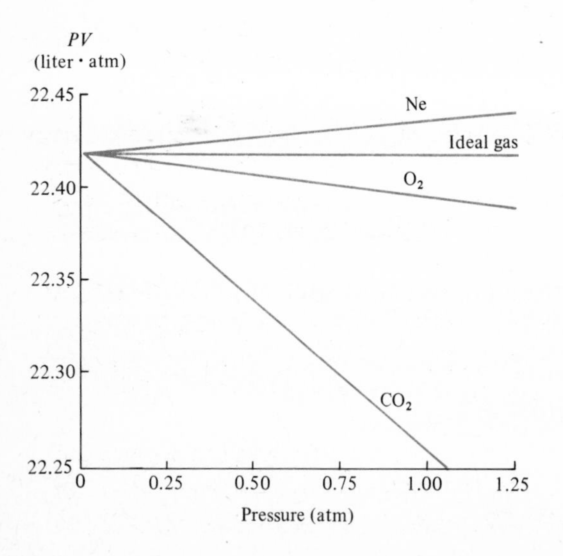

Figure 4-15 How *PV* varies with pressure for a number of gases at ordinary pressures. [Redrawn from *Physical Chemistry*, 3rd ed., by Gordon M. Barrow. Copyright 1973 by McGraw-Hill, Inc. Used with permission of McGraw-Hill Book Company.]

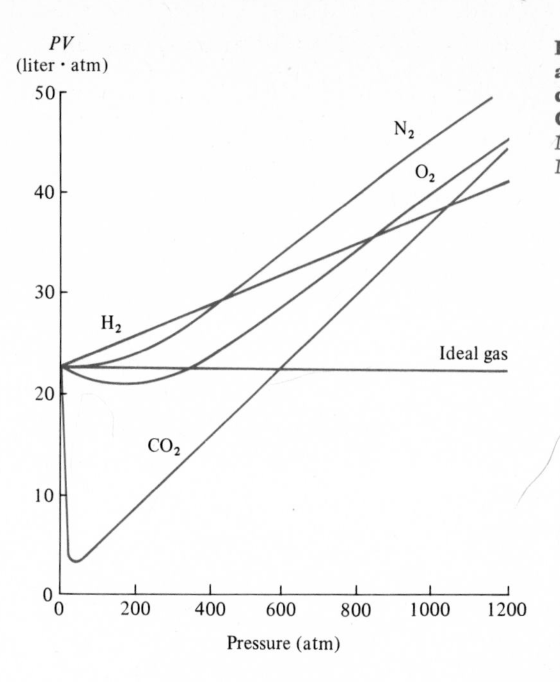

Figure 4-16 How *PV* varies with pressure for a number of gases at high pressures. [Redrawn from *Physical Chemistry*, 3rd ed., by Gordon M. Barrow. Copyright 1973 by McGraw-Hill, Inc. Used with permission of McGraw-Hill Book Company.]

product *PV* changes with pressure and with the identity of the gas. Similar deviations are found when the temperature changes. There are two main reasons why the behavior of real gases deviates from ideality, both of them connected with the assumptions of the kinetic molecular theory.

1. Although the volume of a molecule of a gas is very small compared to the distance between molecules, the molecules *do occupy some space*. When the volume of the container decreases, the volume of the molecules begins to occupy a sizable proportion of the container. The ideal gas equations are actually based on the assumption that molecules occupy *no* space at all.
2. There really *are* attractive and repulsive forces between molecules. When the molecules are far apart, these interactions are negligible. However, as the volume of the container becomes smaller and the distances between molecules decrease, these interactions become important and cause the behavior of the gases to deviate from ideality. A discussion of the various types of interactions between molecules is given in Section 6-6.

If real gases deviate from ideal behavior because they occupy space and because they attract and repel each other, it should be possible for us to predict what type of circumstances will permit gases to behave most ideally.

1. *Low pressures*. At high pressures the molecules of a gas are crowded close to each other. Under these conditions, the forces between the molecules become most important, and the actual volume of the molecules themselves becomes a larger fraction of the total volume of the gas. Therefore, real gases behave most ideally at *low pressures*. Figure 4-15 shows this graphically.
2. *High temperatures*. Again, because the volume of a gas contracts with decreasing temperature, the molecules are closest together at low temperatures. Therefore, real gases behave most ideally at *high temperatures*. When the temperature becomes low enough, and the pressure high enough, the gas liquifies and ceases to obey any kind of gas law.
3. *Nonpolar molecules*. In Section 6-6, we shall see that some molecules are dipolar and behave like little magnets. Molecules of this type (HCl, for example) are most likely to attract each other, and so such gases will deviate most

from ideality. Real gases whose behavior is closest to ideality are those whose molecules are nonpolar and not easily polarized (examples include Ne, Ar, and N_2).

4-13 Equations of State

An **equation of state** describes completely the variables of a gas (P, V, T) in a single equation. The ideal gas law ($PV = nRT$) is an equation of state, but of course, it does not work exactly for real gases, especially at high pressures and low temperatures. Many attempts have been made to find other equations of state that would work better over wide ranges of T, P, and V. All of them suffer from one or another type of disadvantage. The empirical ones introduce additional constants which must be experimentally determined for each gas. The theoretically based equations also have additional constants, although in this case, these constants are related to specific properties of molecules which again may be determined experimentally or may be calculated from empirically proposed equations.

The best-known empirical equation of state for gases is the *van der Waals* equation:

$$\left(P + \frac{n^2a}{V^2}\right)(V - nb) = nRT \qquad (5)$$

where n = number of moles, and a and b are two new empirical constants which have to be determined experimentally for each gas under consideration. Some values of a and b are given in Table 4-3. The constant b is related to the volume of the molecule, and takes into account the fact that the space actually occupied by the molecules themselves is unavailable for the molecules to move in and hence must be subtracted from the total volume of the gas, V. The constant a is related to the forces interacting between molecules and modifies the pressure term.

The van der Waals equation provides a reasonably accurate description of gases under moderate pressures. In fact, of all the equations of state, the van der Waals equation is the only one which predicts that a gas will liquify under specific conditions. It is because of this that the van der Waals equation is the best-known equation of state for real gases.

To represent the behavior of real gases with greater accuracy, one needs more than two adjustable constants. In particular, two-parameter equations fail at high pressure. The *virial equation of state* has as many terms as are necessary to represent the experimental PVT data. For 1 mole, the virial equation has the form

$$PV = RT + BP + CP^2 + DP^3 + \cdots \qquad (6)$$

The constants B, C, D, and so on, are different for each gas.

Table 4-3 Values of a and b for Some Gases, for Use in the van der Waals Equation

	a (liter^2 · atm/mole^2)	b (liter/mole)
H_2	0.244	0.0266
He	0.0341	0.0237
CH_4	2.25	0.0428
NH_3	4.17	0.0371
H_2O	5.46	0.0305
CO	1.49	0.0399
N_2	1.39	0.0391
O_2	1.36	0.0318
CO_2	3.59	0.0427
Cl_2	6.49	0.0562

Those readers with a mathematical background will note that equation (6) has the form of a power series in the pressure P. A power series gives a very accurate description of any curve since we have as many adjustable constants as are needed to assure the best fit to the curve. It is also interesting that the virial equation has a substantial theoretical justification. The term B is called the second virial coefficient, C the third virial coefficient, and so on. For gases at low pressure, the virial equation takes the form

$$PV = RT + BP$$

end

4-14 The Critical Region

Figure 4-17 shows the isotherms of carbon dioxide near 31°C. An **isotherm** is a graph showing measurements made at a constant temperature. In this case the graphs show what happens to the volume of a sample of CO_2 as we increase the pressure while the temperature is constant. In the figure, CO_2 is a gas in the region marked G, and a liquid in the region marked L. The area below the dashed line represents those temperatures and pressures at which gas and liquid coexist. Consider the isotherm at 29.9°C. At point A, carbon dioxide is a gas. Keeping the temperature constant, the gas is compressed, which, of course, decreases the volume.

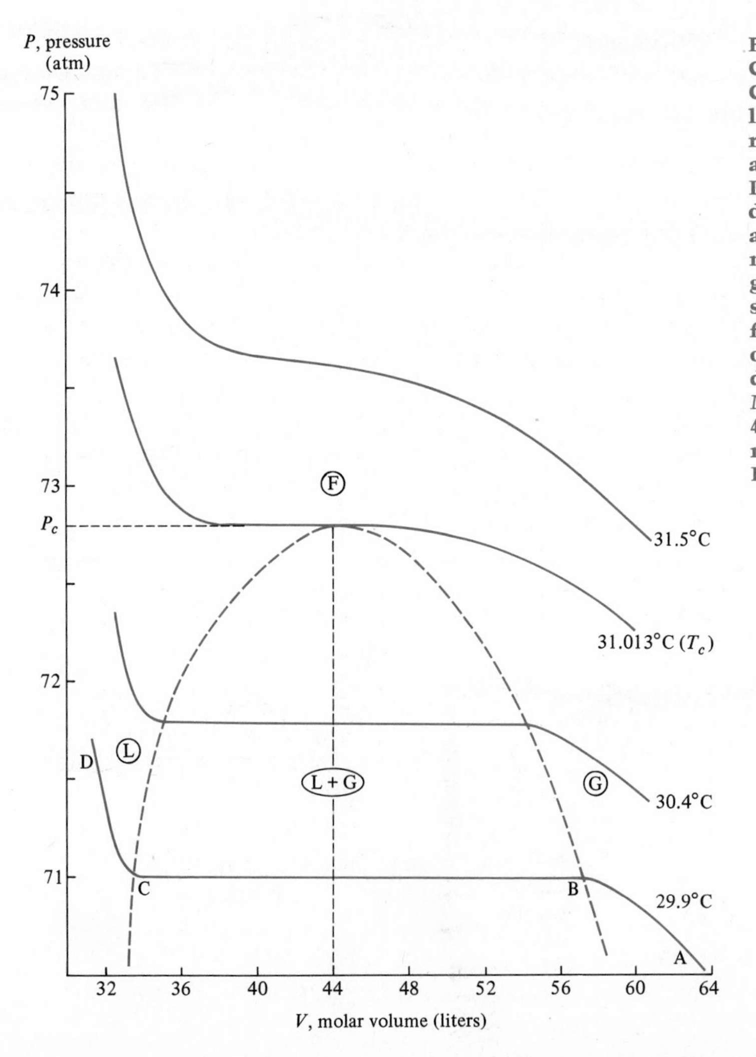

Figure 4-17 Isotherms of Carbon Dioxide near the Critical Temperature. The letters G and L show the regions where CO_2 is a gas and a liquid, respectively. In the region below the dashed line (L + G) liquid and gas coexist. The region marked F is the fluid region. Note that there is no sharp line between the fluid region and the gaseous and liquid regions. [Redrawn from Walter J. Moore, *Physical Chemistry*, 4th ed., © 1972, p. 24. By permission of Prentice-Hall, Inc., Englewood Cliffs, N.J.]

Table 4-4 Critical Constants of Some Common Substances

Substance	T_c (°K)	P_c (atm)	V_c (cm³/mole)
He	5.3	2.26	57.6
H_2	33.3	12.8	65.0
N_2	126.1	33.5	90.0
CO	134.0	35.0	90.0
O_2	153.4	49.7	74.4
CH_4	190.2	45.6	98.8
C_2H_4	282.9	50.9	127.5
CO_2	304.2	73.0	95.7
NH_3	405.6	111.5	72.4
H_2O	647.2	217.7	45.0
Hg	1823.0	200.0	45.0

At point *B*, a drop of liquid begins to form. If we continue to compress the gas, more liquid will form, while the amount of the gaseous portion decreases. All of this takes place at practically no change in pressure. This is shown in the figure as that portion of the curve between *B* and *C*. At point *C* all the gas has been converted to liquid, and further increasing of the pressure slightly (because liquids are not very compressible) decreases the volume of the liquid—that portion of the curve between *C* and *D*. When both liquid and gas are present together, we have a two-phase liquid–gas region (the area below the dashed line).

As we repeat the process at slightly higher temperatures, the two-phase region becomes smaller and smaller. Finally, when we reach 31.013°C, no second phase appears. The temperature at which this occurs is termed the critical temperature of carbon dioxide. At any temperature above this critical temperature, no liquid can be formed no matter how much pressure is applied to the gas. We may define the **critical temperature** as **that temperature above which, no matter how much pressure is applied to a gas, a liquid cannot be formed.** The symbol for critical temperature is T_c. It may at first sight seem surprising that, say at 40°C, it is impossible to liquify CO_2 by pressure alone, but it is nevertheless true. Related to the critical temperature are the critical pressure (P_c) and the critical volume (V_c). The **critical pressure** is the vapor pressure at the critical temperature. The **critical volume** is the volume occupied by one mole of a substance at its critical temperature and critical pressure.

Every gas has a critical temperature. Table 4-4 lists the critical temperatures, pressures, and volumes for some common gases.

Although we cannot compress a gas above its critical temperature to the point where it liquifies, it is possible to compress it so much that its density is comparable to that of a liquid. How then does one distinguish between a liquid and a vapor under these conditions? There is really no reason to distinguish between these two states at these conditions! Chemists have recognized this and generally refer to gases under these conditions as *fluids*. It is important to note that although there is a critical temperature for a gas–liquid system, there are *no* critical temperatures for liquid–solid systems. That is, given a liquid at any temperature, a pressure can be found which when applied to the liquid will convert the liquid into a solid.

Example: By referring to Figure 4-17, show how one may convert a gas completely into a liquid without going through a region where liquid and gas are together.

Answer: (refer to Figure 4-18) Start at (1) in Figure 4-18, where the system is a gas. Keeping the volume constant, increase the pressure and temperature above the critical temperature to point (2). Keeping the pressure constant, decrease the volume and temperature to point (3). From point (3), keep the volume constant and decrease the pressure and temperature to point (4). The system is now a liquid and yet has never passed through a region where gas and liquid exist together as two phases! This phenomenon is termed the *continuity of states* for gases and liquids. There is no continuity of states for liquids and solids.

Returning to Table 4-4, we see that those gases which have low critical temperatures generally behave as ideal gases at normal temperatures and pressures while those gases that have high critical temperatures behave nonideally. This is not difficult to understand once we realize that the critical temperatures directly relate to the attractive forces between molecules. If the attractive forces between molecules are weak, liquefaction cannot occur at elevated temperatures, since the molecules have a large amount of kinetic energy compared to the forces of attraction between them. They therefore have low critical temperatures. We conclude that the critical temperatures as well as boiling points are good yardsticks for comparing attractive forces between molecules.

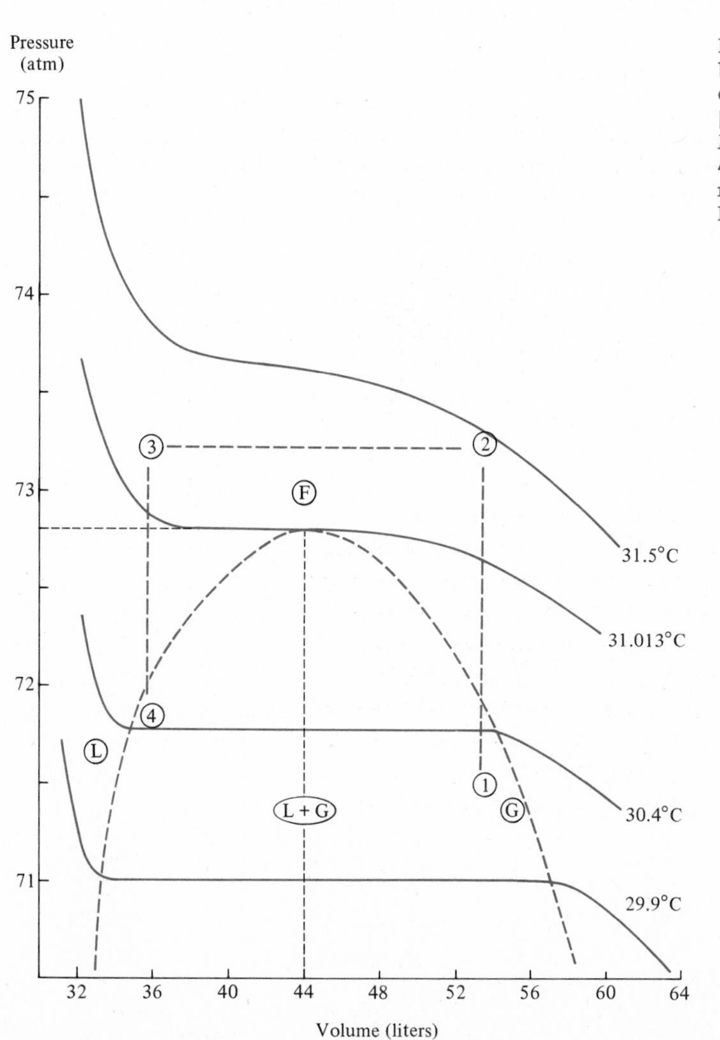

Figure 4-18 Isotherms of carbon dioxide, illustrating the example given in the text. [Redrawn from Walter J. Moore, *Physical Chemistry*, 4th ed., © 1972, p. 24. By permission of Prentice-Hall, Inc., Englewood Cliffs, N.J.]

Problems

Assume that all gases are ideal unless otherwise stated.

4-1 Define each of the following terms. **(a)** volume of a gas **(b)** atmospheric pressure **(c)** barometer **(d)** manometer **(e)** absolute zero **(f)** STP **(g)** ideal gas **(h)** partial pressure **(i)** diffusion **(j)** effusion **(k)** elastic collision **(l)** translational energy **(m)** mean-square velocity **(n)** root-mean-square velocity **(o)** equation of state **(p)** virial equation **(q)** critical temperature **(r)** critical pressure **(s)** critical volume **(t)** isotherm

4-2 State: **(a)** Boyle's law **(b)** Charles's law **(c)** the combined gas law **(d)** Avogadro's law **(e)** Gay-Lussac's law of combining volumes **(f)** the ideal gas law **(g)** Dalton's law of partial pressures **(h)** Graham's law **(i)** the van der Waals equation

4-3 At 5 miles above the Earth's surface, the pressure is reduced to approximately 0.30 atm. What is this pressure in each of the following units? **(a)** torr **(b)** dyn/cm^2 **(c)** lb/in^2

4-4 Convert each of the following as noted.
(a) 0.9 dyn/cm^2 to torr and to atm
(b) 3.2 atm to dyn/cm^2 and to torr
(c) 185 torr to atm and to dyn/cm^2

4-5 What is the pressure, in torr and in atm, of the gas inside the container in the diagram:

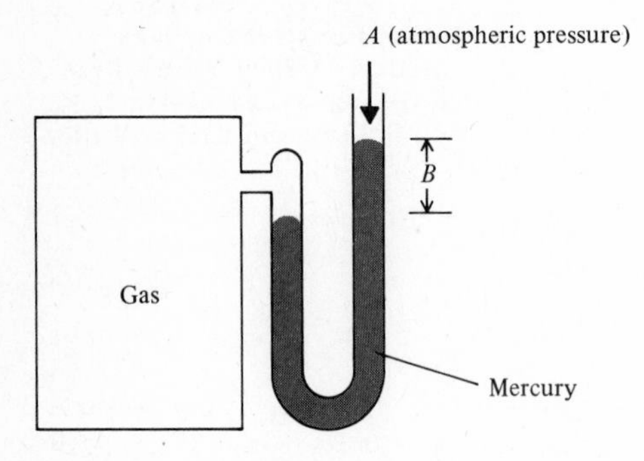

(a) When $B = 275$ mm and $A = 755$ torr? **(b)** When $B = 83$ mm and $A = 728$ torr?
(c) When $B = -475$ mm (the right side is lower) and $A = 629$ torr?

4-6 Fill in the table.

°K	°F	°C
273	_____	_____
_____	40	_____
_____	_____	0
0	_____	_____
233	_____	_____
_____	212	_____
_____	_____	−150

4-7 One mole of an ideal gas occupies 7.22 liters at 520 torr. Calculate the Boyle's law constant in liter · atm.

4-8 An ideal gas at 700°K occupies 0.710 liter at a pressure of 605 torr. If this gas is compressed to 0.120 liter at the same temperature, what is the resultant pressure?

4-9 A gas at 710 cm^3 and 25°C is heated at constant pressure to 90°C. What volume would it now occupy?

4-10 In each case an ideal gas is converted to a new T and P. Calculate the new volume.
(a) 32.7 ml is at 728 torr and 25°C. New conditions: 895 torr and −15°C.
(b) 4.71 liters is at 440 torr and 150°C. New conditions: 1.35 atm and 320°C.
(c) 17.86 liters is at 18 torr and 300°K. New conditions: 0.50 atm and 0°C.
(d) 44.7 cm^3 is at STP. New conditions: 475 torr and 100°C.
(e) 2.886 ml is at 43.5 atm and −185°C. New conditions: STP.
(f) 21.44 liters is at 5×10^{-2} torr and 550°C. New conditions: 0.10 atm and 550°K.

4-11 An ideal gas occupies 83.72 liters at 850 torr and 35°C. If the gas is compressed to 46.1 liters, and the temperature is raised to 75°C, what is the new pressure in atm?

4-12 An ideal gas occupies 26.4 ml at 2.50 atm and 125°C. If the volume is increased to 36.2 ml and the pressure decreased to 1155 torr, what is the new temperature?

4-13 How many liters (at STP) of CO gas are present in 4.2 moles of CO?

4-14 What is the weight of the following gases? **(a)** 1.0 liter of CO_2 at STP **(b)** 2.5 liters of Ne at 25°C and 1.25 atm **(c)** 8.37 ml of B_2H_6 at 273°C and 455 torr **(d)** 3.0×10^{-4} liter of NH_3 at −10°C and 1.2×10^{-2} torr

4-15 Calculate the weight, the number of molecules, and the volume at STP of 1.35 moles of formaldehyde gas, H_2CO.

4-16 Calculate the number of molecules of O_2 in 1.00 liter of air at 27°C and 745 torr. (Assume that air = 79% N_2 and 21% O_2 by volume.)

4-17 When 4.011 g of an unknown liquid was vaporized at −129°C, 2.00 liters of the gas exerted 1.477 atm pressure. Calculate the number of moles of the unknown gas in the 4.011 g and its molecular weight.

4-18 6.02×10^{21} molecules of an ideal gas will occupy what volume at STP?

4-19 How many moles does 75.0 g of the following gases contain? What volume would each occupy at STP? **(a)** H_2 **(b)** H_2S **(c)** NO_2 **(d)** Ar

4-20 What are the units of each of the following? **(a)** molecular weight **(b)** atomic weight **(c)** 6.02×10^{23} **(d)** 22.4

4-21 How many milliliters will 0.675 g of UF_6 gas occupy at 375°C and 475 torr?

4-22 When chemists first discovered hydrazine gas, they determined its empirical formula to be NH_2. How did they find out that the molecular formula was N_2H_4?

4-23 Calculate the molecular weights of the following gases, given the densities.
(a) D = 1.339 g/liter at STP
(b) D = 1.24 g/liter at 17°C and 700 torr
(c) D = 3.17 g/liter at −20°C and 2.35 atm
(d) D = 12.9 g/liter at 75°C and 4.4 atm
(e) D = 1.597 g/liter at 135°C and 635 torr

(f) $D = 0.893$ g/liter at 525°C and 175 torr
(g) $D = 5.639 \times 10^{-4}$ g/ml at 50°C and 710 torr

4-24 What is the density in g/liter of each of the following gases at the conditions given?
(a) HCN at STP **(b)** PH_3 at 75°C and 0.75 atm **(c)** CCl_4 at 275°C and 655 torr
(d) Cl_2 at −10°C and 955 torr

4-25 At STP, 36.9 liters of a certain gas weighs 88.96 g. What is the density of this gas in g/liter at 20°C and 555 torr?

4-26 At a pressure of 800.0 torr, at what temperature (°C) will the density of C_6H_6 vapor be 1.50 g/liter?

4-27 The gas constant R is, in different units:

$$\begin{aligned} R &= 8.314 \text{ joules/°K} \cdot \text{mole} \\ &= 1.987 \text{ cal/°K} \cdot \text{mole} \\ &= 82.06 \text{ ml} \cdot \text{atm/°K} \cdot \text{mole} \\ &= 0.08206 \text{ liter} \cdot \text{atm/°K} \cdot \text{mole} \\ &= 8.314 \times 10^7 \text{ ergs/°K} \cdot \text{mole} \end{aligned}$$

Using the values above as conversion factors, fill in the following table for values of R.

joules	*calories*	*ml · atm*	*liter · atm*	*ergs*
1.00	____	____	____	____
____	____	270	____	____
____	____	____	16.9	____
____	____	____	____	4.318×10^5
____	1.600×10^4	____	____	____
____	1.000×10^5	____	____	____
____	____	____	____	1.00
____	____	____	1.00	____

4-28 In the reaction of phosphorus with gaseous bromine,

$$2P(s) + 3Br_2(g) \longrightarrow 2PBr_3(l)$$

how many liters of Br_2 gas at 135°C and 525 torr are necessary to react completely with 175.0 g of P?

4-29 In the oxidation of HCl with $KMnO_4$,

$$2KMnO_4 + 16HCl \longrightarrow 2KCl + 2MnCl_2 + 5Cl_2(g) + 8H_2O$$

how many liters of Cl_2 at 50°C and 0.855 atm will be produced by the treatment of 85.7 g of $KMnO_4$ with excess HCl?

4-30 In the Haber process for the production of ammonia, the chemical equation is

$$N_2(g) + 3H_2(g) \longrightarrow 2NH_3(g)$$

If 100 g of N_2 were reacted with 50 g of H_2, what volume of NH_3 would be formed at 27°C and 700 torr, assuming that the reaction goes to completion?

4-31 In the reaction

$$2NO_2(g) + O_3(g) \longrightarrow N_2O_5(s) + O_2(g)$$

if 13.5 liters of NO_2 at 30.0°C and 900.0 torr is treated with 56.0 g of O_3 and the reaction allowed to proceed until one reagent is used up, which reagent will remain and how much?

4-32 The important gas acetylene, C_2H_2, is prepared by the treatment of calcium carbide, CaC_2, with water:

$$CaC_2(s) + 2H_2O(l) \longrightarrow C_2H_2(g) + Ca(OH)_2(s)$$

how many kg of CaC_2 are necessary to produce 827 liters of C_2H_2 at STP, assuming excess water is present?

4-33 In the burning of pentane,

$$C_5H_{12}(l) + 8O_2(g) \longrightarrow 5CO_2(g) + 6H_2O(g)$$

how many liters of CO_2 at 750°C and 740 torr will be produced if 18.75 liters of H_2O vapor are also produced at the same conditions?

4-34 In the reaction

$$3NO_2(g) + H_2O(l) \longrightarrow 2HNO_3(l) + NO(g)$$

if it is desired to produce 173 liters of NO at 48°C and 633 torr, how many liters of NO_2 at 20°C and 12.5 atm must be used, assuming complete reaction?

4-35 If 17.32 g Ar, 21.44 g C_2H_6, and 14.4 g N_2O are placed in a 7.75-liter container at 30°C, what is the total pressure in the container and the partial pressure and mole fraction of each gas? (Assume that the three gases do not react with each other.)

4-36 If 835 ml of H_2 is collected over water at a total pressure of 755 torr and a temperature of 21.5°C, how many moles and grams of hydrogen have been collected?

4-37 If 2.77 liters of CH_4 is collected over water at a total pressure of 747 torr and a temperature of 29.5°, how many moles and grams of CH_4 have been collected?

4-38 **(a)** If 6 moles of N_2, 3 moles of O_2, and 17 moles of H_2 are mixed together at STP, what would be the final volume? **(b)** If 0.200 liter of N_2, 700 ml of O_2, and 6000 ml of H_2, each at 0.00°C and 2 atm, were mixed together and the final volume adjusted to 900 ml, what would be the final pressure at 0.00°C and at 25°C?

4-39 The following gases are placed into a 600-ml container at 55°C: 6.0 moles of O_2, 9.73×10^{24} molecules of N_2, and 80 g of CO_2. **(a)** Calculate the mole fraction of each of the gases. **(b)** Calculate the total pressure. **(c)** Calculate the partial pressure of each of the gases. **(d)** If the container was heated to 600°K at constant volume, what new pressure would be exerted by the gas mixture?

4-40 To a good approximation air consists of 79.0% N_2 and 21.0% O_2. Calculate the mean molecular weight of air. What volume does 100 g of air occupy at STP?

4-41 What are the comparative rates of diffusion of the following pairs of gases at the same temperature and pressure? **(a)** H_2 vs. Ne **(b)** SO_3 vs. He **(c)** CO_2 vs. CO **(d)** C_2H_4 vs. NO **(e)** UF_6 vs. He

4-42 State the assumptions of the kinetic-molecular theory.

4-43 Explain Boyle's law, Charles's law, and Dalton's law of partial pressures according to the kinetic-molecular theory.

4-44 **(a)** Calculate the root-mean-square velocity $(\overline{v^2})^{1/2}$ of an oxygen molecule and a carbon tetrachloride (CCl_4) molecule at 300°K and 600°K. Express the velocity in both cm/s and miles/h. **(b)** Compare their kinetic energies at the two temperatures, in erg/mole.

4-45 At what temperature will the *average* H_2 molecule have enough velocity (10^6 cm/s) to escape from the earth's gravity if it is moving upward? Repeat the calculation for He, O_2, and CO_2.

4-46 **(a)** Why do real gases deviate from the ideal gas laws? **(b)** Under which conditions and for what kind of molecules are the deviations greatest? **(c)** Under which conditions do real gases behave most like ideal gases?

4-47 Use the van der Waals equation to calculate the pressure of 2.75 g of CO_2 if it occupies 1.28 liters at 30°C. What would the pressure be if CO_2 were an ideal gas?

4-48 The critical temperature for ethane is 32.3°C and the critical pressure is 48.2 atm. Calculate the critical volume using: **(a)** The ideal gas law. **(b)** The van der Waals equation, noting that for a gas following this equation, $P_cV_c/RT_c = \frac{3}{8}$. **(c)** Compare the results in parts (a) and (b) with the experimental result 0.139 liter/mole.

Additional Problems

4-49 If $H_2(g)$, $O_2(g)$, and $H_2O(g)$ were confined in a 1000-liter container at 1000°K, how many grams of oxygen would you need along with 1000 g of H_2 and 1000 g of H_2O to exert 1000 atm pressure? Assume that no chemical reaction occurs between the species.

4-50 A compound contained 59.96% C, 13.42% H, and 26.62% O. The compound is a gas, and a sample weighing 0.5850 g was found to occupy 275 ml at 100°C and 825 torr. What is the molecular formula?

4-51 A 1.50-liter vessel contains N_2 saturated with water vapor at 70°C. The total pressure is 740 torr. The gas is transferred to another vessel, whose volume is 780 ml, and the temperature is lowered to 20°C, whereupon most of the water vapor condenses and is removed. Find the number of moles of N_2 and the total pressure of gas in the new container.

4-52 Explain why gases are transparent.

4-53 At a temperature of 25°C, at what pressure will the density of CO_2 be 1.00 g/liter?

4-54 How many liters will 13.75 g of C_2H_6 gas occupy at −25°C and 2.44 atm?

4-55 **(a)** Using the mean molecular weight of air calculated in Problem 4-40, what is the total weight of the air in a room 15 ft long, 12 ft wide, and 7.0 ft high, at 1.0 atm and 25°C? **(b)** What is the total number of molecules of N_2 and O_2 in the room? **(c)** What is the number of O_2 molecules?

4-56 In the burning of carbon,

$$C(s) + O_2(g) \longrightarrow CO_2(g)$$

how many grams of C could be completely burned by 375 liters of air at 800°C and 985 torr? (Assume that air contains 21% O_2.)

4-57 In the reaction of hydrogen and iodine,

$$H_2(g) + I_2(s) \longrightarrow 2HI(g)$$

how many liters of HI at 35°C and 1.25 atm will be produced from 8.45 liters of H_2 at the same conditions?

4-58 If 2.4×10^{-2} g He, 1.4×10^{-2} g H_2, 3.5×10^{-3} g C_3H_8, and 0.077 g NH_3 are placed in a

27.0-ml container at 10°C, what is the total pressure in the container, and the partial pressure and mole fraction of each gas? (Assume that the four gases do not react with each other.)

4-59 Calculate $\overline{v^2}$ and $(\overline{v^2})^{1/2}$ for each of the following. Express $(\overline{v^2})^{1/2}$ in both cm/s and miles/h. **(a)** gaseous H_2 at 0°C **(b)** gaseous H_2 at 500°C **(c)** gaseous CO_2 at 500°C

4-60 In the reaction

$$PCl_5(s) + 4H_2O(l) \longrightarrow H_3PO_4(l) + 5HCl(g)$$

how many grams of H_2O are required to react with excess PCl_5 in order to produce 275 ml of HCl at 275 torr and 175°C?

4-61 A container whose volume is 650 ml contains argon at a pressure of 255 torr. Another container, with a volume of 475 ml, contains CO_2 at a pressure of 900 torr. If the gases are mixed at a total volume of 905 ml, what is the final pressure and the percentage of each gas? (Assume that the temperature is constant throughout.)

4-62 At 400°C, 1 volume of an element (symbol E), which is a gas at that temperature, reacts with 3 volumes of fluorine gas (F_2) at the same temperature and pressure, to give 1 volume of compound A. What is the *simplest* formula for compound A?

4-63 A sample of cyanogen gas weighing 6.802 g occupies 3.74 liters at 25.0°C and 650.0 torr. Calculate the molecular weight of cyanogen.

4-64 A mixture of argon and oxygen at 5.50 atm pressure is stored in a 15.25-ml iron container at constant temperature. The iron eventually reacts with all the oxygen, converting it a solid oxide whose volume is negligible. The final pressure is 1275 torr. **(a)** What is the final volume of argon? **(b)** What was the initial pressure of argon? **(c)** What was the initial pressure of oxygen?

4-65 Into a specially reinforced glass tank that can withstand up to 200 atm pressure is placed 700.0 g of dry ice (frozen CO_2), which is allowed to sublime (turn into a gas) at 25°C. If the volume of the tank is 2.22 liters, what is the pressure exerted by the CO_2? How high can the temperature of the gas in the tank be raised without breaking the tank?

4-66 An ideal gas occupies 786 ml at 0.275 atm and -75°C. If the volume is increased to 1275 ml and the pressure is increased to 575 torr, what is the new temperature?

4-67 **(a)** One mole of an ideal gas at 0.0°C and 45 torr occupies what volume? **(b)** If the same gas is compressed to 10.0 liters at 100°C, what pressure will it exert?

4-68 An unknown gas occupies a volume of 322 ml at 29°C and 616 torr. **(a)** What volume would this gas occupy at STP? **(b)** How many molecules are in this gas?

4-69 **(a)** If an ideal gas occupies 720 ml at STP, how many molecules are there in this volume? **(b)** If the volume was doubled, keeping the temperature and number of moles constant, what would happen to the pressure?

4-70 **(a)** The escape velocity from any planet's surface is given by $v_e = \sqrt{2gR}$. For the planet earth $g = 980$ cm/s² (gravitational acceleration) and $R = 6.37 \times 10^8$ cm (the radius). Calculate the temperature that hydrogen must have so that the root-mean-square speed of the average H_2 molecule will be equal to the escape velocity. **(b)** Repeat the calculation for oxygen. **(c)** On the moon, $g = 167$ cm/s² and $R = 1.74 \times 10^8$ cm. Repeat the calculations in parts (a) and (b) for the moon.

4-71 Calculate the pressure exerted by 100.0 g of N_2 confined to a 10.00-liter vessel at 25.0°C using: **(a)** The ideal gas equation. **(b)** The van der Waals equation.

4-72 If 10.0 g of a gas occupies 45.0 liters at 20.0°C and 800.0 torr, what is the density of the gas at STP, in g/liter?

5 Energy Changes and Heats of Reaction

Joseph Black, 1728–1799

Edgar Fahs Smith Collection

The subject of energy is a very important one in chemistry and in the world. Almost all chemical reactions are accompanied either by a loss or a gain in energy. Throughout the history of civilization, human beings have used and depended upon the energy derived from chemical reactions to improve their lot. Indeed, an important part of the history of civilization is our increasing capacity to develop and use sources of energy. In the past 200 years, great advances have been made in the understanding of these energy changes, and it is no coincidence that these years have also seen the greatest technical advances in the history of mankind. We are able to use fuels to heat a home, run a car, drive a rocket to the moon, or to power a pacemaker, a hand calculator, a giant computer, or a child's toy. These are but a few examples of how the energy changes in chemical and electrochemical reactions have been harnessed. Additionally, we are able to explain how the human body stores its energy and then, by a systematic breakdown of certain chemicals, releases that energy (Section 21-3). Plants use light energy in combination with the carbon dioxide (CO_2) in the air and water in the soil and chlorophyll in their leaves to produce glucose, a sugar.

$$6CO_2(g) + 6H_2O(l) \xrightarrow[\text{chlorophyll}]{\text{light}} C_6H_{12}O_6(s) + 6O_2(g)$$

We depend upon glucose in plants as a primary source of our energy. Glucose in the human body is broken down by a sequence of over 25 separate complicated chemical reactions producing carbon dioxide, water, and energy. All life processes as we know them depend on energy for survival of the individual and the species.

Apart from the food we eat, the combustion of fossil fuels is our major source of energy. Hydroelectric power and nuclear power are now responsible for less than 10% of all energy production. Unfortunately, the availability of cheap and plentiful energy sources (mostly fossil fuels: petroleum, coal, and natural gas) has encouraged the people of developed countries to be profligate and wasteful in their use of these sources. We now realize that the supply of fossil fuels in the ground is limited, and that when they are gone, no more will be left (this subject is discussed at greater length in the Interlude between Chapters 8 and 9).

The study of the energetics of chemical reactions is but a small part of a discipline called thermodynamics. **Chemical thermodynamics** is concerned with the transfer of heat and the production of work in physical and chemical systems. During its early history, thermodynamic investigators were primarily interested in heat engines (such as the steam or diesel engine), because they wanted to learn the necessary conditions to obtain maximum efficiencies. The principles that came out of such studies are quite general and have universal applicability.*

*The principles of classical thermodynamics are applicable only to systems containing large numbers of particles. It is for this reason that thermodynamics is applied only to macroscopic phenomena.

In this chapter, we shall study the first law of thermodynamics and the application of this law to heats of chemical reactions. In Chapter 8, we shall discuss the second and third laws of thermodynamics and their applications to chemical equilibrium.

5-1 The First Law of Thermodynamics

The **first law of thermodynamics** is stated very simply: **Energy can neither be created nor destroyed.** As most students will know, this statement is also called the law of conservation of energy. Energy may change from one form to another, but it can neither be created nor destroyed. There is no mathematical derivation of the first law, either simple or complex. Rather, it is based on our experiences with nature. That is, in all the experiments accumulated in scientific books and journals, no one has ever shown that he or she has been able either to create or destroy energy.* The first law is now assumed to be true and will be so assumed until such time as someone can show by repeatable experiment that it is invalid.

In Section 1-7, it was mentioned that heat and work are different forms of energy. It was precisely an understanding of the fact that heat and work are interconvertible forms of energy which was responsible for the universal acceptance of the law of conservation of energy, when it was first clearly stated by Hermann von Helmholtz (1821–1894) in 1847. Figure 5-1 illustrates this interconvertibility. The energy of a system† is given the symbol E. When changes (for example, changes in temperature, volume, or pressure) take place in a system, its energy is changed. This change is

*It is possible to "destroy" energy by changing it to matter; and to "create" energy from matter. (These processes, which take place in atomic bombs and in nuclear reactions in the sun, will be discussed in Chapter 23.) Mass may be considered a form of potential energy that can be released under appropriate conditions. Therefore, a true conservation law combines the laws of conservation of matter and energy and might be stated: Matter–energy can neither be created nor destroyed. However, since mass changes in *ordinary* chemical reactions are too small to be detected by present methods, chemical thermodynamics concerns itself only with the energy conservation law.

†The term **system** may have different meaning in different experiments. If we are considering a gas at a certain temperature, pressure, and volume in a container with a frictionless piston, and wish to study the changes in these properties when the piston compresses the gas to, say, one-tenth of its initial volume, the system is the gas in the container. However, if we are looking at the conservation of energy in such an experiment, the system must include the surroundings of the container, or else the law of conservation of energy would be contradicted. In fact, it was just such subtle changes in the understanding of the term *system* which finally led to the conservation of energy law only about 150 years ago.

(a)

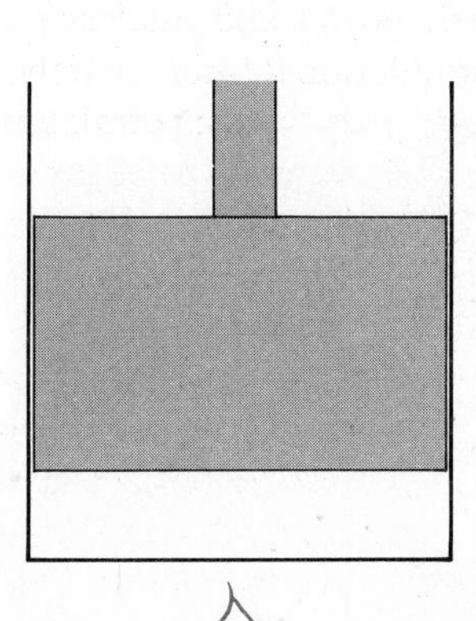

(b)

Figure 5-1 Two Situations in Which Work and Heat are Interconverted. (a) The work done in boring a cannon is changed to heat, raising the temperature of the metal. (b) A gas is heated. The heat expands the gas, raising the piston. Thus work has been done by the gas on the piston, and heat has been converted to work.

labeled ΔE, where the Greek letter delta (Δ) stands for "difference" or "change."

$$\Delta E = \text{difference in the energy of the two states of the system}$$
$$\Delta E = \text{energy of the final state} - \text{energy of the initial state}$$
$$\Delta E = E_2 - E_1 \qquad E_2 = \text{energy of final state of the system}$$
$$E_1 = \text{energy of initial state of the system}$$

In thermodynamics, we are not concerned with the absolute energy of systems (E), only with the *changes* in energy (ΔE). In fact, absolute energies cannot be calculated thermodynamically or by any other technique.

Recalling that heat and work are just forms of energy, the first law can be restated as: *If a certain quantity of heat, q, is added to a system, the added heat can change the energy of the system E and also perform a certain amount of work, w.* In symbols, we write this

$$q = \Delta E + w$$

To show this experimentally, we can visualize a gas in a container with a frictionless piston at a certain temperature T, pressure P, mass m, and volume V (see Figure 5-2a). By adding a quantity of heat to the system, the system changes to a new temperature, T'; a new pressure, P'; and a new volume, V' (but the same mass, m; see Figure 5-2b). The energy of the system has changed (which changes T), and it has performed work (a change in V) by moving the piston against an opposing pressure. We shall discuss separately the concepts of work, heat, and energy.

WORK. **Work** is defined* as the product of a force (F) moving through a distance (Δx),

$$w = F \cdot \Delta x$$

Pressure is defined as force per unit area:

$$P = \frac{F}{A}$$

We may divide and multiply the right-hand side of $w = F \cdot \Delta x$ by A (area). This yields

$$w = \frac{F}{A} \cdot A\,\Delta x \qquad (1)$$

*There are a number of other types of work—electrical work, magnetic work, gravitational work, and so on. Here we shall be solely concerned with pressure–volume work.

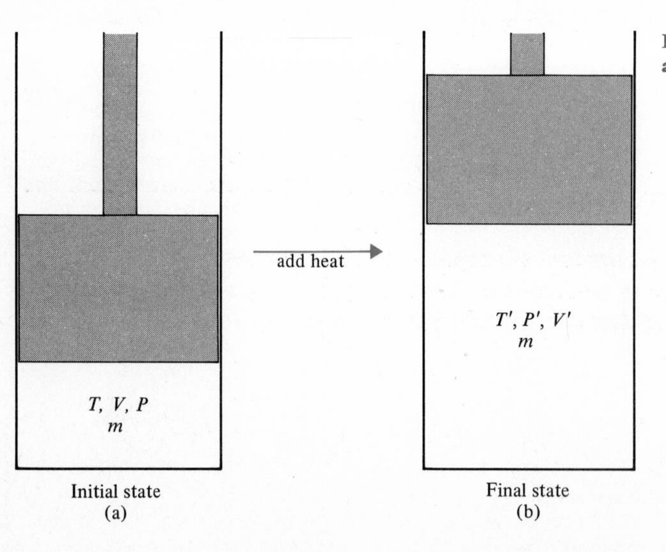

Figure 5-2 The addition of heat to a gaseous system.

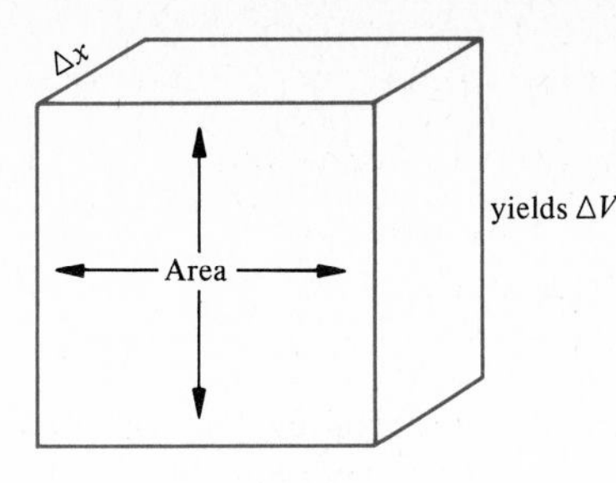

Figure 5-3 $A \cdot \Delta x = \Delta V$.

Noting that

$$P = \frac{F}{A}$$

and

$$\Delta V = A \cdot \Delta x \qquad \text{(a change in volume; see Figure 5-3)}$$

the definition of work becomes

$$\text{work} = w = P\,\Delta V$$

In words, work is a pressure times a change in volume. Note that the equation tells us that if there is no change in volume, the system performs no work. When this is the case, $\Delta V = 0$ and the first law takes the form

$$q_V = \Delta E \qquad \text{(at constant volume)} \tag{2}$$

where the subscript V shows that heat is added at constant volume. Under these conditions, all the heat added is used to change the energy E of the system, since none goes to create work.

When ΔV is greater than zero, the system expands by performing a certain amount of work. The work that the system performs is done at the expense of its energy so that when the system performs work (by expanding), its energy drops. The following example shows this.

Example: A system performs 37 cal of work by expansion without allowing heat to enter or leave the system ($q = 0$). Calculate the change in energy of the system.
Answer: From the first law,

$$q = \Delta E + w$$

For this problem $q = 0$, so

$$\Delta E = -w = -37 \text{ cal}$$

Therefore, the energy of the system has decreased by 37 cal.

What happens when the system is compressed (ΔV is less than zero)? In this case, work is done *on* the system (someone or something compresses the system; it does not compress itself). The sign of the calculated work is negative and the energy of the system is increased.

Example: Work was performed *on* a system by compressing it. A total of 59 cal of work was done on the system without allowing heat to enter or leave the system. Calculate the change in energy of the system for this process.

Answer:

$$q = \Delta E + w \qquad (q = 0 \text{ for this example})$$
$$\Delta E = -w = -(-59 \text{ cal})$$
$$= +59 \text{ cal}$$

The energy of the system is increased when the system is compressed.

HEAT. To give an exact definition of heat is very difficult. We do know intuitively what we mean when we discuss the transfer of heat and the ways in which it is measured. When a hot body is placed next to a cold body, we know that heat is transferred from the hot body to the cold body until the two bodies reach the same temperature (see Figure 5-4). This does not mean that when the process is over, the two bodies have the same quantity of heat—only that they have the same temperature.*

Heat may be expressed in various units, but the most common is the calorie. A **calorie** is defined as the amount of heat necessary to raise the temperature of 1 gram of water 1 degree Celsius.† The interactions between a system and its environment in transferring heat are quite complex and will not be discussed here since it would serve no useful purpose. It will be sufficient for our purposes to consider the concept of heat in terms of the first law. That is, for a system at constant volume, the heat absorbed or evolved is measured in terms of energy change in the system (because, then, $q = \Delta E$).

When a process occurs without heat entering or leaving the system, $q = 0$, and the process is called **adiabatic.** For an adiabatic process, the first law takes the form

$$\Delta E = -w$$

because $q = 0$. In an adiabatic process, work done on or by the system is completely used to change the energy of the system. A constant-volume process which is also adiabatic occurs without changing the energy of the system.

$$\Delta E = 0 \qquad (\Delta V = 0;\ \text{hence, } w = 0) \qquad (q = 0) \tag{3}$$

As might be expected, heat added to a system is taken to be positive, while heat lost by a system is negative. In summary, the signs of heat and work are:

work done *by* the system is + $\quad w > 0$
work done *on* the system is − $\quad w < 0$
heat *gained* by the system is + $\quad q > 0$
heat *lost* by the system is − $\quad q < 0$

Table 5-1 shows relationships involving the first law when either q or w is zero.

*The great scientist Joseph Black (1728–1799) was the first to make this important distinction. He also discovered the concepts of specific heat (Section 5-2), heat of vaporization (Section 5-3), and heat of fusion (Section 5-3) and made crude measurements of these quantities for various substances. He also made fundamental discoveries in the chemistry of gases. For example, it was only through the work of Black that gases were recognized as individual chemical substances, essentially no different in this respect from liquids or solids. Before his work, all gases were generally regarded as various forms of air and, furthermore, were thought to be chemically inert.

†Strictly speaking, the amount of heat necessary to raise the temperature of 1 gram of water by 1 degree Celsius depends on the temperature of the water; for example, it is not exactly the same for 1° → 2° as for 85° → 86°. Therefore, a strict definition of the calorie is the amount of heat necessary to raise the temperature of 1 gram of water from 14.5 to 15.5°C. This value was arbitrarily chosen because some standard was needed. However, the differences are slight, and we will use the definition quoted in the text.

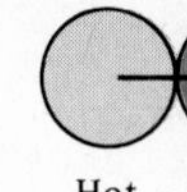
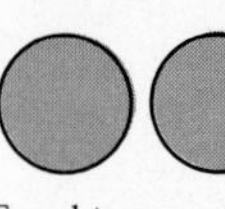
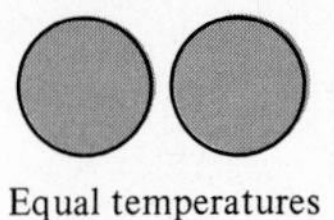

Figure 5-4 Heat flows from a hot body to a cold body.

What happens when both q and w are not zero? The calculated value of ΔE depends on the relative values of q and w, as the following examples illustrate.

Example: During the addition of 100 cal of heat to a system, the system was simultaneously compressed and 190 cal of work was done on the system. Calculate ΔE.

Answer:

$$\Delta E = q - w$$
$$q = +100 \text{ cal}$$

For a compression, $\Delta V < 0$ and w is negative:

$$\Delta E = (+100 \text{ cal}) - (-190 \text{ cal}) = +290 \text{ cal}$$

Example: While a system expands and performs 400 cal of work, 79 cal of heat are added to the system. Calculate ΔE for this process.

Answer:

$$\Delta E = q - w$$
$$q = +79 \text{ cal}$$

For an expansion, $\Delta V > 0$ and w is positive:

$$\Delta E = (+79 \text{ cal}) - (+400 \text{ cal}) = -321 \text{ cal}$$

Example: A system is subjected to the following series of changes. Calculate q, w, and ΔE for each step.

(a) 820 cal of work is done on the system adiabatically.

(b) 605 cal of heat is withdrawn from the system while it expands and does 120 cal of work.

(c) 200 cal of heat is added to the system at constant volume.

Table 5-1 The First Law of Thermodynamics and Some General Calculations

$$\Delta E = q - w$$

Process	ΔE	q	w	*Remarks*
Adiabatic	$-w$	0	$-\Delta E$	$\Delta E = -w$
Adiabatic compression ($\Delta V < 0$)	+	0	−	Energy increases
Adiabatic expansion ($\Delta V > 0$)	−	0	+	Energy decreases
$\Delta V = 0$ (no work done)	q	ΔE	0	$q = \Delta E$
$\Delta V = 0$ (heat added to system; no work done)	+	+	0	Added heat increases ΔE
$\Delta V = 0$ (heat subtracted from system; no work done)	−	−	0	Heat subtracted decreases ΔE

Answer:

(a) $q = 0$	$w = -820$ cal	$\Delta E = -w = +820$ cal
(b) $q = -605$ cal	$w = +120$ cal	$\Delta E = (-605 \text{ cal}) - (120 \text{ cal}) = -725$ cal
(c) $q = +200$ cal	$w = 0$	$\Delta E = q = 200$ cal

How does a system gain or release energy? A system (solid, liquid, or gas) is composed of molecules. It is the molecules comprising the system that store or release energy. Molecules can translate (move through space), rotate, vibrate, and change their electronic structures. In all these actions, energy can be absorbed or released. For example, additional heat can be absorbed by molecules causing them to translate faster. The same is true for rotation and vibration (Figure 5-5). In Section 12-10, we will discuss how atoms and molecules can absorb or release energy by changing the positions of their electrons. These things occur *microscopically* in a system when a thermodynamic change in energy takes place on a *macroscopic* scale.

We can deduce a very important property of the energy of a system by appealing once again to our experiences with nature. These experiences concern what is termed a "perpetual motion machine." A perpetual-motion machine is a machine that operates constantly without using any fuel. The running of the machine generates energy which is then applied to continue the operation of the machine. It is our experience that although many ingeneous devices have been proposed by people who are either swindlers or who have deluded themselves, a true perpetual-motion machine has never yet been built. Nor do we expect that any such machine ever will be built because its operation would constitute a violation of the first law of thermodynamics (Figure 5-6). If anyone devises a thought process that would lead to the construction of this kind of machine, then, by necessity, the process must be incorrect, or some assumption in the process must be false. Noting this, consider the following processes (see Figure 5-7).

A system is in state I with a pressure, volume, temperature, and energy which we will call P_1, V_1, T_1, and E_1; and by expansion (adding heat) proceeds to state II, with P_2, V_2, T_2, and E_2. From state II, the system is returned to state I, with P_1, V_1 and T_1. Let us *assume* that the final energy of the system E_1' is different from the initial energy E_1. The energy of the system follows the path, $E_1 \rightarrow E_2 \rightarrow E_1'$. The system is now in a higher energy state than it was previously, and $E_1' > E_1$. If this is true, then the extra energy of the system can be applied to the system to continue the process I $\rightarrow$ II $\rightarrow$ I, always yielding energy that is used to drive the system. This constitutes a perpetual-motion machine because work is created from nothing. From what was just discussed, we must conclude that the assumption that the two energies (E_1, E_1')

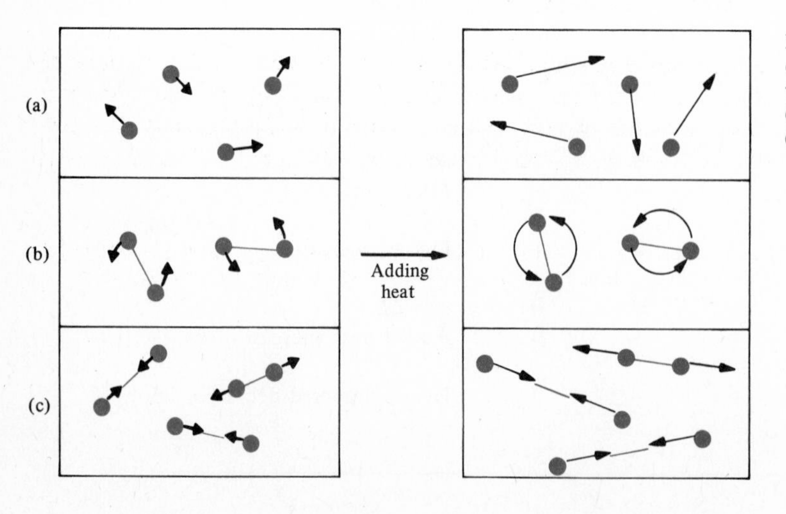

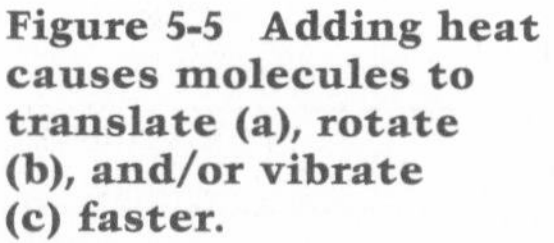
Figure 5-5 Adding heat causes molecules to translate (a), rotate (b), and/or vibrate (c) faster.

Figure 5-6 It would be nice if we could get energy from the operation of a waterwheel in a system such as this. Can you see why it would not work? The drawing is by the artist M. C. Escher, who made many fascinating drawings involving paradoxes and optical illusions. [Escher Foundation, Haags Gemeentemuseum, The Hague]

are different is incorrect. That is, E_1 *must* equal E_1'. The energy E of a system at a particular T, P, and V has a specific and fixed value. It *does not matter* how the system reaches this state (defined by the values of P, V, and T); once the system has reached it, it must have this specific value, E. Putting it another way, the *prior history* of the system plays no role in determining its energy. When the system

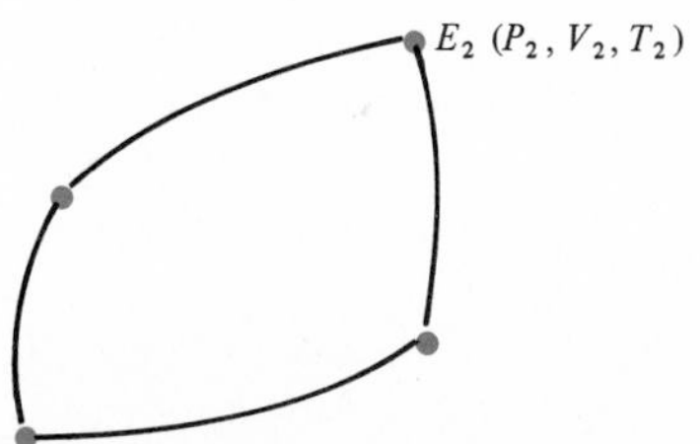

Figure 5-7 The energy of the system E_2 is independent of the path by means of which the system reached this state.

reaches state II, it can only have the energy E_2 and no other. It is of no concern how it arrives at state II. At state II, it will always have the energy value E_2.

Variables, such as the energy of a system, which possess the property discussed above are termed **state functions.** Other state functions are the pressure, temperature, and volume. Consider the volume. When a system goes through a series of expansions and compressions and finally reaches a state where the volume is say 50 liters, does it matter how the volume of the system reached the value of 50 liters? The volume is 50 liters, independently of how it arrived at that state. This is an important concept and we will rely on it heavily in discussing heats of reaction.

ENTHALPY. Although the state function E is highly useful in many ways, it does not fully describe what happens in chemical reactions, for a reason we shall now discuss. Almost all chemical reactions take place at constant pressure. When a system absorbs heat at constant pressure, the first law states that

$$\Delta q_P = \Delta E + P\,\Delta V \qquad (4)$$

(recall that the subscript P means "at constant pressure"). Although the system has absorbed all the heat (Δq_P), often only part of this heat has been used to change the energy of the system, E. The rest of the heat has been used to perform work: to cause a volume change (ΔV). Obviously, the function ΔE does not *fully* describe what is happening to the system because the heat absorbed at constant pressure has done more than just change the energy of the system E; it has also been used to change the volume at constant pressure, $P\,\Delta V$.

Most chemical reactions are performed in laboratory vessels that are open to the atmosphere, and the work that is done is performed against the confining atmosphere as the volume of the system is increased or decreased (a decrease in volume means that work has a negative sign, but still, Δq is not equal to ΔE). For example, consider the combustion of ethane (C_2H_6) at constant atmospheric pressure:

$$\underset{\text{ethane}}{C_2H_6(g)} + \tfrac{7}{2}O_2(g) \longrightarrow 2CO_2(g) + 3H_2O(l)$$

The heat evolved from this reaction is 372,820 cal/mole of ethane used. All this heat came from the chemical reaction, but not all of it has been used to increase the total amount of energy E possessed by the products (CO_2 and H_2O). Some of the energy has been used to expand the system, since the total volume of the products is greater than that of the reactants (see Section 4-8).

The heat of reaction (372,820 cal/mole of ethane) has not only changed the energy of the system, but has also expanded it. Although this expansion (or compression) effect is greatest when gases are involved, it is also present, to a smaller extent, even in reactions of liquids, as long as they take place at constant pressure. Chemists would like to have a function that would show all the energy lost by reactants as being used to increase the energy of the products. Because E does not do this, a new thermodynamic state function was invented, called enthalpy, and given the symbol, H. **Enthalpy** is defined as

$$H = E + PV \qquad (5)$$

As was the case with E, we are interested only in *changes* in H, not in the actual values of H. At constant pressure, P, equation 5 becomes

$$\Delta H = \Delta E + P\,\Delta V = q_P \qquad (6)$$

That is, the change in enthalpy equals the change in heat at constant pressure.

The new state function, H, describes a multitude of processes. Not only can it account for heats of reaction at constant pressure in which no useful work is performed, but it also describes processes in which the system simultaneously changes its energy *and* performs work. Consider the process that occurs when a solid expands on melting. Part of the heat of melting is used to overcome the attractive forces in the solid state, while the rest of the heat is used to perform work of expansion against the atmosphere when a solid changes to a liquid. The heat of melting is accounted for by an enthalpy change. From the above we have

$$\Delta q_P = \Delta H \qquad \text{(constant } P\text{)} \tag{7}$$

$$\Delta q_V = \Delta E \qquad \text{(constant } V\text{)} \tag{8}$$

A synonym for enthalpy is "heat content."

5-2 Heat Capacities

As we have previously mentioned, if we add 1 calorie of heat to 1 gram of water the temperature will increase by 1 degree. However, this is not true for other substances. For example, if we add 1 calorie of heat to 1 gram of copper at about room temperature, the temperature will increase by 10.9°. That is, the same amount of heat that will raise the temperature of 1 gram of water by only 1 degree will raise the temperature of 1 gram of copper nearly 11 times as much. The amount of heat it takes to raise the temperature of 1 gram of a substance by 1 degree is called the specific heat or the **heat capacity.***

The heat capacity of water is 1.000 cal/g · °C; of copper, 0.092 cal/g · °C. Heat capacities can also be expressed in cal/mole · deg. This is termed the *molar heat capacity.* For water the molar heat capacity is 18 cal/mole · °C; for copper, 5.8 cal/mole · °C. The heat capacity of a substance is an innate property of the substance. The value is different for different substances; it is never zero, and generally changes with changing temperature. It is also different for different phases. Thus, liquid water, ice, and steam all have different heat capacities.

As we have said, the molar heat capacity is the amount of heat necessary to raise the temperature of 1 mole of a substance by 1°C. Experimentally, the heat can be added at constant volume, Δq_V, or at constant pressure Δq_P. For this reason we must define two heat capacities, the *molar heat capacity at constant volume,* C_V, and the *molar heat capacity at constant pressure,* C_P. By definition,

$$C_P = \frac{\Delta q_P}{\Delta T} \qquad \text{(change in heat per unit temperature change at constant pressure)}$$

and

$$C_V = \frac{\Delta q_V}{\Delta T} \qquad \text{(change in heat per unit temperature change at constant volume)}$$

From equations (7) and (8), we know that $\Delta q_P = \Delta H$ and $\Delta q_V = \Delta E$ so that C_P and C_V take the form

$$C_P = \frac{\Delta H}{\Delta T} \tag{9}$$

$$C_V = \frac{\Delta E}{\Delta T} \tag{10}$$

*Technically speaking, specific heat has no units since it is a ratio of heat capacity of the substance to the heat capacity of water.

$$\text{specific heat of substance A} = \frac{\text{heat capacity of substance A (cal/g}\cdot\text{deg)}}{\text{heat capacity of water (cal/g}\cdot\text{deg)}}$$

Since we will use the value of 1.000 cal/g · deg for the heat capacity of water, the specific heat of a substance will numerically equal its heat capacity. In this book, we shall refer only to heat capacity.

Table 5-2 Molar Heat Capacities at Constant Pressure at 25°C[a]

Substance	C_P (cal/mole · °K)
$H_2(g)$	6.52
$O_2(g)$	7.16
$N_2(g)$	6.83
$CO(g)$	6.97
$Cl_2(g)$	8.85
$CO_2(g)$	10.57
$CH_4(g)$	8.60
$C_2H_6(g)$	12.71
$NH_3(g)$	7.11
$H_2O(g)$	7.30
$H_2O(l)$	18.04
$I_2(l)$	19.20
NaCl(l)	16.0
Hg(l)	6.62
C(s)(graphite)	4.03
Al(s)	4.94
Cu(s)	5.41
Na(s)	6.71
Pb(s)	5.29
$H_2O(s)$	9.0
$I_2(s)$	9.59
NaCl(s)	10.98

[a] These values change with temperature.

From the definitions of energy and enthalpy ΔH is almost always equal to or greater than ΔE, which also means that $C_P > C_V$. In words, it takes more energy to raise the temperature of a substance 1°C at constant pressure than it does at constant volume.

Molar heat capacities are easily measurable quantities and tables of their values are readily available (see Table 5-2).

Example: A small copper vessel at 25°C weighs 213.5 g. How many calories of heat are necessary to raise the temperature of the vessel to 100°C, the boiling point of water? Compare this amount of heat with the amount necessary to raise 213.5 g of water from 25 to 100°C.

Answer:

heat (calories) = molar heat capacity (cal/mole · °K) × (number of moles) × (change in temperature)

From Table 5-2, C_p for copper is 5.41 cal/mole · °K.

For the copper:

$$\text{heat} = 5.41 \frac{\text{cal}}{\text{mole} \cdot {}^\circ\text{K}} \times \frac{213.5 \text{ g}}{63.546 \text{ g/mole}} \times (373^\circ - 298^\circ)$$

$$= 1360 \text{ cal of heat for the copper vessel}$$

For the water:

$$\text{heat} = 18 \frac{\text{cal}}{\text{mole} \cdot {}^\circ\text{K}} \times \frac{213.5 \text{ g}}{18 \text{ g}} \times (373^\circ - 298^\circ)$$

$$= 16{,}000 \text{ cal of heat for the water}$$

Example: A pail contains 1250 ml of water at 15.0°C. To this pail we add 200 ml of boiling water. What is the resultant temperature of the water? Neglect the heat changes of the pail. Take the density of water as 1.000 g/ml.

Answer:

$$\text{heat gained} = \text{heat lost}$$

$$\text{heat capacity} \times \text{grams of water} \times \text{temperature change} = \text{heat capacity} \times \text{grams of water} \times \text{temperature change}$$

$$1.00 \times 1250 \times (T_2 - 288.0°\text{K}) = 1.00 \times 200(100 - T_2)$$

$$T_2 = 299°\text{K} \quad \text{or} \quad 26°\text{C}$$

Copper is not the only substance with a heat capacity much less than that of water. In fact, water has one of the highest heat capacities of any substance known. This fact is responsible for the differences in climate generally found between coastal areas and areas that are well inland. For example, the east and west coasts of the United States have comparatively moderate climates, while the climate of inland regions like Kansas and Oklahoma is more extreme. The summers are hotter and the winters are colder. The reason for this is that the coastal areas are adjacent to vast bodies of water. When the hot sun beats down in the summer, the water is able to absorb much of this heat. Because the heat capacity of the water is so large, the temperature of the water (and consequently of the air above it, which is warmed by the heat coming from the water) rises comparatively little. In the inland areas where there are no large bodies of water, the heat of the sun must be absorbed by the land. The heat capacity of the land is considerably less, and the temperature of the land (and of the air above it) rises more than it does in the coastal regions, even when the intensity of the heat sent by the sun is the same in both places. In the winter these factors are reversed. The oceans, which have been warmed by the sun for the entire summer, take months to lose their heat, because each time a calorie is lost to the air, the temperature of 1 g of ocean water goes down only 1°C. The inland areas lose their heat more rapidly. The heat capacity of land varies, but an average value is about 0.2 cal/g, and loss of 1 cal of heat decreases the temperature of 1 g of land by about 5°C. An additional factor is that land cannot *mix*. When the upper few meters of ocean lose heat to the air, the temperature of this portion of the ocean drops. This, in turn, causes its density to increase,* which makes it sink below the surface, exposing new water, which in turn can lose *its* heat to the air. The land cannot do this. When the upper part loses heat it cannot sink, and the ground several meters below the surface maintains a temperature that changes much less with the seasons than does the land at the surface. Thus, not only is the heat-capacity factor important but in any given area of ocean, there is a much greater quantity of water available to be heated or cooled than there is land in an equivalent area of inland territory.

5-3 Heat of Vaporization and Heat of Fusion

Almost all substances can exist either as gases, liquids, or solids. A gas, a liquid, and a solid are examples of different **phases.** If we are dealing with a system that contains only gases, then only *one* phase exists. Liquid water and ice together is an example of a two-phase system; and a system composed of a gas, liquid, and solid together is a three-phase system. Whether a substance exists as a gas, a liquid, or a

*As long as the temperature is above 4°C, the density of water decreases with increasing temperature.

solid depends on the nature of the substance and on the conditions of temperature and pressure. For example, a sample of water at 1 atm pressure and 90°C is a liquid. As we saw in Section 5-2, the addition of heat to this sample causes the temperature to rise. Since the heat capacity of water (liquid) is 18 cal/mole · °C, it will take 180 cal to bring 1 mole of water from 90 to 100°C. However, at this point, the temperature increase stops. If we add more heat to the sample now, the temperature does not rise. Instead, the water begins to boil. It is impossible for liquid water at 1 atm pressure to reach a temperature higher than its boiling point,* 100°C. The addition of more heat causes more of the water to evaporate until it has all turned to steam. At this point, the temperature of the steam is still at 100°C. A large quantity of heat has gone into the water, but it has *not* changed the temperature. What it has done is turn the water into steam. The fact that heat could be added to any substance whatever without increasing its temperature was very surprising to Joseph Black, who discovered it, and to the other chemists of that time. It still seems surprising at first sight today, but it is nevertheless true. Without this heat, which is called the **heat of vaporization** (ΔH_{vap}), the water could not boil. Also surprising to Black was the *magnitude* of the quantity. The heat of vaporization for water is 540 cal/g or 9720 cal/mole. Thus, it takes only 100 cal to bring 1 g of water from 0 to 100°C, but 540 *more* calories to change the water to 1 g of steam at 100°C. It is only at this point, if we add still more heat, that the temperature of the steam will rise *above* 100°C.† Incidentally, this explains why steam at 100°C will cause a worse burn than water at the same temperature.

Black also discovered that a similar phenomenon takes place when ice melts. There is a **heat of fusion,** amounting to 80 cal/g or 1440 cal/mole necessary to be added to ice before it can be completely converted to water. In fact, we may generalize this and make the statement that **every phase change is accompanied by a heat change.** This is true not only for changes from liquid to and from gas and for solid to and from liquid, but also for changes from solid directly to gas (this is called **sublimation**; typical substances which behave in this way are iodine and carbon dioxide), and from one solid phase to another (many substances crystallize in more than one way; for example, sulfur, carbon, and phosphorus):

$$\begin{aligned}
\text{liquid} &\rightarrow \text{gas} & \Delta H &= \text{molar heat of vaporization}\\
\text{gas} &\rightarrow \text{liquid} & \Delta H &= \text{molar heat of condensation}\\
\text{solid} &\rightarrow \text{liquid} & \Delta H &= \text{molar heat of fusion}\\
\text{liquid} &\rightarrow \text{solid} & \Delta H &= \text{molar heat of freezing}\\
\text{solid} &\rightarrow \text{gas} & \Delta H &= \text{molar heat of sublimation}
\end{aligned}$$

It is obvious that the law of conservation of energy requires that the heat of vaporization must be equal in magnitude to the heat of condensation for any given substance at a given temperature and pressure, and similarly, the heat of fusion must equal the heat of freezing. Otherwise, we could have a perpetual-motion machine, which we have seen to be impossible. Heats of fusion and of vaporization for several substances are listed in Table 5-3.

Example: How many calories are required to convert 7.50 g of ice at −15°C to steam at 137°C?

*For a definition of boiling point, see Section 6-3.

†In practice, things seldom go so smoothly. A portion of the water may change to steam, and additional heat may reach the steam part before it reaches the remaining water; so the temperature of the steam may for this reason rise before the water has completely boiled. The more intimately the two-phase steam–water system is mixed, the more will the real behavior approach the behavior described in the text.

Answer:

$$7.50 \text{ g} \times \frac{1 \text{ mole}}{18.0 \text{ g}} = 0.417 \text{ mole}$$

From Table 5-2, the heat capacities are:

$H_2O(g)$	7.30 cal/mole · °C
$H_2O(l)$	18.04 cal/mole · °C
$H_2O(s)$	9.0 cal/mole · °C

Thus, it requires for each step:

$H_2O(s)$ at $-15°C \longrightarrow H_2O(s)$ at 0°C

$$0.417 \text{ mole} \times 9.0 \text{ cal/mole} \cdot °C \times 15°C = 56 \text{ cal}$$

$H_2O(s) \longrightarrow H_2O(l)$ $\quad 0.417 \text{ mole} \times 1440 \text{ cal/mole} = 600 \text{ cal}$

$H_2O(l)$ at 0°C $\longrightarrow H_2O(l)$ at 100°C

$$0.417 \text{ mole} \times 18.04 \text{ cal/mole} \cdot °C \times 100°C = 752 \text{ cal}$$

$H_2O(l) \longrightarrow H_2O(g)$ $\quad 0.417 \text{ mole} \times 9720 \text{ cal/mole} = 4.05 \times 10^3 \text{ cal}$

$H_2O(g)$ at 100°C $\longrightarrow H_2O(g)$ at 137°C

$$0.417 \text{ mole} \times 7.30 \text{ cal/mole} \cdot °C \times 37°C = 113 \text{ cal}$$

$$\text{total heat required} = 5.57 \times 10^3 \text{ cal}$$

Example: Three ounces of scotch is considered a very stiff drink. Let us assume that the scotch has a heat capacity of 1.4000 cal/g and a density of 0.94000 g/ml. One ounce is equivalent to 29.570 ml. If we drop an ice cube weighing 30.000 g into the scotch at 25°C and wait for the ice to melt completely, what will be the resultant temperature? Neglect the change in temperature of the glass and the loss of heat to the air.

Table 5-3 Heats of Fusion and of Vaporization for Various Substances

Conversion	*Heat of fusion* ΔH_{fus} *(kcal/mole)*	*Melting point (°K)*
$H_2O(s) \longrightarrow H_2O(l)$	1.435	273
$NaCl(s) \longrightarrow NaCl(l)$	7.22	1073
$KCl(s) \longrightarrow KCl(l)$	6.41	1043
$KNO_3(s) \longrightarrow KNO_3(l)$	2.58	581
$K_2Cr_2O_7(s) \longrightarrow K_2Cr_2O_7(l)$	8.77	671
$Na(s) \longrightarrow Na(l)$	0.631	371
$K(s) \longrightarrow K(l)$	0.581	336
$Ag(s) \longrightarrow Ag(l)$	2.70	1234
$Hg(s) \longrightarrow Hg(l)$	0.581	234
Conversion	*Heat of vaporization* ΔH_{vap} *(kcal/mole)*	*Normal boiling point (°K)*
$H_2O(l) \longrightarrow H_2O(g)$	9.72	373
$NH_3(l) \longrightarrow NH_3(g)$	5.57	239.8
$(CH_3)_2CO(l) \longrightarrow (CH_3)_2CO(g)$	7.23	329.4
$CHCl_3(l) \longrightarrow CHCl_3(g)$	7.04	334.4
$C_6H_6(l) \longrightarrow C_6H_6(g)$	7.35	353.3
$Hg(l) \longrightarrow Hg(g)$	14.20	629.8
$Zn(l) \longrightarrow Zn(g)$	23.70	1180
$H_2(l) \longrightarrow H_2(g)$	0.216	20.5

Answer: The problem is done in two stages. First, one considers the heat required to melt 30.000 g of ice. The heat necessary to do this comes from the scotch and lowers its temperature. After the ice melts, we are confronted with a mixing problem of two liquids at different temperatures. From Table 5-3,

$$H_2O(s) \longrightarrow H_2O(l) \qquad \Delta H_{273} = 1435.0 \text{ cal/mole}$$

Stage 1

Ice:

$$30.000 \text{ g ice} \times \frac{1 \text{ mole ice}}{18.0152 \text{ g ice}} \times \frac{1435.0 \text{ cal}}{\text{mole}} = 2.3896 \times 10^3 \text{ cal heat required to melt the ice}$$

Scotch:

$$3.0000 \text{ oz scotch} \times \frac{29.570 \text{ ml}}{1 \text{ oz}} \times \frac{0.94000 \text{ g}}{\text{ml}} = 8.3387 \times 10^1 \text{ g scotch in } 3.0000 \text{ oz}$$

Heat needed for melting = heat released by scotch.

$$2.3896 \times 10^3 \text{ cal} = 8.3387 \times 10^1 \text{ g} \times \text{heat capacity} \times \text{change in temperature}$$

$$2.3896 \times 10^3 \text{ cal} = 8.3387 \times 10^1 \text{ g} \times 1.4000 \frac{\text{cal}}{\text{g} \cdot {}^\circ\text{K}} \times (298 - T)$$

$$298 - T = 2.0469 \times 10^1 \, {}^\circ\text{K}$$

$$T = 277.53^\circ\text{K}$$

Stage 2. The ice is now melted and has turned to water at a temperature of 273.00°K. The scotch is at 277.53°K. They are mixed. We now calculate the final temperature.

$$\text{heat gained by water} = \text{heat lost by scotch}$$

$$\text{g water} \times \text{heat capacity} \times \text{change in temperature} = \text{g scotch} \times \text{heat capacity} \times \text{change in temperature}$$

$$30.000 \text{ g} \times 1.0000 \frac{\text{cal}}{\text{g} \cdot {}^\circ\text{K}} \times (T_2 - 273.00) = 83.397 \text{ g} \times 1.4000 \frac{\text{cal}}{\text{g} \cdot {}^\circ\text{K}} \times (277.53 - T_2)$$

$$T_2 = 276.60^\circ\text{K} = 3.60^\circ\text{C}$$

Although it requires much less heat to melt 1 g of ice (80 cal) than to boil 1 g of water (540 cal), it is still a sizable quantity and has important consequences for the world's climate. Consider the polar ice caps. Both the arctic and antarctic regions are covered with vast fields of ice. It has been estimated that the total amount of polar ice in the world (mostly sea ice in the arctic, except for the Greenland glacier, but land ice in the antarctic) is about 26.25 million km³ (6.3 million cubic miles). During the summer months the temperatures frequently rise above 0°C, especially in the arctic. Yet despite the fact that the temperature of the air is above the melting point of ice, the ice fields do not melt away. Some of this ice has remained in place without melting for more than 10,000 years. The reason for this is that each gram of ice at 0°C requires 80 cal in order to melt, and this quantity of heat must come from the air and the sunshine. The very act of giving this heat to the ice causes the temperature of the air to drop, and soon it reaches 0°C once again. Consequently, only a small part of the ice is able to melt each summer, and of course, it freezes again the following winter. Now, if the heat of fusion of ice were much lower; say 1 cal/g, then much more ice would melt, perhaps the whole of both ice caps over the course of several years. If this happened, the additional quantity of water would, of course, enter the oceans, and it can be calculated that it would raise sea level by about 65 m (210 ft). This would be more than enough to flood many coastal cities and other coastal areas. Many whole islands would be completely submerged beneath the sea. The resulting damage to the world's economy would be enormous, even if the flooding were slow enough so that there would be time for people to prepare for it.

The fact that the addition of heat to a solid at its melting point does not change the temperature, but is only used to melt the solid, is often used in laboratories for the purpose of maintaining a constant temperature. For example, a beaker that contains both ice and water, intimately mixed, will remain at 0°C, and so will anything that is immersed in it. If a small amount of heat is added, some of the ice will melt; if a small amount of heat is removed, some of the water will freeze, but in neither case will the temperature change.

5-4 Heats of Reaction: Thermochemistry

Almost all chemical reactions are accompanied by a gain or a loss of heat. **Thermochemistry** is the branch of thermodynamics devoted to a study of heat changes in chemical reactions. A reaction that produces heat is called an **exothermic** reaction, while one that requires heat is called an **endothermic** reaction. When 1 mole of liquid water is decomposed at 25°C to yield hydrogen and oxygen, 68,320 cal of heat are required, and we may write

$$H_2O(l) \longrightarrow H_2(g) + \tfrac{1}{2}O_2(g) - 68{,}320 \text{ cal (endothermic)} \qquad (11)$$

The enthalpy change for the reaction at 298°K (written as ΔH_{298}) = +68,320 cal. For the opposite reaction, the production of 1 mole of liquid water from hydrogen and oxygen, 68,320 cal are produced:

$$H_2(g) + \tfrac{1}{2}O_2(g) \longrightarrow H_2O(l) + 68{,}320 \text{ cal (exothermic)} \qquad (12)$$

ΔH_{298} for this reaction is −68,320 cal.

A word concerning sign conventions is appropriate here. In discussing the enthalpy changes in a reaction, the convention is that

an exothermic reaction means ΔH is −

an endothermic reaction means ΔH is +

However, if we incorporate the heat in the balanced equations [see equations (11) and (12)], then the sign convention is different. For example, equation (12) reads: 1 mole of hydrogen plus $\frac{1}{2}$ mole of oxygen yields 1 mole of liquid water *plus* 68,320 cal of heat. This is obviously an exothermic reaction; heat is evolved. By convention, ΔH for this reaction is taken to be $\Delta H_{298} = -68{,}320$ cal. Since the heat of reaction (ΔH) describes the heat changes which take place in a chemical reaction, a balanced equation is necessary. If the stoichiometric coefficients in a balanced equation are multiplied or divided by a common factor, ΔH for the reaction changes accordingly. For example, if we write

$$2H_2(g) + O_2(g) \longrightarrow 2H_2O(l)$$

then ΔH_{298} for this reaction is −136,460 cal (2 × −68,320 cal) [see equation (12)].

We have noted that the enthalpy H is a state function. From the properties of state functions discussed on page 124, it follows that if there are two or more ways to convert a given amount of a given substance at a given temperature and pressure to another substance at a given temperature and pressure, then **the net ΔH for each of these pathways must be identical.** This statement, first made in 1840, is known as **Hess's law** (Germain Hess, 1802–1850). For example, it is possible to combine the elements oxygen and carbon (in the form of graphite) to

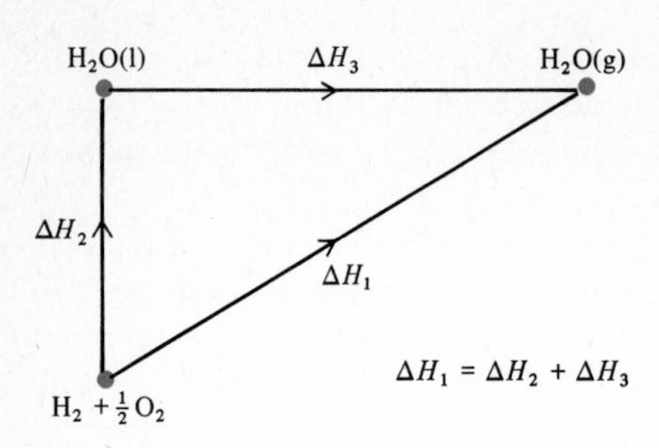

Figure 5-8 The enthalpy difference between $H_2 + \frac{1}{2}O_2$ on the one hand, and $H_2O(g)$ on the other, is the same, no matter what the pathway between them.

produce carbon dioxide. It is also possible to combine these elements to form carbon monoxide; and then to combine this with oxygen to give carbon dioxide.

$$\text{C (graphite)} + O_2 \longrightarrow CO_2 \qquad \Delta H = \Delta H_1$$
$$\text{C (graphite)} + \tfrac{1}{2}O_2 \longrightarrow CO \qquad \Delta H = \Delta H_2$$
$$CO + \tfrac{1}{2}O_2 \longrightarrow CO_2 \qquad \Delta H = \Delta H_3$$

Hess's law tells us that the *total* change in H for the two procedures must be the same ($\Delta H_1 = \Delta H_2 + \Delta H_3$) provided that (1) equal amounts are involved, (2) the starting materials were at the same temperature and pressure, and (3) the products are at the same temperature and pressure. Since E is also a state function, the same law holds with respect to ΔE.

Part of the value of Hess's law is that it enables us in many cases to calculate ΔH for reactions even if it is difficult or impossible to carry them out experimentally, provided that we have data for related reactions. For example, assume that the following are known:

$$H_2(g) + \tfrac{1}{2}O_2(g) \longrightarrow H_2O(l) \qquad \Delta H_{298} = -68{,}320 \text{ cal}$$

and

$$H_2O(g) \longrightarrow H_2O(l) \qquad \Delta H_{298} = -10{,}522 \text{ cal}$$

and it is necessary to determine ΔH_{298} for the following reaction:

$$H_2(g) + \tfrac{1}{2}O_2(g) \longrightarrow H_2O(g) \qquad \Delta H_{298} = ?$$

The above reactions can be rewritten as

$$H_2(g) + \tfrac{1}{2}O_2(g) \longrightarrow H_2O(l) \qquad \Delta H_{298} = -68{,}320 \text{ cal}$$
$$H_2O(l) \longrightarrow H_2O(g) \qquad \Delta H_{298} = +10{,}522 \text{ cal}$$

Simply adding the two equations yields

$$H_2O(l) + H_2(g) + \tfrac{1}{2}O_2(g) \longrightarrow H_2O(g) + H_2O(l) \qquad \Delta H_{298} = -57{,}798 \text{ cal}$$

(see Figure 5-8)

The addition of two or more equations is a simple algebraic procedure. We add everything on the left of the arrow, then everything on the right, and cancel whatever appears on both sides. The ΔH values are also added algebraically.

In order to apply the method shown above, *the reactions must all be at the same temperature*. In addition, one must be alert in matching the proper stoichiometric coefficients of the species in the different reactions, so that the wanted equation will be obtained. This is illustrated by the following example.

Example: Given the following reactions:

$$C(s) + \tfrac{1}{2}O_2(g) \longrightarrow CO(g) \qquad \Delta H_{298} = -26{,}416 \text{ cal}$$
$$C(s) + O_2(g) \longrightarrow CO_2(g) \qquad \Delta H_{298} = -94{,}014 \text{ cal}$$
$$2H_2(g) + O_2(g) \longrightarrow 2H_2O(g) \qquad \Delta H_{298} = -115{,}596 \text{ cal}$$

Calculate the heat of reaction of

$$H_2(g) + CO_2(g) \longrightarrow H_2O(g) + CO(g) \qquad \Delta H_{298} = ?$$

Answer: By rearranging the known reactions, we obtain

$$(1) \quad C(s) + \tfrac{1}{2}O_2(g) \longrightarrow CO(g) \qquad \Delta H_{298} = -26{,}416$$

$$(2) \quad CO_2(g) \longrightarrow C(s) + O_2(g) \qquad \Delta H_{298} = +94{,}014$$

$$(3) \quad H_2(g) + \tfrac{1}{2}O_2(g) \longrightarrow H_2O(g) \qquad \Delta H_{298} = -57{,}798$$

Adding the three yields the desired equation:

$$H_2(g) + CO_2(g) \longrightarrow CO(g) + H_2O(g) \qquad \Delta H_{298} = +9800 \text{ cal}$$

Note that when we reversed the direction of equation (2) it was necessary to change the sign of ΔH.

5-5 Heats of Formation: Standard States

The heat of reaction, ΔH_{reac}, may be thought of as being equal to the sum of the enthalpy of the products minus the sum of the enthalpy of the reactants.

$$\Delta H_{reac} = \sum H_{products} - \sum H_{reactants}$$

[where the Greek symbol Σ (sigma) stands for sum].

We can only know (measure) ΔH. We have no information concerning the *absolute* enthalpies (H) of either the reactants or products. Recalling the reaction of the formation of $H_2O(g)$ from the elements $H_2(g)$ and $O_2(g)$.

$$H_2(g) + \tfrac{1}{2}O_2(g) \longrightarrow H_2O(g) \qquad \Delta H_{298} = -57{,}798 \text{ cal}$$

and

$$\Delta H_{reac} = \sum H_{products} - \sum H_{reactants}$$
$$= H(H_2O)(g) - H(H_2)(g) - H(\tfrac{1}{2}O_2)(g)$$

If the enthalpies of hydrogen and oxygen at 298°K were zero (of course, their enthalpies are not zero and in fact we have *no* information concerning their absolute values), then we could say that

$$\Delta H_{298} = H_{products} - H_{reactants}$$
$$\Delta H_{298} = H(H_2O) - \text{zero}$$
$$\Delta H_{298} = H(H_2O)$$

The heat of formation per mole of $H_2O(g)$ at 298°K would be equal to the absolute enthalpy per mole of $H_2O(g)$ at 298°K. This idea forms the basis of **standard states.**

STANDARD STATES FOR ENTHALPY. By convention, *all elements* in the states in which they are most stable at 1 atm pressure and 298°K are *assigned* an enthalpy value of zero.

Let us discuss this definition term by term.

By convention: This signifies that what follows is *agreed* to by chemists; it is not a law of nature, merely an agreement.

All elements: This definition applies to *elements* and not to compounds.

The states in which they are most stable: This phrase takes into account that some elements occur in more than one form (allotropism). Carbon occurs as graphite and diamond, and sulfur occurs in two crystal forms, called monoclinic and rhombic, to name just two. When an element can exist in more than one form at 298°K and 1 atm pressure, the definition applies only to the form of the element which is the most stable under these conditions.

At 298°K: The definition can only be applied at this specific temperature.

Assigned an enthalpy value of zero: This phrase takes note of the fact that the value of the enthalpy is an artificial value and not the true enthalpy value, which of course we do not know.

1 atm pressure: This is taken to be the standard pressure.

The ΔH is labeled with a superscript zero, $\Delta H°$, to show that the reaction is measured at 1 atm pressure. The heat change that occurs when a compound is formed from its elements in their standard states is called the **standard heat of formation,** or **standard enthalpy of formation,** and given the symbol $\Delta H_f°$. Note that this convention allows us very simply to determine heats of formation for many compounds. All we have to do is measure $\Delta H°$ for the reaction in which 1 mole of the compound is formed from its elements in their standard states. Although this is not always possible, it can be done in many cases.

Why have chemists gone through all the trouble of defining such a standard state and, in particular, one that defines an artificial enthalpy value for elements? The reason is that given these standard states, chemists can calculate heats of formation for compounds, and these values are *not* zero. The *absolute* values of these heats of formation are of course incorrect, but their *relative* values are correct. Since there is no way for us to get the absolute values, we make do with the relative ones. Actually, this is not much of a handicap. For most purposes, the relative values do just as well. Table 5-4 is a list of standard heats of formation ($\Delta H_f°$) for various substances.

Note that the values are given for 1 mole of each substance formed. For example, the heat given off when 1 mole of Na(s) and $\frac{1}{2}$ mole of $Cl_2(g)$ react at 298°K and 1 atm to give 1 mole of NaCl(s) is 98,232 cal; and the heat required for 1 mole of rhombic sulfur to be converted to 1 mole of monoclinic sulfur is 71 cal. Two moles would involve twice as many calories. The values in Table 5-4 may be used to calculate heats of reaction for a great many reactions at 298°K. In order to do this, we apply the meaning of $\Delta H°_{reac}$.

Table 5-4 Standard Enthalpies of Formation at 298°K

Substance	$\Delta H_f°$ *(cal/mole)*	Substance	$\Delta H_f°$ *(cal/mole)*
$H_2O(g)$	−57,789	$C_6H_6(g)$	19,820
HCl(g)	−22,063	$C_2H_2(g)$	54,190
HBr(g)	−8,660	$C_2H_4(g)$	12,500
HI(g)	6,200	$C_2H_6(g)$	−20,240
$NH_3(g)$	−11,040	AgCl(s)	−30,360
NO(g)	21,600	C(graphite)	0
$NO_2(g)$	8,091	C(diamond)	453.2
$N_2O_4(g)$	2,580	$CaCO_3(s)$	−228,450
$O_3(g)$	34,000	CaO(s)	−151,900
$PCl_3(g)$	−73,220	$Fe_2O_3(s)$	−196,500
$PCl_5(g)$	−95,350	KCl(s)	−104,175
S(rhombic)	0	NaCl(s)	−98,232
S(monoclinic)	71	$CH_3OH(l)$	−57,020
$SO_2(g)$	−70,960	$C_2H_5OH(l)$	−66,350
$SO_3(g)$	−94,450	$CH_3COOH(l)$	−116,400
CO(g)	−26,416	$C_6H_6(l)$	11,720
$CO_2(g)$	−94,052	$H_2O(l)$	−68,320
$CH_4(g)$	−17,890		

$$\Delta H^\circ_{\text{reac}} = \sum \Delta H^\circ_{\text{products}} - \sum \Delta H^\circ_{\text{reactants}}$$
$$= \sum \Delta H^\circ_{f(\text{products})} - \sum \Delta H^\circ_{f(\text{reactants})}$$

Example: Calculate ΔH°_{298} for the reaction

$$C_6H_6(g) + 7\tfrac{1}{2}O_2(g) \longrightarrow 6CO_2(g) + 3H_2O(g)$$

using the data in Table 5-4.

Answer: Table 5-4 yields ΔH°_f values as follows:

$$\begin{aligned} H_2O(g) &= -57{,}789 \text{ cal} & 3H_2O(g) &= -173{,}367 \text{ cal} \\ CO_2(g) &= -94{,}052 \text{ cal} & 6CO_2(g) &= -564{,}312 \text{ cal} \\ C_6H_6(g) &= 19{,}820 \text{ cal} & & \\ O_2(g) &= 0 \text{ cal} & &\text{(definition from standard state)} \end{aligned}$$

$$\Delta H^\circ_{f(\text{reac})} = \sum \Delta H^\circ_{f(\text{products})} - \sum \Delta H^\circ_{f(\text{reactants})}$$

$$\begin{matrix} C_6H_6(g) & + & 7\tfrac{1}{2}O_2(g) & \longrightarrow & 6CO_2(g) & + & 3H_2O(g) \\ 19{,}820 & & 0 & & -564{,}312 & & -173{,}367 \end{matrix}$$

$$\sum \Delta H^\circ_{\text{products}} = (-564{,}312) + (-173{,}367) = -737{,}679 \text{ cal}$$

$$\sum \Delta H^\circ_{\text{reactants}} = 19{,}820 \text{ cal}$$
$$\Delta H^\circ_{\text{reac}} = -737{,}679 \text{ cal} - 19{,}820 \text{ cal} = -757{,}499 \text{ cal}$$

We have thus calculated that combustion of 1 mole of benzene (C_6H_6) would give off 757,499 cal at 298°K and 1 atm, and by using standard heats of formation it was not necessary actually to carry out the combustion.

The reader may be wondering what would happen if instead of defining a standard-state enthalpy of zero for elements at 298°K a set of different enthalpy values were assigned. The answer to this question is that the values of the heats of formation as given in Table 5-4 would be correspondingly different, but the ΔH°_{298} of reactions as calculated in the example above would have the *same* values. Most important, it makes no difference what value is assigned to the elements. As long as the same value was always used for each element, the ΔH°_{298} of the reaction would not change. A chemical reaction is balanced in terms of masses. The heats of formation of the compounds on the right and left sides of the equation are in terms of the elements that make up the compounds. Since the equation is balanced, the masses of the elements occur equally on both sides and their enthalpy values cancel, regardless of the assigned values.

From tables of heats of formation more extensive than Table 5-4, it is possible to determine ΔH°_{298} for a wide variety of chemical reactions. Even though the experimentalist thus possesses a powerful tool for predicting the heat of reactions, he or she is constrained to a single temperature, 298°K. It very often happens that we know ΔH°_{298} but need to know the ΔH° of the same reaction at another temperature. An equation that accomplishes this result (which we shall not derive) is

$$\Delta H^\circ_T = \Delta H^\circ_{298} + \Delta C_P(T - 298) \tag{13}$$

where ΔH°_T = standard heat of reaction at a temperature T
ΔH°_{298} = standard heat of reaction at 298°K
ΔC_P = difference in the heat capacities between the reactions and products
$\Delta C_P = \Sigma\, C_P(\text{products}) - \Sigma\, C_P(\text{reactants})$

Equation (13) is valid only for the case when the heat capacities of all the reactants and products are constant* over the temperature range given.

Example: Using tables of heat capacities and heats of formation, calculate the heat of reaction at 750°K for the reaction

$$N_2(g) + 3H_2(g) \longrightarrow 2NH_3(g)$$

Answer: From the tables, we have (at 298°K)

	N_2	$+ 3H_2$	$\longrightarrow 2NH_3$
Heat of formation: (from Table 5-4)	0	0	2(−11,040) = −22,080
Molar heat capacities: (from Table 5-2)	6.83	3(6.52)	2(7.11)

$$\Delta H^\circ_{298} = -22{,}080 \text{ cal}$$
$$\Delta C_p = 2(7.11) - 3(6.52) - 6.83 = -12.17 \text{ cal/mole} \cdot {}^\circ\text{K}$$

Applying equation (13), we obtain

$$\Delta H^\circ_{750} = (-22{,}080) + (-12.17)(750 - 298)$$
$$\Delta H^\circ_{750} = -27{,}580 \text{ cal}$$

5-6 Experimental Determination of Heats of Chemical Reactions

The measurement of heat is quite old compared to other aspects of chemistry. It was in 1780, when Lavoisier and Pierre Laplace (1749–1827) published their work on the heat of combustion of carbon that experimental thermochemistry really began. These scientists obtained a value of −98,848 cal/mole for the heat of the reaction:

$$C(s) + O_2(g) \longrightarrow CO_2(g)$$

This value is astonishingly accurate when one examines their ice calorimeter shown in Figure 5-9 and compares their value to the best modern value of −94,044 cal/mole. It is no accident that Lavoisier and Laplace carried out their studies on a combustion reaction. **Combustion,** which is defined as the reaction of substances with oxygen, is mankind's oldest known chemical reaction, as well as one of the most important. It is from combustion that the human race has always derived most of its energy, and this of course is still true today. For these reasons, chemists have always been interested in all aspects of combustion, especially the energy derived from it.

The experimental study of the heat changes that occur in chemical reactions is known as **calorimetry.** When the reaction involved is a combustion reaction, the heat evolved is called the **heat of combustion.** The heats of combustion of a number of typical compounds are given in Table 5-5.

The basic principle used in measuring heats of combustion today is the same as that used by Lavoisier and Laplace, though of course the apparatus is more modern and less subject to experimental errors. Figure 5-10 is a schematic representation of a bomb calorimeter used for the experimental determination of heats of combustion. The inner vessel is known as the reaction vessel. It is within this vessel that the

*Generally, C_p is a function of temperature. This means that the value of C_p changes with changing temperature [for example, the relationship between C_p and T might be $C_p = 4.3T$; then at 298°K, $C_p = 4.3(298) = 1281.4$, and at 500°K, $C_p = 4.3(500) = 2150$]. When C_p = a constant, it has the same value at all temperatures considered.

Figure 5-9 The ice calorimeter of Lavoisier and Laplace.

material to be burned is placed. The reaction vessel is then filled with pure oxygen under pressure and placed within the calorimeter. The chamber outside the calorimeter contains a known amount of water or some other suitable fluid and a precision thermometer and stirrer. The insulation around the calorimeter is constructed so that the heat evolved in the reaction is not lost from the water. This is an *adiabatic calorimeter.* The combustion process begins when a spark of electricity is transmitted from a battery outside the calorimeter to the reaction vessel. What is measured is the rise in temperature of the *water.* Of course, the heat given off by the reaction is absorbed not only by the water, but also by the reaction chamber, the stirrer, and the thermometer, and each of these things has its own heat capacity. In order to translate this rise in the temperature of the water to the amount of heat absorbed by all the component parts of the calorimeter, one must introduce into the calorimeter a *known* quantity of heat and then measure the temperature rise caused by the introduction of this heat. Electrical energy is always used for this, since it can be measured with the highest precision. This process is known as *calibrating* the calorimeter. The *water equivalent* of a calorimeter is the quantity of water required

Table 5-5 Heats of Combustion at 298°K

Substance	*Formula*	*Equation*	H°_{298} *(cal/mole)*
Methane (g)	CH_4	$CH_4(g) + 2O_2(g) \rightarrow CO_2(g) + 2H_2O(l)$	−212,800
Ethane (g)	C_2H_6	$C_2H_6(g) + \frac{7}{2}O_2(g) \rightarrow 2CO_2(g) + 3H_2O(l)$	−372,820
Ethylene (g)	C_2H_4	$C_2H_4(g) + 3O_2(g) \rightarrow 2CO_2(g) + 2H_2O(l)$	−337,230
Propane (g)	C_3H_8	$C_3H_8(g) + 5O_2(g) \rightarrow 3CO_2(g) + 4H_2O(l)$	−530,600
Propene (g)	C_3H_6	$C_3H_6(g) + \frac{9}{2}O_2(g) \rightarrow 3CO_2(g) + 3H_2O(l)$	−491,800
Sucrose (s)	$C_{12}H_{22}O_{11}$	$C_{12}H_{22}O_{11}(s) + 12O_2(g) \rightarrow 12CO_2(g) + 11H_2O(l)$	−1,348,900
Diamond (s)	C	$C(s) + O_2(g) \rightarrow CO_2(g)$	−94,500
Graphite (s)	C	$C(s) + O_2(g) \rightarrow CO_2(g)$	−94,050
Hydrogen (g)	H_2	$H_2(g) + \frac{1}{2}O_2(g) \rightarrow H_2O(l)$	−68,317

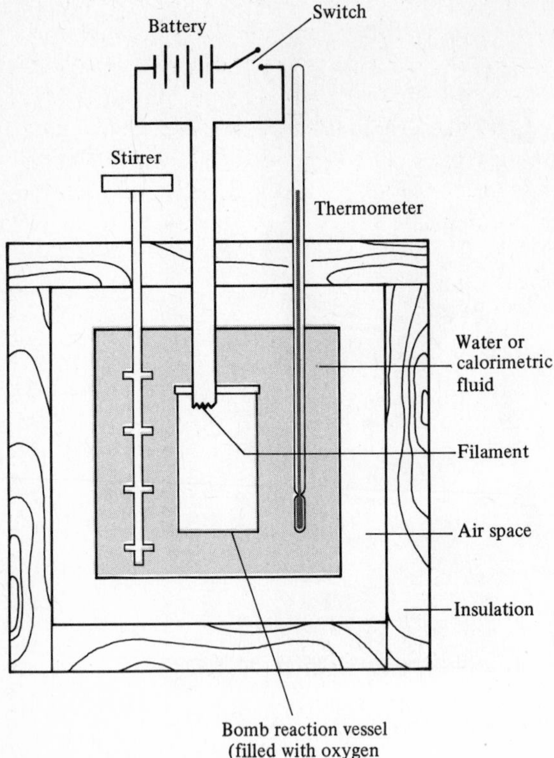

Figure 5-10 Schematic representation of a bomb calorimeter used for the measurements of heats of combustion.

to absorb the same quantity of heat as the calorimeter with its components—that is, water, reaction vessel, thermometer, and stirrer.

Everyone is familiar with the fact that the energy values of foods are expressed in calories, and tables of these values are readily available (for example, see Table 5-6). How are these energy values determined? The body gets its energy by making the food combine with the oxygen that is breathed in, turning these ingredients into the ultimate products, CO_2 and H_2O. The reactions are quite complicated, and are

Table 5-6 Energy Content of Some Foods

	Chemist's kilocalories[a]		*Chemist's kilocalories*[a]
Milk, 8 oz	165	Apple, 5 oz	70
Swiss cheese, 1 oz	105	Banana, 5 oz	85
Egg, 1 boiled, large	80	Grapefruit, $\frac{1}{2}$ medium	50
Hamburger patty, 3 oz	245	Orange juice, 8 oz	100
Beef, lean, 2 oz	115	Bread, rye, 1 slice	55
Lamb, shoulder, lean, 2 oz	125	Bread, white, 1 slice	60
Haddock, fried, 3 oz	135	Fudge, 1 oz	115
Shrimp, 3 oz	110	Honey, 1 tablespoon	60
Peanuts, roasted, 2 oz	210	Cookie, 3-in-diameter	110
Carrot, raw, 8 oz	45	Cornflakes, 1 oz	110
Corn, 5-inch ear	65	Noodles, 8 oz	200
Potato, 5 oz	90	Butter, 1 tablespoon	100
Tomato, 5 oz	30	Margarine, 1 tablespoon	100

[a] A word must be said about the units. The "calorie" used by nutritionists is not the same as the calorie used by chemists. The nutritionist's calorie is actually a kilocalorie, equal to 1000 of the calories that chemists use. This table gives food values in nutritionist's calories, or chemist's kilocalories. For example, the caloric content of 8 oz of milk is 165 kilocalories (165,000 calories) to a chemist, but 165 calories to a nutritionist.

carried out at the temperature of the body (about 37°C), which is far below the temperatures of combustion. Nevertheless, Hess's law (Section 5-4) tells us that the amount of energy given off in the conversion of 1 g of a given food to CO_2 and H_2O, by reaction with oxygen, *must be constant, regardless of the path* (provided that comparisons are made at the same temperature). Therefore, if we burn 1 g of a given food in a bomb calorimeter, and measure the energy given off, we know that this is the same amount of energy we obtain when we eat 1 g of this food, if we make the temperature correction as given by equation 13.

Example: Nutritionists are concerned about the dietary habits of the American public. We would like to compare the caloric values of fats and sugars. Glucose ($C_6H_{12}O_6$) is a sugar that is metabolized in the body and palmitic acid ($C_{15}H_{31}COOH$) is a typical constituent of fat (see Section 21-7). One gram each of glucose and palmitic acid was separately burned in a calorimeter whose water equivalent is 450.0 g. The temperature rise for glucose was determined to be 8.31°C, while that for palmitic acid was found to be 20.67°C. The heat capacity of water is 1.000 cal/g · °K. Calculate the heat of combustion (a) of glucose and (b) of palmitic acid per mole of each compound.

Answer: For our calorimeter we have

$$\begin{aligned}\text{heat capacity of water} \times \text{water equivalent of calorimeter} &= 1.000\ \text{cal/g}\cdot{}^\circ\text{K} \times 450.0\ \text{g}\\ &= 450.0\ \text{cal/}{}^\circ\text{K}\end{aligned}$$

This tells us that we need 450.0 cal to raise the temperature of the thermometer 1°C.
(a) For glucose:

$$450.0\,\frac{\text{cal}}{{}^\circ\text{K}} \times 8.31\,\frac{{}^\circ\text{K}}{\text{g glucose}} = 3740\ \text{cal/g glucose}$$

Glucose has a molecular weight of 180 g/mole.

$$3740\,\frac{\text{cal}}{\text{g}} \times \frac{180\ \text{g}}{\text{mole}} = 673{,}000\,\frac{\text{cal}}{\text{mole glucose}} = 673\,\frac{\text{kcal}}{\text{mole glucose}}$$

(b) For palmitic acid:

$$450.0\,\frac{\text{cal}}{{}^\circ\text{K}} \times 20.67\,\frac{{}^\circ\text{K}}{\text{g palmitic acid}} = 9302\,\frac{\text{cal}}{\text{g palmitic acid}}$$

Palmitic acid has a molecular weight of 256.4 g/mole.

$$9302\,\frac{\text{cal}}{\text{g}} \times \frac{256.4\ \text{g}}{\text{mole}} = 2{,}385{,}000\,\frac{\text{cal}}{\text{mole palmitic acid}} = 2385\,\frac{\text{kcal}}{\text{mole palmitic acid}}$$

We therefore see, as nutritionists know, that 1 g of fat produces approximately $2\frac{1}{2}$ times as much heat as 1 g of sugar. One obtains many more calories from eating fats than from eating equal weights of sugars.

Chemical reactions are carried out either at constant volume or at constant pressure. An example of a reaction vessel that is used for constant volume processes is the bomb calorimeter shown in Figure 5-10. As we have noted, this method is used experimentally to determine heats of reaction and especially heats of combustion. A constant-pressure apparatus is shown in Figure 5-11. Figure 5-11 represents a typical chemical reaction performed in a chemistry laboratory. The difference in the two methods which concerns us is that the bomb calorimeter carries out chemical reactions at constant *volume*, but most reactions (including the one shown in Figure 5-11) are carried out at constant *pressure*. This is an important distinction.

$\Delta E = q_v$
$\Delta H = q_p$

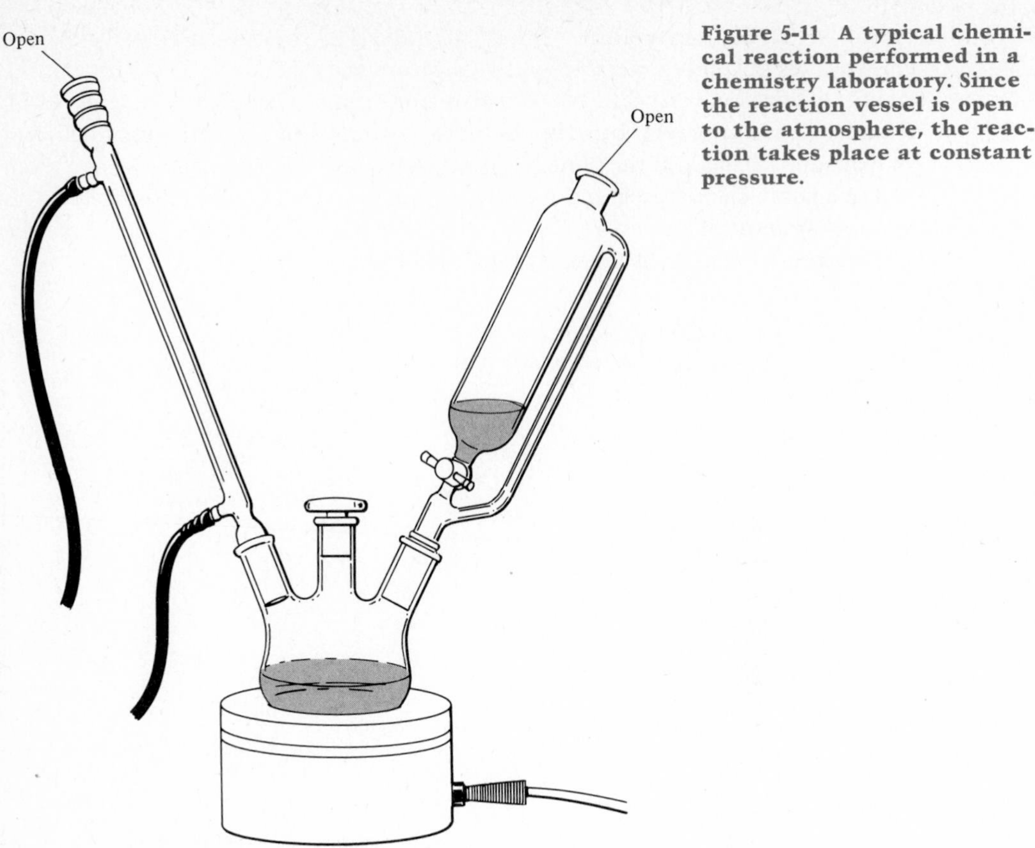

Figure 5-11 A typical chemical reaction performed in a chemistry laboratory. Since the reaction vessel is open to the atmosphere, the reaction takes place at constant pressure.

From equations (7) and (8) (page 125), we see that the heat changes in the different processes are given by $q_V = \Delta E$, and $q_P = \Delta H$. Therefore, when one makes an experimental determination of the heat of a reaction in a bomb calorimeter, one measures ΔE and not ΔH. However, one can obtain the value of ΔH of the reaction from the observed ΔE value rather easily. From equation (5), we have

$$\Delta H = \Delta E + \Delta(PV)$$

We are interested in the $\Delta(PV)$ term. For solids and liquids, $\Delta(PV)$ can, to a very good approximation, be taken as zero. The reason is that the volume of solids and liquids does not appreciably change in a chemical reaction. If all the reactants and products in the reaction are in the solid and/or in the liquid form, then for all practical purposes we can write $\Delta H \approx \Delta E$. However, if any of the substances in the reaction is in the gaseous state, $\Delta(PV)$ may not be negligible. To calculate $\Delta(PV)$ when gases are involved in the reaction, we appeal to the ideal gas law. We have, from Section 4-7,

$$PV = nRT$$

and

$$\Delta(PV) = (\Delta n)RT$$

where Δn is the number of moles of gaseous products minus the number of moles of gaseous reactants. This equation says that, for a reaction at constant temperature, the change in the pressure–volume product is equal to the change in the number of gaseous moles in the reaction times the gas constant and temperature.

Example: The combustion of methane in a bomb calorimeter at 25°C yields −13.30 kcal/g. Estimate ΔH°_{298} for this reaction.

Answer: Since the process was carried out at constant volume, $\Delta E = -13.30$ kcal/g of methane. The equation for the combustion of methane is

$$CH_4(g) + 5O_2(g) \longrightarrow 4CO_2(g) + 2H_2O(g)$$

Since all the components are gases, $\Delta n = (4 + 2) - (5 + 1) = 0$; therefore, $\Delta(PV) = (\Delta n)RT = 0$, $\Delta E = \Delta H$, and

$$-13.30 \text{ kcal/g methane} \times \frac{16.043 \text{ g methane}}{\text{mole methane}} = -213.4 \text{ kcal/mole} = \Delta E^\circ_{298} = \Delta H^\circ_{298}$$

Example: The combustion of 1 mole of sucrose, $C_{12}H_{22}O_{11}(s)$, in a bomb calorimeter generated 1348.9 kcal at 25°C. Estimate ΔH°_{298} for the reaction.

Answer:

$$\Delta E^\circ_{298} = -1348.9 \text{ kcal/mole}$$
$$\Delta H^\circ_{298} = \Delta E^\circ_{298} + \Delta nRT$$

The equation for the combustion of sucrose is

$$C_{12}H_{22}O_{11}(s) + 12O_2(g) \longrightarrow 12CO_2(g) + 11H_2O(g)$$

$$\Delta n = \text{number of moles of gaseous products} - \text{number of moles of gaseous reactants}$$
$$= (12 + 11) - 12 = 11$$
$$\Delta H^\circ_{298} = -1{,}348{,}900 \text{ cal} + (11 \text{ moles})(1.99 \text{ cal/°K} \cdot \text{mole})(298\text{°K})$$
$$= -1{,}342{,}400 \text{ cal} \quad \text{or} \quad -1342.4 \text{ kcal/mole of sucrose}$$

Note that it was necessary to convert ΔE from kilocalories to calories, because we took ΔnRT in calories.

In the above example there are more moles of gas in the products than in the reactants. The additional number of moles had to expand against the atmosphere and hence perform work. Some of the heat of this exothermic reaction was used to perform this work (Figure 5-12).

Example: The combustion of sulfur dioxide, SO_2, yields sulfur trioxide. When this reaction was carried out in a bomb calorimeter at 25°C, it generated 23.485 kcal/mole of SO_2. What is ΔH°_{298} for this reaction?

Answer: For 1 mole of SO_2, the equation is

$$SO_2(g) + \tfrac{1}{2}O_2(g) \longrightarrow SO_3(g)$$

$$\Delta H^\circ_{298} = \Delta E^\circ_{298} + \Delta nRT$$
$$\Delta n = 1 - \tfrac{3}{2} = -\tfrac{1}{2}$$
$$\Delta H^\circ_{298} = -23{,}485 \text{ cal} - (\tfrac{1}{2} \text{ mole})(1.99 \text{ cal/mole} \cdot \text{°K})(298\text{°K})$$
$$= -23{,}800 \text{ cal} \quad \text{or} \quad -23.8 \text{ kcal/mole of } SO_2$$

In order to perform these calculations it is obviously necessary to write the exact chemical equations for the reactions and to specify the states of all reactants and products.

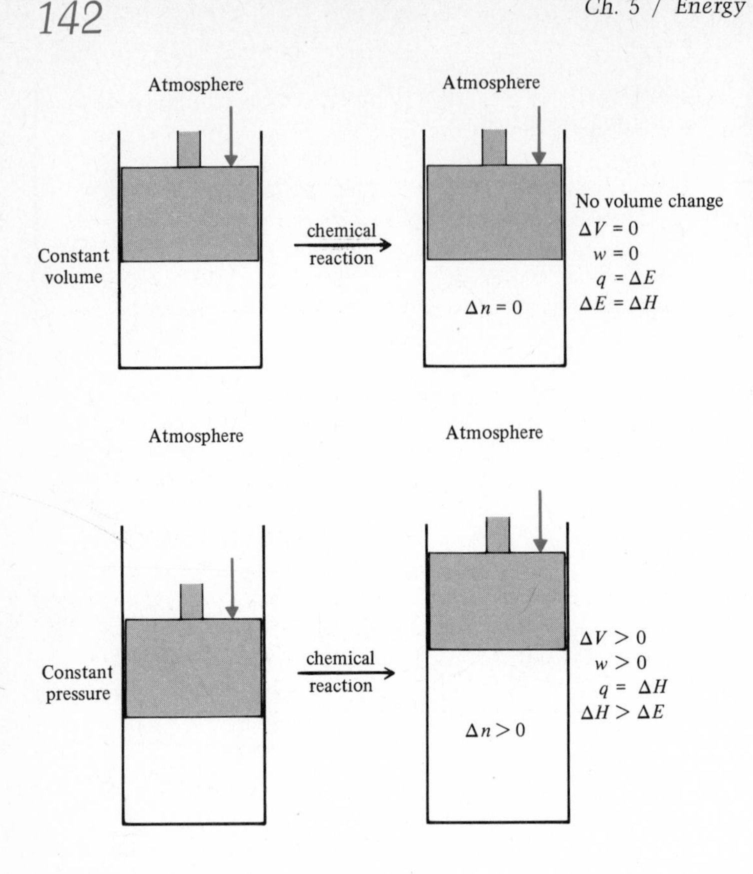

Figure 5-12 When a reaction takes place at constant volume, no work is done, but when it takes place at constant pressure, and $\Delta n > 0$, the system expands, performing work against the atmosphere.

Problems

5-1 Define each of the following terms. **(a)** work **(b)** calorie **(c)** adiabatic process **(d)** state function **(e)** enthalpy **(f)** heat capacity **(g)** phase **(h)** heat of vaporization **(i)** heat of fusion **(j)** sublimation **(k)** exothermic **(l)** endothermic **(m)** Hess's law **(n)** standard state **(o)** heat of formation **(p)** combustion **(q)** calorimetry

5-2 Define each of the following symbols. **(a)** Δ **(b)** E **(c)** q **(d)** w **(e)** q_V **(f)** H **(g)** C_P **(h)** ΔH_{298} **(i)** Σ **(j)** $\Delta H°$ **(k)** ΔH_f°

5-3 What makes us believe that the first law of thermodynamics is valid?

5-4 State the first law of thermodynamics in algebraic form.

5-5 Calculate the change in energy ΔE for each of the following processes. **(a)** 100.0 cal of work is done on a system; no heat is added to or subtracted from the system. **(b)** 100.0 cal of work is done by a system; 100.0 cal of heat is added to the system. **(c)** 100.0 cal of work is done by a system; 100.0 cal of heat is removed from the system. **(d)** No work is done, and no heat is transferred in or out of a system. **(e)** No work is done; 100.0 cal of heat is abstracted from a system.

5-6 How much work in calories is gained or lost when neon gas expands from 350 ml to 550 ml at a constant pressure of 825 torr?

5-7 A system is taken through the following steps: (1) expanded adiabatically with $w = 1000$ cal; (2) expanded with $w = 225$ cal and 195 cal of heat withdrawn from the system; (3) 110 cal of heat added to the system, no volume change taking place. Calculate ΔE for the process.

5-8 A sample of C_2H_6 gas is in a closed 2.55-liter container at 125°C and 750 torr. The container is cooled to 50°C. Assume that C_2H_6 is an ideal gas and for this problem $C_P = C_V$. **(a)** What is the new pressure? **(b)** What is the change in energy ΔE?

5-9 If 487 g of water at 85°C is added to 855 g of water at 20°C, what is the final temperature of the water, ignoring any heat lost to or gained from the surroundings?

5-10 If 574 g of Hg at 175°C is thoroughly mixed with 375 g of H_2O at 20°C, what is the final temperature of the mixture, ignoring any heat lost to or gained from the surroundings?

5-11 183 g of water at 25°C is poured into a flask containing 275 g of water at 75°C. The heat capacity of the flask is 350 cal/deg. What is the final temperature of the flask containing the total amount of water?

5-12 Joseph Black measured the heat capacity of various substances by mixing them with a liquid of known heat capacity and noting the final temperature. If 100.0 g of water at 95.0°C is mixed with 200.0 g of ethyl alcohol C_2H_6O at 15.0°C, the final temperature is 51.6°C. What is the heat capacity of ethyl alcohol in cal/g and cal/mole, ignoring heat gained by or lost to the surroundings?

5-13 Another way in which Black measured heat capacities was by the method of constant heat supply. He placed two substances at an equal distance from a strong fire and measured how high the temperature was raised. The idea is that both substances have received equal quantities of heat. If a beaker containing 125 g of water and another containing 425 g of acetic acid, $C_2H_4O_2$, both initially at 20.0°C, are exposed to a constant heat source, and after a period of time the temperature of the water is 85.0°C and that of the acetic acid is 58.9°C, what is the heat capacity of acetic acid in cal/g and cal/mole, ignoring heat lost to, or gained by, the surroundings?

5-14 Calculate the amount of heat necessary to convert 1.0 mole of ice at 0°C to a gas at 100°C.

5-15 Calculate the amount of heat necessary to take 100.0 g of NaCl(s) at 0°C to NaCl(l) at 1500°C.

5-16 If 1500.0 cal of heat is added to 10.0 g of water at 75.0°C, what will happen to the water?

5-17 How many calories are required for each of the following processes? (Assume that the values in Table 5-1 do not change with temperature.)
(a) Heating 1.0 qt of water from 25°C to 80°C.
(b) Heating 57.5 g of Cl_2(g) from −30°C to 120°C.
(c) Heating 2.0 lb of NaCl(s) from 25°C to 750°C.
(d) Converting 575 g of ice at −20°C to steam at 125°C.
(e) Melting 25.5 kg of Na(s).
(f) Boiling 5.7 g of NH_3(l) at its boiling point, 240°K, and then heating the NH_3(g) to 75°C.

5-18 Some people like to cool a cup of hot coffee by adding an ice cube. If an ice cube weighing 28.000 g is added to a cup containing 135 ml of coffee at 92°C, what will the final temperature be when all the ice melts, ignoring heat lost to or gained from the surroundings? Assume that for coffee the heat capacity is 0.950 cal/g · °K, and the density 1.05 g/ml, and that neither of these values is changed by the melting ice.

5-19 Given

$$CO_2(g) \longrightarrow C(s) + O_2(g) \qquad \Delta H^\circ_{298} = 94{,}014 \text{ cal}$$

calculate ΔH°_{298} for each of the following.
(a) $2CO_2(g) \rightarrow 2C(s) + 2O_2(g)$
(b) $2C(s) + 2O_2(g) \rightarrow 2CO_2(g)$ = 188,028 K cal

5-20 Given

$$H_2O(l) + \tfrac{1}{2}O_2(g) \longrightarrow H_2O_2 \qquad \Delta H^\circ_{298} = 22.64 \text{ kcal}$$
$$H_2O(l) \longrightarrow H_2O(g) \qquad \Delta H^\circ_{298} = 10.52 \text{ kcal}$$

calculate ΔH°_{298} for the reaction

$$H_2O_2 \longrightarrow H_2O(g) + \tfrac{1}{2}O_2(g)$$

5-21 Given

$$H_2(g) + F_2(g) \longrightarrow 2HF(g) \quad \Delta H^\circ = -124 \text{ kcal}$$
$$H_2(g) \longrightarrow 2H(g) \quad \Delta H^\circ = 104 \text{ kcal}$$
$$F_2(g) \longrightarrow 2F(g) \quad \Delta H^\circ = 37.8 \text{ kcal}$$

calculate ΔH° for

$$H(g) + F(g) \longrightarrow HF(g)$$

5-22 From the table of standard enthalpies of formation, calculate ΔH°_{298} for each of the following reactions.

(a) $Fe_2O_3(s) + 3CO(g) \rightarrow 2Fe(s) + 3CO_2(g)$
(b) $Cl_2(g) + 2HBr(g) \rightarrow 2HCl(g) + Br_2(l)$
(c) $2NH_3(g) \rightarrow N_2(g) + 3H_2(g)$
(d) $S(\text{monoclinic}) + O_2(g) \rightarrow SO_2(g)$
(e) $S(\text{rhombic}) + O_2(g) \rightarrow SO_2(g)$
(f) $2NO_2(g) \rightarrow N_2O_4(g)$
(g) $NO(g) + CO_2(g) \rightarrow NO_2(g) + CO(g)$

5-23 From the table of standard enthalpies of formation, calculate ΔH°_{298} for each of the following reactions.

(a) $C_2H_4(g) + H_2(g) \rightarrow C_2H_6(g)$
(b) $C(\text{graphite}) + O_2(g) \rightarrow CO_2(g)$
(c) $CaO(s) + CO_2(g) \rightarrow CaCO_3(s)$
(d) $C_2H_6(g) + \frac{7}{2}O_2(g) \rightarrow 2CO_2(g) + 3H_2O(l)$
(e) $S(\text{rhombic}) \rightarrow S(\text{monoclinic})$
(f) $SO_2(g) + \frac{1}{2}O_2(g) \rightarrow SO_3(g)$
(g) $2C(\text{graphite}) + H_2(g) \rightarrow C_2H_2(g)$
(h) $PCl_3(g) + Cl_2(g) \rightarrow PCl_5(g)$

5-24 Using the tables of heat capacities and heats of formation, calculate ΔH° at the temperatures given. (Assume that the C_p values in Table 5-2 do not change with temperature.)

(a) 80°C: $2Na(s) + Cl_2(g) \rightarrow 2NaCl(s)$
(b) −25°C: $C(\text{graphite}) + 2H_2(g) \rightarrow CH_4(g)$
(c) 275°C: $2CO(g) + O_2(g) \rightarrow 2CO_2(g)$

5-25 From the data given in the chapter, calculate each of the following, at 298°K.

(a) The heat of formation of propane, C_3H_8.
(b) The heat of combustion of methanol, CH_3OH.
(c) The heat of combustion of ethanol, C_2H_5OH.
(d) The heat of formation of sucrose. $C_{12}H_{22}O_{11}$.

5-26 In Section 5-6, we said that for an ideal gas at constant temperature $\Delta(PV) = (\Delta n)RT$. Why is this true?

5-27 Given

$$H_2(g) + \tfrac{1}{2}O_2(g) \longrightarrow H_2O(l) \quad \Delta H^\circ_{298} = -68.3 \text{ kcal}$$
$$C(\text{graphite}) + O_2(g) \longrightarrow CO_2(g) \quad \Delta H^\circ_{298} = -94.0 \text{ kcal}$$
$$12C(\text{graphite}) + 11H_2(g) + 5\tfrac{1}{2}O_2(g) \longrightarrow C_{12}H_{22}O_{11}(s) \quad \Delta H^\circ_{298} = -531 \text{ kcal}$$

calculate each of the following.

(a) ΔH°_{298} for the combustion of sucrose, $C_{12}H_{22}O_{11}$:

$$C_{12}H_{22}O_{11}(s) + 12O_2(g) \longrightarrow 12CO_2(g) + 11H_2O(l)$$

(b) ΔH°_{298} for the decomposition of sucrose into carbon and water:

$$C_{12}H_{22}O_{11}(s) \longrightarrow 12C(\text{graphite}) + 11H_2O(l)$$

5-28 From the values of the heats of combustion in Table 5-5, calculate H°_{298} for each of the following reactions.

(a) $C(\text{diamond}) \rightarrow C(\text{graphite})$
(b) $C_2H_4 + H_2 \rightarrow C_2H_6$
(c) $C_2H_6 + C(\text{graphite}) \rightarrow C_3H_6$

5-29 One of the rocket fuels in the Apollo moon explorations was liquid hydrazine, N_2H_4. From the following set of equations, calculate the heat of combustion of liquid N_2H_4.

$$N_2O(g) + 3H_2 \longrightarrow H_2O(l) + N_2H_4(l) \quad \Delta H = -75.76 \text{ kcal}$$
$$4N_2(g) + 3H_2O(l) \longrightarrow 2NH_3(g) + 3N_2O(g) \quad \Delta H = +241.35 \text{ kcal}$$
$$N_2H_4(l) + H_2O(l) \longrightarrow 2NH_3(g) + \tfrac{1}{2}O_2(g) \quad \Delta H = +34.18 \text{ kcal}$$
$$H_2O(l) \longrightarrow H_2(g) + \tfrac{1}{2}O_2(g) \quad \Delta H = +68.32 \text{ kcal}$$

5-30 As a primary source of energy the body "burns" glucose, $C_6H_{12}O_6$, yielding CO_2 and H_2O. **(a)** Write a balanced equation for the combustion of glucose. **(b)** The combustion of 1.00 g of glucose generates 3.74 kcal of heat. How much heat is produced per mole of glucose? **(c)** A popular breakfast cereal contains 40% sugar, which for our purposes here can be considered to be glucose. How many ounces of this breakfast cereal will yield 1000.0 cal of heat (that is, 1 nutritionist's calorie)? (1 lb = 16 oz.)

5-31 The combustion of methane was carried out in a bomb calorimeter (constant volume) at 300°K. The reaction is

$$CH_4(g) + 2O_2(g) \longrightarrow CO_2(g) + 2H_2O(l)$$

If 212.8 kcal of heat are evolved for each mole of CH_4 which has been burned: **(a)** Calculate ΔE, q, and w for the process. **(b)** Calculate ΔE, ΔH, q, and w for the same process carried out at constant pressure instead of constant volume. **(c)** Comment on the values of ΔE calculated in parts (a) and (b). **(d)** How much heat would be evolved for each gram of water that was formed?

5-32 For the following reactions at 298°K [aq means aqueous (that is, water) solution]:
(1) $\frac{1}{2}H_2(g) + \frac{1}{2}Cl_2(g) \rightarrow HCl(g)$
(2) $2H^+(aq) + CO_3^{2-}(aq) \rightarrow CO_2(g) + H_2O(l)$
(3) $Ag^+(aq) + Cl^-(aq) \rightarrow AgCl(s)$
(4) $CH_4(g) + O_2(g) \rightarrow CO_2(g) + H_2O(l)$
(5) $C_2H_6(g) + \frac{7}{2}O_2(g) \rightarrow 2CO_2(g) + 3H_2O(l)$
(a) Calculate Δn ($n_{\text{products}} - n_{\text{reactants}}$) for each reaction. **(b)** Calculate $(\Delta n)RT$ for each reaction. For which reactions does $\Delta E = \Delta H$?

5-33 When propane gas, C_3H_8, was burned in a bomb calorimeter at 25°C, the heat given off was 11.99 kcal/g of propane. What is ΔH°_{298} for this reaction?

Additional Problems

5-34 The total volume of the sea is 3.30×10^8 cubic miles. The total volume of polar ice is 6.3×10^6 cubic miles. If the average temperature of the ice is −15°C and of the sea is 10.0°C, what would be the temperature of the sea if all the ice melted? (Density of ice = 0.917 g/ml; density of sea water = 1.025 g/ml; heat capacity of seawater = 1.00 cal/g·°K; assume thorough mixing.)

5-35 **(a)** From the ΔH°_{298} of the following reaction and the standard enthalpies of formation given in Table 5-4, calculate the heat of formation of naphthalene, $C_{10}H_8$.

$$C_{10}H_8(s) + 12O_2 \longrightarrow 10CO_2(g) + 4H_2O(l) \qquad \Delta H^\circ_{298} = -1228 \text{ kcal}$$

(b) What is the ΔH°_{298} for this reaction per gram of naphthalene burned?

5-36 If 173 g of Hg at 325°C is added to 411 g of Hg at −10°C, what is the final temperature of the Hg, ignoring any heat lost to or gained from the surroundings?

5-37 The reaction for ethane, C_2H_6, as a rocket fuel at 298°K is

$$C_2H_6(g) + \tfrac{7}{2}O_2(g) \longrightarrow 2CO_2(g) + 3H_2O(l) \qquad \Delta H = -372.8 \text{ kcal}$$

(a) Calculate the ΔH of this reaction per gram of ethane burned. **(b)** Calculate the ΔE of the reaction. **(c)** If only 100.0 kcal of heat is evolved, what volume of CO_2 is produced at a pressure of 720 torr?

5-38 Four moles of an ideal gas at 25°C are held in a cylinder by a piston at a pressure of 100 atm. The pressure is reduced in three stages, first to 50 atm, then to 20 atm, and finally to 2 atm. Calculate the work done by the gas during these expansions, and compare it with the total work done when the gas expands from 100 atm to 2 atm in a single step. What can you conclude as to whether or not work is a state function?

5-39 If 10.0 g of ice at 0°C is introduced into a beaker containing 80.0 g of water at 40°C, what is the temperature of the water after the ice has completely melted?

5-40 The combustion of diethyl ether, $C_4H_{10}O(l)$, in a bomb calorimeter at 25°C gave off 8.81 kcal/g of diethyl ether. Calculate ΔH°_{298} for this reaction.

5-41 The heat of combustion of gasoline, a mixture whose average formula may be taken as $C_8H_{18}(l)$, is $\Delta H^\circ_{298} = -1303$ kcal/mole of gasoline. **(a)** Write a balanced equation for the combustion of gasoline. **(b)** From these data and the heats of formation given in Table 5-4, calculate the heat of formation of C_8H_{18}. **(c)** Calculate the heat evolved per gram of gasoline burned.

5-42 A sample of NH_3 gas is in a closed 850-ml container at 25°C and 450 torr. The container is heated to 50°C. (Assume that NH_3 is an ideal gas and that $C_P = C_V$.) **(a)** What is the new pressure? **(b)** What is the change in energy ΔE?

5-43 From the table of standard enthalpies of formation, calculate ΔH°_{298} for each of the following reactions.

(a) $NO(g) + \frac{1}{2}O_2(g) \rightarrow NO_2(g)$
(b) $N_2O_4(g) \rightarrow 2NO_2(g)$
(c) $C_2H_2(g) + H_2(g) \rightarrow C_2H_4(g)$
(d) $CO(g) + \frac{1}{2}O_2(g) \rightarrow CO_2(g)$
(e) $SO_3(g) \rightarrow S(monoclinic) + \frac{3}{2}O_2(g)$
(f) $Fe_2O_3(s) + 3C(graphite) \rightarrow 2Fe(s) + 3CO_2(g)$
(g) $C_6H_6(l) + \frac{15}{2}O_2(g) \rightarrow 6CO_2(g) + 3H_2O(l)$
(h) $PCl_5(g) \rightarrow PCl_3(g) + Cl_2(g)$

5-44 In a Carnot cycle, a gas is taken through the following four-stage cyclic process:
Step (i) Start with a gas at T_1, P_1, V_1 and go to T_1, P_2, V_2.
Step (ii) From T_1, P_2, $V_2 \rightarrow T_2$, P_3, V_3.
Step (iii) From T_2, P_3, $V_3 \rightarrow T_2$, P_4, V_4.
Step (iv) From T_2, P_4, $V_4 \rightarrow T_1$, P_1, V_1.
What is ΔE for the entire process?

6 Liquids

Ludwig Boltzmann, 1844–1906

6-1 Characteristics of Liquids

In many ways the liquid state occupies a middle ground between the gaseous state and the solid state. To consider a liquid in this way, recall that one of the assumptions of the kinetic-molecular theory of a gas is that the molecules of the gas are in continuous *random* motion. At the other extreme, a solid (as we shall see in Chapter 15) is a state of matter where the molecules are in an almost perfect order; each molecule of the solid is confined to a small specific region in space. The molecules of a liquid represent a compromise situation between the disorder of gases and the almost perfect harmony of a solid.

We can examine these differences experimentally by considering how the three states of matter are changed into one another. Let us begin with a gas at a high temperature and a constant low pressure. We want to observe the behavior of this gas as the temperature is lowered. Initially, the temperature of the gas is high; the molecules are moving rapidly and possess high kinetic energies. Therefore, any interactions between the molecules of the gas are negligible. Under these conditions (high temperature and low pressure) the gas behaves as an ideal gas and experimentally obeys the ideal gas law. As the temperature of the gas is lowered (at constant pressure), the total kinetic energy of the molecules correspondingly decreases and the interactions between the molecules are no longer negligible. Measurements on the gas will begin to show deviations from ideal behavior. By lowering the temperature still more, at constant pressure, the condensation temperature will be reached and the gas will turn into a liquid.

At any specific pressure, molecules in a liquid are much closer to each other than the same molecules would be in the gaseous state. Because of this closeness, the interactions between molecules in the liquid are much greater than in the gas and it is these overall attractive interactions which hold the molecules together in the liquid.

The completely random characteristics of the gaseous state are lost when it becomes a liquid. A liquid possesses some order as well as some disorder. If the temperature of the liquid is lowered further, the liquid will freeze, and a solid will eventually form. The molecules in a solid no longer possess enough kinetic energy to roam throughout the entire phase but are forced to stay in specific regions in space. Molecules in the three different phases are shown in Figure 6-1.

Mathematically, we are able to treat gases and solids with a moderate amount of success. This is because it is relatively easy, in a mathematical sense, to treat a system that is in complete disorder or in almost complete order. However, it is much more difficult to describe a system that exhibits only some order and some disorder. It is not surprising, therefore, that we have no comprehensive mathematical theory of liquids, although experimentally we know as much about liquids as we do about solids and gases.

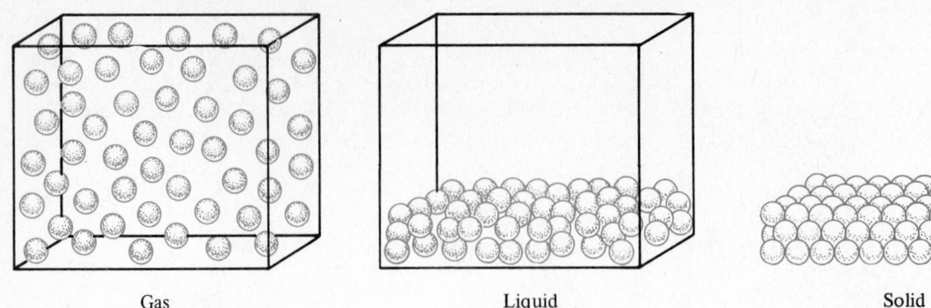

Figure 6-1 Molecules in the three phases—gas, liquid and solid. The molecules in the gas phase show complete disorder, the liquid shows some order and some disorder, and the solid shows almost complete order.

6-2 Vapor Pressure

If we put a liquid into a container (Figure 6-2) and, then, a short time later, examine the air above the surface of the liquid, we will find molecules from the liquid mixed in with the air molecules. Obviously, some molecules have escaped the liquid and entered the gas phase. How does this happen? A molecule in the interior of the liquid phase is surrounded on all sides by other molecules and is held in the liquid by forces of attraction between it and the other nearby molecules. When a molecule reaches the surface of the liquid these forces are unbalanced (Figure 6-3). If that particular molecule possesses enough kinetic energy, it can escape from the surface of the liquid and enter the gas phase. This is called **evaporation.** Two conditions must be present for a molecule to evaporate. They are: (1) the molecule must reach the surface of the liquid, and (2) the molecule must possess enough kinetic energy to overcome the attractive forces of the other molecules on the surface of the liquid.

We shall refer to those molecules of liquid which have escaped into the gas phase as **vapor molecules.** What happens to the vapor molecules after they escape into the gas phase? If the container is open, they will diffuse away, but if the container is closed (Figure 6-4), they cannot get away, and must remain in the gas phase, which is in contact with the surface of the liquid. Like all molecules in gas phases, the vapor molecules are in constant random motion and continually strike the surface of the liquid. A large number of them possess so much kinetic energy that when they strike the surface, they bounce off. However, some of the molecules in the vapor phase lose

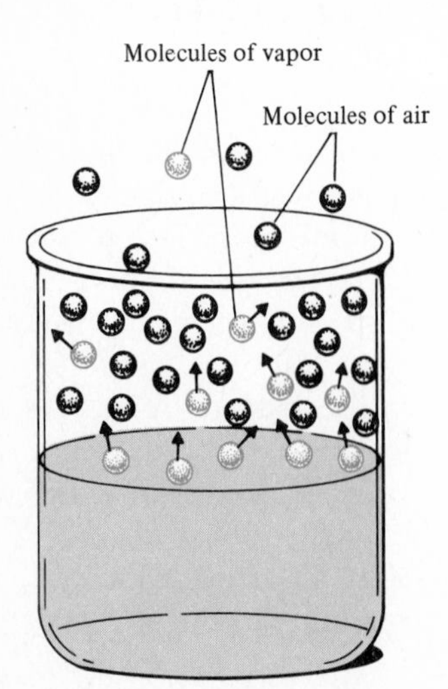

Figure 6-2 Evaporation of molecules of a liquid.

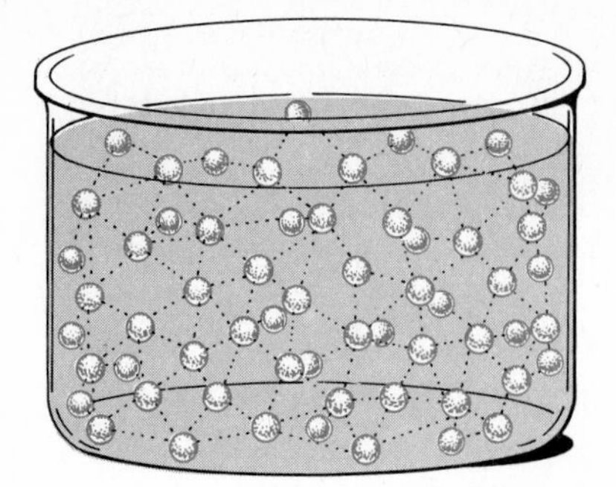

Figure 6-3 Most of the molecules on the surface of a liquid are held on the surface by the attractive forces of the other molecules on the surface and immediately below.

some of their high kinetic energy by colliding with air molecules and with other vapor molecules, and do not have enough kinetic energy remaining to escape the attractive forces of the molecules on the surface of the liquid, so that when they strike the surface they are trapped there and return to the liquid phase. This is **condensation.** In a closed container both processes, evaporation and condensation, are constantly occurring. After a while, the rate at which molecules evaporate from the surface equals the rate at which vapor molecules condense onto the surface. Once the two opposing rates become equal (assuming that nothing is done to change the experimental setup), the partial pressure of the vapor above the liquid remains constant. The partial pressure of the vapor above the surface of the liquid when the two opposing rates are equal is called the **vapor pressure of the liquid** (Figure 6-4). In Section 4-9, we saw that each gas in a mixture of gases has a partial pressure. In this case, we measure the partial pressure *of the vapor,* but we *call* it the vapor pressure *of the liquid.* This is reasonable when we remember that the two opposing pressures are now equal: the pressure that the liquid exerts on the gas phase is equal to the pressure that the vapor (but not the air) exerts on the liquid phase. Next, we will see how to measure vapor pressures.

6-3 Measurement of Vapor Pressure. Effect of Temperature

The vapor pressure of liquids can easily be measured with a type of manometer of a somewhat different design from the one we looked at in Section 4-1. This manometer is shown in Figure 6-5a. As shown in the figure, the space above the mercury contains only mercury vapor (no air). The mercury vapor, of course, is produced by the vapor pressure of the mercury. Figure 6-5b shows a few drops of

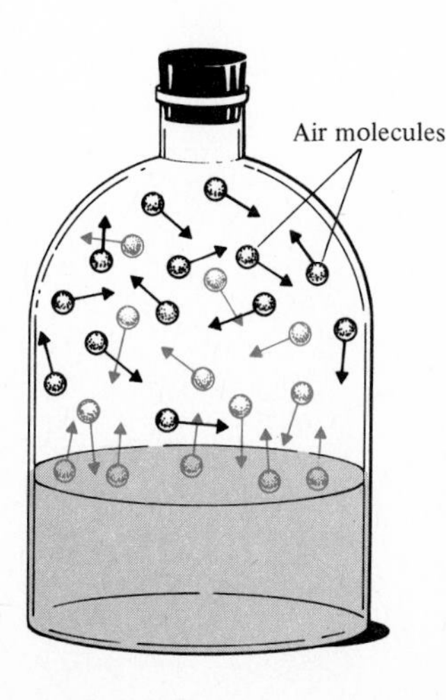

Figure 6-4 Molecules evaporating from and condensing on a liquid surface.

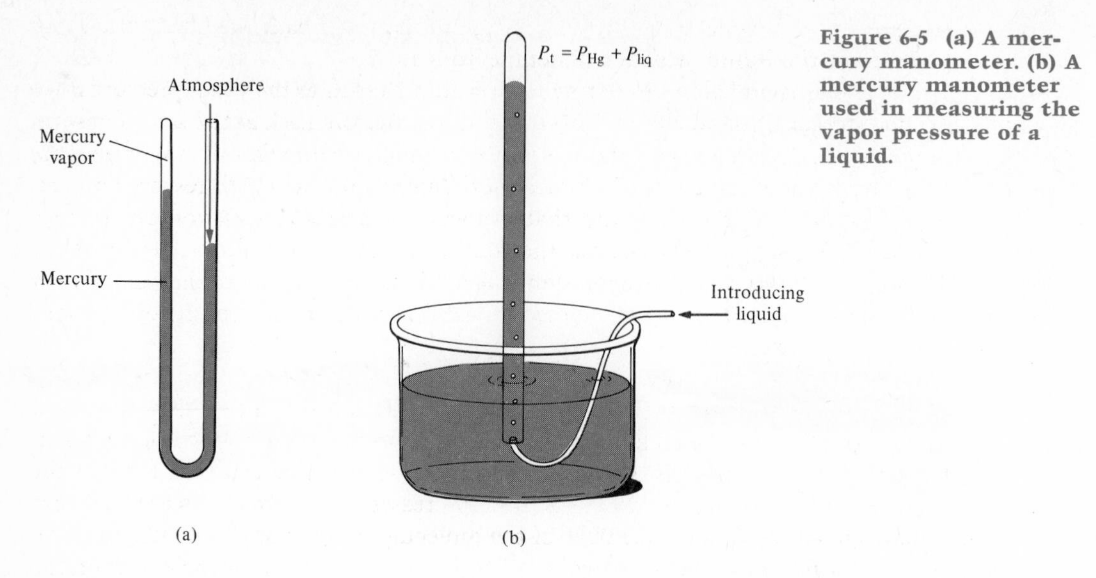

Figure 6-5 (a) A mercury manometer. (b) A mercury manometer used in measuring the vapor pressure of a liquid.

liquid being introduced into the mercury column and the liquid rising to the top of mercury level. The total pressure above the liquid increases, forcing the mercury level to drop. The difference between the total pressure and that of the initial pressure due to the mercury vapor alone is the vapor pressure of the liquid.

It is important in all experiments in which vapor pressures are measured in a manometer that some residual liquid remain. If not, we are no longer measuring the vapor pressure of the liquid but are measuring the pressure of the vapor.

What happens to the vapor pressure of a liquid when the temperature is increased? To answer this question experimentally, we need another apparatus which is somewhat more elaborate than a manometer. This apparatus is shown in Figure 6-6. Initially, the flask is empty and the system is evacuated through stopcock B. The mercury levels in both arms are now equal because both sides contain only a very small amount (trace) of mercury vapor. The liquid whose vapor pressure is to be measured is introduced into the flask by opening stopcock A. The difference in mercury levels of the two arms measures the vapor pressure of the liquid at the bath

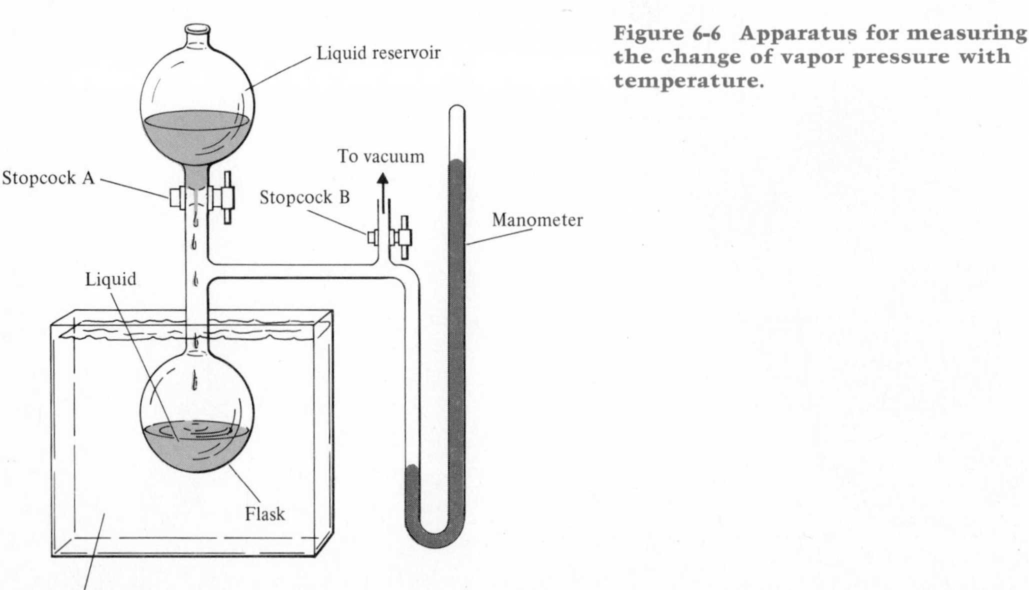

Figure 6-6 Apparatus for measuring the change of vapor pressure with temperature.

temperature *T*. By changing the temperature of the water bath, we obtain the vapor pressure of the liquid at different temperatures.

The reader is probably wondering what would happen to the vapor pressure if we opened stopcock A and allowed more liquid to enter the flask at the same constant temperature. By actually performing this experiment, we find that the vapor pressure of the liquid does not change as long as both liquid and vapor are present. This can be understood by remembering that evaporation and condensation are surface effects. The larger the surface area, the greater the rate of evaporation as well as the rate of condensation. Both evaporation and condensation depend *to the same extent* on the surface area. Increasing or decreasing the volume of liquid or the surface area does not change the vapor pressure.

The results of many such experiments tell us that *the vapor pressure of a liquid always increases with increasing temperature.* Figure 6-7 shows this for a few liquids, but it is true for all liquids. What happens to the molecules of a liquid when the temperature is raised? As we know, an increase in temperature means that the average kinetic energy of the molecules increases. With this increased average kinetic energy, a higher percentage of the molecules have enough kinetic energy to escape into the gas phase by overcoming the attractive forces of the molecules on the surface. Continuing to increase the temperature will produce a continuing increase in vapor pressure. This explains the curves in Figure 6-7. If the temperature continues to rise, a temperature will finally be reached when the vapor pressure of the liquid is equal to the pressure of the air in contact with the surface of the liquid. This temperature is called the boiling point of the liquid.

The **boiling point** of a liquid is defined as **the temperature at which the vapor pressure of the liquid equals the pressure of the atmosphere in contact with its surface.** The boiling point of a liquid is not a constant, but is different depending on the pressure of the atmosphere it is in contact with. For example, let us look at the boiling point of water. From Figure 6-7 we can see that when the atmospheric pressure is 400 torr, water boils at 83°C; when it is 200 torr, it boils at 66.5°C; and at 100 torr, it boils at 51.5°C. The boiling point of any liquid decreases when the atmospheric pressure in contact with its surface

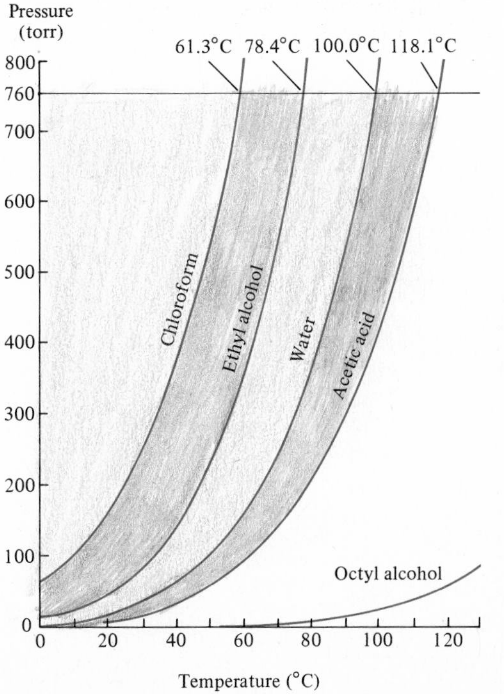

Figure 6-7 Variation of vapor pressure with temperature of a number of different liquids.

decreases. Since our standard pressure (Section 4-1) is 760 torr (1 atm), we define the **normal boiling point** of a liquid as the boiling point at that pressure. The normal boiling point of water is 100°C, but at lower pressures, such as at the top of a mountain, the boiling point is less than the normal boiling point, while at higher pressures the boiling point is greater than 100°C. At 2 atm pressure, water boils at 121°C. A pressure cooker operates on this principle; it confines the water vapor inside the cooker and does not allow it to escape. By use of the adapter on the pressure cooker, the confining pressure can be regulated, to say 3 atm. In order to boil, water at 3 atm must be heated to a temperature of 134°C. Because the water is at a higher temperature, food in pressure cookers cooks faster.

The same principle, in reverse, is responsible for the fact that people at high altitudes must cook their foods *longer*. For example, it takes more time to boil an egg at an altitude of 1 km (3280 ft) than it does at sea level. At 1 km the average atmospheric pressure is 730 torr. At this pressure, water boils at about 97°C, compared to 100°C at sea level. Since the egg is being heated at a lower temperature, it takes more time for it to be cooked.

Table 6-1 lists a number of liquids (including some that are gases at room temperature) and their normal boiling points in degrees Kelvin.

An interesting phenomenon occurs during boiling. When the molecules of a nonboiling liquid evaporate, they must first reach the surface of the liquid. When a liquid boils, its vapor pressure is equal to the confining pressure and can push the atmosphere from the surface of the liquid, which allows the molecules to evaporate from the center of the liquid as well as from its surface. This is what occurs during a "rolling boil" and is responsible for the fact that bubbles of gas in a boiling liquid form at the bottom and in the middle as well as on top.

In Section 6-2 we pointed out that if a liquid is in an open container, the molecules of vapor can diffuse into the atmosphere well away from the surface of the liquid. Those that do so are then unable to return to the liquid. This means that more molecules are escaping from the liquid than are returning. If the process continues long enough, the entire liquid will evaporate to dryness. *Thus, even at atmospheric pressure, a liquid can change into a gas not only at its normal boiling point, but at any temperature* (of course, the *rate* of evaporation increases with increasing temperature). Indeed, we have all seen puddles of water in the street dry up shortly after the end of a rainstorm, although the temperature remains well below the normal boiling point of water.

Table 6-1 Normal Boiling Points and Heats of Vaporization of Some Liquids

Substance	*Normal boiling point (°K)*	ΔH°_{vap} *(cal/mole)*[a]
Ethylene glycol, $C_2H_6O_2$	471	
Water, H_2O	373	9720
Carbon tetrachloride, CCl_4	349.8	7170
Ethyl alcohol, C_2H_5OH	351.3	9970
Benzene, C_6H_6	353	7350
Bromine, Br_2	332	7280
Diethyl ether, $C_2H_5OC_2H_5$	308	6210
Ethane, C_2H_6	184	3610
Oxygen, O_2	90	1630
Ammonia, NH_3	240	5570
Helium, He	4	
Hydrogen, H_2	20	220
Krypton, Kr	121	2320
Nitrogen, N_2	77	1330

[a] ΔH°_{vap} is the heat of vaporization at 1 atm pressure.

In Section 5-3 we learned that to boil a liquid a certain amount of heat (ΔH_{vap}) was necessary. Does this also apply to evaporation at temperatures other than the normal boiling point? The answer is yes, *but*. There is still a heat of vaporization required to evaporate a liquid at any temperature, but it is not exactly the same quantity as it is at the boiling point. For example, ΔH_{vap} for water at 100°C is 540 cal/g; while at 25°C it is around 580 cal/g. As this example shows, the difference is not very great, and for most purposes we can ignore it. Values of ΔH_{vap} at the boiling point are given for some liquids in Table 6-1.

We have seen that a molecule which evaporates from the surface of a liquid does so because it possesses a high-enough kinetic energy to pull away from the attraction of the other molecules. This fact allows us to see why there is such a quantity as ΔH_{vap}. Most of the molecules at the surface do not evaporate since they do not possess enough kinetic energy to pull away from the surface. The molecules that do escape must possess higher kinetic energies than the average kinetic energy of the molecules in the liquid. The process of evaporation therefore lowers the average kinetic energy of the liquid because the molecules that evaporate are the "hot" molecules (the ones with higher-than-average kinetic energies). Since the average kinetic energy of the remaining molecules is now less, the temperature of the liquid falls; the amount of heat lost is ΔH_{vap}.

It is by this mechanism that human beings and many animals cool themselves when they become overheated. When you become overheated, you sweat. The evaporation of the sweat cools the body. A dog has no sweat glands so that to cool itself it pants and evaporates the liquid on its tongue.

Solids as well as liquids have a vapor pressure. This means that molecules in the solid can escape from the solid and enter the gas phase directly. It happens because the molecules in a solid are always vibrating and colliding with their nearest neighbors. Through these collisions, energy is transferred from one molecule in a solid to another, so that there are some molecules in the solid with very little energy while others have a great deal of energy. Some molecules at the surface of the solid have gained enough energy through this collision process to break away from the surface of the solid and enter the atmosphere. When molecules go directly from a solid into the gaseous phase without passing through a liquid phase, the process is called **sublimation.** The energy necessary to transform 1 mole of a solid to a gas is called the heat of **sublimation,** ΔH_{subl}.

If we place the solid in a closed container (review Section 6-2), the number of vapor molecules directly above the surface of the solid will begin to increase. Concurrently, energetic molecules in the vapor which came from the solid will lose some of their energy by colliding with other molecules and the walls of the container. These slower-moving (less energetic) molecules will be able to return to the solid phase after striking the surface of the solid. After a period of time the rates (leaving the solid and returning to the solid) will become equal and the partial pressure of the vapor above the solid will remain constant. This partial vapor pressure of the solid is called the *vapor pressure of the solid.*

The vapor pressure of a solid, like that of a liquid, increases with increasing temperature. When a solid is heated, its vapor pressure rises. If this is continued, the solid will finally melt, and a liquid will form. At this temperature the vapor pressure of the solid and liquid are equal. This temperature is defined as the **melting point** (or the **freezing point**) of the solid. This process can take place over a wide range of external pressures. The temperature at which a solid melts (or a liquid freezes) when the external pressure is 1 atm is called the **normal melting point** of the solid or the **normal freezing point** of the liquid. A liquid and its corresponding solid can coexist for an extended period of time only when their vapor pressures are the same. Since the vapor pressure changes with temperature, there are different combinations of temperatures and pressures where solids and liquids can coexist. As we saw above, if the temperature is not high enough to melt the solid, sublimation can occur.

end

6-4 Calculation of Vapor Pressure at Different Temperatures

We can, of course, *measure* vapor pressures of a given liquid at various temperatures, by the method described in Section 6-3. But it is also possible to *calculate* them without making experimental measurements, provided that we know two things: (1) the heat of vaporization (ΔH°_{vap}) of the liquid and (2) the vapor pressure of the liquid at any single temperature. To make these calculations, we use the following equation, which we present without proof (although it can be derived by the use of advanced thermodynamic arguments):

$$\log P = \frac{-\Delta H^\circ_{vap}}{(2.303)RT} + A \qquad (1)$$

In this equation, P is the vapor pressure of the liquid at temperature T (in degrees Kelvin), ΔH°_{vap} is the *molar* heat of vaporization of the liquid (the ° means that the air pressure is 1 atm), A is a constant whose units depend upon the units of P (for example, atm, torr), 2.303 is the conversion factor of natural logarithms to logarithms to the base 10 (Section 1-5), and R is the gas constant whose units must be the same as those of ΔH°_{vap}. To use equation (1) we need to know the vapor pressure of the liquid at any single temperature. In most cases, the most convenient temperature to choose is simply the normal boiling point of the liquid, since the vapor pressure at this temperature is 1 atm or 760 torr. If we then substitute into equation (1) the known values for P, T, and ΔH°_{vap}, we can calculate A for that liquid. Once we have A, we can then use equation (1) to calculate P for any other value of T.

Example: From the data in Table 6-1, calculate the vapor pressure of ethyl alcohol at 300°K.

Answer: From Table 6-1,

$$\Delta H^\circ_{vap} = +9970 \text{ cal/mole}$$

$$\text{boiling point, } T_b = 351.3°\text{K}$$

The vapor pressure at its boiling point is, of course, 760 torr. Putting this information into equation (1), we have

$$\log 760 = \frac{-9970 \text{ cal/mole}}{(2.303)(1.99 \text{ cal/°K} \cdot \text{mole})(351.3°\text{K})} + A$$

$$A = 9.07$$

Applying equation (1) again, we have

$$\log P = \frac{-9970 \text{ cal/mole}}{(2.303)(1.99 \text{ cal/°K} \cdot \text{mole})(300°\text{K})} + 9.07$$

$$= 1.82$$

$$P = 66.1 \text{ torr}$$

Equation (1) is quite general and can be used to calculate the change in the vapor pressure with temperature for almost all liquids. The only restriction is that ΔH°_{vap}

must be constant over the temperature range considered. When ΔH°_{vap} varies with temperature, the exact form of the equation changes depending on just how ΔH°_{vap} changes. For our purposes, equation (1) as given above will be quite sufficient, since for most liquids ΔH°_{vap} is fairly constant over temperature changes up to 50°C. Equation (1) is one form of a general equation known as the *Clausius–Clapeyron equation.*

Equation (1) can also be used to calculate ΔH°_{vap} for any liquid, if we know (or can measure) the vapor pressures of that liquid at several temperatures. We can do this by plotting log P vs. $1/T$. Equation (1),

$$\log P = \frac{\Delta H^\circ_{vap}}{2.303RT} + A$$

is of the form $y = mx + b$, where

$$y = \log P \qquad m = \frac{-\Delta H^\circ_{vap}}{2.303R} \qquad x = \frac{1}{T} \qquad b = A$$

Therefore, a plot of log P vs $1/T$ will give a straight line with a slope $-\Delta H^\circ_{vap}/2.303R$ and an intercept A.

Example: From the following data, calculate the molar heat of vaporization of water.

Vapor pressure (torr)	*Temperature (°K)*
760	373
740	372
700	371
600	366
500	361
300	348
100	322
10	280

Answer: In order to obtain the H°_{vap} we apply equation (1) and plot log P vs. $1/T$.

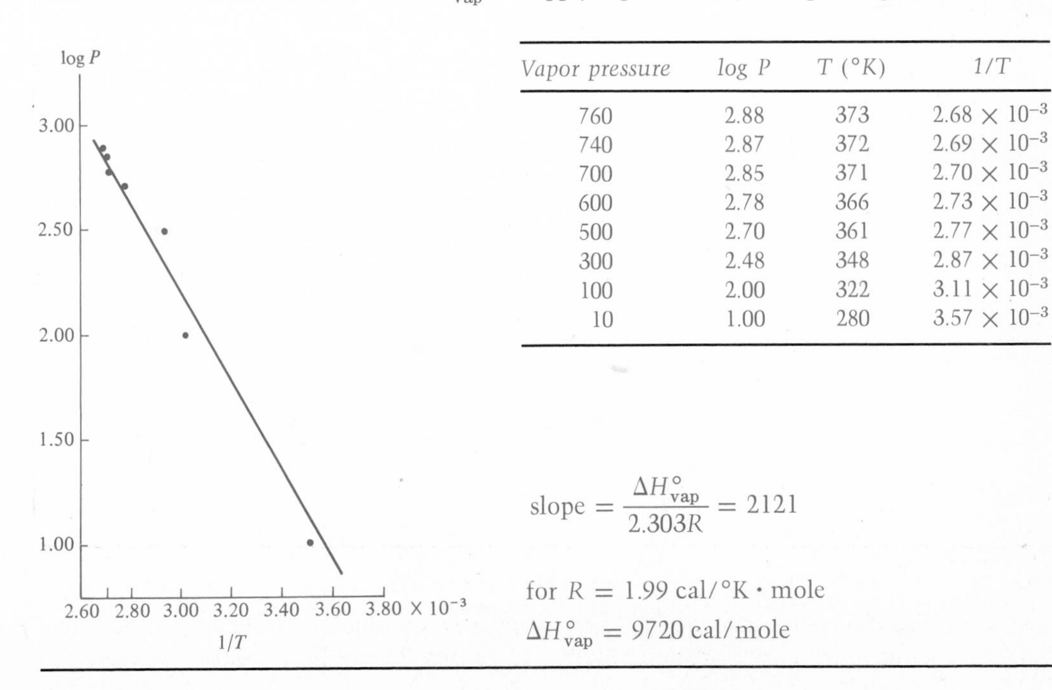

Vapor pressure	*log P*	*T (°K)*	*1/T*
760	2.88	373	2.68×10^{-3}
740	2.87	372	2.69×10^{-3}
700	2.85	371	2.70×10^{-3}
600	2.78	366	2.73×10^{-3}
500	2.70	361	2.77×10^{-3}
300	2.48	348	2.87×10^{-3}
100	2.00	322	3.11×10^{-3}
10	1.00	280	3.57×10^{-3}

$$\text{slope} = \frac{\Delta H^\circ_{vap}}{2.303R} = 2121$$

for $R = 1.99$ cal/°K · mole

$\Delta H^\circ_{vap} = 9720$ cal/mole

6-5 Factors Affecting Boiling Points

Table 6-1 shows that liquids vary greatly in their normal boiling points. At one extreme we have liquid helium, with a boiling point of 4°K, while the highest-boiling liquid in the table, ethylene glycol, boils at 471°K. There are other liquids known with boiling points that are higher still. What are the reasons for such wide variations? There are two factors that determine boiling points: molecular weights and intermolecular forces.

1. *Molecular weight.* Where intermolecular forces are similar, *boiling points approximately increase with increasing molecular weight.* For example, Table 6-2 lists two series of liquids* in which the intermolecular forces are very similar. In both series, the boiling points increase smoothly with increasing molecular weight. To understand the reason for this, let us compare two liquids with similar intermolecular forces but different molecular weights: we will call them A and B, and say that A has a higher molecular weight than B. In order to escape from the surface of the liquid, the molecules of A must be traveling at about the same average velocity as those of B, since the intermolecular forces are about the same. But if both have about the same velocity, then A must have a higher kinetic energy, since kinetic energy $= \frac{1}{2}mv^2$, and a molecule of A has a larger mass. Since we know that temperature is proportional to average kinetic energy, it follows that A will boil at a higher temperature.

2. *Intermolecular forces.* It is obvious from Table 6-1 that molecular weight cannot be the only factor. For example, note that nitrogen (mol wt = 28) has a normal boiling point of 77°K, while water (mol wt = 18), boils at 373°K. In such cases, the explanation lies in vastly different attractive forces *between* molecules. (These are the forces that we first encountered in Section 4-12, the ones that were in part responsible for the nonideal behavior of gases.) The molecules of water are much more strongly attracted to each other than are those of nitrogen. Consequently, more energy, and a higher temperature, are required for water molecules to break away from each other than for nitrogen molecules to do the same thing. Forces between molecules (*intermolecular forces*) also affect many other properties of matter besides boiling point and nonideality of gases. Among other things they are responsible for solubility behavior (Section 7-5), and for the presence of life on earth (see Chapter 21). In the next section we shall discuss these forces and see what it is

*All the substances in this table are gases at room temperature and atmospheric pressure, but the principle is equally valid for any series of liquids with similar intermolecular forces.

Table 6-2 Two Series of Liquids, with Molecular Weights and Boiling Points

Liquid[a]	Molecular weight (to the nearest integer)	Normal boiling point (°K)
Methane, CH_4	16	112
Ethane, C_2H_6	30	184
Propane, C_3H_8	44	231
n-Butane, C_4H_{10}	58	273
Helium, He	4	4
Neon, Ne	20	27
Argon, Ar	40	87
Krypton, Kr	84	121
Xenon, Xe	131	166

[a] All of these are gases at room temperature and atmospheric pressure, but the principle of boiling point increasing with molecular weight holds for other liquids too.

that makes the molecules of one compound strongly attract each other while the molecules of another compound do not.

6-6 Intermolecular Forces

Forces that exist between a molecule or ion and another molecule are called **intermolecular forces,** in contrast to forces between parts of the same molecule, which are termed intramolecular forces (the latter are discussed in Chapter 14). All intermolecular forces are electrical in origin;* that is, they are all based on Coulomb's law, which states that unlike charges attract; like charges repel; and the strength of the attraction or repulsion is directly proportional to the magnitude of the charges and inversely proportional to the square of the distance between the charges (Section 1-8).

In order to obey Coulomb's law, molecules must possess a charge. There are three kinds of species which do so: *ions, dipoles,* and *induced dipoles.* Let us examine each type.

Ions. We have already met these in Section 2-7. All ions are fully charged particles (a full charge is a charge of at least $+1$ or -1). The charge is equal to the difference between the total number of electrons and protons, and may be $+1$, $+2$, -1, -2, or some other positive or negative *integer.*

Dipoles. Many molecules are neutral (that is, they have the same total number of electrons as protons), and yet possess permanent electrical charges. To understand this property, we must backtrack a bit. If we consider any mass, there is a point in the body called the center of gravity. In mechanics, the center of gravity acts as if all the mass has been concentrated at that point. In atoms that make up molecules, we have positively charged parts (nuclei) and negatively charged parts (electrons). In each molecule, then, we can think of the center of positive charge and the center of negative charge, analogous to a center of gravity. In some molecules, these two points coincide in space. Such molecules are *not* dipoles. If the centers of positive and negative charges do not occur at the same point in space, the molecule is a **dipole** or a **dipolar molecule** (sometimes just a **polar molecule**). The extent of charge separation is measured by a number called the **dipole moment.** The greater the charge separation, the larger the value of the dipole moment (in Section 14-7 we will see how dipole moments are measured). A molecule that is a dipole has one end which is positively charged and another which is negatively charged, making it behave like a little magnet. The negative end will attract a positive ion or the positive end of another dipole, and vice versa for the positive end (Figure 6-8). Dipoles are neutral overall because the magnitude of the positive charge is equal to the magnitude of the negative charge (why?), and the net result is no total charge. These magnitudes are usually less than $+1$ and -1 (which means that dipoles do not have a full charge). Dipoles are often represented in this way:

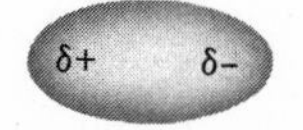

where the Greek lowercase letter δ (delta) is used to indicate a partial charge.

Induced Dipoles. Although many molecules are permanent dipoles, many others are not. (In Section 14-6 we will see how one can tell, by looking at the formula of a molecule, whether it is a permanent dipole or not). As we saw in the preceding paragraph, molecules that are not permanent dipoles have their centers of positive and negative charges located in the same place (Figure 6-9). We know that the positive parts of molecules are the nuclei, which are small, hard, and relatively

*We shall see in Chapter 14 that intramolecular forces are also electrical in origin.

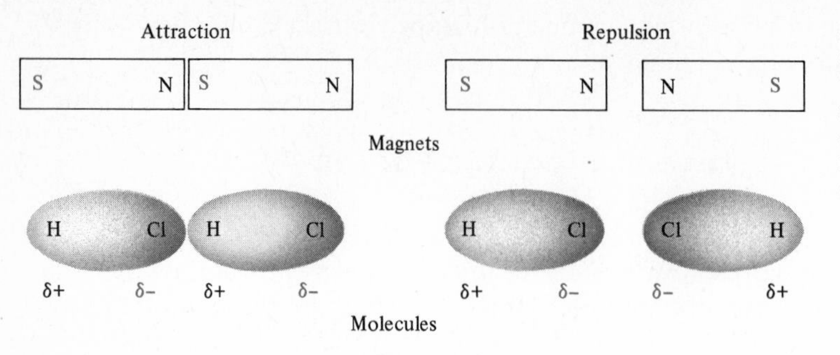

Figure 6-8 Dipoles in magnets and molecules. They can attract or repel each other depending upon their alignment with respect to each other.

immobile (with respect to the rest of the molecule); but the negative parts are electrons which are in the form of clouds. These are much more mobile. In a typical nonpolar molecule, H_2, the center of positive charge coincides with the center of negative charge, as long as there are no neighbors to influence it, but if a positive ion (or the positive end of a permanent dipole) comes nearby (close enough for coulombic forces to work), the electrons of the molecule will be attracted toward the positive ion. The result will be that as long as the positive ion is nearby, the center of negative charge will not coincide with the center of positive charge, and the molecule will not coincide with the center of positive charge, and the molecule will now be a dipole. We say that the positive ion has *induced* a dipole in the H_2 molecule, and the H_2 molecule is now an **induced dipole.** As soon as the positive ion leaves, the H_2 will be nonpolar again. Induced dipoles have no permanent dipole moment. However, for as long as the dipole moment exists, the molecule behaves just like a molecule with a permanent dipole moment; for example, the negative end is attracted to a positive ion (Figure 6-9).

We can now explain how gases that are made of ions* or dipolar molecules form liquids. As the temperature is lowered at constant pressure, the particles are confined to a smaller volume. Being closer, they have greater chances of interacting with one another. Since their kinetic energy is also being decreased (by lowering the temperature), a point is reached where their kinetic energies can no longer overcome the attractive forces acting between them. They remain close to each other and begin "sticking" to one another. A liquid is produced.

But what about gases whose molecules have no permanent dipole moments. How do they liquify? In a pure sample of such a gas there are no ions (or dipolar molecules) present to induce dipole moments in these molecules. Yet every gas can be liquified. The answer was given in 1930 by Fritz London (1900–1954). Although the center of negative charge coincides with the center of positive charge over a period of time, London showed that this was not necessarily true at any given instant. At some particular instant in time, the electron cloud may have a greater density on one side of the molecule than on the other (Figure 6-10 shows this for the gas neon). This is called an instantaneous dipole. At another instant of time the electron density may be greater on the other side of the nucleus, and the instantaneous negative charge will appear on the other side of the molecule. An instantaneous

*At ordinary temperatures, there are no gases made up of ions; however, such gases can exist at high temperatures.

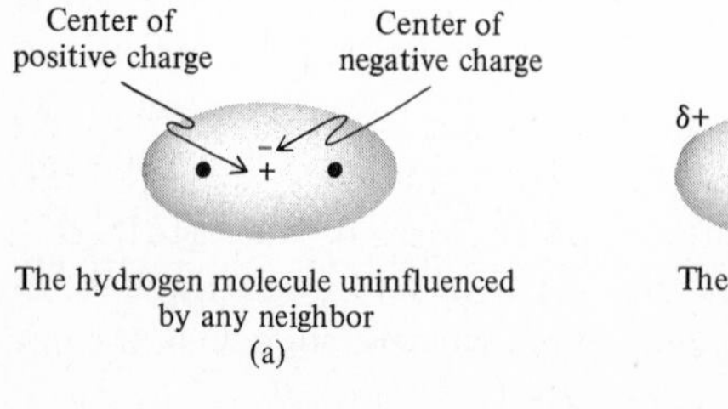

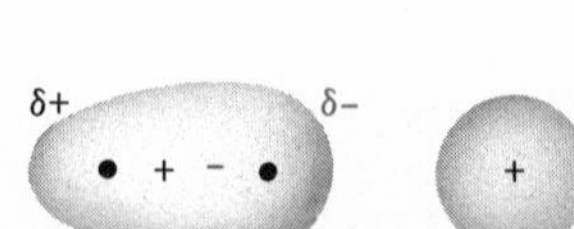

Figure 6-9 The hydrogen molecule is not a permanent dipole, but it can be an induced dipole.

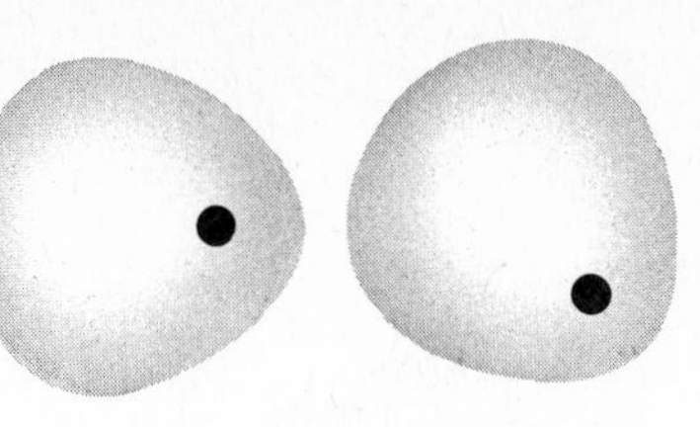

Figure 6-10 **A molecule of neon, showing its neutral character over a long time period and instantaneous, short-lived dipole.**

dipole can induce another instantaneous dipole in a neighboring molecule. These induced dipoles continue to induce other dipoles in molecules, and the process goes on and on from one molecule to the next. As the temperature gets lower and lower, the molecules can no longer "shake off" this interaction and they finally "stick" together and condense. This process of induction of dipoles into neutral molecules explains how noble gases and other nonpolar gases condense. The interaction between two such molecules is termed **induced dipole-induced dipole.** We may also note that molecules which possess permanent dipoles can be induced to yield larger dipole moments than they originally had. The magnitude of an induced dipole moment does not depend on whether or not the molecule starts with a permanent dipole moment.

We are now ready to classify the different types of interactions between ions and molecules. The force (F) of interaction between species is always inversely proportional to some power of the distance r between the species; that is,

$$F \propto \frac{1}{r^n}$$

where n is usually an integer. When n is small, say 2, $F \propto 1/r^2$ and the force of interaction is *long range*. When n is, say, 9, $F \propto 1/r^9$, the force is *short range*. The different types of forces between species are classified by the polar nature of the ions and molecules involved. They are, in order of decreasing magnitude:

1. *Ion–ion*: These are the forces that hold ions together in ionic bonds (Section 14-2). For this case, $F \propto 1/r^2$ (Coulomb's law).
2. *Ion–dipole:* $F \propto 1/r^5$.
3. *Ion–induced dipole:* $F \propto 1/r^7$.
4. *Dipole–dipole:* $F \propto 1/r^7$.
5. *Dipole–induced dipole:* $F \propto 1/r^7$.
6. *Induced dipole–induced dipole:* $F \propto 1/r^7$.

Forces arising in cases 4, 5, and 6 are termed **van der Waals forces.** The forces of case 6 are also known as **London forces** or **dispersion forces.**

Chemists are mostly concerned with attractive interactions. However, there also are repulsive forces between atoms, molecules, and ions. They arise from the repulsive interactions of the positive nuclei in different molecules as well as repulsive interactions of negative electronic clouds. The forces of repulsion are very short range, with $F \propto 1/r^{10}$ and $F \propto 1/r^{13}$.

6-7 The Structure of Liquids

At the beginning of this chapter we mentioned that liquids exhibit neither the complete disorder of a gas or the almost regular order of a solid. We can "see" this from x-ray diffraction patterns of liquids. The method and uses of x-ray diffraction are discussed in Section 15-1 in conjunction with our study of the solid state. A typical pattern for a liquid is shown in Figure 6-11. What we observe is that at some

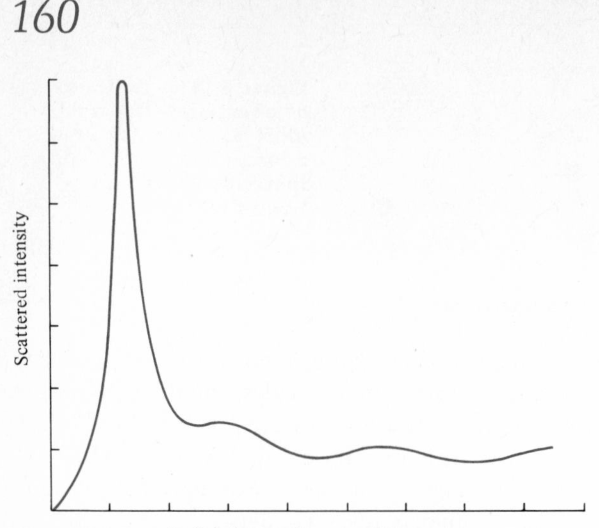

Figure 6-11 Scattering curve for a monatomic liquid.

short distance r, the pattern shows a large peak which then dampens out as the distance increases. Such a pattern means that each molecule in the liquid is surrounded by a certain number of nearest-neighbor molecules. This number, known as the **coordination number,** is fixed for each liquid and every liquid has one. Past the first layer of molecules surrounding the central molecule, there are no longer any fixed numbers of molecules. Whatever structure we observed at close range has died out quickly. This is typical of nearly all liquids. That is, a liquid has short-range order but not long-range order. In comparison, a gas shows neither short nor long-range order, whereas a solid shows both short-and long-range order. Figure 6-1 shows these differences in gases, liquids, and solids. An additional aspect of liquid structure is that there are *holes* in the structure (regions where there are no molecules). This occurs because of the lesser order of a liquid as compared to a solid. The nature of these holes (that is, their size, their movements, their changing structures) are challenges for theorists in constructing a model theory of liquids.

6-8 Properties of Liquid Water. Hydrogen Bonding

Water is a clear transparent liquid that appears bluish-green in thick layers. It is the most abundant liquid on our planet and we encounter it constantly in our everyday experiences. A cursory examination of the physical properties of water immediately shows that it is quite unusual when compared with other liquids. First let us examine the boiling points of the hydrides of the group VIA elements, H_2O, H_2S, H_2Se, and H_2Te. Figure 6-12 compares the boiling points of the group VIA hydrides with those of the group IVA (CH_4, SiH_4, GeH_4, SnH_4) hydrides. The group IVA hydrides show the expected increase in boiling points with increasing molecular weights (Section 6-5). The group VIA hydrides, except for H_2O, follow the same trend. Why does H_2O have such an abnormal boiling point? If water followed the trends of the others we might expect water to have a boiling point of about $-80°C$ instead of $100°C$. In the same way, Figure 6-13 shows the variation of the melting points of the hydrides of the same two families. Again water behaves abnormally. But these are not the only unusual physical properties of water. Most liquids steadily decrease in volume with decreasing temperature. Water has the unusual property of having its maximum density (smallest volume) at 4°C (not 0°C) and also shows the rather abnormal property of *increasing* its volume upon freezing. Figure 6-14 shows how the molar volumes (volumes per mole) of liquid water and ice change with temperature.

The unusual properties of water extend to its surface tension, which is unusually

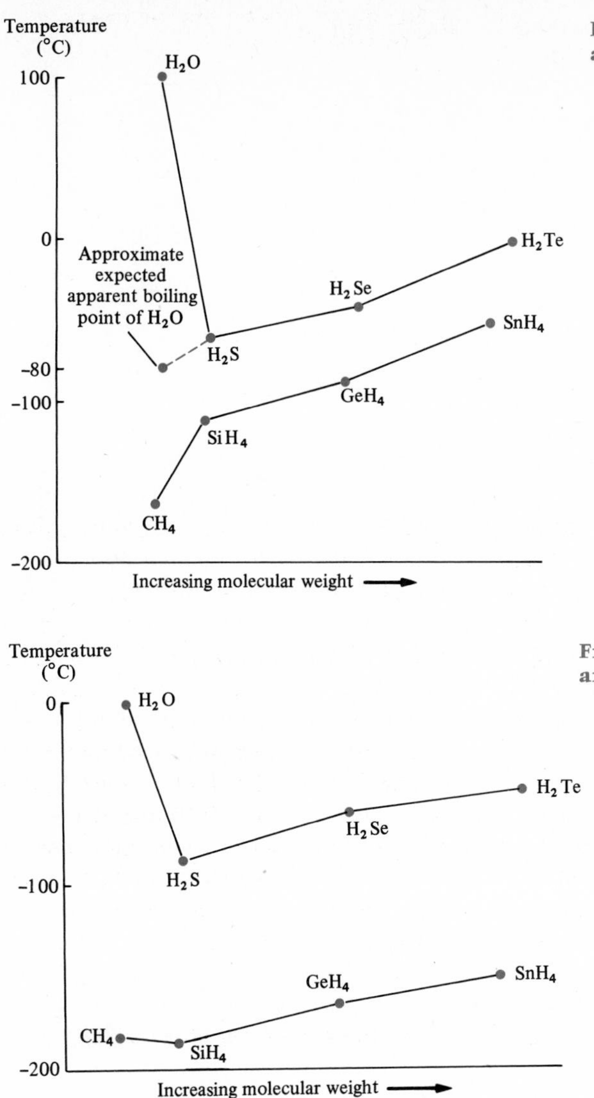

Figure 6-12 Boiling points of groups IVA and VIA hydrides.

Figure 6-13 Melting points of groups IVA and VIA hydrides.

high for a substance made up of small molecules; and to its heats of vaporization, fusion, and heat capacity, all of which are very large when compared to other liquids. Table 6-3 lists some properties of water and of some other liquids.

These abnormal properties of water have pronounced effects upon our everyday lives and on the environment. Since liquid water is denser than ice, ice does not sink

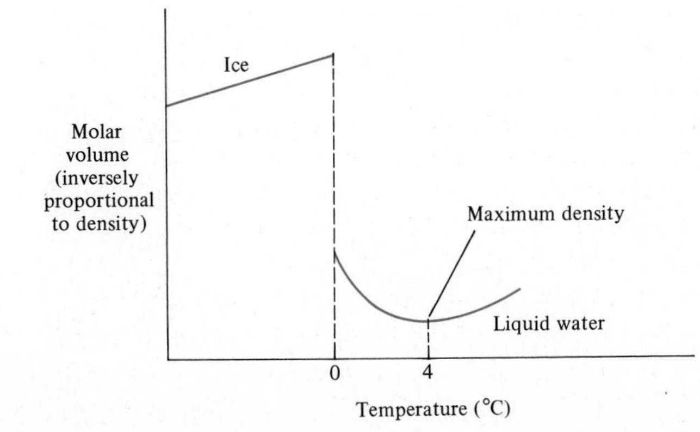

Figure 6-14 The change in the density of liquid and solid water with temperature. Molar volume = volume/mole, which is inversely proportional to density.

Table 6-3 Comparison of Some Properties of Water and Substances of Similar Molecular Weight

Property	Liquid			
	Water	CH_4	NH_3	*Ne*
Mol. wt. to the nearest integer	18	16	17	20
Boiling point (°C)	100	−161	−33	−246
Melting point (°C)	0	−183	−78	−249
Heat capacity (cal/g)	1.000	0.527	0.493	0.246
ΔH_{vap} (cal/g)	540	138	330	22
ΔH_{fus} (cal/g)	80	14.5	108.1	4.005

but floats on the surface of the water. If the densities were reversed, the ice would sink to the bottom of a lake or ocean instead of floating on the top. This would have various consequences, among them that shallow rivers and lakes in cold climates would freeze solidly from bottom to top every winter, thus killing all the aquatic life (when a body of water freezes from the top down, the layer of ice serves to insulate the water below from further freezing, making it much harder for it to freeze all the way).

Some of the other effects of the properties of water on the environment have been discussed in Sections 5-2 and 5-3. See also Chapter 17.

In Section 6-5 we noted that abnormally high boiling points are caused by intermolecular attractive forces. Water is a dipole (as we shall see in Section 14-7), and we might expect that the unusual properties of water might be caused by dipole–dipole interactions among the water molecules. However, this is only part of the explanation (other dipolar compounds do not have such unusual properties). The unusual properties of water are mostly due to a particularly strong type of intermolecular force, called a **hydrogen bond,** which causes water molecules to have strong attractions for each other.

As we shall see, water is not the only compound that forms hydrogen bonds, but the effect in water is particularly strong. The hydrogen bond in water is *the result of an attractive force between a hydrogen atom in one molecule and the oxygen atom*

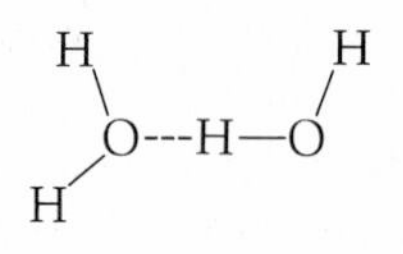

of another (a hydrogen bond is usually represented by a dashed line). A hydrogen bond is much weaker than an ordinary covalent bond (Chapter 14). For example, the O—H bond in water has a bond energy (the amount of energy necessary to pull the atoms apart) of about 110 kcal/mole, while the O---H bond has an energy of about 6 kcal/mole. However, a hydrogen bond is much stronger than the ordinary van der Waals forces discussed in Section 6-6. As with all bonding and intermolecular forces, hydrogen bonds are electrostatic in origin.

What other compounds have hydrogen bonds? Obviously, in order to form a hydrogen bond, a molecule must have hydrogen. But is that enough? From Figures 6-12 and 6-13, we might guess that it is not enough. For example, methane (CH_4) has typical boiling and melting points, and laboratory examinations of its structure show that hydrogen bonding is not present. Two molecules are required to make up a hydrogen bond: one to supply the hydrogen and the other to accept it. These molecules may be the same or different. Many studies have shown that with only a few exceptions, the only atoms that can act as acceptors are O, N, and F, while the only hydrogens capable of forming hydrogen bonds are those covalently bonded to

O, N, or F. We therefore expect hydrogen bonding, for example, in ammonia (NH_3) and ethyl alcohol (C_2H_5OH) but not in CH_4 or H_2S.

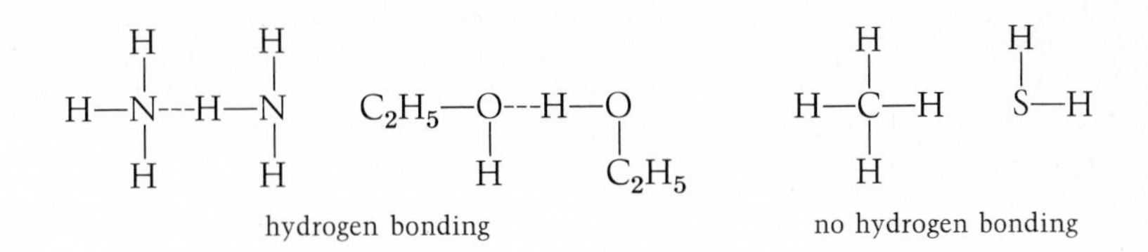

A molecule such as ether, $C_2H_5OC_2H_5$, which has no O—H bond, cannot hydrogen-bond with another ether molecule but can and does form hydrogen bonds with water, ammonia, or HF:

$$\underset{\textstyle C_2H_5}{C_2H_5—\overset{}{O}}\text{---}\underset{}{\overset{\textstyle O—H}{H}} \qquad \underset{\textstyle C_2H_5}{C_2H_5—O}\text{---}H—\overset{\textstyle H}{N}—H \qquad \underset{\textstyle C_2H_5}{C_2H_5—O}\text{---}H—F$$

The strongest hydrogen bond known is that between HF and F^-. When the solid potassium fluoride, KF, is added to liquid hydrogen fluoride, HF, the anion found is not the expected fluoride ion, F^-, but the hydrogen difluoride ion, HF_2^-. The hydrogen difluoride ion exists in large concentrations in acidic fluoride solutions and in salts such as KHF_2. The salts KHF_2, $NaHF_2$, etc, are stable crystalline solids, the main anion being HF_2^-. The HF_2^- anion is actually an HF molecule hydrogen-bonded to an F^- ion,

$$F—H\text{---}F^-$$

but the hydrogen bond in this case is so strong that HF_2^- behaves as a single particle.

It is easy to see how hydrogen bonding may be responsible for abnormally high boiling points. When a liquid is boiled, the molecules leave the surface and go into the gas phase. We have already seen that this requires more energy as the molecular weight increases. But if the molecules are attracted to each other by hydrogen bonds, then either still more energy must be added to break the hydrogen bonds, or if the hydrogen bonds are still there in the gas phase (as they are in some cases), then the molecular weight is effectively doubled, and more energy is required for this reason. In either case, boiling points are abnormally high. This effect is especially strong in the case of water, because each molecule has two hydrogens, so that hydrogen bonding results in clusters of molecules (see Section 6-10).

In order to understand how hydrogen bonding is responsible for some of the other abnormal properties of water, we must consider the structure of ice and of liquid water. Water in the gas phase shows negligible amounts of hydrogen bonding. We know this because gas density measurements (Section 4-6) show that the molecular weight is 18.

6-9 Structure of Ice

The currently accepted theory of the structure of ice is that of Linus Pauling. Figure 6-15 shows the three-dimensional structure of H_2O molecules in the solid state (ice). There is one hydrogen atom between each O——O axis and each oxygen atom is surrounded by four hydrogen atoms. Two of these are connected to the oxygen by covalent bonds and the other two by hydrogen bonds. We see that in this case each oxygen atom can form *two* hydrogen bonds, and thus create this three-dimensional

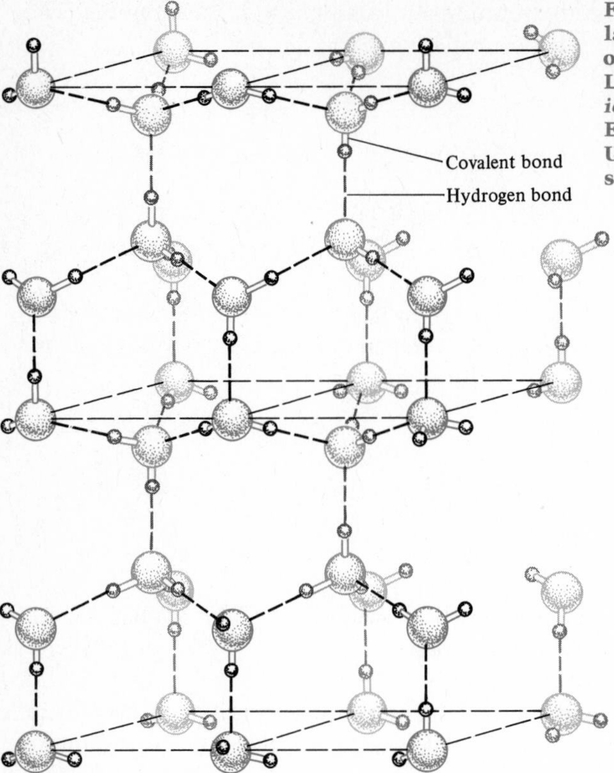

Figure 6-15 The Structure of Ice. The large atoms are oxygen; the small ones are hydrogen. [Redrawn from Linus Pauling: *The Nature of the Chemical Bond.* Copyright © 1939, 1940, Third Edition © 1960 by Cornell University. Used by permission of Cornell University Press.]

network. The two covalent O—H distances are 0.97 Å, while the two hydrogen-bonded O---H distances are 1.79 Å. Electron diffraction studies have shown that this structure proposed by Pauling is essentially correct.

The structure of ice, as shown in the figure, is a large open network, accounting for its low density. When ice melts, the open structure is partially destroyed. It is the collapse of some of the open structure that causes liquid water to have a greater density than ice. At the freezing point, a large portion of the hydrogen bonding still exists, which means that in the liquid at 0°C large groups of open structures still exist. As the temperature is raised from 0°C, these open structures continue to collapse, causing a further increase in density. But this is not the only thing that happens as the temperature is raised. A liquid normally expands with increasing temperature. This is due to the increased kinetic energy which shows itself in increased vibrations or molecular agitation. This expansion effect will finally overcome the density increase, giving water a maximum density at 4°C. Above this temperature, water shows the usual decrease in density with increasing temperature.

6-10 Structure of Liquid Water

When we began this chapter, we pointed out the difficulty theorists have in describing the liquid state. Water, with all its hydrogen bonding and anomalous properties, severely complicates this already difficult situation. For this reason there is not one completely acceptable picture of liquid water. Many theorists believe that water is composed of varying amounts of clusters of water molecules. There may be clusters composed of a very large number of water molecules, which are held together by hydrogen bonds, and of course there may be small clusters. An impor-

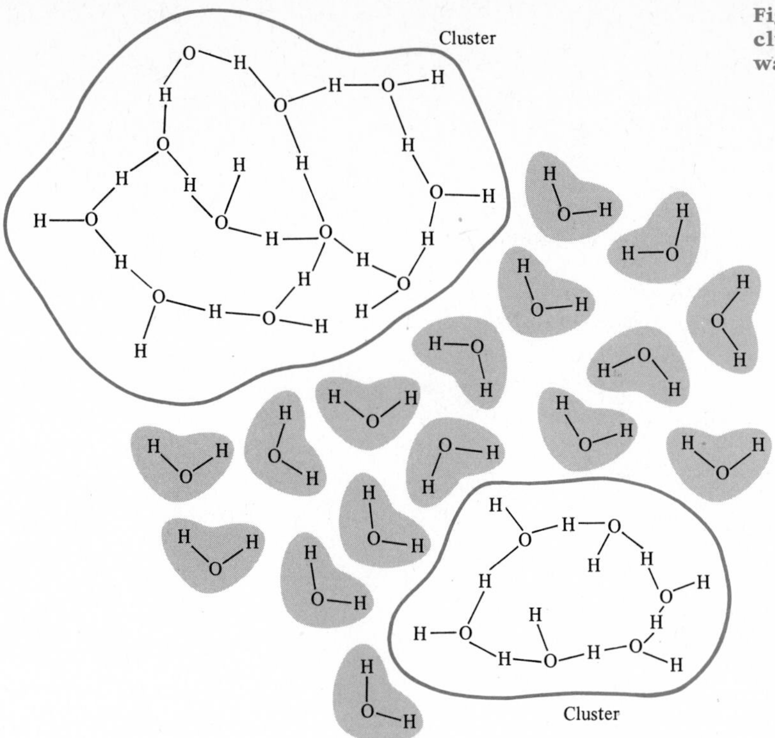

Figure 6-16 The flickering cluster model for liquid water.

tant aspect of all cluster theories is that the clusters change in size continually. Some people picture liquid water as being formed by ice–water molecules and gas–water molecules together. The strength of the interactions and the average size of the clusters change with temperature. Figure 6-16 shows such an instantaneous picture of water.

6-11 Comparison of Some Hydrogen-Bonded Liquids

It is important to point out that the properties of water are unique even when they are compared to other hydrogen-bonded liquids. Some properties of H_2O, HF, and NH_3 are given in Table 6-4.

If we compare the hydrogen-bonded structures of H_2O, NH_3, and HF, we find that only H_2O can form a three-dimensional structure. Figure 6-17 shows the one-dimensional structure possible for NH_3, the one and two-dimensional structures possible for HF, and the three-dimensional structure of H_2O. Oxygen in water is singular in being able to form three-dimensional networks. It is this ability that distinguishes water from other hydrogen-bonded liquids. In Section 14-13 we shall see why of these three molecules, only water can form such three-dimensional networks.

Table 6-4 Comparison of Hydrogen-Bonded Liquids

	NH_3	HF	H_2O
Boiling point (°C)	−33	+19	+100
Melting point (°C)	−78	−92	0
ΔH_{vap} (cal/g)	330	360	540
ΔH_{fus} (cal/g)	108.1	54.7	80

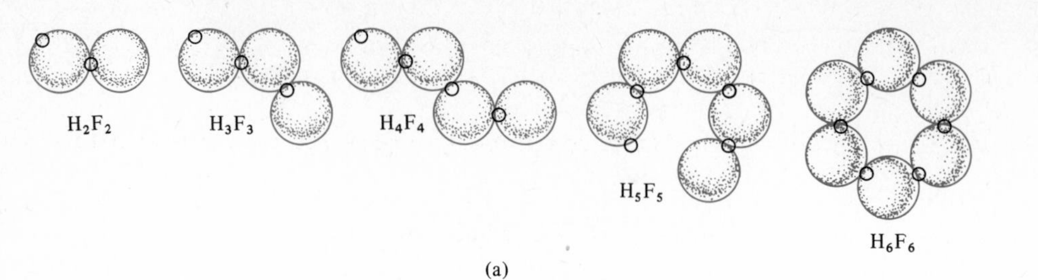

(b) (c)

Figure 6-17 A comparison of the hydrogen-bonded structure in HF (a), NH_3 (b), and H_2O (c). [(c) redrawn from Linus Pauling: *The Nature of the Chemical Bond.* Copyright © 1939, 1940, Third Edition © 1960 by Cornell University. Used by permission of Cornell University Press.]

Problems

6-1 Define each of the following terms. **(a)** evaporation **(b)** vapor molecules **(c)** vapor pressure **(d)** condensation **(e)** boiling point **(f)** normal boiling point **(g)** melting point **(h)** freezing point **(i)** intermolecular forces **(j)** dipole **(k)** induced dipole **(l)** dipole moment **(m)** van der Waals forces **(n)** London forces **(o)** dispersion forces **(p)** coordination number **(q)** molar volume **(r)** hydrogen bond **(s)** sublimation

6-2 At 25°C a mercury manometer like that shown in Figure 6-5a read 758.2 torr. Water was introduced into the manometer and the reading dropped to 734.4 torr. Calculate the vapor pressure of water at this temperature.

6-3 **(a)** If 0.00100 g of water is introduced into a mercury manometer which read 758.2 torr as in Problem 6-2 and the water completely evaporated, what would be the final reading of the manometer? (Assume that the density of the water vapor in the manometer at 25°C is 0.020 g/ml.)
(b) If you subtract the final reading of the manometer from 758.2 torr, would that be the vapor pressure of water at that temperature? Explain.

6-4 From Figure 6-7, estimate the vapor pressure of chloroform, ethyl alcohol, acetic acid, and octyl alcohol at 20°C and at 50°C. By what percentage has each increased?

6-5 From Figure 6-7, estimate the boiling points of ethyl alcohol and acetic acid at 200 torr and at 500 torr.

6-6 Explain why food cooks faster in a pressure cooker than in an open pot.

6-7 Explain why snow often remains on the ground even when the atmospheric temperature rises considerably above 0°C.

6-8 **(a)** Which liquid in Table 6-1 has the highest ΔH°_{vap} per gram? **(b)** Which has the lowest?

6-9 Explain how the vapor pressure of a liquid is affected by each of the following variables. **(a)** The volume of the liquid. **(b)** The temperature. **(c)** The external pressure on the surface of the liquid. **(d)** The surface area of the liquid. **(e)** The magnitude of intermolecular forces.

6-10 **(a)** How many calories are required to evaporate 7.5 moles of water at 25°C (ΔH_{vap} at 25°C = 580 cal/g)? **(b)** How many calories are required first to raise the temperature of this amount of water to 100°C and then to boil it at that temperature? **(c)** If the answer to part **(a)** does not equal the answer to part **(b),** does that violate Hess's law?

6-11 Using the data in Table 6-1, calculate the amount of heat necessary to evaporate each of the following at its normal boiling point. **(a)** 2.1 moles of benzene **(b)** 58.5 g of carbon tetrachloride **(c)** 110 g of krypton **(d)** 2.55 kg of ethyl alcohol

6-12 The vapor pressure of water varies with the temperature as follows:

T (°C)	*P* (torr)	*T* (°C)	*P* (torr)
0	4.6	40	55.3
10	9.2	60	149.4
20	17.5	80	355.1
30	31.8	100	760.0

Plot P in torr vs. T in °C. See if you can obtain a graph similar to that given in Figure 6-7. Would your curve look any different if you plotted P in torr vs T in °K?

6-13 Using the data in Problem 6-12, plot $\log P$ (in torr) vs. $1/T$ (in °K) and from the graph obtain ΔH_{vap}. Compare your result with that given in Table 6-1.

6-14 Using the data in Table 6-1, calculate the vapor pressure of benzene at 65°C by the method of Section 6-4. What assumption is necessary for this calculation?

6-15 Repeat the calculation of Problem 6-14 for carbon tetrachloride at 50.5°C.

6-16 The normal boiling point of chlorobenzene is 132°C and its heat of varporization is 8.730 kcal/mole. Calculate the temperature at which chlorobenzene will boil under a pressure of **(a)** 0.750 atm **(b)** 0.500 atm

6-17 The normal boiling point of *n*-hexane (C_6H_{12}) is 98°C. If its vapor pressure at 30°C is 189 torr, estimate its heat of vaporization.

6-18 Select the substance in each of the following pairs which you think has the higher boiling point and give your reason. **(a)** argon, Ar, and xenon, Xe **(b)** ethanol, C_2H_5OH, and dimethyl ether, $CH_3—O—CH_3$ **(c)** water, H_2O, and hydrogen selenide, H_2Se **(d)** nitrogen,

N_2, and nitric oxide, NO **(e)** methane, CH_4 and silane, SiH_4 **(f)** hydrogen fluoride, HF, and hydrogen iodide, HI

6-19 Tell whether each of the following properties would be increased, decreased, or unchanged by an increase in intermolecular forces. **(a)** normal boiling point **(b)** vapor pressure **(c)** molecular weight **(d)** heat of vaporization **(e)** heat of fusion

6-20 Why is the heat of vaporization generally much higher than the heat of fusion of the same substance?

6-21 In a dipolar molecule, why is the magnitude of the partial positive charge ($\delta+$) always exactly equal to that of the partial negative charge ($\delta-$)?

6-22 For each of the following, predict whether the intermolecular forces (between two identical molecules) would include dipole interactions, hydrogen bonds, and/or London forces.
(a) HF **(b)** Ar **(c)** H_3CCl **(d)** H_3COH **(e)** CH_4
(f) $CH_3-\underset{\underset{O}{\|}}{C}-OH$ **(g)** CCl_4 **(h)** H_2Te

6-23 The boiling point of diethyl ether $C_2H_5-O-C_2H_5$ is 35°C. The boiling point of ethanol CH_3CH_2-OH, which has a similar structure but a much lower molecular weight, is 78°C. Explain why the boiling point of ethanol is *higher* than that of diethyl ether.

6-24 Tell whether hydrogen bonding is likely between the two molecules or ions in each of the following pairs.
(a) H_2S and H_2S **(b)** HF and HF
(c) H_3O^+ and HCl **(d)** CH_3-O-CH_3 and NH_3
(e) CH_3-O-H and C_6H_6 **(f)** H_2Se and PH_3
(g) H_2O and $CH_3-\underset{\underset{CH_3}{|}}{N}-CH_3$ **(h)** NH_3 and CH_3-F
(i) CH_3OH and F^- **(j)** H_3O^+ and CH_3-O-CH_3

6-25 Acetic acid has the molecular formula $C_2H_4O_2$ and a molecular weight of 60.048. However, when one determines its molecular weight by gas-density methods, it seems to have a molecular weight almost twice what its molecular formula tells us. If its structure is $CH_3-\underset{\underset{O}{\|}}{C}-O-H$, show how hydrogen bonding between two acetic acid molecules can account for the anomalous molecular weight.

6-26 To break a single hydrogen bond in ice requires 8.3×10^{-21} cal. **(a)** How many calories are necessary to break 1 mole of hydrogen bonds in ice? **(b)** If the heat of fusion of ice is 80 cal/g and if we consider that all the energy in the heat of fusion is used to break hydrogen bonds, calculate the number of hydrogen bonds broken when 100.0 g of ice is converted to liquid water.

Additional Problems

6-27 The vapor pressure of benzene at various temperatures is

T (°C)	20	30	40	50
P (torr)	75	118	181	269

(a) Use these data to calculate the heat of vaporization of benzene.
(b) Assuming that ΔH°_{vap} does not change with temperature, calculate the normal boiling point of benzene.

6-28 Explain why the gas NH_3 is soluble in water, but the gases CH_4 and PH_3 are not, although all three of these molecules are very similar in structure.

6-29 **(a)** If the average kinetic energy of liquid benzene C_6H_6 molecules is 4.33×10^{10} ergs/mole, what is the average velocity? **(b)** If a sample of liquid carbon tetrachloride (CCl_4) is at the same temperature, then its molecules have the same average kinetic energy. What is their average velocity? **(c)** What is the ratio of the velocities?

6-30 From the data in Table 6-2, plot the molecular weights of CH_4, C_2H_6, C_3H_8, and C_4H_{10} vs. their boiling points. What would you predict for the boiling point of *n*-pentane, C_5H_{12}?

6-31 If water is raised to the boiling point at 1 atm pressure and then allowed to cool, will it cool faster if it is covered or left uncovered? Explain.

6-32 Explain why most liquids contract in volume when they freeze.

6-33 Assume that 1000 g of water is mixed with 100 g of diethyl ether, and the mixture brought to 35°C, the normal boiling point of ether. If all the ether evaporated, calculate the final temperature of the 100 g of water, assuming that the heat capacity of water is 1.00 cal/g · °K and the ΔH°_{vap} of ether remains constant during the temperature change.

6-34 The packages of some foods, such as cake mixes, have special directions for cooking or baking at high altitudes (look for them in your supermarket). Why are such directions necessary?

6-35 The vapor pressure of *n*-octane is 228 torr at 85°C, and its heat of vaporization is 8.36 kcal/mole. Estimate the normal boiling point of *n*-octane.

6-36 Ammonia has a normal boiling point of 240°K and a ΔH°_{vap} of 5560 cal/mole at 240°K. Assuming that the ΔH°_{vap} does not change with temperature, calculate the temperature at which the vapor pressure of ammonia would be $\frac{1}{760}$ atm.

6-37 The vapor pressure of an unknown liquid is 400 torr at 301°K and 130 torr at 273°K. If the liquid has a molecular weight of approximately 76 g/mole, calculate its ΔH°_{vap} in cal/g.

6-38 Explain how a small patch of snow on the ground can disappear without leaving a puddle of water behind.

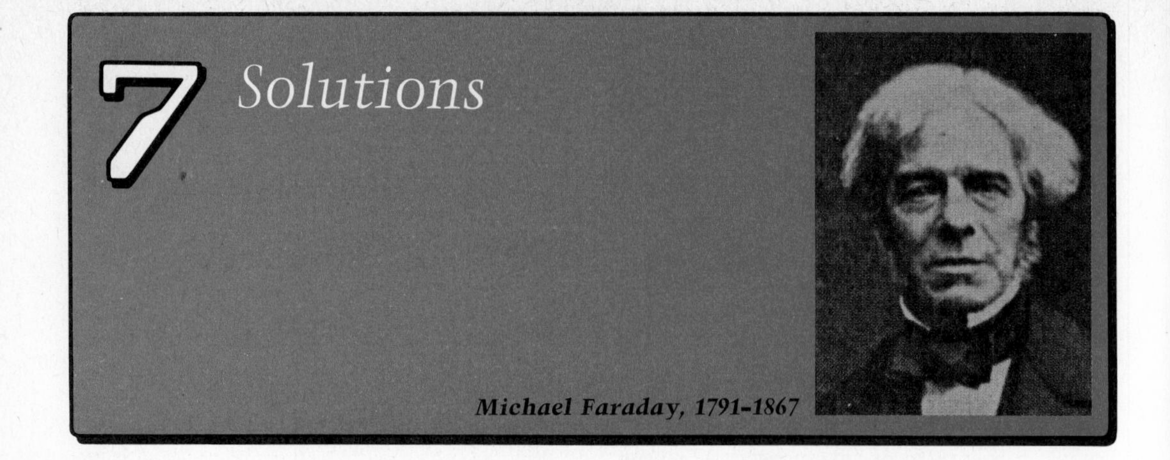

The great majority of all chemical reactions ever studied take place in solution. The chemicals that are to react are mixed in a medium (solvent) and the reaction between the chemicals themselves takes place in the medium. We can think of the solvent as an environment that permits the substances to mix intimately and to react. As an illustration, consider the reaction between solid sodium chloride (NaCl) and solid silver nitrate ($AgNO_3$). If we mix the two powders, even if we crush them together very well with a mortar and pestle or with a mechanical mixer, the best we can expect is that very small particles will come together so that their surfaces touch. The reaction ($Ag^+ + Cl^- \rightarrow AgCl$) will take place very, very slowly since the number of silver and chloride ions that do touch is still very small in comparison to the total number present. If, instead of this procedure, we dissolve both substances in water (the medium or solvent), the gross crystals in the solid will dissolve in the solution, and the ions will separate from each other. The number of contacts between the Ag^+ ions and the Cl^- ions is now about 10^{15} times as great as that found in the solid state, and a white precipitate of silver chloride forms immediately. Hence the reason for the solvent: *It permits the ultimate particles (ions, atoms, or molecules) of the reacting materials to move about and to mix intimately, resulting in many contacts* (Figure 7-1). To appreciate further the importance of solvents

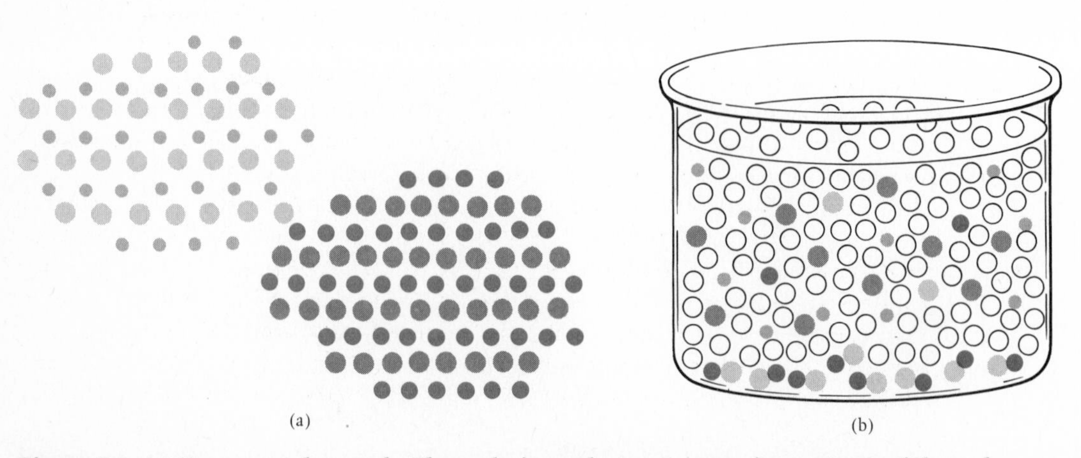

Figure 7-1 (a) Two crystals touch. The only ions that can come in contact with each other are those at the surface. (b) When the ions are dissolved in solution, their new freedom of motion permits them to come into frequent contact with each other.

consider that all the thousands upon thousands of chemical reactions that take place every day in the body take place in a solvent.

7-1 Types of Solutions

Usually, when we think of a solution we think of a liquid, and indeed most solutions we will encounter are liquids, but there are solid and gaseous solutions as well. A liquid solution may be made up of a liquid dissolved in a liquid (example: wine, which is mostly alcohol dissolved in water); a solid dissolved in a liquid (example: a detergent dissolved in a wash liquid), or a gas dissolved in a liquid (example: club soda, which is carbon dioxide dissolved in water); or there may be several compounds (solids, liquids, and/or gases) simultaneously dissolved in the same liquid (example: beer, which contains water, alcohol, carbon dioxide, and solid carbohydrates). Solutions that are gases always consist of mixtures of gases, or of liquids or solids that have evaporated. A good example is air, which is a gaseous solution made up chiefly of the gases nitrogen, oxygen, carbon dioxide, and argon (Section 17-1) but which also contains some water vapor. Good examples of solid solutions are alloys of various metals (for example, steel, which contains not only iron, but also carbon, chromium, nickel, and often other metals). To prepare a solid solution of this kind, it is necessary to melt at least one of the components, converting it to a liquid. The other components may then be dissolved or melted in this liquid. When the mixture has cooled and solidified (provided that it has been thoroughly stirred and the components are soluble), it is a solid solution. Only in this way can the intimate mixing be obtained without which a true solution is impossible.

7-2 Terminology

There are various ways of describing solutions. In this section, we shall discuss some of the common terms used. The material used as the medium is called the **solvent.** Any material dissolved in the medium is called a **solute.** When the medium is a liquid and the other component or components are solids or gases, there is never any doubt about which component is the solvent and which the solute. The liquid is always the solvent, and the others are solutes. With other types of solutions, however, the definitions are not always so clear-cut. If both components are liquids, the one present in excess is generally, though not always, called the solvent.* In cases where the two liquids are present in comparable amounts, there are no set rules. One often calls the solute the species whose properties are being studied. In the same solution, when the other species is being studied, *it* is called the solute. For these solutions, then, the question as to which is the solute and which is the solvent is arbitrary and depends on the particular investigation being performed.

As a thought experiment, let us take 100 ml of water at 25°C. We begin to introduce potassium chloride into the water. Continuing to add the salt, and stirring as we add until each added amount has dissolved, we shall reach a point where no more KCl will dissolve and we begin to see crystals of undissolved salt at the bottom of the water. We say that the solution is now **saturated.** Up to the time of saturation, the solution is said to be **unsaturated.** If we begin to heat the solution, we can dissolve more salt in the water. This happens because the solubility of KCl in

*The same rule can apply where both components are solids or both gases, but in these cases the terms "solvent" and "solute" are not often used.

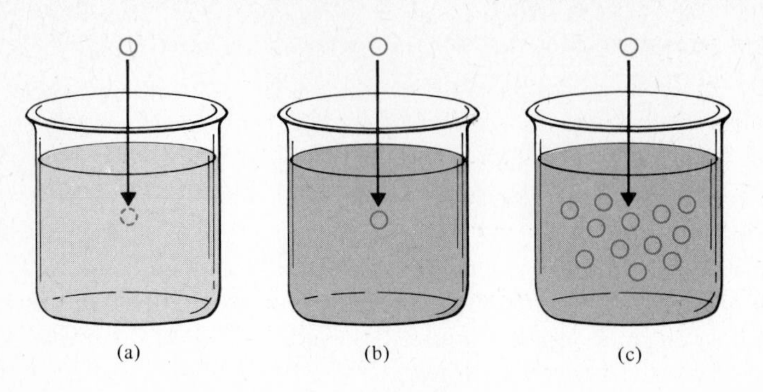

Figure 7-2 Unsaturated (a), saturated (b), and supersaturated (c) solutions.

H_2O increases with increasing temperature.* We cool the heated solution slowly, and if we are careful, it is possible to reach 25°C with more salt dissolved in the water than we had in the saturated solution. The solution is now called **supersaturated.**

We may distinguish between unsaturated, saturated, and supersaturated solutions, by dropping into the beaker a crystal of the salt in question and stirring (see Figure 7-2). In beaker (a), the crystal dissolves in the water. This solution is unsaturated. In beaker (b), the crystal drops to the bottom of the beaker. Here the solution is saturated. In beaker (c), dropping a crystal into the solution causes precipitation of some of the salt dissolved in the water. This solution was supersaturated, containing more salt than it should; adding the crystal to this solution provided the dissolved salt crystal with a surface on which it could crystallize. This is called **seeding.** A supersaturated solution is an unbalanced (metastable) system. The reason it can exist is that in the solution the K^+ ions and the Cl^- ions do not have a crystalline face on which to grow. It is very difficult for the ions to form a microcrystal initially, but when a seed crystal is added, rapid precipitation can take place.

The words **concentrated** and **dilute** are often used to describe solutions. These terms are useful but very imprecise. A solution that is called concentrated in one laboratory may be called dilute in a different laboratory, where it is being compared with another solution that is more concentrated. The terms "concentrated" and "dilute" have no relation to the concept of saturation or unsaturation. Thus, the solubility of sugar in water is very high, and it is possible to have a solution that most people would regard as concentrated but still far from being saturated. On the other hand, the solubility of lead chloride in water is very low, and in this case even a saturated solution must be dilute; here one can only have a concentrated solution if it is supersaturated. The words "concentrated" and "dilute" are often used, but it must be remembered that their meaning is always subjective.

7-3 Concentration Units

The amount of solute dissolved in a given amount of solvent is called the **concentration.** This can be expressed in a variety of ways. Which method one chooses depends on the situation. No one method is intrinsically superior to any other.

PERCENTAGE COMPOSITION. For some purposes we express the concentration simply by giving the percentage (by weight or by volume) of the components present. For example, a solution of 10 g of sugar in 90 g of water may be called a 10% solution by weight of sugar in water. A solution made up of 5 ml of carbon tetrachloride and 95 ml of benzene would be a 5% solution by volume of CCl_4 in benzene. Note that the final volume of this solution is *not* exactly 100 ml. When different liquids are mixed, volumes are not in general additive! Students may try

*This is the case for most solids dissolved in a liquid, but not all. In some cases the solubility of a solid in a liquid decreases with increasing temperature.

the experiment themselves in the laboratory. Mix measured volumes of 95% alcohol with water, and compare the final volume with the sum of the volumes before mixing. When two different liquids are mixed, the final volumes may be greater or less than the sum of the individual volumes. Volume percentage also has another meaning: volume of solute per 100 volumes of total *solution*. Under this definition, the solution of carbon tetrachloride in benzene referred to above would not be exactly 5%, since the total volume is not 100 ml.

The nonadditivity of volumes arises for the following reason. In a liquid *A*, all the molecules of *A* maintain a certain average distance from each other, depending on the attraction the molecules have for each other (Section 6-6). When *B* is dissolved in *A*, a molecule of *A* may be more (or less) attracted to a molecule of *B* than to another molecule of *A*. If it is more attracted to *B*, it will be closer than it would have been to *A*, and the combined total volume will shrink. If it is less attracted to *B*, the total volume will expand.

In the United States, bottles of wine and stronger alcoholic beverages are required to show on the label the amount of alcohol present. This is expressed either directly, as percentage of alcohol *by volume*, or as **proof**, which in the United States means twice the percentage of alcohol by volume. That is, a whiskey that is 86 proof has 43% alcohol by volume.*

When the concentration of a solution of a solid in a liquid is expressed as a percentage, it is almost always a percentage by weight, and it is, therefore, usually unnecessary to indicate this. However, when one wishes to express as a percentage the concentration of a liquid in a liquid, one must specify whether this is a weight or a volume percentage.

MOLE FRACTION. The **mole fraction** is a ratio of the number of moles of the species we are talking about (solute or solvent) to the total number of moles present. For example, let us say we have a solution containing 100 g of water and 60 g of potassium chloride. (The symbol for mole fraction is χ.)

$$\chi_{KCl} = \frac{\text{number of moles of KCl}}{\text{total number of moles}}$$

$$\chi_{H_2O} = \frac{\text{number of moles of } H_2O}{\text{total number of moles}}$$

We must first convert the number of grams of each component into moles.

$$\text{moles of KCl} = \frac{\text{grams of KCl}}{\text{mol wt of KCl}} = \frac{60 \text{ g}}{74.55 \text{ g/mole}} = 0.81 \text{ mole}$$

$$\text{moles of } H_2O = \frac{\text{grams of } H_2O}{\text{mol wt of } H_2O} = \frac{100 \text{ g}}{18.0 \text{ g/mole}} = 5.55 \text{ moles}$$

Now we can compute the mole fraction of each component:

$$\chi_{KCl} = \frac{0.81}{5.55 + 0.81} = \frac{0.81}{6.36} = 0.127$$

$$\chi_{H_2O} = \frac{5.55}{6.36} = 0.873$$

It is very easy to check one's result in calculating mole fractions. First, each mole fraction must be a number less than 1 (why?), and second, the sum of the mole fractions of all the components of the mixture must equal 1.

*The origin of the term "proof" is interesting. Before modern methods of analysis, people still wanted to know how much alcohol there was in the whiskey they were buying. The test they used was to soak some gunpowder in the whiskey and then to see if it could be ignited. The most dilute mixture that would allow the gunpowder to burn was about 50% alcohol. This was called 100 proof.

Let us prove this using the example above.

$$\text{sum of mole fractions} = \text{mole fraction of KCl} + \text{mole fraction of } H_2O$$

$$\frac{\text{moles of KCl}}{\text{total number of moles}} + \frac{\text{moles of } H_2O}{\text{total number of moles}} = \frac{\text{total number of moles}}{\text{total number of moles}} = 1$$

Example: Say that we have a three-component system (a system containing three different species). It contains 6.0 moles of component A, 4.0 moles of component B, and 2.0 moles of component C. Calculate the mole fraction of each.

Answer: From the definition:

$$X_A = \frac{\text{moles of } A}{\text{total number of moles}} = \frac{6.0}{6.0 + 4.0 + 2.0} = \frac{6.0}{12.0} = 0.50$$

$$X_B = \frac{\text{moles of } B}{\text{total number of moles}} = \frac{4.0}{6.0 + 4.0 + 2.0} = \frac{4.0}{12.0} = 0.33$$

$$X_C = \frac{\text{moles of } C}{\text{total number of moles}} = \frac{2.0}{6.0 + 4.0 + 2.0} = \frac{2.0}{12.0} = 0.17$$

The sum of the mole fractions is $0.50 + 0.33 + 0.17 = 1.00$.

MOLARITY. Consider another situation. Dissolve 1 mole of NaCl in enough water to make the final volume of the solution 500 ml. If we take 100 ml of this solution, the number of moles of NaCl present is 0.2 mole of NaCl. Each milliliter of the 500-ml solution contains 0.002 mole ($\frac{1}{500}$) of NaCl. Now let us take the original 500 ml of solution and add enough water to it to give us a total volume of 1 liter (approximately 500 ml). Even though we have not changed the total amount of NaCl in solution, each milliliter of the solution now contains 0.001 mole ($\frac{1}{1000}$) of NaCl. We need a unit that expresses the number of moles of solute per volume of solution. Such a unit is *molarity*, the symbol for which is M. Among chemists, this is the most common method of expressing concentration.

Molarity is defined as the **number of moles of solute per liter of solution.** Note that this is not the same as saying "the number of moles of solute per liter of *solvent*." If we take, say, 1 mole of NaCl and add it to 1 liter of water, the volume of the final solution will be more than 1 liter, because the 1 mole of salt (58.44 g) takes up some space, too. In order to prepare a solution of salt in water which has a concentration of 1 M, we use a *volumetric flask* (Figure 7-3), which is a flask that has been precisely calibrated to hold exactly 1.000 liter when filled to the mark. The carefully weighed-out salt (58.44 g) is placed in the flask along with about 500 ml of water, and the salt is completely dissolved by stirring. Then additional water is added to bring the total volume to the mark on the flask. In this way, we are assured that the flask contains exactly 1.000 liter of total solution (water plus salt). In most cases, we neither know nor care how much water has been added. It is enough to know that the volume of the total solution is exactly 1.000 liter.

Molarity may be expressed as

$$M = \frac{n}{\ell}$$

where n is the number of moles of solute, and ℓ the volume of the solution, in liters.

Figure 7-3 A 500-ml volumetric flask. Volumetric flasks also come in other sizes.

It follows that

$$n = M\ell$$

The units of molarity are, therefore, *moles per liter*. Note that from the definition of moles (page 49), we have

$$\text{number of moles of solute} = \frac{\text{number of grams of solute}}{\text{mol wt of solute}}$$

so that

$$n = M\ell = \frac{\text{grams of solute}}{\text{mol wt}}$$

Example: Calculate the number of grams of $AgNO_3$ in 325 ml of a 0.140 M solution.

Answer:

$$\begin{aligned} n &= M\ell = 0.140 \text{ mole/liter} \times 0.325 \text{ liter} \\ &= 0.0455 \text{ mole } AgNO_3 \\ \text{number of grams} &= \text{number of moles} \times \text{molecular weight} \\ &= 0.0455 \text{ mole} \times 169.87 \text{ g/mole} = 7.73 \text{ g } AgNO_3 \end{aligned}$$

Example: A 0.165 M solution of H_2SO_4 is diluted from 700.0 ml to 900.0 ml. Calculate the new molarity of H_2SO_4.
Answer: Since only water is being added to the original solution of H_2SO_4, the total number of moles of H_2SO_4 in the initial and final solutions must be the same. We can therefore write

$$\begin{matrix}\text{number of moles of } H_2SO_4 \\ \text{in initial solution}\end{matrix} = \begin{matrix}\text{number of moles of } H_2SO_4 \\ \text{in diluted solution}\end{matrix}$$

$$M\ell = M\ell$$

$$0.165 \text{ mole/liter} \times 0.700 \text{ liter} = x \times 0.900 \text{ liter}$$

so that

$$x = \frac{(0.165 \text{ mole/liter})(0.700 \text{ liter})}{0.900 \text{ liter}} = 0.128 \text{ mole/liter} = 0.128\ M$$

Note that it is not necessary in doing this problem to calculate the number of moles of sulfuric acid present, although one can, of course, easily obtain this number if desired (it is 0.165×0.700).

The equation

$$M_A \ell_A = M_B \ell_B \quad \text{or} \quad M_A \text{ml}_A = M_B \text{ml}_B$$

where A refers to the solution before it is diluted and B after it is diluted, is a general useful equation that can be used to solve dilution problems. At least in some cases, the same equation (except that A stands for acid and B for base) can be used to solve problems involving neutralization of an acid by a base, and vice versa.

Example: We wish to neutralize completely 433 ml of a 0.147 M NaOH solution with HCl. The HCl solution we have has a molarity of 1.37 M. How many milliliters of the HCl solution will be needed to neutralize completely the NaOH?

Answer: The net ionic reaction between HCl and NaOH may be written $H_3O^+ + OH^- \rightarrow 2H_2O$ (Section 3-2). Hence, for every mole of OH^- present, it will be necessary to add 1 mole of H_3O^+. Therefore, at equivalence (complete neutralization), we have

number of moles of acid added = number of moles of base initially present

and

$$M_{\text{acid}} \times \ell_{\text{acid}} = M_{\text{base}} \times \ell_{\text{base}}$$

$$1.37 \times x = 0.147 \times 0.433$$

$$x = \frac{0.147 \times 0.433}{1.37} = 0.0465 \text{ liter} \quad \text{or} \quad 46.5 \text{ ml of HCl}$$

Example: How many grams of KCl must be placed in a 400-ml volumetric flask to produce 400 ml of a 0.560 M solution of KCl in water?

Answer:

0.560 mole/liter × 0.400 liter = 0.224 mole of KCl

number of grams of KCl = number of moles of KCl × mol wt KCl

number of grams of KCl = 0.224 mole × 74.551 g/mole = 16.7 g

NORMALITY. It would, of course, be most convenient if it were possible to have just one concentration unit, namely molarity. Unfortunately, the molarity unit does not cover all situations and, if used improperly, will give fallacious results. To illustrate this, consider the following reaction. We wish to neutralize a sodium hydroxide solution, not with HCl as we did earlier, but with sulfuric acid, H_2SO_4. Say that we have 100 ml of a 1 M NaOH solution and an unlimited amount of 1 M H_2SO_4 solution. How many ml of the 1 M solution will neutralize the NaOH solution? Proceeding as we did previously, we would write

number of moles of acid = number of moles of base

$$M\ell = M\ell$$

$$x \times 1.0 = 0.1 \times 1.0$$

$$x = 0.1 \text{ liter} \quad \text{or} \quad 100 \text{ ml}$$

This answer is *wrong*. Let us see why.

The formula for sodium hydroxide is NaOH, so 100 ml of a 1 M solution of NaOH contains 0.1 mole of Na^+ ions and 0.1 mole of OH^- ions. But the formula for sulfuric acid is H_2SO_4, so 100 ml of 1 M H_2SO_4 contains 0.1 mole of SO_4^{2-} but *0.2 mole of H_3O^+ ions*. This is so because sulfuric acid dissociates as $2H_2O + H_2SO_4 \rightarrow 2H_3O^+ + SO_4^{2-}$. That is, each molecule of H_2SO_4 gives rise to *two* H_3O^+ ions. Clearly to neutralize 0.1 mole of OH^- we only need 0.1 mole of H_3O^+, not 0.2 mole of H_3O^+. We can get around this difficulty by devising another concentration unit (*normality*), which takes into account the fact that certain substances yield more than one hydronium ion or hydroxide ion per molecule. Let us look at some acids and bases that show this behavior.

Barium hydroxide, $Ba(OH)_2$:

$$Ba(OH)_2 \longrightarrow Ba^{2+} + 2OH^-$$

Calcium hydroxide, $Ca(OH)_2$:

$$Ca(OH)_2 \longrightarrow Ca^{2+} + 2OH^-$$

Aluminum hydroxide, $Al(OH)_3$:

$$Al(OH)_3 \longrightarrow Al^{3+} + 3OH^-$$

Sulfuric acid, H_2SO_4:

$$H_2SO_4 + 2H_2O \longrightarrow 2H_3O^+ + SO_4^{2-}$$

Phosphoric acid, H_3PO_4:

$$H_3PO_4 + 3H_2O \longrightarrow 3H_3O^+ + PO_4^{3-}$$

Before we can define normality, we must define two other terms, *equivalent weight* and *equivalent*.

The **equivalent weight** of a substance, when applied to acid–base reactions, is the weight of the substance that will produce or react with 1 mole of H_3O^+ or of OH^-.

Example: Barium hydroxide, $Ba(OH)_2$ has a molecular weight of 171. What is the equivalent weight?

Answer: Since barium hydroxide will yield two OH^- ions per molecule of barium hydroxide, we must determine what weight of barium hydroxide will produce 1 mole of hydroxide ions. It should be clear that 1 mole of $Ba(OH)_2$ yields 2 moles of OH^-, so $\frac{1}{2}$ mole (or 171 g/2 = 85.5 g) of barium hydroxide will produce 1 mole of hydroxide ions. Hence, the equivalent weight of $Ba(OH)_2$ is 85.5 g.

Note that for an acid, the equivalent weight may be obtained by dividing the molecular weight by the number of hydrogens that will be removed by a base, while for a base the equivalent weight is obtained by dividing the molecular weight by the number of OH groups it contains. We may now define equivalent in a manner similar to the way we described a mole (Section 2-8); that is, one **equivalent** of a substance is the equivalent weight in grams. Since the equivalent weight of $Ba(OH)_2$ is 85.5 g, then 85.5 g of $Ba(OH)_2$ is one equivalent of this substance. The units of equivalent weight are, therefore, grams per equivalent (note the parallel with the units of molecular weight: grams per mole).

Example: What is the number of equivalents of $Ba(OH)_2$ in 500 g of $Ba(OH)_2$?

Answer:

$$\text{Number of equivalents of } Ba(OH)_2 = \frac{\text{grams of } Ba(OH)_2}{\text{equiv wt of } Ba(OH)_2}$$

$$= \frac{500\text{ g}}{85.5\text{ g/equivalent}} = 5.85 \text{ equivalents}$$

Example: Compare the number of equivalents in 225 g each of HCl, H_2SO_4, and H_3PO_4.

Answer: First we calculate the equivalent weight for each of the above.

Equivalent weight of HCl:

$$OH^- + HCl \longrightarrow H_2O + Cl^-$$

Hence 1 mole of HCl neutralizes 1 mole of OH^-, so that the equivalent weight of HCl is numerically equal to its molecular weight, that is, 36.5 g/equivalent.

Equivalent weight of H_2SO_4:

$$2OH^- + H_2SO_4 \longrightarrow 2H_2O + SO_4^{2-}$$

We have seen that 1 mole of H_2SO_4 neutralizes 2 moles of OH^-; or $\frac{1}{2}$ mole of H_2SO_4 neutralizes 1 mole of OH^-. The equivalent weight in this instance is one half the molecular weight or 98.0/2 = 49.0 g/equivalent.

Equivalent weight of H_3PO_4:

$$H_3PO_4 + 3OH^- \longrightarrow 3H_2O + PO_4^{3-}$$

Here 1 mole of H_3PO_4 neutralizes 3 moles of OH^- so that we need only $\frac{1}{3}$ mole of H_3PO_4 in order to neutralize 1 mole of OH^-. The equivalent weight of H_3PO_4 is, therefore, molecular weight/3 or 98.0/3 = 32.7 g/equivalent.

To calculate the number of equivalents in each, we write

$$\text{HCl:} \quad \frac{\text{grams of solute}}{\text{equiv wt}} = \frac{225\text{ g}}{36.5\text{ g/equiv}} = 6.16\text{ equiv}$$

$$H_2SO_4\text{:} \quad \frac{\text{grams of solute}}{\text{equiv wt}} = \frac{225\text{ g}}{49.0\text{ g/equiv}} = 4.59\text{ equiv}$$

$$H_3PO_4\text{:} \quad \frac{\text{grams of solute}}{\text{equiv wt}} = \frac{225\text{ g}}{32.7\text{ g/equiv}} = 6.87\text{ equiv}$$

We are now ready to define **normality** (symbol *N*) as **equivalents of solute per liter of solution** (note the parallel with the definition of molarity). From the definition of normality, we have

$$N = \frac{\text{number of equivalents}}{\text{liter}}$$

or

$$\text{number of equivalents} = (N)(\text{liters})$$

When an acid neutralizes a base, or vice versa, we have

$$\text{number of equivalents of acid} = \text{number of equivalents of base}$$

(Note that the very word *equivalent* comes from this fact.)

$$N_{\text{acid}} \times \ell_{\text{acid}} = N_{\text{base}} \times \ell_{\text{base}}$$

and from

$$1\text{ ml} = \frac{1}{1000}\text{ liter}$$

we have

$$N_{\text{acid}} \times \text{ml}_{\text{acid}} = N_{\text{base}} \times \text{ml}_{\text{base}}$$

Example: Calculate the normality of a NaOH solution for which 27.3 ml will completely neutralize 50.0 ml of a 0.107 *N* H_2SO_4 solution.

Answer: Since the sulfuric acid concentration is already expressed as normality, we need simply apply the concept that equivalents of acid = equivalents of base (for complete neutralization), or

$$\text{milliequivalents of acid} = \text{milliequivalents of base}$$

$$\text{ml}_{\text{acid}} \times N_{\text{acid}} = \text{ml}_{\text{base}} \times N_{\text{base}}$$

$$50.0\ \text{ml} \times 0.107\ N = 27.3\ \text{ml} \times \text{x}$$

$$\text{x} = \frac{50.0 \times 0.107}{27.3} = 0.196\ N\ \text{NaOH}$$

Note that the *normality* of a solution of an acid or a base in water is simply related to the *molarity* of the same solution by

$$N = aM$$

where a is equal to the number of hydrogens removed by a base (for a solution of an acid) or to the number of OH groups present (for a solution of a base). For example, if a solution of $Ba(OH)_2$ is 0.200 *M*, then it must be 0.400 *N*, since $Ba(OH)_2$ has two OH groups. Quantitative aspects of acid–base reactions will be considered further in Chapter 10.

The concepts of equivalents and normality are not limited to acid–base reactions, but can also be applied to other types, most notably oxidation–reduction reactions. In such cases a different definition of equivalent weight must be used: it is now the weight of the substance that will produce or react with 1 mole of some other standard substance, depending on the type of reaction being run (see Section 16-4).

The concepts of equivalents and normality are not absolutely necessary: it is possible to perform all these calculations on the basis of moles and molarity alone if we are careful to allow for the fact that some acids give rise to more than one H_3O^+ ion and some bases to more than one OH^- ion. Indeed, many chemists prefer not to use equivalents or normality at all. Historically, they were very important in the first half of the nineteenth century, before there was an accurate knowledge of atomic weights (footnote on page 40) and equivalent weights were all that chemists could be sure of. We have discussed them because they do make acid–base and oxidation–reduction calculations more convenient, and there are many chemists who do use them.

MOLALITY. Molality is a concentration term that is used much less than molarity or normality (mostly for colligative property studies; see Section 7-9). The **molality** of a solution (symbol *m*) is defined as **the number of moles of solute per 1000 grams of solvent.** There are two important differences between molality and molarity. First, whereas molarity is a ratio of moles of solute to *liters* (moles to volume), molality is a ratio of moles to *grams* (moles to weight). Second, the ratio in molality is a weight of solute to a *weight of solvent,* not to a *volume of solution.* We must emphasize these differences since molarity deals with liters of *solution,* and students become very accustomed to dealing with solutions (volumes or weights).

We may illustrate the difference by the following example, which really shows the difference between these units.

Example: Describe how you would prepare a 6 molal (6 *m*) and a 6 molar (6 *M*) solution of HCl in water.

Answer: You would prepare a 6 *m* solution of HCl by adding 6 moles of HCl (218.766 g of HCl) to 1000 g of water. You would prepare a 6 *M* solution of HCl by taking 6 moles of HCl and adding enough water to the HCl to get a final volume of 1 liter (in a volumetric flask).

Example: Calculate the molality of a solution prepared by dissolving 40.0 g of KCl in 190 g of water.

Answer:

$$\text{molality} = \frac{\text{moles of solute}}{1000 \text{ g } H_2O}$$

We must first calculate the number of moles of KCl in 40.0 g of KCl.

$$\text{number of moles of KCl} = \frac{40.0 \text{ g KCl}}{74.55 \text{ g/mole KCl}} = 0.537 \text{ mole KCl}$$

If we have 0.537 mole of KCl in 190 g of water, how many moles of KCl would we have in 1000 g of H_2O?

$$\frac{x}{1000 \text{ g } H_2O} = \frac{0.537 \text{ mole KCl}}{190 \text{ g } H_2O}$$

$$x = \frac{0.537}{190} \times 1000 = 2.81 \text{ moles of KCl}$$

This 2.81 moles of KCl would be dissolved in 1000 g of water, hence the molality of the solution is 2.81 *m*.

7-4 Solubility

We may define the **solubility** of a solute in a solvent as the amount of solute which, when dissolved in a given amount of solvent, will make a saturated solution. The units of solubility can be grams of solute per liter of solution, moles of solute per liter of solution, or similar units. That is, they are concentration units. The solubility of a given solute in a given solvent at any particular temperature is constant, but it does change when the temperature changes.

7-5 Properties of Solvents

Whenever we speak of a solvent, we immediately think of water, because of our intimate relationship with it. It is the solvent in our life processes. It is also the most abundant liquid on our planet. The study of water as a solvent has become so important that there are whole treatises devoted to its properties and structure. Water is, indeed, our most important solvent—but it is not the only one. We deal daily with other solvents: for example, cleaning solvents to remove grease and wax stains and solvents to dissolve oil base paints are just a few of many common **nonaqueous** (nonwater) solvents.

One of the most distinguishing characteristics or properties of a solvent is its *dielectric constant*. Recall from Section 1-8 that the electrical force in a vacuum between charged bodies is given by Coulomb's law:

$$F = \frac{kq_1q_2}{r^2}$$

where q_1 and q_2 are the charges on bodies 1 and 2. When any material is placed between the bodies, the electrical force decreases and is given by

$$F = \frac{kq_1q_2}{\varepsilon r^2}$$

where ε is a number, called the **dielectric constant,** which is characteristic of the material placed between the bodies. ε always has a value greater than unity.

For H_2O, $\varepsilon \approx 78$. This is a very high value, much higher than most other solvents. Water has one of the highest dielectric constants of any liquid known. We can say, therefore, that the energy necessary to separate two charged particles (oppositely charged) will be much less in an aqueous medium than in a vacuum.

Suppose that we wish to dissolve potassium chloride, KCl, which is an ionic solid. That is, in the solid state, KCl occurs as K^+ ions and Cl^- ions. Since a positive ion is strongly attracted to a negative ion, a great deal of energy will be required to separate them. When dissolved, K^+ and Cl^- will be in solution, separated from each other by solvent molecules. What kind of solvent will be best for dissolving this type of solid? It is clear that we want a solvent with a high dielectric constant, so that the energy which the solvent has to expend in separating the ions is small. Water, with its high dielectric constant, fills the need nicely.

Table 7-1 lists different solvents and their dielectric constants. What is it about the molecular structure of these solvents that gives rise to the different dielectric constants? A dielectric constant (which is a pure number) is a result of several different properties, the most important being the *dipole moment* (Section 6-6). Molecules with a dipole moment can be turned in an electric field or (more important to us) attracted to ions or to other molecules that possess dipole moments. The negative end of a dipolar molecule will be attracted to a positive ion or to the positive end of another dipolar molecule, and vice versa for the positive end. It is the possibility of such interactions between the solute and the solvent that plays so important a role in solubility. When one substance is dissolved in another, we can mentally construct an artificial series of events that take place. A certain amount of energy is expended in order to separate the solute molecules. The system then retrieves energy by the interaction of solute and solvent molecules. Note that the solvent molecules themselves have some structure and the introduction of solute molecules into their midst expends energy in disrupting the solvent structure. Depending upon the *net* energy obtained (expended plus returned) one can qualitatively understand solubility.

Water is a polar molecule. When KCl dissolves in water, each K^+ ion is surrounded by a shell of water molecules: the negative end of each water molecule is attracted to the positive K^+. In a similar manner, each Cl^- is surrounded by a shell of water molecules, but in this case, it is the positive ends of the water dipoles that are attracted to the Cl^- (Figure 7-4). As we mentioned before, a great deal of energy was necessary to separate the positive K^+ ions from the negative Cl^- ions in the crystal

Table 7-1 Dielectric Constants and Dipole Moments of Certain Molecules

Species	*Dielectric constant*	*Dipole moment (Debye units)*
H_2O	81.1 (18°C)	1.86
NH_3	22 (−34°C)	1.47
HCN	114.9 (20°C)	2.93
CH_3OH	32.63 (25°C)	1.69
CH_3COCH_3	20.70 (25°C)	2.85
SO_2	14.1 (20°C)	1.61

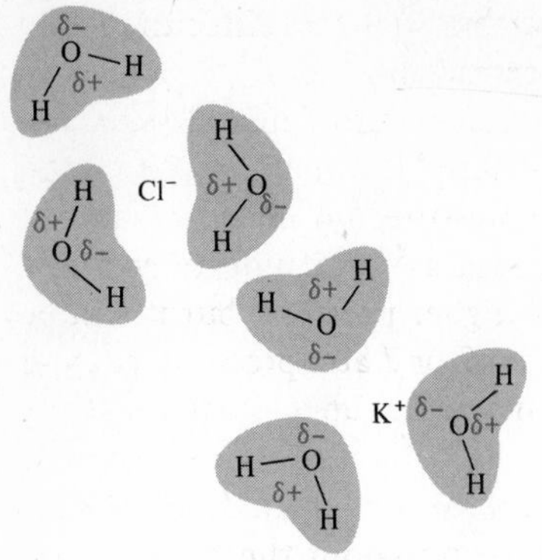

Figure 7-4 Solvation of ions by water molecules.

lattice of solid KCl. However, this energy is more than compensated for by the attractive forces between each ion (K^+ or Cl^-) and its shell of water molecules. The K^+ and Cl^- are said to be **solvated** by the water molecules, and the process is known as **solvation.** On the other hand, KCl will not dissolve to any appreciable extent in a solvent like benzene. Benzene is not a polar molecule (Section 20-5) and cannot solvate either K^+ or Cl^-. Therefore, there is no energy available to separate the ions, and KCl remains undissolved. This does not necessarily mean that *all* ionic solids dissolve in water. An example of one that does not is barium sulfate. In this case, the amount of energy gained by solvation of Ba^{2+} and SO_4^{2-} is less than the energy required to separate the ions in the lattice, and $BaSO_4$ is practically insoluble in water (there is, however, an extremely tiny solubility; see Section 11-1).

A general rule in solubility is that **like dissolves like.** By this we mean that substances which are chemically similar to each other are more mutually soluble than are dissimilar substances. Therefore, we expect solvents with dipole moments to dissolve ionic solutes and solutes possessing dipole moments. Here we have solvation or other strong attractions between the solute and solvent. Substances that have no dipole moments (or small dipole moments) do not readily dissolve in water. In this case, very little energy is required to separate the molecules, nor is there much solvation energy, so this aspect is simply not a factor here. What is important in this case is the semiordered structure of liquid water (Section 6-10). The introduction of molecules that it cannot solvate disturbs this structure without supplying any compensating gain. Another way to look at it is that benzene (an example of a nonpolar substance) is insoluble (or practically insoluble) in water because a water molecule would rather be next to another water molecule (because the positive end of the water dipole attracts the negative end of the next) than it would to a benzene molecule.

Similarly, solvents with no or very small dipole moments will not dissolve ions or dipolar solutes but will dissolve substances with no or very small dipole moments. In this case, energy expended in structure breaking is small, since the solvent–solvent interactions are small. However, note that they are not zero. The solute–solvent interactions are also small, but again, not zero.

7-6 Solubility Factors in Different Types of Solutions

GAS–GAS. Gases are soluble in each other in all proportions. None of the preceding discussion holds for gas–gas solubility. A gas is mostly empty space and the introduction of one gas into another involves no structure breaking and very little net energy gain or loss due to interactions (see Section 4-11) unless the molecules react with each other chemically.

GAS–LIQUID. Gases are most soluble in liquids with similar polarities. For example, the noble gases are more soluble in benzene than in water since the noble gases and benzene possess no dipole moments, whereas water does.

Apart from polarities, the solubility of gases in liquids is also dependent on the pressure and the temperature. For pressure the rule is simple: The solubility of any gas in any liquid *always* increases with increasing pressure, no matter what the polarity of the gas or the liquid. Of course, if a gas has a low solubility at a low pressure, it may still have a fairly low solubility at a higher pressure, but it will be higher than it was before. People who must work under 2 or 3 atm pressure, such as deep-sea divers, can get a disease called the bends. This comes about because at the higher pressures a higher concentration of atmospheric N_2 dissolves in the blood. If the rise to the surface is too sudden, the nitrogen comes out of solution faster than it can be carried to the lungs, and bubbles of N_2 form throughout the body.

For most solutions of liquids in liquids or solids in liquids, solubility increases with increasing temperatures. For solutions of gases in liquids, however, the situation is reversed. For most (though not all) such solutions, solubility *decreases* with increasing temperature.

LIQUID–LIQUID. Liquid–liquid systems generally follow the like-dissolves-like rule. For example, water and methanol are soluble in each other in all proportions. However, the solubility of water in gasoline, a nonpolar liquid, is quite limited. Carbon tetrachloride, which is a nonpolar molecule (no dipole moment), is completely soluble in benzene (a nonpolar liquid).

The discussion of solubility that we have just presented is necessarily somewhat oversimplified. The picture is actually very complex, involving many kinds of effects. It is sometimes necessary to invoke special explanations for certain solubility phenomena. These are done after the effects are noted; that is, they are not generally predictable in advance.

Let us summarize some of our discussion. A general rule of solubility is that like substances dissolve like. This means that, in general, polar solvents dissolve ionic and polar solutes, and that nonpolar solvents dissolve nonpolar solutes. Solubility involves separating solute molecules or ions and breaking up solvent structure. This costs energy. However, the system gets back energy from the interaction of solute and solvent molecules, so that one must look at three types of interaction to get the whole picture: solute–solute interactions, solvent–solvent interactions, and solute–solvent interactions.

7-7 Ions in Solution

One general way of classifying solutes dissolved in water is according to whether or not the solution will conduct an electric current. If a solution of a solute in H_2O does not conduct a current, it is referred to as a **nonelectrolyte.** If the solution conducts weakly, the solute is called a **weak electrolyte;** strong conductors are called **strong electrolytes.** Table 7-2 lists some of the different types of electrolytes.

When an aqueous solution does conduct a current, it is because ions are dissolved in the water. The study of ions in solution and the whole area of the action of ions in an electric field is part of the field of electrochemistry. See Chapter 16.

7-8 Raoult's Law

When a compound that is a nonvolatile nonelectrolyte is dissolved in a solvent, one finds that the vapor pressure of the solvent is lower than that of the pure solvent at the same temperature. It is a very interesting and useful fact that in dilute solutions

Table 7-2 Some Strong Electrolytes, Weak Electrolytes, and Nonelectrolytes

Species	*Electrical behavior*
HCl	Strong
NaCl	Strong
KBr	Strong
BaI_2	Strong
$Cu(NO_3)_2$	Strong
Acetic acid, HOAC	Weak
HF	Weak
NH_3	Weak
Sugar, $C_{12}H_{22}O_{11}$	Nonelectrolyte
Urea, $CO(NH_2)_2$	Nonelectrolyte
Methyl alcohol, CH_3OH	Nonelectrolyte

the amount of this lowering of the vapor pressure depends only on the number of solute molecules present (per liter of solvent) and does not depend on the nature of the solute. That is, it makes no difference if in a liter of water is dissolved 0.1 mole of glucose or 0.1 mole of aspirin or 0.1 mole of LSD. The vapor pressure of the water will be lowered by the same amount in all three cases. The rule also applies to electrolytes, but there we must take into account the *number* of ions present. Thus, 0.05 mole of NaCl dissolved in a liter of water will produce 0.05 mole of Na^+ and 0.05 mole of Cl^-; so in this case, too, there is 0.1 mole of total particles, and the vapor pressure of the water will be lowered by the same amount as in the three solutions mentioned previously.

We also find that when more solute is added to the solvent, the vapor pressure is lowered in proportion to the *amount* of solute added. Qualitatively, the reason for this is easy to understand. The vapor pressure of a liquid depends on its volatility. Some of the molecules at the surface of the liquid have enough energy (that is, they are moving fast enough), and are traveling in the right direction, to escape completely from the other molecules which make up the liquid and to enter the gas phase. However, if a solute is added, the number of molecules of solvent at the surface must be diminished because they are replaced by solute particles. If 10% of the molecules in the entire solution are solute molecules, then 10% of the molecules at the surface are also solute molecules, and 90% are molecules of solvent (Figure 7-5).

If the solute is not volatile, it is clear that (1) the vapor pressure of the pure solvent will be diminished by about 10%, and (2) as long as the solute is nonvolatile, it does not matter what it is, only how many particles there are. A quantitative statement of these facts is called **Raoult's law.** Consider a solution composed of a solvent A and a nonvolatile solute B. If

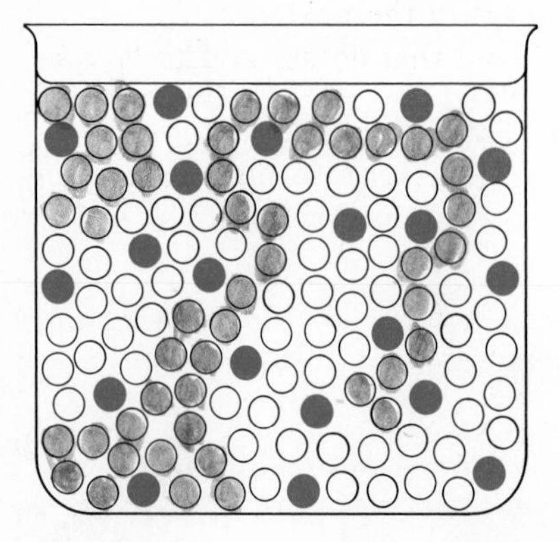

Figure 7-5 In a solution in which 10% of the molecules are a nonvolatile solute, only 90% of the molecules at the surface can vaporize.

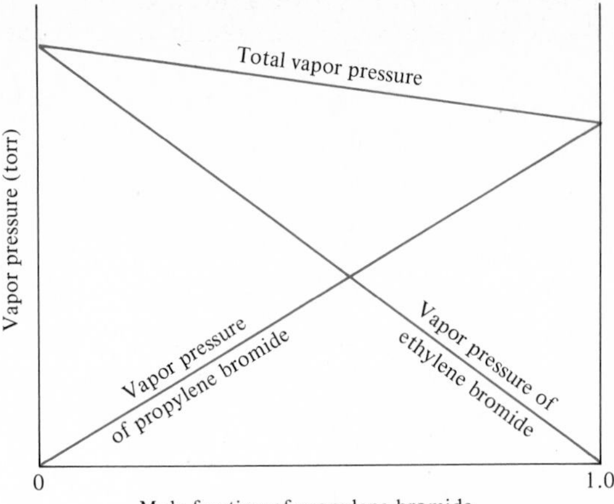

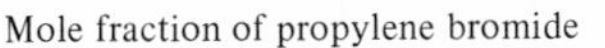

Figure 7-6 The variation of vapor pressure of a solution containing only propylene bromide and ethylene bromide at 85°C. The solutions follow Raoult's law.

P_A = vapor pressure of component A above the solution
χ_A = mole fraction of component A in solution
P_A^0 = vapor pressure of pure component A

Raoult's law states that

$$P_A = \chi_A P_A^0$$

This equation may be manipulated to yield the following:

$$P_A^0 - P_A = P_A^0 - \chi_A P_A^0$$
$$= (1 - \chi_A) P_A^0$$
$$P_A^0 - P_A = \chi_B P_A^0$$

(remember that in a two-component solution, $\chi_A + \chi_B = 1$). These equations indicate that the vapor-pressure lowering of component A is directly proportional to the mole fraction of component B. When the solute B is also volatile, Raoult's law holds for both solute and solvent, and we can write

$$P_A = \chi_A P_A^0$$
$$P_B = \chi_B P_B^0$$

A typical graph showing Raoult's law for two volatile components is shown in Figure 7-6.

The fact that the vapor pressure of a solvent is lowered by the addition of a nonvolatile solute has important consequences. Among other things, it causes an elevation of the boiling point and a lowering of the freezing point of the solution and gives rise to the phenomenon of osmotic pressure. Just how it does these things will be explained when we discuss colligative properties in the next section.

Not all solutions strictly obey Raoult's law. Those that do are said to be *ideal* solutions.

Calculations using Raoult's law are generally simple. In order to calculate the lowering of the vapor pressure one must be given the vapor pressure of the pure material.

Example: The vapor pressure of pure water at 25°C is 23.76 torr. What is the vapor pressure of 100.0 g of water to which 100.0 g of $C_6H_{12}O_6$ (glucose) has been added?

Answer: Using Raoult's law,

$$P_{H_2O} = \chi_{H_2O} P_{H_2O}^0$$

$P^0_{H_2O} = 23.76$ torr. It is necessary to calculate the mole fraction of H_2O. The molecular weight of $C_6H_{12}O_6$ is 180.2; of H_2O, is 18.0.

$$\text{moles of } H_2O = \frac{100.0 \text{ g } H_2O}{18.0 \text{ g/mole}} = 5.55 \text{ moles}$$

$$\text{moles of glucose} = \frac{100.0 \text{ g}}{180.2 \text{ g/mole}} = 0.555 \text{ mole}$$

$$\chi_A(H_2O) = \frac{5.55}{5.55 + 0.56} = 0.907$$

$$P_A = \chi_A P^0_A = 0.907 \times 23.76 = 21.4 \text{ torr}$$

If both components of a two-component solution are volatile, the total vapor pressure P_T is given by $P_T = P_A + P_B$, so we have

$$P_T = P_A + P_B = \chi_A P^0_A + \chi_B P^0_B$$

(since $P_A = \chi_A P^0_A$ and $P_B = \chi_B P^0_B$). The mole fractions of material in solution we labeled χ_A and χ_B. The mole fractions of material in the vapor, which we shall call χ^v_A and χ^v_B, are simply calculated by using Dalton's law of partial pressures (Section 4-9).

$$\chi^v_A = \frac{P_A}{P_T}, \qquad \chi^v_B = \frac{P_B}{P_T}$$

Summarizing, for a two-component system A and B,

$$P_A = \chi_A P^0_A$$
$$P_B = \chi_B P^0_B$$

The method for the calculation of χ_A and χ_B was presented in Section 7-3.

$$P_T = P_A + P_B$$

$$\chi^v_A = \frac{P_A}{P_T} \qquad \chi^v_B = \frac{P_B}{P_T}$$

Note that in general χ_A is not equal to χ^v_A and χ_B is not in general equal to χ^v_B. That is, the mole fractions of A and B in the vapor are generally not equal to the mole fractions of A and B in solution. They are only equal when $P^0_A = P^0_B$.

7-9 Colligative Properties. Freezing-Point Depression and Boiling-Point Elevation

Colligative properties of solutions are properties that depend solely on the *number of particles* of solute dissolved in the solvent and have nothing to do with the chemical behavior of the solute. We have already met one colligative property: the lowering of the vapor pressure. According to Roault's law, 0.1 mole of any solute particles, regardless of the chemical nature of the solute, lowers the vapor pressure of 1000 g of a given solvent by a constant amount. This amount is not the same for all solvents, but once we have chosen a solvent, we can know that the vapor pressure of a given quantity of the solvent will be lowered by the same amount no matter what solute is dissolved in it, provided that the number of moles of solute is the same. For example, at 30°C, 0.1 mole of any solute lowers the vapor pressure of 1000 g of water by 0.057 torr and of 1000 g of carbon tetrachloride by 2.2 torr.

However, as we saw in Section 7-8, we must be careful that we know what we mean by the number of moles of solute particles. Thus, 0.1 mole of sugar, when dissolved in water, really does consist of 0.1 mole of sugar particles; but 0.1 mole of NaCl, dissolved in water, results in *0.2 mole* of particles, because each "NaCl unit" is actually one Na^+ ion and one Cl^- ion. Similarly, 0.1 mole of $BaCl_2$ gives 0.3 mole of particles, and 0.1 mole of $La_2(SO_4)_3$ gives 0.5 mole of particles. Just as some substances "dissociate" in certain solvents, by giving rise to ions or to other kinds of dissociated particles, there are other substances that *associate* in solution, and here the number of moles of solute particles will be less than we might otherwise expect them to be. However, Raoult's law and other colligative properties depend only on the number of particles present; how they got that way is irrelevant. It is evident that if we know how much to expect the vapor pressure of a given solvent to be reduced, and we know the formula of a given solute, we can use Raoult's law (and similarly for the other colligative properties) to help us find out the degree of dissociation or association of a given solute in a given solvent.

We shall discuss three additional colligative properties of solutions: boiling-point elevation, freezing-point lowering, and osmotic pressure. We have learned (Section 6-3) that the vapor pressure of a solvent increases with increasing temperature, and that the vapor pressure is lowered by the addition of a nonvolatile solute. Figure 7-7 puts these two facts together: the vapor pressure of water, and the vapor pressure of a solution made up of a nonvolatile solute dissolved in water, are plotted against the temperature. Looking at this figure, we see that the vapor pressure of the pure water is higher than that of the solution at all temperatures, just as we would expect. We also remember that the boiling point of water at 760 torr (the *normal* boiling point, Section 6-3) is 100°C. The curve representing the vapor pressure of pure water thus

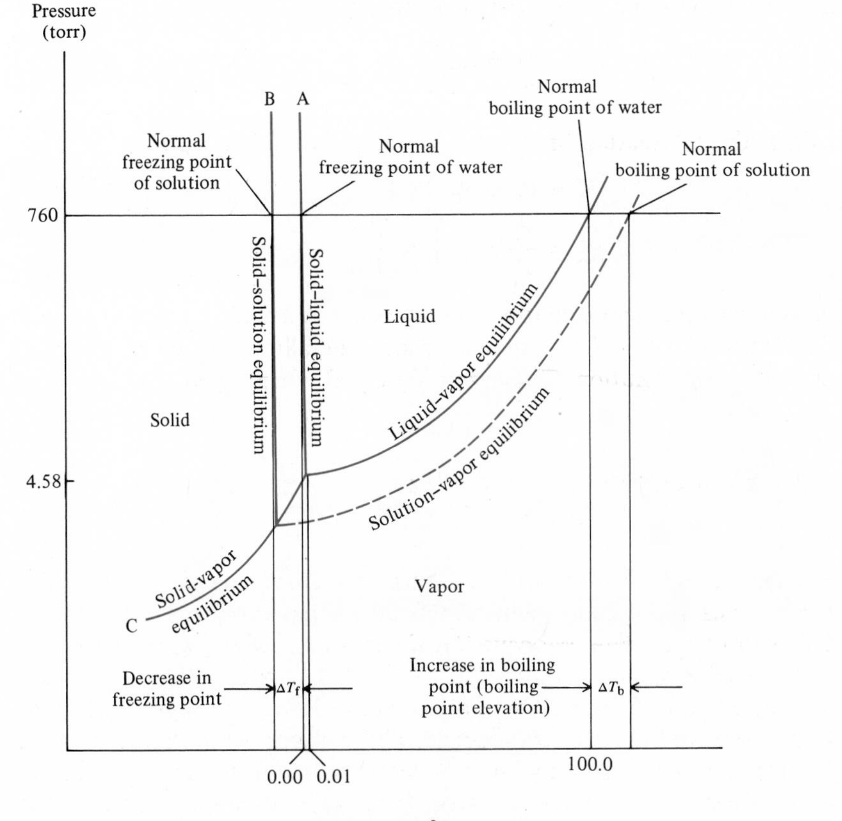

Figure 7-7 The lowering of the vapor pressure of water upon addition of a nonvolatile solute. This gives rise to the elevation of the boiling point and depression of the freezing point as shown.

crosses the line marked 760 torr at 100°C. However, when this temperature is reached, the other curve, representing the solution, must be *below* 760 torr. Consequently, at 100°C, the solution is not yet at the boiling point. If we look at Figure 7-7, we see that the temperature must be *higher* than 100°C for the solution to boil. Thus, *the addition of a nonvolatile solute raises the boiling point of the solvent,* or, putting it another way, the boiling point of the solution is higher than that of the pure solvent. This is true not only for water but for all other solvents as well. Of course, as we have seen, the raising of the boiling point is actually the result of the lowering of the vapor pressure.

We may now look at the left side of Figure 7-7 to see what happens to the freezing point of the solution. We have defined the normal freezing point of a liquid as that temperature at which the vapor pressure of the solid and the vapor pressure of the liquid are equal (see Section 6-3) at normal atmospheric pressure. The line labeled A is the line that shows equal liquid and solid vapor pressures at different temperatures.

For pure ice to melt at normal atmospheric pressure, it must be heated to 0°C, at which temperature the vapor pressure of the ice is equal to that of the liquid at 1 atm. But for the *solution,* the vapor pressure of the liquid at 0°C is less than that of pure water (Figure 7-7), and the solid–solution equilibrium (line B) reaches atmospheric pressure at a lower temperature than that of pure ice. Hence, the freezing point of the solution is less than that of the pure liquid. The line showing the vapor pressure of ice (line C) is unaffected because in most cases the solid that freezes from solution is the pure solvent.

Note that in its effect on the freezing point of the solvent, it does not matter whether the solute is volatile or nonvolatile, since evaporation is not taking place here. Therefore, all solutes, whether volatile or not, have the same effect in lowering the freezing point of a solvent.

It is not difficult to calculate, for a given solvent, how much the freezing point will be lowered, or how much the boiling point will be raised, by the addition of a solute (remember that for boiling point, but not for freezing point, it must be a nonvolatile solute). Because only the number of solute particles matter, not the nature of these particles, every solvent has a **freezing-point-depression constant** and a **boiling-point-elevation constant,** both of which are independent of the nature of the solute. It is necessary to know two things: (1) the *molality* of the solution, and (2) the appropriate constant for that solvent.

The two equations used for these calculations are quite similar. They are

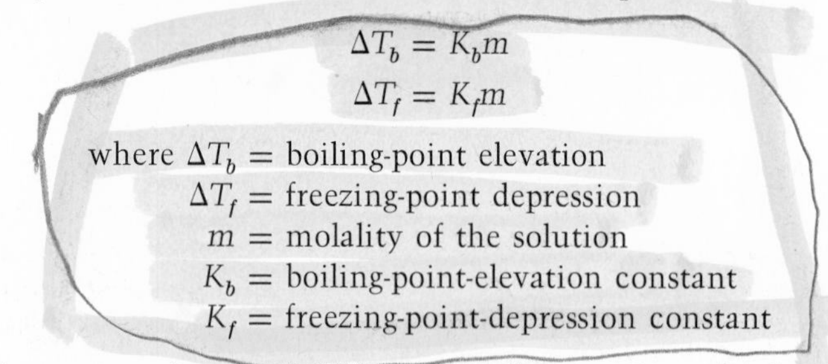

$$\Delta T_b = K_b m$$

$$\Delta T_f = K_f m$$

where ΔT_b = boiling-point elevation
ΔT_f = freezing-point depression
m = molality of the solution
K_b = boiling-point-elevation constant
K_f = freezing-point-depression constant

Table 7-3 lists some values of these constants for a variety of solvents.

Example: Calculate the freezing-point depression and boiling-point elevation of a solution of 1.00 g of urea (mol wt = 60.06) in 50.0 g of water at atmospheric pressure.

Answer: From Table 7-3,

K_b for H_2O = 0.512

K_f for H_2O = 1.86

Table 7-3 Boiling-Point-Elevation Constants and Freezing-Point-Depression Constants for a Variety of Solvents

Solvent	*Freezing point (°C)*	*Freezing-point constant, K_f (°C/mole)*	*Boiling point (°C)*	*Boiling-point constant, K_b (°C/mole)*
Water	0	1.85	100	0.512
Acetic acid	16.6	3.90	118	3.07
Benzene	5.48	4.90	80.1	2.53
Camphor	178	40.0	208	5.98
Cyclohexane	6.5	20.5	81.4	2.79
Nitrobenzene	5.7	7.00	211	5.24
Chloroform	—	—	61.2	3.63
Naphthalene	80	6.8	—	—

To use the equations previously given for these calculations, we must calculate the molality of the solutions:

$$1.00 \text{ g urea} \times \frac{1.00 \text{ mole urea}}{60.06 \text{ g urea}} = 0.0167 \text{ mole urea}$$

Hence, we have 0.0167 mole of urea in 50 g of water, so

$$\frac{x \text{ moles urea}}{1000 \text{ g } H_2O} = \frac{0.0167 \text{ mole urea}}{50 \text{ g } H_2O}$$

$$x \text{ moles urea} = \frac{0.0167}{50} \times 1000 = 0.334 \text{ mole urea}$$

Therefore, $m = 0.334$ (we have 0.334 mole of urea per 1000 g of water).

$$\Delta T_b = K_b m$$
$$= (0.512)(0.334)$$
$$\Delta T_b = 0.171$$

The *elevation* of the boiling point is 0.171°C. Therefore, the solution boils at 100.171°C.

$$\Delta T_f = K_f m$$
$$= (1.86)(0.334) = 0.621°\text{C}$$

The freezing-point depression is 0.621°C, so the solution freezes at −0.621°C.

It is possible to use the freezing-point depression and boiling-point elevation to calculate the molecular weight of solutes. We can do this because we know that for 1000 g of water, 1 mole of any solute will elevate the boiling point 0.512°C and depress the freezing point 1.86°C. If we can calculate the number of *grams* that will do either of these things, this will be 1 mole of substance or its molecular weight in grams. The following example will illustrate this.

Example: The boiling point of a solution made by dissolving 12.0 g of glucose in 100 g of water is 100.340°C. Calculate the molecular weight of glucose.

Answer: 12.0 g of glucose in 100 g of H_2O is in the same proportion as 120 g of glucose in 1000 g of water. Now 1 mole of solute in 1000 g of water will increase the boiling point 0.512°C. Therefore,

$$\frac{120 \text{ g glucose}}{0.340} = \frac{x \text{ g glucose}}{0.512}$$

$$x = 120 \times \frac{0.512}{0.340} = 181 \text{ g}$$

Therefore, 181 g of glucose in 1000 g of water will elevate the boiling point 0.512°C. Hence, this is 1 mole of glucose and its molecular weight. We can also calculate the number of moles directly from the formula $\Delta T_b = K_b m$

$$m = \frac{\Delta T_b}{K_b} = \frac{0.340°\text{C}}{0.512°\text{C/mole}} = 0.664 \text{ mole glucose}$$

Since this amount of glucose weighs 120 g, we get the molecular weight by

$$\frac{120 \text{ g}}{0.664 \text{ mole}} = 181 \text{ g/mole}$$

In practice, freezing-point depressions are more often used to determine molecular weights than are boiling-point elevations because (1) they are easier to measure accurately, (2) they are much less dependent on pressure, and (3) the question of volatility of solute does not enter.

As we pointed out earlier, boiling-point elevation and freezing-point depression measurements can be used to measure the degree of association or dissociation of a solute in a solvent, if we know the formula of the solute.

Example: When dissolved in benzene, a compound whose formula is $C_{38}H_{30}$ partially dissociates by the following equation:

$$C_{38}H_{30} \longrightarrow 2C_{19}H_{15}$$

When 25.6 g of $C_{38}H_{30}$ is dissolved in 400 g of benzene, the freezing point is lowered by 0.680°C. What percentage of the $C_{38}H_{30}$ molecules have dissociated?

Answer: First we convert to 1000 g of benzene:

$$\frac{25.6 \text{ g solute}}{400 \text{ g benzene}} = \frac{x}{1000 \text{ g benzene}}$$

$$x = 64.0 \text{ g of solute per 1000 g benzene}$$

Next, we find out how many moles of solute are present in the solution, by remembering that 1 mole depresses the freezing point by 4.90°C (Table 7-3).

$$\frac{0.680°\text{C}}{4.90°\text{C/mole}} = 0.139 \text{ mole}$$

If the $C_{38}H_{30}$ were entirely undissociated, there would have been

$$\frac{64.0 \text{ g}}{486.655 \text{ g/mole}} = 0.132 \text{ mole of } C_{38}H_{30}$$

present because the molecular weight of $C_{38}H_{30}$ is 486.655. Actually, there is less than this because some of it has dissociated. Let us call the number of moles of $C_{38}H_{30}$ that has dissociated y. Therefore, the amount that remains is $0.132 - y$, and the number of moles of $C_{19}H_{15}$ that is formed is $2y$. But the freezing-point depression has told us that the *total* number of moles of both species in the solution is 0.139 mole. So we have

$$(0.132 - y) + 2y = 0.139$$

Solving for y, we obtain

$$y = 0.007 \text{ mole}$$

Thus, the number of moles of $C_{38}H_{30}$ in the solution is 0.132 − 0.007 = 0.125 mole and the percentage of dissociation is

$$\frac{0.007}{0.132} \times 100 = 5\%$$

(Note that in this calculation we have lost significant figures.)

The fact that freezing points of solutions are lower than those of pure solvents has important applications. The most common example is the use of antifreeze in automobile engines. The water used to cool automobile engines would freeze when the temperature falls below 0°C (32°F), as it often does in winter in many parts of the world. Since ice occupies a greater volume than an equal weight of water, the pressure of ice forming in a confined space is often great enough to crack the metal parts of the engine. In order to prevent the water from freezing, a solute is added to the water, lowering the freezing point. The most common solute used as an antifreeze is ethylene glycol (Section 20-6). Enough must be added so that the freezing point of the water is lowered below the lowest temperature that may reasonably occur in the area in which the automobile will be used. Ethylene glycol is a "permanent" antifreeze because its boiling point (198°C) is considerably higher than that of water, and so when liquid evaporates from a hot automobile engine, that liquid will be water and not ethylene glycol.

7-10 Osmotic Pressure

The final colligative property that we will discuss is osmotic pressure. For there to be an osmotic pressure, there must be a system composed of (1) a solution consisting of a solute dissolved in a particular solvent separated by (2) a semipermeable membrane from (3) the pure solvent (Figure 7-8). A semipermeable membrane is a material with extremely small holes in it which allows solvent molecules but not solute molecules to pass through. Of course, these holes are much too tiny for us to see. Cellophane is a typical semipermeable membrane used in laboratory experiments. When the solvent is water, what we observe is that the water molecules from the pure solvent side pass through the membrane into the solution. This happens for the following reason. On one side of the membrane, we have pure solvent; on the

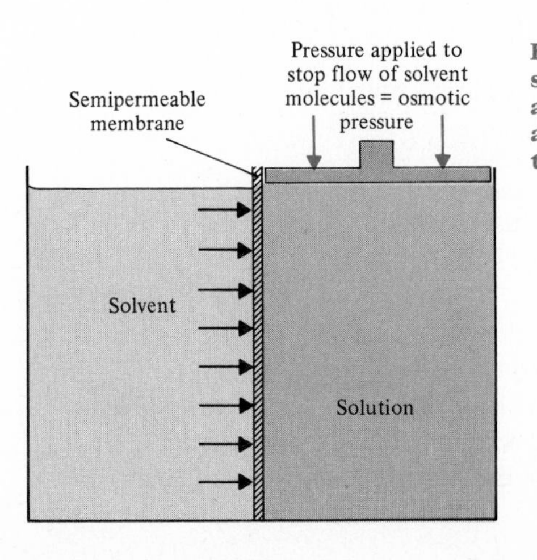

Figure 7-8 The flow of solvent molecules through a semipermeable membrane into the solution. The amount of external pressure necessary to counteract this flow is equal to the osmotic pressure of the system.

other side, a solution; but only solvent molecules, not solute molecules, can pass through. With respect to the solvent, the barrier is not there because solvent molecules go through easily. Therefore, there is a flow of solvent molecules *from the solvent compartment to the solution compartment in an effort to dilute the solution compartment* until both sides have the same concentration. Of course, this goal is impossible because solute molecules are in the solution compartment and have no way to get to the solvent compartment. Nevertheless, solvent molecules keep trying, and by passing from one compartment to the other, they lower the height of the solvent side and raise the height of the solution side. Of course, this process cannot continue indefinitely, because there is a countereffect, that of gravity (a liquid seeks its own level). At the end, the result of the two effects is that the level of the liquid in the solution compartment is higher than that on the solvent side. This pressure which the solvent has exerted is called **osmotic pressure.** Osmotic pressure (which has the symbol Π) can be measured in the following way: An external pressure is applied to the solution compartment (see Figure 7-8) until the two levels are equal. The amount of external pressure necessary to do this is equal to the osmotic pressure.

Osmotic pressure can develop in other ways. For example, if we enclose a solution in a semipermeable membrane and then place the solution into a container of the solvent, the membrane will swell.

As an example we find that if red blood corpuscles are placed in a beaker of pure water, they are observed to swell and finally burst. This happens because the cell membranes of the red corpuscles are permeable to water but not to the other materials dissolved in the red cells.

It was found experimentally by Jacobus H. van't Hoff (1852–1911) in 1887 that the osmotic pressure of a dilute solution satisfies the equation

$$\Pi V = nRT$$

where Π = osmotic pressure of the solution
V = volume of the total solution in liters
n = number of moles of solute in solution
R = gas constant (the same one we saw in Chapter 4)
T = absolute temperature

The form of this equation is remarkably similar to the ideal gas law equation (see Section 4-7). Consider a problem similar to the two problems we did previously (p. 188). We wish to calculate the osmotic pressure of 1.00 g of urea in 50 ml of H_2O. Our previous calculations show that 1.00 g of urea is 0.0167 mole. Using the above, we can now calculate Π at, say, 25°C:

$$\Pi = \frac{nRT}{V}$$

$$= \frac{(0.0167 \text{ mole})(0.08206 \text{ liter} \cdot \text{atm/mole} \cdot {}^\circ\text{K})(298{}^\circ\text{K})}{0.050 \text{ liter}}$$

$$= 8.2 \text{ atm}$$

This value of 8.2 atm is equivalent to a column of water approximately 280 ft high or a column of mercury 6200 mm high. It is plain that this is a huge value for such a simple solution. Compare this result with the boiling-point elevation of 0.171°C and freezing-point depression of 0.621°C calculated previously for the same solution. It is this rather large effect found in osmotic pressure compared to boiling-point elevation and freezing-point depression which gives us the clue to the major use of osmotic pressure. That is, osmotic pressure is an important tool for determining the molecular weight of polymers. (A polymer is a giant molecule with an extremely high molecular weight compared to the relatively small molecules that we have been studying.)

Example: An important protein has recently been synthesized. Only a few grams of the material are available and by certain other experiments it has been determined that the molecular weight of the material is approximately 10,000. Calculate the osmotic pressure, freezing-point depression, and boiling-point elevation for 1 g of this protein dissolved in 100 g of water at 25°C. Which method would you use to check the molecular weight? (Assume that a solution of 1 g of protein in 100 g of H_2O has a volume of approximately 100 ml.)

Answer: The number of moles of the protein is approximately

$$\frac{1\ \text{g}}{10{,}000\ \text{g/mole}} = 10^{-4}\ \text{mole}$$

$$\Pi = \frac{nRT}{V} = \frac{(10^{-4}\ \text{mole})(0.08206\ \text{liter}\cdot\text{atm/mole}\cdot{}^\circ\text{K})(298^\circ\text{K})}{0.100\ \text{liter}}$$

$$\approx 0.025\ \text{atm} = 19\ \text{mm of Hg (19 torr)}$$

$$\Delta T_b = K_b m = (0.512)(10^{-3}) = 0.512 \times 10^{-3}\ {}^\circ\text{C} = 0.000512^\circ\text{C}$$

$$\Delta T_f = K_f m = (1.86)(10^{-3}) = 1.86 \times 10^{-3}\ {}^\circ\text{C} = 0.00186^\circ\text{C}$$

Certainly, the osmotic pressure could be measured much more accurately than such tiny temperature differences.

Problems

7-1 Define each of the following terms. **(a)** solvent **(b)** solute **(c)** saturated **(d)** unsaturated **(e)** supersaturated **(f)** seeding **(g)** concentrated **(h)** dilute **(i)** concentration **(j)** proof **(k)** mole fraction **(l)** molarity **(m)** equivalent weight **(n)** equivalent **(o)** normality **(p)** molality **(q)** solubility **(r)** aqueous **(s)** dielectric constant **(t)** solvation **(u)** electrolyte **(v)** colligative property **(w)** freezing-point-depression constant **(x)** dissociation **(y)** association **(z)** osmotic pressure

7-2 What is meant by each of the following symbols? **(a)** χ **(b)** M **(c)** N **(d)** m **(e)** ε **(f)** K_b **(g)** K_f **(h)** Π

7-3 Give an example, other than those given in the book, of each of the following solutions. **(a)** a liquid in a liquid **(b)** a solid in a liquid **(c)** a gas in a liquid

7-4 For solutions made up of the following components, tell, where possible, which is the solute and which the solvent (all of these are liquid solutions).
(a) 25 g of $CH_3OH(l)$ and 175 g of $H_2O(l)$ **(b)** 175 g of $CH_3OH(l)$ and 25 g of $H_2O(l)$
(c) 10 g of sucrose(s) and 20 g of $H_2O(l)$ **(d)** 20 g of sucrose(s) and 10 g of $H_2O(l)$
(e) 75 g of AcOH(l) and 75 g of $H_2O(l)$ **(f)** 25 g of $SO_2(g)$ and 15 g of $H_2O(l)$

7-5 How much water must you add to 65.6 g of NaCl to prepare a 7.00% NaCl solution by weight?

7-6 **(a)** A bottle of rum is 120 proof. What is the percent alcohol by volume? **(b)** A bottle of cordial has 23.5% alcohol by volume. What is its proof? **(c)** Vodka is a mixture of pure alcohol and water. If a vodka is made up by adding 150 ml of water to 125 ml of pure alcohol, what is its proof?

7-7 An analysis of a sample of mixed amino acids showed that it contained 0.300 g of glycine, 0.018 g of alanine, and 0.330 g of phenylalanine. If the approximate molecular weights of each are glycine, 75 g/mole; alanine, 89 g/mole; and phenylalanine, 165 g/mole, calculate the mole fraction of each in the mixture.

7-8 In a mixture of liquid hydrocarbons, the mole fractions are as follows: hexane (C_6H_{14}), 0.270;

octane (C_8H_{18}), 0.420; the rest is benzene (C_6H_6). How many grams of each component are present in 475 g of the mixture?

7-9 How many grams of solute are present in each of the following aqueous solutions? **(a)** 783 ml of 0.35 *M* KOH **(b)** 21.6 ml of 1.27 *M* $Ba(NO_3)_2$ **(c)** 31.8 liters of 3.1×10^{-3} *M* $Ca_3(PO_4)_2$ **(d)** 3.5 ml of 6.1×10^{-2} *M* $C_{12}H_{22}O_{11}$

7-10 How many millimoles and milligrams of solute are present in each of the following aqueous solutions **(a)** 3.70 ml of 1.100 *M* H_2S **(b)** 60.7 ml of 0.0362 *M* HNO_3 **(c)** 4.65 liters of 1.38 *M* $Ba(OH)_2$ **(d)** 0.41 ml of 1.1×10^{-3} *M* $Al(OH)_3$

7-11 Calculate the number of milliequivalents of solute for each solution in Problem 7-10.

7-12 Calculate the molarity of each of the following aqueous solutions.
(a) 7.12 mg of H_3PO_4 per milliliter of solution.
(b) A concentrated HCl solution that is 37.0% HCl and has a density of 1.20 g/ml.
(c) 74.0 ml of pure H_2SO_4 (density 1.84 g/ml) diluted with water to a final volume of 148 ml.

7-13 Calculate the normality of the solutions in Problem 7-12.

7-14 You have 2.00 liters of a 0.750 *M* solution of HCl. How many milliliters of this solution would you need to prepare 250 ml of a 0.150 *M* solution?

7-15 Calculate the equivalent weight of each of the following acids or bases. **(a)** HNO_3 **(b)** oxalic acid, $H_2C_2O_4$ **(c)** $Al(OH)_3$ **(d)** citric acid, $H_3(C_6H_5O_7)$ **(e)** $Fe(OH)_2$ **(f)** $Fe(OH)_3$ **(g)** LiOH

7-16 How many equivalents are present in each of the following pure compounds? **(a)** 27.5 g of H_2CO_3 **(b)** 16.71 g of HCl **(c)** 1.52 g of $H_2Cr_2O_7$ **(d)** 44.5 g of $Mn(OH)_2$ **(e)** 1.7×10^{-2} g of $Y(OH)_3$ **(f)** 14.6 kg of $H_4As_2O_7$

7-17 How many grams of solute are present in each of the following aqueous solutions? **(a)** 575 ml of 2.0 *N* HBr **(b)** 11.7 ml of 0.135 *N* H_2MnO_4 **(c)** 8.42 ml of 3.75 *N* $Sr(OH)_2$ **(d)** 42.61 liters of 1.52 *N* H_3AsO_4

7-18 Calculate the normality of each of the following solutions. **(a)** 85.7 g of HNO_3 in 4.377 liters of solution **(b)** 9.755 g of $Ba(OH)_2$ in 133 ml of solution **(c)** 45 g of H_3PO_4 in 635 ml of solution **(d)** 2.50 moles of H_2SO_3 in 8755 ml of solution **(e)** 0.1015 g of $Al(OH)_3$ in 1.000 liter of solution **(f)** 71 g of NaOH in 927 ml of solution **(g)** 6.6 mg of H_2CO_3/ml of solution

7-19 Calculate the number of moles, the number of equivalents, and the number of grams of solute in each of the following aqueous solutions. **(a)** 45.5 ml of 2.5 *M* $Ca(OH)_2$ **(b)** 73.3 ml of 1.2 *N* HNO_3 **(c)** 105 ml of 0.0100 *M* H_3PO_4 **(d)** 50.5 ml of 4.0 *N* $Al(OH)_3$

7-20 Fill in the following table. The volume of each of the solutions is 250 ml.

Solute	*Grams*	*Moles*	*Molarity*	*Normality*
KCl	10			–×–
NH_4Cl		3.2		–×–
HCl			0.250	
NaOH				0.110
$Ca(OH)_2$		1.5		
NH_3			6.0	–×–
H_3PO_4	43.5			
H_2SO_4	0.980			
$NaHCO_3$		1.25		–×–
$Al(OH)_3$			2.5	
LiOH				0.0200
HNO_3				0.0100

7-21 Tell exactly how you would make up the following aqueous solutions **(a)** 2.50 liters of 0.25 *N* H_2SO_4 (your only source of H_2SO_4 is a bottle labeled 2.00 *N* H_2SO_4) **(b)** 50.0 ml of 1.50 *N* $Ba(OH)_2$ [your only source of $Ba(OH)_2$ is solid $Ba(OH)_2$] **(c)** 275 ml of 0.85 *N* H_3PO_4 (your only source of H_3PO_4 is a bottle labeled 14.6 *M* H_3PO_4) **(d)** 830 ml of 3.75 *N* H_2SO_4 (your only source of H_2SO_4 is a bottle labeled 96.3% H_2SO_4, density = 1.84 g/ml)

7-22 What volume of a 2.50 *M* solution of H_3PO_4 would be required to neutralize completely 2.50 liters of 0.350 *M* KOH?

7-23 What is the normality of a $Mg(OH)_2$ solution if 18.7 ml of it are completely neutralized by 29.7 ml of an 0.434 *N* solution of HCl?

7-24 5.22 g of sucrose ($C_{12}H_{22}O_{11}$) is dissolved in 120 g of water. Assuming that the volume of the solution is 120.0 ml, calculate each of the following. **(a)** the weight percent of sucrose **(b)** the mole fraction of sucrose **(c)** the molarity of the solution **(d)** the molality of the solution

7-25 What is the molality of each of the following solutions? **(a)** 27.5 g of NaCl dissolved in 850.0 ml of water **(b)** 152 g of ethyl alcohol, C_2H_5OH, dissolved in 2.75 liters of water **(c)** 1.77 g of acetamide, C_2H_5NO, dissolved in 475 g of benzene, C_6H_6 **(d)** 35.5 g of NaCl dissolved in 1.15 liters of liquid ammonia (density of NH_3, 0.771 g/ml) **(e)** 27.5 g of benzophenone, $C_{13}H_{10}O$, dissolved in 935 ml of carbon tetrachloride, CCl_4 (density of CCl_4, 1.594 g/ml) **(f)** 3.25 g of succinic acid, $H_2(C_4H_4O_4)$, dissolved in 748 ml of diethyl ether, $C_4H_{10}O$ (density of $C_4H_{10}O$, 0.714 g/ml)

7-26 A bottle of concentrated sulfuric acid, H_2SO_4, has the following information printed on the label: "Assay" 96.3%; density, 1.84 g/ml. The assay is the percentage of H_2SO_4 by mass (the rest is water). Calculate each of the following. **(a)** the number of moles of water and sulfuric acid in 100.0 ml of the solution **(b)** the number of moles of water and sulfuric acid in 100.0 g of the solution **(c)** the mole fraction of H_2SO_4 **(d)** the molarity **(e)** the molality **(f)** the normality

7-27 State exactly how you would prepare the following solutions. **(a)** 100.0 ml of a 0.00100 *M* solution of HCl starting with 5.00 *M* HCl **(b)** 1.0 liter of a 7.2 *M* solution of KOH, starting with pure solid KOH pellets **(c)** 25.0 ml of a 1.50 *M* H_2SO_4 solution, starting with a 7.00 *M* H_2SO_4 solution **(d)** an 0.0100 *M* solution of KCl, starting with 100.0 g of KCl and using enough water to dissolve all the KCl **(e)** 1.00 liter of a 4.5% (by weight) solution of NaCl in water (assume the density of this solution is 1.00 g/ml) **(f)** an 0.250 molal solution of glucose $C_6H_{12}O_6$ in water, starting with 25.0 g of pure glucose

7-28 Why does a freshly opened bottle of beer or soda foam up much more when opened at room temperature than it does when opened at refrigerator temperature?

7-29 Predict whether each of the following substances is more likely to be soluble in water or in benzene. **(a)** NaCl **(b)** Ar **(c)** carbon tetrachloride, CCl_4 **(d)** I_2 **(e)** NaOH **(f)** gasoline **(g)** ethyl alcohol **(h)** methane, CH_4 **(i)** ammonia, NH_3 **(j)** cane sugar, $C_{12}H_{22}O_{11}$ **(k)** $Cu(NO_3)_2$ **(l)** urea

7-30 The vapor pressure of pure toluene, C_7H_8, an organic solvent, is 100.0 torr at 52°C. What is the vapor pressure of 75.0 g of toluene in which 35.0 g of naphthalene $C_{10}H_8$, a nonvolatile solute, has been dissolved?

7-31 At 50°C the vapor pressure of pure benzene, C_6H_6, is 268 torr and that of ethylene chloride, $C_2H_4Cl_2$, is 236 torr. These liquids behave ideally when mixed. If 100 g of benzene is mixed with 200 g of ethylene chloride, calculate each of the following. **(a)** the mole fraction of each in the liquid **(b)** the vapor pressure of each in the mixture **(c)** the total vapor pressure of the mixture **(d)** the mole fraction of benzene and ethylene chloride in the vapor

7-32 The vapor pressure of pure water at 25°C is 23.756 torr. When 12.0 g of a nonvolatile substance was dissolved in 80.000 g of water, the vapor pressure was found to be 23.332 torr. Calculate the molecular weight of the unknown solute.

7-33 The boiling point of a solution made by dissolving 8.9 g of benzophenone, a nonvolatile solute, in 16.2 g of acetic acid is 127°C. Find the molecular weight of benzophenone.

7-34 A solution contains 0.170 g of an unknown solute dissolved in 85.0 g of water. It freezes at −0.050°C. Find the molecular weight of the solute.

7-35 An antifreeze mixture used in car radiators contains ethylene glycol $C_2H_6O_2$ and water. The density of pure ethylene glycol is 1.113 g/ml at 20°C. Calculate the number of milliliters of ethylene glycol that will lower the freezing point of 500 g of water to −20.0°C, and the molality of the resulting solution.

7-36 When 7.50 g of an unknown substance is dissolved in 100 g of water, its freezing point is lowered by 1.395°C. The unknown substance completely dissociates into three ions in water while it remains undissociated in benzene. If 7.50 g of the unknown substance is added to 100.0 g of benzene, how many degrees will the freezing point be lowered?

7-37 If an automobile radiator contains 2.0 gal of water, how many quarts of ethylene glycol ($C_2H_6O_2$) antifreeze must be added to protect the engine from freezing down to temperatures as low as −25°C (density of ethylene glycol = 1.113 g/ml).

7-38 The freezing point of a solution of glycerol, $C_3H_8O_3$, in water is −2.55°C. What is the molality?

7-39 When 300.0 mg of potassium ferricyanide, $K_3Fe(CN)_6$ are dissolved in 10.0 g of water, the freezing point is lowered 0.68°C. Determine the number of ions produced by 1 mole of $K_3Fe(CN)_6$ when it dissolves in water.

7-40 What is the boiling point of a solution made by dissolving 2.77 g of the nonvolatile solute caffeine, $C_8H_{10}N_4O_2$, in 12.8 ml of chloroform (density of chloroform = 1.483 g/ml)?

7-41 Explain how sprinkling with NaCl or $CaCl_2$ helps to melt snow.

7-42 Predict which side, if any, will rise in each case and tell why. The solvent is water.

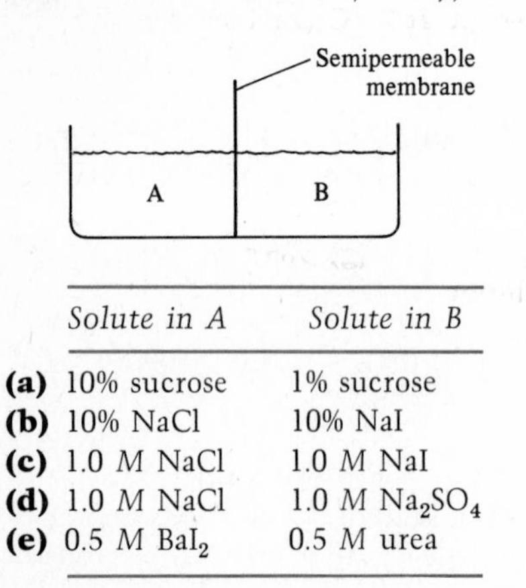

	Solute in A	Solute in B
(a)	10% sucrose	1% sucrose
(b)	10% NaCl	10% NaI
(c)	1.0 *M* NaCl	1.0 *M* NaI
(d)	1.0 *M* NaCl	1.0 *M* Na_2SO_4
(e)	0.5 *M* BaI_2	0.5 *M* urea

7-43 A solution of an unknown polymer contains 1.37 g of polymer dissolved in 75 ml of solution. The osmotic pressure is found to be 53 torr at 25°C. What is the approximate molecular weight of the polymer?

7-44 Cytochrome *c*, a protein, has a molecular weight of 12,400. If 0.100 g of the protein is dissolved in 1.00 ml (1.00 g) of water, calculate **(a)** the expected freezing point of the solution **(b)** the expected boiling point of the solution **(c)** the osmotic pressure of the solution at 25°C in torr and in cm H_2O (1 atm = 34 ft of H_2O) **(d)** What information do we get from this problem concerning methods for determining the molecular weight of proteins (or any very high molecular weight substances)?

7-45 The osmotic pressure of blood at body temperature (37°C) is 7.65 atm. If a solution of glucose, $C_6H_{12}O_6$, in water is to be prepared for intravenous feeding with the same osmotic pressure as blood, what should the concentration of glucose be, in g/liter?

7-46 Explain why osmotic pressure measurements provide a better means for determining molecular weights of high-molecular-weight molecules than freezing-point-depression or boiling-point-elevation measurements.

Additional Problems

7-47 A solution is made by dissolving 245 g of H_2SO_4 in enough water to make a total of 800 ml of solution. Calculate each of the following. **(a)** the number of grams of H_2SO_4 per ml of solution **(b)** the number of moles of H_2SO_4 per ml of solution **(c)** the molarity of the solution **(d)** the volume of solution which will yield 1.5 moles of H_2SO_4 **(e)** the normality of the solution

7-48 A solution of vitamin C, ascorbic acid, $C_6H_8O_6$, contains 1.00 g of the acid in 3.00 ml of water. If the density of water is 1.00 g/ml and volume of water does not change upon addition of ascorbic acid, calculate each of the following. **(a)** the mole fraction of ascorbic acid in the solution **(b)** the weight percent of ascorbic acid in the solution **(c)** the molarity of the solution **(d)** the molality of solution

7-49 If ethylene glycol, $C_2H_6O_2$, and methanol, CH_3OH, sold at the same price per kilogram, which would be the cheaper antifreeze?

7-50 The solute can often be induced to precipitate from a supersaturated solution in a glass container by scratching the side of the container with a glass rod. Explain why this happens.

7-51 One gram (1.00 g) of common table salt, NaCl, along with 1.00 g of household sugar (sucrose, $C_{12}H_{22}O_{11}$) are dissolved in 10.00 g of acetic acid. (Assume that NaCl and sucrose behave the same way in acetic acid as in water.) Calculate each of the following. **(a)** the expected boiling point **(b)** the expected feezing point **(c)** the osmotic pressure at 25°C (Assume that the density of the solution is 1.06 g/ml.)

7-52 Calculate the molarity of each of the following solutions. **(a)** 179 g of CsCl in 1.57 liters of solution **(b)** 0.0110 mole of NH_3 in 111 ml of solution **(c)** 22.22 g of HNO_3 in 2222 ml of solution **(d)** 100.0 g of I_2 in 150.00 ml of carbon tetrachloride solution **(e)** 25.0 g of $Ba(OH)_2$ in 250 ml of solution **(f)** 175 g of H_3PO_4 in 1.050 liters of solution **(g)** 985 g of HCl in 1.100 liters of solution **(h)** 1.015 g of $Al(OH)_3$ in 1.000 liter of solution

7-53 The freezing point of a solution made by dissolving 4.18 g of cocaine in 27.5 g of liquid camphor is 157 °C. Find the molecular weight of cocaine.

7-54 What is the normality of each of the following solutions? **(a)** 3.55 *M* HOAc **(b)** 1.62 *M* LiOH **(c)** 0.57 *M* H_2CO_3 **(d)** 2.10 *M* $Al(OH)_3$ **(e)** 3.0×10^{-3} *M* $Bi(OH)_3$ **(f)** 4.7×10^{-2} *M* H_3PO_4

7-55 Explain with the aid of a diagram why red blood corpuscles swell and burst when they are placed in a beaker of pure water.

7-56 Two unknown liquids, A and B, form ideal solutions. At 30°C the total vapor pressure of a solution containing 2.0 moles of A and 3.0 moles of B is 300 torr. On adding 2.0 more moles of A to the solution the total vapor pressure increased to 400 torr. Calculate the pure vapor pressures of the two unknown liquids.

7-57 Describe how you would prepare each of the following. **(a)** a 2.50 molal and **(b)** a 2.50 molar solution of $Ba(OH)_2$ in water

7-58 When 3.76×10^{-3} g of an unknown compound is dissolved in 0.0503 g of melted naphthalene, the freezing point is lowered from 80.113 to 78.971°C. What is the molecular weight of the solid?

7-59 What are the units of each of the following? **(a)** molarity **(b)** normality **(c)** mole fraction **(d)** equivalent weight

7-60 A 40.0-ml solution is 0.105 *M* in H_2SO_4. **(a)** Calculate the normality of the solution. **(b)** Calculate the number of moles of H_2SO_4 in this sample. **(c)** Calculate the number of equivalents of H_2SO_4 in this sample. **(d)** Calculate the number of grams of H_2SO_4 per milliliter of solution.

7-61 The vapor pressure of pure benzene C_6H_6 is 182 torr at 40°C. What is the vapor pressure at 40°C of a 275-g sample of benzene in which are dissolved both 57.0 g of biphenyl $C_{12}H_{10}$ and 28.5 g of camphor $C_{10}H_{16}O$, both nonvolatile solutes?

7-62 The density of an acetic acid ($HC_2H_3O_2$) solution is 1.070 g/ml. It contains 60.0% acetic acid by weight, the rest being water (consider water to be the solvent). Calculate each of the following-ing. **(a)** the number of grams of each component in 100 ml of solution **(b)** the number of moles of each component in 100 ml of solution **(c)** the mole fraction of each **(d)** the molarity of the solution **(e)** the molality of the solution **(f)** the normality of the solution

7-63 A solution made by dissolving 11.39 g of iodic acid HIO_3 in 175 g of water has a freezing point of −1.17°C. What is the percentage of dissociation $HIO_3 \rightarrow H^+ + IO_3^-$?

7-64 37 g of $Ca(OH)_2$ are put into a 500-ml volumetric flask, dissolved, and enough additional water is added to fill it to the mark. **(a)** How many g/ml are in this solution? **(b)** Calculate the number of moles of $Ca(OH)_2$ in 150 ml of this solution. **(c)** Calculate the molarity of this solution. **(d)** How many milliliters of this solution will neutralize 75 ml of a 2.00 *M* acetic acid (HOAc) solution?

7-65 Fill in the table.

Acid						*Base*				
Species	*ml*	*M*	*N*	*Number of milli-moles*	*Number of milli-equivs*	*Number of milli-moles*	*N*	*M*	*ml*	*Species*
HCl	25	0.15					2.5			NaOH
H_2SO_4			0.00100		0.125				75.0	KOH
HCl	40.0			2.10				0.110		$Ba(OH)_2$

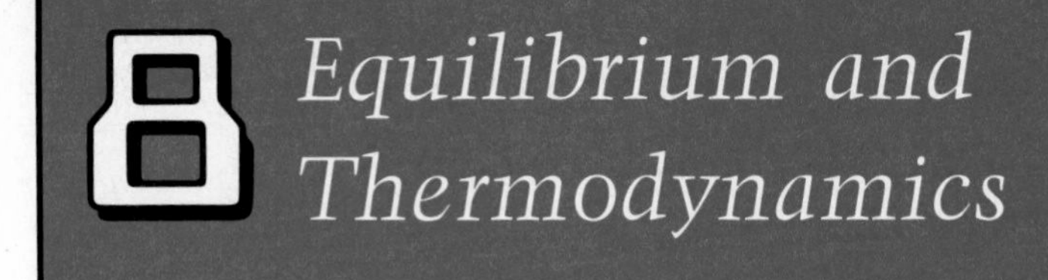

8 Equilibrium and Thermodynamics

Josiah Willard Gibbs, 1839–1903

American Institute of Physics

Part I: EQUILIBRIUM

Since its earliest beginnings, the practitioners of that branch of chemistry called physical chemistry have concerned themselves with two basic questions. The first of these is: Why do certain chemical reactions occur spontaneously while for others energy must be supplied before the reactants will form products? This is the problem of **chemical equilibrium.** The second question concerns the speed or *rate* of chemical reactions: Why are some reactions fast, and others slow? This chapter will be devoted to chemical equilibrium while in Chapter 9 we will look at the second general question, which deals with the area called **chemical kinetics.**

8-1 Reversible Reactions

A start was made in studying chemical reactions in Chapter 3. In that chapter, we calculated the quantities of reactants and products in a balanced equation. In performing these calculations, it was assumed (without explicitly saying so) that chemical reactions go to completion. That is, reactants continue to form products until at least *one of them is used up,* and no more of that reactant remains.

This will indeed happen if the reaction is *irreversible*. For example, the reaction between NaOH and HCl is in practice irreversible. If, say, 1 mole of NaOH is added to 0.3 mole of HCl, then all the HCl and 0.3 mole of the NaOH would be used up, and although 0.7 mole of NaOH would remain, the reaction would end because there would be no more HCl present. Putting NaCl and H_2O together *never* gives HCl and NaOH.

	OH^- +	HCl ⟶	H_2O +	Cl^-
Moles at start:	1	0.3	0	0
Moles at end:	0.7	0	0.3	0.3

A more everyday example of an essentially irreversible reaction is the complete combustion of paper, wood, or coal (to give CO_2 and H_2O). However, although we have just given examples of irreversible reactions, most reactions are **reversible.** In fact, all reactions are theoretically reversible, although for many the reversibility remains theoretical and impossible to put into practice. We cannot, for example, put CO_2 and H_2O together and get wood (of course, a tree does just this, but not in one simple reaction; rather by a series of reactions).

In a reversible reaction, as soon as the products are formed (and before the reactants are used up), they react with each other to re-produce the reactants. So the reaction *never* goes to completion (unless pushed in some way, as we shall see later in this chapter). An example is the HCl-catalyzed reaction between acetic acid and ethanol.

$$\underset{\text{acetic acid}}{CH_3COOH} + \underset{\text{ethanol}}{C_2H_5OH} \overset{HCl}{\rightleftharpoons} \underset{\text{ethyl acetate}}{CH_3COOC_2H_5} + \underset{\text{water}}{H_2O}$$

Note that the double-arrow symbol $\rightleftharpoons$ is used for a reversible reaction, instead of the usual single arrow. In this case, if we mix acetic acid, ethanol, and HCl, we will get ethyl acetate and water, but the reaction will not go all the way, because ethyl acetate and water react with each other (in the presence of HCl) to give acetic acid and ethanol. In practice, as soon as we mix acetic acid, ethanol, and HCl, the concentrations of acetic acid and ethanol will decrease, and the concentrations of ethyl acetate and water will increase until a point is reached at which there is no further change. At this point, all five substances will be present in the mixture, and the concentrations will no longer change. The mixture is said to be at **equilibrium.** (Chemical equilibrium is a special case of a more general phenomenon known as **dynamic equilibrium.**)

Part II of this chapter will deal with the discipline of thermodynamics. It is through a study of thermodynamic principles that we get insights into the factors governing chemical equilibrium.

8-2 Dynamic Equilibrium

Our study of chemical equilibrium will begin with a discussion of dynamic equilibrium since the principles underlying dynamic equilibria are directly applicable to chemical equilibria. To this end, we shall consider three examples of dynamic equilibria.

1. Assume that you are a demographic expert studying population changes. A country has a population of 200 million, and from your study it is observed that the rate of birth equals the death rate. Barring any catastrophe, such as earthquakes, wars, epidemics, etc., you can predict that over a period of time the population will remain constant.
2. Assume that you are a book dealer and it is your desire to keep 10 copies of Philip Greene's *The Jane Castle Manuscript* on hand. If you know the rate at which this book is bought, you can adjust your purchases of the book from the publisher to equal the buying rate. Doing this, you can be confident that your inventory of this book will be approximately constant.
3. A bathtub is half full of water. Simultaneously release the stopper and open the faucet to the extent that the flow of water entering the bathtub is the same as the flow out. Over a period of time the level of the water will remain constant.

If an observer just focused on the population level, the inventory of a particular book and the water level in the bathtub, it would appear that nothing was changing in all three examples. The three systems *would appear* to be *static*. The point which is being made is that although a system may appear to be static, there may be forces in operation. In each of the examples, there are two opposing rates (birth–death, selling books–buying books, water entering–water leaving), and *they are equal*. Systems such as those above are said to be in a state of *dynamic equilibrium*. *Dynamic*, because the systems are not static; in each there are two opposing processes. *Equilibrium*, because the *systems do not change with time*.

In a chemical equilibrium the process is somewhat more subtle. A chemical reaction always begins with the rates of the forward and reverse reactions being unequal (the rate of the reverse reaction is zero at the beginning because there are no

at time $t = 0$

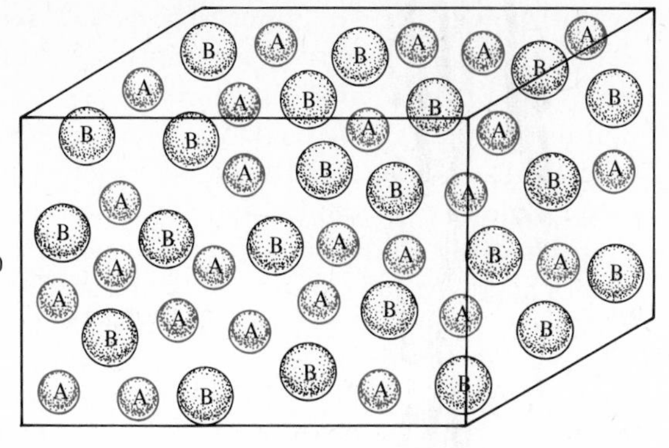

Figure 8-1 Molecules at the start of a reaction.

molecules present to react), and after a period of time, it is observed that the rates become equal. To see how this occurs, let us consider a simple chemical reaction in the gas phase: A + B → C + D, and also the reverse reaction C + D → A + B. We assume the reaction begins with only A and B molecules in the container (Figure 8-1). We also assume that molecules must collide in order for a reaction to take place (see Figure 8-2), and that every collision leads to a reaction.*

We have said that at the start only A and B are present. Therefore, they collide and hence react at a rate that is certainly much greater than the rate of the reverse reaction C + D → A + B; obviously at this point, the rate of C + D → A + B is zero because there are as yet no molecules of C or D present. As the reaction proceeds, C and D molecules are produced and soon start to encounter each other as their concentrations increase. As this happens, the rate of C + D → A + B increases; while at the same time, the rate of the forward reaction A + B → C + D *decreases,* since fewer A and B molecules are available (because they have reacted to form C and D). After a period of time, the rate of A + B → C + D has *decreased,* and the rate of C + D → A + B has *increased,* until the two rates are equal. From this point on, **both reactions are taking place, but there is no further change in the concentration of A, B, C, or D** because A and B are being formed (by C + D → A + B) as fast as they are being used up (by A + B → C + D), and the same of course is true for C and D.

A state of **chemical dynamic equilibrium** has been attained. Note that the concentrations of A and B do not have to be equal to those of C and D; in fact, in

*In most cases, not every collision leads to a reaction, but the argument is the same whether this is true or not.

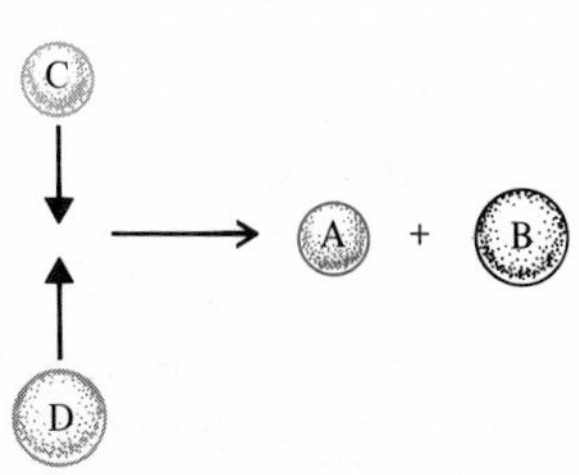

Figure 8-2 Molecules colliding and reacting.

most cases, they will not be. For example, we would not be surprised if we started with 1 mole of A and 1 mole of B and found that at equilibrium we had 0.9 mole of A, 0.9 mole of B, 0.1 mole of C, and 0.1 mole of D. Furthermore, we did not even have to start with equal amounts of A and B. We could have started, for example, with 1 mole of A and 2 moles of B, and we could still come to an equilibrium mixture of, say, 0.8 mole of A, 1.8 moles of B, and 0.2 mole each of C and D. Note also that we could have approached the equilibrium *from the opposite side:* by starting with C and D instead of A and B. At equilibrium, we would still get the same mixture of A, B, C, and D, and the system would not "remember" from which direction we started.

At equilibrium, the rates that are equal are the rate of formation of products and the rate of formation of reactants.

In 1863, two Norwegian chemists, Cato Guldberg (1836–1902) and Peter Waage (1833–1900), showed that the concentrations of all the species present at equilibrium can be related to each other by a mathematical equation called the **equilibrium expression** or the **law of mass action.**

The reaction

$$aA + bB + \cdots \rightleftharpoons cC + dD + \cdots$$

where the capital letters A, B, C, D, . . . , stand for the different chemical species and the small letters a, b, c, d, . . . , signify the stoichiometric coefficients in the balanced equation, is the most general reversible reaction we can write, and stands for all possible reversible reactions. For this general reaction, the mathematical equation is

$$K = \frac{[C]^c[D]^d \cdots}{[A]^a[B]^b \cdots}$$

The brackets stand for concentration in moles/liter. For example, [C] means concentration of C in moles/liter. The equilibrium expression K is a constant (the **equilibrium constant**) for any particular reaction at a given temperature. In Section 9-9, after we have studied reaction rates, we will show how this equation is derived.

The equilibrium expression is a general equation that holds for all reversible reactions once they have reached equilibrium. The forward reaction and reverse reaction are still taking place, but at equilibrium the rates of these opposing reactions are exactly equal.

Figure 8-3 shows how the concentrations in a reaction change with time. As the figure shows, at the beginning of the reaction there are only reactants. As time proceeds, the concentration of reactants decreases while the concentration of

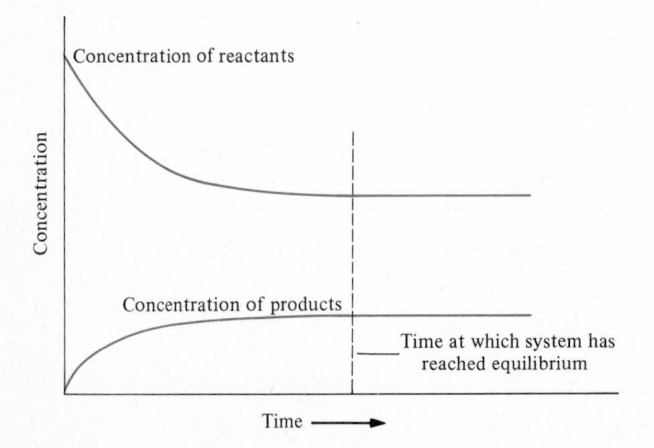

Figure 8-3 A plot of the concentration of reactants and products versus time. In this case, the equilibrium concentration of products is less than that of reactants, but the reverse is also possible, of course.

products increases. After a period of time the concentrations of both species remain constant and no longer change. It is at this time that the system has reached equilibrium.

8-3 The H_2, I_2, HI Equilibrium

Hydrogen gas and iodine vapor react to form hydrogen iodide.

$$H_2(g) + I_2(g) \rightleftharpoons 2HI(g)$$

The equilibrium expression for this reaction is (from the general equation above)

$$K = \frac{[HI]^2}{[H_2][I_2]}$$

Let us start with 1.0 mole of $H_2(g)$ and 2.0 moles of $I_2(g)$ in a volume of 4.0 liters at a temperature of 700°K. After a period of time, the concentrations of material were found to be as given in Table 8-1, and from that point on, no longer changed with time.

If one calculates the ratio $[HI]^2/[H_2][I_2]$ at equilibrium, the value obtained is 55.6. This number, 55.6, is the equilibrium constant (K) for this reaction at 700°K. As long as the temperature does not change, the equilibrium ratio $[HI]^2/[H_2][I_2]$ will be 55.6, no matter what the initial concentrations of H_2, I_2, or of HI. If we started with different concentrations of reactants, or if we had HI present initially, the *equilibrium concentrations* of the reactants and products would be different, but the *ratio* as expressed above, $[HI]^2/[H_2][I_2]$, would be 55.6, the value of the equilibrium constant for this reaction at 700°K.

Table 8-2 shows how it is possible to start with a variety of different initial concentrations and obtain different equilibrium concentrations. Note that we can begin with just H_2 and I_2,* just HI, or all three, but when equilibrium is reached, the value of the equilibrium constant remains constant.

Some points which are worth reviewing:

1. Up to now, we have been dealing with gaseous systems. However, the general equilibrium-constant expression holds for any equilibrium in any phase.
2. The equilibrium constant is a ratio of *concentrations,* not masses. The equilibrium expression is valid only for concentrations. If only the masses are given, it is necessary to divide by the volume to determine the concentrations.
3. The initial concentrations are generally different from the equilibrium concentrations (except in the rare case where, by chance, the initial concentrations happen to be the equilibrium concentrations; see Table 8-2, experiment 5).

* We could not start with just H_2 or I_2 alone, since no reaction could then take place, but we could start with H_2 and HI, or I_2 and HI.

Table 8-1 Concentrations in the Reaction $H_2 + I_2 \rightleftharpoons 2HI$

	Initial number of moles	*Volume of container*	*Initial concentration*	*Equilibrium concentration*
H_2	1.0	4.0 liters	$\frac{1.0}{4.0}$ mole/liter	0.0600 mole/liter
I_2	2.0	4.0 liters	$\frac{2.0}{4.0}$ mole/liter	1.06 moles/liter
HI	0.0	4.0 liters	$\frac{0.0}{4.0}$ mole/liter	1.88 moles/liter

Experiment[a]	Initial number of moles			Initial concentration (moles/liter)		
	H_2	I_2	HI	H_2	I_2	HI
1	1	1	0	0.2	0.2	0
2	5	5	5	1	1	1
3	2	2	3	0.4	0.4	0.6
4	0	0	2	0	0	0.4
5	0.32215	0.32215	2.4105	0.06443	0.06443	0.4821

[a] All experiments were performed in a 5-liter flask.

4. The equilibrium expression holds only *after* equilibrium has been reached. It does not hold before equilibrium is reached. Many reversible reactions reach equilibrium almost instantaneously, as soon as the reactants are thoroughly mixed. Other reactions may take hours, days, or even millions of years to reach equilibrium. The laws of equilibrium and of thermodynamics (see Part II of this chapter) say *nothing* about how long it takes a given reaction to reach equilibrium. They only say what the ratios of concentrations will be after equilibrium has been attained.
5. Once an equilibrium has been reached, the concentrations of all the substances present (products and reactants) will not change with time unless there is a change in conditions: for example, a change in temperature, pressure, or the removal of one of the components (see Le Châtelier's principle, Section 8-7). If the conditions do not change, the concentrations will remain unchanged forever. This does not mean that the chemical reactions within the system have ceased. It *does* signify that the rates of formation of products and of reactants are equal.
6. Although the equilibrium constant does not change as long as the temperature remains the same—in fact, this is the reason why the equilibrium constant is valuable—it *does* change when the temperature changes (see Section 8-7). Therefore, one must be careful to note the temperature when comparing equilibrium values.
7. An equilibrium can be approached from both sides of the equation.

8-4 Writing Equilibrium Constants for Chemical Reactions

Recall the general chemical reaction we presented in Section 8-2:

$$aA + bB + \cdots \rightleftharpoons cC + dD + \cdots$$

The equilibrium constant is

$$K = \frac{[C]^c[D]^d \cdots}{[A]^a[B]^b \cdots}$$

This general form should be memorized since it is so useful. To write an equilibrium constant for any reaction:

1. It is necessary to have a balanced equation for the reaction.
2. One multiplies the concentrations of the products (where each of the concentration terms is raised to the power of its stoichiometric coefficient, $[C]^c[D]^d$), and then divides this by the product of the concentrations of the reactants (where each of its terms is raised to the power of its stoichiometric coefficients, $[A]^a[B]^b$).

Table 8-2 Initial and Equilibrium Concentrations for the $H_2(g) + I_2(g) \rightleftharpoons 2HI(g)$ Equilibrium at 700°K

Equilibrium concentration			
H_2	I_2	HI	$K = \frac{[HI]^2}{[H_2][I_2]}$
0.0423	0.0423	0.3154	55.6
0.3175	0.3175	2.3650	55.49
0.1481	0.1481	1.1038	55.55
0.0423	0.0423	0.316	55.8
0.06443	0.06443	0.4821	55.16

3. Since the same equilibrium can be approached from either direction, it would be just as valid mathematically to put the concentrations of the reactants in the numerator and of the products in the denominator. However, it is useful for all chemists to speak the same language, and custom has decreed that as a reaction is written, for example,

$$H_2 + I_2 \rightleftharpoons 2HI$$

we write the equilibrium expression with the products in the numerator and the reactants in the denominator:

$$K = \frac{[HI]^2}{[H_2][I_2]}$$

The points above should be reviewed as the examples given below are studied.

Example: Write the equilibrium expression for the reaction $I_2(g) \rightleftharpoons 2I(g)$.

Answer: The equation as written is balanced, so that we can write

$$K = \frac{[I]^2}{[I_2]}$$

Example: Write the equilibrium expression for the reaction

$$2NH_3(g) \rightleftharpoons N_2(g) + 3H_2(g)$$

Answer: The equation as written is balanced so that we can write

$$K = \frac{[N_2][H_2]^3}{[NH_3]^2}$$

Example: Write the equilibrium expression for the reaction

$$N_2O_5(g) \rightleftharpoons NO_2(g) + O_2(g)$$

Answer: This equation as written is not balanced. It is necessary to balance the equation before writing the equilibrium expression. Balancing yields

$$2N_2O_5 \rightleftharpoons 4NO_2 + O_2$$

The equilibrium expression is

$$K = \frac{[NO_2]^4[O_2]}{[N_2O_5]^2}$$

We have previously seen (Table 8-2) that the equilibrium constant for the reaction $H_2 + I_2 \rightleftharpoons 2HI$ at 700°K is 55.6. That is,

$$K_1 = \frac{[HI]^2}{[H_2]\,[I_2]} = 55.6$$

It is possible to write equations for this same reaction in a number of other ways and still have chemically balanced equations.

$$2HI \rightleftharpoons H_2 + I_2 \qquad K_2 = \frac{[H_2][I_2]}{[HI]^2}$$

$$\tfrac{1}{2}H_2 + \tfrac{1}{2}I_2 \rightleftharpoons HI \qquad K_3 = \frac{[HI]}{[H_2]^{1/2}[I_2]^{1/2}}$$

$$HI \rightleftharpoons \tfrac{1}{2}H_2 + \tfrac{1}{2}I_2 \qquad K_4 = \frac{[H_2]^{1/2}[I_2]^{1/2}}{[HI]}$$

All these expressions are *valid* chemical equations showing the reaction between hydrogen, iodine, and hydrogen iodide. We have to be prepared to write all these expressions and to know the relationships between them. Compare K_1 and K_2. Clearly, K_2 is the reciprocal of K_1, so that $K_1 = 1/K_2$. Since $K_1 = 55.6$, $K_2 = 1/55.6 = 0.0180$. An examination of K_1 and K_3 leads to the conclusion that K_3 is the square root of K_1; thus $K_1^{1/2} = K_3 = 7.46$. For K_4, we have

$$K_4 = K_2^{1/2} = \frac{1}{K_1^{1/2}} = 0.134$$

To summarize, a chemical reaction can be balanced in a number of different ways, each of them correct. Each different form has its own equilibrium-constant expression and value. Knowing the value of one of the equilibrium expressions enables us to calculate the remaining ones, from the relationships between them. It is necessary that the balanced equation be given along with the equilibrium constant value, or else it could be very misleading.

Example: The equilibrium constant for the reaction

$$H_2S(g) \rightleftharpoons H_2(g) + \tfrac{1}{2}S_2(g) \qquad \text{at 1400°K is } K_1 = 23.90$$

Calculate the equilibrium constant for the following at 1400°K:

(a) K_2 for $2H_2S \rightleftharpoons 2H_2(g) + S_2(g)$
(b) K_3 for $H_2(g) + \tfrac{1}{2}S_2(g) \rightleftharpoons H_2S(g)$
(c) K_4 for $nH_2(g) + \frac{n}{2}S_2(g) \rightleftharpoons nH_2S(g)$ where n is any positive integer (not zero).

Answer:

$$K_1 = 23.90 = \frac{[H_2][S_2]^{1/2}}{[H_2S]}$$

(a) The equilibrium expression for (a) is

$$K_2 = \frac{[H_2]^2[S_2]}{[H_2S]^2}$$

Comparing with the above, we see that

$$K_2 = (K_1)^2 = 571.2$$

(b) The equilibrium expression for (b) is

$$K_3 = \frac{[H_2S]}{[H_2][S_2]^{1/2}}$$

and $K_3 = 1/K_1 = 0.04184$.

(c) The equilibrium expression is

$$K_4 = \frac{[H_2S]^n}{[H_2]^n[S_2]^{n/2}}$$

comparing with K_1, we obtain

$$K_4 = \left(\frac{1}{K_1}\right)^n = \left(\frac{1}{23.90}\right)^n = 0.04184^n$$

8-5 Uses of Equilibrium Constants

This section discusses ways in which the equilibrium constant gives insight into chemical reactions and some other applications.

If a reaction, as written, has a very large value of K (say, 10^{15} or higher), then, at equilibrium, the concentrations of products will be high, and the concentrations of reactants low. We can say that the equilibrium lies well to right and that the reaction will proceed to produce a higher concentration of products than of reactants (remember, though, that the value of K does not tell us how long it will take to reach equilibrium). An example of such a reaction is

$$NO_2(g) + CO(g) \rightleftharpoons NO(g) + CO_2(g) \qquad K = 7.7 \times 10^{38}$$

On the other hand, if the reaction has a very small value of K (say 10^{-15}), then we can say that at equilibrium the concentrations of the *reactants* are very high and those of the *products* very low. An example is

$$H_2(g) \rightleftharpoons 2H(g) \qquad K = \frac{[H]^2}{[H_2]} = 1.2 \times 10^{-71}$$

In this case the equilibrium lies well to the left, and the reaction does not proceed (the passage of time will not help us here!). The concentration of hydrogen atoms is vanishingly small. If the equilibrium constant is neither very large nor very small, the reaction partially proceeds and significant concentrations of both reactants and products are present at equilibrium (it is these reactions that would generally be called reversible). An example is

$$CO_2(g) + H_2(g) \rightleftharpoons CO(g) + H_2O(g) \qquad K = 0.1$$

Example: Consider the reaction

$$SO_2Cl_2(g) \rightleftharpoons SO_2(g) + Cl_2(g)$$

At 375°K, the value of this equilibrium constant is 0.0032. Into a flask we put 0.50 mole/liter of SO_2Cl_2 and 0.50 mole/liter of SO_2. Discuss what will occur in the reaction vessel.

Answer: The equilibrium constant expression is given by

$$K = \frac{[Cl_2][SO_2]}{[SO_2Cl_2]} = 0.0032$$

The experimental data above show that Cl_2 is not yet present in the reaction vessel. The reaction has not reached equilibrium. For this to occur, some of the SO_2Cl_2 must decompose to form SO_2 and Cl_2. The reaction goes to the right. The final (equilibrium) concentration of SO_2Cl_2 will be less than 0.50 mole/liter and the final concentration of SO_2 will be more than 0.50 mole/liter. (Additional SO_2 will form when SO_2Cl_2 decomposes.)

Example: Consider the same reaction again, and assume that the concentration of each of the three species is 0.050 mole/liter. Discuss what will occur in the reaction vessel.

Answer: The equilibrium-constant expression is given in the previous example. The concentrations in the reaction flask are 0.050 mole/liter. Inserting these values into the equilibrium expressions yields

$$\frac{[Cl_2][SO_2]}{[SO_2Cl_2]} = \frac{(0.050)(0.050)}{0.050} = 0.050$$

Clearly, $0.050 > K$, the equilibrium value. Therefore, the system is not in equilibrium and the concentrations must adjust until $K = 0.0032$. This can only happen if the products react to form more reactants (the numerator values must get smaller while the denominator must become larger). It can thus be stated that the reaction will go to the left. The final concentrations of Cl_2 and SO_2 will be less than 0.050 mole/liter while the concentration of SO_2Cl_2 will reach a value greater than 0.05 mole/liter.

The preceding description can be summarized as follows: Given a reaction, $aA + bB + \cdots \rightleftharpoons cC + dD + \cdots$, with

$$K = \frac{[C]^c[D]^d \cdots}{[A]^a[B]^b \cdots}$$

at any time during the reaction process, if we obtain the concentrations of the different species and substitute them into the equilibrium expression, then the three possibilities are

If $\dfrac{[C]^c[D]^d \cdots}{[A]^a[B]^b \cdots} < K$ — Equilibrium has not yet been reached. The reaction will proceed to the right, increasing the product concentrations while simultaneously decreasing the reactant concentrations.

If $\dfrac{[C]^c[D]^d \cdots}{[A]^a[B]^b \cdots} = K$ — The reaction is at equilibrium and the concentrations of reactants and products will remain constant.

If $\dfrac{[C]^c[D]^d \cdots}{[A]^a[B]^b \cdots} > K$ — Equilibrium has not yet been reached. The reaction will proceed to the left, increasing the reactant concentrations while simultaneously decreasing the product concentrations.

It is possible to calculate the equilibrium constant for a given reaction from a knowledge of equilibrium expressions for related reactions. To show this, assume that we wish to calculate the equilibrium constant for the reaction

$$H_2(g) + CO_2(g) \rightleftharpoons H_2O(g) + CO(g) \qquad K = \frac{[CO][H_2O]}{[H_2][CO_2]}$$

and we know the equilibrium constants for two other reactions:

$$H_2(g) + \tfrac{1}{2}O_2(g) \rightleftharpoons H_2O(g) \qquad K_1 = \frac{[H_2O]}{[H_2][O_2]^{1/2}}$$

and

$$CO_2(g) \rightleftharpoons CO(g) + \tfrac{1}{2}O_2(g) \qquad K_2 = \frac{[O_2]^{1/2}[CO]}{[CO_2]}$$

Adding the two known equations yields

$$H_2(g) + CO_2(g) + \cancel{\tfrac{1}{2}O_2(g)} \rightleftharpoons H_2O(g) + CO(g) + \cancel{\tfrac{1}{2}O_2(g)}$$

The $\frac{1}{2}O_2$ cancels, and we have the wanted equation. *Multiplying* the equilibrium expressions, we obtain

$$K_1 \times K_2 = \frac{[H_2O]}{[H_2][O_2]^{1/2}} \times \frac{[O_2]^{1/2}[CO]}{[CO_2]} = \frac{[H_2O][CO]}{[H_2][CO_2]} = K$$

We may thus obtain K by multiplying K_1 by K_2:

$$K = K_1K_2$$

There are two points to keep in mind in applying this technique.

1. All the equilibrium constants must be at the same temperature. If the K to be found is at 25°C, the values of K_1 and K_2 must also be the values at 25°C. This is necessary because equilibrium constants change with changing temperatures.
2. We multiply the equilibrium constants when it is necessary to *add* the equations. When we *subtract* equations we must *divide* equilibrium constants.

8-6 Calculations Involving Equilibrium Constants

In Section 8-3, we saw how to calculate the equilibrium constant if we knew the equilibrium concentrations of all the components. In this section, we shall perform the reverse calculations: Given the equilibrium constant and the amounts of materials we begin with, what will be the equilibrium concentrations of all the components?* Let us look again at the equilibrium

$$H_2 + I_2 \rightleftharpoons 2HI$$

for which, as we have seen, $K = 55.6$ at 700°K. If we put 2.00 moles of H_2 and 2.00 moles of I_2 into a 2.00-liter flask, the reaction will begin; HI will be formed and some of the H_2 and I_2 will be used up. What will be the concentrations of the three species at equilibrium? In handling this type of calculation, it is advantageous to construct a table of initial concentrations and of equilibrium concentrations (see Table 8-3). The initial concentrations are (before any reaction has taken place):

$$[H_2] = \frac{2.00 \text{ moles}}{2.00 \text{ liters}} = 1.00 \text{ mole/liter}$$

$$[I_2] = \frac{2.00 \text{ moles}}{2.00 \text{ liters}} = 1.00 \text{ mole/liter}$$

$$[HI] = \frac{0 \text{ mole}}{2.00 \text{ liters}} = 0 \text{ mole/liter}$$

As we have seen, at equilibrium the concentration of H_2 must be less than 1.0 mole/liter. At the moment we do not know how much less, so let us call that part of the concentration which is used up x mole/liter, so that the concentration of H_2

*This kind of calculation is much more common than the former. Once we calculate K for a reaction at a given temperature, we know it for all time and do not have to calculate it again.

Table 8-3 Initial and Equilibrium Concentrations for the Reaction $H_2 + I_2 \rightleftharpoons 2HI$

Species	*Initial concentration (moles/liter)*	*Equilibrium concentration (moles/liter)*
$[H_2]$	1.00	$1.00 - x$
$[I_2]$	1.00	$1.00 - x$
[HI]	0	2x

which is present at equilibrium is (1.00 − x) mole/liter. Now that we have a value for the concentration of H_2 at equilibrium, we need values for the other two species. At this point, we turn to the balanced equation, $H_2 + I_2 \rightleftharpoons 2HI$. This equation tells us that the number of moles of H_2 that is used up in the reaction must be equal to the number of moles of I_2 that is used up, since the stoichiometric ratio is 1:1. Similarly, the *concentration* of H_2 that is used up must equal the *concentration* of I_2 that is used up because the volume is the same for both. We have already called this x, so the concentration of I_2 at equilibrium is also (1.00 − x) mole/liter. Also, the equation tells us that for every mole of H_2 or I_2 that is used up, *2 moles of HI* are formed. The concentration of HI at equilibrium is therefore 2x. The three values are listed in Table 8-3. Since these numbers (1.00 − x, 1.00 − x, and 2x) are concentrations, we may insert them directly into the equilibrium expression and solve for x. Note that the equilibrium expression only works for concentrations. If, for example, we inserted the number of moles instead of the concentrations, we would risk getting the wrong answer.

The equilibrium expression is

$$K = \frac{[HI]^2}{[H_2][I_2]} = 55.6$$

and using the equilibrium concentrations obtained above, we have

$$55.6 = \frac{(2x)^2}{(1.00 - x)(1.00 - x)}$$

We must now solve the equation for x and then determine the equilibrium concentrations. This can be done by taking the square root of both sides, obtaining

$$7.46 = \frac{2x}{1.00 - x}$$

and solving yields

$$x = 0.789 \text{ mole/liter}$$

This number, of course, is the concentration of H_2 or of I_2 that was *used up*. We may now calculate the equilibrium concentrations as follows:

$$[H_2] = 1.00 - x = 1.00 - 0.789 = 0.21 \text{ mole/liter } H_2$$
$$[I_2] = 1.00 - x = 1.00 - 0.789 = 0.21 \text{ mole/liter } I_2$$
$$[HI] = 2x = 2(0.789) = 1.58 \text{ moles/liter HI}$$

The calculations, once the equilibrium concentrations have been written, are generally not difficult. In the example we have just given, the numbers of moles of H_2 and I_2 at the start were equal, and that is why we were able to solve the mathematical equation by taking the square root of both sides. In a more general case, we cannot do this, and it is necessary to solve a quadratic equation ($ax^2 + bx + c = 0$). The best way to solve such an equation is to use the quadratic formula (Section 1-6). Solving the quadratic usually gives two solutions. Of the two solutions, one is physically impossible, and the correct chemical answer is therefore the other one. The following example illustrates this.

Example: For the reaction $H_2 + I_2 \rightleftharpoons 2HI$, we place 2.00 moles of H_2 and 4.00 moles of I_2 in a 5.00-liter flask at 700°K. Calculate the equilibrium concentration of all species.

Answer:

Species	*Initial concentration (moles/liter)*	*Equilibrium concentration (moles/liter)*
$[H_2]$	0.400	$0.400 - x$
$[I_2]$	0.800	$0.800 - x$
$[HI]$	0.0	$2x$

The equilibrium expression is

$$55.6 = \frac{(2x)^2}{(0.400 - x)(0.800 - x)}$$

$$55.6 = \frac{4x^2}{0.320 - 1.20x + x^2}$$

Rearranging yields

$$x^2 - 1.29x + 0.345 = 0$$

Using the quadratic expression, we get

$$x = \frac{1.29 \pm \sqrt{(-1.29)^2 - 4 \times 1 \times 0.345}}{2}$$

$$= \frac{1.29 \pm \sqrt{0.284}}{2} \qquad x = \begin{cases} 0.91 \\ 0.38 \end{cases}$$

Equilibrium concentrations (if $x = 0.91$):

$$[H_2] = 0.400 - x = -0.51$$
$$[I_2] = 0.800 - x = -0.11$$
$$[HI] = 2x = 1.82$$

It is impossible to have a negative concentration of a species, so that this is a physically unacceptable solution.

Equilibrium concentrations (if $x = 0.38$):

$$[H_2] = 0.400 - 0.38 = 0.02 \text{ mole/liter}$$
$$[I_2] = 0.800 - 0.38 = 0.42 \text{ mole/liter}$$
$$[HI] = 2(0.38) = 0.76 \text{ mole/liter}$$

This answer must therefore be the correct one.

We must be extremely careful in setting up these calculations to take into account the numbers of moles in the balanced equation. Otherwise, the unknown may be introduced in an incorrect fashion. To illustrate this, consider the reaction of hydrogen and nitrogen to form ammonia:

$$3H_2(g) + N_2(g) \rightleftharpoons 2NH_3(g)$$

The balanced equation tells us that 3 moles of hydrogen react with 1 mole of nitrogen to produce 2 moles of ammonia. If we let x be the number of moles/liter of nitrogen which are used up at equilibrium, and the initial concentration of H_2 is 4 moles/liter and N_2 is 2 moles/liter, the concentrations are

Species	*Initial concentration (moles/liter)*	*Equilibrium concentration (moles/liter)*
$[H_2]$	4	$4 - 3x$
$[N_2]$	2	$2 - x$
$[NH_3]$	0	$2x$

The equilibrium concentration of H_2 is $4 - 3x$ because, if x moles of N_2 are used up at equilibrium, then 3x moles of H_2 must be used up. We chose x to be the concentration of N_2 that was used up at equilibrium. We can also set up the problem differently, by letting x equal the concentration of *hydrogen* that is used up at equilibrium. In this case, the equilibrium concentrations are

$$[H_2] = 4 - x$$

$$[N_2] = 2 - \frac{x}{3}$$

$$[NH_3] = \frac{2x}{3}$$

The equilibrium expressions may appear different, depending upon how x is originally defined. Nevertheless, they are mathematically equivalent, and each will give the correct answer when they are solved, although once the solution is obtained, it is necessary to remember how x was originally defined. There are still other ways of setting up this problem. For example, the student may wish to try setting it up so that x equals the concentration of NH_3 at equilibrium. The examples below illustrate these principles.

Example: Given the reaction

$$2H_2(g) + O_2(g) \rightleftharpoons 2H_2O(g)$$

Set up the equilibrium expression if 3 moles/liter of hydrogen is mixed with 5 moles/liter of oxygen.

Answer: Let x = concentration of *hydrogen* that is used up.

Species	*Equilibrium concentration (moles/liter)*
$[H_2]$	$3 - x$
$[O_2]$	$5 - x/2$
$[H_2O]$	x

$$K = \frac{x^2}{(3 - x)^2(5 - x/2)}$$

For the same reaction, if x = concentration of *oxygen* that is used up, then

Species	*Equilibrium concentration (moles/liter)*
$[H_2]$	$3 - 2x$
$[O_2]$	$5 - x$
$[H_2O]$	$2x$

$$K = \frac{(2x)^2}{(3 - 2x)^2(5 - x)}$$

Example: Given the reaction

$$2Cl_2(g) + 2H_2O(g) \rightleftharpoons 4HCl(g) + O_2(g)$$

Set up the equilibrium expression if 3 moles of chlorine are mixed with 1 mole of water vapor and 6 moles of oxygen in a 1-liter container.

Answer: Let x = concentration of oxygen formed by the reaction.

Species	*Initial concentration (moles/liter)*	*Equilibrium concentration (moles/liter)*
$[Cl_2]$	3	$3 - 2x$
$[H_2O]$	1	$1 - 2x$
$[HCl]$	0	$4x$
$[O_2]$	6	$6 + x$

$$K = \frac{(4x)^4(6 + x)}{(3 - 2x)^2(1 - 2x)^2}$$

For the same reaction, if x = concentration of HCl formed:

Species	*Initial concentration (moles/liter)*	*Equilibrium concentration (moles/liter)*
$[Cl_2]$	3	$3 - \frac{x}{2}$
$[H_2O]$	1	$1 - \frac{x}{2}$
$[HCl]$	0	x
$[O_2]$	6	$6 + \frac{x}{4}$

$$K = \frac{(x)^4(6 + x/4)}{(3 - x/4)^2(1 - x/2)^2}$$

For all the examples given above, if K is known, then x can be calculated. In these examples the calculation of x would be exceedingly difficult for most students (we do not recommend attempting to solve fourth-degree equations), although they could be solved with the aid of a computer and in practice such equations often are solved in this manner. Once x is calculated then the equilibrium concentrations are readily obtainable.

8-7 The Principle of Le Châtelier

As we have mentioned before, once a system has reached equilibrium, it will stay there, and the concentrations will no longer change, *unless something happens to the system from outside.* If an outside stress is applied to the system, the system *will* change. **Le Châtelier's principle** (Henri Louis Le Châtelier, 1850–1936) enables us to predict in what way the system will change. This famous principle, first stated in 1888, says that **if an external stress is applied to a system in equilibrium, the system will react in such a way as to relieve the stress.** In a chemical system, an external stress means a change in the volume, pressure, concentration, or temperature of the system. The only way a chemical system can change is to shift the equilibrium to the right (produce more products) or shift to the left (produce more reactants).

In the rest of this section, we will see how Le Châtelier's principle allows us to predict in which direction an equilibrium will shift if we make specific changes in one of the reaction conditions. Incidentally, it may be noted that Le Châtelier's principle is not confined to chemistry, and applies to many areas of natural and human behavior. For example, the law of supply and demand, from economics, is a special case of Le Châtelier's principle (the price of a commodity is an equilibrium that goes up and down in response to external stresses: in this case, changes in the supply or demand).

1. CHANGES IN CONCENTRATION. One way we can put a stress on a system is to change the concentration of one of the components. There are two ways to accomplish this: (1) we can remove some of a component, or (2) we can add some more of a component.

Consider the equilibrium between oxygen (O_2) and ozone (O_3):

$$2O_3(g) \rightleftharpoons 3O_2(g)$$

We can disturb the equilibrium by introducing some additional oxygen gas (O_2) into the system. Le Châtelier's principle states that the system must shift so as to relieve the stress. The addition of more O_2 means that the equilibrium has been disturbed as a result of the increase in O_2 concentration. To relieve the stress, the system has to use up some O_2, and it can only do this by shifting the equilibrium position to the left so as to produce more O_3. On the other hand, if we introduce ozone into the equilibrium system, the reverse occurs; more O_2 is formed and some of the O_3 is decomposed. The equilibrium shifts to the right.

It is usually easy to add some more of a compound to a system at equilibrium. The removal of all or part of a component is not always easy. Nevertheless, it can often be accomplished, and when it happens, the effect on the system is the reverse. For example, the removal of O_2 from the above equilibrium drives the equilibrium *to the right:* some of the O_3 will decompose to form more O_2. Similarly, if part or all of the O_3 were to be removed, some of the O_2 would react to form more O_3.

The rule about changes in concentration is, therefore: *The addition of any component to a system in equilibrium drives the equilibrium in the direction away from that side; the removal of any component drives the equilibrium in the direction toward that side.* Note that this rule holds for any component, no matter whether it appears on the left or right as the equation is written.

Example: Given the reaction

$$2Cl_2(g) + 2H_2O(g) \rightleftharpoons 4HCl(g) + O_2(g)$$

at equilibrium, discuss what happens when (a) O_2 is added, (b) O_2 is removed, (c) H_2O is added, (d) Cl_2 is removed

Answer: (a) The equilibrium shifts to the left producing more moles of Cl_2 and H_2O while decreasing the concentration of HCl. The concentration of O_2 will be more than it was before we disturbed the equilibrium (after all, we *added* O_2 from the outside), but less than the sum of the original quantity plus the amount we added.

(b) The equilibrium shifts to the right, producing more moles of HCl and O_2. The concentrations of Cl_2 and H_2O decrease. The new concentration of O_2 will be less than it was before we disturbed the equilibrium (after all, we took some out) but more than the concentration that remained after we removed some.

(c) The equilibrium shifts to the right [see part (b)].
(d) The equilibrium shifts to the left [see part (a)].

We have discussed in a qualitative fashion what happens to an equilibrium when we change the concentration of one of the components. Let us now look at it quantitatively. Remember that the value of the equilibrium constant K does not change when the concentrations are changed. We have seen (page 210) that a system containing 0.21 mole/liter of H_2, 0.21 mole/liter of I_2, and 1.58 moles/liter of HI at 700°K is at equilibrium, because it satisfies the equilibrium expression, with $K = 55.6$:

$$H_2 + I_2 \rightleftharpoons 2HI$$

$$K = \frac{[HI]^2}{[H_2][I_2]} = \frac{(1.58)^2}{(0.21)(0.21)} = 55.6$$

Example: What will be the new equilibrium concentrations of HI, H_2, and I_2 if we add 0.35 mole/liter of H_2 without changing the total volume?

Answer: We use a technique similar to that we have used before. Let x = the concentration of H_2 that is used up.

	Concentration (moles/liter)		
	$[H_2]$	$[I_2]$	$[HI]$
(a) At the old equilibrium	0.21	0.21	1.58
(b) As soon as 0.35 mole/liter of H_2 is added	0.56	0.21	1.58
(c) At the new equilibrium	0.56 − x	0.21 − x	1.58 + 2x

$$K = 55.6 = \frac{(1.58 + 2x)^2}{(0.56 - x)(0.21 - x)}$$

Solving by the quadratic formula, we obtain

$$x = \begin{cases} 0.78 \\ 0.12 \end{cases}$$

The first answer is physically impossible (why?), and we obtain for x = 0.12:

	$[H_2]$	$[I_2]$	$[HI]$
(c) At the new equilibrium	0.56 − 0.12 = 0.44	0.21 − 0.12 = 0.09	1.58 + 2(0.12) = 1.82

The new concentrations are, therefore, H_2, 0.44; I_2, 0.09; and HI, 1.82 moles/liter. The addition of H_2 from the outside has driven the equilibrium to the right; it has increased the concentration of HI and decreased the concentration of I_2. The new concentration of H_2 is greater than it was originally (because we *added* H_2), but less than the sum of the original concentration and the concentration we added. Note that though we added a large quantity of H_2, it did not use up *all* the I_2. The law of equilibrium ensures that at least a little bit of the I_2 remains (to be sure, if we added enough H_2, this would be a *very* little bit).

We emphasize again that the equilibrium constant K does not change when we change the concentrations. Both our qualitative and our quantitative conclusions rest on this fact.

2. Changes in Volume or Pressure. A change in volume or pressure can shift an equilibrium involving gases. Consider once again the reaction

$$2O_3(g) \rightleftharpoons 3O_2(g)$$

The equation states that every time 2 molecules of O_3 are used up, 3 molecules of O_2 are formed. Therefore, shifting the equilibrium to the right produces more molecules, and shifting the equilibrium to left decreases the number of molecules in the system. From the ideal gas law we know (Section 4-7) that $PV = nRT$. This equation tells us that if we keep the pressure and temperature (PT) constant, increasing the number of moles increases the volume and decreasing the number of moles decreases the volume. We now are able to understand how changing the volume at a constant temperature and pressure can shift the equilibrium. For $2O_3 \rightleftharpoons 3O_2$, if the volume is decreased, the system will move in such a way as to relieve the stress; that is, it will shift in the direction that produces fewer molecules (the equilibrium shifts to the left). Conversely, increasing the volume will cause the equilibrium to shift to the right. If a reaction has an equal number of moles of gas on both sides of the equation [$H_2(g) + I_2(g) \rightleftharpoons 2HI(g)$], then a shift in equilibrium cannot relieve the stress in the same way, since whether the reaction moves to the left or to the right, the total number of moles present cannot change. In this case, it is impossible, at constant temperature, to change the volume without also changing the pressure. However, if we do change the volume, with a corresponding change in pressure, the position of equilibrium will still not change.

Changes in pressure at constant temperature are generally accompanied by changes in volume. Boyle's law ($PV = k$) tells us that increasing the pressure decreases the volume and decreasing the pressure increases the volume. Changes in pressure can be analyzed in terms of changes in volume. For $2O_3 \rightleftharpoons 3O_2$, decreasing the pressure increases the volume and shifts the equilibrium to the right. Increasing the pressure decreases the volume and shifts the equilibrium to the left. It is also possible to increase the pressure at constant temperature without changing the volume. This can be accomplished by adding some nonreactive gas to the reaction vessel. In this instance, no shift in equilibrium will take place. (Why?)

Changes in pressure or volume only affect equilibria in which at least one of the components is a gas; and even for these equilibria, only the gaseous components are counted. It is then very easy to predict in which direction the equilibrium is shifted. *A decrease in volume, or an increase in pressure accompanied by a decrease in volume* drives the equilibrium toward the side with the smaller number of gaseous moles.* Moles of liquids or solids are ignored. For example, look at the equilibrium

$$C(s) + H_2O(g) \rightleftharpoons CO(g) + H_2(g)$$

There are 2 moles of gas on the right and only one on the left. Thus, a decrease in volume or an increase in pressure drives the equilibrium to the left. An increase in volume or a decrease in pressure drives it to the right.

Example: Predict the consequences of an increase in pressure on these equilibria.

(a) $CO_2(g) + H_2(g) \rightleftharpoons CO(g) + H_2O(g)$
(b) $ZnO(s) + H_2(g) \rightleftharpoons Zn(s) + H_2O(l)$
(c) $C_6H_{12}(l) + Br_2(l) \rightleftharpoons C_6H_{12}Br_2(l)$

*That is, not resulting from the addition of a nonreactive component.

Answer:

(a) There are 2 moles of gas on each side; therefore, changing the pressure has no effect on the equilibrium.

(b) There is 1 mole of gas on the left, none on the right; increasing the pressure drives the equilibrium to the right.

(c) There are no moles of gas on either side; increasing the pressure has no effect.

Note that, as in the case of changes in concentration, changes of volume or pressure do not affect the value of the equilibrium constant, K.

3. CHANGES IN TEMPERATURE. The situation differs here from what it was in the two preceding sections. The equilibrium constant K is *not* constant when the temperature changes. A change in temperature changes the value of K. Nevertheless, we can predict which way an equilibrium will shift if we know in which direction the reaction is exothermic. In most reversible reactions, heat is given off either in the forward or reverse direction (Section 5-4). Of course, if heat is given off in the forward direction, then, if energy is to be conserved, the same amount of heat must be absorbed in the reverse direction. According to LeChâtelier's principle, adding heat to the system (raising the temperature) will shift the equilibrium in such a way as to favor the system absorbing the heat, and lowering the temperature will shift the equilibrium in such a way as to favor the production of heat.

For example, the following reaction is endothermic (when it goes from left to right) by 58,000 cal/mole of H_2O.

$$H_2O(g) \rightleftharpoons H_2(g) + \tfrac{1}{2}O_2(g) - 58{,}000 \text{ cal}$$

One must put 58,000 cal into the system to decompose 1 mole of H_2O into hydrogen and oxygen. Adding heat (raising the temperature) will shift the equilibrium to the right, since by shifting to the right, the system will consume the heat in producing more H_2O and O_2, relieving the stress. Conversely, cooling the system (lowering the temperature) will shift the equilibrium to the left, since this will offset the stress by producing the heat. We can look upon heat as a reactant or product. For example, the preceding equation could be written

$$H_2O(g) + 58{,}000 \text{ cal} \rightleftharpoons H_2(g) + \tfrac{1}{2}O_2(g)$$

Writing it in this way, we can easily see that adding heat (raising the temperature) is analogous to adding additional H_2O; either of these will drive the equilibrium to the right.

We can predict not only which way the equilibrium will shift when the temperature changes but also whether the value of K will increase or decrease. In the example above, we saw that the reaction

$$H_2O(g) \rightleftharpoons H_2(g) + \tfrac{1}{2}O_2(g) - 58{,}000 \text{ cal}$$

is shifted to the right when the temperature is raised (introducing heat). This shift will produce more H_2 and O_2, and less H_2O, so that the value of K in

$$K = \frac{[H_2][O_2]^{1/2}}{[H_2O]}$$

must increase. If the temperature of the same reaction is lowered, K will decrease. This is true for all endothermic reactions, and the opposite is true for exothermic reactions.

4. ADDITION OF A CATALYST. A catalyst is a substance that increases the *rate* of a reaction without itself being used up

(Chapter 9). *The addition of a catalyst has no effect on the position of equilibrium, since it accelerates the forward and reverse reactions to the same extent.* However, if a system is not yet at equilibrium, a catalyst will help it to reach equilibrium faster.

Example: For the following reaction at equilibrium at 25°C:

$$H_2O(g) \rightleftharpoons H_2(g) + \tfrac{1}{2}O_2(g) - 58{,}000 \text{ cal}$$

discuss what happens when (a) the volume decreases, (b) O_2 is withdrawn from the system, (c) the pressure increases, (d) the pressure decreases, (e) H_2 is added to the system, (f) Ar gas is added to the system (Ar does not react with any of the species present), increasing the total pressure at constant volume, (g) the temperature is lowered, (h) a catalyst is added

Answer: For all these questions, we apply Le Châtelier's principle.

A. Shifting the equilibrium to the right
   1. increases the total number of moles of gases
   2. produces more H_2 and O_2
   3. decreases the concentration of H_2O
   4. absorbs more heat from the surroundings

B. Shifting the equilibrium to the left
   1. decreases the total number of moles of gases
   2. produces more H_2O
   3. decreases the concentration of H_2 and O_2
   4. generates heat

Stress		*Equilibrium shifts*
(a) The volume decreases	B1	to the left
(b) O_2 is withdrawn	A2	to the right
(c) The pressure increases (volume decreases)	B1	to the left
(d) The pressure decreases (volume increases)	A1	to the right
(e) H_2 is added	B3	to the left
(f) Ar is added		no change
(g) The temperature is lowered	B4	to the left
(h) A catalyst is added		no change

8-8 Equilibrium Constants in Terms of Pressures

We have emphasized that the numbers which go into an equilibrium-constant expression must be concentrations—that they may *not* be numbers of grams or moles. However, when gaseous compounds are involved, it is also possible to write valid equilibrium-constant expressions using partial *pressures*. For example, for the reaction

$$2SO_2(g) + O_2(g) \rightleftharpoons 2SO_3(g)$$

we can write *two* valid expressions:

$$K_c = \frac{[SO_3]^2}{[SO_2]^2[O_2]}$$

$$K_p = \frac{(P_{SO_3})^2}{(P_{SO_2})^2 P_{O_2}}$$

The second of these is an equilibrium-constant expression that is written using partial pressures instead of concentrations, but it is equally valid and useful. It obeys all the rules of equilibrium that we have been discussing, and all the types of equilibrium calculations we have done could have been done in a similar manner using K_p. We have called the first equilibrium constant K_c, to distinguish it from K_p, but it is the same type of constant we have been using throughout this chapter ($K_c = K$)*.

For a given reaction the values of K_p and K_c are in general numerically different. However, $K_p = K_c$ when the number of moles of gaseous products is equal to the number of moles of gaseous reactants, as, for example, in $H_2(g) + I_2(g) \rightleftharpoons 2HI(g)$.

To summarize, we can always use K_c for *any* reversible reaction, including those involving gases; but in the latter case, we can *also* use K_p if we so desire.

Example: For the reaction

$$N_2(g) + 3H_2(g) \rightleftharpoons 2NH_3(g)$$

at a temperature of 450°C, when the total pressure of N_2, H_2, and NH_3 is 300 atm, the equilibrium partial pressures are N_2, 48.4 atm; H_2, 145 atm; and NH_3, 106.5 atm. Calculate K_p.

Answer:

$$K_p = \frac{(P_{NH_3})^2}{P_{N_2}(P_{H_2})^3} = \frac{(106.5)^2}{(48.4)\,(145)^3} = 7.69 \times 10^{-5}$$

Part II: The Thermodynamics of Chemical Equilibrium

Energy, Entropy, and Free Energy. In Chapter 5, we discussed the heat changes occurring in a chemical reaction in terms of thermodynamic principles. In this section, we shall continue our study of thermodynamics and in particular will be concerned with a fundamental result of thermodynamics, the ability to predict whether a chemical reaction will take place spontaneously. We shall find that the thermodynamic state functions defined in Chapter 5 are insufficient for solving this problem. It will be necessary to define new state functions that can be used to determine when a reaction will occur spontaneously, and when it will not. Furthermore, these new functions will allow us to tell from a thermodynamic point of view when a chemical reaction has reached equilibrium.

8-9 Energy

In order to understand why some reactions occur spontaneously and others do not, we must consider the factors that force reactants to form products. Our understanding of these factors is based on certain experiences that we have had with nature. The first factor we shall discuss is the potential energy of a system. **Every system in the universe tends toward a minimum in its potential energy.** This statement, which is one of the most fundamental laws of nature, will be illustrated by a few examples:

*In general, the symbol K, when written without a subscript, means that concentrations are used. If partial pressures are being referred to, the symbol is always K_p.

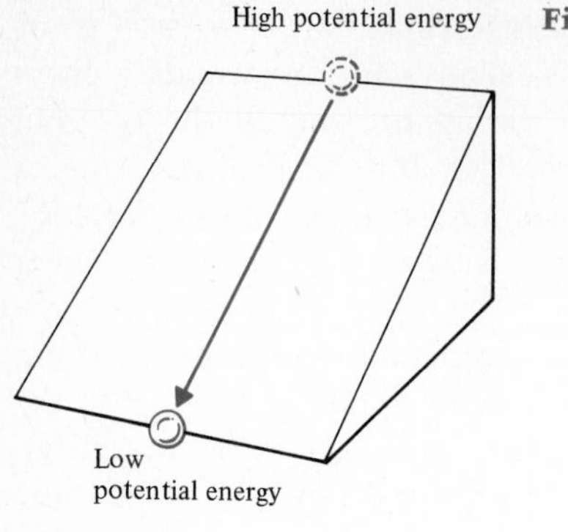

Figure 8-4 A ball rolls *down* an inclined plane, not up.

1. A ball on an inclined plane (see Figure 8-4) will spontaneously roll down the plane. In doing so, the potential energy of the ball has decreased. It is our experience with nature that the ball never spontaneously rolls up the inclined plane.
2. A book balanced on its edge is less stable than the book laid flat (see Figure 8-5). The book when it is flat has a lower potential energy than the book on its edge.
3. A spring is more stable when it is relaxed than when it is pulled taut (Figure 8-6). The relaxed spring has a lower potential energy.

The student can no doubt supply many additional examples.

It is tempting to try to make a similar statement about chemical reactions: Reactions that are exothermic proceed spontaneously. After giving off the energy of the reaction, these systems possess a lower potential energy than they had before the reaction took place. If this statement were true, we could use the knowledge of whether a reaction is exothermic or endothermic as the sole criterion in determining whether a reaction will occur spontaneously. Simply stated, this would mean that all reactions that are exothermic would occur spontaneously, whereas endothermic reactions would not occur spontaneously. This statement is simple, appealing, and easy to use. However, it is incorrect. For one thing, there are reactions that are endothermic and yet proceed spontaneously. For example, when NH_4Cl is dissolved in water, the solution spontaneously gets colder, demonstrating that heat is absorbed. Even more important, if only reactions that give off heat proceeded spontaneously, then all exothermic reactions would proceed to completion. The reverse reactions would not occur. Again, this is contrary to our experience and knowledge (see the discussion of reversible reactions in Section 8-1).

The water gas reaction

$$H_2(g) + CO_2(g) \longrightarrow CO(g) + H_2O(g) \qquad \Delta H_{298} = -9800 \text{ cal}$$

is an exothermic reaction. However, its equilibrium constant at 298°K is 9550. Since the reaction *has* an equilibrium constant, we realize that reactions are *proceeding spontaneously in both directions* even though one direction is exothermic and the other must be endothermic. This reaction (and many other reversible reactions) is highly exothermic. There is a tremendous drive for product formation. However, before the reaction can reach completion, the products begin to form reactants, even though *this* process is highly endothermic. The question then becomes: What drives the reverse reaction?

Although the energy gives us some insight into how certain processes proceed spontaneously, it is clearly incomplete in fully describing all factors affecting equilibrium. Another condition is operating, which must offset the energy factor. This other condition is called the *entropy*. It is a thermodynamic state function and is discussed in the following section.

Figure 8-5 A book is most stable when it is lying flat.

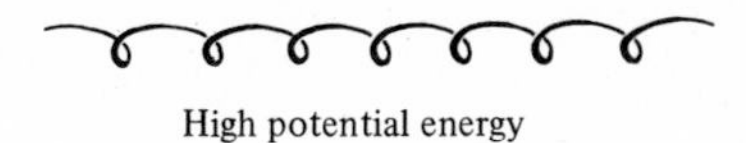

Figure 8-6 A spring is most stable when it is relaxed.

8-10 Entropy and Free Energy

Entropy is a quantity that measures randomness. Its symbol is S. The higher the entropy of a system, the greater the randomness, and vice versa. The reader is probably wondering how the factor of randomness is related to this whole discussion, and in particular to chemical equilibria. To see this, consider the following examples:

1. A hot brick is placed on top of a cold brick. Our experience tells us that heat will spontaneously flow from the hot body into the cold body until they reach the same temperature. Why does not heat spontaneously flow from the cold body (making it colder) to the hot body (making it hotter)? This process does not violate the law of conservation of energy, and yet it never happens.
2. We are in a room with an open window. All the air molecules in the room spontaneously go out the window, leaving a vacuum. It is not impossible for this to happen. If it did, it would not violate the first law of thermodynamics. However, it never occurs.
3. A book on a desk spontaneously cools (using its own energy in overcoming gravity) and rises into the air. Again, this would not violate the first law, and yet all our experience tells us that it will not happen.
4. We have a plate filled with 10,000 coins, all heads. The plate is shaken repeatedly. Again, our experience tells us that with so many coins we should not expect to get all heads again, even though the possibility exists.

The preceding examples illustrate that there are certain processes which occur spontaneously and others which do not, although the latter are not forbidden by energy considerations. The last example, with the coins, is probably the easiest to understand. The *probability* of obtaining all heads with 10,000 coins is 1×10^{-3010}, an extremely small number. *For all practical purposes,* it will not occur. We are not saying that the process *cannot* occur, only that the probability is so small that it does not occur. Similarly, all the other examples are also illustrations of processes for which the probability is so negligibly small that they never happen. *The entropy is a measure of this probability of occurrence,* because entropy is a measure of randomness, and what all our examples have in common is that for any of them to happen, the system would *spontaneously* have to become *much less random—much more ordered.* Our entire experience tells us this does not take place. That is, all other things being equal, changes do not spontaneously take place in a system if these changes mean that the entropy of the system has to decrease. The greater the probability of occurrence, the greater the entropy; and the less the probability, the lower the entropy. If we shake 50 dice very hard and throw them on the floor 20

Table 8-4 Probability of the Outcome of Tossing a Coin Eight Times and the Number of Ways Each Particular Configuration Can Occur

Outcome of experiment	*Percent probability*	*Number of ways*
All heads	0.390625	1
7 heads, 1 tail	3.125	8
6 heads, 2 tails	10.9375	28
5 heads, 3 tails	21.875	56
4 heads, 4 tails	27.34375	70
3 heads, 5 tails	21.875	56
2 heads, 6 tails	10.9375	28
1 head, 7 tails	3.125	8
All tails	0.390625	1

times, we do not expect to see 50 sixes come up every time. It is not impossible, merely extremely improbable.*

The other three examples can also be explained in terms of probability (see Problem 8-29).

We can glean another important feature from the example about the coins. The higher probabilities are associated with greater randomness (disorder) and the lower probabilities with lesser randomness (more order). From Table 8-4, we observe that a mixture of heads and tails (more disorder) is more likely to occur than all heads or all tails (highly ordered), just as a mixture of numbers on 50 dice is more likely to turn up than 50 sixes. *There are more ways of achieving disorder than order.* The terminology we have been using may be grouped as follows:

Probability	*Order*	*Entropy*
High	Disordered	High
Low	Ordered	Low

From the illustrations above and countless other observations, scientists are able to state: **All naturally occurring (spontaneous) processes in nature take place with an increase in entropy (disorder) of the universe.**

We previously saw that the change in energy was one driving force in chemical equilibria. The change in entropy is the other driving force. **The entropy of a system tends toward a maximum (disorder), while the energy tends toward a minimum.** Given both of these state functions, we can consider two different definitions of equilibrium.

1. *If the energy of a system is kept constant, the system† will attain equilibrium when its entropy reaches a maximum.* The system will no longer spontaneously change, since any spontaneous change can only occur with an increase in total entropy. If the entropy is already at a maximum, there is no higher value that it can attain and, therefore, no spontaneous change can occur.
2. *If the entropy of a system is kept constant, the system will attain equilibrium when its energy reaches a minimum.* The system can no longer lose energy once the energy has already reached its smallest value and it will not spontaneously change with time.

*If it did happen, we might begin to suspect that the dice were loaded.

†To be rigorous, the system we are referring to is the universe. The *universe* is defined as the system plus the surroundings. There are process that occur spontaneously with a decrease in entropy in a system contained within a relatively small volume. However, when the total entropy change of the universe (system plus surroundings) is calculated, we find that the entropy of the universe has increased even though the entropy of the system has decreased. The entropy of the surroundings has increased to a greater extent than the entropy of the system has decreased.

The two definitions given above are rigorous and exact, but they have very limited applicability because experiments are not usually conducted at either constant energy or constant entropy. What, then, is the condition for equilibrium when the energy and entropy of a system are both allowed to change? To achieve this condition for equilibrium, we define a *new state function* that combines tendencies of the energy and entropy functions. This new state function is defined as

$$G = H - TS$$

where G is the **"Gibbs free energy,"** named after one of the fathers of thermodynamics—Josiah Willard Gibbs (1839–1903). Since we are interested only in changes in state functions, the change in G at constant temperature is given as

$$\Delta G = \Delta H - T\,\Delta S$$

From an analysis of this equation we can answer some of the fundamental questions asked at the beginning of this chapter. Consider the signs of each of the terms:

ΔH = heat of reaction—the change in enthalpy is usually negative for spontaneous reactions (exothermic); the system has lost a certain number of calories as heat

$T\,\Delta S$ = change in entropy times the temperature (the temperature is necessary to keep the units correct since ΔS has the units of cal/deg · mole); the entropy change is positive for a spontaneous change; in the equation the term is multiplied by a minus sign

Hence both terms (ΔH and $T\,\Delta S$) will be negative for a spontaneous change.

We can now state: For a chemical reaction,

if ΔG is negative, the reaction can occur spontaneously;

if ΔG is positive, the reaction cannot occur spontaneously, *no matter how long we wait.* If ΔG for a reaction is positive, energy (heat) must be applied in order to make the reaction take place.

What happens when $\Delta G = 0$? At this point in the reaction the energy factor (ΔH) is just offset by the entropy factor ($T\,\Delta S$); $\Delta H = T\,\Delta S$ and the two antagonistic tendencies are in balance. When this occurs, the system is at equilibrium.

In summary:

1. When we use the term "universe," we mean the system under consideration plus all the surroundings.
2. By equilibrium we mean a system that does not change with time.
3. The energy of a system tends toward a minimum. However, that the energy is at a minimum cannot be the sole criterion for a system to be at equilibrium.
4. The entropy is a term used to describe probability. Entropy is a measure of randomness. The greater the randomness, the greater the entropy and the higher the probability of occurrence. All spontaneous changes occur with an increase in the entropy of the *universe.*

The development of the concept of entropy came about as a necessary factor in the understanding of certain processes relating to the efficiency of heat engines. Scientists were interested in knowing what were the controlling factors in producing maximum efficiency in an engine. For example, is it possible for an engine to convert heat from the environment completely into useful work? Such an extraction of heat from the environment and complete conversion of it into useful work does not violate the first law of thermodynamics. However, in all human experience, no one has ever been able to accomplish this. This process is termed a "perpetual-motion machine of the second kind," whereas the production of work from nothing is called "a perpetual-motion machine of the first kind" (see Section 5-1). The impossibility of producing a perpetual-motion machine of the second kind is a consequence of the **second law of thermodynamics,** which may be stated: **It is impossible in any cyclic process completely to convert heat into work.** The link with the entropy comes about through a more detailed study. It turns out that when one tries *completely* to convert heat into work by a cyclic process (that is, a process that returns its variables to their initial conditions, which is necessary for the operation of an engine), a certain amount of the heat is lost and is therefore unavailable for conversion to work. This amount of wasted heat is measured by the expression q/T. This term has all the properties of a state function and indeed *is* the entropy, so that $\Delta S = q/T$. Thus, the *reason* we cannot convert heat *completely* into work is that the wasted heat ends up as increased randomness of the molecules.

Up to this point, we have discussed the concepts of entropy (S) and free energy (G), but we have not shown how to perform calculations involving these functions. The next two sections are devoted to this.

8-11 Calculations Involving Entropy

The entropy is thermodynamically* defined as

$$\Delta S = \frac{q_e}{T}$$

where ΔS = entropy change for the process studied
T = temperature in °K at which the process is conducted
q_e = heat change for the process, which must be carried out through a series of equilibrium steps (the e stands for equilibrium)†

Entropy Calculations in Phase Changes. It follows from the definition of entropy that the change in entropy for a change in phase is‡

$$\Delta S = \frac{q_e}{T} = \frac{\Delta H_{\text{pc}}}{T_{\text{pc}}}$$

where ΔH_{pc} = heat of the phase change
T_{pc} = temperature of the phase change

Example: The heat of vaporization of water is 9720 cal/mole. Calculate the entropy change when 1 mole of liquid water is completely vaporized at 373°K.

Answer:

$$H_2O(l) \longrightarrow H_2O(g) - 9720 \text{ cal}$$

9720 cal has to be added to the system (endothermic) in order to vaporize 1 mole of $H_2O(l)$.

$$\Delta S = \frac{9720}{373} = 26.1 \text{ cal/°K} \cdot \text{mole}$$

*This equation, which defines a change in entropy, also turns out to be the thermodynamic definition of temperature. Temperature is that quantity which when divided into the heat change in an equilibrium process generates a new state function, the entropy.

†A complete statement of q_e is well beyond the scope of this book. For our purposes the qs we shall consider will always be proper in a thermodynamic sense.

‡This equation can be obtained from $\Delta G = \Delta H - T\Delta S$. For an equilibrium phase change, $\Delta G = 0$ and $\Delta S = \Delta H/T$.

For the reverse process at 373°K,

$$H_2O(g) \longrightarrow H_2O(l) + 9720 \text{ cal}$$

the system loses 9720 cal (exothermic), so that

$$\Delta S = \frac{-9720}{373} = -26.1 \text{ cal/°K} \cdot \text{mole}$$

Note that in this calculation we have assumed that the water is vaporized at 373°K (the boiling point when the pressure is 1 atm). At this point we do not have enough information to calculate ΔS for vaporization of water at any other temperature. In the rest of our discussions and calculations, we shall *always* consider the pressure on the system under discussion to be 1 atm unless otherwise specified. As we saw in Section 5-5, the superscript ° is used to indicate 1 atm. Thus, $\Delta H°$, $\Delta S°$, $\Delta G°$, and $\Delta E°$ mean that all these changes take place at a pressure of 1 atm.

The preceding example shows that when a liquid becomes a gas, the entropy increases. This is certainly in accord with our discussion of the meaning of entropy. The molecules of a liquid are in a less random condiiton than those of a gas, since the gas molecules are free to roam throughout the container, while the liquid molecules cannot. The entropy increases since the system has become more random. We would also predict that a similar statement would apply to a phase change between a solid and a liquid. This prediction is borne out: when a solid melts, the entropy of the system increases (Figure 8-7). The entropy change for the reaction

$$H_2O(l) \longrightarrow H_2O(g)$$

can be viewed as

$$\Delta S° = \sum S°_{\text{(products)}} - \sum S°_{\text{(reactants)}}$$
$$\Delta S° \text{ positive means } S(H_2O(g)) > S(H_2O(l))$$

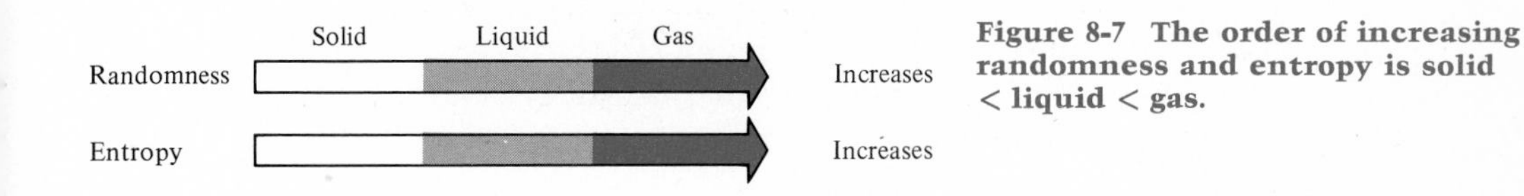

Figure 8-7 The order of increasing randomness and entropy is solid < liquid < gas.

ENTROPY CALCULATIONS IN CHEMICAL REACTIONS. We stated in Section 5-1 that we cannot measure the individual values of state functions, only the changes in these values (for example, ΔH, but not H). *The entropy is an exception to this rule.* The ability to calculate absolute entropies came about as a result of the achievement by certain scientists of extremely low temperatures. At this time the lowest temperature on record is 0.00002°K.*

Given such a low temperature, it would *appear* that the attainment of absolute zero would be forthcoming shortly. However, studies of all processes which lead to such low temperatures have shown that absolute zero could only be attained after an *infinite* number of cooling cycles. Obviously, it is impossible to carry out an infinite number of cycles. A generalized statement of this inability to reach absolute zero is called the **third law of thermodynamics: It is impossible by any procedure, no matter how idealized, to reduce the temperature of any system to absolute zero in a finite number of operations.**

What is the entropy of substances at absolute zero? If at absolute zero a substance is a perfect crystal; that is, all the atoms, molecules, or ions are arranged in an absolutely regular order, and this is the only possible arrangement, we have a system with *perfect order*. Since entropy is a measure of randomness, it must be zero for a

*D. deKlerk, M. J. Steenland, and C. J. Gortner, *Physica* **16,** 571 (1950).

Table 8-5 Third-Law Entropies at 298°K, S°_{298} (cal/°K · mole)

Solids		Gases		Liquids	
Substance	S°_{298}	Substance	S°_{298}	Substance	S°_{298}
Ag	10.20	CH_4	44.47	Br_2	36.4
AgCl	23.00	CO	46.20	H_2O	16.73
Al	6.77	CO_2	51.08	Hg	18.17
C (graphite)	1.37	Cl_2	53.29	CH_3OH	30.3
C (diamond)	2.44	H_2	31.21	C_2H_5OH	38.4
Ca	9.95	HCl	44.64	C_6H_6	41.3
CaO	9.5	HBr	47.6		
Cd	12.37	N_2	45.90		
Cu	7.97	NO	50.34		
S (rhombic)	7.62	O_2	49.01		
S (monoclinic)	7.78	D_2	34.60		
Fe	6.49	H_2O	45.11		
Na	12.2	NH_3	46.01		
I_2	27.76	SO_2	59.40		
NaCl	17.30	C_2H_2	47.50		
AgCl	23.00	C_2H_4	52.45		
AgBr	25.60	C_2H_6	54.85		

perfectly ordered system. This type of analysis gives us another form of the third law of thermodynamics: **At absolute zero, the entropy of perfectly crystalline substances is zero.**

We are still one step away from being able to make use of these laws in entropy calculations. The problem is: What do we do if a substance at absolute zero is *not* a perfect crystal? The answer to this question is that for such substances we use *changes* in entropy. We will now demonstrate that even if the entropy of a substance at absolute zero is not zero, we can *for convenience* take this value to be zero without affecting our entropy calculations.

Consider any particular substance, A. We shall label its entropy value at 0°K as S^*_A and its entropy value at some other temperature, say 300°K, as S_A. We may think of S_A as being composed of two parts:

$$S_A = S^*_A + S_T$$

where S_T is the gain in entropy of substance A in going from 0°K to 300°K. Since only changes or differences in entropy have any physical meaning in chemical thermodynamics, we must ask what happens to S^*_A when substance A is involved in a chemical reaction at, say, 300°K. We know from Chapter 3 that matter is neither created nor destroyed in any chemical reaction. Therefore, all the atoms of A must also appear in some form on the product side of the equation. When we measure or calculate differences in entropy, S^*_A must cancel out, since it appears on both sides of the equation. This is true for all substances. We can therefore assign to the entropy a substance possesses at absolute zero any value we choose, since it will in no way affect any ΔS calculations. It is certainly convenient to *choose the value of* S°_0 (the standard entropy at 0°K) *for all elements to be zero.* Using thermodynamic techniques not discussed here, it is possible to calculate the changes in entropy of substances on going from 0°K to any other temperature *T*. Table 8-5 lists entropies of substances at 298°K and 1 atm pressure. They are labeled **third-law entropies,** since use was made of the third law in calculating them.

Example: From Table 8-5, calculate the change in entropy (ΔS°) for the following reaction at 298°K.

$$CH_4(g) + 2O_2(g) \longrightarrow CO_2(g) + 2H_2O(l)$$

Answer:

$$\Delta S^\circ_{298} = \sum S^\circ_{\text{products}} - \sum S^\circ_{\text{reactants}}$$
$$= 51.08 + 2(16.73) - 44.47 - 2(49.01)$$
$$= 84.54 - 142.49$$
$$= -57.95 \text{ cal/deg}$$

8-12 Standard Free Energies

In Section 5-5, we found it convenient to define standard *heats* of formation. By using the tables of standard heats of formation, we saw that it is possible to calculate heats of reaction at 298°K. We may apply the same reasoning in defining a standard state for the *free energies* of formation of compounds. By convention, *elements* (not compounds) in the states in which they are most stable at 1 atm pressure and 298°K are assigned a free-energy value of zero. The **standard free energy of formation** *of a compound* is the free energy of the reaction by which it is formed from its elements at 1 atm pressure and 298°K, the standard state. In an equation similar to the one we wrote for ΔH°_{298} of formation, we can write for the free energy of formation ΔG°_{298}:

$$\Delta G^\circ_{298} = \sum G^\circ_{298\text{(products)}} - \sum G^\circ_{298\text{(reactants)}} \tag{1}$$

The free energy of formation is equal to the sum of the free energies of the products minus the sum of the free energies of reactants. Table 8-6 lists standard free energies of formation for various compounds.

Example: Calculate the standard free energy change (ΔG°) for the reaction

$$CO(g) + H_2O(l) \longrightarrow CO_2(g) + H_2(g) \quad \text{at 298°K and 1 atm pressure}$$

Answer: From equation 1, ΔG°_{298} for the reaction equals

$$\sum \Delta G^\circ_{298\text{(products)}} = \Delta G^\circ_{298}(CO_2(g) + \Delta G^\circ_{298}(H_2(g))$$
$$\text{minus} \sum \Delta G^\circ_{298\text{(reactants)}} = \Delta G^\circ_{298}(CO(g)) + \Delta G^\circ_{298}(H_2O(l))$$

From Table 8-6, the values are

$$CO_2(g) = -94{,}260 \text{ cal/mole}$$
$$H_2(g) = 0.0 \text{ (element in its standard state)}$$

Table 8-6 Standard Free Energies of Formation at 25°C and 1 atm Pressure

Compound	ΔG°_{298} (*cal/mole*)	*Compound*	ΔG°_{298} (*cal/mole*)
$H_2O(g)$	−54,635	$C_6H_6(g)$	+30,989
$H_2O(l)$	−56,690	$C_6H_6(l)$	+29,756
$H_2S(g)$	−7,892	$AgCl(s)$	−26,224
$NH_3(g)$	−3,976	$AgBr(s)$	−22,930
$N_2O(g)$	+24,760	$AgI(s)$	−15,850
$NO(g)$	+20,719	$CaCO_3(s)$	−269,780
$NO_2(g)$	+12,390	$CaSO_4(s)$	−315,560
$N_2O_4(g)$	+23,440	$CuO(s)$	−30,400
$SO_2(g)$	−71,790	$Fe_2O_3(s)$	−177,100
$SO_3(g)$	−88,520	$HBr(g)$	−12,700
$CH_4(g)$	−12,140	$HCl(g)$	−22,800
$CO(g)$	−32,808	$HF(g)$	−64,700
$CO_2(g)$	−94,260	$HI(g)$	+300

$$CO(g) = -32{,}808 \text{ cal/mole}$$
$$H_2O(l) = -56{,}690 \text{ cal/mole}$$
$$\Delta G^\circ_{298(\text{reaction})} = (-94{,}260 + 0.0) - (-32{,}808 - 56{,}690)$$
$$\Delta G^\circ_{298} = -4762 \text{ cal}$$

CALCULATIONS OF ΔG°_{298} VALUES. The reader may be wondering how the values of ΔG°_{298} presented in Table 8-6 were obtained. One method makes use of third-law entropies. By using the equation $\Delta G^\circ_{298} = \Delta H^\circ_{298} - T\,\Delta S^\circ_{298}$ and applying the values of ΔH°_{298} and ΔS°_{298}, ΔG°_{298} is readily calculable. We have already seen that ΔH°_{298} values are obtained from tables of heats of formation, which come from experimental measurements (Section 5-6). ΔS°_{298} values are obtained from third-law entropies.

Example: In the previous section, we calculated ΔS°_{298} for the reaction

$$CH_4(g) + 2O_2(g) \longrightarrow CO_2(g) + 2H_2O(l)$$

From Tables 8-6 and 5-4 (page 134), calculate ΔG°_{298}, for this reaction.

Answer: First we calculate ΔH°_{298}, using values from Table 5-4.

$$\Delta H^\circ_{298} = \sum \Delta H^\circ_{\text{products}} - \sum \Delta H^\circ_{\text{reactants}}$$
$$= (-94{,}052) + 2(-68{,}320) - (-17{,}890) = -212{,}802 \text{ cal}$$

We have already calculated ΔS°_{298} to be -57.95 cal/°K.

Therefore,

$$\Delta G^\circ_{298} = \Delta H^\circ_{298} - T\,\Delta S^\circ_{298}$$
$$= (-212{,}802 \text{ cal}) - 298\text{°K}(-57.95 \text{ cal/°K})$$
$$= -195{,}533 \text{ cal}$$

However, this method is not the one most often used in practice. A more general method for obtaining ΔG°_{298} values for compounds is through the use of electrochemical cells. This method is discussed in Section 16-11.

8-13 Relationships Between ΔG, ΔG°, and the Equilibrium Constant

In this section, we will show that ΔG° is related to the equilibrium constant by the equation: $\Delta G^\circ = -RT \ln K_p$. K_p is the equilibrium constant expressed in pressure units (Section 8-8). When we discussed ΔG°_{298} and the construction of the table of free energies, it was emphasized that the elements and the corresponding compounds must be in their standard states, that is, at one atmosphere pressure. But what happens to the free energies when the substances are not at 1 atm pressure? This is a very important question, since free-energy changes with changing pressure even if the temperature is constant. More important, once we determine ΔG (not ΔG°) for a given reaction, we can state *with certainty* whether that reaction will occur spontaneously.

For an ideal gas, thermodynamics tells us that the free-energy changes with changing pressure in the following manner:

$$\Delta G = G_2 - G_1 = nRT \ln\left(\frac{P_2}{P_1}\right) \tag{2}$$

where G_2 = free energy of n moles of a gas at pressure P_2
G_1 = free energy of the same amount of the same gas at pressure P_1
ΔG = change in free energy in going from P_1 to P_2
(at a constant temperature T)

Since we are going to relate this to ΔG_{reac} and $\Delta G^\circ_{\text{reac}}$, we will choose the pressure P_1 to be the standard-state pressure, 1 atm. Since G_1 is the free energy of the substance at 1 atm, it is the standard free energy and we label it as G_1°. Equation (2) now takes the form

$$G_2 - G_1^\circ = nRT\ln\left(\frac{P_2}{P_1}\right) \qquad (P_1 = 1 \text{ atm}) \tag{3}$$

Note that P_2 must also be expressed in atmospheres, since the ratio P_2/P_1 must be a dimensionless number. The reason is purely mathematical: we can only take the logarithm of a dimensionless number. We can rewrite equation (3) as

$$G_2 - G_1^\circ = nRT\ln\left(\frac{P_2 \text{ (atm)}}{1 \text{ atm}}\right)$$

or

$$G_2 - G_1^\circ = nRT\ln P_2 \tag{4}$$

In the rest of the derivation relating $\Delta G^\circ_{\text{reac}}$ to K_p, we shall apply equation (4) to *all* the different species in a chemical reaction.

Let us consider a general chemical reaction, where all the species are in the gaseous state.

$$a\text{A} + b\text{B} \rightleftharpoons c\text{C} + d\text{D}$$

where A, B, C, and D are the various gaseous chemical species and a, b, c, and d are their corresponding stoichiometric coefficients in the balanced equation.

For each of the species in the chemical equation, we can write an equation of the form of equation (4):

$$G_2 - G_1^\circ = nRT\ln P$$

For species A,

$$G_{2,\text{A}} - G^\circ_{1,\text{A}} = aRT\ln P_\text{A} = RT\ln(P_\text{A})^{a}* \tag{5}$$

For species B,

$$G_{2,\text{B}} - G^\circ_{1,\text{B}} = bRT\ln P_\text{B} = RT\ln(P_\text{B})^{b} \tag{6}$$

For species C,

$$G_{2,\text{C}} - G^\circ_{1,\text{C}} = cRT\ln P_\text{C} = RT\ln(P_\text{C})^{c} \tag{7}$$

For species D,

$$G_{2,\text{D}} - G^\circ_{1,\text{D}} = dRT\ln P_\text{D} = RT\ln(P_\text{D})^{d} \tag{8}$$

Since we are only interested in the *change* in free energy (ΔG) for the reaction, we have

$$\Delta G_{\text{reac}} = \sum G \text{ products} - \sum G \text{ reactants} \tag{9}$$

Applying this to species A, B, C, and D, we get

$$(G_{2,\text{D}} + G_{2,\text{C}} - G_{2,\text{B}} - G_{2,\text{A}}) - (G^\circ_{1,\text{D}} + G^\circ_{1,\text{C}} - G^\circ_{1,\text{B}} - G^\circ_{1,\text{A}}) = RT\ln\left(\frac{P_\text{C}^c P_\text{D}^d}{P_\text{A}^a P_\text{B}^b}\right) \tag{10}$$

*This form is obtained by making use of the properties of logarithms, $a\ln P_\text{A} = \ln(P_\text{A})^a$.

or

$$\Delta G_{\text{reac}} - \Delta G^\circ_{\text{reac}} = RT \ln\left(\frac{P_C^c P_D^d}{P_A^a P_B^b}\right) \tag{11}$$

Let us pause here and review what we wish to obtain and also discuss the individual terms in equation (11). First we wish to show when, in a thermodynamic sense, a chemical reaction will *spontaneously* occur. Second, we wish to relate $\Delta G^\circ_{\text{reac}}$ to the equilibrium constant, K_p. ΔG_{reac} is the difference in free energies of products and reactants when the species are at any particular partial pressures P_D, P_C, P_A, and P_B. ΔG_{reac} will change with changing pressure or temperature. $\Delta G^\circ_{\text{reac}}$ is the free-energy difference of products and reactants *when the species are in their standard states* and its value is calculable from the data in Table 8-6. $\Delta G^\circ_{\text{reac}}$ also changes with changing temperature.

$$\Delta G = \Delta G^\circ + RT \ln\left(\frac{P_D^d P_C^c}{P_B^b P_A^a}\right) \tag{11}$$

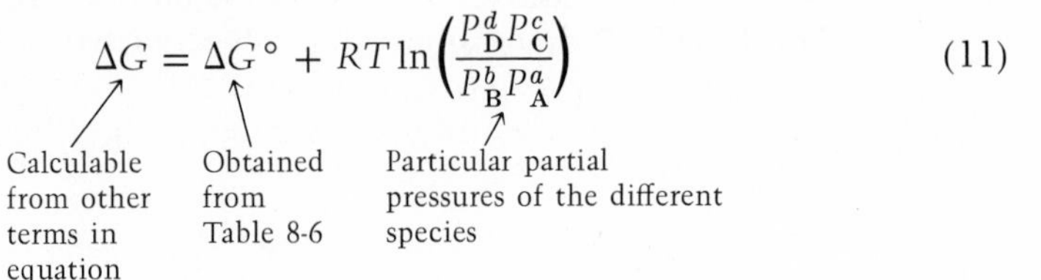

The sign of ΔG_{reac} gives us the answer to our first question. From the thermodynamic principles stated on page 223 we have:

If ΔG_{reac} is negative, the reaction will proceed spontaneously to equilibrium.

If ΔG_{reac} is positive, the reaction will not proceed spontaneously and energy must be supplied in order for the reaction to take place.

If ΔG_{reac} is zero, the reaction *is* at equilibrium and, of course, will not proceed in either direction.

Let us illustrate all of this by an example. Consider the following reaction at 298°K:

$$H_2(g) + \tfrac{1}{2}O_2(g) \rightleftharpoons H_2O(g)$$

We shall assume that at the beginning of the reaction, all the species in the reaction are in their standard states, 1 atm pressure. From equation (11), we have

$$\Delta G = \Delta G^\circ + RT \ln\left(\frac{P_{H_2O}}{P_{H_2} P_{O_2}^{1/2}}\right) = \Delta G^\circ + RT \ln\left[\frac{1\ \text{atm}}{(1\ \text{atm})(1\ \text{atm})^{1/2}}\right] \tag{12}$$

Since ln 1 equals zero, we obtain

$$\Delta G = \Delta G^\circ \text{ for the above conditions}$$

ΔG° is calculable from the standard free energies of formation given in Table 8-6:

$$\Delta G^\circ = G_{H_2O}(g) - G_{H_2}(g) - \tfrac{1}{2}G_{O_2}(g)$$
$$\Delta G^\circ = [(-54{,}635) - 0 - 0] = -54{,}635 \text{ cal/mole}$$

The free energies of both hydrogen and oxygen are zero by definition of the standard state. Now, since

$$\Delta G = \Delta G^\circ = -54{,}635 \text{ cal/mole}$$

we can say that *the reaction will spontaneously occur, and proceed to equilibrium.* However, as the reaction proceeds, *the initial values of the pressure will be changing,* and hence ΔG will change accordingly. As we stated before, when the pressures reach their equilibrium values, ΔG will equal zero and we have:

At equilibrium, $\Delta G_{\text{reac}} = 0$, so

$$\Delta G° = -RT \ln \left[\frac{P_{H_2O}(\text{equilibrium value})}{P_{H_2}(\text{equilibrium value})\ P_{O_2}^{1/2}(\text{equilibrium value})} \right] \tag{13}$$

The ratio of pressures at equilibrium is K_p, the *pressure equilibrium constant,* because at equilibrium, the partial pressures of all the species are the equilibrium pressures.

$$K_p = \frac{P_{H_2O}(\text{equilibrium})}{P_{H_2}(\text{equilibrium}) P_{O_2}^{1/2}(\text{equilibrium})} \tag{14}$$

so equation (13) may be rewritten as

$$\Delta G° = -RT \ln K_p \tag{15}$$

This relationship between K_p and $\Delta G°$ is one of the most useful in chemistry. It allows us to calculate the value of the equilibrium constant for a reaction by simply looking up the $G°$ values of formation of the different species.

Whenever the pressures have values different from their equilibrium values, the system is no longer at equilibrium and ΔG_{reac} is not zero.

Example: Estimate the value of the equilibrium constant, K_p, for the reaction $H_2(g) + \frac{1}{2}O_2(g) \rightleftharpoons H_2O(g)$ at 298°K.

Answer: From the above we have that $\Delta G° = -54{,}635$ cal/mole and $\Delta G° = -RT \ln K_p$. Therefore,

$$\ln K_p = \frac{-\Delta G°}{RT} = \frac{-(-54{,}635)}{(1.99)(298)} = 92.1$$

$$K_p \approx 10^{40}$$

Because the value of K_p is so high, we can say that, for all practical purposes, this reaction will go to completion. We may also note that one consequence of equation (15) is that for any reaction for which $\Delta G°$ is negative, $\ln K_p$ is positive, so $K_p > 1$. (Readers can work this out for themselves using the properties of logarithms.) Conversely, when $\Delta G°$ is positive, $\ln K_p$ is negative and K_p is less than 1. In the former case ($K_p > 1$), this means that at equilibrium there will be more products than reactants, and in the latter case there will be more reactants than products. It is instructive to see how the value of K_p changes with different values of $\Delta G°$. Table 8-7 presents these data for a temperature of 298°K. Note that this table applies to *any* reaction at 298°K. Table 8-7 shows that K_p values change much faster than $\Delta G°$ values. For example, when $\Delta G°$ varies from -5000 to $-10{,}000$ cal/mole, K_p varies from $\sim 5 \times 10^3$ to 2×10^7. As a rough rule, if the $\Delta G°$ values are equal to or greater

Table 8-7 How K_p Changes with $\Delta G°$ at 298°K

$\Delta G°$ (*cal/mole*)	$\ln K_p = \frac{-\Delta G°}{RT} = \frac{-\Delta G°}{(1.99)(298)}$	K_p
−10,000	16.9	2.19×10^7
−5,000	8.43	4.58×10^3
−2,000	3.37	29.1
−1,000	1.69	5.42
0.0	0.0	1
+1,000	−1.69	0.185
+2,000	−3.37	0.0344
+5,000	−8.43	2.18×10^{-4}
+10,000	−16.9	4.58×10^{-8}

than 10,000 cal/mole, we can use *them* to predict the spontaneity of a reaction instead of ΔG. This works because the actual ratio $P_C^c P_D^d / P_A^a P_B^b$ would have to be greater than 10^7 to overcome the $\Delta G°$ value.

As we have mentioned, whenever the pressures have values different from their equilibrium values, the system is no longer at equilibrium. Therefore, ΔG_{reac} is not zero and can be calculated from equation (11). This is illustrated by the following example.

Example: Given that the partial pressures of the species in the reaction at 298°K,

$$CO(g) + H_2O(g) \longrightarrow H_2(g) + CO_2(g)$$

are

$$CO(g) = 5 \text{ atm} \qquad H_2(g) = 3 \text{ atm}$$
$$H_2O(g) = 2 \text{ atm} \qquad CO_2(g) = 3 \text{ atm}$$

and the fact that the reaction is not at equilibrium, using appropriate tables and equations, calculate ΔG_{reac}, $\Delta G°_{reac}$, and K_p and discuss in which direction the reaction will proceed.

Answer: From Table 8-6, $\Delta G°_{298}$ for the reaction is

$$\Delta G°_{298} = -94{,}260 - (-32{,}808 - 54{,}635) = -6817 \text{ cal}$$
$$-\Delta G°_{298} = RT \ln K_p$$
$$\ln K_p = \frac{6817}{(1.99)(298)} = 11.5$$
$$K_p = 1 \times 10^5$$

Note that K_p for the reaction depends only on $\Delta G°$ and is thus independent of the values of partial pressures.

For such a large value of the equilibrium constant, we would expect to have mostly products and very little reactants.

$$\Delta G_{reac} - \Delta G° = RT \ln \left(\frac{P_C^c P_D^d}{P_A^a P_B^b} \right)$$

Remember that the ratio $P_C^c P_D^d / P_B^b P_A^a$ in the equation above is *not* the equilibrium constant since the pressures are not the equilibrium pressures.

$$\Delta G_{reac} + 6817 = (1.99)(298) \ln \left[\frac{(3)(3)}{(5)(2)} \right]$$
$$\Delta G_{reac} = -6879 \text{ cal}$$

Thus, the reaction will spontaneously proceed to the right (as written) until the pressures of the products and reactants adjust themselves to obtain a ΔG_{reac} equal to zero.

Example: For the reaction at 298°K,

$$H_2(g) + Cl_2(g) \rightleftharpoons 2HCl(g)$$

Calculate the equilibrium constant using the data in Table 8-6.

Answer: From Table 8-6, we have

$$\Delta G°_{298} = 2(-22{,}800) = -45{,}600$$

From $\Delta G°_{298} = -RT \ln K_p$, we get

$$\ln K_p = \frac{\Delta G^\circ_{298}}{RT} = \frac{+45{,}600}{(1.99)(298)} = 76.9$$

$$K_p = 2 \times 10^{33} = \frac{P^2_{HCl}}{P_{H_2}P_{Cl_2}}$$

This is a huge equilibrium constant.

Before ending this section, we would like to state once again that a chemical reaction with a negative free energy proceeds spontaneously, while a reaction whose free energy is positive does not proceed. However, these signs tell us nothing about how long it takes for these spontaneous processes to occur. In some cases, it could take billions of years for a reaction to reach equilibrium, even though the free-energy change for that reaction is negative.

This is why, for example, it is possible to keep N_2O (laughing gas) in a container without it decomposing to give nitrogen and oxygen. It is also the reason why a mixture of hydrogen and oxygen can be kept at 298°K and 1 atm indefinitely without any water being formed, although ΔG° is negative for water in both the liquid and gaseous phases. The *rate* at which a chemical reaction proceeds is *not* controlled by its free energy and is often not even remotely related to its free energy. The rate is determined by other factors. Chemical kinetics (the study of reaction rates) is the subject of Chapter 9.

8-14 Calculation of ΔG° at Any Temperature

We have shown how a knowledge of the free energy of formation of compounds may be used to calculate equilibrium constants. In this calculation we were, however, restricted to the standard state temperature, 298°K. If we were able to calculate ΔG° for a reaction at some other temperature, we could apply equation (15) and calculate the equilibrium constant at that other temperature.

The relevant equation which enables us to do this is called the **Gibbs–Helmholtz equation** and gives the free energy at any temperature T, if we know ΔG°_{298}. The Gibbs–Helmholtz equation is

$$\frac{\Delta G^\circ_T}{T} - \frac{\Delta G^\circ_{298}}{298} = \Delta H^\circ_{298}\left(\frac{1}{T} - \frac{1}{298}\right) \qquad (16)$$

where ΔG°_T = free-energy change of the reaction at a temperature T
ΔG°_{298} = free-energy change of the reaction at 298°K
ΔH°_{298} = heat of the reaction at 298°K

Since both ΔG°_{298} and ΔH°_{298} can be obtained from tables, we can make this calculation on reactions without even having to run them. However, equation (16) is applicable only when ΔH°_{298} is *constant* over the temperature range considered. When ΔH°_{298} is not constant, the Gibbs–Helmholtz equation becomes quite complex and will not be pursued here. Generally, ΔH°_{298} is constant over a range of about 50°K or less. The Gibbs–Helmholtz equation is not just valid for 298°. If necessary, we can find ΔG° at any temperature if we know ΔG° at any other temperature.

Problems

8-1 Define each of the following terms. **(a)** reversible reaction **(b)** dynamic equilibrium **(c)** chemical equilibrium **(d)** equilibrium constant **(e)** catalyst **(f)** entropy **(g)** free energy **(h)** standard free energy of formation

8-2 What is meant by each of the following symbols? **(a)** $\rightleftharpoons$ **(b)** [A] **(c)** K **(d)** K_p **(e)** S **(f)** $\Delta S°$ **(g)** G

8-3 Write the equilibrium expression for each of the following reactions.
(a) $N_2(g) + 3H_2(g) \rightleftharpoons 2NH_3(g)$
(b) $PCl_3(g) + Cl_2(g) \rightleftharpoons PCl_5(g)$
(c) $CO(g) + H_2O(g) \rightleftharpoons CO_2(g) + H_2(g)$
(d) $N_2O_4(g) \rightleftharpoons 2NO_2(g)$
(e) $\frac{1}{2}N_2O_4(g) \rightleftharpoons NO_2(g)$
(f) $D_2O(g) + H_2O(g) \rightleftharpoons 2HDO(g)$
(g) $H_2(g) \rightleftharpoons 2H(g)$
(h) $2C_2H_6(g) + 7O_2(g) \rightleftharpoons 4CO_2(g) + 6H_2O(l)$
(i) $PH_3(g) + 3F_2(g) \rightleftharpoons PF_3(g) + 3HF(g)$
(j) $S_8(s) \rightleftharpoons 8S(g)$
(k) $2Cl_2O_5(g) \rightleftharpoons 4ClO_2(g) + O_2(g)$
(l) $PO_4^{3-} + 3HCl \rightleftharpoons H_3PO_4 + 3Cl^-$

8-4 In Section 6-2, we discussed vapor pressure. Explain how a dynamic equilibrium must be present if the vapor pressure of a liquid is to be measured.

8-5 At 1225°C the equilibrium constant for the reaction

$$2HBr(g) \rightleftharpoons H_2(g) + Br_2(g)$$

is $K = 2.86 \times 10^{-5}$. Write equilibrium expressions and calculate the equilibrium constants for each of the following.
(a) $H_2(g) + Br_2(g) \rightleftharpoons 2HBr(g)$
(b) $HBr(g) \rightleftharpoons \frac{1}{2}H_2(g) + \frac{1}{2}Br_2(g)$
(c) $\frac{1}{2}H_2(g) + \frac{1}{2}Br_2(g) \rightleftharpoons HBr(g)$
(d) $2H_2(g) + 2Br_2(g) \rightleftharpoons 4HBr(g)$

8-6 Given these equilibrium constants at 25°C:

$$S + S^{2-} \rightleftharpoons S_2^{2-} \qquad K = 1.7$$
$$S + S_2^{2-} \rightleftharpoons S_3^{2-} \qquad K = 3.1$$

Calculate K for $2S + S^{2-} \rightleftharpoons S_3^{2-}$ at 25°C.

8-7 Given the reaction

$$2H_2S(g) + 3O_2(g) \rightleftharpoons 2H_2O(g) + 2SO_2(g)$$

if 8.6 g of $H_2S(g)$ are mixed with 4.7 g of $O_2(g)$ and 0.50 g of $H_2O(g)$ in a 3.50-liter container, what are the concentrations of all four components at equilibrium. (Since you are not given K, set up the calculations, but do not solve them.)

8-8 In a closed 700-ml flask, there are initially 8.0 g of Cl_2O_5, 3.0 g ClO_2, and 12 g of O_2 (all gases); calculate the initial concentration of each of the species in moles/liter. If the size of the flask was 900 ml, what would be the initial concentrations?

8-9 The $N_2O_4(g) \rightleftharpoons 2NO_2(g)$ equilibrium constant at 25°C is $K = 5.85 \times 10^{-3}$. **(a)** Write the equilibrium expression. **(b)** Fill in the table.

Size of container (liters)	*Initial amounts (g)*		*Equilibrium concentration (moles/liter)*		*Percent dissociation of N_2O_4*
	$N_2O_4(g)$	$NO_2(g)$	N_2O_4	NO_2	
2.00	200.0	0	____	____	____
3.00	0	18.0	____	____	– × –
1.50	18.5	18.5	____	____	– × –

8-10 For the reaction

$$PCl_3(g) + Cl_2(g) \rightleftharpoons PCl_5(g)$$

$K = 17.5$ at 575°K.
(a) If 17.0 g each of PCl_5 and Cl_2 are initially introduced into a 6.00-liter vessel at 575°K (no PCl_3) calculate the equilibrium concentration of each species. **(b)** If the experiment was repeated and just 17.0 g of PCl_5 was introduced into the 6.00-liter vessel, calculate the equilibrium concentration of each species. **(c)** If the experiment was repeated again and just 17.0 g of PCl_3 was initially introduced into the 6.00-liter vessel, calculate the equilibrium concentration of each species.

8-11 The equilibrium constant for the water–gas reaction

$$CO(g) + H_2O(g) \rightleftharpoons CO_2(g) + H_2(g) \text{ is } 4.00$$

(a) If 100.0 g each of CO, H_2O, CO_2, and H_2 are introduced into a 1.500-liter vessel and allowed to come to equilibrium, calculate the equilibrium concentration of all the species **(b)** What effect would doubling the volume have on the equilibrium concentrations of each of the species?

8-12 Given the water–gas equilibrium

$$H_2(g) + CO_2(g) \rightleftharpoons H_2O(g) + CO(g)$$

at 1260°K and 1 atm in a 1-liter vessel, and $K = 1.60$, fill in the table.

	Initial composition (moles/liter)				*Equilibrium composition (moles/liter)*			
	CO_2	H_2	H_2O	CO	CO_2	H_2	H_2O	CO
(a)	1	9	0	0	___	___	___	___
(b)	0	0	1	9	___	___	___	___
(c)	1	1	1	0	___	___	___	___
(d)	0	1	9	9	___	___	___	___

8-13 For a hypothetical gas reaction

$$2AB_3(g) \rightleftharpoons A_2(g) + 3B_2(g)$$

2.00 moles of AB_3 are introduced into an evacuated 1.00-liter flask. At equilibrium, 1.00 mole of AB_3 remains. Calculate the equilibrium constant, K.

8-14 For the reaction

$$2NOCl(g) \rightleftharpoons 2NO(g) + Cl_2(g)$$

the equilibrium constant is 0.10. This reaction was studied in a 3.00-liter vessel, where at equilibrium the numbers of moles of NOCl and Cl_2 were 6.00 and 0.60, respectively. Calculate the number of moles of NO which are present at equilibrium.

8-15 For the reaction

$$COCl_2(g) \rightleftharpoons CO(g) + Cl_2(g)$$

the equilibrium constant at 1000°K is 0.320. This reaction was studied in a 4.00-liter vessel, where at equilibrium there were present 1.40 moles of $COCl_2$ and 1.16 moles of Cl_2. If an additional 1.00 mole of CO was added to the vessel and the system allowed to reestablish equilibrium, calculate the equilibrium concentrations of all species.

8-16 CO_2 and H_2 were mixed in a container kept at 959°K. Their initial concentrations were 0.0200 M for CO_2 and 0.0100 M for H_2. The reaction

$$CO_2(g) + H_2(g) \rightleftharpoons CO(g) + H_2O(g)$$

then occurred. At equilibrium the concentration of $H_2O(g)$ was found to be 8.5×10^{-3} M. Calculate the concentrations of CO_2, H_2, and CO at equilibrium, and K for the reaction at 959°K.

8-17 For the reaction

$$CH_3COOH + CH_3OH \rightleftharpoons CH_3COOCH_3 + H_2O$$

$K = 2.00$ at 50°C. If 1.00 mole of CH_3COOH is added to 0.500 mole of CH_3OH in enough solvent to make up a total volume of 1.00 liter, what will be the concentrations in moles/liter of all four substances at equilibrium? (The solvent is not any of the above four substances.)

8-18 The gaseous reaction

$$3O_2 \rightleftharpoons 2O_3 - \text{heat}$$

is allowed to come to equilibrium. Explain what happens to the concentrations of O_2 and O_3 when the following stresses are applied to the system. **(a)** adding O_3 **(b)** Increasing the pressure by a piston **(c)** lowering the temperature **(d)** removing O_2 **(e)** increasing the volume of the container **(f)** adding a catalyst **(g)** adding heat **(h)** adding an inert gas such as Ar without changing the volume **(i)** raising the temperature

8-19 If the reaction

$$C(s) + CO_2(g) \rightleftharpoons 2CO(g) - \text{heat}$$

is at equilibrium, what would happen *to the concentration of* CO_2 and *to the equilibrium constant K* if each of the following things were done? **(a)** the temperature was increased **(b)** the pressure was decreased by raising a piston **(c)** more CO_2 was added **(d)** a catalyst was added **(e)** the volume of the system was increased **(f)** the pressure was increased by adding the inert gas helium **(g)** some, but not all, of the C(s) was removed **(h)** more CO was added

8-20 Given the reaction

$$N_2(g) + O_2(g) \rightleftharpoons 2NO(g) - 43.2 \text{ kcal}$$

At 2300°K, the measured equilibrium constant was found to be 6.2×10^{-2}. Discuss the effect on the equilibrium constant when **(a)** the pressure is increased by a piston **(b)** the temperature is lowered **(c)** a catalyst is introduced into the reaction vessel **(d)** after equilibrium is obtained, more $O_2(g)$ is introduced into the reaction vessel **(e)** after equilibrium is obtained, 2 moles of the inert gas argon are introduced into the reaction vessel, without changing the volume.

8-21 In Problem 8-20, **(a)** if 10.0 g of NO(g) are introduced into a 4.00-liter reaction vessel at 2300°K, calculate the equilibrium concentrations of each of the species. **(b)** if after equilibrium is attained, another 10.0 g of NO is introduced, calculate the new equilibrium concentrations.

8-22 Predict, using Le Châtelier's principle, the effect of (1) lowering the temperature, and (2) raising the pressure for each of the following equilibria.
(a) $2NH_3(g) \rightleftharpoons 3H_2(g) + N_2(g) - \text{heat}$
(b) $2O_3(g) \rightleftharpoons 3O_2(g) + \text{heat}$
(c) $2NO_2(g) \rightleftharpoons N_2O_4(g) + \text{heat}$
(d) $Fe_3O_4(s) + 4H_2(g) \rightleftharpoons 3Fe(s) + 4H_2O(l) + \text{heat}$
(e) $H_2O(s) \rightleftharpoons H_2O(g) - \text{heat}$
(f) $CO(g) + H_2O(g) \rightleftharpoons CO_2(g) + H_2(g) - \text{heat}$

8-23 For the gaseous reaction

$$C_2H_2(g) + D_2O(g) \rightleftharpoons C_2D_2(g) + H_2O(g) + 520 \text{ cal}$$

at 25°C, the equilibrium constant is 0.80.
(a) Calculate the number of moles of C_2D_2 formed when 3.0 moles of C_2H_2 is mixed with 1.0 mole of D_2O in a 7.50-liter vessel.
(b) If the reaction was carried out at 100°C instead of 250°C, would the value of C_2D_2 obtained in part (a) be larger, smaller, or the same?

8-24 For the reaction

$$NO_2(g) \rightleftharpoons NO(g) + \tfrac{1}{2}O_2(g)$$

at 500°K the concentrations at equilibrium were found to be $[NO_2] = 0.180\ M$, $[NO] = 0.0200\ M$, and $[O_2] = 0.0100\ M$. Calculate K.

8-25 For the equilibrium in Problem 8-24, calculate K_p.

8-26 Carbon monoxide and water react as follows:

$$CO(g) + H_2O(g) \rightleftharpoons CO_2(g) + H_2(g)$$

48.1 g of CO and 72.8 g of H_2O were placed in a 3.00-liter container at 986°C and the mixture allowed to come to equilibrium. 47.8 g of CO_2 was found to be present at equilibrium. Calculate K at 986°C.

8-27 For the equilibrium in Problem 8-26, calculate K_p.

8-28 State for the system whether each of the following processes occurs with an increase in entropy, a decrease in entropy, no entropy change, or you cannot tell. **(a)** ice melting **(b)** frying an egg **(c)** a gas condensing into a liquid **(d)** mixing of a container of helium with a container of argon **(e)** mixing of a container of helium with another container of helium **(f)** a chemical reaction **(g)** an endothermic reaction **(h)** an exothermic reaction **(i)** the burning of a book

8-29 Explain in terms of probability the first three examples given at the beginning of Section 8-10. **(a)** Why does heat not flow spontaneously from a cold body to a warm body? **(b)** Why do all the molecules of air in a room not spontaneously go out the window at once? **(c)** Why does a book on a desk not spontaneously rise into the air, cooling itself in the process?

8-30 Is this statement true? Mixing of two different substances always increases the total entropy of the substances, assuming that no reaction takes place between them. Explain.

8-31 Using the third-law entropies in Table 8-5, calculate the change in entropy ΔS^0 for the following reactions at 25°C.
(a) $H_2(g) + Cl_2(g) \rightleftharpoons 2HCl(g)$
(b) $2Ag(s) + Cl_2(g) \rightleftharpoons 2AgCl(s)$
(c) $C(graphite) \rightleftharpoons C(diamond)$
(d) $N_2(g) + O_2(g) \rightleftharpoons 2NO(g)$
(e) $3H_2(g) + N_2(g) \rightleftharpoons 2NH_3(g)$

8-32 For the reaction

$$N_2O_4(g) \rightleftharpoons 2NO_2(g)$$

$\Delta H = 13.9$ kcal and $\Delta S = 42.5$ cal/deg.
(a) Calculate ΔG at 25°C and at 500°C. **(b)** In which direction will the reaction proceed at each temperature? **(c)** At what temperature (°C) will $\Delta G = 0$? **(d)** Describe the state of the system at that temperature. (Assume that ΔH and ΔS are independent of temperature.)

8-33 For the reaction

$$BaO(s) + SO_3(g) \longrightarrow BaSO_4(s) \qquad \text{at } 298°K$$

ΔG°_{298} is -135.000 kcal/mole. If at 298°K, the standard free energy of formation of BaO(s) is -126.300 kcal/mole and of $BaSO_4(s)$ is 350.200 kcal/mole, calculate the standard free energy of formation of $SO_3(g)$ at 298°K.

8-34 For the reaction

$$2H_2O_2(g) \rightleftharpoons 2H_2O(g) + O_2(g)$$

at 25°C, $\Delta G^\circ = -59.8$ kcal and $\Delta H^\circ = -50.4$ kcal. Calculate ΔS°.

8-35 Using the data in Table 8-6, calculate ΔG°_{298} for the following reactions.

(a) $H_2(g) + \frac{1}{2}O_2(g) \rightleftharpoons H_2O(g)$
(b) $H_2O(g) \rightleftharpoons H_2O(l)$
(c) $H_2(g) + Br_2(l) \rightleftharpoons 2HBr(g)$
(d) $CO(g) + \frac{1}{2}O_2(g) \rightleftharpoons CO_2(g)$
(e) $\frac{1}{2}N_2(g) + \frac{1}{2}O_2(g) \rightleftharpoons NO(g)$
(f) $2NO(g) + O_2(g) \rightleftharpoons 2NO_2(g)$

8-36 Calculate K_p at 25°C for each reaction in Problem 8-35.

8-37 Using the data in Table 8-6, calculate ΔG°_{298} for the following reactions.
(a) $N_2O(g) + 4H_2(g) \rightleftharpoons 2NH_3(g) + H_2O(g)$
(b) $CO(g) + H_2O(l) \rightleftharpoons CO_2(g) + H_2(g)$
(c) $CH_4(g) + 2O_2(g) \rightleftharpoons CO_2(g) + 2H_2O(g)$
(d) $2Ag(s) + Cl_2(g) \rightleftharpoons 2AgCl(s)$
(e) $N_2O_4(g) \rightleftharpoons 2NO_2(g)$
(f) $CH_4(g) + CO_2(g) \rightleftharpoons 2CO(g) + 2H_2(g)$

8-38 Calculate K_p at 25°C for each reaction in Problem 8-37.

8-39 **(a)** If ΔG°_{298} fo a reaction is −10 kcal, what is the K_p of the reaction? **(b)** If K_p for a reaction at 298°K is 10, what is the ΔG°_{298} of the reaction?

8-40 For the reaction

$$PCl_5(g) \longrightarrow PCl_3(g) + Cl_2(g)$$

at 25°C, ΔG°_{298} is 9.35 kcal. **(a)** If initially only PCl_5 was present and its equilibrium partial pressure is 0.00100 atm, calculate the equilibrium partial pressures of $PCl_3(g)$ and $Cl_2(g)$. **(b)** If at equilibrium the partial pressures of PCl_5 and PCl_3 are equal, what must be the equilibrium partial pressure of Cl_2?

8-41 Given the reaction

$$SbCl_5(g) \rightleftharpoons SbCl_3(g) + Cl_2(g)$$

(a) If K_p for this reaction is 1.532×10^{-6} at 298°K, calculate ΔG°_{298}.
(b) Using equation (11) at 298°K, fill in the table for the following values of the partial pressures.

P_{SbCl_5} (atm)	P_{SbCl_3} (atm)	P_{Cl_2} (atm)	ΔG_{298}	ΔG°_{298}
1	1	1	____	____
0.1	10	0.001	____	____
1×10^{-6}	1.237×10^{-1}	1.237×10^{-1}	____	____
1.532×10^{-6}	1.532×10^{-6}	1.532×10^{-6}	____	____
0.001	0.001	0.001	____	____

8-42 A vessel contains SO_2, O_2, and SO_3 in equilibrium:

$$2SO_2(g) + O_2(g) \rightleftharpoons 2SO_3(g)$$

The equilibrium partial pressures are $[O_2] = 0.119$ atm, $[SO_2] = 0.238$ atm, and $[SO_3] = 0.138$ atm. Calculate ΔG° for this reaction at 25°C.

8-43 **(a)** Using the Gibbs–Helmholtz equation, calculate K_p at 100°C for the reaction $SbCl_5(g) \rightleftharpoons SbCl_3(g) + Cl_2(g)$ for which K_p is 1.532×10^{-6} at 25°C and $\Delta H^\circ_{298} = 19.3$ kcal and is independent of temperature. **(b)** Repeat the calculation to obtain K_p at 500°K still assuming that ΔH° is independent of temperature.

8-44 Assume that the standard enthalpies of formation of NO(g) and $NO_2(g)$ are 22.00 and 8.00 kcal/mole, respectively. Assume the standard entropy of formation of NO_2 is 59.00 cal/mole · °K.

(a) Use these data and the data in Table 8-5 to calculate ΔG°_{298} for the reaction

$$2NO(g) + O_2(g) \longrightarrow 2NO_2(g)$$

and K_p at 298°K. **(b)** Repeat the calculations in (a) at 0°C, 35°C, 50°C, and 1000°K, assuming that enthalpy and entropy values do not change with temperature.

Additional Problems

8-45 Given the reaction

$$H_2(g) + I_2(g) \rightleftharpoons 2HI(g)$$

initially 4.00 moles of H_2 and 2.00 moles of I_2 are introduced into a 2.00-liter vessel. After equilibrium is attained, the number of moles of hydrogen was found to be 2.10 moles. Calculate the equilibrium concentrations of I_2 and HI and the equilibrium constant.

8-46 If we compare 1 g of water at 100°C and 1 g of steam at 100°C, we see that both have the same kinetic energy, since kinetic energy is proportional to temperature, and both are at the same temperature. Yet 1 g of steam obviously has 540 cal more energy than 1 g of water, since it requires 540 cal to turn 1 g of water to steam. Explain how this can be so.

8-47 Show how a thermostat in a house is a device for maintaining a dynamic equilibrium.

8-48 For the gas reaction

$$4NH_3(g) + 5O_2(g) \rightleftharpoons 4NO(g) + 6H_2O(g) + \text{heat}$$

explain what happens to the concentration of each of the species when each of the following stresses is applied to the system at equilibrium. **(a)** heat is withdrawn from the system **(b)** oxygen is added to the system **(c)** the temperature of the system is increased **(d)** NO is withdrawn **(e)** the pressure is increased by adding the inert gas helium **(f)** the pressure is increased by a piston **(g)** the volume is increased **(h)** a catalyst is added **(i)** water vapor is added to the system

8-49 What is K_p for any reaction that has $\Delta G^\circ_{298} = -3000$ cal/mole at 25°C?

8-50 For the reaction

$$PCl_5(g) \rightleftharpoons PCl_3(g) + Cl_2(g)$$

the equilibrium constant is 5.00×10^{-2}. Calculate the number of grams of Cl_2 in a 3.0-liter flask containing this reaction at equilibrium if $[PCl_5] = [PCl_3]$.

8-51 In a certain experiment $H_2(g)$ and $I_2(g)$ were mixed at 700°K, and after a period of time the concentrations were found to be $[H_2] = 0.072\ M$, $[I_2] = 0.096\ M$, and $[HI] = 0.55\ M$. **(a)** Insert these values into the equilibrium expression and calculate K. **(b)** If you look up K for $H_2(g) + I_2(g) \rightleftharpoons 2HI(g)$ at 700°K, you will find that $K = 55.6$. Explain how it is possible that the value you calculated in (a) does not equal 55.6.

8-52 The Haber process for the production of ammonia is

$$N_2(g) + 3H_2(g) \rightleftharpoons 2NH_3(g) + 22.0\ \text{kcal}$$

Suggest four ways in which the concentration of ammonia can be increased.

8-53 The equilibrium constant for the reaction

$$H_2(g) + Br_2(g) \rightleftharpoons 2HBr(g)$$

is 1.6×10^5 at 1300°K and 3.5×10^4 at 1500°K. **(a)** Calculate the ΔH° of this reaction assuming it does not change with temperature between 1300°K and 1500°K. **(b)** What are the equilibrium constants for the reaction

$$HBr(g) \rightleftharpoons \tfrac{1}{2}H_2(g) + \tfrac{1}{2}Br_2(g)$$

at 1300°K and at 1500°K?

8-54 State: **(a)** Le Châtelier's principle **(b)** the second law of thermodynamics **(c)** the third law of thermodynamics

8-55 For the reaction

$$N_2(g) + 2O_2 \rightleftharpoons 2NO_2(g)$$

$\Delta H = 16.3$ kcal and $\Delta S = -28.8$ cal/deg. **(a)** Calculate ΔG at 25°C. **(b)** Will the reaction proceed to the left or right?

8-56 For the reaction

$$C(\text{graphite}) + 2H_2(g) \rightleftharpoons CH_4(g)$$

the heat of reaction at 600°K (ΔH^0) was determined to be $-21{,}050$ cal/mole of $CH_4(g)$ formed. Third-law entropies at 600°K are: 4.8 cal/mole · °K for C(graphite), 39.0 cal/mole · °K for H_2, and 51.6 cal/mole · °K for CH_4. From this information, calculate the equilibrium constant for the reaction at 600°K.

8-57 For the reaction

$$PCl_5(g) \rightleftharpoons PCl_3(g) + Cl_2(g)$$

$K = 0.0041$ at 250°C. **(a)** If 75.0 g of PCl_5 are heated at 250°C in a 5.00-liter container, calculate the concentrations of all three species at equilibrium. **(b)** If, equilibrium having been reached, 25.0 g of Cl_2 is now added, calculate the new concentrations of all three species.

8-58 Calculate the entropy change when 1.00 mole of each of the following substances undergoes the given phase change, given ΔH and the temperature in each case.
(a) freezing of water at 0°C, $\Delta H_{fus} = -80.0$ cal/g
(b) boiling of ammonia at -33°C, $\Delta H_{vap} = +328$ cal/g
(c) conversion of monoclinic to rhombic sulfur at 96°C, $\Delta H = -88$ cal/mole
(d) sublimation of iodine (I_2) at 18°C, $\Delta H_{sub} = +58.7$ cal/g

55.6

8-59 At 700°K the equilibrium constant for the reaction $H_2(g) + I_2(g) \rightleftharpoons 2HI(g)$ is 5.56×10^1. Calculate the number of grams of HI in equilibrium with 0.80 moles of H_2 and 0.50 moles of I_2 in a 2.5-liter flask at 700°K.

Interlude
Energy Resources

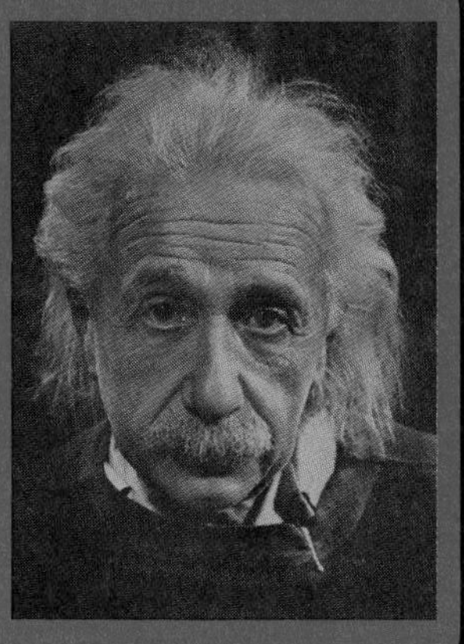

Albert Einstein, 1879–1955

E-1 Present Sources of Energy

Life for the average person around 1750 was not very different than it was 1000 or 2000 years earlier. Since 1750 there has been an enormous revolution in the way people live, at least in the developed countries. The amazing technological advances and the vast increases in the standard of living all depend on energy. Before 1750 the principal sources of energy were the muscles of men, women, and domestic animals, although windmills and waterwheels did supply some energy. Fires for cooking and heating were fueled with wood. Today our energy comes chiefly from the fossil fuels oil (petroleum), coal, and natural gas. Figure E-1 shows the world's sources of energy for approximately the last 75 years. Note that about three-fourths of our energy today is derived from oil and natural gas.

We have only to look around us to see how much our life-styles depend on these fuels. About 26% of the energy in the United States is used for transportation—about half of this for automobiles, and the rest for trucks, airplanes, trains, farm vehicles, etc. Without this energy we could not live very far from where we work or go to school, and factories and farms would be unable to ship their products rapidly and over great distances. The factories and farms themselves would produce much less (37% of U.S. energy is used in factories), since all labor would have to be done by hand or with the help of animals. About 20% of U.S. energy is used for heating and cooling of buildings. With only firewood, the heating would be much less efficient, and we probably would not be able to cool our buildings at all.

In short, without our modern sources of energy—chiefly the fossil fuels—we would have to go back to living the old way, and with the world population today much larger than it was in 1750, it is unlikely that we could even live as well as they did then. That is why we must be concerned with the fact that oil and natural gas are running out.

E-2 Oil and Natural Gas

The amount of oil in the ground is finite. When a barrel of oil is removed from the ground, it will not be replaced. As Figure E-1 shows, oil is our principal fuel, so its decline is important. Right now, on a world basis, we can still produce oil as fast as we can use it. The recent (since about 1974) higher prices are due to political factors rather than scarcity: Most of the producing countries simply got together and demanded higher prices. But this cannot continue indefinitely. Estimates vary about how much oil is still in the ground; it is likely that more oil fields will be discovered; nobody knows how many. Table E-1 shows some of the most recent estimates of the amount of recoverable oil in the earth, but these figures may be too high or too low. They are only guesses. From the values in the table it might seem that we have about

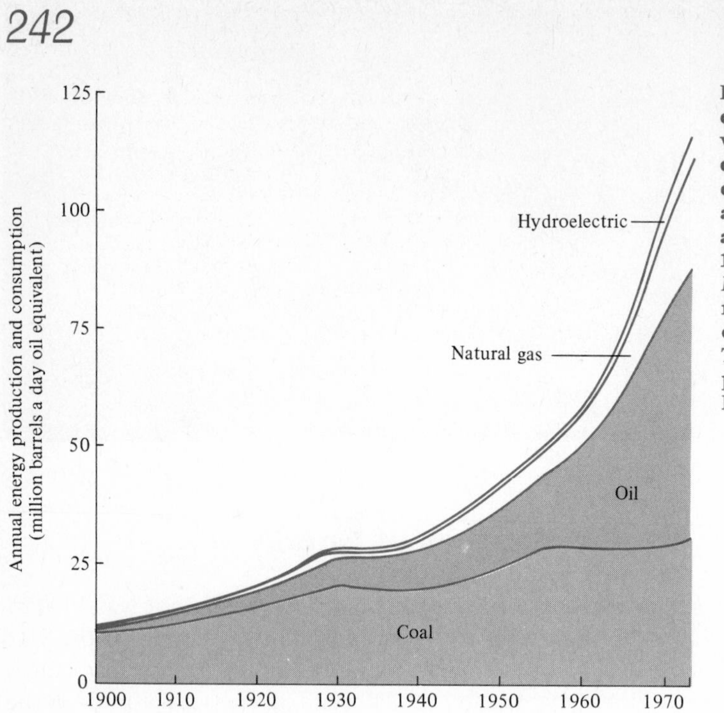

Figure E-1 Sources of the energy consumed in the world from 1900. The graph does not show nuclear energy. A small but growing amount has been produced and consumed since about 1970. [From *Energy: Global Prospects 1985–2000.* Copyright 1977 by the Massachusetts Institute of Technology. Used with permission of McGraw-Hill Book Company.]

Table E-1 World's Estimated Recoverable Oil and Gas Supply[a]

	Recoverable oil (billions of barrels)	*Recoverable natural gas (billions of barrels–oil equivalent)*
Total "proven" resources on earth	1000	520
Already taken out	− 350	− 130
"Proven" resources remaining	650	390
Undiscovered resources	1300	1000
Total "proven" and undiscovered remaining	1950	1390
Used per year (1977)	20	8

[a]"Proven" resources are not actually proven, but educated guesses. The figures given for undiscovered resources are even more uncertain. The units for oil are billions of barrels (1 barrel in the petroleum industry is 42 U.S. gallons or 158.98 liters). The units for natural gas are billions of barrels–oil equivalent; that is, the number of billions of barrels of oil that would have to be burned to get the same amount of energy.

1950/20 = 98 years before all our oil is used up, but this depends on (1) the rate of consumption remaining constant at 20 billion barrels per year, which is extremely unlikely, and (2) all the undiscovered oil being discovered and used as we need it. (Note that a *barrel* in the petroleum industry is defined as 42 U.S. gallons, equal to 158.98 liters.)

Not only is the amount of recoverable oil uncertain, but also how much we will use. Figure E-1 shows that the amount used per year has steadily increased, and this is likely to continue, although nobody can be certain about the rate of increase. It depends on economic conditions (a recession means less demand for oil), on new technologies that might be invented either to use more oil or to conserve it, etc. A report published in 1977 by the Workshop on Alternative Energy Strategies (a project sponsored by the Massachusetts Institute of Technology) estimates that it is likely that sometime between 1985 and 1995 the world will no longer produce oil as fast as it can use it. After that, the difference between the oil available and the oil needed will increase each year. One possible projection is shown in Figure E-2. The gap

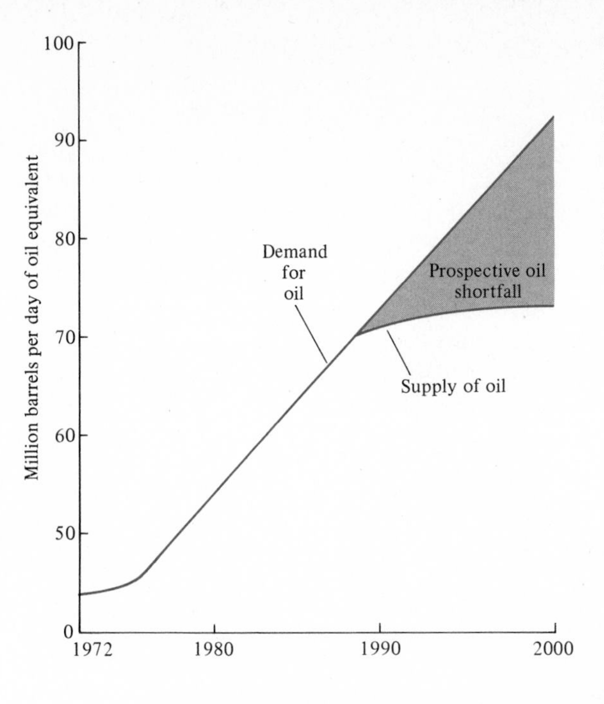

Figure E-2 One possible projection of the future supply and demand for oil. In this projection, the supply fails to keep up with demand in 1988. After that, the shortfall increases each year. [From *Energy: Global Prospects 1985-2000.* Copyright 1977 by the Massachusetts Institute of Technology. Used with permission of McGraw-Hill Book Company.]

between supply and demand will have to be filled by other energy sources or reduced by energy conservation if we are to *maintain* our standard of living, let alone increase it. Remember that more than half of the world's population has a very low standard of living now and would like to better it. Eventually, of course, the recoverable oil will all be gone; perhaps by 2050. How long it will last depends on how successful we are in conservation and in the development of other energy sources.

In Section 20-2 we shall see that petroleum has another major use besides burning: it is used as a raw material for plastics, synthetic rubbers, artificial fibers, etc. It would seem that we would be well advised to hasten the development of other energy sources so that less oil would have to be burned and more could be used as a raw material.

From Table E-1 it would appear that the supply picture for natural gas is more favorable than for oil (1390/8 = ~175 years). However, this is misleading for two reasons: (1) as the oil declines, larger amounts of natural gas will be used as a substitute; and (2) the gas is mostly in the wrong places. The places it is needed are far away from the places where supplies are plentiful. Natural gas is easy to transport by pipeline but not in any other way. In the United States we have been fortunate in that we have had large amounts of gas, but supplies are now diminishing, and we can no longer produce as much as we can use. On a world basis, the areas where natural gas is in large supply are separated by oceans from the areas that need it, so pipelines are impractical. The other alternative is to turn it into a liquid and ship it that way. There are two main ways of doing this, but both are costly and inefficient. One of these involves liquifying the gas under pressure. Natural gas is mostly methane (CH_4), whose critical temperature is −83°C. Therefore it can only be liquified below this temperature. In practice, it is cooled to −161°C and shipped as a liquid under pressure at that temperature (it is called liquified natural gas: LNG). Special refrigerated ships are required, and about 25% of the gas is unavoidably lost in handling and shipping. The other method involves chemical conversion of CH_4 to methanol (CH_3OH), which is a liquid at room temperature and can be shipped without refrigeration. However, it is impractical to convert methanol back to methane, and the methanol is burned directly. Since it gives off less energy on burning than methane does, the process is inefficient.

E-3 What Can We Do: The Near Future

It is only recently that many people in and out of national governments have begun to realize that a serious problem exists and that we will have to do something about it if we wish to retain our civilization. Many other people still do not realize it, and continue to use up the oil and natural gas as if they would last forever. There are two major approaches to the problem, and we will have to use both: (1) conservation, and (2) development of alternative energy resources. We will discuss these possibilities in this section and in Section E-4. Most of the alternative possibilities are not used much now, if at all, because they cost too much. Oil and natural gas are cheaper. However, we have all seen the price of oil rise sharply. If this continues, and it seems certain to do so, the cost of other energy sources will become competitive.

Here are the main possibilities for the near future.

1. Conservation. There is plenty of room for improvement here. Because until recently energy was very cheap, we have tended to use it inefficiently. Our automobiles have been mostly "gas guzzlers"—heavy and powerful with lots of optional power equipment, all of which uses up large amounts of gasoline. We have built excellent highways, which encourages people to live in ways that require a lot of driving. One thing we can do is to make our cars lighter, smaller, and less powerful, to increase the average number of miles per gallon. Another is to provide more and better mass transportation—buses and trains—and encourage people to leave their cars at home.

Many of our buildings—homes, offices, stores, and factories—are also energy-inefficient, especially those built recently. Architects did not have to have energy conservation in mind, so buildings are lighted, heated, and cooled in an inefficient manner. For example, the use of large expanses of glass walls without windows that can be opened means that air conditioning must be used virtually every day in the summer time, and when the sun comes in, the glare is so bright that shades must be drawn and the lights turned on in the daytime. The World Trade Center in New York City is a pair of buildings that uses as much electricity as a city of 100,000 people. These buildings were designed with only one light switch per floor, so that if someone works late in one office, all the lights on that floor must remain on. To make our buildings more energy-efficient will be much harder than for automobiles, and require many more years. One thing that can be done relatively quickly is to add insulation to houses, as well as caulking, weatherstripping, etc., so that leakage of heat is reduced, and less fuel is required for heating and cooling. Other changes, such as the design of buildings that are inherently more energy efficient, will have to be put into effect only as new buildings slowly replace old ones.

Another place for energy conservation is in industrial processes, many of which could be redesigned to use much less energy. Factories in the United States are particularly energy wasteful compared to those in other industrial countries. Of course, these changes will also have to be very slow, as large amounts of money are required to build new factory equipment.

Energy conservation is the most important thing we can do right now to meet the energy crisis, because if properly done it can substantially lengthen the time before the oil and natural gas run out. However, it cannot be the only thing, since all it can do is postpone the inevitable. New energy sources must also be developed.

2. Increase the Recoverable Oil. When an oil well is drilled, natural pressure forces the oil out, but this does not continue until it is all out. On the average, only about 25% of the oil in the reservoir comes out before the pressure stops. Techniques are known that will increase this amount: for example, by pumping water into the reservoir to maintain the pressure for a longer time. By such techniques it is possible to increase the percentage of oil that comes out to about 30 to 35%. This is *recoverable oil,* and is what is shown in Table E-1. The other 65 to 70% remains in the ground, and at present we do not know any economical way to get it out. If a good method could be discovered, we could greatly add to our supplies of oil. A similar statement can be made for natural gas.

A related possibility is to use *oil shale* and *oil sands.* These are rocks and sands interpenetrated with small amounts of oil. The oil can be extracted by heating or by other methods. The known deposits of oil shale and oil sands are much more abundant than ordinary petroleum, but the oil is costly to extract, and in most cases the oil shale and oil sands lie too deep for surface mining, so that still more costly methods must be used. As long as ordinary petroleum was cheap and plentiful it did not pay to exploit the sands and shale, but when the price goes high enough, perhaps it will pay. This is a good example of how an increase in price can lead to the development of a new energy source.

3. COAL. There is a vast amount of coal still underground. The *proven* economically recoverable amount is equivalent to at least 3000 billion barrels of oil, and there is probably several times that amount still undiscovered (Figure E-3). Most of the known reserves are in the United States, the USSR, and China. Many other potential coal areas, particularly in the southern hemisphere, have not been investigated, because oil has been cheaper and more convenient. It is probable that there is enough coal to last for several centuries, even if it completely replaces oil and natural gas. However, while coal will be of great help in the next few decades, it is unlikely to be the exclusive answer to the energy problem, because it has several major disadvantages.

a. Coal mining is relatively slow. It is unlikely that *the rate of getting it out of the ground* will be sufficient to meet our needs. It has been calculated that if the United States were to rely exclusively on coal by the year 2000, it would have to open five new mines every day (of the size of the average present mine) from now until the end of the century!

b. Coal mining in the past has despoiled the environment. Much of our coal lies deep under the ground, and this kind of mining need not greatly harm the environment, although it can be very hazardous to the miners. But there are vast amounts of coal near the surface, and this must be removed by *strip mining.* Huge earth-moving machines remove the entire surface of the land (they can shear off the tops of mountains) and leave large areas of desolation—land that is ugly and cannot be used

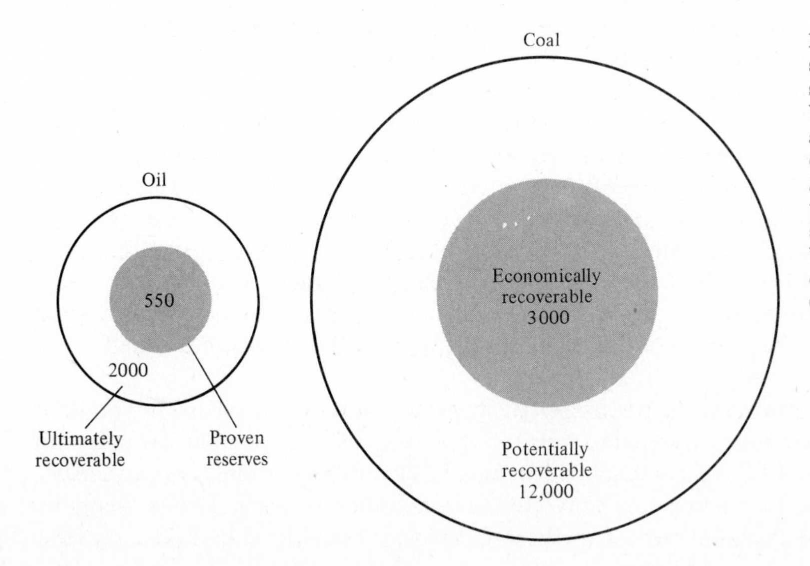

Figure E-3 A comparison of oil and coal reserves. Units are billion barrels of oil-equivalent. [From *Energy: Global Prospects 1985–2000.* Copyright 1977 by the Massachusetts Institute of Technology. Used with permission of McGraw-Hill Book Company.]

Figure E-4 A strip mine. [TVA photograph]

for anything else without undergoing extensive reclamation (Figure E-4). The reclamation will, of course, increase the cost of the coal. However, strip mining is more efficient in terms of recovery. About 25 to 75% of the coal can be economically recovered from deep mines, compared with 85 to 95% from strip mines.

c. Coal is a solid and thus expensive to ship. Much of it lies in areas far from where it is needed. It is generally shipped in railroad cars. A newer method is to pulverize the coal, mix it with water, and send the mixture through a pipeline. This method, called the coal slurry pipeline method, costs only about one-third as much as railroad shipping, but it depends on the availability of water, which is scarce in some areas.

d. Coal is a solid fuel, inconvenient to handle, dirty, and leaves ashes that must be disposed of. As a solid it can be burned to generate electricity and to heat buildings. However, it is not a direct substitute for oil in automobiles and most other means of transportation. Liquid and gaseous fuels are cleaner and much more convenient to ship and store. One possible solution here would be to turn coal into liquid or gaseous fuels. This can be done (in fact, Germany got much of its oil in this way during World War II), but it is costly and inefficient. The possibility even exists of converting coal to a liquid or gaseous fuel while it is still in the ground, but efforts in this direction have been disappointing.

e. Much of our coal contains relatively large amounts of sulfur. When it is burned, the sulfur is converted to sulfur dioxide SO_2, which is a major air pollutant (Section 17-7). This can be removed (Section 17-9), but the technology is expensive.

f. When coal is burned, it is converted to carbon dioxide, CO_2. This is the major product of coal combustion, and there is no way to remove it. Ever-increasing amounts of CO_2 in the atmosphere cause the greenhouse effect (Section 17-6), which may raise the average temperature of the atmosphere to unacceptable levels. This has not been a problem so far, but may become one in the future.

4. Nuclear Energy. Energy derived from nuclear fission already accounts for a small but significant amount of the world's energy (see Section 23-8). It is likely that this amount will increase, although not rapidly, because of a number of problems involved, the main one of which is the problem of dealing with radioactive waste. Another problem is that all nuclear fuel is either uranium or made from uranium or thorium, and the supplies of natural uranium and thorium are limited, though how limited they are is not certain; some estimates are that they will not last past A.D. 2000. This problem might be solved by the *breeder reactor*, which makes more nuclear fuel than it uses up, but breeder reactors have their own disadvantages, and whether they will ever be used on a large scale is problematical. All these points are discussed further in Section 23-8, but here we can say that while it is likely that nuclear fission will supply a part of the energy needed to close the gap left by diminishing oil and natural gas, it is unlikely that it will ever close it completely, and almost certainly not in the next 50 years.

5. Hydroelectric Energy. Figure E-1 shows that about 2% of the world's energy is hydroelectric, that is, derived from the flow of water (see Figure 1-6). Although small in relative terms, this is a fairly large amount of energy, and it differs from the others previously mentioned in that it is inexhaustible. As long as the sun evaporates water from lakes and oceans, and the water returns as rain that runs downhill, this source of energy will be present. Most of the possible sources of hydroelectric energy in the developed countries are already in use, so not much expansion can be looked for here, but there is room for much more expansion in the underdeveloped countries. All told, the amount of energy worldwide from hydroelectric power might double by 2000.

It should be clear from the above that there is no single magic solution to the energy crisis in the near future. We will have to use conservation as well as all the other possibilities mentioned. Furthermore, none of the above is a long-term solution. Although we have large supplies of coal, oil shale, and oil sands (if we can solve all the problems in getting them out and using them), they, too, are finite and will eventually run out. The same for uranium, even if breeder reactors lengthen the life of this fuel. Of the energy sources mentioned so far, only hydroelectric power is inexhaustible, and there is not nearly enough of this to meet the needs of the world even today, to say nothing of the larger needs expected for the future.

E-4 What Can We Do: The Long Run

In this section we shall consider some possible sources of energy for the more distant future. All of them are inexhaustible or virtually inexhaustible. None is yet practical for meeting a significant portion of our needs, but eventually we will have to get our energy from one of these or from a combination of two or more of them.

1. Nuclear Fusion. This process, which is discussed in Section 23-9, is where the sun gets its energy. If we can solve the problems involved in duplicating it on earth, we would have a virtually endless supply of energy, since the raw material is deuterium, which is found in enormous quantities in the ocean. Nuclear fusion has few of the problems associated with nuclear fission. The main problem at present is that we do not know how to do it yet.

2. Solar Energy. All of us feel solar energy when we stand out in the hot sun in the summertime. Most of the other sources of energy (all but nuclear fission and fusion) come indirectly from the sun, but there are many advantages in getting energy directly from the sun. Not only is this an inexhaustible source, but such a vast amount of it falls on the Earth. In the climate of the United States, Europe, and Japan, enough solar energy falls on the roof of an average house to supply all its energy needs, if this energy could be tapped. There are several ways to utilize solar energy. One of these is to allow the sun to heat water or some other liquid and then to transfer the heat to where it can be used or stored. This type of system can be used for small applications such as single buildings. A number of practical devices for doing this are available right now. Initial installation costs are high, though the long-run savings in other types of fuel can be great. The principle can also be applied on a large scale. Three experimental solar power plants are being built right now; one each in the United States, Japan, and France, all of which use mirrors to focus the sun. Another type of device converts sunlight directly into electricity. These devices, too, are available right now but are expensive. They have proven highly useful in space vessels, where the cost is not important, but are not yet in widespread use on earth. They, too, can be made small for use in individual buildings, or large enough to generate electricity on a fairly large scale.

All the types we have discussed thus far depend on direct sunlight, which disappears at night, is reduced on cloudy days, and is not greatly available in some areas. Another method makes use of the fact that sunlight warms the surface of oceans, to the point where the temperature 200 meters below might be about 15 to 20°C cooler than on the surface. A liquid (for example, ammonia) is evaporated on the surface, and then cooled and condensed by water brought up from below. The ammonia would work a heat engine on a principle similar to that of a refrigerator (see Figure 17-6). The advantage of this method is that it does not depend on direct sunlight. However, it is not yet practical.

3. Trees and Other Plants. It has been suggested that large supplies of energy can be obtained by burning wood. We could plant large plantations of fast-growing trees and harvest a percentage of them every year for use as fuel. This, too, would be an inexhaustible supply of energy, although we would have to contend with air pollution problems. The CO_2 would not be a problem (see page 247), because the CO_2 given off by the burning wood would be no more than that used up by the growing trees in the first place. Another suggestion is that we grow large plantations of grains and/or fruits and ferment them to obtain ethyl alcohol, which we would then burn. Still another possibility is the burning of garbage and other municipal solid waste. Some cities are already doing this, but it creates an air pollution problem. The amount of energy obtained by burning municipal waste could never be enough to serve as our only source of energy, but if enough cities do it, it could, like hydroelectric power, meet a small but significant portion of our energy needs. It has the added advantage that it supplies a way to get rid of the wastes.

4. Geothermal and Wind Energy. In many parts of the world, there are substantial amounts of heat either at the surface of the earth or just below. There may be hot water, or underground steam, or just hot rocks. In some places, especially in the western part of the United States and in certain other countries, it is possible to use this heat either to generate electricity or for the heating of buildings. Several installations of this type are in use right now. The Geysers field in California supplies enough electricity for a city of half a million people. Iceland, which is located on top of a large geothermal field, obtains most of its energy from this source. Many other geothermal locations are known, and it is likely that a substantial amount of energy could be obtained from them, although there are problems. For example, geothermal hot water often contains high concentrations of salts, which corrode pipes, as well as gaseous impurities such as hydrogen sulfide (H_2S) and ammonia (NH_3), which are poisonous and must be removed.

One of our oldest sources of energy, wind energy, is free, clean, and inexhaustible. In many parts of the world the wind is strong and fairly steady. Windmills connected to generators of electricity can be practical in many places: for example, on an individual farm or for a small city. The electricity would be stored in batteries for times when the wind velocity is too low to generate power.

Eventually all the exhaustible energy sources will run out, and for the long run we will have to get our energy from inexhaustible sources. We cannot foretell just which ones they will be because that depends on whatever technological advances are made in the next 50, 100, or 200 years. But at this time it appears likely that, while sources such as hydroelectric power and winds may supply a part of our needs, ultimately the bulk of our energy will come from one or both of two sources: *nuclear fusion* and *direct solar energy*. These two are inexhaustible, clean, and nonpolluting. Eventually, one or both will also be made cheap, and when that happens the costs of energy (which at present include the costs of pollution as well as the costs of the fuel) will no longer be a major restraint on the world's standard of living.

One problem with most of the forms of energy mentioned in this section and in Section E-3 is that the energy will be generated in the form of electricity, or in some other manner that makes it difficult to transport from its source. A major advantage of coal, oil, and natural gas is that they can easily be transported to wherever they are needed. For example, how would we drive cars if all our energy came from fusion and direct solar energy? So far, electrical cars are not very practical because they cannot go far on a single charge. If energy supplies become cheap and abundant, this problem can be solved by the use of hydrogen (H_2). Hydrogen is a gas that burns easily to give off energy. The only product is water,

$$2H_2 + O_2 \longrightarrow 2H_2O$$

so that, unlike coal and oil, its burning presents no pollution problem, not even a CO_2 problem. Devices which at present burn coal, oil, or natural gas can be modified to burn hydrogen. Hydrogen can also be combined with oxygen in a different way (though thermodynamics tells us that the same amount of energy is given off) in a device called a fuel cell (Section 16-9).

Where would we get the hydrogen from? Very simple! From the *reverse* of the preceding equation—that is, from the electrolysis of water! (Section 16-12).

$$2H_2O \xrightarrow{\text{electrolysis}} 2H_2 + O_2$$

We would use electricity from fusion or solar energy to electrolyze water, and then burn the hydrogen wherever it is needed. The cycle is unending. Of course, we know from the second law of thermodynamics that the amount of *useful* energy we get from buring the hydrogen must be less than the energy we put into the electrolysis, but if energy is cheap and plentiful, that would not matter. The situation is not quite as utopian as it sounds. Hydrogen can be a dangerous fuel (see Figure 18-1), and it will have to be handled carefully.

E-5 Implications of Cheap and Plentiful Energy

If it turns out that we are able to solve the energy crisis with fusion, solar energy, or some other cheap and plentiful source, then not only will the energy crisis be solved, but also any potential matter crisis! We would no longer have to worry about anything running out (except space on the Earth's surface). For example, consider iron, which is certainly one of the world's most important raw materials. The day will come when all the iron in the iron mines is used up. But that will not mean that we will have to do without iron. Very little of the matter on earth leaves the earth to go into space. Practically all of the iron that has been removed from mines over the centuries is still on the Earth. However, most of it is not in a directly usable form, because of enthalpy and entropy.

Enthalpy. Much of the iron has been oxidized—it has been converted into iron oxides. This is an exothermic process, which means that the oxides have a lower enthalpy than the original iron.

Entropy. The iron, both in the metallic form and as the oxide, has scattered with the four winds. It is dispersed all over, meaning that its concentration in any one place is low.

A source of cheap and plentiful energy enables us to get around both these factors. If the enthalpy of iron is greater than that of iron oxides, we can introduce enough energy to overcome the difference and convert the iron oxides to iron (this is exactly what is done in iron refining; see Section 19-1). Entropy is also a form of energy, and we can collect and concentrate scattered iron, iron oxides, or anything else, provided that we use enough energy.

The same situation holds for any kind of matter. As long as energy is cheap and plentiful we can move matter from one place to another, and convert it into any form we wish.

This does not mean that with a source of abundant and inexpensive energy we can expand our population without limit. No matter how much energy we have, the total amount of matter and space on the earth is finite, and there is a limit to the number of people that can be supported on the planet. Remember also that we are talking about the distant future. We do not yet have cheap and plentiful energy, so that for now, the energy costs of converting one form of matter to another are often prohibitively high. An important example is the lack of fresh water in many parts of the world (Section 17-10). Here is one matter crisis that we cannot yet solve, although it would be easily solved with cheap energy.

9 Chemical Kinetics

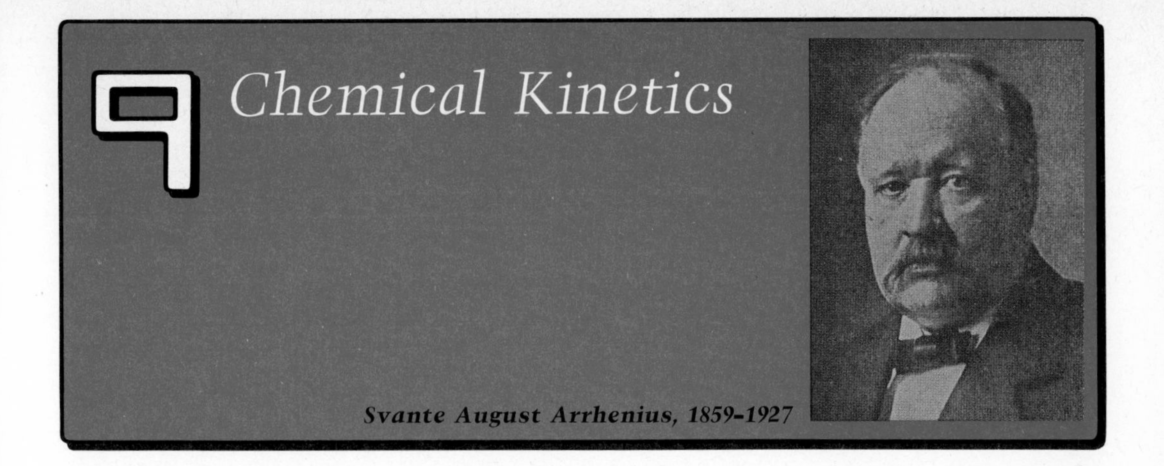

Svante August Arrhenius, 1859–1927

In Chapter 8, we discussed one of the fundamental questions in chemistry: Will a chemical reaction proceed spontaneously? As we saw there, that question is answered as soon as we know whether ΔG for the reaction is positive or negative. Thus, *whether* a given reaction will proceed is the subject matter of chemical thermodynamics. The second fundamental question of chemistry is concerned with the *speed* of chemical reactions. How fast the reaction will proceed comprises the subject matter of **chemical kinetics.** As an illustration, the reaction of hydrogen and oxygen in the gas phase is

$$H_2(g) + \tfrac{1}{2}O_2(g) \longrightarrow H_2O(g)$$

ΔG for this reaction is -54.6 kcal at 25°C. This is a huge negative free-energy change, and we might expect that the reaction would proceed spontaneously and quickly. However, as we have noted before, though ΔG is negative and the reaction therefore proceeds spontaneously (according to thermodynamic principles), in actual fact no appreciable reaction occurs even after long time periods. How can this apparent contradiction be explained? Thermodynamic principles state that the reaction will proceed, but thermodynamics does *not* tell us how long it will take to happen. That answer is in another subject area, namely, chemical kinetics, which is the subject of this chapter. If we introduce a lighted match or a platinum catalyst, the reaction will take place immediately and explosively. Again, although thermodynamics tells us whether a reaction will proceed spontaneously or not, kinetics explains why certain reactions take place slowly and others quickly and what happens when a catalyst is introduced into the reaction.

9-1 Rates of Chemical Reactions and Rate Orders

By the **rate of a reaction,** we mean the speed or velocity of the reaction. The rate may be defined as the **change in the concentration of material per unit time.** This can be written

$$\left.\begin{matrix}\text{rate}\\ \text{speed}\\ \text{velocity}\end{matrix}\right\} \text{of reaction} = \frac{\text{loss of concentration of reactants}}{\text{time interval}}$$

or

$$\left.\begin{matrix}\text{rate}\\ \text{speed}\\ \text{velocity}\end{matrix}\right\} \text{of reaction} = \frac{\text{gain in concentration of products}}{\text{time interval}}$$

If we let Δ = change (loss or gain) and t = time, then

$$\text{rate of reaction} = \frac{\Delta(\text{concentration of reactants or products})}{\Delta t}$$

We can write an expression for the rate of a reaction in terms of a decrease in concentration of any reactant or an increase in concentration of any product. For example, consider the simple reaction

$$A + B \longrightarrow C + D$$

where all the stoichiometric coefficients are unity. For this reaction, the rate can be expressed in any of the following ways:

$$\text{rate of reaction} = \frac{-\Delta[A]}{\Delta t}$$ decrease in concentration of species A per unit time

or

$$= \frac{-\Delta[B]}{\Delta t}$$ decrease in concentration of species B per unit time

or

$$= \frac{+\Delta[C]}{\Delta t}$$ increase in concentration of species C per unit time

or, finally,

$$= \frac{+\Delta[D]}{\Delta t}$$ increase in concentration of species D per unit time

For many thousands of reactions studied, it has been found that the rate of reaction depends in some way upon the concentration of some or all of the species undergoing the reaction.

Examples

(a) Decomposition of nitrogen pentoxide: $2N_2O_5 \longrightarrow 4NO_2 + O_2$

$$\text{rate} = \frac{-\Delta[N_2O_5]}{\Delta t} = k[N_2O_5]$$

(b) Reaction of hydrogen and iodine: $H_2 + I_2 \longrightarrow 2HI$

$$\text{rate} = \frac{\Delta[HI]}{\Delta t} = k[H_2][I_2]$$

(c) Decomposition of acetaldehyde: $CH_3CHO \longrightarrow CH_4 + CO$

$$\text{rate} = \frac{-\Delta[CH_3CHO]}{\Delta t} = k[CH_3CHO]^{3/2}$$

(d) Decomposition of ozone: $2O_3 \longrightarrow 3O_2$

$$\text{rate} = \frac{-\Delta[O_3]}{\Delta t} = k\frac{[O_3]^2}{[O_2]}$$

The expressions above, in which the rate of reaction is given in terms of the concentrations of certain species taking part in the reaction, are called **rate laws.** Each reaction has its own rate law, which must be determined experimentally. Note that not all of the species taking part in the reaction are necessarily found in the rate law. For example, both O_3 and O_2 appear in the rate law for the decomposition of ozone, but only N_2O_5 (and not NO_2 or O_2) is found in the rate law for the decomposition of N_2O_5. There is no general way to predict rate laws. Each must be separately determined by laboratory measurements. The proportionality constants k which appear in all the rate-law expressions are called the **rate constants** of each reaction. The rate constants k are specific characteristics of each reaction and are

1. Different for different reactions.
2. Independent (they do not change) of changes in concentration of the species in the reaction.
3. Dependent (they do change) on changes in the temperature at which the reaction is carried out.

Note that the rate laws tell us that the rate of each reaction *increases* as the concentrations of the species in the rate laws *increase** and *decreases* when these concentrations *decrease*. Furthermore, since the species in the rate law are being used up as the reaction proceeds, the rate laws predict that the reaction rates will *decrease with time*. For example, both $[H_2]$ and $[I_2]$ appear in the rate law for the reaction between H_2 and I_2. Therefore, if the concentration of either or both of those species is increased (say by adding some from the outside), the rate will increase. Conversely, as they are converted to HI, *their* concentrations decrease, and the rate of the reaction therefore decreases with time.

Let us now consider the general reaction.

$$a\text{A} + b\text{B} + c\text{C} + \cdots \longrightarrow \text{products}$$

where A, B, C, . . . are the various species and $a, b, c, \ldots$ are the stoichiometric coefficients in the balanced equation. A general expression for the rate law is

$$\frac{-\Delta[\text{A}]}{\Delta t} = k[\text{A}]^l[\text{B}]^m[\text{C}]^n \cdots$$

where $\dfrac{-\Delta[\text{A}]}{\Delta t}$ = rate of reaction

k = the rate constant for this reaction, which, as mentioned above, is a constant if the temperature is constant

$l, m, n, \ldots$ = the *exponents* to which the concentration terms must be raised to fit the *experimentally determined* rate expression. Note that one or more of these exponents may well be zero, in which case that substance does not appear in the rate expression at all. Note also that the numbers $l, m, n, \ldots$ may or may not be equal to the stoichiometric coefficients, $a, b, c, \ldots$. Whether they are or not cannot be predicted, but only determined by experiment. The *sum of the exponents* ($l + m + n + \cdots$) is called the **order of the reaction.** The order of a reaction may be a whole number (positive or negative), zero, or fractional.

Let us return to the four examples given on page 252. The orders of the reactions are (obtained by adding the exponents)

(a) First order.
(b) Overall second order; first order with respect to hydrogen and first order with respect to iodine.

*These statements, of course, do not apply to species that are written in the denominator of a particular rate law (for example, O_2 in the fourth example given). For these species, the effects are the reverse of those noted in the statements.

(c) Three-halves order.

(d) Overall first order; second order with respect to ozone and *minus* first order with respect to oxygen.

Observe the orders of the reactions and compare them with the stoichiometry in the balanced equations.

Examples	*Stoichiometric coefficient of reactants*	*Order*
$2N_2O_5 \longrightarrow 4NO_2 + O_2$	2	1
$H_2 + I_2 \longrightarrow 2HI$	1 + 1	2
$CH_3CHO \longrightarrow CH_4 + CO$	1	$\frac{3}{2}$
$2O_3 \longrightarrow 3O_2$	2	1

It should be evident that the order of the reaction bears *no* necessary relationship to the stoichiometry of the equation. We might have intuitively supposed that there would be such a relationship. For example, we might have predicted that a reaction whose stoichiometry was

$$A + B \longrightarrow C + D$$

would be first order each in A and in B, second order overall, because it is tempting to think that the rate should decrease with a decrease in concentration of each reactant. However, *one cannot predict the order of a reaction from a knowledge of the stoichiometric coefficients of the balanced equation.* Many reactions whose stoichiometry is $A + B \rightarrow C + D$ *do* obey our intuitive prediction, and are first order each in A and B, second order overall. However, many others do not. Some are first order only in A, and changes in the concentration of B *do not affect the rate.* Others are overall third order, or even have fractional or negative orders. We thus repeat: One cannot predict the order of a reaction from a knowledge of its stoichiometry.

9-2 Reaction-Rate Constants and Their Units

The rate of reaction is most generally expressed in terms of changes in concentration (generally moles/liter) per unit time (generally seconds). The units of the *rate constants k* depend on the order of the reaction, because the left-hand and right-hand sides of the rate-law expressions must have the same units. The following examples should clarify this.

Order of reaction	*Rate law*	*Units of k*
1	$\text{rate} = \frac{-\Delta[A]}{\Delta t} = k[A]$	$\frac{1}{s}$
2	$\text{rate} = \frac{-\Delta[A]}{\Delta t} = k[A]^2$	$\frac{\text{liters}}{\text{mole} \cdot s}$
$\frac{3}{2}$	$\text{rate} = \frac{-\Delta[A]}{\Delta t} = k[A]^{3/2}$	$\left(\frac{\text{liters}}{\text{mole}}\right)^{1/2} \frac{1}{s}$
n	$\text{rate} = \frac{-\Delta[A]}{\Delta t} = k[A]^n$	$\left(\frac{\text{liters}}{\text{mole}}\right)^{n-1} \frac{1}{s}$

Experimental data in kinetic experiments are expressed in terms of concentrations of reactants or products at different time intervals. We have been expressing our rates in terms of the loss of reactants, but we could just as easily write them in terms of the formation of products. Which way to do it is usually a matter of personal choice, since the different expressions for the rate of reaction are related by the balanced equation. For example, if we are considering the reaction

$$2HI \longrightarrow H_2 + I_2$$

we could write the rate law as

$$\text{rate} = \frac{-\Delta[HI]}{\Delta t} = k_1[HI]^2$$

or as

$$\text{rate} = \frac{\Delta[H_2]}{\Delta t} = k_2[HI]^2$$

The stoichiometry of the equation tells us that 1 mole of H_2 is formed for each 2 moles of HI that disappear. The rate of disappearance of HI is therefore *twice* the rate of appearance of H_2, so we can write

$$\frac{-\Delta[HI]}{\Delta t} = \frac{2\Delta[H_2]}{\Delta t}$$

A comparison of the rate laws shows that $k_1 = 2k_2$.

As a second example, consider the decomposition of nitrogen pentoxide,

$$2N_2O_5 \longrightarrow 4NO_2 + O_2$$

Going through the same kind of reasoning, we find

$$\frac{-2\Delta[N_2O_5]}{\Delta t} = \frac{\Delta[NO_2]}{\Delta t} = \frac{4\Delta[O_2]}{\Delta t}$$

The rate law for this equation may be written

$$\text{rate} = \frac{-\Delta[N_2O_5]}{\Delta t} = k_1[N_2O_5]$$

or

$$\text{rate} = \frac{\Delta[NO_2]}{\Delta t} = k_2[N_2O_5]$$

or

$$\text{rate} = \frac{\Delta[O_2]}{\Delta t} = k_3[N_2O_5]$$

From the material above, we have

$$2k_1 = k_2 = 4k_3$$

This shows that one should not hesitate to write different rate expressions for the same reaction. The rate constants are related by the stoichiometry of the equation. On the other hand, we must be aware which rate law is being used to express the experimental results, or else we may obtain erroneous values of k.

9-3 Experimental Methods

The experimental determination of the rate law for a reaction depends on the measurement of three variables; temperature, time, and concentration. The temperature must be kept as constant as possible since both the rates and the rate constants change with temperature. This can be done by the use of constant-temperature baths. The technology of temperature control has advanced to the stage where this variable now ordinarily offers no difficulty. An accurate measurement of time is necessary since the concentration changes are measured over a time interval. Experimental time measurements of high accuracy are not difficult to obtain, except for very fast reactions—those which are over in small fractions of a second. It is the experimental measurements of the concentrations of reactants and/or products that are the most difficult to carry out. The chemist has many ways to determine the concentration of species in a solution or a mixture of gases (some of which we shall discuss later), but the difficulty here is that the concentrations of reactants and products are constantly changing as the reaction proceeds. It is necessary to catch them on the run, so to speak. Any ordinary measurement takes time to carry out, and by the time the measurement is finished, the concentrations have changed. There are three basic approaches to this problem.

1. *The method of quenching*. In this method we start a series of parallel identical reactions in a number of reaction vessels all at precisely the same time. Then after a certain time interval (say 10 minutes), we stop one of them (this is called **quenching**) by some method that we know in advance will stop that particular reaction; for example, we might add a compound that will (practically) instantaneously remove one of the reactants, or we might suddenly lower the temperature by 50°C or so, which will effectively stop most reactions. Meanwhile, in the other vessels, the reaction is continuing. Then after a second interval of 10 minutes, we quench a second vessel; and so on. We can now analyze each vessel at our leisure, since the reactions have been effectively stopped.

2. *The method of aliquots*. In this method we start only one large reaction vessel, and after an interval of time remove a small portion (called an **aliquot**) from the reaction mixture. The aliquot is immediately quenched, and analyzed when convenient. The rest of the reaction continues unabated. After a second time interval, another aliquot is withdrawn; etc.

3. *The method of reaction monitoring*. In this method only one reaction vessel is used, and no aliquots are taken, but the vessel is constantly monitored by some physical process. The process most commonly used is the measurement of some type of spectrum. For example, the concentration of certain types of compounds can be determined by measuring the height of a peak in an infrared spectrum. An infrared spectrophotometer sends a beam of infrared light through a sample. If the molecules in the sample absorb some of this light, then the light emitted has a lower intensity (Figure 9-1). The difference between the intensity of light sent in and the light coming out is measured electronically, and is shown graphically on the chart paper. The height of the peak on the graph is proportional to the concentration of the light-absorbing species. The reaction vessel is placed inside the infrared instrument

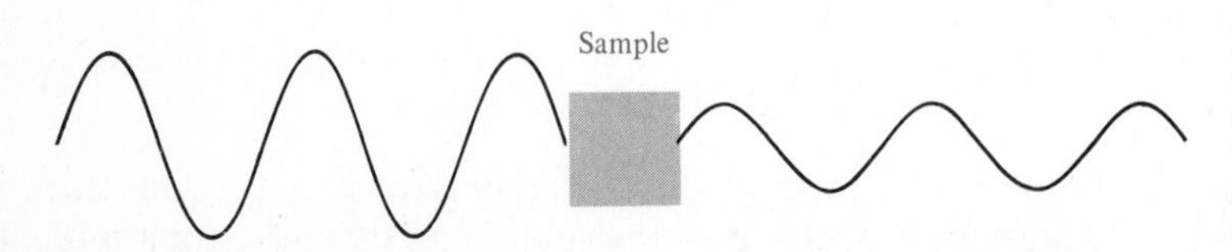

Figure 9-1 Infrared light of a given frequency is sent through a sample. The sample may absorb light at that frequency; if so, the light that comes out has a lower intensity than the light that went in.

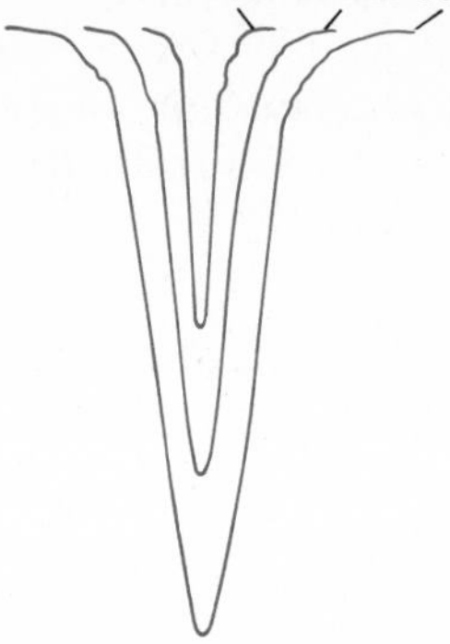

Figure 9-2 Peaks heights on an infrared spectrum.

and the instrument is manipulated in such a way as to draw this peak, say every 10 minutes (Figure 9-2). The concentrations can then be determined by measuring the heights of the peaks. Where it can be used, this method is the best of the three, because it requires only one reaction vessel, and no aliquots need to be removed.

Optimally, the measurements should yield the concentration of the reactants and/or products at specific time intervals while not changing the concentrations of materials in the reaction. Any change in the concentration of the species alters the reaction rate.

9-4 Determination of the Order of a Reaction

To determine the rate law for a reaction, we must determine two quantities: the rate constant and the reaction order. For the reaction $a\text{A} + b\text{B} + c\text{C} + \cdots \rightarrow$ products, we have seen that the general rate law is written

$$\text{rate of reaction} = k[\text{A}]^l[\text{B}]^m[\text{C}]^n \cdots$$

Determination of the order is tantamount to determination of the values of the exponents $l, m, n, \ldots$. This task is not as easy as it sounds, because accurate measurements of rate are complicated by the fact, mentioned before, that the rate of the reaction is constantly changing with time. One way to get around this difficulty is to measure the rate only at the very beginning of the reaction (the initial reaction rate method), when the decrease in concentration of the reactants has taken place to only a small extent, and we can ignore the fact that both the concentrations and the rate are changing. This can be done with a fair amount of precision (except for very fast reactions). What is actually done is the following: we prepare solutions of A, B, C, . . . of known concentrations; mix them; and then after a short period of time* measure the concentration of A, B, or C, or of one of the products, by one of the methods previously discussed. From this we can of course calculate, from the stoichiometric equation, the change in concentration of all the other reactants and products. The initial rate of the reaction is then the change in concentration that has taken place divided by the amount of time that has elapsed between the mixing of the reactants and the measurement of the concentration; that is,

$$\frac{-\Delta[\text{A}]}{\Delta t}$$

*The short period of time may be a few seconds, a few minutes, or a few hours, depending on how long the reaction takes to go to completion. We want the reaction to have proceeded long enough for us to be able to detect (and carefully measure) a change in concentration, but not long enough so that the rate of the reaction has significantly changed because of changing concentrations of the reactants and products.

However, this information is only a starting point. We do not yet know whether or in what way the rate is proportional to the concentrations of A, B, or C.

The next step is to run a series of reactions in which the initial concentration of A is varied while those of B and C are held constant. In each case we measure the initial rate in the same way. Typical data may appear as follows:

Concentration of A (moles/liter)	*Concentration of B (moles/liter)*	*Concentration of C (moles/liter)*	*Rate of reaction (moles/liter · s)*
0.100	0.100	0.100	0.05
0.200	0.100	0.100	0.10
0.300	0.100	0.100	0.15

Since we have kept the concentrations of B and C fixed, the change in the rate of the reaction must be due solely to the change in the concentration of A. We observe that doubling the concentration of A doubles the rate, tripling the concentration of A triples the rate, etc. We therefore conclude that A appears in the rate equation, and the exponent l has the value of unity.

Let us look at some typical data for B and C (this time keeping constant the concentration of A).

Concentration of A (moles/liter)	*Concentration of B (moles/liter)*	*Concentration of C (moles/liter)*	*Rate of reaction (moles/liter · s)*
0.001	0.001	0.001	0.05
0.001	0.002	0.001	0.05
0.001	0.005	0.001	0.05
0.001	0.005	0.002	0.20
0.001	0.005	0.003	0.45

We see that a change in the concentration of B has no effect on the rate, so that $m = 0$, and B does not appear in the rate equation at all, while a doubling of the concentration of C results in a fourfold increase in the rate, so that $n = 2$ and the rate is proportional to the *square* of the concentration of C. The rate law is thus

$$\text{rate of reaction} = k[\text{A}][\text{C}]^2$$

This is therefore an overall third-order reaction, first order in A and second order in C. The reaction is zero order in B.

Now that we know the initial rate of the reaction and the order, it is a simple matter to determine the rate constant, for which we may use any of the rates measured above. For example, for the rate of reaction $= k[\text{A}][\text{C}]^2$, using the top line of the preceding table,

$$0.05\,\frac{\text{mole}}{\text{liter}\cdot\text{s}} = k\left(0.001\,\frac{\text{mole}}{\text{liter}}\right)\left(0.001\,\frac{\text{mole}}{\text{liter}}\right)^2 = k \times 10^{-9}\left(\frac{\text{mole}}{\text{liter}}\right)^3$$

$$k = \frac{5 \times 10^{-2}}{10^{-9}}\left(\frac{\text{liters}}{\text{mole}}\right)^2\frac{1}{\text{s}}$$

$$= 5 \times 10^{7}\left(\frac{\text{liters}}{\text{mole}}\right)^2\frac{1}{\text{s}}$$

Example: The gas-phase reaction

$$SO_2Cl_2(g) \longrightarrow SO_2(g) + Cl_2(g)$$

was studied by the initial reaction-rate method and the following data were obtained at 320°C.

Initial concentration of SO_2Cl_2 (moles/liter)	*Rate of reaction (moles/liter · s)*
0.1	2.2×10^{-6}
0.2	4.4×10^{-6}
0.3	6.6×10^{-6}
1.0	2.2×10^{-5}

From these data, determine (a) the order of the reaction (b) the rate constant (c) the rate of the reaction at an initial concentration of 10 moles/liter.

Answer: From

$$\text{rate of reaction} = k[SO_2Cl_2]^\alpha \tag{1}$$

and from the data, we have

$$2.2 \times 10^{-6} = k[0.1]^\alpha$$
$$4.4 \times 10^{-6} = k[0.2]^\alpha$$
$$6.6 \times 10^{-6} = k[0.3]^\alpha$$
$$2.2 \times 10^{-5} = k[1.0]^\alpha$$

We observe that doubling the initial concentration doubles the rate, tripling the concentration triples the rate, etc. The order of the reaction is therefore unity, that is, $\alpha = 1$.

This calculation by observation is all well and good for orders that are simple integers, but it is plain that it will not do for fractional orders. If the order is fractional, it is more difficult to obtain the exact result by the initial reaction-rate method. It can be done, but it takes a bit more work. The technique we must use is quite powerful and could be used for any problem in which we wish to obtain orders of reaction given rates of reaction and initial concentrations, including the ones we could also solve by observation. To do this, we take logarithms of both sides of equation (1):

$$\log(\text{rate of reaction}) = \log k + \alpha \log(\text{initial concentration})$$

This equation, when plotted on a graph, will give a straight line. We recall from algebra that $y = mx + b$ is the equation of a straight line. In this case, $y = \log(\text{rate of reaction})$, $x = \log(\text{initial concentration})$, $b = \log k$, $m = \alpha$.

If we plot log (rate of reaction) vs. log (initial concentration) we will obtain a straight line whose slope is the order of the reaction α and whose intercept is the log of the rate constant k. Taking the log of our previous values:

Rate of reaction (moles/liter · s)	*log (rate of reaction)*	*Initial concentration (moles/liter)*	*log (initial concentration)*
2.2×10^{-6}	−5.6576	0.1	−1.0000
4.4×10^{-6}	−5.3565	0.2	−0.6990
6.6×10^{-6}	−5.1805	0.3	−0.5229
2.2×10^{-5}	−4.6576	1.0	0.0000

and plotting the graph (Figure 9-3), we get $\alpha = 1$ and $k = 2.2 \times 10^{-5}\ s^{-1}$. To calculate the initial rate at a concentration of 10 moles/liter, we have

$$\text{rate} = (2.2 \times 10^{-5}\ s^{-1})(10\ \text{moles/liter})$$
$$\text{rate} = 2.2 \times 10^{-4}\ \text{mole/liter} \cdot s^{-1}$$

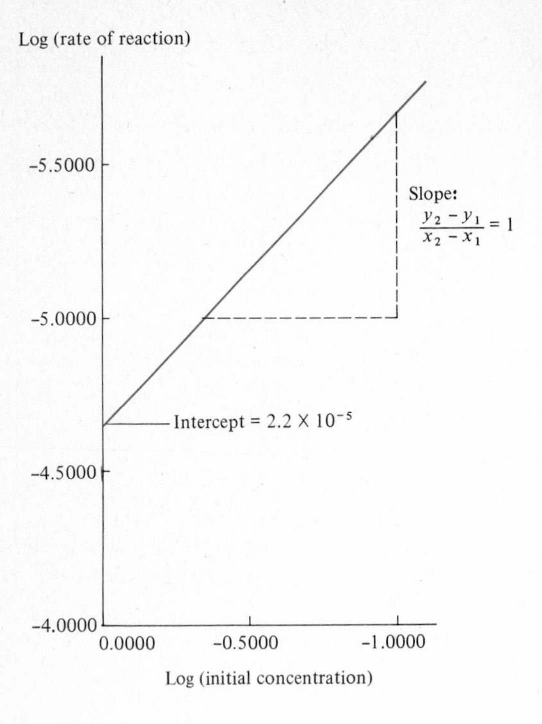

Figure 9-3 A plot of log (reaction rate) vs. log (initial concentration) for the reaction given in the text.

INTEGRATED RATE LAWS. The rate laws we have discussed can be converted by the mathematical technique called *integration** to another set of rate laws which are often easier to work with because the actual *rate* does not appear in them. They are expressions that show us how the concentration changes with *time*. We shall convert the first- and second-order rate laws to the integrated expressions. In each case,

$$t = \text{time}$$
$$[\text{A}]_0 = \text{concentration of A at time} = 0, \text{ that is, before any reaction takes place}$$
$$[\text{A}] = \text{concentration of A at time} = t$$
$$k = \text{rate constant}$$

In each case we will obtain an expression which shows us how the *concentration* changes with *time*.

1. *First-order reactions.* Our previous rate expression was rate $= k[\text{A}]$. The integrated expression† is

$$\log \frac{[\text{A}]_0}{[\text{A}]} = \frac{kt}{2.303} \quad \text{or} \quad \log [\text{A}]_0 - \log [\text{A}] = \frac{kt}{2.303}$$

In most cases, we know the value of $[\text{A}]_0$, because this is the concentration of A before any of the reaction takes place, and since we can generally prepare any solution we wish, this value can be controlled as we see fit. Furthermore, in any given individual experiment, the value of $[\text{A}]_0$ cannot possibly change once the

*The method of integration is taught in courses on calculus. We will merely content ourselves with the results of the integrations and omit the process.

†The factor 2.303 arises when one changes natural logarithms to common logarithms. That is, 2.303 $\log x = \ln x$.

reaction has begun (it is the *initial* concentration), so we may treat $[A]_0$ as a constant. We may then use the integrated expression in two ways.

1. Note that a plot of log $[A]_0/[A]$ vs. t gives a straight line whose slope is equal to $k/2.303$. This is therefore another way to calculate k for first-order reactions, in addition to the method we saw before.
2. If we know the values of k and $[A]_0$, we can find out how much A will remain at any given time t, merely by substituting the values of $[A]_0$, k, and t into the equation and solving for $[A]$.

2. *Second-order reactions.* Our previous rate expressions were

$$\text{rate} = k[A]^2 \tag{2}$$

$$\text{rate} = k[A][B] \tag{3}$$

where equation (2) was for a reaction second order in A, and equation (3) for a reaction which is first order each in A and B, second order overall. In the latter case, integration is difficult unless we choose to make the initial concentration of A equal to the initial concentration of B. If we do so, then mathematically we get $[A] = [B]$. Consequently, if we integrate only equation (2), the result will apply not only to that equation, but also to equation (3), provided that $[A] = [B]$. The integrated expression is

$$\frac{1}{[A]} - \frac{1}{[A]_0} = kt$$

Once again, we have an equation that relates $[A]_0$ (a constant for any particular run), $[A]$, k, and t. This time a plot of $1/[A]$ vs. t gives a straight line whose slope is k and whose intercept is $-1/[A]_0$.

3. *Zero-order reactions.* Up to now we have only discussed first- and second-order reactions, but as we mentioned before, there are some reactions that are *zero order*. A zero-order reaction is a reaction whose rate is completely *independent* of the concentrations of reactants. This means that the reaction rate (speed of the reaction) does not change as the concentrations of reactants decrease. For example, there are reactions which require that the molecules of gaseous reactants all be on a solid surface in order to react. In such cases, the reaction takes place on the walls of the container. Almost immediately after the reaction begins, the walls of the container are completely covered by products and the reaction can only proceed when some product molecules fall from the walls in order to leave room for further reactant molecules to meet. The rate of the reaction is now essentially equal to the rate at which product molecules fall off the walls, and so is independent of how much material is present. This is an example of a zero-order reaction. For this case,

$$\text{the rate law is:} \quad \text{rate} = k$$

$$\text{the integrated expression is:} \quad [A]_0 - [A] = kt$$

If we plot $([A]_0 - [A])$ vs. t, a straight line will be obtained whose slope equals the rate constant k.

9-5 Half-life

The **half-life** of a reaction (the symbol is τ) is the time it takes for *half* of the initial concentration of reactants to form products. In terms of the symbols we have been using, the concentration of the reactants at time $= \tau$ is half the initial concentration; that is, $[A]_0/2$. We shall now derive expressions for the half-lives of reactions of the various orders.

1. *First-order reactions.* The integrated expression is $\log \frac{[A]_0}{[A]} = \frac{kt}{2.303}$.

At $t = \tau$, $[A] = \frac{[A]_0}{2}$, so

$$\log \frac{[A]_0}{[A]_0/2} = \frac{k\tau}{2.303}$$
$$2.303 \log 2 = k\tau$$
$$\tau = \frac{0.693}{k}$$

It is obvious that if we know the rate constant k, we can obtain the half-life, and vice versa. It is interesting to note that for a first-order reaction, *the half-life is independent of initial concentration*. This means that it will take the same amount of time to convert 10 moles of reactants to 5 moles as it will to convert 0.02 mole of reactant to 0.01 mole. This relationship between half-life and rate constant holds for all first-order reactions. If we know that the half-life of a given reaction is independent of initial concentration, then it must be a first-order reaction. In Section 23-4, we will see how the concept of half-life applies to radioactive decay—a first-order reaction.

2. *Second-order reactions.* The integrated expression is $\frac{1}{[A]} - \frac{1}{[A]_0} = kt$.

At $t = \tau$, $[A] = \frac{[A]_0}{2}$, so

$$\frac{1}{[A]_0/2} - \frac{1}{[A]_0} = k\tau$$
$$\frac{2}{[A]_0} - \frac{1}{[A]_0} = k\tau$$
$$\tau = \frac{1}{k[A]_0}$$

We can thus obtain the half-life if we know the initial concentration and the rate constant. For a second-order reaction, the half-life is *inversely* proportional to the initial concentration. This relationship is characteristic of second-order reactions.

3. *Zero-order reactions.* The integrated expression is $[A]_0 - [A] = kt$.

At $t = \tau$, $[A] = \frac{[A]_0}{2}$, so

$$[A]_0 - \frac{[A]_0}{2} = k\tau$$
$$\frac{[A]_0}{2} = k\tau$$
$$\tau = \frac{[A]_0}{2k}$$

In this case, we see that the half-life for a zero-order reaction is *directly* proportional to the initial concentration.

If we look at the three half-lives obtained above, it becomes clear that there is a simple relationship between half-lives and orders of reactions which can be seen if we observe the dependence of the half-lives on initial concentration. This can be generalized as follows: For an nth-order reaction in which all the initial concentrations are equal, the half-life τ is inversely proportional to the initial concentration to the $(n - 1)$st power:

$$n\text{th-order reaction} \quad \tau \propto \frac{1}{[A]_0^{n-1}}$$

The reader should verify the results we obtained previously for zero-, first-, and second-order reactions.

Example: The half-life τ of a first-order reaction is 81 minutes. Calculate the rate constant for this reaction.

Answer: The equation is

$$\tau = \frac{0.693}{k}$$

$$k = \frac{0.693}{(81\ \text{min})(60\ \text{s/min})} = 1.43 \times 10^{-4}/\text{s}$$

For convenience, we list in Table 9-1 all the pertinent equations derived in the previous sections. We have also indirectly obtained two new methods for obtaining order of reactions:

1. Given a certain reaction, we assume an order, say first. Using the integrated relationship, a graph of the data is constructed and if found to be linear, then we have a first-order reaction. If not, another order is attempted. Of course, this method is limited to reactions whose order is a simple integer.
2. Data relating to half-lives and initial concentrations will allow us to determine the order of the reaction (see Problem 9-14).

9-6 Effects of Temperature on the Rate Constant

Up to now we have been concerned with the dependence of the reaction rate on the *concentration* of the compounds involved. We have looked at the concept of reaction order and have seen how the order is experimentally determined. In this section, we shall discuss how reaction rates are affected by changes in *temperature*. For almost all reactions, **an increase in temperature means an increase in rate.** This is in accord with what our intuitions would predict, because, as we

Table 9-1 Summary of Pertinent Equations for Reactions of Different Orders

Order	*Sum of exponents*	*Rate law*	*Concentration–time relationship*	*Linear plots to determine order*	*Half-life–rate constant relationship*
0	0	rate = k	$[A]_0 - [A] = kt$	$([A]_0 - [A])$ vs. t	$\tau = \frac{[A]_0}{2k}$
1	1	rate = $k[A]$	$\log\frac{[A]_0}{[A]} = \frac{kt}{2.303}$	$\log\frac{[A]_0}{[A]}$ vs. t	$\tau = \frac{0.693}{k}$
2	2	rate = $k[A]^2$	$\frac{1}{[A]} - \frac{1}{[A]_0} = kt$	$\frac{1}{[A]}$ vs. t	$\tau = \frac{1}{k[A]_0}$
2	2	rate = $k[A]\,[B]$	(complicated)	—	—
		(set [A] = [B] initially so that relationships above will hold)			
n	n	rate = $k[A]^n$	—	—	$\tau \propto \frac{1}{[A]_0^{n-1}}$

have seen (Section 4-11), increasing the temperature causes the molecules to move faster, and faster molecules collide more often and with greater force than do slower molecules. (Collision theory is discussed more fully in the next section.) Chemists have a general rule of thumb which says that *the rate of any reaction approximately doubles for any 10°C rise in temperature*. Because there are so many widely different kinds of reactions, it may seem surprising that all of them should follow the same simple rule, and indeed the rule is only approximate and there are many exceptions. However, the student will not go far wrong to assume that in most cases the rule is approximately valid.

Not only does the *rate* of a reaction change with temperature, but so does the *rate constant* (which is therefore not a constant with respect to temperature). Fortunately, the variation of the rate constant with changing temperature is not unpredictable but follows a regular law. In 1889, a Swedish scientist, Svante Arrhenius (1859–1927), observed that the rate constant k for any given reaction varied with the absolute temperature in the following manner:

$$k = Ae^{-E_a/RT}$$

where k = rate constant
A = a constant (called the **frequency factor**)
E_a = a constant (called the **activation energy**)
R = the gas constant (1.99 cal/deg · mole)
T = absolute temperature

This equation is known as the **Arrhenius equation.**

If we take the logarithm of both sides, we get

$$\log k = \log A - \frac{E_a}{2.303RT}$$

This is the equation of another straight line ($y = mx + b$), so that we can determine the values of E_a and A by plotting $\log k$ vs. $1/T$. (We already know R, of course; it is the familiar gas constant.) The slope of this line will be $-E_a/R(2.303)$ and the intercept will be $\log A$. To get the data necessary to make such a plot, we must run the reaction at several temperatures and experimentally determine the value of k (as in the preceding section) at each of these temperatures. It might be thought that only two temperatures would be necessary, because that would give us two points on the graph, and two points are all that is needed to determine a straight line. Strictly speaking, this is true, but in practice our measurements are always subject to a certain degree of experimental error, so that we can improve the accuracy of our slope by plotting more than two points. The method is illustrated in the following example.

Example: For the decomposition of N_2O_5, the experimentally determined values of k at several temperatures are:

k	T (°K)
1.72×10^{-5}	298
6.65×10^{-5}	308
24.95×10^{-5}	318
75×10^{-5}	328
240×10^{-5}	338

Calculate A and E_a for this reaction from the Arrhenius equation, $k = Ae^{-E_a/RT}$.

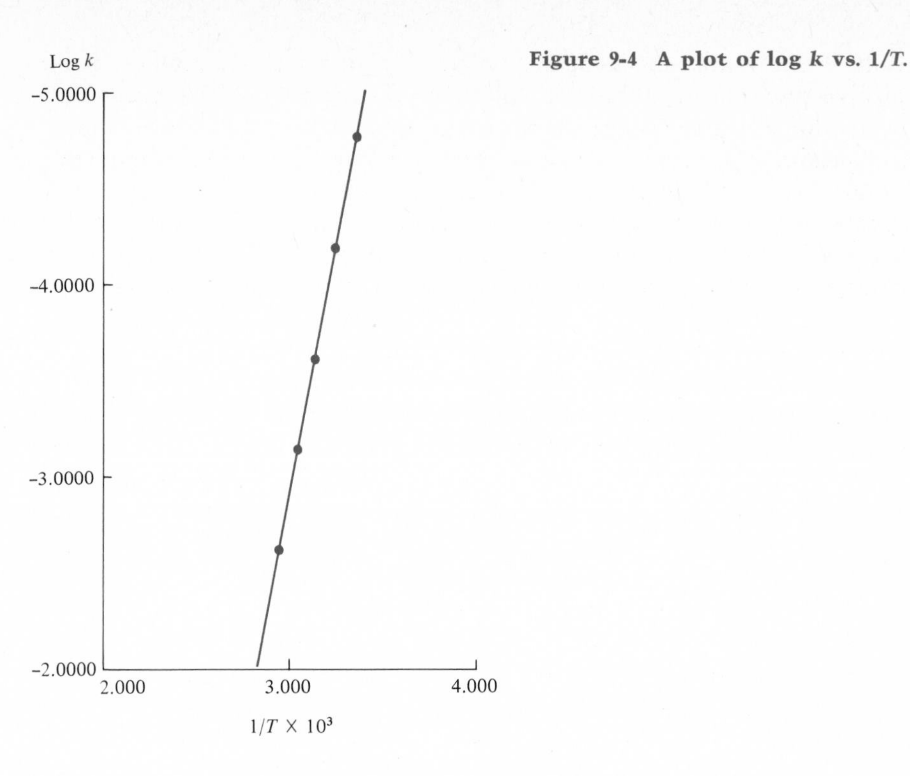

Figure 9-4 A plot of log k vs. 1/T.

Answer: To obtain A and E_a, we plot $\log k$ vs. $1/T$. From the data given, we get

$\log k$	$1/T$
−4.7645	3.356×10^{-3}
−4.1772	3.247×10^{-3}
−3.6030	3.147×10^{-3}
−3.1249	3.059×10^{-3}
−2.6198	2.959×10^{-3}

Plotting these data (see Figure 9-4) we draw the best straight line through the five points, and from the slope we find that $E_a = 24{,}800$ cal, while A is calculated to be 2.84×10^{13} s.

Now that we know mechanically how to compute these values, let us look into their physical meanings.

9-7 Collision Theory of Gas Reactions

One theory that chemists have used to explain the kinetics of reactions which take place in the gas phase is the *collision theory*. The first postulate of this theory is that in order for two reactant molecules to change into one or more product molecules, they must approach each other closely enough for us to say that they have collided. This seems eminently reasonable, since we would not expect a reaction to occur between two gas molecules which are separated from each other. They must collide, and some rearrangements must take place (old bonds must break and/or new bonds must form) in order for new compounds to form. Whether enough collisions can take place to account for the rate of a given gaseous reaction can be tested mathematically if one knows the temperature, the pressure, and some other factors. Calculations show that an average gas molecule experiences about 10^{28} collisions per second at ordinary temperatures and pressures. Under some conditions of temperature and pressure this number may vary somewhat, but it is hardly ever more than 10^{30} or less than 10^{26}. Under any conditions, it is an extremely large number. From this fact we might conclude that all reactions involving about 1 mole ($\sim 10^{23}$ molecules) of a gas should be completed in a fraction of a second. While this is, in fact, the case for some reactions, it is not true for many others. Some reactions take many seconds, still others minutes, and some take much longer. Other factors must be involved in a reaction besides the mere collision of molecules.

One of these factors is the energy involved when two molecules collide. This also appears quite reasonable. In order for most reactions to begin, a chemical bond must be broken and energy is required to do this. In Section 4-11, we learned that in a container containing gas molecules, some molecules are traveling slowly (little energy) while others are traveling with moderate speed and still others are traveling at great speeds (very high energies). If a collision takes place between two molecules both of which are traveling slowly, there may not be enough energy to break the bond.

The second postulate of the collision theory is that in order to have an *effective* collision (one that leads to a reaction), the molecules must possess, between them, at least a certain amount of energy. This amount of energy is the activation energy (E_a); we have already met it in the previous section and have seen how it is calculated. The activation energy is different for every reaction.

In Figure 4-14 (page 104) we saw how the energies of gas molecules are distributed at different temperatures. Increasing the temperature increases the number of molecules with high energies (that is, those that are moving very fast). Thus, reaction rates increase with increasing temperature for two reasons: (1) the molecules are moving faster so that they collide more often (there are more collisions per second), and (2) because of the larger number of high-energy molecules, the average collision takes place with more force, so that a larger percentage of collisions have energies greater than the activation energy and so can lead to reaction.

Summarizing, the collision theory of gas reactions postulates that for a reaction to take place between two molecules:

1. The molecules must collide.
2. The molecules must possess an amount of energy equal to or greater than the activation energy of the particular reaction under investigation.

The interpretation of the Arrhenius equation $k = Ae^{-E_a/RT}$ is that A, the frequency factor, is related to the collision frequency (the number of collisions per second), and the exponential factor, $e^{-E_a/RT}$, is related to the probability that any given collision is an effective collision (that is, one which leads to a chemical reaction).

RESULTS OF THE COLLISION THEORY. In order for a theory to be valid, it must be in accord with experimental facts. When the collision theory was tested (by comparing experimentally measured rate constants with calculated rate constants), large discrepancies were obtained. For example, the calculated rate constant for the reaction

$$NO + O_3 \longrightarrow NO_2 + O_2$$

differs from the measured rate constant by a factor of a thousand. Other reactions show smaller discrepancies, while others show discrepancies up to 10^6.

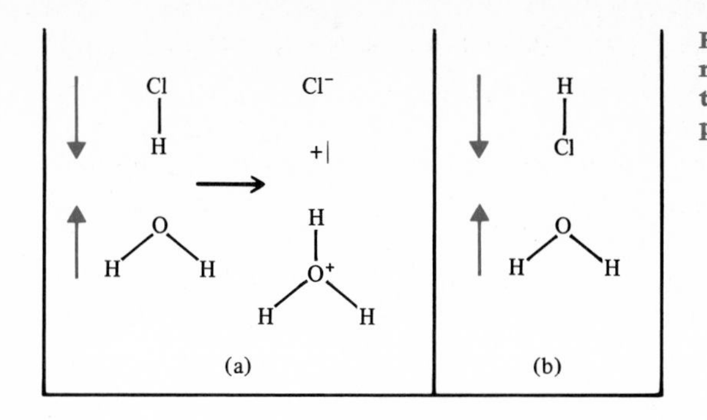

Figure 9-5 (a) A collision that can lead to reaction. (b) A collision that cannot lead to reaction. The colored arrows show the paths of motion of the molecules.

Theorists are often confronted with this kind of problem. What they wish to do is to describe the physical universe in terms of some model. The collision theory gives a picture of how gas reactions occur which is approximately correct but has discrepancies. The question that arises is whether it is worthwhile to abandon this model totally and begin again with a different theory or to look for additional factors that may influence the rate of the reaction but are still within the model of the collision theory. In the collision theory we treat molecules as if they were billiard balls which simply collide and bounce away again as either product or reactant molecules, depending on whether or not they react. In reality, molecules are not simply hard spheres, but have a certain structure. This means that there may be certain geometric requirements in the collisions which would facilitate effective collisions. As a hypothetical example, a reaction between a molecule of H_2O and one of HCl may take place if, when the molecules approach each other, the hydrogen of the HCl is pointing toward the oxygen of the H_2O, but not if the chlorine of the HCl is pointing toward the oxygen (Figure 9-5). We may label this geometric or steric factor p and introduce it in the Arrhenius equation, which now becomes $k = pAe^{-E_a/RT}$. Unfortunately, the problems of calculating p theoretically have not been surmounted; furthermore, some theorists have stated that there are probably other factors involved also. Because of this, the steric factor p has become a number by which one measures the differences in the calculated and experimentally determined rate constant. Of course, this is not a satisfactory situation, and eventually the collision theory has given way to a more modern theory, discussed in the next paragraph.

We have seen that the activation energy is the amount of energy the molecules must possess in order to yield an effective collision. During the collision process this energy is needed in order to change the reactants to products. The reaction itself cannot proceed instantaneously (that is, by some hypothetical magic process in which at one time we have reactants and an instant later we have products). The reaction requires a finite (though very short) time period, and during that time period there must be species which are neither reactants nor products but something in-between. The reactant molecule (or molecules) have started to break up, but the product molecule (or molecules) are not yet fully formed. We call this intermediate species the **activated complex.** It must be a species of higher energy than either the reactants or products, since an activation energy is needed to form it. We can describe the process graphically (Figure 9-6). In the figure we are looking at the differences in energies of the reactants, the products, and the activated complex. The energy necessary for the reactants to form an activated complex is the activation energy.

The horizontal axis, called the **reaction coordinate,** signifies the progression of the reaction. For the reaction between HCl and NH_3, the activated complex may look like this:

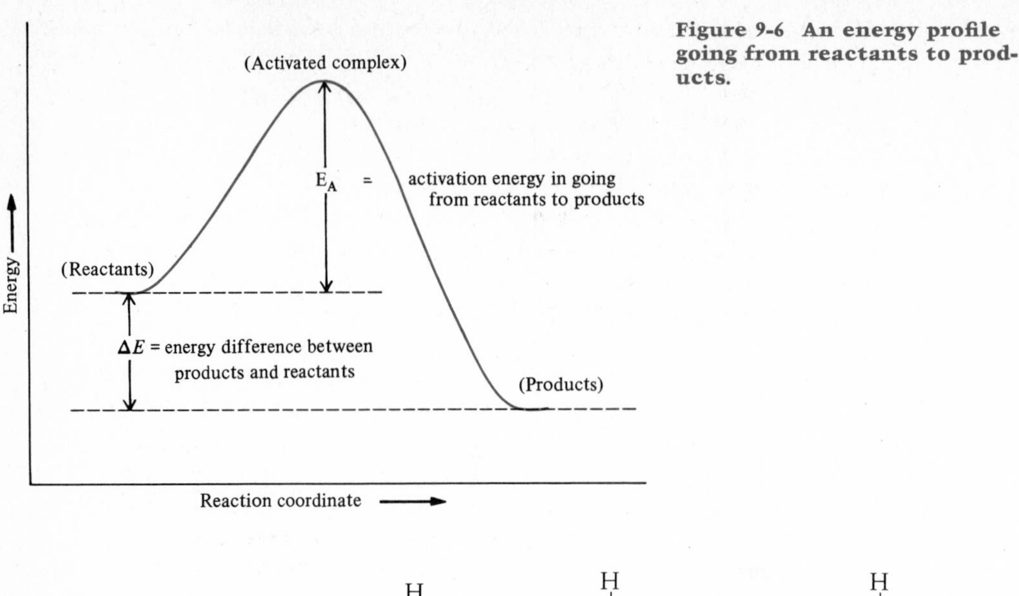

Figure 9-6 An energy profile going from reactants to products.

$$\text{Cl—H} \quad \text{H}_3\text{N} \longrightarrow \text{Cl}\cdots\text{H}\cdots\text{NH}_3 \longrightarrow \text{Cl}^{\ominus} + \text{H—NH}_3^{\oplus}$$

reactants — activated complex — products

where the dotted lines represent bonds that are in the process of breaking or being made.

9-8 Reaction Mechanisms

Consider the reaction

$$H_2 + I_2 \longrightarrow 2HI$$

The rate law for this reaction is rate $= k[H_2][I_2]$. This is a simple second-order rate law and is physically interpretable in terms of a collision mechanism. That is, if we assume that a reaction takes place when a molecule of I_2 collides with a molecule of H_2, then the rate of the reaction should depend, as the rate equation tells us, on the concentrations of H_2 and of I_2. Since bromine Br_2 is a substance that in many ways greatly resembles iodine, we might logically assume that the reaction

$$H_2 + Br_2 \longrightarrow 2HBr$$

would obey a similar rate law. But chemistry is full of surprises, and when M. Bodenstein and S. C. Lind studied the kinetics of this reaction in 1906, they found that the experimental rate law was

$$\text{rate} = \frac{k[H_2][Br_2]^{1/2}}{k' + [HBr]/[Br_2]}$$

A very surprising result. Can this complicated rate law be interpreted in terms of simple collision theory? No! The reason is that this reaction, like a great many others, does not take place in one step but goes through a *sequence* of reactions. The individual reactions (called **elementary reactions**), which make up such a sequence, take place so rapidly that it is not obvious that they are going on. All we *see* is that H_2 and Br_2 are being used up, and that HBr is the product. The fact that several elementary reactions are actually taking place (these are described later in this section) can only be detected by indirect evidence, such as the rate law determined above. An outline of such a sequence of reactions is called the **mechanism** of a reaction. A reaction mechanism shows exactly which elementary reactions are taking place, which bonds are breaking and forming, and tells which species are **reaction intermediates**; that is, substances that are formed, have an extremely short lifetime, and then are converted to other intermediates or to the reaction products. What are observed are the reactants and the products. It is up to the chemist to unravel the series of elementary reactions that constitutes the mechanism of the reaction. The reaction mechanism gives us the intimate details concerning the reaction itself. The kinetics of a reaction are an important tool, although not the only one, in helping the chemist unravel the mechanism.

However, the determination of the mechanism of a given reaction is not always easy, and the mechanisms of many reactions are still unknown, although in many other cases mechanisms are fairly well understood.

In order to use kinetics as an aid to the understanding of mechanism, we make use of the following fact: **The rate of an elementary step is proportional to the concentrations of the reactants which take part in that step.** Therefore, there *is* a relationship between the rate and stoichiometry *of each individual step*. This satisfies our intuition, since each elementary step takes place through a collision. We can now understand why we cannot in general tell the overall order of a reaction by appealing to the stoichiometry. It is because a reaction may take place through a sequence of elementary reactions that are not knowable from the stoichiometry. It is this reaction mechanism that finally yields the order of the reaction.

When a reaction involves two or more elementary steps, then in the usual case, one step will be slower than all the others. We can then make the statement that the **overall rate of the entire reaction is equal to the rate of the slowest step,** which we therefore call the **rate-determinating step.** That this statement is true can be demonstrated by the following analogy. Imagine a room filled with white marbles. Two goblins (not much bigger than the marbles) are engaged in painting the marbles. The first goblin leaps about at random; when he lands on a white marble, he paints it blue. The second goblin also leaps about, but he does not paint the white marbles. His job is to repaint the blue marbles black, and he leaps directly from one blue marble to another. First, let us suppose that the first goblin is much slower than the second (Figure 9-7a). It is easy to see that the overall rate (that is, the rate in marbles per minute at which the marbles are converted from white to black) is the same as that of the first goblin. No marble can get from white to black without first becoming blue; and the rate of this is the rate set by the first goblin. Once each marble has been painted blue, it is rapidly repainted black. Now let us suppose that the second goblin is the slow one (Figure 9-7b). This time the rate of the overall change is equal to the rate at which the second goblin works. The first goblin paints a great many marbles blue, but the marbles only become black when they have been repainted by the second goblin. In this case, the blue marbles pile up. Thus, we see that the rate of the overall reaction is equal to the rate of the slow step, *whichever step that may be*. Of course, some reactions *are* single-step overall reactions, and in those cases, the overall rate law *does* correspond directly to the stoichiometry. The difficulty is that we do not in general know which reactions these are without a great deal of evidence, from kinetics and other sources.

Example: Consider a reaction whose stoichiometry is

$$A + 2B \longrightarrow 2C$$

The experimental rate law is found to be

$$\text{rate} = k[A][B]$$

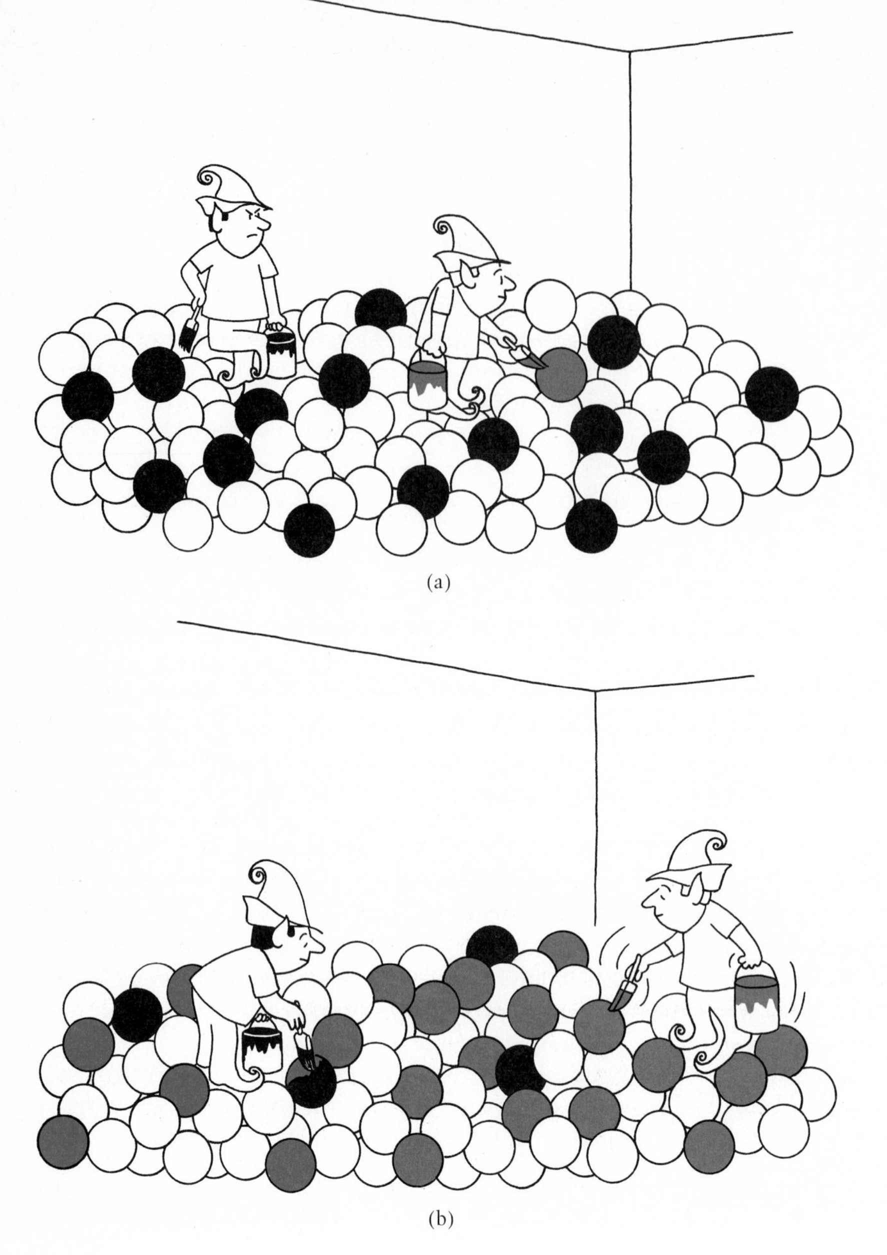

Figure 9-7 Goblins painting marbles blue, then black. In (a), the first goblin is the rate-determining goblin. In (b), it is the second goblin.

A plausible mechanism is

(1) $$A + B \longrightarrow C + D \qquad \text{(slow)}$$

(2) $$D + B \longrightarrow C \qquad \text{(fast)}$$

The mechanism as written involves two elementary reactions. The first is slow and the second fast. As soon as species D is formed, it reacts with a second molecule of B and forms products. Step 1 is the rate-determining step and whatever happens in step 2 has no effect on either the rate or the rate law. We can therefore (going by the fact that the rate law of a single step is related to the stoichiometry of that step) write the rate law as

$$\text{rate} = k[A][B]$$

We see that this conforms with the experimentally determined rate law.

Example: Consider the decomposition of ozone,

$$2O_3 \longrightarrow 3O_2$$

We have already mentioned (Section 9-1) that the rate law for this reaction is experimentally found to be

$$\text{rate} = k\frac{[O_3]^2}{[O_2]}$$

Clearly, the rate law cannot be predicted from the stoichiometry of the balanced equation. A reasonable mechanism is

$$(1) \qquad O_3 \underset{k_{-1}}{\overset{k_1}{\rightleftharpoons}} O_2 + O \qquad \text{(fast)} \qquad (4)$$

$$(2) \qquad O + O_3 \xrightarrow{k_2} 2O_2 \qquad \text{(slow)} \qquad (5)$$

The symbol O signifies an individual atom of oxygen, which is an intermediate in this reaction. Since it has a very short lifetime, and is always present in very small concentrations, it is difficult or impossible to measure its concentration at any given time.

Step 1 is an elementary reaction that proceeds very quickly and comes to a rapid equilibrium. Step 2 is the slow step and hence the rate-determining step. We can write for step 2

$$\text{rate} = k_2[O_3][O] \qquad (6)$$

This equation expresses the rate in terms of the concentrations of ozone and of atomic oxygen. Since atomic oxygen does not appear in the experimentally determined rate expression and we cannot directly measure its concentration, we must find a relationship between the concentration of atomic oxygen and quantities we *can* measure, namely the experimentally determinable concentrations of O_3 and O_2. Since step 1 is assumed to come to a rapid equilibrium, we can write an equilibrium expression for step 1.

$$K = \frac{[O_2][O]}{[O_3]} \qquad K = \text{an equilibrium constant, not a rate constant}$$

Solving for [O], the concentration of atomic oxygen, yields

$$[O] = \frac{K[O_3]}{[O_2]}$$

Substituting the concentration of atomic oxygen into the rate law for step 2, equation (6), gives

$$\text{rate} = \frac{k_2[O_3]^2K}{[O_2]} \qquad (7)$$

and, since k_2K is also a constant (which we can call k),

$$\text{rate} = k\frac{[O_3]^2}{[O_2]} \qquad (8)$$

the experimentally determined rate law. Thus, the mechanism we have outlined (steps 1 and 2) is consistent with the experimentally determined rate law. This does not mean that we have proven that this is indeed the mechanism (because other mechanisms might also be consistent with this rate law), but at least we have suggested a plausible mechanism that does not conflict with the rate law. This mechanism now becomes a hypothesis which we would next try to prove or disprove by other experiments. For example, the mechanism predicts that atomic oxygen is an intermediate, and we might try to "trap" this by adding something to the reaction mixture that would preferentially react with this species.

Summarizing, we have shown that chemical reactions often are not simple one-step processes. Frequently, they occur as a sequence of steps. Such sequences are referred to as *reaction mechanisms*. Given a reaction, we would like to unravel the reaction mechanism so that we can better understand and perhaps control it. A mechanism usually has a slow step, which is the rate-determining step. Many times, one finds in the rate-determining step a concentration term involving a species that is neither reactant nor product—an *intermediate*. One then must find a way of expressing the concentration of the intermediate in terms of the concentrations of reactants and products. Our example illustrated one method. The following example illustrates another useful technique.

Example: As we mentioned before, the reaction between hydrogen and bromine,

$$H_2 + Br_2 \longrightarrow 2HBr$$

yields the following rate law;

$$\text{rate of formation of HBr} = \frac{k[H_2][Br_2]^{1/2}}{k' + [HBr]/[Br_2]} \tag{9}$$

After the experimentally determined rate law was published, it took 13 years to work out an acceptable mechanism. This mechanism is a classic and is generally used to illustrate what is termed a **chain mechanism.** In a chain mechanism we have a series of intermediates being formed which are in very low concentration and extremely reactive. The first step in the mechanism forms an intermediate and is called the **chain-initiating step;** it starts the chain.

(1) $Br_2 \xrightarrow{k_1} 2Br\cdot$ — The chain starts with the formation of two bromine atoms from a bromine molecule.

(2) $Br\cdot + H_2 \xrightarrow{k_2} HBr + H\cdot$ — Steps 2 and 3 are termed **chain-propagating steps.**

(3) $H\cdot + Br_2 \xrightarrow{k_3} HBr + Br\cdot$ — Two molecules of HBr and another atom of bromine have been formed. Since each $H\cdot$ could react with another Br_2 and each $Br\cdot$ with another H_2, this sequence, if nothing interfered, could continue until all the H_2 and Br_2 were used up.

(4) $H\cdot + HBr \xrightarrow{k_4} H_2 + Br\cdot$ — **Chain-inhibiting step**—uses up a molecule of product. This is necessary in this mechanism since the experimentally determined rate law shows that the rate of formation of HBr is inhibited by HBr ([HBr] appears in the denominator so that an increase in [HBr] *decreases* the rate).

(5) $2Br\cdot \xrightarrow{k_5} Br_2$ — **Chain-terminating step**—this step breaks the sequence of the chain reaction. Such a step is necessary for all chain mechanisms unless experiments show they proceed without termination.

From the foregoing sequence, we can write the rate of formation of HBr as

$$\frac{\Delta[\mathrm{HBr}]}{\Delta t} = k_2[\mathrm{Br\cdot}][\mathrm{H_2}] + k_3[\mathrm{H\cdot}][\mathrm{Br_2}] - k_4[\mathrm{H\cdot}][\mathrm{HBr}] \qquad (10)$$

(HBr is formed in the second and third steps and is lost in the fourth step.)

Since all these steps are rapid, and none is much slower than any other, we have no rate-determining step and it is necessary to find another method for expressing the concentrations of the intermediates $[\mathrm{H\cdot}]$ and $[\mathrm{Br\cdot}]$ in terms of the measurable $[\mathrm{H_2}]$, $[\mathrm{Br_2}]$, and $[\mathrm{HBr}]$: the reactant and/or product species. An approximation is used here that is often made in chemical kinetics when the going (mathematically) becomes too difficult. The approximation is termed **the steady-state approximation.** We assume that after the reaction has proceeded for a while, **the rate of formation and the rate of disappearance of the intermediate species are equal,** so that the *net* rate of formation or the *net* rate of disappearance is zero. Another way of saying this is that the concentration of the intermediates during the period of the reaction is constant. By using this assumption, we can write

$$\text{rate of formation of H}\cdot = \text{rate of disappearance of H}\cdot$$
$$\text{rate of formation of Br}\cdot = \text{rate of disappearance of Br}\cdot$$
$$\text{net rate of formation of H}\cdot = 0$$
$$\text{net rate of formation of Br}\cdot = 0$$

But the net rate of formation of H· and of Br· can be expressed as follows (from the equations above; H· appears in steps 2, 3, and 4; Br· appears in all five steps):

$$0 = \text{net rate of formation of H}\cdot = k_2[\mathrm{H_2}][\mathrm{Br\cdot}] - k_3[\mathrm{H\cdot}][\mathrm{Br_2}] - k_4[\mathrm{H\cdot}][\mathrm{HBr}]$$

$$0 = \text{net rate of formation of Br}\cdot$$
$$= k_1[\mathrm{Br_2}] - k_2[\mathrm{H_2}][\mathrm{Br\cdot}] + k_3[\mathrm{H\cdot}][\mathrm{Br_2}] + k_4[\mathrm{H\cdot}][\mathrm{HBr}] - k_5[\mathrm{Br\cdot}]^2$$

We may treat these as simultaneous equations with $[\mathrm{H\cdot}]$ and $[\mathrm{Br\cdot}]$ as the two unknowns. Solving for $[\mathrm{H\cdot}]$ and $[\mathrm{Br\cdot}]$, we get

$$[\mathrm{Br\cdot}] = \frac{k_1}{k_5}[\mathrm{Br_2}]^{1/2}$$

$$[\mathrm{H\cdot}] = \frac{k_2(k_1/k_5)^{1/2}[\mathrm{H_2}][\mathrm{Br_2}]^{1/2}}{k_3[\mathrm{Br_2}] + k_4[\mathrm{HBr}]}$$

These look like complicated equations, but all they consist of are constants plus the measurable quantities $[\mathrm{H_2}]$, $[\mathrm{HBr}]$, and $[\mathrm{Br_2}]$. We may now insert these values into equation (10) to get (after rearranging)

$$\frac{\Delta[\mathrm{HBr}]}{\Delta t} = \frac{k_3k_2k_4^{-1}k_1^{1/2}k_5^{-1/2}[\mathrm{H_2}][\mathrm{Br_2}]^{1/2}}{k_3k_4^{-1} + [\mathrm{HBr}]/[\mathrm{Br_2}]}$$

which is the same as the experimentally determined rate law, equation (9). Of course, we do not know the individual values of k_3, k_2, etc. (these can be determined, but more experimental work is required), but the experimentally measured k is equal to $k_3k_2k_4^{-1}k_1^{1/2}k_5^{-1/2}$, and k' is equal to $k_3k_4^{-1}$.

Without using the steady-state approximation we would have been hopelessly lost in a maze of mathematics, but using it, we have verified the rate law. This is a *good* approximation: reasonable, logical, and used widely.

When we have described a mechanism, we always use the term "reasonable." As we have previously indicated, the fact that a mechanism agrees with a given rate law does not, in fact, prove that the mechanism is correct, since other mechanisms might also agree with this rate law. Other kinds of experimental evidence are needed before we can say that a mechanism is "proven." In fact, the history of mechanistic investigation has shown us that even mechanisms regarded as proven have been changed as new data have been obtained. This is an area where we cannot see directly into what is happening, and so must be guided by inference. Although we cannot, therefore, use a rate law to prove a mechanism, we can use rate laws in a negative sense: *If a rate law conflicts with a proposed mechanism, the mechanism must be wrong.* The rate law is a *fact* (we assume that no laboratory mistake was made in obtaining the rate law); the mechanism is a *theory.* When a fact conflicts with a theory, it is the theory that must give way.

9-9 Derivation of the Equilibrium Expression

In Section 8-2, we presented the equilibrium expression

$$K = \frac{[C]^c[D]^d \cdots}{[A]^a[B]^b \cdots}$$

for the general equilibrium reaction

$$aA + bB + \cdots \rightleftharpoons cC + dD + \cdots$$

We are now ready to derive the equilibrium expression.

We have learned that the rate law for any reaction is not necessarily related to the stoichiometry, but if the reaction consists of two or more elementary steps, then the rate law for each elementary step *is* related to the stoichiometry of that step. If the overall reaction is reversible, each elementary step must also be reversible. (Why?) We will now derive an equilibrium expression for any particular elementary step, for example for

$$2A + B \rightleftharpoons 2C + D$$

Because this is an elementary step, we can write for the forward reaction,

$$\text{rate} = k_f[A]^2[B]$$

and for the reverse reaction,

$$\text{rate} = k_r[C]^2[D]$$

At equilibrium the rates must be equal, so at equilibrium

$$k_f[A]^2[B] = k_r[C]^2[D]$$

Rearranging, we get

$$\frac{k_f}{k_r} = \frac{[C]^2[D]}{[A]^2[B]}$$

Since k_f and k_r are both constants, their quotient (K) is also a constant,

$$\frac{k_f}{k_r} = K = \frac{[C]^2[D]}{[A]^2[B]}$$

K is, in fact, the equilibrium constant for the elementary step $2A + B \rightleftharpoons 2C + D$. It is obvious that we could follow the same procedure for any elementary step.

Having shown that a correct equilibrium expression is derivable from any elementary step, we will now show that the equilibrium expression is also valid for any overall reaction consisting of two or more elementary steps. In Section 8-5, we learned that if two chemical equations are *added,* their equilibrium expressions and equilibrium constants are *multiplied.* Assume that the reaction

$$A + 2B \rightleftharpoons 2C$$

involves the elementary steps (pages 269–270)

$$(1)\ A + B \rightleftharpoons C + D$$

$$(2)\ D + B \rightleftharpoons C$$

Then K for each step is

$$(1)\ K_1 = \frac{[C][D]}{[A][B]}$$

$$(2)\ K_2 = \frac{[C]}{[D][B]}$$

Adding the equations for the steps, we get the overall equation

$$\begin{array}{l} A + B \rightleftharpoons C + \not{D} \\ \not{D} + B \rightleftharpoons C \\ \hline A + 2B \rightleftharpoons 2C \end{array}$$

Multiplying the equilibrium expressions, we get

$$K_1K_2 = K = \frac{[C][D]}{[A][B]} \cdot \frac{[C]}{[D][B]} = \frac{[C]^2}{[A][B]^2}$$

Note that $K = [C]^2/[A][B]^2$ is the equilibrium expression corresponding to the overall equation.

The same result would be obtained for any reversible reaction, no matter how many or how complicated the elementary steps. The sum of all the equilibrium equations must equal the overall equation (why?), and the product of all the individual equilibrium expressions will always be the same as the equilibrium expression for the overall equation.

9-10 Catalysts

Many major industrial processes are possible only because of catalysis. Among these are the Haber process for the production of ammonia (Section 17-3) and the cracking of petroleum (Section 20-2).

A **catalyst** is a **substance that increases the rate of a reaction without being used up in the reaction.** This definition is due to Friedrich Wilhelm Ostwald (1853–1932), who also showed that although a catalyst affects the *rate* of a reaction, it cannot affect the *position of equilibrium* of the reaction or the value of the equilibrium constant. Since a catalyst changes the rate but not the equilibrium, it must also change the rate of the reverse reaction to the same degree as the rate of forward reaction. For a reversible reaction,

$$A \underset{k_2}{\overset{k_1}{\rightleftharpoons}} B$$

we have seen that

$$K = \frac{k_1}{k_2}$$

and since K is not affected by a catalyst, k_1 and k_2 (the rate constants for the forward and reverse reactions) must be affected to the same extent by any catalyst.

The question that naturally arises is: How does a catalyst increase the rate of a

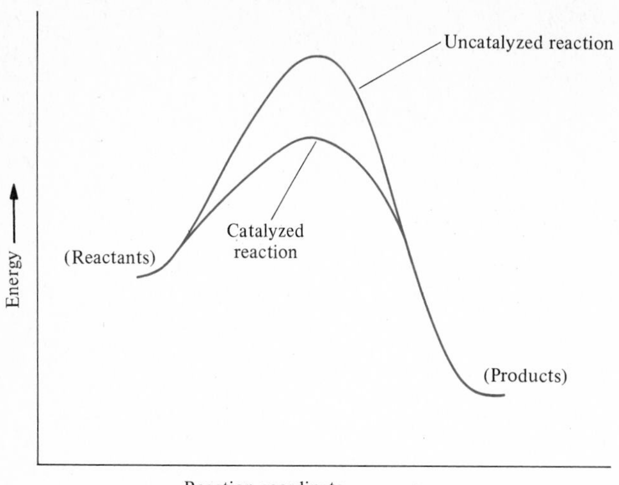

Figure 9-8 A catalyst makes a reaction go faster by lowering the activation energy. A different mechanism is always involved.

reaction? It does this by allowing the reaction to proceed by a *different mechanism*. This lowers the activation energy of the reaction so that it proceeds at a faster rate (see Figure 9-8). **Inhibitors** or *poisons* increase the activation energy and hence slow the reaction. Catalysts may be classified as **homogeneous** or **heterogeneous.** When the entire reaction takes place in a single phase, it is termed homogeneous. Heterogeneous catalysis must take place at an interface, between two phases. Heterogeneous catalysis is also termed *surface catalysis*. Most known examples of homogeneous catalysis are those that take place in the liquid phase.

Many reactions that proceed immeasurably slowly in a single phase (gaseous or liquid) take place rather quickly upon the introduction of a particular solid surface. The catalytic action of the surface depends on the **adsorption** (when particles go into the body of a substance it is called absorption, when they go onto the surface it is called adsorption) of the reactant molecules onto the surface of the catalyst.

Enzymes are the most remarkable catalysts known, since they catalyze reactions in the body in a highly specific way and with extreme rapidity compared to the reactions run by chemists in laboratories. The way they do this is discussed in Section 21-4. Enzymes acting as catalysts have been studied as both homogeneous and as heterogeneous systems.

Problems

9-1 Define each of the following terms **(a)** chemical kinetics **(b)** reaction rate **(c)** rate law **(d)** rate constant **(e)** reaction order **(f)** first order **(g)** second order **(h)** zero order **(i)** quenching **(j)** aliquot **(k)** half-life **(l)** activation energy **(m)** frequency factor **(n)** activated complex **(o)** reaction coordinate **(p)** elementary reaction **(q)** reaction mechanism **(r)** intermediate **(s)** rate-determining step **(t)** chain mechanism **(u)** chain-initiating step **(v)** chain-propagating step **(w)** chain-terminating step **(x)** steady-state approximation **(y)** inhibitor **(z)** homogeneous catalyst **(aa)** heterogeneous catalyst **(bb)** adsorption

9-2 Determine the orders of the corresponding reactions, given the rate laws.

(a) $-\dfrac{\Delta NO}{\Delta t} = k_a[NO]^2[O_2]$

(b) $-\dfrac{\Delta NO_2}{\Delta t} = k_b[NO_2]^2$

(c) $-\dfrac{\Delta COCl_2}{\Delta t} = k_c\dfrac{[COCl_2]^2}{[CO]}$

(d) $+\frac{\Delta I_2}{\Delta t} = k_d[C_2H_4Br_2][KI]$

(e) $-\frac{\Delta AsCl_5}{\Delta t} = k_e$

(f) $-\frac{\Delta CH_3COOC_2H_5}{\Delta t} = k_f[CH_3COOC_2H_5]$

9-3 Work out the units of k for each of the rate laws in Problem 9-2.

9-4 The reaction

$$2NOCl \longrightarrow 2NO + Cl_2$$

was found to be a second-order reaction with respect to NOCl. Write the rate law for this reaction.

9-5 For the reaction

$$H_2 + I_2 \longrightarrow 2HI$$

the rate laws are

$$\frac{\Delta[HI]}{\Delta t} = k_1[H_2][I_2]$$

$$-\frac{\Delta[H_2]}{\Delta t} = k_2[H_2][I_2]$$

$$-\frac{\Delta[I_2]}{\Delta t} = k_3[H_2][I_2]$$

Work out the relationships among k_1, k_2, and k_3.

9-6 The typical exchange reaction $H + D_2 \rightarrow HD + D$ has a simple second-order rate law,

$$-\frac{\Delta[D_2]}{\Delta t} = k[H][D_2]$$

(a) If all concentrations are given in moles/liter and time is in seconds, calculate the units of the rate constant k. **(b)** Fill in the table at 25°C.

$\frac{-\Delta[D_2]}{\Delta t}$ (*moles/liter · s*)	k	[H] (*moles/liter*)	$[D_2]$ (*moles/liter*)
5×10^{-4}	______	5×10^{-4}	1×10^{-3}
______	______	1×10^{-3}	1×10^{-3}
1×10^{-2}	______	______	1.0
3.5×10^{-2}	______	______	2.5×10^{-4}

9-7 For the reaction

$$2H_2 + 2NO \longrightarrow 2H_2O + N_2$$

the rate law was found to be

$$\text{rate} = k[H_2][NO]^2$$

At a certain temperature k was found to be 27 (liters/mole)^2/s. Calculate the rate when $[H_2] = 3.6 \times 10^{-2}\ M$ and $[NO] = 4.5 \times 10^{-3}\ M$.

9-8 The reaction

$$2NO(g) + 2H_2(g) \longrightarrow N_2(g) + 2H_2O(g)$$

was found to be second order in NO and first order in H_2. **(a)** Write rate laws for this reaction in terms of changing concentrations of each of the four species in the reaction. **(b)** Show the relationships among the four rate constants.

9-9 Under special conditions, the rate of the reaction

$$2H_2(g) + O_2(g) \longrightarrow 2H_2O(g)$$

can be represented by

$$\frac{\Delta[H_2O]}{\Delta t} = k[H_2]^m[O_2]^n$$

From the following data, find the values of m, n, and k.

$[H_2]$ *(moles/liter)*	$[O_2]$ *(moles/liter)*	$\frac{\Delta[H_2O]}{\Delta t}$ *(moles/liter · s)*
1.00	1.00	1.00
1.00	2.00	2.00
1.00	4.00	4.00
3.00	1.00	9.00
4.00	2.00	32.00

9-10 The reaction

$$CO + Cl_2 \longrightarrow COCl + Cl$$

was studied by the initial rate method and the following data were obtained:

$[CO]$ *(moles/liter)*	$[Cl_2]$ *(moles/liter)*	$-\frac{\Delta[CO]}{\Delta t}$ *(moles/liter · s)*
1.00×10^2	1.00×10^2	6.60×10^3
2.00×10^2	1.00×10^2	1.32×10^4
3.00×10^2	1.00×10^2	1.98×10^4
1.00×10^2	2.00×10^2	2.64×10^4
2.00×10^2	3.00×10^2	1.19×10^5

From the data, obtain the values of m, n, and k in the rate law:

$$-\frac{\Delta[CO]}{\Delta t} = k[CO]^m[Cl_2]^n$$

9-11 The reaction

$$A + B \longrightarrow C + D$$

is first order each in A and in B, second order overall. The rate was studied by measuring the concentration of A at time intervals of 200 s, with the following results:

Time (s)	200	400	600	800	1000
[A] (moles/liter)	0.219	0.195	0.176	0.160	0.147

The initial concentrations of A and B were equal. Use the graphical method to calculate **(a)** the rate constant k **(b)** the initial concentrations of A and B

9-12 A certain compound X spontaneously converts itself to another compound, Y. The rate of disappearance of X is measured and its half-life is found to be 1×10^{-8} s at 100°K. Calculate the first-order rate constant for this conversion.

9-13 A compound A is converted at a temperature of 125°C by a first-order reaction to another compound B. The reaction has a half-life of 4.42 h. If we bring a 7.50-g sample of A to a temperature of 125°C, how long will it take for 7.15 g of the sample to be converted to B?

9-14 The half-life τ of a reaction is related to the initial concentration of reactant species A as follows:

$$\tau = \frac{C}{[A]_0^{n-1}}$$

where C is a constant and n is the order of the reaction. Use the following data to calculate the order of the reaction A → products:

$[A]_0$ (*moles/liter*)	0.5000	1.000	2.000	3.000	5.000
τ (*s*)	5.557×10^3	1.111×10^4	2.223×10^4	3.334×10^4	5.557×10^4

9-15 Assume that a reaction A → B can be zero, first, or second order. If the reaction begins with 0.500 mole/liter of A, and the rate constant is 1.00×10^{-2} (with appropriate units), calculate the half-lives if the reaction is **(a)** zero order **(b)** first order **(c)** second order

9-16 A compound A decomposes with a first-order half-life of 36.5 days. If we have a sample of pure A weighing 850.0 g, how much will be left at the end of 219 days?

9-17 For a reaction

$$M(g) \longrightarrow N(g) + P(g)$$

the frequency factor A is 9.0×10^{12}/s and the activation energy E_a is 25500 cal/mole. What is the value of k at 400.0°K?

9-18 A reaction has an activation energy $E_a = 15.5$ kcal/mole. What is the effect on the rate if: **(a)** The temperature increases from 35.0°C to 45.0°C? **(b)** The temperature increases from 100.0°C to 110.0°C? (Assume that the rate law and frequency factor A do not change with temperature.)

9-19 Give possible reasons why, in a given reaction vessel, two molecules A and B react with each other when they collide, whereas two other molecules A and B collide but do not react.

9-20 For the reaction

$$X(g) \underset{k_2}{\overset{k_1}{\rightleftharpoons}} Y(g) + Z(g)$$

where k_1 and k_2 have the values 0.050 s^{-1} and 2.5×10^{-5} liter/mole · s at 298°K and 0.100 s^{-1} and 4.0×10^{-5} liter/mole · s at 308°K, calculate the equilibrium constants at 298°K and at 308°K.

9-21 If we can show that a proposed mechanism is consistent with the experimentally determined rate law for that reaction, does that prove the mechanism? Explain.

9-22 One plausible mechanism for the reaction $H_2 + I_2 \rightarrow 2HI$ is

(1)	$I_2 \rightleftharpoons 2I$	(fast, reversible)
(2)	$I + H_2 \rightleftharpoons IH_2$	(fast, reversible)
(3)	$IH_2 + I \longrightarrow 2HI$	(slow)

Show that this mechanism is consistent with the rate law

$$\frac{\Delta[\mathrm{HI}]}{\Delta t} = k[\mathrm{H_2}][\mathrm{I_2}]$$

9-23 The following are some other possible mechanisms for the reaction given in Problem 9-22.
(a) (1) $H_2 \longrightarrow 2H$ (slow)
(2) $H_2 + I_2 \longrightarrow HI + I$ (fast)
(3) $H + I \longrightarrow HI$ (fast)
(b) (1) $H_2 + I_2 \longrightarrow H_2I_2$ (slow)
(2) $H_2I_2 \longrightarrow 2HI$ (fast)
(c) (1) $H_2 \rightleftharpoons 2H$ (fast, reversible)
(2) $I_2 \rightleftharpoons 2I$ (fast, reversible)
(3) $H + I \longrightarrow HI$ (slow)
Calculate the predicted rate law for each. Which, if any, of these mechanisms is consistent with the rate law given in Problem 9-22?

9-24 Write the intermediates that appear in each of the four mechanisms shown in Problems 9-22 and 9-23.

9-25 The reaction

$$CO(g) + Cl_2(g) \longrightarrow COCl_2(g)$$

has the experimentally determined rate law

$$\frac{\Delta[\mathrm{COCl_2}]}{\Delta t} = k[\mathrm{CO}][\mathrm{Cl_2}]^{3/2}$$

Show that the following mechanism is consistent with this rate law.

(1) $Cl_2 \rightleftharpoons 2Cl$ (fast, reversible)
(2) $Cl + CO \rightleftharpoons COCl$ (fast, reversible)
(3) $Cl_2 + COCl \longrightarrow COCl_2 + Cl$ (slow)

9-26 If an overall reaction consisting of several elementary steps is reversible, why can we conclude that all the elementary steps are also reversible?

Additional Problems

9-27 A first-order gas reaction has $k = 1.5 \times 10^{-6}\ \mathrm{s^{-1}}$ at 200°C. **(a)** Calculate the half-life of this reaction. **(b)** If the reaction is allowed to run for 10.0 h, what percentage of the initial concentration has changed to product?

9-28 Under some conditions, the reaction $H_2 + I_2 \rightleftharpoons 2HI$ follows the rate law $\Delta[\mathrm{HI}]/\Delta t = k[\mathrm{H_2}]^m[\mathrm{I_2}]^n[\mathrm{HI}]^p$. At a particular temperature T, the following data were collected:

$[HI]$ (moles/liter)	$[H_2]$ (moles/liter)	$[I_2]$ (moles/liter)	$\frac{\Delta[HI]}{\Delta t}$ (moles/liter · s)
2.0	0.10	0.10	10.0
2.0	0.10	0.40	40.0
3.0	0.10	0.20	20.0
2.0	0.20	0.40	80.0

Find m, n, and p in the rate expression and calculate the value of k.

9-29 Calculate the activation energy E_a for a reaction whose rate exactly doubles when the temperature rises from 27°C to 40°C (assume that the rate law and frequency factor A do not change with temperature).

9-30 For the reaction

$$2NH_3 \xrightarrow[\text{catalyst}]{\text{tungsten}} N_2 + 3H_2$$

the rate laws were determined to be

$$-\frac{\Delta[NH_3]}{\Delta t} = k_1 \quad \text{(constant)}$$

$$\frac{\Delta[N_2]}{\Delta t} = k_2 \quad \text{(constant)}$$

$$\frac{\Delta[H_2]}{\Delta t} = k_3 \quad \text{(constant)}$$

What is the relationship among k_1, k_2, and k_3?

9-31 The inversion of sucrose $C_{12}H_{22}O_{11}$, a first-order reaction, proceeds as follows at 25°C:

$$\underset{\text{sucrose}}{C_{12}H_{22}O_{11}} + H_2O \longrightarrow \underset{\text{glucose}}{C_6H_{12}O_6} + \underset{\text{fructose}}{C_6H_{12}O_6}$$

Time (s)	0	1800	3600	5400	7800	10,800
Sucrose inverted (moles/liter)	0	0.100	0.195	0.280	0.370	0.470

The initial concentration of sucrose is 1.000 mole/liter. By plotting the appropriate variables, calculate the first-order rate constant k and the half-life of the reaction.

9-32 The reaction

$$C_4H_9Br + 2H_2O \longrightarrow C_4H_9OH + Br^- + H_3O^+$$

has the following mechanism:

$$\begin{array}{lll} (1) & C_4H_9Br \longrightarrow C_4H_9^+ + Br^- & \text{(slow)} \\ (2) & C_4H_9^+ + H_2O \longrightarrow C_4H_9OH_2^+ & \text{(fast)} \\ (3) & C_4H_9OH_2^+ + H_2O \longrightarrow C_4H_9OH + H_3O^+ & \text{(fast)} \end{array}$$

Write the rate law for this reaction.

9-33 Determine the order of the reaction in each case.

$$\frac{\Delta[X]}{\Delta t} = [A]^m[B]^n$$

(a) $m = n = 0$ **(b)** $m = n = 2$ **(c)** $m = 1, n = 0$ **(d)** $m = -1, n = 1$
(e) $m = 1, n = -2$ **(f)** $m = n = 1$

9-34 Work out the units of k for each of the rate laws in Problem 9-33.

9-35 For the reaction

$$2NO(g) + 2H_2(g) \longrightarrow N_2(g) + 2H_2O(g)$$

the rate law at 25°C was experimentally determined to be

$$-\frac{\Delta[NO]}{\Delta t} = k[NO]^2[H_2]$$

(a) If all the concentration terms are in moles/liter and the time is in seconds, calculate the units of the rate constant k. **(b)** Fill in the table at 25°C.

$\frac{-\Delta[NO]}{\Delta t}$ (moles/liter · s)	k	$[NO]$ (moles/liter)	$[H_2]$ (moles/liter)
1.0×10^6	______	1.0	0.038
5.2×10^4	______	0.20	______
______	______	0.010	0.010
6.2×10^4	______	______	0.025

9-36 For the reaction A → B, the rate varied with initial concentration of A as follows:

Initial concentration of A (moles/liter)	*Rate* (moles/liter · s)
0.20	3.5×10^{-5}
0.28	4.1×10^{-5}
0.36	4.7×10^{-5}
0.44	5.2×10^{-5}

From these data, use the graphical method to determine each of the following.
(a) the reaction order **(b)** the rate constant k **(c)** the rate of the reaction at initial concentrations of A of 0.050 *M* and 1.00 *M*

9-37 For the reaction A + 2B → products, if the concentrations are in moles/liter and the rate of reaction in moles/liter · s, determine the units of k for each of the following rate laws.
(a) rate = k_a **(b)** rate = k_b[A] **(c)** rate = k_c[B] **(d)** rate = k_d[A][B]
(e) rate = k_e[A][B]^2 **(f)** rate = k_f[A]^2[B]^2 **(g)** rate = k_g[B]^2

9-38 The rate constant for the second-order reaction

$$CH_3CHO \xrightarrow{I_2} CH_4 + CO$$

varies with temperature as follows:

T (°K)	630	645	660	675	695
k (liters/mole · s)	12	22	41	77	140

Calculate *A* and E_a in the Arrhenius equation.

10 Acids and Bases

Johannes Nicolaus Brønsted, 1879–1947

In Chapter 8, we discussed the laws of chemical equilibrium. In this chapter and the next we will see how these laws apply to reactions that take place among ions dissolved in water. It may be recalled that we have emphasized that it takes a long period of time for some chemical reactions to reach equilibrium. Fortunately, this consideration will not worry us here. Reactions that take place among ions dissolved in water are almost always extremely fast. If the solutions are properly stirred, equilibrium is almost always reached in less than 1 second. In this chapter, we shall discuss acids and bases: in the next, equilibria involving solubility.

10-1 Early Theories of Acids and Bases

Acids and bases have been known for a great many years. The strong mineral acids hydrochloric, sulfuric, and nitric were discovered by the alchemists in the years between 1100 and 1600 (actually nobody knows exactly when they were discovered nor who their discoverers were, because alchemists generally worked in secrecy and were deliberately unclear in what they wrote down). Bases like potassium hydroxide were known even earlier (potassium hydroxide can be obtained from wood ashes). It was soon learned that although both acids and bases are corrosive to human skin and to many other materials, they have the ability to neutralize each other to give water and harmless salts. Because of this, acids and bases have always been associated with each other. Although these compounds were known, there was no clear definition of acids or bases, though the chemists who worked with them could generally recognize an acid or a base. One of the earliest attempts to find something that the various known acids had in common was that of Lavoisier, whose experiments led him to believe that all acids contained oxygen, and that it was this element alone that gave them their acidic properties (the very name *oxygen*, coined by Lavoisier, means "acid-former"), though he also recognized that some compounds that were not acids also contained oxygen. This was one of the few areas of chemistry in which Lavoisier was wrong. A theory of this kind (that all of a given class of compounds contain a given element) is very hard to prove but very easy to disprove, if it is indeed wrong—all it takes is the discovery of one acid that does not contain oxygen. Two such acids were discovered in the early nineteenth century. Gay-Lussac proved that hydrocyanic acid (HCN; at that time called prussic acid) contained no oxygen, and Humphry Davy (1778–1829) did the same for hydrochloric acid. After that, overwhelming evidence showed that *hydrogen*, not oxygen, is the element that gives acids their acidic properties (although many compounds that are not acids also contain hydrogen).

Without a satisfactory theory to explain or even define acids and bases, nineteenth-century chemists fell back on descriptions. To them acids were substances that tasted sour, changed the color of litmus (a substance derived from a vegetable) from blue to red, reacted with certain metals (such as iron and zinc) to give off hydrogen gas and with a group of compounds called carbonates to give off carbon dioxide, and, above all, *were able to neutralize bases*. Bases (originally called alkalies) were substances that tasted bitter, felt slippery, changed the color of litmus back from red to blue, and, above all, *were able to neutralize acids*. These descriptions were the same ones that chemists and alchemists had been using for centuries to recognize acids and bases.

It was in 1884 that Svante Arrhenius, as part of his theory of ionization, proposed a new theory of acids and bases. According to the Arrhenius theory, acids were substances that produced hydrogen ions (H^+) when mixed with water; bases were substances that, when mixed with water, produced hydroxide ions (OH^-). This explained why all acids had to have hydrogen, but, more important, it also explained how acids neutralized bases to give water:

$$H^+ + OH^- \longrightarrow H_2O$$

One of the facts that led Arrhenius to his theory was that the *heat of neutralization*, ΔH of a neutralization reaction, of aqueous solutions of a strong acid and a strong base was the same (per mole of water formed) no matter which acid and base were used. Thus, $HCl + NaOH$, $HCl + KOH$, and $HNO_3 + NaOH$ all gave the same heat of neutralization. Arrhenius said that HNO_3 and HCl in water both consisted of H^+ (along with NO_3^- or Cl^-) and that KOH and NaOH both consisted of OH^- (along with K^+ or Na^+); so that when an aqueous solution of a strong acid and a strong base are mixed, the only reaction that takes place is $H^+ + OH^- \rightarrow H_2O$. Therefore, it is not surprising that the heat of the reaction should be the same.

Arrhenius's theory of acids and bases was a good beginning, but it required a modification and an extension. The modification was in the hydrogen ion. We now know that H^+ cannot exist by itself in water or in any other solvent. When in water it always combines with a molecule of water to give the *hydronium ion*, H_3O^+.* Thus, a more correct representation for the neutralization of a strong acid and a strong base in aqueous solution is

$$H_3O^+ + OH^- \longrightarrow 2H_2O$$

With this modification, the Arrhenius picture is still valid for acid and base behavior in water solutions.

Around 1923, Johannes Brønsted (1879–1947) and Gilbert N. Lewis (1875–1946) each realized that acid–base behavior was not confined to aqueous solutions, and each significantly extended Arrhenius's theory, in different ways. In the next section, we shall look at Brønsted's theory. We will postpone consideration of the Lewis acid–base theory to Section 15-5.

10-2 The Brønsted Acid-Base Theory

In the **Brønsted theory, an acid is defined as a proton donor; a base is a proton acceptor.** These definitions significantly enlarge the Arrhenius definition of acids and bases. Acids and bases are no longer restricted to water solutions. An acid is any hydrogen-containing compound that donates its proton to any base, in any solvent, or in the absence of a solvent. A base is no longer just a substance that produces OH^- when added to water, but any substance that takes a

*The representation of hydronium ion in water as H_3O^+ is a simplification. Actually, several molecules of water (somewhere between 3 and 11) attach themselves to the H_3O^+ ion, completely surrounding it with a shell of water molecules (Section 7-5).

proton (as does OH^-) in any solvent or in the absence of a solvent. An acid–base reaction is simply a reaction in which an acid transfers a proton to a base:

$$AH + B \longrightarrow A^- + BH^+$$

We use the general formulas AH and B to stand for all uncharged acids and bases, respectively. When the transfer has been completed, B now has the proton (it has become BH^+). But having a proton means that at least potentially, BH^+ can give it up again. Therefore, according to the Brønsted theory, **BH^+ is an acid.** Similarly, AH, having lost the proton, has now become A^-; since it could, at least potentially, take the proton back, **A^- is a base.** The key to the Brønsted picture is the concept that in all acid–base reactions, an acid gives a proton to a base; and that in doing so, the acid becomes a base and the base, in accepting a proton, becomes an acid. We may write the general reaction this way:

$$\underset{acid_1}{AH} + \underset{base_2}{B} \rightleftharpoons \underset{base_1}{A^-} + \underset{acid_2}{BH^+}$$

BH^+ is called the **conjugate acid** of the base B; A^- is the **conjugate base** of the acid AH. AH and A^- are called a *conjugate acid–base pair,* as are BH^+ and B. In the Brønsted theory, every acid has a conjugate base (we may write it by simply removing a proton); every base has a conjugate acid (just add a proton). Some examples are:

Conjugate acid	*Conjugate base*
HCl	Cl^-
H_2SO_4	HSO_4^-
NH_4^+	NH_3
HSO_4^-	SO_4^{2-}
HPO_4^{2-}	PO_4^{3-}

In the Brønsted picture, then, the products of every reaction between an acid and a base are an acid and a base. Therefore, every acid–base reaction is at least potentially reversible (note that we have used equilibrium arrows above), since the new acid could give its newly acquired proton back to the new base; and if this happened the old acid and base would be regenerated. How can we tell in any given case in which direction the equilibrium lies? For example, when we add hydrofluoric acid (HF) to the base ammonia (NH_3),

$$\underset{acid_1}{HF} + \underset{base_2}{NH_3} \rightleftharpoons \underset{base_1}{F^-} + \underset{acid_2}{NH_4^+}$$

will the reaction go to completion so that at equilibrium we will have practically only F^- and NH_4^+; or will the equilibrium lie in the other direction so that at equilibrium we will have only HF and NH_3, and essentially no reaction taking place; or something in between? Questions like this can be answered by use of the concept of *acid strength.* Acid strength is defined as the tendency of an acid to give up its proton. A **strong acid** gives up its proton more readily than does a weak acid. Similarly, a **strong base** accepts a proton more readily than does a weak one.

If we have two acids, how can we tell which is stronger? One way to do this is to allow the conjugate bases of the two acids to compete for the proton. For example, suppose our two acids are acetic acid (HOAc) and phenol (C_6H_5OH). We could take some acetic acid and add it to the conjugate base of phenol ($C_6H_5O^-$, in the form of its sodium salt, C_6H_5ONa) and then observe if a reaction takes place.

$$HOAc + C_6H_5O^- \rightleftharpoons OAc^- + C_6H_5OH$$

If we carried out the experiment, we would find that a reaction does indeed take place; at equilibrium we would have essentially only OAc^- and C_6H_5OH; virtually no HOAc or $C_6H_5O^-$ would remain. As a check, we could carry out the reaction the other way also: add C_6H_5OH to NaOAc. This time we would observe essentially no reaction. Practically all of the C_6H_5OH and NaOAc would remain as they are. What is happening here is that OAc^- and $C_6H_5O^-$ are competing for the proton. *The one that gets it must be the stronger base,* since that is how we defined the concept of base strength. The experiment showed that it was $C_6H_5O^-$ that got the proton, not OAc^-. Therefore, $C_6H_5O^-$ is a stronger base than OAc^-. We could also look at it the other way: HOAc is a stronger acid than C_6H_5OH because, as the experiments showed us, it is more eager to give up its proton, and that is exactly what we mean by acid strength.

By the use of competition experiments of this kind, we could test a large number of acids against each other, making use of the following relationships, which directly follow from the arguments we have just given.

1. The stronger an acid, the weaker is its conjugate base. This must be so because the more eager an acid is to give up its proton, the less eager will be the resulting base to take it back. Therefore, the conjugate base of a strong acid must be a weak base, and vice versa; and the conjugate base of a weak acid must be a strong base, and vice versa.
2. If an acid HA is stronger than another acid HB; then it must also be stronger than any acid that is weaker than HB.

Table 10-1 is a list of some acids (only a few of the thousands of known acids are listed) *in order of acid strength,* along with their conjugate bases. The strongest acid listed is at the top; the weakest at the bottom; it follows that the strongest base must be at the bottom, the weakest at the top. Once we have made a list of acids and bases

Table 10-1 Acids and Bases

	Conjugate acid	*Conjugate base*	
strongest acid on this list	$HClO_4$	ClO_4^-	weakest base on this list
	HI	I^-	
	HBr	Br^-	
strong acids	H_2SO_4	HSO_4^-	feeble bases
	HCl	Cl^-	
	HNO_3	NO_3^-	
	HSO_4^-	SO_4^{2-}	
	H_3O^+	H_2O	
	HF	F^-	
	HNO_2	NO_2^-	
	HOAc	OAc^-	
intermediate or weak acids	H_2CO_3	HCO_3^-	intermediate or weak bases
	H_2S	HS^-	
	NH_4^+	NH_3	
	C_6H_5OH	$C_6H_5O^-$	
	HCO_3^-	CO_3^{2-}	
	H_2O	OH^-	
	CH_3OH	CH_3O^-	
feeble acids	CH_3CH_2OH	$CH_3CH_2O^-$	strong bases
	NH_3	NH_2^-	
weakest acid on this list	CH_4	CH_3^-	strongest base on this list

in order of their strengths, we may then use this list to answer the question we asked before: How can we predict whether a reaction will or will not take place between a given acid and a given base? The rule is: **Any acid on the list will react with the conjugate base of any acid on the list *below it*, but will not react with the conjugate base of any acid on the list *above it**** (this rule obviously follows from our previous discussion).

Thus, H_2CO_3 will undergo a reaction with OH^- but not with NO_3^-; while HI will undergo a reaction with both of them. A simple inspection of Table 10-1 is also sufficient to answer our earlier question. HF will react with NH_3 to give F^- and NH_4^+; the reverse reaction will essentially not take place.

Table 10-1 illustrates the following points:

1. Acids and bases may be uncharged (neutral) or may carry positive or negative charges. In any case, the conjugate base of an acid always has a charge one positive unit less (or one negative unit more) than the acid itself.

2. Certain acids can give up two or three protons. These are called **diprotic** (for example, H_2SO_4) or **triprotic** (for example, H_3PO_4) acids, respectively. The general term for such acids is **polyprotic.** However, in the Brønsted picture *each acid is regarded as giving up only one proton*. Thus, H_2SO_4 is treated as if it gives up a single proton to become HSO_4^-; and HSO_4^- is then regarded as a separate acid in its own right (note that it has its own place in the table), which gives up the other proton. In such cases, the second acid is always weaker than the first. Thus, H_3PO_4 is a stronger acid than $H_2PO_4^-$; which, in turn, is stronger than HPO_4^{2-}. Similarly, each base is regarded as accepting only one proton, and bases become weaker as they accept protons (for example, NH_2^- is a stronger base than NH_3).

3. Note that HSO_4^- appears in the table as an acid and also as a base. A species that can act as either an acid or a base is called **amphoteric.** Many such substances are known. Whether an amphoteric substance will be an acid or a base at any given time depends on what other substance is present. Thus, if NO_2^- is added to HSO_4^-, the latter will behave as an acid (because NO_2^- is below the acid HSO_4^-); but in the presence of HI, HSO_4^- will function as a base.

4. *Water* is an acid and also a base, and is therefore amphoteric. When water is a base, its conjugate acid is H_3O^+. Any acid stronger than H_3O^+ is called a *strong acid* and its conjugate base is called a *feeble base*. When water is an acid, its conjugate base is OH^-. Any base stronger than OH^- is called a *strong base*, and its conjugate acid is a *feeble acid*.

The Brønsted acid–base picture is a simple and convenient way to look at acids and bases. With the use of tables like Table 10-1, it is possible to predict whether or not thousands of possible acid–base reactions will actually take place.

Example: Which of the following reactions will take place?

$$(1)\quad HCl + OAc^- \longrightarrow HOAc + Cl^-$$

$$(2)\quad H_2O + C_6H_5O^- \longrightarrow C_6H_5OH + OH^-$$

$$(3)\quad CH_4 + NH_2^- \longrightarrow NH_3 + CH_3^-$$

Answer: Reaction (1) will take place, because HCl is above OAc^- in the table (that is, HCl is a stronger acid than HOAc). Reactions (2) and (3) will not take place, because H_2O as an acid is below $C_6H_5O^-$, and CH_4 is below NH_2^-. In fact, CH_4 will not react with *any* base in Table 10-1, because it is the weakest acid in the table.

*What the rule actually says is that the equilibrium will lie in the predicted direction; the reaction does not necessarily have to go all the way. For example, if an acid HA is only slightly stronger than another acid HB, then a reaction between HA and B^- might go only 60 or 70% at equilibrium and not the 99% or more that happens in most cases.

The strengths of thousands of acids and bases have now been determined and published in tables like Table 10-1. A study of such tables shows that there are certain correlations between acidity or basicity and the structure of molecules, which allow us in certain cases to make predictions about acidity or basicity merely by inspecting the formulas. Among these are the following:

1. The strengths of binary acids H_nA (that is, acids in which the hydrogen or hydrogens are connected to only one other atom) *increase in going from left to right* across any row of the periodic table. Thus acidities increase in the order:

$$CH_4 < NH_3 < H_2O < HF$$

and

$$SiH_4 < PH_3 < H_2S < HCl$$

Of course, this mean that base strengths of the conjugate bases *decrease* in the order $CH_3^- > NH_2^- > OH^- > F^-$ and $SiH_3^- > PH_2^- > SH^- > Cl^-$.

2. The strengths of binary acids H_nA *increase in going down* any column of the periodic table. Thus, acidities increase in the order

$$HF < HCl < HBr < HI$$

and

$$H_2O < H_2S < H_2Se < H_2Te$$

3. In most ternary acids of the form H_nAO_m, each hydrogen is connected to an oxygen, not to the atom A. In many of these compounds, acid strength *increases with an increase in the number of oxygen atoms.** Thus, acidities increase in the order

$$H_2SO_3 < H_2SO_4$$

and

$$HNO_2 < HNO_3$$

and

$$HOCl < HClO_2 < HClO_3 < HClO_4$$

Note that HCl does not belong in the latter sequence. In HCl the H is obviously directly connected to the Cl, while in the other compounds listed, each hydrogen is connected to an oxygen atom, for example

```
        O
        |
H—O—Cl—O      HClO4
        |
        O
```

4. Compounds in which OH groups are attached to a central atom $A_n(OH)_m$ usually show acidic or basic properties (in some cases both; these are amphoteric). For these compounds, *acid strength increases, and base strength decreases, in going from left to right across a given row of the periodic table.* For example, in the third row of the periodic table, we can write:

NaOH	$Mg(OH)_2$	$Al(OH)_3$	$Si(OH)_4$	$P(OH)_5$	$S(OH)_6$	$Cl(OH)_7$
		H_3AlO_3	H_2SiO_3	H_3PO_4	H_2SO_4	$HClO_4$
strong base	weak base	amphoteric	very weak acid	weak acid	strong acid	very strong acid

*An important class of acids to which this generalization does *not* apply are acids in which A is carbon (organic acids, see Section 20-6). There are thousands of known organic acids, among them acetic and formic, and such simple generalizations cannot apply in these cases.

Note that the hydroxides $Si(OH)_4$, $P(OH)_5$, $S(OH)_6$, and $Cl(OH)_7$ are unstable; they lose one or more molecules of H_2O to give, respectively, H_2SiO_3, H_3PO_4, H_2SO_4, and $HClO_4$. In water solution, NaOH and $Mg(OH)_2$ are basic and produce OH^- (NaOH is stronger than $Mg(OH)_2$ and produces a higher concentration of OH^-). The hydroxides of Si, P, S, and Cl are acidic and produce H_3O^+ in water solution; acid strength increases from left to right in the periodic table. $Al(OH)_3$ is amphoteric, and in water solution exhibits both acidic and basic properties.

Although the generalizations given above are useful as far as they go, they do not give us guidance in all cases (not even in most cases). For most acids and bases such simple generalizations are lacking, and the degree of acid or base strength must be determined in laboratory experiments.

In the rest of this chapter (except for Section 10-14), while recognizing the validity of the Brønsted picture, which defines acids and bases in all solvents, we shall confine ourselves to a study of acids and bases in water solutions.

10-3 Acid-Base Reactions in Water Solution. K_w

Most acid-base reactions are carried out in a solvent. By far the most common such solvent is water, which, as we shall see, offers many advantages for the study of most acid–base reactions. We have seen in the previous section that water itself is amphoteric: it can be either an acid or a base. When an acid stronger than water is added to water, the water behaves as a base; it takes a proton and is converted to its conjugate acid, H_3O^+:

$$\underset{\text{acid}}{HA} + \underset{\text{base}}{H_2O} \rightleftharpoons H_3O^+ + A^-$$

Thus, the addition of an acid to water produces H_3O^+. When a base stronger than water is added to water, the water behaves as an acid. It gives up a proton and is converted to its conjugate base OH^-:

$$\underset{\text{base}}{B} + \underset{\text{acid}}{H_2O} \rightleftharpoons BH^+ + OH^-$$

Thus, the addition of a base to water produces OH^-. (Some bases, for example NaOH and KOH, produce OH^- directly on being added to water. No reaction with the water is required in this case.)

In the reactions we are talking about it is rare for *all* the water to react in the way described. Water is the solvent, and only a relatively small amount of HA or B is added. For example, 1 liter of water contains about 55.5 moles, so the addition of 0.1 mole of HCl (about 3.65 g) to 1 liter of water will use up only 0.1 mole of H_2O (and produce 0.1 mole of H_3O^+), so about 55.4 moles of water will still remain.

The addition of an acid to water produces H_3O^+; the addition of a base produces OH^- (note that this is essentially the Arrhenius definition of acids and bases). However, *even if no acid or base is added, there is still some H_3O^+ and OH^- present in pure water.* Some of the water molecules react with each other; one molecule of water acting as an acid, another as a base, to produce OH^- and H_3O^+.

$$\underset{\text{acid}}{H_2O} + \underset{\text{base}}{H_2O} \rightleftharpoons OH^- + H_3O^+ \qquad (1)$$

Pure water is neutral (it is neither acidic nor basic), not because there is no H_3O^+ or OH^- present, but because they are present in equal concentrations. In pure water, $[H_3O^+]$ must equal $[OH^-]$, since the two species are formed by the same process. This reaction of water with itself is called **autoprotolysis.**

Equation (1) is an equilibrium, and we can write an equilibrium expression for it:

$$K = \frac{[OH^-][H_3O^+]}{[H_2O]^2} \qquad (2)$$

Rather than use this expression, it will be more convenient for us to change it into another. We may easily do this by noting that the concentration of water is not essentially changed by reaction (1), because only a very small amount of the water undergoes the reaction: water is the solvent and is present in very large amounts compared to the amount of H_3O^+ and OH^-. [Only about 2 out of every 10 million water molecules undergo reaction (1). The decrease from 10,000,000 to 9,999,998 is not a significant one.] Since $[H_2O]$ is essentially unchanged, it is constant, and so is its square, $[H_2O]^2$. Since $[H_2O]^2$ is constant, we may multiply it by K, which is also a constant, giving us another constant. This changes equation (2) to

$$K[H_2O]^2 = K_w = [OH^-][H_3O^+] \qquad (3)$$

Equation (3) is a modified equilibrium expression, and K_w is a *modified equilibrium constant*. We call it K_w, not K, because we do not wish to confuse a modified equilibrium constant with a real equilibrium constant. Although the deviation from reality as expressed by equation (3) is insignificant, it is there nevertheless, and we want to make sure we keep things clear. That is why we use a subscript, in this case w. $\mathbf{K_w}$ is also known as the **ion product constant for water.** Like other equilibrium constants, K_w is constant only at a given temperature and varies with changing temperature. At 25°C, $K_w = 1.0 \times 10^{-14}$ mole²/liter²; at 40°C, it equals 3.0×10^{-14} mole²/liter².

Equation (3) is a very important equation, because it does not only apply to *pure* water. It applies to *any* water solution, no matter what else is in it. It tells us that the product of the H_3O^+ concentration and the OH^- concentration in *any* water solution at 25°C is 10^{-14}.

$$K_w = [OH^-][H_3O^+] = 10^{-14} \text{ mole}^2/\text{liter}^2 \qquad (3)$$

Of course, in pure water, and in any other neutral water solution, this means that $[OH^-]$ and $[H_3O^+]$ must each be equal to 10^{-7} mole/liter. But if, for example, we have a solution where $[H_3O^+]$ is 10^{-3} mole/liter,* then equation (3) tells us that OH^- *must* equal $10^{-14}/10^{-3}$ or 10^{-11} mole/liter.

Example: The $[OH^-]$ of a water solution is 10^{-5} mole/liter. What is the $[H_3O^+]$?

Answer:

$$K_w = 10^{-14} = [OH^-][H_3O^+]$$
$$10^{-14} = 10^{-5}[H_3O^+]$$
$$[H_3O^+] = \frac{10^{-14}}{10^{-5}}$$
$$[H_3O^+] = 10^{-9} \text{ mole/liter}$$

Example: The $[H_3O^+]$ of a water solution is 4.3×10^{-2} mole/liter. What is the $[OH^-]$?

Answer:

$$(4.3 \times 10^{-2})[OH^-] = 10^{-14}$$
$$[OH^-] = \frac{10^{-14}}{4.3 \times 10^{-2}} = 2.3 \times 10^{-13} \text{ mole/liter}$$

*Such a solution might be obtained, for example, by adding hydronium ions (in the form of HCl) to pure water.

We may define an *acidic aqueous solution* as one in which the concentration of H_3O^+ exceeds that of OH^-. A *basic aqueous solution* is the opposite: $[OH^-]$ exceeds $[H_3O^+]$. In a neutral solution, the concentrations of the two species are the same: they are each 10^{-7} mole/liter.

10-4 pH and pOH

The acidity or basicity of an aqueous solution can be expressed by the hydronium ion concentration; for example, a solution with $[H_3O^+] = 5 \times 10^{-3}$ mole/liter is more acidic and less basic than one in which $[H_3O^+]$ is 6×10^{-4} mole/liter. However, numbers of this kind are cumbersome to work with, and for this reason, another way to express $[H_3O^+]$ was devised by Sören Sörensen (1868–1939) in 1909. For this purpose we define a quantity **pH** to be the **negative logarithm of the hydronium ion concentration.**

$$pH = -\log [H_3O^+]$$

It is very easy to convert $[H_3O^+]$ to pH, and vice versa.

Example: The hydronium ion concentration of an aqueous solution is 10^{-5} mole/liter. What is the pH?

Answer:

$$\begin{aligned} pH &= -\log [H_3O^+] \\ &= -\log [10^{-5}] \end{aligned}$$

Note that the log of any number 10^x is x, so $\log 10^{-5}$ is -5:

$$\begin{aligned} pH &= -(-5) \\ &= 5 \end{aligned}$$

Thus, the pH of a solution whose $[H_3O^+]$ is 10^{-5} is 5.

Example: The hydronium ion concentration of an aqueous solution is 6.2×10^{-9} mole/liter. What is the pH?

Answer:

$$\begin{aligned} pH &= -\log [H_3O^+] \\ &= -\log [6.2 \times 10^{-9}] \\ &= -(\log 6.2 + \log 10^{-9}) \end{aligned}$$

We were able to do the last step because *the logarithm of the product of two numbers is equal to the sum of the individual logarithms of the two numbers.* $\log 10^{-9} = -9$. (We can find log 6.2 in a table of logarithms.)

$$\begin{aligned} pH &= -(0.79 - 9) \\ &= -(-8.21) \\ &= 8.21 \end{aligned}$$

The pH of a solution whose $[H_3O^+] = 6.2 \times 10^{-9}$ is 8.21.

In order to go from pH to $[H_3O^+]$ we follow the same procedure backward:

Example: The pH of a solution is 11.74. What is the $[H_3O^+]$?

Answer:

$$\log [H_3O^+] = -\text{pH}$$
$$= -11.74$$

At this point we convert −11.74 to −12 + 0.26. (We choose −12 because it is the next negative integer after −11.)

$$\log [H_3O^+] = 0.26 - 12$$

Now we take the *antilog* of −12 (antilog −12 is, of course, 10^{-12}) and of 0.26 (which is 1.8):

$$[H_3O^+] = 1.8 \times 10^{-12} \text{ mole/liter}$$

Example: The pH of a solution is 3.16. What is the $[H_3O^+]$?

Answer:

$$\log [H_3O^+] = -3.16$$
$$= 0.84 - 4$$
$$[H_3O^+] = 6.9 \times 10^{-4}\ M$$

In a similar manner we may define pOH as $-\log [OH^-]$. Now we can express the acidity or basicity of an aqueous solution by using these numbers. Thus, a solution whose pH is 7 is neutral. A solution with a pH less than 7 is acidic; greater that 7 it is basic. The lower the pH, the more acidic the solution; the higher the pH, the more basic the solution. Since $[H_3O^+][OH^-] = 10^{-14}$, pH + pOH must equal 14. Consequently, a solution with a pH of 4.7 has a pOH of 9.3. Remember, the only reason we use pH and pOH instead of $[H_3O^+]$ and $[OH^-]$ is that the former numbers are more convenient to work with.

10-5 Indicators

Fortunately, it is very easy to determine the pH of an aqueous solution in a laboratory. There are two main methods for doing this. The older of the two methods uses certain chemical compounds called **acid-base indicators.** An acid–base indicator is a compound that changes color over a narrow pH range. For example, if a drop of the compound *methyl red* is added to an aqueous solution, it will turn the solution red if the pH is below 4.8, but yellow if the pH is above 6.0 (the solution will take on an intermediate color if the pH is between 4.8 and 6.0). Note that methyl red cannot tell the difference between a pH of 6.0 and one of 8.0 or 10.0. It will be yellow in all three cases. But it can tell if the pH is greater than 6.0 or less than 4.8. If we wish to specify the pH more precisely, we may take a fresh sample of the aqueous solution and add a drop of a different indicator. For example, let us say that the solution turned yellow with methyl red, so we know the pH is greater than 6. We may then use the indicator *thymolphthalein*. This indicator is colorless below pH 9.4 but blue above pH 10.6. If the fresh solution remains colorless when a drop of thymolphthalein is added, we now know the pH is between 6.0 and 9.4. We may specify it still more precisely by using other indicators. An indicator is itself a weak acid or base. When the pH changes, it is converted to its conjugate base, or acid,

Table 10-2 Some Acid-Base Indicators

Indicator	pH Range	Color[a]	
		Acid	Base
Methyl violet	0.0–1.6	yellow	blue
Quinaldine red	1.0–2.2	colorless	red
Thymol blue	1.2–2.8	red	yellow
	8.0–9.6	yellow	blue
Bromphenol blue	3.0–4.6	yellow	blue
Congo red	3.0–5.0	blue	red
Methyl orange	3.2–4.4	red	yellow
Methyl red	4.8–6.0	red	yellow
Bromcresol purple	5.2–6.8	yellow	purple
Bromthymol blue	6.0–7.6	yellow	blue
Cresol red	0.0–1.0	red	yellow
	7.0–8.8	yellow	red
Phenolphthalein	8.2–10.0	colorless	pink
Thymolphthalein	9.4–10.6	colorless	blue
Alizarin yellow R	10.1–12.0	yellow	red
Clayton yellow	12.2–13.2	yellow	amber

[a]Note that some of the indicators have *two* color changes.

which has a different color. A list of some indicators is given in Table 10-2.* The indicator method is slow and not precise, but it is inexpensive and has been used for many years. A more rapid procedure involves the use of pH paper, which is commercially available. pH paper has been soaked in a blend of indicators, and a not-very-precise value of the pH may be obtained merely by dipping the paper into the solution, noting the color, and comparing the color with a color code that is supplied with the paper.

When precise values of pH are desired, the best method is the use of an instrument called a pH meter. This instrument contains electrodes which are dipped into the unknown solution. The pH of the solution appears on a dial which is part of the meter. Although the pH meter is more expensive than the indicator method, it is both more rapid and more precise.

Many ordinary foods and other substances commonly found in households and factories are acidic or basic aqueous solutions. For example, vinegar is approximately a 5% solution of acetic acid in water; lemon juice consists mostly of citric acid dissolved in water, and household ammonia is approximately a 5% solution of NH_3 in water. Table 10-3 gives approximate pH values of these and some other common materials and body fluids.

10-6 Acidity Constants. pK_a

In Section 10-2, it was mentioned that the relative acidities of a group of acids could be determined by a series of competition experiments, where two bases are allowed to compete for the same proton. In practice, this method often involves experimental difficulties. For acids weaker than H_3O^+ and stronger than H_2O (the ones called

*Note that litmus, which we mentioned earlier, is not listed in this table. This indicator, which changes color at about ph 7, was one of the first indicators known, and is widely used in chemical and biological laboratories for crude determinations of whether a given solution is acidic (litmus turns red) or basic (litmus turns blue). However, litmus is unsuitable as one of the indicators to be used for a precise determination of pH because the pH range over which it changes color is much greater than that of other indicators.

Table 10-3 pH of Some Common Materials

Material	pH
Soft drinks	2.0–4.0
Lemons	2.2–2.4
Vinegar	2.4–3.4
Wines	2.8–3.8
Oranges	3.0–4.0
Sauerkraut	3.4–3.6
Peaches	3.4–3.6
Tomatoes	4.0–4.4
Urine, human	4.8–8.4
Milk	6.3–6.6
Saliva, human	6.5–7.5
Blood, human	7.4
Egg whites	7.6–8.0
0.1 *N* solution of borax	9.2
Household bleach	10.0
Milk of magnesia	10.5
Household ammonia	11.7
1 *M* solution of NaOH (lye)	14.0

weak or *intermediate* in Table 10-1), there is a better method, which depends on the fact that the pH of an aqueous solution can be very accurately determined. First, let us see what happens when a *strong* acid (that is, an acid stronger than H_3O^+) is added to water. Let us imagine that 0.1 mole of nitric acid is added to 1 liter of water (assume that the total volume of the solution remains at 1 liter).

	$HNO_3 + H_2O$	$\rightleftharpoons$	H_3O^+	$+ NO_3^-$
Concentration at start	0.1 *M*		0	0
Concentration at equilibrium	0		0.1 *M*	0.1 *M*

Because HNO_3 is a much stronger acid than H_3O^+, the equilibrium lies far to the right, and we can say that addition of 0.1 mole of HNO_3 to 1 liter of H_2O results in a solution containing essentially *no* HNO_3, but rather 0.1 mole of H_3O^+. Because it is stronger than H_3O^+, essentially all the HNO_3 is converted to H_3O^+. The pH of the solution is 1. The same is true when 0.1 mole of any acid stronger than H_3O^+ is dissolved in 1 liter of water. There are a number of such acids known, but six of them are fairly common, and they are the ones shown in Tables 10-1 and 10-4. For the purposes of this book we shall assume that these are the six strong acids, and that all other acids (stronger than water) are weak (some of these are sometimes called intermediate acids).

When an acid weaker than H_3O^+ is added to water, the equilibrium lies in the other direction. For example, assume that 0.1 mole of HA, a weak acid, is added to 1 liter of H_2O (again, assume the total volume is 1 liter).

Table 10-4

Strong acids		*Strong bases*	
$HClO_4$	perchloric	NaOH	sodium hydroxide
HI	hydroiodic	KOH	potassium hydroxide
HBr	hydrobromic	LiOH	lithium hydroxide
HCl	hydrochloric	$Ba(OH)_2$	barium hydroxide
H_2SO_4	sulfuric		
HNO_3	nitric		

	$HA + H_2O \rightleftharpoons$	H_3O^+	$+ A^-$
Concentration at start	$0.1\ M$	0	0
Concentration at equilibrium	$0.1\ M - x$	x	x

In this case, only some of the HA reacts with the water (we have written this quantity as x); most of it remains as HA. We may write the equilibrium expression

$$K = \frac{[H_3O^+][A^-]}{[HA][H_2O]}$$

Once again, the concentration of H_2O remains essentially unchanged, and it is convenient to use the modified form

$$K[H_2O] = K_a = \frac{[H_3O^+][A^-]}{[HA]}$$

This is the general equilibrium expression for the reaction of a weak acid with excess water. $\mathbf{K_a}$, a modified equilibrium constant, is often referred to as an **ionization constant,** an **acidity constant,** or a **dissociation constant.** The value of K_a for a given weak acid HA can easily be determined in the laboratory by making up a solution of HA (in H_2O) of known molarity, and measuring the pH with a pH meter or with indicators.

Example: The pH of an 0.0100 M solution of formic acid (HCOOH) in H_2O is 2.90. Calculate K_a for this acid (note that formic acid gives up only one proton).

Answer: First we convert pH to $[H_3O^+]$:

$$\log [H_3O^+] = -2.90$$
$$= 0.10 - 3$$
$$[H_3O^+] = 1.26 \times 10^{-3}\ M$$

The equilibrium expression is

$$K_a = \frac{[H_3O^+][HCOO^-]}{[HCOOH]}$$

We need to substitute the values for $[H_3O^+]$, $[HCOO^-]$, and [HCOOH] into this equation and solve for K_a. We know $[H_3O^+]$; it is $1.26 \times 10^{-3}\ M$. This means that $[HCOO^-]$ is also $1.26 \times 10^{-3}\ M$.

$$\underset{(0.0100 - 1.26 \times 10^{-3})}{HCOOH} + H_2O \rightleftharpoons \underset{(1.26 \times 10^{-3})}{H_3O^+} + \underset{(1.26 \times 10^{-3})}{HCOO^-}$$

In this case, these two numbers must be equal, because there is no way an $HCOO^-$ ion could form without an H_3O^+ being formed at the same time. That is, $HCOO^-$ and H_3O^+ are formed by the same reaction, and the stoichiometry shows that they are formed in equal amounts. Therefore, if $[H_3O^+] = 1.26 \times 10^{-3}\ M$, then $[HCOO^-]$ must also be $1.26 \times 10^{-3}\ M$. The remaining quantity we need is [HCOOH]. At the start this was 0.0100 M. A small amount of this reacted with H_2O to form $HCOO^-$ and H_3O^+. This amount must obviously also be $1.26 \times 10^{-3}\ M$, so the concentration of HCOOH at equilibrium is $0.0100\ M - 0.00126\ M = 0.0087\ M$. Inserting these numbers into the equilibrium expression, we get

$$K_a = \frac{[H_3O^+][HCOO^-]}{[HCOOH]} = \frac{(1.26 \times 10^{-3})(1.26 \times 10^{-3})}{8.7 \times 10^{-3}}$$
$$= 1.8 \times 10^{-4}$$

Therefore, K_a for HCOOH is 1.8×10^{-4}.

Table 10-5 Some Intermediate or Weak Acids and Their Dissociation Constants in Aqueous Solutions

	Acid	K_a	pK_a	*Conjugate base*
H_2SO_3	sulfurous acid	1.5×10^{-2}	1.81	HSO_3^-
HSO_3^-		1.0×10^{-7}	6.91	SO_3^{2-}
HSO_4^-		1.2×10^{-2}	1.92	SO_4^{2-}
H_3PO_4	phosphoric acid	7.5×10^{-3}	2.12	$H_2PO_4^-$
$H_2PO_4^-$		6.2×10^{-8}	7.21	HPO_4^{2-}
HPO_4^{2-}		2.2×10^{-13}	12.66	PO_4^{3-}
HONO	nitrous acid	4.6×10^{-4}	3.37	ONO^-
HF	hydrofluoric acid	3.5×10^{-4}	3.45	F^-
HCOOH	formic acid	1.8×10^{-4}	3.75	$HCOO^-$
C_6H_5COOH	benzoic acid	6.5×10^{-5}	4.19	$C_6H_5COO^-$
HOAc	acetic acid	1.8×10^{-5}	4.75	OAc^-
H_2CO_3	carbonic acid	4.3×10^{-7}	6.37	HCO_3^-
HCO_3^-		5.61×10^{-11}	10.25	CO_3^{2-}
H_2S	hydrosulfuric acid	9.1×10^{-8}	7.04	HS^-
HS^-		1.0×10^{-14}	14.00	S^{2-}
HOCl	hypochlorous acid	2.95×10^{-8}	7.53	ClO^-
H_3BO_3	boric acid	7.3×10^{-10}	9.14	$H_2BO_3^-$
NH_4^+	ammonium ion	5.6×10^{-10}	9.25	NH_3
HCN	hydrocyanic acid	4.93×10^{-10}	9.31	CN^-
C_6H_5OH	phenol	1.3×10^{-10}	9.89	$C_6H_5O^-$

The K_a for any acid stronger than water and weaker than H_3O^+ can easily be determined just by measuring the pH of a known concentration and making the simple calculation shown above. Table 10-5 contains a list of some of the more common weak acids, along with their K_a values. Note that a diprotic acid such as H_2CO_3 is listed twice, once under H_2CO_3 and then again under HCO_3^-. H_2SO_4 is not listed because it is a strong acid, but HSO_4^- is weak enough to be listed. Since K_a values are generally of the form $a \times 10^{-b}$, they are often expressed as pK_a values instead. **pK_a** is defined as

$$pK_a = -\log K_a$$

so that pK_a bears the same relationship to K_a as pH does to $[H_3O^+]$, and is calculated in the same way. The numerical values of K_a and of pK_a tell us how strong an acid is. K_a values increase, and pK_a values decrease, with increasing acid strength: **The stronger the acid, the higher the K_a value and the lower the pK_a value.**

It is easy to see why K_a values cannot be obtained for strong acids. When a strong acid HA is added to water, the reaction with water is virtually complete, making the value of [HA] extremely small.

$$K_a = \frac{[H_3O^+][A^-]}{[HA]}$$

Since [HA] is in the denominator, the value of K_a approaches infinity. For all practical purposes, the reaction goes to completion.

If we know K_a or pK_a for a given acid, we can calculate the hydronium ion concentration of any aqueous solution of this acid, provided that we know the overall concentration of the acid.

Example: What is the pH of a solution that is made up to be 1.00 *M* in HOAc (K_a for HOAc is 1.8×10^{-5})?

Answer:

	$HOAc + H_2O \rightleftharpoons$	H_3O^+	$+ OAc^-$
Concentration at start	1.00	0	0
Concentration at equilibrium	1.00 − x	x	x

Let x be the concentration of HOAc which is *used up* at equilibrium. Then x is the concentration of H_3O^+ *and* of OAc^- at equilibrium, since one H_3O^+ ion *and* one OAc^- ion are formed every time one HOAc molecule reacts with water. If x is the concentration of HOAc that is used up, then 1.00 − x must be the concentration remaining. The concentrations at equilibrium of HOAc, H_3O^+, and OAc^- are, respectively, 1.00 − x, x, and x; and we may substitute these into the equilibrium expression,

$$K_a = \frac{[H_3O^+][OAc^-]}{[HOAc]}$$

$$1.80 \times 10^{-5} = \frac{(x)(x)}{1.00 - x}$$

This equation can be solved just the way it stands, but it requires a somewhat tedious solution of a quadratic equation (see below). In this particular problem, we can simplify the solution, without greatly affecting the answer, if we assume that the amount of a weak acid which reacts with water is very small in comparison with the total amount present, so 1.00 − x is essentially the same as 1.00. This assumption can be made with all but the strongest of the weak acids without causing a significant change in the answer provided that the initial concentration of the acid is substantial compared to the quantity of acid that reacts with water. If these conditions are met, our equation simplifies to

$$1.80 \times 10^{-5} = \frac{(x)(x)}{1.00}$$

$$x^2 = 1.80 \times 10^{-5}$$

It is not convenient to take the square root of a number of the form 10^a when a is odd. Therefore, we change it to an even number. This can easily be done by multiplying one part of the number by 10; and dividing the other part by 10 (note that this does not change the value of the number).

$$x^2 = 1.80 \times 10^{-5}$$

$$x^2 = 18.0 \times 10^{-6}$$

$$x = 4.24 \times 10^{-3} \text{ mole/liter} = [H_3O^+] = [OAc^-]$$

$$pH = -\log(4.24 \times 10^{-3})$$

$$pH = -(\log 4.24 + \log 10^{-3})$$

$$pH = -(0.63 - 3)$$

$$pH = -(-2.37)$$

$$pH = 2.37$$

As a quick check, let us see if our simplifying assumption was valid. Remember that we assumed that x is very much smaller than the original concentration. In this case, the original concentration was 1.00 *M*, while x is 0.00424 *M*. The error is about 0.4%. In most cases this error can be safely ignored, but if more accuracy is needed, the complete calculation can be performed as follows.

We will now see how the result would change if we solve the equation exactly, without introducing the assumption that x is very much smaller than 1.00.

$$1.80 \times 10^{-5} = \frac{(x)(x)}{1.00 - x}$$

$$x^2 = 1.80 \times 10^{-5} - (1.80 \times 10^{-5})x$$

$$x^2 + (1.80 \times 10^{-5})x - 1.80 \times 10^{-5} = 0$$

We may solve this quadratic equation with the aid of the quadratic formula for an equation of the form $ax^2 + bx + c = 0$.

$$x = \frac{-b \pm \sqrt{b^2 - 4ac}}{2a} \qquad \begin{aligned} a &= 1 \\ b &= 1.80 \times 10^{-5} \\ c &= -1.80 \times 10^{-5} \end{aligned}$$

$$= \frac{-(1.80 \times 10^{-5}) \pm \sqrt{3.24 \times 10^{-10} + 7.20 \times 10^{-5}}}{2}$$

$$= \frac{-1.80 \times 10^{-5} \pm \sqrt{7.20 \times 10^{-5}}}{2}$$

$$= \frac{8.52 \times 10^{-3}}{2} = 4.26 \times 10^{-3} \text{ mole/liter}$$

This answer, 4.26×10^{-3} mole/liter, differs very little from the previous answer (4.24×10^{-3} mole/liter), justifying our use of the simplifying assumption in this case.

The pH of a 1.00 *M* solution of HOAc is 2.37, and the $[H_3O^+]$ (and the $[OAc^-]$) is about 4.3×10^{-3} mole/liter. Since the original concentration of HOAc was 1.00 mole/liter, the **degree of dissociation** (actually the degree of reaction with water) is

$$\text{degree of dissociation} = \frac{0.0042}{1.00} \times 100 = 0.42\%$$

Thus, at equilibrium we find that about $\frac{1}{2}$ of 1% of the acetic acid molecules have reacted; about 99.5% are still there, unreacted. The ratio $[HOAc]/[OAc^-]$ is about 200.

It is interesting to find out what happens to the degree of dissociation if we make the solution more dilute. Let us calculate $[H_3O^+]$ and pH for a 0.100 *M* solution of HOAc.

	$HOAc + H_2O \rightleftharpoons$	H_3O^+ +	OAc^-
Concentration at start	0.100	0	0
Concentration at equilibrium	0.100 − x	x	x

$$K_a = 1.8 \times 10^{-5} = \frac{(x)(x)}{0.100 - x}$$

Making the same assumption as before, we have

$$1.8 \times 10^{-5} = \frac{x^2}{0.100}$$

$$x^2 = 1.8 \times 10^{-6}$$

$$x = 1.3 \times 10^{-3}\,M = [H_3O^+] = [OAc^-]$$

$$pH = -\log(1.3 \times 10^{-3})$$

$$\text{pH} = -(\log 1.3 + \log 10^{-3})$$
$$= -(0.11 - 3)$$
$$= 2.89$$

$$\text{degree of dissociation} = \frac{0.0013}{0.100} \times 100 = 1.3\%$$

Thus, the hydronium ion concentration is lower in the more dilute solution (1.3×10^{-3} vs. 4.3×10^{-3} mole/liter), and the pH is higher (2.89 vs. 2.38), as we might expect; but the degree of dissociation has *increased*. In the 0.100 M solution, 1.3% of the HOAc molecules have reacted, as compared with only 0.42% in the 1.00 M solution. This is generally true for all weak acids: *The degree of dissociation increases as the solution becomes more dilute.*

If we calculate the error introduced by our simplifying assumption in this case [we assumed that $(0.100 - x) \approx 0.100$], we obtain a 1.3% error. We find that the error using the simplifying assumption increases as the solution becomes more dilute. For very dilute solutions, the quadratic solution must be used.

10-7 Leveling Effects

For weak acids (those stronger than H_2O but weaker than H_3O^+), it is much more convenient and far more accurate to measure the K_a values in the manner described in Section 10-6 than to use the competition method mentioned in Section 10-2. Unfortunately, pK_a values in water solution cannot be measured for acids on either side of this range, because of leveling effects. Let us first look at strong acids. Both $HClO_4$ and HNO_3 are stronger acids than H_3O^+. Therefore, if say 0.1 mole of each is added to 1 liter of water, the reactions will go essentially all the way, and at equilibrium we will have, in each case, a solution whose hydronium ion concentration is 0.1 M (pH = 1).

	$HClO_4 + H_2O \rightleftharpoons$	H_3O^+	$+\ ClO_4^-$
Concentration at start	0.1	0	0
Concentration at equilibrium	~0	0.1	0.1

	$HNO_3 + H_2O \rightleftharpoons$	H_3O^+	$+\ NO_3^-$
Concentration at start	0.1	0	0
Concentration at equilibrium	~0	0.1	0.1

We can measure the pH, of course, but because it is essentially the same in both cases, such a measurement will not tell us whether $HClO_4$ is stronger or weaker than HNO_3. Although we know from direct competition experiments that $HClO_4$ is indeed very much stronger than HNO_3, their reactions with water do not show this difference. The weaker of the two acids, HNO_3, is already strong enough so that virtually all the molecules transfer their protons to the water, and $HClO_4$ cannot do any better than that. This is called a **leveling effect.** When matched against the base H_2O, both HNO_3 and $HClO_4$ are equally strong acids, although against a weaker base (or in competition with each other), $HClO_4$ is much stronger than HNO_3. The situation may be likened to a boxing tournament in which an inexperienced amateur fights against the world's champion in one match, and against a competent professional in another. The amateur will badly lose both matches and we may well be unable to judge which of the two professionals is better, although in a direct contest, the world's champion might have little trouble in handily defeating the merely competent professional.

The pK_a method is also useless for acids weaker than water (feeble acids) because of a different kind of leveling effect. What we did in Section 10-6 was add a known quantity of acid to a known volume of water and then measure the resulting [H_3O^+]. However, this procedure will not work with a feeble acid because *the concentration of H_3O^+ produced by the feeble acid is much less than that produced by the water alone.* We have seen that in pure water, without any acid added, the [H_3O^+] is 10^{-7} *M*. But a feeble acid, by the very fact that it is a weaker acid than water, will be less effective at giving a proton to the water. The [H_3O^+] produced by the reaction

$$HA + H_2O \rightleftharpoons H_3O^+ + A^-$$

must be *less* than 10^{-7}, in most cases a good deal less. Another way of putting it is that the base water would rather take a proton from the acid water than from an acid weaker than water. If we attempt to measure the [H_3O^+] of a solution of a feeble acid in water, we will find it to be 10^{-7} *M*. Any additional H_3O^+ ions generated by the feeble acid are simply not measurable (for example, $10^{-10} + 10^{-7} \approx 10^{-7}$). Thus, to the solvent water all acids weaker than water are equal in strength (the leveling effect), although when compared against each other, or against a stronger base, their acidities may be very different.

The two kinds of leveling effects prevent us from obtaining accurate pK_a values in water for strong acids and for feeble acids. This is unfortunate, because other ways of determining relative acidities are far less accurate as well as being more difficult to carry out (see Section 10-14). Because of this fact, relative strengths of strong acids and of feeble acids are known only approximately.

10-8 Basicity Constants. pK_b

We have seen that a *strong* acid is one that in water is completely converted to H_3O^+, while a *weak* acid also gives H_3O^+, but to a much lesser extent. Similarly, a *strong base* is one that in water is completely converted to OH^-, and a weak base also gives OH^-, but much less. Bases can give rise to OH^- in water solution in two ways. One class of bases are some metallic hydroxides [for example, NaOH, KOH, $Mg(OH)_2$]. When added to water, these simply dissociate. A strong base such as NaOH dissociates *completely:*

	NaOH $\rightleftharpoons$	OH^+ +	Na^+
Concentration at start	1	0	0
Concentration at equilibrium	~0	1	1

while a weak base such as magnesium hydroxide, $Mg(OH)_2$, dissociates to a much smaller extent:

	$Mg(OH_2)$ $\rightleftharpoons$	Mg^{2+} +	$2OH^-$
Concentration at start	1	0	0
Concentration at equilibrium	1	very small	very small

The other kind of base is not a hydroxide. It forms OH^- in water solution by taking a proton from the water. The most important example of this type of base is ammonia, a weak base.

$$NH_3 + H_2O \rightleftharpoons NH_4^+ + OH^-$$

As with acids, there are not many important strong bases. We will consider NaOH, KOH, LiOH, and $Ba(OH)_2$ to be the only strong bases. All others are to be regarded as weak or feeble (the very strong bases listed in Table 10-1 are not usually considered to be important). Note that all four of our strong bases are of the hydroxide type. Very few bases of the other type are strong.

Just as we were able to obtain K_a and pK_a values for weak acids so can we treat weak bases (bases stronger than water but weaker than OH^-). In this case, the reactions are (for our two types of bases)

$$B + H_2O \rightleftharpoons HB^+ + OH^- \qquad MOH \longrightarrow M^+ + OH^-$$

and the modified equilibrium expressions are

$$K_b = \frac{[HB^+][OH^-]}{[B]} \quad \text{or} \quad \frac{[M^+][OH^-]}{[MOH]}$$

Once again, all we need to do is add a known quantity of base (say 0.10 mole) to a known volume of water (say 1 liter), and measure the pH. From this we can calculate $[OH^-]$, which in this situation must be equal to $[HB^+]$. We can then insert these values into the equation and solve for K_b. As before, **pK_b** is defined as $-\log K_b$. Table 10-6 is a list of some weak bases and their K_b and pK_b values.

Example: The pH of an 0.000100 M solution of AgOH is 9.80. Determine K_b and pK_b.

Answer:

$$AgOH \rightleftharpoons OH^- + Ag^+$$

$$K_b = \frac{[OH^-][Ag^+]}{[AgOH]}$$

$$pH = 9.80$$

$$pOH = 14 - 9.80 = 4.20$$

$$\log [OH^-] = -4.20$$

$$= -5 + 0.80$$

$$[OH^-] = 6.3 \times 10^{-5}\ M$$

Thus, $[OH^-]$ (and therefore Ag^+) is equal to $6.3 \times 10^{-5}\ M$.

$$[AgOH] = 1.00 \times 10^{-4} - 6.3 \times 10^{-5} = 3.7 \times 10^{-5}\ M$$

$$K_b = \frac{(6.3 \times 10^{-5})(6.3 \times 10^{-5})}{3.7 \times 10^{-5}}$$

$$= 1.1 \times 10^{-4}$$

$$pK_b = -\log(1.1 \times 10^{-4})$$

$$= -(\log 1.1 + \log 10^{-4})$$

$$= -(0.04 - 4)$$

$$= -(-3.96)$$

$$= 3.96$$

Since every acid has a conjugate base, and vice versa, we will not be surprised to find that there is a mathematical relationship between K_a for an acid and K_b for its conjugate base. This relationship is easily discovered by simple algebraic manipulations.

For a weak acid, HA:

(a) $HA + H_2O \rightleftharpoons H_3O^+ + A^-$

$$K_a = \frac{[H_3O^+][A^-]}{[HA]} \quad \text{and} \quad \frac{[HA]}{[A^-]} = \frac{[H_3O^+]}{K_a} \tag{3}$$

For its conjugate base, A^-:

(b) $A^- + H_2O \rightleftharpoons HA + OH^-$

$$K_b = \frac{[HA][OH^-]}{[A]} \quad \text{and} \quad \frac{[HA]}{[A]} = \frac{K_b}{[OH^-]} \tag{4}$$

Since both equilibria, (a) and (b), occur simultaneously in the same solution, we may equate the two expressions involving $[HA]/[A^-]$.

$$\frac{[H_3O^+]}{K_a} = \frac{K_b}{[OH^-]}$$

and

$$K_aK_b = [OH^-][H_3O^+]$$

But

$$[OH^-][H_3O^+] = K_w$$

So

$$K_aK_b = K_w = 10^{-14}$$

Thus, if we know K_a for any acid, we may obtain K_b for its conjugate base simply by dividing 10^{-14} by K_a, and vice versa. Similarly, $pK_a + pK_b = 14$ (see the values in Table 10-6).

Example: K_b for the base piperidine $C_5H_{11}N$ is 1.3×10^{-3}. Calculate pK_a for its conjugate acid $C_5H_{11}NH^+$.

Answer:

$$\begin{aligned} K_b &= 1.33 \times 10^{-3} \\ pK_b &= 2.88 \\ pK_a &= 14 - 2.88 \\ &= 11.12 \end{aligned}$$

Table 10-6 Some Weak Bases, Their Conjugate Acids, and Their Dissociation Constants in Aqueous Solution

	Base	K_b	pK_b	*Conjugate acid*	K_a	pK_a
$Pb(OH)_2$	lead hydroxide	9.6×10^{-4}	3.02	—		
CH_3NH_2	methylamine	3.7×10^{-4}	3.34	$CH_3NH_3^+$	2.7×10^{-11}	10.66
AgOH	silver hydroxide	1.1×10^{-4}	3.96	—		
NH_3	ammonia	1.8×10^{-5}	4.75	NH_4^+	5.6×10^{-10}	9.25
NH_2NH_2	hydrazine	1.7×10^{-6}	5.77	$NH_2NH_3^+$	5.9×10^{-9}	8.23
C_5H_5N	pyridine	1.8×10^{-9}	8.75	$C_5H_5NH^+$	5.6×10^{-6}	5.25
$C_6H_5NH_2$	aniline	4.3×10^{-10}	9.37	$C_6H_5NH_3^+$	2.3×10^{-5}	4.63

10-9 Dissociation of Polyprotic Acids

When a diprotic weak acid, H_2A, is dissolved in water, we must write separate equilibrium expressions for H_2A and for HA^-. For example, for H_2CO_3 we have

$$H_2CO_3 + H_2O \rightleftharpoons H_3O^+ + HCO_3^-$$
$$HCO_3^- + H_2O \rightleftharpoons H_3O^+ + CO_3^{2-}$$
$$K_{a_1} = \frac{[H_3O^+][HCO_3^-]}{[H_2CO_3]} = 4.3 \times 10^{-7}$$
$$K_{a_2} = \frac{[H_3O^+][CO_3^{2-}]}{[HCO_3^-]} = 5.61 \times 10^{-11}$$

Similarly, for H_3PO_4:

$$H_3PO_4 + H_2O \rightleftharpoons H_3O^+ + H_2PO_4^-$$
$$H_2PO_4^- + H_2O \rightleftharpoons H_3O^+ + HPO_4^{2-}$$
$$HPO_4^{2-} + H_2O \rightleftharpoons H_3O^+ + PO_4^{3-}$$
$$K_{a_1} = \frac{[H_3O^+][H_2PO_4^-]}{[H_3PO_4]} = 7.5 \times 10^{-3}$$
$$K_{a_2} = \frac{[H_3O^+][HPO_4^{2-}]}{[H_2PO_4^-]} = 6.2 \times 10^{-8}$$
$$K_{a_3} = \frac{[H_3O^+][PO_4^{3-}]}{[HPO_4^{2-}]} = 2.2 \times 10^{-13}$$

Each species has its own value of K_a (see Table 10-5). It is important to realize that in a given solution of such an acid, *all* the equilibria must be simultaneously satisfied. Furthermore, while $[H_3PO_4]$ and $[PO_4^{3-}]$ each appears in only one of the three expressions, the other quantities appear in more than one expression, *and a given quantity must have the same value in each expression* for a given solution. For example, if in a solution made by dissolving a certain quantity of H_3PO_4 in water, $[H_2PO_4^-]$ is 1.0×10^{-4} in the expression for K_{a_1}, it will also be 1.0×10^{-4} in the expression for K_{a_2}.

These simultaneous equations become complicated when we try to solve them exactly, and it is customary to apply simplifying assumptions.

Example: Calculate the concentrations of all solute species present in a solution made up of 0.10 mole of H_2S in 1.00 liter of water ($K_{a_1} = 9.1 \times 10^{-8}$; $K_{a_2} = 1.0 \times 10^{-14}$).

Answer:

$$H_2S + H_2O \rightleftharpoons H_3O^+ + HS^- \qquad \text{step 1}$$
$$HS^- + H_2O \rightleftharpoons H_3O^+ + S^{2-} \qquad \text{step 2}$$
$$K_{a_1} = 9.1 \times 10^{-8} = \frac{[H_3O^+][HS^-]}{[H_2S]}$$
$$K_{a_2} = 1.0 \times 10^{-14} = \frac{[H_3O^+][S^{2-}]}{[HS^-]}$$

In this solution, the $[H_3O^+]$ present comes both from dissociation of H_2S and of HS^-, but we will assume that the latter contributes very little (because it is so much weaker); and that essentially all the $[H_3O^+]$ comes from dissociation of H_2S. We will also assume that the amount of $[HS^-]$ that is lost by step 2 is only a very small fraction of that which is formed by step 1. With these assumptions, we may now solve the first equation, in the same way as in Section 10-6.

	H_2S	$+ H_2O \rightleftharpoons$	H_3O^+	$+ HS^-$
Concentration at start	0.10		0	0
Concentration at equilibrium	$0.10 - x$		x	x

Once again, we assume that x is very small compared to 0.10.

$$9.1 \times 10^{-8} = \frac{x^2}{0.10}$$

$$x^2 = 9.1 \times 10^{-9} = 91 \times 10^{-10}$$

$$x = 9.5 \times 10^{-5}\ M$$

Thus the concentrations of three of the four species present are

$$H_2S: \quad (0.10\ M) - (9.5 \times 10^{-5}\ M) = 0.10\ M$$

$$H_3O^+: \quad 9.5 \times 10^{-5}\ M$$

$$HS^-: \quad 9.5 \times 10^{-5}\ M$$

The remaining solute species present is S^{2-}. We may calculate the concentration of this by using the equilibrium expression for step 2, inserting the values for $[H_3O^+]$ and $[HS^-]$ that we have just computed.

$$1.0 \times 10^{-14} = \frac{[H_3O^+][S^{2-}]}{[HS^-]} = \frac{(9.5 \times 10^{-5})[S^{2-}]}{9.5 \times 10^{-5}}$$

$$[S^{2-}] = 1.0 \times 10^{-14}\ M$$

The values of $[H_3O^+]$ and $[HS^-]$ cancel, and the concentration of $[S^{2-}]$ is numerically equal to the value of K_{a_2}. It is obvious that this will be true for *any* diprotic acid dissolved in water at *any* concentration (provided that our assumption that we could neglect x with respect to the initial concentration is valid, and that $K_{a_1} >> K_{a_2}$ as is usually the case). If these assumptions hold, then for any acid H_2A dissolved in pure water, the concentration of A^{2-} is always numerically equal to the value of K_{a_2} for the acid H_2A.

10-10 Hydrolysis of Ions in Aqueous Solution

In previous sections, we have seen that the pH of pure water is 7, and that we can make it fall below 7 by adding an acid, or rise above 7 by adding a base. But what happens to the pH of pure water when we add a salt? At first sight it might seem that nothing would happen: the pH would remain at 7. However, this is true only for certain salts. Each salt may be considered as the reaction product of an acid and a base. For example, the solid salt potassium bromide (KBr) may be considered to be derived from potassium hydroxide (KOH) and hydrobromic acid (HBr). In fact, solid KBr is usually made by allowing KOH to react with HBr in aqueous solution, and then evaporating the water:

$$K^+ + OH^- + H_3O^+ + Br^- \rightleftharpoons 2H_2O + K^+ + Br^-$$

although it can also be made in other ways. What happens to the pH of pure water when a salt is added depends on the relative strengths of the acid and base from which the salt is derived. If the salt is composed of a *strong acid* and a *strong base*, then the pH does in fact remain at 7. But if the salt is composed of a *strong acid* and a *weak base*, the pH drops below 7, while if it is made up of a *weak acid* and a *strong base,* it rises above 7. For salts composed of a weak acid and a weak base, the pH behavior depends on the K_a and K_b values of the acid and base, in a manner that we will discuss later in this section.

Type of salt	*pH of aqueous solution*
strong acid–strong base	7 neutral
strong acid–weak base	<7 acidic
weak acid–strong base	>7 basic
weak acid–weak base	varies

The reason for the behavior we have just described is that when a salt is dissolved in water, the positive and negative ions that make up the salt *react with the water.* This reaction, which is known as **hydrolysis,** is nothing new, but simply another example of an acid-base reaction such as those we have already discussed. A negative ion A^- produces the acid HA along with hydroxide ions OH^-

$$A^- + H_2O \rightleftharpoons HA + OH^-$$

while a positive ion such as Mg^{2+} or Fe^{3+} gives hydronium ions and undissociated hydroxides.

$$Mg^{2+} + 4H_2O \rightleftharpoons Mg(OH)_2 + 2H_3O^+$$
$$Fe^{3+} + 6H_2O \rightleftharpoons Fe(OH)_3 + 3H_3O^+$$

Positively charged ions that contain hydrogen (for example, NH_4^+) can give a proton directly to the water:

$$NH_4^+ + H_2O \rightleftharpoons NH_3 + H_3O^+$$

However, like all acid-base reactions, these are equilibrium reactions and *the extent of the reactions depends on the identity of the ions.* If A^- is derived from a strong acid (for example, HBr), the equilibrium lies so far to the left that for all practical purposes we can say that the reaction does not occur at all.

$$Br^- + H_2O \rightleftharpoons HBr + OH^- \qquad \text{(essentially no reaction)}$$

That is, the addition of Br^- to water produces essentially *no* HBr and *no* OH^-. On the other hand, if A^- is derived from a weak acid (for example, HOAc), the equilibrium may still lie predominantly to the left, but there will now be a considerable amount of HOAc and of OH^- (later in this section we will see just how much).

$$OAc^- + H_2O \rightleftharpoons HOAc + OH^- \qquad \text{(some reaction takes place)}$$

We can say that *the weaker the acid HA, the more HA and OH^- will be produced when A^- is added to water.* This principle directly follows from our discussions of acid strength in Section 10-2. If hydrolysis of A^- produces a strong acid, such as HBr, this acid, being strong, will immediately give the proton up again (we have seen that strong acids cannot exist in water solution but are converted essentially entirely to their conjugate A^- ions). A weaker acid has less tendency to do so, and the weaker the acid, the greater the percentage of it that can remain as undissociated HA in water solution.

Similar reasoning applies to positive ions. Ions derived from a strong base (see, for example, the four strong bases in Table 10-4) undergo essentially no reaction in water, because the base takes a proton back from the water as fast as it can be formed, while ions derived from weak bases undergo hydrolysis that increases in extent as the bases become weaker.

$$Na^+ + 2H_2O \rightleftharpoons NaOH + H_3O^+ \quad \text{(essentially no reaction)}$$
$$NH_4^+ + H_2O \rightleftharpoons NH_3 + H_3O^+ \quad \text{(some reaction takes place)}$$

We are now ready to understand why salts dissolved in water cause the pH to behave as it does. First, let us consider a typical salt of a strong acid and a weak base: ammonium nitrate NH_4NO_3. The NO_3^-, being derived from a strong acid, undergoes essentially no hydrolysis, and produces no HNO_3 and no OH^-. However, the NH_4^+ does undergo hydrolysis, giving measurable amounts of NH_3 and of H_3O^+.

$$NH_4^+ + H_2O \rightleftharpoons NH_3 + H_3O^+ \quad \text{(some reaction takes place)}$$
$$NO_3^- + H_2O \rightleftharpoons HNO_3 + OH^- \quad \text{(essentially no reaction)}$$

It is this production of H_3O^+ that results in the lowering of the pH. It is true that the base NH_3 is also produced in this reaction, but the pH of a solution is not a measure of the amount of base present, but *only of the concentration of H_3O^+.* Since hydrolysis of NH_4NO_3 increases the concentration of H_3O^+, the pH of pure water must fall when NH_4NO_3 is added. Similar reasoning applies to a salt made up of a weak acid and a strong base, for example, potassium fluoride (KF). In this case the K^+ does not react with the water, while hydrolysis of the F^- produces some OH^-, resulting in an increase of the pH.

$$K^+ + 2H_2O \rightleftharpoons KOH + H_3O^+ \quad \text{(essentially no reaction)}$$
$$F^- + H_2O \rightleftharpoons HF + OH^- \quad \text{(some reaction takes place)}$$

The undissociated acid HF is also produced, but once again this does not affect the pH, since pH is only a measure of $[H_3O^+]$ (which of course is related to $[OH^-]$ by $[OH^-][H_3O^+] = K_w$). For a salt made up of a strong acid and a strong base (for example, Na_2SO_4), *neither* ion reacts with water, so in this case the pH remains at 7.

When a salt is composed of a weak acid and a weak base, *both* ions react with water; whether the pH rises or falls depends upon the dissociation constants of the acid and base. If the numerical value of K_a for the acid is greater than that of K_b for the base, the pH will be less than 7 and the salt will be an acidic salt; if the contrary is true, the pH will be greater than 7 and the salt will be a basic salt. For example, consider the salt ammonium fluoride, NH_4F. From Tables 10-5 and 10-6 we find that K_a for HF is 3.5×10^{-4}; while K_b for NH_3 is 1.8×10^{-5}. Since the numerical value of K_a is greater than that of K_b, the pH of this solution will be less than 7. Both ions hydrolyze, but the equilibrium

$$NH_4^+ + H_2O \rightleftharpoons NH_3 + H_3O^+$$

lies farther to the right than does the equilibrium

$$F^- + H_2O \rightleftharpoons HF + OH^-$$

so there is an excess of H_3O^+ and the pH is less than 7. On the other hand, the salt hydrazinium cyanide (NH_2NH_3CN) is a basic salt, because K_b for hydrazine 1.7×10^{-6} is numerically greater than K_a for hydrocyanic acid, 4.93×10^{-10}. It is important to realize that the fact that both ions are hydrolyzing in these cases does *not* mean that the concentrations of *both* $[OH^-]$ and $[H_3O^+]$ are rising above 10^{-7}. This is impossible, because $[H_3O^+][OH^-] = 10^{-14}$, and if one of these concentrations goes above 10^{-7}, the other must fall below this value. What is happening here is that both OH^- and H_3O^- are produced by the hydrolysis reactions, but that they then react with each other to form water. The concentration of the one that is in excess remains above 10^{-7}, so the concentration of the other falls below this value.

In a few cases it happens that a weak acid and a weak base which make up a salt have K_a and K_b values that are virtually identical. The most important such salt is ammonium acetate (NH_4OAc), where the pK_a for acetic acid = the pK_b for ammonia = 1.8×10^{-5}. In such cases, the solution is neutral, and the pH is 7.

So far our discussion of hydrolysis has been qualitative. We will now see that it is possible in many cases to calculate the pH values obtained when a salt is added to water. For the salt of a strong acid and a strong base, of course, no calculation is necessary; the pH is 7. For the salt of a weak acid and a strong base, only the negative ion hydrolyzes,

$$A^- + H_2O \rightleftharpoons HA + OH^-$$

and the equilibrium expression is

$$K_h = \frac{[HA][OH^-]}{[A^-]} = K_b \text{ for the base } A^-$$

K_h is the **hydrolysis constant** for the ion A^-. Note that this expression is exactly the same as the equilibrium expression for a weak base shown in equation (4) on page 302. Therefore, the hydrolysis constant for any ion A^- is equal to the value of K_b for the same ion. This is not surprising, since what is happening here is that A^- is acting as a base toward the acid H_2O, and that is just what was happening when we derived K_b in equation (4). We have already seen that for a given conjugate acid-base pair,

$$K_a K_b = K_w$$

Thus, for any salt of a weak acid HA and a strong base B, we can obtain the value of K_h from the value of K_a of the acid HA, by use of the simple relationship

$$K_h = K_b = \frac{K_w}{K_a}$$

This value can then be used to calculate the degree of hydrolysis and pH of the solution, provided that we know the concentration.

Example: Calculate the percentage of hydrolysis and the pH of a 0.0100 *M* solution of potassium butyrate (C_3H_7COOK). The K_a for butyric acid (C_3H_7COOH) is 1.54×10^{-5}.

Answer: First we note that K^+ does not react with water, so we consider only the anion, $C_3H_7COO^-$.

$$K_h = K_b = \frac{1.00 \times 10^{-14}}{1.54 \times 10^{-5}} = 6.49 \times 10^{-10}$$

	$C_3H_7COO^- + H_2O$	$\rightleftharpoons C_3H_7COOH$	$+ OH^-$
Concentration at start	0.0100 *M*	0	0
Concentration at end	0.0100 − x	x	x

$$K_h = \frac{[C_3H_7COOH][OH^-]}{[C_3H_7COO^-]}$$

$$6.49 \times 10^{-10} = \frac{x \cdot x}{0.0100 - x}$$

Once again, we assume that x is small compared to 0.0100:

$$6.49 \times 10^{-10} = \frac{x^2}{0.0100}$$

$$x^2 = 6.49 \times 10^{-12}$$

$$x = 2.55 \times 10^{-6}\ M$$

Our assumption is borne out: 2.55×10^{-6} is very much smaller than 0.0100. The $[OH^-]$ is therefore 2.55×10^{-6} *M*, and the pOH is 5.6. Since pH + pOH = 14, the pH of this solution is 8.4. Thus, it is basic, in accord with our qualitative predictions.

percentage of hydrolysis is the percentage of $C_3H_7COO^-$ ions that have been converted to C_3H_7COOH ions. This number is

$$\% \text{ hydrolysis} = \frac{[C_3H_7COOH] \times 100}{[C_3H_7COO^-] \text{ at the start}} = \frac{2.55 \times 10^{-6}}{0.0100} \times 100 = 0.0255\%$$

Thus, only 0.0255% of the original butrate ions (or about 1 of every 4000) have reacted with water to produce butyric acid and OH^- ions at equilibrium, indicating that the hydrolysis reaction does not have to go very far in order to produce a basic solution. The corresponding percentage for a 0.01 *M* solution of sodium acetate is 0.0236%.

Similar calculations can be made for the salt of a strong acid and a weak base, where only the positive ion hydrolyzes:

$$BH^+ + H_2O \rightleftharpoons B + H_3O^+$$

The equilibrium expression is

$$K_h = \frac{[B][H_3O^+]}{[BH^+]} = K_a \text{ for the acid } BH^+$$

This time the hydrolysis constant is equal to the value of K_a for the acid BH^+, and we can obtain this value from the value of K_b for the base B:

$$K_h = K_a = \frac{K_w}{K_b}$$

Example: Calculate the percentage of hydrolysis, and the pH of a 0.100 *M* solution, of ammonium chloride (NH_4Cl). K_b for $NH_3 = 1.80 \times 10^{-5}$.

Answer:

$$K_h = K_a = \frac{1.00 \times 10^{-14}}{1.80 \times 10^{-5}} = 5.56 \times 10^{-10}$$

	NH_4^+	$+ H_2O \rightleftharpoons$	NH_3	$+ H_3O^+$
Concentration at start	0.100 *M*		0	0
Concentration at end	0.100 − x		x	x

$$K_h = \frac{[NH_3][H_3O^+]}{[NH_4^+]}$$

$$5.56 \times 10^{-10} = \frac{x^2}{0.100 - x}$$

Once more we assume that x is small compared to 0.100:

$$5.56 \times 10^{-10} = \frac{x^2}{0.100}$$

$$x^2 = 5.56 \times 10^{-11} = 55.6 \times 10^{-12}$$

$$x = 7.46 \times 10^{-6}\ M$$

The $[H_3O^+]$ is thus 7.46×10^{-6} *M*, and the pH is 5.1. As predicted, the solution is acidic. The percentage of hydrolysis is

$$\% \text{ hydrolysis} = \frac{[NH_3] \times 100}{[NH_4^+] \text{ at the start}} = \frac{7.46 \times 10^{-6} \times 100}{0.100} = 0.00746\%$$

In this case at equilibrium only 0.00746% of the original NH_4^+ ions are present as NH_3 molecules (about 1 out of every 13,000), but this is enough to lower the pH of pure water from 7.0 to 5.1.

For a salt of a weak acid and a weak base, calculations can become quite complicated, since there are three equilibrium expressions involved, all of which must be satisfied; for example, for NH_4CN:

$$K_{h_{NH_4^+}} = \frac{[NH_3][H_3O^+]}{[NH_4^+]} \qquad K_{h_{CN^-}} = \frac{[HCN][OH^-]}{[CN^-]} \qquad K_w = [H_3O^+][OH^-]$$

10-11 Titration

Suppose that we have a solution of an acid in water and we wish to know its concentration. This type of problem is frequently met with when working in a chemical or biological laboratory. (Note that we are *not* referring here to the H_3O^+ concentration; the latter can always be measured by the use of indicators or a pH meter. What we are talking about here is the concentration of HA, and, as we have seen, if HA is a weak acid, only a small amount of it will be converted to hydronium ions.) We can easily determine the concentration of any acid dissolved in water, whether strong or weak (but not feeble), by the use of a simple laboratory technique called **titration.** Furthermore, if we know the concentration of an unknown acid, we can use titration to determine its equivalent weight (see page 177 for a definition of equivalent weight).

To determine the concentration of an acid in water solution, the technique of titration is carried out as follows. We fill a *buret* (Figure 10-1) with a solution of a strong base (usually KOH or NaOH) whose concentration is known. We then place a measured volume of the acid solution into a flask. The base is added to the acid in the flask, drop by drop, until *exactly* the right amount of base has been added to neutralize *completely* the acid in the flask (but not one drop more). This point is called the **equivalence point,** and is always accompanied by a sharp change in pH (we shall see how sharp later in the section). We can tell when the equivalence point is reached in one of two ways. Either we can add to the acid solution (before we begin the titration) a drop or two of an appropriate indicator, or we can use a pH meter to measure the pH after each drop has been added (the solution must be thoroughly stirred before each measurement). In this way we can locate the sharp pH change that signals the equivalence point. The indicator procedure is generally simpler to carry out, but it is necessary to use an *appropriate* indicator. The point at which the indicator changes color is called the **end point** of the titration, and the end point must coincide with the equivalence point if the correct result is to be obtained. For example, if at the equivalence point the final drop of added base changes the pH from about 6 to about 13, it would not be appropriate to use the indicator methyl red, which changes color at 4.8 to 6.0 (Table 10-2). An appropriate indicator for this titration would be phenolphthalein. In the next section we shall see how to select an appropriate indicator for a given titration.

When the equivalence point has been reached, we know (by reading the buret) how many milliliters of base solution have been added, and of course we know the concentration of the base solution (or we wouldn't have done the titration to begin with). For the purposes of titration, this concentration is generally expressed as *normality* (for the reason discussed in Section 7-3), though for NaOH and KOH,

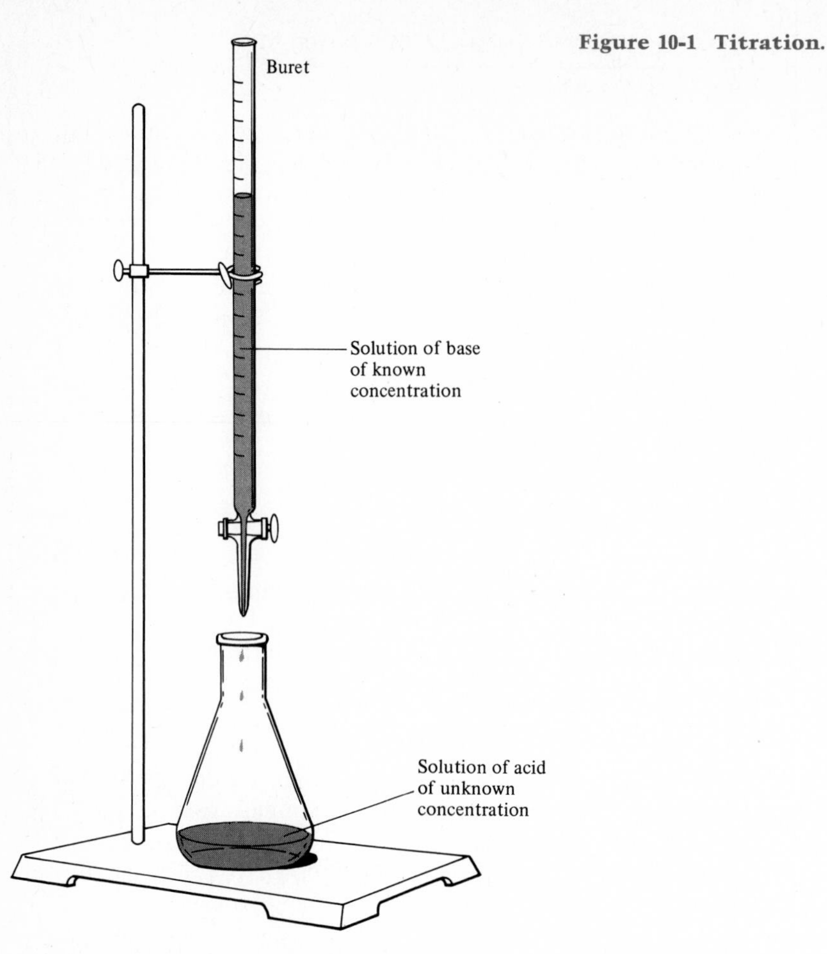

Figure 10-1 Titration.

normality is numerically equal to molarity. Since we carefully measured the volume of acid solution before we began the titration, we may use the equations from page 178 to solve for the normality of the acid.

$$\text{millequivalents of acid} = \text{milliequivalents of base}$$

$$\underset{\text{measured before titration}}{\text{ml}_{\text{acid}}} \times N_{\text{acid}} = \underset{\text{taken from buret reading}}{\text{ml}_{\text{base}}} \times \underset{\text{previously known}}{N_{\text{base}}}$$

Note that this procedure does not give us the *molarity* of the acid solution, only the normality. In order to determine the molarity of an unknown solution of acid, we must have additional information. For example, if we know the *identity* of the acid, then we know whether it is monoprotic, diprotic, or triprotic, and we can convert normality to molarity by

$$N = aM$$

where a is equal to the number of protons donated by the acid.

The titration procedure described above can also be used to determine the *equivalent weight of an unknown acid.* In this case, it is necessary to weigh a sample of the pure acid before dissolving it in the water (the amount of water used is irrelevant in this case—why?) and then carrying out the titration. Once again, we know the normality of the base and the number of milliliters of base added. The *product* of these numbers is equal to the number of milliequivalents of base:

$$\text{ml}_{\text{base}} \times N_{\text{base}} = \text{meq}_{\text{base}}$$

The titration procedure ensures that at the equivalence point

$$meq_{base} = meq_{acid}$$

so we know how many milliequivalents of acid are present in our sample. But we also know the *weight* of the acid, since we carefully weighed it before we dissolved it in the water. We can, therefore, calculate the equivalent weight by

$$E.W._{acid} = \frac{g_{acid}}{eq_{acid}} = \frac{mg_{acid}}{meq_{acid}}$$

Example: If 0.441 g of an unknown acid required 36.44 ml of an 0.100 N NaOH solution for complete neutralization, what is the equivalent weight of the acid?

Answer:

$$E.W. = \frac{mg_{acid}}{meq_{acid}} = \frac{mg_{acid}}{ml_{base} \times N_{base}}$$

$$= \frac{441 \text{ mg}}{36.44 \text{ ml} \times 0.100 \text{ meq/ml}}$$

$$= 121 \text{ mg/meq}$$

The equivalent weight is 121 mg/meq or 121 g/equiv.

It is true that this technique does not directly give us the *molecular weight* of the acid, only the equivalent weight, but the molecular weight of an acid is very simply related to the equivalent weight by

$$\text{mol wt} = a(\text{E.W.})$$

where once again a is the number of protons donated by the acid, and it is often possible to determine a in some independent manner (from the shape of the titration curve, for example; see below). In other cases, we can determine the *molecular weight* in an independent manner, and then the titration technique gives us a means of determining a, that is, whether the acid is monoprotic, diprotic, or triprotic.

We have shown how the technique of titration can be used to find the concentration of a solution of an acid in water or the equivalent weight of an unknown acid. In a similar manner, it can also be used to determine the concentration of a solution of a base in water or the equivalent weight of an unknown base. For this purpose, of course, the basic solution can be placed in the flask, and the buret filled with a solution of a strong acid of known concentration. The technique and the calculations are otherwise the same.

Example: If 0.863 g of an unknown base required 21.06 ml of an 0.500 N HCl solution for complete neutralization, what is the E.W. of the base?

Answer:

$$E.W. = \frac{mg_{base}}{ml_{acid} \times N_{acid}}$$

$$= \frac{863 \text{ mg}}{21.06 \text{ ml} \times 0.0500 \text{ meq/ml}} = 82.0 \text{ mg/meq}$$

The equivalent weight of the base is 82.0 g/equiv.

Titrations are not limited to acid–base reactions. They can be carried out for any kind of reaction provided that (1) there is some means available, such as a color change, to determine when the equivalence point has been reached, and (2) the rate of the reaction is relatively rapid, so that the reaction goes to completion within a few minutes or less. In Section 16-4, we shall see how titrations are applied to oxidation–reduction reactions.

10-12 Titration Curves

Valuable information can be gained by looking at a graph of how the pH of a solution changes during the course of an acid–base titration. Such a graph, in which we plot the pH of the solution vs. the number of ml of base added, is called a **titration curve.** The shape of the curve will be different depending on which acid and base are involved. We shall discuss several typical examples.

1. Strong Acid–Strong Base. Since the actual acid present in all strong acid aqueous solutions is H_3O^+ and the actual base present in all strong base aqueous solutions is OH^-, all titrations of a monoprotic strong acid vs. a strong base will have the same titration curve, shown in Figure 10-2. The points of the curve are determined by calculations such as the following.

Example: We begin with 10.00 ml of a solution of 1.000 *M* HCl. Calculate the pH after we add the following volumes of a 1.000 *M* solution of NaOH: (a)before any NaOH is added, (b) 9.00 ml, (c) 9.90 ml, (d) 9.99 ml, (e) 10.00 ml. (f) 10.01 ml, (g) 10.10 ml, (h) 11.00 ml.

Answer:

(a) Before any NaOH is added, the concentration of HCl is 1.000 *M*. Since HCl is a strong acid, this means that $[H_3O^+] = 1.000\ M$, so the pH is 0.

(b) 10.00 ml of 1.000 *M* HCl contains

$$10.00 \text{ ml} \times 1.000 \frac{\text{mmole}}{\text{ml}} = 10.00 \text{ mmoles of } H_3O^+$$

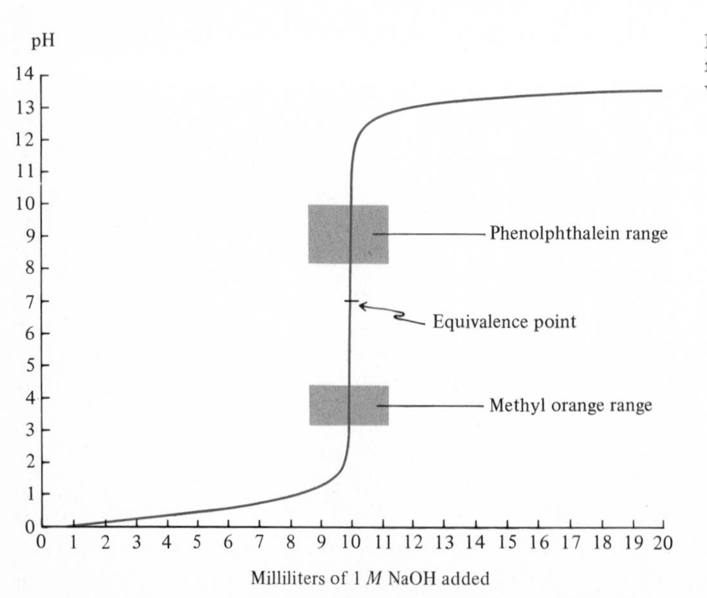

Figure 10-2 Titration curve for 10 ml of 1.00 *M* HCl titrated with 1.00 *M* NaOH solution.

9.00 ml of 1.000 M NaOH contains

$$9.00 \text{ ml} \times 1.000 \frac{\text{mmole}}{\text{ml}} = 9.00 \text{ mmoles of } OH^-$$

9.00 mmoles of OH^- neutralizes 9.00 mmoles of H_3O^+, so

$$\underset{\text{at start}}{10.00 \text{ mmoles of } H_3O^+} - \underset{\text{neutralized}}{9.00 \text{ mmoles of } H_3O^+} = \underset{\text{present}}{1.00 \text{ mmole of } H_3O^+}$$

The solution now contains 1.00 mmole of H_3O^+ in a total volume of 10.00 ml + 9.00 ml = 19.00 ml. The concentration of H_3O^+ is, therefore, 1.00/19.00 mmole/ml = 5.26×10^{-2} M and the pH is 1.28.

(c) 9.90 mmoles of OH^- neutralizes 9.90 mmoles of H_3O^+, leaving 0.10 mmoles of H_3O^+. The total volume is now 19.90 ml, so the concentration of H_3O^+ *is* 0.10/19.90 mmole/ml = 5.0×10^{-3} M, and the pH is 2.30.

(d) 9.99 mmoles of OH^- neutralizes 9.99 mmoles of H_3O^+, leaving 0.01 mmoles of H_3O^+. The total volume is 19.99 ml, so $[H_3O^+]$ = 0.01/19.99 mmoles/ml = 5×10^{-4} M. pH = 3.30.

(e) We are now exactly at the equivalent point. 10.00 mmoles of OH^- completely neutralizes 10.00 mmoles of H_3O^+. The pH is thus the same as that of pure water; pH = 7.00.

(f) 10.00 mmoles of H_3O^+ have now been neutralized by 10.00 mmoles of OH^-. Therefore, 0.01 mmole of OH^- is now present, in a total volume of 20.01 ml. $[OH^-]$ is thus 0.01/20.01 mmole/ml = 5×10^{-4} M. pOH = 3.30 and pH = 10.70.

(g) The OH^- concentration is now 0.10/20.10 mmole/ml = 5.0×10^{-3} M. pOH = 2.30 and pH = 11.70.

(h) The OH^- concentration is now 1.00/21.00 mmole/ml = 4.76×10^{-2} M. pOH = 1.32 and pH = 12.68.

The results are summed up in the following table.

ml of 1.00 M NaOH added	*pH*
0	0
9.00	1.28
9.90	2.30
9.99	3.30
10.00	7.00
10.01	10.70
10.10	11.70
11.00	12.68

Figure 10-2 is plotted from such data. Note that at first the pH changes very slowly. It takes 9.00 ml of the NaOH solution to raise the pH from 0 to 1.28. Then the rise becomes more rapid, until, near the equivalence point, the addition of only 0.02 ml of the basic solution causes the pH to increase from 3.30 to 10.70, a rise of 7.40 pH units! It is this sharp rise that makes titration a very accurate technique for solutions of acids and bases. Once the equivalence point is passed, the rate of pH increase once again diminishes, and the pH levels off. In this part of the curve, the pH behavior is the same as if the NaOH solution were being added to pure water.

When we use a pH meter in a titration, the equivalence point is determined by the position of the sharp pH increase. When we use an indicator, it is necessary to choose one for which the color change falls on the steep part of the curve. As is shown in Figure 10-2, any of several indicators, including, for example, phenolphthalein and methyl orange, will be satisfactory for a titration of a strong acid vs. a

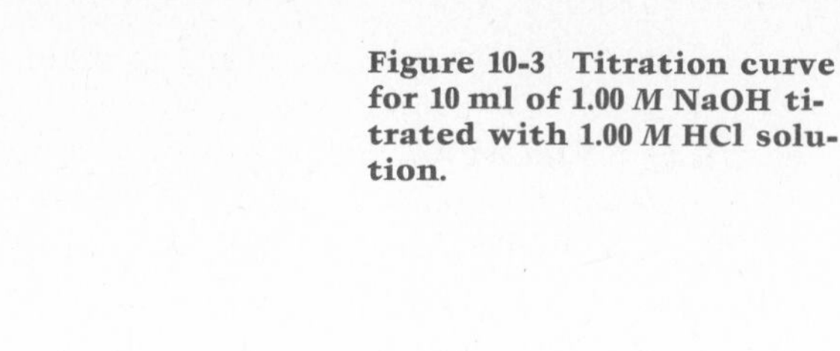

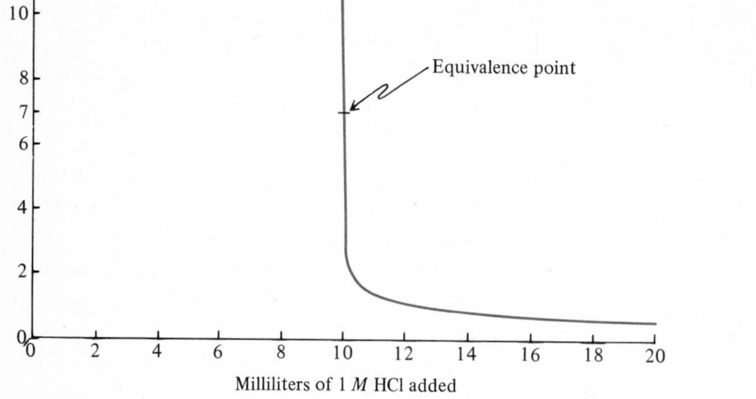

Figure 10-3 Titration curve for 10 ml of 1.00 *M* NaOH titrated with 1.00 *M* HCl solution.

strong base. However, indicators that change color at the extreme ends of the scale (for example, methyl violet, Clayton yellow; see Table 10-2) would not be satisfactory.

In a similar manner, we can also draw a titration curve for a titration in which the strong base is in the flask and the strong acid is added from a buret. Such a curve (Figure 10-3) is essentially the mirror image of Figure 10-2.

2. Weak Acid–Strong Base. The calculations necessary to draw a titration curve for the titration of a weak acid by a strong base are similar to those shown above for the strong acid–strong base case. Figure 10-4 shows a titration curve for a typical weak acid—acetic acid ($K_a = 1.8 \times 10^{-5}$). We immediately see that the shape of the curve is different. Because acetic acid is a weak acid, the pH at the start of the titration (which is simply the pH of acetic acid dissolved in water) is higher than in the previous case (see Section 10-6). The pH rises more rapidly than it did before, but soon forms a level section approximately halfway to the equivalence point (in this case, 5.0 ml). The sharp pH rise at the equivalence point is smaller

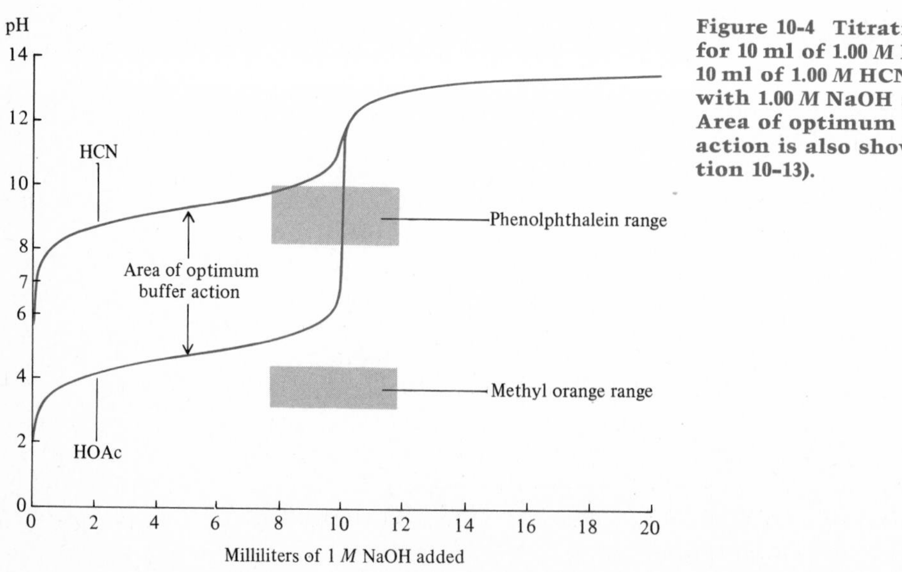

Figure 10-4 Titration curve for 10 ml of 1.00 *M* HOAc or 10 ml of 1.00 *M* HCN titrated with 1.00 *M* NaOH solution. Area of optimum buffer action is also shown (see Section 10-13).

than in the previous case (a rise of about 5 pH units rather than 11 units). This is understandable when we remember that at the equivalence point, we have a solution of sodium acetate in water, which is a basic solution (Section 10-10) rather than the neutral NaCl solution we had in the previous titration. This means, as Figure 10-4 clearly shows, that methyl orange is totally unsuitable as an indicator for this titration. If we did use methyl orange, it would change color long before the equivalence point was reached. On the other hand, phenolphthalein is entirely suitable. As we proceed beyond the equivalence point, the titration curve resembles the earlier one, which is not surprising, since in both titrations there is now excess NaOH present.

For acids very much weaker than acetic, the left-hand portion of the curve is much higher, and the length of the steep section at the equivalence point greatly diminishes. For example, note the titration curve for HCN ($K_a = 4.93 \times 10^{-10}$) shown in Figure 10-4. For this acid, even phenolphthalein is not a satisfactory indicator, and in fact, no indicator is satisfactory because the pH at the equivalence point does not change sharply enough. Acids as weak as this can only be titrated with a pH meter, and even here the accuracy is less than it is with stronger acids.

3. Strong Acid–Weak Base. Figure 10-5 shows a titration curve for a typical weak base—ammonia—against a strong acid—HCl. Comparing this with Figure 10-3, we see that the initial portion of the curve is lower here (because NH_3 is a weaker base than NaOH) and that the steep portion of the curve begins only at about pH 7. Therefore, phenolphthalein is not satisfactory for this titration, whereas methyl orange does make a satisfactory indicator. After the equivalent amount of HCl has been added, the curve is essentially the same as in Figure 10-3, which again is understandable, since both solutions now have excess HCl.

4. Weak Acid–Weak Base. When a weak acid is titrated with a weak base, the titration curve has essentially no steep portion at all (Figure 10-6). Because of this, such titrations are generally not feasible at all.

5. Polyprotic Acids. If a polyprotic acid is titrated with a strong base, it may be possible to find *two* steep portions. Figure 10-7 shows a titration curve for 10 ml of 1.00 *M* phosphoric acid (H_3PO_4) titrated with 1.00 *M* NaOH. The figure shows a steep pH rise when 10 ml of NaOH have been added, and a second steep rise

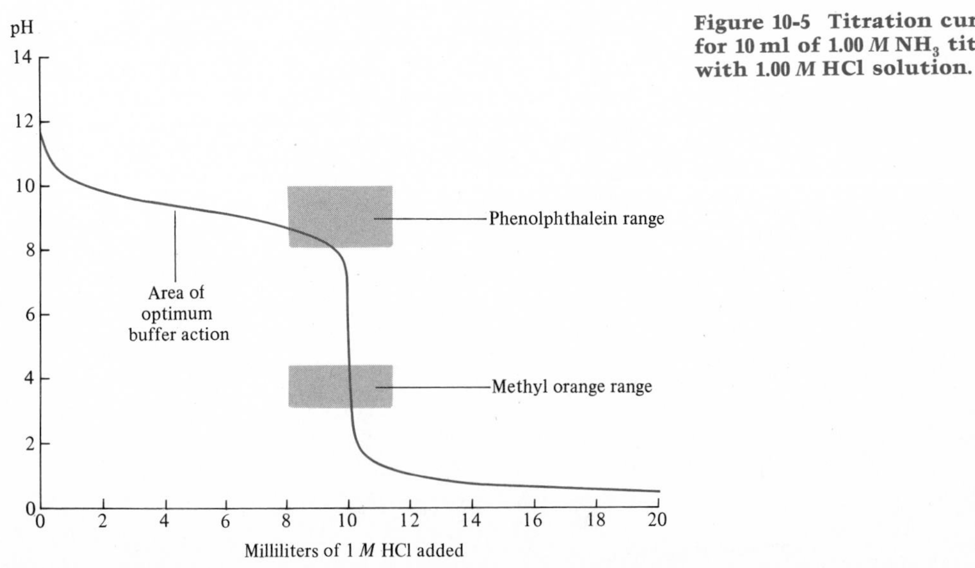

Figure 10-5 Titration curve for 10 ml of 1.00 *M* NH_3 titrated with 1.00 *M* HCl solution.

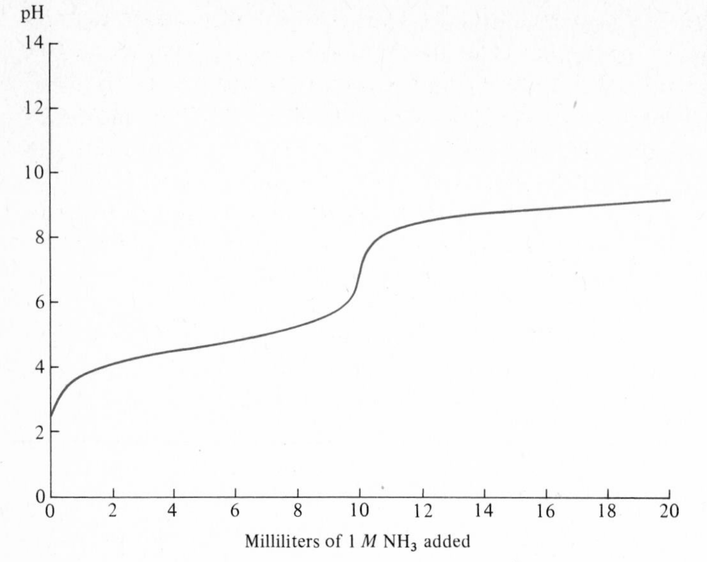

Figure 10-6 Titration curve for 10 ml of 1.00 *M* HOAc titrated with 1.00 *M* NH_3.

after the addition of another 10 ml of NaOH. This type of behavior comes about because the reaction of H_3PO_4 with base occurs in three distinct stages, and *each stage is essentially complete before the next one begins.*

$$\text{Stage 1:}\quad H_3PO_4 + OH^- \rightleftharpoons H_2PO_4^- + H_2O$$
$$\text{Stage 2:}\quad H_2PO_4^- + OH^- \rightleftharpoons HPO_4^{2-} + H_2O$$
$$\text{Stage 3:}\quad HPO_4^{2-} + OH^- \rightleftharpoons PO_4^{3-} + H_2O$$

At the first equivalence point, stage 1 is essentially complete; virtually all the H_3PO_4 has been converted to $H_2PO_4^-$; and there is a relatively sharp pH increase, as sharp as we would expect for a weak acid. The solution now contains 0.01 mole of $H_2PO_4^-$; and additional NaOH now reacts with this (stage 2), converting it to HPO_4^{2-}. When this reaction is complete, we have the second equivalence point and another steep rise (though smaller, because $H_2PO_4^-$ is a weaker acid than H_3PO_4) in the pH. It is quite possible to titrate such a solution using two different indicators, one for the first pH change and another for the second. However, because the indicator colors may interfere with each other, it is generally more convenient to use a pH meter.

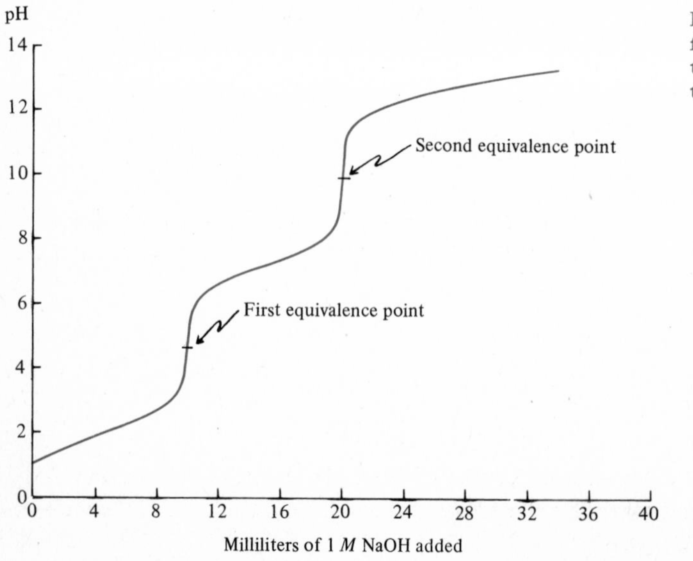

Figure 10-7 Titration curve for 10 ml of 1.00 *M* H_3PO_4 titrated with 1.00 *M* NaOH solution.

Not all diprotic and triprotic acids exhibit this behavior. In many cases there is no second steep rise, simply because the second acid is too weak. We have previously seen that as acids become weaker, the steep portion of the titration curve decreases in length, and for very weak acids disappears entirely. Thus, many diprotic and triprotic acids show only one steep rise and can be titrated to only one equivalence point. It is for this reason that phosphoric acid does not show a third steep rise, corresponding to stage 3. At the end of the second stage, all the $H_2PO_4^-$ has been converted to HPO_4^{2-}, and it might be expected that we could titrate this acid by adding 10 additional ml of 1.00 *M* NaOH. We can, of course, add the 10 ml of NaOH solution, and this will convert the HPO_4^{2-} to PO_4^{3-}, but there is no sharp pH increase at this point and there is no way to tell, either by an indicator or a pH meter, that we have arrived at an equivalence point.

10-13 Buffers

The pH of normal human blood is 7.4, and it is very important for the health of the individual that it maintain this value. One reason is that many of the enzyme-catalyzed reactions that take place in blood go with optimum speeds at this pH. If the pH of blood goes up as high as 7.8 or down as low as 7.0, serious illness or death can result. Every day we eat and drink foods whose pH is a great deal higher or lower than 7.4, and this eventually adds considerable quantities of H_3O^+ or OH^- ions to the blood. How, then, is the body able to maintain the pH of the blood at 7.4? It is able to do this because it is *buffered*. A **buffer solution** may be defined as *a solution whose pH does not change very much when H_3O^+ or OH^- ions are added to it*. Besides being present in blood and in certain other biological fluids, buffer solutions have many useful applications in chemical and biological laboratories, and it is fortunate that they are so easy to prepare. To make a buffer solution, all we have to do is to make up a solution that contains *approximately equal quantities of a weak acid and its conjugate base*. A typical buffer solution might consist of 0.1 mole of acetic acid and 0.1 mole of sodium acetate dissolved in 1 liter of water. This solution behaves as a buffer because it is able to neutralize any H_3O^+ or OH^- ions which may be added to the solution from the outside (up to a certain limit; see the discussion of buffer capacity later in this section).

$$HOAc + OH^- \rightleftharpoons OAc^- + H_2O$$

$$OAc^- + H_3O^+ \rightleftharpoons HOAc + H_2O$$

Any H_3O^+ ions that are added are neutralized by the OAc^- ions, while added OH^- ions are neutralized by the HOAc molecules.

We will get a clearer picture of buffer action if we look at it quantitatively. We will see what happens to the pH if we add H_3O^+ or OH^- to an unbuffered solution, and then to a buffered solution.

An unbuffered solution: The pH of pure water is 7.00. How will this change if we add (a) 0.10 mole of HCl? (b) 0.10 mole of NaOH? In both cases, assume that the final volume of the solution is 1.00 liter.

1. A solution that contains 0.10 mole of H_3O^+ in 1.00 liter of solution has $[H_3O^+] = 1 \times 10^{-1}\,M$. pH = 1.00.
2. A solution that contains 0.10 mole of OH^- in 1.00 liter of solution has $[OH^-] = 1 \times 10^{-1}\,M$. pOH = 1.00 and pH = 13.00.

Thus, the addition of 0.10 mole of H_3O^+ lowers the pH of 1.00 liter of water from 7 to 1, while the addition of 0.10 mole of OH^- raises the pH from 7 to 13—in either case a change of 6 pH units.

A buffered solution: As we shall see below, the pH of a 1.00-liter solution containing 0.50 mole of HOAc and 0.50 mole of NaOAc is 4.75. How will this change if we add (a) 0.10 mole of HCl? (b) 0.10 mole of NaOH? In both cases assume that the final volume is still 1.00 liter.

1. In any aqueous solution in which HOAc is present, the familiar equilibrium expression must be valid:

$$K_a = \frac{[OAc^-][H_3O^+]}{[HOAc]}$$

In this case we began with 0.50 mole of HOAc and 0.50 mole of OAc^-, so that $[H_3O^+] = K_a$ and pH $= pK_a$ for acetic acid, which is 4.75. When we added 0.10 mole of H_3O^+, it immediately reacted with OAc^- to form HOAc.

	$OAc^- + H_3O^+$	$\rightleftharpoons$	$HOAc + H_2O$
Number of moles at start	0.50		0.50
Number of moles after 0.10 mole of H_3O^+ added	0.40		0.60

After the addition of 0.10 mole of H_3O^+, we have 0.40 mole of OAc^- and 0.60 mole of HOAc, in 1.00 liter of solution. We can now calculate $[H_3O^+]$ by

$$K_a = 1.8 \times 10^{-5} = \frac{0.40[H_3O^+]}{0.60}$$

$$[H_3O^+] = 2.7 \times 10^{-5}\ M$$

$$\text{pH} = 4.57$$

2. When we added 0.1 mole of NaOH to the original buffer solution, it reacted with HOAc to form OAc^-.

	$HOAc + OH^-$	$\rightleftharpoons$	$OAc^- + H_2O$
Number of moles at start	0.50		0.50
Number of moles after 0.1 mole of OH^- added	0.40		0.60

After the addition of 0.10 mole of OH^-, we now have 0.60 mole of OAc^- and 0.40 mole of HOAc in 1.00 liter of solution. Thus

$$1.8 \times 10^{-5} = \frac{0.60[H_3O^+]}{0.40}$$

$$[H_3O^+] = 1.2 \times 10^{-5}\ M$$

$$\text{pH} = 4.92$$

Thus, the addition of 0.10 mole of H_3O^+ lowers the pH of this buffer solution from 4.75 to 4.57, while the addition of 0.10 mole of NaOH raises it to 4.92, a change in either case of less than 0.2 pH unit.

These calculations and results present a graphic illustration of the pH stability provided by a buffer solution, and also illustrate more clearly just *how* this stability is provided. For any buffer solution, we have

$$K_a = \frac{[A^-][H_3O^+]}{[HA]} \quad \text{or} \quad [H_3O^+] = K_a\frac{[HA]}{[A^-]}$$

and

$$A^- + H_3O^+ \rightleftharpoons HA + H_2O$$

$$HA + OH^- \rightleftharpoons A^- + H_2O$$

When $[H_3O^+]$ is added from the outside, it is neutralized by A^-, resulting in an *increase* in [HA] and a *decrease* in $[A^-]$. This will increase the value of the ratio $[HA]/[A^-]$ and therefore change $[H_3O^+]$. As long as the amount of external H_3O^+ added is not too great, the value of $[HA]/[A^-]$ will remain less than, say, 20 and the effect on the pH is relatively small. When $[OH^-]$ is added from the outside, the opposite behavior takes place. The external OH^- is neutralized by the HA, resulting in a *decrease* in [HA] and an increase in $[A^-]$. Thus, $[HA]/[A^-]$ decreases, but again, as long as the value remains more than about $\frac{1}{20}$, the pH will not be greatly affected.

A graphical picture of the buffering effect may be seen in the titration curves of a weak acid against a strong base, or a strong acid against a weak base. For example, consider Figure 10-4. At the start of the titration we have HOAc dissolved in water, along with a very small amount of OAc^-, formed by ionization of the weak acid HOAc. The value of $[HOAc]/[OAc^-]$ is about 200 (we calculated this on page 298), and the solution is not a buffer. As NaOH is added from the buret, the ratio $[HOAc]/[OAc^-]$ decreases and soon reaches a point where buffer action begins. This is shown in the graph as the approximately level portion at about pH 5. The graph clearly shows that in this area, addition of either NaOH or of HCl changes the pH very little. Maximum buffer action is at the point when 5.0 ml of the NaOH solution has been added. At this point half of the HOAc has been converted to OAc^-, and the $[HOAc]/[OAc^-]$ ratio is 1.

Every buffer solution has two major characteristics: its pH value and its capacity.

1. *The pH value of a buffer solution.* If we make up a buffer solution by dissolving in water equimolar quantities of a weak acid and its conjugate base, the pH of the solution must be equal to the pK_a value of the weak acid. This follows from the equation

$$[H_3O^+] = K_a \frac{[HA]}{[A^-]}$$

If $[HA] = [A^-]$, then $[H_3O^+] = K_a$ and $pH = pK_a$. This fact enables us to prepare buffer solutions for almost any pH we desire. In the laboratory the need frequently arises for a buffer solution that has to maintain a given pH. When we are confronted with this problem, all we have to do is to consult a chart of weak acids (such as Table 10-5), select an acid with a pK_a value close to the desired pH value, and make up a solution consisting of equimolar quantities of the acid and its conjugate base dissolved in water. This can be done in two ways: either we can directly add to the water both the acid and its conjugate base, or, if the conjugate base is not available, we can add only the acid, and then add enough NaOH to half-neutralize the acid (if only the conjugate base is available, we could proceed in the other direction: add the base and enough HCl to half-neutralize the base). For example, suppose that we wanted to prepare a buffer solution that would maintain a pH of 10.00. Table 10-5 shows us that the acid phenol has a pK_a value of 9.89 ($K_a = 1.3 \times 10^{-10}$). A solution made up of equimolar quantities of phenol and its conjugate base (phenoxide ion) would thus have a pH of 9.89, which is close to 10. In some cases, however, just being close will not do. We might need a buffer solution whose pH is *exactly* 10.00. Such a solution can be prepared by adjusting the relative amounts of phenol and phenoxide ion.

Example: Prepare a buffer solution of pH 10.00 consisting of phenol and phenoxide dissolved in water.

Answer: We need to find which ratio of $[C_6H_5OH]/[C_6H_5O^-]$ will result in a $[H_3O^+]$ of 1.0×10^{-10}.

$$[H_3O^+] = K_a \frac{[HA]}{[A^-]}$$

$$1.0 \times 10^{-10} = 1.3 \times 10^{-10} \times \frac{[C_6H_5OH]}{[C_6H_5O^-]}$$

$$\frac{[C_6H_5OH]}{[C_6H_5O^-]} = \frac{1.0 \times 10^{-10}}{1.3 \times 10^{-10}} = \frac{10}{13}$$

Thus, a solution made up of phenol and phenoxide ion in a molar ratio of 10:13 will have $[H_3O^+] = 1.0 \times 10^{-10}$ *M* and a pH of 10.00. By calculations such as these, and a sufficiently long list of weak acids, we can prepare buffer solutions to maintain virtually any pH value we choose.

2. *Buffer capacity.* We have already pointed out that a buffer solution cannot neutralize very large amounts of acid or base. Every buffer solution has a capacity, and when that capacity is exceeded, the buffering action is swamped out. As we have just seen, the pH of a buffer solution depends on the pK_a value of the acid and on the ratio of $[HA]/[A^-]$. The buffer *capacity,* however, depends on the *total concentrations* of HA and of A^-. For example, compare a solution that is 1.00 *M* in HOAc and 1.00 *M* in OAc^- with another solution which is 0.10 *M* in HOAc and 0.10 *M* in OAc^-. Both solutions have the same pH (4.75), but the first solution has a capacity 10 times that of the second solution. This can be illustrated by seeing what happens to the pH of each solution when we add, say, 0.20 mole of HCl to 1.0 liter of each solution.

Example: Calculate the pH that results when 0.20 mole of HCl is added to 1.00 liter of (a) a solution that is 1.00 *M* in both HOAc and OAc^- and (b) a solution that is 0.10 *M* in both HOAc and OAc^-. K_a for HOAc is 1.8×10^{-5}. Assume that the total volume in each case remains at 1 liter.

Answer:

(a) We began with 1.00 mole each of HOAc and OAc^-. The addition of 0.20 mole of HCl converts 0.20 mole of OAc^- to HOAc, and we now have 0.80 mole of OAc^- and 1.20 mole of HOAc.

$$[H_3O^+] = 1.8 \times 10^{-5} \times \frac{1.2}{0.8}$$

$$[H_3O^+] = 2.7 \times 10^{-5}\ M$$

$$\text{pH} = 4.57$$

(b) We began here with 0.10 mole each of HOAc and OAc^-. When we add 0.20 mole of HCl, the first 0.10 mole of this virtually completely converts the OAc^- to HOAc. The second 0.10 mole of HCl is left over. So at the end of this reaction we have a solution consisting of 0.20 mole of HOAc and 0.10 mole of HCl. The HCl, of course, completely reacts with water to give H_3O^+, while the HOAc does so only to a very small extent. We can assume that the concentration of H_3O^+ coming from the HOAc is very small compared to that coming from the HCl. Thus,

$$[H_3O^+] = 0.10\ M \quad \text{and} \quad \text{pH} = 1.0$$

Thus, the first buffer solution, with a much larger capacity, had its pH lowered only from 4.75 to 4.57 by the 0.20 mole of HCl added, while the buffering action of the second solution was completely destroyed and the pH went all the way down to 1.0.

Living organisms, of course, do not have available a long list of weak acids such as that given in Table 10-5, and only a relatively few of these systems serve as buffers in maintaining the pH of bodily fluids. The most important of these are the H_2CO_3–HCO_3^- buffer and the $H_2PO_4^-$–HPO_4^{2-} buffer. Both of these systems are present in blood and help to maintain the pH of blood at 7.4.

10-14 Acid-Base Behavior in Nonaqueous Solvents

We have seen that water is an excellent solvent for determining acid and base ionization constants, but it does have its limitations. Ionization constants can be determined in water only for acids stronger than H_2O and weaker than H_3O^+ (or bases stronger than H_2O and weaker than OH^-), because of leveling effects. Given these limitations, it was only natural that chemists should turn to other solvents to try to determine if they could be used instead of water for acids and bases outside these limits. Several such solvents have been found, and acid–base behavior in nonaqueous solvents has been extensively studied. Unfortunately, experimental difficulties with such solvents are greater than they are with water, and ionization constants are generally a good deal less precise than those determined in water. Furthermore, many of these solvents have low dielectric constants (Section 7-5), which make them inhospitable to free ions. Because of this, ions formed by normal reactions of an acid or a base with the solvent

$$HA + S \rightleftharpoons HS^+ \cdots A^- \qquad S = \text{a basic solvent}$$

$$B + SH \rightleftharpoons BH^+ \cdots S^- \qquad SH = \text{an acidic solvent}$$

ion pairs

do not float freely in the solution, but rather remain closely associated with each other as **ion pairs.** Such behavior greatly complicates the measurement of the concentrations of these ions and hence affects the calculation of ionization constants. Nevertheless, nonaqueous solvents have proved useful for this purpose, since without them we would have almost no information at all. We shall mention two nonaqueous solvents: ammonia and acetic acid.

AMMONIA. Ammonia, NH_3, is a gas at room temperature, and in order to use it as a solvent, it is necessary to liquify it. At atmospheric pressure, this means that we must cool it below its boiling point (−33°C) before we can use it as a solvent. Although this is experimentally inconvenient, ammonia is still an important nonaqueous solvent for acid-base reactions because it is so similar to water in its properties. Like water, NH_3 is amphoteric and undergoes autoprotolysis, although to a lesser extent than water:

$$NH_3 + NH_3 \rightleftharpoons NH_4^+ + NH_2^-$$

$$K_{ion} = [NH_4^+][NH_2^-] = 10^{-33} \text{ at } -50°C$$

When an acid is dissolved in liquid NH_3, it reacts with it to form ammonium ion, NH_4^+, analogous to the reaction of acids with water to form H_3O^+:

$$HA + NH_3 \rightleftharpoons NH_4^+ + A^- \qquad (a)$$

$$HA + H_2O \rightleftharpoons H_3O^+ + A^- \qquad (b)$$

Similarly, bases when added to liquid ammonia take a proton to give amide ion, NH_2^-, analogous to OH^-:

$$B + NH_3 \rightleftharpoons NH_2^- + BH^+ \qquad (c)$$

$$B + H_2O \rightleftharpoons OH^- + BH^+ \qquad (d)$$

The important difference is that NH_3 is a stronger base than H_2O and a weaker acid (Table 10-1). Therefore, equilibrium (a) lies farther to the *right* than does equilibrium (b), while equilibrium (c) lies farther to the *left* than does equilibrium (d). The consequences of this are that the leveling effects of ammonia appear in a different place on the pK scale than do those of water. In water we could not distinguish the acidities of ethanol (CH_3CH_2OH) and methanol (CH_3OH) (Table 10-1) because both are weaker than water, and the amount of H_3O^+ produced by either of them is less than that produced by water alone. But both of these compounds are stronger acids than ammonia, so that in ammonia, each of them will produce concentrations of NH_4^+ [by reaction (a)] which are *greater* than that produced by ammonia itself. These concentrations can be measured by methods analogous to those used to determine $[H_3O^+]$ in water (though less accurately), and the relative acidities of methanol and ethanol can thus be determined (methanol is stronger). Of course, the equilibrium constants we obtain in NH_3 are not the same as the K_a values we would have obtained in water, since the equilibrium expressions are different:

In water: $$K_a = \frac{[A^-][H_3O^+]}{[HA]}$$

In ammonia: $$K_a^{NH_3} = \frac{[A^-][NH_4^+]}{[HA]}$$

However, methods are known for conversion of $K_a^{NH_3}$ values to the more familiar K_a scale, and we can say, for example, that ethanol would have $K_a \approx 1 \times 10^{-18}$ if we were able to measure it in water. By the use of ammonia as a solvent, we can distinguish acidities of acids (and basicities of their conjugate bases), which in water would have pK_a values of about 10 to 25 (Figure 10-8). Liquid ammonia is not suitable for acids stronger or weaker than this, because the two kinds of leveling effect come into play. The strongest acid that can exist in NH_3 solution is NH_4^+ (just as the strongest acid that can exist in H_2O is H_3O^+), and all acids stronger than NH_4^+ behave as strong acids in ammonia and are essentially completely converted to NH_4^+ and A^-. For example, HF and HOAc are strong acids in liquid ammonia, though they are weak acids in water. Similarly, the strongest base that can exist in ammonia solution is NH_2^-. Of course, we do not need ammonia for the study of acids with pK_a values from 10 to about 15, since for these acids we can obtain better

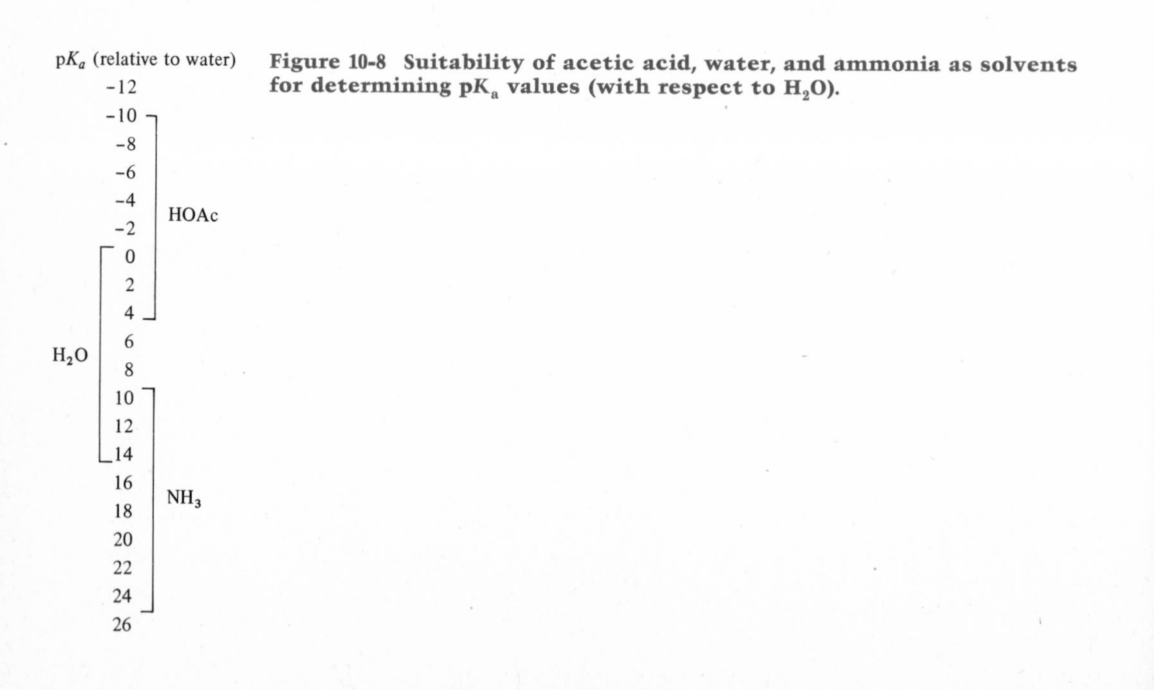

Figure 10-8 Suitability of acetic acid, water, and ammonia as solvents for determining pK_a values (with respect to H_2O).

and more convenient results in water, so that in practice, liquid ammonia is useful for acids that in water would have pK_a values of about 15 to 25.

Acetic Acid. This is an acidic solvent and the leveling effects are in the opposite direction from ammonia. In this solvent, such acids as NH_4^+, C_6H_5OH, and H_2S are feeble acids, but acidities of acids like HCl and HBr can be told apart, since they are weak acids in this solvent, and thus have measurable (and different) values of K_a^{HOAc}.

$$HA + HOAc \rightleftharpoons H_2OAc^+ + A^-$$

$$K_a^{HOAc} = \frac{[A^-][H_2OAc^+]}{[HA]}$$

Of course, we must convert K_a^{HOAc} values to K_a values if we wish to compare these values with K_a values for other acids determined in water solutions. Acetic acid can be used for acids which in water would have pK_a values between about −10 and +4, and for their conjugate bases (Figure 10-8).

Problems

10-1 Define each of the following terms. **(a)** Brønsted acid **(b)** Brønsted base **(c)** Arrhenius acid **(d)** Arrhenius base **(e)** diprotic acid **(f)** amphoteric **(g)** autoprotolysis **(h)** pH **(i)** indicator **(j)** strong acid **(k)** weak base **(l)** equivalence point **(m)** buffer **(n)** leveling effect **(o)** hydrolysis **(p)** titration **(q)** ion pair

10-2 What is the conjugate acid of each of the following? **(a)** NH_3 **(b)** NH_2^- **(c)** OAc^- **(d)** OH^- **(e)** H_2O **(f)** F^- **(g)** HF **(h)** PO_4^{3-} **(i)** $H_2PO_4^-$ **(j)** HPO_4^{2-} **(k)** Cl^- **(l)** CN^- **(m)** SO_4^{2-} **(n)** HSO_4^-

10-3 What is the conjugate base of each of the following? **(a)** HNO_3 **(b)** H_3PO_4 **(c)** $H_2PO_4^-$ **(d)** HPO_4^{2-} **(e)** NH_4^+ **(f)** NH_3 **(g)** HF **(h)** H_2S **(i)** HS^- **(j)** H_3BO_3 **(k)** H_2O **(l)** H_3O^+ **(m)** HOAc **(n)** H_2F^+ **(o)** H_2CO_3 **(p)** HCO_3^-

10-4 Indicate the two conjugate acid–base pairs in the following reactions.

Example: $\underset{base_1}{NH_3} + \underset{acid_2}{HCl} \rightleftharpoons \underset{acid_1}{NH_4^+} + \underset{base_2}{Cl^-}$

(a) $OAc^- + H_2O \rightleftharpoons HOAc + OH^-$
(b) $CO_3^{2-} + H_2O \rightleftharpoons HCO_3^- + OH^-$
(c) $HCO_3^- + OH^- \rightleftharpoons CO_3^{2-} + H_2O$
(d) $HS^- + H_3O^+ \rightleftharpoons H_2S + H_2O$
(e) $2H_2O \rightleftharpoons H_3O^+ + OH^-$
(f) $H_3O^+ + H_2PO_4^- \rightleftharpoons H_3PO_4 + H_2O$
(g) $Al(H_2O)_3(OH)_3 + OH^- \rightleftharpoons [Al(H_2O)_2(OH)_4]^- + H_2O$
(h) $H_2O + NH_3 \rightleftharpoons NH_4^+ + OH^-$
(i) $H_2O + HCl \rightleftharpoons H_3O^+ + Cl^-$
(j) $CN^- + HBr \rightleftharpoons HCN + Br^-$
(k) $NH_4^+ + NH_2^- \rightleftharpoons 2NH_3$

10-5 For each of the following, tell whether the equilibrium will lie to the right or to the left.
(a) $H_2O + HOAc \rightleftharpoons H_3O^+ + OAc^-$
(b) $C_6H_5OH + F^- \rightleftharpoons C_6H_5O^- + HF$
(c) $HNO_2 + NH_3 \rightleftharpoons NO_2^- + NH_4^+$
(d) $H_2S + OH^- \rightleftharpoons H_2O + HS^-$
(e) $HCl + Br^- \rightleftharpoons HBr + Cl^-$
(f) $H_2O + CH_3O^- \rightleftharpoons OH^- + CH_3OH$
(g) $NH_4^+ + OAc^- \rightleftharpoons NH_3 + HOAc$

10-6 In each case predict whether a reaction will take place. If a reaction does take place, write the formulas for the products. If a reaction does not take place, tell why.

(a) $HCl + NO_2^- \rightarrow$ **(b)** $HSO_4^- + NH_3 \rightarrow$
(c) $HCO_3^- + HS^- \rightarrow$ **(d)** $HCO_3^- + OH^- \rightarrow$
(e) $HCO_3^- + H_2O \rightarrow$ **(f)** $OAc^- + HCl \rightarrow$
(g) $OAc^- + F^- \rightarrow$ **(h)** $OAc^- + H_2S \rightarrow$
(i) $NH_4^+ + HNO_2 \rightarrow$ **(j)** $CH_3OH + NH_2^- \rightarrow$
(k) $CH_3^- + H_2O \rightarrow$

10-7 Write balanced equations to show how each of the following substances can react with an acid HA and with a base B^-. Indicate the conjugate acid–base pairs.

Example (for H_2O):

$$\underset{\text{base}_1}{H_2O} + \underset{\text{acid}_2}{HA} \rightleftharpoons \underset{\text{acid}_1}{H_3O^+} + \underset{\text{base}_2}{A^-}$$

$$\underset{\text{acid}_2}{H_2O} + \underset{\text{base}_1}{B^-} \rightleftharpoons \underset{\text{acid}_1}{HB} + \underset{\text{base}_2}{OH^-}$$

(a) HCO_3^- **(b)** $H_2PO_4^-$
(c) HS^- **(d)** H_2S
(e) HPO_4^{2-} **(f)** HSO_4^-

10-8 Arrange the following bases in increasing order of base strength: HSe^-, Br^-, GeH_3^-, H_2As^-.

10-9 Arrange the following acids in increasing order of acid strength: H_3As, H_4Ge, HBr, H_2Se.

10-10 Arrange the following acids in increasing order of acid strength: H_2Se, H_2O, H_2S, H_2Te.

10-11 Arrange the following acids in increasing order of acid strength: GeH_4, CH_4, SiH_4, SnH_4.

10-12 For each of the following pairs, tell which is the stronger acid.

(a) H_2S, H_2Te **(b)** HBr, HF
(c) HBr, H_2Se **(d)** H_2O, HCl
(e) HI, HBr **(f)** PH_3, SiH_4
(g) HF, CH_4 **(h)** H_2Te, HI
(i) CH_4, HCl **(j)** H_2CO_3, HCO_3^-
(k) $H_2BO_3^-$, H_3BO_3

10-13 For each of the following pairs, tell which is the stronger acid.

(a) H_2SO_4, H_2SO_3 **(b)** HNO_2, HNO_3
(c) $HClO_3$, HOCl **(d)** HOCl, $HClO_4$
(e) HOI, HIO_4 **(f)** $HClO_2$, $HClO_3$
(g) H_3AsO_4, H_2SeO_4 **(h)** H_3PO_4, H_3AlO_3
(i) H_2TeO_4, HIO_4 **(j)** HF, HCO_3^-
(k) C_6H_5OH, H_2S **(l)** NH_4^+, H_2O

10-14 For each of the following pairs, tell which is the stronger base.

(a) Br^-, F^- **(b)** PH_2^-, AsH_2^-
(c) OH^-, NH_2^- **(d)** HTe^-, I^-
(e) CO_3^{2-}, HCO_3^- **(f)** $H_2PO_4^-$, HPO_4^{2-}
(g) Cl^-, OH^- **(h)** KOH, $Ca(OH)_2$
(i) CsOH, $Ba(OH)_2$ **(j)** Br^-, HS^-
(k) ClO_3^-, ClO_2^- **(l)** HS^-, OAc^-
(m) F^-, NO_2^- **(n)** ClO_4^-, CO_3^{2-}

10-15 In each case, calculate the $[OH^-]$ at 25°C.

(a) $[H_3O^+] = 10^{-11}\,M$ **(b)** $[H_3O^+] = 10^{-7}\,M$
(c) $[H_3O^+] = 10^{-15}\,M$ **(d)** $[H_3O^+] = 3.5 \times 10^{-9}\,M$
(e) $[H_3O^+] = 8.7 \times 10^{-13}\,M$ **(f)** pOH = 12.4
(g) pH = 6.4

10-16 Calculate the pH and the pOH of each of the following solutions.
(a) $[H_3O^+] = 10^{-5}\ M$ **(b)** $[OH^-] = 10^{-10}\ M$
(c) $[H_3O^+] = 7 \times 10^{-7}\ M$ **(d)** $[OH^-] = 7 \times 10^{-7}\ M$
(e) $[H_3O^+] = 4 \times 10^{-4}\ M$ **(f)** $[H_3O^+] = 10\ M$
(g) $[OH^-] = 1 \times 10^{-14}\ M$ **(h)** $[H_3O^+] = 2.5 \times 10^{-11}\ M$
(i) $[OH^-] = 6.1 \times 10^{-3}\ M$ **(j)** $[H_3O^+] = 4.4 \times 10^{-5}\ M$
(k) $[OH^-] = 7.2 \times 10^{-12}\ M$

10-17 Calculate the $[H_3O^+]$ and $[OH^-]$ in solutions having the following.
(a) pH = 4.74 **(b)** pOH = 4.74
(c) 0.250 *M* HCl **(d)** pH = 0.1
(e) pH = −1.00 **(f)** pOH = 1.00
(g) pOH = 7.35 **(h)** pH = 14.00

10-18 Calculate the $[H_3O^+]$ and $[OH^-]$ in solutions of the following: **(a)** 25.00 g of NaOH dissolved in 400 ml of water. **(b)** pH = 9.25 **(c)** 50.0 ml of a 0.500 *M* HCl solution diluted to 175 ml **(d)** 0.75 *M* HCl **(e)** 0.69 *M* KCl **(f)** 3.0 *M* NaOH **(g)** 30 ml of 2.5 *M* KOH diluted to 120 ml **(h)** pOH = 3.81 **(i)** pH = 8.66 **(j)** $3.4 \times 10^{-2}\ M\ HNO_3$ **(k)** $0.27\ M\ H_2SO_4$ **(l)** $4.5 \times 10^{-4}\ M\ Ba(OH)_2$

10-19 The K_w for water at 25°C is 1.0×10^{-14}. At 37°C (98.6°F) K_w is 2.42×10^{-14}. What is the pH of this neutral solution at 37°C?

10-20 A solution turns yellow when a drop of the indicator methyl orange is added. What can you tell about the pH of the solution? Another sample of the same solution turns red when a drop of methyl red is added. What can you tell now?

10-21 Calculate the pH of the following solutions: **(a)** 0.200 *M* HCl **(b)** 0.200 *M* HOAc **(c)** 0.200 *M* HCN **(d)** 0.200 *M* HOCl

10-22 Chlorous acid, $HClO_2$, ionizes as follows:

$$HClO_2 + H_2O \rightleftharpoons H_3O^+ + ClO_2^-$$

with a K_a of 1.1×10^{-2}.
(a) Calculate the pK_a of this acid.
(b) Calculate the $[H_3O^+]$ and pH of a 0.500 *M* $HClO_2$ solution.
(c) Calculate the pOH of a 0.125 *M* solution of $HClO_2$.

10-23 Dichloroacetic acid ionizes as follows:

$$Cl_2CHCO_2H + H_2O \rightleftharpoons H_3O^+ + Cl_2CHCO_2^-$$

If the concentration of $Cl_2CHO_2^-$ in a 1.00 *M* Cl_2CHCO_2H solution is 0.199 *M*, calculate the K_a of Cl_2CHCO_2H and the pH of the solution.

10-24 Hydrazine, N_2H_4, is a weak base with a $pK_b = 5.77$. Calculate the pH of a 0.150 *M* hydrazine solution, and the degree of dissociation.

10-25 The pH of a 1.00 *M* solution of barbituric acid (a monoprotic acid) is 2.02. Calculate K_a and pK_a for this acid.

10-26 The pH of a $1.00 \times 10^{-2}\ M$ solution of saccharin (a monoprotic acid) is 6.76. Calculate K_a and pK_a for this acid.

10-27 The pH of a 0.00100 *M* solution of morphine (this base takes only 1 proton) is 8.40. Calculate K_b and pK_b for this base.

10-28 What is the molarity of a solution of benzoic acid whose pH is 3.0?

10-29 Given K_a for the following acids, write the formulas and calculate K_b for their conjugate bases.

(a) *ortho*-cresol, HC_7H_7O, $K_a = 6.3 \times 10^{-11}$
(b) hypobromous acid, HOBr, $K_a = 2.06 \times 10^{-9}$
(c) quinolinium ion, $HC_9H_7N^+$, $K_a = 1.25 \times 10^{-5}$

10-30 Given K_b for the following bases, write the formulas and calculate K_a for their conjugate acids.
(a) urea, CH_4N_2O, $K_b = 1.26 \times 10^{-14}$
(b) dimethylamine, C_2H_7N, $K_b = 5.40 \times 10^{-4}$
(c) furoate ion, $C_5H_3O_3^-$, $K_b = 1.48 \times 10^{-11}$

10-31 Assume that you have discovered two new acids, both of them stronger than water, but weaker than H_3O^+. Give two different ways in which you could tell which of the two is a stronger acid.

10-32 Acetone is a weaker base than water. Explain why it is impossible to measure K_b for acetone in water solution.

10-33 Phosphoric acid is a polyprotic acid and dissociates stepwise as follows:

$$H_2O + H_3PO_4 \rightleftharpoons H_3O^+ + H_2PO_4^- \qquad K_{a_1} = 7.5 \times 10^{-3}$$
$$H_2O + H_2PO_4^- \rightleftharpoons H_3O^+ + HPO_4^{2-} \qquad K_{a_2} = 6.2 \times 10^{-8}$$
$$H_2O + HPO_4^- \rightleftharpoons H_3O^+ + PO_4^{3-} \qquad K_{a_3} = 2.2 \times 10^{-13}$$

(a) Calculate pK_{a_1}, pK_{a_2}, and pK_{a_3}. **(b)** If we consider that the major source of H_3O^+ is from the first ionization, calculate the pH of a 0.200 *M* H_3PO_4 solution.

10-34 Sulfurous acid dissociates stepwise as follows:

$$H_2O + H_2SO_3 \rightleftharpoons H_3O^+ + HSO_3^- \qquad K_{a_1} = 1.5 \times 10^{-2}$$
$$H_2O + HSO_3^- \rightleftharpoons H_3O^+ + SO_3^{2-} \qquad K_{a_2} = 1.0 \times 10^{-7}$$

(a) If we start with a 0.0100 *M* solution of H_2SO_3, calculate the pH of the solution and the SO_3^{2-} concentration. **(b)** If the pH is fixed at 1.00, calculate the SO_3^{2-} concentration for a 0.100 *M* H_2SO_3 solution.

10-35 In each case, if the given salt was added to pure water (pH 7), would the water become acidic, basic, or remain neutral? **(a)** KBr **(b)** LiOAc **(c)** $Mg(ClO_4)_2$ **(d)** FeI_3 **(e)** Na_2CO_3 **(f)** $CaSO_4$ **(g)** $Ba(ClO_4)_2$ **(h)** $AlCl_3$ **(i)** $Sn(NO_3)_2$ **(j)** NH_4Cl **(k)** NH_4CN **(l)** NH_4OAc

10-36 Find the $[H_3O^+]$, the $[OH^-]$, the pH, the pOH, and the percent dissociation or hydrolysis of each of the following solutions. **(a)** 0.250 *M* HOCl **(b)** 0.125 *M* HCN **(c)** 0.123 *M* NaCN **(d)** 0.250 *M* NH_2NH_3Cl (hydrazinium chloride)

10-37 Calculate the pH of each of the following solutions. **(a)** 0.0200 *M* NaCl **(b)** 0.0200 *M* NaOAc **(c)** 0.0200 *M* NaF **(d)** 0.0200 *M* NaCN **(e)** 0.0200 *M* NH_4Cl **(f)** 0.0200 *M* $Ba(OH)_2$

10-38 Calculate the $[H_3O^+]$, $[OH^-]$, pH, pOH, and percent hydrolysis of each of the following solutions. **(a)** 0.125 *M* $(NH_4)_2SO_4$ **(b)** 0.100 *M* NaCl **(c)** 0.400 *M* $NaNO_2$ (the sodium salt of nitrous acid) **(d)** 0.250 *M* $NaCHO_2$ (sodium formate) **(e)** 0.750 *M* $C_5H_5NH^+ Br^-$ (pyridinium bromide) **(f)** 3.0×10^{-3} *M* $CH_3NH_3^+ NO_3^-$ (methylammonium nitrate)

10-39 When 37.50 ml of an acidic solution is titrated with 0.3255 *N* base, 25.55 ml of base are needed to reach the end point. What is the normality of the acid solution?

10-40 A 1.48-g sample of a pure unknown acid was dissolved in 25.00 ml of H_2O and the solution titrated with 0.527 *N* NaOH solution. The indicator changed color when 43.65 ml of the NaOH solution had been added. What is the equivalent weight of the unknown acid? Can we determine the molecular weight by this procedure alone?

10-41 A sample of 0.715 g of a pure unknown acid was dissolved in 30.00 ml of water and the solution titrated with 0.6154 *N* NaOH solution. One end point was reached after 22.34 ml of the NaOH solution had been added. The addition of 22.34 ml more of the standard solution gave a second end point. Calculate the molecular weight of the unknown acid, assuming that additional base no longer reacts.

10-42 Choose a suitable indicator from Table 10-2 for each of the following titrations. (Do not calculate end points; just estimate.) **(a)** H_2SO_4 vs. NaOH **(b)** Formic acid vs. NaOH **(c)** H_2S vs. NaOH **(d)** HCl vs. hydrazine **(e)** HNO_3 vs. pyridine **(f)** HCl vs. sodium acetate

10-43 Assume that you need 500 ml of a buffer solution of pH 5.5. Explain how you would prepare such a solution, using any acids or bases in Table 10-5 or 10-6.

10-44 If 50.00 ml of a 0.1000 *M* KOH solution is titrated with 0.1000 *M* HNO_3, calculate the pH of the solution after addition of each of the following volumes of HNO_3.
(a) 00.00 ml **(b)** 5.00 ml **(c)** 10.00 ml
(d) 25.00 ml **(e)** 45.00 ml **(f)** 49.00 ml
(g) 49.90 ml **(h)** 49.99 ml **(i)** 50.01 ml
(j) 51.00 ml **(k)** 75.00 ml

10-45 Calculate the pH of a solution which is 0.070 *M* in HCl and 0.090 *M* in NaCl.

10-46 Calculate the pH of a solution that is 0.030 *M* in formic acid and 0.060 *M* in sodium formate.

10-47 What molar ratio of HOAc to NaOAc does one need to produce buffer solutions of **(a)** pH 4.70 **(b)** pH 7 **(c)** pH 5.10?

10-48 A buffer solution is made by adding equal volumes of 0.040 *M* NaF and 0.020 *M* HF. **(a)** Calculate the pH of the solution. **(b)** If the solution is now made 0.010 *M* in HCl (no volume change), calculate the new pH.

10-49 How many grams of NaOAc ($C_2H_3O_2Na$) must be added to 0.150 mole of HOAc in order to prepare 0.500 liter of a buffer whose pH equals 5.25?

10-50 Blood is mainly buffered by the H_2CO_3–HCO_3^- system in which the ratio $[HCO_3^-]/[H_2CO_3]$ is 20:1.
(a) Calculate the pH of this system. (Note that blood has other buffering systems also.)
(b) What would the pH be if the ratio was 10:1? **(c)** If it was 1:10?

10-51 Calculate the pH that results when 0.150 mole of NaOH is added to: **(a)** 1.000 liter of a solution that is 1.000 *M* in both pyridine and pyridinium ion; and **(b)** 1.000 liter of a solution that is 0.100 *M* in both pyridine and pyridinium ion (K_b for pyridine is 1.8×10^{-9}). Assume that the total volume of each solution remains at 1.000 liter. **(c)** What was the pH of each solution before any NaOH was added?

10-52 Calculate the pH of solutions obtained by mixing each of the following.
(a) 11.6 ml of 0.25 *M* H_2SO_4 with 25.1 ml of 0.20 *M* KOH
(b) 35.0 ml of 0.065 *M* KOH with 25.0 ml of 0.30 *M* HOAc
(c) 525 ml of 0.0200 *M* HCl with 350 ml of 0.550 *M* hydrazine N_2H_4

10-53 Hydrazine, N_2H_4, is sometimes used as a nonaqueous solvent for acid–base reactions. What are the strongest acid and the strongest base that can exist in hydrazine? What are the strongest acid and the strongest base that can exist in acetic acid?

10-54 When sulfuric acid is the solvent, nitric acid behaves as a base. Write an equation for this reaction.

Additional Problems

10-55 If K_a for the acid phenol, HC_6H_5O, is 1.3×10^{-10}, what is K_h for the hydrolysis of sodium phenoxide, NaC_6H_5O?

10-56 When 21.55 ml of a solution of NaOH was titrated with a standard H_2SO_4 solution whose normality was 0.457 *N*, 32.51 ml of the acid solution was required to reach the end point. What was the normality and molarity of the NaOH solution?

10-57 In a 0.010 *M* NH_3 solution the NH_3 is a weak base that ionizes 4.2%. Calculate K_b and pK_b for NH_3.

10-58 You have a solution that is 0.200 *M* in NH_3 and also 0.100 *M* in NH_4Cl.
(a) Calculate the pH of this solution.
(b) To this solution you add 0.050 *M* NaOH (keeping the volume constant). Calculate the new pH.
(c) To the solution prepared in part (a) you add 0.050 *M* HCl (keeping the volume constant). Calculate the new pH.

10-59 A solution is made by combining 0.50 mole of HOAc and 0.30 mole of NaOH made up to a liter of solution.
(a) What would the pH of the solution be if there had been only HOAc initially?
(b) Find the pH of the buffer solution.
(c) What would the pH be if 0.10 mole of NaOH is added?
(d) What would the pH be if 0.10 mole of HCl is added to the original solution?

10-60 The pH of a 0.100 *M* solution of arsenious acid (a monoprotic acid) is 5.11. Calculate K_a and pK_a for this acid.

10-61 A 1.728-g sample of the base nicotine (isolated from tobacco) was dissolved in 250 ml of water and titrated with a standard acid solution of normality 0.284 *N*. The end point was reached when 37.52 ml of acid solution had been added. What is the equivalent weight of nicotine?

10-62 If possible, write the formula for each of the following. **(a)** an Arrhenius acid that is not a Brønsted acid **(b)** a Brønsted acid that is not an Arrhenius acid **(c)** an Arrhenius base that is not a Brønsted base **(d)** a Brønsted base that is not an Arrhenius base

10-63 Calculate the S^{2-} concentration in a saturated solution of H_2S (= 0.100 *M*) whose pH is fixed at 2.85.

10-64 Is it possible to prepare a buffer solution from a solution of Na_3PO_4 and a solution of HCl? If so, tell how you would do it and why it would be a buffer.

10-65 What is the concentration of H_3O^+ in pure water at 40°C?

10-66 Why is it not possible to tell whether HI is stronger or weaker than H_2SO_4 by adding them to separate samples of water and measuring the resulting pH values? How could we tell?

10-67 Anisic acid, $HC_8H_7O_2$, has a pK_a of 4.47. Calculate the pH and degree of dissociation of: **(a)** 0.150 *M* $HC_8H_7O_2$ solution **(b)** 0.0001 *M* $HC_8H_7O_2$ solution

10-68 Calculate the pH and the pOH of a 0.350 *M* solution of hydrazium chloride, NH_2NH_3Cl. The K_b for hydrazine = 1.7×10^{-6}.

10-69 The pH of a 0.100 *M* solution of aniline (this base takes only one proton) is 8.82. Calculate K_b and pK_b for this base.

10-70 How many moles of HCl must be added to 2.0 liters of a mixture containing 0.0100 *M* HOAc and 0.0200 *M* NaOAc in order to produce a pH of 3.80?

10-71 Normal urine has a pH of approximately 6.2. If a patient eliminates 1500 ml of urine per day, how many moles of H_3O^+ have been eliminated?

10-72 What is the pH of a solution obtained by making the concentration of HCl in pure water equal to 10^{-8} *M*? Does the pH = 8? Can we make water basic by adding HCl?

In this chapter, we shall continue our discussion of the behavior of ions in aqueous solution. In particular, we will look at those aspects having to do with solubility.

11-1 Reactions Between Ions in Solution. Solubility of Salts in Water

In Section 3-2 we saw that reactions between ions in aqueous solution only take place when (a) an undissociated soluble species, (b) an insoluble compound, or (c) a gas is formed, or when (d) charges on the ions change. We would like to be able to predict when such reactions will, in fact, take place, and to do this it is necessary to know which ions will behave in one of these four ways. We will now discuss three of these ways. Reactions in which charges on ions change are called oxidation–reduction reactions. It is inconvenient to discuss them at this point, and such discussion will be postponed to Chapter 16.

A. FORMATION OF AN UNDISSOCIATED SOLUBLE SPECIES. The chief example here is formation of water, which, as we saw in Chapter 10, is what happens every time an acid is neutralized by hydroxide ion or a base neutralized by hydronium ion in water solution.

$$H_3O^+ + OH^- \rightleftharpoons 2H_2O$$

or

$$HA + OH^- \rightleftharpoons H_2O + A^-$$

or

$$H_3O^+ + B \rightleftharpoons BH^+ + H_2O$$

These equilibria lie far to the right because the ionization constant of water, K_w, is 10^{-14} at 25°C. Apart from water, the only important water-soluble undissociated species that are formed by the reactions of ions in aqueous solution are certain ions called *complex ions*, examples being $Ag(NH_3)_2^+$, $Cd(CN)_4^{2-}$, and $Zn(OH)_4^{2-}$. These will be discussed in Chapter 19. Practically all simple salts can be regarded as 100% dissociated when dissolved in water. There are only a few exceptions, two notable ones being mercuric chloride, $HgCl_2$, and lead acetate, $Pb(OAc)_2$. These two salts are only partially dissociated into ions when dissolved in water.

B. FORMATION OF A GAS. When a reaction between ions in aqueous solution gives rise to a gas as one of the products, the gas is lost from the solution. This is a case of an equilibrium that is shifted by removal of a product, thus driving the reaction to completion. We can mention two important examples: formation of the gases carbon dioxide and sulfur dioxide by the reaction of hydronium ions with, respectively, bicarbonate ion, HCO_3^-, and bisulfite ion, HSO_3^-.

$$\underset{\text{bicarbonate}}{HCO_3^-} + H_3O^+ \rightleftharpoons CO_2\uparrow + 2H_2O$$

$$\underset{\text{carbonate}}{CO_3^{2-}} + 2H_3O^+ \rightleftharpoons CO_2\uparrow + 3H_2O$$

$$\underset{\text{bisulfite}}{HSO_3^-} + H_3O^+ \rightleftharpoons SO_2\uparrow + 2H_2O$$

$$\underset{\text{sulfite}}{SO_3^{2-}} + 2H_3O^+ \rightleftharpoons SO_2\uparrow + 3H_2O$$

The symbol ↑ means a gas is formed.

Since carbonate ions (CO_3^{2-}) react with H_3O^+ to give HCO_3^- ions, it is clear that CO_3^{2-} ions will also give CO_2, if enough H_3O^+ is present. Similarly, sulfite ions (SO_3^{2-}) will give SO_2 if treated with enough H_3O^+.

An important consequence of these reactions is that *it is impossible to have a significant concentration of CO_3^{2-}, HCO_3^-, SO_3^{2-}, or HSO_3^- ions present in an acidic solution.* Under these conditions, these ions are converted to the gases CO_2 and SO_2, which are lost from the solution.

Baking powder is an important material, used for household and industrial baking, whose effectiveness results from one of the preceding reactions. It is formation of the gas CO_2 that causes bread and cakes to rise during the baking process. The traditional method of accomplishing this is the use of yeast, which ferments the sugars present, forming CO_2 in the process. Yeast is still used today for the baking of bread, but for cakes the yeast is usually replaced by baking powder, which consists of sodium bicarbonate ($NaHCO_3$, also called *baking soda*) and an acidic compound such as monocalcium phosphate or sodium aluminum sulfate. The acidic compound reacts with the HCO_3^- ions in the presence of water to give CO_2. The earliest acidic compound used for this purpose was potassium acid tartrate (cream of tartar), and it is still one of the best, but it has been little used in recent years because of its high cost.

C. FORMATION OF AN INSOLUBLE COMPOUND. When two ions in water solution can react to form an insoluble salt, the result is the same as in the previous case: The reacion goes almost to completion because the equilibrium is driven by removal from the solution of the insoluble salt, which precipitates out and falls to the bottom of the container. For example, sodium sulfate and barium nitrate are each soluble in water. If we mix aqueous solutions of these two salts, we get the reaction

$$Ba^{2+} + SO_4^{2-} \rightleftharpoons BaSO_4\downarrow$$

This reaction goes almost to completion because barium sulfate, $BaSO_4$, is insoluble in water and precipitates out. The Na^+ and NO_3^- ions are spectator ions in this case.

It is clear that in any given case we can only predict whether or not a reaction of this kind will, in fact, take place if we know which salts are soluble in water and which are not. [At this point it is necessary to mention that our use of the words "soluble" and "insoluble" in this context is not very precise. No salt is *completely* insoluble in water, and no salt is *totally* soluble in water in all concentrations. When we say that a salt is insoluble, we really mean that its solubility is extremely low, and when we say that another salt is soluble, we mean that a convenient amount of it will dissolve in a given quantity of water. Of course, the solubility of each salt can be quantitatively determined, and each has a different value, which changes with changes in temperature. Some quantitative aspects of the solubility of salts in water will be discussed in the next section.] There are so many possible salts that few people if any take the trouble to remember the relative solubilities of all of them. If needed, they can be looked up in tables such as those found in the *Handbook of Chemistry and Physics* (CRC Press, Cleveland, Ohio). However, there are a few generalizations that are not difficult to remember.

1. Virtually all sodium, potassium, and ammonium salts are soluble in water, no matter which anion is present.
2. Virtually all nitrates and acetates are soluble in water, no matter which cation is present.
3. Most chlorides are soluble in water. Significant exceptions are silver chloride ($AgCl$), mercurous chloride (Hg_2Cl_2), and lead chloride ($PbCl_2$) (the last is soluble in hot water).
4. Most sulfates are soluble in water. Significant exceptions are the alkaline earth sulfates: barium, calcium, and strontium sulfates ($BaSO_4$, $CaSO_4$, and $SrSO_4$); silver, mercurous, and lead sulfates (Ag_2SO_4, Hg_2SO_4, and $PbSO_4$); and ferric sulfate ($Fe_2(SO_4)_3$).
5. Most sulfides are *insoluble* in water. Significant exceptions are sodium, potassium, and ammonium sulfides (Na_2S, K_2S, and $(NH_4)_2S$), and the alkaline earth sulfides: BaS, CaS, and SrS. Aluminum sulfide decomposes when dissolved in water because hydrolysis of the two ions is practically complete when both are present:

$$Al^{3+} + 6H_2O \rightleftharpoons Al(OH)_3 + 3H_3O^+$$

$$S^{2-} + 2H_2O \rightleftharpoons H_2S + 2OH^-$$

$$OH^- + H_3O^+ \rightleftharpoons 2H_2O$$

A list of some insoluble salts and hydroxides is given in Table 11-1.

Table 11-1 Some Solubility Products in Water

Compound	*Temperature (°C)*	*Solubility product*	*Compound*	*Temperature (°C)*	*Solubility product*
$AgBr$	25	7.7×10^{-13}	Hg_2Cl_2	25	2×10^{-18}
Ag_2CO_3	25	6.2×10^{-12}	HgS	—	1×10^{-52}
$AgCl$	25	1.56×10^{-10}	$MgCO_3$	12	2.6×10^{-5}
Ag_2CrO_4	25	9×10^{-12}	$Mg(OH)_2$	18	1.2×10^{-11}
AgI	25	1.5×10^{-16}	MgC_2O_4	18	8.6×10^{-5}
Ag_2S	18	1.6×10^{-49}	$Mn(OH)_2$	18	4×10^{-14}
$Al(OH)_3$	—	2×10^{-33}	MnS	18	1.4×10^{-15}
$BaCO_3$	25	8.1×10^{-9}	NiS	18	1.4×10^{-24}
$BaCrO_4$	18	1.6×10^{-10}	$PbCO_3$	18	3.3×10^{-14}
BaF_2	25	1.7×10^{-6}	$PbCl_2$	25	1.7×10^{-5}
$BaSO_4$	25	1.08×10^{-10}	$PbCrO_4$	18	1.8×10^{-14}
$CaCO_3$	25	8.7×10^{-9}	PbF_2	27	3.7×10^{-8}
CaF_2	18	3.4×10^{-11}	PbI_2	25	1.4×10^{-8}
$CaSO_4$	10	2.0×10^{-4}	$PbSO_4$	18	1.06×10^{-8}
CdS	18	3.6×10^{-29}	PbS	18	3.4×10^{-28}
CoS	25	4×10^{-21}	SnS	25	1×10^{-26}
CuS	18	8.5×10^{-45}	SrC_2O_4	18	5.6×10^{-8}
$Fe(OH)_2$	18	1.6×10^{-14}	$SrSO_4$	17	3.8×10^{-7}
$Fe(OH)_3$	18	1.1×10^{-36}	$Zn(OH)_2$	—	1×10^{-17}
FeS	18	3.7×10^{-19}	ZnS	18	1.2×10^{-23}

11-2 Solubility Product

We have pointed out that salts generally regarded as insoluble in water are not *completely* insoluble; they have a very small solubility. Imagine a beaker of water into which a fair amount of the insoluble salt Ag_2CO_3 is added. A very small amount of the salt dissolves, and because salts are composed of ions, this very small amount is 100% ionized. The situation, however, is not static. There are always some of the dissolved ions resolidifying, while an equal quantity of fresh solid is dissolving. Thus, we have a true equilibrium, in which solidifying and dissolving take place at equal rates, and the concentrations of the dissolved Ag^+ and CO_3^{2-} ions do not change with time.

$$Ag_2CO_3(s) \rightleftharpoons 2Ag^+ + CO_3^{2-}$$

Since this is an equilibrium, we may write an equilibrium-constant expression for it:

$$K = \frac{[Ag^+]^2[CO_3^{2-}]}{[Ag_2CO_3]}$$

The concentration of Ag_2CO_3 is constant at a given temperature because Ag_2CO_3 is a solid and is not dissolved in the water. The concentration of any solid is equal to its density (note that both density and concentration can be expressed in moles/liter), and the density of a solid is of course constant at a given temperature. We can therefore combine $[Ag_2CO_3]$ with K to obtain the modified equilibrium expression (see Section 10-3).

$$K[Ag_2CO_3] = K_{sp} = [Ag^+]^2[CO_3^{2-}]$$

$\mathbf{K_{sp}}$ is called a **solubility product.** For Ag_2CO_3 it has the value 6.2×10^{-12} at 25°C. Solubility products for a number of other insoluble salts and hydroxides are given in Table 11-1. The solubility-product concept is only valid for insoluble salts and hydroxides. *It cannot be applied to soluble salts.*

It is important to keep clear the difference between the *solubility* and the *solubility product.* The solubility of a salt in water depends on other substances that may be present in the water. The solubility product is an equilibrium constant, and as such is independent of the presence of any other substance. Like any equilibrium constant, K_{sp} can only change with temperature.

For any given insoluble salt or hydroxide, the solubility product may easily be calculated from a knowledge of the solubility, and vice versa. First we shall give examples of these calculations and then some uses of K_{sp}.

Example: The solubility of calcium iodate $Ca(IO_3)_2$ in water at 18°C is 2.1 g/liter. Calculate K_{sp}.

Answer: The solubility is given in g/liter. K_{sp} is an equilibrium constant, and all concentrations must be expressed in moles/liter. The first step is convert g/liter to molarity. The molecular weight of $Ca(IO_3)_2$ is 389.89.

$$\frac{2.1 \text{ g/liter}}{389.89 \text{ g/mole}} = 5.4 \times 10^{-3} \text{ mole/liter}$$

The equation is

$$Ca(IO_3)_2(s) \rightleftharpoons Ca^{2+} + 2IO_3^-$$

We have calculated that 5.4×10^{-3} mole of $Ca(IO_3)_2$ is dissolved in each liter of solution. Since the salt is 100% ionized, *each mole* of dissolved $Ca(IO_3)_2$ gives rise to *one mole* of Ca^{2+} ions and *two moles* of IO_3^- ions. Thus, the concentration of Ca^{2+} ions at equilibrium is 5.4×10^{-3} mole/liter, while the concentration of IO_3^- is twice this amount:

$$Ca(IO_3)_2(s) \rightleftharpoons Ca^{2+} + 2IO_3^-$$

Concentrations at equilibrium (moles/liter) 5.4×10^{-3} $2(5.4 \times 10^{-3})$

We may substitute these values into the solubility product expression

$$K_{sp} = [Ca^{2+}][IO_3^-]^2$$
$$= (5.4 \times 10^{-3})(10.8 \times 10^{-3})^2 = 6.3 \times 10^{-7}$$

K_{sp} for $Ca(IO_3)_2$ at 18°C is 6.3×10^{-7}.

Example: K_{sp} for silver sulfide, Ag_2S, at 18°C is 1.6×10^{-49}. Calculate the solubility of Ag_2S in water at 18°C in g/liter.

Answer:

$$Ag_2S(s) \rightleftharpoons 2Ag^+ + S^{2-}$$

When Ag_2S ionizes in water each mole of dissolved Ag_2S produces one mole of S^{2-} ions and two moles of Ag^+ ions. The concentration of Ag^+ ions in any solution made up by dissolving Ag_2S in water (where no Ag^+ or S^{2-} ions are present from any other source) must be exactly twice the concentration of S^{2-} ions. If we let x be the concentration of S^{2-} ions, then the concentration of Ag^+ ions will be 2x. Inserting these values into the solubility-product expression, we have

$$K_{sp} = [Ag^+]^2[S^{2-}]$$

Equilibrium concentrations (moles/liter) $(2x)^2$ (x)

$$1.6 \times 10^{-49} = (2x)^2x$$
$$4x^3 = 1.6 \times 10^{-49}$$
$$x^3 = 4.0 \times 10^{-50}$$

To solve this, we must find the cube root of 4.0×10^{-50}, which is done more easily by converting it to 40×10^{-51}.

$$x^3 = 40 \times 10^{-51}$$
$$x = 3.4 \times 10^{-17} \text{ mole/liter}$$

The solubility in water of Ag_2S at 18°C is 3.4×10^{-17} mole/liter. We convert moles/liter to g/liter by multiplying by the molecular weight of Ag_2S (247.80 g/mole).

$$3.4 \times 10^{-17} \text{ mole/liter} \times 247.80 \text{ g/mole} = 8.4 \times 10^{-15} \text{ g/liter}$$

This is a very small solubility, indeed.

In writing solubility products we must be careful to insert the proper exponents, which are the same as the subscripts in the formula. For example, the solubility products for BaF_2 and Bi_2S_3 are

$$BaF_2 \rightleftharpoons Ba^{2+} + 2F^- \qquad K_{sp} = [Ba^{2+}][F^-]^2$$
$$Bi_2S_3 \rightleftharpoons 2Bi^{3+} + 3S^{2-} \qquad K_{sp} = [Bi^{3+}]^2[S^{2-}]^3$$

The chief use of solubility product is that it allows us to determine whether or not a precipitate of an insoluble salt or hydroxide will form when we mix known concentrations of ions, and also allows us to determine the concentrations of ions remaining in the solution. For example, we can answer such questions as: If 500 ml of an 0.0050 M solution of $AgNO_3$ is added to 500 ml of an 0.0010 M solution of KCl, will a precipitate of AgCl form? And if so, what will be the final concentrations of Ag^+ and Cl^- still dissolved in the water? What we do first is to calculate a number called the **ion product** of AgCl, and then compare this with the solubility product. The ion product for AgCl (for which we may use the symbol Q) is calculated by inserting the concentration values into the equation

$$Q = [Ag^+][Cl^-]$$

Note that this is *not* an equilibrium expression, although it has a similar appearance. In this case the concentration of Ag^+ is 0.0025 M (it is half of 0.0050 M because we diluted the solution in half when we mixed 500 ml each of the two solutions), while the concentration of Cl^- is 0.00050 M:

$$Q = (2.5 \times 10^{-3})(5.0 \times 10^{-4})$$
$$= 1.3 \times 10^{-6}$$

The rule is:

1. If Q is larger than K_{sp}, a precipitate will form.
2. If Q is smaller than K_{sp}, a precipitate will not form and the solution will be unsaturated in that salt.
3. If Q is equal to K_{sp}, a precipitate will not form and the solution will be saturated in that salt.

In this case, Q is 1.3×10^{-6} and hence larger than K_{sp}, which for AgCl is 1.56×10^{-10}. Thus, a precipitate will form.

Let us now turn to the second question: What are the final concentrations of Ag^+ and Cl^- still present in the solution? In this case there was a large excess of Ag^+: we added 0.0025 mole of Ag^+ to 0.00050 mole of Cl^-. Therefore, if virtually all the Cl^- reacted, we would have $0.0025 - 0.00050 = 0.0020$ mole of Ag^+ remaining in our liter of solution. It is a good assumption that this in fact is the case, and that we can assume that the final concentration of Ag^+ is 0.0020 M. Since an equilibium expression is valid no matter what concentrations are present, we can obtain the $[Cl^-]$ by substituting into the K_{sp} formula:

$$K_{sp} = [Ag^+][Cl^-]$$
$$1.56 \times 10^{-10} = (2.0 \times 10^{-3})[Cl^-]$$
$$[Cl^-] = \frac{1.56 \times 10^{-10}}{2.0 \times 10^{-3}} = 7.8 \times 10^{-8} \text{ mole/liter}$$

The final concentration of Cl^- is 7.8×10^{-8} mole/liter, which justifies our assumption. Of the original 5.0×10^{-3} mole/liter of Cl^-, all but 7.8×10^{-8} mole/liter has precipitated with the Ag^+ to form the solid AgCl.

In writing the ion product, we use the same exponents as in the corresponding solubility product. For example, the ion product for BaF_2 is

$$Q = [Ba^{2+}][F^-]^2$$

The solubility-product concept has other uses as well. For example, we can use K_{sp} to tell us (1) the solubility of an insoluble salt or hydroxide in water that already contains one of the ions and (2) what concentration of a negative ion must be added to a mixture of two positive ions in order to precipitate one without precipitating the other (or vice versa). Examples of these uses follow.

Example: What is the solubility of $Ca(IO_3)_2$ in an 0.10 *M* solution of $Ca(NO_3)_2$? [That is, if we add solid $Ca(IO_3)_2$ to the solution in question, how many moles of $Ca(IO_3)_2$ will dissolve in each liter of solution?] K_{sp} for $Ca(IO_3)_2$ is 6.3×10^{-7} (from page 334).

Answer: The equilibrium expression

$$K_{sp} = [Ca^{2+}][IO_3^-]^2$$

must be fulfilled, even though the Ca^{2+} can now come from two sources: from the $Ca(NO_3)_2$ as well as the $Ca(IO_3)_2$. The $Ca(NO_3)_2$ produces concentrations of 0.10 *M* Ca^{2+} and 0.20 *M* NO_3^-. The $Ca(IO_3)_2$ only dissolves to the extent necessary to satisfy the solubility-product expression:

$$K_{sp} = 6.3 \times 10^{-7} = [Ca^{2+}][IO_3^-]^2$$

	$[Ca^{2+}]$	$[IO_3^-]$	
	0.10	0	from $Ca(NO_3)_2$
	x	2x	from $Ca(IO_3)_2$

$$6.3 \times 10^{-7} = (0.10 \times x)(2x)^2$$

At this point we assume that the concentration of Ca^{2+} ions produced in solution from the $Ca(IO_3)_2$ is much smaller than the 0.10 *M* derived from the $Ca(NO_3)_2$. If that is true (and we will check the assumption at the end), we can put $0.10 + x \approx 0.10$. Therefore,

$$[IO_3^-]^2 = (2x)^2 = \frac{6.3 \times 10^{-7}}{0.10} = 6.3 \times 10^{-6}$$

$$[IO_3^-] = 2x = 2.5 \times 10^{-3} \text{ mole/liter}$$

The concentration of $[IO_3^-]$ in the final solution is 2.5×10^{-3} mole/liter, so that the solubility of $Ca(IO_3)_2$ in an 0.10 *M* solution of $Ca(NO_3)_2$ is 1.3×10^{-3} mole/liter.

Checking our assumption, we see that 0.10 — 0.0013 is indeed very close to 0.10 (it does not change the two significant figures at all). The error amounts to 1.39%. If the assumption could not be made, we would have to solve a quadratic equation.

Note that the solubility of $Ca(IO_3)_2$ in the above solution is smaller than the solubility of $Ca(IO_3)_2$ in pure water, which we saw on page 333 to be 5.4×10^{-3} mole/liter. This is what we would expect from Le Châtelier's principle, since the presence of the excess Ca^{2+} must drive the equilibrium to the left, so that less $Ca(IO_3)_2$ can dissolve.

$$Ca(IO_3)_2(s) \rightleftharpoons Ca^{2+} + 2IO_3^-$$

This is an example of the **common-ion effect,** which may be defined as the driving of an equilibrium involving ions to the right or to the left by the addition of an additional quantity of one of the ions. It is a general rule that any salt is less soluble in a solution containing a common ion than it is in pure water.

Example: An aqueous solution contains 0.10 *M* Cd^{2+} and 0.10 *M* Ni^{2+}. What range of concentrations of S^{2-} will precipitate one of these ions and leave the other in solution? K_{sp} for $CdS = 3.6 \times 10^{-29}$; K_{sp} for $NiS = 1.4 \times 10^{-24}$.

Answer: We know that when the ion product equals the solubility product a saturated solution results. Consequently, the maximum concentration of S^{2-} that can be present in a solution containing 0.10 *M* Cd^{2+} *without precipitating CdS* may be found by solving the solubility-product expression for $[S^{2-}]$:

$$K_{sp} = [Cd^{2+}][S^{2-}]$$
$$3.6 \times 10^{-29} = (0.10)[S^{2-}]$$
$$[S^{2-}] = 3.6 \times 10^{-28} \text{ mole/liter}$$

Any concentration of S^{2-} above this value will cause CdS to precipitate because the ion product would then be greater than the solubility product.

Making the same calculation for NiS, we get

$$K_{sp} = [Ni^{2+}][S^{2-}]$$
$$1.4 \times 10^{-24} = (0.10)[S^{2-}]$$
$$[S^{2-}] = 1.4 \times 10^{-23} \text{ mole/liter}$$

The maximum concentration of S^{2-} that can be present without precipitating NiS is 1.4×10^{-23} mole/liter. It follows that any concentration of S^{2-} greater than 3.6×10^{-28} mole/liter but less than 1.4×10^{-23} mole/liter will precipitate CdS, but not NiS. For maximum efficiency, it would be better to make the concentration of S^{2-} closer to 10^{-23} than to 10^{-28}, since the amount of CdS that precipitates increases with increasing concentration of S^{2-}, thus achieving a greater degree of separation of the Cd^{2+} ions from the Ni^{2+} ions. A S^{2-} concentration of 1.4×10^{-23} mole/liter will precipitate the maximum amount of Cd^{2+} that can be precipitated without precipitating any of the Ni^{2+}.

11-3 The Effect of One Equilibrium on Another. Simultaneous Equilibria

It often happens that in a given solution some of the ions present can take part in two or more reversible reactions. In such cases there are two or more equilibria going on at the same time. Since the equilibrium laws are equally valid for all equilibria, **all the corresponding equilibrium expressions must be simultaneously satisfied.** We have already seen at least two examples: (1) multiple equilibria in the dissociation of polyprotic acids (Section 10-9), and (2) hydrolysis of a salt of a weak acid and a weak base (Section 10-10). Another phenomenon often regarded as an example of the same principle is the common-ion effect. The principle can often be turned to practical purposes, by using one equilibrium to drive another. We will discuss three examples.

1. In Section 11-2, we calculated that it is possible to keep Ni^{2+} ions completely in solution while precipitating the maximum amount of Cd^{2+} ions, if we make the S^{2-} ion concentration equal to 1.4×10^{-23} mole/liter. But how can we achieve this concentration of S^{2-} ions? One conceivable way we might try to do it would be by adding a salt such as Na_2S directly to the solution. In practice, this would be extremely difficult, because 1.4×10^{-23} mole/liter is such a low concentration—the solution has to contain only about eight S^{2-} ions in a whole liter! We would either have to weigh out 1.1×10^{-22} g of Na_2S for a 1-liter solution (and there are no balances sensitive enough to detect such a small quantity) or else make a more concentrated solution and then dilute it, but the amount of dilution necessary would be so great that we could no longer say just how much, if any, S^{2-} ions were still present.

A solution to the difficulty is found when we remember (Section 10-9) that an aqueous solution of H_2S contains a very low concentration of S^{2-} ions. The concentration of S^{2-} ions can be controlled by controlling the pH. We can make the concentration of S^{2-} ions any value we like (within certain limits) by adding a known concentration of H_2S and then adjusting the pH (by the addition of HCl or NaOH) to whatever value we calculate. In this case, we have the following equilibria:

$$CdS \rightleftharpoons Cd^{2+} + S^{2-} \qquad K_{sp} = [Cd^{2+}][S^{2-}] = 3.6 \times 10^{-29} \quad (1)$$

$$NiS \rightleftharpoons Ni^{2+} + S^{2-} \qquad K_{sp} = [Ni^{2+}][S^{2-}] = 1.4 \times 10^{-24} \quad (2)$$

$$H_2S + H_2O \rightleftharpoons H_3O^+ + HS^- \qquad K_{a_1} = \frac{[H_3O^+][HS^-]}{[H_2S]} = 9.1 \times 10^{-8} \quad (3)$$

$$HS^- + H_2O \rightleftharpoons H_3O^+ + S^{2-} \qquad K_{a_2} = \frac{[H_3O^+][S^{2-}]}{[HS^-]} = 1.0 \times 10^{-14} \quad (4)$$

all of which must be simultaneously satisfied. The method we use to control them is the adjustment of the pH. In the absence of added HCl or NaOH, the pH of an 0.10 *M* H_2S solution is about 4 and the S^{2-} concentration about 1.0×10^{-14} *M* (page 304). When we lower the pH (by adding HCl), we drive equation (4) to the *left*, reducing the concentration of S^{2-}. This, in turn, causes equations (1) and (2) to move to the *right*, because we are reducing the concentration of S^{2-}, which is a product of the reaction. Conversely, we can increase the concentration of S^{2-} by adding NaOH. An S^{2-} ion concentation of 1.0×10^{-14} *M* is obviously much too high for the present purpose (we need an S^{2-} concentration of 1.4×10^{-23} *M*); and we will therefore have to add HCl and not NaOH. But how much?

Example: In order to achieve as complete a separation as possible of 0.10 *M* Cd^{2+} ion from 0.10 *M* Ni^{2+} ion, we begin by adding enough H_2S to make the solution 0.10 *M* in H_2S (this is approximately a saturated solution of H_2S in water). To what value must we then adjust the $[H_3O^+]$ in order to make the $[S^{2-}]$ equal to 1.4×10^{-23} *M*?

Answer: We can obtain the value of $[H_3O^+]$ by using the equilibrium expressions (3) and (4). Since we do not know the value of $[HS^-]$, it is best to combine (3) and (4) as follows, so that $[HS^-]$ cancels out:

$$K_a = K_{a_1}K_{a_2} = \frac{[H_3O^+][HS^-]}{[H_2S]} \times \frac{[H_3O^+][S^{2-}]}{[HS^-]}$$

$$K_a = (9.1 \times 10^{-8})(1.0 \times 10^{-14}) = \frac{[H_3O^+]^2[S^{2-}]}{[H_2S]}$$

$$9.1 \times 10^{-22} = \frac{[H_3O^+]^2[S^{2-}]}{[H_2S]} \quad (5)$$

Equation (5) is valid if we know two of the concentrations and wish to solve for the third. It would not be valid, for example, if we wished to calculate both $[H_3O^+]$ and $[S^{2-}]$ in a solution of H_2S in water (note that we calculated this in Section 10-9). In this case, we can assume that $[H_2S]$ is 0.10 *M*, since very little of it dissociates. We can then find out what $[H_3O^+]$ would have to be in order for $[S^{2-}]$ to be 1.4×10^{-23} *M* by substituting in the equation

$$\frac{[H_3O^+]^2(1.4 \times 10^{-23})}{0.10} = 9.1 \times 10^{-22}$$

$$[H_3O^+]^2 = \frac{9.1 \times 10^{-23}}{1.4 \times 10^{-23}} = 6.5$$

$$[H_3O] = \sqrt{6.5} = 2.5 \text{ moles/liter}$$

An H_3O^+ concentration of 2.5 *M* will ensure that the maximum amount of CdS precipitates without precipitating any of the Ni^{2+}. This corresponds to a pH of −0.4, which is a very strongly acidic solution.

2. Another practical application of the principle involves the use of a strong acid to dissolve a salt of a weak acid. For example, $BaSO_3$ is insoluble in water, which

means, of course, that a very small amount of it dissolves and ionizes 100%. One of the ions is SO_3^{2-}. When a strong acid is added to the mixture, the hydronium ions react with SO_3^{2-} to produce the weak acid HSO_3^-. Removal of SO_3^{2-} in turn causes the $BaSO_3$ equilibrium to move to the right, and the $BaSO_3$ dissolves.

$$BaSO_3 \rightleftharpoons Ba^{2+} + SO_3^{2-}$$
$$SO_3^{2-} + H_3O^+ \rightleftharpoons HSO_3^- + H_2O$$

In general, we may say that many salts of weak acids can be dissolved by the addition of a strong acid. This fact is used in analyzing an unknown solution for negative ions, to tell the difference between SO_4^{2-} and SO_3^{2-} ions. Both SO_4^{2-} and SO_3^{2-} ions will form white precipitates when treated with Ba^{2+} ions, but only $BaSO_3$, and not $BaSO_4$, will dissolve if a strong acid such as HCl is added. $BaSO_4$ will not dissolve because SO_4^{2-} is the conjugate base of a relatively strong acid (HSO_4^-), which reacts with hydronium ions to a much smaller extent.

There are salts of weak acids that cannot be dissolved by strong acids, simply because a high-enough concentration of H_3O^+ cannot be achieved.

Example: A liter of water contains 0.10 mole of solid FeS. Another contains 0.10 mole of solid HgS. What concentration of H_3O^+ is necessary to dissolve completely the precipitate in each case? Assume the total volumes remain at 1 liter. K_{sp} for FeS is 3.7×10^{-19}; K_{sp} for HgS is 1×10^{-52}.

Answer: If the precipitates dissolve completely, it will be because S^{2-} ions are converted by the H_3O^+ to HS^- and H_2S. Therefore, the $[S^{2-}]$ will not be 0.10 *M*. But the metallic ion concentrations will be 0.10 *M*, since nothing will happen to these ions. When we have added just enough H_3O^+ to dissolve FeS completely (and no more), a saturated solution will be present and K_{sp} will be valid. We can then solve for $[S^{2-}]$.

$$K_{sp} = [Fe^{2+}][S^{2-}]$$
$$3.7 \times 10^{-19} = (0.10)[S^{2-}]$$
$$[S^{2-}] = 3.7 \times 10^{-18}\ M$$

If we can reduce $[S^{2-}]$ to 3.7×10^{-18} *M*, the FeS will completely dissolve. The next step is to find out what concentration of H_3O^+ is necessary to reduce $[S^{2-}]$ to 3.7×10^{-18} *M*. For this we can use equation (5), provided that we have a value for the concentration of H_2S. We will assume that the concentration of H_2S is 0.10 *M*. This is reasonable, because we started with 0.10 mole of S^{2-} ion (in the form of solid FeS), and after the precipitate is completely dissolved, the $[S^{2-}]$ is only 3.7×10^{-18} *M*. The rest of the S^{2-} was essentially completely converted to H_2S. Putting it another way, $(0.10) - (3.7 \times 10^{-18}) = 0.10$. We may now substitute the concentrations of S^{2-} and H_2S into equation (5) and solve for $[H_3O^+]$:

$$9.1 \times 10^{-22} = \frac{[H_3O^+]^2(3.7 \times 10^{-18})}{0.10}$$
$$[H_3O^+]^2 = 2.5 \times 10^{-5} = 25 \times 10^{-6}$$
$$[H_3O^+] = 5 \times 10^{-3}\ M$$

We now make a similar calculation for HgS:

$$K_{sp} = [Hg^{2+}][S^{2-}]$$
$$1 \times 10^{-52} = (0.10)[S^{2-}]$$
$$[S^{2-}] = 1 \times 10^{-51}\ M$$
$$9.1 \times 10^{-22} = \frac{[H_3O^+]^2(1 \times 10^{-51})}{0.10}$$

$$[H_3O^+]^2 = 9.1 \times 10^{28}$$
$$[H_3O^+] = 3.0 \times 10^{14}\ M$$

These calculations show that FeS can easily be dissolved by the addition of a strong acid, since a $[H_3O^+]$ of $5 \times 10^{-3}\ M$ is readily obtainable, but that HgS cannot be dissolved in this way, despite the fact that it is a salt of a weak acid, because a $[H_3O^+]$ of $3.0 \times 10^{14}\ M$ is impossible to achieve.

3. Our final application of the principle also involves the dissolving of precipitates, but in a different way. Some precipitates can be dissolved by the addition of compounds or ions that cause formation of a complex ion salt that is soluble in water. An important example is silver chloride, which can be dissolved by the addition of ammonia. In this case, Ag^+ ions react with ammonia to form the soluble silver–ammonia complex ion salt $Ag(NH_3)_2Cl$. When ammonia is added to water that contains solid AgCl, it reacts with the small amount of Ag^+ ions present in the solution to form $Ag(NH_3)_2{}^+$.

$$AgCl(s) \rightleftharpoons Ag^+ + Cl^- \qquad (6)$$
$$Ag^+ + 2NH_3 \rightleftharpoons Ag(NH_3)_2{}^+ \qquad (7)$$

This removal of Ag^+ ions causes equilibrium (6) to shift to the right, and if enough ammonia is added, the AgCl can be completely dissolved.

Other examples of solids that can be dissolved by conversion to complex ions include $Zn(OH)_2$ and $Al(OH)_3$, both of which form complex hydroxide ions when treated with excess OH^-.

$$Zn(OH)_2(s) + 2OH^- \rightleftharpoons Zn(OH)_4{}^{2-} \quad \text{soluble in water}$$
$$Al(OH)_3(s) + 3OH^- \rightleftharpoons Al(OH)_6{}^{3-} \quad \text{soluble in water}$$

Quantitative aspects of complex-ion formation will be discussed in Section 19-11.

11-4 Qualitative Analysis of Cations

Suppose we are given an aqueous solution containing a number of positive metallic ions. How could we go about analyzing this solution to determine which ions are present and which absent? More than a hundred years ago a method was devised for solving this problem. This method, called the *qualitative analysis scheme,* was used in chemical laboratories all over the world, and was taught to every chemistry major (and to many other students) in colleges and universities. In recent years the importance of the scheme has diminished, owing to the advent of instrumental methods, but it is still valuable, because many of the procedures provide quick and easy tests for the presence or absence of ions.

Over the years many variations of the scheme have been developed and it has been applied to almost every metallic positive ion, but the standard basic qualitative analysis scheme includes only the 24 ions shown in Figure 11-1. The principles used in the scheme are largely those we have discussed in this chapter: formation of precipitates, the use of one equilibrium to drive another, the dissolving of precipitates by their conversion to complex ions, etc. The basic method consists of *separation* and *confirmation*. Each ion is separated from all the others, and its presence or absence is then confirmed in some way (for example, by the color of a precipitate or the fact that a given precipitate dissolves when treated with a given reagent). To make the process more manageable, the 24 ions are first separated into five groups, as shown in Figure 11-1 (note that these groups are *not* the same as the groups of the

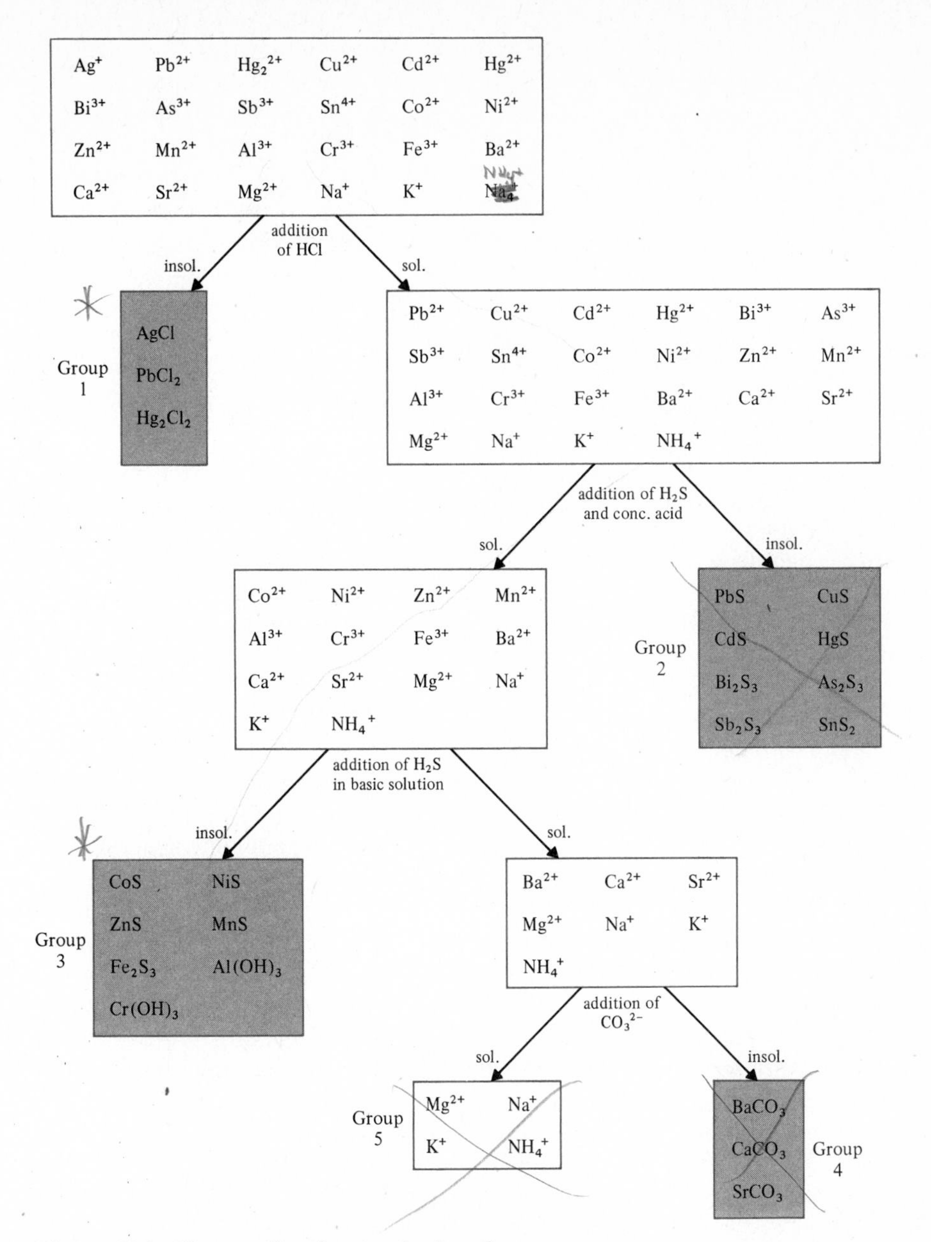

Figure 11-1 The qualitative analysis scheme.

periodic table), and then each group is analyzed separately. It is important that each separation be essentially complete, because if an ion is not completely removed when it is supposed to be, it may interfere with later tests. The separation into groups is done as follows (see Figure 11-1).

Group 1. The solution is treated with dilute HCl. As mentioned in Section 11-1, only three of these ions (Pb^{2+}, Ag^+, Hg_2^{2+}) form insoluble chlorides. If one or more of these three ions are present, they will be precipitated at this point. Any of the other 21 ions that are present will remain in solution. The test tube is then placed in a centrifuge, a device that causes the precipitate to drop to the bottom of the test tube, after which it can be separated from the remaining solution by decantation. If no precipitate forms when dilute HCl is added, then none of the three group 1 ions is present. (Pb^{2+} is a special case. If it is present, only part of it will precipitate as $PbCl_2$, because this compound is only partially insoluble. Therefore, some of the Pb^{2+} will remain in solution and be carried over into group 2.)

Group 2. The pH of the solution is now lowered by the addition of more acid and the solution is saturated with H_2S. Most of the 22 ions form insoluble sulfides, but, as we have seen in Section 11-3, some of them are soluble under highly acidic conditions and others are not. Eight of the 22 ions (including the remaining Pb^{2+} ions) form insoluble sulfides under these conditions and any of these which are present will now precipitate as the sulfides. The precipitate, if any, is then separated from the remaining solution by centrifugation and decantation.

Group 3. The pH is raised by the addition of a base and the solution again saturated with H_2S. The concentration of S^{2-} is now much greater (Section 11-3) and five more sulfides precipitate. The increased concentration of OH^- causes two additional ions to precipitate as hydroxides [$Al(OH)_3$ and $Cr(OH)_3$]. Once again, the precipitate containing any of these seven ions that may be present is separated from the rest of the solution.

Group 4. There are only seven ions that may still be present in the solution. A solution of $(NH_4)_2CO_3$ is now added, and three of these ions will precipitate as insoluble carbonates.

Group 5. The ions remaining now (Na^+, K^+, Mg^{2+}, NH_4^+) are generally tested for by individual tests. For example, K^+ gives a violet flame test, while NH_4^+ can be detected by adding NaOH and heating. The NH_3 given off is detected by holding a piece of wet red litmus paper at a short distance above the heated test tube. If the litmus paper turns blue, it means that NH_4^+ was present. The test for NH_4^+ is usually done at the very beginning of the analysis (before the separation of group 1), so that NH_4^+ ions or NH_3 can then be added as reagents later in the scheme.

After the ions have been separated into groups, it is necessary to test each group for the specific ions present in it. The methods for doing this are similar in principle to the methods already discussed (they involve separation and confirmation) and would require too much space for us to discuss completely (the details are available in books on qualitative analysis). We will confine ourselves to describing the analysis of group 1 (see Figure 11-2).

Recall that the group 1 precipitate might contain any or all of the three chlorides $PbCl_2$, AgCl, Hg_2Cl_2. After removal of the solution that contains the ions of groups 2 to 5, the group 1 precipitate is treated with hot water. If $PbCl_2$ is present, it will dissolve in the hot water, but AgCl and Hg_2Cl_2 are not soluble under these conditions. The precipitate containing AgCl and/or Hg_2Cl_2 is removed from the solution. It is usually not easy to tell if any of the precipitate has dissolved. Therefore, it is necessary to test the resulting solution for the presence of Pb^{2+}. This is done by treating the solution with K_2CrO_4. A yellow precipitate of $PbCrO_4$ indicates the presence of Pb^{2+}.

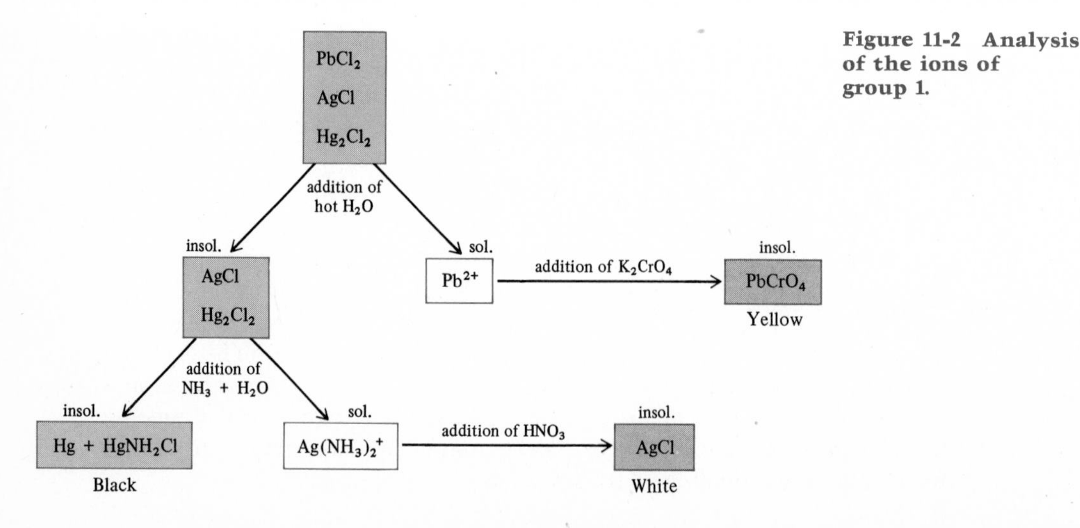

Figure 11-2 Analysis of the ions of group 1.

To the precipitate remaining after treatment with hot water is now added a solution of NH_3. As we saw on page 340, the NH_3 dissolves AgCl (if present) to give the soluble $Ag(NH_3)_2^+$ ion. The Hg_2Cl_2 also reacts with NH_3, but in this case it is converted to a mixture of two insoluble substances: Hg and $HgNH_2Cl$, which appear as a black precipitate. This black precipitate, if formed, confirms the presence of Hg_2^{2+}. In order to confirm the presence of Ag^+, the black precipitate (if any) resulting from the NH_3 treatment is removed, and the remaining solution is treated with HNO_3. Because this acid reacts with NH_3, it drives equilibrium (7) to the left, which in turn drives equilibrium (6) to the left, resulting in a reprecipitation of AgCl.

$$AgCl(s) \rightleftharpoons Ag^+ + Cl^- \quad (6)$$

$$Ag^+ + 2NH_3 \rightleftharpoons Ag(NH_3)_2^+ \quad (7)$$

Qualitative analysis also includes the analysis of anions. However, attempts to devise formal schemes, similar to the scheme for cations, have been mostly unproductive, and in most cases separate tests are used for each anion. An example of such a test (for SO_4^{2-} and SO_3^{2-} ions) was given on page 339.

Problems

11-1 Define each of the following terms. **(a)** solubility product **(b)** solubility **(c)** ion product **(d)** common-ion effect.

11-2 Which of the following salts would you expect to be soluble in water? NaCl, KOAc, AgS, $AgNO_3$, $FeCl_3$, $BaSO_4$, $Mn(NO_3)_2$, AgCl, $As(OAc)_3$, NH_4Br, K_2SO_4, $Fe_2(SO_4)_3$, Na_2S, MoS_2, $Ba(OAc)_2$.

11-3 Complete and balance those reactions that go to completion in water solution. If they do not go to completion, write NR.

(a) $Ba^{2+} + SO_4^{2-} \rightarrow$
(b) $K^+ + Cl^- \rightarrow$
(c) $Pb^{2+} + CrO_4^{2-} \rightarrow$
(d) $CO_3^{2-} + H_3O^+ \rightarrow$
(e) $SO_3^{2-} + H_3O^+ \rightarrow$
(f) $SO_4^{2-} + H_3O^+ \rightarrow$
(g) $Ag^+ + Cl^- \rightarrow$
(h) $Ba^{2+} + S^{2-} \rightarrow$
(i) $Ag^+ + NH_3 \rightarrow$
(j) $Pb^{2+} + NO_3^- \rightarrow$
(k) $HOAc + OH^- \rightarrow$
(l) $NH_4^+ + CO_3^{2-} \rightarrow$
(m) $NO_3^- + H_3O^+ \rightarrow$
(n) $Hg^{2+} + Cl^- \rightarrow$
(o) $S^{2-} + H_3O^+ \rightarrow$

11-4 On page 331 it is stated that monocalcium phosphate, $CaHPO_4$, and sodium aluminum sulfate, $NaAl(SO_4)_2$, are acidic compounds. Write equations to show how they are acidic in water solution.

11-5 Write the solubility-product and ion-product expressions for the following salts. **(a)** Ag_2S **(b)** $PbCrO_4$ **(c)** $Fe(OH)_3$ **(d)** SrC_2O_4

11-6 The solubility of PbI_2 at 25°C is 0.063 g/100 ml of solution. Calculate the solubility product of PbI_2 at 25°C.

11-7 The solubility of Bi_2S_3 is 2.55×10^{-12} g/liter. What is the K_{sp} for Bi_2S_3?

11-8 The solubility product of $PbCO_3$ is 3.3×10^{-14} at 18°C. Calculate the solubility of $PbCO_3$ in moles/liter and g/100 ml of solution at this temperature.

11-9 The solubility product of BaF_2 is 1.7×10^{-6} at 25°C. Calculate the solubility of BaF_2 in moles/liter and g/100 ml of solution at this temperature.

11-10 A 50-ml solution at 25°C contains 5.0×10^{-4} M $AgNO_3$. How many grams of KBr must be added to the solution to just begin precipitation of AgBr? (Assume no volume change.)

11-11 In each case, calculate whether or not a precipitate will form. **(a)** If we mix equal volumes of an 0.0350 M $Ca(NO_3)_2$ solution and an 0.0280 M Na_2SO_4 solution. **(b)** If we mix equal volumes of a 2.50×10^{-4} M $AgNO_3$ solution and a 2.30×10^{-4} M K_2CO_3 solution. **(c)** If 400 ml of 0.035 M $Pb(NO_3)_2$ solution is mixed with 100 ml of 0.00200 M CaI_2 solution. **(d)** If 200 ml of 0.00450 M $Al_2(SO_4)_3$ solution is mixed with 300 ml of 0.00200 M $Ni(OH)_2$ solution.

11-12 1 liter of a 1.0×10^{-1} M $Ba(OAc)_2$ solution is mixed with 1 liter of a 1.1×10^{-4} M K_2SO_4 solution. Calculate the concentration of Ba^{2+} and SO_4^{2-} in the resulting solution.

11-13 500 ml of a 2.50×10^{-3} M $AgNO_3$ solution is mixed with 200 ml of a 6.80×10^{-2} M solution of K_2S. Calculate the concentrations of Ag^+ and S^{2-} ions in the resulting solution (neglect hydrolysis).

11-14 The solubility product of ZnS at 18°C is 1.2×10^{-23}. **(a)** Calculate its solubility in g/ml at this temperature. **(b)** Calculate its solubility in g/ml when the solution also contains 0.025 mole/liter of $ZnCl_2$.

11-15 The solubility of $PbCl_2$ in a solution of 0.130 M $Pb(OAc)_2$ is 5.7×10^{-3} mole/liter. Calculate the solubility product of $PbCl_2$ at this temperature.

11-16 Calculate the concentration of SO_4^{2-} in a saturated solution of $BaSO_4$ under each of the following conditions. **(a)** No other ions are present. **(b)** The solution contains 1.0×10^{-2} M $Ba(OAc)_2$.

11-17 A solution contains the following compounds in the concentrations given:

1.5×10^{-2} M KCl

3.6×10^{-1} M KBr

7.2×10^{-3} M K_2CrO_4

A solution of $AgNO_3$ is added to this solution drop by drop. Calculate which silver salt will precipitate first.

11-18 Why is the solubility of $PbCl_2$ smaller in a solution of $Pb(NO_3)_2$ than in pure water?

11-19 A solution saturated with H_2S has an $[H_2S]$ of 0.10 M. What is the concentration of S^{2-} under each of the following conditions? **(a)** pH = 7 **(b)** pH = 3.5

11-20 The K_{sp} of AgCl at 25°C is 1.56×10^{-10}. If HCl is added to a saturated solution of AgCl, what must the final pH of the solution be if the concentration of Ag^+ is to be 1.25×10^{-8} M.

11-21 An aqueous solution contains 0.0100 M F^- and 0.0100 M SO_4^{2-}. It is desired to separate the two by adding Ba^{2+}. What concentration of Ba^{2+} will precipitate the maximum amount of one without precipitating any of the other?

11-22 An aqueous solution contains 0.0500 M Ag^+ and 3.75×10^{-3} M Pb^{2+}. What concentration of I^- will precipitate the maximum amount of one without precipitating any of the other?

11-23 Which of the following salts would be expected to dissolve in a strong acid solution? **(a)** $MgSO_4$ **(b)** $MgSO_3$ **(c)** $CaCO_3$ **(d)** AgCl **(e)** HgS **(f)** MnS

11-24 Calculate K_{sp} for $Ni(OH)_2$ given that the pH of a saturated solution of this compound is 8.83.

11-25 How many moles of HCl must be added to 2.5 liters of water that has been saturated with H_2S ($[H_2S] = 0.10$ M) in order to dissolve completely 0.125 mole of solid CoS?

11-26 A solution that contains 0.10 *M* $ZnCl_2$ and 0.10 *M* $FeCl_2$ is saturated with H_2S ($[H_2S] = 0.10\ M$). What must the pH of the solution be to precipitate selectively the ZnS? What is the concentration of Zn^{2+} in solution when FeS first begins precipitating?

11-27 It is desired to dissolve completely 10.0 g of $Mg(OH)_2$, which is in contact with 1.00 liter of water, by adding a strong acid. At what pH will the solid $Mg(OH)_2$ completely dissolve?

11-28 Assume that you have an unknown solution which might contain any or all of the following ions: Ag^+, Pb^{2+}, Cd^{2+}, Co^{2+}, Ba^{2+}, and NH_4^+. Describe the tests you would carry out to determine which ions are present.

11-29 A solution contains all the following ions, each at a concentration of 0.200 *M*: Mn^{2+}, Ni^{2+}, Pb^{2+}, and Sn^{2+}. If the solution is to be saturated with H_2S, devise a scheme for selectively precipitating these ions by controlling the pH of the solution. Is it physically possible to separate all these ions by this procedure?

11-30 Repeat the calculations of Problem 11-29 for the following ions: Ag^+, Cd^{2+}, Co^{2+}, and Cu^{2+}.

Additional Problems

11-31 Calculate the concentration of all species present in solution when 100 ml of 0.120 *M* $AgNO_3$ is mixed with 100 ml of an HCl solution whose pH is 1.5.

11-32 The solubility product of MgC_2O_4 at 18°C is 8.6×10^{-5}. If 220 ml of a 0.12 *M* $MgCl_2$ solution is mixed with 200 ml of a 0.230 *M* $Na_2C_2O_4$ solution, calculate how much MgC_2O_4 is dissolved in the resulting solution.

11-33 The solubility product of $Fe(OH)_3$ is 1.1×10^{-36} at 18°C. Calculate its solubility in moles/liter and mg/ml at this temperature.

11-34 What is the solubility in g/100 ml of $Mg(OH)_2$ in a 0.20 *M* NH_3 solution? (Don't forget that an aqueous solution of NH_3 is basic.)

11-35 A liter of H_2O contains 0.20 mole of solid $CaCO_3$. What concentration of H_3O^+ is necessary to dissolve it completely? (Assume that the total volume remains at 1 liter.)

11-36 Would you expect the following solids to dissolve in the solvents given?

(a) NaCl in concentrated H_2SO_4
(b) $BaSO_3$ in concentrated H_2SO_4
(c) $AgNO_3$ in concentrated HCl
(d) $Al(OH)_3$ in concentrated HCl
(e) AgCl in aqueous NH_3
(f) $PbCl_2$ in hot water

11-37 How many grams of Ag_2CrO_4 will dissolve in 150 ml of a 0.11 *M* $AgNO_3$ solution?

11-38 Assume that you add base to an acidic solution of 0.01 *M* $Mg(NO_3)_2$. At what pH will $Mg(OH)_2$ precipitate?

11-39 At 25°C, the solubility of $BaCO_3$ is 9.0×10^{-5} mole/liter. Calculate the solubility product of $BaCO_3$.

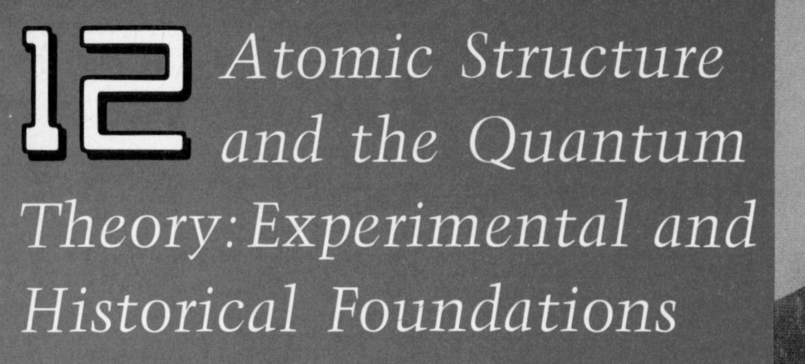

12 Atomic Structure and the Quantum Theory: Experimental and Historical Foundations

Niels Bohr, 1885–1962

Edgar Fahs Smith Collection

Up to this point we have largely discussed the properties of matter in bulk. The observations made, and the theories and laws described, were obtained from experiments carried out on materials that we could actually see and work with in the laboratory. Although we know that bulk matter actually consists of very large numbers of very tiny particles, we were able to calculate and predict many things involving chemical reactions and other properties without worrying too much about the microscopic properties of individual atoms and molecules. In this and the following two chapters we will begin to study in detail the actual structures of atoms and molecules. Our aim will be to correlate the *macroscopic* behavior of bulk matter with the *microscopic* structure of the individual particles.

In Chapter 13, we will deal with the structure of *atoms,* in Chapter 14 with that of *molecules.* In this chapter, we will see that our modern theories of atomic and molecular structure did not suddenly spring into bloom, but are based on a firm historical background. We will begin the chapter by examining the way nineteenth-century physicists looked at the concepts of *particles* and *waves.* We will see that these physicists thought they had theories that completely explained particles and waves and the differences between them. These theories (which we now refer to as *classical*) were mainly Newton's laws of motion and Maxwell's laws of electromagnetic radiation. We will then show that (as has happened so often in the history of science) a number of new facts were discovered that simply could not be explained by the classical theories. Among these were (1) the distribution of wavelengths of light given off by a blackbody (Section 12-4), (2) the photoelectric effect (Section 12-6), and (3) the presence of lines in spectra given off by the elements (Section 12-7). In order to explain these experimental facts, new and (at that time) revolutionary ideas were proposed that struck at the heart of the classical theories of Newton and Maxwell. We will see that one is almost forcibly led to the conclusion that light waves and particles are manifestations of the same thing, namely energy.

The beginning of the twentieth century has been labeled the quantum revolution. It may more aptly be called the quantum evolution, since scientists were forced step by step to reject the classical ideas of physics. The aim of this chapter is to present the experimental evidence and ideas that led to the evolution of the quantum theory and to its application to atomic structure. The unraveling of the structure of atoms and molecules is one of the great success stories in the history of science.

12-1 Wave Properties of Light

The idea that light is a wave, similar in some of its properties to ocean waves and sound waves, was first proposed by the Dutch astronomer Christiaan Huygens (1629–1695), and confirmed by the experiments of Thomas Young (1773–1829).

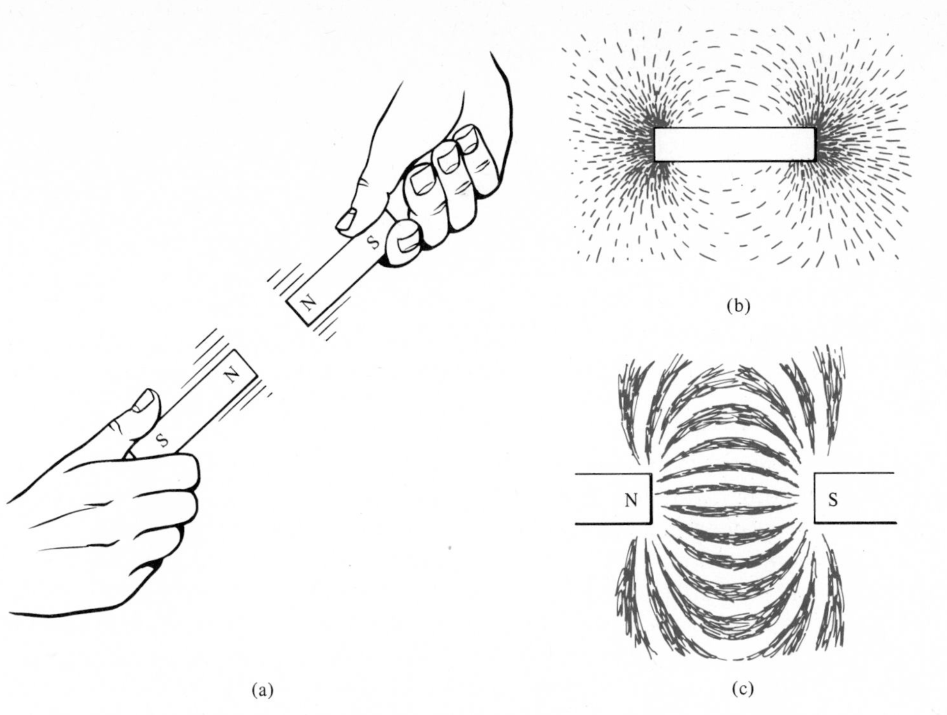

Figure 12-1 (a) When two magnets are brought close together without touching, they interact with one another through a distance. (b) Lines of force coming from a magnet. (c) Interaction of two magnetic fields.

However, it was not until the 1860s that James Clerk Maxwell (1831–1879) made two extremely important discoveries about the nature of light: (1) light is related to electricity and magnetism (Maxwell had previously shown that electricity and magnetism always existed together), and (2) the light that we can see (visible light) is just a very small part of what is called the **electromagnetic spectrum.** Very briefly, the ideas leading up to these conclusions are as follows.

Almost everyone is familiar with magnets and their properties. If you hold one end of a magnet lightly with one hand on a table and bring another magnet close to it, you will feel the magnets quivering, showing the interactions of the two magnets upon each other (Figure 12-1a). What is particularly important to us is that the interactions are taking place through space *without the magnets touching each other.* If we place a sheet of paper over the magnets and sprinkle iron filings on them, we see the filings assume a pattern caused by lines of force emanating from each of the poles of the magnets (Figure 12-1b). This type of experiment, in which one actually sees two bodies acting upon each other through space, gave rise to the concept of **fields.** That is, the magnets have magnetic fields emanating from them and these fields interact with other magnetic fields (Figure 12-1c). Consider what will happen to the field if we take one of the magnets and begin gently to move it up and down (an oscillating motion). From Figure 12-2, we see that the field begins to pulse. What we have done by moving the magnet up and down is to generate a *wave* that will now travel away from the magnet in all directions. The pulsating type of action that takes place when the magnet is moved in this manner accounts for the wave type of motion of the moving field. Electrical fields act in the same manner. **It is pulsating electrical and magnetic fields that constitute light.** Maxwell's equations showed that light is generated by the oscillating motions of either a magnet or an electrically charged body. Consider a typical far-off galaxy, perhaps 5 billion light-years away. Within the stars of this galaxy, the oscillating motions of charged particles inside the atoms generate this type of pulsating electric and magnetic field, which now begins to travel away from the source. The pulsating

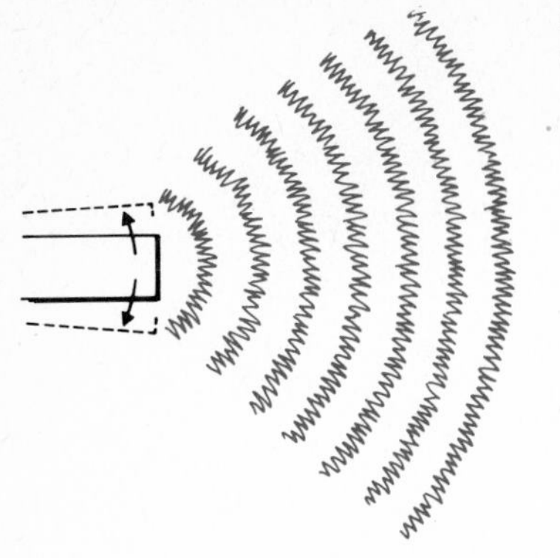

Figure 12-2 A pulsating magnetic field obtained by a slight up-and-down movement of a magnet.

wave continues from the distant galaxy, for millions or billions of years, and finally arrives at the earth, where we can detect it by telescopes or radio telescopes in astronomical observatories. It is because of this that we see the stars and galaxies of our universe. Indeed, light from our own sun comes to us in the same way.

Light is composed of waves, made up of electric and magnetic fields. It travels through space, having been generated by an initial pulsating mechanism. The first of Maxwell's two discoveries gave rise to the second. If the light that we can see is caused by very rapid oscillations of charged particles, then there could be other charged particles that oscillate faster or slower. Such particles would then give off light that we *cannot* see [this prediction was confirmed by Heinrich Hertz (1857–1894)]. Thus, Maxwell recognized that there is a broad spectrum of light—not only the visible light that we can see but also ultraviolet and infrared light,* microwaves, radio waves, x rays, and gamma rays, which we cannot see (Figure 12-3). All these types of light are waves and are composed of pulsating electrical and magnetic fields. All of them can be generated in the same way: by oscillating motion of an object that carries a magnetic or an electrical field. All of this light, visible or invisible, is called **radiation**—more specifically, **electromagnetic radiation.**

Figure 12-4a shows a string tied to a piece of wood and held taut. If we now pulse the string (Figure 12-4b) and follow the pulse along the string, we can draw an outline of the pulse (Figure 12-4c). This outline is simply a wave traveling along the string. Electromagnetic radiation travels in the same manner, except that of course there is no string. Figure 12-4d shows two different wave forms.

The basic nature of all electromagnetic waves is the same. All electromagnetic waves travel in a vacuum at the same speed. This speed is called the **speed of light** and has the symbol *c*. It is one of the fundamental constants of nature.

$$c = \text{speed of light in vacuum} = 3.00 \times 10^{10}\ \text{cm/s}$$

A number of characteristics can help us distinguish between different electromagnetic waves. They are:

1. What makes one kind of radiation different from another is the wavelength (symbol λ). The **wavelength,** λ, is defined as the distance between any point on a cycle and the corresponding point on the next cycle (Figure 12-4d).
2. Another important characteristic of a wave is the **frequency** (symbol ν). This is defined as the number of crests that pass a given point in 1 second. The unit of frequency is a number of cycles or vibrations per second (also called hertz, Hz). The units of frequency are often given as s^{-1}, the word "cycles" being understood.

*Both infrared and ultraviolet light had already been discovered: infrared by William Herschel in 1800 and ultraviolet by Johann Ritter in 1801.

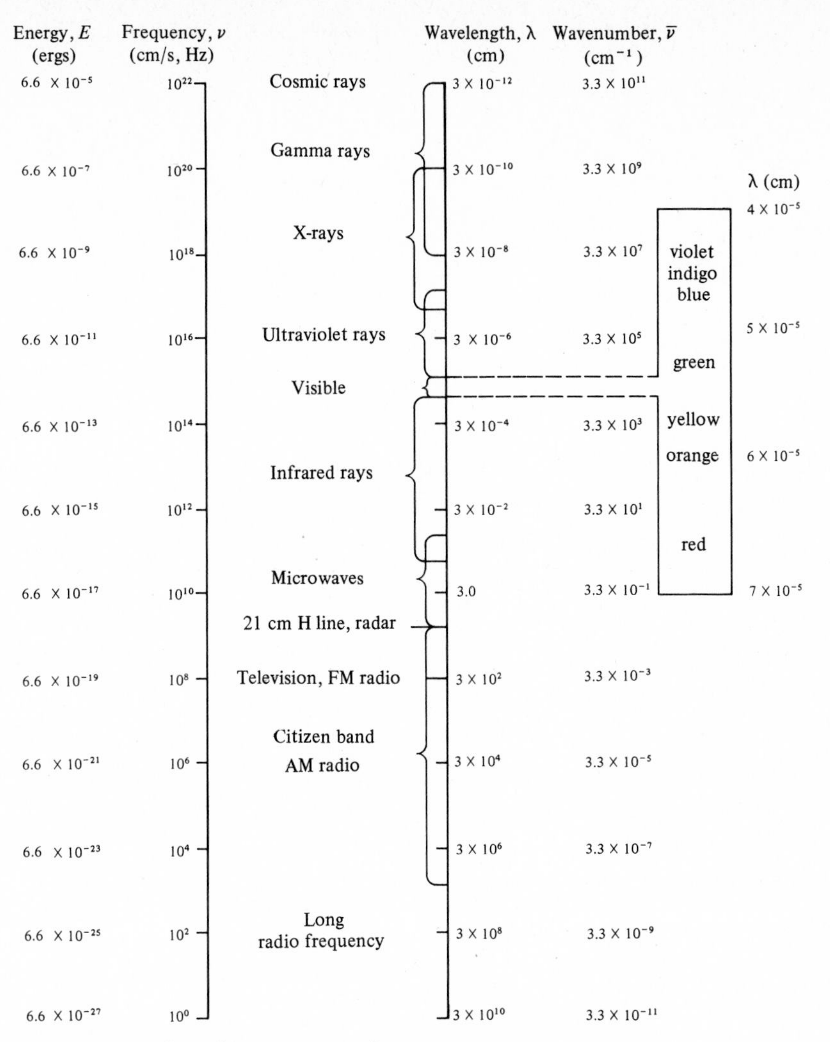

Figure 12-3 The electromagnetic spectrum.

Because all electromagnetic waves travel at the speed of light in a vacuum, there is a relationship between these quantities. The speed of light equals the wavelength times the frequency. In symbols,

$$c = \lambda\nu \tag{1}$$

$$c\ (\text{cm/s}) = \lambda\ (cm) \cdot \nu\ (1/\text{s}) \tag{2}$$

Equation (2) allows us to calculate the frequency of a wave given its wavelength, or vice versa.

Example: The wavelength of the light coming from an infrared heating lamp is 5.00×10^{-4} cm. Calculate the frequency ν of this emitted light.

Answer: From equation (2),

$$\nu = \frac{c}{\lambda} = \frac{3.00 \times 10^{10}\ \text{cm/s}}{5.00 \times 10^{-4}\ \text{cm}} = 6.00 \times 10^{13}\ \text{s}^{-1} = 6.00 \times 10^{13}\ \text{Hz}$$

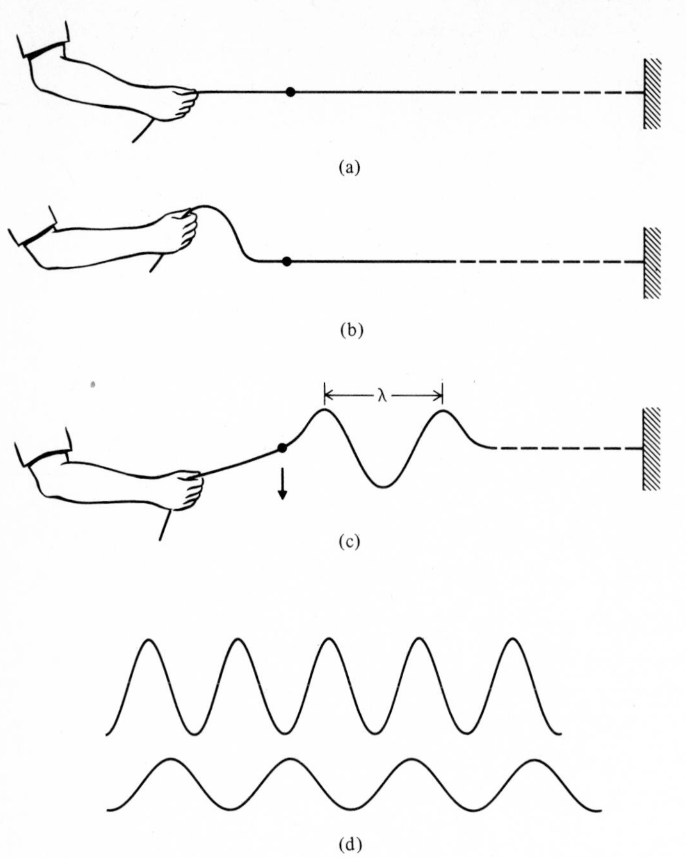

Figure 12-4 A description of how one may generate a wave motion in a taut piece of string. (a) Cord tied to a mounting. (b) Cord pulsed. (c) Outline of pulse. (d) Different wave forms.

Example: The frequency of x rays used by dentists for examination is 3.0×10^{16} Hz. Calculate the wavelength of this electromagnetic radiation.

Answer:

$$\lambda = \frac{c}{\nu} = \frac{3.00 \times 10^{10}\ \text{cm/s}}{3.0 \times 10^{16}\ \text{s}^{-1}} = 1.0 \times 10^{-6}\ \text{cm}$$

Note that the kinds of radiation shown in Figure 12-4d have different wavelengths and hence different frequencies. Also note that the different colors of the visible spectrum also differ in wavelength as well as in frequency. All electromagnetic waves have the following properties in common:

1. As previously mentioned, all electromagnetic radiation travels at the speed of light in a vacuum.
2. Electromagnetic radiation shows the property of **reflection.** This is shown in Figure 12-5a. We say that light is reflected and that the angles of reflection are equal, $\theta_1 = \theta_2$.
3. Electromagnetic radiation shows the property of **refraction.** This is shown in Figure 12-5b. When light travels from one medium into another, the path of the light is bent, $\theta_1 \neq \theta_2$. If one takes a pencil and sticks it in a glass of water, the pencil appears to be bent because the light path is bent. This is due to the phenomenon of refraction, which is exhibited by all light.
4. Electromagnetic radiation shows the property of constructive and destructive interference. When two waves encounter each other, the resulting wave is the sum of the two. When crests are aligned with crests, the sum is larger than the

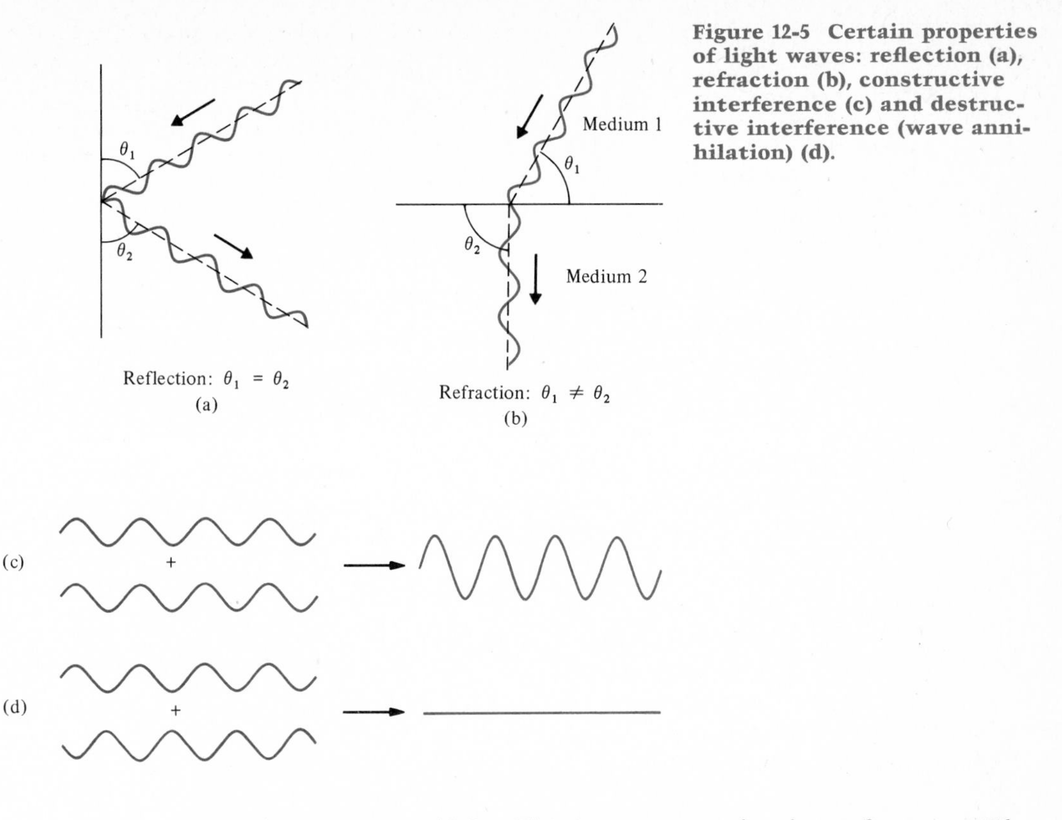

Figure 12-5 Certain properties of light waves: reflection (a), refraction (b), constructive interference (c) and destructive interference (wave annihilation) (d).

original waves (Figure 12-5c). This is **constructive interference.** When crests are aligned with troughs, **destructive interference** takes place. The resulting wave is smaller than the original, and in the case where the two original waves had equal wave heights, the waves are completely annihilated (Figure 12-5d).

12-2 Particle Properties of Matter

Late-nineteenth-century scientists believed that they completely understood not only the nature of light but also the nature of matter. According to them, matter was made up of tiny indestructible particles (atoms), and all matter, whether atoms, ordinary-sized objects, or very large bodies such as planets or stars, obeyed the same laws of motion. The first of these laws states that if a moving object is isolated so that there are no forces to disturb it, it will move with its constant velocity in a straight line along its original path. If the object was not moving to start with, it will remain still. This law, which is called the *principle of inertia* (Figure 12-6), was discovered by Galileo Galilei (1564–1642) and adopted as the first law of motion by Isaac Newton (1642–1727). The property of inertia is one possessed by all matter. It makes no difference where the object comes from; this property of matter is valid throughout the universe. We use the term *mass* as a measure of the amount of inertia possessed by a body. The difference between mass and weight is discussed in Section 1-1.

Now that we know what happens to an object in the absence of outside forces, we can next inquire what happens to its velocity if an outside force does intervene. This question is answered by Newton's second law of motion, which states that if an object is acted upon by an outside force, a property of the object called its *momentum* will change with time. The rate of change of the momentum with time will be

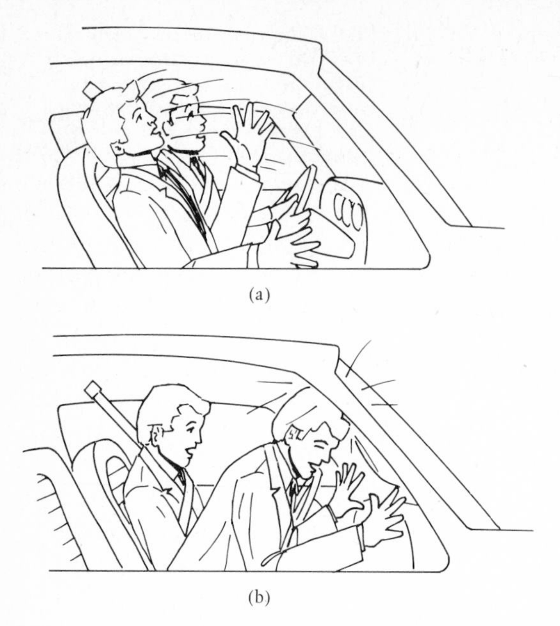

Figure 12-6 Properties of Inertia. (a) If a car makes a sudden start, a passenger, not holding the wheel, is thrown backward. (b) If a car makes a sudden stop, a person without a seat belt continues to move forward.

in proportion to the force acting upon it. The **momentum** of an object (symbol p) is the product of the mass of the object, m, and the velocity of the object, v.

$$\text{momentum, } p = mv \tag{3}$$

When objects collide and they transfer momentum during a collision process, one must take into account not only the velocity of the colliding objects but also their mass. Figure 12-7 shows how momentum is exchanged during collisions of billiard balls.

In summary, scientists of the late nineteenth century were very clear as to the distinction between particles and waves. There was no ambiguity in their minds, and each of these things had its own characteristic properties. If you were a scientist of that era, and you wanted to find out if something was a particle or a wave, you would examine its properties. If it had mass and inertia, it was a particle. If it showed reflection, refraction, and had a wavelength and frequency, it was a wave. To them, nothing could be clearer.

This was the state of the art in physics in the late nineteenth century. Albert A. Michelson (1852–1931), a distinguished physicist in the late 1890s and the first Nobel Laureate in science from the United States, suggested that the really exciting times in physics were over; that almost all the known laws of the universe had already been discovered. The only job left for physicists to do was to measure the various physical constants to a higher degree of accuracy than they had previously been measured ("find the next significant figure").

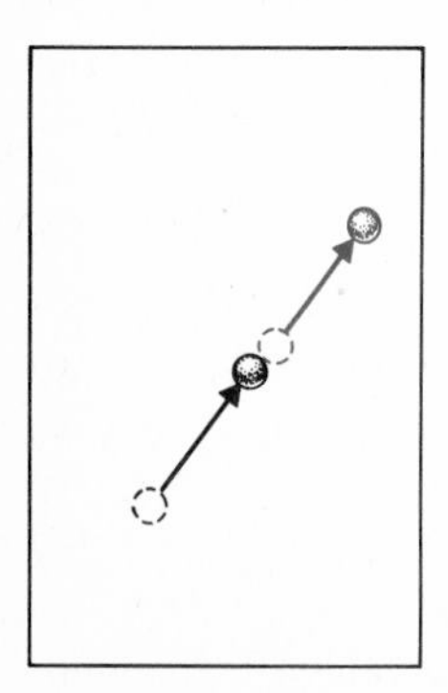

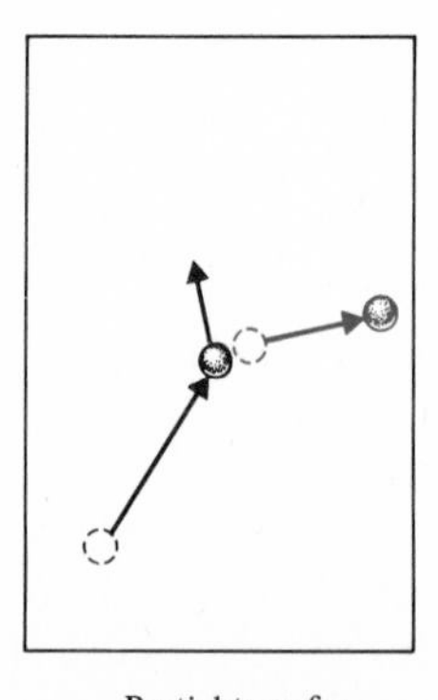

Figure 12-7 Collision of billiard balls, illustrating the transfer of momentum.

12-3 The Properties of Electrons

While he was in his early twenties, William Crookes (1832–1919) was investigating electrical discharges in different gases. In order to do this type of research, he began to conduct experiments using glass tubes of the type shown in Figure 12-8a (these are now called *Crookes tubes*, although several other scientists had used them earlier). A high voltage (10,000 volts) is applied to the system, which consists of two electrodes connected to the voltage source but not to each other. The positive electrode is called the **anode** and the negative one is the **cathode.** When the tube contains air (or any other gas) at normal atmospheric pressure, nothing happens. No current flows, and the ammeter registers zero. This is not surprising, since the electrical circuit is not complete. However, when the pressure is lowered by evacuating the gas, we soon observe that the remaining gas within the tube begins to glow. When the pressure is further lowered, the glow moves toward the anode end of the tube, and a dark space appears near the cathode. As you may have guessed, this glow is the same as the one seen in neon lights (a neon light is simply a Crookes tube containing neon).

In order to identify and characterize what has actually happened in the tube, further experiments were conducted, as follows:

1. A screen coated with fluorescent material, such as zinc sulfide, is placed between the cathode and the anode. Upon evacuation, the screen is seen to glow on the side facing the cathode, showing that something is coming from the cathode (these are called **cathode rays**) and striking the screen.
2. If a pinwheel is placed within the tube, the pinwheel rotates (Figure 12-8b).
3. If the anode is coated with zinc sulfide, and a maltese cross is placed within the tube, an outline of the cross is observed on the anode screen (Figure 12-8c).
4. If a magnetic field is placed around the tube, the glow from the fluorescent screen is displaced toward the north pole of the magnet (Figure 12-8d).
5. All the preceding results are found to occur regardless of the identity of the gas or of the material used as a cathode.

We can now make some definitive statements concerning these cathode rays.

a. *They are constituents of all matter (experiment 5 above).*
b. *They travel in straight lines when emitted (experiment 3).*

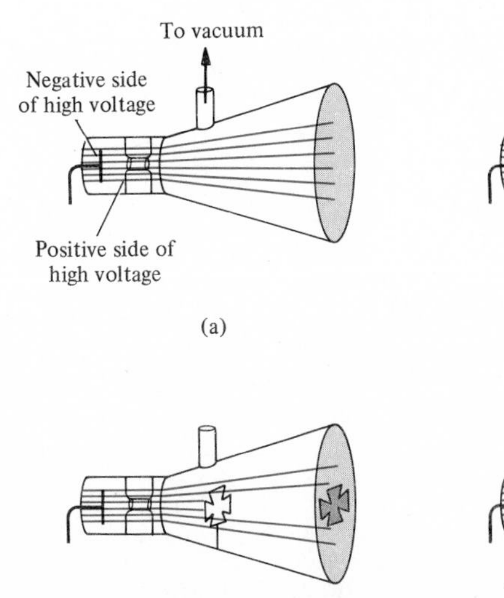

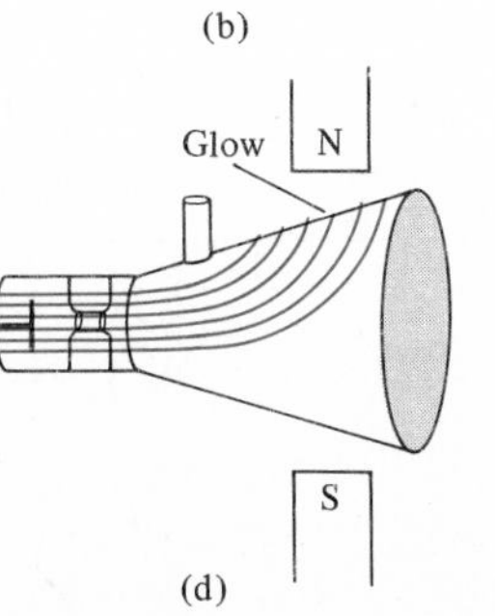

Figure 12-8 Cathode ray tube and screen (a), with pinwheel (b), with maltese cross (c), and with magnetic field (d).

c. *They are negatively charged (experiment 4).*
d. *They have mass (experiment 2), which means that they are particles.*

It was J. J. Thomson (1856–1940) who first realized that these negatively charged particles are constituents of all matter and who gave them the name **electrons,** a name that George Stoney (1826–1911) had previously invented for the minimum quantity of electrical charge. They are, of course, the same electrons which, as we have already seen in Chapter 2, are constituents of all atoms, and to J. J. Thomson belongs the credit for being the first to demonstrate that Dalton's atoms were not indivisible.

We can now explain what happens in the cathode ray tube. The voltage applied accelerates electrons which are generated from the cathode, which is heated to the point where electrons are "boiled off." The electrons collide with atoms of the gas, knocking electrons out of them, and turning them into positive ions. It is the ionized gas atoms that glow. When the pressure is reduced, the electrons can travel farther before striking a gaseous atom, accounting for the fact that the glow moves farther and farther toward the anode with reduced pressure. The dark spot appears near the cathode, since no ionization is taking place there.

After the discovery that the electron is a constituent of all matter, a large number of experiments were conducted with the aim of accurately measuring its properties. Two of these properties, the *mass* and the *charge,* are most important, and it is not surprising that many attempts were made to measure them. The first successful experiments along these lines were those of J. J. Thomson, who, in 1897, measured the *ratio* of charge (e) to mass (m) of the electron. This value is -1.76×10^8 coulombs/gram.

$$e/m \text{ for the electron} = -1.76 \times 10^8 \text{ C/g} \tag{4}$$

Thomson was not able to measure either the electrical charge on the electron or its mass. What he was able to measure was the *ratio* of the charge to the mass. It was Robert A. Milliken (1863–1953), who, in 1911, was able to measure the charge on an electron. This value is

$$e = -1.602 \times 10^{-19} \text{ C} \tag{5}$$

Since e/m was already known, it was a simple matter to calculate the mass of the electron:

$$\begin{aligned} \text{mass of an electron} &= \frac{\text{charge of the electron } (e)}{e/m} \\ &= \frac{-1.602 \times 10^{-19} \text{ C}}{-1.76 \times 10^8 \text{ C/g}} \\ &= 9.10 \times 10^{-28} \text{ g} \end{aligned} \tag{6}$$

Example: What is the total number of coulombs carried by a collection of chloride ions (Cl^-) that weighs 20.1 g?

Answer: Each chloride ion contains one excess electron charge or -1.602×10^{-19} C.

$$20.1 \text{ g chloride ions} \times \frac{6.02 \times 10^{23} \text{ chloride ions}}{35.453 \text{ g of chloride ions}} \times \frac{-1.602 \times 10^{-19} \text{ C}}{1 \text{ chloride ion}} = -5.47 \times 10^4 \text{ C}$$

12-4 Blackbody Radiation

Scientific activities may be divided into two kinds: pure and applied, although they overlap many times. The applied investigator devotes himself to work that is directly useful to society: curing a disease, inventing a better plastic, finding a way to make some drug cheaper, or some other mission-oriented goal. The pure scientist focuses his energies on very basic questions such as proving that matter is stable or explaining why a solid melts. Some people may think that all scientists ought to work only on applied problems, since it is these which directly improve our lot, but, apart from the goal of satisfying the intellectual curiosity of mankind, it has often happened that a particular practical problem could not be solved until certain basic questions were answered first. But even more than this, applied science is only successful because it is based on knowledge gained by pure science.

In the late nineteenth century, a fundamental question that occupied the attention of some theoretical physicists and chemists (pure scientists) was the color of hot objects. An iron rod when heated glows first dull red, then orange, then white as its temperature is raised. It is not difficult to measure the light being emitted from an iron rod or any other body and to determine what wavelengths (or frequencies) are being emitted and how much energy and intensity belong to each wavelength. Figure 12-9 shows the wavelength distribution of radiation from a so-called blackbody at three different temperatures. A **blackbody** is an idealized system that will absorb *all* the light shining upon it and then reemit all the radiation. Actually, no real body can perform in this ideal way. In the working laboratory, the best blackbody is an empty container with a little peephole that emits light after the container is heated. It is called black since the orifice appears to be black when the container is cold. As Figure 12-9 shows, as the temperature is increased, the maximum in the distribution curves moves to smaller wavelengths (or higher frequencies).

However, another aspect of these curves caused great difficulty for late nineteenth-century physicists. The trouble was that the presence of maxima in the curves was incompatible with their theories. When they attempted to predict the distribution of frequencies using the wave theory of light and classical thermodynamics based on Newton's laws of motion, what they were able to show is that the intensity emitted from a blackbody should rise continuously with increasing frequency (dashed line in Figure 12-9). *There should be no maxima at all.* Physicists tried time and time again to predict the correct behavior without success. From what we know today, we can state that the classical theories of light and matter are absolutely incapable of correctly describing the distribution of the frequencies of the light emitted from a blackbody.

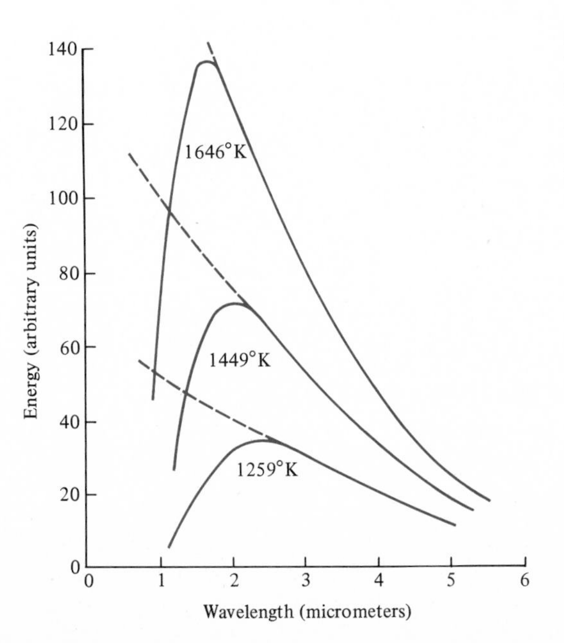

Figure 12-9 Experimental measurement on the spectral distribution of radiation from a blackbody at three different temperatures. [From Walter J. Moore, *Physical Chemistry*, Fourth Edition, © 1972, p. 581. Reprinted by permission of Prentice-Hall, Inc., Englewood Cliffs, N.J.]

The Bettmann Archive

Max Planck, 1858–1947

12-5 Planck's Quantum of Light

It was on October 19, 1900, that Max Planck (1858–1947), a professor of physics at Berlin, presented to the Berlin Physical Society the explanation of the distribution of radiation from a blackbody and in the process shook the science of physics to its very foundations. Planck's explanation was revolutionary because it rejected what was considered an ancient truth. All scientists at that time believed that change takes place in a continuous fashion. Nature does not make jumps. If a body is heated, it can absorb all possible energies sent to it and can reemit all possible energies.

Planck's explanation broke with this idea by postulating that electromagnetic radiation (including visible light) is emitted in small packages called **quanta.** The energy of the quantum of radiation is directly proportional to the frequency (ν) of the radiation. The proportionality constant is called **Planck's constant,** h. We have then

$$E = h\nu \tag{7}$$

where E = energy of radiation, ergs
h = Planck's constant = 6.626×10^{-27} erg · s
ν = frequency of radiation, s^{-1}

Using this new concept, Planck was able to reproduce the experimental spectral distribution curves from a blackbody.

Equation (7) is very interesting. It tells us that the amount of energy of light (or any other electromagnetic radiation) is proportional to its frequency. **The higher the frequency, the greater the energy.** Thus, ultraviolet light has more energy than visible light, which in turn has more energy than infrared light. Also, blue light has a higher energy than green light, which in turn has a higher energy than red light. Equation (7) also tells us how energetic x rays are. From equations (1) and (7) we can calculate how the energy depends on the wavelength of light.

$$E = h\nu \tag{7}$$

$$c = \lambda\nu \tag{1}$$

Therefore,

$$E = \frac{hc}{\lambda} \tag{8}$$

Example: Dentists use what is termed "soft" x rays as a tool for discovering cavities in teeth. Their wavelength is approximately 10^{-6} cm. Calculate the frequency of soft x rays and their energy and compare them with the frequency and energy emitted from an infrared lamp with a wavelength of approximately 7×10^{-4} cm.

Answer: For x rays,

$$\lambda = 10^{-6}\ \text{cm}$$

$$c = \nu\lambda$$

$$\nu = \frac{c}{\lambda} = \frac{3 \times 10^{10}\ \text{cm/s}}{10^{-6}\ \text{cm}} = 3 \times 10^{16}\ \text{s}^{-1}$$

$$E = h\nu = (6.6 \times 10^{-27}\ \text{erg} \cdot \text{s})(3 \times 10^{16}\ \text{s}) = 2 \times 10^{-10}\ \text{erg}$$

For infrared,

$$\lambda = 7 \times 10^{-4}\ \text{cm}$$

$$\nu = \frac{c}{\lambda} = \frac{3 \times 10^{10}\ \text{cm/s}}{7 \times 10^{-4}\ \text{cm}} = 4 \times 10^{13}\ \text{s}^{-1}$$

$$E = h\nu = (6.6 \times 10^{-27}\ \text{erg} \cdot \text{s})(4 \times 10^{3}\ \text{s}^{-1})$$
$$= 3 \times 10^{-13}\ \text{erg}$$

Clearly x rays have much higher energies than infrared radiation.

The energy of visible light, and some radiation with lower frequencies than that, is enough to cause some harm to human beings. Skin cancers have been produced by overexposure to sunlight. This effect may take many years of prolonged exposure. High-frequency radiation has much greater energies and these, for example x rays and gamma rays, can cause genetic damage, radiation sickness, and even death (see Section 23-2).

There is another unit which is frequently used to describe electromagnetic radiation. This is called the **wave number** (symbol $\bar{\nu}$), and it is the number of crests (or troughs) in 1 cm (Figure 12-10). The wave number is obviously the reciprocal of the wavelength and can be calculated by

$$\bar{\nu} = \frac{1}{\lambda} \tag{9}$$

From the facts above, we have

$$E = h\nu = \frac{hc}{\lambda} = hc\bar{\nu} \tag{10}$$

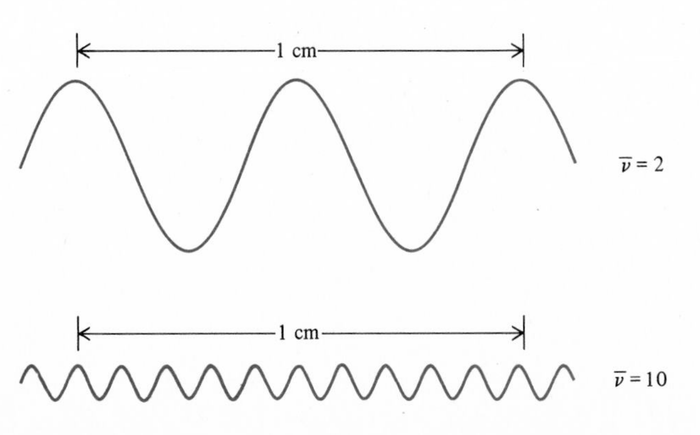

Figure 12-10 Wave numbers.

Example: Calculate the wave number, wavelength, and frequency of a quantum of light with an energy of 2×10^{-12} erg.

Answer:

$$E = h\nu$$

$$\nu = \frac{E}{h} = \frac{2 \times 10^{-12}\ \text{erg}}{6.6 \times 10^{-27}\ \text{erg} \cdot \text{s}} = 3 \times 10^{14}\ \text{s}^{-1}$$

$$\lambda = \frac{c}{\nu} = \frac{3 \times 10^{10}\ \text{cm/s}}{3 \times 10^{14}\ \text{s}^{-1}} = 1 \times 10^{-4}\ \text{cm}$$

$$\bar{\nu} = \frac{1}{\lambda} = \frac{1}{1 \times 10^{-4}\ \text{cm}} = 1 \times 10^{4}\ \text{cm}^{-1}$$

12-6 The Photoelectric Effect

At about the same time that blackbody radiation was causing such a problem for physicists, another phenomenon was emerging. This was the **photoelectric effect,** discovered by Heinrich Hertz and Philipp Lenard (1862–1947) around 1890. When light of the proper frequency range is allowed to shine on certain metal surfaces, electrons are emitted. The following two observations can be made using a typical photoelectric apparatus as shown in Figure 12-11:

1. When light of low frequency is allowed to shine on the metal surface, no effect occurs. As the frequency is raised, a point is reached when current flows. Any higher frequency will also produce a current.
2. One can determine the kinetic energy of the emitted electrons. This kinetic energy increases with increasing *frequency* of the light shining on the metal surface. Keeping the *frequency* of the light constant while increasing the *amount* (or intensity) of the light shining on the metal surface increases the number of electrons being emitted from the electrode but does *not* increase the kinetic energy of the emitted electrons. This, too, was contrary to the laws of nineteenth-century physics, since these laws predicted exactly the opposite: that the energy of the electrons emitted should depend on the intensity of the light and not on the frequency.

In 1905, Albert Einstein (1879–1955) applied Planck's theory of the quantization of light to explain the photoelectric effect. Einstein reasoned that electrons are constituents of all matter. When a quantum of light (also called a **photon**) strikes

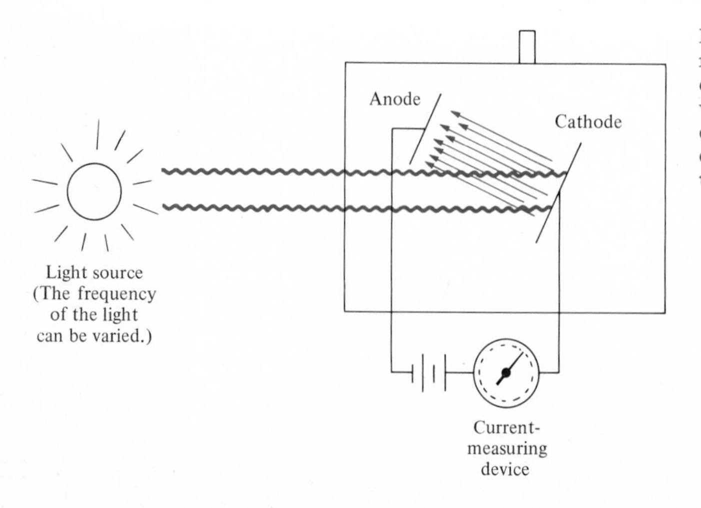

Figure 12-11 Apparatus for measuring the photoelectric effect. The circuit is completed when light of the proper frequency shines on the cathode, emitting electrons that jump to the anode.

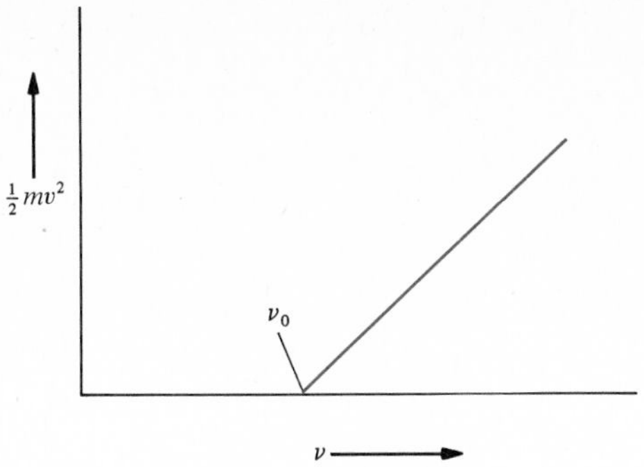

Figure 12-12 A plot of the frequency of impinging radiation on a metal surface vs. the kinetic energy of the ejected electrons.

a metal surface, it imparts its energy to the electrons in the metal. In order for an electron to escape from the surface of the metal, it must overcome the attractive force exerted on it by the positive ions in the metal. Since, for example, a photon of blue light contains more energy than a photon of red light, it would not be surprising to find electrons emitted when blue light shines on the metal surface, whereas red light may not produce emission. Upon being struck by the blue light, the electron would have enough energy to overcome the attractive potential and escape. The energy absorbed by the electron would be equal to the attractive potential of the surface plus the kinetic energy of the emitted electron:

$$E = h\nu = \tfrac{1}{2}mv^2 + \phi \tag{11}$$

energy of photon ($h\nu$); kinetic energy of emitted electron ($\tfrac{1}{2}mv^2$); attractive potential of the surface; called the "work function" of the surface (ϕ)

If the energy of the photons is plotted against the kinetic energy of the emitted electrons, one obtains the curve shown in Figure 12-12. Electrons are emitted only after the frequency of light has passed a critical frequency, labeled ν_0 on the graph. This frequency is proportional to the work function, ϕ, of the metal.

$$\phi = h\nu_0 \tag{12}$$

For higher frequencies $\nu > \nu_0$, the kinetic energy of the electrons increases directly with the increased frequency of the photons. From equations (11) and (12), we have

$$h\nu = \tfrac{1}{2}mv^2 + h\nu_0$$

or

$$h\nu - h\nu_0 = \tfrac{1}{2}mv^2 \tag{13}$$

This equation was confirmed by many carefully performed experiments. Not only did Einstein's explanation of the photoelectric effect lend major support to Planck's hypothesis concerning the nature of light, it also provided physicists a novel method for determining Planck's constant, h.

Example: Calculate the kinetic energy of an electron emitted from a surface of potassium metal (work function $= 3.62 \times 10^{-12}$ erg) by light of the following wavelengths: (a) 7.0×10^{-5} cm (b) 5.5×10^{-8} cm

Answer:

$$h\nu = \tfrac{1}{2}mv^2 + \phi$$
$$\tfrac{1}{2}mv^2 = h\nu - \phi$$

(a) For $\lambda = 7.0 \times 10^{-5}$ cm:

$$\nu = \frac{c}{\lambda} = \frac{3.0 \times 10^{10}\ \text{cm/s}}{7.0 \times 10^{-5}\ \text{cm}} = 4.3 \times 10^{14}\ \text{s}^{-1}$$

$$\tfrac{1}{2}mv^2 = (6.6 \times 10^{-27}\ \text{erg} \cdot \text{s})(4.3 \times 10^{14}\ \text{s}^{-1}) - 3.62 \times 10^{-12}\ \text{erg}$$
$$= 28.3 \times 10^{-13} - 3.62 \times 10^{-12} = -0.79 \times 10^{-12}\ \text{erg}$$

It is physically impossible to have a negative kinetic energy. The work function is greater than the energy of the photon, so the electron will not be emitted and its kinetic energy will be zero.

(b) For $\lambda = 5.5 \times 10^{-8}$ cm

$$\nu = \frac{c}{\lambda} = \frac{3.0 \times 10^{10}\ \text{cm/s}}{5.5 \times 10^{-8}\ \text{cm}} = 5.5 \times 10^{17}\ \text{s}^{-1}$$

$$\tfrac{1}{2}mv^2 = (6.6 \times 10^{-27}\ \text{erg} \cdot \text{s})(5.5 \times 10^{17}\ \text{s}^{-1}) - 3.62 \times 10^{-12}\ \text{erg}$$
$$= 3.63 \times 10^{-9} - 3.62 \times 10^{-12}\ \text{erg}$$
$$= 3.63 \times 10^{-9}\ \text{erg}$$

In this case, an electron will be emitted, with an energy of 3.63×10^{-9} erg.

12-7 The Atomic Spectrum of Hydrogen

As we saw in Section 12-4, when any object is heated to a high enough temperature, it gives off light. In 1860, Robert Bunsen (1811–1899) and Gustav Kirchhoff (1824–1887) decided to analyze the light given off by hot objects by passing it through a prism. Isaac Newton had shown, almost 200 years earlier (around 1665), that when ordinary sunlight is passed through a prism, we get the now-familiar ROYGBIV spectrum. Bunsen and Kirchhoff made the surprising discovery that when they volatilized an element (for example, H_2, Cu, Zn) by heating it in a flame, and passed the light given off through a prism (the device they invented for doing this is called a *spectroscope*), what they got was not a continuous spectrum, but rather a series of lines (a **line spectrum**; Figure 12-13). Even more surprising, *each element gave a different pattern of lines.* This discovery had at least three major consequences for the progress of science:

1. It greatly facilitated the discovery of new elements, since if one suspected that a certain rock, say, contained a new element, all one had to do was to analyze its spectrum, and look for unknown lines.
2. It provided an excellent analytical tool, which is still used today to analyze for the presence of various elements.
3. However, the third consequence was the most important of all: the discovery of spectral lines was to lead (as we shall see) to our present knowledge of atomic structure.

As with blackbody radiation and the photoelectric effect, nineteenth-century physics had no explanation for these spectral lines.

It was soon discovered that atoms and molecules will also *absorb* light as well as emit light. In fact, it was Kirchhoff who stated the general law governing emission and absorption of electromagnetic radiation. This law states that *if a body emits light of a certain wavelength, it will also absorb light of the same wavelength.* Figure 12-14 shows simple experimental setups for obtaining the absorption and emission patterns of hydrogen. The pattern obtained during absorption is called the **absorption spectrum** of the material and the emission pattern is termed the **emission spectrum** of the material. Each substance, not only elements, has its

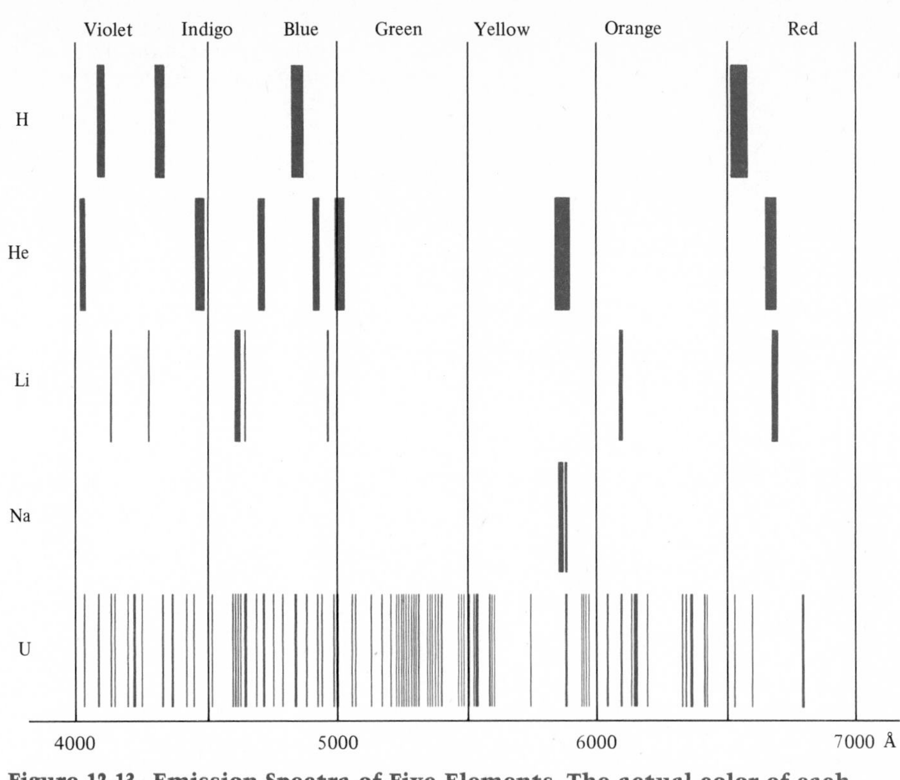

Figure 12-13 Emission Spectra of Five Elements. The actual color of each line corresponds to its position in the spectrum. For example, lines near 4000 Å are violet; those near 7000 Å are red.

own *characteristic* absorption and emission spectrum. A spectrum of a substance is a fingerprint of that substance and can often be used to identify it and to learn something of its structure.

Because it is the simplest substance, it is not surprising that a great deal of attention was paid to the emission spectrum of atomic hydrogen (Figure 12-13b).

In 1885, J. J. Balmer (1825–1898) discovered that the lines in the visible region of the spectrum of atomic hydrogen were not just randomly placed. There was a mathematical relationship which governed the position of these lines: This relationship is

$$\bar{\nu} = R_H\left(\frac{1}{2^2} - \frac{1}{n^2}\right) \tag{14}$$

where $n = 3, 4, 5, \ldots$. That is, if we let $n = 3$, the expression in parentheses in equation (14) becomes $\frac{1}{4} - \frac{1}{9}$, or 0.139. If this is multiplied by an appropriate constant (R_H), we will get the wave number corresponding to the first line in the spectrum; if we let $n = 4$ and repeat the procedure, *using the same constant,* we will get the second line; and so on. This proved that nature did not put the lines in at random, although the reason for the pattern could not be guessed at that time.

Other series were found in different regions of the spectrum. Johannes Rydberg (1854–1919) finally put all the different series together into one general expression, called the *Rydberg formula:*

$$\bar{\nu} = R_H\left(\frac{1}{n_1^2} - \frac{1}{n_2^2}\right) \tag{15}$$

In the Rydberg formula, n_1 and n_2 are integers with n_2 always greater than n_1. R_H is

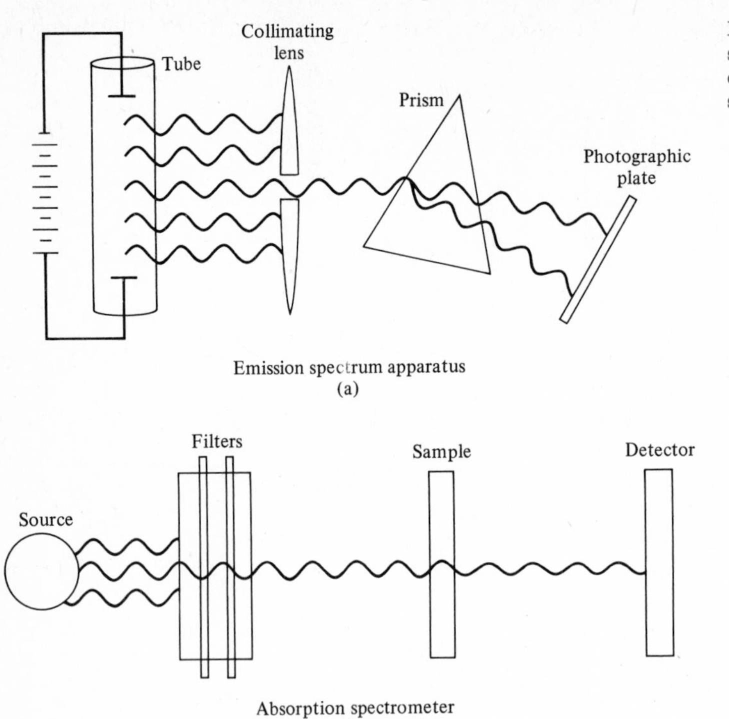

Figure 12-14 Experimental setups for obtaining emission and absorption spectra.

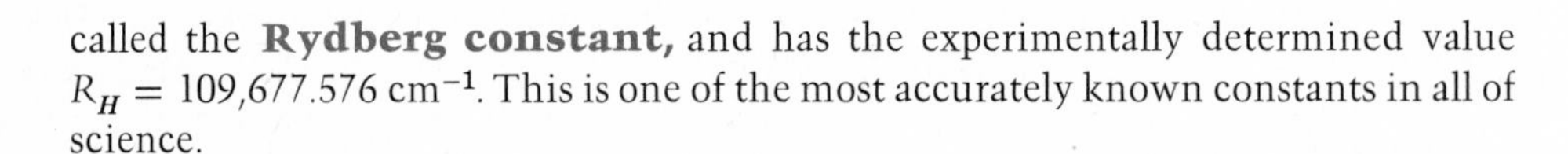

called the **Rydberg constant,** and has the experimentally determined value $R_H = 109{,}677.576\ \text{cm}^{-1}$. This is one of the most accurately known constants in all of science.

Example: In the Balmer series, $n_1 = 2$. Calculate $\bar{\nu}$ for the lines for $n_2 = 3$, 4, and 7.

Answer: From equation (15), $\bar{\nu} = R_H(1/n_1^2 - 1/n_2^2)$.

$$n_2 = 3: \quad \bar{\nu} = 109{,}677.576\left(\frac{1}{2^2} - \frac{1}{3^2}\right) = 15{,}232.997\ \text{cm}^{-1}$$

$$n_2 = 4: \quad \bar{\nu} = 109{,}677.576\left(\frac{1}{2^2} - \frac{1}{4^2}\right) = 20{,}564.546\ \text{cm}^{-1}$$

$$n_2 = 7: \quad \bar{\nu} = 109{,}677.576\left(\frac{1}{2^2} - \frac{1}{7^2}\right) = 25{,}181.074\ \text{cm}^{-1}$$

Example: In the Lyman series, $n_1 = 1$, calculate $\bar{\nu}$ for the line $n_2 = \infty$. This is the highest emission line in this series.

Answer: From the equation, $\bar{\nu} = R_H(1/n_1^2 - 1/n_2^2)$.

$$n_2 = \infty: \quad \bar{\nu} = 109{,}677.576\left(\frac{1}{1^2} - \frac{1}{\infty^2}\right)$$
$$= 109{,}677.576(1 - 0) = 109{,}677.576$$

12-8 The Interpretation of Atomic Spectra

Many similar series were subsequently observed for other elements. The discovery of mathematical relationships like equations (14) and (15) only deepened the mystery of the spectral lines. A number of questions naturally arose:

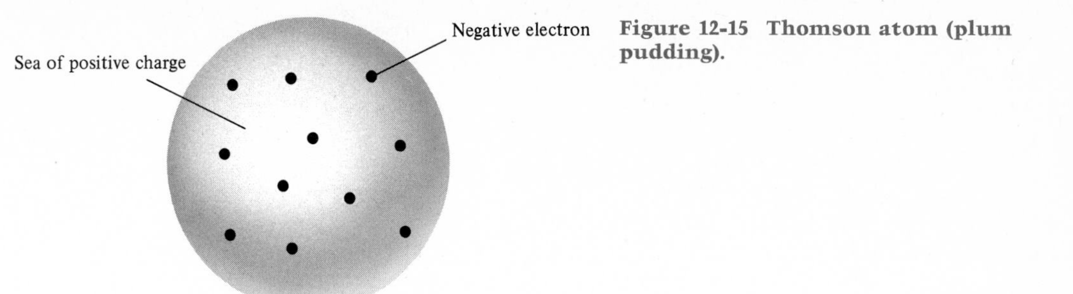

Figure 12-15 Thomson atom (plum pudding).

1. In the spectrum of hydrogen, what do the values of $n = 1, n = 2, \ldots, n = \infty$ signify in a physical sense?
2. Why do the lines arise only when the values of n are integers? Why don't fractional values of n give rise to lines?
3. Since other elements show the same behavior, what structures do atoms have in common that produce this similarity?

We will turn now to the structure of atoms to see how the above questions were finally resolved.

12-9 The Thomson and Rutherford Atoms

In 1898, J. J. Thomson proposed that atoms may be considered as being composed of a ball of positive electricity in which negative electrons are stuck. This became known as the "plum pudding" model of the atom (Figure 12-15). It was an excellent proposal since it conformed to all the observations known at that time.

In 1911, Ernest Rutherford (1871–1937) had a graduate student named Marsden, who was performing experiments on the scattering of alpha particles (alpha particles are helium nuclei*) by thin metal sheets. The sheet of metal was vertically mounted and a stream of alpha particles was directed at it (Figure 12-16). Since the Thomson atom was the accepted model of atoms, both Rutherford and Marsden expected that the very highly energetic particles would pass through the thin sheet without major deflections. They observed that almost all of the alpha particles did indeed pass through the sheet. However, to their great astonishment, every now and then an alpha particle came flying back at them. In his description of this, Rutherford wrote that it was incredible to find an alpha particle being returned; it was as if you were firing a 15-in artillery shell at a piece of tissue paper and the tissue paper knocked the shell back at you. Rutherford quickly realized that the Thomson atom could in no way account for this experimental observation. The Thomson atom had its positive and negative electricity smeared out in such a manner that the positively charged

*Rutherford, of course, did not know that alpha particles were helium nuclei, since the fact that an atom had a nucleus was not yet known, but he did know that they were very dense positively charged particles.

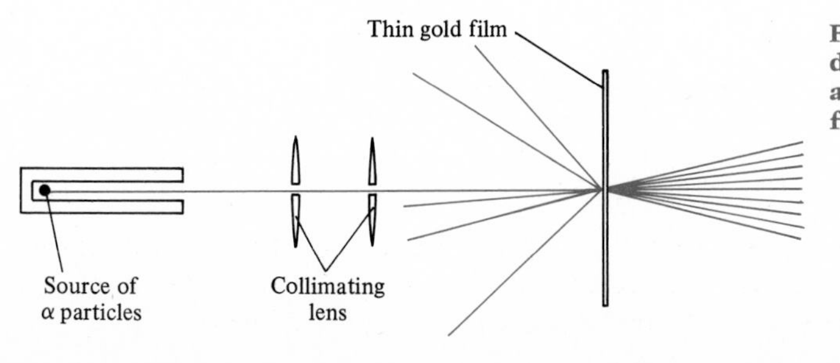

Figure 12-16 Rutherford-Marsden experiment, in which alpha particles are shot at thin foils.

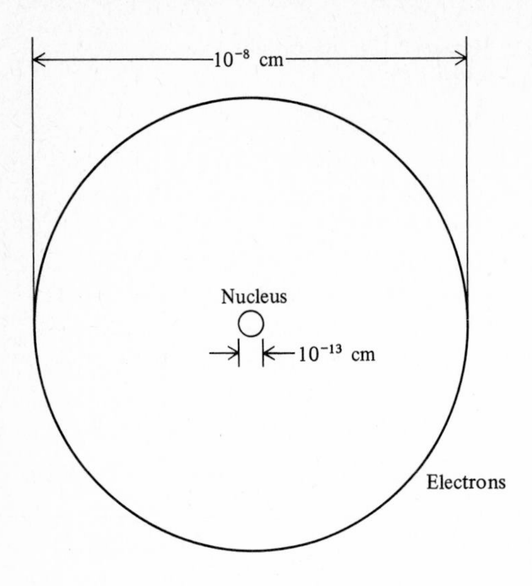

Figure 12-17 The Rutherford atom.

alpha particles would never encounter a repulsion strong enough to deflect them 180°.

To account for the large deflections observed in his alpha particle experiments, Rutherford proposed that an atom is composed of a dense positive nucleus, thinly surrounded by negative electrons. The mass and positive charge of the nucleus are concentrated in a very small region of space. The electrons are outside the nucleus, so that an atom is almost entirely empty space (Figure 12-17). We now know that the diameter of an atom is of the order of 10^{-8} cm, while the nucleus has a diameter of approximately 10^{-13} cm. Thus, the ratio of the diameter of an entire atom to that of the nucleus is about 100,000 : 1. If the nucleus were the size of a baseball, the entire atom would have a diameter of about 5 miles.

Rutherford used his model of an atom to explain the scattering experiments. Since atoms are mostly empty space, almost all the alpha particles pass through the thin metal sheet without being deflected. Every once in a while, an alpha particle will directly approach a nucleus. The nuclei are very dense, highly positively charged species. Since the alpha particle is also positively charged, it will be repelled very strongly upon approaching a nucleus (Figure 12-16).

This was a very satisfactory explanation of the scattering of particles by thin metal sheets. There was, however, a problem with the Rutherford model. Certain scientists of that time thought that the electrons might be in orbits outside the nuclei. Classical electromagnetic theory says that if a charged particle accelerates around an oppositely charged particle, the former will radiate energy. If it radiates energy, it will not be able to stay in a fixed orbit about the nucleus (because its speed must decrease) but should go into a spiral motion, finally falling into the nucleus (Figure 12-18). This, of course, does not happen for electrons in atoms and the discrepancy could not be explained at that time.

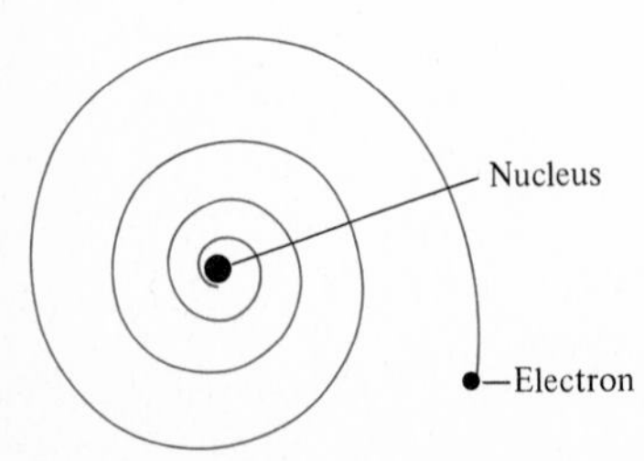

Figure 12-18 An accelerating electron was supposed to radiate energy and spiral into the nucleus.

12-10 The Bohr Atom

Rutherford's picture of the structure of atoms was a big step forward, but it did not say anything about where the electrons were (except that they were outside the nucleus) and, as we saw, it could not explain why the electrons did not fall into the nucleus. In 1913, a Danish genius named Niels Bohr (1885–1962) proposed a structure for atoms that explained some of these things and furthermore explained the lines in the hydrogen spectrum, which had remained a mystery for more than 25 years. As we shall see, Bohr's theory was not the last word in answering all the fundamental questions posed in the early 1900s, and, in fact, it can only explain the emission spectrum of hydrogen; it fails miserably when applied to other atoms. Nevertheless, it was important in its time, and the most important of its basic ideas are still valid. Bohr's theory was based on Planck's quantum theory and was built on two fundamental postulates.

1. *The electron of the hydrogen atom moves in a circular orbit about the positive nucleus. In moving in this orbit, the electron does not radiate any energy.*

In this first postulate, Bohr never explained *why* an electron does not radiate energy, although, as we have seen (Figure 12-18), it was a consequence of classical physics that an accelerating charged particle must do so. He simply stated that this is the way nature behaves. An electron in orbit around a nucleus does not radiate energy. Of course, he was also telling his fellow physicists that their physics of "classical electromagnetism" cannot be applied to atomic particles.

2. *An electron in a hydrogen atom may occupy orbits for which the* **angular momentum** *(mvr) of the electron is an integral multiple of Planck's constant divided by 2π:*

$$\text{angular momentum} = mvr = n\frac{h}{2\pi} \qquad (16)$$

where m = mass of the electron
v = velocity of the electron
r = radius of the orbit
n = a positive integer
h = Planck's constant

Figure 12-19 shows how to calculate the angular momentum of a particle in an orbit. If no external forces are acting on a particle, its angular momentum does not change.

Note that the right-hand side of equation (16) incorporates Planck's constant h into the theorem concerning the angular momentum of an electron. It is important to realize that Planck's constant does not occur in any way in Newtonian mechanics. It is unique to the new mechanics of elementary particles.

What Bohr's second postulate says is that there are an infinite number of orbits possible for the electron in a hydrogen atom. The radius of one of these orbits can be

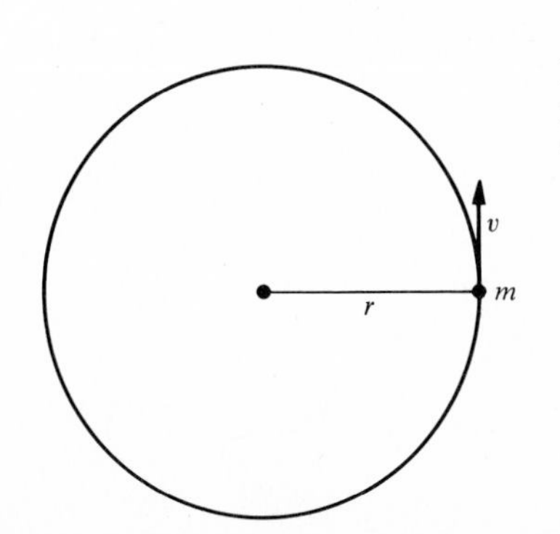

Figure 12-19 The calculation of angular momentum of a particle of mass *m* traveling with a velocity *v* in a circular orbit of radius *r*.

obtained by letting $n = 1$ and solving equation (16) for r. Similarly, if we let n = other integers, we get the radii of other orbits.

The value of n can be 1, 2, 3, or any positive integer up to infinity. This means that the value of the angular momentum can be

$$mvr = \frac{h}{2\pi} \quad \text{or}$$

$$mvr = \frac{2h}{2\pi} \quad \text{or}$$

$$\vdots$$

$$mvr = \frac{17h}{2\pi}$$

and so on

but despite the fact that there are an infinite number of possible values, **there are no values in between any two consecutive solutions of equation (16).** The angular momentum is said to be **quantized;** that is, it can have only certain specific values and *no others*. The integer n in equation (16) is called a **quantum number.**

This part of Bohr's theory has proven to be perfectly correct, and required a very profound change in the thinking of scientists. If we are in a car going 35 miles/h, we know that it is also possible to go 34 or 36 miles/h, or any speed in between, for example, 34.73 or 35.21 miles/h. If we were told that it was impossible for the car to travel at 34.84 miles/h, that it could travel at 34.81 or 34.88 but never at any speed in between these numbers, we would not believe it. *Yet this is just what Bohr said about the behavior of electrons in an atom.* They may travel only with angular momenta that are equal to an integral multiple of $h/2\pi$, but not with any angular momentum in between two adjacent allowed numbers. Only certain energies were allowed; the electrons could never have any others. We might suppose that the scientists of the time would not accept such a revolutionary theory. Yet most of them did accept it, because with these two postulates, Bohr was able to calculate the Rydberg constant (Section 12-7), and thus explain the spectral lines of hydrogen. The conclusion that the angular momentum (and consequently the energy) possessed by the electron of a hydrogen atom could only take on certain values, but could never take on a value in between any two consecutive values, thus forced itself upon the scientific thinking of the time. (Another reason that the theory was so quickly accepted was that previous revolutionary theories and discoveries—those of Thomson, Planck, Einstein, and others—had paved the way; classical physics was breaking down.)

By use of equation (16), along with the laws of classical physics and electromagnetism, Bohr was able to show that the energy of a hydrogen electron is quantized and given by the expression

$$E = -\frac{J}{n^2} \tag{17}$$

where E is the energy of the electron and n is the same quantum number found in equation (15). J is a constant composed of known universal constants:

$$J = \frac{2\pi^2 m e^4}{h^2} \tag{18}$$

$$J = 313.5 \text{ kcal/mole}$$

$$J = 2.179 \times 10^{-11} \text{ erg/atom}$$

where m = mass of the electron
$= 9.10953 \times 10^{-28}$ g

$$e = \text{charge of the electron} = 4.8030 \times 10^{-10}\ \text{esu}$$
$$h = \text{Planck's constant} = 6.6262 \times 10^{-27}\ \text{erg} \cdot \text{s}$$

Bohr was also able to calculate the radii of the electron orbits in a hydrogen atom. They fit the expression

$$r = n^2\left(\frac{h^2}{4\pi^2 me^2}\right) = n^2 \times 0.529 \times 10^{-8}\ \text{cm} \tag{19}$$

where n is the quantum number. For $n = 1$,

$$r = 0.529 \times 10^{-8}\ \text{cm} = a_0 \tag{20}$$

This last quantity, a_0, is called the first **Bohr radius,** and was taken by Bohr to be the radius of the orbit of lowest possible energy of the electron in a hydrogen atom.

Example: Calculate the first five Bohr radii.

Answer: From equation (19), $r = n^2 \times 0.529 \times 10^{-8}$ cm

$$n = 1: \quad r = 1^2 \times 0.529 \times 10^{-8} = 0.529 \times 10^{-8}\ \text{cm}$$
$$n = 2: \quad r = 2^2 \times 0.529 \times 10^{-8} = 2.12 \times 10^{-8}\ \text{cm}$$
$$n = 3: \quad r = 3^2 \times 0.529 \times 10^{-8} = 4.76 \times 10^{-8}\ \text{cm}$$
$$n = 4: \quad r = 4^2 \times 0.529 \times 10^{-8} = 8.46 \times 10^{-8}\ \text{cm}$$
$$n = 5: \quad r = 5^2 \times 0.529 \times 10^{-8} = 13.2 \times 10^{-8}\ \text{cm}$$

Example: Calculate the five lowest energy levels of the hydrogen atom.

Answer: From equation (17), $E = -J/n^2 = \dfrac{-2.179 \times 10\ \text{erg/atom}}{n^2}$.

$$n = 1: \quad E = \frac{-2.179 \times 10^{-11}}{1^2}\ \text{erg/atom} = -2.179 \times 10^{-11}\ \text{erg/atom}$$
$$n = 2: \quad E = \frac{-2.179 \times 10^{-11}}{2^2} = -0.5448 \times 10^{-11}\ \text{erg/atom}$$
$$n = 3: \quad E = \frac{-2.179 \times 10^{-11}}{3^2} = -0.2421 \times 10^{-11}\ \text{erg/atom}$$
$$n = 4: \quad E = \frac{-2.179 \times 10^{-11}}{4^2} = -0.1362 \times 10^{-11}\ \text{erg/atom}$$
$$n = 5: \quad E = \frac{-2.179 \times 10^{-11}}{5^2} = -0.08716 \times 10^{-11}\ \text{erg/atom}$$

Bohr's theory was thus able to explain why the absorption and emission spectra of hydrogen atoms consisted of lines whose positions in the spectra fitted equation (15). The electron in each hydrogen atom has an infinite number of possible energy values, but at ordinary temperatures, practically all of them are in the lowest possible level ($n = 1$), called the **ground state.** Figure 12-20 shows the possible energy levels. Remember that each frequency of light corresponds to a different energy [equation (7)]. White light consists of many frequencies, each with a corresponding energy. When white light shines on a sample of hydrogen, one of these frequencies (and only one) has *exactly* the right amount of energy necessary to promote an electron from level 1 to level 2. This frequency will thus be absorbed by the sample, which will use it to promote some of its electrons to level 2.

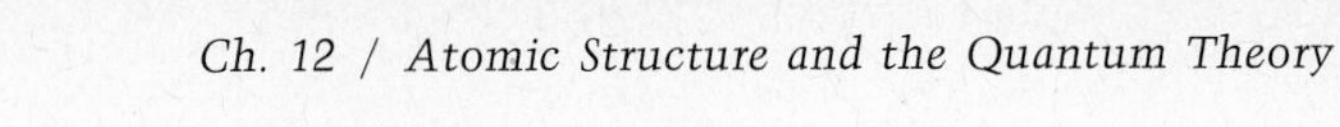

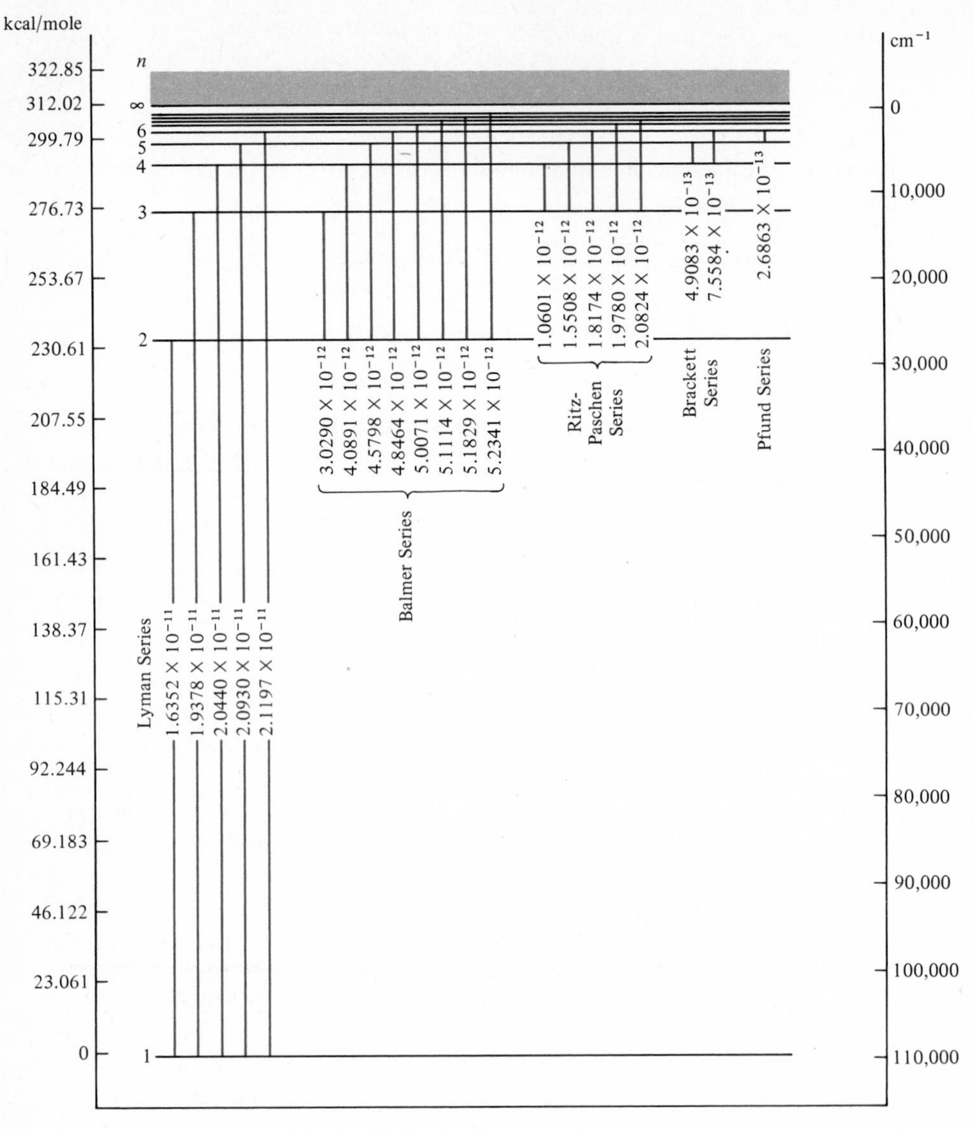

Figure 12-20 The energy levels of the hydrogen atom.

Another frequency has *exactly* the right amount of energy to promote an electron from level 1 to level 3. This frequency will also be absorbed by the sample. The same will be true for frequencies that promote to levels 4, 5, etc. The other frequencies, the ones whose energies do not correspond to any ΔE value, will not be absorbed by the sample; they will pass through. Thus, what will pass through the sample will be white light containing black lines at just those frequencies which were absorbed (since they were absorbed, they do not pass through). If the amount of energy necessary to promote the electron to level 2 (ΔE_1) corresponds to light of $\bar{\nu} = 8.20 \times 10^4$ cm^{-1}, it is obvious that light of lower energy, say $\bar{\nu} = 8.0 \times 10^4$ cm^{-1}, will not be able to promote the electron, and will simply pass through. But the same also applies to light of higher energy. For example, light of 8.50×10^4 cm^{-1} will also pass through. The electron is unable to take out 8.20×10^4 cm^{-1} of this and give back 0.30×10^4 cm^{-1} in change. *It can only absorb the exact amount of energy.*

Emission spectra are formed in a somewhat different way. In this case, the sample must be heated or irradiated, and the added energy of heat or light causes the electrons to be promoted to higher levels. However, they do not stay up there, but drop back down. The law of conservation of energy says that when they drop down to lower energy levels, they must give back this energy. When they do, it is often in the form of light, and this light constitutes the emission spectrum. It is clear that when an electron falls from level 2 to level 1, it must give out light of $\bar{\nu} = 8.20 \times 10^4\ \text{cm}^{-1}$ (since that is the wave number corresponding to ΔE_1). That is the reason why the lines of an emission spectrum correspond to the lines in the absorption spectrum of the same substance.

Example: Calculate the wave number of a photon of light that will excite an electron from the $n = 1$ to $n = 4$ energy level.

Answer: From the example on page 367,

$$\Delta E(n_1 \longrightarrow n_4) = (-2.179 \times 10^{-11} + 0.1362 \times 10^{-11})\ \text{erg}$$
$$= -2.043 \times 10^{-11}\ \text{erg/atom}$$

Energy of the photon must be equal to 2.043×10^{-11} erg/atom. From equation (10),

$$\Delta E = hc\bar{\nu}$$

$$\bar{\nu} = \frac{2.043 \times 10^{-11}\ \text{erg}}{6.626 \times 10^{-27}\ \text{erg} \cdot \text{s} \times 3.000 \times 10^{10}\ \text{cm/s}} = 1.028 \times 10^5\ \text{cm}^{-1}$$

Bohr's theory predicted exactly the known emission spectrum of hydrogen. At that time no series was known where $n_1 = 1$ and $n_2 = 2, 3, 4, \ldots$. Bohr predicted that such a series did exist, and a year later, it was observed by Lyman, in the ultraviolet spectrum of hydrogen. Furthermore, as we have already mentioned, Bohr showed that if one starts with his two postulates and the laws of classical mechanics, one can derive equation (15) with a value for the Rydberg constant R_H of 109,733 cm^{-1} (compare the experimental value of 109,677.576 cm^{-1}). The Bohr theory is capable of explaining the emission spectra of all one-electron systems. He^+, Li^{2+}, and so on, are one-electron ions and their emission spectra can be reproduced by this theory. However, despite the remarkable success of the Bohr theory in explaining the spectra of one-electron systems, when scientists examined the emission spectra of other elements and ions, the Bohr theory was found wanting. It simply failed to explain what was observed experimentally for systems containing more than one electron. There were many attempts to salvage Bohr's theory, but they also failed.

12-11 Matter Waves

Louis Victor de Broglie (1892–) in 1923, a year before he was granted his doctorate in theoretical physics, showed that with any particle, it was possible to associate a wave. These waves are *not* electromagnetic waves and hence have been called **matter waves.** De Broglie reasoned that the wavelength λ of a particle is inversely proportional to its momentum, p:

$$\lambda \propto \frac{1}{p}$$

The proportionality constant is Planck's constant h.*

*This relationship can be simply derived from Planck's ($E = h\nu$) and Einstein's ($E = mc^2$; see Section 23-7) equations:

$$E = h\nu$$
$$E = mc^2$$
$$h\nu = mc^2 = \frac{hc}{\lambda}$$
$$\lambda = \frac{h}{mc} = \frac{h}{p}$$

$$\lambda = \frac{h}{p} \tag{21}$$

where $p = mv$ = momentum of the particle
h = Planck's constant
λ = wavelength associated with a particle of momentum p

Example: Calculate the wavelength of a 5-oz baseball traveling at 110 miles/h.

Answer: First calculate the momentum of the baseball:

$$5 \text{ oz} \times \left(\frac{1 \text{ lb}}{16 \text{ oz}}\right) \times \left(\frac{464 \text{ g}}{1 \text{ lb}}\right) = 145 \text{ g}$$

$$110 \frac{\text{miles}}{\text{h}} \times \left(\frac{1 \text{ h}}{60 \text{ min}}\right)\left(\frac{1 \text{ min}}{60 \text{ s}}\right)\left(\frac{5280 \text{ ft}}{1 \text{ mile}}\right)\left(\frac{12 \text{ in}}{1 \text{ ft}}\right)\left(\frac{2.54 \text{ cm}}{1 \text{ in}}\right) = 4.92 \times 10^3 \frac{\text{cm}}{\text{s}}$$

$$p = mv = (145 \text{ g})\left(4.92 \times 10^3 \frac{\text{cm}}{\text{s}}\right) = 7.13 \times 10^5 \frac{\text{g} \cdot \text{cm}}{\text{s}} \text{ (momentum)}$$

$$\lambda = \frac{h}{p} = \frac{6.63 \times 10^{-27} \text{ erg} \cdot \text{s}}{7.13 \times 10^5 \text{ g} \cdot \text{cm/s}} = 9.30 \times 10^{-33} \text{ cm} = 9.30 \times 10^{-25} \text{ Å}$$

This is an incredibly small wavelength and would be impossible to observe.

Example: Calculate the wavelength associated with an electron traveling at one-tenth the speed of light.

Answer:

$$\lambda = \frac{h}{p}$$

$$m = 9.1 \times 10^{-28} \text{ g}$$

$$v = 0.30 \times 10^{10} \text{ cm/s}$$

$$p = mv = (9.1 \times 10^{28} \text{ g})(3.0 \times 10^9 \text{ cm/s}) = 27 \times 10^{-19} \text{ g} \cdot \text{cm/s}$$

$$\lambda = \frac{h}{p} = \frac{6.626 \times 10^{-27} \text{ erg} \cdot \text{s}}{2.7 \times 10^{-18} \text{ g} \cdot \text{cm/s}} = 2.5 \times 10^{-9} \text{ cm} = 0.25 \times 10^{-8} \text{ cm} = 0.25 \text{ Å}$$

High-speed electrons have the same wavelength as high-energy x rays. In Section 15-1 we shall see that x rays can be diffracted by crystals (ordinary light can be diffracted; x rays are a high-energy form of light). If that is the case, why not electrons? In 1927, C. J. Davisson and L. H. Germer studied the scattering reflection of electrons from a metallic nickel target. They found that not only were the electrons reflected from the target as expected, but they were also diffracted. Diffraction is a property of waves, not particles, and in this way they confirmed de Broglie's theory of matter waves. In 1937, they were awarded the Nobel Prize in Physics. Electron diffraction is today a standard technique used for determining the structure of gaseous molecules.

The energy of an electron that is accelerated through a potential difference of one volt is called an **electron volt,** eV. Electron volts, being a unit of energy, can be converted to ergs. The conversion factor is

$$1 \text{ eV} = 1.6021 \times 10^{-12} \text{ erg} \tag{22}$$

Example: Calculate the energy in ergs, the velocity, the momentum, and the wavelength of an electron that has been accelerated through a potential difference of 50,000 volts. (The mass of an electron is 9.11×10^{-28} g.)

Answer: Accelerating an electron through a potential difference of 50,000 volts yields 50,000 eV.

$$50{,}000 \text{ eV} \times \frac{1.6021 \times 10^{-12} \text{ erg}}{1 \text{ eV}} = 8.01 \times 10^{-8} \text{ erg}$$

All of this energy is kinetic energy, $\frac{1}{2}mv^2$:

$$\frac{1}{2}mv^2 = 8.01 \times 10^{-8} \text{ erg}$$

$$v = \left[\frac{(2)8.01 \times 10^{-8} \text{ erg}}{9.11 \times 10^{-28} \text{ g}}\right]^{1/2}$$

$$= 1.33 \times 10^{10} \text{ cm/s}$$

$$p = mv = (9.11 \times 10^{-28} \text{ g})(1.33 \times 10^{10} \text{ cm/s})$$

$$= 12.1 \times 10^{-18} \text{ g} \cdot \text{cm/s}$$

$$\lambda = \frac{h}{mv} = \frac{6.63 \times 10^{-27} \text{ erg} \cdot \text{s}}{12.1 \times 10^{-18} \text{ g} \cdot \text{cm/s}} = 5.48 \times 10^{-10} \text{ cm}$$

12-12 The Compton Effect

As we have seen, Davisson and Germer experimentally confirmed that particles have wave properties associated with them. A few years before their work, Arthur Holly Compton (1892–1962) observed that when x rays were scattered by matter, some of the x rays had, in the act of scattering, lengthened their wavelength. Compton was able to explain this result by postulating that when a photon of light struck an electron, the photon lost some energy, thereby increasing its wavelength (Figure 12-21). This process makes it appear that a photon acts as a particle, under certain circumstances. This completes the picture of a wave acting as a particle which was earlier foreshadowed in the work of Planck and Einstein (Section 12-6), and also allows us to go the other way in the de Broglie theory. That is, given the wavelength of a photon, one can calculate its associated particle momentum. For this discovery, called the **Compton effect,** Arthur Compton was awarded the Nobel Prize in Physics in 1927.

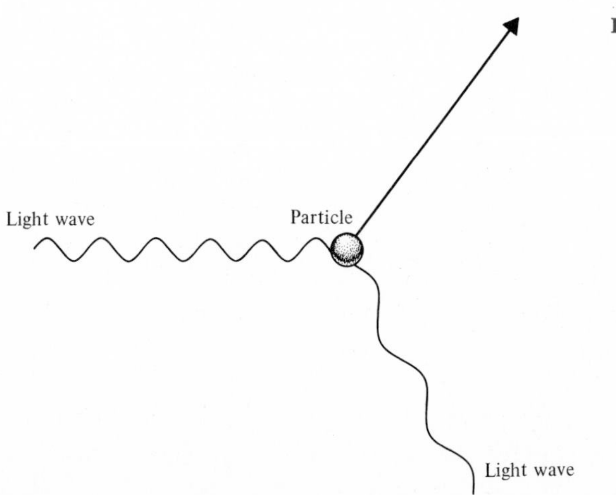

Figure 12-21 The Compton effect.

Example: Calculate the momentum of a photon whose frequency is $6.00 \times 10^{14}\ s^{-1}$.

Answer: $\lambda = h/p$ or $p = h/\lambda$. To calculate λ:

$$\lambda = \frac{c}{\nu} = \frac{3.00 \times 10^{10}\ cm/s}{6.00 \times 10^{14}\ s^{-1}} = 5.00 \times 10^{-5}\ cm$$

$$p = \frac{6.63 \times 10^{-27}\ erg \cdot s}{5.00 \times 10^{-5}\ cm} = 1.33 \times 10^{-22}\ g \cdot cm/s$$

Thus, the classical physics of the late nineteenth century is now confounded. Those physicists thought there was a clear distinction between particles and waves; but now we see that particles can behave as waves and waves can behave as particles. In the next section we shall discuss still another blow to the thinking of the nineteenth-century physicists.

12-13 Heisenberg's Uncertainty Principle

Newtonian mechanics, which we briefly mentioned at the beginning of this chapter, is based on a purely *deterministic* philosophy of science. It was maintained by many that the entire history of the whole universe, *both its past and its future,* could be predicted from calculations if the *position, velocity,* and forces acting on every particle in the universe were known at a *single instant of time.* This philosophy, that everything that happens is determined by what happened before, is called **determinism.** The past would be known because from the known position and momentum of all particles, we could extrapolate back to where these particles had previously been; and the future would be known because given the position, velocity, and forces of all particles, we could calculate all future positions and velocities that these particles would have to assume. Of course, the nineteenth-century physicists knew that there were two extremely huge tasks involved here: learning the position, velocity, and forces of all particles simultaneously, and then making the necessary calculations. They did not really expect to be able to do it in practice. But in *principle,* they believed it was possible to do it.

Werner K. Heisenberg (1901–1976) destroyed this approach to science by his **uncertainty principle.** He stated that it was impossible to know exactly both the position and momentum of any particle at the same instant of time. If we let

Δx = uncertainty in the measurement of position
Δp = uncertainty in the measurement of momentum

then, if we attempt *simultaneously* to measure the position and momentum of any particle, the following is true:

$$\Delta x \cdot \Delta p \geq \frac{h}{2\pi} \qquad (23)$$

Equation (23) says that if one attempts simultaneously to measure the position and momentum of any particle, the error in the measurement of the position, Δx, times the error in the measurement of the momentum, Δp, must be equal to or greater than Planck's constant divided by 2π.

Equation (23) also says that if you knew the position of a particle exactly (no error in its measurement), you would have no information concerning its momentum; and if you knew the momentum exactly, you would have no information concerning its position.

Example: An electron is traveling at a speed of 1.00×10^8 cm/s. The Bohr theory predicts that an electron in a hydrogen atom at its lowest energy will have a radius of 0.529×10^{-8} cm. Assume the error in this measurement is 0.0010×10^{-8} cm. Calculate the uncertainty (error) in the momentum of the electron. The mass of an electron is 9.11×10^{-28} g.

Answer:

$$p = mv = (9.11 \times 10^{-28}\ \text{g})(1.00 \times 10^8\ \text{cm/s}) = 9.11 \times 10^{-20}\ \text{g/cm} \cdot \text{s}$$

$$\Delta p \geq \frac{h}{2\pi\, \Delta x} = \frac{6.63 \times 10^{-27}\ \text{erg} \cdot \text{s}}{(2\pi)\,(1.0 \times 10^{-11}\ \text{cm})} = 6.7 \times 10^{-17}\ \text{g} \cdot \text{cm/s}$$

The error in the momentum measurement is approximately 1000 times the value of the momentum itself, which means that we cannot measure the momentum at all. This means that we cannot measure the velocity either, so our initial statement (in the example) that an electron is traveling at a speed of 1.00×10^8 cm/s is not an experimental statement, only hypothetical.

Example: A green squash ball weighs 75 g and comes traveling toward you at 400 cm/s. The photon of light emitted from the green ball has a wavelength of 5×10^{-5} cm. Assume that the error in position of the ball is the same as the wavelength itself, namely 5×10^{-5} cm. Calculate the error in momentum of the squash ball.

Answer:

$$p = mv = (75\ \text{g})(400\ \text{cm/s}) = 3.0 \times 10^4\ \text{g} \cdot \text{cm/s}$$

$$\Delta p \geq \frac{h}{2\pi\, \Delta x} = \frac{6.63 \times 10^{-27}\ \text{erg} \cdot \text{s}}{(2\pi)(5 \times 10^{-5}\ \text{cm})} = 2 \times 10^{-23}\ \text{g} \cdot \text{cm/s}$$

This value is so small as to be undetectable.

The preceding two examples illustrate an important feature of Heisenberg's uncertainty principle. The consequences of the principle show up very clearly when one is looking at particles of atomic dimensions moving near the speed of light. But when one performs calculations on macroscopic particles, the errors in position and/or momentum are too small to be detected. Even though the uncertainty principle exists for large-scale bodies, it does not manifest itself in any meaningful way, and under these conditions we safely ignore it.

It is important to understand that the uncertainty principle is a theoretical principle and not based on deficiencies in experimental techniques. The limitation on knowing simultaneously both the position and the momentum of small particles traveling at high velocity is not one that can be overcome by devising new experimental methods. The principle says that no matter how fine our experimental tools become, we will never be able to measure simultaneously both position and momentum to any desired accuracy. Of course, it may turn out that the theory will be modified or proven wrong, as has happened to many theories in the past, but if the theory is true (and it is accepted by all physicists today), it provides a severe limitation on our knowledge of electrons and other atomic particles.

12-14 What Is an Electron?

Some place in this chapter or the next, the reader is probably going to ask himself the question: What is an electron? Is it a wave or is it a particle? This is an old question; it has been asked many times before and the phenomenon is known as the **wave-**

particle duality. In one sense the question arises because in our everyday world we observe particles and we observe waves. In our experience, they are never mixed or confused.

Consider this analogy: There is a man who has been blind since birth and has never encountered a cat. Now he is interested in knowing what a cat is and wants to be able to recognize one if he meets one. We describe a cat to him in the following way. A cat is a small four-legged animal with a tail. It is furry all over. When you pet it, it will purr, and if you scare it, its spine will contract and hunch up. The blind man goes out to seek a cat. When he meets an animal he applies the tests we gave him—four legs, a tail, furry, purring when petted, a spine that contracts when frightened. If it passes all the tests, it is a cat.

We are like the blind man in our understanding of an electron. We cannot see it or feel it, but we can ask it certain questions by experimenting with it. We ask the electron—are you a particle? Do you have momentum? The electron responds yes. Then we ask it: Are you a wave? Can you be refracted? Again the electron responds yes. It turns out that whether the electron tells us it is a particle or a wave depends on which question we ask (that is, which experiment we perform). Nothing in our everyday experience reacts this way. We have no label for it. To the question: Is an electron a particle or a wave? we can only say that it behaves as a particle or as a wave, depending on the experiment.

Problems

12-1 Define each of the following terms. **(a)** electromagnetic spectrum **(b)** magnetic field **(c)** electromagnetic radiation **(d)** wavelength **(e)** frequency **(f)** hertz **(g)** reflection **(h)** refraction **(i)** constructive and destructive interference **(j)** momentum **(k)** cathode ray **(l)** blackbody **(m)** quantum **(n)** wave number **(o)** photoelectric effect **(p)** photon **(q)** work function **(r)** line spectrum **(s)** Rydberg constant **(t)** Balmer series **(u)** angular momentum **(v)** quantum number **(w)** Bohr radius **(x)** ground state **(y)** matter waves **(z)** electron volt **(aa)** determinism **(bb)** wave–particle duality

12-2 Define each of the following symbols. **(a)** c **(b)** λ **(c)** ν **(d)** p **(e)** e/m **(f)** h **(g)** $\bar{\nu}$ **(h)** R_H

12-3 Calculate the wavelength (λ) corresponding to each of the following frequencies. **(a)** 7.14×10^{14} cycles/s **(b)** $1.56 \times 10^{6}\ s^{-1}$ **(c)** $3.00 \times 10^{22}\ s^{-1}$ **(d)** 1.11×10^{15} Hz **(e)** 2.50×10^{13} Hz

12-4 Tell what kind of radiation corresponds to each of the frequencies in Problem 12-3 (for example, visible, ultraviolet, radio waves).

12-5 For each frequency in Problem 12-3, calculate the corresponding wave number $\bar{\nu}$ in cm^{-1}.

12-6 For each frequency in Problem 12-3, calculate the energy in ergs and in calories.

12-7 Calculate the frequency (ν) corresponding to each of the following wavelengths. **(a)** 6800 Å **(b)** 3×10^{-3} nm **(c)** 288 cm **(d)** 3.7×10^{-4} m **(e)** 0.75 Å

12-8 Tell what kind of radiation corresponds to each of the wavelengths in Problem 12-7 (for example, visible, ultraviolet, radio waves).

12-9 For each wavelength in Problem 12-7, calculate the corresponding wave number $\bar{\nu}$ in cm^{-1}.

12-10 For each wavelength in Problem 12-7, calculate the energy in ergs and in calories.

12-11 In Figure 12-5d, two waves of the same frequency are shown canceling each other (complete destructive interference). Could two waves of different frequencies completely cancel each other? Explain.

12-12 Arrange the following kinds of radiation in order of increasing energy.
(a) the light from an infrared oven **(b)** yellow visible light **(c)** a physician's x rays **(d)** radio waves from radio station WABC **(e)** television waves from Channel 5 **(f)** citizen-band radio waves **(g)** light from an ultraviolet lamp **(h)** gamma rays
(i) green visible light

12-13 The first regular radio broadcasting station in the United States was KDKA in Pittsburgh, which in 1920 was licensed to operate for 1 h a day on 300 meters. What frequency was this, in kilohertz?

12-14 Calculate the kinetic energy and momentum of a bullet of mass 18.8 g moving at a velocity of 2.5×10^4 cm/s. Compare the results with the kinetic energy and momentum of a 200-lb man moving with a velocity of 10 yards/s.

12-15 What is the total number of coulombs carried by each of the following?
(a) a collection of Br^- ions that weighs 4.27 g
(b) a collection of S^{2-} ions that weighs 0.386 g
(c) a collection of SO_4^{2-} ions that weighs 2.62×10^{-4} g
(d) a collection of K^+ ions that weighs 16.55 g
(e) a collection of Al^{3+} ions that weighs 8.31×10^{-2} g

12-16 The experimentally determined charge-to-mass ratio (e/m) of the electron is -1.76×10^8 coulomb/g. **(a)** Calculate the mass of an electron if its measured charge is -1.602×10^{-19} coulomb **(b)** If the charge/mass ratio of a proton is 9.65×10^4 coulombs/g, calculate the mass ratio of a proton to an electron. **(c)** From the results obtained in parts (a) and (b), calculate the mass of a proton. **(d)** Calculate the mass of a mole of protons and a mole of electrons.

12-17 One of the methods used in the treatment of cancer is to irradiate the cancer with a gamma-ray source. What is the energy of a radioactive cobalt-60 gamma ray source that emits radiation with a wavelength of 9.3×10^{-3} Å?

12-18 In a photoelectric experiment using the surface of chromium (work function = 7.00×10^{-12} erg), incident light of wavelengths 2000 and 3500 Å was used. Calculate the kinetic energy and associated wavelength of the electron driven from the surface of chromium by light of these two wavelengths. What wavelength of light should be used just to begin photoelectric emission from this surface?

12-19 Calculate $\bar{\nu}$ for the lines in the Balmer series for which **(a)** $n_2 = 12$ **(b)** $n_2 = 13$ **(c)** Subtract the two values and compare with the value obtained by subtracting the value for $n_2 = 3$ from that for $n_2 = 4$. Do the lines get closer together as n_2 increases?

12-20 Calculate $\bar{\nu}$ for the lines in the Lyman series for which **(a)** $n_2 = 2$ **(b)** $n_2 = 3$ **(c)** $n_2 = 8$

12-21 How do astronomers find out which elements are present in distant stars?

12-22 Why did the results of Rutherford and Marsden's thin-metal-sheets experiment destroy Thomsons's plum pudding model of the atom?

12-23 How did Bohr explain the lines in the hydrogen emission spectrum? Was his explanation right or wrong?

12-24 Calculate the energy necessary to take an electron in a hydrogen atom from the lowest energy level ($n = 1$) to $n = \infty$, and from the fifth energy level ($n = 5$) to $n = \infty$.

12-25 The Lyman series of lines in the hydrogen spectrum results from electronic transitions from higher energy levels to the $n = 1$ level. Calculate the energies of photons that are emitted for transitions **(a)** $n_2 = 2$ to $n_1 = 1$ **(b)** $n_2 = 4$ to $n_1 = 1$ **(c)** $n_2 = 7$ to $n_1 = 1$

12-26 What are the wavelengths of each of the following? **(a)** a photon **(b)** an electron **(c)** a proton, each having a kinetic energy of 1 eV? (The mass of an electron = 9.11×10^{-28} g; the mass of a proton = 1.67×10^{-24} g.)

12-27 A certain star in space emits a total energy of 3.17×10^{33} ergs/s. If all this energy was encompassed in a single photon, calculate its wavelength, frequency, and associated momentum.

12-28 Calculate the energy in ergs, the velocity, the momentum, and the wavelength of a proton that has an energy of 50,000 electron volts. The mass of a proton is 1.67×10^{-24} g.

12-29 Calculate the wavelength and frequency of a 3750-lb automobile traveling at 55 miles/h.

12-30 Calculate the energy and associated momentum of a mole of photons under each of the following conditions. **(a)** in the microwave region $\bar{\nu} = 10 \text{ cm}^{-1}$ **(b)** in the x-ray region, $\bar{\nu} = 1.0 \times 10^7 \text{ cm}^{-1}$ **(c)** in the gamma-ray region, $\bar{\nu} = 1.0 \times 10^{10} \text{ cm}^{-1}$

12-31 By use of Heisenberg's uncertainty principle, determine Δx for **(a)** a car of mass 1.5 tons moving at 60 miles/h **(b)** a bullet of mass 10 g moving at 700 yards/s **(c)** a hydrogen molecule moving at a speed of 2.0×10^6 cm/s (Use reasonable error estimates to calculate Δp.)

12-32 Is an electron a particle or a wave?

Additional Problems

12-33 In a photoelectric experiment, electrons are first emitted when light of wavelength 7098 Å shines on the surface of the metal. Calculate the maximum kinetic energy and associated wavelength of the emitted electron when light of the following wave numbers shines on the metal. (The mass of an electron = 9.11×10^{-28} g.) **(a)** 1 cm^{-1} **(b)** 10^7 cm^{-1}

12-34 Calculate the wavelength of the absorption line that corresponds to an energy difference of 4.5 eV and one for an energy difference of 79 kcal/mole.

12-35 The Balmer series of lines occurs in the hydrogen spectrum when there are electronic transitions from higher levels to the $n = 2$ level. Calculate the energies of photons that are emitted for transitions **(a)** from $n_2 = 4$ to $n_1 = 2$ **(b)** from $n_2 = 7$ to $n_1 = 2$

12-36 In the electromagnetic spectrum, the region of visible light is from approximately 400 to 800 nm. Calculate the frequencies corresponding to 400 and to 800 nm.

12-37 Calculate the associated wavelength of an electron (mass = 9.1×10^{-28} g) traveling at $0.010\,c$ (one hundredth the speed of light). If the size of a nucleus that consists of a single proton is of the order of 10^{-13} cm, what is the possibility of an electron residing in the nucleus?

12-38 While the great majority of solar radiation in the wavelength range 200 to 300 nm is absorbed by ozone (O_3) in the outer atmosphere, some of it gets through, resulting in sunburn and probably some skin cancers. Calculate the energy of a photon of light with a wavelength of 250 nm and compare it with that of photon of wavelength 5×10^{-5} cm, which is the maximum wavelength for a blackbody heated to 6000°K.

12-39 The lines in the hydrogen spectrum for which $n_1 = 3$ are called the Paschen series. Calculate $\bar{\nu}$ for the lines in this series for which $n_2 = 4$ and 5.

12-40 When and if astronauts reach Mars, it will be possible to talk to them by radio (even now we communicate with two Viking spacecraft on Mars). If we ask a question and they reply immediately on hearing it, how much time will elapse between asking the question and hearing the reply when Earth and Mars are at their closest distance, 49 million miles? At their farthest distance, 235 million miles?

12-41 Assume that you could move a bar magnet up and down at the rate of 15 motions per second. What would be the wavelength of the pulsating field that you generate in this manner?

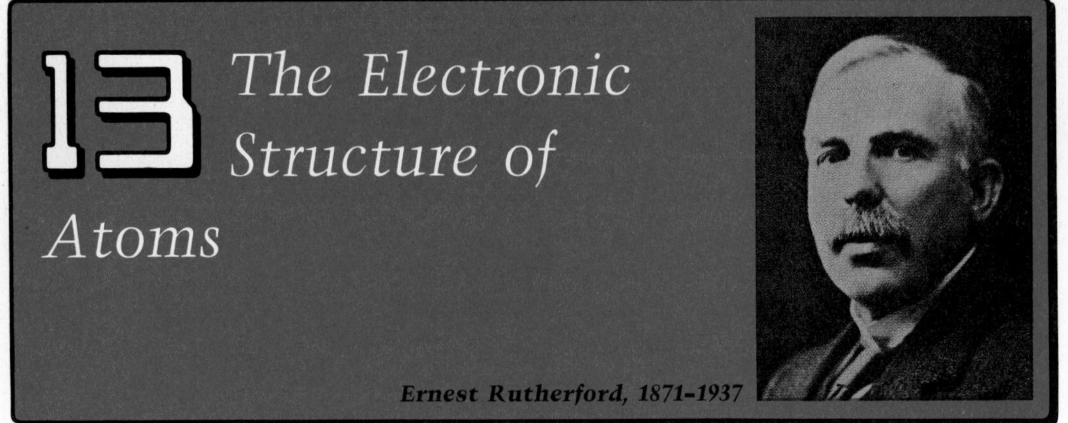

In Chapter 12, we reviewed what we called the quantum evolution. The old theories of Newton and Maxwell were shown to be inadequate to explain certain aspects of the behavior of atomic particles and light. A host of unexpected experimental results created the need for new explanations and theories which taken together bring an entirely new viewpoint to the atomic world. We have seen that particles can act as if they were waves and waves as if they were particles. The uncertainty principle leads to a philosophy of science that can no longer be termed deterministic. The breakdown of the old theories did not occur all at once. A continuing erosion was taking place. Scientists like Planck, Einstein, de Broglie, Bohr, Heisenberg, etc., were forced to conclude that there was something fundamentally wrong with the view of nineteenth-century physics. Each of their contributions was startling, imaginative, and productive. They forever changed the way scientists view the world.

At the close of Chapter 12, we saw that Bohr's theory, while adequate to explain the structure of the hydrogen atom, failed when applied to larger atoms. The structures of larger atoms (and molecules—see Chapter 14) could be understood only by taking into account their *wave properties*. This was done by Schrödinger in 1926.

13-1 The Schrödinger Wave Equation

Erwin Schrödinger (1887–1961) felt that the structure of atoms could be explained if the Bohr model of the atom were united with de Broglie's concept of matter waves (Section 12-11). Bohr's theory suffered from the fact that it could only explain the emission spectrum of hydrogen. Bohr additionally did not explain why an electron in an orbit did not radiate energy, but simply postulated that it did not do so. Schrödinger was able to combine the two ideas by assuming that an electron in an orbit around a nucleus behaves as a wave. These matter waves of the electron extend throughout the orbit and *contain an integral number of wavelengths* (that is, the number of wavelengths is an exact integer and not a fraction). Figure 13-1a shows the Schrödinger representation of two such matter waves for circular orbits, one with four and one with five wavelengths. Any circulating wave which completely fills its orbit so that the number of crests and troughs remain constant is called a **standing wave.** Because the number of wavelengths is an exact integer, there is neither constructive nor destructive interference (Figure 12–5c, d). The electron is not accelerating and therefore will not radiate energy. In Figure 13-1b, we show a wave in the same orbit but which does not have an integral number of wavelengths. In this case the wave will show the property of destructive interference and the wave will be annihilated. Since real electrons neither spiral into the nucleus nor are annihilated, the standing-wave explanation is quite acceptable. *It explains the experimental results.*

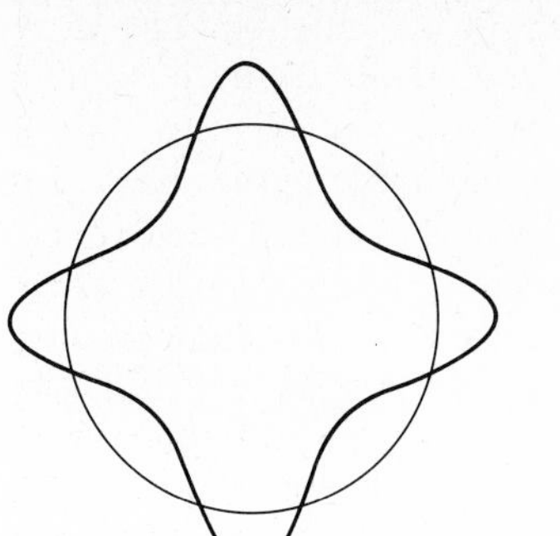

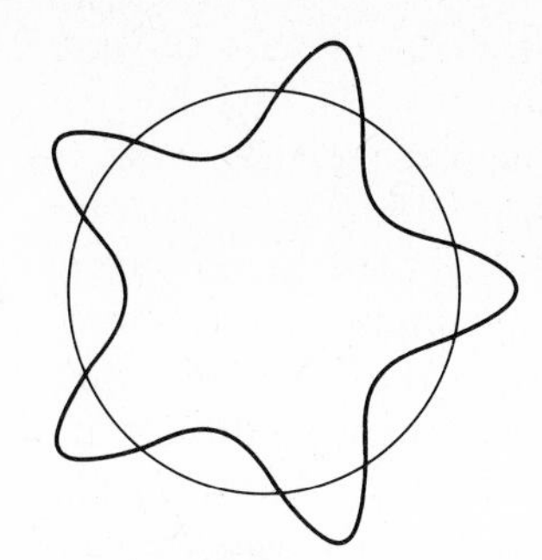

(a)

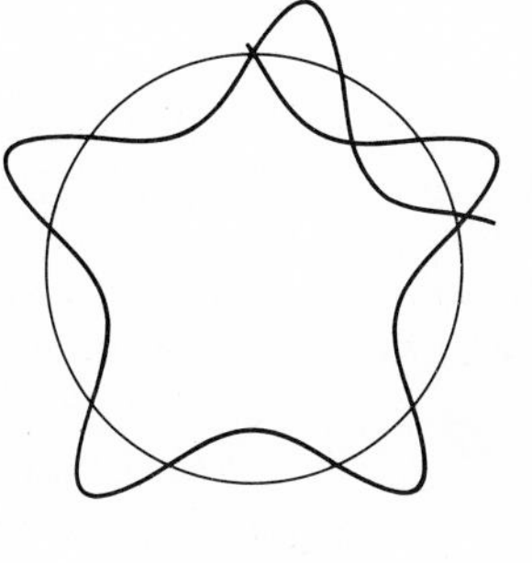

(b)

Figure 13-1 (a) Standing waves in circular orbits with four and five crests. There is no interference. (b) A standing wave in circular orbit with a nonintegral number of crests. The wave destroys itself through interference.

Schrödinger also developed the necessary mathematical framework for analyzing these concepts. This framework is embodied in an equation known as the **Schrödinger wave equation.*** The Schrödinger equation is a differential equation, the detailed mathematics of which are well beyond us. Therefore, we will not develop it here; however, we can gain quite a bit of insight by qualitatively looking at the equation and at the results derived from it. With this in mind, we will list a number of its features and characteristics.

1. The Schrödinger equation is a mathematical description of a matter wave. The equation is similar in structure to other mathematical wave equations found in mechanics, such as the equation describing the vibration of a violin string or the motions of a membrane on a drum after it is struck.

2. The Schrödinger wave equation is applied to atomic particles. We could, if we wished, apply it to everyday macroscopic objects. However, the results we would get would be identical to those obtained from Newton's equations. Since Newton's equations are easier to use, we reserve the Schrödinger equation for atomic particles, and Newton's equations for macroscopic objects. Does this mean that there are two different laws of physics—one dealing with atomic particles and the other with macroscopic particles? The answer is *no!* There is one set of physical laws that governs all particles. If we review the examples on pages 370 and 373, we see that although the quantum theory can be applied to macroscopic bodies, the effects are so infinitesimally small as to be undetectable. When dealing with macroscopic bodies, we use Newton's laws as a practical matter, since the calculations are easier to carry out, and the results are essentially indistinguishable from those obtained by

*The Schrödinger equation, for an atom containing one electron, looks like this:

$$\frac{\partial^2\psi}{\partial x^2} + \frac{\partial^2\psi}{\partial y^2} + \frac{\partial^2\psi}{\partial z^2} + \frac{8\pi^2 m}{h^2}(E - V)\psi = 0$$

where m is the mass of the electron; E its total energy; V its potential energy; h is Planck's constant; x, y, and z are the three rectangular coordinates: and ψ is the wave function, which we shall shortly discuss. For atoms containing more than one electron, the Schrödinger equation is more complicated.

the more exact wave equations; that is, wave mechanics reduces to Newtonian mechanics for large bodies.

3. The wave equation, when solved, gives us the wave motions *and the energies* of electrons. Since the equation explicity takes account of de Broglie's and Heisenberg's principles, it cannot describe simultaneously the exact position and exact momentum of an electron. It could tell the exact position of the electron if the uncertainty in the momentum, Δp, was infinite and vice versa (remember $\Delta x \cdot \Delta p \geq h/2\pi$). However,these are very extreme conditions. If, in general, we cannot say *where* an electron is, what can we say?

We can say where an electron would most probably be found. We can also say where an electron will very rarely be found. Note that our language has now changed. We are talking now in terms of *probabilities* instead of exact locations.

For example, we can say that an electron at its lowest energy in the hydrogen atom will have a very high probability of being found at 0.529×10^{-8} cm (one Bohr radius) from the nucleus, while the same electron will have a small but finite probability of being found at 2.17×10^{-8} cm from the nucleus. We believe and accept that our knowledge about electrons in atoms is not perfect. We get probabilities instead of exact locations. We also believe that *this is the best we can do.*

4. The Schrödinger equation is a differential equation. The solutions to a differential equation are themselves equations. The Schrödinger equation for a given atom has a large number of solutions, each of which is an equation, called a ψ **function.** Each ψ function corresponds to a given energy value for the electrons of that atom. Thus, when we solve the Schrödinger equation for the hydrogen atom, for example, we get a series of equations, each of which describes a state that has its own energy value. The state of lowest energy is called the **ground state.** The electron in a given hydrogen atom resides in only one of these possible energy states at any given time, although it can be promoted to a higher energy state, or drop to a lower energy state (unless it is already in the ground state). If this picture of possible energy states sounds familiar, it is because these are the same states postulated by the Bohr treatment of atomic structure (Section 12-10).

13-2 The Wave Functions ψ

The wave functions ψ obtained by solving the Schrödinger equation are quite similar to functions obtained by solving other kinds of wave equations. They also have some rather unique features.

The functions ψ are solutions to a mathematical equation; however *we cannot observe them in nature.* Only the square of the absolute value of these functions, $|\psi|^2$, can be measured in experiments. The reason for this is that ψ may be a complex function. By this we mean it may contain the quantity $\sqrt{-1}$. This quantity $\sqrt{-1}$ does not naturally occur in the solution of classical mechanical problems; however, it is an integral part of some quantum-mechanical wave functions. Because of this it is impossible to assign any *physical* meaning to the wave function ψ. It is $|\psi|^2$, the square of the absolute value of the wave function, which has a physical interpretation. $|\psi|^2$ is the *relative* probability density (or probability distribution) of a particle in space.

Since we will be discussing plots of $|\psi|^2$, it is necessary to describe the coordinate systems we will be using. In the most familiar system of coordinates (called the **Cartesian** system), a point in space is located by specifying the coordinates x, y, and z with respect to the origin (Figure 13-2). However, there are other coordinate systems which are more convenient for certain purposes. For plotting $|\psi|^2$ for the hydrogen atom, the system of choice is the **spherical polar coordinate**

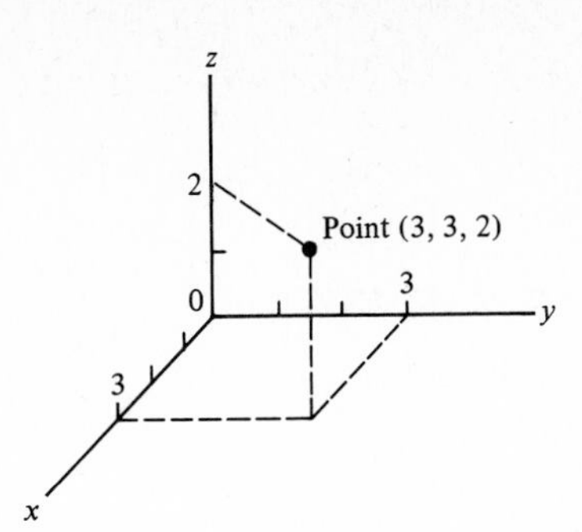

Figure 13-2 A point in space is located by specifying its x, y, and z coordinates in the Cartesian coordinate system.

system. As shown in Figure 13-3, a point in this system is represented by three numbers: r, θ, and ϕ:

r = distance of the point from the origin
θ = angle that r makes with the z axis
ϕ = angle that the line segment OA makes with the x axis

As we have already pointed out, each ψ function for the hydrogen atom corresponds to a given energy state in which the electron may reside. When we plot $|\psi|^2$ for a single energy state against the three coordinates r, θ, and ϕ (putting the nucleus at the origin), we get a three-dimensional picture which looks like a cloud (see, for example, Figures 13-9 and 13-10) because it is essentially a map of the distribution in space of the probability of finding the electron at any given point r, θ, ϕ. We get a cloud instead of a sharp outline, because the electron has some probability of being found almost anywhere. These electron clouds are called **orbitals.** The terms **orbital, electron cloud, energy state,** and **energy level** are often used almost interchangably, to mean essentially the same thing. Note the essential difference between the Bohr and Schrödinger pictures of the energy levels of the hydrogen atom. In the Bohr picture, each orbit is sharp; for example, recall that an electron in the ground state of the hydrogen atom was said to have a radius of 0.529×10^{-8} cm. In the Schrödinger picture, which takes account of the uncertainty principle, there is no sharp orbit (which is why we use the term *orbital* instead of *orbit*); the best we can say is that an electron in the ground state of the hydrogen atom is a spherical cloud whose maximum density is 0.529×10^{-8} cm from the nucleus, but which also has finite electron density both closer in and farther out. We use the term **electron density** to mean the same thing as the probability of finding the electron. A region of space with a high probability of finding an electron is said to have a high electron density.

Although the lowest energy level (ground state) of the hydrogen atom is an electron cloud whose shape is spherical, not all orbitals have a spherical shape. As we

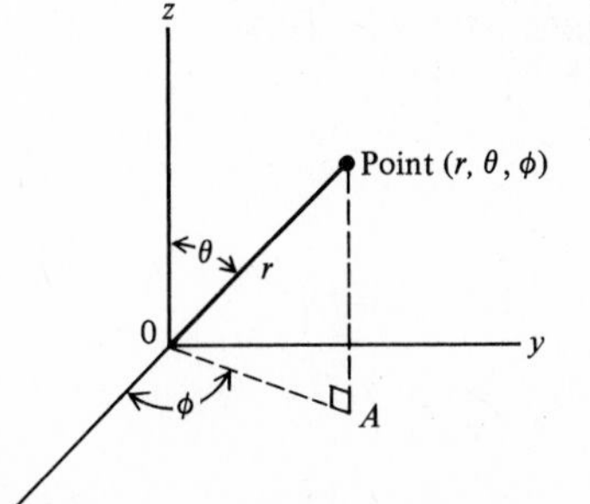

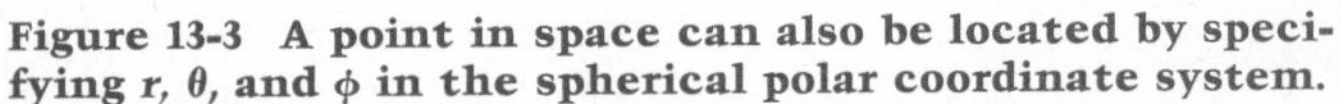
Figure 13-3 A point in space can also be located by specifying r, θ, and ϕ in the spherical polar coordinate system.

shall see, the shape of an orbital (which, of course, is obtained by plotting $|\psi|^2$ for that orbital) is one of its most important characteristics.

13-3 The Quantum Numbers n, ℓ, m

As we stated previously, it is the square of the absolute value of the wave function $|\psi|^2$ which has a physical meaning. It turns out that when one solves the wave equation for the hydrogen atom, certain restrictions must be placed on ψ itself, so that $|\psi|^2$ can make physical sense. The restrictions take the form of three new numbers, n, ℓ, and m, which are related to the coordinates r, θ, and ϕ. The numbers n, ℓ, and m are called **quantum numbers** since they may only assume certain specific integral values.

Each orbital has a different set of quantum numbers n, ℓ, and m, so that if we specify a set of numbers n, ℓ, and m, we are specifying one particular orbital. In the following paragraphs we will discuss what values are possible for n, ℓ, and m.

n is termed the **principal quantum number** and is directly related to the *energy* of the electron. The possible values of n are 1, 2, 3, 4, ... That is, n may be any positive integer greater than zero.

$$n = 1, 2, 3, 4, \ldots$$

The lower the value of n, the more stable is any orbital corresponding to it. This means that the lower the n value, the more energy one must use to eject the electron from the atom. For $n = 1$ (the lowest possible n value, the ground state), the electron is in its most stable orbital. For an electron in a hydrogen atom, the energy depends solely on the n value. This is also true for other species containing one electron: He^+, Li^{2+}, and so on. For systems that contain two or more electrons, the energy depends on the value of ℓ as well as n.

ℓ is termed the **azimuthal quantum number** or **orbital quantum number** and is related to the *shape* of the orbital. The possible values of ℓ are 0, 1, 2, . . . , $n - 1$. That is, ℓ may be any positive integer *including* zero, but it must be less than n (Table 13-1).

Table 13-1 Possible Values of ℓ for a Given n Value

n	ℓ
1	0
2	0, 1
3	0, 1, 2
4	0, 1, 2, 3
5	0, 1, 2, 3, 4

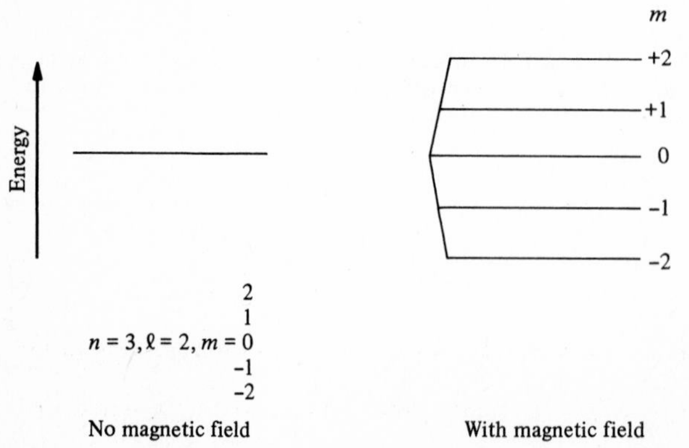

Figure 13-4 The relative energy levels for an electron with $n = 3$, $\ell = 2$ with and without a magnetic field.

Example: For $n = 4$, what are the possible values of ℓ?

Answer: For $n = 4$, ℓ may have the value 0 or 1 or 2 or 3.

Orbitals are given labels depending on the values of n and ℓ associated with them. The first part of the label is simply the quantum number n. The second part of the label is a letter corresponding to the value of the quantum number ℓ. These letters are:

ℓ	*Orbital label*
0	*s*
1	*p*
2	*d*
3	*f*
4	*g*

Thus, an orbital for which $n = 3$ and $\ell = 2$ is called a 3*d* orbital. Other examples are

n	ℓ	*Orbital label*
1	0	1*s*
1	1	Cannot exist— ℓ cannot equal n
2	1	2*p*
5	3	5*f*
6	2	6*d*

The letters *s*, *p*, *d*, etc., are derived from an old spectroscopic notation* that has little or no relevance to our knowledge of orbitals today.

m is termed the **magnetic quantum number** and is related to the different *orientations* of the orbitals in space. One way of distinguishing these different orientations is to place atoms in a magnetic field. Orbitals with the same n and ℓ values but different orientations possess different energies in the magnetic field: hence the term "magnetic quantum number." When an atom is not in a magnetic field, electrons with the same n and ℓ values but different m values have the same energy (Figure 13-4).

The possible values of m are directly related to the value of ℓ. For a given ℓ value, the m values may be any integer from $-\ell$ to $+\ell$ including zero (Table 13-2).

Example: For $\ell = 3$, what are the possible m values?

Answer: If $\ell = 3$, m may be any integer from $-\ell$ to $+\ell$; $m = -3, -2, -1, 0, 1, 2, 3$.

Example: For $n = 2$, what are the possible values of ℓ and m?

Answer:

$n = 2$ $\quad \ell = 1 \quad m = -1, 0, +1$

$\ell = 0 \quad m = 0$

*The letters were taken from certain words used to describe spectral series: *s* for sharp, *p* for principal, *d* for diffuse, *f* for fundamental.

Table 13-2 m Values Corresponding to a Given ℓ Value

ℓ	m
0	0
1	−1, 0, +1
2	−2, −1, 0, +1, +2
3	−3, −2, −1, 0, +1, +2, +3
4	−4, −3, −2, −1, 0, +1, +2, +3, +4

Table 13-3 shows all the quantum states for the hydrogen atom up to $n = 4$. The wave function ψ can now be written as $\psi_{n,l,m}$ since it depends on the different values of n, ℓ, and m. The next section will deal with the different ψ functions as the values of n, ℓ, and m are changed.

Table 13-3 The Possible Quantum States for Hydrogen up to $n = 4$

n	ℓ	m	Orbital label
1	0	0	1s
2	0	0	2s
	1	−1	2p
		0	2p
		+1	2p
3	0	0	3s
	1	−1	3p
		0	3p
		+1	3p
	2	−2	3d
		−1	3d
		0	3d
		+1	3d
		+2	3d
4	0	0	4s
	1	−1	4p
		0	4p
		+1	4p
	2	−2	4d
		−1	4d
		0	4d
		+1	4d
		+2	4d
	3	−3	4f
		−2	4f
		−1	4f
		0	4f
		+1	4f
		+2	4f
		+3	4f

13-4 The Shapes of Atomic Orbitals

Table 13-4 lists the first few wave functions n, ℓ, m obtained by solving the Schrödinger equation for a hydrogen atom. We are going to study the graphs of these functions in some detail, since this information is necessary to understand the shapes of molecules, as well as the concepts involved in chemical bonding.

Let us consider the first wave function shown in Table 13-4, $\psi_{100} = [1/(\pi a_0^3)^{1/2}]e^{-\rho}$. The letter ρ is used to stand for r/a_0, where a_0 is the first Bohr radius, 0.529 Å. e is the base of natural logarithms. This function is for the 1s orbital. Note that the coordinates θ and ϕ do not appear in this function, because the 1s orbital is spherically symmetrical and is therefore only a function of the radius r. We can plot the function ψ_{100} by tabulating the different values that ψ_{100} would have as r is changed. This is done most conveniently by changing the units of r. Instead of measuring r in Å, we define a new unit, called an **atomic unit.**

$$1 \text{ atomic unit} = 0.529 \text{ Å} = a_0 \quad \text{(the first Bohr radius)}$$

When $r = 0.529$ Å, the term $e^{-r/a_0} = e^{-1}$, which simplifies the calculations. If r is measured in atomic units, ψ_{100} takes the form

$$\psi_{100}(\text{atomic units}) = \frac{1}{\sqrt{\pi}}e^{-r} \quad (r \text{ in atomic units})$$

Table 13-5 shows values of the function ψ_{100} in atomic units as the value of r is increased from 0 to 10.0 atomic units (5.29 Å), which is a considerable distance from the nucleus.

If in atomic units, $\psi_{100} = (1/\sqrt{\pi})e^{-r}$, then $|\psi|^2_{100}$ is equal to

$$|\psi|^2_{100} = \frac{1}{\pi}e^{-2r}$$

In Table 13-6 we have tabulated $|\psi|^2$ against r for the ψ_{100} function. As noted previously, $|\psi|^2$ gives us the probability density of finding the electron at a distance r from the nucleus. If we plot the data from Table 13-6 for $|\psi|^2$ as a function of r and draw a smooth curve through the calculated points, we obtain Figure 13-5. This

Table 13-4 The First Few Hydrogen-Atom Wave Functions, $\psi_{n,l,m}(r, \theta, \varphi)(\rho = r/a_0)^a$

$n = 1, \ell = 0, m = 0$:	$\psi_{100} = \left(\frac{1}{\pi a_0^3}\right)^{1/2} e^{-\rho}$
$n = 2, \ell = 0, m = 0$:	$\psi_{200} = \frac{1}{8}\left(\frac{2}{\pi a_0^3}\right)^{1/2}(2 - \rho)e^{-\rho/2}$
$n = 2, \ell = 1, m = 0$:	$\psi_{210} = \frac{1}{8}\left(\frac{2}{\pi a_0^3}\right)^{1/2}\rho e^{-\rho/2}\cos\theta$
$n = 2, \ell = 1, m = +1$:	$\psi_{211} = \frac{1}{8}\left(\frac{1}{\pi a_0^3}\right)^{1/2}\rho e^{-\rho/2}\sin\theta\, e^{i\varphi}$
$n = 2, \ell = 1, m = -1$:	$\psi_{21-1} = \frac{1}{8}\left(\frac{1}{\pi a_0^3}\right)^{1/2}\rho e^{-\rho/2}\sin\theta\, e^{-i\varphi}$
$n = 3, \ell = 0, m = 0$:	$\psi_{300} = \frac{1}{243}\left(\frac{3}{\pi a_0^3}\right)^{1/2}(27 - 18\rho + 2\rho^2)e^{-\rho/3}$
$n = 3, \ell = 1, m = 0$:	$\psi_{310} = \frac{1}{81}\left(\frac{2}{\pi a_0^3}\right)^{1/2}\rho(6 - \rho)e^{-\rho/3}\cos\theta$
$n = 3, \ell = 1, m = +1$:	$\psi_{311} = \frac{1}{81}\left(\frac{1}{\pi a_0^3}\right)^{1/2}\rho(6 - \rho)e^{-\rho/3}\sin\theta\, e^{i\varphi}$
$n = 3, \ell = 1, m = -1$:	$\psi_{31-1} = \frac{1}{81}\left(\frac{1}{\pi a_0^3}\right)^{1/2}\rho(6 - \rho)e^{-\rho/3}\sin\theta\, e^{-i\varphi}$

[a]For explanation of the symbols, see the text.

Table 13-5 Values of the function ψ_{100} as a Function of r

$$\left(\psi_{100} = \frac{1}{(\pi a_0^3)^{1/2}} e^{-r/a_0}\right)$$

in atomic units $\psi_{100} = \frac{1}{\sqrt{\pi}} e^{-r}$

r (atomic units)	ψ_{100} (atomic units)
0.0	0.564
0.1	0.510
0.2	0.462
0.5	0.342
0.8	0.254
1.0	0.208
1.5	0.126
2.0	0.076
3.0	0.028
4.0	0.010
5.0	0.004
10.0	2.56×10^{-5}

Table 13-6 Values of the Functions $|\psi|^2_{100}$ and $4\pi r^2|\psi|^2_{100}$ as a Function of r

$$|\psi|^2 = \frac{1}{\pi} e^{-2r}$$

r (atomic units)	$|\psi|^2$ (atomic units)	$4\pi r^2|\psi|^2$ (atomic units)
0.0	0.318	0.0
0.1	0.260	0.0327
0.2	0.213	0.107
0.5	0.117	0.368
0.8	0.065	0.523
1.0	0.043	0.540
1.5	0.016	0.452
2.0	0.006	0.302
3.0	7.8×10^{-4}	0.088

curve has some interesting features. We note that the electron has the highest probability density of being found close to the nucleus. Also, the curve of $|\psi|^2$ vs. r approaches but never reaches zero even at $r = \infty$. One can say that the probability density of the electron is spread throughout the universe. In principle this is correct, but from the values of $|\psi|^2$ we can see that the probability density of finding the electron even as far as 3.0 atomic units (1.6 Å) from the nucleus is extremely small.

Figure 13-5 can be divided into a series of spherical shells as shown in Figure 13-6. One may now inquire as to the probability of finding an electron in one of these shells. Note that this is a different question from the one we discussed previously. When we looked at $|\psi|^2$, we were looking at probability *per unit volume of space.* This means that we were drawing little cubes of identical size (see Figure 13-6) and asking how the electron density changed as we moved from cube to cube, beginning

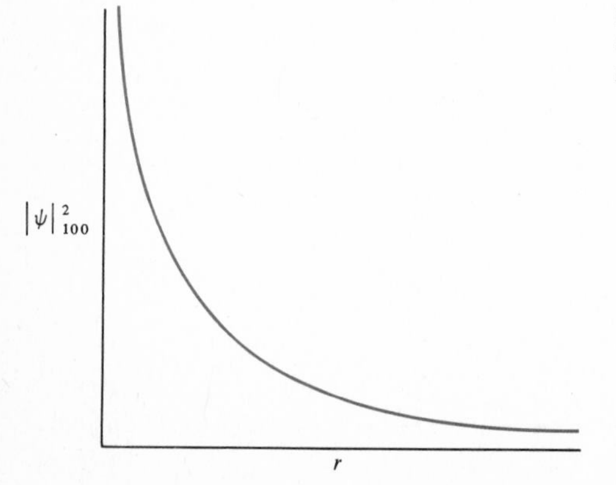

Figure 13-5 A plot of the $|\psi|^2_{100}$ wave function of the hydrogen electron vs. r. Data obtained from Table 13-6.

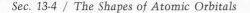

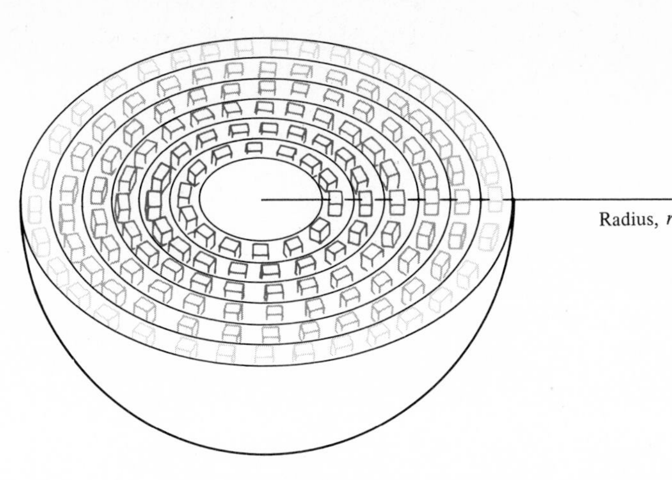

Figure 13-6 Probability density of a $|\psi|^2_{100}$ hydrogen atom wave function divided into a series of spherical shells.

at the nucleus and going outward. As we saw from the column marked $|\psi|^2$ in Table 13-6, this density was highest at the nucleus and then consistently decreased. What we are asking now is: How does the probability of finding the electron *in each shell* in Figure 13-6 change as we begin at the nucleus and move outward? Obviously, the number of identically sized cubes per shell increases as we get farther away from the nucleus. Physically, what we want to do is gather up all the cubes in each spherical shell and look at the total probability as r varies. Mathematically, this is accomplished by plotting $4\pi r^2 |\psi|^2$ vs. r ($4\pi r^2$ is the area of a sphere of radius r). Values are given in Table 13-6 for ψ_{100}. The plot is shown in Figure 13-7. This probability is called the **radial probability function.**

For the radial probability function, the highest value occurs at $r = 0.529$ Å, the first Bohr radius, a_0. As previously mentioned, while Bohr confined the electron to $r = 0.529$ Å, the wave-mechanical picture does not. It may be close to the nucleus or 100 times a_0 away from the nucleus. However, as with the probability density, $|\psi|^2$, the radial probability function becomes extremely small for large r. Because of this, it is convenient to represent orbitals by enclosing shells around 90 or 99% of the total probability region. Such shells are shown in Figure 13-8 for the ψ_{100}, ψ_{200}, and ψ_{300} functions. These figures show that these functions are spherically symmetrical, which means that the probability of finding the electron at any given distance from the nucleus is the same in all directions for a given value of r. These three orbitals are s orbitals since $\ell = 0$.

All s orbitals are spherically symmetrical. The $2s$ orbital contains a feature not found in the $1s$ orbital. This is a region of *zero electron density,* a region in which the probability of finding the electron is zero. Such a region in space is called a **node.** The node in the $2s$ orbital is found at $r = 2$ atomic units. The $3s$ orbital contains two nodes. For any s orbital the number of nodes is always one less than its n value.

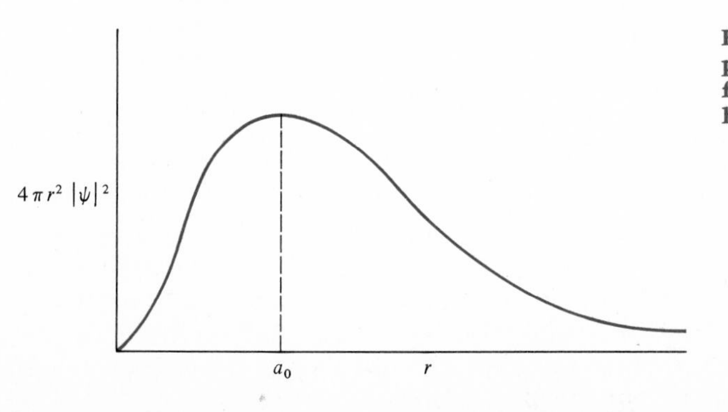

Figure 13-7 A plot of the radial probability function $4\pi r^2|\psi|^2$ vs. r for the ψ_{100} wave function of the hydrogen atom.

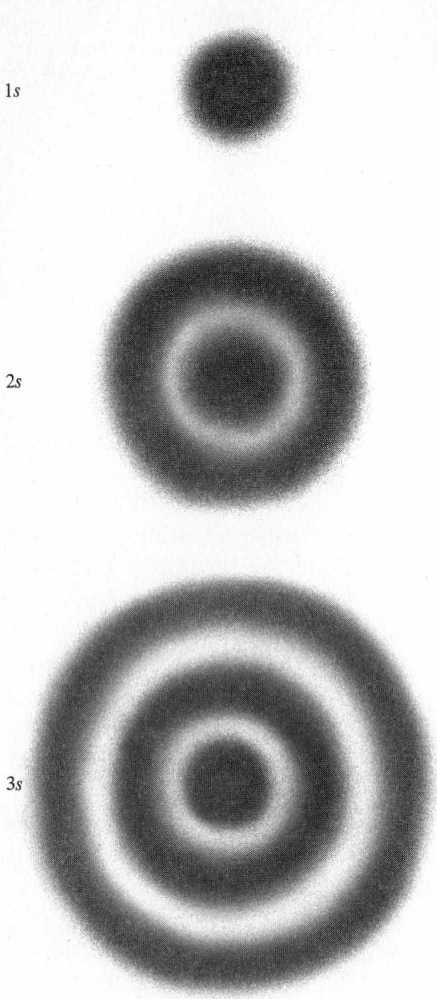

Figure 13-8 The 90% contours of probability densities for the ψ_{100}, ψ_{200}, and ψ_{300} functions (1*s*, 2*s*, and 3*s*, orbitals).

1*s*

2*s*

3*s*

The first orbital with an ℓ value greater than zero ($\ell = 1$) occurs when $n = 2$. As mentioned above, orbitals for which $\ell = 0$ are spherically symmetrical. This is not the case for orbitals with ℓ greater than zero. We have previously seen (Table 14-3) that orbitals with $\ell = 1$ are called *p* orbitals, and that there are three of them for each value of *n*, since *m* can be $+1$, 0, or -1. Thus, there are three 2*p* orbitals, three 3*p* orbitals, and so on. We can determine the shape of the *p* orbitals by plotting the *p* wave functions as we did for the *s* functions. When this is done, we obtain the shapes shown in Figure 13-9. As we can see from the graphs, the three *p* orbitals are identical in shape but have different orientations in space. They are mutually perpendicular. This is clearly shown in the figure if we introduce an *arbitrary* Cartesian coordinate system with the origin at the nucleus. We also note that each of the three *p* orbitals contains a node. The p_x orbital has a node in the *yz* plane; the p_y orbital has a node in the xz plane, and the p_z orbital has a node in the *xy* plane. As with all orbitals, the size of the *p* orbitals increases as the principal quantum number *n* increases.

The three 3*p* orbitals resemble the 2*p* orbitals in shape but are larger in size. The + and − signs in the lobes shown in Figure 13-9 come from signs of the wave function. They have no physical significance but are very important in bonding theory. It is important to understand that *they do not indicate charge.* Since an orbital is an electron cloud, *both* lobes of any particular *p* orbital must be negatively charged. The + sign in one lobe, therefore, does not indicate positive charge, merely a positive sign for the wave function in that region. *It is a rule that the sign of an orbital is always different on either side of a node.*

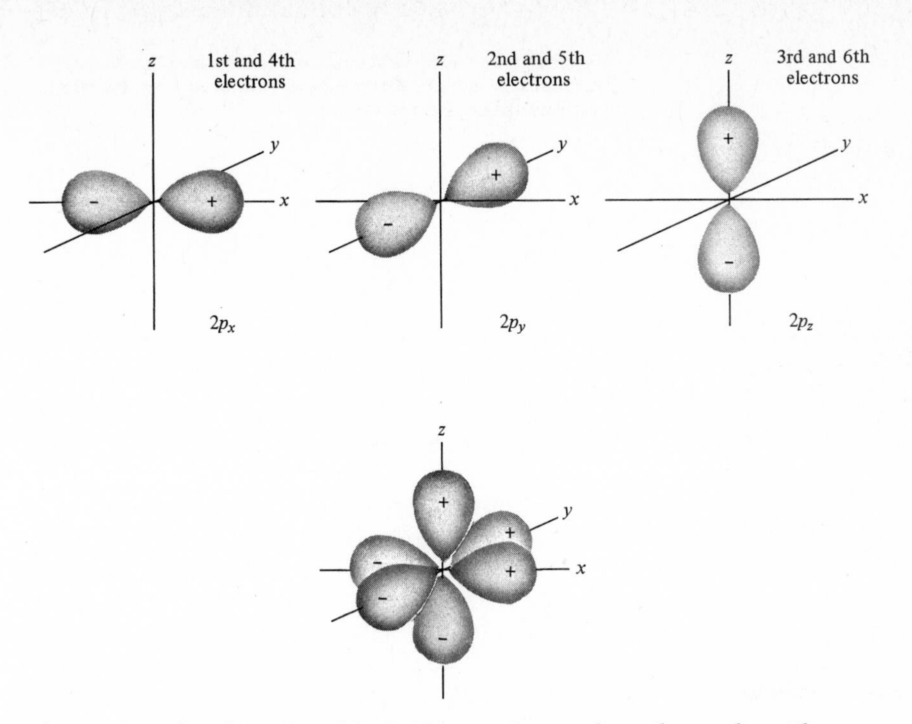

Figure 13-9 The three 2*p* orbitals shown separately and together. These are surfaces enclosing 90% probabilities. The signs (+ and −) in the different lobes come from the signs of the wave function, ψ. They have no physical significance in terms of calculating probability densities, $|\psi|^2$, but are very important in bonding theory.

The lowest value of n for $\ell = 2$ is $n = 3$. Therefore, the first d orbitals are the 3d orbitals. Since m can be $-2, -1, 0, +1$, or $+2$, there are five 3d orbitals, and five d orbitals for each higher value of n. The five 3d orbitals are shown in Figure 13-10. Note that each has two nodes. The five d orbitals are designated xy, xz, yz, z^2, and $x^2 - y^2$. The higher d orbitals (4d, 5d, etc.) are similar in appearance, but larger.

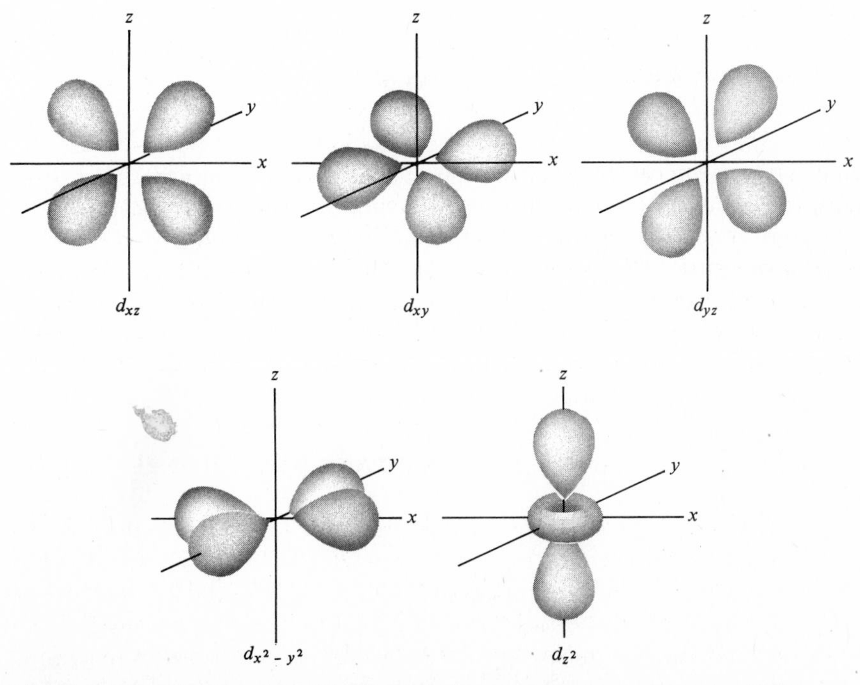

Figure 13-10 The five 3*d* orbitals of hydrogen shown separately. Each surface encloses an orbital of 90% probability.

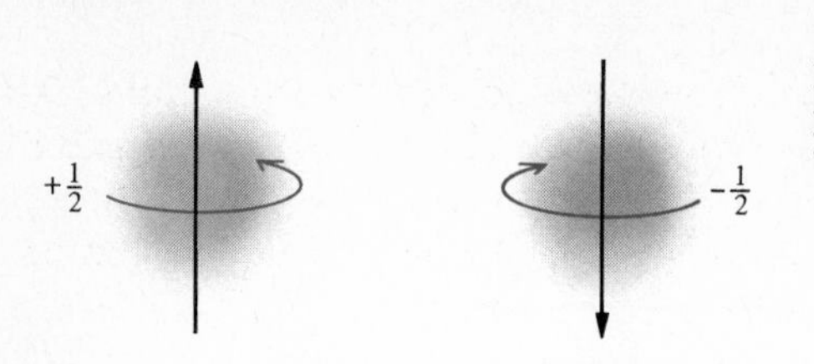

Figure 13-11 The intrinsic spin of the electron, pictorially represented as a ball of negative electricity spinning on its axis.

When $\ell = 3$, we have the f orbitals. There are seven of these for each value of n. Consequently, we have seven $4f$, seven $5f$ orbitals, and so on. We do not show these, but as we might expect, each of the seven $4f$ orbitals has three nodes.

13-5 Electron Spin

Even with the Schrödinger solution to the hydrogen atom in hand, scientists were still troubled by a feature occurring in the atomic spectra of alkali metals. Lines in these spectra are not simply single lines as predicted, but are actually made up of two closely spaced lines, called *doublets*. G. E. Uhlenbeck (1900–) and S. Goudsmit (1902–1978) identified these lines as two different states of intrinsic angular momentum or **spin** possessed by an electron. If we do not take too seriously the idea of trying to picture an electron, we may imagine an electron as a little ball of negative electricity spinning on its axis (Figure 13-11). We introduce a new quantum number, the **spin quantum number** s, which may have only one of two values, $s = +\frac{1}{2}$ or $s = -\frac{1}{2}$. With this assumption, the experimental data are explained.

Schrödinger's equation as generally solved gives us no knowledge concerning this additional feature of the electron. Paul A. M. Dirac (1902–), an English theoretical physicist, solved a relativistic form of the wave equation (that is, a form consistent with relativity theory) and showed that this feature of spin was a natural consequence of the theory. Therefore, we accept this new quantum number, s, not as a hypothesis that merely explains certain experimental results, but as a coherent part of our fundamental theory. From this point, we will use four quantum numbers, n, ℓ, m, and s, to describe an electron in an atom.

13-6 Many-Electron Atoms

Up to now, we have confined our discussions to the hydrogen atom, which has one proton and one electron. What happens when we apply the Schrödinger equation to other atoms: those with more than one proton and more than one electron? The answer is simply that the Schrödinger equation cannot be solved *exactly* for any atom more complex than hydrogen. However, all is not lost. Theoreticians have worked hard to develop approximate solutions to describe more complex systems. We have at our disposal a series of rules that will enable us to write the ground-state electronic structure of atoms more complex than hydrogen. First, we shall state the rules and then we shall apply them.

Rules for Determining Ground-State Electronic Structures

1. Each atom has available to it a series of orbitals that are similar to those of the hydrogen atom. Not only hydrogen, but every atom has one $1s$, one $2s$, three $2p$, five $3d$ orbitals, etc. These orbitals have the same shapes as those of the hydrogen atom (for example, s orbitals are spherical, etc.), although their *sizes* and *energies* are different. (In general, the size and energy of a given orbital decreases with increasing atomic number.)

2. No two electrons in the same atom can have the same four quantum numbers. That is, they cannot have the same n, ℓ, m, and s values. This is a famous principle known as the **Pauli exclusion principle.** Since each orbital has a different set of n, ℓ, and m values which determines its energy level, shape, and orientation in space, it follows that **any orbital can hold a maximum of two electrons, provided that the two electrons have opposite spins** ($s = +\frac{1}{2}$ and $-\frac{1}{2}$). This is another way of stating the Pauli exclusion principle.

3. We shall construct the atoms by adding electrons to the bare nucleus one at a time. Each added electron will be placed in the *lowest energy* orbital that is available. This is known as the **aufbau** (German for "building up") process. Doing this will generate the ground-state electronic structure for each atom.

4. The p, d, and f orbitals come in groups of three, five, and seven equivalent orbitals, respectively. These equivalent orbitals have the same energy but differ only in spatial orientation. Since the aufbau principle requires us to add electrons to the available orbitals in order of increasing energy, we need a rule for this case. The rule is the following: When we add electrons to a group of equivalent orbitals, **each orbital becomes half-filled before any are filled.** This is known as **Hund's rule.** For example, there are three $2p$ orbitals ($2p_x$, $2p_y$, and $2p_z$). As with all other orbitals, these three $2p$ orbitals can accomodate two electrons each. When electrons are added to them (Figure 13-12), the first goes into, say, the $2p_x$, the second into the $2p_y$, and the third into the $2p_z$ orbital. At this point each of the three equivalent $2p$ orbitals has one electron (is half-filled). After this the fourth electron goes into the $2p_x$ orbital again, filling it completely with two electrons. The fifth and sixth electrons fill the other two $2p$ orbitals.

By doing this, the electrons stay as far away from each other as possible. This is reasonable if we consider the effect of electron–electron repulsion. Say that we are dealing with two electrons that go into p orbitals. Our choices are either to put both into the same orbital, say p_y, or to put the two electrons into different orbitals: for example, the p_y and the p_x. The first choice, putting them both into the orbital p_y, will have the electrons in closer proximity to each other so that there is a greater degree of electron–electron repulsion. This would be a state of higher energy than if they were separated with less chance of repulsion. The system will tend toward its lowest energy, hence the separation and Hund's rule.

A supplement to Hund's rule is that a particularly stable system is obtained when a set of equivalent orbitals is either filled or half-filled, each containing exactly one electron or a pair of electrons. We shall see examples of this when we discuss the periodic table.

5. We are filling orbitals in order of increasing energy and need to know what the order is. For the hydrogen atom the energy of an orbital depends solely on the value of n. The one $3s$, three $3p$, and five $3d$ orbitals all have the same energy. For atoms larger than hydrogen, this is not the case. For these atoms, orbital energies depend not only on n, but also on ℓ. The rule is that when comparing two orbitals, *the one*

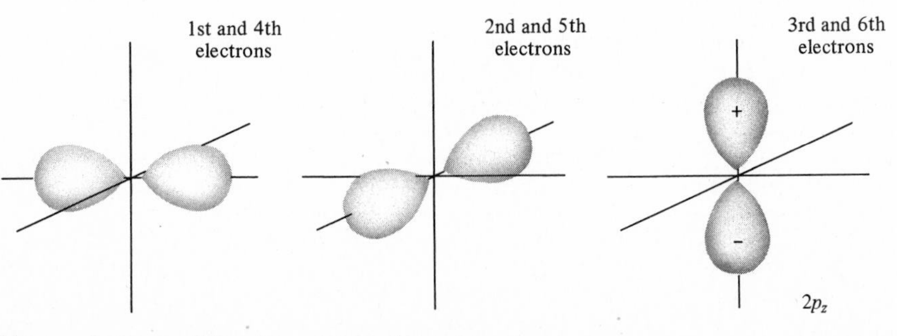

Figure 13-12 Application of Hund's rule.

with lower energy is the one with the lower value of $(n + \ell)$, and consequently this one will fill first. For example, the $3s$ orbital will be filled before the $3d$ orbitals.

Example: In the $n = 2$ energy level, ℓ may be either 0 or 1. Which ℓ value fills first?

Answer: From rule 5, $(n + \ell)$ must be a minimum. Our choices are

$$2 + 0 = 2 \quad \text{or} \quad 2 + 1 = 3$$

Therefore, $n = 2$, $\ell = 0$ fills first, then $n = 2$, $\ell = 1$.

Example: Which orbital should fill first: $n = 3$, $\ell = 2$ or $n = 4$, $\ell = 0$?

Answer: $(n + \ell)$ must be a minimum. Our choices are

$$3 + 2 = 5 \quad \text{or} \quad 4 + 0 = 4$$

Therefore, the $n = 4$, $\ell = 0$ fills before any of the $n = 3$, $\ell = 2$ orbitals.

6. If the $n + \ell$ values of different orbitals are equal, the one with the lowest n value fills first.

Example: Which orbital fills first: $n = 3$, $\ell = 2$ or $n = 4$, $\ell = 1$?

Answer: $n + \ell = 5$ for both, so the one with the lower n value fills first; that is, $n = 3$, $\ell = 2$ fills before $n = 4$, $\ell = 1$.

7. A **shell** of electrons is defined as all of the electrons with the same principal quantum number. These shells are given alphabetical names beginning with the letter *K*. The first shell (the *K* shell) consists only of the two electrons in the $1s$ orbital. The *L* shell contains eight electrons. Subsequent shells contain 18 and 32 electrons. A group of equivalent orbitals within a shell is called a **subshell.** The *M* shell consists of three subshells: (1) a subshell of one $3s$ orbital, (2) a subshell of three $3p$ orbitals, and (3) a subshell of five $3d$ orbitals. Table 13–7 lists the principal quantum numbers, their shell designations, the maximum number of electrons in each shell, and the maximum number of subshells.

13-7 Paramagnetism, Diamagnetism, and Ferromagnetism

In this section, we shall examine one of the consequences of the Pauli exclusion principle, which we stated as: Any orbital can hold a maximum of two electrons provided that the two electrons have opposite spins. This means that for any orbital

Table 13-7 Principal Quantum Numbers, Their Shell Designations, Maximum Number of Electrons in Each Shell, and Maximum Number of Subshells

Principal quantum number, n	*Shell designation*	*Maximum number of electrons*	*Maximum number of subshells*
1	*K*	2	1 (*s*)
2	*L*	8	2 (*s*, *p*)
3	*M*	18	3 (*s*, *p*, *d*)
4	*N*	32	4 (*s*, *p*, *d*, *f*)
5	*O*	50	5 (*s*, *p*, *d*, *f*, *g*)

there are three possible situations: either it contains zero electrons, one electron, or two electrons. If it has zero electrons, it is unoccupied and does not concern us here, but there is an important difference in properties between an orbital that contains one electron and an orbital that contains two electrons.

Any electron in an orbital is considered to have two types of motion. One is its motion about the nucleus and the other is the spin on its own axis. (This is the same idea as the moon revolving around the earth and, simultaneously, rotating on its axis while it revolves.) A single electron spinning on its own axis generates a magnetic field. In an orbital that contains two electrons, the spins are opposed so that the fields cancel each other. Consequently, an orbital that has one electron generates a magnetic field while an orbital with two electrons does not. This type of magnetism is called **paramagnetism. Any atom, ion, or molecule that contains one or more unpaired electrons will be paramagnetic.** Any atom, ion, or molecule in which all the electrons are paired will not be paramagnetic.

Paramagnetism is one type of magnetism; there are several others, of which we shall mention two. One is **diamagnetism.** We can imagine the orbital motions of the electrons as being like a current in a wire. Diamagnetic behavior occurs when an external magnetic field is applied to the substance. The current in the wire induces small magnetic fields in opposition to the applied field. Since all electrons revolve, all matter possesses diamagnetism. However, diamagnetism is much weaker than paramagnetism, and in any substance that is paramagnetic, the diamagnetism is swamped out.

It is relatively easy to distinguish experimentally between paramagnetic and diamagnetic substances. A paramagnetic substance is attracted to an external magnetic field while a diamagnetic substance is repelled by one (see Figure 13-13).

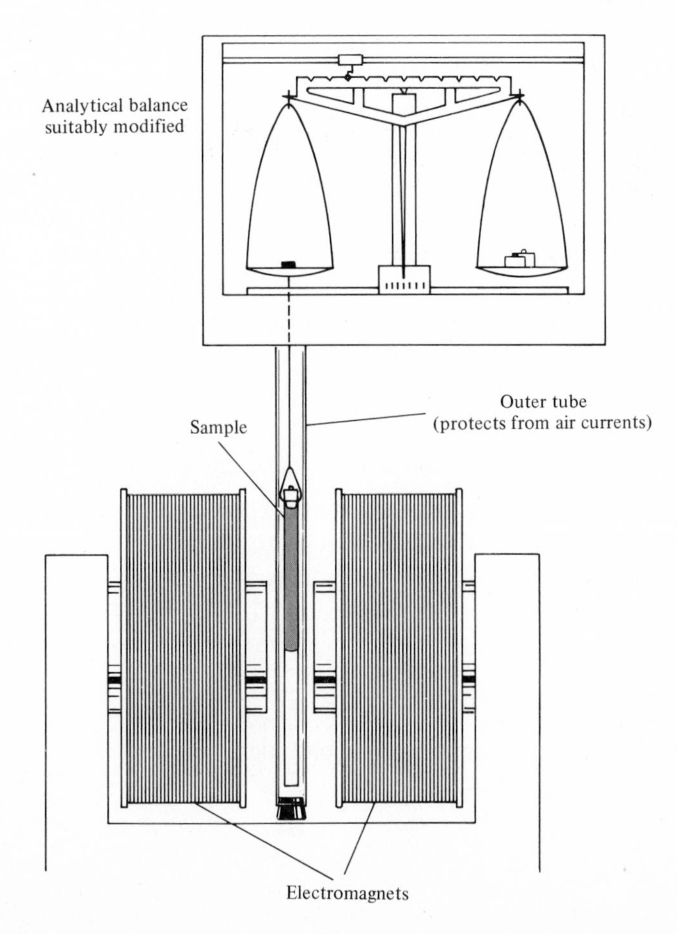

Figure 13-13 A Gouy balance, used to detect paramagnetic material. [Redrawn from *Experiments in Physical Chemistry* by Shoemaker and Garland. Copyright 1967 by McGraw-Hill, Inc. Used with permission of McGraw-Hill Book Company.]

Neither of the above types of magnetism is the everyday magnetism we are all familiar with. This type of magnetism is called **ferromagnetism.** When a paramagnetic substance is placed in a magnetic field, the field will align the unpaired spins and magnetize the substance. If the substance keeps its magnetism when the field is removed, it is called ferromagnetic. Ferromagnetism is much stronger than paramagnetism. The most familiar ferromagnetic substance is iron. Cobalt and nickel are also ferromagnetic.

13-8 The Ground-State Electronic Structures of Atoms

In this section, we shall use the above rules to determine the electronic ground states of the first 36 elements. (The electronic ground states of all the elements are summed up in Table 13-8.) The reader will find it helpful to consult the periodic table shown on the inside front cover of this book while reading this section.

Hydrogen ($_1H$). Hydrogen has one electron, so that in the ground state that electron must occupy the orbital of lowest energy, which is, of course, the 1*s* orbital. The electronic configuration for hydrogen is, therefore,

$1s^1$ the superscript signifies that there is one electron in the 1*s* orbital

The quantum numbers of this electron are

$$n = 1,\ \ell = 0,\ m = 0,\ s = +\tfrac{1}{2} \text{ or } -\tfrac{1}{2}$$

The choice of the *s* quantum number here is arbitrary.

Another representation is

1*s*

(↑)

The arrow in the circle symbolizes the direction of electron spin. It is a matter of convention whether one draws the arrow ↑ up or ↓ down. We shall follow the general convention of drawing all first electrons in an orbital ↑ up and the second electrons ↓ down.

Helium ($_2He$). Here we have two electrons. The first electron is characterized by $n = 1$, $\ell = 0$, $m = 0$, $s = +\frac{1}{2}$; the second by $n = 1$, $\ell = 0$, $m = 0$, $s = -\frac{1}{2}$. The ground-state configuration for helium is $1s^2$. The circle representation is

1*s*

(↑↓)

The helium electrons just complete a shell of electrons (the *K* shell).

Lithium ($_3Li$). The first two electrons in lithium are like those in helium $1s^2$. The 1*s* orbital is now completely filled, and so we must go to the next principal quantum number $n = 2$ for the third electron. $n = 2$, $\ell = 0$, $m = 0$, $s = +\frac{1}{2}$. $1s^2 2s^1$

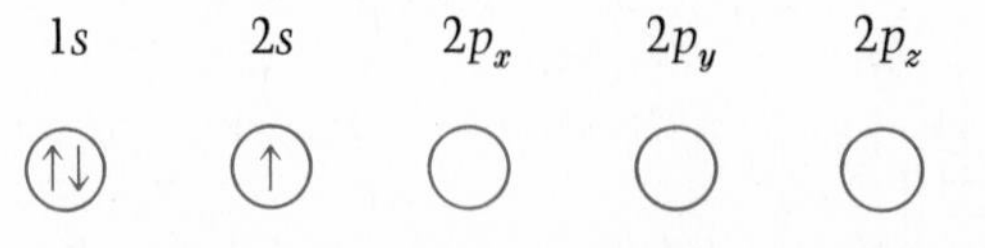

Beryllium ($_4Be$). The fourth electron completes the 2*s* orbital of beryllium. $1s^2 2s^2$

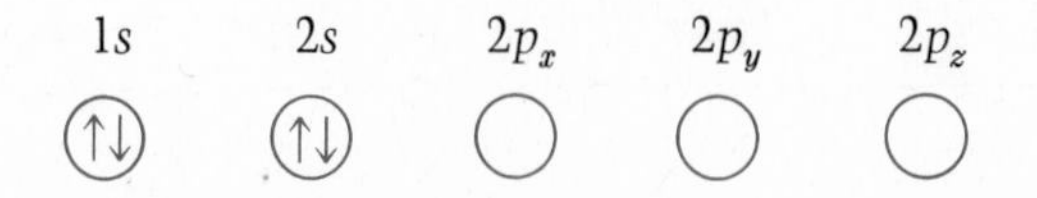

From Section 13-7, we would expect lithium to be paramagnetic, but beryllium should not be. This is indeed the case.

Boron ($_5B$). The first four electrons in boron are like those in beryllium. The fifth electron goes into a p orbital since the 1s and 2s orbitals are filled. $1s^22s^22p_x^{\ 1}$

1s	2s	$2p_x$	$2p_y$	$2p_z$
(↑↓)	(↑↓)	(↑)	○	○

The quantum numbers for the fifth electron are $n = 2$, $\ell = 1$, $m = 1$, $s = +\frac{1}{2}$.
The choice of which p orbital we place the fifth electron in is arbitrary. We could have chosen with equal validity the $2p_y$ or the $2p_z$ orbital.

Carbon ($_6C$). Here is the first application of Hund's rule. The sixth electron does not go into an already occupied 2p orbital because there are 2p orbitals that are vacant. $1s^22s^22p_x^{\ 1}2p_y^{\ 1}$

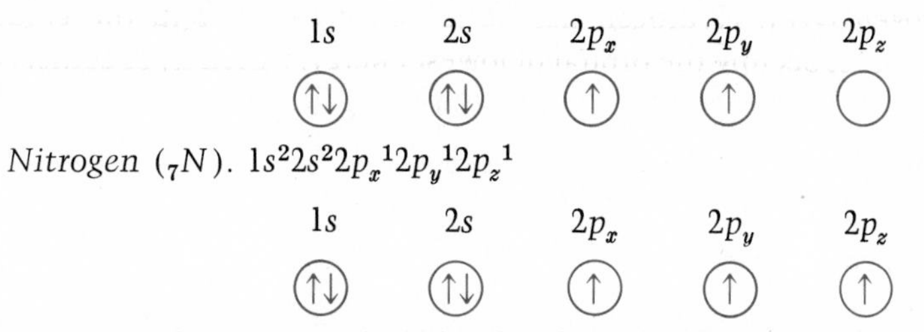

Nitrogen ($_7N$). $1s^22s^22p_x^{\ 1}2p_y^{\ 1}2p_z^{\ 1}$

Again, according to Hund's rule, the three 2p electrons are each in a different p orbital. Nitrogen has a half-filled 2p subshell.

Oxygen ($_8O$). The eighth electron of oxygen must go into one of the half-filled 2p orbitals. $1s^22s^22p_x^{\ 2}2p_y^{\ 1}2p_z^{\ 1}$

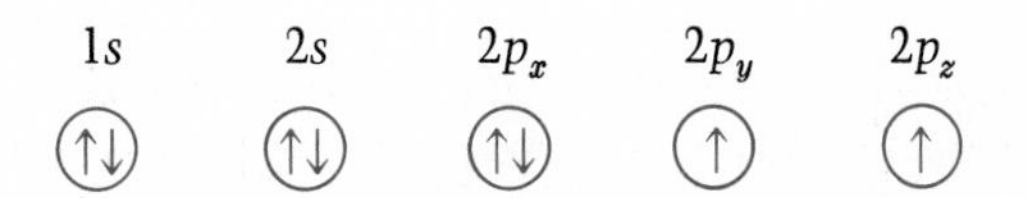

Fluorine ($_9F$). The ninth electron will complete another 2p orbital. $1s^22s^22p_x^{\ 2}2p_y^{\ 2}2p_z^{\ 1}$

1s	2s	$2p_x$	$2p_y$	$2p_z$
(↑↓)	(↑↓)	(↑↓)	(↑↓)	(↑)

Neon ($_{10}Ne$). The tenth electron completes the occupation of all the 2p orbitals. $1s^22s^22p_x^{\ 2}2p_y^{\ 2}2p_z^{\ 2}$

1s	2s	$2p_x$	$2p_y$	$2p_z$
(↑↓)	(↑↓)	(↑↓)	(↑↓)	(↑↓)

With neon we have now filled the second shell (the L shell), so that like helium, neon has a complete outer shell of electrons. The next element, sodium, has 11 electrons. Ten of these will fill the K and L shells, so these 10 will have the same configuration as the 10 electrons of neon. The eleventh electron of sodium must have a principal quantum number of $n = 3$, since the $n = 1$ and $n = 2$ levels are completely filled. In fact, *all* the remaining elements must have complete K and L shells. This being the case, we may use the symbol [Ne] to stand as a shorthand representation of $1s^22s^22p_x^{\ 2}2p_y^{\ 2}2p_z^{\ 2}$. The next eight elements will fill in the 3s and 3p orbitals in a manner similar to the way elements with atomic numbers of 3 to 10 filled in the 2s and 2p orbitals.

Table 13-8 Ground-State Electronic Configurations of the Elements

Atomic Number	*Symbol*	1*s*	2*s*	2*p*	3*s*	3*p*	3*d*	4*s*	4*p*	4*d*	4*f*	5*s*	5*p*	5*d*	5*f*	6*s*	6*p*	6*d*	7*s*
1	H	1																	
2	He	2																	
3	Li	2	1																
4	Be	2	2																
5	B	2	2	1															
6	C	2	2	2															
7	N	2	2	3															
8	O	2	2	4															
9	F	2	2	5															
10	Ne	2	2	6															
11	Na	2	2	6	1														
12	Mg	2	2	6	2														
13	Al	2	2	6	2	1													
14	Si	2	2	6	2	2													
15	P	2	2	6	2	3													
16	S	2	2	6	2	4													
17	Cl	2	2	6	2	5													
18	Ar	2	2	6	2	6													
19	K	2	2	6	2	6		1											
20	Ca	2	2	6	2	6		2											
21	Sc	2	2	6	2	6	1	2											
22	Ti	2	2	6	2	6	2	2											
23	V	2	2	6	2	6	3	2											
24	Cr	2	2	6	2	6	5	1											
25	Mn	2	2	6	2	6	5	2											
26	Fe	2	2	6	2	6	6	2											
27	Co	2	2	6	2	6	7	2											
28	Ni	2	2	6	2	6	8	2											
29	Cu	2	2	6	2	6	10	1											
30	Zn	2	2	6	2	6	10	2											
31	Ga	2	2	6	2	6	10	2	1										
32	Ge	2	2	6	2	6	10	2	2										
33	As	2	2	6	2	6	10	2	3										
34	Se	2	2	6	2	6	10	2	4										
35	Br	2	2	6	2	6	10	2	5										
36	Kr	2	2	6	2	6	10	2	6										
37	Rb	2	2	6	2	6	10	2	6			1							
38	Sr	2	2	6	2	6	10	2	6			2							
39	Y	2	2	6	2	6	10	2	6	1		2							
40	Zr	2	2	6	2	6	10	2	6	2		2							
41	Nb	2	2	6	2	6	10	2	6	4		1							
42	Mo	2	2	6	2	6	10	2	6	5		1							
43	Tc	2	2	6	2	6	10	2	6	6		1							
44	Ru	2	2	6	2	6	10	2	6	7		1							
45	Rh	2	2	6	2	6	10	2	6	8		1							
46	Pd	2	2	6	2	6	10	2	6	10									
47	Ag	2	2	6	2	6	10	2	6	10		1							
48	Cd	2	2	6	2	6	10	2	6	10		2							
49	In	2	2	6	2	6	10	2	6	10		2	1						
50	Sn	2	2	6	2	6	10	2	6	10		2	2						
51	Sb	2	2	6	2	6	10	2	6	10		2	3						
52	Te	2	2	6	2	6	10	2	6	10		2	4						
53	I	2	2	6	2	6	10	2	6	10		2	5						
54	Xe	2	2	6	2	6	10	2	6	10		2	6						

Table 13-8 (continued)

Atomic Number	Symbol	1s	2s	2p	3s	3p	3d	4s	4p	4d	4f	5s	5p	5d	5f	6s	6p	6d	7s
55	Cs	2	2	6	2	6	10	2	6	10		2	6			1			
56	Ba	2	2	6	2	6	10	2	6	10		2	6			2			
57	La	2	2	6	2	6	10	2	6	10		2	6	1		2			
58	Ce	2	2	6	2	6	10	2	6	10	2	2	6			2			
59	Pr	2	2	6	2	6	10	2	6	10	3	2	6			2			
60	Nd	2	2	6	2	6	10	2	6	10	4	2	6			2			
61	Pm	2	2	6	2	6	10	2	6	10	5	2	6			2			
62	Sm	2	2	6	2	6	10	2	6	10	6	2	6			2			
63	Eu	2	2	6	2	6	10	2	6	10	7	2	6			2			
64	Gd	2	2	6	2	6	10	2	6	10	7	2	6	1		2			
65	Tb	2	2	6	2	6	10	2	6	10	9	2	6			2			
66	Dy	2	2	6	2	6	10	2	6	10	10	2	6			2			
67	Ho	2	2	6	2	6	10	2	6	10	11	2	6			2			
68	Er	2	2	6	2	6	10	2	6	10	12	2	6			2			
69	Tm	2	2	6	2	6	10	2	6	10	13	2	6			2			
70	Yb	2	2	6	2	6	10	2	6	10	14	2	6			2			
71	Lu	2	2	6	2	6	10	2	6	10	14	2	6	1		2			
72	Hf	2	2	6	2	6	10	2	6	10	14	2	6	2		2			
73	Ta	2	2	6	2	6	10	2	6	10	14	2	6	3		2			
74	W	2	2	6	2	6	10	2	6	10	14	2	6	4		2			
75	Re	2	2	6	2	6	10	2	6	10	14	2	6	5		2			
76	Os	2	2	6	2	6	10	2	6	10	14	2	6	6		2			
77	Ir	2	2	6	2	6	10	2	6	10	14	2	6	7		2			
78	Pt	2	2	6	2	6	10	2	6	10	14	2	6	9		1			
79	Au	2	2	6	2	6	10	2	6	10	14	2	6	10		1			
80	Hg	2	2	6	2	6	10	2	6	10	14	2	6	10		2			
81	Tl	2	2	6	2	6	10	2	6	10	14	2	6	10		2	1		
82	Pb	2	2	6	2	6	10	2	6	10	14	2	6	10		2	2		
83	Bi	2	2	6	2	6	10	2	6	10	14	2	6	10		2	3		
84	Po	2	2	6	2	6	10	2	6	10	14	2	6	10		2	4		
85	At	2	2	6	2	6	10	2	6	10	14	2	6	10		2	5		
86	Rn	2	2	6	2	6	10	2	6	10	14	2	6	10		2	6		
87	Fr	2	2	6	2	6	10	2	6	10	14	2	6	10		2	6		1
88	Ra	2	2	6	2	6	10	2	6	10	14	2	6	10		2	6		2
89	Ac	2	2	6	2	6	10	2	6	10	14	2	6	10		2	6	1	2
90	Th	2	2	6	2	6	10	2	6	10	14	2	6	10		2	6	2	2
91	Pa	2	2	6	2	6	10	2	6	10	14	2	6	10	2	2	6	1	2
92	U	2	2	6	2	6	10	2	6	10	14	2	6	10	3	2	6	1	2
93	Np	2	2	6	2	6	10	2	6	10	14	2	6	10	5	2	6		2
94	Pu	2	2	6	2	6	10	2	6	10	14	2	6	10	6	2	6		2
95	Am	2	2	6	2	6	10	2	6	10	14	2	6	10	7	2	6		2
96	Cm	2	2	6	2	6	10	2	6	10	14	2	6	10	7	2	6	1	2
97	Bk	2	2	6	2	6	10	2	6	10	14	2	6	10	8	2	6	1	2
98	Cf	2	2	6	2	6	10	2	6	10	14	2	6	10	10	2	6		2
99	Es	2	2	6	2	6	10	2	6	10	14	2	6	10	11	2	6		2
100	Fm	2	2	6	2	6	10	2	6	10	14	2	6	10	12	2	6		2
101	Md	2	2	6	2	6	10	2	6	10	14	2	6	10	13	2	6		2
102	No	2	2	6	2	6	10	2	6	10	14	2	6	10	14	2	6		2
103	Lr	2	2	6	2	6	10	2	6	10	14	2	6	10	14	2	6	1	2

Sodium ($_{11}$Na) $[Ne]3s^1$

Magnesium ($_{12}$Mg) $[Ne]3s^2$

Aluminum ($_{13}$Al) $[Ne]3s^2 3p_x{}^1$

Silicon ($_{14}$Si) $[Ne]3s^2 3p_x{}^1 3p_y{}^1$

Phosphorus ($_{15}$P) $[Ne]3s^2 3p_x{}^1 3p_y{}^1 3p_z{}^1$

Sulfur ($_{16}$S) $[Ne]3s^2 3p_x{}^2 3p_y{}^1 3p_z{}^1$

Chlorine ($_{17}$Cl) $[Ne]3s^2 3p_x{}^2 3p_y{}^2 3p_z{}^1$

Argon ($_{18}$Ar) $[Ne]3s^2 3p_x{}^2 3p_y{}^2 3p_z{}^2$

In terms of shells, the electronic configuration of argon is 2, 8, 8. Remember (Table 13-7) that a completed *M* shell has a total of 18 electrons. Argon has only 8 in this shell. The five 3*d* orbitals in argon are vacant. Nevertheless, argon behaves as a noble gas and as an element that possesses a completed shell. In terms of energy, the 3*d* orbitals are much higher than the 3*p* orbitals, so the element argon behaves as if it had a completed shell. This is also true for the other noble gases: krypton, xenon, and radon.

The next element is potassium, with 19 electrons. The first 18 electrons will, of course, assume the argon configuration (which we may represent as [Ar]). At first sight, it might seem that the nineteenth electron would go into a 3*d* orbital, since the principal quantum number of this orbital is 3. However, rule 6 (page 392) tells us that the 4*s* orbital becomes filled *before* the 3*d* orbitals, and the next two elements are

Potassium ($_{19}$K) $[Ar]4s^1$

Calcium ($_{20}$Ca) $[Ar]4s^2$

Now that the 4*s* orbital is filled, the 3*d* orbitals become the lowest-energy orbitals available, and the next 10 elements will fill these five orbitals, one electron at a time, except for two irregularities to be noted below.

Since the five 3*d* orbitals are equivalent, Hund's rule will be followed, and each of the five will be half-filled before any are filled.

Scandium ($_{21}$Sc) $[Ar]4s^2 3d_{xy}{}^1$

Titanium ($_{22}$Ti) $[Ar]4s^2 3d_{xy}{}^1 3d_{xz}{}^1$

Vanadium ($_{23}$V) $[Ar]4s^2 3d_{xy}{}^1 3d_{xz}{}^1 3d_{yz}{}^1$

Chromium ($_{24}$Cr) $[Ar]4s^1 3d_{xy}{}^1 3d_{xz}{}^1 3d_{yz}{}^1 3d_{z^2}{}^1 3d_{x^2-y^2}{}^1$ *irregularity*

Manganese (25 Mn) $[Ar]4s^2 3d_{xy}{}^1 3d_{xz}{}^1 3d_{yz}{}^1 3d_{z^2}{}^1 3d_{x^2-y^2}{}^1$

Iron ($_{26}$Fe) $[Ar]4s^2 3d_{xy}{}^2 3d_{xz}{}^1 3d_{yz}{}^1 3d_{z^2}{}^1 3d_{x^2-y^2}{}^1$

Cobalt ($_{27}$Co) $[Ar]4s^2 3d_{xy}{}^2 3d_{xz}{}^2 3d_{yz}{}^1 3d_{z^2}{}^1 3d_{x^2-y^2}{}^1$

Nickel ($_{28}$Ni) $[Ar]4s^2 3d_{xy}{}^2 3d_{xz}{}^2 3d_{yz}{}^2 3d_{z^2}{}^1 3d_{x^2-y^2}{}^1$

Copper ($_{29}$Co) $[Ar]4s^1 3d_{xy}{}^2 3d_{xz}{}^2 3d_{yz}{}^2 3d_{z^2}{}^2 3d_{x^2-y^2}{}^2$ *irregularity*

Zinc ($_{30}$Zn) $[Ar]4s^2 3d_{xy}{}^2 3d_{xz}{}^2 3d_{yz}{}^2 3d_{z^2}{}^2 3d_{x^2-y^2}{}^2$

The two irregularities, at chromium and copper, are easily explained on the basis of our discussion of rule 4, where we saw that a filled or half-filled subshell was particularly stable. Vanadium ($_{23}$V) has two electrons in the $4s^2$ orbital, and one each in three 3*d* orbitals. For chromium ($_{24}$Cr) we would expect the next electron to enter a fourth 3*d* orbital, and indeed it does, but at the same time one of the $4s^2$ electrons leaves the 4*s* orbital and occupies the fifth 3*d* orbital, because this produces an atom with a half-filled subshell. Similarly, copper has five filled 3*d* orbitals (a complete subshell) and a half-filled 4*s* orbital, rather than the less stable situation of a filled 4*s* orbital and nine electrons in the 3*d* subshell.

Note that with scandium ($_{21}Sc$), we encountered a situation we had not met earlier. This element is the first to have an electron in the $3d$ subshell, which means that the third shell (the *M* shell), which had behaved chemically as if it were complete when we reached argon ($_{18}Ar$) has now been reopened. Of course, the fourth shell (the *N* shell) is also occupied in scandium, since there are two electrons in the $4s$ orbital. Scandium, therefore, has *two* incomplete shells. The fourth shell is the *outer* shell, and the third shell is an incomplete *inner* shell. We may now redefine a **transition element** (Section 2-10) as **an element with one or more incomplete inner shells.** Scandium is thus the first transition element. In argon, and also in potassium ($_{19}K$) and calcium ($_{20}Ca$), the third shell behaves as a complete shell, but in scandium, it is now incomplete. Similarly, the elements titanium ($_{22}Ti$) through nickel ($_{28}Ni$) are also transition elements, with incomplete third and fourth shells. In copper and zinc, however, the third shell is complete, and these two are technically not transition elements, although copper (and less often zinc) is frequently classified in that way, since its properties resemble those of the transition elements.

After zinc, the next six elements simply fill in the $4p$ subshell (we will write the d electrons as $3d^{10}$ to save space).

Gallium ($_{31}Ga$)	$[Ar]4s^2 3d^{10} 4p^1$
Germanium ($_{32}Ge$)	$[Ar]4s^2 3d^{10} 4p^2$
Arsenic ($_{33}As$)	$[Ar]4s^2 3d^{10} 4p^3$
Selenium ($_{34}Se$)	$[Ar]4s^2 3d^{10} 4p^4$
Bromine ($_{35}Br$)	$[Ar]4s^2 3d^{10} 4p^5$
Krypton ($_{36}Kr$)	$[Ar]4s^2 3d^{10} 4p^6$

With krypton we complete another shell (the fourth or *N* shell), although the $4d$ and $4f$ orbitals are still vacant. As with argon, these orbitals are too far away to be considered part of the fourth shell (although they are occupied in atoms of higher atomic number). In terms of shells, the electronic configuration of krypton is 2, 8, 18, 8. Note that it is the *M* shell that has 18 electrons. The outer shell (the *N* shell) has only eight. In higher elements, when the $4d$ and $4f$ orbitals start to fill, the *N* shell will be the inner shell, and a shell of higher principal quantum number will be the outer shell. It is a general rule that **the outer shell of an atom can hold a maximum of eight electrons.**

We have completed the detailed electronic configurations for the first 36 elements. The electronic structures of most of the other elements will be discussed in Chapter 19, which deals with the transition and inner transition elements, as well as the other metals.

As noted, the complete electronic configurations of all the elements are given in Table 13-8. A quick perusal of this table shows that there are many irregularities in the electronic structure of the heavier elements. At present there is no theoretical explanation as to why these occur in the exact way in which they do; they will therefore not be explored further in this text.

Ions have the same electronic structures as the corresponding atoms except that they have gained or lost electrons. For example, the electronic structure of K is $[Ar]4s^1$, while that of K^+ is [Ar], the $4s$ electron having been lost. For representative elements, the electrons lost by positive ions are the p electrons, if any, and then the s electrons. In forming negative ions, atoms generally gain p electrons. For transition elements, electrons lost first are normally those of the highest principal quantum number, so $4s$ electrons are lost before $3d$ electrons. For example, the electronic structure of Mn is $[Ar]4s^2 3d^5$, while that of Mn^{2+} is $[Ar]3d^5$. Note that this is contrary to the way these levels were built up. The $4s$ level was filled before the $3d$. The reason for this is that in ions the $3d$ orbitals have a lower energy than the $4s$.

13-9 Electronic Structure and the Periodic Table

We saw in Chapter 2 that Mendeleev and the chemists who came after him were able to use the periodic table, even though they had no idea why it worked. Now that we have seen the ground-state electronic configurations of the atoms, it is possible to understand why the table works: it works primarily because atoms within a single column of the table have the same electron configurations in their outer shells and, for transition elements, in their incomplete inner shells (although there are a few irregularities for the latter). For example, some outer-shell and incomplete inner-shell configurations are:

Group IIA		*Group IVA*		*Group IVB*	
Be	$2s^2$	C	$2s^2 2p^2$	Ti	$4s^2 3d^2$
Mg	$3s^2$	Si	$3s^2 3p^2$	Zr	$5s^2 4d^2$
Ca	$4s^2$	Ge	$4s^2 4p^2$	Hf	$6s^2 5d^2$
Sr	$5s^2$	Sn	$5s^2 5p^2$		
Ba	$6s^2$	Pb	$6s^2 6p^2$		
Ra	$7s^2$				

required

For representative elements (defined in Section 2-10), outer-shell electrons are the ones that take part in chemical bonding (we will discuss this in Chapter 14), and these electrons are, therefore, called **valence electrons.** Since the outer shell of each element in group IIA, for example, consists of two electrons in an *s* orbital, it is not surprising that all of them should be able to give up two electrons to form ions with a charge of +2. Because they form ions by giving up two electrons, each of these elements is said to have a valence of +2 (each charge unit is assumed to represent one bond). In a similar manner, all the elements within any other column of the periodic table would be expected to have the same valence, because in each case the outer-shell configurations are the same. If we know the valence of one element in a column, it is a good bet that the other elements in the same column have the same valence. We say a good bet because this is not an absolute rule—there are exceptions. For example, the compounds PbF_2, SnF_2, and GeF_2 are all known stable compounds at room temperature, from which it might be predicted that SiF_2 and CF_2 could also exist at room temperature. However, the latter two are not stable compounds; and, indeed, stable compounds in which carbon or silicon has a valence of 2 are unknown. However, in most cases we can use the above rule with some degree of confidence: if we know the valence of one element in a column, it is a good bet that the other elements in the same column have the same valence.

It is also easy to see why properties other than valence should change in a regular way as we proceed down a column. Since the outer-shell configuration remains the same, the only differences in the atoms as we go down the table are that the nuclear charge and the number of inner shells increase, and, of course, this happens in a regular manner, each atom being larger than the one above it. Such properties as density, melting point, heat of formation, and heat of fusion change in a fairly regular way from top to bottom of a column (though there are some exceptions). Table 13-9 presents some examples and several others will be discussed in Sections 13-10 to 13-12.

In many cases, properties also change in a regular way as we proceed *across* the periodic table, from left to right, because the configuration of each element is the same as the one before, except that an additional electron has been added, in most cases to the same subshell. For the representative elements, this applies especially to valence and is the reason valences change in the regular manner shown in Table 2-13

Table 13-9 Some Properties of Elements in Individual Columns of the Periodic Table[a]

Density (g/ml)[b]				*Melting point (°C)*			
IA		*IIB*		*VIIA*		*VB*	
Li	0.53	Zn	7.14	F	−219.6	V	1917
Na	0.97	Cd	8.6	Cl	−101	Nb	2487
K	0.86	Hg	13.55	Br	−7.2	Ta	2997
Rb	1.53			I	113.6		
Cs	1.90			At	—		

Heat of formation as monatomic gas (kcal/mole)				*Heat of fusion (cal/mole)*[c]			
IVA		*IVB*		*IIIA*		*VIB*	
C	171.7	Ti	112.6	B	5.3	Cr	3.3
Si	105	Zr	146	Al	2.55	Mo	6.6
Ge	90.0	Hf	168	Ga	1.336	W	8.42
Sn	72			In	0.78		
Pb	46.8			Tl	1.02		

[a] *Source:* R. T. Sanderson, *Chemical Periodicity*, Reinhold Publishing Corp., New York, 1960
[b] Note the irregularity in the values of K and Na.
[c] Note the irregularity in the values of In and Tl.

(page 58). However, many other properties also change in a regular manner as we proceed across the table. In Sections 13-10 to 13-12, we shall examine some of these properties (see also page 288).

13-10 Sizes of Atoms and Ions

In this section, we shall see how the sizes of atoms and ions of the representative elements vary with their positions in the periodic table. Unfortunately, it is not meaningful to measure directly the size of an atom (or an ion) because, as we saw in Section 13-2, the outer part of an atom is composed of electron clouds which have no sharp boundaries, but in fact go out to infinity. Therefore, we cannot say exactly where an atom ends. Because of this, it is necessary to measure the sizes indirectly. One technique is to begin with molecules that are made up of two atoms of the same element (for example, Cl_2, H_2, F_2). These atoms are held together by a covalent bond (Section 14-2) to form a molecule (Figure 13-14), where the two nuclei remain at a fixed average distance (d) from each other.* This distance can be measured, and we can define the **atomic radius** (also called the **covalent radius**) in this case to be one-half of d. For example, d for the Cl_2 molecule is 1.99 Å, so that the atomic radius of chlorine is taken to be 0.99 Å.

*It is necessary to say *average distance* because the molecules are constantly vibrating (Figure 5-5) so that the nuclei get closer together and farther apart many times each second. However, the *average* distance is a constant for the ground state of a given molecule.

d

Figure 13-14 For the Cl_2 molecule, d = 1.99 Å.

This method cannot be used for an element like carbon, which does not form a molecule C_2. In a case like this what we do is to measure the C—Cl bond distance in a compound like CCl_4, and then subtract the value of the chlorine radius to obtain the carbon radius.

Example: The C—Cl bond distance in CCl_4 is 1.77 Å. What is the atomic radius of carbon?

Answer: The atomic radius of chlorine is 0.99 Å. So the atomic radius of carbon is 1.77 − 0.99 Å = 0.78 Å

Using this method we can calculate atomic radii for most of the elements. There are a number of other techniques for calculating atomic radii, but, as expected, the values obtained for the same atom using different methods are different. This is because the atom does not have a fixed radius and different measuring techniques will squeeze the electron cloud to different extents yielding different values. Because helium, neon, and argon do not form compounds, the covalent radius method cannot be used for them. These values are obtained by other methods. Ionic radii are measured by a totally different method which will be discussed in Section 15-2 when we take up ionic crystal lattices.

When we have obtained all these values, what do we find? We find that we can make several generalizations about how the sizes of atoms and ions vary with their positions in the table:

1. The sizes of *atoms* increase as we go down a column of the periodic table (some values are given in Table 13-10). This is understandable, since each time we go down one step, the outer-shell configuration is the same, but there is an additional inner shell. Consequently, the atom is larger.

2. The sizes of *ions* increase as we go down a column of the table (Table 13-11),

Table 13-10 Covalent Radii in Ångstrom Units for Some Representative Families of Elements

Li 1.33	Be 0.90	O 0.74	F 0.71
Na 1.54	Mg 1.36	S 1.02	Cl 0.99
K 1.96	Ca 1.74	Se 1.16	Br 1.14
Rb 2.16	Sr 1.92	Te 1.35	I 1.33
Cs 2.35	Ba 1.98		

Table 13-11 Some Ionic Radii in Å

Li^+ 0.60	Be^{2+} 0.31	F^- 1.36
Na^+ 0.95	Mg^{2+} 0.65	Cl^- 1.81
K^+ 1.33	Ca^{2+} 0.99	Br^- 1.95
Rb^+ 1.48	Sr^{2+} 1.13	I^- 2.16
Cs 1.69	Ba^{2+} 1.35	

Li	Be	B	C	N	O	F
1.33	0.90	0.80	0.78	0.73	0.74	0.71
Na	Mg	Al	Si	P	S	Cl
1.54	1.36	1.25	1.15	1.1	1.02	0.99

Table 13-12 Covalent Radii in Ångstrom Units of Elements of the Second and Third Rows of the Periodic Table

provided that we are comparing ions of the same charge (for example, Be^{2+}, Mg^{2+}, Ca^{2+}, etc.). The reasons are similar to those given above for atomic radii.

3. In general, the sizes of *atoms* decrease as we go from left to right along a row of the table (Table 13-12), except for column 0. This seems at first sight surprising, but we can explain it in the following way.

All electrons in an atom to a good approximation (neglecting quantum effects) are under two general influences. One is the attraction of the negative electrons for the positive nucleus and the other is the mutual repulsion of the electrons for each other. Both of these forces are governed by Coulomb's law (force of attraction = $-Ze^2/r^2$; force of repulsion = e^2/r^2).

It is important to remember that these forces are inversely proportional to the square of the distance, so that increasing the distance between an electron and the positive nucleus considerably decreases the force of attraction between the two. Let us compare Li (1.35 Å) and Be (0.90 Å). A lithium atom has three protons and three electrons. The first two electrons occupy a $1s^2$ orbital and the third electron is in the $2s^1$ orbital. (There are also four neutrons in the most common isotope, but these are irrelevant to our discussion, because they are uncharged and therefore do not influence the attraction, and hence the size of the ion.)

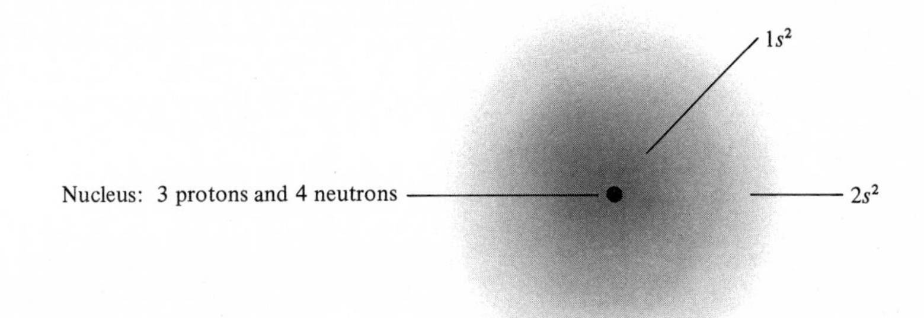

We shall assume (and it is a *good* assumption) that the $1s^2$ electrons in the inner shell effectively are **shielding** the $2s^1$ electron, so that for practical purposes the $2s^1$ electron in lithium *sees* only one proton in the nucleus. The force of attraction then between the $2s^1$ electron and the nucleus would be ($Z = 1$) $F_a = -e^2/r^2$. If the $2s^1$ electron were able to feel or be under the influence of all three protons, then F_a would equal $-3e^2/r^2$, a much greater force of attraction (three times as much).

The situation for beryllium is somewhat similar but also somewhat different. There are four protons in the nucleus of a beryllium atom and four electrons. The first two electrons occupy a $1s^2$ orbital and the remaining two are in a $2s^2$ orbital. The inner two electrons shield the outer two, so the latter see an effective $Z = 2$ instead of $Z = 4$. What of the outer two electrons? They are both in the same shell and one might think that they shield each other so that each electron would only see one proton. This is not so. They are both negatively charged and tend to *repel* each other. Therefore, they are not very effective in shielding each other, although some shielding does occur. The electrons in the $2s^2$ orbital on the average experience a $Z > 1$ and are attracted to the nucleus to a greater extent than the $2s^1$ electron in lithium. Because of this, the atomic size decreases. The same argument can be extended across the rest of the row. When we go across a row of the periodic table, we are adding protons and electrons in the same principal shell. Shielding of the outer electrons by each other becomes less effective as we go across the table and the electrons experience a larger and larger Z factor. The force of attraction becomes stronger and the atoms become smaller.

4. A positive ion is always *smaller* than the corresponding neutral atom (compare the values in Tables 13-10 and 13-11). A positive ion contains a smaller number of electrons, compared to the neutral atom, while both have the same number of protons. The electrons in the positive ion cannot shield each other as well as those in the neutral atom, and experience a greater force of attraction (than the corresponding neutral atoms) toward the nucleus, making the ion smaller.

5. A negative ion is always *larger* than the corresponding neutral atom (compare the values in Tables 13-10 and 13-11). A negative ion has more electrons than the neutral atom. These electrons can shield each other more effectively and therefore experience less attraction for the nucleus. Additionally, the electrons in the negative ions repel each other to a greater extent and the outer shell expands.

6. Atoms or ions with the same electronic configurations are said to be **isoelectronic.** If we consider a series of isoelectronic atoms or ions, the size *decreases* with increasing atomic number. Thus, the following are values (in Å) for ionic and atomic radii of five members of an isoelectronic series.

O^{2-}	F^-	Ne	Na^+	Mg^{2+}
1.40	1.36	1.31	0.95	0.65 Å

The reasons are similar to those discussed under generalization 3. As we move from left to right, the electronic configuration remains the same, but the charge on the nucleus has increased by +1. Therefore, there is a greater attraction between the nucleus and the outer-shell electrons, and the atom or ion becomes smaller.

13-11 Ionization Energy

The **first ionization energy** is defined as the energy necessary to remove a single electron from a neutral gaseous atom. Symbolically, we have

$$E(g) \longrightarrow E^+(g) + e^- \quad \text{where E stands for any element}$$

These energies can be measured with a fair amount of precision. Table 13-13 shows the ionization energies for some of the elements in the periodic table. The values in Table 13-13 have a positive sign because energy is absorbed by the system.

As we would expect from our previous discussions, the values of the first ionization energies vary in a regular way across or down the periodic table. First, let us consider the groups of elements. Table 13-13 shows that the first ionization energy *decreases* as we go down the table. This is exactly what we would expect, and is consistent with our previous statement that the size of atoms increases as we go down the table. The first ionization energy measures the ease of removing an electron from the outer shell, and as we go down the table this shell is farther away from the nucleus. There is a smaller attraction between the electrons in this shell and the positive nucleus. It is true that the positive charge of the nucleus also increases as we go down the table, but this of course is offset by the shielding supplied by the inner electrons.

As we go from left to right, the atomic size decreases; the outer electrons are more greatly attracted to the nucleus and are more difficult to remove. Consequently, the first ionization energy *increases* as we go from left to right across the table. Table 13-13 shows that there are some irregularities here, and most of these can be explained by the stability of filled and half-filled subshells. Thus, beryllium has a higher ionization energy than boron. When beryllium gives up an electron, it must

Table 13-13 First (top) and Second (bottom) Ionization Energies for the Representative Elements from Hydrogen to Krypton in Electron Volts[a]

IA	IIA	IIIA	IVA	VA	IVA	VIIA	VIIIA
H 13.60 —							He 24.59 54.42
Li 5.39 75.64	Be 9.32 18.21	B 8.30 25.15	C 11.26 24.38	N 14.53 29.60	O 13.62 35.12	F 17.42 34.97	Ne 21.56 40.96
Na 5.14 47.29	Mg 7.65 15.04	Al 5.99 18.83	Si 8.15 16.35	P 10.49 19.73	S 10.36 23.33	Cl 12.97 23.81	Ar 15.76 27.63
K 4.34 31.63	Ca 6.11 11.87	Ga 6.00 20.51	Ge 7.90 15.93	As 9.81 18.63	Se 9.75 21.19	Br 11.81 21.8	Kr 14.00 24.36

[a]*Source:* J. F. Liebman, *J. Chem. Educ.* **50,** 831 (1973).

come from a filled 2*s* subshell, while boron need only give up its lone 2*p* electron. Another example is oxygen, which has a lower ionization energy than nitrogen. Nitrogen has a half-filled 2*p* subshell that will, of course, no longer be intact when an electron is given up, but oxygen will have a half-filled 2*p* subshell *after* it has given up one electron.

All atoms other than hydrogen have two or more electrons. We can define the **second ionization energy** as the amount of energy necessary to remove the second electron from an atom that has already lost one (third and higher ionization energies are analogously defined):

$$\text{Second ionization energy:}\quad E^+(g) \longrightarrow E^{2+}(g) + e^-$$
$$\text{Third ionization energy:}\quad E^{2+}(g) \longrightarrow E^{3+}(g) + e^-$$

Table 13-13 shows second ionization energies for some of the elements. We immediately see from the table that the second ionization energy for any element is always numerically larger than the first ionization energy for the same element (the third ionization energy is larger still, etc). The explanation for this is quite simple. After the first electron has been removed from a neutral atom, the number of protons in the remaining ion (E^+) has not changed, but the number of electrons has diminished by one. Thus, the remaining electrons in E^+ are held more closely to the nucleus than they were in the original atom E. It is more difficult to remove the second electron than it was the first, and the second ionization energy is higher. The same reasoning holds for third and higher ionization energies.

Another fact that strikes us as we look at the second ionization energies in Table 13-13 is that they approximately increase in going from left to right, *except for column IA*. The first ionization energy for each element in column IA is the lowest for any element in that row, but the second ionization *energy is the highest*. There is a much greater difference between the second and the first ionization energies for the group IA elements than for any other in Table 13-13. For example, the second ionization energy for lithium (group IA) is about 14 times the first, while for beryllium (group IIA), it is only about 2.

	Li	Be
First ionization energy:	5.39	9.32
Second ionization energy:	75.64	18.21

To account for the large jump in ionization energies for lithium compared to beryllium, consider what the remaining configurations of both of them are once the first electron is removed.

Li^+	Be^+
$1s^2$	$1s^22s^1$

Li^+ has lost its only outer electron and now has only inner electrons left. The inner electrons occupy a much smaller space than the valence electrons and are held very tightly to the nucleus. Conversely, beryllium still has one outer electron left at a much greater distance from the nucleus than the inner electrons of Li^+. The second ionization energy of Li^+ will be much, much greater than the first because of this stripping of its outer electrons.

In summary, we can state that:

1. Second ionization energies are always greater than first ionization energies (and third and fourth are higher still).
2. Whenever all the outer-shell electrons are removed, the next ionization energy is *much* greater than the previous values for the same element.

If we compare the first ionization energies of elements going across a row, the noble gas always has the highest value in its respective row. A noble-gas configuration is extraordinarily stable. It therefore takes more energy to disrupt this particular electronic configuration compared to a non-noble-gas configuration.

13-12 Electron Affinity

In Section 13-11, we defined the ionization energy as the amount of energy necessary to remove a single electron from a neutral gaseous atom to produce a positive ion. The values were given a positive sign because energy is absorbed by the system when an electron is given up. A neutral gaseous atom can also *add* an electron to give a negative ion. The energy given up or absorbed by this process is called the **electron affinity.**

$$E(g) + e^- \longrightarrow E^-(g)$$

Table 13-14 lists the values of the electron affinity for some of the representative elements. In this case the sign convention used by most chemists is opposite to that normally employed for other reactions. When energy is given off, we use the + sign; when it is absorbed, we use the − sign. As Table 13-14 shows, the electron affinity in general increases in going from left to right across a row of the table, although there are several irregularities. Some of these irregularities are explained by the stability of filled and half-filled subshells, but others are unexplained.

Table 13-14 Electron Affinities of Some Elements in Electron Volts[a]

H 0.75							He −0.22
Li 0.62	Be −2.5	B 0.24	C 1.27	N 0.0	O 1.47	F 3.34	Ne −0.30
Na 0.55	Mg −2.4	Al 0.46	Si 1.24	P 0.77	S 2.08	Cl 3.61	Ar −0.36
						Br 3.36	Kr −0.40
						I 3.06	Xe −0.42

[a] *Source:* E. C. M. Chen and W. E. Wentworth, *J. Chem. Educ.* **52,** 486 (1975).

We would expect electron affinities to decrease in going down a column of the table, because as the atom gets larger, the attraction of the positive nucleus for an outside electron becomes less. This is generally true, but here, too, there are unexplained irregularities, most notably at fluorine, which has a *lower* value than chlorine.

Part of the explanation for these irregularities may lie in the fact that electron affinity values are difficult to measure; much more so than ionization energies. Therefore, there are many gaps in the table, and the values given are not generally known with high accuracy.

13-13 Classification of the Elements

We can divide all the elements into five classes, based on their electronic structures and positions in the periodic table (see Figure 13-15).

1. METALS. This is the largest class of elements. There are more metals than all the other elements combined. In addition to their shiny metallic luster, a characteristic property of metals is that they readily give up electrons (their ability to carry an electric current is one aspect of this). We saw in Section 2-10 that metallic character *increases in going down* a column and *decreases in going from left to right* across a row of the periodic table. This, of course, is precisely the order of ionization energy, and we now see that metals are elements with low ionization energies. In general, the higher the ionization energy, the lower the degree of metallic character. As shown in Figure 13-15, all the transition and inner transition elements are metals, as well as all the elements of groups IA and IIA. For the elements of groups IIIA to VA, those near the bottom are metals.

2. NONMETALS. These elements do not readily give up electrons. They either take electrons to form negative ions, or form covalent bonds (as we shall see in Chapter 14). They have high ionization energies and do not carry an electric current. We shall use the word "nonmetal" to refer to 10 elements; these are shown in Figure 13-15.

3. METALLOIDS. Since metallic character *gradually* increases as we go from right to left across a row, or down a column of the table, it is not possible to draw a sharp line between metals and nonmetals. For example, in column VA, nitrogen and phosphorus are clearly nonmetals, although phosphorus is more metallic than

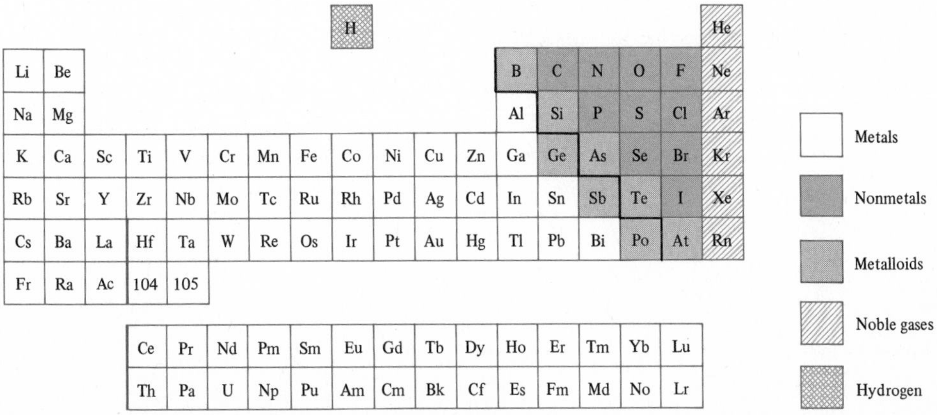

Figure 13-15 Classification of the elements.

nitrogen. If we go down to the bottom of the column, bismuth is clearly a metal. But the two intermediate elements, arsenic and antimony, have both metallic and nonmetallic properties (as expected, antimony is more metallic than arsenic). Such elements are called **metalloids.** We recognize eight of these (B, Si, Ge, As, Sb, Te, Po, and At), although other chemists may have somewhat different classifications. The heavy zigzag line shown in Figure 13-15 and in most periodic tables divides the metals from the nonmetals. Most of the elements on either side of this line are usually classified as metalloids.

4. Noble Gases. The elements in column 0 do not behave like metals, nor do they behave like nonmetals. Thus, they clearly belong in a different category, the chief characteristic of which is that they are very reluctant to form compounds (see Section 18-7).

5. Hydrogen. The element hydrogen is different from any other element and cannot be easily classified in any of the preceding categories. It is in a class by itself, as we shall see in Section 18-1.

Problems

13-1 Define each of the following terms. **(a)** standing wave **(b)** ψ function **(c)** wave function **(d)** Cartesian coordinates **(e)** spherical polar coordinates **(f)** orbital **(g)** energy state **(h)** electron cloud **(i)** energy level **(j)** electron density **(k)** quantum number **(l)** atomic unit **(m)** radial probability function **(n)** node **(o)** electron spin **(p)** Pauli exclusion principle **(q)** aufbau process **(r)** Hund's rule **(s)** half-filled orbital **(t)** shell **(u)** subshell **(v)** paramagnetism **(w)** diamagnetism **(x)** ferromagnetism **(y)** transition element **(z)** valence electrons **(aa)** atomic radius **(bb)** shielding **(cc)** isoelectronic **(dd)** ionization energy **(ee)** electron affinity **(ff)** metalloid

13-2 Discuss the picture of the electron in the ground state of the hydrogen atom as proposed by **(a)** Bohr and **(b)** Schrödinger. What are the chief differences between these pictures?

13-3 **(a)** What are the permitted values for the azimuthal quantum number ℓ for an electron with a principal quantum number $n = 5$?
How many electrons can be accommodated **(b)** in all the 3*d* orbitals? **(c)** in all the 4*d* orbitals? **(d)** in all the 3*f* orbitals? **(e)** in all the 4*f* orbitals?
How many different values of the magnetic quantum number *m* are permitted for an electron with an azimuthal quantum number **(f)** $\ell = 5$? **(g)** $\ell = 0$?

13-4 Arrange in order of increasing energy electrons with the following sets of quantum numbers.

	n	ℓ	m	s		n	ℓ	m	s
(a)	1	0	0	$\frac{1}{2}$	**(f)**	3	2	2	$\frac{1}{2}$
(b)	2	1	-1	$-\frac{1}{2}$	**(g)**	3	1	-1	$\frac{1}{2}$
(c)	2	0	0	$-\frac{1}{2}$	**(h)**	3	0	0	$\frac{1}{2}$
(d)	3	2	-1	$\frac{1}{2}$	**(i)**	1	0	0	$-\frac{1}{2}$
(e)	3	1	-1	$-\frac{1}{2}$					

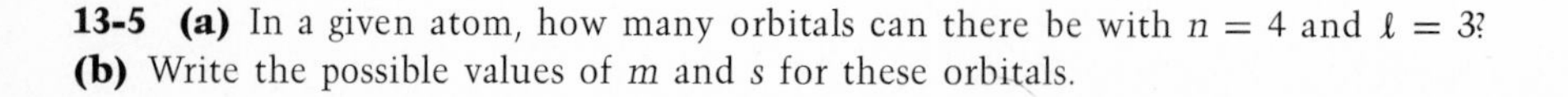

13-5 **(a)** In a given atom, how many orbitals can there be with $n = 4$ and $\ell = 3$?
(b) Write the possible values of *m* and *s* for these orbitals.

13-6 Give the label of the orbital corresponding to each of the following cases, if there is one. **(a)** $n = 3$, $\ell = 0$ **(b)** $n = 5$, $\ell = 5$ **(c)** $n = 4$, $\ell = 1$ **(d)** $n = 5$, $\ell = 4$ **(e)** $n = 3$, $\ell = 5$ **(f)** $n = 6$, $\ell = 3$ **(g)** $n = 5$, $\ell = 3$, $m = -2$

13-7 What are the n and ℓ quantum numbers for each of the following orbitals? (Where the designation given does not refer to any orbital, say so.) **(a)** $2s$ **(b)** $3d$ **(c)** $2d$ **(d)** $7p$ **(e)** $6g$ **(f)** $3f$ **(g)** $5f$ **(h)** $1p$

13-8 In each of the following pairs, predict which orbital will fill first. **(a)** $n = 3$, $\ell = 1$ vs. $n = 3$, $\ell = 2$ **(b)** $n = 5$, $\ell = 3$ vs. $n = 6$, $\ell = 1$ **(c)** $n = 5$, $\ell = 3$ vs. $n = 6$, $\ell = 5$ **(d)** $n = 6$, $\ell = 2$ vs. $n = 5$, $\ell = 0$

13-9 **(a)** What is the total number of electrons that can fit into the $5s$, $5p$, $5d$, and $5f$ orbitals? **(b)** How many orbitals are there in a $5f$ subshell; a $6f$ subshell?

13-10 Arrange the following orbitals in order of increasing energy: $5s$, $3p_x$, $4f$, $3s$, $4p_y$, $3p_z$, $5d$, $2p$.

13-11 Write out the electronic configuration for each of the following atoms. **(a)** Ca **(b)** Al **(c)** Y **(d)** Cr **(e)** Cd **(f)** Te **(g)** P **(h)** Na

13-12 Write out the electronic configuration for each of the following atoms. **(a)** Sc **(b)** Ag **(c)** Zn **(d)** As **(e)** Cu **(f)** Mg **(g)** Ar **(h)** Rn

13-13 Write out the electronic configuration for each of the following ions. **(a)** Sc^{2+} **(b)** Zn^{2+} **(c)** Cr^{3+} **(d)** K^{2+} **(e)** S^{2-} **(f)** I^- **(g)** Sr^{2+} **(h)** Al^{3+}

13-14 Which of the ions in Problem 13-13 are paramagnetic, and how many unpaired electrons does each have?

13-15 What is wrong with each of the following attempts to write ground-state electronic configurations? **(a)** $1s^22s^3$ **(b)** $1s^21p^62s^22p^5$ **(c)** $1s^22s^22p^62d^{10}3s^2$ **(d)** $1s^22s^23s^24s^2$ **(e)** $1s^22s^22p^63s^23d^{10}4s^24p^6$

13-16 **(a)** How many electrons does the N shell contain in the krypton atom? Is it complete? **(b)** How many electrons does the N shell contain in the Pd atom? Is it complete? **(c)** How many electrons does the N shell contain in the Yb atom? Is it complete? **(d)** Which of these three atoms behaves like an atom with a complete outer shell?

13-17 What is the maximum number of electrons **(a)** that any orbital can hold? **(b)** that any outer shell can hold? **(c)** that any inner shell can hold?

13-18 Which of the following are paramagnetic? **(a)** F^- **(b)** Na **(c)** S^{2-} **(d)** S^- **(e)** Ti **(f)** Fe **(g)** Zn **(h)** Zn^+ **(i)** Zn^{2+} **(j)** Fe^{2+} **(k)** Fe^{3+} **(l)** Pb **(m)** Xe **(n)** Te^{2+} **(o)** Ca **(p)** Ca^{2+}

13-19 Write the four quantum numbers for each electron in the ground state of each of the following. **(a)** the carbon atom **(b)** the phosphorus atom **(c)** the chromium atom **(d)** the nickel atom

13-20 For each electronic state: (1) decide whether a neutral atom, a positive ion, or a negative ion is represented; and (2) decide whether the atom is in a ground state, an excited state, or an impossible state. If the state is impossible, explain why. **(a)** H $1s^1$ **(b)** H $1s^2$ **(c)** Ne $1s^22s^22p^5$ **(d)** Ne $1s^22s^32p^4$ **(e)** O $1s^22s^22p^4$ **(f)** Sc $1s^22s^22p^63s^23p^64s^2$ **(g)** Cl $1s^22s^22p^63s^23p^54s^1$ **(h)** C $1s^22s^22p^2$ **(i)** Li $1s^22s^22p^1$ **(j)** Na $1s^22s^22p^63d^1$

13-21 For the first 36 elements in the periodic table, list those whose atoms are paramagnetic in the ground state. List the number of unpaired electrons in each paramagnetic element.

13-22 Why do the halogens have similar properties but not identical properties?

13-23 Select the atom or ion in the following pairs which you would expect to be larger. **(a)** S, Cl **(b)** Se, Te **(c)** Na, Na^+ **(d)** Al, P **(e)** Cl^-, S^{2-} **(f)** K^+, Cl^- **(g)** K^-, K **(h)** Na, S **(i)** O, O^{2-} **(j)** Ar, Kr **(k)** Cl^-, I^- **(l)** Ge, Ga **(m)** H^+, Li^+ **(n)** Be, Na **(o)** Pb, Si **(p)** Se^{2-}, Sr^{2+} **(q)** Si, N

13-24 **(a)** The B—Cl bond distance in BCl_3 is 1.72 Å. Calculate the atomic radius of boron. **(b)** The B—Br distance in BBr_3 is 1.87 Å. Calculate the atomic radius of bromine.

13-25 A hydrogen atom and a helium atom both consist of a nucleus and a 1*s* orbital. The only difference in electronic structure is that H has only one electron in the 1*s* orbital, while He has two. Yet the helium atom is *smaller*. Explain why.

13-26 The ionization energies for the carbon atom are:

First	11.3 eV	Fourth	64.5 eV
Second	24.4 eV	Fifth	392.1 eV
Third	47.9 eV	Sixth	490.0 eV

Explain the very large increase in going from the fourth to the fifth ionization energy. Based on your explanation, where would you predict that the large difference would occur for B, Mg, P, and F?

13-27 Without consulting Table 13-13, select the atom or ion in each of the following pairs which has the higher ionization energy (energy necessary to remove one electron from the species given) and explain why. **(a)** S, Cl **(b)** Se, Te **(c)** Na, Na^+ **(d)** Al, P **(e)** Na, S **(f)** Ar, Kr **(g)** Ge, Ga **(h)** Be, Na **(i)** Pb, Si **(j)** Si, N **(k)** S, S^+ **(l)** C^{3+}, C^{2+} **(m)** Be^{2+}, B^{2+} **(n)** C^{3+}, B^{3+}

13-28 Arrange each group of atoms or ions in order of increasing ionization energy and decreasing size. **(a)** Ca^+, Sc^{2+}, K **(b)** Ge, C, Si

13-29 Give a possible explanation for the sharp dip in electron affinity in column IIA of the periodic table (note from Table 13-14 that Be and Mg have much lower electron affinities than the atoms on either side).

Additional Problems

13-30 Write out the electronic configurations for each of the following ions. **(a)** V^{3+} **(b)** Co^{3+} **(c)** Fe^{3+} **(d)** As^{3+} **(e)** Se^{3-} **(f)** F^- **(g)** H^-

13-31 Which of the ions in Problem 13-30 are paramagnetic, and how many unpaired electrons does each have?

13-32 **(a)** How many electrons can be accommodated in the $n = 5$ energy level? **(b)** How many 4*f* orbitals does an atom have? **(c)** How many 6*f* orbitals does an atom have? **(d)** What is the total number of electrons that can be accommodated in an atom in which the highest principal quantum number is $n = 4$?

13-33 Predict the C—N bond distance in methylamine,

```
   H
   |
H—C—N—H.
   |  |
   H  H
```

13-34 Titanium, which has the electronic configuration $[Ar]4s^2 3d^2$, loses two of these electrons to form the Ti^{2+} ion. What would be a quick and easy way to tell, in the laboratory, whether the two electrons lost are the $4s^2$ or the $3d^2$ electrons?

13-35 Give the name of an atom that has each of the following properties. **(a)** 59 electrons **(b)** 17 protons **(c)** an electronic structure ending with $4s^1 3d^5$

(d) an electronic structure ending with $4s^23d^8$ **(e)** a valence electron structure of $6s^26p^4$ **(f)** a valence electron structure of $2s^22p^5$

13-36 Which of the following atoms has completely filled inner subshells? **(a)** K **(b)** Sc **(c)** Zn **(d)** Ge **(e)** Ba **(f)** Cl **(g)** Xe **(h)** He **(i)** Mn

13-37 **(a)** What is the lowest principal quantum number for which there are g orbitals?
(b) How many g orbitals are there for that quantum number?
(c) How many electrons can each g subshell hold?
(d) How many nodes are there in each g orbital of the lowest energy?

13-38 Arrange each group of atoms or ions in order of increasing size and decreasing ionization energy. **(a)** S^-, S, S^{2-} **(b)** Na, S, Si

13-39 In the ground state of the arsenic atom, how many electrons do each of the following have? **(a)** $\ell = 1$ **(b)** $m = -1$ **(c)** $m = -2$ **(d)** $\ell = 0$ and $m = 0$ **(e)** $\ell = 2$ and $m = -1$ **(f)** $\ell = 1$ and $m = +2$ **(g)** $n = 2$, $\ell = 1$, $m = 0$, and $s = -\frac{1}{2}$

13-40 Name four ions that have this electronic configuration: $1s^22s^22p^6$.

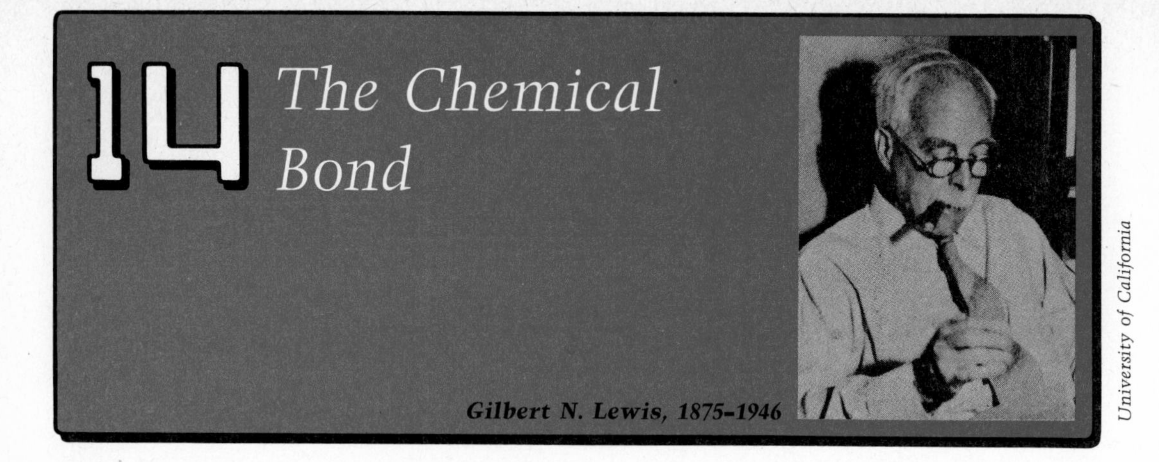

University of California

In Chapter 13, we learned about the electronic structures of atoms. In this chapter, we will see how atoms are attached to each other. The forces that hold atoms or ions together are called **chemical bonds.**

In Section 2-3, we defined a *molecule* as an uncharged entity in which atoms are held together by chemical bonds. We find that atoms link together to form an almost infinite variety of molecules. A molecule of hydrogen gas, for example, is composed of two atoms of hydrogen held together by a single chemical bond. The molecules of sucrose, which is ordinary household sugar, are much larger. Each is made up of 12 atoms of carbon, 22 atoms of hydrogen, and 11 atoms of oxygen. These 45 atoms are linked together by chemical bonds in a particular arrangement in space which gives the sucrose molecule its own individual shape (Figure 14-1). A molecule may be composed of a single atom if that atom is unconnected to any other atom by chemical bonds; examples are helium and mercury vapor, each of which consists of monatomic molecules. In sharp contrast, there are other molecules which are composed of thousands and even millions of atoms. (Such large molecules, which are called giant molecules or polymers, are discussed in Chapters 21 and 22). The largest molecules of all are certain crystals in which every atom in the crystal is part of one gigantic molecule. A good example is diamond. Diamond is one form of the element carbon and each individual perfect diamond is a single molecule, all of whose carbon atoms are joined by chemical bonds in a vast array (see Figure 14-1). For example, a perfect diamond weighing 2 carats (400 mg) is a single molecule containing 2×10^{22} carbon atoms.

Not all matter is made up of molecules. The solid salts we discussed in Chapter 11 are composed of ions. For example, in common table salt (NaCl) we cannot find a simple unit consisting of one sodium and one chlorine atom. Instead, there is a tremendous network of sodium ions (Na^+) and chloride ions (Cl^-), arranged in a definite and particular geometric array (see Figure 15-8). The same is true for other salts, and each salt has its own type of array (this topic is discussed in Section 15-4). We cannot speak of a sodium chloride molecule or a molecule of any other solid salt. Solid metals are another kind of matter that is not composed of molecules. The atoms in solid metals are connected to each other by a special kind of bond, called a metallic bond (Section 19-2). In general, we can say that gases and liquids under ordinary conditions are always composed of molecules (and relatively small molecules—usually less than about 40 or 50 atoms), but that solids may be made up of small or large molecules, or of ions, or of metallic atoms held together by metallic bonds.

With the sophisticated instruments that are available today, we have been able to measure, calculate, and accumulate a great deal of information concerning molecules. For example, we know that the water molecule, H_2O, is composed of two atoms of hydrogen and one atom of oxygen. We also know that the hydrogens are

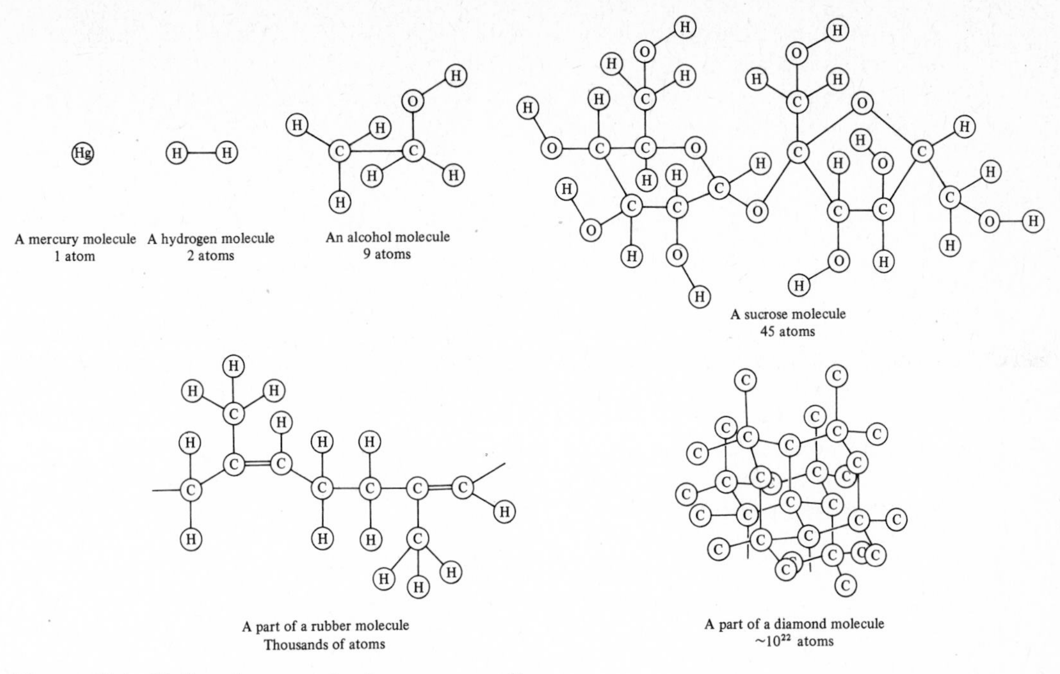

Figure 14-1 Molecules may be large or small.

connected not to each other but through the oxygen. The angle H—O—H is 105°. It is proper to ask why is the water molecule H_2O and not H_3O, H_4O, or HO_2? Also, why is the water molecule nonlinear (that is, why is the H—O—H angle not 180°)? It is the object of this chapter to show that questions of this kind, namely, why atoms combine to form molecules, the way they link with one another (H—O—H not H—H—O), and the particular arrangements of their atoms in space can be answered with a great deal of accuracy and certainty.

We saw in Chapter 13 that electrons in atoms occupy orbitals that have definite energies and shapes. When molecules form, these electrons will occupy different energy levels within the molecule and the shapes of their orbitals will change correspondingly. It is through our knowledge of the electronic structure of atoms that we will understand molecules and chemical bonding.

Part I Properties of Bonds and Shapes of Molecules

14-1 The Ionic Bond

Richard Abegg (1869–1910) was one of the first chemists to be concerned with the electronic structure of atoms. He observed that the noble gases were extraordinarily stable and thought that if he studied the reasons for their unique stability, he might find some of the answers to chemical bonding. It soon became evident that the stability of the noble gases was related to their electronic structures. (Helium possesses two electrons in its valence shell while all other noble gases have a valence shell containing eight electrons.) The noble gases are so stable that in Abegg's time, they were called "inert" gases, since it was believed that they were incapable of forming compounds with themselves or with any other elements. (In Section 18-7 we will discuss some of the noble-gas compounds which are known today.) It was soon concluded that if the number of electrons in the valence shell of atoms was the key, then such atoms as the halogens, with seven valence electrons and high electron affinity, could accept an electron (the electron would have to come from some outside source, possibly another atom), complete a noble-gas configuration, and become a negative ion.

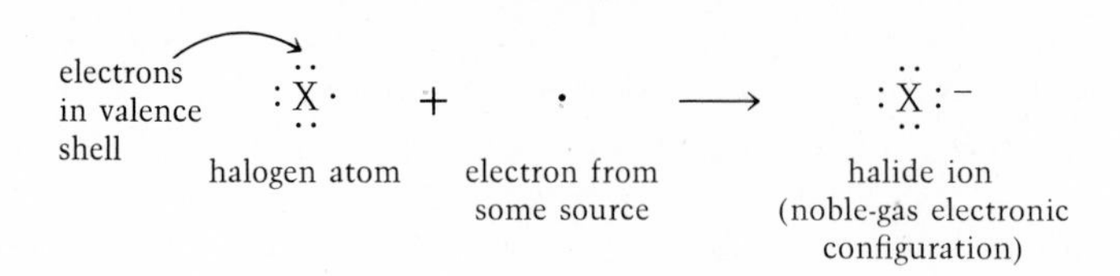

An alkali metal atom, having a low ionization energy and one valence electron, could give up that valence electron (possibly to a halogen atom). Loss of the valence electron would mean that the next inner shell, which is, of course, completely filled, would now become the outer shell, so that it would also have a noble-gas configuration and be a positive ion.

$$\mathrm{E}\cdot \longrightarrow \mathrm{E}^{+} + \cdot$$

alkali atom	alkali metal ion	electron donated to some species, possibly a halogen atom

If both processes occurred simultaneously, we would have

$$\mathrm{E}\cdot + \cdot\overline{\underline{\mathrm{X}}}| \longrightarrow \mathrm{E}^{+} + |\overline{\underline{\mathrm{X}}}|^{-}$$

(Here we have used a single line to represent two electrons. We will continue to use this convention when it is clear that there will be no confusion.)

The type of bond associated with the process occurring in the preceding equation is called an **ionic bond.** The distinguishing feature in an ionic bond is the **complete transfer of one or more electrons from one atom to another.** When the transfer is complete, the atoms have been converted to ions. What holds the ions together is the electrostatic attraction between a positive and a negative species. *The bonding force in an ionic bond is this electrostatic attraction.*

Alkali metals have only one electron in their valence shells, and so each alkali metal atom gives up just one electron. Other metal atoms may have to give up two or more electrons to reach a complete inner shell. Similarly, halogen atoms, with seven valence-shell electrons, need just one more electron to complete the valence shell, but in Chapter 13 we saw that other atoms have fewer than seven outer-shell electrons, and these atoms require more than one electron to complete their outer shells.

Table 14-1 gives a number of examples of atoms reacting to form ionic compounds. **The number of electrons that an atom gives up or gains when it becomes an ion is called its ionic valence, or electrovalence.** This number is equal to the charge on the ion. For example, as shown in Table 14-1, the

Table 14-1 Some Representative Examples of Atoms That React by Transferring Electrons[a]

Reactants		Products	Formula	Name
$\cdot\dot{\mathrm{Al}}\cdot + 3\cdot\overline{\underline{\mathrm{Cl}}}\|$	$\longrightarrow$	$\mathrm{Al}^{3+} + 3\|\overline{\underline{\mathrm{Cl}}}\|^{-}$	$AlCl_3$	aluminum chloride
$\mathrm{K}\cdot + \cdot\overline{\underline{\mathrm{I}}}\|$	$\longrightarrow$	$\mathrm{K}^{+} + \|\overline{\underline{\mathrm{I}}}\|^{-}$	KI	potassium iodide
$\cdot\mathrm{Ca}\cdot + \cdot\dot{\underline{\mathrm{S}}}\|$	$\longrightarrow$	$\mathrm{Ca}^{2+} + \|\overline{\underline{\mathrm{S}}}\|^{2-}$	CaS	calcium sulfide
$\cdot\mathrm{Mg}\cdot + 2\cdot\overline{\underline{\mathrm{Br}}}\|$	$\longrightarrow$	$\mathrm{Mg}^{2+} + 2\|\overline{\underline{\mathrm{Br}}}\|^{-}$	$MgBr_2$	magnesium bromide
$2\mathrm{Rb}\cdot + \cdot\overline{\mathrm{O}}\|$	$\longrightarrow$	$2\,\mathrm{Rb}^{+} + \|\overline{\underline{\mathrm{O}}}\|^{2-}$	Rb_2O	rubidium oxide

[a]Only outer-shell electrons are shown. All the ions have noble-gas configurations.

ionic valence of aluminum is +3, and that of sulfur is −2. By convention, an ionic valence must have a sign: + if the original atom lost electrons, − if the atom gained electrons.

What happens when the number of electrons given up by the donor atom does not equal the number gained by the acceptor atom? Table 14-1 provides several illustrations. For example, an aluminum atom (valence +3) gives up three electrons, but a chlorine atom (valence −1) takes only one. In this case, three chlorine atoms are needed for each aluminum atom, and the formula for aluminum chloride is $AlCl_3$. In each case the total number of electrons lost is equal to the total number gained, as it must be, *because in each reaction there must be a conservation of charge.* Electrons are neither produced nor destroyed, only transferred from one species to another.

The fact that the formula for aluminum chloride is $AlCl_3$ does not mean that there exist discrete $AlCl_3$ molecules. As we mentioned on page 412, solid ionic compounds (and virtually all ionic compounds are solids at room temperature) are composed of a vast number of positive and negative ions held together (by ionic bonds) in a regular geometric array. The formula $AlCl_3$ simply means that each crystal of $AlCl_3$ contains three times as many Cl^- ions as Al^{3+} ions. It is clear that this ratio must be present or else the crystals would have to have an overall positive or negative charge.

Not all atoms form ionic compounds. The formation of an ionic bond most often takes place between metallic atoms (group IA–group IIA and certain transition elements) of low ionization energy and reactive nonmetals (group VIA–group VIIA) with high electron affinities.

14-2 The Covalent Bond

The explanation given in Section 14-1 accounted very well for the bonding in simple inorganic salts. However, it soon became evident that most other kinds of matter could not contain this kind of bond. For example, it was well known that the molecules of such elements as hydrogen, oxygen, nitrogen, and fluorine each contain two atoms. Could the two atoms of a fluorine molecule be connected by an ionic bond? Each fluorine atom has seven electrons in its outer shell. If one fluorine atom gave an electron to the other, the latter would have a complete outer shell, but the donor atom would only have six electrons in its outer shell.

$$|\overline{\underline{F}}\cdot + \cdot\overline{\underline{F}}| \not\longrightarrow |\overline{\underline{F}}|^- + \overline{\underline{F}}|^+$$

Some other kind of bond must be present here.

The problem was solved by Gilbert N. Lewis (1875–1946), who extended the previous ideas to organic compounds and to those inorganic compounds that could not be explained in terms of a *complete* transfer of electrons. Lewis proposed that chemical bonds could also result when two atoms *share* a pair of electrons in such a way that both atoms form complete outer shells.

To illustrate this principle, we shall consider the valence electrons of the second-row elements and label them with dots or straight lines (which stand for a pair of electrons).

$$\mathrm{Li}\cdot \quad \dot{\mathrm{Be}}\cdot \quad \cdot\dot{\mathrm{B}}\cdot \quad \cdot\dot{\underset{\cdot}{\mathrm{C}}}\cdot \quad \cdot\overline{\underset{\cdot}{\mathrm{N}}}\cdot \quad |\dot{\underset{\cdot}{\mathrm{O}}}| \quad |\overline{\underset{\cdot}{\mathrm{F}}}| \quad |\overline{\underline{\mathrm{Ne}}}|$$

These formulas are known as **Lewis formulas.**

The fluorine molecule, F_2, is formed by the combination of two fluorine atoms. Its structure is

$$|\overline{\underline{F}}\cdot + \cdot\overline{\underline{F}}| \longrightarrow |\overline{\underline{F}}-\overline{\underline{F}}|$$

(this line signifies a pair of electrons shared by two atoms (a covalent bond))
(this line signifies a pair of electrons belonging to one atom alone (an unshared pair))

From the diagram one can see that both of the fluorine atoms in F_2 have complete outer shells. For this to happen it is necessary that both atoms in the molecule *share a pair of electrons*. There is no transfer of electrons here. **A bond that is formed when two atoms share a pair of electrons is called a covalent bond.**

Example: Methane has the chemical formula CH_4. Draw its structure showing how each atom in methane has a noble-gas structure.

Answer:

$$4H\cdot + \cdot\dot{\underset{\cdot}{C}}\cdot \longrightarrow H-\underset{\underset{H}{|}}{\overset{\overset{H}{|}}{C}}-H$$

Each of the hydrogen atoms in methane has two electrons in its outer shell (the helium configuration) while the carbon atom has a neon configuration.

Example: Write out the structure of ammonia, NH_3, drawing the noble-gas configurations for each atom.

Answer:

$$3H\cdot + \cdot\overline{\underset{\cdot}{N}}\cdot \longrightarrow H-\overline{\underset{\underset{H}{|}}{N}}-H$$

(unshared pair)
(covalent bond)

Note that each element forms as many covalent bonds as is necessary to give it a complete outer shell. Since a complete outer shell for most atoms contains eight electrons (Chapter 13) this is known as the **octet rule.** (Of course, a complete outer shell for hydrogen contains only two electrons, but this is not regarded as an exception to the octet rule.)

The number of covalent bonds formed by a given atom is called its valence. For example, the illustrations given above show the following:

in methane CH_4, the valence of carbon is 4
in ammonia NH_3, the valence of nitrogen is 3
in both CH_4 and NH_3 the valence of hydrogen is 1

Unlike ionic valences, covalences are always written *without* + or − signs, because electrons are neither given nor taken, but shared, and ions are not formed. Values for the valences of various atoms will be discussed in Section 14-11.

Two different types of bonding have been discussed. In the first there is an essentially *complete transfer of electrons* from one atom to another yielding two ions. This is called an *ionic bond.* The second kind of bonding occurs when two atoms in the molecule *share a pair of electrons.* Here, a complete transfer of electrons does not take place. This kind of bond is called a *covalent bond.*

In Section 13-2 we saw that the electrons of atoms exist in clouds called orbitals. A covalent bond may be thought of as arising from overlap of two electron clouds (one from each atom) to form a new electron cloud, called a *molecular orbital.* An atomic orbital surrounds one nucleus; a molecular orbital surrounds two or more nuclei. The greater the overlap between the two atomic orbitals, the more stable the covalent bond. Like atomic orbitals, molecular orbitals can hold a maximum of two electrons. Molecular orbitals will be further discussed in Part II of this chapter.

It is quite possible for a given compound to contain both ionic and covalent bonds, and many such compounds are known. A common example is ammonium chloride, NH_4Cl. This compound consists of ammonium ions (NH_4^+) and chloride ions (Cl^-), which in the solid are held together by ionic bonds in much the same way as sodium chloride or any other ionic salt (Figure 14-2). The difference here is that the sodium ion results from loss of an electron by a single atom (the sodium atom), whereas the ammonium ion cannot be derived in this way. The ammonium ion consists of a group of five atoms held together by covalent bonds.

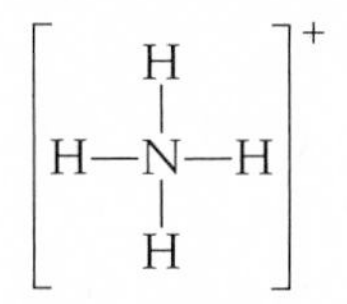

The positive charge results from the fact that NH_4^+ contains a total of 11 protons and only 10 electrons. Other common ions which are made up of atoms connected by covalent bonds (they are called **polyatomic ions**) are sulfate (SO_4^{2-}), nitrate (NO_3^-), and phosphate (PO_4^{3-}) ions.

A compound containing both ionic and covalent bonds is considered to be an ionic compound no matter how many of each kind of bond is present. For example, compounds such as NH_4Cl, Na_2SO_4, and $NaNO_3$ are ionic compounds.

14-3 Exceptions to the Octet Rule

For a time it was believed that *all* compounds obeyed the octet rule, and in fact the rule does apply to the vast majority of known compounds. However, it gradually became apparent that there were quite a few exceptions. Atoms in perfectly stable

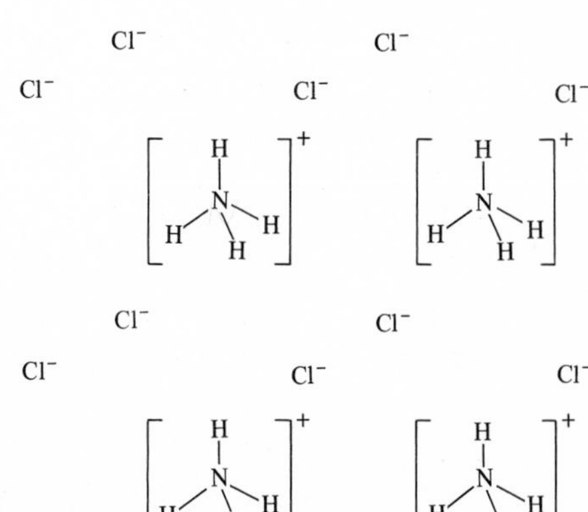

Figure 14-2 The structure of ammonium chloride in the solid state.

molecules could exist with some number of electrons in their valence shells other than eight. Some important examples are:

FOUR OR SIX ELECTRONS. The compound boron trifluoride, a gas at room temperature, has the formula BF_3. The electronic configuration is

$$\cdot\dot{\mathrm{B}}\cdot + 3\cdot\overline{\underline{\mathrm{F}}}| \longrightarrow |\overline{\underline{\mathrm{F}}}-\underset{}{\overset{|\overline{\mathrm{F}}|}{\overset{|}{\mathrm{B}}}}-\overline{\underline{\mathrm{F}}}|$$

The boron atom has only six electrons in its outer shell. Clearly, the octet rule is not obeyed in this compound. The boron atom has six electrons in its outer shell in many other of its compounds, as do aluminum and other elements in group IIIA of the periodic table. An example of a stable molecule that contains an atom with four electrons in its outer shell is beryllium chloride, $BeCl_2$:

$$\cdot\dot{\mathrm{Be}} + 2\cdot\overline{\underline{\mathrm{Cl}}}| \longrightarrow |\overline{\underline{\mathrm{Cl}}}-\mathrm{Be}-\overline{\underline{\mathrm{Cl}}}|$$

SEVEN ELECTRONS. In the overwhelming majority of known stable compounds, every atom, even if it does not have an octet, at least has an even number of electrons in its outer shell. However, there are some compounds in which this is clearly impossible. A simple example is the compound chlorine dioxide, ClO_2, a stable gas, which, however, has the tendency to explode violently when mixed with air. Since a chlorine atom has seven electrons in its valence shell, and each oxygen atom has six, the total number of valence electrons in ClO_2 is 19, so that one of the atoms must have an odd number of electrons in its outer shell. The bonding may be written as follows:

$$|\overline{\underset{\cdot}{\mathrm{O}}}\cdot + |\overline{\underline{\mathrm{Cl}}}\cdot + \cdot\overline{\underset{\cdot}{\mathrm{O}}}| \longrightarrow |\overline{\underline{\mathrm{O}}}-\dot{\underline{\mathrm{Cl}}}-\overline{\underline{\mathrm{O}}}|$$

(In this case all four of the electrons in the two bonds come from the chlorine atom and none from the oxygen; this kind of covalent bonding is discussed in Section 14-5). The chlorine atom in ClO_2 has seven electrons in its outer shell, so one electron is unpaired. **Any species with an unpaired electron is called a free radical.** As we saw in Section 13-7, all free radicals must be paramagnetic, which is one way to detect them. There are not many known stable free radicals. Besides ClO_2, a few examples are two oxides of nitrogen NO and NO_2 (count the total number of valence electrons), and, surprising as it may seem, oxygen gas O_2 (why O_2 is a free radical will be explained in Section 14-20). Most free radicals cannot be isolated as stable compounds because they react with each other to give *dimers.* A dimer is a species made up of two identical units. For example, chlorine atoms $|\overline{\underline{\mathrm{Cl}}}\cdot$ can be generated but do not exist as a stable species because they dimerize in a very short time (much less than 1 second) to give chlorine molecules $|\overline{\underline{\mathrm{Cl}}}-\overline{\underline{\mathrm{Cl}}}|$. In the conditions of outer space, where gases exist at extremely low pressures and molecules are far apart from each other, free radicals may exist indefinitely. For example, the hydroxy free radical $\cdot\overline{\underline{\mathrm{O}}}-\mathrm{H}$ is not a stable species under normal conditions on Earth, because it dimerizes to hydrogen peroxide $\mathrm{H}-\overline{\underline{\mathrm{O}}}-\overline{\underline{\mathrm{O}}}-\mathrm{H}$, but in outer space it exists in large quantities, in vast clouds, where the rate of OH dimerization is very low, simply because the OH radicals are so far apart that they seldom encounter each other.

TEN OR MORE ELECTRONS. For elements in the second row of the periodic table (up to neon), *eight is the maximum number of electrons that can exist in the* *outer shell.* There are no compounds known in which an atom of any of these elements has more than eight electrons in its valence shell. However, this restriction does not apply to elements in the third row or lower. Many compounds are known in which atoms such as chlorine, sulfur, arsenic, etc., have 10, 12, or even more electrons. Some typical examples are PCl_5 and SF_6. The compounds formed by the noble gases krypton, xenon, and radon also fall into this category.

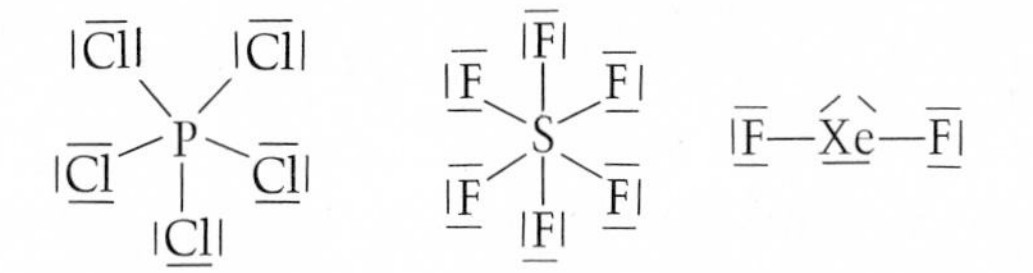

It is obvious that if the octet rule had no exceptions, the noble gases would form no compounds at all, since these atoms already have an octet. Whenever an atom in a molecule has more than eight electrons in its valence shell, the electrons are said to be in an **expanded shell.**

14-4 The Energetics of Covalent Bonding

At this point it would not be surprising if the reader began wondering if it is possible to predict whether the bonding between two atoms will be ionic or covalent or whether this must always be experimentally determined. Another question may be: Why do atoms combine to form covalent bonds? Are there some strange forces at work which hold atoms together to form molecules? We shall address the latter question first and then in Section 14-6 look at the question of predicting the type of chemical bond that is formed.

As we have seen from the Lewis octet rule, when bonding occurs, we can, for an overwhelmingly large number of compounds, draw structures in which electrons in a noble-gas configuration surround each of the atoms in the molecule. Nevertheless, the existence of many stable compounds in which atoms are not surrounded by a noble-gas configuration shows that we cannot simply use the octet rule as the whole answer to covalent bonding. What must and does happen is that the total energy of the system *must be lower* in the bonded state than in the nonbonded state. For both ionic and covalent bonds, bonding can occur only when the system goes to a lower energy state. If the system goes to a higher energy state or to one that is equal in energy to the nonbonded state, then bonding will not occur. These statements are simple, valid, and there are no exceptions. In Part II of this chapter we will discuss some of the theories of chemical bonds and show that the energy of the bonded state is lower than that of the nonbonded state.

When two atoms combine to form a covalent bond, the total energy of the system must be lowered. Therefore, energy must have been given off when the bonding took place: *Bonding is an exothermic process* (Figure 14-3). But if energy was given off, then the separated atoms have more energy than the molecule. It is obvious that the

Figure 14-3 The total energy of the hydrogen molecule is lower than that of the two hydrogen atoms. The difference is called the bond energy.

law of conservation of energy requires that in order to break the bond, the same amount of energy must be added. This, then, is what holds atoms together in covalent bonds. *They cannot separate unless they are given the required amount of energy.* In fact, covalent bonds can be broken if enough energy, in the form of heat or light, is added.

The explanation we have just given is correct as far as it goes, but it leads to a new question. It is true that atoms are held together in covalent bonds because energy was given off when the bonding took place, and that energy must be resupplied if the bond is to be broken. But the next question is: Why does the bonded state have a lower energy than the unbonded state? What are the forces that hold atoms together in a chemical bond? Physical chemists working on bonding theory have shown that it is not necessary to invoke any new or mysterious quantum-mechanical forces to explain covalent bonding. Covalent bonding is completely explained by electrostatic interactions. The forces that hold atoms together in molecules are all *electrical* in origin. The electrons are electrically negative and the protons are electrically positive, and it is the electrical coulombic interactions between all these particles that give rise to the covalent bond. Coulombic forces are thousands of times stronger than gravitational forces so that in all calculations involving bonding, gravitational forces are considered negligible and are completely ignored. Since we have seen (Section 14-1) that ionic bonds also result from electrical attraction, this means that all chemical bonding can be accounted for by coulombic forces in some sense or other.

14-5 The Coordinate Covalent Bond. Lewis Acids and Bases

In the original Lewis picture of the covalent bond, two atoms each supply one electron to form the bond. Some years later, Irving Langmuir (1881–1957) realized that in some cases two atoms could form a covalent bond if one atom supplied *both* electrons and the other atom supplied *neither*. Such a bond is called a **coordinate covalent bond.** An example is the compound formed by reaction of boron trifluoride with ammonia

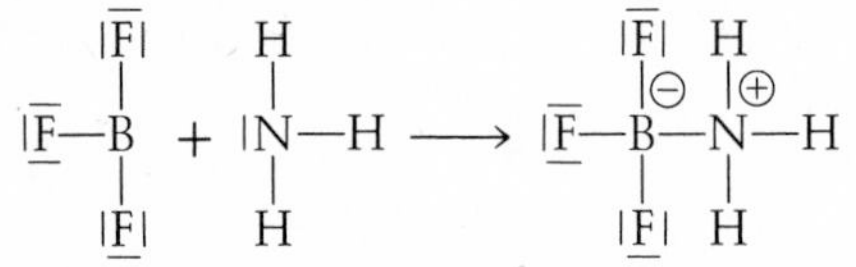

```
    |F̅|       H              |F̅|  H
     |        |               |⊖  |⊕
|F̅—B    +  |N—H   ⟶   |F̅—B—N—H
     |        |               |    |
    |F̲|       H              |F̲|  H
```

The use of the positive and negative charges will be explained in Section 14-9. They can be ignored for now.

Once the bond has been formed, a coordinate covalent bond is identical to an ordinary covalent bond, and there is no way to distinguish between them.

Since the octet rule is important (even though there are exceptions), it is not surprising that a major class of compounds which can form coordinate covalent bonds by accepting a pair of electrons are those compounds (Section 14-3) in which an atom has six electrons in the outer shell.

In 1923, G. N. Lewis used this fact to develop a theory of acids and bases which is broader than the Brønsted theory (Section 10-2). According to the Lewis acid–base theory, a **Lewis acid** is defined as **any species that forms a covalent bond by accepting a pair of electrons.** A **Lewis base** is **a species that supplies the electron pair.** A Lewis acid–base reaction is the formation of a coordinate covalent bond. The reaction between BF_3 and NH_3 shown above is a Lewis acid–base reaction. Note that electrons are not transferred from the base to the acid. The entire base, including the electron pair, becomes bonded to the Lewis acid.

How can we recognize a Lewis acid? It has to be a species that has room for a pair of electrons. For example, we would not expect hydrogen, methane, ammonia, or water to be Lewis acids. All the atoms in these molecules have complete octets (two for hydrogen),

$$\mathrm{H{-}H} \qquad \mathrm{H{-}\underset{H}{\overset{H}{C}}{-}H} \qquad \mathrm{H{-}\underset{H}{\bar{N}}{-}H} \qquad \mathrm{H{-}\underset{H}{\bar{O}}|}$$

and none of these atoms can take more than eight electrons in its valence shell. Lewis acids are chiefly of three types.

1. Molecules with atoms that have only four or six electrons in their outer shells. Most of these are compounds of the group IIIA elements. Some important examples are BF_3, $AlCl_3$, AlH_3, and $GaCl_3$. Compounds of beryllium which have four electrons in their valence shells are also Lewis acids. In these cases, two coordinate covalent bonds can form; for example, in the reaction between beryllium difluoride (BeF_2) and fluoride ions:

$$|\underline{\overline{F}}{-}Be{-}\underline{\overline{F}}| + 2|\underline{\overline{F}}|^- \longrightarrow \left[|\underline{\overline{F}}{-}\underset{|\underline{F}|}{\overset{|\overline{F}|}{Be}}{-}\underline{\overline{F}}| \right]^{2-}$$

2. Metallic positive ions. Many metallic positive ions function as Lewis acids when in the presence of suitable bases. Typically, each ion can accept a certain number of electron pairs. In most cases this number, called the **coordination number,** is six or four, but other numbers are also possible. The compounds formed in this manner are called *coordination compounds.* A typical example is the Cr^{3+} ion, which usually has a coordination number of six. The Cr^{3+} ion forms thousands of coordination compounds, with various combinations of Lewis bases. A few examples are

$$Cr(H_2O)_6{}^{3+} \quad Cr(NH_3)_6{}^{3+} \quad Cr(H_2O)_3(NH_3)_3{}^{3+} \quad Cr(H_2O)_4(NH_3)_2{}^{3+} \quad CrCl_6{}^{3-} \quad Cr(H_2O)_3Cl_3$$

Positive ions of group IA and IIA metals show little or no tendency to act as Lewis acids, but such behavior is important for most other metallic ions, especially for ions of the transition elements. Coordination compounds will be further discussed in Chapter 19.

3. Molecules with atoms that can expand their outer shells. We have seen that atoms in the third row of the periodic table can have 10, 12, or even more electrons in their outer shells. Some atoms with less than this number can react with bases to expand their outer shells. Typical examples are stannic chloride ($SnCl_4$) and silicon tetrafluoride (SiF_4):

$$SnCl_4 + 2Cl^- \longrightarrow SnCl_6{}^{2-}$$
$$SiF_4 + 2F^- \longrightarrow SiF_6{}^{2-}$$

How can we recognize a Lewis base? Obviously, a Lewis base must be a species that has available a pair of electrons which it can share with a Lewis acid. In nearly all cases this will be an unshared pair, and we can say that any species with one or more unshared pairs is a potential Lewis base, although some are stronger Lewis bases than others. Lewis bases may be neutral, like water or ammonia, or negatively charged, like chloride, hydroxide, or cyanide ions.

$$\mathrm{H{-}\overset{H}{\underline{O}}|} \qquad \mathrm{H{-}\overset{H}{\underline{N}}{-}H} \qquad |\underline{\overline{Cl}}|^{\ominus} \qquad {}^{\ominus}|\underline{\overline{O}}{-}H \qquad {}^{\ominus}\bar{C}{\equiv}\bar{N}$$

Some Lewis bases

Comparison Between the Brønsted and Lewis Acid-Base Theories. In the Brønsted theory, a base is a proton acceptor. But protons cannot exist in a free state. Thus, when a Brønsted base accepts a proton, it can only do so by forming a bond with the proton, using an unshared pair of electrons for this purpose, for example:

$$H—\overline{\underline{O}}|^- + H—Cl \longrightarrow H—\overline{\underline{O}}—H + Cl^-$$

Consequently, a Brønsted base is also a Lewis base, since an unshared pair of electrons is required for both purposes. However, Brønsted restricts acids to species that can donate protons. In the Lewis picture the "real" acid in a case like HCl or HOAc is regarded to be the proton itself, and this is the species that forms the covalent bond with the base. (This, of course, is only conceptual, since protons cannot exist in the free state.) In the Lewis picture, all Brønsted acids are looked upon as precursors of a single "real" acid, H^+, which of course has room for an electron pair. The advantage of the Lewis picture is that it ties together much chemical behavior, since it also takes into account many Lewis acids other than H^+.

Example: How do the Lewis and Brønsted theories account for the reaction between HBr and NH_3?

Answer:

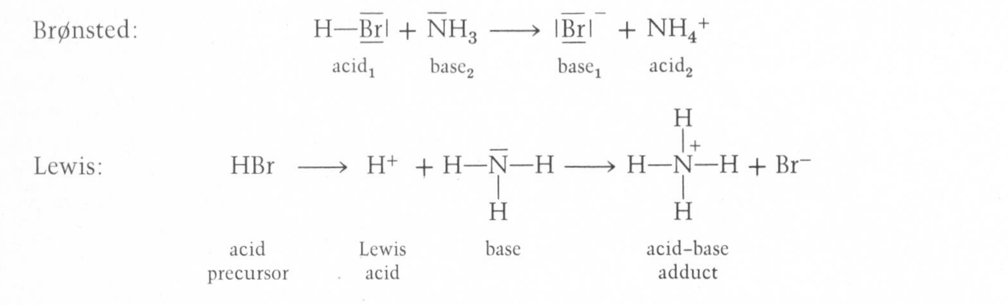

Example: How do the Lewis and Brønsted theories account for the reaction between $AlCl_3$ and Cl^-?

Answer:

Lewis:

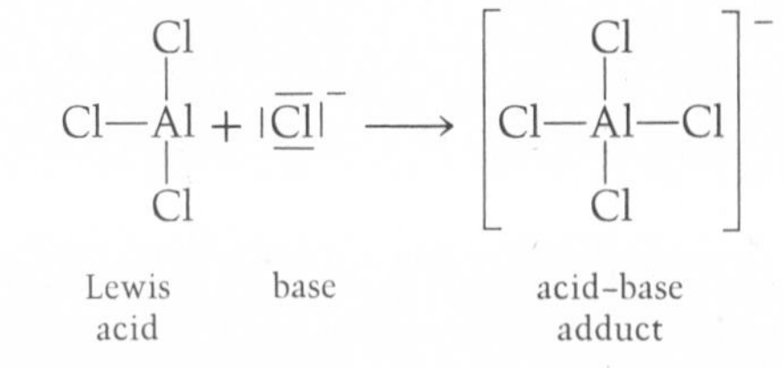

Brønsted: The Brønsted theory does not account for this reaction.

It is obvious that there is no conflict between the two acid–base theories. They are both highly useful and chemists use each one where it seems most applicable.

14-6 Electronegativity

How can we tell whether a given compound is made up of ionic or covalent bonds? We have seen that in a simple salt such as KI, which is made up of elements from opposite ends of the periodic table (groups IA and VIIA), the bonding is ionic. On the

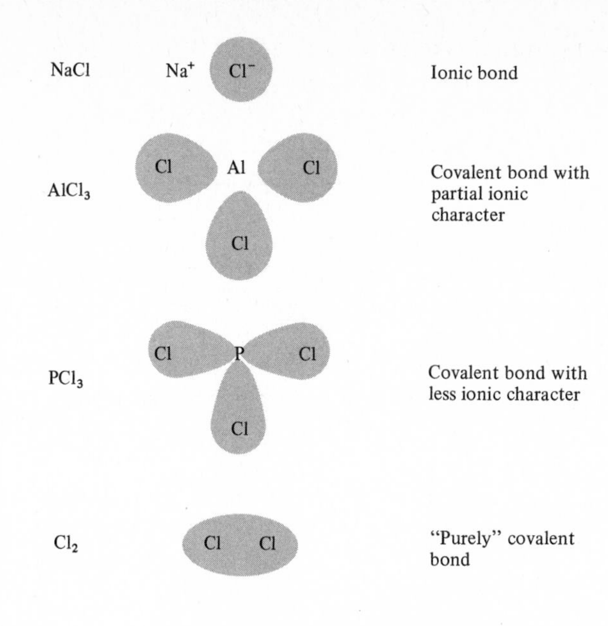

Figure 14-4 Some Compounds of Chlorine with Thrid-Row Elements. The shaded areas represent the positions of the two electrons of each bond.

other hand, the bond between Br and Cl in the compound BrCl (where both elements are in group VIIA) is covalent. It would appear that the answer to the question as to whether a given compound is covalent or ionic lies hidden in the periodic table. This conjecture is correct. It is certain properties of atoms, namely their ionization energies and electron affinities that govern whether a bond between them will be ionic or covalent. We saw in Sections 13-11 and 13-12 that the values of IE and EA within the periodic table change *gradually*. As a consequence, the transition from ionic to covalent bonding, as one goes across a row in the periodic table also occurs gradually, for a series of similar compounds (for example, $Na^+Cl^- \rightarrow Cl—Cl$; Figure 14-4). Because of this fact, chemists talk about the **partial ionic character** of a covalent bond.

Consider a bond formed between two different atoms, A and B:

$$A—B$$

Whether the bond will be ionic, "purely" covalent, or covalent with partial ionic character depends upon the strength of attraction for the pair of electrons exerted by atoms A and B. It also depends upon how readily each of them is willing to release the electrons to the other. These releases and attractions are, of course, ionization energy and electron affinity.

To illustrate, let us look at atom A. The power that atom A exerts to pull the pair of electrons completely to it is measured by the electron affinity of A and also by the ionization energy of B (which measures how readily B will give up the pair of electrons). The same is true for atom B. The power that B exerts to pull the electrons completely to itself depends on the electron affinity of B and on the ionization energy of A.

The measure of the ability of an atom in a molecule to attract to itself the electrons within a bond is called **electronegativity.** Since there are no simple and direct ways of measuring electronegativities, a number of different scales have been proposed. Robert S. Mulliken (1896–   ) proposed that the electronegativity of an element is proportional to the average of its first ionization energy and its electron affinity. However, another scale that is more widely used is due to Linus Pauling (1901–   ). Pauling assigned the most electronegative element, fluorine, an arbitrary value of 4.0 and calculated the other values from that. Values of the electronegativity of the elements according to the Pauling scale are given in Table 14-2.

Electronegativities are used to predict the type of bonding that will occur between

Table 14-2 Electronegativities of the Elements

1 H 2.1																	2 He —
3 Li 1.0	4 Be 1.5											5 B 2.0	6 C 2.5	7 N 3.0	8 O 3.5	9 F 4.0	10 Ne —
11 Na 0.9	12 Mg 1.2											13 Al 1.5	14 Si 1.8	15 P 2.1	16 S 2.5	17 Cl 3.0	18 Ar —
19 K 0.8	20 Ca 1.0	21 Sc 1.3	22 Ti 1.5	23 V 1.6	24 Cr 1.6	25 Mn 1.5	26 Fe 1.8	27 Co 1.8	28 Ni 1.8	29 Cu 1.9	30 Zn 1.6	31 Ga 1.6	32 Ge 1.8	33 As 2.0	34 Se 2.4	35 Br 2.8	36 Kr —
37 Rb 0.8	38 Sr 1.0	39 Y 1.2	40 Zr 1.4	41 Nb 1.6	42 Mo 1.8	43 Tc 1.9	44 Ru 2.2	45 Rh 2.2	46 Pd 2.2	47 Ag 1.9	48 Cd 1.7	49 In 1.7	50 Sn 1.8	51 Sb 1.9	52 Te 2.1	53 I 2.5	54 Xe —
55 Cs 0.7	56 Ba 0.9	57–71 La–Lu 1.1–1.2	72 Hf 1.3	73 Ta 1.5	74 W 1.7	75 Re 1.9	76 Os 2.2	77 Ir 2.2	78 Pt 2.2	79 Au 2.4	80 Hg 1.9	81 Tl 1.8	82 Pb 1.8	83 Bi 1.9	84 Po 2.0	85 At 2.2	86 Rn —
87 Fr 0.7	88 Ra 0.9	89– Ac– 1.1–1.7															

Source: Linus Pauling, *The Nature of the Chemical Bond,* 3rd ed. (Ithaca, N.Y.: Cornell University Press, 1960).

any two atoms. The important thing is the difference between the electronegativities of the two atoms. When this *difference* is zero or very small, the bond is essentially covalent, and the electron density is more or less equally shared by the two atoms. When the difference is large, one atom more-or-less completely relinquishes the electron pair to the other, and the bonding is ionic. In practice, we find all possible situations between these extremes. Nature does not draw a sharp boundary line between ionic and covalent bonding. However, such a boundary line is useful, and chemists have found it convenient to select the arbitrary value of 1.7 as the approximate borderline between ionic and covalent bonding. When the electronegativity difference between A and B is greater than 1.7, we say the bonding is ionic; when it is less than 1.7, it is covalent.

What happens to the electron cloud in a molecule when there is a covalent bond formed whose electronegativity difference lies between zero and 1.7? We can get a clear understanding of this by taking an example. Consider the molecule HBr.

	H	Br
Electronegativities:	2.1	2.8

In HBr the electronegativity difference is 0.7, so the bond is covalent. Since the bromine atom has the higher electronegativity value, it attracts the pair of electrons more strongly than the hydrogen atom. The electron cloud of the covalent bond will be denser in the vicinity of the bromine atom than in that of the hydrogen atom.

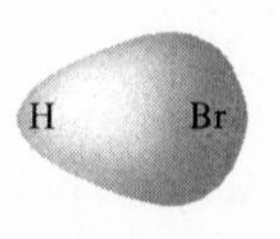

Because of this, the bromine end of the bond is partially negatively charged while the hydrogen end is partially positively charged.

$$\overset{\delta+}{H}-\overset{\delta-}{Br}$$ The partial charges are indicated by the symbols $\delta-$ and $\delta+$.

The molecule as a whole is electrically neutral (as it must be since the total number of electrons is equal to the total number of protons). Any covalent bond in a molecule with a separation of positive and negative partial electrical charges is called a **polar covalent bond.** We can now see that the greater the electronegativity difference, the greater will be the *difference* between the electron attracting abilities of the two atoms and the more polar will be the bond. At the extreme situation when the electronegativity difference becomes greater than 1.7, there is a complete transfer of electrons and an ionic bond results. The atoms no longer carry partial electrical charges but full charges. They are now ions.

Example: Describe the type of bonding for the following molecules: H_2, HgF_2, $HgCl_2$, $HgBr_2$, HgI_2.

Answer:

Compound	*Electronegativity difference*	*Bond*
H_2	0	"Pure" covalent
HgF_2	$4.0 - 1.9 = 2.1$	Ionic
$HgCl_2$	$3.0 - 1.9 = 1.1$	Polar covalent
$HgBr_2$	$2.8 - 1.9 = 0.9$	Polar covalent
HgI_2	$2.5 - 1.9 = 0.6$	Polar covalent

14-7 Dipole Moments

As discussed in Section 6-6, an electric dipole exists in a neutral molecule when there is a separation between the positive and negative centers of electrical charge. As we saw in Section 14-6, a separation of partial positive and negative charges results when there are differences in electronegativities of the bonding atoms. A heterodiatomic molecule* such as HBr has an *electric dipole moment* caused by the separation of charge. The dipole moment μ is defined as the product of the distance, r, separating equal charges of opposite sign and the magnitude of the charge q:

$$\mu = qr$$

The charge q on an electron is 4.8×10^{-10} esu, so two charges of this magnitude separated by 1 Å will have a dipole moment of

$$\mu = (4.8 \times 10^{-10}\ \text{esu})\ (1 \times 10^{-8}\ \text{cm}) = 4.8 \times 10^{-18}\ \text{esu} \cdot \text{cm}$$

Dipole moments are usually expressed in terms of a defined unit, the Debye (symbol D; named for Peter Debye, 1884–1966); $1\ \text{D} = 1 \times 10^{-18}$ esu · cm, so for the previous example, we have

$$4.8 \times 10^{-18}\ \text{esu} \cdot \text{cm} = 4.8\ \text{D}$$

Molecules that have dipole moments orient themselves in electric fields. This can be accomplished by placing the material between the plates of a condenser (Figure

*A molecule made up of two different atoms.

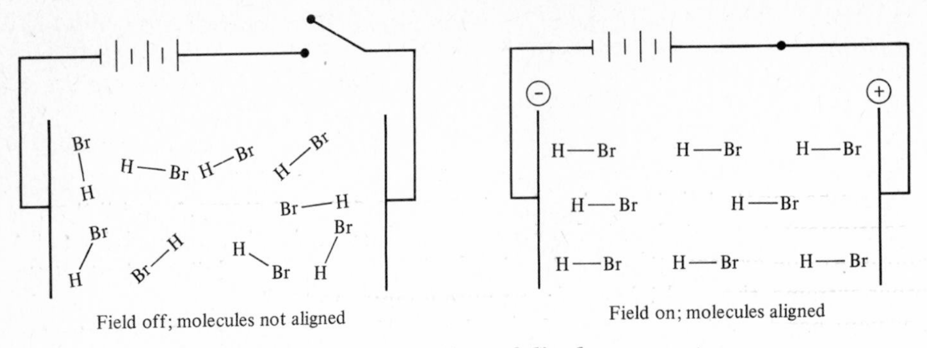

Figure 14-5 Experimental determination of dipole moments.

14-5). When the field is off, the molecules are randomly oriented. Upon applying an electric field, the negative end of the molecule orients itself toward the positive plate of the condenser while the positive end of the molecule is oriented toward the negative plate. Because of the orientation of the molecules, the amount of electricity the plates can hold is increased, and this increase can be accurately measured. As a result of measurements of this type, the actual dipole moment of molecules can be calculated. Table 14-3 contains a list of some diatomic molecules and their measured dipole moments.

If the molecules were completely ionic, each of the ions would bear a full electrical charge (4.8×10^{-10} esu). From the bond distance of the molecule, a *hypothetical* dipole moment can be calculated. For example, HBr has a measured bond length of 1.41 Å. Its hypothetical dipole moment would be

$$\mu = rq = (1.41 \times 10^{-8}\ \text{cm})(4.8 \times 10^{-10}\ \text{esu}) = 6.77\ \text{D}$$

The experimentally observed dipole moment for HBr is 0.79 D. The ratio $\mu_{\text{observed}}/\mu_{\text{hypothetical}} \times 100$ has been termed the **percent ionic character** of the bond in the molecule. For HBr this would be

$$\text{HBr:}\quad \frac{\mu_{\text{obs}}}{\mu_{\text{hypo}}} \times 100 = \frac{0.79\ \text{D}}{6.77\ \text{D}} \times 100 = 0.116 \times 100 = 12\%\ \text{ionic character}$$

Table 14-3 Dipole Moments of Some Diatomic Molecules (all in the gas phase) and Their Percent Ionic Character

Molecules (in the gas phase)	μ *(Debye units)*	*Percent ionic character*
H_2	0	0
HF	1.8	41
HCl	1.07	17.6
HBr	0.79	12
HI	0.38	5
HLi	5.9	81
F_2	0	0
FCl	0.88	11.2
FBr	1.29	15.3
Cl_2	0	0
ClBr	0.57	5.6
ClI	0.6	5.4
N_2	0	0
O_2	0	0
NO	0.1	1.8
CO	0.13	2.4
NaCl	9	82
NaI	4.9	37.7

From this calculation the covalent bond of HBr has 12% ionic character. The percent ionic character of the other diatomic molecules is also given in Table 14-3. This is a rather simple method for calculating the percent ionic character of bonds. Additional factors may also contribute to the ionic character. For example, it has been shown that in some cases the nonbonding electrons play a role in determining the dipole moment and hence would affect the percent ionic character of the bond.

It can be concluded that in bonds between two different atoms A and B there is a continuous range of electrical polarity, depending on the electronegativities of A and B. At one extreme there are bonds which are almost "purely" covalent; at the other extreme, those which are almost purely ionic; and in between there are polar covalent bonds with a smaller or larger degree of ionic character. Despite the fact that this continuous range exists and that any borderline between ionic and covalent bonds (including the one we have used) must be arbitrarily placed, chemists generally have little trouble in classifying most bonds either as ionic, nonpolar covalent, or polar covalent, based on the electronegativity difference between A and B, and on the properties of the compound.

A molecule with more than two atoms may also have a dipole moment, but in this case the value of the dipole moment depends not only on the bond polarities, but also on the shape of the molecule. For example, let us consider the water molecule, which has a bond angle of 105°.

O
H 105° H

From the electronegativity values of oxygen (3.5) and hydrogen (2.1), we know that each of the two O—H bonds is polar-covalent, with the oxygen as the negative and the hydrogen the positive end. We do not know how to measure the individual dipole moment of a single bond. We can only measure the moment for the entire molecule, by the method we just described. When the molecule lines itself up in the condenser, it behaves as if it has only one positive pole and one negative pole, no matter how many atoms in the molecule or how complicated the shape.

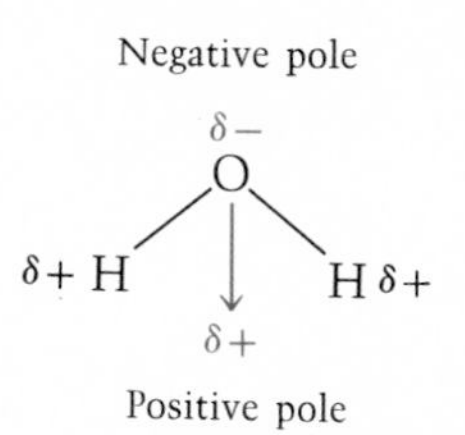

In this case, the negative pole of the molecule is close to the oxygen atom, and the positive pole is on a line that bisects the angle of 105°.* Thus, the positive pole is not on any atom but at a position in space halfway between the two hydrogen atoms. The dipole moment of water is 1.85 D.

In some cases the geometry of the molecule is such that the individual bond moments cancel each other, and the molecule is left with no moment at all, even though the individual bonds may be polar covalent. An example here is gaseous mercuric chloride ($HgCl_2$), a linear molecule.

$$\overset{\delta-}{\text{Cl}}\text{—}\overset{\delta+}{\text{Hg}}\text{—}\overset{\delta-}{\text{Cl}}$$

All three atoms are on a straight line. The electronegativity of Cl(3.0) and of Hg(1.9) tell us that both bonds are polar covalent, but there can be no positive pole and

*The two O—H bonds act as vectors, and the line between the positive and negative poles is the resultant of these vectors.

hence no negative pole (the average position of the two partial negative charges is right on the positive mercury atom). Therefore, $HgCl_2$ has no dipole moment.

14-8 Multiple Bonds, Bond Distances, and Bond Energies

In Section 14-2, we presented the Lewis dot-structure method so that we could easily picture the octet rule in chemical bonding. In many molecules, we find that in order to satisfy the octet rule, it is necessary to have more than one covalent bond between two atoms. A good example of this is carbon dioxide, CO_2. From analysis of its structure we know that the carbon is bonded to both oxygens, so that we can start by drawing the structure as

O C O

Next we place the valence electrons around each atom:

:Ö· ·Ċ· ·Ö:

Since CO_2 is a discrete molecule, we know that there must be at least one bond between the carbon atom and each of the oxygen atoms. If we allow one bond to form in each direction (by the sharing of one electron from the carbon atom with one electron from each oxygen atom), we have

:Ö··Ċ··Ö:

This structure is unsatisfactory as it stands, because none of the atoms obey the octet rule. Each oxygen atom has seven valence electrons, and the carbon atom has six. Although we have seen that there are molecules in which the octet rule is disobeyed, in general this only happens when the molecules have no way to obey the rule. In the vast majority of cases, when there is a way for the octet rule to be obeyed, it is obeyed. In the case of CO_2, the rule can be obeyed if each oxygen atom uses its seventh electron to form a second bond, by joining with one of the two unpaired electrons on the carbon.

:Ö··C··Ö: ⟶ :Ö::C::Ö: ≡ |Ō=C=Ō|

Now each of the atoms in CO_2 fulfills the octet rule. In doing this we had to place *four* bonding electrons between each of the oxygens and the central carbon atom. A single covalent bond between two atoms takes two electrons. Four bonding electrons between two atoms is termed a **double bond,** and for molecules where there are six bonding electrons between two atoms we have a **triple bond.**

Example: Using the Lewis dot structure, determine the bonding in nitrogen gas, N_2.

Answer: Drawing the atoms with their valence electrons, we have

·N̈· ·N̈·

To achieve octets, the atoms must share electrons as follows:

N̈:::N̈ ≡ |N≡N|

Therefore, nitrogen gas, N_2, has one triple bond between the nitrogen atoms.

Example: Given the structure of ethylene, C_2H_4, as

H H
C C
H H

determine the bonding.

Answer:

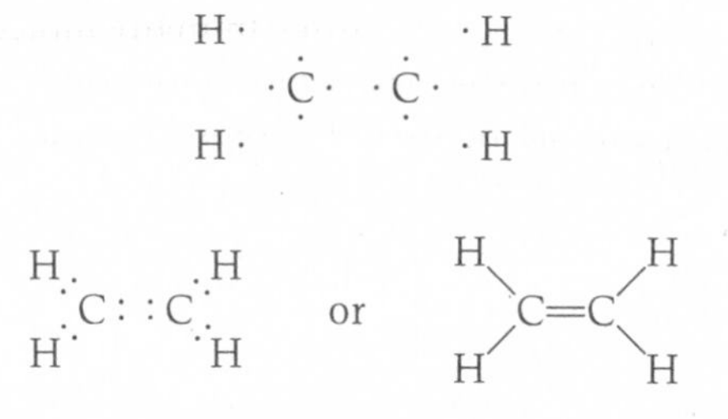

In ethylene each carbon is joined to two hydrogens by a single covalent bond and the carbons are linked together by a double bond.

We have just developed a nice scheme or model which helps to satisfy the octet rule in certain molecules which could not be drawn with single bonds alone. Does nature really operate in this fashion? That is, is there any experimental proof we can find that will give us confidence in using the multiple-bond model? The answer is definitely *yes*. We have very good evidence, even though the evidence is indirect, that multiple bonding occurs in CO_2, N_2, C_2H_4, and many other molecules. Some of the evidence comes from the experimental measurements of bond distances in these molecules. As shown in Table 14-4, the average bond distance for C—C bonds is 1.54 Å, for C=C bonds, it is 1.34 Å and for C≡C bonds, it is 1.21 Å. These results come from a large variety of molecules and are generally consistent for single, double, and triple bonds worked out according to the octet rule. These bond distances are evidence for the existence of multiple bonding because a shorter bond is a stronger bond, and we would predict that multiple bonds should be stronger than single bonds because they contain more electrons. We are squeezing the positive nuclei of the two atoms together while simultaneously putting more electrons (negative charges) between them. The electrons interact favorably (attractive interactions) with the protons, which offsets the additional repulsion of bringing the protons closer.

The strengths of bonds can also be obtained more directly by measuring bond energies. The **bond energy** is defined as the energy necessary to break a molecule into two pieces in such a way that each piece gets one electron from each bond, and the two pieces separate far enough away from each other so that they do not interact. For example, ethane and ethylene could cleave as follows:

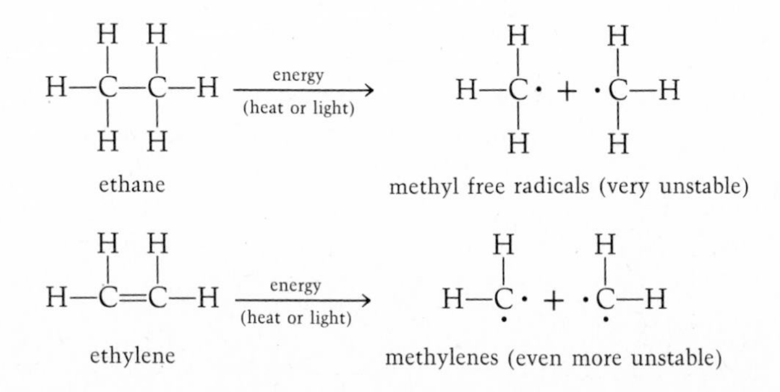

Table 14-4 Average Distances and Energies for Some Bonds

Bond	*Approximate bond distance (Å)*	*Approximate bond energy (kcal/mole)*
H—H	0.74	104
H—F	0.92	135
H—Cl	1.27	103
H—Br	1.41	87
H—I	1.61	71
H—O	0.96	110
H—N	1.0	93
H—C	1.11	96–99
C—C	1.54	84
C═C	1.34	146–151
C≡C	1.21	200
C—N	1.47	69–75
C═N	1.28	143
C≡N	1.16	204
C—O	1.41	85–91
C═O	1.20	173–181
C—S	1.8	66
C—F	1.38	115
C—Cl	1.78	79
C—Br	1.94	66
C—I	2.14	52
O—O	1.48	50
O═O	1.21	118
N—N	1.45	—
N═N	1.25	—
N≡N	1.10	227
F—F	1.42	37
Cl—Cl	1.99	58
Br—Br	2.28	46
I—I	2.67	37

For single covalent bonds, the bond energy is equal to the amount of energy lost when the bond was originally formed (Section 14-4).

Table 14-4 lists bond energies for a number of bonds. As expected, a triple bond is stronger than a double bond, which is stronger than a single bond.

$$\text{Strength of chemical bonds:} \quad \equiv \; > \; = \; > \; -$$
$$\text{Length of chemical bonds:} \quad - \; > \; = \; > \; \equiv$$

14-9 Lewis Structures and Formal Charges

We can write formulas for compounds in several different ways. A **molecular formula** tells us which elements are in a molecule, and how many of each (for example, CO_2). A **structural formula** shows all covalent bonds (O═C═O). A **Lewis structure** shows not only all covalent bonds, but all other valence electrons as well (|Ō═C═Ō|). In this section we will give some rules that will allow us to write Lewis structures for compounds and for polyatomic ions if we are given the molecular formulas.

In order to write a Lewis structure, we must not only know the molecular formula, but also which atoms are connected to which. In some cases we can figure this out for ourselves, by the use of the valence rules, but in most cases we cannot, and we must obtain this information before we can draw a Lewis structure. For example, in the compound ethane C_2H_6, the connections must be

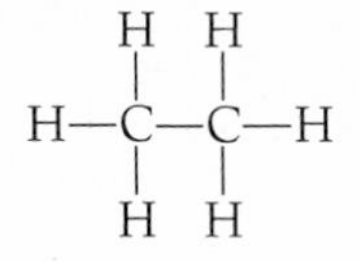

because no other connections are possible for these eight atoms, but this is not the case, for example, for N_2O, where the atoms might be connected N—N—O or N—O—N; and valid Lewis structures, obeying all the rules, can be drawn for both situations. The only general way to tell what are the connections in an actual molecule or ion under discussion is by laboratory experiments. Since most students do not have the means to perform such experiments, they must be given the information before they can write Lewis structures.

RULES FOR WRITING LEWIS STRUCTURES

Rule 1. The total number of valence electrons in any molecule or ion is equal to the sum of the number of valence electrons that each atom had before it bonded (which can be obtained from its position in the periodic table), plus or minus the charge in the case of ions.

Example: What are the total numbers of valence electrons in CO_2, C_2H_2, CH_3NH_2, NH_4^+, and SO_4^{2-}?

Answer:

	Number of valence electrons		
CO_2:	C has 4 each O has 6	Total	16 valence electrons
C_2H_2:	each C has 4 each H has 1	Total	10 valence electrons
CH_3NH_2:	C has 4 N has 5 each H has 1	Total	14 valence electrons
NH_4^+:	N has 5 each H has 1	Subtotal	9 valence electrons − 1 for the positive charge
		Total	8 valence electrons
SO_4^{2-}:	S has 6 each O has 6	Subtotal	30 valence electrons + 2 for the negative charge
		Total	32 valence electrons

Rule 2. All of the valence electrons calculated by rule 1 are inserted by pairs into the molecule or ion, in such a way as to give every atom an octet (two for hydrogen) if possible. No distinction is made between ordinary covalent and coordinate covalent bonds.

Example: Draw Lewis structures for (a) nitrous acid (HNO_2) and (b) acetylene (C_2H_2). The atoms are connected as follows: H O N O and H C C H.

Answer:

(a) The total number of valence electrons in HNO_2 is 18. We can put nine pairs of electrons into H O N O in this way:

$$\mathrm{H{-}\overline{\underline{O}}{-}\underline{N}{=}\underline{O}|}$$

Each atom has a complete octet. Note that we had to use a double bond. If we had written

$$\mathrm{H{-}\overline{\underline{O}}{-}\underline{N}{-}\overline{\underline{O}}|} \quad \text{or} \quad \mathrm{H{-}\overline{\underline{O}}{-}\overline{\underline{N}}{-}\underline{O}|}$$

It would not be a satisfactory answer because in each case there is one atom that lacks an octet.

(b) The total number of valence electrons in C_2H_2 is 10. We put in five pairs as follows:

$$\mathrm{H{-}C{\equiv}C{-}H}$$

Here we had to use a triple bond.

Rule 3. If an octet is not possible, an outer shell of six or seven electrons is permissible. No atom in the second row of the periodic table (up to neon) can hold more than eight electrons, but atoms in the third row or below can hold 10 or more.

Example: Draw a Lewis structure for $SbCl_5$ (the antimony atom is connected to each of the chlorines).

Answer: The total number of valence electrons in $SbCl_5$ is 40. Twenty pairs can be inserted only if the antimony atom holds 10 electrons:

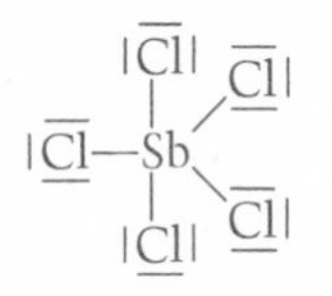

Rule 4. When all the valence electrons have been inserted, it is necessary to inspect each atom to see if it must be given a **formal charge.** A Lewis structure is not correct if proper formal charges are omitted.

Formal charge is an artificial concept that is needed to keep our charges straight. The formal charge for each atom in a Lewis structure is calculated by the following formula:

$$\text{formal charge on a given atom} = \text{number of valence electrons in the neutral atom before bonding} - \text{number of nonbonding valence electrons on the atom} - \text{number of covalent bonds on the atom}$$

Note that the number of valence electrons is always constant for a given neutral atom. For example, this number is always six for oxygen and sulfur, five for nitrogen and phosphorus, four for carbon, and seven for the halogens.

Example: Draw Lewis structures, including formal charges, for (a) H_3O^+, (b) NH_3, (c) NH_4^+, (d) ClO^-, (e) ClO_2^- (f) ClO_3^- [In parts (e) and (f), all the oxygen atoms are individually connected to the chlorine atom and not to each other.]

Answer:

(a) H_3O^+ has eight valence electrons. Its Lewis structure before formal charges is

```
H—Ō—H
  |
  H
```

For each hydrogen, the number of valence electrons in the neutral atom before bonding is 1, the number of nonbonding electrons is 0, and the number of bonds is 1. Thus, the formal charge is

$$1 - 0 - 1 = 0$$

For the oxygen, the formal charge is

$$6 - 2 - 3 = +1$$

The complete Lewis structure for H_3O^+ is

```
   _⊕
H—O—H
  |
  H
```

It is obvious that a hydrogen atom with a single bond and no nonbonding electrons must always have a formal charge of 0. Therefore, we can confine our attention to other atoms.

(b) NH_3

```
H—N̄—H
  |
  H
```

For nitrogen, the formal charge = $5 - 2 - 3 = 0$ (it is not necessary to indicate a formal charge of 0)

(c) NH_4^+

```
  H
  |⊕
H—N—H
  |
  H
```

For nitrogen, the formal charge = $5 - 0 - 4 = +1$

(d) ClO^-

```
 __  _⊖
|Cl—O|
 ‾‾  ‾
```

The formal charge for chlorine = $7 - 6 - 1 = 0$
for oxygen = $6 - 6 - 1 = -1$

(e) ClO_2^-

```
⊖_   _⊕  _⊖
|O—Cl—O|
 ‾   ‾   ‾
```

The formal charge for chlorine = $7 - 4 - 2 = +1$
for each oxygen = $6 - 6 - 1 = -1$

(f) ClO_3^-

```
⊖_   (2+)  _⊖
|O—Cl—O|
 ‾   |    ‾
    |O|
     ‾ ⊖
```

The formal charge for chlorine = $7 - 2 - 3 = +2$
for each oxygen = $6 - 6 - 1 = -1$

Rule 5. The sum of all the formal charges in any species must equal the charge on the species (positive, negative, or neutral). A glance at the previous examples will show that this is always the case. (For example, in ClO_3^- there are two + and three − charges, which add up to −1, which is the charge on the ClO_3^- ion.) It is quite possible for a neutral molecule to have formal charges on two or more of its atoms, but the sum of such charges must be zero.

Example: Draw a Lewis structure for carbon monoxide, CO.

Answer: The total number of valence electrons is 10, so we must put five pairs into CO. The only way to do this and give each atom an octet is

$$^{\ominus}\bar{\text{C}}{\equiv}\bar{\text{O}}^{\oplus}$$

The formal charge

$$\text{for carbon} = 4 - 2 - 3 = -1$$
$$\text{for oxygen} = 6 - 2 - 3 = +1$$

The total charge is zero, and carbon monoxide is a neutral molecule, not an ion (although it has a dipole moment).

14-10 Resonance

Most molecules and polyatomic ions are satisfactorily described by a single Lewis structure of the type we saw in Section 14-9. However, there are some molecules and ions that *cannot* be correctly described in this way. For example, let us draw a Lewis structure for nitric acid, HNO_3. The connections are H O N O (with the second O attached to N). The total number of valence electrons is 24, and we can put 12 pairs in as follows, giving each atom an octet:

$$\text{H—}\bar{\underline{\text{O}}}\text{—}\overset{\oplus}{\text{N}}\text{—}\bar{\underline{\text{O}}}|^{\ominus}$$
$$\text{N}{=}\text{O}|$$

By our formal-charge rule, the nitrogen atom has a formal charge of $+1$ $(5 - 0 - 4)$, while one of the oxygens has a formal charge of -1 $(6 - 6 - 1)$. The other two oxygens are uncharged (each has $6 - 4 - 2$). The trouble with this Lewis structure is that it fails to describe the real HNO_3 molecule! This structure says that the two N—O bonds are different: that one is a single bond and the other a double bond. If so, they should have different bond distances. But in the real HNO_3 molecule, these distances are equal and there is no experimental evidence to show that the two N—O bonds are different in any way.

In fact, there is no single Lewis structure which correctly describes the real HNO_3 molecule. We can, however, correctly describe the real molecule by drawing *two* Lewis structures, the one given above and a second one, in which the electrons have been put in in a different way.

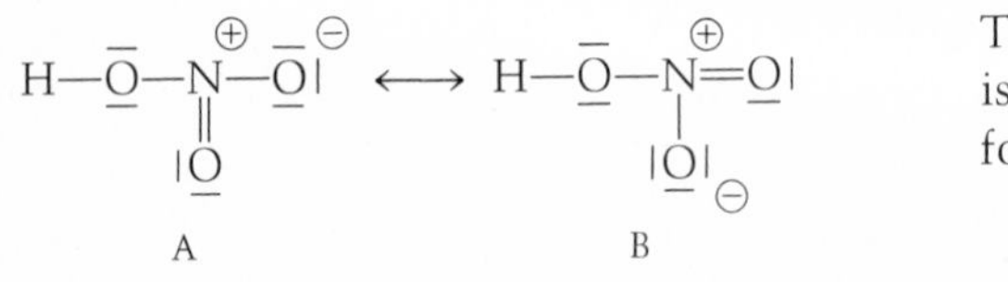

The arrow with two heads is used to indicate canonical forms. *It has no other use.*

The real HNO_3 molecule has *neither* of the two Lewis structures A or B. These two structures are fictitious. The structure of the real molecule is an *average of the two fictitious structures.* When it is necessary to use an average of two or more Lewis structures to describe a real molecule, we call the phenomenon **resonance.*** The fictitious structures are the *canonical forms,* and the real molecule is called a **resonance hybrid.** The use of the word hybrid is quite appropriate since there is an analogy here with the biological use of this word; for example, a mule is a hybrid

*The use of the word "resonance" for this purpose has no connection with its use in physics for quite a different phenomenon.

between a horse and a donkey. However, there is a difference between the chemical and biological usage of the word hybrid: Horses and donkeys really exist, but canonical forms do not. If the real molecule is an average of the two canonical forms, then the two N—O bonds are each intermediate between a single and a double bond, and the fact that they have equal distances is accounted for. As we would expect, the measured N—O bond distance in the real HNO_3 molecule is longer than an ordinary N=O double bond, but shorter than an ordinary N—O single bond.

Note that it is not true that a given HNO_3 molecule spends part of the time as canonical form A and the rest of the time as canonical form B. Nor is it true that some HNO_3 molecules have structure A and others have structure B. Canonical forms A and B are both *totally* fictitious, and no real HNO_3 molecule has either structure at any time. All the *real* HNO_3 molecules have the same structure, which does not change with time and which is an average, or hybrid, of A and B.

Let us compare the positions of the electrons in A and B. Note that most of the electrons (printed in black) are in the same positions in the two structures.

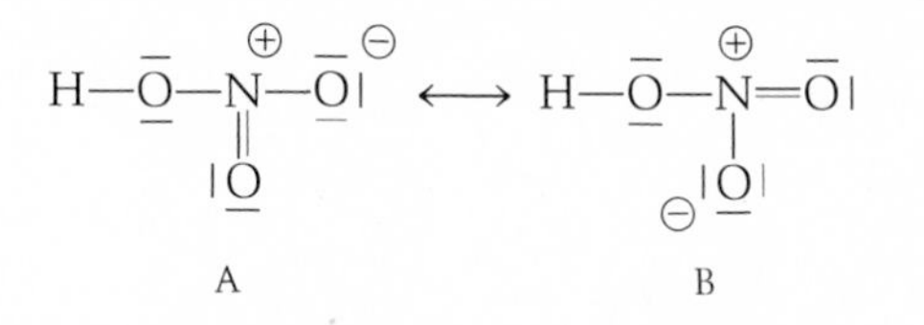

The only difference between A and B is in the positions of the two electron pairs shown in color. Since the real molecule is an average of A and B, the positions of all the electrons shown in black are the same in the real molecule as in A and B. These electrons are called **localized.** They are located either on a single atom as an unshared pair or as a bond between two atoms. The four electrons shown in color, however, are not localized. They are called **delocalized,** and are spread out over a greater distance (in this case, three atoms). Some chemists prefer not to use A and B (which are, after all, fictitious structures) to represent HNO_3, but use the following style instead (note that this is not a Lewis structure).

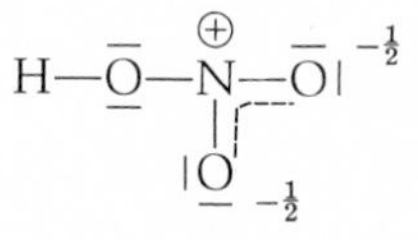

Why is each oxygen atom given a charge of $-\frac{1}{2}$?

In this style, all localized electrons are shown in the usual way, but delocalized electrons are indicated by dashed lines which are spread out to cover all the positions at which they are actually found. *We can make the statement that resonance is found only in molecules or ions with delocalized electrons.* Molecules or ions in which all electrons are localized can always be satisfactorily represented by a single Lewis structure.

The following rules will help us to recognize when resonance is involved and to write the canonical forms.

1. Resonance exists whenever, in attempting to draw a Lewis structure for a molecule or ion, we can put in the electrons in two or more ways to give two or more reasonable Lewis structures. If so, the real molecule will not have any of these structures, but will be a hybrid of them all.
2. All canonical forms must be valid Lewis structures that obey the rules for writing Lewis structures. For example, the following is *not* a canonical form for nitric acid:

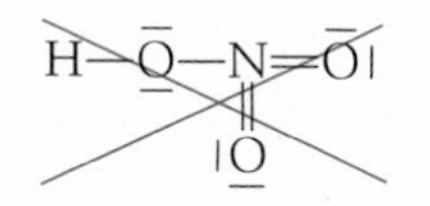

because it has 10 electrons in the outer shell of the nitrogen atom, which is impossible for nitrogen.

3. The positions of the nuclei must be the same in all the canonical forms for a given molecule or ion. Only electrons can be in different places.
4. All the atoms covered by the delocalized electrons must be in the same plane. In the case of nitric acid this rule is unimportant, because the delocalized electrons cover only three atoms, and plane geometry tells us that any three atoms must lie in a plane in any case, but the rule is important for most cases of resonance.
5. Canonical forms do not necessarily contribute *equally* to the real molecule. In the case of nitric acid, A and B do contribute equally, because these two forms are equivalent to each other, but in cases where forms are not equivalent, reasonable forms contribute more than unreasonable ones. For example, the following four valid Lewis structures can be drawn for hydrazoic acid, HN_3:

$$\mathrm{H{-}\overset{\ominus}{\underline{N}}{-}\overset{\oplus}{N}{\equiv}\overline{N}} \longleftrightarrow \mathrm{H{-}\overline{N}{=}\overset{\oplus}{N}{=}\overset{\ominus}{\underline{\overline{N}}}} \longleftrightarrow \mathrm{H{-}\overset{\ominus}{\underline{\overline{N}}}{-}\overset{2+}{N}{=}\overset{\ominus}{\underline{\overline{N}}}} \longleftrightarrow \mathrm{H{-}\overset{\oplus}{N}{\equiv}\overset{\oplus}{N}{-}\overset{2-}{\underline{N}}|}$$

C	D	E	F
important	important	not important	not important

Of these four, only C and D contribute substantially to the structure of the real molecule. E is an unreasonable form because it has only six electrons around the central nitrogen; it has an atom with a formal charge of +2 (such a high charge on a small atom is unstable); and it has twice as much charge separation as A or B (a total of two + and two − charges). F is unreasonable because of the charge of −2 on a single atom, because of the high charge separation, and because there are two adjacent nitrogen atoms with the same charge. Although C and D are both important canonical forms, they do not contribute *equally* to the structure of the real molecule, because they are not equivalent to each other. This means that the real molecule resembles one of them more than it does the other (although it is not known which of them contributes more).

Resonance is a stabilizing phenomenon. That is, **the energy of the real molecule is lower than any of the canonical forms.** It is obvious that this must be true, because if it were not, then whichever canonical form had the lowest energy would *be* the real molecule. Canonical forms contribute to the real molecule in proportion to their stability—stable forms contribute more than unstable ones (note that the stability of canonical forms can only be estimated and not measured, because they do not exist). The stabilizing nature of resonance has many important chemical consequences (an example is given in Section 20-5).

14-11 More about Ionic Valence and Covalence

WHICH ELEMENTS FORM WHICH KINDS OF BONDS. In Section 14-6 we learned that we could predict the type of bonding in a compound formed between two elements by considering the electronegativity difference between them. We used the somewhat arbitrary electronegativity difference of 1.7 as the borderline between an ionic and a covalent bond. At this point we can make some general statements about which kinds of elements form which kinds of bonds. (We shall not explicitly discuss metalloids here. Their bonding behavior is intermediate between that of metals and nonmetals.)

1. Metals form ionic bonds by losing electrons (they give *positive* ions). Many metals do not form covalent bonds at all. As is apparent from the electronegativity values in Table 14-2, the tendency for a representative metal to form covalent bonds

increases in going from left to right across the periodic table. For example, cesium and barium form only ionic bonds, while lead and bismuth form covalent bonds in some compounds and ionic bonds in others. Metals in general cannot form ionic bonds with other metals because both positive and negative ions are needed for an ionic bond and metals do not generally form negative ions. Metals in general also do not form covalent bonds with other metals (those metals which do form covalent bonds generally do so with nonmetals or with hydrogen). It follows that *metals seldom form compounds with metals.*

2. Most nonmetals can form ionic bonds by gaining electrons (they give *negative* ions), but all nonmetals can form covalent bonds as well. The tendency to form negative ions *increases* in going from left to right in the periodic table and in going down the table. For example, fluorine forms ionic bonds (as well as covalent bonds), while carbon forms only covalent bonds. Oxygen forms O^{2-} ions in a few compounds, but sulfur forms S^{2-} ions in a great many more compounds. *Most compounds of metals with nonmetals are ionic. Most compounds of nonmetals with nonmetals are covalent.*

3. Some nonmetals are very reluctant to be completely converted to negative ions. As we would expect from the preceding paragraph, these elements are carbon, silicon, germanium, nitrogen, phosphorus, and oxygen. For carbon, silicon, and germanium, we can assume that all bonding is covalent. Nitrogen, phosphorus, and oxygen atoms can form ionic bonds (negative charges), but in most cases only if the same atom also forms one or more covalent bonds at the same time. The most common example is the OH^- ion. In a compound such as NaOH, the oxygen is connected to sodium by ionic bonding, but to hydrogen by a covalent bond.

What Valences Are Possible. In Section 2-11, we defined *valence* as the number of bonds formed, and saw that the valence of an element is related to its position in the periodic table. We are now ready to return to this subject, this time taking note of the difference between ionic and covalent bonding.

1. *Covalence.* Each element in the second row of the periodic table has a covalence that corresponds to its position in the periodic table.

B	C	N	O	F
3	4	3	2	1

Covalences of second-row elements

Fluorine has seven outer-shell electrons, so it completes its octet by forming one covalent bond. Similarly, oxygen, nitrogen, and carbon have covalences of 2, 3, and 4, respectively. These three elements can use any combination of single, double, and triple bonds to make up the total number (see Figure 14-6). Boron, which has three outer-shell electrons, can form only three bonds, and has a covalence of 3. For second-row elements, these are the *only* covalences possible, except when the atom has a formal charge or an unpaired electron. As shown in Figure 14-7, second-row elements can have a higher covalence than normal *if they have a positive formal charge,* or a lower covalence than normal if they have a *negative formal charge* or *an unpaired electron.* In some compounds of boron, the atom can have a valence higher than three if it carries a negative charge: for example,

Elements below the second row also form many compounds in which their covalences correspond to their positions in the periodic table, but these elements can

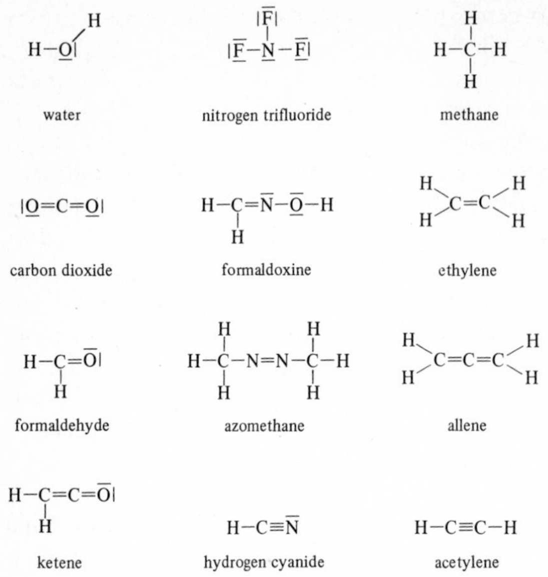

Figure 14-6 Some compounds of oxygen, nitrogen, and carbon. The total valence of O, N, and C is 2, 3, and 4, respectively, though any combination of single, double, and triple bonds can be used to achieve this number.

also form compounds with higher valences, *because they can expand their octets* (Table 14-5). Like the second-row elements, these elements can also have valences higher or lower than normal if they have formal charges or unpaired electrons.

2. *Ionic valence.* When an atom is converted to an ion, the ionic valence is equal to the number of electrons lost or gained. For elements in groups IA, IIA, VIA (except oxygen, which seldom forms O^{2-} ions), and VIIA, this number is $+1$, $+2$, -2, and -1, respectively, because this is the number of electrons that are lost or gained to give a complete octet.

K^+	Sr^{2+}	Te^{2-}	I^-
Valence $+1$	Valence $+2$	Valence -2	Valence -1

Some of the representative elements of the other four groups (especially those near the bottom of the table) also form ions whose valences correspond to their positions in the table. For transition elements, matters are more complex. These elements often form ions with different valences (for example, Fe^{2+}, Fe^{3+}), and many of their ions do not have complete outer shells. Transition elements will be discussed in Chapter 19.

For polyatomic ions, the ionic valence of the ion is equal to the total charge on the ion. For example, the valence of SO_4^{2-} is -2, of NO_3^- is -1, and of NH_4^+ is $+1$. **Remember that ionic valences always carry a + or − sign, but covalences never do.**

Physical Properties of Ionic and Covalent Compounds. Ionic and covalent compounds generally have very different physical properties, though in order to discuss them we must distinguish between compounds made up of small covalent molecules (fewer than about 50 to 100 atoms) and *polymeric*

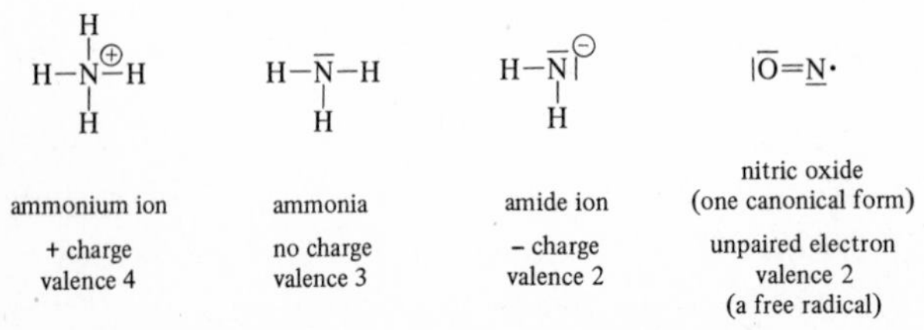

Figure 14-7 Ions and molecules showing possible valences of nitrogen.

Table 14-5 Some Compounds of As, S, and Br Showing Different Covalences

	H_2S	HBr
AsH_3	SCl_4	BrF_3
AsF_5	SF_6	BrF_5

compounds, which are made up of larger covalent molecules (thousands, millions, or even $\sim 10^{22}$ atoms—see page 412). We will discuss three properties: physical states, solubility, and the ability to conduct an electric current. They are summarized in Table 14-6. Remember that we are talking about a vast number of compounds here—there are millions of known compounds—and that with such a large variety of compounds there are bound to be exceptions to almost any rule, but the statements given here are generally true.

1. *Physical state.* All ionic compounds are solids at room temperature, as are almost all polymeric covalent compounds. Ordinary covalent compounds may be solids, liquids, or gases, but if they are solids, they have comparatively low melting points (less than ~300°C). Ionic compounds and polymeric covalent compounds typically have melting points of 1000°C or more, and many polymeric compounds cannot be melted at all because their bonds break down before a high-enough temperature can be reached. We can be safe in assuming that almost all gases, liquids, and low-melting solids are either elements or ordinary covalent compounds.

2. *Solubility.* In Section 7-5, we saw that the rule for solubility is: *like dissolves like*. The most important aspect of a solvent for the purpose of classifying solubility is its polarity (*the dielectric constant—Section* 7-5*—is good measure of this.*) We shall compare water, a highly polar solvent, with benzene, a typical nonpolar solvent (benzene and water are insoluble in each other).

a. Most ionic compounds are soluble in water, because the polar water molecules surround the negative and positive ions (solvation); although some ionic compounds are insoluble (Section 11-1) because the solvation energy is not large enough to overcome the lattice energy of the crystal. Ionic compounds are insoluble in benzene and other nonpolar solvents because the solvent molecules, being nonpolar, are not attracted to and cannot solvate the ions.

b. Small covalent compounds are soluble in water if they can form hydrogen bonds (Section 6-8) with the water, and if they have relatively low molecular weights (below about 70). For example, ethyl alcohol (CH_3CH_2OH), acetone (CH_3COCH_3), ammonia (NH_3), and acetic acid (CH_3COOH) all form hydrogen bonds with water and are all soluble; propane (C_3H_8), chloromethane (CH_3Cl), benzene (C_6H_6), and phosphine (PH_3) do not form hydrogen bonds with water and are insoluble; while ether ($C_2H_5OC_2H_5$), butyl alcohol (C_4H_9OH), and benzoic acid (C_6H_5COOH) form hydrogen bonds with water but are insoluble because their molecular weights are too high. Some covalent

Table 14-6 Some Physical Properties of Ionic and Covalent Compounds

Ionic compounds	*Covalent compounds*	*Polymeric covalent compounds*
Solids	Solids, liquids, or gases	Solids
Very high melting points	If solids, then low melting points	Very high melting points if can be melted at all
Often soluble in water, insoluble in benzene	Rarely soluble in water, often soluble in benzene	Most are insoluble in any solvent, although some are soluble in benzene
Water solutions always conduct electricity	Water solutions do not conduct electricity unless they react with water to form ions	—

compounds are soluble in water even though they have relatively high molecular weights, because each molecule forms hydrogen bonds with several water molecules. A good example is sucrose ($C_{12}H_{22}O_{11}$; cane sugar), which is very soluble in water despite its high molecular weight. Some covalent compounds are soluble in water because they react with water. We have already seen some examples:

$$HCl + H_2O \rightleftharpoons H_3O^+ + Cl^-$$
$$CO_2 + H_2O \rightleftharpoons H_2CO_3$$

Most small covalent compounds are soluble in nonpolar solvents like benzene, unless the compounds are highly polar (like sucrose). The benzene molecules are not strongly attracted toward each other, and molecules of small covalent compounds can easily get between the benzene molecules without creating much of a disturbance, unless the molecules are very polar, in which case they are too much attracted to each other for the benzene to be able to separate them. Water can separate them because water molecules solvate each solute molecule, but benzene, being nonpolar, is not attracted to the solute molecules and hence cannot solvate them.

c. Most polymeric molecules are insoluble in any solvent. They are too big to be effectively solvated by a polar solvent like water. However, some polymers can be dissolved in nonpolar solvents such as benzene.

3. *The ability to conduct an electric current.* We have said that ionic compounds are always solids at room temperature. In this state they cannot conduct an electric current. However, ionic compounds do conduct an electric current when molten, or when dissolved in water. *Covalent compounds do not conduct electricity in any state.* This is one of the most important distinctions between ionic and covalent compounds, and is often used in laboratories in order to tell them apart. A compound that, when dissolved in water, conducts a current, is called an **electrolyte.** Compounds that are soluble in water and ionize 100% are *strong electrolytes* (this group consists of the six strong acids, the four strong bases, and all soluble salts). Compounds that produce a small number of ions in water (weak acids, weak bases, insoluble salts) are *weak electrolytes.* Covalent compounds can be electrolytes (strong or weak) if they react with water to produce ions. For example, HCl is a gaseous covalent compound but when dissolved in water, it reacts to give the ions H_3O^+ and Cl^- and is thus a strong electrolyte.

14-12 Electron Pair Repulsion Theory (EPRT) and the Shapes of Molecules

So far we have discussed the electronic structures of molecules and polyatomic ions, but have said little about the *shapes* of molecules—the three-dimensional arrangement of the atoms within the molecules. Chemists have available today several kinds of highly sophisticated (and expensive) instruments which are used for determining molecular shapes. It is important to realize that measurements made by these instruments can only detect the *nuclei* of the atoms. We ourselves must decide on the basis of other experiments where the electrons go. The information we get tells us which atoms are next to one another as well as all the bond distances and bond angles.

A rather simple, but surprisingly accurate theory used to *predict* the geometry of molecules is the **electron-pair repulsion theory (EPRT),** also called the valence-shell electron-pair repulsion model (VSEPR). When we use this theory, we focus on the central atom of the molecule.* The theory states that *all valence-shell*

*If a molecule has more than one "central atom," each of them is treated separately in the same way.

electron pairs surrounding the central atom locate themselves in such a manner as to be as far away from each other as possible. Two distinct types of electron pairs may be around the central atom:

1. Covalent bonds—bonding electron pairs.
2. Nonbonding electron pairs (unshared pairs).

Look at H_2O. Its Lewis structure is

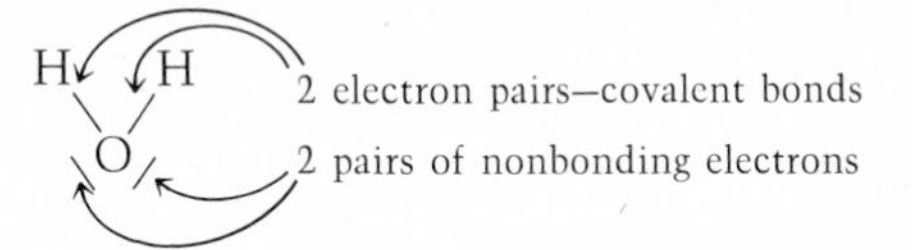

The rationale for the theory is simple. Electrons are negatively charged and repel each other. By separating the electrons (bonding as well as nonbonding) as far from each other as possible, the electrostatic repulsion that raises the energy is minimized. Each molecule tends toward a state of lowest energy. Table 14-7 shows the predicted shapes and angles (geometry) for different types of molecules (water is an $|\overline{A}B_2$ type).

Table 14-7 Geometry for Different Numbers of Electron Pairs About the Central Atom in a Molecule

General type	*Number of electron pairs*	*Examples*	*Shape of the valence-shell orbitals, bonding and nonbonding*	*Bond angle*	*Drawing*
AB_2	2	$BeCl_2$	Linear	180°	
AB_3, $\overline{A}B_2$	3	BF_3, $\overline{Ge}F_2$	Triangular	120°	
AB_4, $|AB_3$, $|\overline{A}B_2$	4	CH_4, $|NH_3$, $H_2\overline{O}|$	Tetrahedral	109°28′	
AB_5, $|AB_4$, $|\overline{A}B_3$, $|\underline{\overline{A}}B_2$	5	PCl_5, $\overline{Te}Cl_4$, $|\overline{Cl}F_3$, $|\underline{\overline{Xe}}F_2$	Trigonal bipyramid	120°, 90°	
AB_6, $|AB_5$, $|\overline{A}B_4$	6	SF_6, $|BrF_5$, $|\overline{Xe}F_4$	Octahedral	90°	

In the next section we shall briefly review the various types of molecules and their geometries. When possible we will compare the predicted angles with experiment. Anticipating what we will find experimentally in the next section, we state the following rule: *Nonbonding pairs of electrons repel adjacent electron pairs more strongly than do bonding pairs.* Two nonbonding electron pairs repel each other to a greater extent than two bonding pairs. The strength of the electron pair repulsion is in this order: **two nonbonding pairs** repel each other more than **one nonbonding pair-one covalent pair** which repel each other more than **two covalent pairs.**

14-13 Molecules and Their Geometry

AB_2 MOLECULES (two electron pairs). An example of a AB_2 molecule is BeF_2. The Lewis structure is

$$|\overline{\underline{F}}\text{—Be—}\overline{\underline{F}}|$$

The central atom, Be, has only two pairs of electrons surrounding it, both of which are bonding pairs. (Remember that BeF_2 does not follow the octet rule.) Consequently, according to the EPRT the angle F—Be—F should be 180° and measurements confirm that it is. Note also that since the angle is 180°, BeF_2 has no permanent dipole moment, even though each of the bonds Be—F is a polar bond.

AB_3 MOLECULES (three electron pairs). BF_3 is an AB_3 molecule which is planar and each angle (F—B—F) in the molecule is 120°. The Lewis structure for BF_3 is

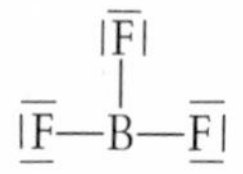

The central atom boron has three pairs of bonding electrons and *no* pairs of nonbonding electrons. Consequently, its shape is planar.

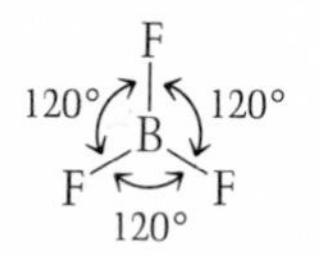

This arrangement of the fluorine atoms provides the greatest possible separation of the electron pairs. Experiments confirm that this is the structure of BF_3.

AB_4, $\overline{A}B_3$, AND $|\overline{A}B_2$ MOLECULES (four electron pairs). A typical AB_4 molecule is methane, CH_4. The Lewis structure is

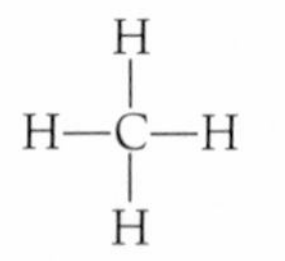

This structure shows four pairs of bonding electrons. Consequently, the geometry of methane is tetrahedral; that is, the four bonds of carbon are directed to the corners of a regular tetrahedron. The H—C—H bond angles are those of a regular tetrahedron: 109°28′.

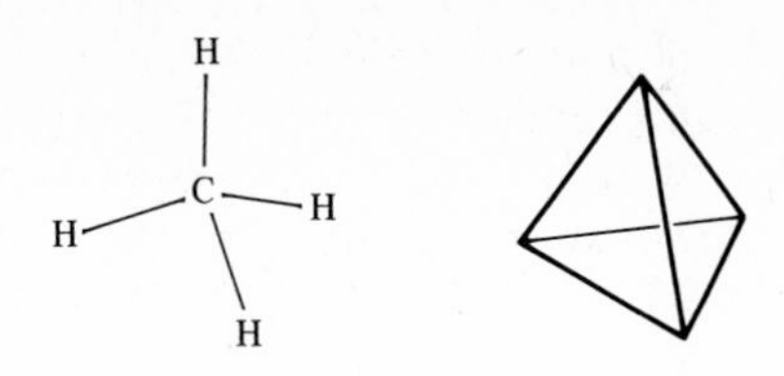

This is the structure that provides the greatest possible separation for the four pairs of bonding electrons.

Example: Describe the geometry of the ammonia molecule, NH_3 (an $\overline{A}B_3$ molecule).

Answer: To determine the structure of any molecule using EPRT, we first draw the Lewis structure.

H—$\overline{N}$—H
|
H

From the Lewis structure we find that ammonia has three pairs of bonding electrons and one pair of nonbonding electrons. The electrons will generate a tetrahedral structure.

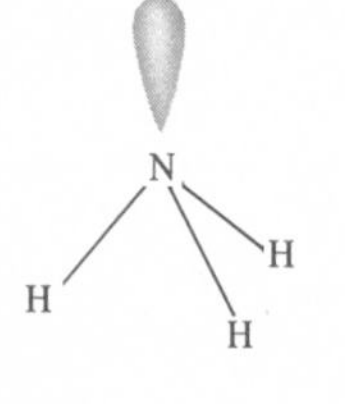

We would expect the H—N—H angles to be slightly less than the tetrahedral angle (109°) since the nonbonding pair repels to a greater extent than the bonding pairs and will squeeze the hydrogens closer to each other. Experimentally, we find that each of the angles H—N—H is around 106° rather than 109°. Looking only at the nuclei, we see that the three hydrogens form an equilateral triangle, while the nitrogen sits above them. This shape is called a *trigonal pyramid.*

Example: Draw the structure of H_2O (an $|\overline{A}B_2$ molecule) according to EPRT.

Answer: The Lewis structure for H_2O is

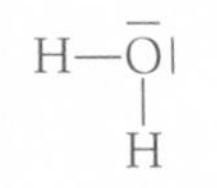

This Lewis structure shows two pairs of bonding electrons and two pairs of nonbonding electrons. Consequently, the geometry of the electron pairs will be tetrahedral.

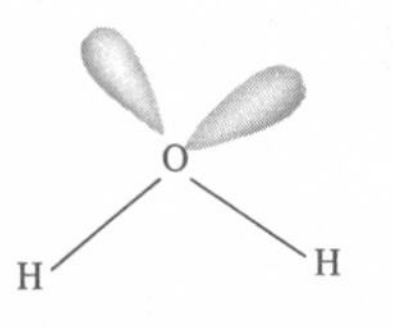

Looking only at the nuclei, we have an angular structure.

O
H H

Again based on EPRT, the H—O—H angle should be less than 109°. There are two pairs of nonbonding electrons, and these repel more strongly than the bonding electrons. The bond angle in H_2O is found to be 105°. Note that the angle in H_2O is less than in NH_3 as we would predict (one pair of nonbonding electrons in NH_3 vs. two in H_2O).

AB_5 and $\bar{A}B_4$ Molecules (five electron pairs). $SbCl_5$ is an AB_5-type molecule. There are 10 electrons around the antimony atom which make up five covalent bonds. Rule 3 states that atoms in the third row or lower in the periodic table can accommodate more than eight electrons. Since $SbCl_5$ has five pairs of bonding electrons and zero nonbonding pairs of electrons, its shape will be a trigonal bipyramid. Experiments confirm that this is so. Three of the five chlorine atoms are located

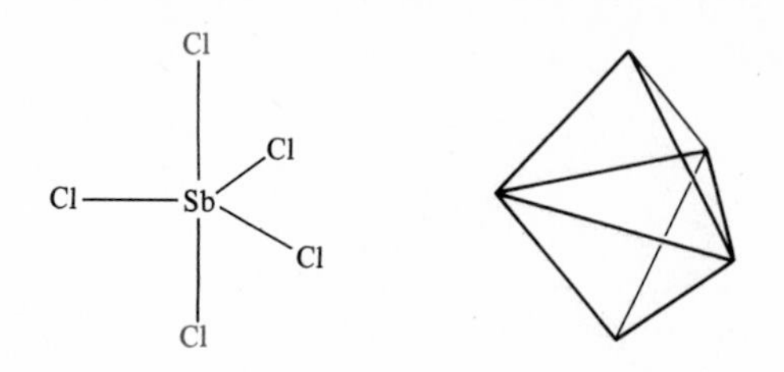

at the corners of an equilateral triangle (these are called *equatorial* positions), while the other two (called *axial*) are above and below the plane of the equilateral triangle.

A good test of EPRT is the compound tellurium tetrachloride, $TeCl_4$. Without EPRT we might have predicted a tetrahedral shape for this molecule, similar to that of methane CH_4. However, the Lewis structure for $TeCl_4$ is

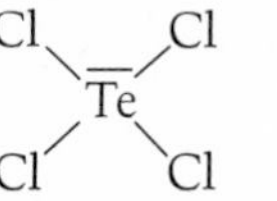

10 outer-shell electrons on Te

$TeCl_4$ is an $\bar{A}B_4$ molecule: four pairs of bonding and one pair of nonbonding electrons. The EPRT predicts a trigonal bipyramid structure (Table 14-7) similar to that of $SbCl_5$, except that one of the five positions contains an unshared pair of electrons. Experiments show that in the gas phase $TeCl_4$ does have a trigonal bipyramid shape.

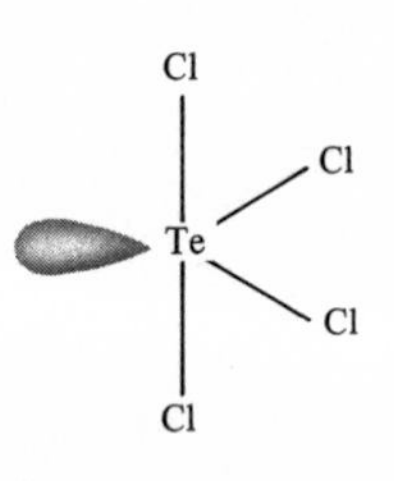

The unshared pair of electrons occupies one of the three equatorial positions.

AB_6 Molecules. SF_6 is an AB_6 molecule. Its Lewis structure is

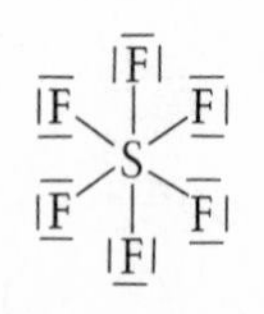

Sulfur obviously has an expanded shell with 12 electrons in six bonding pairs. From Table 14-7 we see that its structure is octahedral.

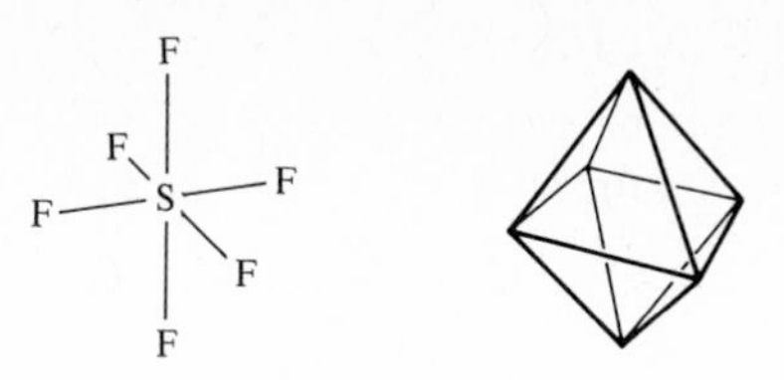

Experiments confirm that this is the shape.

Now that we have seen their shapes, we can understand why only H_2O, and not NH_3 or HF, can form three-dimensional networks of molecules (Figure 6-16). Of the three molecules, only H_2O has two unshared pairs and two hydrogens and can thus form four hydrogen bonds with other water molecules in a three-dimensional array. The NH_3 molecule has three hydrogens but only one unshared pair; the HF molecule has three unshared pairs but only one hydrogen.

14-14 Multiple Bonding and the EPRT

The EPRT is easily extended to accommodate multiple bonding in covalent compounds. All one has to remember is that as far as the EPRT is concerned, *a double or triple bond counts simply as a single bond.* For example, if we consider the bonding in ethylene C_2H_4, we get for its Lewis structure

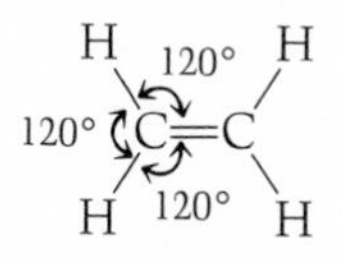

The geometry at each carbon atom is triangular (three bonds; table 14-7), and the entire molecule is planar (all six atoms lie in one plane).

Example: Predict the structure of CO_2.

Answer: The Lewis structure for CO_2 is

$$|\underline{O}=C=\underline{O}|$$

This structure shows two sets of bonding electrons emanating from the carbon (remember that in EPRT a double bond only counts as a single bond). Consequently, EPRT predicts that the structure of CO_2 will be linear. Each bond in CO_2 is polar; however, the linear structure

$$\overset{\delta-}{O}=\overset{\delta+}{C}=\overset{\delta-}{O}$$

results in a *zero* dipole moment. Experimentally, CO_2 is a linear molecule with no dipole moment.

The EPRT helps us to understand why molecules have the shape they have. CO_2, BeF_2, and H_2O are all triatomic molecules. The first two are linear, while the third, H_2O, has an angle of 105°. The EPRT shows us that it is the number of electron pairs that is important, not the number of atoms.

Example: Predict the shape of stannous chloride ($SnCl_2$).

Answer: The tin atom has four electrons in its outer-shell (group IVA). Because we can draw two reasonable Lewis structures,

$$|\overline{\underline{Cl}}-\overset{\ominus}{\overline{Sn}}=\overset{\oplus}{\overline{Cl}}| \longleftrightarrow |\overset{\oplus}{\overline{Cl}}=\overset{\ominus}{\overline{Sn}}-\overline{\underline{Cl}}|$$

the $SnCl_2$ molecule is a resonance hybrid. In each canonical form we count two pairs of bonding electrons (counting a double bond as one pair) and one pair of nonbonding electrons. Consequently, according to the EPRT the structure of $SnCl_2$ is

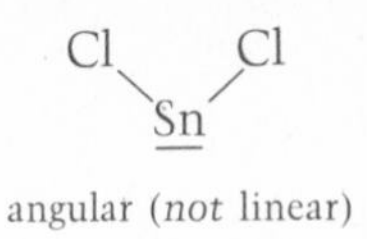

angular (*not* linear)

14-15 Hybridization and Hybrid Orbitals

A theory that is to account for the geometry of molecules must, of course, be consistent with observed experimental facts. Let us use the typical carbon compound methane, CH_4, as our example. From the EPRT we now know that the methane molecule has a tetrahedral structure.

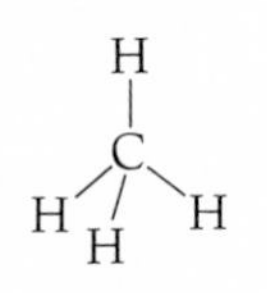

From experimental data on methane, we know that a molecule of methane is composed of one carbon atom and 4 hydrogen atoms (CH_4). All the C—H bond distances are equal (1.11 Å) and there are no chemical or physical experiments we can perform that will distinguish any one of the hydrogens from the other three. For all chemical purposes, the four C—H bonds are identical.

Let us reconsider this matter from a different viewpoint. Assume that we did not have all the experimental information concerning methane. Knowing the electronic structure of the hydrogen and carbon atoms, what would we predict for the formula of the simplest compound made up of these two elements? We know that a carbon atom has the outer-shell electronic configuration $2s^2 2p_x^1 2p_y^1$. It has only two unpaired electrons. It would be logical to conclude that the *simplest* compound of carbon and hydrogen would have the formula CH_2, where each of the 1*s* atomic orbitals of hydrogen combines with one of the 2*p* orbitals of carbon. From the electronic structures of the atoms, we might conclude that the compound CH_2 would look as follows:

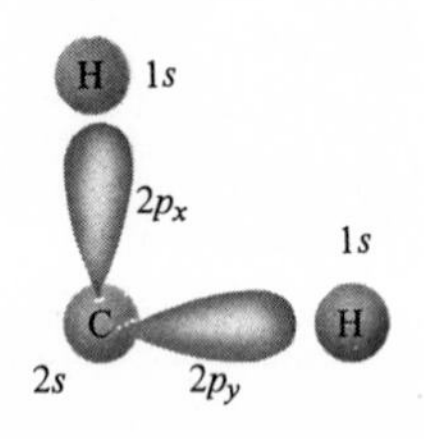

From the EPRT we would predict an angle (H—C—H) of somewhat less than 120°. This would be a logical prediction based on the known electronic structures of carbon and hydrogen. Nevertheless, CH_2 is not a stable compound (it has a lifetime of much less than one-thousandth of a second). The simplest stable compound of carbon and hydrogen is methane CH_4. Thus, our naive prediction that CH_2 should be a stable compound is wrong. The experimental facts are that carbon has a valence of 4, and does not have a valence of 2.

This paradox and the basic geometry and bonding in almost *all* carbon compounds were solved in 1931 by Linus Pauling. He showed that the $2s^2$, $2p_x^1$, $2p_y^1$, and the empty $2p_z$ atomic orbitals in carbon are mixed in such a manner as to *produce four equivalent orbitals,* which are oriented to the corners of a tetrahedron, in order to maximize the distance between them.

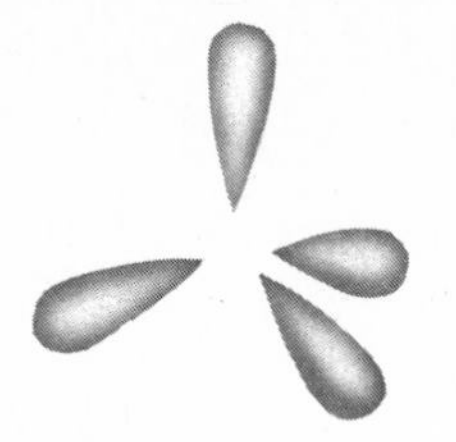

This mixing of atomic orbitals in order to produce new orbitals is called **hybridization.**

sp^3 Hybrid Orbitals. The outer-shell electronic configuration of carbon is $2s^2 2p_x^1 2p_y^1$. This is the lowest energy configuration—the ground state. Imagine that one of the ground-state electrons in the $2s^2$ orbital is "promoted" into the empty $2p_z$ orbital. This would give us

Ground state		*Excited state*
		$2p_z^1$
$2p_y^1$	one electron promoted to a $2p_z$ orbital	$2p_y^1$
$2p_x^1$	$\longrightarrow$ (this requires energy)	$2p_x^1$
$2s^2$		$2s^1$

After the electron has been promoted, all four orbitals are "mixed" ($2s^1$, $2p_x^1$, $2p_y^1$, and $2p_z^1$), yielding four equivalent hybrid orbitals called sp^3 orbitals.

$2p_z^1$		sp^3
$2p_y^1$	mixed or hybridized $\longrightarrow$	sp^3
$2p_x^1$		sp^3
$2s^1$		sp^3

There are four hybrid orbitals here, each of which is called an **sp^3 orbital.** They all have the same energy and are equivalent in every other way. They are directed in space toward the corners of a tetrahedron. This is the best configuration they can have in order to keep out of each other's way and to minimize their electron–electron repulsion. Like all orbitals, each can accommodate a maximum of two electrons. Each of the four sp^3 orbitals in carbon can now combine with a 1*s* orbital

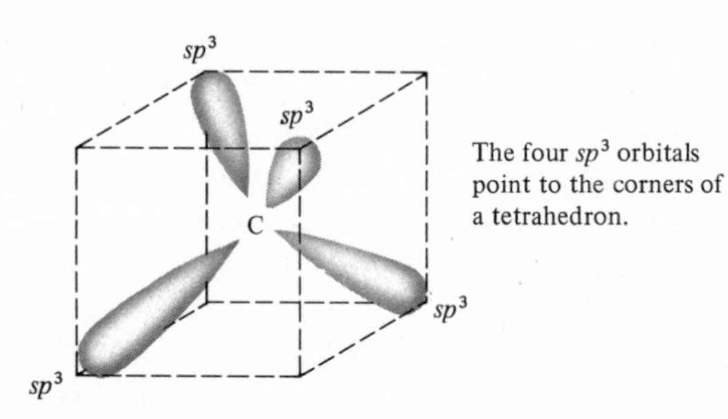

The four sp^3 orbitals point to the corners of a tetrahedron.

of a hydrogen atom, forming four equivalent C—H bonds. A word about the energetics of this procedure. Even though it takes energy to promote a 2*s* electron into a 2*p* orbital, the energy that the system gets back when the bonding occurs more than compensates for this initial expenditure. The reason is that hybrid orbitals are so strongly oriented in space that the overlap with other orbitals (in this case, the 1*s* of hydrogen) is much stronger than the *s* or *p* bonds formed without hybridization.

The oxygen and nitrogen atoms also form four sp^3 orbitals, but here only two or three of them, respectively, are available for bonding, since the other orbitals are already filled. For example, the structures of water and ammonia are:

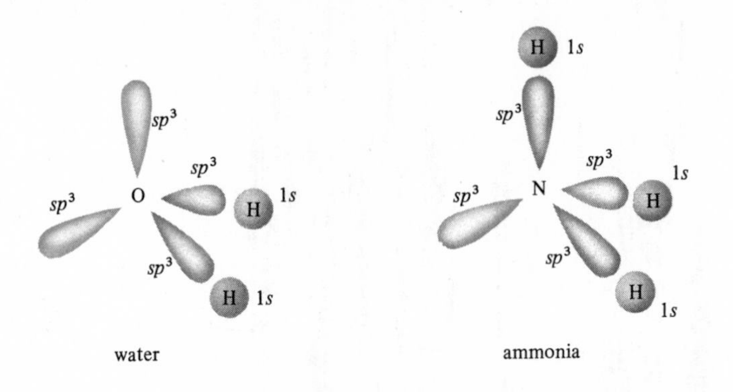

The orbitals printed in color do not form bonds because they are already filled with two electrons each.

sp^2 Hybrid Orbitals. In deducing the geometry of BF_3 using the EPRT, we concluded that BF_3 is a planar triangular molecule. The valence electrons of boron are $2s^22p^1$. However the three B—F bonds are equivalent. In order to account for the three equivalent atomic orbitals in boron, we consider a hybridization scheme similar to the sp^3 in CH_4.

Ground state		*Excited state*
		$2p_y^1$
$2p_x^1$	one electron promoted to a $2p_y^1$ orbital from a 2*s* orbital →	$2p_x^1$
$2s^2$		$2s^1$

The three orbitals in the excited state are now mixed or hybridized to form three **sp^2 orbitals.** The sp^2 orbitals emanate from the central atom in a triangular orientation (once again this is the farthest apart they can get). The three fluorine

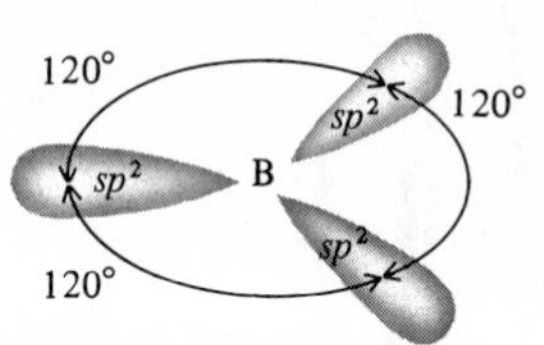

The three sp^2 orbitals are all in the same plane.

atoms now bond to the three sp^2 hybrid orbitals, giving us BF_3. sp^2 hybridization is formed by hybridizing one *s* and two *p* orbitals to form three equivalent sp^2 orbitals.

***sp* Orbitals.** An ***sp* orbital** is formed when an atom hybridizes one *s* and one *p* orbital to form *two* equivalent *sp* orbitals. Two equivalent orbitals emanating from a single atom can get farthest away from each other by adopting an angle of 180° so that the two *sp* orbitals are oriented in space 180° from each other. As an example, BeF_2 is a linear molecule. The valence electrons of Be are $1s^2$ electrons. Using the hybridization scheme, we have:

Ground state		*Excited state*
	promotion of one *s* electron to a p_x orbital →	$2p_x^1$
$2s^2$		$2s^1$

The $2s^1$ and $2p^1$ electrons are now mixed or hybridized, yielding two *sp* orbitals.

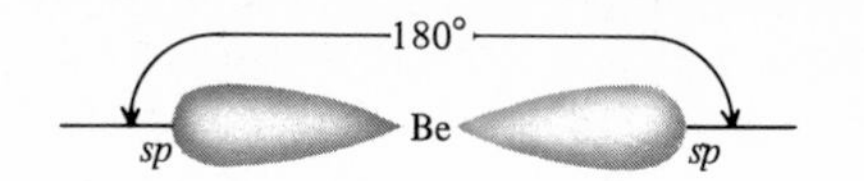

The two *sp* orbitals are in a straight line.

Bonding of the fluorine atoms to the Be atom yields the compound BeF_2.

14-16 Hybridization and Multiple Bonds

We can make very good guesses concerning the hybridization of compounds containing multiple bonds using the EPRT and Table 14-8. Consider ethylene C_2H_4 (discussed in Section 14-14). Since the carbon bonds are triangular, we would expect its hybridization to be sp^2. For carbon this can occur as follows:

Ground-state (carbon) valence electrons		*Excited state (carbon)*
		$2p_z^1$
$2p_y^1$	promotion of one 2*s* electron to the $2p_z$ orbital →	$2p_y^1$
$2p_x^1$		$2p_x^1$
$2s^2$		$2s^1$

To get sp^2 hybridization we mix one 2*s* electron and two 2*p* electrons, say $2p_x{}^1$ and $2p_y{}^1$. This leaves the $2p_z{}^1$ electron in an unhybridized state.

The geometry now for each of the carbons is

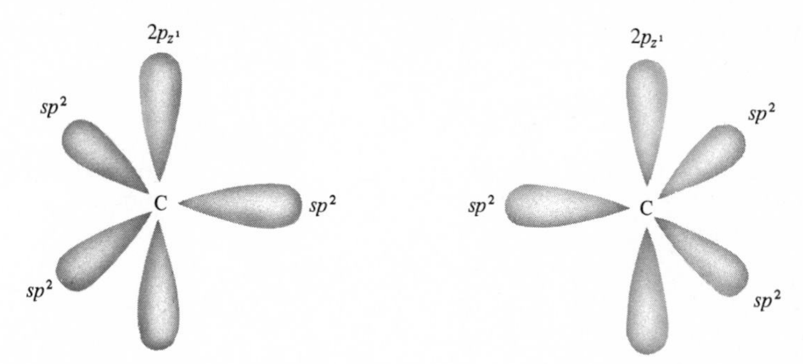

It is very important to recognize that the $2p_z$ orbital is perpendicular to the plane of three sp^2 orbitals. (Why?) Bringing the carbons together and adding the hydrogens, we get

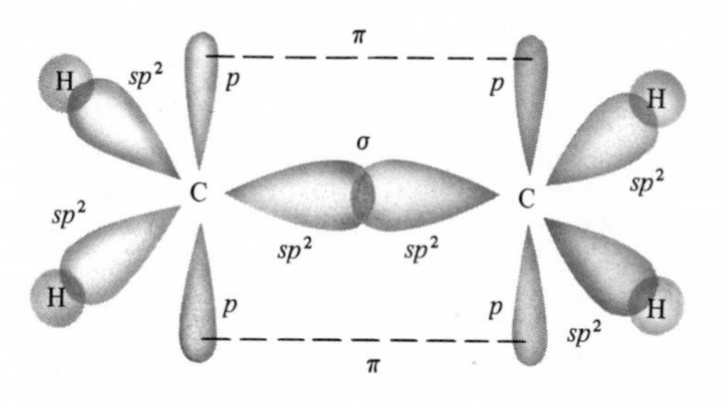

Table 14-8 Hybridization

Example	*Hybridization*	*Number of electron pairs*	*Orientation*	*Bond angles*
CH_4	sp^3	4	Tetrahedral	109°28′
BF_3	sp^2	3	Triangular	120°
BeF_2	*sp*	2	Linear	180°

We can see from the diagram that the two covalent bonds that make up the double bond between the carbon atoms *are not equivalent—they are different*. To see the difference between the two bonds, consider an imaginary line between the nuclei of the two carbon atoms. One of the bonds is formed by overlap of two sp^2 orbitals. In this bond the maximum electron density accumulates between the centers of the atoms being bonded and lies on the imaginary line. This is termed a **σ (sigma) bond.** The other bond is formed by overlap between the two $2p_z$ orbitals. In this bond, the electron density lies above and below the imaginary line between the centers of the atoms being bonded. This type of bond is termed a **π (pi) bond.** For C_2H_4 we have

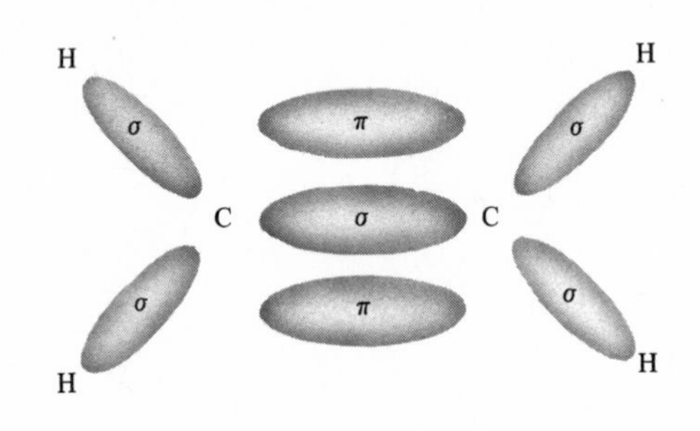

The π bond is almost always weaker than the σ bond because the sp^2 orbitals can overlap to a greater extent than the p orbitals (which must overlap sideways) (Figure 14-8). In almost all cases *single* covalent bonds are σ bonds, while *double* covalent bonds are composed of one σ and one π bond. Only one σ bond can exist between two atoms.

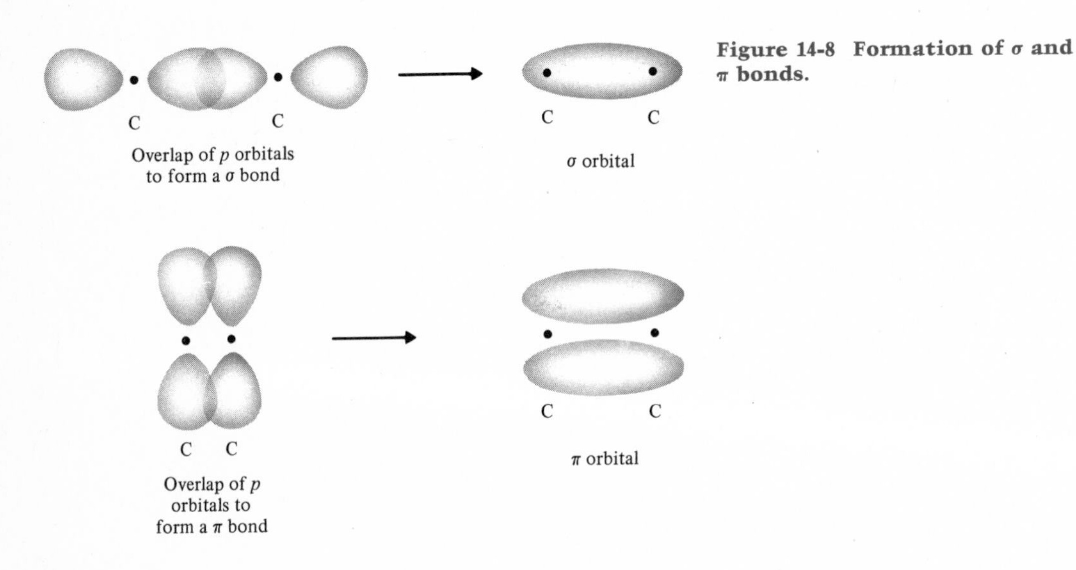

Figure 14-8 Formation of σ and π bonds.

Example: Discuss the hybridization and bonding in acetylene, C_2H_2.

Answer: The Lewis structure of acetylene C_2H_2 is

$$H—C{\equiv}C—H$$

From EPRT, we conclude the molecule is linear and the hybridization in the carbons must be sp. To achieve this, we have

Ground state (carbon) valence electrons		*Excited state (carbon)*
		$2p_z^1$
$2p_y^1$	promotion of one $2s$ electron to the $2p_z$ orbital →	$2p_y^1$
$2p_x^1$		$2p_x^1$
$2s^2$		$2s^1$

For *sp* hybridization we mix an *s* and a *p* orbital, say the $2p_x$ orbital. The $2p_y$ and $2p_z$ orbitals are not hybridized, and they are both perpendicular to the *sp* hybrid orbitals. Drawing the carbons in acetylene, we have

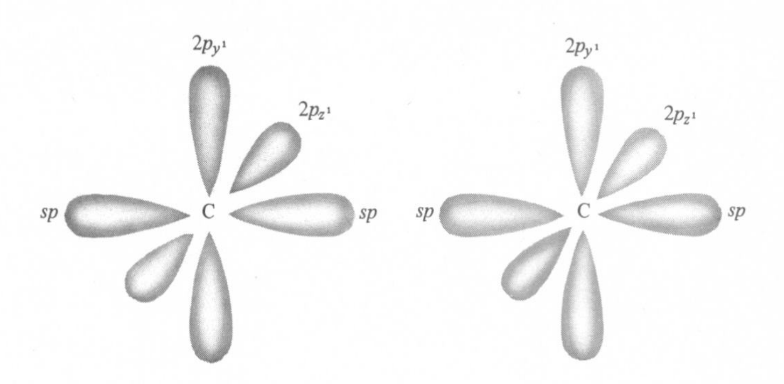

Bringing the two together gives us

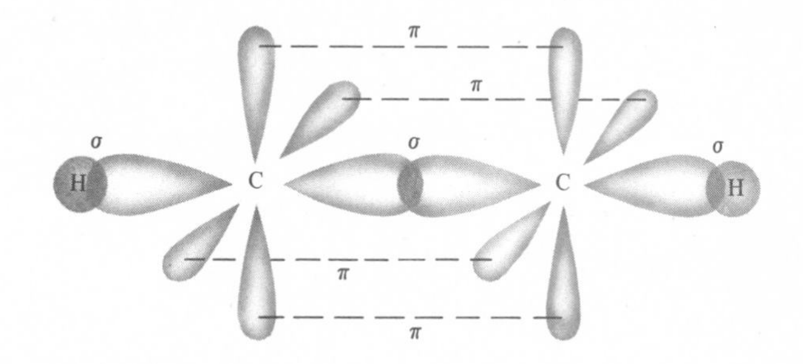

In terms of σ and π bonds, we get

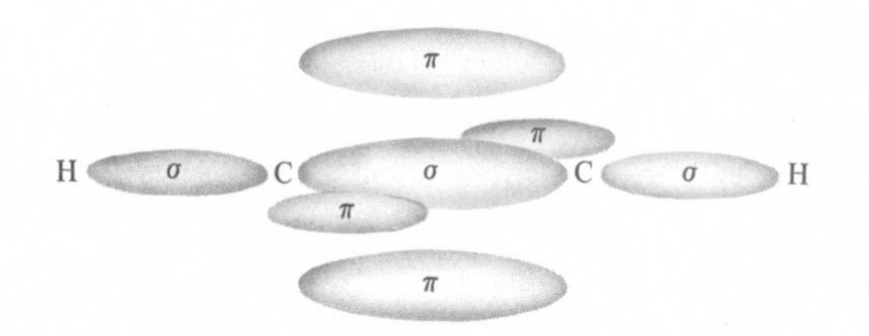

The triple bond in acetylene and in any other molecule with a triple bond is composed of one σ bond and two π bonds.

Part II MOLECULAR ORBITAL THEORY

In Section 14-2 we stated that a covalent bond could be thought of as arising from the overlap of two atomic orbitals. For example, each of the two covalent bonds in the H_2O molecule is formed by overlap of the 1*s* orbital of a hydrogen atom with one of the four sp^3 orbitals of the oxygen atom (Section 14-15). This way of describing covalent bonding is called the **valence-bond method.** The method is quite successful for many compounds (including water), and explains many structural features found in covalently bonded molecules and polyatomic ions; however, the method is not successful in certain cases. For example, it fails to predict the magnetic properties of some simple molecules. The prime example is the oxygen molecule. From the valence-bond method, we would predict that the two oxygen atoms in O_2 form a double bond (since each atom has two half-filled $2p$ orbitals) and that the Lewis structure ought to be $|\overline{\text{O}}=\overline{\text{O}}|$, in which each oxygen atom obeys the octet rule. However, the properties of real oxygen are not in accord with this structure. Laboratory experiments show that oxygen is paramagnetic, which means that it must have unpaired electrons, and a better representation of the structure is $|\underset{\cdot}{\overline{\text{O}}}-\underset{\cdot}{\overline{\text{O}}}|$. The valence-bond method cannot explain why the oxygen molecule keeps two of its electrons unpaired, giving each atom only seven electrons in its valence shell, when it would seem that the two orbitals (one from each atom) could easily overlap to give the double bond and a complete octet for each atom.

In the rest of this chapter we will look at another way of describing covalent bonding, called **molecular orbital theory.** The development of this theory began around 1932 with the work of Erich Hückel (1896–) and Robert Mulliken (1896–) and is still being refined today. As we shall see in Section 14-20 the molecular orbital theory successfully predicts the true structure of the O_2 molecule.

14-17 Comparison Between Valence-Bond and Molecular Orbital Methods

In Section 13-6, we saw that the Schrödinger equation could not be solved exactly for atoms or molecules containing more than one electron. Both the valence-bond and molecular orbital methods are based on *approximate* solutions to the Schrödinger equation. Each uses different assumptions, and each is more successful with some molecules than with others. In this section we shall examine some of these assumptions.

In order to contrast the differences between the valence-bond theory and the molecular orbital theory, let us take the simple molecule H_2 as an example. The valence-bond treatment begins with the individual atoms and then considers the interactions between them. In order to form the bond we bring the two atoms close enough together so that the two $1s$ orbitals (one from each atom) overlap. The covalent bond is formed by overlap of these two orbitals. The overlap cannot be complete, because the two nuclei repel each other. The actual bond distance (0.75 Å) is the result of the optimum balance between two opposing forces: (1) the desire for maximum possible overlap of the two $1s$ orbitals, resulting from the attraction of the bonding electrons and both nuclei; and (2) repulsion between the two nuclei.

The molecular orbital treatment of the molecule *starts with the two nuclei already in place. All* the electrons in the molecule are considered to be under the influence of both nuclei. The orbitals in the molecular orbital theory *span both nuclei.* These orbitals are not atomic orbitals (they do not belong to any individual atom in the molecule); their electron density is spread over both nuclei (Figure 14-9). They are called *molecular orbitals.*

It may be recalled (Section 13-4) that we originally derived atomic orbitals by making graphs of mathematical functions that are solutions to the Schrödinger equation. Molecular orbitals are derived in the same way, but in this case the functions are obtained by making linear combinations of the atomic orbital func-

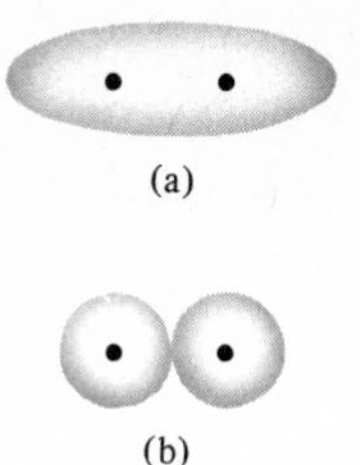

Figure 14-9 (a) The bonding molecular orbital of hydrogen. Note that the electron density is spread around both nuclei, in contrast to (b), a simple overlapping of two atomic orbitals.

tions. In this book, we will not go into the mathematics of this (which get very complicated and normally require the use of computers), but will merely give a qualitative description.

14-18 Molecular Orbitals

We saw in Section 13-6 that even though the Schrödinger equation can be solved exactly only for the hydrogen atom, we have a series of rules that enables us to write the ground-state electronic structure of atoms more complex than hydrogen. For each atom there is available a number of orbitals of varying energy, and all we had to do was to add the proper number of electrons to the available orbitals of lowest energies until all the electrons were used up (the aufbau principle). In a similar manner, we have available to us the results of quantum-mechanical calculations for molecules, which even though not exact are quite helpful. Along with this, there are extensive experimental spectral and magnetic measurements on a large variety of molecules. From all this information we can calculate the energies of molecular orbitals. Once we know them, the method is the same. All we do is insert the available electrons in the orbitals of lowest energies until all the electrons are used up.

The reader will soon be aware of the similarities between the rules for constructing molecular orbitals and the rules for determining the ground-state electronic structure of atoms (Section 13-6).

RULES FOR DETERMINING MOLECULAR ORBITALS OF MOLECULES

1. Molecular orbitals are formed by combining atomic orbitals. When atomic orbitals are combined to form molecular orbitals, **the number of molecular orbitals formed equals the number of atomic orbitals used to form them.** For example, if we combine the 1*s* orbital of a hydrogen atom with the 1*s* orbital of another hydrogen atom, two different molecular orbitals are obtained. For the general case,

$$\begin{matrix} n \text{ atomic orbitals} \\ \text{of one atom} \end{matrix} + \begin{matrix} n \text{ atomic orbitals} \\ \text{of one atom} \end{matrix} \longrightarrow \begin{matrix} 2n \text{ molecular orbitals} \\ \text{for the molecule} \end{matrix}$$

(n stands for some small integer).

2. When two atomic orbitals combine to form two molecular orbitals, *the energies of the two molecular orbitals are different.* One of the molecular orbitals has a lower energy than the sum of the individual atomic orbitals while the other molecular orbital has a higher energy (it is as much higher as the other one is lower) (Figure 14-10). The same kind of symmetry is found whenever molecular orbitals are formed. Half of the molecular orbitals are lower in energy than the sum of the individual energies of the atomic orbitals; the other half are higher in energy.

The molecular orbitals that are lower in energy (compared to the atomic orbitals) are favorable for bonding and are called **bonding molecular orbitals;** the higher-energy molecular orbitals are unfavorable for bonding and are termed **antibonding molecular orbitals.**

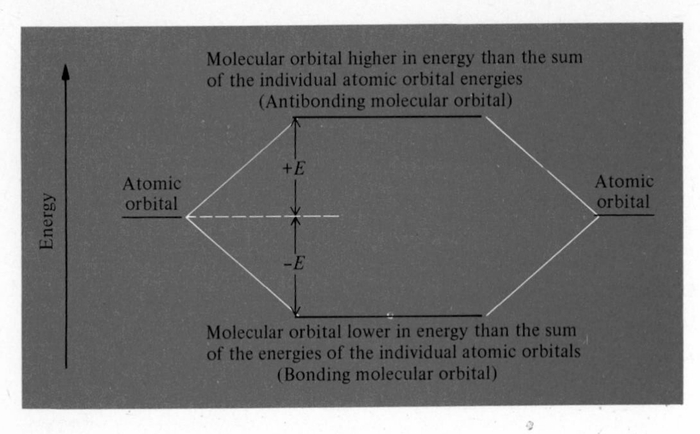

Figure 14-10 The energy of the molecular orbitals formed by the combination of two atomic orbitals. The absolute value of $+E$ is equal to that of $-E$.

If we look at the electron densities of bonding and antibonding molecular orbitals as we did for atomic orbitals, we find that in the bonding orbitals the electron density is much greater between the two atoms than it would be if the atoms existed separately (Figure 14-11). Of course, the electron density is not just located between the atoms, but surrounds the whole molecule. For antibonding orbitals, on the other hand, the electron density between the atoms is less when compared to the individual atoms.

A corollary to rule 2 is that *the number of bonding orbitals in a molecule is equal to the number of antibonding orbitals*. This is so even if some of the bonding and/or antibonding orbitals are not occupied by electrons. The situation is similar to the hydrogen atom, where there are unoccupied $2s$, $2p$, etc. orbitals. The orbitals (atomic and molecular) exist even though they may be vacant.

3. As with atomic orbitals, any single molecular orbital can accommodate a maximum of two electrons. For a completely occupied molecular orbital, the intrinsic spins of the two electrons must be opposed ($s = +\frac{1}{2}$, $s = -\frac{1}{2}$). This is the *Pauli principle for molecules*.

4. When filling the molecular orbitals, we follow the same rule as with atomic orbitals. The first electron is placed in the molecular orbital of lowest energy. The second electron goes into the same orbital (with opposite spin). The next electron goes into the next higher molecular orbital, and so forth. As we discovered in Section

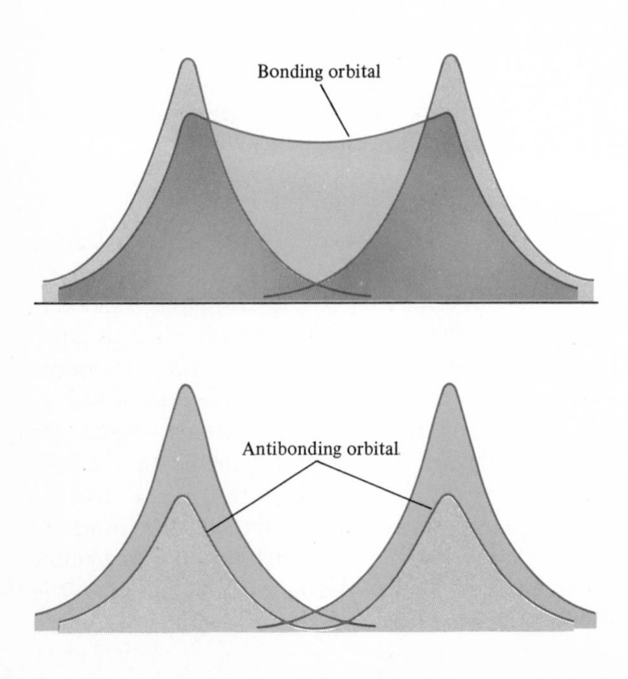

Figure 14-11 Electron-density maps of bonding and antibonding molecular orbitals.

13-6, there are atomic orbitals with the same energy (three *p*, five *d*, etc.); there also are different molecular orbitals possessing the same energy (equivalent orbitals). In this case (as for atoms), when we add electrons to a group of equivalent molecular orbitals, each molecular orbital becomes half-filled before any are filled. This is **Hund's rule for molecules.**

5. Some atoms combine with each other to form molecules; others do not. Molecular orbital theory gives us a rule to determine whether two atoms will combine with each other to form a molecule. To apply this rule, we set up the molecular orbital energy levels and feed in the available electrons from the two atoms according to rules 3 and 4. Next, we add up the number of electrons we have put into bonding orbitals and the number in antibonding orbitals. The rule is: **If the number of electrons in bonding orbitals is greater than the number in antibonding orbitals, a chemical bond will form. Otherwise, no bond will form.** Note that the number of electrons in antibonding orbitals can never exceed the number in bonding orbitals because the bonding orbitals fill first.

When the number of electrons in bonding orbitals is *equal* to the number in antibonding orbitals, no bond will form and the two species remain as atoms. The reason for this is the following: We saw in Section 14-4 that atoms connected to each other by a covalent bond have a lower energy than the unbonded atoms, and they cannot separate unless this energy is supplied to them from an outside source. This lowered energy is what holds the bond together. When bonding and antibonding orbitals contain equal numbers of electrons, the stability gained by having *n* electrons in bonding orbitals is *exactly* offset by the instability of *n* electrons in antibonding orbitals. If there is no stability to be gained, the atoms would rather separate because the *entropy* is greater when they are separated.

In the next section, we describe the energies of molecular orbitals that arise from overlap of 1*s*, 2*s*, and 2*p* orbitals and show the electron configurations for possible molecules formed from the first-row elements H and He. In Section 14-20 we will discuss homonuclear diatomic molecules formed from the second-row elements. We will not explicitly discuss diatomic heteronuclear molecules (those in which the atoms are different) or molecules with three or more atoms, but the principles in these cases are very similar.

14-19 Molecular Orbital Energy Levels

Figure 14-12 is a *relative* energy-level diagram of the molecular orbitals that arise from overlap of 1*s*, 2*s*, and 2*p* orbitals. This is a relative energy diagram because different species have different energy-level values. The important thing in the figure is the *order* of the molecular orbital energy levels.

From Figure 14-12 we see that the two 1*s* orbitals combine to form two molecular orbitals which are labeled σ_{1s} and σ^*_{1s}. The two 2*s* orbitals form the σ_{2s} and σ^*_{2s} molecular orbitals, whereas the two groups of three 2*p* orbitals form six molecular orbitals. The star (*) in the label of an orbital signifies that the orbital is an antibonding orbital. As we previously noted and now find in the figure, the antibonding orbitals are always higher in energy than the corresponding bonding orbitals.

Molecular orbitals are labeled σ or π. The σ molecular orbitals are formed by overlap of two atomic orbitals where the resulting electron density in the bonding orbital is concentrated between the centers of the two nuclei forming the molecule (Figure 14-13a). The π molecular orbitals are formed from two atomic orbitals where the resulting electron density in the bonding orbital is concentrated above and below the centers of the two nuclei forming the molecule (Figure 14-13b). These are the same definitions we used in Section 14-16. Note that *s* atomic orbitals always form σ

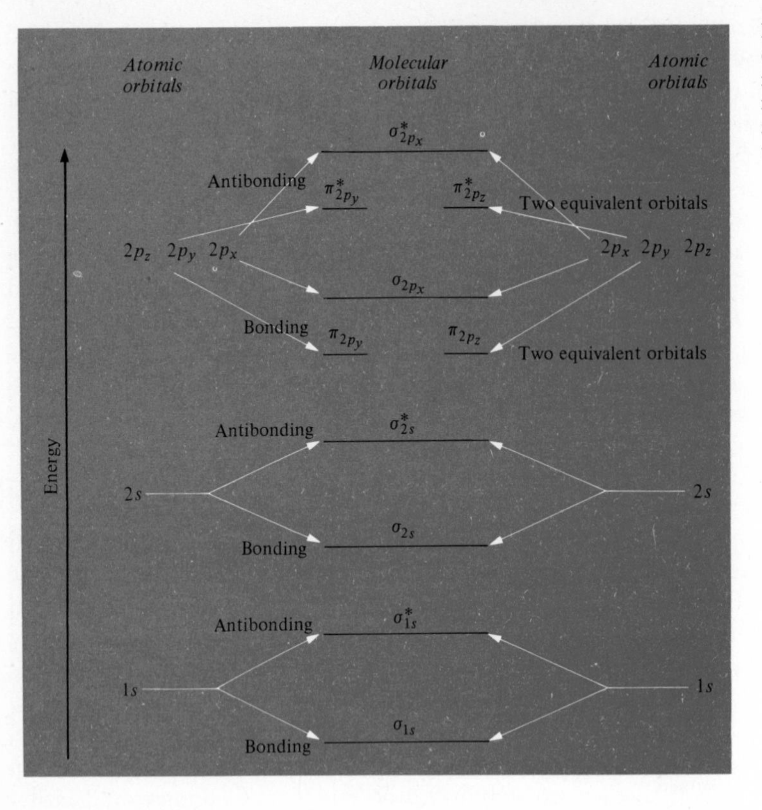

Figure 14-12 Molecular orbitals in order of increasing energy. The molecular orbitals are formed from combinations of atomic orbitals.

molecular orbitals; however, p orbitals may form σ as well as π orbitals (see Figure 14-8).

Each molecular orbital has a subscript that signifies the atomic orbitals from which it was formed. For example, $\sigma^*_{2p_x}$ means a σ antibonding orbital formed from $2p_x$ atomic orbitals.

From Figure 14-12, we see that it is possible to have one, two, three, four, five, or six *more* electrons in bonding orbitals than in antibonding orbitals. The bonds resulting from these situations are called respectively, one electron bonds, two electron bonds, three electron bonds, etc. In valence-bond theory, a single covalent bond contains two electrons, a double bond contains four electrons, and a triple bond six electrons. Molecular orbital theory extends this so that one-electron, three-electron, and five-electron bonds are also possible. A one-electron bond should

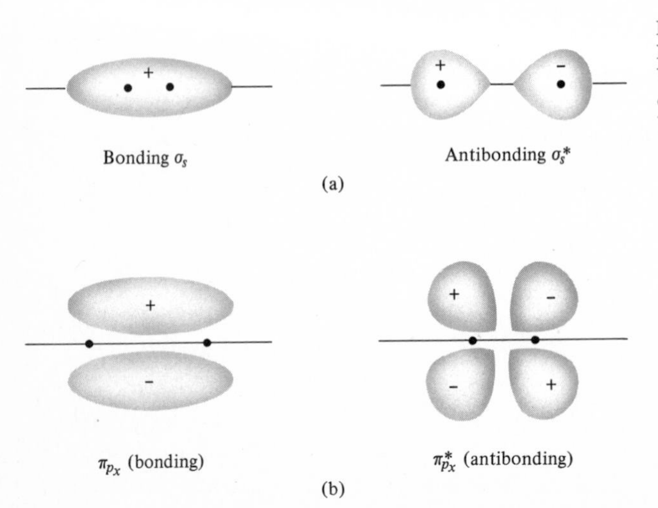

Figure 14-13 (a) Sigma (σ) molecular orbitals (bonding and antibonding). (b) Pi (π) molecular orbitals (bonding and antibonding).

have approximately half the energy of a full covalent bond (two electrons). The **bond order** of a bond is defined as one-half the difference between the number of bonding electrons and antibonding electrons. Thus, a one-electron bond has a bond order of 0.5, a two-electron bond of 1.0, a three-electron bond of 1.5, etc.

With the order of the energy levels given in Figure 14-12, we can consider the formation of homonuclear diatomic molecules from the first two elements.

Example: Describe the bonding in the ground states of the following species: H_2, H_2^+, H_2^-, He_2, He_2^+.

Answer: Since no more than four electrons are present, we will need to consider only the two orbitals of lowest energy, σ_{1s} and σ_{1s}^*.

H_2: 2 electrons — σ_{1s}^*: (empty); σ_{1s}: ↑↓

Hydrogen has two bonding electrons. They are in a σ_{1s} orbital and described as σ_{1s^2}; the superscript shows there are two electrons in this bonding orbital. The bond order is 1.

H_2^+: 1 electron — σ_{1s}^*: (empty); σ_{1s}: ↑ — bond order $= \frac{1}{2}$ — σ_{1s}^1

H_2^-: 3 electrons — σ_{1s}^*: ↑; σ_{1s}: ↑↓ — bond order $= \frac{1}{2}$ — $\sigma_{1s}^2\ \sigma_{1s}^{*1}$

He_2: 4 electrons — σ_{1s}^*: ↑↓; σ_{1s}: ↑↓ — bond order $= 0$ — no bond or molecule is formed

He_2^+: 3 electrons — σ_{1s}^*: ↑; σ_{1s}: ↑↓ — bond order $= \frac{1}{2}$ — $\sigma_{1s}^2\sigma_{1s}^{*1}$

From these results we would conclude that H_2 should possess a full covalent bond (bond order = 1); H_2^+, H_2^-, and He_2^+ should exist, but with weak covalent bonds (bond order $= \frac{1}{2}$); and He_2 should not exist. The results are summarized and compared with the actual data in Table 14-9, which shows that these predictions are borne out, except for H_2^-, for which the prediction has not yet been tested.

Table 14-9 Comparison of Homonuclear Diatomic Molecules Formed from the First-Row Elements

Molecule	*Bonding electrons*	*Antibonding electrons*	*Bond order*	*Bond length*	*Bond energy (kcal/mole)*
H_2	2	0	1	0.74	103
H_2^+	1	0	$\frac{1}{2}$	1.06	61
H_2^-	2	1	$\frac{1}{2}$	data unavailable	data unavailable
He_2	2	2	0	—	—
He_2^+	2	1	$\frac{1}{2}$	1.08	60

Molecular orbital theory gives us a reasonable explanation for the fact that helium does not form a homonuclear diatomic molecule. As shown, the bond order would be zero and the system (He_2) would not have a lower energy when this molecule is formed; therefore, no molecule forms and the system remains as two helium atoms.

14-20 Molecular Orbital Theory of Homonuclear Diatomic Molecules of the Second-Row Elements

We are now ready to apply molecular orbital theory to homonuclear molecules of the second-row elements. For these molecules we will be primarily dealing with the $2s$, and later the $2p$, electrons. In each case, the $1s$ orbitals from both atoms contain four electrons. These are the inner-shell electrons. Two of them enter the σ_{1s} orbital, and the other two, the σ^*_{1s} orbital. This never produces an excess of bonding or of antibonding electrons, so that, even though they enter molecular orbitals, these four electrons will not contribute to nor detract from the bonding properties of the molecules.

Returning to Figure 14-12, we see that the manner of combination of $2s$ atomic orbitals is similar to that of $1s$ atomic orbitals. They both produce a σ and σ^* molecular orbital. The σ_{2s} orbital, being lower in energy than the σ^*_{2s} orbital, will fill first.

The six $2p$ orbitals produce two different types of molecular orbitals. Two of them (say, the $2p_x$ orbitals from each atom) combine to generate two σ_{p_x} orbitals. The other $2p$ orbitals, p_y and p_z, generate four π molecular orbitals (two bonding and two antibonding). These π orbitals, π_{2p_y} and π_{2p_z}, are energetically equivalent and when electrons fill them, Hund's rule for molecular orbitals (rule 4, Section 14-18) will apply. The first electron will go into the π_{2p_y} orbital and the very next electron will enter the π_{2p_z} orbital.

The energy order of the molecular orbitals undergoes a significant change after the N_2 molecule. There is experimental evidence which verifies that the two π_{2p} bonding orbitals are lower in energy than the bonding σ_{2p} orbital up through and including N_2. This statement is a revision of some older ideas concerning molecular-orbital energy levels. The evidence comes from recent spectroscopic and magnetic measurements on B_2 and N_2^+. After N_2 (that is, in O_2 and F_2), there is a *crossover* in energy levels so that the σ_{2p_x} orbital is more stable (lower in energy) than the π_{2p_y} and π_{2p_z} orbitals. Since these levels are completely filled in O_2 and F_2, the crossover will play no role in our further discussions.

We are now ready to examine the homonuclear diatomic molecules Li_2 through Ne_2 (the results are summarized in Table 14-10).

Table 14-10 Molecular-Orbital Energy Levels for the Second-Row Homonuclear Diatomic Molecules

(crossover of molecular-orbital energy levels — between N_2 and O_2)

	Li_2	Be_2	B_2	C_2	N_2	O_2	F_2	Ne_2	
σ^*_{2p}								↑↓	σ^*_{2p}
π^*_{2p}						↑ ↑	↑↓ ↑↓	↑↓ ↑↓	π^*_{2p}
σ_{2p}					↑↓	↑↓ ↑↓	↑↓ ↑↓	↑↓ ↑↓	π_{2p}
π_{2p}			↑ ↑	↑↓ ↑↓	↑↓ ↑↓	↑↓	↑↓	↑↓	σ_{2p}
σ^*_{2s}		↑↓	↑↓	↑↓	↑↓	↑↓	↑↓	↑↓	σ^*_{2s}
σ_{2s}	↑↓	↑↓	↑↓	↑↓	↑↓	↑↓	↑↓	↑↓	σ_{2s}
σ^*_{1s}	↑↓	↑↓	↑↓	↑↓	↑↓	↑↓	↑↓	↑↓	σ^*_{1s}
σ_{1s}	↑↓	↑↓	↑↓	↑↓	↑↓	↑↓	↑↓	↑↓	σ_{1s}

Li_2: 6 electrons $\sigma_{1s}^2\sigma_{1s}^{*2}\sigma_{2s}^2$

The lithium molecule has one single covalent bond (bond order = 1). The Li—Li bond is longer than the H—H bond, which is expected since it is the $n = 2$ electrons in lithium which form the covalent bond rather than the $n = 1$ electrons. For this reason, the bond in Li_2 is much weaker than the H_2 bond.

Be_2: 8 electrons $\sigma_{1s}^2\sigma_{1s}^{*2}\sigma_{2s}^2\sigma_{2s}^{*2}$

There are *no* net bonding electrons for Be_2, and therefore the molecule Be_2 does not exist.

B_2: 10 electrons $\sigma_{1s}^2\sigma_{1s}^{*2}\sigma_{2s}^2\sigma_{2s}^{*2}\pi_{2p_y}^1\pi_{2p_z}^1$

This is the first example of Hund's rule for molecules. The last two electrons go into the π_{2p_y} and π_{2p_z} orbitals. The bond order for this molecule is 1. Additionally, this configuration predicts that the B_2 molecule is paramagnetic (why?), which is borne out by magnetic measurements. If the σ_{2p_x} orbital were more stable than the π_{2p_y} and π_{2p_z} orbitals, the two electrons would be paired in the σ_{2p_x} orbital and the B_2 molecule would be diamagnetic (recall—Section 13-7—that all substances which are not paramagnetic are diamagnetic).

C_2: 12 electrons $\sigma_{1s}^2\sigma_{1s}^{*2}\sigma_{2s}^2\sigma_{2s}^{*2}\pi_{2p_y}^2\pi_{2p_z}^2$

The C_2 molecule has four net bonding electrons. Its bond order is 2. C_2 is diamagnetic, as predicted.

N_2: 14 electrons $\sigma_{1s}^2\sigma_{1s}^{*2}\sigma_{2s}^2\sigma_{2s}^{*2}\pi_{2p_y}^2\pi_{2p_z}^2\sigma_{2px}^2$

N_2 has six net bonding electrons. Its bond order is 3 and it is also diamagnetic. As expected, N_2 has the largest bond energy and the shortest bond length of any second-row homonuclear diatomic molecule.

O_2: 16 electrons $\sigma_{1s}^2\sigma_{1s}^{*2}\sigma_{2s}^2\sigma_{2s}^{*2}\sigma_{2px}^2\pi_{2p_y}^2\pi_{2p_z}^2\pi_{2p_y}^{*1}\pi_{2p_z}^{*1}$

The last two electrons go into the antibonding orbitals $\pi_{2p_y}^*$ and $\pi_{2p_z}^*$, one in each. O_2 has four net bonding electrons and a bond order of 2. Yet it only has a single bond. From its energy-level configuration, we predict that O_2 is paramagnetic since the antibonding electrons are unpaired in the $\pi_{2p_y}^*$ and $\pi_{2p_z}^*$ orbitals. The fact that O_2 is experimentally found to be paramagnetic led to wide acceptance of the molecular orbital theory in the early 1930s, since the valence-bond theory cannot explain this fact.

Note that we have used the crossover of energy levels in writing out the structure for O_2, as previously discussed.

F_2: 18 electrons $\sigma_{1s}^2\sigma_{1s}^{*2}\sigma_{2s}^2\sigma_{2s}^{*2}\sigma_{2px}^2\pi_{2p_y}^2\pi_{2p_z}^2\pi_{2p_y}^{*2}\pi_{2p_z}^{*2}$

Fluorine has a total of two net bonding electrons and a bond order of 1. Bond length and bond energies for the F_2 molecule are as expected for a single bond. F_2 is also diamagnetic.

Ne_2: 20 electrons $\sigma_{1s}^2\sigma_{1s}^{*2}\sigma_{2s}^2\sigma_{2s}^{*2}\sigma_{2px}^2\pi_{2p_y}^2\pi_{2p_z}^2\pi_{2p_y}^{*2}\pi_{2p_z}^{*2}\sigma_{2px}^{*2}$

Ne_2 has no net bonding electrons and a zero bond order. As predicted, there is no Ne_2 molecule.

Table 14-10 lists the molecular orbital energy levels for the second-row homonuclear diatomic molecules. Table 14-11 lists some bond energies and bond lengths for these and additional species.

In this chapter we have discussed ionic and covalent bonds, which are the bonds that hold together atoms in molecules and ions in crystal lattices. Let us remember

Table 14-11 Bond Energies and Bond Lengths of Some Homonuclear Diatomic Molecules and Ions of the Second-Row Elements

Molecule or ion	Bond order	Bond length (Å)	Bond dissociation energy (kcal/mole) ($A_2(g) \rightarrow 2A(g)$; $\Delta E°$)
Li_2	1	2.672	26.3
Be_2	0	does not exist	
B_2	1	1.589	65.5 ± 5
C_2	2	1.2425	144
N_2	3	1.0976	225.07
N_2^+	2.5	1.116	201.28
O_2	2	1.20741	117.96
O_2^+	2.5	1.1227	—
O_2^-	1.5	1.26	93.9
O_2^{2-}	1	1.49	—
F_2	1	1.417	33.2 ± 1.6
Ne_2	0	does not exist	

that there are also weaker forces—those between one molecule and another, between an ion and a dipolar molecule, etc. These forces, which were discussed in Section 6-6, are much weaker than covalent or ionic bonds, but are important because of the effect they have on the properties of matter. For example, they are responsible for the fact that gases liquify under pressure, that sugar is soluble in water, and many other physical and chemical phenomena.

Problems

14-1 Define each of the following terms. **(a)** chemical bond **(b)** molecule **(c)** ionic bond **(d)** covalent bond **(e)** unshared pair **(f)** Lewis structure **(g)** octet rule **(h)** valence **(i)** molecular orbital **(j)** polyatomic ion **(k)** free radical **(l)** dimer **(m)** expanded shell **(n)** coordinate covalent bond **(o)** Lewis acid **(p)** Lewis base **(q)** coordination compound **(r)** partial ionic character **(s)** electronegativity **(t)** polar covalent bond **(u)** percent ionic character **(v)** double bond **(w)** triple bond **(x)** bond energy **(y)** structural formula **(z)** formal charge **(aa)** resonance **(bb)** canonical form **(cc)** localized electrons **(dd)** delocalized electrons **(ee)** electrolyte **(ff)** EPRT **(gg)** hybrid orbitals **(hh)** sp^3 orbital **(ii)** σ (sigma) bond **(jj)** π (pi) bond **(kk)** molecular orbital **(ll)** bonding orbital **(mm)** antibonding orbital **(nn)** Hund's rule for molecules **(oo)** bond order

14-2 What is the ionic valence of each of the following elements? **(a)** F **(b)** Cs **(c)** Se **(d)** Sr **(e)** Ca **(f)** Te **(g)** At **(h)** Fr

14-3 Why do we always write ionic valences with + or − signs, but covalences without any sign?

14-4 Which of the following compounds do not obey the octet rule? For each such compound, tell how many electrons are in the valence shell of each atom. **(a)** CBr_4 **(b)** BH_3 **(c)** H_2Se **(d)** $AlCl_3$ **(e)** IF_7 **(f)** CO **(g)** $BeCl_2$ **(h)** $AsBr_5$ **(i)** $\cdot CCl_3$ **(j)** O_2 **(k)** TeF_6

14-5 Why do the two atoms in the H_2 molecule stay together? Why don't they fly apart?

14-6 Which of the following contain coordinate covalent bonds? **(a)** $BeCl_4^{2-}$ **(b)** CCl_4 **(c)** BH_4^- **(d)** Cl_2 **(e)** O_2^{2-} **(f)** CO_2 **(g)** PH_4^+ **(h)** $Co(NH_3)_6^{3+}$ **(i)** H_3O^+ **(j)** NaCl **(k)** N_2H_4

14-7 If possible, write the formula for each of the following.
(a) a Lewis acid that is not a Brønsted acid
(b) a Brønsted acid that is not a Lewis acid
(c) a Lewis base that is not a Brønsted base
(d) a Brønsted base that is not a Lewis base

14-8 Which of the following can be (1) Lewis acids, (2) Lewis bases, or (3) neither?
(a) $AlCl_3$ **(b)** PH_3 **(c)** H_2O **(d)** CH_4 **(e)** BH_3 **(f)** H^+ **(g)** Ni^{2+} **(h)** NCl_3
(i) SnF_4 **(j)** H_2 **(k)** SO_4^{2-}

14-9 For each of the following compounds, indicate whether the bonds in the molecule are ionic, polar, or "pure" covalent. Indicate partial charges on the atoms if there are any. **(a)** HI
(b) CCl_4 **(c)** H_2S **(d)** CaO **(e)** K_2O **(f)** CO_2 **(g)** OF_2 **(h)** NH_4NO_3
(i) $KClO_4$ **(j)** $AlCl_3$ **(k)** I_2

14-10 From the following data, calculate the percent ionic character for the following:

Species	μ *(D)*	*Bond distance (Å)*
(a) HF	1.8	0.917
(b) HI	0.38	1.61
(c) NO	0.1	1.15
(d) NaCl (molecule in the vapor phase)	9	2.36

14-11 Which of the following molecules would have a dipole moment? In each case where there is a dipole moment, show the positive and negative ends of the molecule. **(a)** AgI
(b) CH_4 **(c)** CH_3OH **(d)** $HCCl_3$ **(e)** H_2O **(f)** I_2 **(g)** CO_2 **(h)** NH_3
(i) BF_3 **(j)** $BeCl_2$

14-12 Account for the fact that the average energy necessary to break the CN bond in H_3CNH_2 is 4.96×10^{-12} erg/bond, while for the bond in HCN it is 1.22×10^{-11} erg/bond.

14-13 If the average energy necessary to break the CC bond in H_2CCH_2 is 1.15×10^{-11} erg/bond, what would you predict for HCCH and H_3CCH_3?

14-14 Judging by the values in Table 14-4, arrange these bonds in predicted order of bond distance and bond energy: N—Cl, N—Br, N—F, N—I.

14-15 What are the total numbers of valence electrons in the following? **(a)** H_2O
(b) C_3H_8 **(c)** N_2H_4 **(d)** NO_2 **(e)** NO_2^- **(f)** PO_4^{3-} **(g)** $CH_3NH_3^+$

14-16 Write Lewis structures for each of the following. (Don't forget the formal charges, if any.) In each case, there is only one way the atoms can be connected. **(a)** I_2 **(b)** N_2
(c) Cl^- **(d)** CN^- **(e)** C_3H_8 **(f)** N_2H_4 **(g)** CH_2Cl_2 **(h)** BF_3 **(i)** BF_4^-
(j) CH_4O **(k)** O_2^{2-} **(l)** CH_3 (free radical)

14-17 Write Lewis structures for each of the following. (Don't forget the formal charges, if any.) Connections are given where needed.
(a) ICl
(b) C_2H_4
(c) $POCl_3$ (O above P; Cl P Cl; Cl below P)
(d) SiO_3^{2-} (O Si O; O below Si)
(e) CCl_4
(f) HCN (H C N)

(g) HNC (H N C)

(h) H_2O_2

(i) CH_4O (H C C H; H above first C; H O below)

(j) $C_2H_4O_2$ (H C C O H; H above first C; H O below)

(k) $H_2BO_3^-$ (H O B O H; O below B)

(l) $Cr_2O_7^{2-}$ (O Cr O Cr O; O above and below each Cr)

(each Cr atom has six electrons available for bonding)

14-18 In each case, tell what, if anything, is wrong with the Lewis structure given.

(a) HCl $\overset{\ominus}{H}=\overset{\oplus}{\underline{\overline{Cl}}}$

(b) HNO_2 $H-\overset{\oplus}{\underline{O}}=\overset{\ominus}{\overline{N}}=\underline{\overline{O}}$

(c) CH_2N_2 $H-\overset{\oplus}{C}(H)-\overset{\ominus}{N}\equiv\overline{N}$

(d) SO_3^{2-} $|\underline{\overline{O}}-\overline{S}(=O|)-\underline{\overline{O}}|$

(e) SO_3 $|\overline{O}=S(=O|)=\underline{O}|$

(f) HNO_2 $H-\underline{\overline{O}}-\overset{\ominus}{\underline{\overline{N}}}-\overset{\ominus}{\underline{\overline{O}}}|$

(g) SO_2 $\overset{\ominus}{|\underline{O}}-\overset{2+}{\underline{S}}-\overset{\ominus}{\underline{\overline{O}}}|$

(h) HOBr $H-\overset{\ominus}{\underline{O}}=\overset{\oplus}{\underline{\overline{Br}}}$

(i) OCl^- $|\overline{O}=\overset{\ominus}{\underline{\overline{Cl}}}|$

(j) SeO_3^{2-} $\overset{\ominus}{|\underline{O}}-\overset{2+}{Se}(=O|)-\overset{\ominus}{\underline{\overline{O}}}|$

(k) AsOCl $|\underline{\overline{Cl}}-\overset{\ominus}{\underline{\overline{As}}}-\overset{\ominus}{\underline{\overline{O}}}|$

(l) SiO_4^{2-} $\overset{\ominus}{|\underline{\overline{O}}}-\overset{\oplus}{Si}(-\overset{\ominus}{\overline{O}}|)(=O|)-\overset{\ominus}{\underline{\overline{O}}}|$

14-19 Draw all the reasonable valid canonical forms for each of the following. (Don't forget the formal charges, if any.) Connections are given in each case.

(a) CHO_2^- (H C O; O below C)

(b) CO_3^{2-} (O C O; O below C)

(c) O_3 (O O O)

(d) N_2O_4 (O N N O; O O below the N atoms)

(e) SO_3 (O S O; O below S)

(f) O_2NF (O N F; O below N)

(g) $C_4O_4^{2-}$ (O at four corners; C C / C C in a square at center)

14-20 Draw Lewis structures for two possible N_2O molecules (N N O and N O N). Do either or both display resonance?

14-21 The following are canonical forms written by a not-so-perfect student for diazomethane, CH_2N_2. Which are (1) reasonable and correct canonical forms which contribute substantially to the real molecule; (2) correct forms, but unreasonable, and so contribute little; (3) incorrect forms which violate the rules and do not contribute at all? Give reasons for your answers.

(a) $H-\overset{\ominus}{C}(H)-\overset{2+}{N}=\overset{\ominus}{N}$ **(b)** $H-\overset{\ominus}{C}(H)-\overset{\oplus}{N}\equiv N$ **(c)** $H-\overset{\ominus}{C}(H)-N=\overset{\ominus}{N}$

(d) $H-\overset{\ominus}{C}(H)=N=\overset{\oplus}{N}$ **(e)** $H-C(H)=\overset{\oplus}{N}=\overset{\ominus}{N}$ **(f)** $\overset{\oplus}{H}\ \ C(H)\equiv\overset{\oplus}{N}-\overset{2-}{N}$

14-22 The following are canonical forms written by another not-so-perfect student for the bicarbonate ion, HCO_3^-. Place them into the same three categories as in Problem 14-20. Give reasons for your answers.

(a) $H-O-\overset{\oplus}{C}(-\overset{\ominus}{O})-\overset{\ominus}{O}$ **(b)** $H-O-\overset{\ominus}{C}(=O)=O$ **(c)** $H-\overset{\oplus}{O}=C(-\overset{\ominus}{O})-\overset{\ominus}{O}$ **(d)** $H-O-C(-\overset{\ominus}{O})=O$

(e) $\overset{\ominus}{O}-C(H)(-O)-O$ **(f)** $H-\overset{2+}{O}-C(-\overset{\ominus}{O})=O$ **(g)** $H-O-C(=O)-\overset{\ominus}{O}$

14-23 **(a)** Draw two canonical forms for the acetate ion, $C_2H_3O_2^-$ $\left(\begin{matrix} & H & & \\ H & C & C & O \\ & H & O & \end{matrix}\right)$. **(b)** Which of these two is the structure of the *real* acetate ion?

14-24 Which, if any, of these formulas disobey the covalent valence rules given in Section 14-11? **(a)** $TeBr_4$ **(b)** PF_5 **(c)** NCl_5 **(d)** CCl_2 **(e)** ICl_3 **(f)** CBr_3^- **(g)** OF_2 **(h)** NBr_4^+ **(i)** NBr_2^- **(j)** SeO_3 **(k)** OF_4

14-25 Some of the following formulas are incorrect (that is, do not represent any real compound) because they violate the valence rules. Which are they, and which atoms have incorrect valences?

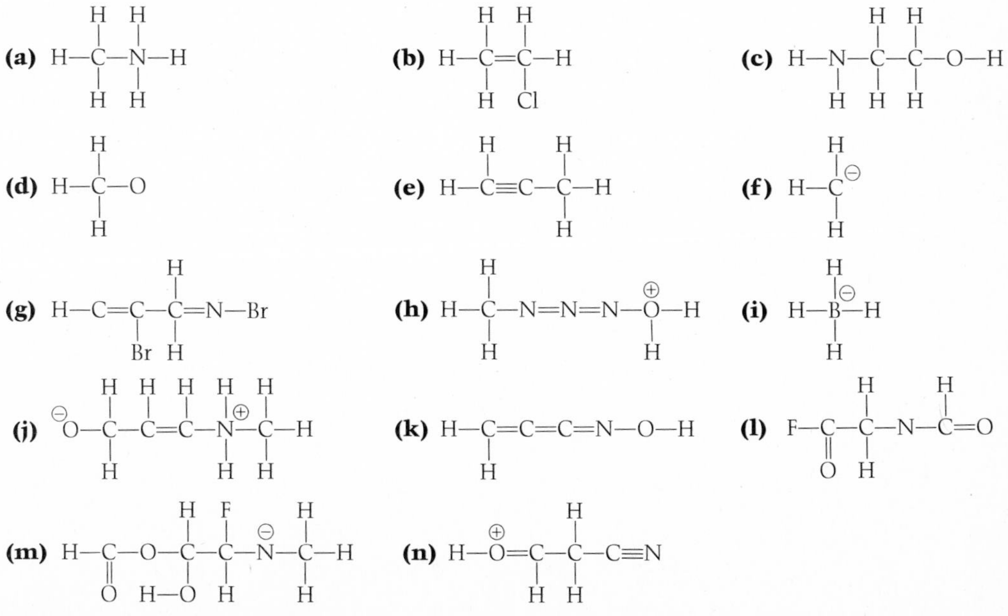

14-26 Predict which of the following is likely to be soluble in (1) water, (2) benzene, (3) both, or (4) neither. **(a)** KCl **(b)** C_2H_6 **(c)** diamond **(d)** formaldehyde, $H_2C{=}O$ **(e)** $BaSO_4$ **(f)** CCl_4 **(g)** methanol, CH_3OH **(h)** glucose, $C_6H_{12}O_6$ **(i)** Dacron (a polymeric molecule) **(j)** P_2S_5 **(k)** NO **(l)** Na_2S **(m)** chloroform, $CHCl_3$

14-27 Classify each of the following as (1) strong electrolyte, (2) weak electrolyte, or (3) nonelectrolyte. **(a)** HBr **(b)** $Mg(OH)_2$ **(c)** KNO_3 **(d)** methanol, CH_3OH **(e)** AgCl **(f)** $HClO_4$ **(g)** HOAc **(h)** LiOH **(i)** acetone CH_3COCH_3 **(j)** HF **(k)** NH_3 **(l)** CCl_4

14-28 Use the EPRT to predict the geometry of each of the following molecules and ions. **(a)** BCl_3 **(b)** PCl_5 **(c)** CO_2 **(d)** SbF_6^- **(e)** CH_2Cl_2 **(f)** BeH_2 **(g)** TeF_6 **(h)** H_3O^+ **(i)** XeF_2 **(j)** ClF_3 **(k)** BrF_5 **(l)** ICl_4^- **(m)** XeF_4 **(n)** ClO_4^-

14-29 Use the EPRT to predict the geometry of each of the following molecules or ions. **(a)** NF_3 **(b)** SO_3 **(c)** $AlCl_3$ **(d)** O_3 (O O O) **(e)** $HC{\equiv}CH$ **(f)** $GeCl_2$ **(g)** $CH_3{-}O{-}CH_3$ **(h)** NO_3^- **(i)** SO_4^{2-} **(j)** $CH_2{=}NH$ **(k)** Cl_2CO (Cl C Cl / O)

14-30 What are the differences between a σ bond and a π bond?

14-31 What angles exist between the hybrid orbitals in each of the following hybrid orbitals. **(a)** sp^3 **(b)** sp^2 **(c)** sp

14-32 State which kind of hybridization the carbons use in each of the following molecules. **(a)** $CH_2{=}CH_2$ **(b)** $CH_3{-}CH_3$ **(c)** $HC{\equiv}CH$ **(d)** CO_2

14-33 The ground-state electronic configuration of silicon is $[Ne]3s^23p^2$. Account for the fact that silicon has a valence of 4 and that all four bonds of a molecule such as silane (SiH_4) are equivalent.

14-34 In the compound ethylene (Section 14-16), why is the $2p_z$ orbital perpendicular to the plane of the three sp^2 orbitals?

14-35 Discuss the hybridization and bonding in CO_2, in the same way as we discussed it for acetylene in Section 14-16.

14-36 What are the similarities and differences between bonding and antibonding orbitals?

14-37 According to the molecular orbital theory, what condition is necessary to form a covalent bond?

14-38 What is the Pauli principle for atoms? What is the Pauli principle for molecules?

14-39 Write molecular orbital electronic configurations for **(a)** O_2^+ **(b)** O_2^-, and **(c)** O_2^{2-}, and show how they verify the bond orders given in Table 14-11. Which, if any, are paramagnetic?

14-40 Using molecular orbital theory, describe the bonding in the following species, and predict the bond orders and whether each species will be paramagnetic. **(a)** Be_2^- **(b)** Be_2^+ **(c)** N_2^{2+} **(d)** F_2^{2-} **(e)** C_2^+ **(f)** C_2^- **(g)** O_2^{2+}

Additional Problems

14-41 In each case tell whether the bonds in these elements or binary compounds are ionic, "pure" covalent, or polar covalent. If the latter, tell which atom(s) has the partial positive charge. **(a)** HCl **(b)** KF **(c)** Br_2 **(d)** IBr **(e)** BeO **(f)** BeS **(g)** PH_3 **(h)** MgI_2 **(i)** $MgCl_2$ **(j)** H_2O **(k)** N_2 **(l)** SF_6 **(m)** CsI

14-42 In the formula $\mathrm{H{-}\bar{\underline{O}}{-}\overset{\oplus}{N}({=}\bar{O}|^{-\frac{1}{2}}){-}\underline{O}|^{-\frac{1}{2}}}$, why are two oxygen atoms each given a charge of $-\frac{1}{2}$?

14-43 State which kind of hybridization the nitrogens use in each of the following molecules. **(a)** HCN **(b)** NH_3 **(c)** HN=NH **(d)** N_2H_4 **(e)** $CH_3{-}N(CH_3){-}CH_3$

14-44 There are a few compounds known which contain C=S bonds. Predict the average bond distance and bond energy of this bond.

14-45 The nitrite ion is NO_2^-. Nitrogen dioxide is NO_2. **(a)** Draw Lewis structures for these two species. (In both cases the connections are O N O) **(b)** Why does NO_2^- have a negative charge, while NO_2 is neutral?

14-46 How do the Lewis and Brønsted theories account for: **(a)** the reaction between HCN and OH^-? **(b)** the reaction between Ag^+ and NH_3 to give $Ag(NH_3)_2^+$?

14-47 Explain why CCl_4 is tetrahedral but $SeCl_4$ is not.

14-48 Write molecular orbital electronic configurations for the heteronuclear molecules and ions given below and tell which species would not exist. For each species, give the bond order and tell whether you expect it to be paramagnetic. **(a)** HHe **(b)** HHe^+ **(c)** HHe^- **(d)** HLi **(e)** LiBe **(f)** BBe^{2+} **(g)** BBe^{2-}

14-49 Would you expect NaI or I_2 to have a higher melting point? Explain.

14-50 Is the valence of carbon in carbon monoxide (CO) in accord with the valence rules given in Section 14-11 (that is, the valence of carbon is 4 unless it is charged or has an unpaired electron)?

14-51 Use the EPRT to predict the geometry of each of the following molecules or ions. **(a)** NH_4^+ **(b)** NH_3 **(c)** NH_2^+ **(d)** NH_2^- **(e)** $TeCl_4$ **(f)** $TeCl_3^+$ **(g)** I_3^-

14-52 Draw all the reasonable valid canonical forms for each of the following species. (Don't forget the formal charges, if any.) In some cases there may be only one reasonable form. Connections are given in each case.

(a) SO_2 (O S O) **(b)** SO_4^{2-} (O S O, with O above and O below S) **(c)** SO_3^{2-} (O S O, with O below S)

(d) N_3^- (N N N) **(e)** CNO^- (C N O) **(f)** OCN^- (O C N)

(g) NO_3^- (O N O, with O below N) **(h)** H_2SO_4 (H O S O H, with O above and O below S)

14-53 There are six chlorinated ethylenes each of which, like ethylene itself, is a flat planar molecule. Which of these have dipole moments and which do not?

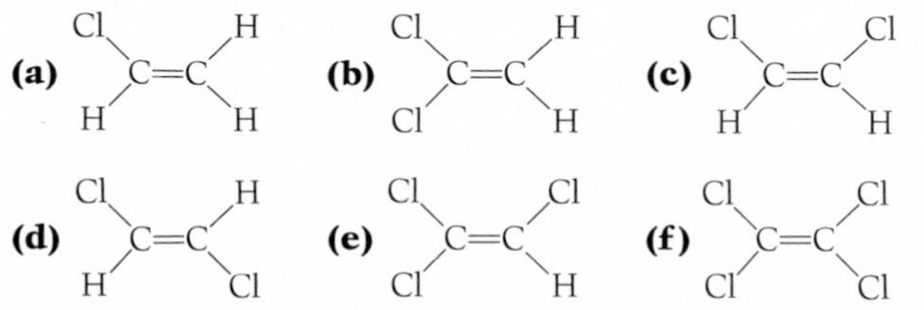

14-54 Predict which of the following elements or compounds will be high-melting solids at room temperature. **(a)** H_2 **(b)** CCl_4 **(c)** NaBr **(d)** boron carbide (which has a diamond-like structure) **(e)** K_2SO_4 **(f)** NH_3 **(g)** IBr **(h)** PH_3 **(i)** CsF

14-55 Write Lewis structures for each of the following. (Don't forget the formal charges, if any.) Connections are given in each case.

(a) CO_2 (O C O)

(b) PO_4^{3-} (O P O, with O above and O below P)

(c) H_3PO_4 (H O P O H, with O above P and O–H below P)

(d) ClO_4^- (O Cl O, with O above and O below Cl)

(e) $COCl_2$ (Cl C Cl, with O below C)

(f) HNO_2 (H O N O)

(g) CH_2O (H C H, with O above C)

(h) C_2H_6O (H C C O H, with H above and H below each C)

(i) C_2H_6O (H C O C H, with H above and H below each C)

14-56 The ground-state electron configuration of the chromium atom is [Ar] $3d^5 4s^1$. **(a)** Show how chromium can promote three of its $3d$ electrons to the $4p$ level and then hybridize all six orbitals to form six equivalent hybrid orbitals, called d^2sp^3. **(b)** What orientation in space can you predict for these six hybrid orbitals?

15 Solids

William Lawrence Bragg, 1890–1971
William Henry Bragg, 1862–1942

Edgar Fahs Smith Collection

In Chapter 4, we characterized the gaseous state as the state of matter that shows almost complete *disorder* (or very high entropy). At the other extreme, the solid state is characterized by almost complete *order* (or very low entropy). Solids are orderly because almost all the atoms, molecules, or ions that make up solids are confined to small regions in the solid. They cannot move very far away from their positions without disrupting the whole structure of the solid. The atoms, molecules, or ions can be present in several different three-dimensional arrangements in space, and, as we shall see, these different three-dimensional arrangements of atoms, molecules, or ions result in the variety of geometrical structures we find in solids. It is their different structures, their compositions, as well as the different kinds of bonding found that are responsible for the way solids behave. For example, diamond has a melting point that is higher than 3550°C. It is composed of a single element, carbon, arranged in a three-dimensional network, where each carbon atom is connected to four other carbon atoms by covalent bonds (Figure 15-1a). In contrast, solid argon melts at −189°C. It has a cubic close-packed structure (Figure 15-1b) and the atoms are held together by weak van der Waals attractive forces.

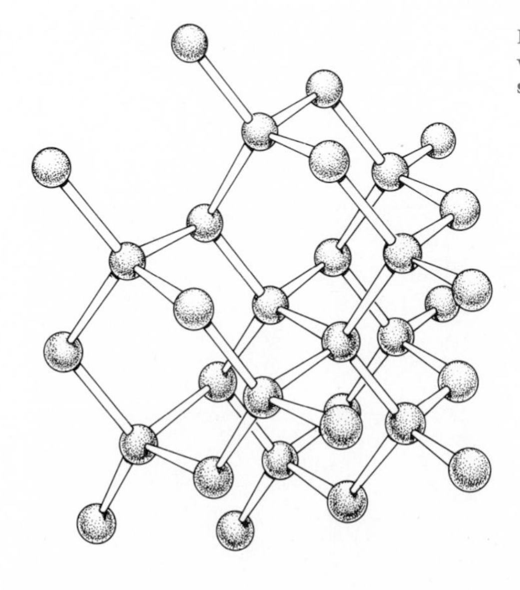

(a)

(b)

Figure 15-1 (a) The three-dimensional network of diamond. (b) Cubic close-packed structure of solid argon.

15-1 X-ray Crystallography

Hundreds of years ago it was noticed that many crystals have regular shapes—hexagonal, cubic, etc. (Figure 15-2). In the eighteenth century a number of scientists proposed that these regular shapes were the result of a regular *internal* arrangement of particles. It was not easy to prove this idea, and science had to wait almost 200 years for the birth of a new experimental tool which allows us to "observe" the internal structure of solids. In this method, called **x-ray crystallography,** x rays are allowed to come into contact with a sample of solid matter. The x rays bounce off the atoms, molecules, or ions at certain angles. From measurements of these angles, and of the intensities of the x rays coming off, x-ray crystallographers gain information about the internal structure of the crystals. X-ray crystallography was developed in a truly serendipitous manner. A scientific conference was held in Munich in 1912. Among the many topics on the program, there were separate discussions of both crystallography and of x rays. One of the participants, Max von Laue (1879–1960), a German physicist, suggested that the interatomic spacings in crystals were of the same order of magnitude as the wavelengths of x rays. If this were so, then if x rays were passed through a crystal, a diffraction pattern would result (Figure 15-3). In the same year, Friedrich and Knipping performed the experiment on a crystal of zinc sulfide and it worked perfectly. Additional important

Figure 15-2 Some crystals that have regular shapes.

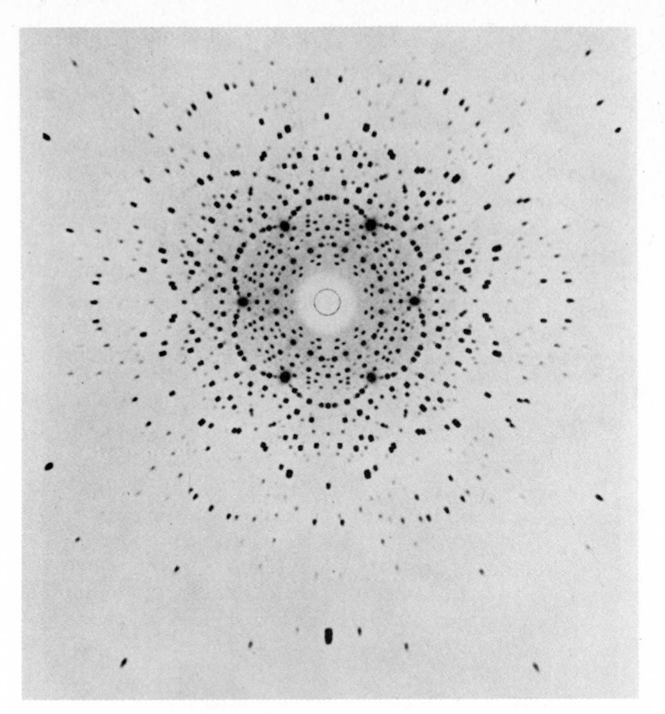

Figure 15-3 X-ray diffraction pattern produced by shining x rays on a crystal of beryl. [Courtesy of Eastman Kodak Research Laboratories]

contributions were made by William H. Bragg (1862–1942) and William L. Bragg (1890–1971) (father and son). A new and very powerful experimental tool was created.

Since that period, x-ray crystallographers have examined thousands of crystals and have been successful in elucidating the structure of a wide variety of solids, ranging from simple inorganic salts to complex proteins. In Chapter 21 we will discuss the structure of proteins and nucleic acids and the contributions made by x-ray crystallography. Figure 15-4 shows a typical modern x-ray diffraction device.

By the use of x-ray crystallography we know that almost all true solids consist of atoms, ions, or molecules arranged in a *regular* three-dimensional array. This regular array continues throughout the entire piece of solid, and shows up in the shape of the crystal itself. That is, the shape of the crystal is dependent on which particles are present and how they are arranged. Nearly all true solids are **crystalline** (that is, made up of crystals). Often the crystals are too tiny to be visible without a microscope or are so imperfect that it is difficult to make out their shapes, but the crystals are nearly always there. A few solids, called **amorphous** solids, are not made up of crystals at all. Some materials seem to be solids, but actually are not. An important example is glass. In spite of its appearance, glass is not a true solid, but actually a very viscous liquid! It has no regular array of atoms, ions, or molecules, and no melting point. You can demonstrate the liquid nature of glass by placing a glass plate onto two supports (Figure 15-5). After some weeks you will notice that the glass has sagged.

The perfect three-dimensional arrangement of atoms, ions, or molecules very seldom continues throughout an entire crystal (nearly all crystals contain some imperfections), but for the most part the stacking of atoms, ions, or molecules in a crystal is very regular.

Figure 15-4 A Modern Automatic X-ray Diffractometer. From left to right: the generator and diffractometer, the computer console, and the disc storage unit. [Courtesy of Syntex Analytical Instruments, Inc.]

15-2 Types of Unit Cells

The primary information obtained from an analysis of x-ray data concerns the unit cell of the crystal. A **unit cell** is the fundamental building block of the crystal. It is the smallest part of the crystal which if repeated by translation (movement) in three-dimensional space would reproduce the entire crystal.

To understand what a unit cell is, we will first look at a piece of wallpaper (Figure 15-6). Like most wallpaper, Figure 15-6 contains a *repeating pattern*. We can draw a parallelogram on one portion of the wallpaper and reproduce the entire pattern by translation of this parallelogram over the entire surface of the paper, provided that we have made our parallelogram large enough to include one complete unit of the pattern. We can use the word **motif** to refer to that part of the pattern that lies within a single parallelogram. The entire wallpaper is made up of this single motif, repeated as many times as is necessary to cover the whole wall.

We can look at a crystal in the same way, except that here the motif is repeated in three dimensions rather than two. In a crystal, the motif may consist of a single atom, ion, or molecule, or two, or three, or many, but it must be large enough to contain a complete unit of the pattern.

In order to picture the structure of crystals, it is helpful to imagine an array of points in space called a **space lattice.** Two such lattices, where the points (called **lattice points**) are connected by straight lines, are shown in Figure 15-7. As

Figure 15-5 A glass plate placed over two supports will slowly sag.

Figure 15-6 Wallpaper. The pattern is repeated by translation of the motif (outlined in color).

shown in the figure, the lattice points are arranged in a regular repeating order. We set up a different space lattice for each kind of crystal, choosing the distances and angles between the lattice points so as to suit the particular crystal involved. There is one lattice point for each motif. We may choose to put a lattice point anywhere we like within a motif, but then we must put all the other lattice points at the same relative place in every motif. The lattice points themselves are geometric points that occupy no volume.

When choosing a lattice we must be careful that there is one and only one lattice point for each motif. For example, consider the NaCl crystal, which has the structure shown in Figure 15-8. It might appear that we could choose single ions of Na^+ and Cl^- to be our lattice points (Figure 15-8a). This is not correct, because Na^+, for

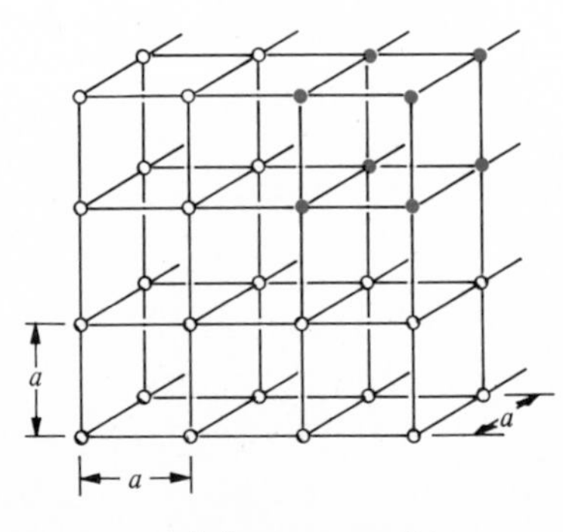

All distances equal
All angles = 90°

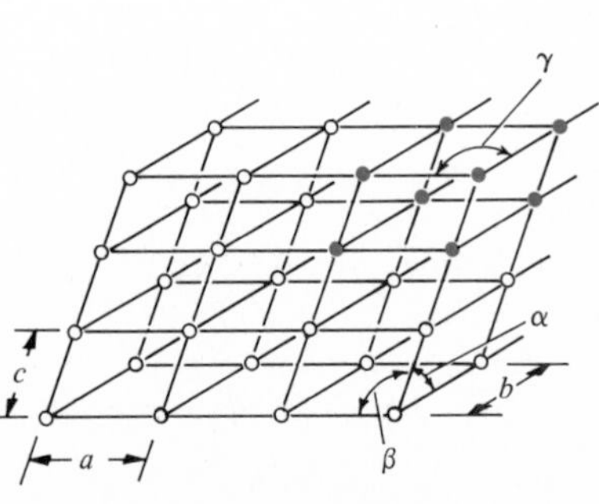

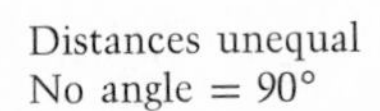
Distances unequal
No angle = 90°

Figure 15-7 Two three-dimensional space lattices. The unit cells are shown in color.

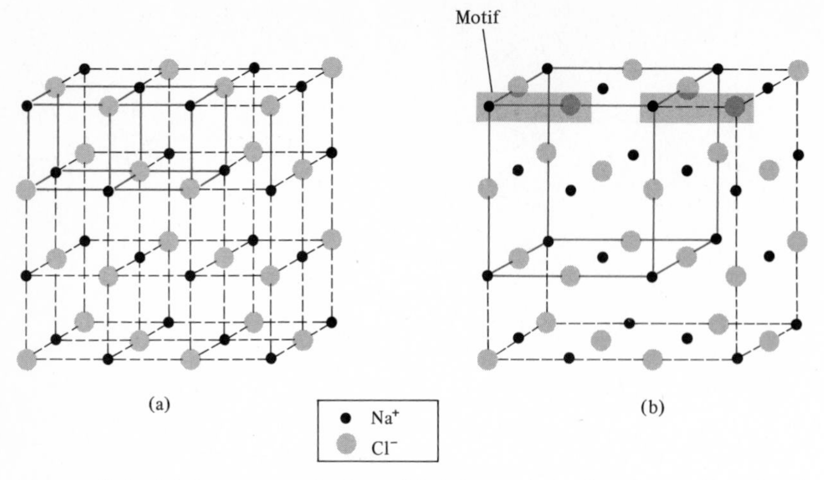

Figure 15-8 (a) Incorrect choice of lattice points for the unit cell of the NaCl crystal. (b) Correct choice of lattice points for the unit cell of the NaCl crystal.

example, is not an entire motif; that is, repetition of this ion will not reproduce the entire crystal. The correct situation is shown in Figure 15-8b. The motif contains *two* ions: one Na^+ and one Cl^-. The lattice point may be chosen anywhere we wish within this motif: for example, in the center of the Na^+, or of the Cl^-, or halfway between them, or anywhere else, as long as we choose the same place in each motif (in Figure 15-8b we have chosen the center of the Na^+ ion).

A *unit cell* consists of the space within eight lattice points which occupy the corners of a cube, or other parallelepiped (Figure 15-7). A. Bravais (1848) showed that all space lattices can be divided into 14 types. Table 15-1 lists the 14 Bravais lattices, and the 14 different structures are depicted in Figure 15-9. From Table 15-1 we see that there are seven basic crystal systems which are identifiable by these unit-cell axes and angles. The most regular is the cubic lattice, with all axes and angles equal, while the triclinic system is the least regular, with no axes or angles equal. As shown in the figure and table, there are three kinds of cubic lattice. In the simplest, called **primitive,** (designated P), the unit cell consists only of the volume defined by the eight lattice points at the corners of the cube. In the other two types the unit cell contains additional lattice points. The **body-centered cubic** lattice (designated I) has a ninth lattice point at the center of the cube, while the **face-centered cubic** lattice (F) has 14 lattice points in the unit cell: eight at the corners and one more at the center of each face of the cube. The monoclinic system occurs in two types: the primitive system, and the **end-centered** (C) system, with an additional lattice point at each end of the unit cell. Similarly, the

Table 15-1 The 14 Bravais Lattices[a]

System	*Unit-cell axes and angles*		*Lattice symbol*			
Cubic	$a = b = c$	$\alpha = \beta = \gamma = 90°$	P	I	F	
Tetragonal	$a = b \neq c$	$\alpha = \beta = \gamma = 90°$	P	I		
Orthorhombic	$a \neq b \neq c$	$\alpha = \beta = \gamma = 90°$	P	I	F	C
Monoclinic	$a \neq b \neq c$	$\alpha = \gamma = 90° \neq \beta$	P	C		
Rhombohedral (trigonal)	$a = b = c$	$\alpha = \beta = \gamma \neq 90°$	P			
Hexagonal	$a = b \neq c$	$\alpha = \beta = 90°;\ \gamma = 120°$	P			
Triclinic	$a \neq b \neq c$	$\alpha \neq \beta \neq \gamma \neq 90°$	P			
				14		

[a]The meaning of the letters a, b, c, α, β, and γ is shown in Figure 15-7.

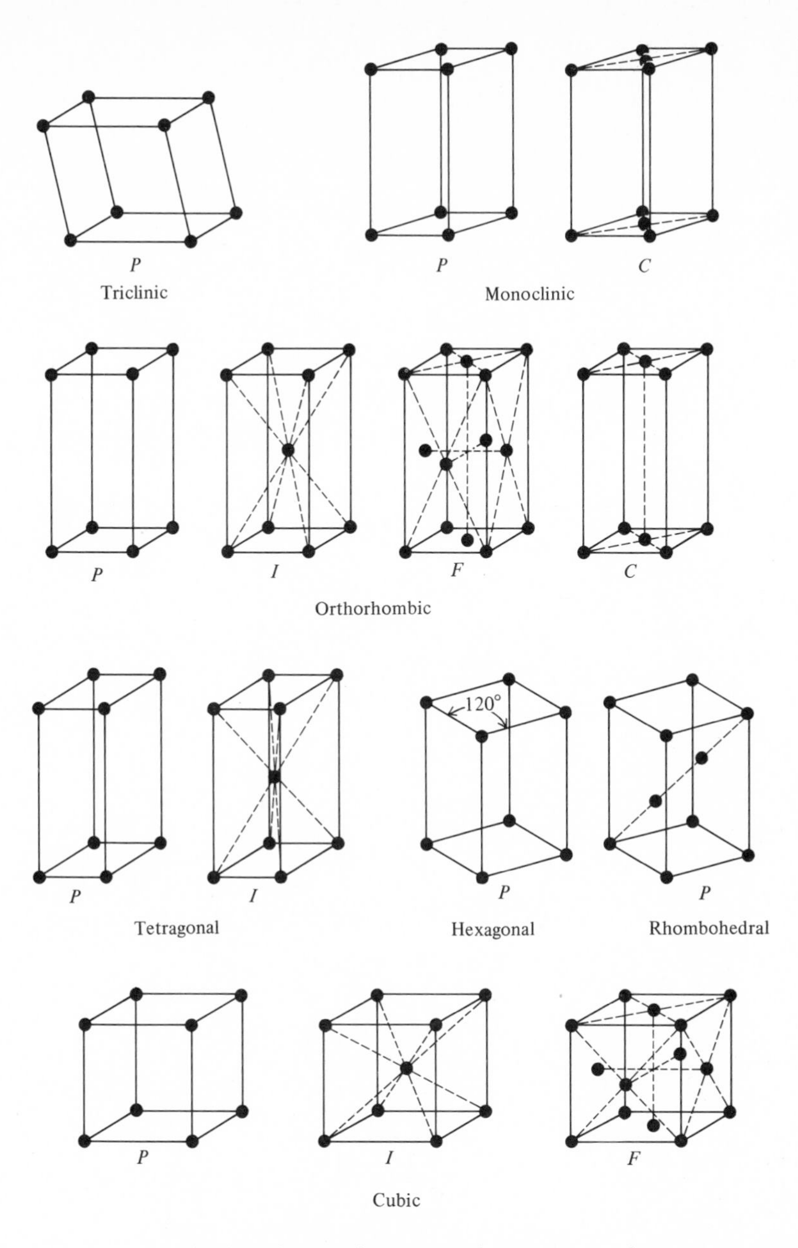

Figure 15-9 The 14 Bravais Lattices. The letters denote the kinds of lattice: *P* (primitive), *I* (body-centered), *F* (face-centered), *C* (end-centered).

orthorhombic and tetragonal systems each occur in more than one type (Figure 15-9). The other three systems, rhombohedral, hexagonal, and triclinic, can only occur as primitive types.

An x-ray study of a given crystal not only tells us which type of lattice is present, but also the dimensions and angles of the unit cell.

On the average, how many lattice points are associated with a given unit cell? Let us look at the simple primitive cubic system (Figure 15-10a). At first glance, we might say that there are eight lattice points in this cell, but a little reflection will show us that this is not so. Each lattice point does not belong exclusively to one unit cell. In fact, every one of these lattice points is shared by eight unit cells, as shown in Figure 15-10d. Therefore, in a primitive cubic system, each unit cell "owns" only $\frac{1}{8}$ of each lattice point. The same is true for the corner points in all the systems: $\frac{1}{8}$ of every corner point belongs to each unit cell. Each lattice point on a face (see Figure 15-10e for the face-centered cubic system) is part of two unit cells (so each cell "owns" $\frac{1}{2}$), while each body-centered lattice point, of course, belongs only to one cell. Figure 15-10b shows a body-centered cubic lattice.

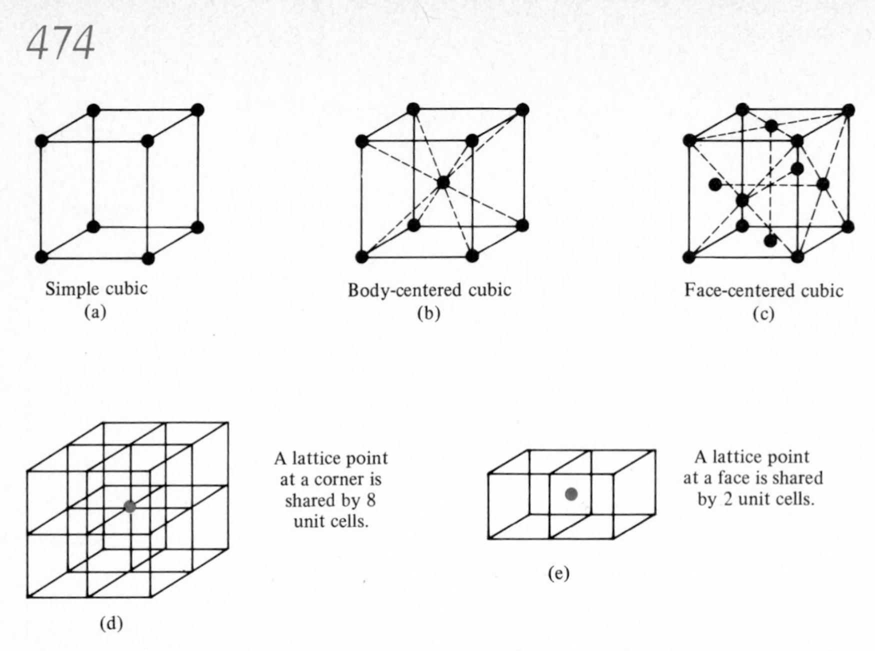

Figure 15-10 (a-c) The three cubic unit cells: primitive, body-centered, and face-centered. (d, e) Corner and edge lattice points are shared by adjacent unit cells.

We are now ready to calculate how many lattice points are associated with the three types of cubic-lattice unit cell.

	Simple cubic	*Face-centered cubic*	*Body-centered cubic*
Corner lattice points	8	8	8
	$\times\frac{1}{8}$	$\times\frac{1}{8}$	$\times\frac{1}{8}$
Number belonging to each unit cell	1	1	1
Face lattice points	0	6	0
		$\times\frac{1}{2}$	
Number belonging to each unit cell	0	3	0
Body lattice points	0	0	1
			$\times 1$
Number belonging to each unit cell	0	0	1
Total lattice points belonging to each unit cell	1	4	2

Example: X-ray analysis shows that nickel crystallizes in a face-centered cubic unit cell with an edge of 3.63 Å. Each motif consists of a single nickel atom, and the lattice point is at the center of each atom.

Calculate: (a) The number of atoms in the unit cell.
(b) The volume of the unit cell, cm^3.
(c) The density of the solid.

Answer:

(a) We have already calculated that the number of atoms in a face-centered cubic lattice unit cell is four atoms/unit cell.

(b) Using Figure 17-10c, we can calculate the volume of the cell as

$$\text{volume} = (3.63\ \text{Å})^3 = 47.8\ \text{Å}^3$$
$$= 47.8 \times 10^{-24}\ \text{cm}^3/\text{unit cell}$$

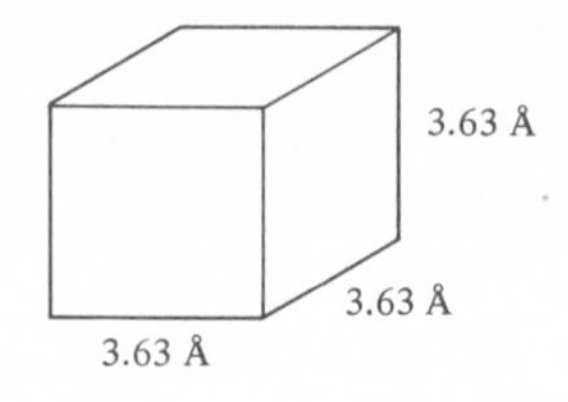

(c) Density = mass/volume. To find the density, we must calculate the mass in the unit cell. The atomic weight of nickel is 58.70. The weight of a single atom is

$$\frac{58.70\ \text{g/mole}}{6.02 \times 10^{23}\ \text{atoms/mole}} = 9.75 \times 10^{-23}\ \text{g/atom}$$

Since there are four atoms in a face-centered unit cell, the mass of the unit cell is

$$4\ \frac{\text{atoms}}{\text{unit cell}} \times 9.75 \times 10^{-23}\ \text{g/atom} = 39.0 \times 10^{-23}\ \text{g/unit cell}$$

The density of the unit cell is

$$\frac{39.0 \times 10^{-23}\ \text{g/unit cell}}{47.8 \times 10^{-24}\ \text{cm}^3/\text{unit cell}} = 8.15\ \text{g/cm}^3$$

At this point the reader may be wondering why a given substance crystallizes in a particular lattice type and not in another. There are a number of factors that determine this. For example, we must take account of the forces within the crystal, the relative size of the atoms that make up the crystal, and finally the different kinds of atoms, molecules, or ions that make up the crystal.

15-3 Close Packing of Spheres

In many cases, substances made up of atoms or ions form crystals in such a way that the atoms or ions get as close together as possible. This maximizes any attractive forces between them and results in the greatest possible stability for the crystal. If we assume that atoms and ions are spheres,* we can see for ourselves which arrangements lead to the closest possible packing in three dimensions.

We will start with a layer of spheres laid out on a flat surface. Figure 15-11 shows

*In most cases, this is a pretty good approximation of the shape of atoms and ions.

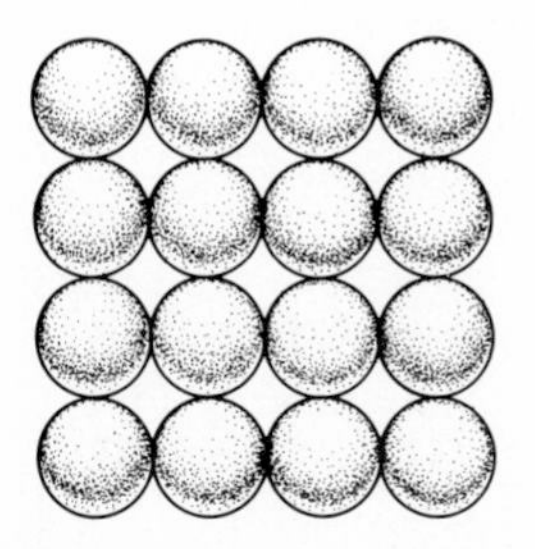

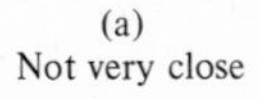

(a)
Not very close

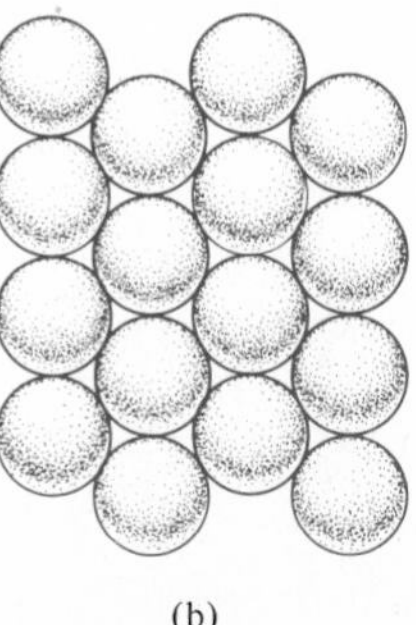

(b)
Much closer

Figure 15-11 Two ways to form a regular layer of spheres on a flat surface. The spheres in (b) are more closely packed than those in (a).

Figure 15-12 Two layers of closely packed spheres. The lower layer is in black; the upper layer is in color. Note that some of the holes in the lower layer (two of them are marked ▨) are completely covered by the upper layer, but that others (marked ■) are uncovered. Note that half of the holes in the upper layer (two of them are marked ▩) cover atoms; the others cover holes in the lower layer.

two possible regular patterns. The one on the right (in color) allows much closer packing than the square pattern on the left and is the one adopted by most atoms or ions. In a three-dimensional structure, a second layer is placed on top of the first. We could put the spheres of the second layer directly over those in the first layer, but, like the square arrangement in Figure 15-11a, this is not the closest packing. A much more efficient arrangement is for each sphere in the second layer to be placed over a hole in the first layer (Figure 15-12). Note that we cannot cover *all* the holes in the first layer; the size of the spheres will not allow this. We cover every other hole. Note also that the second layer has the same triangular arrangement as the first layer.

When it comes to putting on a third layer, there are two ways, both of which are equally efficient, and both of which cover holes in the second layer. We can cover those holes in the second layer which cover *holes* in the first layer (two of them are marked ■ in Figure 15-12) or we can cover those holes in the second layer which cover *spheres* in the first layer (two of them are marked ▩ in Figure 15-12). The two possibilities are shown in Figure 15-13; as shown, if the third-layer spheres are directly over first-layer *spheres*, then the third layer has the same position as the first layer. Additional layers will continue in the pattern *abababab. . .* , where *a* stands for the first-layer arrangement and *b* for the second layer. This arrangement is called **hexagonal close packing (hcp).** In the other arrangement, the third layer does not have the same position as the first layer; however, the fourth layer will have the same arrangement as the first, so the pattern will be *abcabcabcabc. . . .* This is called **cubic close packing (ccp).** Figure 15-14 shows three-dimensional views of the two types of packing. You can see them even better if you make models of them yourself, using plastic balls, or even gumdrops.

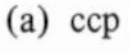

(a) ccp

(b) hcp

Figure 15-13 Two ways to obtain three layers of closely packed spheres. (a) Each third-layer sphere is over a first-layer hole (ccp). (b) Each third-layer sphere is over a first-layer sphere (hcp). Note that in (b) the first layer cannot be seen because each sphere in it is directly under a sphere in the third layer.

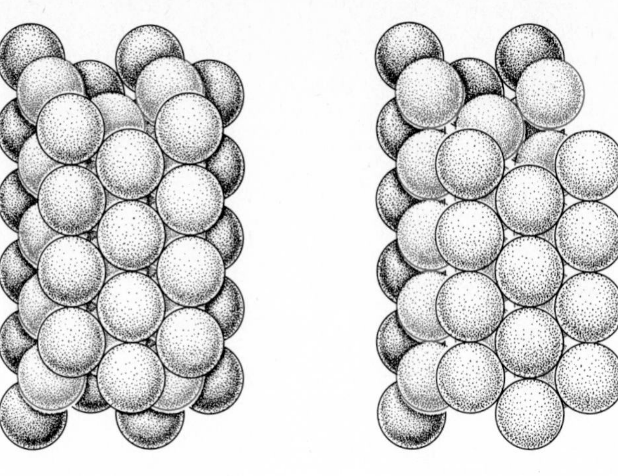

Figure 15-14 The ccp and hcp structures. The second layer is in color.

As previously noted, the ccp and hcp are equally efficient; in each, the spheres cover about 74% of the total volume, leaving about 26% for the holes. This is the minimum amount of empty space that can be achieved when spheres are packed into a volume. Note that the ccp structure is equivalent to the face-centered cubic lattice discussed in Section 15-2.

Most metals crystallize in the ccp, hcp, or body-centered cubic (bcc) structure. The inert gases crystallize in ccp structures, while solid H_2, O_2, and N_2 have hcp structures. Figure 19-5 shows the crystal structures of most of the metals.

15-4 Ionic Solids

Ionic solids are composed of positive and negative ions. We must suppose that in order to form such a crystal, the ions will so arrange themselves as to achieve the lowest possible energy. This can be accomplished if ions of the same sign are at a maximum distance from each other while ions of opposite sign are at a minimum distance. If this is to happen in three dimensions, each ion will be surrounded by a number of oppositely charged ions. This number is called the **coordination number** of the ion.* The most common coordination numbers in ionic crystals are 4, 6, and 8. We will look at an example of each.

1. *Coordination Number 8. Cubic Structure.* Figure 15-15 shows a portion of the CsCl crystal. Note that every Cs^+ ion is surrounded by 8 Cl^- ions, each an equal distance away; and each Cl^- is similarly surrounded by Cs^+ ions. This type of structure is called **cubic.** Note that the lattice type here is simple cubic. A motif consists of one Cs^+ ion and one Cl^- ion, and we may choose our lattice

*We have already seen the term "coordination number" used to express a similar concept (Section 6-7).

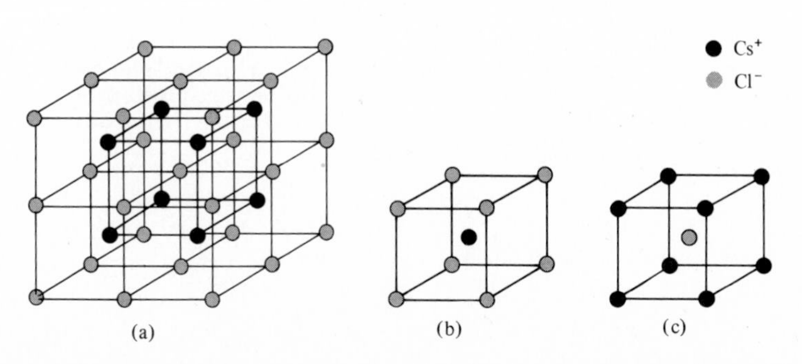

Figure 15-15 The CsCl Crystal. (a) The entire crystal. (b) A small portion of the crystal. Each Cs^+ ion is surrounded by eight Cl^- ions. This is a unit cell if the lattice point is the center of a Cs^+ ion. (c) Each Cl^- ion is surrounded by eight Cs^+ ions.

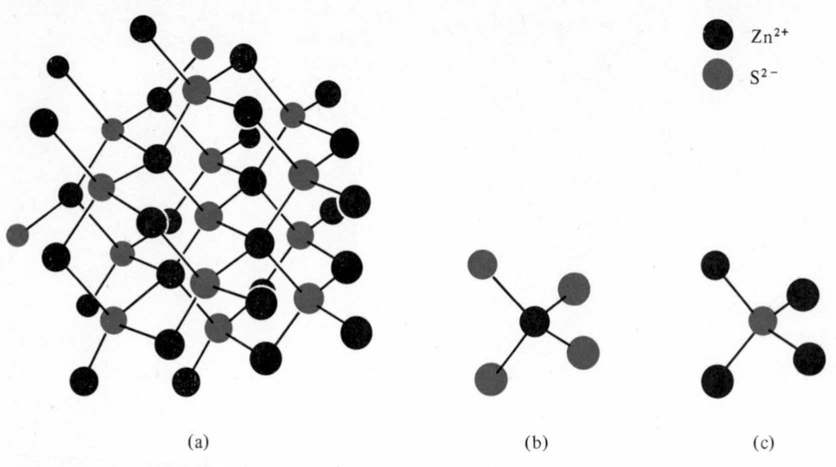

Figure 15-16 The Zincblende Crystal. (a) The entire crystal. (b) A small portion of the crystal. Each Zn^{2+} ion is surrounded by four S^{2-} ions. (c) A small portion of the crystal. Each S^{2-} ion is surrounded by four Zn^{2+} ions.

point to be at the center of a Cl^- ion. The cubic structure of the crystal is then easily seen. However, not every ionic solid in which the coordination number is 8 has a simple cubic lattice.

2. *Coordination Number 4. Tetrahedral Structure.* Figure 15-16 shows the structure of zincblende, a form of ZnS. In this crystal each Zn^{2+} ion is equidistant from 4 S^{2-} ions, and each S^{2-} ion is equidistant from 4 Zn^{2+} ions. In each case, the four equidistant ions are located at the corners of a regular tetrahedron, and this type of structure is called **tetrahedral.** The lattice here is face-centered cubic. It is not easy to see this from Figure 15-16, and it is helpful to look at a model of this system.
3. *Coordination Number 6. Octahedral Structure.* An example of this system is NaCl (Figure 15-8). If we examine this figure carefully, we see (Figure 15-17) that each Na^+ is surrounded by 6 equidistant Cl^- ions, and each Cl^- ion is surrounded by 6 equidistant Na^+ ions. In each case the six equidistant ions are located at the corners of a regular octahedron, and this type of structure is called **octahedral.** We have already seen that the lattice here is face-centered cubic.

In each of our examples, the total number of positive ions is equal to the total number of negative ions, because the formulas are of the type AB. This, of course, is not true for ionic solids such as CaF_2, Na_2S, K_3N, etc. In these cases, the coordination numbers of the two ions are different. For example, in CaF_2, each Ca^{2+} ion has a coordination number of 8 and each F^- has a coordination number of 4.

We might expect that the cubic structure would be the structure of choice for all ionic solids of the form AB, since it produces a coordination number of 8, which is the highest of the common coordination numbers. However, another important factor is involved—the *size* of the ions. The type of structure formed is almost

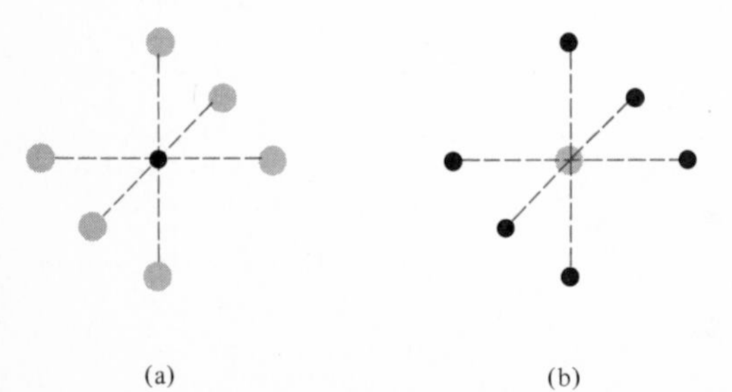

Figure 15-17 Two small portions of the NaCl crystal. (a) Each Na^+ ion is surrounded by six Cl^- ions. (b) Each Cl^- ion is surrounded by six Na^+ ions.

Table 15-2 Crystal Structure Determined by the Limiting Values of r_+/r_-

Limiting values of r_+/r_-	*Crystal structure*	*Coordination number of cation*
>0.732	Cubic	8
0.414–0.732	Octahedral	6
0.225–0.414	Tetrahedral	4

exclusively determined by the larger of the ions, and usually this is the anion. The anions tend to form close-packed structures with the smaller cations fitting into the voids. Since the tetrahedral structure has a coordination number of 4, it has the smallest voids while the cubic structure has the largest voids. Each of the possible structures is limited in how closely the anions can approach each other with the cations in the voids. If we try to push the anions closer together than they already are, we would experience a repulsive force originating from overlapping electron clouds. Since the structures depend on the relative sizes of anion and cation, we will define a term that will help us to decide what structure we might expect given the respective anion and cation sizes. This term, called the **radius ratio,** is defined as

$$\text{radius ratio} = \frac{r_+}{r_-}$$

where r_+ is the radius of the cation and r_- the radius of the anion (see Section 13-10). It has been determined that there are limiting values of the radius ratio which determine the type of crystal structure formed by ionic solids. Table 15-2 gives these values. Table 15-3 lists a number of ionic solids and their radius ratios.

Remember that in ionic solids there are no molecules. In a crystal of NaCl, we find Na^+ ions and Cl^- ions but never molecules of NaCl. This is in contrast to molecular solids, in which molecules are indeed present. Because of this, an ionic crystal can be imagined as a single giant molecule whose molecular weight is enormous. This explains the very high melting points and heats of fusion found for ionic crystals. Since there are no molecules present in the ionic solids the forces holding the solid together are the electrostatic interactions found between ions (ionic bonds). In order to melt an ionic crystal, enough energy must be supplied to overcome the attractive forces between the ions, which are quite large.

Table 15-3 Ionic Solids, r_+/r_- Ratios, and Coordination Numbers of Cations

Solid	r_+/r_-	*Coordination number of cation*
MgF_2	0.48	6
NiF_2	0.51	6
CoF_2	0.53	6
ZnF_2	0.54	6
MnF_2	0.59	6
CdF_2	0.71	8
CaF_2	0.73	8
SrF_2	0.83	8
PbF_2	0.88	8
BaF_2	0.99	8
CsCl	0.93	8
NaCl	0.52	6
ZnS	0.40	4

15-5 Covalent Crystals

At the beginning of Chapter 14, it was pointed out that covalent molecules may consist of a comparatively small number of atoms, or of a very large number, so large that an entire crystal is essentially a single molecule made up of perhaps 10^{22} or 10^{23} atoms. We may refer to such crystals as **giant molecules.** A good example of such a crystal is the diamond structure shown in Figure 15-1a. The bonding between the carbon atoms is purely covalent. Figure 15-1a illustrates a number of interesting features. Each carbon atom is surrounded by four equidistant carbon atoms. The angle between *any* two bonded atoms is the tetrahedral angle, 109°28′. The basic unit cell of the diamond structure is a cubic cell; the covalent bonding permeates throughout the complete crystal. In order to melt such a structure, one must rupture a great many of these strong covalent bonds, which gives diamond its very high melting point. Germanium, gray tin, and silicon also crystallize in the diamond structure. All of them are in group IVA of the periodic table, so that all have four outer-shell electrons. It is not surprising then that all bond in the same way. Other systems that assume the diamond structure are ZnS, AgI, AlP, and SiC. We find in these structures that each atom is tetrahedrally surrounded by four unlike atoms; the bonding is almost exclusively covalent. The covalent structure, as in diamond, permeates the complete crystal.

In small covalent molecules, the atoms are connected to each other by strong covalent bonds within the molecules, but there are no covalent bonds *between* the molecules. The molecules are held together in the crystal by van der Waals forces. The molecules tend to pack within the crystals as closely as their sizes allow; we may call these **molecular crystals.** Typical examples of compounds or elements that form molecular crystals are ethane, carbon tetrachloride, nitrogen, argon, benzene, and napthalene. It is relatively easy to melt molecular crystals, precisely because covalent bonds do not have to be broken. When this type of crystal melts, it is only the weak van der Waals forces that must be overcome. The molecules are still intact in the liquid state, and even in the gaseous state (Figure 15-18). This is the reason that compounds made up of small covalent compounds are usually liquids, gases, or relatively low melting solids. Most nonionic organic substances crystallize as molecular crystals except those in which hydrogen bonding is present.

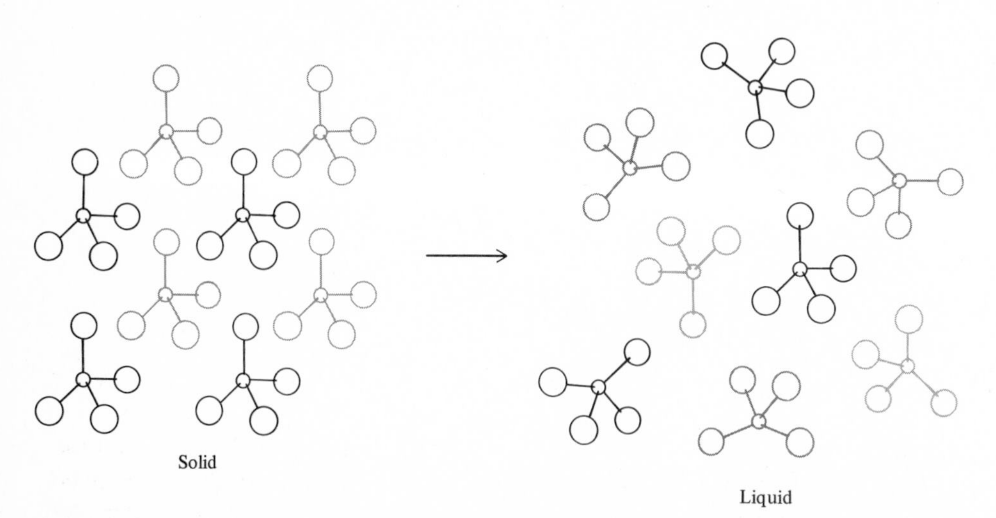

Figure 15-18 **The molecules of solid CCl_4 are rigidly held in a crystal lattice by weak van der Waals forces. When the solid melts, the molecules are still intact (no covalent bonds have been broken) but are no longer held in a rigid array.**

The hydrogen bond is very important in a number of crystal structures. Certain organic and inorganic acids, as well as many hydrated salts, owe their structures to hydrogen bonds. The ice structure is shown in Figure 6-16. All these crystals exhibit the open structure expected of hydrogen-bonded systems.

15-6 Phase Diagrams

The three states of matter, gas, liquid, and solid, are known as **phases.** Most substances can exist in any of these three phases, but only under certain conditions of temperature and pressure. For example, at 1 atm pressure, water is a liquid between 0°C and 100°C, but at 5 atm pressure, water is a liquid between about −0.1°C and 152°C.

As the temperature of a gas is lowered at constant pressure, a point is reached where liquid begins to form. As we know, this is the condensation temperature of the gas (or the boiling point of the liquid) at this particular pressure. The gas and liquid are in equilibrium at this T and P and the system is now a two-phase system. We say that the system is at equilibrium since if it was completely isolated it would not change with time. The gas and liquid would remain as such indefinitely. If we continue to withdraw heat from the system, the temperature remains constant while more gas is converted into liquid. Eventually, a point is reached where all the gas is converted into liquid and the system again becomes a one-phase system. Further withdrawal of heat from the liquid system lowers the temperature until a point is reached when solid begins to form. This is, of course, the freezing temperature of the liquid (or the melting point of the solid) at this particular pressure. The two-phase system (liquid and solid) is in equilibrium. Further withdrawal of heat does not lower the temperature but first converts more liquid into solid. Finally, all the liquid has been converted into solid; further withdrawal of heat now would lower the temperature of the solid. If heat is supplied to the system at the same pressure, the whole process can be reversed. This is shown as follows. Note that the terminology changes, depending on which way the process is proceeding.

$$\text{gas} \xrightarrow[\Delta H_{\text{condensation}}]{\text{heat withdrawn}} \text{liquid} \xrightarrow[\Delta H_{\text{fusion}}]{\text{heat withdrawn}} \text{solid}$$

$$\text{gas} \xleftarrow[\Delta H_{\text{vaporization}}]{\text{heat added}} \text{liquid} \xleftarrow[\Delta H_{\text{melting}}]{\text{heat added}} \text{solid}$$

(solid → gas: heat of sublimation, $\Delta H_{\text{sublimation}}$)

Note that

$$\Delta H_{\text{condensation}} \text{ (exothermic)} = -\Delta H_{\text{vaporization}} \text{ (endothermic)}$$
$$\Delta H_{\text{fusion}} \text{ (exothermic)} = -\Delta H_{\text{melting}} \text{ (endothermic)}$$
$$\Delta H_{\text{sublimation}} = \Delta H_{\text{melting}} + \Delta H_{\text{vaporization}} \text{ (endothermic)}$$

As we saw in Section 6-3, boiling points and melting points are dependent on the external pressure. How a substance behaves under different conditions of temperature and pressure can be represented on a graph called a **phase diagram.** A phase diagram condenses a great of information into a small figure. To see this, let us consider the phase diagram for water given in Figure 15-19. The various phases are labeled on the graph. The line BC is the liquid–vapor line, which tells us the

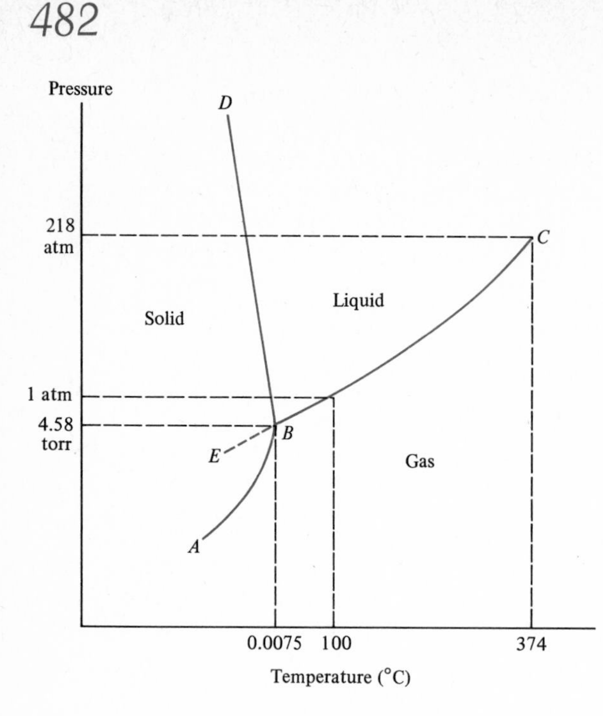

Figure 15-19 The phase diagram of water (not drawn to scale). The letters are explained in the text.

temperature and pressure conditions necessary to condense a gas or to vaporize a liquid. For example, at 1 atm pressure, the temperature must be 100°C, which is the normal boiling point. If the atmospheric pressure is increased to 218 atm, the boiling point of water increases to 374°C. This is also the critical point of water (Section 4-14). *DB* is the melting-point line; it tells us how the equilibrium melting temperatures change with external pressures. Note that there is *no* critical temperature for a solid–liquid system. At point *B*, gas, liquid, and solid are in equilibrium. This is known as the **triple point,** since there are *three* phases in equilibrium. For water it occurs at 0.0075°C and 4.58 torr. The phase diagram shows that if the temperature or pressure even slightly changes from the triple-point values, the three phases are no longer in equilibrium. *AB* is the sublimation curve for water. Solid and gas are in equilibrium along this curve. At the temperatures and pressures along this curve one may convert the solid directly into the gas, or vice versa, without going through a liquid phase.

EB, which is shown as a dashed line in the phase diagram, is a nonequilibrium extension of the liquid–gas equilibrium curve. Water can be **supercooled** below 0°C, where it would normally freeze. This may occur, since for any liquid to form a solid, crystals of ice or other small solid particles or a rough surface must be present so that other crystals can grow upon them. It is quite similar to the phenomenon of supersaturation discussed in Section 7-2. The *EB* line is known as a nonequilibrium or metastable equilibrium state. If we drop the temperature low enough, the water will freeze spontaneously even though impurities are absent.

There is one important aspect of the phase diagram of water that is not true of most other substances. The *BD* line slopes to the left, whereas for most other species, the solid–liquid line slopes to the right. A thermodynamic analysis shows that this occurs when the liquid is more dense than the solid. Bismuth and antimony behave similarly.

Phase diagrams have been experimentally determined for thousands of pure substances as well as for mixtures of two, three, and more components. They are extremely valuable since, as previously noted, they contain such large quantities of information in a single graph.

Problems

15-1 Define each of the following terms. **(a)** x-ray crystallography **(b)** crystalline solid **(c)** amorphous solid **(d)** unit cell **(e)** motif **(f)** space lattice **(g)** lattice point **(h)** hcp **(i)** ccp **(j)** coordination number **(k)** bcc **(l)** radius ratio **(m)** giant molecule **(n)** phase **(o)** phase diagram **(p)** triple point **(q)** supercooling

15-2 What type of crystal structure would you expect the following compounds to have in the solid state? **(a)** SO_2 **(b)** SiC **(c)** HF **(d)** KCl **(e)** SiO_2 **(f)** KBr

15-3 Calculate the amount of empty space in **(a)** a primitive cubic, **(b)** a body-centered cubic, and **(c)** a face-centered cubic lattice packed with identical spheres touching each other. **(d)** Which of these is the most efficient and the least efficient in filling space (the volume of a sphere of radius r is $\frac{4}{3}\pi r^3$).

15-4 NaCl and CsCl are similar in many ways, but their unit cells differ. **(a)** How many Na^+ and Cl^- ions are necessary to make up one unit cell? **(b)** How many Na^+ and Cl^- ions "belong" to each unit cell? **(c)** How many Cs^+ and Cl^- ions are necessary to make up one unit cell? **(d)** How many Cs^+ and Cl^- ions "belong" to each unit cell **(e)** Why do two such similar compounds have different kinds of unit cell?

15-5 The CaF_2 crystal looks like this:

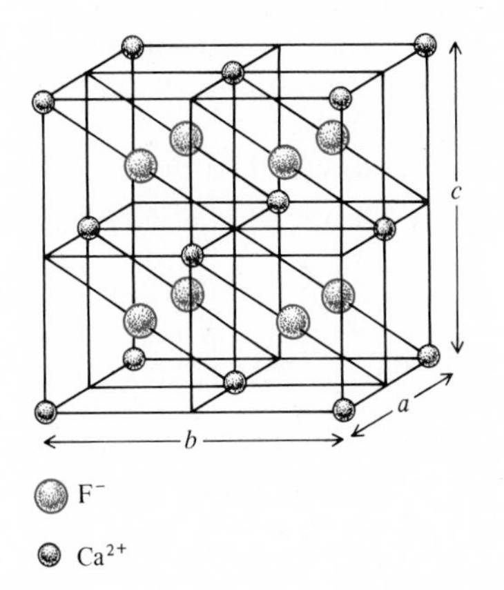

(a) Show the ions that make up a motif. **(b)** If $a = b = c$, which of the 14 Bravais lattice types does this crystal belong to? **(c)** Show that the coordination numbers of Ca^{2+} and F^- are 8 and 4, respectively.

15-6 Which of the following planar figures can serve as a motif to fill up all space on a plane? (For example, if you had floor tiles with these shapes, could you cover the whole floor without leaving any holes?) **(a)** equilateral triangle **(b)** square **(c)** parallelogram **(d)** regular pentagon **(e)** regular hexagon **(f)** regular octagon **(g)** circle

15-7 Describe how spheres would have to be packed to result in a body-centered cubic structure.

15-8 What percentage of the total volume is covered by spheres in the body-centered cubic structure you worked out in Problem 15-7, assuming that the spheres are all close enough to touch their neighbors?

15-9 Strontium crystallizes in the ccp structure. The density of Sr at 20°C is 2.54 g/cm³. Calculate the radius of a strontium atom.

15-10 Vanadium crystallizes in the bcc structure. The density of vanadium is 6.11 g/cm³. Calculate the radius of a vanadium atom.

15-11 An element with a density of 19.3 g/cm^3 crystallizes in the bcc structure. The radius of each atom is 1.37 Å. Assuming that the atoms touch each other in the crystal, what is the element? (*Hint:* Calculate its atomic weight.)

15-12 Platinum crystallizes in the ccp structure. The radius of a Pt atom is 1.39 Å. Calculate: **(a)** the length of one side of the unit cell **(b)** the volume of the unit cell in cm^3 **(c)** the mass of the unit cell **(d)** the density of platinum in g/cm^3.

5-13 Aluminum crystallizes in a face-centered cubic lattice. **(a)** Calculate the number of Al atoms per unit cell **(b)** If the edge of the unit cell is 4.04 Å, what is the density of aluminum?

5-14 Magnesium oxide MgO has the NaCl structure and a density of 3.65 g/cm^3. Calculate the volume of its unit cell.

5-15 Germanium oxide, GeO_2, has a radius ratio of 0.40. GeO_2 crystallizes in two different forms, with one structure having a coordination number of 4 and the other of 6. Explain why.

15-16 The radii of Rb^+ and Br^- are 1.48 Å and 1.95 Å, respectively. Predict whether RbBr will have a cubic, octahedral, or tetrahedral crystal structure.

15-17 Lithium iodide, LiI, which crystallizes in the face-centered cubic system, has a density of 3.49 g/cm^3. Calculate the distance between the centers of a Li^+ ion and an adjacent I^- ion.

15-18 Thallium bromide, TlBr, which crystallizes in the primitive cubic system, has a density of 7.56 g/cm^3. Calculate the size of the unit cell and the distance between the centers of a Tl^+ ion and an adjacent Br^- ion.

15-19 Use the answer to Problem 15-18 to calculate the radius of a Tl^+ ion, if the Tl^+ and Br^- ions are spheres that touch each other and the Br^- has a radius of 1.95 Å.

15-20 Magnesium fluoride, MgF_2, has a radius ratio of 0.48. **(a)** How many fluoride ions would you predict surround each magnesium ion in the crystal? **(b)** How many Mg^{2+} ions surround each F^- ion? **(c)** Does MgF_2 have a low or relatively high melting point?

15-21 Silicon fluoride, SiF_4, forms a solid structure in which each silicon has a coordination number of 4. **(a)** How do the molecules in the crystal compare with those in the gas phase? **(b)** What forces hold the crystal structure together? **(c)** Would you expect a high or a low melting point?

15-22 Shown is the phase diagram for CO_2.

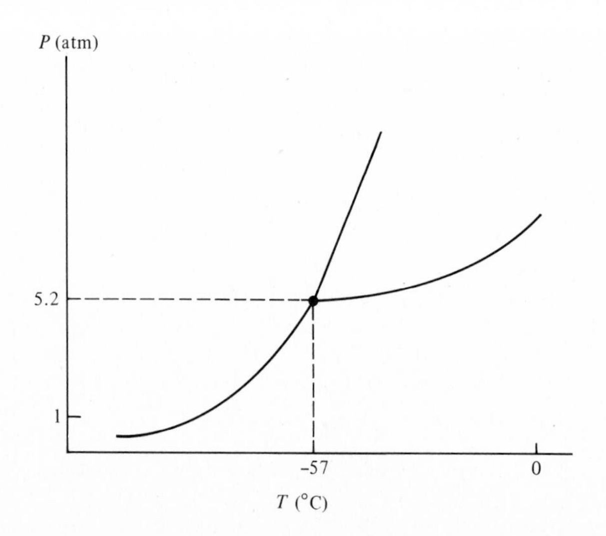

(a) Label the solid, liquid, and vapor regions. **(b)** Label the triple point. **(c)** Explain why CO_2 does not have a normal boiling point. **(d)** When solid CO_2 at 1 atm pressure is dissolved in acetone (mp −95°C) the temperature of the acetone drops until a saturated solution of CO_2 in acetone forms at −78°C. Explain, using the phase diagram of CO_2.

15-23 Referring to the phase diagram for CO_2 given in Problem 15-22, describe the phase changes that occur, and the approximate pressures, when, beginning with CO_2 at a pressure of 7 atm, the pressure is gradually decreased: **(a)** At a constant temperature of −75°C. **(b)** At a constant temperature of −50°C.

15-24 Referring to the phase diagram for CO_2 given in Problem 15-22, describe the phase changes that occur and the approximate temperatures, when, beginning with CO_2 at a temperature of −80°C, the temperature is gradually increased: **(a)** At a constant pressure of 3.0 atm. **(b)** At a constant pressure of 7.0 atm.

Additional Problems

15-25 Barium crystallizes in the bcc structure. The radius of a barium atom is 2.17 Å. Calculate the density of barium.

15-26 The radii of Ba^{2+} and S^{2-} are 1.35 Å and 1.84 Å, respectively. Predict whether BaS will have a cubic, octahedral, or tetrahedral crystal structure.

15-27 **(a)** Show that the ratio of the radius of the inner circle shown below to that of any of the outer circles is 0.414. **(b)** To which of the three geometries in Table 15-2 does this correspond, if the cation is in the middle? **(c)** Why is it that with a larger cation the system can still have this geometry, but not with a smaller cation, assuming that the size of the anion is constant?

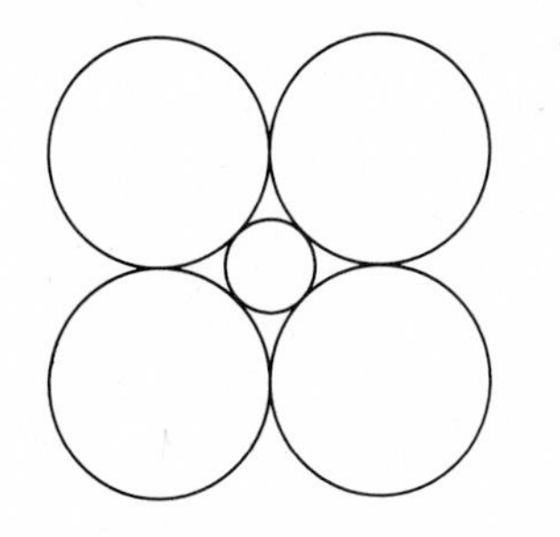

15-28 The density of CaO is 3.35 g/cm³. The oxide crystallizes in one of the cubic systems, with an edge of 4.80 Å. How many Ca^{2+} ions and O^{2-} ions "belong" to each unit cell, and which type of cubic system is present?

15-29 Distinguish between the ccp and hcp arrangements of close-packed spheres.

15-30 In the hcp arrangement of metallic atoms, the unit cell looks like this

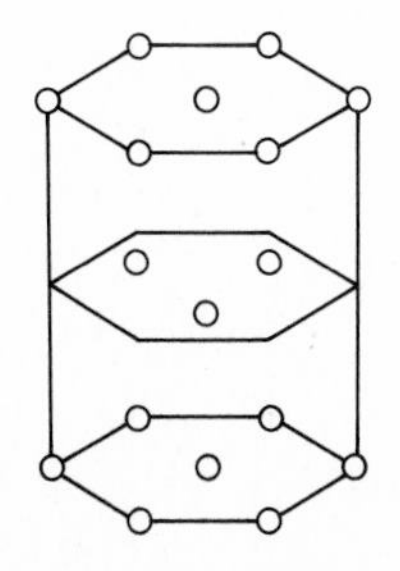

(a) How many atoms "belong" to the unit cell? **(b)** For cadmium, the dimensions of the unit cell are height 5.62 Å, side of the regular hexagon 2.98 Å. Calculate the density of cadmium.

15-31 The zincblende structure is shown in Figure 15-16. **(a)** Show the ions that make up a motif. **(b)** Which of the three cubic lattice types does this crystal belong to?

15-32 The crystals of $AgClO_2$ belong to the orthorhombic system. X-ray analysis shows that the edges of the unit cell are 6.13 Å, 6.68 Å, and 6.07 Å. If there are four Ag^+ ions and four ClO_2^- ions per unit cell, calculate the density of $AgClO_2$.

15-33 How many nearest neighbors does a sphere have if it is packed in each of the following structures? **(a)** body-centered cubic (bcc) **(b)** hcp **(c)** ccp

16 Oxidation-Reduction and Electrochemistry

Alessandro Volta, 1745–1827

In this chapter, we will study one of the most important types of chemical reaction, *oxidation–reduction reactions*. It was Lavoisier who first proposed the concept of oxidation. He defined oxidation as the reaction of any substance with oxygen. It soon became apparent that this definition was too narrow, and gradually, over the course of the nineteenth and early twentieth centuries, the concept of oxidation (and of its opposite, reduction) has grown much broader, until we have reached the modern definitions of oxidation and reduction, which will be given in Section 16-2. The topic of electrochemistry is very closely tied to that of oxidation and reduction, as we shall see beginning in Section 16-5. Before we can define oxidation and reduction, we must introduce a new concept, the concept of *oxidation number*. This concept will also help us to balance oxidation–reduction equations. As we shall see, these equations are often difficult to balance by simple inspection.

16-1 Oxidation Number

There are a number of bookkeeping systems which are used as aids to help us keep track of electrons and of changes in the location of electrons during the course of chemical reactions. Formal charge, which we met in Section 14-9, is one of these. Another very useful concept, which is exceedingly helpful when studying oxidation–reduction reactions, is oxidation number. By means of the following rules, we assign a number, called an **oxidation number,** to every element in any compound.

1. Any element in the free state is assigned an oxidation number of zero. For example H_2, P_4, S_8, O_2, Fe, Br_2, and K have oxidation numbers of zero.
2. The oxidation number of any one-atom (monatomic) ion is equal to its charge. For example, Ca^{2+} has an oxidation number of $+2$, Al^{3+} of $+3$, and Cl^- of -1.
3. *Hydrogen* in almost all its compounds is assigned an oxidation number of $+1$. The exceptions occur when hydrogen forms compounds with metals, called metal hydrides. Examples are KH, MgH_2, CaH_2, and LiH (see Section 18-1). In these compounds hydrogen has an oxidation number of -1.
4. *Oxygen* in almost all its compounds is assigned an oxidation number of -2. Exceptions are a class of compounds called peroxides, which are compounds containing an oxygen–oxygen bond. Some typical examples are sodium peroxide (Na_2O_2), hydrogen peroxide (H_2O_2), and potassium peroxide (K_2O_2). In these compounds the oxidation number of oxygen is -1. Another exception (a very unusual one) is the compound OF_2, in which oxygen has an oxidation number of $+2$.
5. The oxidation number of the alkali metals (group IA) is $+1$ in all their compounds and the oxidation number of the alkaline earth metals (group IIA) is $+2$ in all their compounds.

6. Fluorine is the most electronegative element and is assigned an oxidation number of −1 in all its compounds. For the other halogens the oxidation number is always −1, except when they are bonded to a more electronegative halogen or to oxygen. For example, in ClO^- the oxidation number of chlorine is +1, and in the compound IF_7 the oxidation number of iodine is +7.
7. The algebraic sum of the oxidation numbers of all atoms in a neutral molecule is zero. The algebraic sum of the oxidation numbers of all atoms in an ion equals the charge on the ion. These facts enable us to calculate oxidation numbers of elements not listed above. For example, in H_2S the two hydrogens each have an oxidation number of +1 so the oxidation number of the sulfur is −2. In the sulfate ion, SO_4^{2-}, each oxygen has an oxidation number of −2, so that counting all the oxygens we get $4 \times (-2) = -8$. The oxidation number of sulfur must be +6 in order that the algebraic sum of the five atoms in SO_4^{2-} be equal to $+6 - 8 = -2$, the charge on the ion.

With this set of rules in hand, we can calculate almost all the oxidation numbers of elements in compounds and ions.

Example: Using the above rules calculate the oxidation numbers of all the atoms in the following compounds and ions: SiO_2, ClO_4^-, CH_4, BO_3^{3-}, OF_2, Hg_2O, $Cr_2O_7^{2-}$, C_2H_2, MgS.

Answer:

SiO_2:	each oxygen −2	total charge 0	O	−2
	total oxygen −4	−total oxygen −4	Si	+4
		silicon +4		
ClO_4^-:	each oxygen −2	total charge −1	O	−2
	total oxygen −8	−total oxygen − 8	Cl	+7
		chlorine + 7		
CH_4:	each hydrogen +1	total charge 0	H	+1
	total hydrogen +4	−total hydrogen +4	C	−4
		carbon −4		
BO_3^{3-}:	each oxygen −2	total charge −3	O	−2
	total oxygen −6	−total oxygen −6	B	+3
		boron +3		
OF_2:	given directly by the rules		O	+2
			F	−1
Hg_2O:	total oxygen −2	total charge 0	O	−2
		−total oxygen −2	Hg	+1
		total mercury +2		
		each mercury +1		
$Cr_2O_7^{2-}$:	each oxygen − 2	total charge − 2	O	−2
	total oxygen −14	−total oxygen −14	Cr	+6
		total chromium +12		
		each chromium + 6		
C_2H_2:	each hydrogen +1	total charge 0	H	+1
	total hydrogen +2	−total hydrogen +2	C	−1
		total carbon −2		
		each carbon −1		
MgS:		total charge 0	Mg	+2
		−total magnesium +2	S	−2
		sulfur −2		

Example: Calculate the oxidation numbers of all the elements in the following compounds: $HCCl_3$, Na_3PO_4, $C_3H_6Cl_2$, NH_4NO_3, $BrCl$, Na_2O_2, PbO_2, Pb_3O_4.

Answer: $HCCl_3$: In this case we calculate the oxidation number of carbon from the known values for hydrogen (+1) and chlorine (−1).

total hydrogen	+1	total charge	0	H	+1
total chlorine	−3	−total H + Cl	−2	Cl	−1
total H + Cl	−2	carbon	+2	C	+2

Na_3PO_4: Here we use a similar process to calculate the oxidation number of phosphorus.

total sodium	+3	total charge	0	Na	+1
total oxygen	−8	−total Na + O	−5	O	−2
total Na + O	−5	phosphorus	+5	P	+5

$C_3H_6Cl_2$: Again we have two known (H = +1, Cl = −1) and one unknown oxidation number.

total hydrogen	+6	total charge	0	H	+1
total chlorine	−2	−total H + Cl	+4	Cl	−1
total H + Cl	+4	total carbon	−4	C	$-\frac{4}{3}$
		each carbon	$-\frac{4}{3}$		

This example illustrates that an oxidation number may be a fractional number. However, it may be possible to avoid fractions by giving each carbon a different number. From the oxidation numbers of H and Cl, the sum of the oxidation numbers of the three carbons must be −4, but this can be achieved, for example, if one carbon has an oxidation number of +4, and others of −4 each.

NH_4NO_3: This compound is made up of NH_4^+ and NO_3^- ions. The nitrogens in these two ions have different oxidation numbers, and each is calculated separately.

NH_4^+:	total charge	+1	H	+1
	−total hydrogen	+4	N	−3
	nitrogen	−3		
NO_3^-:	total charge	−1	O	−2
	−total oxygen	−6	N	+5
	nitrogen	+5		

BrCl: Chlorine is more electronegative than bromine, so the oxidation numbers are Cl = −1 and Br = +1.

Na_2O_2: This is an example of a peroxide.

$$Na = +1$$
$$O = -1$$

PbO_2: This is not a peroxide.

total charge	0	O	−2
total oxygen	−4	Pb	+4
lead	+4		

Pb_3O_4:

total charge	0	O	−2
total oxygen	−8	each Pb	$+\frac{8}{3}$
total lead	+8		

This is another example of a fractional oxidation number, which can be avoided, for example, by assigning the numbers +3, +3, +2, or +4, +4, 0, to the three Pb atoms.

RELATIONSHIP OF OXIDATION NUMBER TO VALENCE. Because the numerical value is often the same, it is easy to confuse oxidation number with valence. However, they are fundamentally different things.

The valence of an element is real. It is the *number of bonds* formed by each element within a given compound or ion.

The oxidation number is artificial, with no reality. It is a formal concept invented by chemists to keep their bookkeeping straight.

The oxidation number always has a sign (+ or −); the valence has a sign only when it is ionic valence, never when it is covalence. The rules for oxidation number have been chosen in such a way that the numerical value often comes out the same as the valence (for example, the valence of the group IA metals is always +1 and so is the oxidation number), but often it does not.

Example: Compare the oxidation numbers and valences of (a) C in C_2H_2 (b) O in H_2O_2 (c) N in N_2H_4.

Answer:

(a) The oxidation number of C in C_2H_2 is −1; the valence is 4.

$$H—C{\equiv}C—H$$

(b) The oxidation number of O in H_2O_2 is −1; the valence is 2.

$$H—O—O—H$$

(c) The oxidation number of N in N_2H_4 is −2; the valence is 3.

```
H—N—N—H
  |  |
  H  H
```

16-2 Electron Transfer

We have already seen (Chapter 10) that the transfer of a *proton* from one species to another is an important reaction. The transfer of one or more *electrons* from one species to another is also very important. Consider the simple reaction

$$Cu^{2+} + Pb \longrightarrow Pb^{2+} + Cu \qquad (1)$$

In this reaction Pb atoms have been converted to Pb^{2+} ions. Obviously, each lead atom has lost two electrons in going from reactant to product.

$$Pb \longrightarrow Pb^{2+} + 2e^-$$

The oxidation number of the lead has also changed (0 to +2) in going from reactant to product. We can now give two definitions for the term **oxidation.**

1. **Oxidation is a loss of electrons.**
2. **Oxidation is an increase in oxidation number.**

In the above equation, lead has been oxidized. It has lost electrons and its oxidation number has increased. Both the loss of electrons and the increase in oxidation number occur together. For copper, the opposite has taken place.

$$Cu^{2+} + 2e^- \longrightarrow Cu$$

The oxidation number of the copper has decreased from +2 to 0; in the process, copper has gained two electrons. We now give two definitions for the term **reduction.**

1. **Reduction is a gain of electrons.**
2. **Reduction is a decrease in oxidation number.**

In the above equation copper has gained two electrons in going from reactant to product while decreasing in oxidation number. Copper has been reduced.

Oxidation and reduction must take place simultaneously. We cannot have electrons floating around without being attached to a species. If one species loses electrons, some other species must gain them. It is not possible for a species simply to throw electrons away. When a species is oxidized (loses electrons), it causes another species to be reduced (gain electrons). When a species is reduced, it causes another species to be oxidized. In equation (1) Pb is oxidized and is the **reducing agent** (causes Cu^{2+} to be reduced) and Cu^{2+} is reduced and is the **oxidizing agent** (causes Pb to be oxidized). The species that is oxidized loses electrons, gains in oxidation number, and is the reducing agent.

There is no oxidation without an accompanying reduction. There is no reduction without an accompanying oxidation. Because oxidation and reduction must occur together, we often use the term **redox reaction** to refer to the total process.

In the example given above, it was easy to see that Pb lost electrons and Cu^{2+} gained electrons, but in many other redox reactions, this process is not immediately obvious. Yet, as we shall see in the next section, we can always show that electrons are lost and gained in any redox reaction. Just as we defined a Brønsted acid–base reaction as the transfer of a proton, we may define a redox reaction as the *transfer of electrons.*

Example: In the equation

$$3MnO_2 + 4Al \longrightarrow 3Mn + 2Al_2O_3$$

label the oxidation numbers of all atoms; show the species oxidized, the species reduced, the oxidizing agent, and the reducing agent.

Answer: The oxidation numbers are:

$$3\overset{+4}{Mn}\overset{-2}{O_2} + 4\overset{0}{Al} \longrightarrow 3\overset{0}{Mn} + 2\overset{+3}{Al_2}\overset{-2}{O_3}$$

Aluminum is oxidized (increase in oxidation number, loss of electrons); manganese is reduced (decrease in oxidation number, gain of electrons). Aluminum is the reducing agent; MnO_2 is the oxidizing agent (actually the Mn in MnO_2 is the oxidizing agent, but it is customary to refer to the entire molecule as the oxidizing agent). The oxygen has neither been oxidized or reduced. Its oxidation number has not changed.

Example: In the equation

$$C_2H_4 + 3O_2 \longrightarrow 2CO_2 + 2H_2O$$

label the oxidation number of all atoms; show the species oxidized, the species reduced, the oxidizing agent, and the reducing agent.

Answer:

$$\overset{-2}{C_2}\overset{+1}{H_4} + 3\overset{0}{O_2} \longrightarrow 2\overset{+4}{C}\overset{-2}{O_2} + 2\overset{+1}{H_2}\overset{-2}{O}$$

The carbon in C_2H_4 is oxidized by losing electrons and increasing in oxidation number.

The oxygen is reduced by gaining electrons and decreasing in oxidation number. The oxidizing agent is O_2; the reducing agent, C_2H_4. The hydrogen has neither been oxidized nor reduced.

Example: In the equation

$$2MnO_4^- + 10Cl^- + 16H^+ \longrightarrow 2Mn^{2+} + 5Cl_2 + 8H_2O$$

label the oxidation number of all atoms; show the species oxidized, the species reduced, the oxidizing agent, and the reducing agent.

Answer: The oxidation numbers are:

$$\overset{+7\,-2}{2MnO_4} + \overset{-1}{10Cl^-} + \overset{+1}{16H^+} \longrightarrow \overset{+2}{2Mn^{2+}} + \overset{0}{5Cl_2} + \overset{+1\,-2}{8H_2O}$$

Cl^- is oxidized by losing electrons and increasing in oxidation number. The manganese in MnO_4^- is reduced by gaining electrons and decreasing in oxidation number. Cl^- is the reducing agent; MnO_4^- the oxidizing agent. Hydrogen and oxygen have neither been oxidized nor reduced.

We now have the necessary tools to balance oxidation–reduction equations, and this topic will be taken up next.

16-3 Balancing Oxidation-Reduction Equations by the Half-Equation Method

We have pointed out that redox equations can be difficult to balance by simple inspection. For example, look at

$$Fe^{2+} + H^+ + Cr_2O_7^{2-} \longrightarrow Fe^{3+} + Cr^{3+} + H_2O$$

Can you balance it? (Remember that in ionic equations, the charges must balance as well as the atoms.) It is likely that you would have a hard time doing it by inspection. In this section, we will introduce a method for balancing redox equations which always works, if properly applied. In using this method, called the **half-equation method,** we pretend that the reaction is made up of two separate half-reactions. One of them is the oxidation reaction and the other is the reduction reaction. *Each half-equation is separately and independently balanced.* Each half-equation is then multiplied by an appropriate coefficient if necessary, so that the number of electrons used in the oxidation part equals the number of electrons used in the reduction part. The two half-equations are then added to produce the final balanced equation.

One reason we use half-equations is that, as we shall see in Section 16-5, reactions that occur in electrochemical cells actually do take place in distinctly separate parts of the cell; the oxidation occurring in one location, the reduction in another. In such cells the half-reactions really do exist, and when we write them we are in accord with reality. But even in nonelectrochemical reactions, where the half-reactions may or may not be real, they are useful, because they enable us to balance equations that would otherwise be difficult to balance.

There are some general rules that must be obeyed if you are to balance redox equations successfully.

1. You must know the formulas of the oxidizing and reducing agents and the formulas of the reduced and oxidized products.

2. You must not violate the conservation of mass. Whatever atoms are on the reactant portion of the reaction must also be on the product side.
3. You must not violate the conservation of charge. All electrons that are produced by the oxidation half-reaction must be used up by the reduction half-reaction.

Oxidation–reduction reactions may occur in acid, basic, or neutral solution. The method of balancing them is a little different for acid and basic solution, as we shall see below. How can we tell if a reaction is taking place in acid or basic solution? In most cases, we look for H^+ or OH^-. If H^+, or any acid, appears *on either side* of the equation, the reaction takes place in acid solution. If OH^- or any base appears on either side, the solution is basic. If neither H^+, OH^-, nor any acid or base is present, the solution is neutral. (Note that it is impossible for both H^+ and OH^- to appear together in *any* equation, whether redox or not, except when they are neutralizing each other, in which case they both appear on the left, and the product, water, on the right.)

Let us now balance the preceding equation:

$$Cr_2O_7^{2-} + H^+ + Fe^{2+} \longrightarrow Cr^{3+} + Fe^{3+} + H_2O$$

which of course takes place in acid solution.

Step 1. Find the elements that change oxidation numbers. The oxidation number of Cr in $Cr_2O_7^{2-}$ is $+6$ while in Cr^{3+} it is $+3$. Fe^{2+} in the reactants changes to Fe^{3+} in the products. The half-equations (unbalanced and incomplete) are therefore

$$\text{Oxidation:} \quad Fe^{2+} \longrightarrow Fe^{3+}$$

$$\text{Reduction:} \quad Cr_2O_7^{2-} \longrightarrow Cr^{3+}$$

Step 2. Balance each half-equation separately. This is done according to the following procedure (we will apply it first to the reduction half-equation).

(a) Balance all atoms other than H and O.

$$Cr_2O_7^{2-} \longrightarrow Cr^{3+}$$

There are two chromiums on the left, one on the right. We balance the chromiums by

$$Cr_2O_7^{2-} \longrightarrow 2Cr^{3+}$$

(b) Calculate the oxidation number on the left and on the right. Add electrons to whichever side is necessary to make up for the difference. The oxidation number of the chromium on the left is $+6$; on the right it is $+3$. Each chromium must gain three electrons. Since there are two chromiums, there is a total of six electrons gained.

$$Cr_2O_7^{2-} + 6e^- \longrightarrow 2Cr^{3+}$$

(c) Balance the half-equation so that both sides will have the same charge. The procedure for this step is different depending on whether the reaction takes place in acidic or basic solution. In acidic solution, add as many *H^+ ions* as are needed to account for the extra *positive* charge on one side or the other. Remember that each H^+ ion carries a single positive charge. (The procedure in basic solution will be given in the next example.) In this case the total charge on the left is -8; on the right $+6$. We need $14H^+$ on the left to balance the charges:

$$Cr_2O_7^{2-} + 14H^+ + 6e^- \longrightarrow 2Cr^{3+}$$

(d) Add water molecules to complete the balancing of the half-equation. In this case, $7H_2O$ is required on the right.

$$Cr_2O_7^{2-} + 14H^+ + 6e^- \longrightarrow 2Cr^{3+} + 7H_2O$$

The reduction half-equation is now complete and balanced. We will now apply the same procedure to balance the oxidation half-equation:

$$Fe^{2+} \longrightarrow Fe^{3+}$$

(a) Balance all atoms other than H and O. This is already the case, so nothing further is needed here.
(b) Add electrons to whichever side is necessary to make up for the difference in oxidation number. The oxidation number of Fe on the left is +2; on the right, +3. One electron is needed.

$$Fe^{2+} \longrightarrow Fe^{3+} + e^-$$

(c) Balance the charges. This is already done. Both sides have a total charge of +2.
(d) Add water. This step is unnecessary here, since no oxygen or hydrogen is present.

Step 3. Checking. At this point each half-equation should be balanced atomically and electrically; and the electrons must appear on *opposite* sides in the two half-equations. (Why?) Check to see that these things are correct.

Step 4. Add the two half-equations together. In doing this we want the electrons to cancel, which they can do only if the same number of electrons appears in both equations. To achieve this, we multiply the equations by coefficients so that the number of electrons produced in oxidation equals the number of electrons used in reduction.

Oxidation: $6 \times (Fe^{2+} \longrightarrow Fe^{3+} + e^-)$
Reduction: $1 \times (6e^- + 14H^+ + Cr_2O_7^{2-} \longrightarrow 2Cr^{3+} + H_2O)$

$$6Fe^{2+} \longrightarrow 6Fe^{3+} + 6e^-$$
$$6e^- + 14H^+ + Cr_2O_7^{2-} \longrightarrow 2Cr^{3+} + 7H_2O$$

Adding,

$$\cancel{6e^-} + 6Fe^{2+} + 14H^+ + Cr_2O_7^{2-} \longrightarrow 6Fe^{3+} + 2Cr^{3+} + 7H_2O + \cancel{6e^-}$$

The final balanced equation is

$$6Fe^{2+} + 14H^+ + Cr_2O_7^{2-} \longrightarrow 6Fe^{3+} + 2Cr^{3+} + 7H_2O$$

Step 5. Checking. At this point, check to see if the final equation is balanced both atomically and electrically. If the procedures have been properly carried out, it will be.

Example: Balance:

$$NO_3^- + Zn \longrightarrow Zn^{2+} + NH_4^+$$

This equation is incomplete (note that O appears only on the left and H only on the right), but with the half-equation method, we can balance it anyway, because all we need are the formulas for the oxidizing and reducing agent and for the oxidized and reduced products, and we have those. This particular reaction can be balanced in either acid or basic solution. We will balance it in basic solution.

Step 1

Oxidation: $Zn \longrightarrow Zn^{2+}$
Reduction: $NO_3^- \longrightarrow NH_4^+$

Step 2. For the oxidation half-equation:

(a) Balance all atoms other than H and O. This is already the case.
(b) Add electrons to make up for the difference in oxidation number.

$$Zn \longrightarrow Zn^{2+} + 2e^-$$

(c) Balance the charges. This is already done.
(d) This step is unnecessary here.

For the reduction half-equation:

$$NO_3^- \longrightarrow NH_4^+$$

(a) Balance all atoms other than H and O. This is already the case.
(b) Add electrons to make up for the difference in oxidation number. The oxidation number of N on the left is +5; on the right it is −3. The difference is eight electrons.

$$NO_3^- + 8e^- \longrightarrow NH_4^+$$

(c) Balance the charges. In a basic solution, we add as many *OH^- ions* as are needed to account for the extra *negative* charge on one side or the other. Of course, each OH^- ion carries a charge of −1. In this case the total charge on the left is −9; on the right it is +1. Ten OH^- ions are needed on the right.

$$NO_3^- + 8e^- \longrightarrow NH_4^+ + 10OH^-$$

(d) Add water molecules to complete the balancing. The left side contains 3O; the right side 10O and 14H. To balance, we add $7H_2O$ to the left side.

$$NO_3^- + 7H_2O + 8e^- \longrightarrow NH_4^+ + 10OH^-$$

Step 3. Checking. Both half-equations are balanced atomically and electrically, and the electrons appear on opposite sides.

Step 4

$$\begin{array}{rcl} 4(Zn & \longrightarrow & Zn^{2+} + 2e^-) \\ 1(8e^- + 7H_2O + NO_3^- & \longrightarrow & NH_4^+ + 10OH^-) \\ \hline 4Zn + \cancel{8e^-} + 7H_2O + NO_3^- & \longrightarrow & 4Zn^{2+} + NH_4^+ + 10OH^- + \cancel{8e^-} \end{array}$$

Example: Complete and balance the equation:

$$Cr(OH)_4^- + H_2O_2 \longrightarrow CrO_4^{2-} + H_2O \text{ (basic solution)}$$

Answer:

Oxidation:

$$Cr(OH)_4^- \longrightarrow CrO_4^{2-}$$

$$\overset{+3}{Cr}(OH)_4^- \longrightarrow \overset{+6}{Cr}O_4^{2-} + 3e^-$$

$$4OH^- + Cr(OH)_4^- \longrightarrow CrO_4^{2-} + 3e^-$$

$$4OH^- + Cr(OH)_4^- \longrightarrow CrO_4^{2-} + 3e^- + 4H_2O$$

Check.

Reduction:

$$H_2O_2 \longrightarrow H_2O$$

$$\overset{-1}{H_2O_2} + 2e^- \longrightarrow \overset{-2}{H_2O}$$

we need two electrons, because there are two oxygens and each needs one to go from −1 to −2

$$H_2O_2 + 2e^- \longrightarrow H_2O + 2OH^-$$

At this point there are 2H and 2O on the left; 4H and 3O on the right. We need 1 H_2O on the left:

$$\cancel{H_2O} + H_2O_2 + 2e^- \longrightarrow \cancel{H_2O} + 2OH^-$$

Because 1 H_2O appears on both sides, it cancels. Check.

$$2(4OH^- + Cr(OH)_4^- \longrightarrow CrO_4^{2-} + 4H_2O + 3e^-)$$
$$3(2e^- + H_2O_2 \longrightarrow 2OH^-)$$

We multiply each half-equation by appropriate factors so that when we add them together, the electrons will cancel.

$$\cancel{6e^-} + 8OH^- + 2Cr(OH)_4^- + 3H_2O_2 \longrightarrow 2CrO_4^{2-} + 8H_2O + 6OH^- + \cancel{6e^-}$$

Removing the $6OH^-$ that occur on both sides of the equation gives us

$$2OH^- + 2Cr(OH)_4^- + 3H_2O_2 \longrightarrow 2CrO_4^{2-} + 8H_2O \qquad \text{(balanced)}$$

In steps 2c and 2d in acid solution, we balanced oxygen and hydrogen by using only H^+ and H_2O. In basic solution, we used only OH^- and H_2O. The reason is that these are the only oxygen and hydrogen-containing species present in these solutions. (Actually, the H^+ is in the form of H_3O^+, but it is customary to write the simplified form when balancing oxidation–reduction equations.) In acid solution we can only use H^+ and not OH^-, because the concentration of OH^- is insignificant (Section 10-3). In basic solution, the situation is reversed, and we can only use OH^-. The rules are very simple and easy to remember: in acid solution we add H^+ ions to the side with the greater negative charge; in basic solution we add OH^- ions to the side with the greater positive charge. How can the student tell if the reaction takes place in acid or basic solution? Look at the given equation. If H^+ or an acid appears in it, on either side, the solution is acidic. If OH^- or a base appears in it, on either side, the solution is basic. If neither an acid nor a base is involved, there are two possibilities. Either the solution is neutral (and step 2c will not have to be used at all), or the student will be told whether to balance the equation in acidic or basic solution.

The redox equations we have balanced so far are ionic equations. They are absolutely correct chemical equations, but for certain purposes they are difficult to work with. For example, suppose that we wish to do a weight–weight problem on the preceding equation: How many grams of $Cr(OH)_4^-$ are required to react completely with 100 g of H_2O_2? We could do the calculation (Section 3-4), but it would be of little use, because we cannot weigh out $Cr(OH)_4^-$. We can isolate ions only in the form of salts, where both positive and negative ions are present. For this reason, and others, it is often necessary to balance molecular redox equations. The procedure is the same, but there is an initial and a final step.

Example: Complete and balance:

$$KMnO_4 + H_2SO_4 + K_2C_2O_4 \longrightarrow MnSO_4 + CO_2$$

Answer: Since sulfuric acid is present, the solution is acidic. A calculation of oxidation numbers will show us that manganese is reduced, and carbon is oxidized. In the half-equation method we use ions where these exist. (In Section 11-1 we saw which compounds are completely ionized in water solution.) Thus, $KMnO_4$ and $MnSO_4$ are completely dissociated in water solution, and for one half-equation we write

$$MnO_4^- \longrightarrow Mn^{2+}$$

Potassium oxalate ($K_2C_2O_4$) is also ionized, but CO_2 is not:

$$C_2O_4^{2-} \longrightarrow CO_2$$

Note that we try to be as accurate as possible in writing our formulas as ions or molecules. From this point on, the procedure is the same as before.

Oxidation:

$$C_2O_4^{2-} \longrightarrow CO_2$$
$$C_2O_4^{2-} \longrightarrow 2CO_2$$
$$\overset{+3}{C_2}O_4^{2-} \longrightarrow 2\overset{+4}{C}O_2 + 2e^-$$

Check.

Reduction:

$$\overset{+7}{Mn}O_4^- \longrightarrow \overset{+2}{Mn^{2+}}$$
$$MnO_4^- + 5e^- \longrightarrow Mn^{2+}$$
$$8H^+ + MnO_4^- + 5e^- \longrightarrow Mn^{2+}$$
$$8H^+ + MnO_4^- + 5e^- \longrightarrow Mn^{2+} + 4H_2O$$

Check.

$$5(C_2O_4^{2-} \longrightarrow 2CO_2 + 2e^-)$$
$$2(8H^+ + MnO_4^- + 5e^- \longrightarrow Mn^{2+} + 4H_2O)$$

$$5C_2O_4^{2-} \longrightarrow 10CO_2 + \cancel{10e^-}$$
$$16H^+ + \cancel{10e^-} + 2MnO_4^- \longrightarrow 2Mn^{2+} + 8H_2O$$
$$16H^+ + 5C_2O_4^{2-} + 2MnO_4^- \longrightarrow 10CO_2 + 2Mn^{2+} + 8H_2O \quad \text{(balanced)}$$

This is a complete and balanced *ionic* equation. We now turn it into a complete and balanced *molecular* equation by adding the missing ions. Note that we must be careful always to add the same species in the same quantities to both sides. Having devoted some care and effort to obtaining a balanced equation, we do not want to do anything to unbalance it now. The oxalate ion was given as $K_2C_2O_4$. We therefore add $10K^+$ ions to each side.

$$5K_2C_2O_4 + 16H^+ + 2MnO_4^- \longrightarrow 10CO_2 + 2Mn^{2+} + 8H_2O + 10K^+$$

We need $8SO_4^{2-}$ ions for the $16H^+$. These will balance all the Mn^{2+} and K^+ on the right and leave one SO_4^{2-} left over.

$$5K_2C_2O_4 + 8H_2SO_4 + 2MnO_4^- \longrightarrow 10CO_2 + 2MnSO_4 + 8H_2O + 5K_2SO_4 + SO_4^{2-}$$

Finally, 2 more K^+ ions will balance the remaining negative ions.

$$5K_2C_2O_4 + 8H_2SO_4 + 2KMnO_4 \longrightarrow 10CO_2 + 2MnSO_4 + 8H_2O + 6K_2SO_4$$
(complete and balanced)

At this point we make a final check to see if the equation is indeed balanced.

Many students will find that, after some practice, the half-equation method will allow virtually any redox equation to be balanced easily and fairly quickly. However, this is not the only method. Another method, the oxidation-number method, is discussed in Appendix D; some students will find this one more comfortable. Still other methods exist which are not discussed in this book. Very simple redox equations, such as $Pb^{2+} + Na \rightarrow Na^+ + Pb$, can be balanced by inspection.

16-4 Oxidation-Reduction Titrations

In Section 7-3 we saw that the equivalent weight of an acid or base in neutralization reactions is the weight of a particular acid or base that will give up or take 1 mole of protons. For HCl the equivalent weight is 36.46 g; for H_2SO_4 it is $\frac{98}{2} = 49$ g. For NaOH the equivalent weight is 40 g; for $Ba(OH)_2$ it is $171.34/2 = 85.670$ g. For oxidation-reduction reactions we define the **equivalent weight of an oxi-**

Table 16-1 Selected List of Oxidation-Reduction Indicators

Indicator	Color	
	Oxidized	*Reduced*
Phenosafranine	Red	Colorless
Indigo tetrasulfonate	Blue	Colorless
Methylene blue	Blue	Colorless
Diphenylamine	Violet	Colorless
Diphenylamine-sulfonic acid	Red-violet	Colorless
Ferroin	Pale blue	Red

dizing agent as the *weight of material that will accept 1 mole of electrons.* The **equivalent weight of a reducing agent** is the *weight of material that will liberate 1 mole of electrons.* Titrations involving oxidation-reduction reactions can be worked in the same way as acid-base titrations. Of course, we will need a suitable indicator to let us know when the oxidation-reduction equivalence point has been reached. Table 16-1 shows some of the indicators used in redox titrations.

The last indicator listed in Table 16-1, ferroin, deserves some special mention. Ferroin undergoes a reversible oxidation-reduction reaction which may be written

$$(Phen)_3Fe^{3+} + e^- \rightleftharpoons (Phen)_3Fe^{2+}$$

The Phen in the equation is a symbol for a class of organic compounds called 1,10-phenanthrolines:

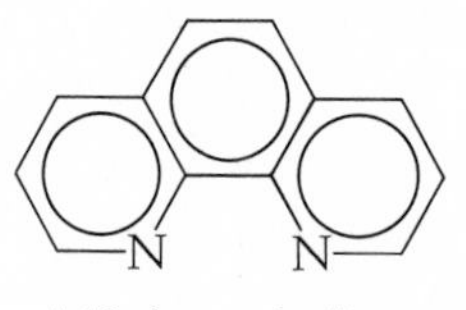

1,10-phenanthroline

Of all oxidation-reduction indicators, ferroin is almost ideal. Its color change is very sharp and its solutions are easily prepared and quite stable. In contrast to many other indicators, the oxidized form of ferroin is quite stable toward strong oxidizing agents. The indicator reaction is rapid and reversible.

Let us consider an example of how to calculate equivalent weights. Potassium permanganate, $KMnO_4$, is used as an oxidizing agent in a number of reactions. In acid solution its half-reaction is (as we have already seen)

$$MnO_4^- + 8H^+ + 5e^- \longrightarrow Mn^{2+} + 4H_2O$$

From this half-reaction we see that *1 mole* of $KMnO_4$ will use up *5 moles* of electrons. Therefore, when $KMnO_4$ undergoes the above reaction, its equivalent weight is 158.03/5 = 31.606 g/equiv. That is, 31.606 g of $KMnO_4$ will use up 1 mole of electrons.

In basic solution $KMnO_4$ acts as an oxidizing agent by the half-reaction

$$MnO_4^- + 2H_2O + 3e^- \longrightarrow MnO_2 + 4OH^-$$

The equivalent weight of $KMnO_4$ in this reaction is 158.03/3 = 52.677 g/equiv. For the half-reaction in basic solution, 52.677 g of $KMnO_4$ will use 1 mole of electrons. Here we have illustrated the important principle that when we are dealing with a redox reaction and wish to calculate the equivalent weight of the oxidizing or reducing agent, we *must* know the half-reaction. As we just saw, the equivalent weight of $KMnO_4$ may be 31.606 g/equiv or 52.677 g/equiv. *It depends on the specific half-equation.*

A 1 normal solution of an oxidizing or reducing agent contains one equivalent of the agent in 1 liter of solution. (Therefore, the normality of the solution depends on which reaction you intend to run.)

$$N \times \text{liters} = \text{equiv} = \frac{\text{g}}{\text{equiv wt}} = \frac{\text{g}}{\text{mol wt/number of electrons}}$$

Example: The following are some common oxidizing agents used for preparing standard solutions. Calculate the weight of each which you will need to prepare 1.000 liter of a 1.000 *N* solution. What is the corresponding molarity of each solution?

Reagent	*Half-reaction*
(a) Potassium permanganate, $KMnO_4$ (in acid)	$MnO_4^- + 8H^+ + 5e^- \rightarrow Mn^{2+} + 4H_2O$
(b) Cerium(IV) ammonium nitrate, $(NH_4)_2Ce(NO_3)_6$	$Ce^{4+} + e^- \rightarrow Ce^{3+}$
(c) Periodic acid, H_5IO_6	$H_5IO_6 + H^+ + 2e^- \rightarrow IO_3^- + 3H_2O$
(d) Potassium dichromate, $K_2Cr_2O_7$	$Cr_2O_7^{2-} + 14H^+ + 6e^- \rightarrow 2Cr^{3+} + 7H_2O$

Answer:

(a) The molecular weight of $KMnO_4$ is 158.03. From the half-reaction its equivalent weight is 158.03/5 = 31.606 g/equiv. Therefore, 31.606 g of $KMnO_4$ in 1.000 liter of solution will yield a 1.000 *N* solution. Its molarity will be

$$\frac{1.000 \text{ equiv/liter}}{5 \text{ equiv/mole}} = 0.2000\ M \text{ (mole/liter)}$$

(b) The molecular weight of cerium(IV) ammonium nitrate, $(NH_4)_2Ce(NO_3)_6$, is 548.2. Since only one electron is transferred in its half-reaction, its equivalent weight is equal to its molecular weight. 548.2 g of cerium(IV) ammonium nitrate in 1.000 liter of solution will produce a 1.000 *N* as well as a 1.000 *M* solution.

(c) The molecular weight of H_5IO_6 is 227.94. Since two electrons are transferred in its half-reaction, its equivalent weight is 227.94/2 = 113.97 g/equiv. It takes 113.97 g of H_5IO_6 in 1.000 liter of solution to make a 1.000 *N* solution. Its molarity is 0.5000 *M*.

(d) The molecular weight of $K_2Cr_2O_7$ is 294.18. From its half-reaction we see that its equivalent weight is 294.18/6 = 49.030 g/equiv. It takes 49.030 g of $K_2Cr_2O_7$ in 1.000 liter of solution to prepare a 1.000 *N* solution. Its molarity is 0.1667 *M*.

Example: Potassium oxalate ($K_2C_2O_4$) reacts with $KMnO_4$ in acid solution to give Mn^{2+} ion, CO_2, and H_2O. We balanced this equation in Section 16-3.

(a) Calculate the equivalent weight for both the oxidizing and reducing agent.

(b) 50.0 ml of an 0.100 *N* solution of $K_2C_2O_4$ is titrated with a 0.200 *N* $KMnO_4$ solution. How many milliliters of this solution will completely oxidize the oxalate?

Answer:

(a) $K_2C_2O_4$ is the reducing agent and $KMnO_4$ is the oxidizing agent. For the reducing agent ($K_2C_2O_4$) the equivalent weight = mol wt/2 = 166.2/2 = 83.1 g/equiv. For $KMnO_4$ the equivalent weight = mol wt/5 = 158.03/5 = 31.606 g/equiv.

(b) The main reason we use equivalents and normality at all is that they allow us to use this very simple formula (see page 178): At equivalence,

$$\text{ml}_{\text{oxid}} \times N_{\text{oxid}} = \text{ml}_{\text{red}} \times N_{\text{red}}$$

so

$$50.0 \text{ ml} \times 0.100\ N = \text{ml} \times 0.200\ N$$

$$\text{ml} = \frac{50.0 \text{ ml} \times 0.100\ N}{0.200\ N} = 25.0 \text{ ml of } 0.200\ N\ KMnO_4$$

One of the principal uses of dichromate ion involves the titration of iron(II).

$$6Fe^{2+} + Cr_2O_7^{2-} + 14H^+ \rightleftharpoons 6Fe^{3+} + 2Cr^{3+} + 7H_2O$$

Solutions of dichromate have an orange hue; the color is not intense enough to give a good end point, so a redox indicator is ordinarily used, most commonly diphenylaminesulfonic acid. The indicator is colorless in its reduced state and red-violet when oxidized.

Example: For the reaction above involving the dichromate ion and Fe(II):

(a) Calculate the equivalent weight for both the oxidizing and reducing agent

(b) 50.0000 g of $K_2Cr_2O_7$ is dissolved in enough water to make 250 ml of solution. 50.000 g of potassium ferrocyanide $K_4Fe(CN)_6$ is also dissolved in enough water to yield 250 ml of solution. How many milliliters of the dichromate solution will completely oxidize 250 ml of the ferrocyanide solution?

Answer:

$$\text{Oxidation:} \quad Fe^{2+} \longrightarrow Fe^{3+} + e^-$$

$$\text{Reduction:} \quad 14H^+ + Cr_2O_7^{2-} + 6e^- \longrightarrow 2Cr^{3+} + 7H_2O$$

(a) Equivalent weight of the oxidizing agent ($K_2Cr_2O_7$) is its mol wt/6 = 294.18/6 = 49.030 g/equiv. The equivalent weight of the reducing agent, $K_4Fe(CN)_6$, is its mol wt/1 = 368.346/1 = 368.346 g/equiv.

(b) We will first calculate the normality of each solution.

$$\text{normality} = \frac{\text{equivs}}{\text{liters}}$$

$$\text{equivs} = \frac{\text{g}}{\text{equiv wt}}$$

$$\text{normality} = \frac{\text{g}}{(\text{equiv wt})(\text{liters})}$$

$$K_2Cr_2O_7: \quad N = \frac{50.0000 \text{ g}}{(49.030 \text{ g/equiv})(0.250 \text{ liter})} = 4.08\ N$$

$$K_4Fe(CN)_6: \quad N = \frac{50.0000 \text{ g}}{(368.346 \text{ g/equiv})(0.250 \text{ liter})} = 0.543\ N$$

The normalities are

$$K_2Cr_2O_7: \quad 4.08\ N$$

$$K_4Fe(CN)_6: \quad 0.543\ N$$

At the equivalence point,

$$\text{liters} \times \text{normality} = \text{liters} \times \text{normality}$$

$$\text{liters} \times 4.08\ N = 0.250 \text{ liter} \times 0.543\ N$$

$$\text{liters} = \frac{0.250 \times 0.543}{4.08} = 0.0333 \text{ liter}$$

Thus, 0.0333 liter or 33.3 ml of the $K_2Cr_2O_7$ solution will completely oxidize the 250 ml of the ferrocyanide solution.

We can also use oxidation-reduction titrations to determine equivalent weights of unknown compounds, in the same way that we did for acid–base titrations (Section 10-11).

16-5 Voltaic Cells

In Section 16-2 we defined an oxidation-reduction reaction as one that involves the transfer of electrons. In Section 16-4 we saw that we can divide any redox reaction into two parts: an oxidation part and a reduction part, and write a half-equation for each part. For example, consider the reaction

$$Zn + Cu^{2+} \longrightarrow Zn^{2+} + Cu$$

The half-equations are

$$Zn \longrightarrow Zn^{2+} + 2e^-$$
$$Cu^{2+} + 2e^- \longrightarrow Cu$$

If we put a piece of zinc into a solution containing Cu^{2+} ions, we can actually see this reaction take place (Figure 16-1). We will see a deposit of metallic copper appear on the surface of the zinc (this is called **plating out**). We cannot see the Zn^{2+} ions as they go into solution, but we can see the blue color of the aqueous Cu^{2+} ions get lighter and eventually disappear as all the Cu^{2+} ions get reduced (Zn^{2+} ions in water are colorless). In this beaker the zinc atoms are *directly* giving electrons to the copper ions.

It was one of the greatest discoveries of mankind that we can make the same reaction take place even if the copper ions and zinc metal are *not* directly in contact. If we put the copper ions and zinc metal into separate containers but connect the two by a metal wire and introduce a salt bridge, then electrons will still be transferred, but **the electrons will now flow through the wire** in order to get from the zinc to the copper ions. Figure 16-2 shows such a system, called an **electrochemical cell.** The oxidation half-reaction $Zn \rightarrow Zn^{2+} + 2e^-$ takes place in the left compartment (and the zinc plate gradually dissolves, turning into Zn^{2+} ions); the reduction half-reaction $Cu^{2+} + 2e^- \rightarrow Cu$ takes place in the right compartment, where we will see copper plating out.

The **salt bridge** is necessary here in order to keep the two solutions electrically neutral. For example, when the reaction $Zn \rightarrow Zn^{2+} + 2e^-$ takes place in the left compartment, additional positive ions are being added to the solution. These must be matched by additional negative ions, which come from the salt bridge. In the other compartment, positive ions (Cu^{2+}) are *leaving* the solution; and these are replaced by positive ions that flow from the salt bridge (thus, the salt bridge remains neutral also).

Alternatively, we could use a porous plug instead of a salt bridge (Figure 16-3). This plug allows the solutions to be in liquid contact with each other and at the same time prevents the two solutions from mixing. But without the salt bridge or the porous plug, the cell will not function. Neither oxidation nor reduction will take place, and no electrons will flow over the wire. Another way of saying it is that a complete circuit is necessary for electrons to flow through the wire. The flow of

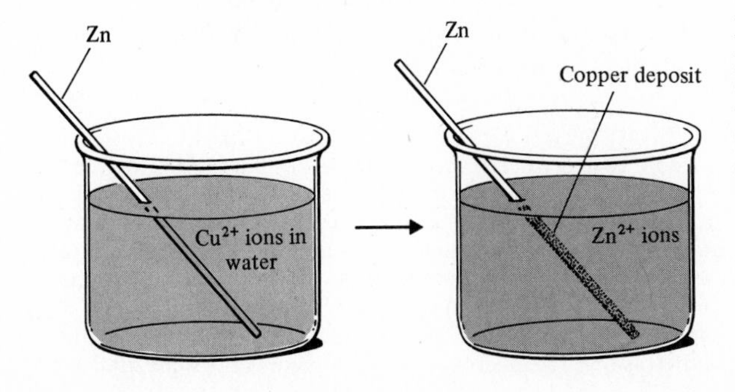

Figure 16-1 If a piece of zinc is put into an aqueous solution containing Cu^{2+} ions, some of the zinc will be converted to Zn^{2+} ions, which will dissolve in the water; the Cu^{2+} ions will plate out as copper metal.

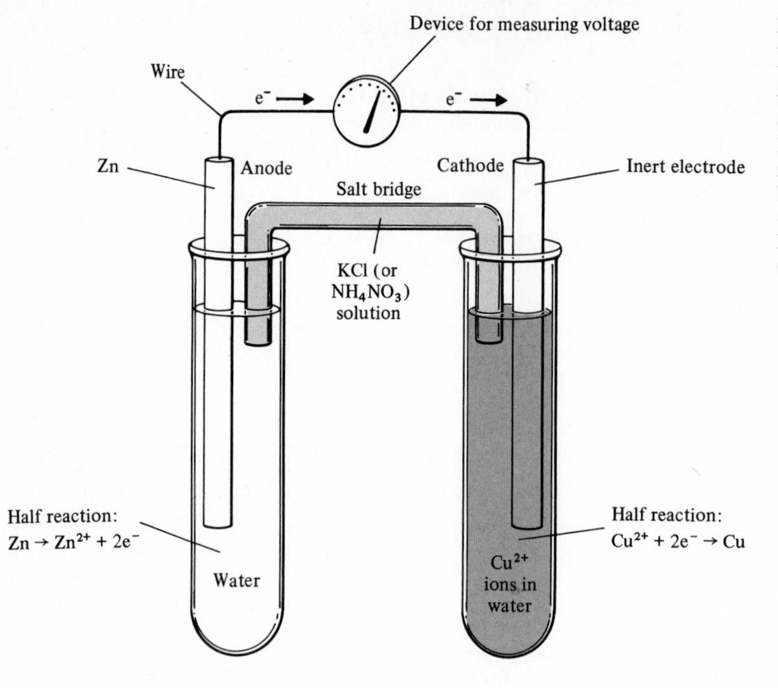

Figure 16-2 A Zn-Cu voltaic cell with a salt bridge. The salt bridge is usually composed of a concentrated solution of KCl or NH_4NO_3. The function of the salt bridge is to keep electrical neutrality in the half-cells as the cell operates.

electrons through the wire constitutes an electrical current, and this type of cell produces such a current.

The two pieces of solid matter in the cell are called **electrodes.** The one where oxidation takes place is the **anode,** while reduction takes place at the **cathode.** In this particular cell, the anode must be made of zinc (or else the cell would have a different reaction), but the cathode may be copper, or platinum, or carbon, or some other solid, as long as electrons can pass through it and as long as it is inert under these conditions. All it does is provide a place for the copper ions to plate out on.

The type of electrochemical cell pictured in Figures 16-2 and 16-3 was invented in 1799 by Alessandro Volta (1745–1827), following some earlier work by Luigi Galvani (1737–1798).* This invention had enormous practical significance, because it was the first source of continuous moving electricity in the history of the world. Before 1799, the only electricity known had been lightning, and the electricity of electroscopes, which is a kind of artificial lightning on a very small scale. In both these cases, a flow of electrons occurs in an instantaneous burst and is immediately over. Volta's invention enabled mankind to utilize continuous electrical currents, with the results with which we are all familiar. Volta may with some justice be called the inventor of electricity.† Electrochemical cells that produce electricity are called **voltaic cells,** in honor of this great man. Voltaic cells produce *direct current* (dc) and today we have electrical generators that work on different principles, and some which provide alternating current, but we still make considerable use of voltaic cells (Section 16-9) and it is unlikely that the other types of electricity-producing devices could have been invented had not the voltaic cell come first.

*Galvani discovered that frogs' legs would twitch when placed in contact with a metal. He proposed that electricity was coming from the frog and called it "animal electricity." Volta investigated this phenomenon, and found that: (1) The frog's leg was not necessary; the electricity was coming from the metal, and the frog was just an indicator that electricity was there; other indicators could also be used. (2) A single pure metal would not do. It had to be two metals in contact. As we know today, it is actually one metal and the ions of a second metal.

†It is a common misconception that Benjamin Franklin (1706–1790) invented electricity. Franklin's great discovery, in which he risked death, was that the electricity in lightning was the same kind as the electricity in electroscopes.

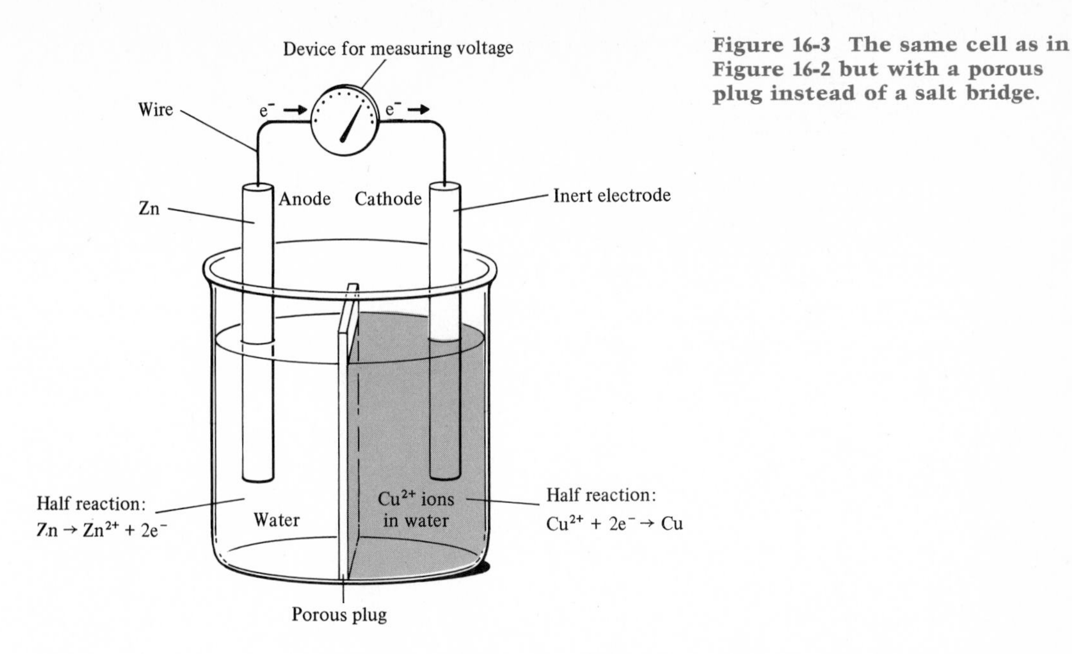

Figure 16-3 The same cell as in Figure 16-2 but with a porous plug instead of a salt bridge.

16-6 Measurement of Electromotive Force in Voltaic Cells

In Figures 16-1, 16-2, and 16-3, we show a reaction taking place between zinc metal and copper ions. What would happen if we tried the opposite situation: a reaction between copper metal and zinc ions? In this case *no reaction would take place,* whether we put a piece of copper directly into a solution containing zinc ions, or we set them up in a cell such as that shown in Figure 16-2 or 16-3. One way of trying this experiment would be to put Zn^{2+} ions into the left compartment and make the cathode out of copper (Figure 16-4). We would then have a cell which is, in principle, reversible. We can imagine the zinc giving electrons to the Cu^{2+} ions (with the electrons flowing from left to right), or the copper giving electrons to the Zn^{2+} ions (with the electrons flowing from right to left). In practice, only the former reaction would take place spontaneously, and the electrons would flow from left to

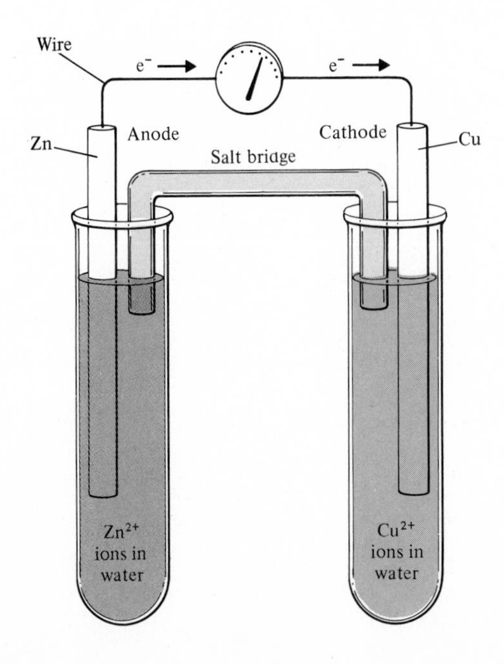

Figure 16-4 The same cell as in Figure 16-2 but this time with a copper cathode and with Zn^{2+} ions in the left compartment. These changes will make no difference in the behavior of the cell, which will operate as before.

right. We may refer to any voltaic cell constructed in this manner as a **reversible voltaic cell.**

Is there any way we could have predicted this—that is, which way the reaction would go—*before* we carried out the experiments? The answer is yes, as we shall shortly see.

What is the reason the reaction proceeds in one direction but not the other way? In Section 8-10, we saw that reactions proceed spontaneously only when the free-energy change (ΔG) is negative. The reaction between copper and zinc ions

$$Cu + Zn^{2+} \not\rightarrow Cu^{2+} + Zn$$

does not happen spontaneously because it would result in an increase in free energy. (We can *make* it happen, but not spontaneously—see Section 16-12.)

In any given voltaic cell consisting of two half-cells, current will flow in one direction only, because in that direction ΔG is negative. But *quantitatively,* all cells do not have the same potential. In some cells the electrons of the reducing agent have a very large tendency to be transferred to the oxidizing agent; in other cells this tendency is smaller (the tendency is related to the magnitude of $-\Delta G$; the larger this value, the greater the tendency). A measure of this tendency is called the **electromotive force (EMF),** the **voltage,** or the **difference in potential**—all three expressions mean the same thing. As an analogy, if we pour a bucket of water from a tenth-story window, it will hit the ground with a greater force than the same water poured from a second-story window. The water on the tenth story has a higher potential energy than the water on the second story.

We can actually measure the EMF in any cell by a device called a *potentiometer.* This device is able to measure the EMF (or voltage) when *no current flows.* It is important to make the measurement in this way, since the potential difference (voltage) between the two half-cells changes as long as current is allowed to flow between them.

To measure the EMF, it will not be necessary to describe the workings of the potentiometer. It is sufficient to know that the EMF between the two half-cells can be measured experimentally with great accuracy. By actually performing the EMF measurement we get an additional piece of information about the voltaic cell—we can tell which half-cell is the cathode and which is the anode. (If our cell is a reversible one such as that in Figure 16-4, we cannot always predict this in advance.) If by mistake we reverse the anode and cathode connections, the potentiometer will not read anything. By getting an EMF reading on the potentiometer, we distinguish between the cathode and the anode.

The EMF value for a cell made up of two specific half-cells is a constant for those two half-cells provided that the concentrations, temperature, and pressure (if gases are involved) are constant. The EMF values do change with concentration, temperature, and pressure. In order to make comparisons between different measurements, chemists have agreed on standards for these conditions. Thus, EMF values are normally reported at a temperature of 25°C, a pressure (if gases are involved) of 1 atm, and at concentrations of 1.0 *M* for all solutes present. It is because EMF values change with changes in concentration that potentiometers must be used only when no current is flowing. It is obvious that concentrations must be changing when the current flows.

We have been discussing the EMF difference between two half-cells. Is it possible to measure the EMF of a single half-cell? This would be the potential difference between the metal electrode and its solution. If it were possible to measure the EMF of each half-cell, we would do it and list these values in a table. By consulting this table we could predict for any reversible voltaic cell the cathode reaction, the anode reaction, and the EMF of the cell. We would do this by writing the equation for each half-cell as a reduction reaction. By consulting the table, the half-cell with the higher reduction EMF value would act as the cathode while the other half-cell is forced to be the anode. This would be a great help in predicting which way redox reactions would go. Unfortunately, it is now accepted that there is no completely satisfactory method for measuring the absolute difference in electrical potential between a metal electrode and a solution containing its ions, that is, the EMF of a half-cell. Therefore, although we can measure the EMF difference between two half-cells, we cannot tell what their individual EMF values are. However let us take heart, all is not lost.

16-7 Standard Reduction Potentials and the Standard Hydrogen Electrode

Even though we cannot measure the absolute EMF of a half-cell, we can still set up a bookkeeping system that will accomplish the same goals. This is done by arbitrarily selecting one half-cell and setting its half-cell EMF value at zero. We can then measure the potentials of all other half-cells by combining them with this standard reference cell. The EMF value we measure on the potentiometer is then the EMF of the unknown half-cell, and we can construct a table of reduction EMFs and use them as we had hoped to do in the previous section. The values we obtain in this manner are called **standard reduction potentials** and are given the symbol $\mathcal{E}°$.

The standard reference cell against which all other half-cells have been measured is the hydrogen electrode half-cell, which is shown in Figure 16-5. Of course, our standard cell will have H^+ (actually H_3O^+) at a concentration of 1.0 *M*, and H_2 at a pressure of 1 atm, and we will make our measurements at 25°C. The standard hydrogen electrode is usually abbreviated as **S.H.E.** The negative ion that accompanies the H^+ may be any anion as long as it allows $[H^+] = 1.0\ M$. The half-cell reaction of S.H.E. written as a reduction is

$$2H^+ + 2e^- \longrightarrow H_2$$

Let us take an example to see how this actually works. Say that we want to calculate the EMF of the reaction

$$Cu^{2+} + 2e^- \longrightarrow Cu$$

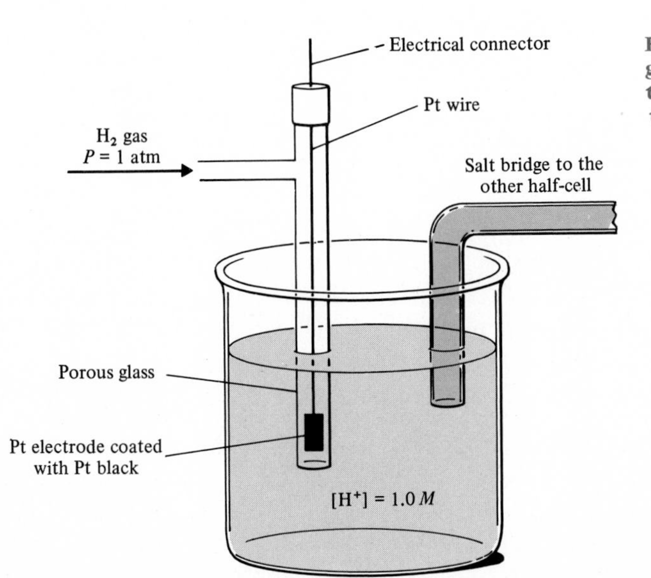

Figure 16-5 The Standard Hydrogen Electrode. Note the concentration of hydrogen ion in solution is 1.0 *M*.

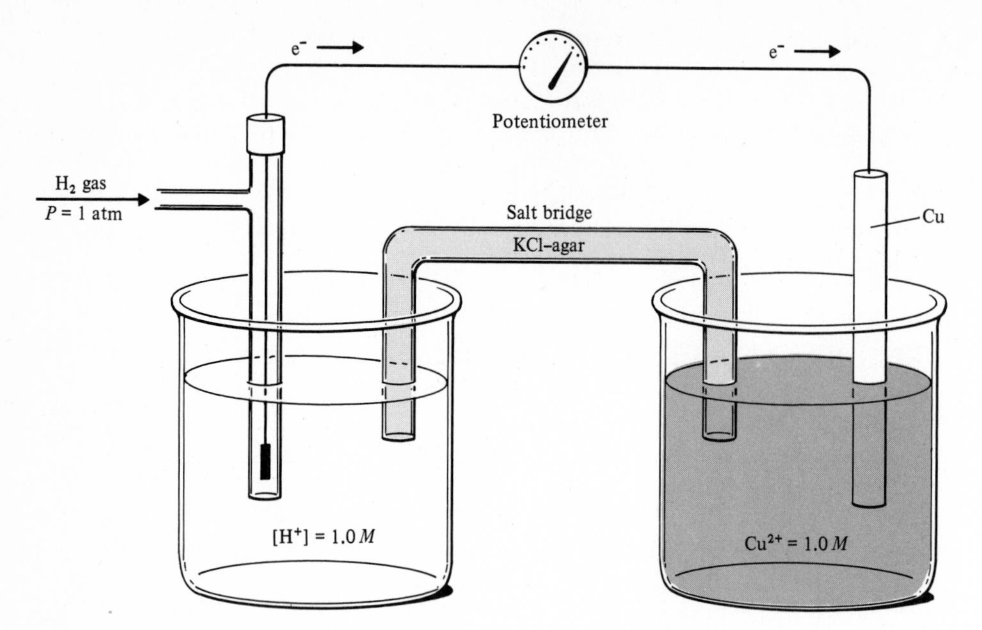

Figure 16-6 The measurement of the standard electrode potential of copper using the standard hydrogen electrode (S.H.E.) as a reference half-cell.

To do this we would experimentally set up a copper half-cell against the S.H.E. (Figure 16-6)*. The potentiometer measurement of the EMF of this cell gives $\mathcal{E}° = +0.340$ volts. Since the S.H.E. is zero, *by definition*, the half-cell EMF of the Cu half-cell is $\mathcal{E}° = +0.340$ V. Let us return to a point we made earlier concerning the experimental setup: How does one tell which half-cell is the anode and which is the cathode? The actual experimental setup was worked out many years ago. What we must do is set up the cell as in Figure 16-6. If the potentiometer shows a positive EMF, then as set up, the left half-cell is the anode (oxidation) and the right half-cell is the cathode (reduction). For the copper-S.H.E. cell, the copper ions are reduced at the cathode while the hydrogen is oxidized at the anode.

		$\mathcal{E}°$
Anode:	$H_2 \longrightarrow 2H^+ + 2e^-$	0.00
Cathode:	$Cu^{2+} + 2e^- \longrightarrow Cu$	+0.340 V

The overall cell reaction is

$$Cu^{2+} + H_2 \rightarrow Cu + 2H^+$$

$\mathcal{E}°$ for the cell = +0.340 V.

If we try a similar experimental setup to measure the EMF of the zinc half-cell ($Zn^{2+} + 2e^- \rightarrow Zn$) against the S.H.E. as the anode (left-half cell), the potentiometer will not show a positive voltage. This means that the reduction potential of the S.H.E. is stronger than the reduction potential of the zinc half-cell. In such a situation the potentiometer will show a positive EMF only if we physically reverse the potentiometer connections. For the case in point, the $\mathcal{E}°$ for the cell will be +0.76 V.

*Note that in this half-cell we have present both Cu^{2+} ions and a copper electrode, so that in theory, the reaction could go either way.

$$Cu \rightleftharpoons Cu^{2+} + 2e^-$$

If we did not know the EMF of this half-cell before we carried out the experiment, we would not know in advance which way the reaction would spontaneously proceed when connected to the hydrogen half-cell.

		$\mathcal{E}°$
Anode:	$Zn \longrightarrow Zn^{2+} + 2e^-$	+0.76 V
Cathode:	$2H^+ + 2e \longrightarrow H_2$	0.00

The overall cell reaction is

$$2H^+ + Zn \longrightarrow Zn^{2+} + H_2$$

$\mathcal{E}°$ for the cell = +0.76 V.

Note that the $\mathcal{E}°$ of the anode reaction is +0.76 V. This reaction is an oxidation. As a reduction, $Zn^{2+} + 2e^- \rightarrow Zn$, the $\mathcal{E}°$ is −0.76 V and will appear as such in a table of standard reduction potentials.

In a similar manner we can set up any half-cell against the S.H.E. under standard conditions, and the measured EMF becomes the standard $\mathcal{E}°$ for that half cell. This has, of course, been done for many half-cells, and a list of standard electrode potentials is given in Table 16-2. (Note that it is not always necessary to use the S.H.E. as our reference cell. For example, once we have determined that $\mathcal{E}°$ for $Zn^{2+} + 2e^- \rightarrow Zn$ is −0.76 V, we can then set up another half-cell against *that* one; etc. In Section 16-9 we shall see some common half-cells which are frequently used as reference cells.)

Every half-reaction in Table 16-2 can be written either as an oxidation or a reduction. The absolute value of $\mathcal{E}°$ is the same either way, *but the sign is reversed*. For example:

$$MnO_4^- + 8H^+ + 5e^- \longrightarrow Mn^{2+} + 4H_2O \qquad \mathcal{E}° = 1.49 \text{ V}$$

$$Mn^{2+} + 4H_2O \longrightarrow MnO_4^- + 8H^+ + 5e^- \qquad \mathcal{E}° = -1.49 \text{ V}$$

The table could be set up either as a list of oxidations or reductions, but it is customary to print such tables as reduction reactions rather than as oxidations. All reactions listed in Table 16-2 which have *positive* $\mathcal{E}°$ values will proceed *in the direction shown in the table* when placed against the S.H.E. under standard conditions. This means, for example, that F_2 will be reduced to F^-, and MnO_2 will be reduced to Mn^{2+}. In the S.H.E., H_2 will be oxidized to H^+. All reactions in Table 16-2 that have *negative* $\mathcal{E}°$ values will proceed *in the opposite direction to that shown in the table,* when compared to the S.H.E., so that, for example, Al will be oxidized to Al^{3+}, and Cr^{2+} to Cr^{3+}. In the S.H.E., H^+ will be reduced to H_2.

We can use Table 16-2 and similar tables for two purposes.

1. We can calculate $\mathcal{E}°$ for a cell made up of any two half-cells, provided both half-cells are at standard conditions. To calculate this value we:

(a) Write the equation for each half cell.
(b) Write the $\mathcal{E}°$ value for each half cell, giving each its proper sign.
(c) Add the two $\mathcal{E}°$ values algebraically.

This gives the $\mathcal{E}°$ value for the whole cell.

Example: Calculate the $\mathcal{E}°$ values ($\mathcal{E}°_{cell}$) for these reactions

(a) $Br_2 + Cu \longrightarrow 2Br^- + Cu^{2+}$
(b) $2La^{3+} + 3S^{2-} \longrightarrow 2La + 3S$

Answer:

(a) $Br_2 + 2e^- \longrightarrow 2Br^- \qquad \mathcal{E}° = +1.06 \text{ V}$

$Cu \longrightarrow Cu^{2+} + 2e^- \qquad \mathcal{E}° = -0.34 \text{ V}$

$\mathcal{E}°_{cell} = +1.06 \text{ V} - 0.34 \text{ V} = +0.72 \text{ V}$

(b) $La^{3+} + 3e^- \longrightarrow La \qquad \mathcal{E}° = -2.37 \text{ V}$

$S^{2-} \longrightarrow S + 2e^- \qquad \mathcal{E}° = +0.51 \text{ V}$

$\mathcal{E}°_{cell} = -2.37 \text{ V} + 0.51 \text{ V} = -1.86 \text{ V}$

Table 16-2 Standard Reduction Potentials at 298°K

	$\mathcal{E}°$ (V)
$F_2 + 2e^- \longrightarrow 2F^-$	2.87
$Ag^{2+} + e^- \longrightarrow Ag^+$	1.99
$H_2O_2 + 2H^+ + 2e^- \longrightarrow 2H_2O$	1.78
$PbO_2 + 4H^+ + SO_4^{2-} + 2e^- \longrightarrow PbSO_4 + 2H_2O$	1.69
$MnO_4^- + 4H^+ + 3e^- \longrightarrow MnO_2 + 2H_2O$	1.68
$Au^+ + e \longrightarrow Au$	1.68
$MnO_4^- + 8H^+ + 5e^- \longrightarrow Mn^{+2} + 4H_2O$	1.49
$PbO_2 + 4H^+ + 2e^- \longrightarrow Pb^{2+} + 2H_2O$	1.46
$Cl_2 + 2e^- \longrightarrow 2Cl^-$	1.36
$Cr_2O_7^{2-} + 14H^+ + 6e^- \longrightarrow 2Cr^{3+} + 7H_2O$	1.33
$Au^{+3} + 2e \longrightarrow Au^+$	1.29
$O_2 + 4H^+ + 4e^- \longrightarrow 2H_2O$	1.23
$MnO_2 + 4H^+ + 2e^- \longrightarrow Mn^{2+} + 2H_2O$	1.21
$Br_2(l) + 2e^- \longrightarrow 2Br^-$	1.06
$AuCl_4^- + 3e^- \longrightarrow Au + 4Cl^-$	0.99
$NO_3^- + 4H^+ + 3e^- \longrightarrow NO + 2H_2O$	0.96
$2Hg^{2+} + 2e^- \longrightarrow Hg_2^{2+}$	0.90
$HO_2^- + H_2O + 2e^- \longrightarrow 3OH^-$	0.87
$Ag^+ + e^- \longrightarrow Ag$	0.80
$Hg_2^{2+} + 2e^- \longrightarrow 2Hg$	0.80
$Fe^{3+} + e^- \longrightarrow Fe^{2+}$	0.77
$O_2 + 2H^+ + 2e^- \longrightarrow H_2O_2$	0.68
$MnO_4^- + 2H_2O + 3e^- \longrightarrow MnO_2 + 4OH^-$	0.59
$MnO_4^- + e^- \longrightarrow MnO_4^{2-}$	0.56
$I_2 + 2e^- \longrightarrow 2I^-$	0.54
$Cu^+ + e^- \longrightarrow Cu$	0.52
$O_2 + 4e^- + 2H_2O \longrightarrow 4OH^-$	0.40
$Cu^{2+} + 2e^- \longrightarrow Cu$	0.34
$Hg_2Cl_2 + 2e^- \longrightarrow 2Hg + 2Cl^-$	0.27
$AgCl + e^- \longrightarrow Ag + Cl^-$	0.22
$SO_4^{2-} + 4H^+ + 2e^- \longrightarrow H_2SO_3 + H_2O$	0.20
$Cu^{2+} + e^- \longrightarrow Cu^+$	0.16
$Co(NH_3)_6^{3+} + e^- \longrightarrow Co(NH_3)_6^{2+}$	0.10
$HgO + H_2O + 2e^- \longrightarrow Hg + 2OH^-$	0.10
$2H^+ + 2e^- \longrightarrow H_2$	0.00

Note that in this procedure we completely ignore the number of electrons being transferred.

2. We can predict, for any possible cell made up of two of the half-cells in Table 16-2, *which way the reaction will proceed spontaneously*, provided the reactants are at standard conditions. One way to do this is to perform the calculation mentioned in 1. If $\mathcal{E}°$ for the whole cell as written comes out positive, the reaction will proceed spontaneously as written. If $\mathcal{E}°$ is negative, the reaction will go the other way. Thus, in the example above, the reaction between Ag^+ and Cu will proceed spontaneously, while the reaction between La^{3+} and S^{2-} will not. However, it is not necessary to carry out the calculation, simple as it is. We can predict which way reactions spontaneously proceed merely by noting the position of the reactants in the table, according to the following rule: **Any oxidizing agent on the table will react spontaneously with any reducing agent below it, but will not react with any reducing agent above it.** This rule holds not only for reactions taking place in half-cells, but also for redox reactions which are carried out by bringing the reactants in direct contact with each other. For example, $Cr_2O_7^{2-}$ in acid solution will oxidize Br^- but will not oxidize Cl^-, whether in a cell or directly.

Table 16-2 (continued)

	$\mathcal{E}°$ (V)
$2H^+ + 2e^- \longrightarrow H_2$	0.00
$MnO_2 + H_2O + 2e^- \longrightarrow Mn(OH)_2 + 2OH^-$	−0.05
$O_2 + H_2O + 2e^- \longrightarrow HO_2^- + OH^-$	−0.08
$Cu(NH_3)_2^+ + e^- \longrightarrow Cu + 2NH_3$	−0.12
$Pb^{2+} + 2e^- \longrightarrow Pb$	−0.13
$Sn^{2+} + 2e^- \longrightarrow Sn$	−0.14
$Ni^{2+} + 2e^- \longrightarrow Ni$	−0.23
$Ag(CN)_2^- + e^- \longrightarrow Ag + 2CN^-$	−0.31
$PbSO_4 + 2e^- \longrightarrow Pb + SO_4^{2-}$	−0.36
$Hg(CN)_4^{2-} + 2e^- \longrightarrow Hg + 4CN^-$	−0.37
$Cd^{2+} + 2e^- \longrightarrow Cd$	−0.40
$Cr^{3+} + e^- \longrightarrow Cr^{2+}$	−0.41
$Fe^{2+} + 2e^- \longrightarrow Fe$	−0.41
$S + 2e^- \longrightarrow S^{2-}$	−0.51
$Pb(OH)_3^- + 2e^- \longrightarrow Pb + 3OH^-$	−0.54
$Fe(OH)_3 + e^- \longrightarrow Fe(OH)_2 + OH^-$	−0.56
$Zn^{2+} + 2e^- \longrightarrow Zn$	−0.76
$Cd(OH)_2 + 2e^- \longrightarrow Cd + 2OH^-$	−0.81
$2H_2O + 2e^- \longrightarrow H_2 + 2OH^-$	−0.83
$SO_4^{2-} + H_2O + 2e^- \longrightarrow SO_3^{2-} + 2OH^-$	−0.92
$Mn^{2+} + 2e^- \longrightarrow Mn$	−1.03
$Zn(NH_3)_4^{2+} + 2e^- \longrightarrow Zn + 4NH_3$	−1.03
$Zn(OH)_4^{2-} + 2e^- \longrightarrow Zn + 4OH^-$	−1.22
$Mn(OH)_2 + 2e^- \longrightarrow Mn + 2OH^-$	−1.47
$Al^{3+} + 3e^- \longrightarrow Al$	−1.66
$H_2 + 2e^- \longrightarrow 2H^-$	−2.23
$Mg^{2+} + 2e^- \longrightarrow Mg$	−2.37
$La^{3+} + 3e^- \longrightarrow La$	−2.37
$Mg(OH)_2 + 2e^- \longrightarrow Mg + 2OH^-$	−2.67
$Na^+ + e^- \longrightarrow Na$	−2.71
$Ca^{2+} + 2e^- \longrightarrow Ca$	−2.76
$Ba^{2+} + 2e^- \longrightarrow Ba$	−2.90
$K^+ + e^- \longrightarrow K$	−2.92
$Ca(OH)_2 + 2e^- \longrightarrow Ca + 2OH^-$	−3.02
$Li^+ + e^- \longrightarrow Li$	−3.05

Another example: A spontaneous reaction will take place between PbO_2 and Fe^{2+} (in acid solution) but not between Pb^{2+} and Fe^{3+}.

There is an interesting parallel here with the Brønsted acid–base theory, which we encountered in Section 10-2. There we had a list of acids (Table 10-1) in order of decreasing acid strength (the ability to donate a proton). Here we have a list of oxidizing agents in order of decreasing oxidizing power (the ability to accept electrons). In both cases, we can predict which way potential reactions will go spontaneously by making use of essentially the same rule. However, in this case we must remember that $\mathcal{E}°$ (hence oxidizing or reducing power) changes with changes in concentration, so that only if standard conditions are used can we be certain that our predictions are valid.

Among the things Table 16-2 tells us is which metals will react with acids to liberate hydrogen. Note that metals always appear on the right (reducing agents). As a corollary of the above rule, we see that all metals *below* the S.H.E. will react with acids (which, of course, supply H^+) to give H_2 and M^+, while all metals *above* the S.H.E. will not react with acids to give hydrogen (some of those metals react with certain acids in other ways, but H_2 is not produced). For example, magnesium reacts with acids,

$$Mg + 2H^+ \longrightarrow Mg^{2+} + H_2$$

but silver does not:

$$Ag + H^+ \longrightarrow \text{no reaction}$$

Since for any cell at standard conditions we can predict which way the reaction will go, we can also predict, for any reversible cell, which side will be the anode and which the cathode. When both half equations are written as reductions, then the half-cell with the higher positive $\mathcal{E}°$ value will be the cathode, and the other half-cell will become the anode.

16-8 Shorthand Notation for Cells

In discussing voltaic cells, it is usually too much trouble to sketch the various cells. Because of this a shorthand notation has been developed which gives the essential features of the cell. Any phase boundary between solid and liquid, solid and gas, or liquid and gas, is indicated by a solid single vertical line |. A liquid junction is shown by a dashed vertical line ¦ and a salt bridge is shown by two parallel vertical lines ||. A comma is used where two layers are not separated by any physical barrier. For example, the Cu half-cell vs. the S.H.E. is shown this way:

$$Pt|H_2|H^+(1.0\ M)||Cu^{2+}(1.0\ M)|Cu$$

In writing the short notation, the concentrations of the ions are always given. In this case each concentration is the standard one, 1.0 *M*. By convention, the anode (oxidation) half-cell is written on the left and the cathode (reduction) on the right. The reaction for the cell as written would be $Cu^{2+} + H_2 \rightarrow Cu + 2H^+$. If we were concerned with the reverse reaction, $Cu + 2H^+ \rightarrow H_2 + Cu^{2+}$, the cell notation would be $Cu|Cu^{2+}(1.0\ M)||H^+(1.0\ M)|H_2|Pt$. Many times we are not sure which way the reaction will go spontaneously. To determine the correct direction for each cell, we apply the principle given in Section 16-7.

Example: The Daniell cell is a zinc–copper cell in which the zinc electrode is in contact with Zn^{2+} ions and the copper electrode is in contact with Cu^{2+} ions. In this version we will keep the two half-cells separated by a salt bridge.

(a) Calculate the standard EMF of the cell ($\mathcal{E}°_{cell}$).
(b) Write the balanced chemical equation.
(c) Show the construction of the cell in the shorthand notation.

Answer:

(a) From Table 16-2 we have the $\mathcal{E}°$ for each of the half-cells.

$$Zn^{2+} + 2e^- \longrightarrow Zn \qquad \mathcal{E}° = -0.76\ V$$

$$Cu^{2+} + 2e^- \longrightarrow Cu \qquad \mathcal{E}° = +0.34\ V$$

At this point we don't know which way the reaction will go. Let us see what $\mathcal{E}°$ is, if the reaction goes in this direction:

$$Cu + Zn^{2+} \longrightarrow Cu^{2+} + Zn$$

For the zinc half-reaction we use the $\mathcal{E}°$ given in the table, -0.76 V, but for the copper half-reaction we must change the sign, since we are writing it the other way.

$$\mathcal{E}°_{cell} = -0.76\ V - 0.34\ V = -1.10\ V$$

Since $\mathcal{E}^\circ_{cell}$ comes out negative, the reaction as written above cannot take place spontaneously, so it must go the other way. The $\mathcal{E}^\circ_{cell}$ will be +1.10 V.

(b) $Cu^{2+} + Zn \rightarrow Cu + Zn^{2+}$

(c) $Zn|Zn^{2+}(1.0\ M)||Cu^{2+}(1.0\ M)|Cu$

Example: The dry cell (usually called the "flashlight battery"),* used to power flashlights, electric clocks, radios, children's toys, and many portable items, is shown in Figure 16-7. Its overall reaction is

$$Zn(s) + 2MnO_2(s) + 8NH_4^+ \longrightarrow Zn^{2+} + 2Mn^{3+} + 8NH_3 + 4H_2O$$

(a) Write the anode reaction.

(b) Write the cathode reaction.

(c) Calculate the standard EMF of the dry cell if the cathode $\mathcal{E}^\circ$ is between +0.49 and +0.74 V.

Answer:

(a) Anode: $Zn(s) \rightarrow Zn^{2+} + 2e^-$ $\quad \mathcal{E}^\circ = 0.76$ V

(b) Cathode:

$$2MnO_2(s) + 8NH_4^+ + 2e^- \longrightarrow 2Mn^{3+} + 4H_2O + 8NH_3 \qquad \mathcal{E}^\circ = +0.49 \text{ to } +0.74 \text{ V}$$

$$\mathcal{E}^\circ_{cell} = +0.49 \text{ V} + 0.76 \text{ V} = +1.25 \text{ V}$$

$$\mathcal{E}^\circ_{cell} = +0.74 \text{ V} + 0.76 \text{ V} = +1.50 \text{ V}$$

The dry cell delivers between 1.25 and 1.50 V.

Example: For each of the following cells:

(a) Write the equation for the cell process.

(b) Using Table 16-2, find $\mathcal{E}^\circ$ for each cell.

(c) Explain the significance of any negative answers in part (b).

(1) $Fe|Fe(NO_3)_2(1.0\ M)||Zn(NO_3)_2(1.0\ M)|Zn$

(2) $Pt|Cl_2(g)|KCl(1.0\ M)|Hg_2Cl_2(s)|Hg(l)$

(3) $Cd|Cd(NO_3)_2(1.0\ M)||AgNO_3(1.0\ M)|Ag$

Answer:

(1) (a) $Fe + Zn^{2+} \rightarrow Fe^{2+} + Zn$

(b) $\mathcal{E}^\circ_{cell} = +0.41 \text{ V} - 0.76 \text{ V} = -0.35 \text{ V}$

(c) The cell reaction as written will not proceed since the $\mathcal{E}^\circ$ is negative. The reaction will proceed in the opposite direction, $Fe^{2+} + Zn \rightarrow Zn^{2+} + Fe$, and $\mathcal{E}^\circ$ for this cell will be $-0.41 \text{ V} + 0.76 \text{ V} = +0.35 \text{ V}$.

(2) (a) $Hg_2Cl_2 \rightarrow Cl_2 + 2Hg$

(b) $\mathcal{E}^\circ_{cell} = 0.27 \text{ V} - 1.36 \text{ V} = -1.09 \text{ V}$

(c) The cell reaction as written will not proceed since the $\mathcal{E}^\circ$ is negative. The reaction will proceed in the opposite direction, $2Hg + Cl_2 \rightarrow Hg_2Cl_2$, with an $\mathcal{E}^\circ = +1.09$ V.

(3) (a) $Cd + 2Ag^+ \rightarrow 2Ag + Cd^{2+}$

(b) $\mathcal{E}^\circ_{cell} = +0.80 \text{ V} + 0.40 \text{ V} = +1.20 \text{ V}$

(c) $\mathcal{E}^\circ_{cell}$ is positive and the reaction will proceed as written.

*The term battery is not strictly correct when referring to a single cell. More properly, a battery consists of two or more voltaic cells connected in series. The automobile lead storage battery (Figure 16-11) is a true battery.

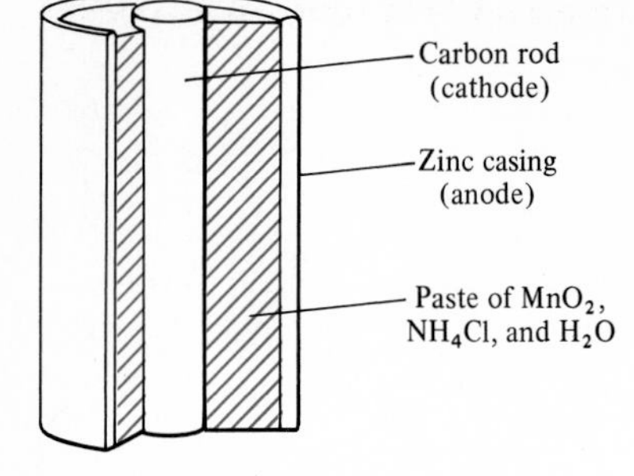

Figure 16-7 The dry cell.

16-9 Some Special Cells and Half-Cells

THE STANDARD WESTON CELL. In electrochemical work it is sometimes necessary to use a cell whose EMF is precisely known. The cell that is almost exclusively used for this purpose is the Weston cell (Figure 16-8). It is represented as

$$Cd(Hg)|CdSO_4 \cdot \tfrac{8}{3}H_2O,\ Hg_2SO_4(sat'd)|Hg$$

A mixture of a metal with mercury is called an **amalgam.** The Weston cell uses a cadmium amalgam.

The half-reactions are

$$\text{Anode:} \quad Cd(Hg) \rightarrow Cd^{2+} + Hg + 2e^-$$
$$\text{Cathode:} \quad Hg_2SO_4 + 2e^- \rightarrow 2Hg + SO_4^{2-}$$

The potential of the Weston cell at 25°C is 1.0183 V. It has a temperature coefficient (change of voltage with temperature) of about −0.04 millivolt/°C.

REFERENCE ELECTRODES. In the measurement of standard reduction potentials of half cells, it is necessary that the unknown half-cell be measured against a half-cell that will remain constant during the course of the analysis. Additionally, the half-cell should be insensitive to changes in composition, easy to put together and certainly reproducible. Such cells are called reference cells. In addition to the S.H.E., already discussed, there are several other electrode systems that meet these requirements. In fact, it is seldom practical to use the S.H.E. We will mention two of them.

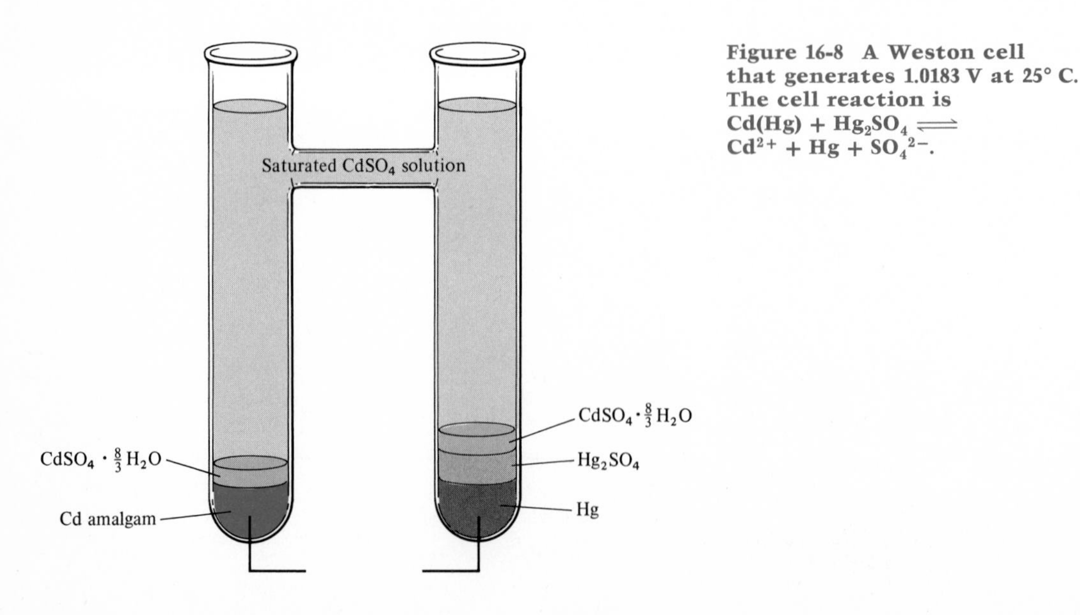

Figure 16-8 A Weston cell that generates 1.0183 V at 25° C. The cell reaction is $Cd(Hg) + Hg_2SO_4 \rightleftharpoons Cd^{2+} + Hg + SO_4^{2-}$.

Table 16-3 Calomel Electrodes

Concentration, KCl	Reduction potentials, $\mathcal{E}°$ (V)
Saturated	$+0.242 - 7.6 \times 10^{-4}(T - 25)$
1.0 *M*	$+0.280 - 2.4 \times 10^{-4}(T - 25)$
0.1 *M*	$+0.334 - 7 \times 10^{-5}(T - 25)$
	T is the temperature in degrees Celsius

1. *Calomel Electrode.* The calomel half-cell may be represented as

$$Hg_2Cl_2(\text{sat'd}),\ KCl(x\ M)|Hg$$

where x represents the molarity of the KCl solution (*calomel* is an old name for mercurous chloride). The electrode reaction is

$$Hg_2Cl_2 + 2e^- \rightleftharpoons 2Hg + 2Cl^-$$

The calomel electrode is a chloride electrode.

The EMF of this half-cell will, of course, change with temperature and concentration of the KCl solution. The specifications of calomel half-cells are given in Table 16-3 for the three most common solutions. Note that all calomel half-cells are saturated with mercurous chloride and therefore the changes are given only with respect to the KCl concentration and the temperature. A typical calomel cell is shown in Figure 16-9.

2. *Silver–silver Chloride Electrode.* Another reference electrode which is also a chloride electrode is the silver–silver chloride electrode. It consists of a silver electrode immersed in a solution of potassium chloride that has been saturated with silver chloride. It is represented as

$$AgCl(\text{sat'd}),\ KCl(x\ M)|Ag$$

The half-cell reaction is given by

$$AgCl + e^- \rightleftharpoons Ag + Cl^-$$

The $\mathcal{E}°$ at 25°C when KCl is saturated is +0.222 V. A typical Ag–AgCl electrode is shown in Figure 16-10.

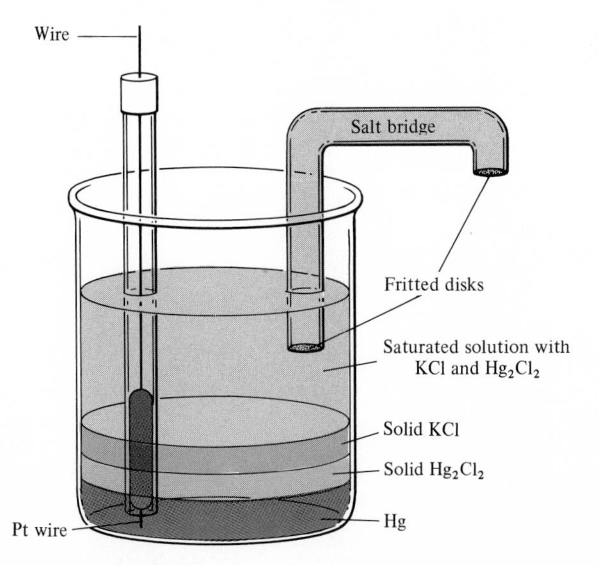

Figure 16-9 A Saturated Calomel Electrode. The half-reaction is $Hg_2Cl_2 + 2e^- \rightleftharpoons 2Hg + Cl^-$ ($\mathcal{E}°$ = 0.268 V). The potential of this cell varies with the concentration of chloride ion (Table 16-3). The saturated calomel electrode (S.C.E.) is most commonly used because it is easy to prepare.

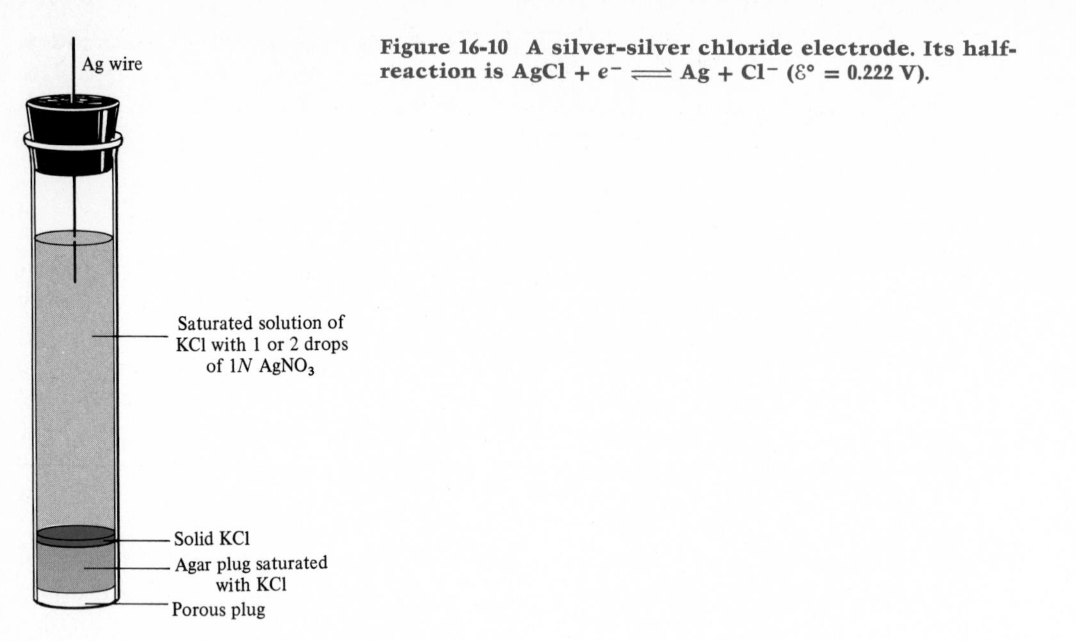

Figure 16-10 A silver-silver chloride electrode. Its half-reaction is $AgCl + e^- \rightleftharpoons Ag + Cl^-$ ($\mathcal{E}° = 0.222$ V).

Commercial Cells

1. *The Dry Cell.* The dry cell was discussed in Section 16-8 and shown in Figure 16-7. As we saw, this cell delivers between 1.25 and 1.50 V.

2. *The Lead Storage Battery.* The lead storage battery consists of a number of cells (three or six) connected in series and capable of delivering 6 or 12 volts. In practice, the cathode plates are joined together and alternated with anode plates, which are also joined.

A typical lead storage battery is shown in Figure 16-11. The anode plates are made

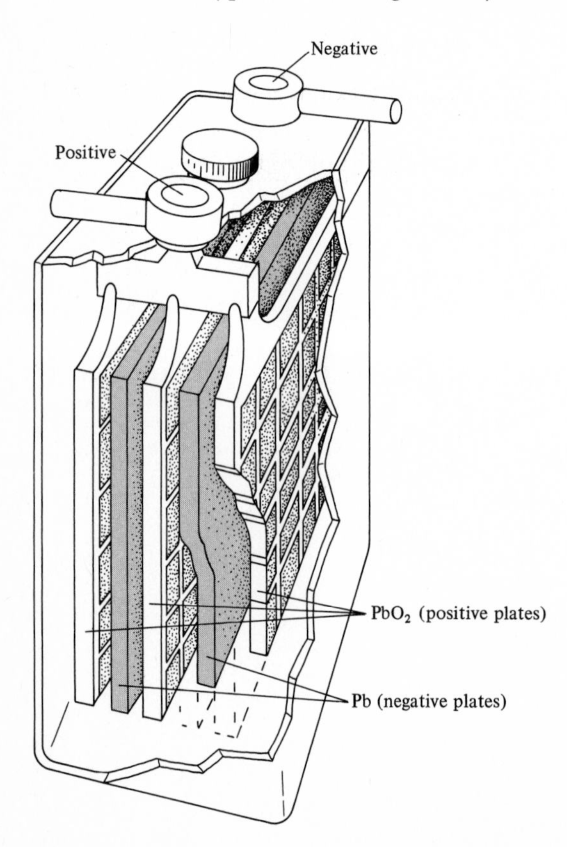

Figure 16-11 A lead storage battery.

of lead, while the cathode plates are grids of lead packed with lead dioxide. A solution of sulfuric acid surrounds the plates. Sulfuric acid is called an *electrolyte* here, since it produces ions in a water solution (Section 14-11). The half-cell reactions may be represented as

$$\text{Anode:} \qquad Pb(s) + SO_4^{2-} \longrightarrow PbSO_4(s) + 2e^-$$

$$\text{Cathode:} \quad 2e^- + PbO_2(s) + SO_4^{2-} + 4H^+ \longrightarrow PbSO_4(s) + 2H_2O$$

This battery is reversible (chargeable) since lead sulfate, which is the product of both the anode and cathode reactions, adheres to the plates instead of moving away as the products in the dry cell do. From the cathode reaction we see that as the battery is used, sulfuric acid is used up, so that one can determine the charge on the battery by measuring the density of the electrolyte (which is proportional to the concentration of H_2SO_4) in the battery cells. Commercial hydrometers are devices which do exactly that. When the density reading is between 1.25 and 1.30 g/ml, the cell is considered full charged. When its value goes below 1.20 g/ml, the battery is in need of a charge.

3. *Fuel Cells.* At the beginning of our space program, scientists quickly realized that the electrical energy needs of space vehicles could not be supplied by conventional batteries since they were much too heavy. Hydrogen–oxygen **fuel cells** were used. Fuel cells use gases that are fed into the cell as electrode materials.

Figure 16-12 shows a hydrogen–oxygen fuel cell. Hydrogen is introduced into the anode, which is a porous carbon chamber whose surface is impregnated with a catalytic material such as finely divided platinum. Oxygen is introduced into the cathode, which is also a porous carbon chamber impregnated with a platinum or silver oxide catalyst. Both electrodes are placed in an electrolytic bath of concentrated KOH or NaOH.

The electrode reactions are

$$\text{Anode:} \qquad H_2 + 2OH^- \longrightarrow 2H_2O + 2e^-$$

$$\text{Cathode:} \quad O_2 + 2H_2O + 4e^- \longrightarrow 4OH^-$$

At the cathode, oxygen is reduced, producing hydroxide ions which at the anode react with hydrogen ions, producing water. The net overall cell reaction is

$$2H_2 + O_2 \longrightarrow 2H_2O$$

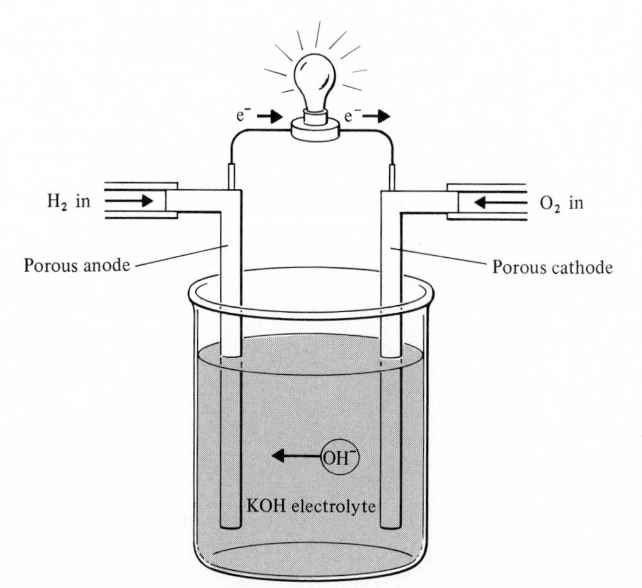

Figure 16-12 A hydrogen-oxygen fuel cell.

In the manned space flights, the fuel cells were operated at high temperatures so that the water produced from the fuel cell could be in the form of steam, which was then condensed and used for drinking.

Hydrogen–oxygen fuel cells offer a number of advantages compared to conventional energy sources. The fuel cell can be used continuously as long as there is a steady supply of hydrogen and oxygen. Conventional batteries run down after a period of time. Additionally, fuel cells offer greater efficiency in their work-producing capacity as well as being nonpolluting, since the output of the cell is harmless steam. Studies are being made to evaluate the feasibility of using fuel cells in automobiles so that gasoline-powered automobiles might be replaced by electric cars, but this is unlikely to happen in the near future, because of engineering problems and high costs. Because of their nonpollution characteristic, fuel cells are being investigated as possible means of generating electricity in large cities.

16-10 The Nernst Equation: How the Cell Voltage Changes with Concentration

We have emphasized that EMF values depend on concentrations. All of our EMF calculations up to now have been for electrolytes having a 1.0 *M* concentration. They have been labeled $\mathcal{E}^\circ_{cell}$. Is it possible to calculate $\mathcal{E}$ values for cells in which the concentrations of the various species are *not* 1.0 *M*? Fortunately, it is. W. F. Nernst* (1864–1941) proposed the relationship between the EMF of a cell and its concentration. The equation is aptly called the **Nernst equation** and is, at 25°C,

$$\mathcal{E}_{cell} = \mathcal{E}^\circ_{cell} - \frac{0.05916}{n} \log Q$$

where $\mathcal{E}^\circ_{cell}$ = standard electrode potential of the cell
n = number of moles of electrons that are transferred in the oxidation–reduction reaction
Q = reaction quotient for the reaction as written: For a generalized reversible reaction†

$$aA + bB + \cdots \rightleftharpoons cC + dD + \cdots$$

$$Q = \frac{[C]^c[D]^d \cdots}{[A]^a[B]^b \cdots}$$

$$\log Q = \log \frac{[C]^c[D]^d \cdots}{[A]^a[B]^b \cdots}$$

The quantity in square brackets may have a number of meanings, depending on the physical state of the substance involved. For example, if A is

a solute	[A] represents concentration, moles/liter
a gas	[A] represents partial pressure, atm
a pure solid or a pure liquid	[A] = 1

The concentration of the solvent (normally water) is generally so large compared to

*This great scientist not only discovered the Nernst equation, but also the third law of thermodynamics, and the concepts of buffer solutions and solubility product, among other things. He was awarded the Nobel prize in chemistry in 1920.

†This looks like an equilibrium expression. The reason we use *Q* instead of *K* is that the system does not have to be at equilibrium.

the other species that its concentration may be considered unchanged and we set $[H_2O] = 1$.

Example: For the cell

$$Cu|Cu^{2+}(x\,M)||(y\,M)Ag^+|Ag$$

write the Nernst equation showing how $\mathcal{E}^\circ_{cell}$ will vary with concentrations of Cu^{2+} and Ag^+.

Answer: For the cell as written:

		$\mathcal{E}^\circ$
Anode:	$Cu \rightleftharpoons Cu^{2+} + 2e^-$	-0.34 V
Cathode:	$2(Ag^+ + e^- \rightleftharpoons Ag)$	$+0.80$ V
	$2Ag^+ + \cancel{2e^-} + Cu \rightleftharpoons Cu^{2+} + 2Ag + \cancel{2e^-}$	$\mathcal{E}^\circ = 0.46$ V

By writing the oxidation–reduction equation, we see that the number of electrons transferred in the balanced equation is 2. Therefore, $n = 2$ in the Nernst equation.

$$\mathcal{E}_{cell} = \mathcal{E}^\circ_{cell} - \frac{0.05916}{n}\log\frac{[Ag(s)]^2[Cu^{2+}]}{[Cu(s)][Ag^+]^2}$$

$$[Ag(s)] = 1 \qquad [Cu^{2+}] = x$$

$$[Cu(s)] = 1 \qquad [Ag^+] = y$$

$$\mathcal{E}_{cell} = +0.46 - \frac{0.05916}{2}\log\frac{x}{y^2}$$

Thus, $\mathcal{E}_{cell}$ will decrease as $[Cu^{2+}]$ increases and as $[Ag^+]$ decreases.

Example: Calculate $\mathcal{E}_{cell}$ for the Daniell cell when $[Zn^{2+}] = 0.5\,M$ and $[Cu^{2+}] = 1.5\,M$. Repeat the calculations for the same cell but with the concentrations reversed.

Answer: As shown in Section 16-8, the $\mathcal{E}^\circ$ for the Daniell cell is $+1.10$ V. The redox reaction is

$$Zn + Cu^{2+} \rightleftharpoons Cu + Zn^{2+}$$

Clearly, $n = 2$ in this reaction. We therefore have

$$\mathcal{E}_{cell} = \mathcal{E}^\circ_{cell} - \frac{0.05916}{2}\log\frac{[Zn^{2+}]}{[Cu^{2+}]}$$

$$\text{(a)}\quad \mathcal{E}_{cell} = 1.10 - \frac{0.05916}{2}\log\frac{0.5}{1.5} = 1.10 - 0.0296(\log 0.333)$$

$$= 1.10 - 0.0296(-0.477)$$

$$= 1.10 + 0.014$$

$$= 1.11 \text{ V}$$

$$\text{(b)}\quad \mathcal{E}_{cell} = 1.10 - \frac{0.05916}{2}\log\frac{1.5}{0.5} = 1.10 - \frac{0.05916}{2}\log 3$$

$$= 1.10 - 0.0296(0.477)$$

$$= 1.10 - 0.014$$

$$= 1.09 \text{ V}$$

Example: In the previous example, what must the ratio of concentrations $[Zn^{2+}]/[Cu^{2+}]$ be so that $\mathcal{E}_{cell}$ will be 1.00 V?

Answer: In this problem we are given $\mathcal{E}_{cell} = 1.00$ V and have to calculate the ratio of concentrations $[Zn^{2+}]/[Cu^{2+}]$. From the previous problem, we have

$$\mathcal{E}_{cell} = \mathcal{E}^\circ_{cell} - \frac{0.05916}{2}\log\frac{[Zn^{2+}]}{[Cu^{2+}]}$$

$$1.00 = 1.10 - 0.0296\log\frac{[Zn^{2+}]}{[Cu^{2+}]}$$

$$\log\frac{[Zn^{2+}]}{[Cu^{2+}]} = \frac{-(1.00 - 1.10)}{0.0296} = 3.38$$

$$\frac{[Zn^{2+}]}{[Cu^{2+}]} = 2.4 \times 10^3$$

Example: Using the data on page 511 on the dry cell, write the Nernst equation for this cell.

Answer: The reaction is

$$2MnO_2(s) + 8NH_4^+ + Zn(s) \rightleftharpoons Zn^{2+} + 2Mn^{3+} + 4H_2O + 8NH_3$$

and $\mathcal{E}^\circ_{cell} \sim 1.25$.

$$\mathcal{E}_{cell} = 1.25 - \frac{0.059}{2}\log\frac{[Zn^{2+}][Mn^{3+}]^2}{[NH_4^+]^8}$$

16-11 EMF, Free Energy, and the Equilibrium Constant

In Section 8-10, we learned that a chemical reaction will occur spontaneously when its free-energy change (ΔG) is negative. In this chapter, we found that oxidation–reduction reactions can take place in electrochemical cells and, if a given reaction occurs spontaneously, the cell will exhibit a positive EMF. This establishes that there is a relationship between a positive EMF and a negative ΔG. The relationship between the free energy of a reaction and the EMF of the corresponding cell has been worked out (the details will not concern us) and takes the mathematical form

$$\Delta G = -23.06n\mathcal{E}$$

and

$$\Delta G^\circ = -23.06n\mathcal{E}^\circ$$

where $\mathcal{E}$ and $\mathcal{E}^\circ$ are measured in volts and ΔG and ΔG° in kcal/mole and n is the same as in the Nernst equation in Section 16-10. Since we previously showed

$$\Delta G^\circ = -2.303\, RT\log K$$

we can combine this with the equation above to get

$$\log K = \frac{23.06n\mathcal{E}^\circ}{2.303RT}$$

At 25°C we arrive at (remember that R is a constant equal to 1.987×10^{-3} kcal/°C)

$$\log K = 16.91\, n\mathcal{E}^\circ$$

To see how to use these relationships, let us consider some examples.

Example: In one of the examples worked out in the last section we calculated that $\mathcal{E}_{cell}$ was 1.11 V for the Daniell cell when $[Zn^{2+}] = 0.5\ M$, and $[Cu^{2+}] = 1.5\ M$. Calculate ΔG, $\Delta G°$, and K for the Daniell cell reaction $Zn + Cu^{2+} \rightleftharpoons Zn^{2+} + Cu$ at 25°C when the concentrations of the zinc ions and copper ions have the values given.

Answer: Using

$$\Delta G = -23.06n\mathcal{E}$$

$\mathcal{E} = 1.11$ V and $n = 2$, so

$$\Delta G = -23.06(2)(1.11) = -51.2 \text{ kcal/mole}$$
$$\Delta G° = -23.06(2)(1.10) = -50.7 \text{ kcal/mole}$$

and

$$\log K = 16.91n\mathcal{E}° = 16.91(2)(1.10) = 37.2$$
$$K = 2 \times 10^{37}$$

Example: Calculate $\Delta G°$ and K for the reaction representing the dry cell whose $\mathcal{E}° = 1.25$ V.

Answer:

$$\begin{aligned} \Delta G° &= -23.06n\mathcal{E}° \\ &= -23.06(2)(1.25) \\ &= -57.65 \text{ kcal/mole} \\ \log K &= 16.91n\mathcal{E}° \\ &= 16.91(2)(1.25) = 42.3 \\ K &= 2 \times 10^{42} \end{aligned}$$

16-12 Electrolytic Cells

Less than two months after Volta announced the invention of the voltaic cell and so made possible the generation of a continuous electric current, William Nicholson (1753–1815) and Anthony Carlisle (1768–1840) in 1800 used a current to electrolyze water. Seven years later, Humphry Davy (1778–1829) isolated the element potassium by electrolyzing molten potassium hydroxide, and followed that up by isolating sodium, barium, strontium, calcium, and magnesium, in a similar manner. Davy is given the credit for discovering all these elements.

Let us first consider the electrolysis of a typical molten salt, NaCl, shown in Figure 16-13. When the current is turned on, Cl^- ions are attracted to the positive electrode (the anode), while Na^+ ions travel to the negative electrode (the cathode). At the anode, the Cl^- ions each give up one electron and are oxidized to Cl_2, which is given off as green bubbles of gas:

$$\text{Anode reaction:} \quad 2Cl^- \longrightarrow Cl_2 + 2e^-$$

Note that, as in voltaic cells, oxidation takes place at the anode. The electrons given up by the Cl^- ions do not react with the electrodes (it is necessary that these be inert to molten NaCl) but are carried over the wires by the electric current (in fact, they *are* the electric current) to the cathode, where they are given to the Na^+ ions:

$$\text{Cathode reaction:} \quad Na^+ + e^- \longrightarrow Na$$

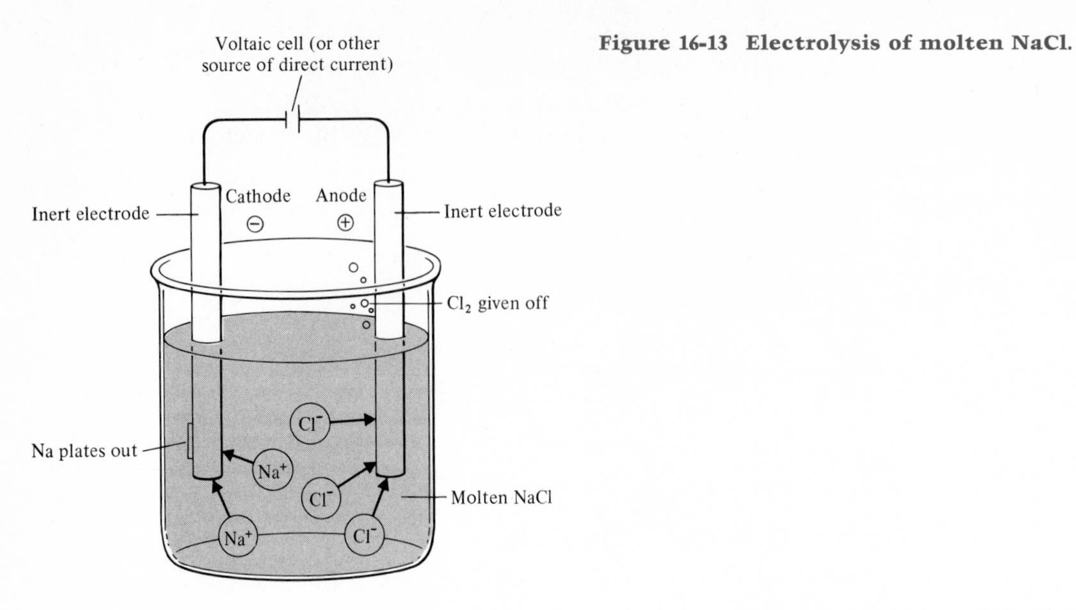

Figure 16-13 Electrolysis of molten NaCl.

Reduction thus takes place at the cathode, and metallic sodium can be seen to plate out there. As usual, the total reaction for the cell is obtained by adding the half-equations:

$$\begin{array}{rcl} 2Cl^- & \longrightarrow & Cl_2 + 2e^- \\ Na^+ + e^- & \longrightarrow & Na \\ \hline 2Cl^- + 2Na^+ & \longrightarrow & Cl_2 + 2Na \end{array}$$

Molten sodium chloride is an electrolyte (carries a current) because the Na^+ and Cl^- ions travel in the manner described, thus allowing the circuit to be completed. The process would not work for solid NaCl since, as we saw on page 412, Na^+ and Cl^- ions in the solid state are not free to travel, so that there is no complete circuit, and no electrolysis will be observed. Positive ions are called **cations** because they are attracted to the cathode, and negative ions are called **anions** because they are attracted to the anode.

This type of cell is called an **electrolytic cell,** and the process is known as **electrolysis.** The essential difference between a voltaic cell and an electrolytic cell is that a voltaic cell *produces* electricity; an electrolytic cell *requires* electricity. In other words, a voltaic cell changes chemical energy into electrical energy, and the products have a *lower* total free energy than the starting materials. An electrolytic cell changes electrical energy into chemical energy, and the products have a *higher* total free energy than the starting materials. The importance of electrolytic cells is that they make it possible for us to isolate and purify elements that have high free energies and so are difficult to separate from their compounds in other ways (we shall see examples in Section 19-1).

Electrolysis is possible not only for molten salts, but also for water, provided that an electrolyte is dissolved in the water. The reactions here may be different. For example, let us look at the electrolysis of sodium sulfate in water (Figure 16-14). This is essentially the experiment of Nicholson and Carlisle. Once again, the circuit is completed by the travels of the electrolyte ions. The Na^+ ions are attracted to the cathode and the SO_4^{2-} ions to the anode, enabling the current to flow. However, this time the reactions are different. It is not the Na^+ ions that take electrons from the cathode, but the H_2O molecules, which react as follows:

$$\text{Cathode reaction:} \quad 2H_2O + 2e^- \longrightarrow H_2 + 2OH^-$$

If we look at the cathode, we will see bubbles of H_2 given off. If we had put a few

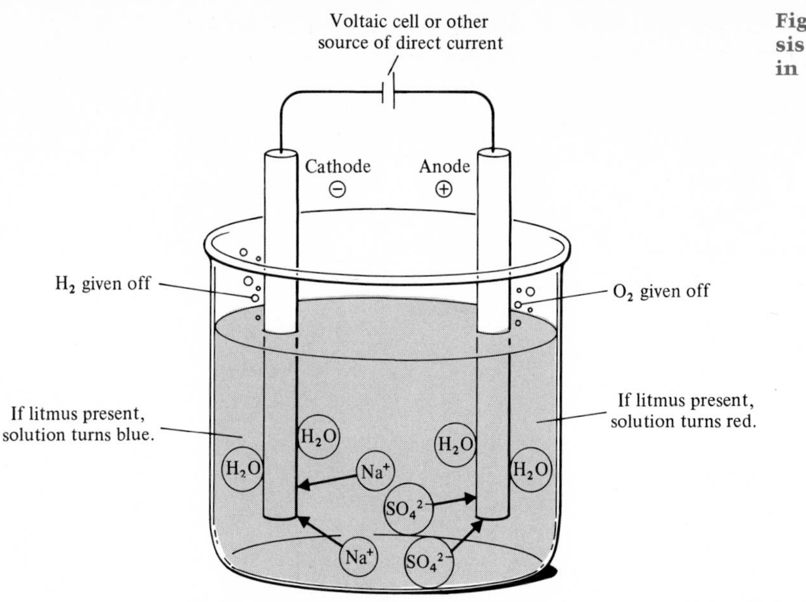

Figure 16-14 Electrolysis of Na_2SO_4 dissolved in water.

drops of litmus solution into the cell, we would also see the solution adjacent to the cathode turn blue from the OH^- ions that are produced simultaneously.

Meanwhile, at the anode, it is not the SO_4^{2-} ions that give up electrons, but once again, the water molecules, by the reaction:

$$\text{Anode reaction:} \quad 2H_2O \longrightarrow O_2 + 4H^+ + 4e^-$$

Here we can see bubbles of O_2; and the litmus solution turning red from the H^+ ions. The overall reaction is

$$\begin{array}{rcl} 2H_2O + 2e^- & \longrightarrow & H_2 + 2OH^- \\ 2H_2O & \longrightarrow & O_2 + 4H^+ + 4e^- \\ \hline 6H_2O & \longrightarrow & O_2 + 2H_2 + 4H^+ + 4OH^- \end{array}$$

or, if allow the H^+ ions to react with the OH^- ions,

$$2H_2O \longrightarrow O_2 + 2H_2$$

Why is it that H_2O molecules react at the anode and not Na^+ ions? The reason is simply that H_2O has a greater affinity for electrons than does Na^+; that is, it is more easily reduced, a fact that can be checked by looking at Table 16-2, where we see that $\mathcal{E}°$ for $Na^+ + e^- \rightarrow Na$ is -2.71 V, while for $2H_2O + 2e^- \rightarrow H_2 + 2OH^-$, it is -0.83 V. Similarly, at the cathode, H_2O molecules are oxidized rather than SO_4^{2-} ions, because they are more easily oxidized.

The Na^+ and SO_4^{2-} ions therefore do not react at all in this electrolysis; only the water molecules do. Yet they, or some other ions, are necessary. If we try to electrolyze pure distilled water, we would find that the cell would not work: no reaction would take place, and no current would flow. The ions perform an essential function: they carry the current and allow the circuit to be complete. Another way of looking at it is that the ions are necessary to maintain electrical neutrality. Before the reaction starts, the Na^+ and SO_4^{2-} ions are uniformly distributed throughout the solution (of course, there are twice as many of the former). When the electrolysis begins, OH^- ions are formed at the cathode, and it is necessary for an equal number of Na^+ ions to migrate to this area. At the same time, SO_4^{2-} ions must migrate to the anode to match the H^+ ions which are being formed there.

Thus, it is the Na^+ and SO_4^{2-} ions that *carry the current,* but it is H_2O molecules that *react.*

When we electrolyze other aqueous solutions, sometimes the ions react, and sometimes the water, depending on which species are more easily oxidized and reduced. For example, electrolysis of a solution of $CuSO_4$ will produce oxygen at the anode and copper at the cathode. Water is more easily oxidized than SO_4^{2-}, but Cu^{2+} is more easily reduced than H_2O ($\mathcal{E}°$ for $Cu^{2+} + 2e^- \rightarrow Cu$ is +0.34 V). Electrolysis of a solution of $CuBr_2$ will produce copper and bromine. In this case, the water takes no part in the reaction at all other than providing a medium, and the reaction is essentially the same as if we electrolyzed molten $CuBr_2$.

$$\begin{array}{rcl} Cu^{2+} + 2e^- & \longrightarrow & Cu \\ 2Br^- & \longrightarrow & Br_2 + 2e^- \\ \hline Cu^{2+} + 2Br^- & \longrightarrow & Cu + Br_2 \end{array}$$

16-13 Faraday's Law

Michael Faraday (1791–1867) showed that the *amount* of the substances that react at the two electrodes in an electrolytic cell is proportional to the *amount* of electricity that flows through the cell. The basic unit of quantity for electricity is the **coulomb.** The coulomb is related to the rate of flow of electricity by the expression

$$1 \text{ coulomb} = 1 \text{ amp} \cdot \text{s}$$

That is, when 1 coulomb of electricity flows past a given point in 1 second, we have a current of 1 **ampere.**

Example: How many coulombs of electricity are passed through a solution in which 6.5 amperes are allowed to run for 1 hour?

Answer:

$$\begin{aligned} C \text{ (coulombs)} &= I \text{ (amperes)} \times t \text{ (seconds)} \\ C &= (6.5 \text{ amperes}) \times \left(1 \text{ h} \times \frac{60 \text{ min}}{1 \text{ h}} \times \frac{60 \text{ s}}{1 \text{ min}}\right) \\ &= 2.34 \times 10^4 \text{ coulombs} \end{aligned}$$

Note that coulombs, amperes, and volts are completely different kinds of units

1. A coulomb is a *quantity* of electricity. In a wire each coulomb represents a certain *number* of flowing electrons.
2. Amperes are the *rate* of flow of electricity; how many coulombs flow past a given point in a unit of time.
3. Volts are the *difference in potential* between the two parts of an electric circuit. The voltage is proportional to the difference in potential energy between one part of the circuit and another.

Faraday showed that the greater the number of coulombs that flowed through an electrolytic cell, the larger the quantity of matter that reacted at the electrodes. In modern terms, **Faraday's law** may be stated: **One equivalent of substance will be produced at each electrode when 96,500 coulombs of electricity are passed through an electrolytic cell.** This quantity, 96,500 coulombs, which is equal to 1 mole of electrons, is called 1 **faraday:**

$$1 \text{ faraday} = 96{,}500 \text{ coulombs} = 1 \text{ mole of electrons}$$

Faraday's law enables us to calculate how much of any substance will be produced in an electrolytic cell if we know how much electricity flows through; and vice versa. We must remember that for oxidation–reduction reactions,

$$\text{equivalent weight} = \frac{\text{molecular weight}}{\text{number of electrons in half-equation}}$$

Example: In the following reduction reactions

$$Zn^{2+} + 2e^- \longrightarrow Zn$$
$$Cu^{2+} + 2e^- \longrightarrow Cu$$
$$Al^{3+} + 3e^- \longrightarrow Al$$
$$Ag^+ + e^- \longrightarrow Ag$$

what weight of metal would be deposited if 1 faraday of electricity were passed through the cell?

Answer:

$Zn^{2+} + 2e^- \longrightarrow Zn$	$\frac{1}{2}$ mole	32.69 g
$Cu^{2+} + 2e^- \longrightarrow Cu$	$\frac{1}{2}$ mole	31.77 g
$Al^{3+} + 3e^- \longrightarrow Al$	$\frac{1}{3}$ mole	8.99 g
$Ag^+ + e^- \longrightarrow Ag$	1 mole	107.868 g

Example: In the series of cathodes given above, how many coulombs are needed to produce 1 g each of (a) Zn, (b) Cu, (c) Al, and (d) Ag?

Answer:

(a) $Zn^{2+} + 2e^- \rightarrow Zn$. The equivalent weight is 65.38/2 = 32.69 g. 96,500 coulombs will produce 32.69 g.

$$1 \text{ g} \times \frac{96{,}500 \text{ coulombs}}{32.69 \text{ g}} = 2950 \text{ coulombs for 1 g Zn}$$

(b) $Cu^{2+} + 2e^- \rightarrow Cu$. The equivalent weight is 63.54/2 = 31.77 g.

$$1 \text{ g} \times \frac{96{,}500 \text{ coulombs}}{31.77 \text{ g}} = 3040 \text{ coulombs for 1 g Cu}$$

(c) $Al^{3+} + 3e^- \rightarrow Al$. The equivalent weight is 26.98/3 = 8.99 g.

$$1 \text{ g} \times \frac{96{,}500 \text{ coulombs}}{8.99 \text{ g}} = 10{,}700 \text{ coulombs for 1 g Al}$$

(d) $Ag^+ + e^- \rightarrow Ag$. The equivalent weight is 107.868/1 = 107.868 g.

$$1 \text{ g} \times \frac{96{,}500 \text{ coulombs}}{107.868 \text{ g}} = 895 \text{ coulombs for 1 g Ag}$$

Example: In the electrolysis of copper sulfate, a current of 2.50 amperes is allowed to flow for exactly 3.00 h. How many g of Cu and liters of O_2 at 25.0°C and 1.00 atm will be produced?

Answer:

$$3.00\ \text{h} \times 60.0\frac{\text{min}}{\text{h}} \times 60.0\frac{\text{s}}{\text{min}} \times 2.50\ \text{amperes} = 27{,}000\ \text{coulombs}$$

$$\frac{27{,}000\ \text{coulombs}}{96{,}500\ \text{coulombs/faraday}} = 0.280\ \text{faraday}$$

Therefore, 0.280 equiv of copper and of oxygen will be produced.

At the cathode: $Cu^{2+} + 2e^- \rightarrow Cu$

For copper, $n = 2$, so

$$\text{equiv wt} = \frac{\text{mol wt}}{2} = \frac{63.55}{2} = 31.8\ \text{g/equiv}$$

$$31.8\frac{\text{g}}{\text{equiv}} \times 0.280\ \text{equiv} = 8.90\ \text{g of copper produced}$$

At the anode: $2H_2O \rightarrow O_2 + 4H^+ + 4e^-$

For oxygen, $n = 4$, so

$$\text{equiv wt} = \frac{\text{mol wt}}{4} = \frac{32.00}{4} = 8.00\ \text{g/equiv}$$

$$8.00\frac{\text{g}}{\text{equiv}} \times 0.280\ \text{equiv} = 2.24\ \text{g of oxygen produced}$$

$$PV = \frac{\text{g}}{\text{mol wt}}RT$$

$$(1.00\ \text{atm})\ V = \frac{2.24\ \text{g}}{32.0\ \text{g/mole}}\left(0.08206\ \frac{\text{liter}\cdot\text{atm}}{\text{mole}\cdot{}^\circ\text{K}}\right)(298^\circ\text{K})$$

$$V = 1.71\ \text{liters of oxygen at } 25^\circ\text{C and } 1.00\ \text{atm}$$

16-14 Reversal of Voltaic Cells

In an electrolytic cell we use electrical energy to make a reaction take place that would otherwise not take place because the starting compound is at a position of low potential energy. We can do the same thing to a voltaic cell, thus turning it into an electrolytic cell! A voltaic cell delivers a certain voltage. By imposing on the cell a voltage higher than the cell can deliver, we can reverse the chemical reactions of the cell. The anode, where oxidation would normally take place spontaneously, now becomes the cathode, since reduction is now taking place there, while the cathode in turn becomes the anode. The reversible lead storage battery (Figure 16-11) is an example. The normal reactions (when the battery is being used) are

Anode: $Pb(s) + SO_4^{2-} \longrightarrow PbSO_4(s) + 2e^-$

Cathode: $PbO_2(s) + SO_4^{2-} + 4H^+ + 2e^- \longrightarrow PbSO_4(s) + 2H_2O$

When the battery is being charged, either during driving, or at the garage where they use a charger, these reactions are reversed:

Cathode: $PbSO_4(s) + 2e^- \longrightarrow Pb(s) + SO_4^{2-}$

Anode: $PbSO_4(s) + 2H_2O \longrightarrow PbO_2(s) + SO_4^{2-} + 4H^+ + 2e^-$

In Figure 16-3 we have an electrochemical cell that will produce 1.10 V when $[Cu^{2+}]$ and $[Zn^{2+}]$ are 1.0 M. If we now impose upon this cell a voltage greater than 1.10 V as in Figure 16-15, we can reverse the reaction.

It is important to remember that when we do this, the cathode of the outside cell

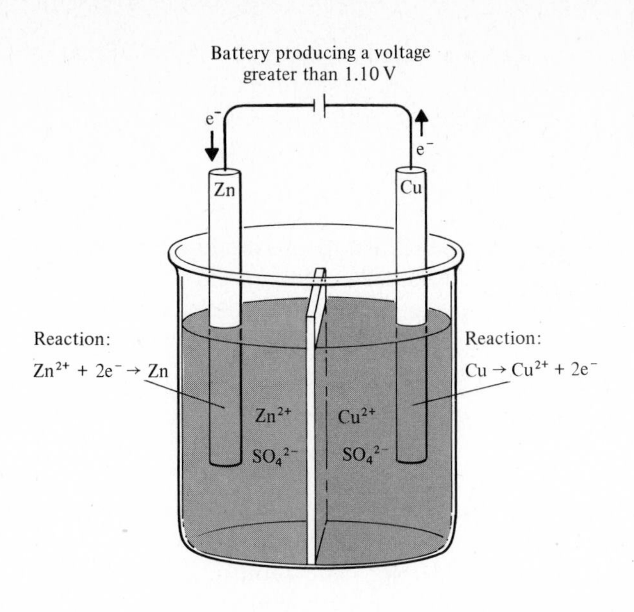

Figure 16-15 The electrochemical cell in Figure 16-3 is reversed by imposing electrical energy from another cell or battery. The reaction is now $Zn^{2+} + Cu \rightarrow Cu^{2+} + Zn$.

or battery is attached to the cathode of the electrochemical cell and the anode to the anode. Only in this way can the electron flow originating from the cell be pushed back, thereby reversing the natural cell reaction.

16-15 Corrosion

Billions of dollars worth of valuable materials are destroyed every year by corrosion, a simple chemical reaction. The corrosion of metals is an oxidation process. Typically, many metals react with the oxygen in the air, converting them to oxides. The reaction of atmospheric oxygen with iron and steel is very slow, but over a period of time these metals eventually oxidize to rust. Rusting is accelerated when the metal is in contact with water (even water vapor accelerates the process), because the water participates in the reaction. The iron is first converted to the +2 oxidation state.

$$\begin{array}{llll} \text{Oxidation:} & Fe \longrightarrow Fe^{2+} + 2e^- & & \mathcal{E}^\circ = +0.41\ V \\ \text{Reduction:} & \underbrace{\tfrac{1}{2}O_2 + H_2O}_{\text{moist air}} + 2e^- \longrightarrow 2OH^- & & \mathcal{E}^\circ = +0.40\ V \\ \hline \text{Overall:} & Fe + \tfrac{1}{2}O_2 + H_2O \longrightarrow Fe(OH)_2 & & \mathcal{E}^\circ = +0.81\ V \end{array}$$

Note that it is the oxygen that oxidizes the iron, not the water, even though the water makes the reaction go faster. The initial product, $Fe(OH)_2$, is further oxidized by moist air to ferric hydroxide, $Fe(OH)_3$:

$$2Fe(OH)_2 + \tfrac{1}{2}O_2 + H_2O \longrightarrow 2Fe(OH)_3 \qquad (\text{or } 2Fe_2O_3 \cdot 3H_2O)$$

The ferric hydroxide rapidly loses water and is converted to Fe_2O_3. Ferric oxide (Fe_2O_3) and pure iron have very dissimilar crystal structures and the iron oxide does not stick to the iron surface. The oxide flakes off the metal, exposing a fresh surface of iron which is then ready for further attack by oxygen and water vapor.

Other metals are also oxidized by atmospheric oxygen, but the process does not always lead to corrosion. In the cases of copper and aluminum, for example, the oxidation products CuO and Al_2O_3 do not flake off but remain attached to the metal. Another layer of fresh metal is not exposed, and the oxide coating prevents further oxidation.

A number of techniques have been developed in order to overcome the destruction of iron and other materials by corrosion. These include (1) coatings, (2) electrochemical methods, and (3) alloying.

COATINGS. The metal to be protected is covered by a layer or layers of paint, grease, or anything else that will adhere to the metal and protect it from corrosive oxygen and water vapor. Generally, these techniques are useful only as temporary solutions since after a period of time, cracks, chips, or holes appear in the coatings, exposing the metal.

ELECTROCHEMICAL METHODS. We shall illustrate this method for iron, but other metals may be protected in a similar manner. Iron will not oxidize or rust as long as there is a more reactive metal (one with a more negative reduction potential) available. Aluminum is generally the metal of choice, so that the iron is surface coated with aluminum. Comparing reduction potentials, we have

$$Fe^{2+} + 2e^- \longrightarrow Fe \qquad \mathcal{E}^\circ = -0.41 \text{ V}$$
$$Al^{3+} + 3e^- \longrightarrow Al \qquad \mathcal{E}^\circ = -1.66 \text{ V}$$

As already mentioned, when aluminum rusts or is oxidized, it forms aluminum oxide, Al_2O_3, which has a crystal structure that is quite similar to the pure aluminum and adheres strongly to the aluminum surface. The aluminum is then protected from further oxidation by the Al_2O_3, its own corrosion product. It is obvious that the iron, being underneath it all, cannot rust either. Aluminum is a good choice for electrochemical protection of iron, but it does have one major drawback. It is expensive. Because of this, industry has used zinc as an alternative. Coating with zinc is called **galvanizing.**

$$Zn^{2+} + 2e^- \longrightarrow Zn \qquad \mathcal{E}^\circ = -0.76 \text{ V}$$

From the $\mathcal{E}^\circ$ value for the zinc half-cell, we see that galvanizing the iron by giving it a coating of zinc protects the iron by the same principle as does aluminum. Galvanized iron is corrosion-free because the zinc protects the iron in two ways. It coats the iron as paint does, and additionally is preferentially oxidized. If a galvanized piece of iron is scratched, the iron will not corrode. When zinc is oxidized, it forms an oxide which is converted by the CO_2 in the air to a carbonate. This oxidized zinc product strongly adheres to the zinc metal.

Both iron cans and copper cookware are protected by coating with tin. What kind of protection is afforded? Let us examine their reduction potentials.

$$Cu^{2+} + 2e^- \longrightarrow Cu \qquad \mathcal{E}^\circ = 0.34 \text{ V}$$
$$Sn^{2+} + 2e^- \longrightarrow Sn \qquad \mathcal{E}^\circ = -0.14 \text{ V}$$
$$Fe^{2+} + 2e^- \longrightarrow Fe \qquad \mathcal{E}^\circ = -0.41 \text{ V}$$

Here we note that the tin plating of iron cans does not protect the cans electrochemically since iron is more easily oxidized. The tin affords the iron a very tight protective coating in the same way as paint would. Scratch a tin can and corrosion will occur. Tin-coated copper cookware, however, is protected the same way as zinc protects iron, both by coating it, and electrochemically.

ALLOYING. Iron can be alloyed with certain other metals to provide steels that are greatly resistant to corrosion (see Section 19-1). Another method consists of adding a rusting inhibitor such as chromate ion or tributylamine.

Problems

16-1 Define each of the following terms. **(a)** oxidation (2 definitions) **(b)** reduction (2 definitions) **(c)** redox reaction **(d)** equivalent weight (for redox reactions) **(e)** plating out **(f)** salt bridge

(g) electrode **(h)** anode **(i)** cathode **(j)** voltaic cell **(k)** reversible voltaic cell
(l) EMF **(m)** voltage **(n)** difference in potential **(o)** S.H.E. **(p)** amalgam
(q) calomel electrode **(r)** fuel cell **(s)** electrolytic cell **(t)** cation **(u)** anion
(v) electrolysis **(w)** coulomb **(x)** ampere **(y)** faraday **(z)** rust
(aa) galvanized iron

16-2 What is the oxidation number of sulfur in the following compounds and ions?
(a) Na_2S **(b)** H_2SO_4 **(c)** SO_2 **(d)** S_8 **(e)** $Na_2S_4O_6$ **(f)** SO_3 **(g)** SO_3^{2-}
(h) $S_2O_3^{2-}$ **(i)** CaS **(j)** SF_2 **(k)** S^{2-}

16-3 What is the oxidation number of each element in the following compounds and ions?
(a) IF_5 **(b)** YBr_3 **(c)** $HOCl$ **(d)** BH_4^- **(e)** OF_2 **(f)** Bi^{3+} **(g)** W **(h)** NH_4NO_3
(i) SiH_4 **(j)** PCl_3 **(k)** PCl_5 **(l)** C_3H_8 **(m)** NaH **(n)** O_3 **(o)** MnO_4^-
(p) $Cr_2O_7^{2-}$ **(q)** CrO_4^{2-} **(r)** K_2O_2 **(s)** $Cr_2(SO_4)_3$ **(t)** IBr **(u)** Fe_3O_4 **(v)** H_2Se
(w) C_2H_6O **(x)** MnO_2

16-4 What is the difference between oxidation number and valence?

16-5 Fill in the table.

Name	*Oxidation number change (increase, decrease, no change)*	*Electron gain or loss (gain, loss, no change)*
Oxidation		
Reduction		
Substance reduced		
Substance oxidized		
Oxidizing agent		
Reducing agent		

16-6 In each of the following reactions, label the oxidation number of all atoms, the species oxidized, the species reduced, the oxidizing agent, and the reducing agent.
(a) $Ni^{2+} + Fe \rightarrow Fe^{2+} + Ni$
(b) $2HI + Cl_2 \rightarrow I_2 + 2HCl$
(c) $3VO_3^- + Al + 12H^+ \rightarrow 3VO^{2+} + Al^{3+} + 6H_2O$
(d) $K_2Cr_2O_7 + 3SnCl_2 + 14HCl \rightarrow 2CrCl_3 + 3SnCl_4 + 2KCl + 7H_2O$
(e) $2MnO_2 + O_2 + 4OH^- \rightarrow 2MnO_4^{2-} + 2H_2O$

16-7 Balance (and complete where necessary) each of the following equations.
(a) $Cr + Cu^{2+} \rightarrow Cr^{3+} + Cu$
(b) $Cu + NO_3^- \rightarrow Cu^{2+} + NO$ (acid solution)
(c) $PbO_2 + Cl^- \rightarrow ClO^- + Pb(OH)_3^-$ (basic solution)
(d) $CO + I_2O_5 \rightarrow CO_2 + I_2$ (acid solution)
(e) $Fe + O_2 \rightarrow Fe_2O_3$ (equation describing the rusting of iron; acid solution)
(f) $S_2O_3^{2-} + Br_2 \rightarrow S_4O_6^{2-} + Br^-$
(g) $H_2O_2 + HI \rightarrow H_2O + I_2$
(h) $Zn + H_3AsO_4 \rightarrow AsH_3 + Zn^{2+}$ (acid solution)
(i) $Zn + As_2O_3 \rightarrow AsH_3 + Zn^{2+}$ (acid solution)
(j) $Pb^{2+} + H_2O \rightarrow Pb + PbO_2$ (acid solution)
(k) $ClO_2 + O_2^{2-} \rightarrow ClO_2^- + O_2$ (basic solution)

16-8 Balance (and complete where necessary) each of the following equations.
(a) $CrO_4^{2-} + I^- \rightarrow Cr^{3+} + IO_3^-$ (basic solution)
(b) $ClO_3^- + I_2 \rightarrow IO_3^- + Cl^-$ (acid solution)
(c) $MnO_4^- + S^{2-} \rightarrow MnO_2 + S$ (basic solution)
(d) $Al + KNO_3 + H_2O \rightarrow Al(OH)_3 + NH_3 + KOH$
(e) $I_2 \rightarrow I^- + IO_3^-$ (basic solution)

(f) $Br_2 + KAsO_2 + KOH \rightarrow K_3AsO_4 + KBr$
(g) $CoSO_4 + Na_2O_2 \rightarrow Co(OH)_3 + Na_2SO_4$
(h) $WO_3 + SnCl_2 + HCl \rightarrow W_3O_8 + H_2SnCl_6 + H_2O$
(i) $NO_2 \rightarrow HNO_3 + NO$
(j) $NaTcO_4 + Ti + H_2SO_4 \rightarrow TcSO_4 + Ti_2(SO_4)_3$

16-9 Calculate equivalent weights of each of the following compounds, given the half-equations.
(a) $PbCl_2$ $Pb \rightarrow Pb^{2+} + 2e^-$
(b) O_2 $O_2 + 4e^- + 2H_2O \rightarrow 4OH^-$
(c) $NaReO_4$ $Re + 8OH^- \rightarrow ReO_4^- + 4H_2O + 7e^-$
(d) Ag $4Ag + Fe(CN)_6^{4-} \rightarrow Ag_4Fe(CN)_6 + 4e^-$

16-10 **(a)** Calculate the weight of UO_2Cl_2 necessary to prepare 1.000 liter of a 1.000 *N* solution in water, given the half-equation

$$UO_2^{2+} + 4H^+ + 2e^- \longrightarrow U^{4+} + 2H_2O$$

(b) What is the molarity of this solution?

16-11 **(a)** Calculate the weight of $NaBrO_3$ necessary to prepare 85.5 ml of a 0.672 *N* solution, given the half-equation

$$BrO_3^- + 6H^+ + 6e^- \longrightarrow Br^- + 3H_2O$$

(b) What would the weight be if this were the half-equation

$$2BrO_3^- + 12H^+ + 10e^- \longrightarrow Br_2 + 6H_2O?$$

(c) What are the molarities of the two solutions?

16-12 Given the equation

$$K_2Cr_2O_7 + 14HCl \longrightarrow 2CrCl_3 + 2KCl + 3Cl_2 + 7H_2O$$

how many milliliters of a 0.325 *N* solution of HCl are necessary to reduce completely 36.7 ml of a 0.526 *N* solution of $K_2Cr_2O_7$?

16-13 **(a)** Calculate the weight of $KClO_3$ necessary to prepare 275 ml of a 0.388 *N* solution, given the half-equation

$$ClO_3^- + 2H^+ + e^- \longrightarrow ClO_2 + H_2O$$

(b) What is the molarity of this solution?

16-14 **(a)** What is the equivalent weight of HNO_3 when it acts as an acid? **(b)** What is its equivalent weight when it acts as an oxidizing agent, being reduced to NO?

16-15 Nitric acid, HNO_3, oxidizes sulfite ion, SO_3^{2-}, to sulfate ion, SO_4^{2-}, in acid solution, and is reduced to NO. **(a)** Write a balanced equation for this reaction. **(b)** What is the molarity and normality of a solution that is prepared by dissolving 65.50 g of HNO_3 in 500.0 ml of water? **(c)** It is found that 35.53 ml of a 0.0775 *M* HNO_3 solution will react exactly with 25.00 ml of a solution of sodium sulfite. Calculate the molarity and normality of the Na_2SO_3 solution and the equivalent weight of Na_2SO_3 in this reaction.

16-16 If a solution made by dissolving 2.17 g of an unknown reducing agent in 75.0 ml of H_2O is titrated with the oxidizing agent Br_2 and requires 37.45 ml of an 0.865 *N* solution of Br_2 to reach the end point, what is the equivalent weight of the unknown reducing agent?

16-17 Draw voltaic cells for each of the following unbalanced reactions. In your drawings, indicate the balanced half-cell reactions, the anode, the cathode, and the flow of electrons.
(a) $Mg(s) + Cl_2(g) \rightarrow Mg^{2+} + 2Cl^-$
(b) $Cu^{2+} + H_2 \rightarrow Cu + 2H^+$
(c) $Al + Fe^{3+} \rightarrow Al^{3+} + Fe^{2+}$
(d) $Zn + AgCl \rightarrow Zn^{2+} + Ag + Cl^-$

16-18 For each of the voltaic cells in Problem 16-17, write the shorthand notation for the cells.

16-19 Calculate the $\mathcal{E}°$ cell at 25°C and 1 atm for each of the following unbalanced equations, considering all species to be at 1.0 *M* concentration. Predict which reactions will go spontaneously in the direction shown.

(a) $I_2 + MnO_2 \rightarrow MnO_4^- + I^- + H^+$
(b) $Na + S \rightarrow Na^+ + S^{2-}$
(c) $NO + Cr_2O_7^{2-} \rightarrow NO_3^- + Cr^{3+}$
(d) $Fe^{2+} + H_2O_2 + H^+ \rightarrow Fe^{3+} + H_2O$
(e) $Zn(s) + H^+ \rightarrow Zn^{2+} + H_2$
(f) $Cu(s) + Zn^{2+} \rightarrow Cu^{2+} + Zn$
(g) $Cu + Ag^+ \rightarrow Cu^{2+} + Ag$
(h) $AuCl_4^- + Hg_2^{2+} \rightarrow Hg^{2+} + Au + Cl^-$
(i) $Pb + O_2 + H^+ \rightarrow Pb^{2+} + H_2O_2$
(j) $Ni^{2+} + Pb + SO_4^{2-} \rightarrow Ni + PbSO_4$
(k) $Co(NH_3)_6^{2+} + Mg(OH)_2 \rightarrow Co(NH_3)_6^{3+} + Mg + OH^-$
(l) $Cu^{2+} + Ag + CN^- \rightarrow Cu^+ + Ag(CN)_2^-$
(m) $Fe^{2+} + MnO_2 + OH^- \rightarrow Fe + MnO_4^-$
(n) $Sn + Ni^{2+} \rightarrow Ni + Sn^{2+}$
(o) $K^+ + Na \rightarrow K + Na^+$

16-20 For each of the reactions in Problem 16-19 that will go spontaneously, write the shorthand notation for the voltaic cell corresponding to the reaction.

16-21 Which of the following unbalanced and incomplete reactions will proceed spontaneously at room temperature?
(a) $AgCl + Hg_2^{2+} \rightarrow$ **(b)** $Cu^{2+} + Mn \rightarrow$ **(c)** $Cu^{2+} + Ag \rightarrow$
(d) $La^{3+} + Ca \rightarrow$ **(e)** $Cu + H^+ \rightarrow$ **(f)** $MnO_4^- + Ag^+ \rightarrow$
(g) $Al + H^+ \rightarrow$ **(h)** $SO_4^{2-} + Br^- \rightarrow$

16-22 Will SO_4^{2-} spontaneously oxidize Sn in basic solution? Will it do so in acid solution?

16-23 In the following pairs, pick out the stronger oxidizing agent. **(a)** F_2, F^-
(b) F^-, MnO_4^- **(c)** O_2, PbO_2 **(d)** Ag, Ag^+ **(e)** Fe^{3+}, Fe^{2+} **(f)** Cu^{2+}, Zn^{2+}
(g) Cu^{2+}, Cu^+ **(h)** H^+, MnO_2 **(i)** Li^+, Li **(j)** Li^+, F_2 **(k)** H_2O, Al^{3+} **(l)** Pb^{2+}, PbO_2

16-24 In the following pairs, pick out the stronger reducing agent. **(a)** F^-, Li **(b)** Na, Zn
(c) Cu, Cu^+ **(d)** Pb, Pb^{2+} **(e)** Cd, La^{3+} **(f)** H_2, Ni **(g)** Cu, Zn **(h)** OH^-, H_2
(i) H_2O_2, I^- **(j)** Cu^{2+}, Cu^+ **(k)** Cl^-, Br^- **(l)** Br^-, F^- **(m)** H_2O, Hg_2^{2+}
(n) Hg_2^{2+}, Hg^{2+} **(o)** Hg^{2+}, Hg

16-25 **(a)** Suppose that the half-cell reaction $AuCl_4^- + 3e^- \rightarrow Au + 4Cl^-$ was taken as the standard half-cell reaction and its $\mathcal{E}°$ was assigned the value zero. What would be the $\mathcal{E}°$ of the following half-cells? (1) $Hg_2^{2+}|Hg$ (2) $H^+|H_2$ (3) $Cu^{2+}|Cu$ (4) $Zn^{2+}|Zn$
(b) Using these $\mathcal{E}°$ values, calculate the $\mathcal{E}°$ of the cell $Zn|Zn^{2+}||Cu^{2+}|Cu$. What can you conclude about the $\mathcal{E}°$ of a cell?

16-26 Write the Nernst equation for each of the following cells. In each case where $H_2(g)$ appears, the pressure is 1 atm.
(a) $Sn(s)|Sn^{2+}(0.050\ M)||H^+(0.20\ M)|H_2(g)|Pt$
(b) $Cu(s)|Cu^{2+}(3.0\ M)||Ag^+(0.30\ M)|Ag(s)$
(c) $Pt|Br_2(l)|Br^-(0.0100\ M)||H^+(0.0300\ M)|H_2(g)|Pt$
(d) $Al|Al^{3+}(0.0150\ M)||Cl^-(0.0250\ M)|AgCl(s)|Ag(s)$
(e) $Pt|Cr^{2+}(0.250\ M)|Cr^{3+}(0.500\ M)||Fe^{3+}(0.0100\ M)|Fe^{2+}(0.200\ M)|Pt$
(f) $Pt|H_2(g)|H^+(0.050\ M)||H^+(0.500\ M)|H_2(g)|Pt$

16-27 Calculate $\mathcal{E}_{cell}$ for each of the cells in Problem 16-26.

16-28 Calculate ΔG, $\Delta G°$, and K for the cells in Problem 16-26.

16-29 The voltage of the following cell $Cu(s)|Cu^+(0.5500\ M)||Ag^+(x\ M)|Ag(s)$ was 0.31 V. Calculate the unknown Ag^+ concentration.

16-30 Given the cell in Problem 16-29, $Cu(s)|Cu^+(0.5500\ M)||Ag^+(x\ M)|Ag(s)$, calculate the equilibrium concentration of Ag^+.

16-31 Given the cell $Ag|AgCl(s)|NaCl(0.05\ M)||AgNO_3(0.30\ M)|Ag$. **(a)** Write the half-reaction occurring at the anode. **(b)** Write the half-reaction at the cathode. **(c)** Write the net ionic equation for the cell. **(d)** Does the cell above go spontaneously as written? **(e)** Draw a diagram of the cell. Label the cathode, anode, direction of electron flow, and salt bridge. **(f)** Calculate $\mathcal{E}^\circ_{cell}$ and $\mathcal{E}_{cell}$ at 25°C.

16-32 For this reaction in aqueous solution at 25°C, $\mathcal{E}^\circ = -1.0181$ V:

$$Cu^{2+} + 2Cl^- \longrightarrow Cu(s) + Cl_2(g)$$

find ΔG° and K.

16-33 For electrolysis in this cell

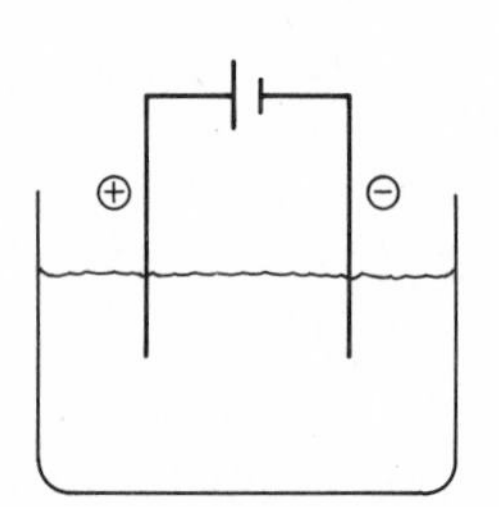

answer the following questions for each of the materials listed. **(a)** molten KI **(b)** aqueous solution of K_2SO_4 **(c)** aqueous solution of FeF_2 **(d)** aqueous solution of $NiBr_2$ **(e)** aqueous solution of CaS (1) Which electrode is the anode (+ or −)? (2) Which species migrates to the + pole? (3) Which species migrates to the − pole? (4) At which electrode (+ or −) does reduction take place? (5) Write the half-equation for the reaction at the + electrode. (6) Write the half-equation for the reaction at the − electrode. (7) Write the overall equation.

16-34 What is the essential difference between a voltaic cell and an electrolytic cell?

16-35 How many coulombs of electricity are passed through a solution in which 37.4 amperes are allowed to run for 30.0 min?

16-36 What weight of product would be produced by electrolysis in each case. **(a)** Mg, if 27,500 coulombs passed through molten $MgCl_2$ **(b)** Pb, if 3.74 faradays passed through molten $PbSO_4$ **(c)** Bi, if 11.5 amperes are passed through molten $BiCl_3$ for 27.5 min.

16-37 How many faradays would be required to reduce 1 mole of each of the following to the indicated product? **(a)** Fe^{3+} to Fe **(b)** MnO_4^- to Mn^{2+} **(c)** Cu^{2+} to Cu **(d)** K^+ to K **(e)** I_2 to I^- **(f)** $Cr_2O_7^{2-}$ to Cr^{3+} **(g)** PbO_2 to Pb^{2+}

16-38 How many seconds would it take to plate out 6.6 g of Ni from a solution of $NiCl_2$ if a current of 4.4 amperes is used? What would your answer be if the solution was $Ni(NO_3)_2$?

16-39 How many faradays of electricity are required to produce the following? **(a)** 5.5 g of Fe from $FeCl_3$. **(b)** 18.8 g of Cd from $CdSO_4$. **(c)** 25.0 g of Na from NaCl. **(d)** 1.00 g of I_2 from Hg_2I_2. **(e)** 10.0 liters of O_2 at STP from $Hg(OH)_2$.

16-40 For 1200.0 min, 25.000 milliamperes are passed through solutions of (a) H_2O (containing Na_2SO_4), (b) $PbSO_4$, (c) $ScCl_3$, and (d) $AgNO_3$. How many grams of **(a)** H_2 **(b)** Pb **(c)** Sc **(d)** Ag are produced?

16-41 When a lead–acid automobile battery is discharged, the anode reaction is

$$Pb(s) + SO_4^{2-} \longrightarrow PbSO_4(s) + 2e^-$$

If the battery is run at a rate of 2.50 amperes for 1.00 h, how many grams of $PbSO_4$ will be deposited on the anode?

16-42 Water is not normally an oxidizing agent (that is, it is not easily reduced). Yet it is common experience that tools and other implements made of iron will rust (oxidize) much faster if left out in the rain. Explain.

Additional Problems

16-43 Chlorine is produced electrolytically from molten NaCl. When a current of 2.30 amperes generates chlorine for 12.32 min, how many milliliters of chlorine gas are produced at 27°C and 740 torr?

16-44 Arrange the following reducing agents in order of increasing reducing ability: Cu, F^-, Mg, Au^+, Pb, Fe^{2+}, OH^- (all except the last in neutral solution).

16-45 Compare the oxidation number and valence of each of the following. **(a)** C in CH_4 **(b)** C in C_2H_6 ($CH_3—CH_3$) **(c)** C in C_2H_4 ($CH_2═CH_2$) **(d)** N in HNO_3 (structure given in Section 14-10) **(e)** S in $S_2O_3^{2-}$ (structure given in Section 18-5).

16-46 In an experiment in acid solution, 73.73 ml of a 0.0125 *N* $K_2C_2O_4$ solution reacted completely with 52.23 ml of a $KMnO_4$ solution to yield Mn^{2+} and CO_2. How many milliliters of this $KMnO_4$ solution will completely react with 0.1550 g of $NaCr(OH)_4$ dissolved in 100 ml of water to yield CrO_4^{2-} and MnO_2?

16-47 If we wish spontaneously to reduce Ba^{2+} to Ba, what are the only three reducing agents in Table 16-2 that could be used?

16-48 In the presence of atmospheric oxygen, Fe^{2+} ions are oxidized to Fe^{3+} ions. Suppose that you had a solution containing Fe^{2+} ions that you wanted to keep free of Fe^{3+}. Which of the following procedures would you recommend and why? (Assume all species at 1.0 *M* concentration.) **(a)** Add Fe. **(b)** Add Au. **(c)** Add Zn^{2+}. **(d)** Add Ag^+. **(e)** Add Ag.

16-49 Given the equation

$$2KMnO_4 + 3H_2SO_4 + 5H_2O_2 \longrightarrow 2MnSO_4 + 5O_2 + K_2SO_4 + 8H_2O$$

if 75.000 g of $KMnO_4$ is dissolved in enough water to make 300.0 ml of solution and 25.000 g of H_2O_2 is dissolved in enough water to make 275.0 ml of solution, how many milliliters of the $KMnO_4$ solution will completely oxidize all the H_2O_2 in the 275.0 ml of the H_2O_2 solution?

16-50 What is the oxidation number of chlorine in each of the following compounds and ions? **(a)** HCl **(b)** $NaClO_4$ **(c)** NaClO **(d)** Cl_2 **(e)** NaCl **(f)** ClO_2^- **(g)** ClO_3^- **(h)** ClO_4^- **(i)** Cl^- **(j)** ICl **(k)** ClF_3 **(l)** CCl_4

16-51 Commercial drain cleaners (such as Drano) contain mixtures of aluminum powder and lye (sodium hydroxide). One reaction that occurs when such a mixture is added to water is

$$Al + 3H_2O(l) \longrightarrow Al(OH)_3 + \tfrac{3}{2}H_2(g)$$

Construct a voltaic cell that represents this reaction and calculate its $\mathcal{E}°$.

16-52 The voltage of the cell $Pt|Fe^{3+}(0.500\ M), Fe^{2+}(0.400\ M)||H^+(n\ M)(g)|H_2|Pt$ was 0.82 V. The pressure of H_2 is 1 atm. Calculate the unknown hydrogen concentration.

16-53 A student taking his final oral examination for the Ph.D. was discussing an oxidation reaction. He was asked "What was reduced in that reaction?" and replied "Nothing was reduced." He was failed on the spot. Why?

16-54 If Cu is spontaneously oxidized by O_2 in basic solution, would the product be Cu^+ or Cu^{2+}?

16-55 Complete and balance this equation in both acidic and basic solution:

$$Cl_2 + IO_3^- \longrightarrow IO_4^- + Cl^-$$

16-56 What is wrong with this equation?

$$6H^+ + 2NO_3^- + 3HgS + 12Cl^- \longrightarrow 2NO + 2H_2O + 3S + 2OH^- + 3HgCl_4^{2-}$$

16-57 An unknown salt contains either Al^{3+} or Ga^{3+} ions. We want to find out which by running an electrolysis experiment on the molten salt. Calculate the number of seconds a current of 10.000 amperes should be sent through the cathode in order to deposit 15.50 g of each metal. If the experiment was carried out, and 15.50 g of the unknown metal (Al or Ga) was deposited, while at the same time 71.948 g of Ag was deposited in another electrolytic cell in series with the first one, which ion was in the unknown salt?

17 Air and Water

Fritz Haber, 1868–1934

Courtesy Morris Goran

So far in this book we have spent most of our time in learning and understanding the *principles* and *laws* of chemistry. In this chapter and in the next five chapters, we shall turn to *descriptive chemistry,* which means that we will be discussing the structures and properties of some of the compounds that make up the world—both living and nonliving things. In doing so, we will be able to see how some of the principles we have already learned help us to understand the behavior of actual materials. Of course, the world contains a vast number of compounds and mixtures, both natural and man-made, and we have room to discuss only a small percentage of them, but still we shall get some idea of the great variety of structures and properties that compounds can have.

The first of our six descriptive chapters is devoted to the two materials on which we depend for life itself—air and water, two of the four original Greek "elements." Without air and water the earth would be like the moon—its surface would consist only of rock and sand, with no weather. During the day, the full heat of the sun would beat down on the ground, with no atmosphere to filter out the most harmful rays (see Section 17-5), and temperatures would climb above 100°C. At night, with no atmosphere to moderate the temperature, it would fall to perhaps −100°C (it would not go near absolute zero, since the interior of the earth produces a good deal of heat, partly from the radioactivity of some of the elements within the earth). Of course, none of this would matter to any living thing, since without air and water there would be no life on earth.

Part I AIR

17-1 The Composition of the Atmosphere

The composition of the atmosphere is remarkably constant from place to place and over relatively long periods of time. Table 17-1 shows the composition of clean dry air at sea level. The table gives percentages by volume (and parts per million for components of low concentration); as we saw in Chapter 4, for an ideal gas, percentage by volume is the same as percentage by moles or percentage by molecules. Therefore, out of every 10,000 molecules in dry air, 7809 are N_2, 2094 are O_2, and 93 are Ar (Figure 17-1). The remainder of the air (0.04%) consists of all of the other gases in Table 17-1, as well as others present in still smaller amounts. Water vapor is not included in the table because its concentration does vary widely; it is normally present in concentrations of 1 to 3%.

Although we have been able to measure the composition of the atmosphere for only a relatively short time (less than 200 years), there is reason to believe that the

Table 17-1 Composition of Clean Dry Air at Sea Level (Percent by Volume; also Parts per Million for Components of Low Concentration)[a]

	% by volume	Parts per million
N_2	78.09	
O_2	20.94	
Ar	0.93	
CO_2	0.0318	318
Ne	0.0018	18
He	5.2×10^{-4}	5.2
CH_4	1.5×10^{-4}	1.5
Kr	1×10^{-4}	1
H_2	5×10^{-5}	0.5
N_2O	2.5×10^{-5}	0.25
CO	1×10^{-5}	0.1
Xe	8×10^{-6}	0.08
O_3	2×10^{-6}	0.02
NH_3	1×10^{-6}	0.01
NO_2	1×10^{-7}	0.001
SO_2	2×10^{-8}	0.0002

[a] Water vapor is not included because its concentration varies greatly.
Source: Cleaning Our Environment: The Chemical Basis for Action (Washington, D.C.: American Chemical Society, 1969).

proportions of the major components N_2, O_2, and Ar have not changed appreciably over much longer time spans, at least thousands of years. Fears have lately been expressed that our modern industrial society, with the vast increase in the amount of combustion taking place, might be causing a decrease in the percentage of O_2 in the atmosphere, but this has so far not happened. A study made in 1970 (by Machta and Hughes) showed that the oxygen concentration in the air changed by less than 0.010% in the 60 years from 1910 to 1970. However, as might be expected, the percentages of some of the minor components of air can change much more rapidly. For example, NO_2 and SO_2 are major pollutants produced by automobile engines and heating furnaces (Section 17-7); because there is so little of these compounds in natural air, it does not take a great deal of pollution to increase their concentrations substantially, especially near well-traveled highways or large factories.

In addition to the components already discussed, the air also usually contains tiny dispersed particles of liquids or solids. These particles, which are too small to be seen, are called **aerosols** or **particulates.** They range in size from clusters of a few molecules to particles with diameters of a few thousandths of a millimeter.

Although the atmosphere now contains about 78% N_2 and 21% O_2, it is likely that

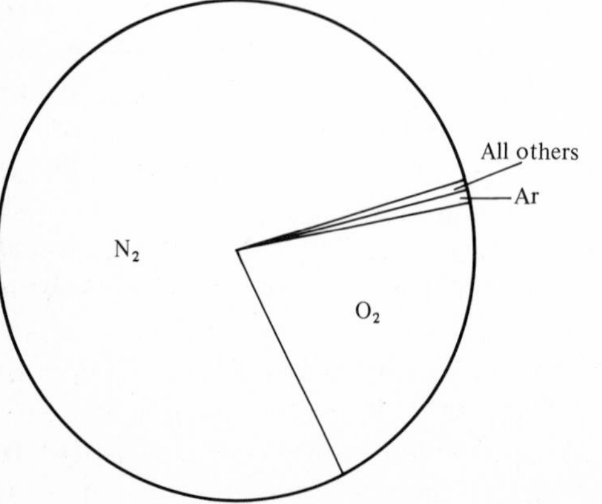

Figure 17-1 Composition of clean dry air.

in the early days of the earth its composition was very different. We are able to say this, even though nobody was there to measure the atmospheric composition several billion years ago, because we know that the O_2 in the air today is produced by plant life (a large proportion from plankton, which is plant life floating on or near the surface of the oceans), by the process called photosynthesis. There was no plant life on the earth several billion years ago, and we know of no other natural process that would lead to large amounts of atmospheric O_2. Thoughts about the early atmosphere can only be conjectures, but it is currently believed that before there was life on earth the atmosphere consisted mostly of H_2, He, CH_4, NH_3, and H_2O, and *no* O_2. The present atmosphere arose from all the natural processes that have taken place in the last few billion years, the most important of which was the establishment of life.

Most of the H_2 and He was lost from the upper reaches of the atmosphere in the following manner. These gases have very low molecular weights. At any given time, a small fraction of the H_2 and He molecules in the upper atmosphere have very high velocities (see the Maxwell distribution—Figure 4-14). Any of these high-velocity molecules which happen to be moving away from the earth will have sufficient energy to overcome earth's gravity, and will be lost from the atmosphere forever. Over a period of billions of years, nearly all of the original H_2 and He was lost in this way.

17-2 Layers of the Atmosphere

About 200 years ago scientists began to measure how the temperature of the air varies with increasing height. Such measurements were impossible before then, because two inventions were required: (1) the thermometer, and (2) a means of ascending higher than the tallest building, tree, or nearby mountain. Practical thermometers were produced by Gabriel Fahrenheit (1686–1736) after 1714,* and the Montgolfier brothers in 1783 invented the hot-air balloon, which was soon improved upon by Jacques Charles (1746–1823), who substituted hydrogen for hot air. This is a good example of how a scientific advance had to wait for improvements in technology.

The early investigators found that the temperature of the air decreased with increasing altitude, with an average decrease of about 6.5°C/km, as high as they could go. They naturally concluded that the temperature continued to decrease all the way up. However, that this conclusion was wrong, was discovered about 1900, when the first high-altitude balloon soundings were made. Beginning at about 10 km, the temperature no longer decreases but begins to level off, and then starts to *increase*.

Today we know that the atmosphere can be divided approximately into four layers, depending on how the temperature varies (Figure 17-2). In the lowest layer, called the **troposphere,** the temperature gradually decreases with increasing height. Then come the **stratosphere,** in which the temperature increases again; the **mesosphere,** where there is a decrease to about −80°; and the **thermosphere,** where the temperature rises once again. The boundaries between the four layers are arbitrary and not sharply defined. They change with latitude and with the season. Virtually all the water vapor is in the troposphere; consequently, so is the weather. Airplanes can escape the weather by flying near the top of the troposphere. In Section 17-5, we shall discuss the reasons for the way the atmospheric temperature changes with altitude.

Although the variation of temperature with altitude is irregular, the same is not true for the pressure. Atmospheric pressure smoothly decreases with increasing

*Fahrenheit did not invent the thermometer, but he was the first to use mercury as the liquid in the thermometer.

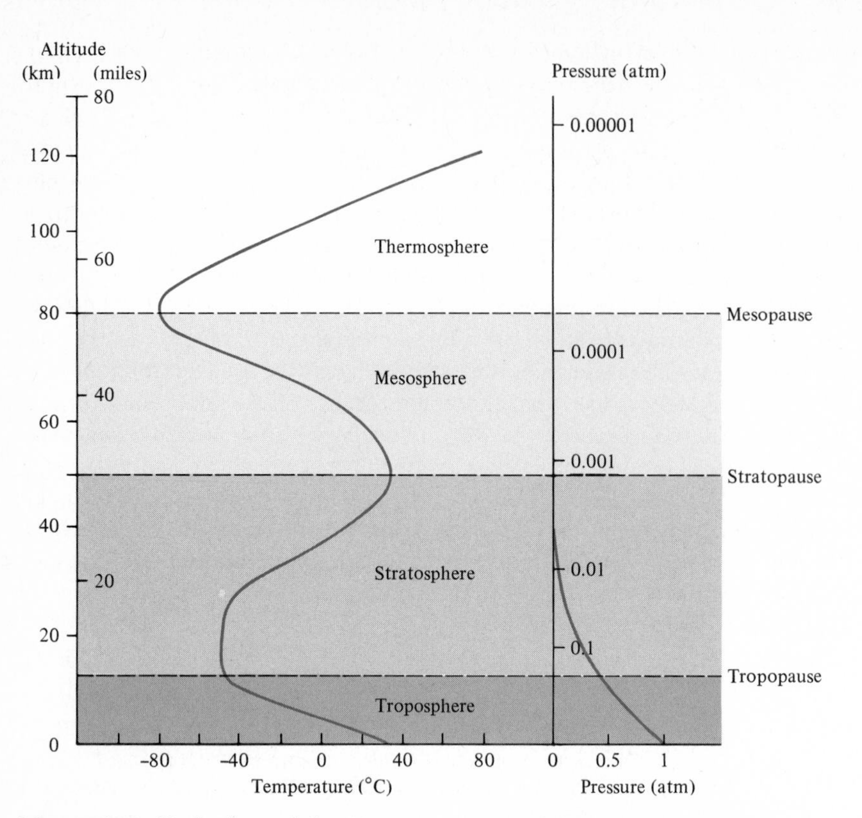

Figure 17-2 Variation of the Temperature and Pressure of the Atmosphere with Increasing Altitude. The four principal layers of the atmosphere are shown. Boundaries between layers are approximate. [Source: A. H. Strahler, *The Earth Sciences*, 2nd ed., Harper & Row, New York, 1971]

height (Figure 17-2). The decrease is fairly sharp, and climbers of high mountains must take oxygen with them, because the air pressure at such heights is not enough to sustain normal breathing. The top of Mount Everest is about 8 km above sea level. At the **mesopause** (the boundary between the mesosphere and the thermosphere), about 80 km high, the atmospheric pressure is only about 10^{-4} atm. About 95% of the atmosphere is within 20 km of sea level.

In the next four sections we shall discuss the properties of the two major constituents of the atmosphere and two of the minor constituents.

17-3 Nitrogen and Nitrogen Fixation

The most important property of N_2 is that *it is inert*. Apart from biological nitrogen fixation (to be discussed below), nitrogen gas at room temperature does not react with any substances at all, except with lithium, to form the ionic compound lithium nitride

$$N_2 + 6Li \longrightarrow 2Li_3N$$

and with a few transition metal complexes. N_2 is more reactive at high temperatures, but even here its reactivity is much less than that of most other gases.

The inertness of nitrogen gas is due to the great stability of the N≡N bond. ΔH for the reaction $N_2 \rightarrow 2 \cdot\overline{\underset{\bullet}{N}}\cdot$ is $+227$ kcal/mole, making this one of the strongest covalent bonds known.

At high temperatures, N_2 reacts with several metals besides Li, including Mg, Ca, Ba, Zn, and Al, and with the metalloid B. Most of the nitrides produced in these reactions are ionic, although BN and AlN are covalent compounds of the giant molecule type with structures similar to graphite and diamond, respectively. At high temperatures N_2 can also react with O_2 and with H_2. These reactions will be discussed below.

The inertness of N_2 posed a major problem to the thinking of the early-nineteenth-century scientists. There are a great many plants in the world, and all of them possess a substantial amount of nitrogen-containing compounds, chiefly amino acids and proteins (see Section 21-4) along with other compounds, including nucleic acids, alkaloids (such as nicotine, quinine, strychnine, morphine, and many others), purines, and certain vitamins and hormones. The problem was: How did the nitrogen get into the plants? Apart from nitrogen, the other major elements that are ingredients of plants are oxygen, carbon, and hydrogen, and Lavoisier had shown that plants obtained these elements from carbon dioxide and water, by the process of photosynthesis. The obvious source of nitrogen for plants was from the N_2 in the atmosphere (and Humphry Davy actually stated this), but it was difficult to establish experimental evidence for this theory, and most nineteenth-century scientists believed that the nitrogen came from nitrogen compounds in the soil—that atmospheric nitrogen was not involved.

In the 1880s, a German scientist, Hermann Hellriegel (1831–1895), and his coworker H. Wilfarth proved that certain biological organisms do have the ability to use N_2 in the atmosphere, and to convert it to the nitrogen-containing compounds found in plants. This process is called **nitrogen fixation.** *Fixed nitrogen* is defined as nitrogen which is present in the form of a compound that can be used by plants. There are basically two types of nitrogen-fixing organisms:

1. Bacteria that live in symbiosis with certain plants, mostly of the type called *legumes* (typical legumes are clover, alfalfa, peas, beans, and peanuts). These bacteria live in nodules attached to the roots of the plants (Figure 17-3). Neither the plants nor the bacteria alone can fix nitrogen; both must be present.
2. Certain organisms that live free in soil and in marshes—mostly certain species of bacteria and of algae. These species can fix nitrogen independently of any plants.

For thousands of years farmers have understood the importance of crop rotation. For two or three years, a given field would be planted with a crop such as wheat or rye, but in the next year it would be a legume such as alfalfa or clover, which was then plowed into the soil, so that wheat or rye could be planted again in the following year. We now understand why this method works. Nonleguminous plants take nitrogen from the soil (in the form of nitrates or ammonium compounds), so that after two or three years the soil is depleted and further nonleguminous crops will yield a much smaller harvest. The occasional leguminous crop fixes atmospheric nitrogen and puts it back into the soil.

The mechanism by which these organisms fix nitrogen is not fully understood, although many details are known. For example, it is fairly certain that the mechanism is the same for all organisms, whether free-living or symbiotic, and that the ultimate product of the reaction is ammonia NH_3. The hydrogen comes from water that is present in all biological systems.

$$N_2 \xrightarrow[\text{organisms}]{\text{nitrogen-fixing}} NH_3$$

The fixation of nitrogen by biological organisms is part of a vast cycle by which nitrogen is taken from the atmosphere, used by plants and animals, and returned to the atmosphere (Figure 17-4). A second important nitrogen-fixation pathway is the

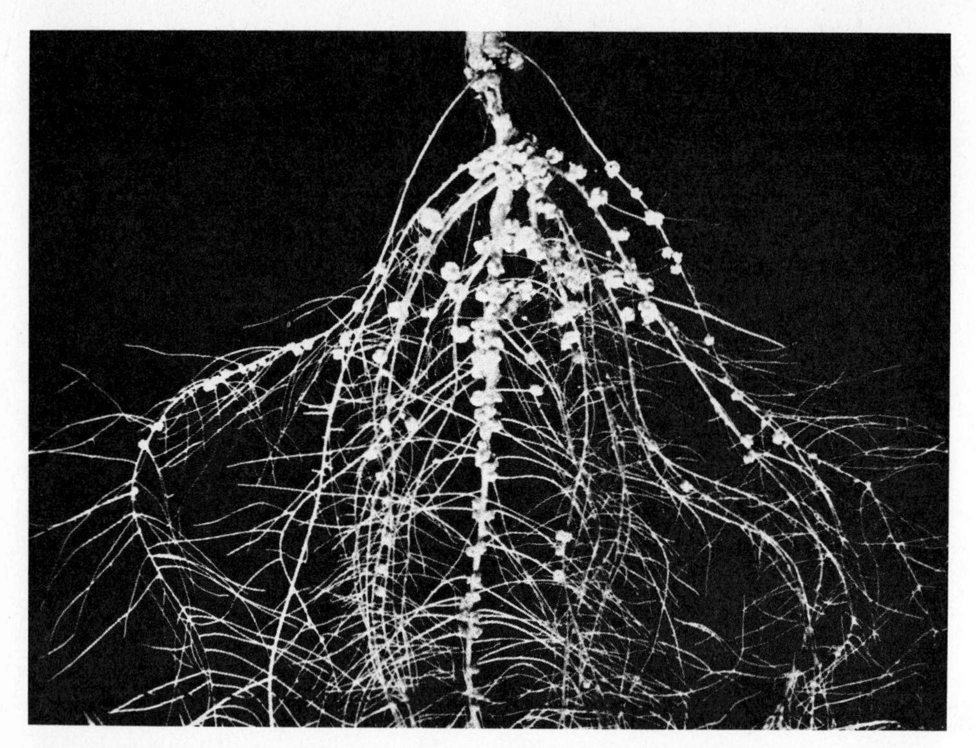

Figure 17-3 Roots of soybeans showing nodules. [Courtesy Professor G. Bond, University of Glasgow]

reaction of the nitrogen and oxygen in the atmosphere to form nitric oxide, NO:

$$N_2 + O_2 \xrightarrow[\text{temperatures}]{\text{high}} 2NO$$

This reaction does not occur at ordinary temperatures (which is a good thing, because if it did, it would use up all the oxygen in the air), but it does occur along the pathways of lightning, cosmic radiation, and meteor trails, all of which produce very high temperatures within a relatively small volume of air. The nitric oxide is oxidized by the oxygen in the air and is washed down by the rain, where it reaches the soil in the form of nitrates (NO_3^-). The amount of nitrogen fixed by this pathway is much less than that fixed by the biological pathway, but it is still important.

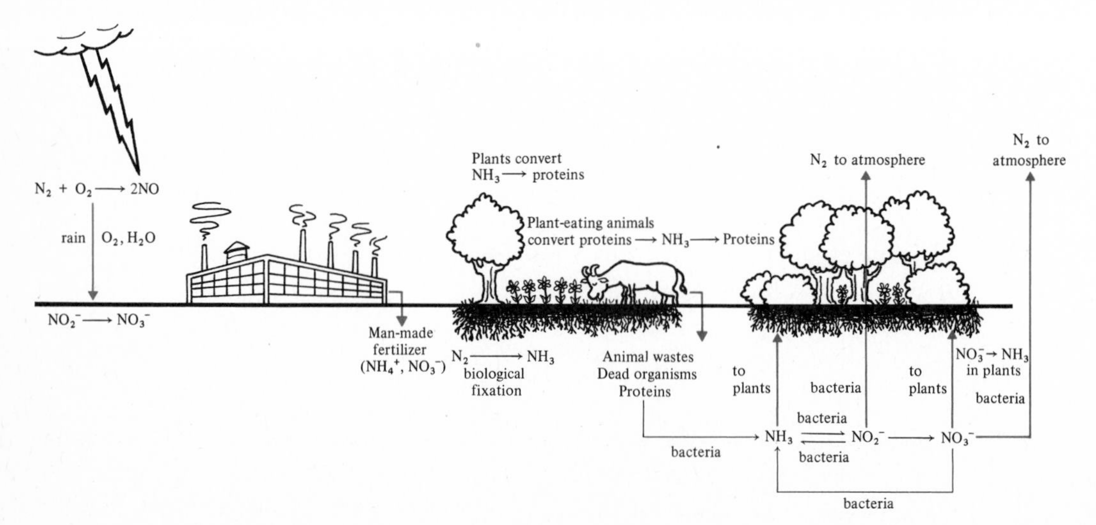

Figure 17-4 The nitrogen cycle.

Virtually all plants (even the legumes) can obtain useful nitrogen from both ammonia (or ammonium ion) and NO_3^-. (The legumes prefer to get their nitrogen from these sources—when these compounds are plentifully supplied in the soil, nitrogen fixation diminishes or stops.) The actual compound used by plants is ammonia, but they are able to reduce NO_3^- to ammonia. Once they have obtained their ammonia (either directly from the soil, by reduction of NO_3^-, or by fixation) the plants convert it to amino acids and then to proteins (and to other nitrogen-containing compounds). Animals get their nitrogen by eating the plants; other animals eat *them*; and so on. Eventually, most of the nitrogen in plants and animals returns to the soil in two forms: (1) animal wastes—mostly manure and urine, both of which contain nitrogen compounds, and (2) dead plants and the dead bodies of animals.

Once in the soil, these wastes and dead organisms are attacked by many bacteria, which decompose amino acids, proteins, and other nitrogen-containing compounds. The nitrogenous product of this reaction is NH_3. In addition, the soil contains bacteria that can carry out the following reactions (NO_2^- is the nitrite ion; N_2O is nitrous oxide):

$$\left.\begin{array}{l} NH_3 \longrightarrow NO_2^- \\ NO_2^- \longrightarrow NO_3^- \end{array}\right\} \textbf{nitrification,} \text{ catalyzed by bacteria}$$

$$\left.\begin{array}{l} NO_3^- \longrightarrow N_2 \text{ or } N_2O \\ NO_2^- \longrightarrow N_2 \text{ or } N_2O \\ NO_3^- \longrightarrow NH_3 \\ NO_2^- \longrightarrow NH_3 \end{array}\right\} \begin{array}{l} \textbf{denitrification,} \text{ catalyzed by} \\ \text{bacteria (anaerobic conditions)} \end{array}$$

The nitrification reactions are important because they cause valuable ammonia to be lost from the soil either by leaching out of NO_3^- or by subsequent denitrification of NO_3^- to N_2.

The denitrification reactions are the principal routes by which N_2 is returned to the atmosphere. These reactions only take place under anaerobic conditions (that is, in the absence of oxygen). When oxygen is present, these bacteria, like almost all other organisms, use oxygen to oxidize their carbohydrates and obtain energy from the process. Unlike most organisms, however, these bacteria, in the absence of oxygen, can use NO_3^- or NO_2^- as oxidizing agents instead. Of course, anaerobic conditions are often present deep within the soil. The product of denitrification can be N_2 or nitrous oxide, N_2O.

Late in the nineteenth century, certain scientists, most notably William Crookes (1832–1919), pointed out that the amount of N_2 being fixed each year by the natural processes mentioned above was fairly constant but the population was not; it was rapidly increasing. Crookes showed that in a relatively short time, the population would increase to the point where there would not be enough food to sustain it. He predicted widespread starvation unless an economical process for artificial nitrogen fixation (that is, conversion of atmospheric N_2 to NH_3 or NO_3^-) could be discovered. Although several artificial processes were known at that time, all were too expensive to be of practical use.

The problem was solved by Fritz Haber (1868–1934), who in 1908 invented the Haber process for the production of ammonia from nitrogen. The process was improved by Karl Bosch (1874–1940). For this work, Haber was awarded the Nobel Prize in Chemistry in 1918* (Bosch won the prize in 1931). In the Haber process,

*Haber's method not only solved the world's food-nitrogen problem, but also supplied Germany with nitrates for munitions in World War I, which was necessary for them, since imported nitrates (mainly from Chile) were cut off by the British blockade. Furthermore, Haber directed poison-gas warfare during the war, and after the war attempted to pay off Germany's postwar debt by isolating gold from seawater (this project did not succeed). Nevertheless, since Haber was Jewish, he was forced by Hitler to leave Germany in 1933, despite his enormous contributions. He died in Switzerland in 1934.

nitrogen is combined with hydrogen at high temperatures and pressures in the presence of a catalyst.

$$N_2(g) + 3H_2(g) \underset{500\text{–}550°C,\ 350\ atm}{\overset{catalyst}{\rightleftharpoons}} 2NH_3(g) \qquad \Delta H = -22.0\ kcal$$

The catalyst used today is iron containing small amounts of K_2O and Al_2O_3.

The Haber reaction is an equilibrium reaction, and at room temperature and atmospheric pressure lies well to the right, so that the formation of NH_3 is strongly favored. Nevertheless, no ammonia forms under these conditions, because the *rate* of the reaction is close to zero. The rate is increased by the addition of the catalyst, and by the high temperature. Since ΔH for the reaction is negative, the equilibrium is driven to the *left* by the higher temperature (Section 8-7), but this has to be done to achieve a reasonable rate. We have the paradoxical situation of finding it necessary to drive the equilibrium in the wrong direction in order to get a decent yield of NH_3 within a reasonable time period! The high pressure does two things: (1) it shifts the equilibrium to the right, in this reaction in which all components are gases (Section 8-7), and (2) it increases the rate of the reaction by increasing the concentration of the components.

The ammonia made by the Haber process is added to the soil in the form of fertilizer. The nitrogen in chemical fertilizers can be supplied in several ways: as NH_3 or NH_4^+; as NO_3^-; as NH_4NO_3; or as urea, NH_2CONH_2. Since all of these can be made from NH_3, the Haber process is the ultimate source of a large part of the world's food-nitrogen. More than 85% of the nitrogen content of the world's chemical fertilizers is derived from NH_3 produced by the Haber process. If all the Haber-process factories were to shut down, widespread starvation would follow within a year or two. Natural fertilizers (manure) have much less nitrogen than chemical fertilizers.

It has been estimated that about 237 million metric tons of atmospheric nitrogen were fixed in 1974. The distribution of this amount is shown in Table 17-2. As the table shows, chemical fertilizers account for about one-fourth of all the fixed nitrogen.

Although the Haber process solved the world's food-nitrogen problem at the time, the problem was not solved forever. The world's population has continued to increase, at an ever-increasing rate. More and more food is required all the time, and the capacity of the Haber process to supply NH_3 is not unlimited. What limits it? The Haber process requires three components: N_2, H_2, and the catalyst. It does not require energy, because it is exothermic and gives off energy. The supply of N_2 is essentially unlimited; it comes from the air. The supply of catalyst is unlimited, since the catalyst is not significantly used up. The limiting factor in the Haber process is the supply of H_2. At present, the H_2 used in the Haber process is largely

Table 17-2 Estimated Percentages of Nitrogen Fixation in 1974 from All Sources[a]

Natural fixation on agriculture land	38%
Natural fixation on forested or unused land	25%
Industrial fixation (mostly Haber process)	24%
Combustion	9%
Lightning	4%
Natural fixation in oceans (by algae or plankton)	?

[a]The total amount was about 237 million metric tons. The percentage of natural fixation in the oceans is unknown, but may be on the order of 1 million metric tons.

derived from methane by the following reactions:

$$CH_4(g) + H_2O(g) \xrightarrow{\Delta} CO(g) + 3H_2(g) \qquad \Delta H^\circ_{298} = 51.5 \text{ kcal}$$

$$CO(g) + H_2O(g) \overset{\Delta}{\rightleftharpoons} CO_2(g) + H_2(g) \qquad \Delta H^\circ_{298} = -9.8 \text{ kcal}$$

How long it will be economical to use these reactions as the source of the H_2 for the Haber process is a matter of conjecture, but it cannot be very long, for two reasons:

1. The supply of methane is limited. Methane is the chief constituent of natural gas, and we saw in Section E-2 that the supply is limited. At present we burn most of our natural gas, rather than saving it for future use as a source of hydrogen.
2. Unlike the Haber process, the conversion of methane to hydrogen is an *endothermic* process: it requires a good deal of energy—more than is given off in the Haber process. As long as energy was cheap, this did not matter, but as the cost of energy goes up, the price of chemical fertilizer must also go up, and eventually may make food too expensive for a large number of the world's poor people. The solution to this problem would be a cheap virtually inexhaustible supply of energy (Section E-4). If we had that, there would be no difficulty in obtaining H_2, because we could then get it from electrolysis of water (Section 16-12), and the supply of water is inexhaustible.

Nitrogen is only one part (although an important part) of the food problem. Plants also need other elements. Apart from C, H, and O, the most important of these are phophorus and potassium, and chemical fertilizers generally contain these elements as well as nitrogen. In addition, a number of other elements are needed in small quantities, among them cobalt, molybdenum, iron, and sulfur.

17-4 Oxygen

The discovery of oxygen was one of the most important events in the history of chemistry because it directly led Lavoisier to the overthrow of the **phlogiston theory,** a theory that had dominated chemistry for more than 100 years.* Like a number of other scientific discoveries, the discovery of oxygen was accomplished independently by two chemists working about the same time: Karl Wilhelm Scheele (1742–1786) and Joseph Priestley (1733–1804).† Scheele actually discovered it first

*According to the phlogiston theory, all combustible substances (such as wood, sulfur, coal) contained an ingredient called phlogiston. When the substance burned, the phlogiston was lost and only the ash remained. A combustible substance such as coal therefore consisted of phlogiston and ash, and the ash, the other product of the combustion, weighed less than the original coal. Phlogiston had never been isolated, although many chemists believed that hydrogen—discovered by Henry Cavendish (1731–1810) around 1766—was pure phlogiston, since it burned without leaving any ash. Lavoisier explained combustion precisely in the opposite way. Instead of losing phlogiston, the combustible portion of a substance such as coal combined with oxygen to give oxides; that is, the total combustion products (the ash and the oxides) weighed *more* than the coal, not less. This fact was not recognized earlier because the most important oxide formed by combustion is CO_2, which is a gas, and most scientists had not thought to test for it, since they believed that gases were unimportant. Lavoisier's correct explanation of the nature of combustion as oxidation allowed chemistry to make enormous strides in the next century, and it is for this, even more than for his many other contributions, that he is often called the father of chemistry.

†Both Scheele and Priestley also discovered other gases: Scheele—Cl_2, HF, H_2S, and HCN (all of these are poisonous, and it is likely that his work with such compounds—it was the custom to taste new compounds at that time—led to his early death); Priestley—HCl, NH_3, N_2O, SO_2, CO, NO, and NO_2. It was Joseph Black's demonstration (footnote on page 120) that a gas was a chemical substance that led the way to these discoveries. In later life Priestley's religious and political opinions became very unpopular (his house was burned down by an angry mob because of his support of the French revolution), and in 1794 (one month before Lavoisier was executed) he was forced to leave England for the United States, where he settled in Northumberland, Pennsylvania. It was in Northumberland that Priestley discovered CO, in 1799.

(1771–1772), but Priestley is generally given the credit because he published his work earlier (1774). Ironically, neither Scheele nor Priestley ever accepted Lavoisier's new theories about combustion. Both continued to believe in phlogiston until they died.

Oxygen is the most abundant element in the earth's crust and constitutes about 50% of its total weight. Not only does the atmosphere contain about 21% oxygen, but about 89% of the earth's vast amount of water is oxygen; and most of the rocks and minerals that constitute the earth's crust are oxides, carbonates, and sulfates, all of which contain oxygen. Furthermore, oxygen is an important constituent of many organic compounds (Chapter 20), and is an essential component of all living things.

In Section 17-3, we learned that N_2 is essentially inert. O_2 in contrast is quite reactive, partly because of its much lower bond energy (ΔH for $O_2 \rightarrow 2\cdot\overline{\underset{\bullet}{O}}|$ is $+119$ kcal/mole), and partly because of its diradical character (Section 14-20). Compounds of oxygen are known with all elements except He, Ne, and Ar. Some of these compounds have to be prepared by indirect methods, but O_2 combines *directly* with all the elements except the noble gases, the halogens, and a few unreactive metals, such as gold, silver, and platinum. Most reactions of O_2 take place very slowly at ordinary temperatures, and high temperatures are usually required. Nevertheless, the slow room-temperature reactions of O_2 are very important and have a profound effect on the earth's economy, and indeed on life itself. Among these reactions are:

1. Breathing. Most animals and plants must continually breathe in order to live. Living organisms use the oxygen to oxidize various organic compounds (mostly carbohydrates and fats—see Chapter 21) to CO_2 and H_2O. *The purpose of this reaction is to provide the energy needed by all living things.* CO_2 and H_2O have lower energies than carbohydrates and fats, so energy is given off when these compounds are used up in this reaction. At one time, it was believed that the reaction took place in one specific area of an animal body (for example, the lungs or the liver), but it is now known that it goes on in every living cell. The blood carries the oxygen (bonded to the iron in the hemoglobin) to each cell and carries away the CO_2.
2. Rusting and other metallic corrosion. The reaction of atmospheric oxygen with iron and other metals was discussed in Section 16-15.
3. Autoxidation. The oxygen in the atmosphere slowly reacts with certain organic materials, causing them to deteriorate over a period of time. This slow room-temperature reaction is called **autoxidation.** Among the materials affected are foods, rubber, paint, and lubricating oils. On the other hand, autoxidation also has its positive side, since it causes the drying of paints and varnishes. Most paints and varnishes do not dry because of evaporation of a solvent, but because small molecules are converted to large molecules by atmospheric oxygen.
4. O_2 reacts with certain laboratory chemicals, sometimes with explosive effects, and these chemicals must be prepared and stored under an inert atmosphere, usually N_2, He, or Ar. Materials that ignite spontaneously in air are called **pyrophoric.**

Besides causing the slow room-temperature reactions just discussed, atmospheric oxygen is of course also essential for combustion reactions, which take place at high temperatures. Most organic substances and many inorganic ones (even metals such as iron) will burn if heated to a high-enough temperature. Combustion reactions are exothermic and give off a great deal of heat, which human beings have used since civilization began,* both for constructive and destructive purposes. These reactions

*Many people count the taming of fire as the beginning of civilization.

illustrate the laws of thermodynamics and of kinetics. Heat is given off because the reactions are exothermic. The products have less energy than the reactants; ΔH and ΔG are negative; the laws of equilibrium say that the reactions should proceed essentially to completion, and they do. Nevertheless, a combustible substance such as paper, wood, coal, or sulfur can be kept in air at room temperature for long periods of time without noticeably reacting, because of kinetics: the reactions, although favorable thermodynamically, are unfavorable kinetically. However, when the materials are heated to a high-enough temperature (the **ignition temperature** or **flash point**), which supplies the energy of activation, the reaction begins, and then continues until all the substance has been burned. In some cases, the energy of activation can be supplied without heating, for example by an electric spark, or by friction, as in striking a match.

What actually goes on within a flame is not completely understood, but the reactions are very complicated and involve unstable free radicals.

17-5 Ozone

A number of elements can exist in more than one molecular form. Different forms of the same element are called **allotropes** or **allotropic forms.** The most common form of the element oxygen is O_2, whose structure we have already discussed, but oxygen also exists in another allotropic form O_3, called *ozone.** We can draw two Lewis structures for ozone, so that the true molecule is a resonance hybrid of the two canonical forms.

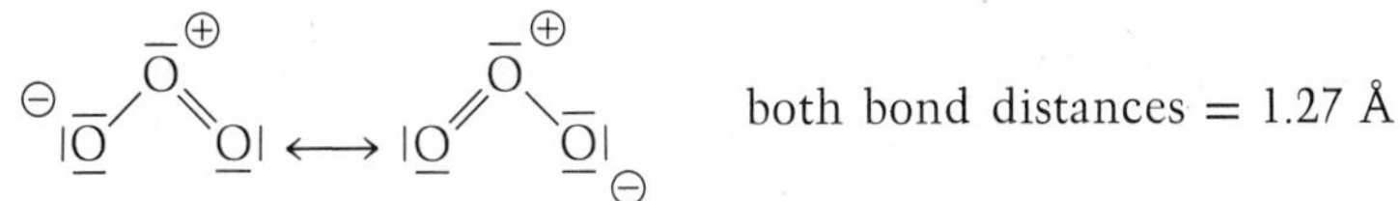

The EPRT tells us to expect an angular molecule, which is what is found. The O—O—O angle is about 117°. Ozone is much more reactive than oxygen and reacts with many substances under conditions where oxygen will not. It is also less stable than O_2 even in the absence of other substances, and slowly decomposes to oxygen at 250°C in the absence of catalysts and ultraviolet light.

$$2O_3 \longrightarrow 3O_2 \qquad \Delta H = -33.8 \text{ kcal/mole of } O_3$$

Ozone is a pale blue poisonous gas with a characteristic odor; it is formed when an electric discharge is passed through air (it is commercially prepared in this manner) and can often be detected near high-tension electric wires and electric railways. For this reason, many people think they are smelling electricity, when they are actually smelling ozone.

Although ozone is only a very minor constituent of the atmosphere (Table 17-1), it has an importance far beyond what might be expected of its low concentration. The ozone in the atmosphere is not uniformly distributed; rather it is concentrated in a layer about 20 to 35 km above the surface of the earth (the **ozone layer**). The exact height of the layer varies with latitude: highest at the equator, lowest at the poles; and with the seasons: lowest in the winter and highest in the summer. Figure 17-5 shows the layer as it was measured above Flagstaff, Arizona, in May in a typical year.

*The word "ozone" also has another meaning, which the dictionary gives as "pure and refreshing air." This meaning has no connection with O_3, which is not refreshing, but poisonous. In chemistry, the word "ozone" always refers to O_3, not to pure and refreshing air.

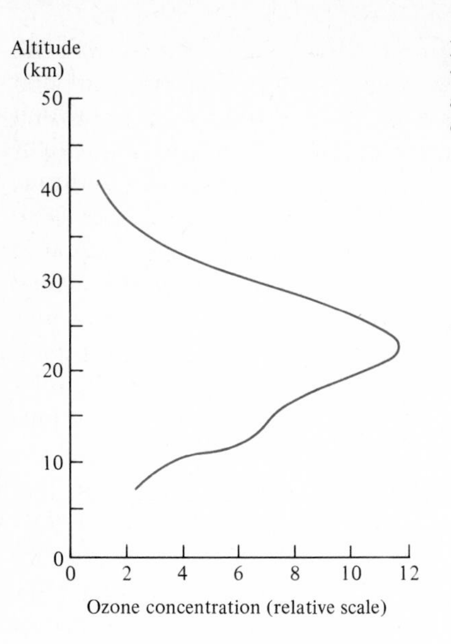

Figure 17-5 Variation of the atmospheric ozone content with altitude, over Flagstaff, Arizona, in May of a typical year. [Source: A. H. Strahler, *The Earth Sciences*, 2nd ed., Harper & Row, New York, 1971]

The ozone layer exists where it does because it is formed up there by the interaction of oxygen molecules with ultraviolet light coming from the sun.

$$O_2 \xrightarrow[1000\text{–}2000\ \text{Å}]{h\nu} 2O \qquad h\nu \text{ signifies the energy of light}$$

$$O_2 + O \longrightarrow O_3$$

The ozone does not reach the surface of the earth because, as mentioned above, it decomposes to O_2 before it can diffuse all the way down (this is fortunate, because it would have a deleterious effect on life if it did). The importance of the ozone layer lies in the fact that *this layer filters out more than 99% of the harmful ultraviolet light that comes from the sun, and prevents it from reaching the surface of the earth.* Ultraviolet light of wavelengths between about 1000 and 2000 Å is used up in the reaction that forms O_3; the ozone itself then absorbs light of higher wavelengths (ultraviolet light from 2000 to 4000 Å; and some visible and infrared light as well).

It is this fact that is responsible for the temperature inversion found in the stratosphere (Section 17-2; Figure 17-2). The absorption of ultraviolet and some other light which takes place in the ozone layer heats the air and causes the temperature of the stratosphere to rise from about −60°C at the tropopause to about 0°C at the stratopause. The temperature inversion at the mesopause (Figure 17-2) is caused by a similar phenomenon. In this case, the air above about 80 km absorbs the very high energy ultraviolet rays (less than 1000 Å in wavelength) and the x rays, producing ions such as N_2^+, N^+, and O^+. For this reason, this region is also called the **ionosphere.** The heat of these reactions causes the temperature increase shown in the thermosphere region of Figure 17-2. Ions such as N_2^+, N^+, and O^+ are relatively stable in the ionosphere simply because the atmospheric pressure is so low that collisions with other molecules are comparatively infrequent. Such ions could not exist near the surface of the earth.

The high temperatures of the thermosphere are somewhat misleading. The temperature of a gas is proportional to the average kinetic energy of the molecules. These molecules are moving very fast, so technically speaking, they have a high temperature. But if you could put your finger into this region, it would not feel hot, but very cold, because there are so few molecules present that the amount of heat reaching your finger from these molecules would be very small. For the same reason, a thermometer not in direct sunlight would register close to absolute zero.

As we have just seen, the ozone layer is very important because it protects us from many of the harmful ultraviolet rays of direct sunlight. If the amount of ozone in the ozone layer were to decrease, there might be several harmful effects. It is fairly certain that the number of cases of skin cancer would rise, with the rise being proportional to the ozone depletion. Furthermore, there could be other effects which cannot be fully predicted; for example, the increased amount of ultraviolet light might affect food crops and other vegetation, and might change weather patterns.

Considering natural sources alone, the amount of ozone in the ozone layer is fairly constant, the rate of ozone formation via $3O_2 \xrightarrow{h\nu} 2O_3$ being approximately equal to the rate of ozone decomposition, $2O_3 \rightarrow 3O_2$; but in recent years two possible man-made threats to the ozone layer have been identified: supersonic airplanes (SSTs) and Freons. It must be emphasized that neither of these is a *proven* threat. Direct research at these altitudes is difficult to carry out and most of what is known is derived from experiments in laboratories and from mathematical calculations.

SUPERSONIC AIRPLANES. The possible threat from SSTs arises because they emit nitrogen oxides (chiefly NO and NO_2). All airplanes (see Section 17-7) emit nitrogen oxides, but most airplanes fly considerably below the ozone layer, and the nitrogen oxides pose no threat to the ozone because they are very reactive and do not live long enough to diffuse that high. But SSTs must fly higher than other airplanes in order to be economical, and therefore release their nitrogen oxides directly into the ozone layer. Once there, they might cause ozone depletion by this sequence of reactions:

$$\begin{array}{rcl} NO + O_3 & \longrightarrow & NO_2 + O_2 \\ O_3 & \xrightarrow{h\nu} & O_2 + O \\ NO_2 + O & \longrightarrow & NO + O_2 \\ \hline \text{net reaction: } 2O_3 & \longrightarrow & 3O_2 \end{array}$$

Of course, this sequence occurs naturally (there is some NO present in the ozone layer through natural processes), but the fear is that large amounts of nitrogen oxides from SSTs might cause an ozone depletion great enough to have deleterious effects on earth.

FREONS. The Freons are a group of organic compounds containing only carbon, fluorine, and chlorine. The most important one, dichlorodifluoromethane CCl_2F_2, is called Freon 12. The Freons have been used for about 50 years as refrigerants. A refrigerant is a gas which is easily liquified by pressure alone. In a refrigerator or an air conditioner, the gas is compressed by a motor until it liquifies, and the liquid is permitted to evaporate again. The evaporation takes heat from the surroundings, resulting in cooling (Figure 17-6). The Freons are very satisfactory for this purpose because they are noncorrosive and nontoxic, and they are used in virtually all refrigerators, air conditioners, ice-making machines, and so on. Although the gas is totally sealed in and does not normally escape (or come in contact with the food stored in a refrigerator or freezer), accidents do happen, and refrigerators occasionally leak. When they do, the Freon that escapes is harmless. Refrigerants that were used before the discovery of the Freons (mostly NH_3 and SO_2) did not have this virtue, and also corroded the insides of the tubes.

In recent years, another important use has been found for the Freons: They are used in many aerosol cans, where they are present as liquids under pressure. When the nozzle is pressed, some of the liquid evaporates and forces some of the contents of the can through the nozzle. Despite the fact that Freons are nontoxic, they are never used for foods; only for nonfoods, such as insect sprays, hair and deodorant sprays, and shaving creams (nitrous oxide is usually used for foods). Unlike the Freons in refrigerators and air conditioners, the Freons in aerosol cans are sprayed into the atmosphere, and this is where the possibility of harm comes in. The Freon molecules are extremely stable and unreactive and will remain in the atmosphere for long periods of time, since there are no natural reactions by means of which they

Figure 17-6 Operation of a Refrigerator. The liquid evaporates in the evaporator, taking in heat, which cools the space. The vapor is compressed in the compressor, turning it back into liquid, which is then reevaporated. The liquid is shown in color.

might decompose. Eventually, many of them (unfortunately, it is hard to predict how many) will diffuse up to the ozone layer. When they get there, it is possible that they might be broken into free radicals by the ultraviolet light; for example

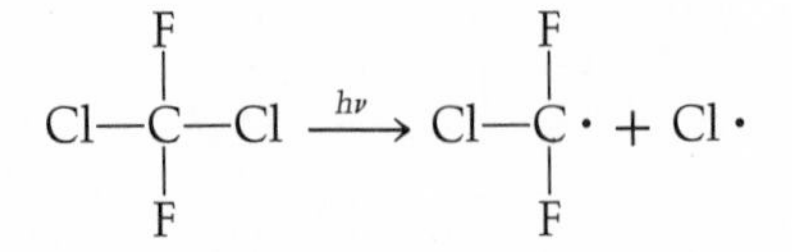

(Again, there is no doubt that this reaction can take place; the question is: How many of the molecules will do so?) Some of the free radicals can then react with ozone as follows:

$$\begin{aligned} Cl\cdot + O_3 &\longrightarrow ClO + O_2 \\ O_3 &\xrightarrow{h\nu} O_2 + O \\ ClO + O &\longrightarrow Cl\cdot + O_2 \\ \hline \text{net reaction: } 2O_3 &\longrightarrow 3O_2 \end{aligned}$$

Once again, some of the ozone might be depleted.

How real are the threats to the ozone layer from SSTs and Freons? Unfortunately, the point is not settled. Some scientists think these are very real threats; others think they are not. However, it would seem that we would do well to be cautious in an area where a substantial depletion of the ozone layer would cause a great deal of harm, much of it in unpredictable ways.

Apart from its presence in the ozone layer, O_3 is also a constituent of photochemical smog. This will be discussed in Section 17-8.

17-6 Atmospheric Carbon Dioxide and the Greenhouse Effect

Carbon dioxide is, after argon, the most abundant of the minor constituents of dry air (Table 17-1). Its presence in the atmosphere is essential to life on earth because it is one of the two raw materials for photosynthesis (Section 21-3). In photosynthesis, plants combine CO_2 and H_2O to form the carbohydrates, which are the principal constituents of all plants. When animals eat the plants, they convert the carbohydrates (by respiration) back to CO_2 and H_2O, gaining energy in the process. When plants die, they are attacked by bacteria, which also convert the carbohydrates back to CO_2 and H_2O. In this way an approximate CO_2 balance is reached. The rate of removal of CO_2 from the atmosphere is about equal to the rate of CO_2 restored, so that the percentage of CO_2 in the atmosphere is fairly constant, or at least it was up to about 100 years ago.

About that time, man began to disturb this equilibrium by burning ever-increasing amounts of fossil fuels: coal, oil, and natural gas. The ultimate products of this combustion are, of course, CO_2 and H_2O; and for the last 100 years the amount of CO_2 in the atmosphere has been slowly increasing (it is now about 10% higher than it was 100 years ago). Theoretically, an increase in the CO_2 content of the atmosphere should have an important effect on the earth's climate, as will now be explained.

In Section 12-4, we learned that a blackbody emits radiation whose wavelength is dependent on its temperature. The sun is a very hot body (surface temperature about 6000°C), and although it emits energy at all wavelengths, the *maximum* amount of its energy is in the visible region of the spectrum. The Earth, on the other hand, is much colder, and emits no visible light at all; its maximum radiation is in the infrared region (wavelength about 1×10^{-4} cm). CO_2 is a colorless substance, which means it is completely transparent to visible light. However, it does absorb infrared radiation, especially of wavelengths about 1×10^{-4} to 2×10^{-4} cm. The consequence of this is that the CO_2 in the atmosphere does nothing to stop solar radiation from reaching the surface of the earth but does stop some radiation from going from the earth back into space. This is called the **greenhouse effect** (Figure 17-7), since a greenhouse was once supposed to work by the same principle.* Each molecule of CO_2 absorbs infrared energy and then reemits it; the reemitted energy is scattered in all directions, some going upward, but some going downward to the surface of the earth. (Water vapor in the atmosphere also absorbs infrared radiation and transmits solar radiation, but does not pose the same kind of problem because the amount of water vapor in the atmosphere is not increasing.)

Before the large-scale burning of fossil fuels, there was an approximate radiation balance: the amount of energy reaching the earth's surface each year was about the same as the amount which went back into space. Calculations show that the net effect of a substantial increase in the CO_2 content of the atmosphere should be a gradual but steady increase in the average surface temperature of the earth, as would

*We know today that greenhouses work by suppressing convection currents.

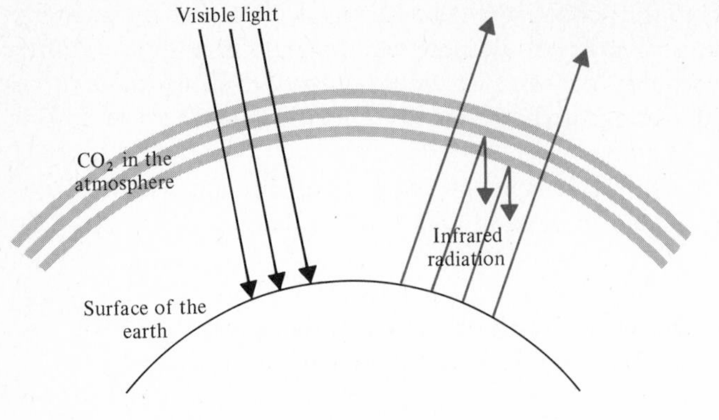

Figure 17-7 The greenhouse effect.

be expected if more energy is coming in than is going out. It would not take a very large increase in the average surface temperature to cause substantial effects on the earth's climate. A rise of about 4°C would probably cause enough of the Antarctic ice to melt so as to flood most of the world's coastal cities.

What, in fact, has been the effect of the 10% increase in atmospheric CO_2 concentration on the average surface temperature? Has it gone up as predicted? Measurements show that the average surface temperature did rise, by about 0.6°C, from 1880 to 1940, but since then it has *fallen* about 0.3°C. The reasons for this are not completely understood and several factors are probably involved. One factor is that the burning of fossil fuels not only produces CO_2 but also considerable quantities of particles (aerosols). These particles vary in size. The large ones quickly settle out over the surrounding neighborhood, causing windows and other surfaces to get dirty, but have little effect on the atmosphere. However, the small particles (diameters about 1×10^{-3} mm or smaller) remain suspended in the atmosphere, and since they, unlike CO_2, are not transparent to visible light, it is likely that they reflect some of the incoming solar radiation, sending it back into space. This reduces the amount of solar energy that reaches the earth's surface and results in a lowering of the average surface temperature. Another possible explanation for the recent reduction in temperature is that such a fall is part of a long-term temperature cycle. There is evidence from a study of polar ice that there has been a succession of regular rises and falls of the average temperature for at least the last 1000 years. These rises and falls are fairly small—less than 1°C each time.

However, the fact that the increase in atmospheric CO_2 concentration has not yet resulted in a steady increase in the average surface temperature should not make us complacent. As more and more fossil fuels are burned at an ever-greater rate, the CO_2 percentage must increase to the point where the magnitude of the greenhouse effect will outweigh normal temperature fluctuations. The possibility of major climatic changes is one that mankind must seriously consider.

AIR POLLUTION

Ever since man appeared, hundreds of thousands of years ago, all human beings have breathed the air on this planet and, with very few exceptions (such as allergy sufferers during the pollen season or people in the vicinity of an erupting volcano), have not had to worry that it would cause them any harm. Since the beginning of the industrial revolution, this situation has changed, and we now have good reason to be concerned about the pollution of the air. There is no universally accepted definition of polluted air. What is a pollutant to one person is not always considered so by another. We shall use a two-part definition and define an **air pollutant** as a substance (1) *which, in the concentrations found in any particular location, causes detectable damage to people, animals, plants, or materials;* and (2) *which is produced by human activities in concentrations significantly greater than those found in natural air.* In this connection a useful word is **anthropogenic,** which refers to substances put into the air by human activities. We will consider a pollutant to be any substance that meets *both* parts of the definition.

We shall find for most of the pollutants that the total amounts entering the atmosphere from anthropogenic sources are much less than from natural sources. Nevertheless, they are pollutants, because of the concentration factor. The natural material is uniformly spread out throughout the entire atmosphere all over the earth; but the anthropogenic material is usually produced in large amounts in a relatively small geographical area. For this reason, the local concentration of a given pollutant often reaches 100 or more times that of the natural concentration.

Sometimes pollution is a very detectable phenomenon: the air *looks* polluted (Figure 17-8) or it *smells* polluted, or both. But this can be deceiving. Many pollutants cannot be so easily detected, and much air that looks and feels clean may actually be very dirty. On the other hand, bad odors or a cloudy appearance do not *always* mean that the air is unhealthy.

Air pollution is very complicated and a difficult subject to study, because the concentrations involved are very low (usually a few parts per million) and often rapidly change with changing time and place; and because the pollutants constantly react with each other and with the principal constituents of air. For these reasons, there is still much that we do not know, and some of the things we think we do know will probably have to be changed as more knowledge is gained. A good illustration is a passage from p. 34 of *Cleaning Our Environment: The Chemical Basis for Action,* a report issued by the American Chemical Society in 1969:

Figure 17-8 Air pollution in New York City.

"Carbon monoxide appears to be almost exclusively a man-made pollutant. The only significant source known is combustion processes. . . . The automobile is estimated to contribute more than 80% of carbon monoxide emissions globally. . . ." These statements were believed in 1969, but today we know that about 90% of the CO in the atmosphere is not anthropogenic, but in fact arises from a natural process (Section 17-7).

In the next section, we will discuss the primary pollutants, where they come from, and the damage that they do. We will cover secondary pollutants and smog in Section 17-8 and pollution control in Section 17-9.

17-7 Primary Pollutants

Primary pollutants are compounds that enter the atmosphere directly. **Secondary pollutants** (Section 17-8) are compounds formed by reactions of primary pollutants with each other and with the natural components of air. There are five principal classes of primary pollutants. The members of these classes altogether account for more than 90% (by weight) of all primary pollutants in the atmosphere. These classes are:

1. Oxides of nitrogen.
2. Oxides of sulfur.
3. Carbon monoxide.
4. Hydrocarbons.
5. Particulates.

We will discuss them one by one.

1. *Oxides of nitrogen.* NO_x. Three oxides of nitrogen are found in air, but one of these, nitrous oxide (N_2O), has no significant anthropogenic sources (it is produced by denitrification bacteria, Section 17-3) and is not considered a pollutant. The two polluting oxides are nitric oxide (NO), which is colorless, and nitrogen dioxide (NO_2), which is brown. They are often jointly referred to as NO_x. NO_2 does not enter the atmosphere directly in significant amounts from any source but is produced by atmospheric oxidation of NO. About 80% of atmospheric NO comes from natural sources: a small amount from lightning (Section 17-3), but more from bacterial action and by atmospheric oxidation of NH_3. The remaining 20% is anthropogenic and arises from the operation of gasoline engines and stationary coal or oil-fired heating furnaces (about equal amounts from the two sources). The nitrogen in the NO does not come from nitrogen compounds in the fuel (neither gasoline, fuel oil, nor coal contain significant amounts of nitrogen-containing compounds) but from the N_2 in the air, which at the high temperatures at which the engines and furnaces operate reacts with O_2 to give NO:

$$N_2 + O_2 \longrightarrow 2NO$$

and a small amount (1 to 2%) of NO_2. Once NO enters the atmosphere from any source, it is rapidly oxidized to NO_2 in a reaction whose mechanism is not understood.

$$2NO + O_2 \longrightarrow 2NO_2$$

Eventually the NO_2 is further oxidized to NO_3^-, which is washed down to the soil.

Although only 20% of the atmospheric NO_x is anthropogenic, it still causes problems because of the concentration effects mentioned earlier. Like the other pollutants, anthropogenic NO_x is produced by combustion, and the problems are most serious where combustion is carried out to a large extent: chiefly large cities and areas where automobile and truck traffic is heavy. NO is poisonous, but at the concentrations found in the atmosphere its toxicity is such that it poses no health hazard, even in polluted air. NO_2 has a higher toxicity and at the high concentrations found in polluted air can cause nasal irritations and breathing discomforts. NO_2 also damages plants, textiles, and some metals. However, the principal disadvantage of NO_2 is not its direct effects, but its role in the production of photochemical smog (next section).

2. *Oxides of sulfur*, SO_x. There are two oxides of sulfur in the atmosphere, sulfur dioxide (SO_2) and sulfur trioxide (SO_3). As in the similar case of NO_x, it is SO_2 which directly enters the atmosphere; the SO_2 is then oxidized to SO_3.

$$2SO_2 + O_2 \longrightarrow 2SO_3$$

Just how this oxidation takes place is unknown: experiments show that SO_2 reacts hardly at all with clean air at room temperature in the absence of sunlight. It has been suggested that the actual oxidizing agent is oxygen atoms O, which are formed by photochemical decomposition of NO_2 (Section 17-8). If so, this illustrates the effect of one pollutant on another. Whatever the mechanism, SO_2 does not remain in the air very long. It is oxidized to SO_3 within a few hours or days. The SO_3 immediately combines with the water in the air to give H_2SO_4, part of which remains as such and part of which combines with other substances. An important example is NH_3, which reacts with H_2SO_4 to give $(NH_4)_2SO_4$. H_2SO_4 is a liquid and $(NH_4)_2SO_4$ is a solid, but a major portion of these substances nevertheless remains in the atmosphere by becoming attached to tiny particles of water or of various forms of dust (aerosols).

$$SO_3 + H_2O \longrightarrow H_2SO_4$$
$$H_2SO_4 + 2NH_3 \longrightarrow (NH_4)_2SO_4$$

About 40% of atmospheric SO_2 comes from anthropogenic sources, and 60% from natural sources (chiefly from oxidation of H_2S, which is produced by decay of dead plants and animals). Most of the anthropogenic SO_2 (about 65%) arises from the burning of coal. For the remaining 35%, the two most important sources are industrial processes in which sulfur plays a part (for example, copper smelting, H_2SO_4 production) and the combustion of fuel oil in heating furnaces. Much less of the SO_2 comes from the burning of gasoline. Both coal and petroleum contain substantial amounts of sulfur compounds (although the amount varies greatly with the origin of the petroleum or the coal), but these can be (and are) more easily removed from the liquid petroleum than from the solid coal.

SO_x is probably the most harmful of the five principal primary pollutants. As stated above, most of the SO_2 in the atmosphere is soon converted to H_2SO_4 and to acid salts such as $(NH_4)_2SO_4$. Eventually, all of these are washed out in rain. In fact, rain falling in industrial countries has become more acidic in recent years. Rain has always been acidic because of the CO_2 it contains, but the pH has been dropping lately because of the increasing amounts of H_2SO_4 and, to a lesser extent, of HNO_3 (from NO_x) which are now present. But before they are washed out, the H_2SO_4 and acid salts have powerful effects on both living and nonliving things. Aerosol particles containing H_2SO_4 damage plants, animals, and humans. The most severe effects are on eyes and the respiratory system. This type of mist is especially bad for asthmatics and others with chronic respiratory problems, and this is probably the most serious air pollution health hazard. SO_2 causes similar damage to plants, animals, and human beings, but to a lesser degree.

Atmospheric H_2SO_4 is also very damaging to many materials. It attacks paints, metals, leather, paper, and even stone, especially carbonates (for example, limestone, marble).

$$H_2SO_4 + CaCO_3 \longrightarrow CaSO_4 + CO_2 + H_2O$$

Over a period of time this is damaging to buildings, stone bridges and walls, outdoor statues, and so on, because the $CaSO_4$ is more soluble in water than $CaCO_3$, and is eventually washed away by the rain. Some outdoor statues that remained in excellent condition for more than 1000 years have become severely corroded within the last 100 years or so (Figure 17-9).

3. *Carbon monoxide,* CO. As mentioned earlier, about 90% of the total CO content of the atmosphere comes from natural sources: chiefly the slow atmospheric oxidation of methane, CH_4, which seeps into the atmosphere from marshes and swamps (page 554). This reaction is probably initiated by O atoms and OH radicals produced by

$$O_3 \xrightarrow{h\nu} O_2 + O$$
$$O + H_2O \longrightarrow 2HO\cdot$$

However, the 10% of the CO that is anthropogenic is still a major air pollutant because it is concentrated in a few local areas, while the 90% that is natural is fairly uniformly spread over the whole earth. Thus, the local concentration of CO in a polluted area is many times higher than the natural concentration in an unpolluted area.

About 75% of the anthropogenic CO is produced by motor vehicles and other forms of motorized transportation (the rest comes mostly from industrial processes and the controlled burning of forest debris, crop residues, and other agricultural material). Complete combustion of the fuel in a gasoline or diesel engine would give CO_2 and H_2O, but unfortunately the combustion is not complete, and a substantial amount of CO is produced. Although CO is oxidized by atmospheric oxygen

$$2CO + O_2 \longrightarrow CO_2$$

this reaction is quite slow and accounts for the removal of only some of the atmospheric CO. The principal pathway for the removal of CO from the atmosphere

1908 1969

Figure 17-9 This statue is at Herten Castle in Westphalia, Germany, which was built in 1702. The picture on the left was taken in 1908; that on the right in 1969. The statue suffered only slight damage in its first 200 years but was almost obliterated in the next 60. [From E. M. Winkler, *Stone: Properties, Durability in Man's Environment* (New York: Springer-Verlag, 1975)]

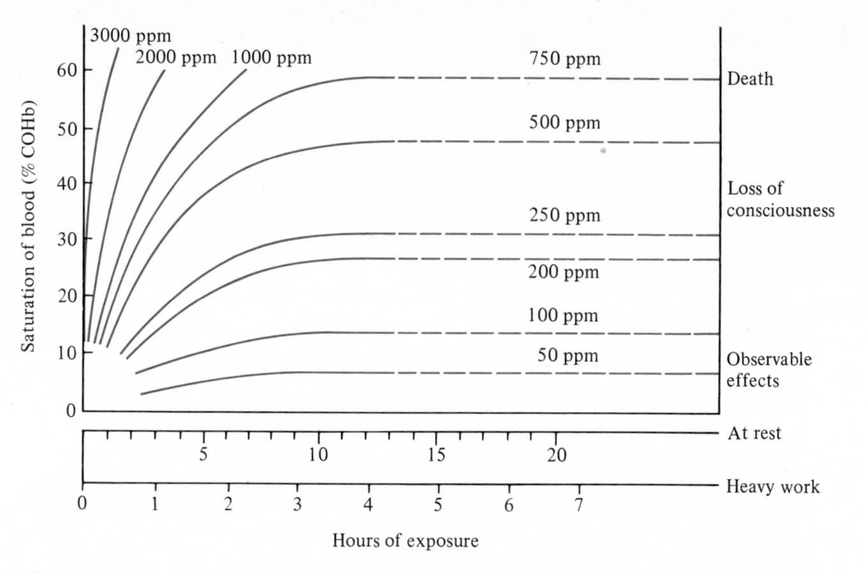

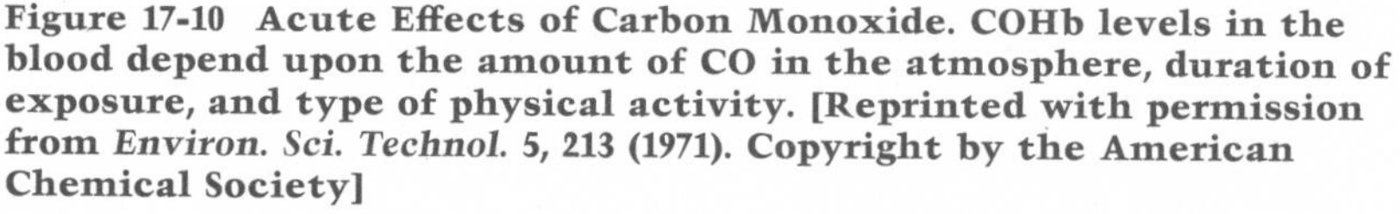
Figure 17-10 Acute Effects of Carbon Monoxide. COHb levels in the blood depend upon the amount of CO in the atmosphere, duration of exposure, and type of physical activity. [Reprinted with permission from *Environ. Sci. Technol.* 5, 213 (1971). Copyright by the American Chemical Society]

is oxidation to CO_2 by soil bacteria. Unfortunately, those parts of the earth with the highest local concentrations of CO (chiefly large cities) are also those with the smallest amounts of open soil available.

In large-enough concentration, CO in the atmosphere is invariably fatal (Figure 17-10). We have mentioned that the hemoglobin in the blood carries O_2 from the lungs to every living cell. It does this by combining with O_2 to form a compound which we may designate as **O_2Hb.** Unfortunately, hemoglobin also forms a compound with CO (**COHb**), and the affinity of hemoglobin for CO is more than 200 times its affinity for O_2. Therefore, a relatively small amount of CO in the lungs will prevent the hemoglobin from performing its normal function of carrying O_2. A percentage of COHb in the blood of about 30 to 50% generally results in death.

Polluted air seldom, if ever, has high-enough CO concentrations to cause death, but even at lower concentrations (those typically found in polluted air) CO causes deleterious effects on human beings and animals: impairment of visual activity, brightness discrimination, cardiac changes, and so on. These effects are proportional to the concentration of COHb in the blood, which in turn is proportional to the CO concentration of the air (Figure 17-10). Incidentally, tobacco smokers usually have between two and four times the COHb levels of nonsmokers; the amount of the increase is directly related to the quantity of tobacco used.

4. *Hydrocarbons.* A **hydrocarbon** is a compound made up entirely of the elements carbon and hydrogen. It may seem at first sight surprising, but thousands of different hydrocarbons are known (in Sections 20-1 to 20-5 we will see how this is possible). Hydrocarbons are classified on the basis of the kinds of bonds they contain:

alkanes contain only single bonds
alkenes contain at least one C=C double bond
alkynes contain at least one C≡C triple bond

Examples of each type are shown in Table 17-3. All members of these three classes are called **aliphatic hydrocarbons.** There is another class of hydrocarbons, called **aromatic hydrocarbons.** The most important member of this class is benzene, C_6H_6.

Table 17-3 Examples of Aliphatic Hydrocarbons[a]

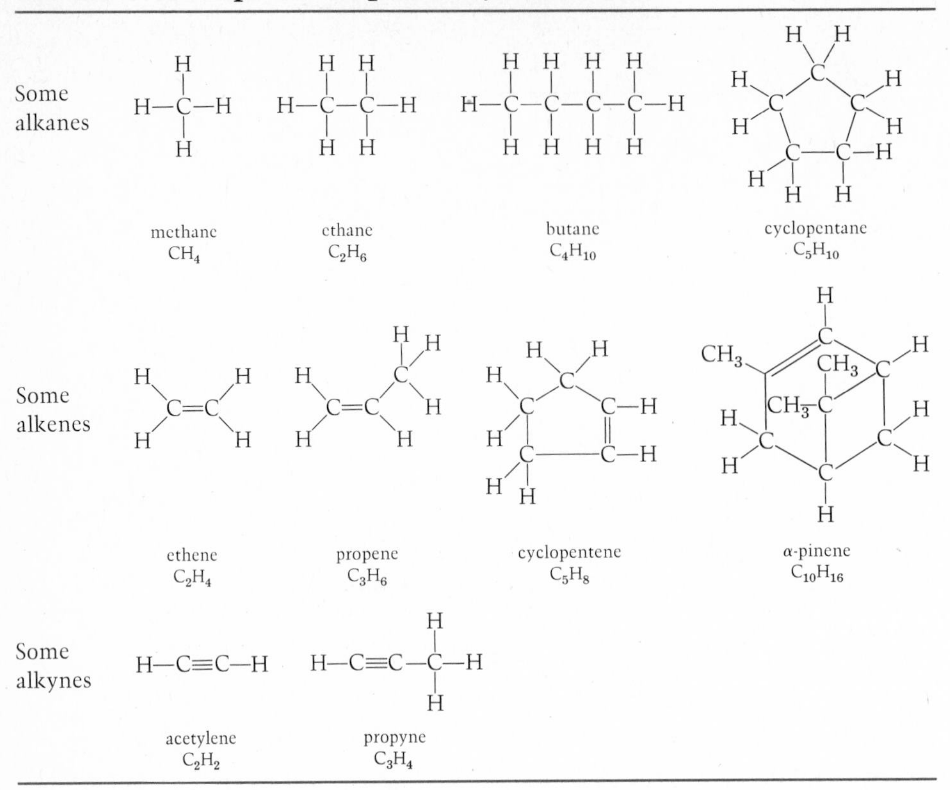

[a]Note that not all these diagrams show the true geometry of the molecules. For example, EPRT tells us that the four CH bonds in methane are directed to the corners of a tetrahedron, not to a square, as shown in the table.

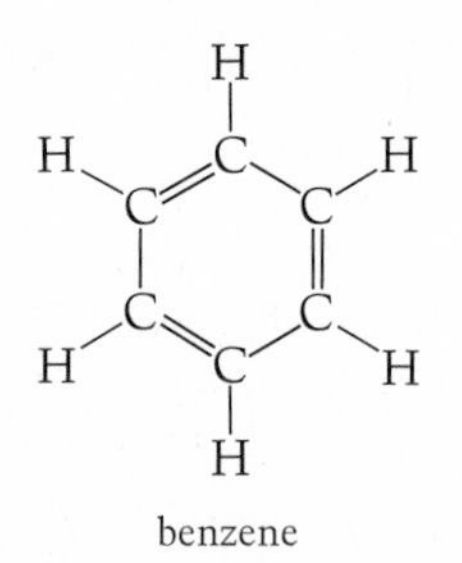

benzene

Note that the six carbon atoms of benzene are arranged in a ring and that each carbon atom in the formula has one C—C single bond and one C═C double bond, and that the total valence of four is achieved by a bond to a hydrogen atom. (In Section 20-5, we shall see that the bonding in benzene is more complicated than this, and that because of resonance, benzene does not actually have C═C double bonds.) Virtually all aromatic hydrocarbons have at least one benzene ring. Some examples are shown in Table 17-4.

About 85% of the hydrocarbons in the atmosphere come from natural sources. There are two important natural sources, each of which gives a different kind of hydrocarbon:

a. Bacterial decay of dead plants—often in swamps and marshes—produces a mixture of gaseous alkanes. The principal component of this mixture is methane (CH_4), but other gaseous alkanes, such as ethane (C_2H_6) and propane (C_3H_8), are also present. It is for this reason that methane is sometimes called *marsh gas*.

Table 17-4 Some Aromatic Hydrocarbons[a]

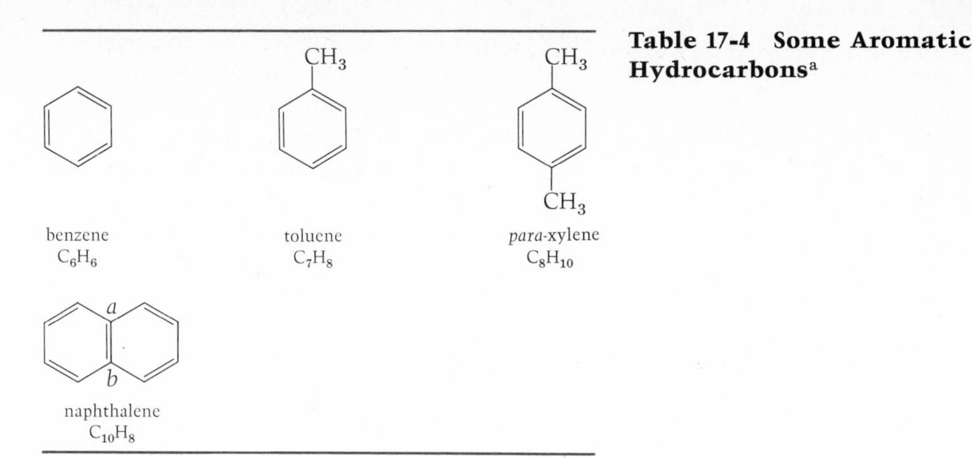

[a]The hydrogens connected to the ring carbons are generally not shown. Note that the carbons marked *a* and *b* in naphthalene are not connected to hydrogens, since they already have valences of 4 through bonds with carbon.

b. The leaves of many plants give off certain alkenes called **terpenes.** Among these are α-pinene (Table 17-3), limonene, and myrcene. These compounds are liquids, but because of their high vapor pressures, there is a substantial amount of these substances in the atmosphere near large areas of vegetation.

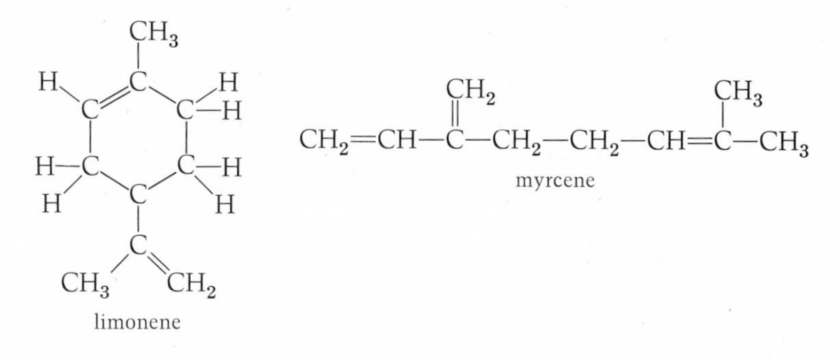

All of us encounter evidence that plants give off organic compounds to the atmosphere every time we smell the aroma of a forest, the scent of a flower, or the stench of rotting vegetation (not all of these are hydrocarbons).

Another source of gaseous alkanes is **natural gas,** which has essentially the same composition as the gas produced in swamps and marshes, consisting largely of methane, with smaller amounts of other gaseous alkanes. (Natural gas was also produced in the same way as the swamp gases—from rotting vegetation—but this happened millions of years ago, and the natural gases have remained in underground caverns ever since.) Most of the natural gas is sealed underground and remains there until tapped by industry, but in certain areas of the world it is near the surface and some of it escapes into the atmosphere from cracks in the ground. Natural gas is flammable (all hydrocarbons burn), and in ancient times some of this escaping gas was discovered by people who accidentally set it afire. Wherever this happened, it was regarded as a great wonder. It was felt that the gods must have had something to do with it, and oracles were frequently set up at these sites, since it was believed that here was where one or more gods were directly talking to us. The only difficulty was interpreting the language, but many people felt they could do that.

Most of the anthropogenic hydrocarbons in the atmosphere come ultimately from gasoline. Gasoline is a mixture of hydrocarbons with low boiling points; it consists mostly of alkanes, but small amounts of aromatic hydrocarbons are also present (Section 20-2). Gasoline has no alkenes or alkynes. The hydrocarbons enter the atmosphere largely in two ways: (1) evaporation of gasoline during transfers: for example, loading of tank trucks, filling automobile tanks; and (2) unburned and incompletely burned fuel emitted from automobile exhausts. The primary reason that the fuel is not burned completely in the cylinder is that a certain amount of it is near the cylinder wall, which is cool and quenches the reaction. In this connection it is important to note that the exhaust is richer in aromatic hydrocarbons than was the original gasoline, and also contains some alkenes, which were not present in the gasoline but were produced in the engine by incomplete combustion.

Hydrocarbons in the atmosphere are not generally harmful in themselves. They are classified as primary pollutants only because some of them (chiefly the alkenes) react with other constituents of the atmosphere to produce secondary pollutants (Section 17-8). Methane, the most abundant atmospheric hydrocarbon, and also the least harmful since it is fairly inert, is eventually oxidized to CO and to CO_2, in a very slow process probably initiated by O atoms and OH radicals.

There is one type of atmospheric hydrocarbon that is directly harmful to human beings. Certain aromatic hydrocarbons (called **polynuclear aromatics**), all of them containing at least four benzene rings, are emitted in small amounts from automobile exhausts. They are produced by the incomplete combustion of gasoline. Examples are the three compounds shown in Figure 17-11. These compounds have been shown to cause cancer when rubbed into the skin of laboratory animals. Compounds that cause cancer are called **carcinogens.** The compounds shown in Figure 17-11 and similar aromatic carcinogens emitted from automobile engines are all solids and remain in the atmosphere only in the form of crystals adhering to small particles. The breathing of air containing these compounds may be an important contributor to lung cancer. Carcinogenic polynuclear aromatics are also produced by combustion of fuel oil and coal and are present in cigarette smoke.

5. *Particulates,* also called *aerosols,* were defined on page 534. They may appear as mist, smoke, fume, dust, and so on (the clouds in the sky are aerosols consisting mostly of tiny droplets of water), but are often invisible. Most particulates (about 90%) are natural—primarily sea-salt aerosols formed by bubbles bursting in sea foam. Most of the anthropogenic particulates come from industrial processes such as stone crushing, grain handling, steel making, and cement and lime making. Another important source is tiny particles of ash from the burning of coal (*fly ash*). Only a small fraction of the total particulates come from motor vehicles, but in large cities it is these which may dominate. As can be seen, the composition of particulates is very diverse.

Eventually, particulates fall to earth in three ways: (1) directly by gravity, (2) washed down by rain (mostly larger particles), and (3) by becoming the nucleus of raindrops (mostly smaller particles).

A high concentration of particulates can interfere with breathing. Small particles can settle in the lungs and remain there for long periods of time. Some particulates are intrinsically poisonous to human beings. An example is particulates containing lead, an important source of which is lead tetraethyl, a gasoline additive (see Section 17-9).

In addition to their effects on human health, particulates also have other disad-

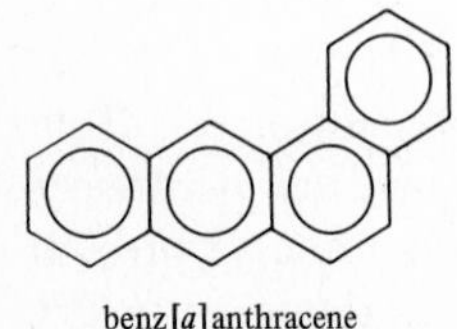
benz[*a*]anthracene

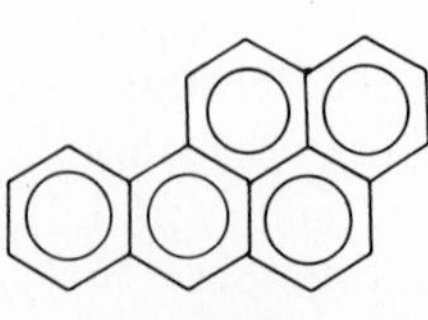
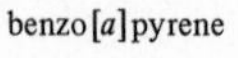
benzo[*a*]pyrene

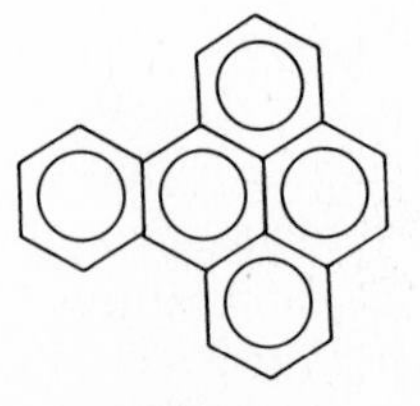
benzo[*e*]pyrene

Figure 17-11 Some carcinogenic polynuclear aromatics.

vantages: (1) they get things dirty so that extra cleaning is required, and things wear out faster; (2) some particulates are corrosive to various materials; (3) particulates in air can decrease visibility—often substantially—since they scatter light; and (4) we have already seen (Section 17-6) how they can decrease the temperature of the earth.

In summary, we can see that combustion is the most important anthropogenic source of all five of the principal types of primary air pollutants (and consequently of the secondary air pollutants). Automobile engines are the major source of NO_x, CO, and hydrocarbons, while stationary combustion produces most of the SO_x and particulates.

17-8 Secondary Pollutants. Smog

The five types of primary pollutants discussed in the previous section are bad enough, but in certain circumstances some of them interact with each other and with other constituents of the atmosphere to produce secondary pollutants, which are often worse. One important example is **photochemical smog,** which is produced by interaction of NO_2, O_2, hydrocarbons, and ultraviolet light from the sun.

The ultraviolet light cleaves the NO_2 molecule, causing the following reactions:

$$NO_2 \xrightarrow{h\nu} NO + O$$
$$O + O_2 \longrightarrow O_3$$
$$O_3 + NO \longrightarrow NO_2 + O_2$$

This series of reactions causes no net change if it goes to completion, but if a sufficiently large concentration of hydrocarbons is present, it does not entirely go to completion because the ozone and oxygen atoms can react with hydrocarbons (principally alkenes and aromatic hydrocarbons rather than alkanes). For example, ozone can react with propene, oxygen, and NO_2 as follows:

$$\underset{\text{propene}}{CH_3(H)C{=}CH_2} + O_3 \longrightarrow \underset{\text{acetaldehyde}}{CH_3{-}\overset{\overset{\displaystyle O}{\|}}{C}{-}H} + \underset{\text{free radicals}}{HO\cdot + H{-}\overset{\overset{\displaystyle O}{\|}}{C}\cdot}$$

$$CH_3{-}\overset{\overset{\displaystyle O}{\|}}{C}{-}H + O \xrightarrow{h\nu} \underset{\text{free radicals}}{CH_3{-}\overset{\overset{\displaystyle O}{\|}}{C}\cdot + HO\cdot}$$

$$CH_3{-}\overset{\overset{\displaystyle O}{\|}}{C}\cdot + O_2 \longrightarrow CH_3{-}\overset{\overset{\displaystyle O}{\|}}{C}{-}O{-}O\cdot$$

$$CH_3{-}\overset{\overset{\displaystyle O}{\|}}{C}{-}O{-}O\cdot + NO_2 \longrightarrow \underset{\text{PAN}}{CH_3{-}\overset{\overset{\displaystyle O}{\|}}{C}{-}O{-}O{-}NO_2}$$

to produce the compound peroxyacetyl nitrate **(PAN).** The free radicals HO·, $CH_3{-}\overset{\overset{O}{\|}}{C}\cdot$, and $H{-}\overset{\overset{O}{\|}}{C}\cdot$ can also interact with other components of the atmosphere.

O_3 and PAN are the two most important of a group of air pollutants known as

photochemical oxidants. Photochemical smog consists of NO_2, hydrocarbons, O_3, PAN, acetaldehyde, and a host of other oxygen-containing compounds, the whole usually in the form of an aerosol. The various free radicals formed are very reactive and react with each other and with all other components of the atmosphere except N_2 and the noble gases. Because of the presence of NO_2, photochemical smog is often brown in color, and is visible as a brown haze. Note that the ozone in photochemical smog is produced near the surface of the earth and is thus entirely distinct from the ozone that occurs naturally in the ozone layer (Section 17-5).

Ozone, PAN, and the other photochemical oxidants have many harmful effects, which become more severe as their concentrations increase. PAN irritates the eyes of human beings and animals. Peroxybenzoyl nitrate, $C_6H_5{-}\underset{\underset{O}{\|}}{C}{-}O{-}O{-}NO_2$, is even more irritating to the eyes but fortunately is present in smaller amounts. Ozone produces fatigue and lack of coordination. Both O_3 and PAN cause damage to plants and to such materials as rubber and textiles.

Because it is dependent on ultraviolet light from the sun, photochemical smog is a *daily* occurrence. Each day an entirely new smog cycle begins in the early morning, builds to maximum intensity, and slowly ends by about the middle of the afternoon. After that the pollutants become dispersed by the settling out of the aerosols, and by winds. The next morning a new cycle begins.

Photochemical smog is one of the two principal types of smog. The other, frequently called **industrial smog,** or **London smog,** mainly consists of SO_2, H_2SO_4, and particulates in the form of fog and smoke. This type of smog does not require ultraviolet light and does not have a daily cycle. Industrial smog occurs mostly in older cities, such as London, New York, Pittsburgh, and Chicago, which have large concentrations of heavy industries that rely on combustion; and in cold and wet weather when heating and electric-power generation are at a maximum. On the other hand, photochemical smog is chiefly found in younger cities with warm dry climates, plenty of sunshine, and large amounts of automobile and truck traffic (for example, Los Angeles, Mexico City, Buenos Aires). Industrial smog is largely caused by stationary combustion (which produces SO_x and particulates); photochemical smog by motor-vehicle combustion (which produces NO_x and hydrocarbons). Table 17-5 summarizes these characteristics.

The effects of both kinds of smog are made much worse when **temperature inversions** take place. Under normal conditions, the temperature of the air steadily decreases with increasing altitude throughout the troposphere (Figure 17-12). This is a very important factor in dispersing polluted air, because the warmest air, being at the very bottom of the atmosphere, rises and carries the

Table 17-5 Characteristics of Industrial and Photochemical Smog

Characteristic	*Industrial smog*	*Photochemical smog*
Typical city	London	Los Angeles
Climate	Cool and humid	Warm and dry
Pollutants	SO_x, particulates	O_3, PAN, NO_x, aldehydes, etc.
Major sources	Industrial and household burning of coal and oil	Motor vehicles
Major effects on humans	Irritation of lungs and throat	Eye irritation
Times when worst episodes occur	Winter months (especially in early mornings)	Summer months (maximum effect around noon)

Source: Adapted from G. T. Miller, Jr., *Chemistry: Principles and Applications* (Belmont, Calif.: Wadsworth Publishing Company, Inc., 1976).

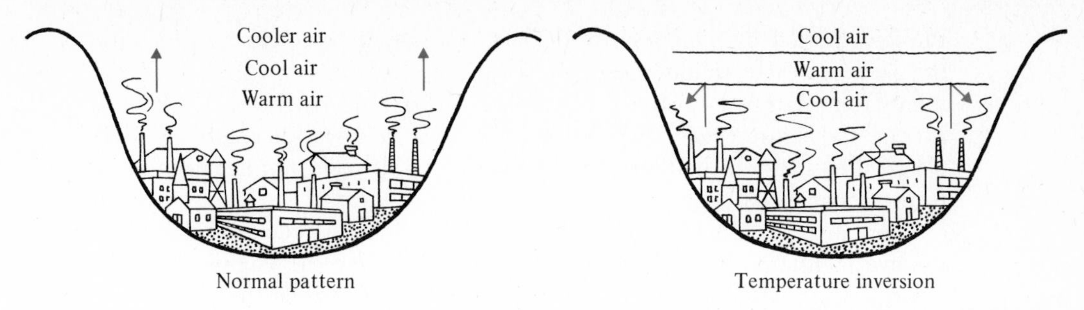

Figure 17-12 Under normal conditions, polluted air is carried away because warm air rises, but in a temperature inversion, a layer of warm air lies above the cool air and the pollutants cannot disperse.

pollutants away with it. In a temperature inversion, however, a layer of warm air rests above a layer of cooler air (Figure 17-12). The cool polluted air at the bottom cannot rise and so remains in place. Factories, homes, and motor vehicles continue to add pollutants to the atmosphere, but as long as the temperature inversion continues, the pollutants cannot escape by rising. Thus, their concentrations steadily increase. The situation can be made worse by other local factors, such as low wind velocities and irregularities of terrain (hills, valleys, buildings).

When smog and temperature inversions combine, the harmful effects of air pollution reach their peak. In October 1948, these conditions existed in the town of Donora, Pennsylvania. At least 17 people died, and more than half the town's population was made ill. A more severe example took place in London in December 1952. At least 4000 people were killed by the smog between December 4 and December 9, when the temperature inversion cleared. The combined effects of smog and temperature inversions have been felt in other cities and other times. Few of these instances make the headlines, because the effects in any single case are generally much less severe than they were in Donora or in London, but the overall deaths, illnesses, and other damage caused by these factors is considerable. In general, the people most affected are those with respiratory illnesses such as emphysema or asthma.

17-9 The Control of Air Pollution

Man has no convenient way to remove pollutants from the atmosphere once they are there, except by letting nature do its normal work of dispersion, washing out, and those reactions discussed in Sections 17-7 and 17-8. Therefore, all attempts at pollution control must focus on *preventing* the pollutants from reaching the atmosphere. It is obvious that we must concentrate on the primary pollutants, since if we can reduce them, the secondary pollutants will also be reduced. At the end of Section 17-7, we noted that the major source of all types of air pollution was combustion: from motor vehicles and from stationary furnaces that produce heat and electricity. Therefore, pollution control primarily consists of finding ways to make these combustions cleaner and/or finding ways to generate energy without combustion.

Motor Vehicles. There is no question that in the near future, developed countries (and, to an ever-increasing extent, underdeveloped countries as well) must continue to rely upon gasoline and diesel engines to meet most of their transportation needs. For this reason, a great deal of work has been done in modifying automobiles and other motor vehicles so that they will emit smaller amounts of pollutants. An important stimulus to this work in the United States has been the laws passed by Congress in 1970 and modified since then. These laws, and the agency created to enforce them, set automobile emission standards which the manufacturers must meet. (For example, 87.5 g/mile of CO was emitted by the average car before 1968. The standard for 1973–1974 was 28 g/mile, and for 1978, 3.4 g/mile.) Most of the research done to reduce air pollution has been in response to these standards.

There are two chief approaches to the problem: (1) redesign engines, and (2) remove pollutants by an afterburner, after they leave the engine. In redesigning engines, the main problem is that the best ways to reduce the three major auto pollutants are to some extent incompatible. The best way to reduce NO production is to operate at a lower flame temperature (Section 17-7), but the best way to reduce CO and hydrocarbon production is to try to achieve more complete combustion of the fuel, and higher flame temperatures are desirable for this! A number of methods for solving these problems have been introduced (for example, adjustment of air-fuel ratios), but we will discuss only two: control of spark timing and exhaust-gas recycling.

An automobile engine contains several cylinders, in each of which a piston moves up and down (Figure 17-13). The path of the piston is so fixed (by the connecting rod) that it must complete its entire stroke (that is, move from the very top to the very bottom) every time. It cannot change direction at any other point in the cycle. When it reaches the top of its stroke, the spark plug is activated, setting on fire the mixture of fuel and air and forcing the piston down. For maximum efficiency, it is necessary that the spark occur just a little bit before the top of the stroke. However, it is possible to lower the amount of NO_x and hydrocarbon emissions by *retarding the spark*, making it come a little later. The emissions of NO_x are lowered because the flame temperature is lowered. Hydrocarbon emissions are lowered because this change increases the temperature of the exhaust gases, and additional combustion of the hydrocarbons takes place in the exhaust. In the early 1970s, control of spark timing was the primary method used in U.S. cars to lower pollution. Unfortunately, there is a substantial cost involved. Retarding the spark decreases engine efficiency, so that fuel economy (that is, miles per gallon) goes down.

In exhaust-gas recycling (EGR), a portion of the exhaust gas is sent back into the engine. The purpose of this is to lower the flame temperature so as to decrease the amount of NO_x formed. EGR has been used in U.S. cars since 1975 as the principal method for achieving the government NO_x emission standards.

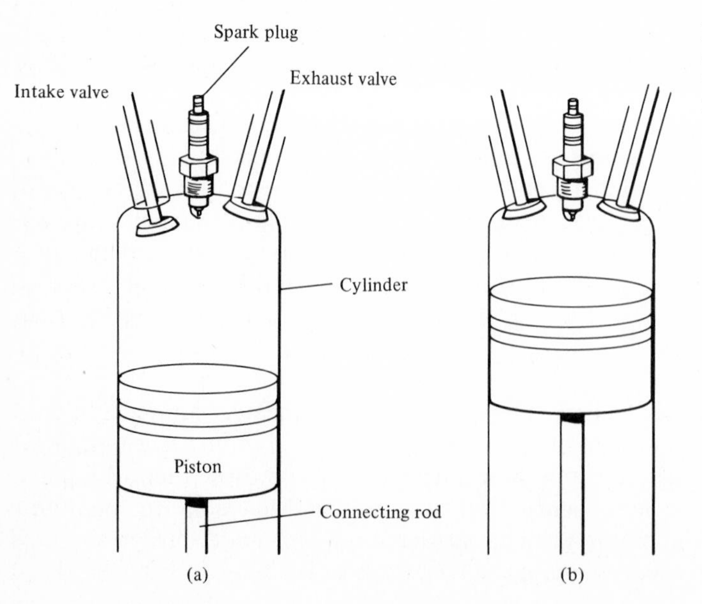

Figure 17-13 Automobile cylinder with piston at the bottom (a) and at the top (b) of its stroke.

Because it causes cars to burn more fuel, the practice of retarding the spark is not satisfactory at a time when the government is trying to conserve gasoline. Because of this, the automobile engineers turned their attention to designing afterburners; the purpose of these devices is to complete the oxidations left incomplete in the engine. They oxidize both CO and unburnt and partially burnt hydrocarbons to CO_2 and H_2O. Afterburners can operate either thermally (that is, by heat alone) or catalytically. The latter have proven more successful, and catalytic converters have been used on almost all U.S. cars since the 1975 model year. The use of catalytic converters has enabled spark timing to be advanced, and the average fuel economy of U.S. cars has significantly increased.

The catalysts used are generally platinum or palladium. These are expensive metals, but each car uses only a small amount, so the cost has not proven prohibitive. However, there was one problem that had to be solved: almost all gasoline sold in the United States before 1975 contained a compound called lead tetraethyl. When

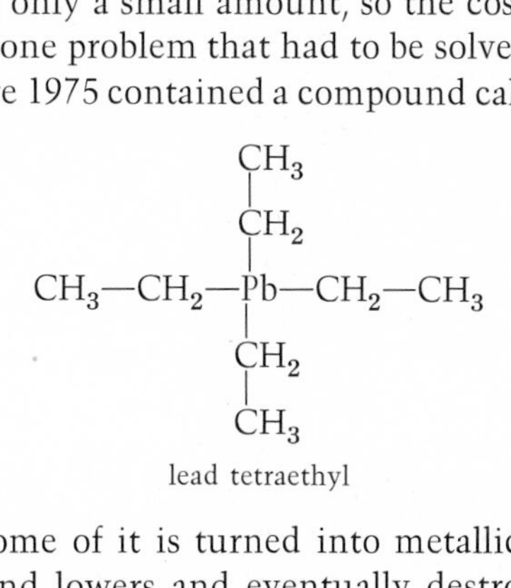

lead tetraethyl

this compound burns, some of it is turned into metallic lead; the lead coats the surface of the catalyst and lowers and eventually destroys its activity. The only solution was to use gasoline without lead tetraethyl. (Although it is possible that lead emissions from automobile engines are harmful to human beings, this was not the reason that lead tetraethyl was removed from some gasolines: the reason is that the lead deactivates the catalysts in the catalytic converter.)

Lead tetraethyl is not a natural constituent of petroleum or of gasoline. It is a synthetic compound that was added in small amounts (about 2 or 3 ml/gal of gasoline) to most gasoline in the United States between about 1922 and 1974. The purpose of the lead tetraethyl was to increase the octane number of the gasoline. We have mentioned that the spark ignites the gasoline-air mixture in the cylinder when the piston has reached approximately the top of its stroke (Figure 17-13). It was discovered more than 50 years ago that with some gasolines, the gasoline-air mixture would ignite spontaneously, *before* the spark went off. This phenomenon is called **knocking.** Since knocking happens when the piston is still on its way up, it lowers efficiency because the piston must still travel up, when the force of the explosion is trying to send it down. Not all gasolines are identical, and the ability of a gasoline to cause knocking can be quantitatively measured. This measure is called the **octane number.** One way to increase octane number is to add lead tetraethyl, although it is not understood how it accomplishes this task. With the advent of catalytic converters, it became necessary to remove lead tetraethyl* and to handle the octane-number problem in other ways. One way is to add aromatic hydrocarbons (which, in general, have higher octane numbers than aliphatic ones); another is to lower compression ratios, since engines with lower compression ratios have lower octane-number requirements.

At this time, the government air pollution standards are being met primarily by catalytic converters, which reduce the amounts of CO and hydrocarbons, and by EGR (page 560), which reduces the amount of NO_x. However, it is likely that the

*The removal of the lead tetraethyl also enabled other gasoline additives to be removed: these had been added to decrease lead deposits in the engine.

standards will become stricter, and these practices will not be enough. Research is now being carried out on catalytic converters which will also remove NO_x. This is difficult, because the NO_x would have to be removed by reduction, so that the converters would have to contain both oxidizing and reducing catalysts. A number of longer-range solutions are also being considered: (1) instead of gasoline, burn other fuels, which pollute less (among those being considered are hydrogen and methanol); (2) replace automobile engines entirely by some other type of device (for example, turbines or electrical cars); and (3) significantly decrease the number of automobiles by increasing mass transportation. All these longer-range possibilities have practical difficulties, and it remains problematical just which of them, if any, will eventually be adopted.

Stationary Combustion. The two major pollutants from stationary combustion are SO_x and particulates. SO_x control has proven difficult, and much work is being done in this field. Since SO_2 is formed by burning the sulfur and sulfur compounds contained in the coal or fuel oil,* one possibility would be to remove these substances before combustion. Unfortunately, this is very difficult for coal, and a satisfactory method is not yet available. Some of the sulfur compounds are removed from fuel oil, but the process is not complete and some of them remain. Another possibility is to use low-sulfur coal and fuel oil, and some cities have regulations that require this (coal and petroleum found in different places vary widely in their compositions, and some have much smaller amounts of sulfur compounds than others). For coal in the United States, this solution also has its difficulties, because most of the low-sulfur coal is near the surface and strip-mining is necessary; and most of it is far from major population centers, so the high cost of transportation must be considered.

If it is not feasible to burn coal and fuel oil that are free of sulfur compounds, then the only way to reduce SO_x pollution is to remove the SO_2 from the exhaust gas. One method suggested for this purpose is to add pulverized limestone ($CaCO_3$) to the fuel. In the flame, the $CaCO_3$ is decomposed to calcium oxide. The calcium oxide then reacts with the SO_2 and oxygen to give calcium sulfate.

$$CaCO_3 \xrightarrow{\Delta} CaO + CO_2$$

$$2CaO + 2SO_2 + O_2 \longrightarrow 2CaSO_4$$

Several other methods have also been suggested, and much research has been done on them, but none of the methods is yet in widespread use.

The methods for controlling particulates are all based on capturing them before they reach the atmosphere. Many devices to do this have been invented, including gravity settling chambers, wet scrubbers, and electrostatic precipitators.

Part II WATER

17-10 Water on the Earth

As everybody knows, the earth contains a great deal of water. It has been estimated that about 1.5 billion cubic kilometers of water are present on the surface of the earth—in the oceans, ice caps, glaciers, lakes, rivers, soils, and in the atmosphere. Water covers more than 75% of the entire surface of the Earth. But of all of this vast

*Coal typically contains sulfur-containing organic and inorganic compounds. The sulfur in petroleum is entirely in the form of organic sulfur compounds.

amount of water, only a small percentage is available for use by human beings. Ninety-seven percent of the earth's water is salt water, and about three-fourths of the rest is in glaciers and polar ice caps. Most of the remaining water lies underground too deep for ready accessibility, or is in the soil, or in other places where it is too inconvenient or expensive to get it out. Thus, we have the paradox poetically expressed by Coleridge in "The Rime of the Ancient Mariner":

Water, water, every where,
And all the boards did shrink;
Water, water, every where,
Nor any drop to drink.

Although in some parts of the earth there is enough water to meet the present needs of the inhabitants, in other parts there is not. Some places have a chronic shortage of usable water; in others there are occasional severe droughts that cause great decreases in the amounts of crops and sometimes result in the deaths of thousands of people and animals. Almost everywhere population is growing, which increases the need for fresh clean water, and at the same time increases the amount of polluted water. In 1976, a United Nations' report stated that at least 20% of the world's urban population and 75% of its rural population lacked reasonably reliable supplies of drinking water. Because of this, the problem of supplying water for drinking, cooking, washing, and other human needs is one that has gained the attention of many governments. Perhaps the answer will come from desalinization of salt water (Section 17-17).

The human body is about 60% water (that is, in a person weighing about 150 lb, about 90 lb is water). Each adult needs to consume on the average about 1.5 liter/day in order to survive,* since this is the amount lost by perspiration, respiration (we breathe out H_2O as well as CO_2), emission of wastes, etc. At the present time, the consumption of fresh water for all purposes in the United States averages about 7000 liters per person per day (this number has been steadily increasing since the beginning of the century), of which about 8%, or 560 liters, is for direct personal use—drinking, cooking, washing, watering of lawns, flushing of toilets, and so on. Agricultural (about 33%) and industrial (about 59%) uses account for the rest. For example, it takes on the average about 150 liters of water to produce one egg; 21,000 liters for 1 kg of beef, 300 liters for 1 kg of steel, and 380,000 liters for one automobile.

We have already seen (Section 6-8) that water is the most unusual liquid known. It has an abnormally high boiling point, melting point, specific heat, heat of fusion, heat of vaporization, thermal conductivity, dielectric constant, and surface tension. Furthermore, it reaches its maximum density at 4°C instead of at its freezing point, and expands on freezing rather than contracting as other liquids do. The high heat of vaporization is highly useful to human beings (and to some animals), since this is the way we keep our bodies cool in hot weather. Each gram of perspiration that evaporates removes 540 cal of heat from our skins. The high value of ΔH_{vap} means that large amounts of heat are removed by evaporation of relatively small quantities of water. The evaporation of perspiration is one of the principal methods used by human beings and most other warm-blooded animals to maintain constant body temperatures. (We have already seen, Section 5-2, how the high ΔH_{vap} moderates the climate of land that is near large bodies of water.)

*It may not seem to you that you regularly consume about 1.5 liters/day of water. However, this figure includes the water present in the other liquids most people normally drink, such as milk, fruit juices, coffee, and beer. It also includes the water present in solid foods, many of which have large amounts. For example, raw carrots have 88% water and lettuce is 95% water (these foods have even more water than whole milk, which is 87% water).

Another unusual property of water—its high ability to dissolve ions—is responsible for the saltiness of the sea. When rain falls on land it flows into streams and rivers (some of them underground), but eventually all flow into the sea. While it is in contact with the ground, it dissolves ionic minerals and rocks. Some of these have fairly highly solubilities in water, but even those with only tiny solubilities eventually become dissolved, little by little, over the course of thousands and millions of years. The flowing water relentlessly carries all these salts to the sea. The sun constantly evaporates water from the oceans, but the dissolved salts remain behind. Since more salts arrive all the time, but none ever leave, the concentration of salts in the sea is ever-increasing. How long will it take for the sea to become saturated? It has been estimated that at the present rate, it will take about 3.5 billion years.

There is a water balance in nature: each year the total amount of water evaporated by the sun is equal to the amount that falls as rain or snow. This amount is approximately 4×10^{17} liters/yr.

17-11 Hard and Soft Water. Soaps and Detergents

We get our household, industrial, and agricultural water supplies basically from two kinds of sources: (1) surface waters (rivers, lakes), and (2) ground waters (wells, which tap underground sources, and springs, where underground water breaks through the soil). In either case, the water has been in contact with rocks and minerals, and so contains dissolved salts (although in much smaller concentrations than in salt water). These salts are almost entirely made up of six ions: the cations Na^+, Ca^{2+}, and Mg^{2+} (mostly Ca^{2+}), and the anions HCO_3^-, SO_4^{2-}, and Cl^- (although small amounts of other ions, such as Fe^{2+} and Mn^{2+}, may also be present). The sodium ion Na^+ causes very little trouble in the concentrations typically found in fresh water, but high concentrations of Ca^{2+} and/or Mg^{2+} do create problems. Water containing high concentrations of these two ions is called **hard water.**

Of course, concentrations of Ca^{2+} and Mg^{2+} vary greatly. Because ground waters generally spend a much longer time in contact with the earth, they are usually much harder than surface waters. A typical soft water might contain 20 ppm of hardness,* while hard waters vary from 80 to 250 ppm or higher.

Spring water, which is usually quite hard, has for centuries been regarded by many people as good for the health, precisely because of the salts it contains, and a number of towns in Europe and the United States (for example, Bath, England; Baden-Baden, Germany; Silver Springs, Fla.), which are located at springs, have prospered because of this.

Most community water supplies in the United States are hard, with varying degrees of hardness. Two exceptions are New York City and San Francisco, which have major supplies of soft water.

Hard water can be divided into two types, depending on the anion. If the anion is chiefly bicarbonate ion, HCO_3^-, then the hardness can be removed by boiling the water. (For this reason, this is often called *temporary* hardness.) Boiling causes these reactions to move to the right:

$$2HCO_3^- \rightleftharpoons CO_3^{2-} + H_2CO_3$$
$$H_2CO_3 \rightleftharpoons H_2O + CO_2(g)$$

because CO_2 is a gas and is thus removed from the solution. In this way, most of the HCO_3^- is converted to CO_2 and CO_3^{2-}. The CO_2 leaves the solution, but the CO_3^{2-}

*This means 20 g of $CaCO_3$ per million grams of water. All the hardness is not present as $CaCO_3$ (in fact, very little of it is, since CO_3^{2-} is not present in significant amounts), but for convenience analytical water chemists express it in this way. In this case, the total number of equivalents of Ca^{2+} and Mg^{2+} present in 1 million grams of water would have a mass of 20 g if it were all in the form of $CaCO_3$.

performs an important function. It causes Mg^{2+} and Ca^{2+} to precipitate in the form of $CaCO_3$ and $MgCO_3$:

$$Ca^{2+} + CO_3^{2-} \longrightarrow CaCO_3\downarrow$$
$$Mg^{2+} + CO_3^{2-} \longrightarrow MgCO_3\downarrow$$

Thus, the boiling not only removes Ca^{2+} and Mg^{2+} but also decreases the total ionic content of the water. In the other type of hard water, the principal anions are SO_4^{2-} or Cl^-. These are not so easily removed, and this type is called *permanent* hardness.

Hard water has several major disadvantages. The most important of these is that it interferes with the action of soap. Soap is a mixture of sodium salts of organic acids, all of which have a long chain of carbon atoms (an alkane-like portion) on one end, and a $—COO^-\ Na^+$ on the other. A typical soap molecule looks like this:

$$CH_3CH_2CH_2CH_2CH_2CH_2CH_2CH_2CH_2CH_2CH_2CH_2CH_2CH_2CH_2CH_2CH_2—\overset{}{\underset{\|}{C}}—O^-Na^+$$
$$\text{sodium stearate} \qquad O$$

The other soap molecules in the mixture are very similar; they differ in the length of the carbon chain (in Section 21-7 we will see how soap is made). The cleaning action of soap is a result of its structure. Whenever we wash anything (hands, dishes, clothing) there are two kinds of dirt we wish to remove: dirt that is soluble in water and dirt that is insoluble. We do not need soap for the soluble dirt; that can be removed with water alone. Soap is used for dirt that is insoluble in water. What kind of compounds are likely to be insoluble in water? In Section 7-5 we saw that *nonpolar* compounds are the ones most likely to be insoluble. A soap molecule has an ionic end ($—COO^-\ Na^+$) and a nonpolar end (the alkane-like portion). The action of soap depends on the fact that the ionic end would like to dissolve in water, while the nonpolar end hates water (we call it **hydrophobic**) and would like to dissolve in the dirt! Without the soap the water cannot remove the dirt because the two layers (water and dirt) cannot penetrate each other (Figure 17-14). The molecules of water are much more strongly attracted to other molecules of water than they are to molecules of dirt, and vice versa. The water cannot *solvate* the dirt molecules. When soap is added, the ionic end of each soap molecule dissolves in water, while the nonpolar end dissolves in the dirt (Figure 17-15). This lowers the interfacial tension and permits the dirt to be pulled off in small particles which are easily washed away by more water.

Hard water interferes with the action of soap because the calcium and magnesium salts of the soap molecules are insoluble in water. The sodium salts (soap) are soluble, but in the presence of Ca^{2+} or Mg^{2+} ions, the solubility products are exceeded, and the calcium and magnesium salts precipitate:

$$2C_{17}H_{35}COO^- + Ca^{2+} \longrightarrow (C_{17}H_{35}COO)_2Ca\downarrow$$

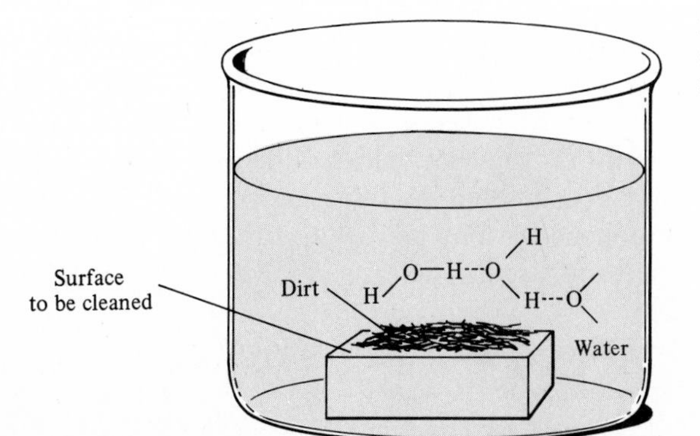

Figure 17-14 Before soap is added, molecules of water cannot penetrate through molecules of dirt.

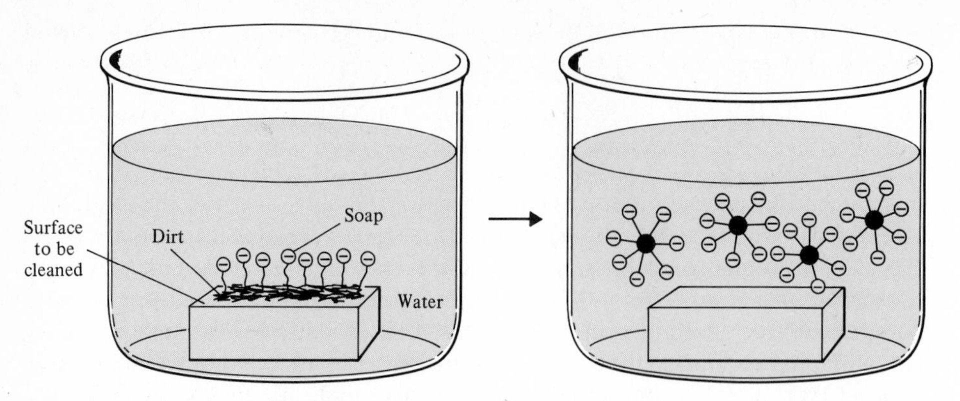

Figure 17-15 The cleaning action of soap. Soap molecules line up at the interface between the dirt and the water, lowering the interfacial tension and breaking the dirt into small particles which can easily be washed away.

This has two unfortunate consequences: (1) a certain amount of soap is wasted, since the anions that precipitate are not available for cleaning; and (2) the precipitate itself is now a kind of dirt (it is in the form of a gray curd). This gray curd deposits on clothes, dishes, and on sinks and bathtubs (it is the "ring" around the bathtub), and must be removed if we want our surfaces to be clean.

Because soap causes so much difficulty in hard water, and because most of the community water supplies in the United States are hard, the use of soap in this country has greatly diminished in the last 30 years or so. Soap has been largely (although not entirely) replaced by **detergents,*** both for household and for industrial cleaning. A typical detergent is a mixture of several types of compounds, but two of these are the most important: (1) a **surfactant,** and (2) a **builder.**

A surfactant is a long-chain molecule with an ionic end and a nonpolar end. Two typical surfactants used in detergents are

$$CH_3CH_2CH_2CH_2CH_2CH_2CH_2CH_2CH_2CH_2CH_2CH_2{-}O{-}SO_2{-}O^-Na^+$$

sodium lauryl sulfate

$$CH_3CH_2CH_2CH_2CH_2CH_2CH_2CH_2CH_2CH_2\underset{\displaystyle CH_3}{CH}{-}C_6H_4{-}SO_2{-}O^-Na^+$$

sodium *para*-(2-dodecyl)benzenesulfonate

It can immediately be seen that these compounds function exactly the same way as do soap molecules. The major difference is that the surfactants used in detergents *do not precipitate with Ca^{2+} or Mg^{2+}* and so work very well in hard water. Another advantage is that they produce a smaller amount of suds, which interfere with the action of many washing machines. One problem that arose with surfactants formerly used is that they were not **biodegradable.** This is an important concept. We saw in the first part of this chapter that bacteria and other organisms that exist in the soil and in the waters of the earth decompose many compounds. Used wash water is eventually sent out into the environment—into the ground, and into lakes, rivers, and oceans, and it is important that the compounds added to the environment in this way be capable of being decomposed by living organisms. If they are not, they will remain in the environment for long periods of time and become a major source of pollution. The surfactants used in detergents today (including the two shown above) are biodegradable.

In most detergents the major constituent is not the surfactant but the builder. A

*Strictly speaking, the dictionary defines a detergent as any cleaning agent, including soap, but modern practice is not to include soap among the detergents.

builder is a compound that removes Ca^{2+} and Mg^{2+} ions. The most common builders in use today are sodium tripolyphosphate ($Na_5P_3O_{10}$) and other polyphosphates.

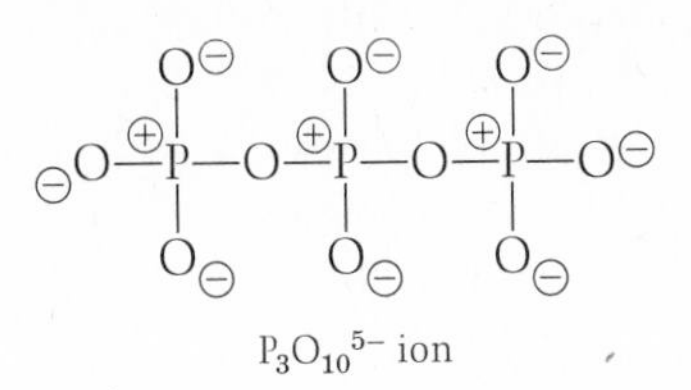

$P_3O_{10}^{5-}$ ion

These ions tie up (this is called **sequestering**) Ca^{2+} and Mg^{2+} ions, converting them to large complex ions, which do not precipitate, but are soluble in water (see chelation, Section 19-7). When they are tied up, they cannot interfere with the action of the surfactant. In previous years, simpler phosphates, such as sodium phosphate (Na_3PO_4), were used in washing powders. These actually precipitate the Mg^{2+} and Ca^{2+} as the solids $Mg_3(PO_4)_2$ and $Ca_3(PO_4)_2$. It is obvious that the polyphosphates are preferable. Since they effectively remove Ca^{2+} and Mg^{2+} and yet avoid a precipitate, polyphosphates are the best builders so far discovered. They have the advantages that they are cheap, nontoxic, noncorrosive, nonflammable, safe for all fabrics and colors, completely biodegradable, and do not interfere with waste-water treatment procedures. Nevertheless, they do have one serious disadvantage, which they share with the phosphates: They can be a major water pollution problem, not because they are not biodegradable, but for the opposite reason! (They can cause eutrophication; see page 574.) The eutrophication problem is so serious that several states and communities (including New York State, Minnesota, Indiana, Michigan, Chicago, Akron, and Dade County, Florida) have banned the sale of detergents containing phosphates or polyphosphates. These and other prospective bans have caused detergent companies to search for other builders which would sequester Ca^{2+} and Mg^{2+} and not cost too much. So far, they have not been successful.

In addition to surfactants and builders, most detergents contain smaller amounts of other ingredients, such as perfumes and brighteners. A brightener is a compound that absorbs ultraviolet light and emits white or blue light, causing a fabric to look brighter.

Besides interfering with washing, another major disadvantage of hard water is that it forms a precipitate, known as **boiler scale,** in boilers and hot-water pipes. This precipitate eventually builds up, and in a pipe may become so thick that it stops the water from flowing (Figure 17-16). What is the nature of this precipitate? If the water is temporary hard water, then, as we have seen, heating precipitates $CaCO_3$. This is all right if we are trying to remove hardness in an open vessel, where the precipitate can be scraped off. But a closed container such as a pipe or a boiler is much less accessible. If the hardness is of the permanent kind, the precipitate is calcium sulfate ($CaSO_4$), which is one of the few inorganic salts whose water solubility *decreases*

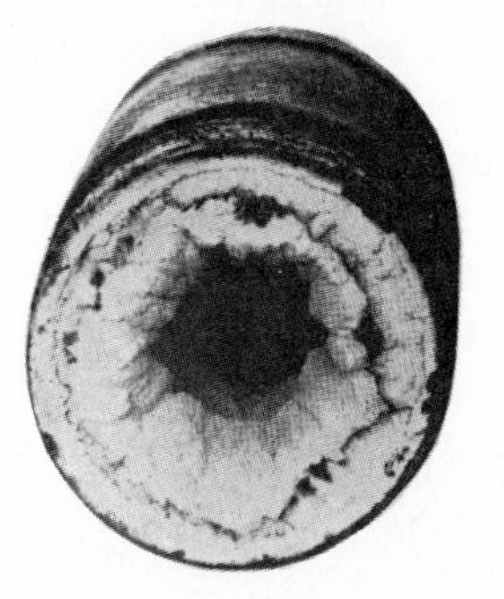

Figure 17-16 Boiler scale in pipes. [From W. F. Ehret, *Smith's College Chemistry* (New York: Appleton-Century-Crofts, 1960)]

with an increase in temperature. Boiler scale is a particularly major problem in industrial processes, where large amounts of water often must be heated. The hardness of water also causes problems with other industrial processes as well; for example, it may interfere with the dying of fabrics.

To many people, another disadvantage of hard water is its taste, which is quite different from that of soft water, although there are many who prefer the taste of hard water. Like the taste of hard water, the taste of soft water results from the salts dissolved in it. Anyone who believes that soft water is tasteless should drink some distilled water and compare.

17-12 Removal of Hardness from Water

Soap itself, and the builders in detergents, remove the hardness from water as the clothes, dishes, etc., are being washed. But suppose that we wish to remove the hardness from water *before* we use it for washing, or from water that is not to be used for washing at all. If the hardness is temporary, we can boil the water, but this is obviously not feasible for large quantities (for example, a community water supply) and, in any case, will not work for permanent hard water. There are two chief methods for removing the hardness from water.

1. *The soda-lime method.* In this method, $Ca(OH)_2$ and/or Na_2CO_3 are added to the water, depending on the type of hardness present. If the hardness is caused by calcium and/or magnesium bicarbonates, then it is removed by adding $Ca(OH)_2$ (this is sometimes called "adding hardness to remove hardness"), which reacts with HCO_3^- as follows:

$$HCO_3^- + OH^- \longrightarrow CO_3^{2-} + H_2O$$

The CO_3^{2-} produced in this way precipitates Ca^{2+}:

$$Ca^{2+} + CO_3^{2-} \longrightarrow CaCO_3\downarrow$$

The Mg^{2+} ions are precipitated by the OH^- ions.

$$Mg^{2+} + 2OH^- \longrightarrow Mg(OH)_2\downarrow$$

If the hardness is caused by $CaSO_4$ or $CaCl_2$, then CO_3^{2-} is added directly, in the form of Na_2CO_3. Once again, the CO_3^{2-} precipitates the Ca^{2+}. To remove $MgSO_4$ or $MgCl_2$, both $Ca(OH)_2$ and Na_2CO_3 are added. The OH^- precipitates the Mg^{2+} as $Mg(OH)_2$, and the Na_2CO_3 removes the Ca^{2+} ions that have just been added. Because of this action, Na_2CO_3 is known as *washing soda.*

When $Ca(OH)_2$ is used alone, then not only the hardness, but the total ionic content of the water, is decreased (why?), but when Na_2CO_3 is used, either alone or in combination with $Ca(OH)_2$, the total ionic content is unchanged and Na^+ ions have been substituted for Ca^{2+} and/or Mg^{2+} ions.

2. *Ion exchange.* This method is entirely different. The water is passed through a tube filled with a polymeric substance called a **resin** (which may be in the form of hard beads, though tiny openings are present that allow the water molecules and small ions to pass through). The resin contains a large number of negative ion sites (Figure 17-17), near each of which is a positive Na^+ ion. (Note that this is not like the structure of typical ionic solids, which are crystal arrays.) When water containing Ca^{2+} or Mg^{2+} is passed through, the Na^+ ions are pushed out, and the Ca^{2+} or Mg^{2+} ions remain attached to the site. This means that the water coming out contains Na^+ instead of Ca^{2+} or Mg^{2+}. The process can be represented:

$$Ca^{2+} + 2\text{ resin–}Na^+ \longrightarrow \text{resin}_2\text{–}Ca^{2+} + 2Na^+$$

When all the sites have been converted from Na^+ to Ca^{2+} or Mg^{2+}, the resin can be

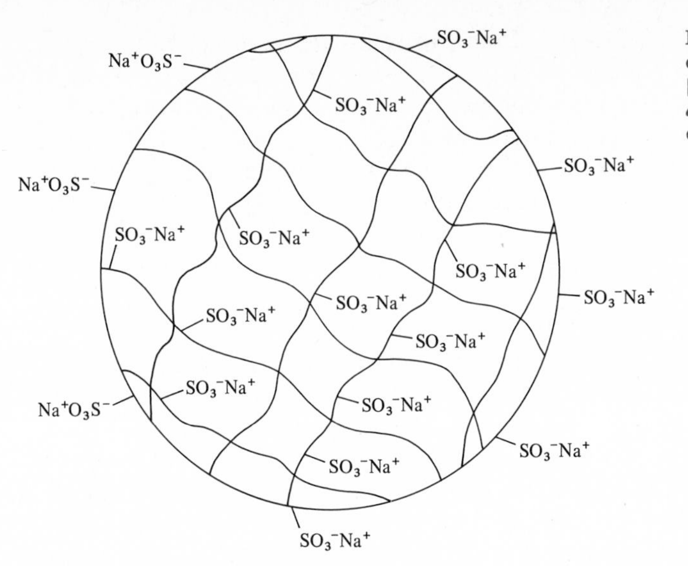

Figure 17-17 Schematic diagram of cation-exchange resin. [After Kirk-Othmer, *Encyclopedia of Chemical Technology*, 2nd ed., Wiley, New York, 1966.]

regenerated by passing a concentrated solution of NaCl through it: this reverses the process; the large concentration of Na^+ pushes out the Ca^{2+} and Mg^{2+} ions, and the resin can be used again. Thus, the only continuing cost is for NaCl, which is very cheap. The water that has been softened in this way is satisfactory for most purposes, although it now contains Na^+ ions.

If necessary, water can be *completely* deionized by using two ion exchanges, first a cation-exchange process in which the resin contains H^+ instead of Na^+, and then an *anion-exchange* resin, where any anions present are converted to OH^-. For example, let us say that a sample of water contains $CaSO_4$. This is first run through a cation-exchange resin containing H^+, which converts the $CaSO_4$ to H_2SO_4, and then through an anion-exchange resin, where the H_2SO_4 is converted to H_2O.

$$1.\ CaSO_4 + 2\ \text{resin–}H^+ \longrightarrow H_2SO_4 + \text{resin}_2\text{–}Ca^{2+}$$

$$2.\ H_2SO_4 + 2\ \text{resin–}OH^- \longrightarrow H_2O + \text{resin}_2\text{–}SO_4^{2-}$$

The soda-lime process is much less efficient than the ion-exchange process. It leaves a substantial amount of Ca^{2+} and Mg^{2+} ions in solution, while the ion-exchange process gets nearly all of them out. On the other hand, the ion-exchange process is much less adaptable to very large scale operation. Thus, the soda-lime process is used by cities and towns to soften large quantities of water in community treatment plants, while the ion-exchange process is used in homes and in some industrial plants.

17-13 Treatment of Community Water Supplies

Above all, we want the water that comes out of our faucets to be safe. Beyond that, we would like it to be aesthetically pleasant—free from suspended particles, and from color, odor, and taste (other than the natural taste of some of the dissolved salts). The water available for community supplies does not always have these characteristics; because of this, most water is treated in community treatment plants before it is piped into our houses. The following are the steps most frequently used, though not all water is put through all the steps. For example, the first three steps are used to get rid of suspended particles. These are much more prevelant in surface waters than in ground waters (where the water has already been filtered through miles of sand or dirt), so that some communities which get their supplies from ground waters omit these steps.

1. *Sedimentation.* The water is allowed to stand for a while, allowing large particles to settle to the bottom.

2. *Coagulation.* Many waters contain particles that are too small to settle out in reasonable periods of time. These particles are coagulated into larger particles, most often by the addition of alum, which is hydrated aluminum sulfate, $Al_2(SO_4)_3$. When alum is added to waters, it reacts with OH^- or HCO_3^- to give aluminum hydroxide, $Al(OH)_3$:

$$Al^{3+} + 3OH^- \longrightarrow Al(OH)_3\downarrow$$
$$Al^{3+} + 3HCO_3^- \longrightarrow Al(OH)_3\downarrow + 3CO_2$$

(most water is naturally alkaline; if it is not, then a base is added). Aluminum hydroxide is a gelatinous (sticky) precipitate, and tiny particles adhere to it. Also commonly used is $Fe_2(SO_4)_3$, which produces gelatinous $Fe(OH)_3$.

3. *Filtration.* Whatever solid particles remain are filtered, usually by allowing the water to pass through a layer of sand, although most of the actual filtration is accomplished by a thin layer of coagulated particles which lies on the surface of the sand particles. Thus, sand filters do a better job when they have been in use for a while than when they are perfectly clean.

4. *Removal of hardness.* In many community water supplies hardness is allowed to remain, but in others it is so great that it is removed. The soda-lime process (previous section) is the chief method used.

5. *Chlorination.* Water can look and taste clean and yet be contaminated by harmful bacteria. The purpose of chlorination is to kill these bacteria. Chlorination can be done in various ways, but by far the most common is the direct addition of Cl_2 gas. The amount of Cl_2 varies, but usually enough is added to give a Cl_2 concentration of about 0.2 to 1.0 ppm. When Cl_2 is added to water, it reacts to give hypochlorous acid, HOCl:

$$Cl_2 + 2H_2O \rightleftharpoons HOCl + H_3O^+ + Cl^-$$

HOCl is even more toxic to bacteria than Cl_2 itself. Chlorination not only kills bacteria but also oxidizes organic matter that is responsible for objectionable colors, tastes, and odors.

The most important disease caused by contaminated water is typhoid. This disease has been almost completely wiped out in the United States and Western Europe by the chlorination of public water supplies and by the pasteurization of milk. And not only typhoid. Studies made on communities before and after chlorination showed that for every person saved from death by typhoid, three other persons were saved from death by other causes, some of which were thought to have no connection with water supplies. An occasional typhoid epidemic still occurs, when water-purification facilities break down (for example, in Zermatt, Switzerland, in January 1963).

Because chlorination can cause problems (for example, it can oxidize organic matter in the water to carcinogenic chlorinated hydrocarbons), some communities disinfect their water with other oxidizing agents. The most important of these is ozone, which is used in many European cities.

6. *Removal of tastes and odors.* If water has a high degree of taste and odor caused by organic compounds, some communities filter the water through a layer of activated charcoal, which removes essentially all of these, by adsorption of the offending molecules onto its surface. The charcoal also removes excess chlorine and organic chlorine compounds resulting from chlorination. In some community water supplies, the bad odor is caused by hydrogen sulfide, H_2S. This problem can be removed by **aeration** of the water, which decreases the solubility of H_2S gas in the water. Aeration can be accomplished by spraying the water into the air and allowing it to fall back; or by blowing air through the water.

7. *Fluoridation*. Several communities in the United States have a high natural concentration of fluoride ions, F^-, in their water supplies. During the 1930s it was noticed that the inhabitants of these communities had an unusually low number of cavities in their teeth. This led to the hypothesis that the presence of F^- caused the decrease in the cavities. A number of studies were undertaken to test this hypothesis; pairs of cities were selected in which the water supply contained little or no F^- (for example, Kingston and Newburgh, N.Y.), and F^-, in a concentration of about 1 ppm, was added to the water in one of these cities for a long period of time, say 10 years. Then the number of cavities in the two cities was compared. The results of these studies were clear and conclusive: in every test, the addition of F^- resulted in a significant decrease in the number of cavities. There is no scientific doubt about it: The presence of F^-, in a concentration of about 1 ppm in community water supplies, reduces cavities by about 60 to 70%, especially in children, and has no deleterious effects on human health. If it had, these effects would surely have been noticed in those communities which have been using natural fluoridated water in concentrations considerably higher than 1 ppm for more than 100 years.

For this reason, many communities in the United States and other countries routinely fluoridate their public water supplies. It is important to keep clear that *this is not fluorination*. It is not the extremely reactive and violent F_2 that is being added, but the ion F^-. (Note the difference from chlorination, where Cl_2 is added, and not Cl^-.) Fluoride is mostly added in the form of sodium silicofluoride (Na_2SiF_6), but NaF and H_2SiF_6 are used to a lesser extent. The concentration of the added F^- is important. The effective concentration is about 1 ppm, but concentrations greater than this (above about 1.5 ppm) cause mottling of the teeth. The inhabitants of communities with a high natural fluoride concentration (which can reach 5 ppm or more) often do have mottled teeth, and some of these communities remove excess F^-. This is done by passing the water through a bed of calcium phosphate, $Ca_3(PO_4)_2$, or alumina, $Al(OH)_3$. These compounds bind the F^-, in a form of ion exchange.

Topical applications of F^-, as in the use of fluoride toothpastes (where the fluoride is usually in the form of stannous fluoride, SnF_2) or fluoride treatments by a dentist, are also valuable but are less effective than the presence of F^- in the drinking water.

WATER POLLUTION

As we have already indicated, water in nature is never completely pure. Even rain contains dissolved CO_2, O_2, N_2, and often suspended dust and other particles. As soon as it reaches the ground, it begins to pick up dissolved salts. How shall we define a water pollutant? We use water for many different purposes: drinking, washing, swimming, boating, fishing, irrigation, industry, etc. Water that is satisfactory for some of these purposes may be unsatisfactory for others. We shall thus define a **water pollutant** as *a substance or condition that interferes with our normal use of the water, for whichever purpose we intend.* Note that this definition is not the same as the one we used for an air pollutant (pages 548–549).

There are nine recognized categories of water pollutants:

1. Oxygen-demanding wastes.
2. Disease-causing agents.
3. Plant nutrients.
4. Synthetic organic compounds.
5. Oil.
6. Inorganic chemicals and mineral substances.
7. Sediments from erosion of land.

Figure 17-18 A paper mill in Canada discharging effluent into a river. Note the white color of the polluted water. [Beverly March]

8. Radioactive materials.
9. Heat.

These pollutants enter our waters chiefly from three sources (in descending order of importance): (1) manufacturing and other industrial processes (see Figure 17-18); (2) domestic wastes (sewage); and (3) agricultural and urban runoff. In the remainder of this chapter, we will discuss the nine categories of water pollutants and the methods used for the treatment of polluted water. In the last section, we will consider the desalination of salt water.

17-14 The Major Water Pollutants

1. *Oxygen-demanding wastes.* Fish and other aquatic organisms need oxygen to survive, just as land animals do, but instead of getting it from the air, they must make use of the oxygen that is dissolved in the water. In general, a concentration of at least 5 to 6 ppm of O_2 in the water is necessary to ensure that a diversified group of organisms will be able to survive. (Note that this does not leave much margin, because the solubility of oxygen in water at 20°C and 1 atm pressure is only about 9 ppm.) Fish require the highest oxygen concentration, invertebrates less, and bacteria the least. Oxygen-demanding wastes are substances which, when added to water, cause the dissolved oxygen concentration to decrease. This happens because bacteria present in the water oxidize these compounds, using the O_2 in the water as oxidizing agent. When the O_2 is consumed in this fashion, its concentration of course decreases.

Oxygen-demanding wastes are primarily wastes from living organisms. The chief pollutant in this group is sewage—both human and animal. In the United States, animals produce more than 10 times as much sewage as humans. Among other oxygen-demanding wastes are industrial wastes from food-processing plants, paper-mill wastes, and effluents from slaughter houses. Most of these wastes are dumped into rivers and lakes. These materials consist chiefly of carbon compounds, and the bacterial oxidation of them (to CO_2) is part of nature's way of maintaining the carbon cycle (see Section 17-6), but when the amount of sewage and other wastes in a limited volume of water becomes too large, the oxygen depletion becomes great enough to kill off large numbers of fish and other aquatic organisms (Figure 17-19). Of course, the dead bodies of the fish themselves now become part of the oxygen-depleting wastes, since they, too, will be oxidized by the bacteria. Besides killing aquatic organisms, oxygen-demanding wastes impart bad odors to the water, both directly, by causing the odor of decaying fish or algae, and indirectly, by increasing the action of anaerobic bacteria, which produce such odorous substances as amines and H_2S. When the problem of oxygen-demanding wastes becomes bad enough, we say that a body of water has been "killed."

It is valuable to have a quantitative measure of just how serious this problem is in a given river or lake. Such a measure is the **biochemical oxygen demand (BOD).** The BOD is determined by taking a sample of known volume and measuring how much oxygen it consumes at 20°C in a period of 5 days. The sample is saturated with O_2, and the dissolved oxygen content is measured at the beginning and at the end. The difference is the BOD. The units of BOD are grams of oxygen per million grams of water (expressed as ppm). Pure water has a low BOD (about 1 ppm). A BOD of about 5 ppm indicates water of doubtful purity, while untreated

Figure 17-19 Dead fish (alewives) in Lake Michigan near Chicago. [Nancy Hays from Monkmeyer]

municipal sewage has a BOD range of about 100 to 400 ppm. In general, when the BOD of any effluent (for example, the runoff from a factory or a sewage treatment plant) entering a river or lake is greater than 20 ppm, it is a cause for concern.

2. *Disease-causing agents.* The bacteria mentioned in the previous section do not cause diseases in human beings or animals. However, as indicated previously, there are other waterborne bacteria and viruses which cause serious diseases such as typhoid, cholera, dysentery, hepatitis, and polio. In developed countries, these are largely controlled (chiefly by chlorination), but in underdeveloped countries they still cause serious problems. Even in developed countries, the control is not total. Thus, hepatitis outbreaks occasionally occur, and many beaches are temporarily or permanently closed because of the danger of disease.

3. *Plant nutrients.* Rivers and lakes have plants growing in them—algae, aquatic weeds, and other plants. They are essential for the presence of aquatic animal life, since they begin the food chain: some fish eat the plants; other fish eat these fish; etc. These plants get their nutrients from compounds dissolved in the water. In general, the total plant growth will be as much as the nutrients allow. When the supply of any single necessary nutrient is essentially used up, total plant growth stops. Plant nutrients become pollutants *when they cause too much plant growth.* Excessive growth of algae and aquatic weeds is called **eutrophication** (Figure 17-20). Eutrophication is bad primarily because the plants eventually die, and the decay of dead plant matter uses up oxygen. Thus, the effects of plant nutrients as pollutants are essentially the same as the oxygen-demanding wastes discussed in Section 1. In addition, eutrophication causes the water to be covered with unsightly masses of living and dead plant matter, which often give rise to bad odors, interfering with the normal use of the water for boating, swimming, and fishing. In general,

Figure 17-20 Stream in northern Louisiana blocked by water hyacinths. [From *Eutrophication: Causes, Consequences, Correctives* (Washington, D.C.: National Academy of Sciences, 1969)]

eutrophication is more of a problem with lakes than with rivers, since the flowing water of a river usually reduces the high concentration of any particular nutrient.

The two plant nutrients that are the chief causes of eutrophication are nitrogen and phosphorus compounds, since these are the ones that usually limit the natural growth of aquatic plants (which get their carbon and oxygen from the atmosphere). Excessive amounts of nitrogen nutrients come from sewage and from excess fertilizer that is washed into rivers and lakes. Phosphorus nutrients also come from fertilizer and sewage, but an important source of excessive phosphorus nutrients is the phosphates and polyphosphates in detergents (Section 17-11).

4. *Synthetic organic compounds.* In some parts of the United States, factories making or using synthetic organic compounds pour their wastes into neighboring rivers or lakes, and this may create a severe local pollution problem, but by far the most important members of this class of pollutant on a national level are detergents and pesticides, which enter the environment on a broad scale, and all over the United States, wherever used wash water is flushed down the drain, or insecticides or weed killers sprayed over a garden or a farmer's field or orchard. We have already discussed detergents in Section 17-11; we will treat pesticides in Section 17-15.

Synthetic organic compounds are pollutants because: (1) many of them, in sufficiently high concentrations, are toxic to fish and other wildlife; (2) some of them cause offensive odors, and also can affect the taste of fish and shellfish; and (3) many of them are not biodegradable and so can persist in the environment for long periods of time. Fortunately, as we have seen in Section 17-11, the compounds in present use as detergents are biodegradable, though the same cannot be said for many pesticides. When detergents *are* found in polluted water, they may make their presence known by producing suds. Some people have had the unpleasant experience of finding that their tap water comes out sudsy. Such pollution is not especially harmful in itself, but it indicates that untreated sewage water (which is where detergent-containing water is usually sent) is getting in the drinking water, and this can be quite harmful. In 1971, Suffolk County in New York State became the first, and so far the only, county in the United States to prohibit the sale of all detergents except dishwater detergents.

Also included in this category are chlorinated hydrocarbons produced by the chlorination treatment of water supplies.

5. *Oil.* In recent years there have been a number of spectacular examples of large oil spills. Three which might be mentioned are those at Cornwall, England, where the tanker *Torrey Canyon* ran aground in 1967, spilling more than 100,000 tons of crude petroleum, at Santa Barbara, California, where a blowout at an offshore oil well in 1969 resulted in large quantities of petroleum fouling up the beaches and other shore areas, and off the coast of France, where in March 1978 the tanker *Amoco Cadiz* crashed on reefs, spilling the largest amount of oil ever yet spilled by a tanker, 220,000 tons, causing enormous pollution of the Brittany coast, and killing many thousands of birds and fish. However, noteworthy examples such as these are only a small part of the problem. Numerous smaller oil spills, which do not make the headlines, and routine flushing of tankers, are responsible for much larger quantities of spilled petroleum. The National Academy of Sciences has estimated that of some 2 billion tons of oil produced worldwide each year, about 6 million tons eventually go into the ocean.

Petroleum is a mixture consisting chiefly of alkanes. These compounds are nonpolar and have low densities, so the spilled petroleum neither dissolves in the water, nor sinks, but instead floats on the surface, where it has the following deleterious effects: (1) by coating the surface, it decreases the amount of O_2 from the atmosphere that can enter the water, which can have harmful effects on aquatic life; (2) it can kill aquatic plants by smothering them; (3) it is toxic to many animals, especially larvae and other immature forms; (4) it coats the feathers of birds, reducing or eliminating their power of flight, and decreasing the insulating effect of the feathers, causing many of them to die; and (5) it decreases the amount of light that can get through the surface of the water. This light is necessary for photosynthesis by marine plants, many of which are under the surface of the water.

6. *Inorganic chemicals and mineral substances.* These kinds of compounds cause three types of water pollution: increased acidity, increased salinity, and direct toxicity.

a. *Increased acidity.* The pH of natural fresh water usually lies between about 6.5 and 8.5. This pH is maintained by the natural buffering action (Section 10-13) of CO_2 and the HCO_3^- ion. (The pH of ocean water is about 8 to 9; many marine animals die if the pH gets lower than about 7.5.) Because of the buffering action, small amounts of added acidity do not affect the pH very much, but large amounts swamp out the buffer and cause the pH to drop. These large amounts primarily come from two sources: acid rain (Section 17-7), and mine drainage. The mine drainage, which comes mostly from exhausted and abandoned coal mines, comes about when iron pyrites (FeS_2) are oxidized by air (in the presence of water) to H_2SO_4, and the rain washes the H_2SO_4 into rivers and lakes. Increased acidity has several harmful effects: (1) it kills much aquatic life; (2) it corrodes boats, piers, plumbing systems, and so on; (3) it dissolves more Ca^{2+} and Mg^{2+} from the rocks and minerals, and thus increases the hardness of the water; and (4) it can cause agricultural crop damage.

b. *Increased salinity.* We have mentioned that 97% of the earth's water is salt water (meaning that it contains all kinds of salts, not just NaCl). But fresh water can also acquire salinity by pollution. The principal ions involved are the cations Na^+, K^+, NH_4^+, Ca^{2+}, Mg^{2+} and the anions Cl^-, HCO_3^-, SO_4^{2-}, PO_4^{3-}, NO_3^-, and NO_2^-. These ions enter the water (in addition to the natural hardness) from industrial effluents, mine drainage, irrigation (which dissolves minerals), and the use of salts such as $CaCl_2$ on highways in winter. The main deleterious effects are that high concentrations of the ions render the water unfit for drinking and for irrigation; and that such concentrations are harmful to animal and plant life. Even if the ions are not toxic themselves, they affect osmosis (Section 7-10), and many fresh-water species cannot survive in water with a high concentration of ions.

c. *Direct toxicity.* Heavy-metal ions (virtually all except those of groups I and II of the periodic table) are directly toxic to man and to many plants and animals. Among the most important pollutants are ions of beryllium, mercury, cadmium, lead, copper, and nickel. These ions reach the water supply from the drainage coming from mining, smelting, and refining operations (see Section 19-1) and from various industrial and other human activities; for example, an important source of Pb^{2+} in the environment is the burning of lead tetraethyl (Section 17-9) in gasolines. We will discuss only one of these toxic metals: mercury. The most important use of mercury is in making electrical equipment: batteries, switches, mercury lamps, etc. Large amounts are also used in the production of Cl_2 and NaOH by electrolysis of NaCl. Mercury compounds are used in paints and fungicides. Some of this escapes into the environment, along with some from mining and smelting, and from the burning of coal, which contains a very small percentage of mercury. Although Hg^{2+} ions are toxic (as are all other forms of mercury), the greatest environmental danger is from dimethylmercury, $(CH_3)_2Hg$, and related compounds (called alkylmercurys), which enter the environment both directly and because certain microorganisms convert Hg^{2+} to these compounds. Dimethylmercury and other alkylmercurys are extremely toxic. Once small amounts get into the body, they stay there a long time (weeks or months), and the damage that they do is permanent, even after they are eliminated. Larger amounts cause death. In 1953, a mysterious neurological disease broke out among Japanese fishermen and their families. In all, 44 people died and many others were paralyzed for life. The mystery was solved when it was discovered that these people had eaten fish and shellfish taken from Minamata Bay, where a plastics factory had discharged dimethylmercury. Of course, human beings must have small amounts of certain heavy metals in order to sustain life (for example, Fe, Co, Cu), but even for these, larger quantities are toxic.

7. *Sediments from erosion of land.* The water in streams and rivers normally contains some suspended solids which are washed into it from rain running off of the land. Erosion is greater from cleared land than it is from forested land, and some soil is lost in this way when forested land is cleared for agricultural purposes. Erosion is enormously increased (by factors of 10 to 100) by such activities as construction and strip mining. The sediments cause problems because they fill streams, channels, harbors, and reservoirs, kill shellfish (which cannot get out of the way) and other aquatic animals, reduce the amount of light reaching aquatic plants, and reduce the visibility of fish. In addition, the sediments can adsorb and bind various substances (for example, insecticides), and serve as sites for bacterial growth.

8. *Radioactive materials.* These reach the water chiefly from the processing of uranium ore. Since uranium ore contains only a small percentage (about 0.1%) of U_2O_3, large amounts of ore must be processed to extract relatively small amounts of uranium oxide. The rest of the ore, called *tailings,* is piled up nearby. Huge mounds of these tailings, which still contain radioactive compounds, exist wherever uranium is mined. The rain slowly washes them into streams and rivers. In the past, radioactive materials also reached the soil from atmospheric nuclear weapons testing. Some of them are still there. In the future, it is likely that nuclear power plants will become a significant source of this type of pollution, through accidental leaks and other kinds of contamination. The effects of radioactivity are discussed in Section 23-2.

9. *Heat.* Excessive heat added to water is often called **thermal pollution.** Many industries use water for cooling purposes. They take water from a river; send it through a factory or power plant, where it is used to keep some industrial process from reaching too high a temperature; and then return it to the river. The second law of thermodynamics ensures that the water coming back to the river has a higher temperature than it had when it left the river. The temperature increase is frequently as high as 10 or 12°C. The added heat raises the temperature of the river, where it often has serious effects on aquatic life. These effects are (1) The higher temperature results in a decrease in the concentration of dissolved oxygen. We have already seen the harm that this can do. In water with a relatively high BOD, the effects will of course be worse, since the amount of oxygen there is already low. (2) The higher temperature increases the rate of chemical reactions. Nearly all body processes are chemical reactions, and when the rates are increased, natural processes are greatly upset. (3) All fish have temperature limits beyond which they cannot survive. Thermal pollution can cause these limits to be exceeded.

17-15 Pesticides

Pesticides are synthetic organic compounds that are used to kill insects, fungi, weeds, rodents, worms, etc. Many of them are sprayed or dropped onto large or small areas. Eventually, those that do not rapidly decompose are washed by the rain into rivers and lakes, or soak through the soil into underground water systems. Pesticides are highly important for agriculture. Without them, crop yields in many places would be drastically lowered. Even with pesticides it has been estimated that one-third of the world's food crops are destroyed by pests each year. Chemically, there are four major types of pesticides.

1. *Chlorinated hydrocarbons.* Certain organic compounds, containing only carbon, hydrogen, and chlorine, have proven to be very effective in killing many

insects. Among them are DDT, chlordane, heptachlor, lindane, dieldrin, aldrin, and endrin. Formulas for some of them are shown in Table 17-6. Use of these compounds has been of major benefit to mankind in two ways: (a) In many places chlorinated hydrocarbons have been so effective in killing crop-eating insects that crop yields have shown enormous increases; and (b) chlorinated hydrocarbons have wiped out from large areas of the world a number of insects which carry such diseases as malaria and typhus. The man who discovered that DDT is an insecticide, Paul Mueller, was awarded the Nobel Prize in Medicine in 1948.

However, chlorinated hydrocarbons are a two-edged sword. Despite their enormous benefits, they also have enormous disadvantages. These disadvantages stem from the facts that (1) chlorinated hydrocarbons are not just toxic to insects, but to *all* animal life, including human beings; and (2) they are not biodegradable, and so are persistent. They remain in the environment (in the soil and the waters) for long periods of time: months and years. Even worse, there is evidence that insects develop tolerances to chlorinated hydrocarbons (through the Darwinian process of natural selection), and after some years, they are no longer effective.

One thing that increases the harmful effect of chlorinated hydrocarbons on animals is that their concentrations increase as they are carried up the food chain, because they are not biodegradable and because they are soluble in the body-fat portion of living organisms and not in water (which would wash them away in body wastes). For example, an insect might have only a small concentration of DDT; but a bird, which eats many insects, has a much higher percentage, since it accumulates all of this DDT, and can eliminate very little of it (this is called **biological amplification**; see Figure 17-21). An actual example is as follows: In Lake

Table 17-6 Some Chlorinated Hydrocarbons That Have Been Used as Insecticides

DDT
*d*ichloro*d*iphenyl*t*richloroethane
1,1,1-trichloro-2,2-bis(*p*-chlorophenyl)ethane

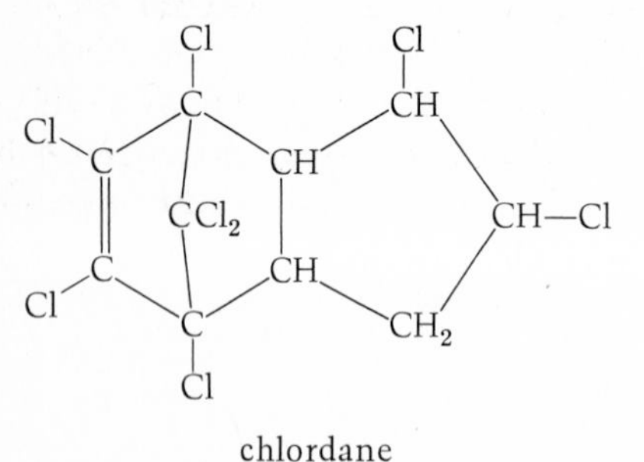

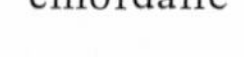
chlordane

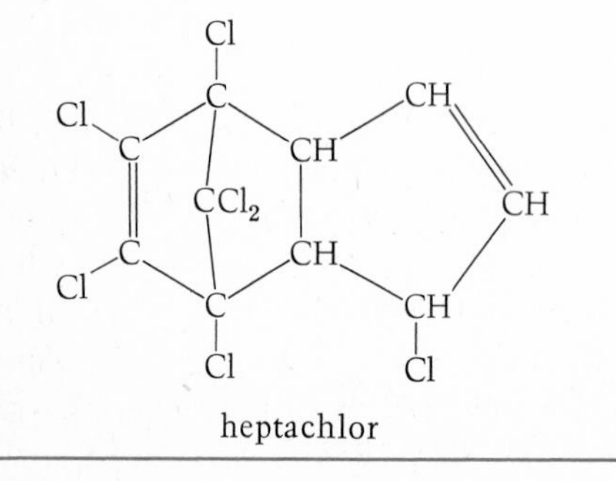

heptachlor

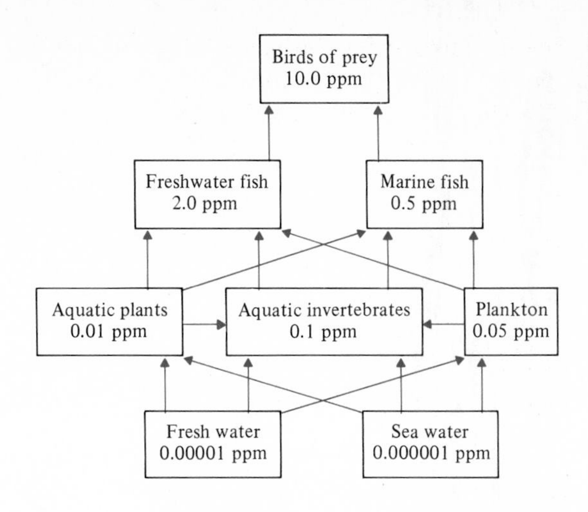

Figure 17-21 A typical food-chain concentration of DDT. [Source: *Persistent Pesticides in the Environment*, 2nd ed., by C. A. Edwards, Copyright CRC Press, Inc., 1973.]

Michigan in the 1960s the DDT level in the mud of the lake bed averaged 0.014 ppm. In the food chain: shrimps, fish, herring gulls, the levels were found to be 0.41, 5.0, and 99 ppm, respectively. This has had devastating effects on several species of wildlife, especially birds of prey, such as eagles, hawks, and falcons, which are at the top of a food chain. Many of these species have vanished from various parts of the world; others show delayed breeding, a thinning of eggshells, much breaking of eggs when laid, and a high mortality rate among young birds. The DDT in the herring gulls of Lake Michigan was sufficient to prevent 30 to 35% of their eggs from hatching.

Because of their persistence, chlorinated hydrocarbons have spread all over the world. Small quantities have been found in penguins in Antarctica and in seals near the North Pole. It is estimated that the tissues of the average human being contain about 5 to 10 ppm of DDT. There is DDT in mother's milk. As a result of all these problems, some governments have banned some or all chlorinated hydrocarbons, and their worldwide use is declining.

A closely related problem is that of polychlorinated biphenyls (PCBs). These compounds are not pesticides but are used in industry in various ways: for example, as insulators and heat-transfer fluids, in fluorescent light fixtures, and as lubricating oils for the motors of drills, air conditioners, washing machines, etc. They are excellent for these purposes because they are nonflammable (unlike the mineral oils previously used) and resistant to acids, bases, and biological degradation. Nevertheless, they are chlorinated hydrocarbons (note the similarity of the structure of a

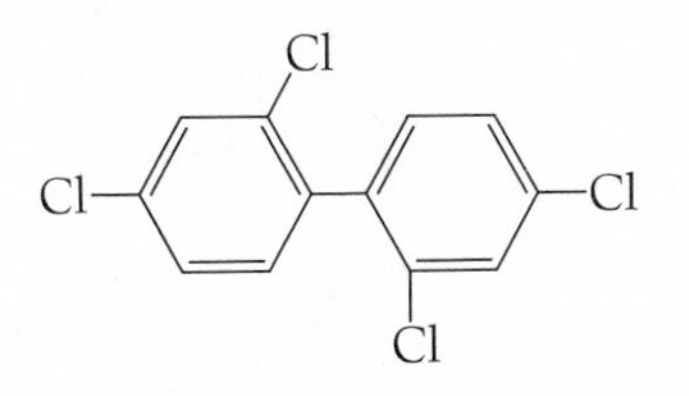

a typical PCB: 2,2′,4,4′-tetrachlorobiphenyl

typical PCB: 2,2′,4,4′-tetrachlorobiphenyl to that of DDT, Table 17-6) and cause the same kind of pollution when they reach the environment. PCBs are even less biodegradable than DDT. In this case, they are not deliberately sprayed out into the environment but reach it through accidental spillages and factory effluents, which over the years have been substantial. Like DDT, PCBs have been found in the environment all over the world and are present in mother's milk. More than 90% of the people of the United States have them in their body tissues. The U.S. government has banned the production of PCBs, effective in 1979.

2. *Organophosphates.* Because of the persistence of the chlorinated hydrocarbons, many people have turned to less persistent insecticides. Among these are the organophosphates, two of the most important of which are *parathion* and *malathion.*

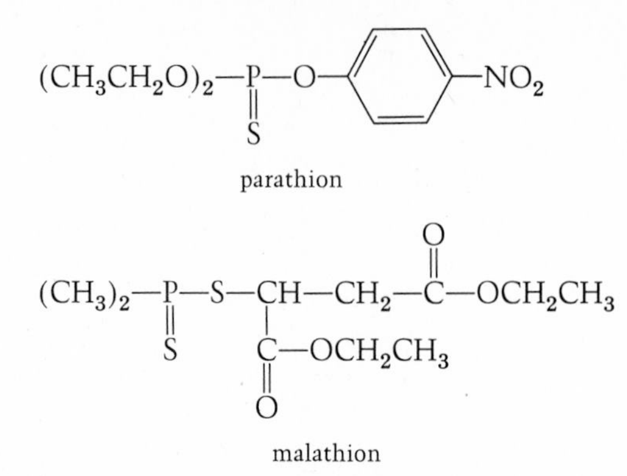

parathion

malathion

Like the chlorinated hydrocarbons, these compounds are toxic to human beings (malathion much less so than parathion), but are much less persistent, and so safer when added to the environment. Their toxic action results from attacks on the nervous system. Certain other organophosphates (these are not used as insecticides!) are much more toxic in this respect, and some of them, which are extremely lethal, have been prepared by the military for possible use as nerve gases. For example, about 1 mg of the compound called Agent GB is enough to kill an average person.

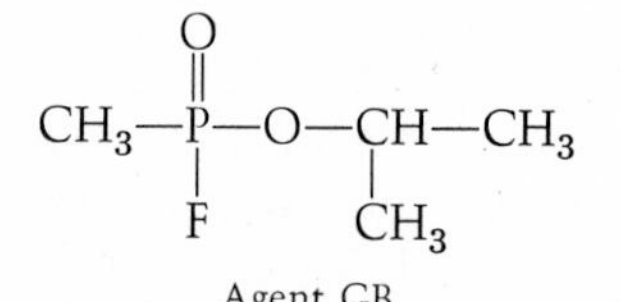

Agent GB
isopropyl methylphosphonofluoridate

3. *Carbamates.* A typical member of this class of insecticides is Sevin (carbaryl). These compounds are rapidly biodegradable.

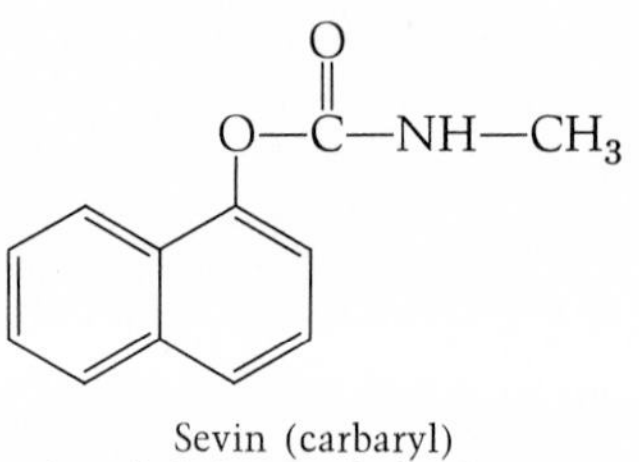

Sevin (carbaryl)
1-naphthyl-*N*-methylcarbamate

Their toxicities vary, some less toxic than DDT, some much more so.

4. *Chlorophenoxy acids.* These compounds (two typical ones are 2,4,5-T and 2,4-D)

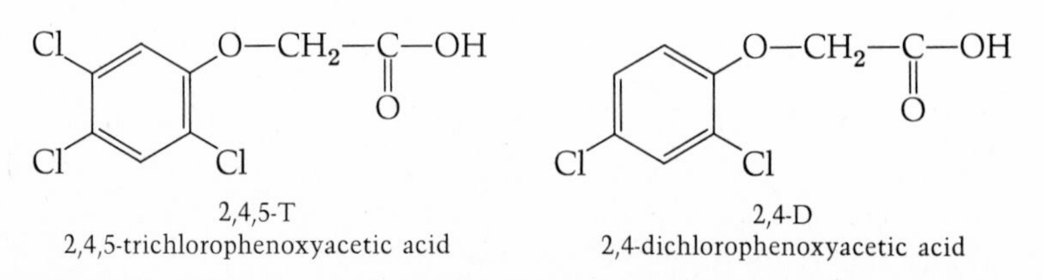

2,4,5-T
2,4,5-trichlorophenoxyacetic acid

2,4-D
2,4-dichlorophenoxyacetic acid

are not insecticides, but herbicides. They are toxic to broad-leaf plants and are used as weed killers for lawns and along the edges of highways, power lines, etc. They were also used by the military as defoliants in the Vietnam war. They kill broad-leaf plants by mimicking the action of natural plant hormones, making plants grow abnormally fast, which causes them to use up their stored energy and die.

The use of 2,4,5-T has one notable disadvantage. The compound as commercially made contains very small quantities (parts per billion) of an impurity called 2,3,7,8-tetrachlorodibenzo-*p*-dioxin (TCDD).

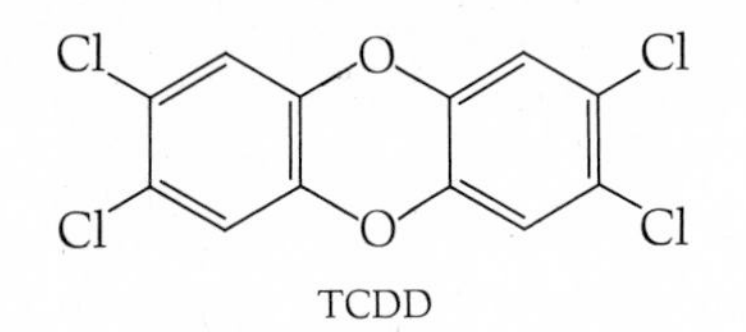

TCDD

TCDD is one of the most toxic compounds known (1 g of it is enough to kill 5 million guinea pigs), and is not biodegradable. Much 2,4,5-T has been sprayed on foliage in many parts of the world, and small amounts of TCDD are widespread in the environment. Tiny amounts of TCDD are also present in *hexachlorophene*, which was once widely used in making soaps and cosmetics, and TCDD from this source is also present in the environment. The TCDD problem was forcibly brought to public attention when a chemical plant near Seveso, Italy, blew up in July 1976 and sprayed considerable amounts of TCDD (more than 10 kg) over the neighboring area, killing more than 1000 animals, causing extensive damage to plants, and hospitalizing more than 60 people. Among other things, TCDD causes vomiting, diarrhea, and a bad skin condition called chloracne, and hundreds of the inhabitants of the area and nearby regions developed these symptoms. More than 30 pregnant women had spontaneous abortions. The entire area was evacuated, and more than 200 acres was sealed off with barbed wire and protected by armed guards. It was estimated that no one would be able to live in the area for more than 10 years. The average concentration of TCDD in the soil of the evacuated area is less than 50 ppm, but this is enough to cause all these problems. Because of the TCDD problem, the use of 2,4,5-T has been banned in the United States.

It is not surprising, with such major disadvantages accompanying the use of pesticides, that people have begun looking in other directions to see if ways can be found to control the pest problem without poisonous chemicals. Several alternative methods have been under study for a number of years, and though none of them has yet reached the point where it can even come close to supplanting the total need for pesticides, enough progress has been made so that we can be quite hopeful that the future will see a much smaller need for poisonous chemicals.

Among these methods are:

1. The use of biological controls, whereby a harmless type of organism is used to kill a harmful one. For example, insects can be killed by birds, by parasites, or by bacteria that cause diseases in insects.
2. The use of genetic controls, whereby male insects are sterilized (usually with radiation) and then allowed to mate with females. Naturally, no offspring are produced. This technique has been used to eradicate a number of insects in certain specific areas, such as the Mediterranean fruit fly (from Florida), the screwworm fly, the Khapra beetle, and the citrus black fly.
3. Quarantine, which prevents the spread of pests across borders (for example, controlling the kinds of plants that can be brought into a country also controls insects and insect larvae that may be clinging to the plants).
4. The use of sex attractants. It is known that many male insects are attracted to females by certain chemicals (called **pheromones**) which are secreted by the females. Many of these have simple structures that can be synthesized in laboratories (for example, bombykol, the sex attractant of the silkworm moth).

$$CH_3CH_2CH_2{-}CH{=}CH{-}CH{=}CH{-}CH_2CH_2CH_2CH_2CH_2CH_2CH_2CH_2CH_2{-}OH$$

bombykol
10,12-hexadecadien-1-ol

The synthetic pheromones can be released and used to attract the insects to places where they can be killed.

5. The use of juvenile hormones. All insects, at certain stages in their lives, secrete a type of compound known as a *juvenile hormone*.

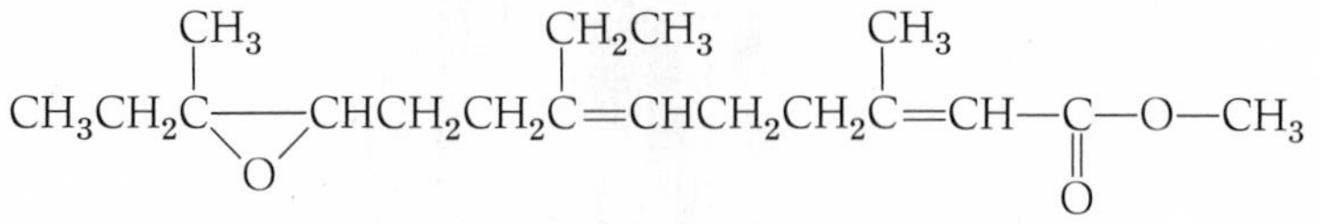

$$CH_3CH_2\overset{CH_3}{\overset{|}{C}}\underset{O}{\diagdown\!\diagup}CHCH_2CH_2\overset{CH_2CH_3}{\overset{|}{C}}{=}CHCH_2CH_2\overset{CH_3}{\overset{|}{C}}{=}CH{-}\underset{\underset{O}{\|}}{C}{-}O{-}CH_3$$

juvenile hormone from the *Cecropia* moth

However, at other stages this hormone must be absent, if normal development is to occur. Therefore, spraying with juvenile hormones means that the insects cannot develop and must die. The advantage over the more common insecticides is that juvenile hormones are not known to have any effect on any other form of life. Even for insects they are not toxic. They just don't allow them to develop.

Still, despite these and other methods, chemical pesticides will remain necessary for the forseeable future.

17-16 The Treatment of Waste Water

Raw sewage and agricultural and factory wastes must be gotten rid of somewhere. In most cases the amounts are so great that the only feasible means of disposal is to dump them into waterways: rivers, lakes, and oceans. In some cases these raw wastes are dumped directly into the waterways, but mostly (at least in the United States and other developed countries) they are subjected to waste-water treatment before they are dumped. In many places, waste-water treatment is essential, because the river or lake into which it flows will be used as a source of drinking water (sometimes, such purified waste water is reused four or five times down the course of a river), but even where this is not true, waste-water treatment is regarded as desirable because the pollutants can cause much harm (see Section 17-14). Some of the waste-water treatments are similar to those used to purify community water supplies (Section 17-13).

About three-fourths of the U.S. population is served by sewers. Almost all the water from this source is treated before being dumped into waterways. About one-third receives only **primary treatment,** which consists of screening to remove large objects; settling, to remove whatever solids settle out; and chlorination, to kill pathogenic bacteria. Primary treatment removes about one-third of the BOD and about 60% of the suspended solids, but none of the dissolved minerals. The rest of the sewage gets not only primary treatment but also **secondary treatment,** which lowers the BOD still more and also removes most of the remaining solids. There are two chief methods of secondary treatment, the *trickling filter* method and the *activated sludge* method.

In the trickling filter method, the effluent from the primary treatment is allowed to pass slowly through a bed of stones and gravel. During this time the organic matter is oxidized by bacteria that grow on the stones. In the activated sludge method, the water is aerated while it is in contact with the bacteria (Figure 17-22). Compressed air has traditionally been used for this purpose, but many treatment plants are now using pure oxygen, which allows more vigorous growth of bacteria and faster processing of the sewage.

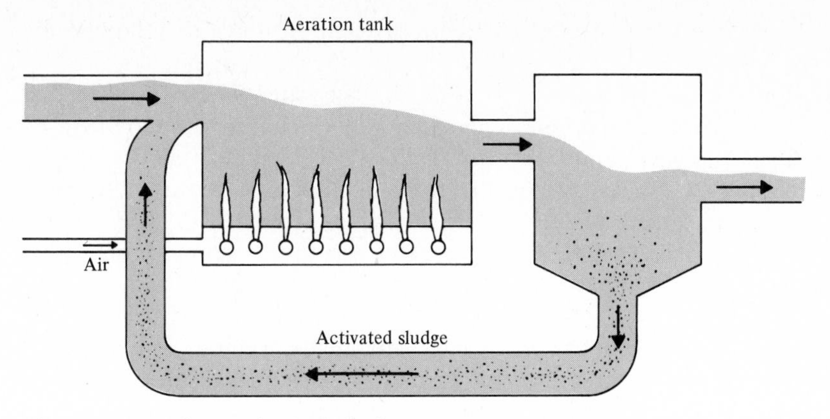

Figure 17-22 The activated sludge process.

The solids that settle out from primary and secondary water treatment are known as *sludge,* and they too present a waste-disposal problem, often the most troublesome problem in the entire waste-water treatment, because they contain large amounts of bacteria. Much of the sludge is dumped in the sea, especially that which arises from coastal cities. However, there are at least two ways in which sludge can be potentially of economic value: (1) it can be burned for energy, or digested by bacteria to produce methane, which can be burned for energy; or (2) it can be used for fertilizer. Neither of these uses is yet widespread.

Water that has undergone secondary treatment is still not very pure (for example, secondary treatment does not remove dissolved minerals), and much research has been carried out on additional treatments (called **tertiary treatment**), which would remove dissolved compounds and the remaining solids. Some of these methods involve adsorption on activated charcoal, oxidation by hydrogen peroxide, microfiltration, ion exchange, electrodialysis (Section 17-17), and coagulation with alum. Only about 1 to 2% of the U.S. waste water currently gets any kind of tertiary treatment, but this percentage will undoubtedly grow.

17-17 Desalination of Salt Water

In Section 17-10 we mentioned that many parts of the world do not have adequate supplies of fresh water, while at the same time vast quantities of salt water are all around us. Salt water is present not only in the oceans, but also in some lakes, such as the Great Salt Lake in Utah and the Dead Sea. The ocean contains about 3.5% by weight of salts, about three-fourths of which is NaCl. Salt lakes can contain more or less than this.

With so much salt water often directly adjacent to areas with short supplies of fresh water, it is not surprising that the possibility of purifying salt water should have been extensively investigated. Technically, it is very easy to purify salt water. Many methods have been discovered. The only difficulty is that all of them have high costs. For example, it might be thought that crops could be grown in arid areas such as the Sahara desert if water from the nearby Mediterranean Sea were desalinated and used for irrigation. Unfortunately, the cost of desalinating the water would be greater than the value of the crops. Nevertheless, research continues in efforts to bring the costs down, and it may be that some day, desalination will supply a significant fraction of the world's fresh water. Already, despite the costs, desalination plants are in operation in some parts of the world where other sources of fresh water are still more expensive. For example, Kuwait, a small country in the Middle East, gets virtually all its fresh water from desalination.

All desalination methods require energy, since salt water is a mixture and entropy is decreased whenever a mixture is separated (see Section 8-10). The following are the principal methods.

1. *Distillation.* This is nature's way of purifying salt water. The heat of the sun evaporates the water but not the salts, since their vapor pressures are too low. The water vapor then condenses and falls back to the earth as rain or snow. Human beings can carry out this process also (a laboratory distillation apparatus is shown in Figure 17-23), but it requires a source of heat (remember that 540 cal of heat is required to vaporize each gram of water), and this gets quite expensive on a large scale. On a small scale, the process is quite feasible; for example, ships at sea have carried water-distilling apparatus for many years. In some parts of the world (chiefly in tropical areas) direct sunlight can be used to evaporate the water without reaching the normal boiling point, but in most cases heating by a flame is needed, and some kind of fuel must be burned. Various technological devices have been designed to lower the cost, for example, by reusing some of the heat gained when the water vapor recondenses. One factor that increases the cost of any distillation process (except where direct sunlight is used) is that salt water has a lower vapor pressure at any given temperature (Section 7-8), and a higher boiling point, than pure water, so more heat is required to evaporate it, although this factor is not great (see Problem 17-37). One possible way to lower the cost of distillation is to use very cheap sources of fuel, such as garbage or sludge from sewage treatment. This solves two problems at once, since it supplies a way of getting rid of these, and a number of cities are studying the feasibility of such methods. However, burning of garbage or sludge does add to the air pollution problem, so even this is not a perfect solution.

2. *Freezing.* This method is the opposite of the previous one. When an aqueous

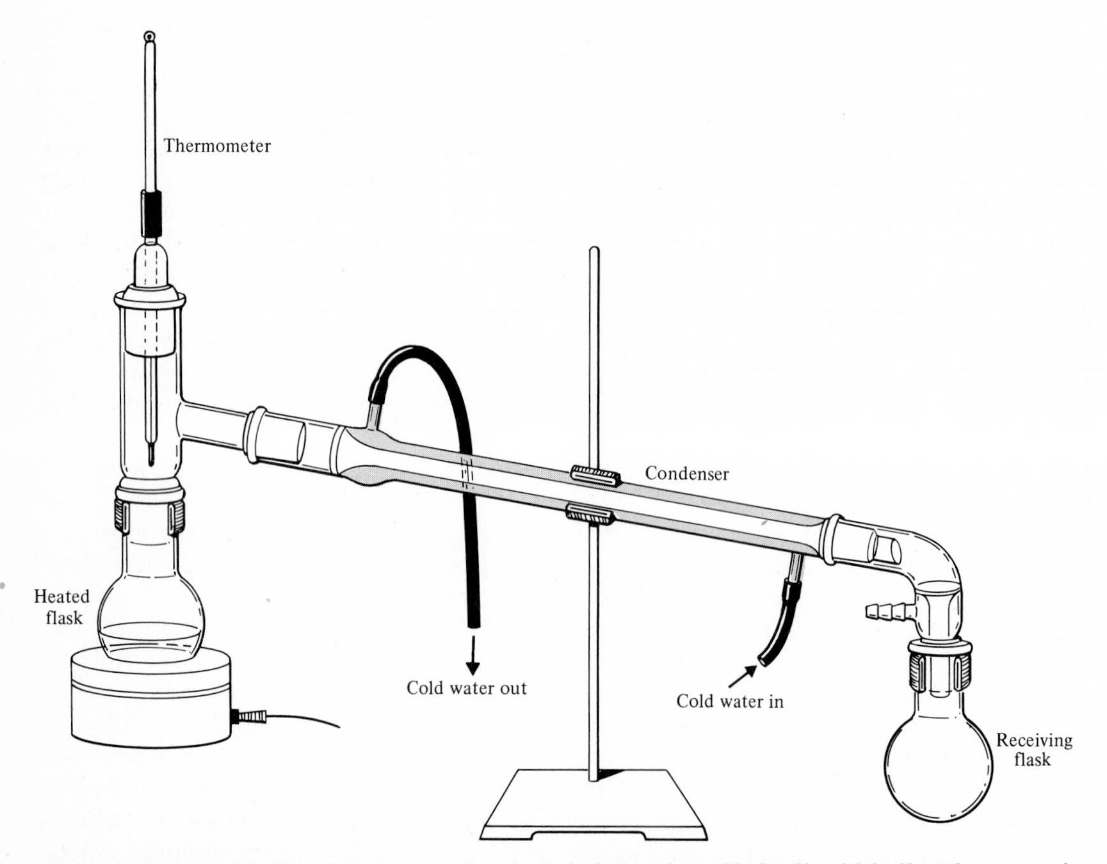

Figure 17-23 Distillation Apparatus. The flask is heated until the liquid boils. The vapor is then condensed in the condenser, which is kept at a low temperature by constantly flowing water. The condensed liquid is collected in the receiving flask.

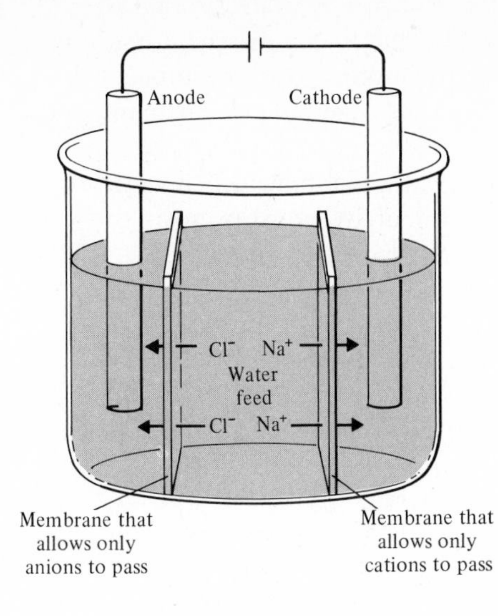

Figure 17-24 The basic electrodialysis process.

solution of salts freezes, the ice comes out almost pure: the salts do not remain in the ice, but separate out. Remelting of the ice thus gives fairly pure water. This method has some advantages over the distillation method, among them that ΔH_{fus} is much lower than ΔH_{vap}; and that salt water is much less corrosive to the apparatus at low temperatures than at high temperatures. But it also has a major disadvantage, to wit, freezing and melting are much slower than boiling and condensation, which reduces the capacity of the method. Also, the separation of the water from the salts is not as complete. However, freezing methods are in use in some places, though on a fairly small scale, about 100,000 liters/day. Freezing of the water is done by a refrigerant (see Figure 17-6), and, of course, energy is required.

3. *Ion exchange*. This process has already been discussed (Section 17-12). It is generally applicable only on a fairly small scale.

4. *Electrodialysis*. This method is not practicable for seawater, since the concentration of salts is too high, but it is quite useful for communities that have fresh water containing concentrations of salts of about 2000 to 3000 ppm (such water is called *brackish*). Many communities in the western United States have such water, and some of them use electrodialysis to purify it. The method simply consists of sending the water through a series of electrolytic cells, which contain two semipermeable membranes, which are generally made of ion-exchange resins (Figure 17-24).

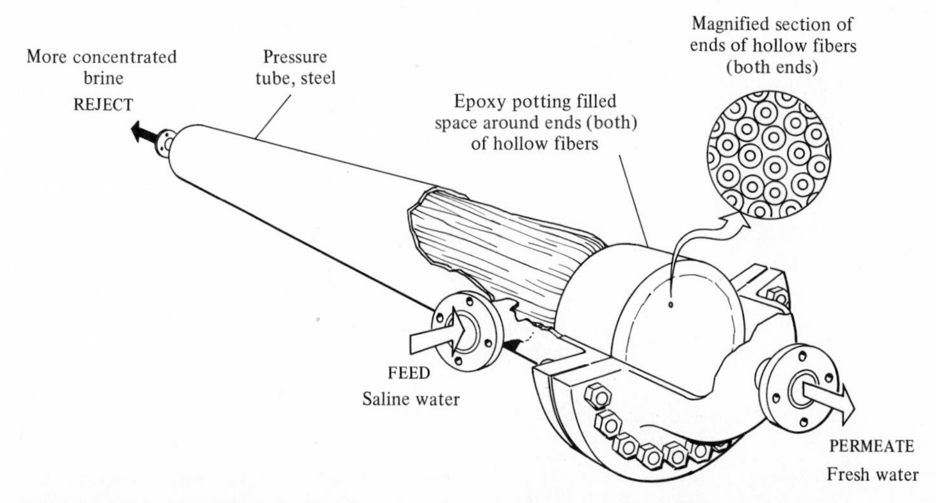

Figure 17-25 Reverse osmosis of brackish water. [Courtesy E. I. du Pont de Nemours & Co., Inc.]

One of the membranes is permeable only to water and to cations, the other only to water and to anions. When the current flows, the anions go to the anode and the cations to the cathode, but neither can get back into the central compartment, so the water in this compartment becomes purer.

5. *Reverse osmosis.* In osmosis (Section 7-10), water spontaneously flows from the dilute compartment to the concentrated compartment. In reverse osmosis, the water is made to flow the other way, by the application of high pressures (Figure 17-25).

Problems

17-1 Define each of the following terms. **(a)** aerosol **(b)** troposphere **(c)** mesopause **(d)** nitrogen fixation **(e)** nitrification **(f)** phlogiston **(g)** autoxidation **(h)** allotropic form **(i)** ozone layer **(j)** ionosphere **(k)** greenhouse effect **(l)** anthropogenic **(m)** air pollutant **(n)** water pollutant **(o)** NO_x **(p)** particulates **(q)** hydrocarbon **(r)** COHb **(s)** carcinogen **(t)** PAN **(u)** temperature inversion **(v)** octane number **(w)** knocking **(x)** hard water **(y)** hydrophobic **(z)** biodegradable **(aa)** eutrophication **(bb)** BOD **(cc)** thermal pollution **(dd)** pheromone **(ee)** electrodialysis

17-2 Compare the advantages and disadvantages of using each of the following for manned balloon flights. **(a)** hot air **(b)** hydrogen **(c)** helium

17-3 On a day when the water vapor content of the atmosphere is 1.50% and the atmospheric pressure 750 torr, what are the partial pressures of O_2, Ne, and water vapor in the atmosphere, in torr? Assuming that the percentage composition is the same, what are the partial pressures of these components at the top of Mount Everest on a day when the atmospheric pressure is 300 torr?

17-4 Of the four regions of the atmosphere (Figure 17-2): **(a)** Which has the largest volume? **(b)** Which has the largest number of molecules?

17-5 Water is a liquid at ordinary temperatures and a solid below 0°C. How is it possible for the atmosphere to contain water in the gaseous state at ordinary temperatures, especially on a cold winter day when the temperature drops below 0°C?

17-6 Write equations for the reactions of N_2 at high temperatures with Mg, Al, Zn, and B. In each case, a binary nitride is formed.

17-7 At 300 atm and 400°C in the Haber process

$$N_2 + 3H_2 \rightleftharpoons 2NH_3$$

the mixture at equilibrium contains 47.0 mole % NH_3, 13.0 mole % N_2, and 40.0 mole % H_2. Calculate K_c and K_p at 400°C, assuming ideal behavior. Will K_c change if the pressure is increased to 450 atm? If the temperature is increased to 500°C?

17-8 An oxygen tank has an approximate volume of 55 liters. Calculate the weight of oxygen that can be put into this tank at a temperature of 25°C and a pressure of 500 lb/in^2, assuming ideal behavior (14.7 lb/in^2 = 1 atm).

17-9 Assume that in the reaction $O_2 \xrightarrow{h\nu} 2O$, the wavelength of light used is 1500 Å. Calculate the amount of energy in calories that would be needed to decompose 1.00 mole of oxygen.

17-10 Explain why the ozone molecule is angular and not linear.

17-11 Before the actual structure of ozone was known, it was proposed that the three oxygen atoms might be connected to each other in a ring. Draw a Lewis structure for this possibility. How do we know that ozone does not actually have this structure?

17-12 If the bond energy of the C—Cl bond in CF_2Cl_2 is 80 kcal/mole, what frequency of ultraviolet light possesses the right amount of energy to cleave it?

17-13 The amount of CO_2 in the atmosphere has increased about 10% in the last 100 years because of the large-scale burning of coal, oil, and natural gas. Water is also a product of this combustion, at least from oil and natural gas. Why has the water-vapor content of the atmosphere not also increased?

17-14 A single reaction is the ultimate source of virtually all the anthropogenic NO_x in the atmosphere, and also some of the naturally occurring NO_x. Write the equation for this reaction.

17-15 What is the ultimate source of the anthropogenic SO_x in the atmosphere?

17-16 If a certain type of coal contains 2.5% sulfur, how much limestone must be added per metric ton of coal to remove the sulfur completely (Section 17-9)? How many grams of $CaSO_4$ will be formed by this process? (1 metric ton = 1000 kg.)

17-17 Two cities with a major smog problem are Los Angeles and London. What kind of legislation would you recommend to reverse this, in each case?

17-18 If a practical electric automobile can be built, would this completely solve the air pollution problem?

17-19 Suppose that you are the manager of a chemical plant. Your laboratory researchers inform you that this compound,

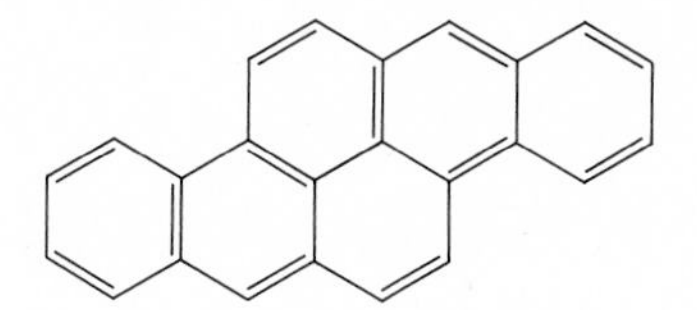

shows promise as a possible preservative for meats. Would you allow them to continue to work on it?

17-20 Why does unleaded gasoline cost *more* than gasoline containing lead tetraethyl?

17-21 What is the difference between temporary and permanent hard water? How may each be softened?

17-22 What is the purest form of natural water? Is it chemically pure?

17-23 If a 100-liter sample of water contains 200 ppm of hardness, caused entirely by Ca^{2+} and HCO_3^- ions, how many grams of $Ca(OH)_2$ should be added to soften it?

17-24 Show how the total ionic content of hard water is decreased when $Ca(OH)_2$ is used to remove the hardness, but not when Na_2CO_3 is used.

17-25 A sample of ground water was found to have a hardness of 135 ppm. Assuming that the hardness was made up of equimolar quantities of $Ca(HCO_3)_2$ and $Mg(HCO_3)_2$, how many grams of $Ca(HCO_3)_2$ and $Mg(HCO_3)_2$ are present in 1 million grams of water? (See the footnote on page 564.)

17-26 What are the advantages and disadvantages of phosphates and polyphosphates in detergents?

17-27 What is the difference between a soap and a detergent?

17-28 Of the processes used for the treatment of community water supplies (Section 17-13), which are physical processes and which are chemical processes? Write equations for the chemical processes.

17-29 Why are Cl_2 and F^- added to many public water supplies?

17-30 Discuss five methods used to control pests other than chemical pesticides.

17-31 Why are oxygen-demanding wastes considered to be a water pollutant?

17-32 The concentration of dissolved oxygen in a sample of waste water at 20°C decreases from 8.8×10^{-3} g/liter to 3.1×10^{-3} g/liter in 5 days. What is the BOD in ppm?

17-33 Explain why wastes from each of the following might cause pollution problems when dumped into rivers or lakes.
(a) a slaughter house
(b) an electricity-generating plant
(c) an area with large dairy and chicken farms
(d) a factory making electrical equipment
(e) an area heavily planted with crops
(f) uranium mines

17-34 A state legislator decides to solve the water pollution problem in his state by introducing a bill which would require that all water coming out of household taps be chemically pure H_2O. What is your opinion of this proposed legislation? Are there other possible laws that might handle the problem better?

17-35 If the concentration of mercury in a sample of tuna fish is 0.75 ppm, how many grams of mercury are present in a $6\frac{1}{2}$-oz can? (16 oz = 1 lb.)

17-36 If the total volume of the earth's oceans is 1.4×10^9 km^3 and the ocean contains 3.5% by weight of salts, what is the weight in kg of all the salts in the oceans? If 75% of it is NaCl, how many moles of NaCl are there in the sea? (The density of seawater is 1.025 g/ml.)

17-37 Suppose that you have a liter of salt water, whose density is 1.025 g/ml and which contains 3.50% by weight of NaCl. Calculate the boiling point of the salt water. Assuming that the heat capacity of the salt water is the same as pure water, 1.000 cal/g · deg, calculate the increased amount of heat required to evaporate the liter of salt water compared to a liter of distilled water.

17-38 Explain, using equations, how water containing a substantial amount of $FeCO_3$ can be deionized with ion-exchange resins.

17-39 What is the freezing point of seawater, assuming that the 3.5% of salts it contains is all NaCl? The density of seawater is 1.025 g/ml.

Additional Problems

17-40 What are the advantages and disadvantages of using chlorinated hydrocarbons as insecticides?

17-41 **(a)** Why does the burning of natural gas produce less air pollution than the burning of oil or coal? **(b)** If natural gas causes less air pollution, why don't we simply ban the burning of oil and coal altogether and substitute natural gas for all combustion uses? (See Section E-3.)

17-42 Assume that you have a clothes washing machine that holds 60 liters of water. The water in your area has a hardness of 175 ppm. If you wash with ordinary soap powder (not a detergent), how many grams of soap are required to precipitate all the Ca^{2+} from one 60-liter wash load before the rest of the soap can be used for washing? (Assume that the formula for soap is $C_{18}H_{35}O_2Na$.)

17-43 A proposed solution to plant nutrients as a water pollutant is to ban the use of fertilizers, detergents containing phosphate and polyphosphate ions, and pesticides. Discuss the consequences of such a ban.

17-44 The ion N_2^+ is found in the ionosphere. Show all the molecular orbitals in this ion, as we did in Section 14-20.

17-45 When distillation is used to desalinate seawater, each gram of water boiled requires 540 cal of heat. However, when the steam is condensed 540 cal/g are given back. Why, then, is any energy required for this process? Why not simply use the 540 cal given back to evaporate the next gram, etc.?

17-46 New York City uses about 7.0×10^9 liters of water per day. If we consider that the maximum permissible salt content of drinking water is 500 mg/liter, calculate the maximum number of moles of NaCl in 1 day's consumption of water in New York City.

17-47 After it was realized that PCBs in the environment caused a pollution problem, the government decided to prohibit the use of these compounds and to retrieve and destroy as much as possible of those which had already been used in the manufacture of products. Among the products containing PCBs are 800 million fluorescent light fixtures in use throughout the United States. How can the government retrieve 800 million fluorescent light fixtures?

17-48 Show how the SSTs and Freons might cause depletion of the ozone layer.

17-49 If mankind ever achieves a situation where we get all or nearly all our energy from sources that do not involve combustion (Section E-4), will we have solved the air pollution problem?

17-50 What are the harmful effects of high atmospheric concentrations of NO_2, SO_2, and CO?

17-51 About 180 million metric tons of N_2 are removed from the atmosphere each year by natural fixation processes. How does the N_2 get back into the atmosphere?

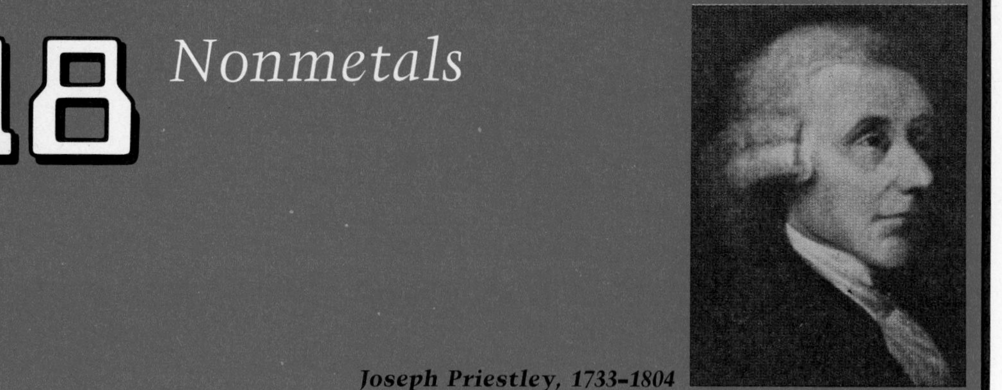

18 Nonmetals

Joseph Priestley, 1733–1804

Portrait by Fuseli; Dr. Williams's Library

In Section 13-13, we divided the elements into five classes: metals, nonmetals, metalloids, noble gases, and hydrogen. This chapter deals with some of the descriptive chemistry of the nonmetallic elements and their compounds, that is, with the four classes of elements that are not metals. In Chapter 19, we shall turn our attention to the metals.

The chief characteristic that these elements have in common is their reluctance to give up electrons, although some of them are more reluctant than others. In general, they tend to form covalent rather than ionic bonds, and when they do form ionic bonds they do so by gaining electrons, not by losing them. Three of these elements (helium, neon, argon) do not form any bonds at all.

Following a custom used by most chemistry books for many years, we will discuss these elements according to their positions in the periodic table. We begin with hydrogen and then move from group IIIA to group 0.

18-1 Hydrogen

Hydrogen is one of the most important elements, as well as the most unusual. There is no other like it. In some of its properties it resembles the metals; in others, the

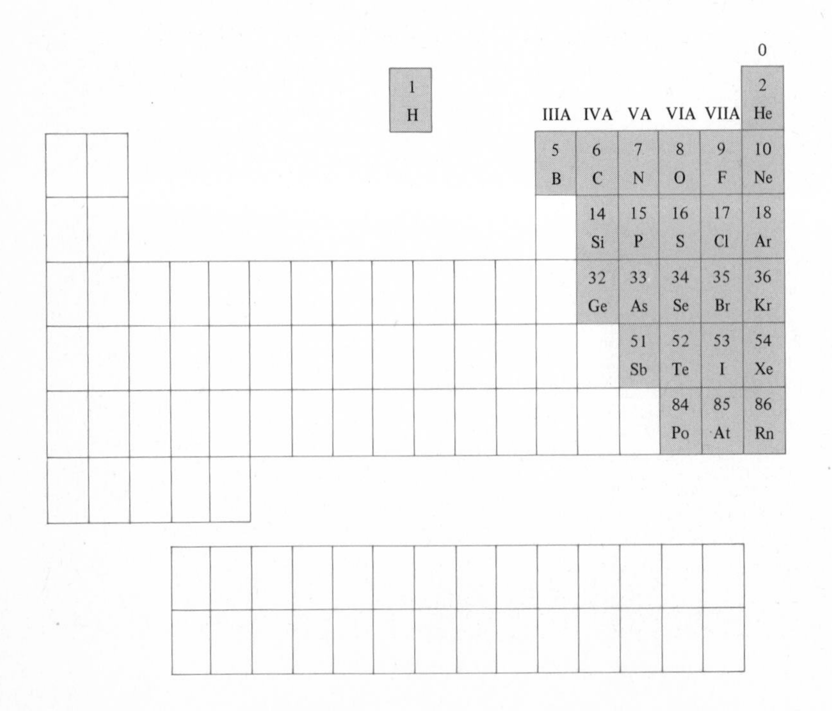

nonmetals, yet unlike the metalloids, it cannot be said of hydrogen that its properties are in between those of metals and nonmetals.

OCCURRENCE. Hydrogen as the free element H_2 is not found on this planet except in very small quantities (Table 17-1). The chief form in which it occurs on Earth is in the compound water. Hydrogen is also an important constituent of the alkanes, which make up petroleum, and as we shall see in Chapter 21, it is one of the major essential ingredients of all living things. There is very little hydrogen in the rocks and minerals that make up most of our planet (the hydrogen here is mostly in the form of water which is chemically bound to some of the rocks). Outside the Earth, hydrogen is the most abundant element in the universe. It makes up most of the sun and of other stars, and there are vast clouds of hydrogen floating in space between stars. It has been estimated that about 90% of all atoms in the universe are hydrogen atoms.

PHYSICAL PROPERTIES. Hydrogen (H_2) is a colorless, odorless gas. Because it has the lowest molecular weight of any gas, it also has the lowest density at any given temperature and pressure. For this reason, it was used in the early balloon experiments (Section 17-2) and indeed in blimps and dirigibles until comparatively recently. However, hydrogen is flammable, and the burning of the dirigible *Hindenburg* in 1937 (Figure 18-1) convinced most balloonists to switch to helium, even though the latter has twice the molecular weight of H_2.

Figure 18-1 Burning of the *Hindenburg*. [The Bettmann Archive]

H_2 is nonpolar (why?); and this, combined with its very low molecular weight, gives it a very low boiling point (−252.5°C) and freezing point (−259°C). Its nonpolarity also causes it to be essentially insoluble in water and other polar solvents.

PREPARATION. In the laboratory hydrogen can be easily prepared by the reaction between acids and metals. In Section 16-7 we learned that strong acids react in this way with all metals that have a lower standard reduction potential than H_2: for example,

$$Zn + 2H_3O^+ \longrightarrow H_2 + Zn^{2+} + 2H_2O$$
$$Sn + 2H_3O^+ \longrightarrow H_2 + Sn^{2+} + 2H_2O$$

For metals with very low standard reduction potentials (near the bottom of the table), water is a strong enough acid for this reaction: for example,

$$2Na + 2H_2O \longrightarrow H_2 + 2Na^+ + 2OH^-$$

Iron does not react with water at room temperature, but hydrogen can be prepared by the reaction of iron with steam:

$$Fe + 2H_2O \xrightarrow{\Delta} H_2 + Fe^{2+} + 2OH^-$$

The most important industrial method for the preparation of hydrogen is the process that begins with methane (Section 17-3). Hydrogen is also prepared by the electrolysis of water and in several other ways.

CHEMICAL PROPERTIES. Because the energy of the H—H bond is high (103 kcal/mol), H_2 is relatively stable, that is, unreactive. It does not react with O_2 at room temperature, but when ignited it readily burns, or explodes if it is mixed with oxygen or air in the right proportions. At high temperatures it is a reducing agent, and will, for example, reduce many metallic oxides to lower-valent oxides or to the metals. Several elements, among them germanium and vanadium, were first prepared by this method. In the presence of the proper catalyst, H_2 reduces the double bonds of organic compounds (Section 20-3).

COVALENT COMPOUNDS OF HYDROGEN. Hydrogen is present in most organic compounds and in a substantial number of inorganic compounds. From the electronic structure of the hydrogen atom $1s^1$ we might expect hydrogen to form bonds in three different ways: (1) by loss of an electron to give H^+, (2) by gain of an electron to give H^-, and (3) by sharing its electron in a covalent bond. In fact, only the last two of these possibilities are actually found. Positive hydrogen ions H^+ (bare protons) do not exist in any compounds (they are present on Earth only in certain gaseous ionic beams, as in a mass spectrometer, and in the ionosphere of the upper atmosphere). In most of its compounds, hydrogen forms covalent bonds, always with a valence of 1. In Section 10-2 we learned that all such compounds can act as Brønsted acids, some strong, such as HCl and HNO_3; some weak, such as HF and H_2CO_3; and some feeble, such as NH_3 and CH_4. In an ion such as H_3O^+, all three of the hydrogen atoms are connected to the oxygen by covalent bonds. In all these compounds the bonding is essentially the same, except that electronegativity differences cause each bond to have a different degree of polarity. In all of them, hydrogen has an oxidation number of +1.

METAL HYDRIDES. Hydrogen forms binary compounds with a number of metals. Because metals have lower electronegativities than hydrogen, such compounds are called **hydrides,** and the hydrogen is written on the right (for example, sodium hydride, NaH). Some of these hydrides are ionic compounds in which hydrogen is present as hydride ions, H^-. Hydrogen has a low electron affinity (Table 13-14), much lower than say bromine or chlorine, and so it forms simple ionic hydrides only with the group IA metals and with four of the group IIA metals (CaH_2, SrH_2, BaH_2, RaH_2). These are ionic salts, and conduct electricity in the molten state (hydrogen, of course, comes off at the anode). Metal hydrides can be prepared by heating the metal directly with hydrogen at a high temperature (for example, 500°C). We expect most ionic compounds to be soluble in water, but in this case they *react instantly* with water: for example,

$$\underset{\text{base}_1}{H^-} + \underset{\text{acid}_2}{H_2O} \longrightarrow \underset{\text{acid}_1}{H_2} + \underset{\text{base}_2}{OH^-}$$

because H_2 is a weaker Brønsted acid than H_2O. Obviously, these hydrides will also react with any acid stronger than water. Several other known binary metal hydrides are covalent compounds (for example, BeH_2, AlH_3, SnH_4) and not ionic compounds. There are also a number of metallic hydrides with more than two elements present; these are usually known as complex metal hydrides. The most important of them, lithium aluminum hydride ($LiAlH_4$) and sodium borohydride ($NaBH_4$), are widely used as reducing agents. The oxidation number of hydrogen in metal hydrides is −1. *Thus, hydrogen can exist in oxidation states of −1, 0, or +1.*

18-2 Boron

The element boron is the only element of group IIIA that is not a metal. It does not occur free in nature, and is obtained chiefly from natural supplies of borates [one of these is *borax*, $Na_2B_4O_5(OH)_4 \cdot 8H_2O$], which are found in large quantities in California. The element occurs in several allotropic forms, one of which (called α-rhombohedral boron) is made up of B_{12} units in which the boron atoms are located at the corners of a regular icosahedron (Figure 18-2). Note that each boron atom is covalently bonded to five other boron atoms, although a boron atom only has three outer-shell electrons. The type of bonding here is similar to that in B_2H_6 (see below). The B_{12} units have a relatively high molecular weight, and there are fairly strong intermolecular forces holding the B_{12} units together in the crystal lattice. Consequently, boron has a high melting point, ~2250°C. Boron is fairly inert. It does not react with acids and is only slowly oxidized by strong oxidizing agents. The element is not a conductor of electricity, like metals, but is a semiconductor, like silicon and germanium (Section 19-3).

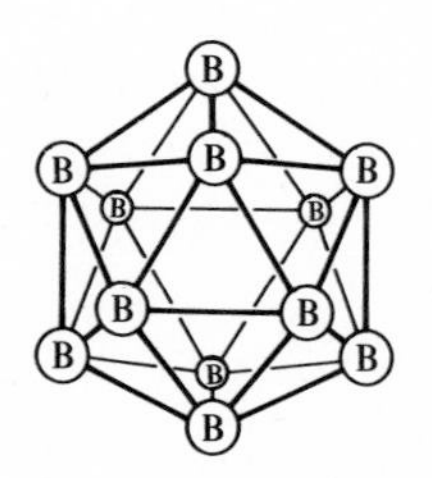

Figure 18-2 A molecule of B_{12}, α-rhombohedral boron.

Compounds of Boron. Because its ionization energy is high and its electron affinity is low, boron forms only covalent bonds, and the valence is always 3 (unless it has a formal charge, as, for example, in BH_4^-). The oxidation number of boron in its compounds can vary from +3 to −3; most often it is +3. As we saw in Section 14-5, when boron forms three covalent bonds, it has only six electrons in its outer shell, so that such compounds are Lewis acids.

The bonds are sp^2-hybridized, and the EPRT tells us to expect planar molecules with bond angles of 120°. All four boron halides are known, and are all Lewis acids. When a Lewis acid such as BF_3 combines with a negative ion, ions such as BF_4^- are obtained.

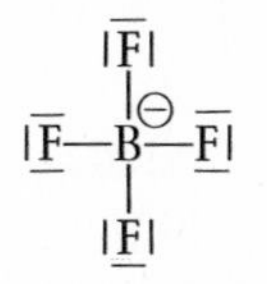

Many of these ions are quite stable. The geometry here is tetrahedral, with sp^3 hybridization. An important example is the borohydride ion, BH_4^-, a component of $NaBH_4$, which was mentioned in Section 18-1.

The most important oxide of boron is B_2O_3 (there is also an oxide BO). When B_2O_3 is added to water, boric acid is obtained. This acid has the structure $B(OH)_3$, though it is often written as H_3BO_3 to emphasize its acidic properties. It is a weak acid, with $K_a = 7 \times 10^{-10}$. Despite its three hydrogens, it is a *monoprotic* acid.

Borane BH_3 is not a stable compound, but exists only in the form of its dimer, B_2H_6 (*diborane*). B_2H_6 is an example of an **electron-deficient** compound, whose structure puzzled chemists for many years. To see what the trouble was, let us try to draw a Lewis structure for diborane. Each boron has three outer-shell electrons, and each hydrogen has one, making a total of 12 electrons, or six pairs. But B_2H_6 would seem to require seven pairs: six for H—B bonds and one for a B—B bond. Consequently, a Lewis structure cannot be drawn for this compound, and yet it exists. Today we explain the bonding in diborane as follows. The two borons and four of the hydrogens lie in a plane (Figure 18-3a). These four B—H bonds are ordinary covalent bonds (x-ray crystallography shows that these B—H distances are 1.19Å), and use up four of the six pairs. The other two electron pairs are each in a **three-center-two-electron bond;** that is, a single bond, with two electrons, covers the three atoms B—H—B. This is a weaker bond and therefore longer (the H—B distance is 1.33Å). The two hydrogen atoms in these bonds lie above and

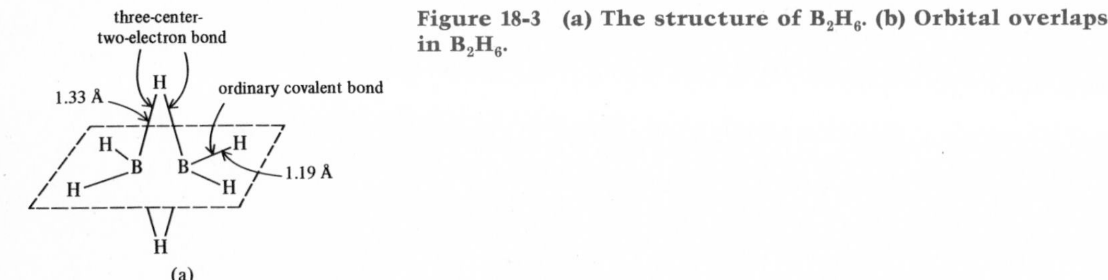

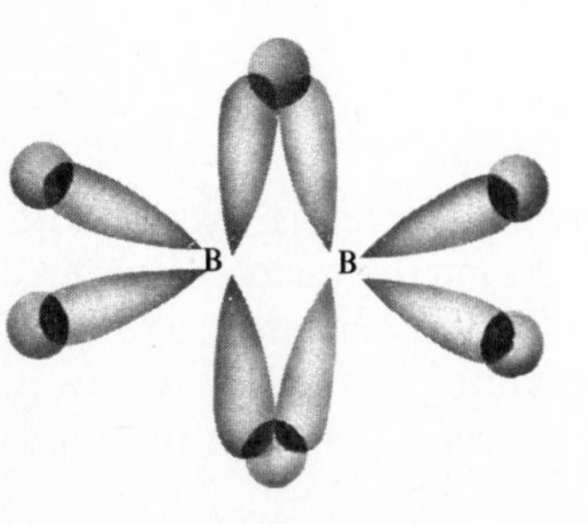

Figure 18-3 (a) The structure of B_2H_6. (b) Orbital overlaps in B_2H_6.

below the plane. Each boron atom uses four sp^3 orbitals, and each hydrogen atom, of course, uses its single 1*s* orbital. Each of the four planar hydrogen atoms overlaps with one of the sp^3 orbitals, but the other two hydrogen 1*s* orbitals each overlap with *two* of the boron sp^3 orbitals (Figure 18-3b). A number of other boranes are known (for example, B_4H_{10}, B_5H_9, B_5H_{11}). All of them are electron deficient and have three center–two electron bonds.

18-3 Carbon, Silicon, and Germanium

Silicon, the second most abundant element by weight on the earth's crust, is found in sand and in silicate rocks. Carbon is much less abundant but is still found in large quantities in carbonate rocks, in coal and petroleum, in the CO_2 in the atmosphere, and in all living things. Germanium is fairly rare. All three of these elements form four covalent bonds per atom, as we would expect from their positions in the periodic table. However, there is one major difference. Carbon can form double and triple bonds as well as single bonds (Section 14-8), but silicon and germanium can only form single bonds. This has an important influence on the types of compounds which these elements can form, as we shall see. The oxidation numbers of each of them can vary from +4 to −4.

THE ELEMENTS. There are two allotropic forms of pure carbon. Both of them have structures which depend on the fact that carbon forms four covalent bonds. The thermodynamically more stable form (lower ΔG) is *graphite*, a soft black solid which is used in "lead" pencils (the "lead" is not lead but graphite mixed with clay). Graphite is made up of carbon atoms connected to each other in flat layers composed of six-membered rings (Figure 18-4). Within each layer, each carbon atom is

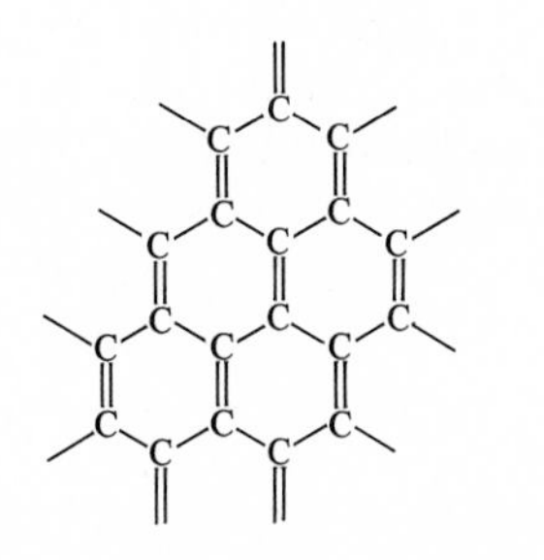

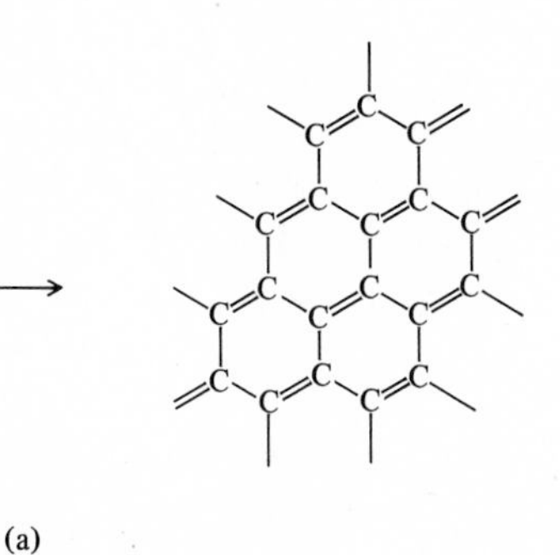

(a)

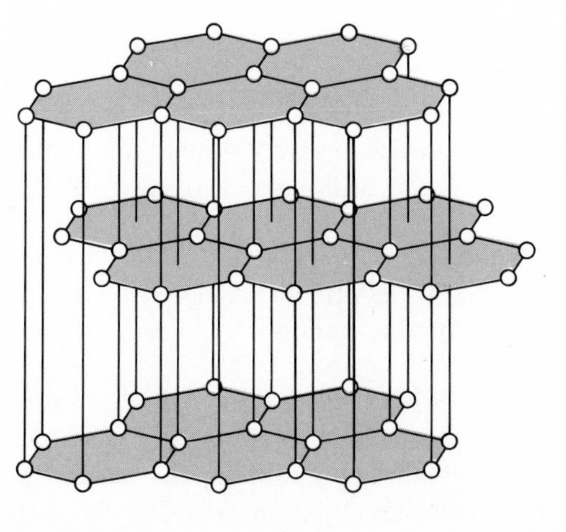

(b)

Figure 18-4 Structure of Graphite. (a) A portion of a single sheet, showing two canonical forms. (b) The stacking of several sheets.

connected to other carbon atoms by four covalent bonds. There are three canonical forms, two of which are shown in Figure 18-4a. In each form, each atom uses sp^2 orbitals to form two single bonds and one double bond, but the resonance makes all the bonds equal, and x-ray crystallography shows that all of the C—C bond distances are 1.415 Å. The fact that all the carbon atoms are using sp^2 orbitals makes the entire layer planar and all the angles 120°. The layers are held together (Figure 18-4b) by van der Waals forces, which of course are not very strong and allow the layers to slide past each other, thus accounting for the softness of graphite, and allowing it to be useful in pencils, since the layers easily flake off. Graphite is also used as a "dry" lubricant.

The other allotropic form of carbon is *diamond*, which has the three dimensional structure shown in Figure 15-1a. The rigidity of this structure, which continues throughout the entire crystal of a diamond, makes diamond the hardest substance known. Unlike graphite, diamond has no resonance and no double bonds. Diamond is thermodynamically less stable than graphite, and can only be produced from it by the application of very high pressures. Consequently, diamond is found in nature only in these few places where natural high-pressure processes have had a chance to work on graphite. It is only recently that human beings have been able to imitate these conditions and produce artificial diamonds. Because of its hardness, diamond is an important industrial abrasive.

There are other forms of carbon. Charcoal and soot were long regarded as "amorphous" forms of carbon, but today it is known that these substances are made up of microcrystals of graphite. Coal is impure carbon, containing varying amounts of organic compounds (mostly aromatic) as impurities.

Silicon and germanium are ordinarily produced by reduction of their oxides. Both of them have structures similar to that of diamond. There is no structure similar to graphite, because that requires double bonds which these two elements cannot form. Silicon and germanium are semiconductors, and are used to make transistors (see Section 19-3).

Silicon carbide, SiC, is a giant three-dimensional molecule with a structure similar to that of diamond. Like diamond, it is very hard and is used industrially as an abrasive, under the name *carborundum*. Another important industrial abrasive, boron carbide (B_4C), also has a giant-molecule structure, but the pattern of the atoms is not similar to that of diamond.

CHAINS OF ATOMS. More than any other elements, C, Si, and Ge can form compounds containing chains of identical atoms. For Si and Ge, this property is limited, and the longest simple chain compounds have no more than nine atoms in a row, but carbon forms chain compounds that vary from two or three carbon atoms (for example, ethanol, propane)

$$\underset{\text{ethanol}}{CH_3—CH_2—OH} \qquad \underset{\text{propane}}{CH_3—CH_2—CH_3}$$

to thousands and even millions of carbon atoms. It is the ability to form such chains that makes carbon such an important element, and makes the chemistry of carbon compounds (called *organic chemistry*) a separate branch of chemistry. We will discuss the compounds of carbon in Chapters 20, 21, and 22. Here we will limit ourselves to the oxides of carbon, which are often considered to be inorganic compounds, despite the fact that they contain carbon.

OXIDES. There are two important oxides of carbon, carbon monoxide and carbon dioxide, as well as several others that are less important. CO and CO_2 are both gases, as expected for low-molecular-weight nonpolar compounds (CO has a low dipole moment; CO_2 has none at all). Both are colorless and odorless. CO is very poisonous, as we saw on page 553, but CO_2 is not, although CO_2 can be fatal if it replaces most of the oxygen in a room. CO is produced by incomplete combustion, CO_2 by complete combustion. CO is insoluble in water; CO_2 has an appreciable solubility. Part of the CO_2 reacts with water to form carbonic acid H_2CO_3:

$$CO_2 + H_2O \rightleftharpoons H_2CO_3$$

Carbonic acid is not a stable compound. It cannot be isolated away from water. If a solution of H_2CO_3 is heated, the equilibrium moves to the left, and CO_2 is driven out. Despite the fact that H_2CO_3 does not have an independent existence, its salts do. It is a diprotic acid and forms two series of salts. Salts that result from loss of one proton are called **bicarbonates***; loss of both protons gives **carbonates.**

$$H_2CO_3 \rightleftharpoons H^+ + \underset{\text{bicarbonate ion}}{HCO_3^-}$$

$$HCO_3^- \rightleftharpoons H^+ + \underset{\text{carbonate ion}}{CO_3^{2-}}$$

Carbonates (chiefly $CaCO_3$ and Na_2CO_3) are important constituents of many rocks. Limestone, marble, and chalk are all forms of $CaCO_3$. The bicarbonate ion is a major constituent of hard water. We have already seen that sodium carbonate is washing soda (Section 17-12) and sodium bicarbonate is baking soda (Section 11-1).

Silicon also forms an oxide, silicon dioxide (SiO_2), also called *silica* (there is no SiO). There are two forms of pure SiO_2 found in nature: the minerals *quartz* and *cristobalite*; sand is an impure form. Silicon dioxide is completely different from CO_2. Though CO_2 is a gas, SiO_2 is a high-melting solid (mp 1600 to 1700°C). Furthermore, CO_2 is soluble in water; SiO_2 is insoluble. Why is there such an enormous difference in physical properties between CO_2 and SiO_2? The answer comes from the fact that carbon forms double bonds and silicon does not. The Lewis structure of carbon dioxide is

$$|\overline{O}{=}C{=}\overline{O}|$$

so CO_2 is a small molecule with a molecular weight of 44. SiO_2 cannot have such a structure because silicon forms no double bonds. Therefore, the silicon must be connected to four oxygen atoms.

$$\begin{array}{c} | \\ O \\ | \\ -O-Si-O- \\ | \\ O \\ | \end{array}$$

Each oxygen has one valence left over, and this can only combine with another silicon atom, so that SiO_2 is a three-dimensional polymeric molecule (Figure 18-5) of very high molecular weight.

When, in the nineteenth century, it was realized that silicon was just below carbon in the periodic table, and that life on earth is based on carbon compounds, it was proposed by some people that a form of life based on silicon rather than carbon might be present on other planets (several science-fiction stories, most notably "A Martian Odyssey," by Stanley G. Weinbaum, have used this idea). The idea seemed plausible at that time, but we now know that it is very unlikely because of the simple fact that carbon forms double bonds and silicon does not. In Chapter 21 we shall see that many compounds important to living organisms contain C=C and

*"Bicarbonate" is a name left over from more than a century ago, when incorrect notions of the structures of these salts were prevalent. For this reason, many chemists call them acid carbonates or hydrogen carbonates rather than bicarbonates.

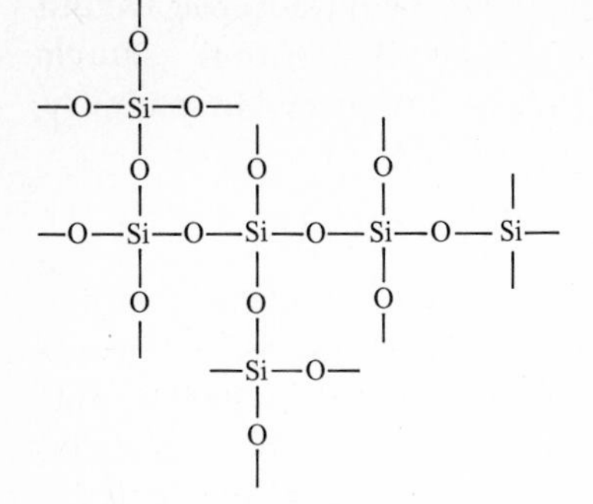

Figure 18-5 Schematic structure of SiO_2.

C=O bonds and there are no silicon analogs of these compounds. Furthermore, CO_2 itself is a crucial compound for both plant and animal metabolism. It can function in the way it does only because it is a gas and soluble in water. It is obvious that SiO_2 cannot perform any similar function. Because double bonds are necessary for aromatic rings (Section 17-7), there are no silicon analogs of aromatic compounds.

Silicate rocks, which make up the majority of the rocks on earth, vary greatly in their compositions. Some contain simple silicate ions, SiO_4^{4-}, and others contain

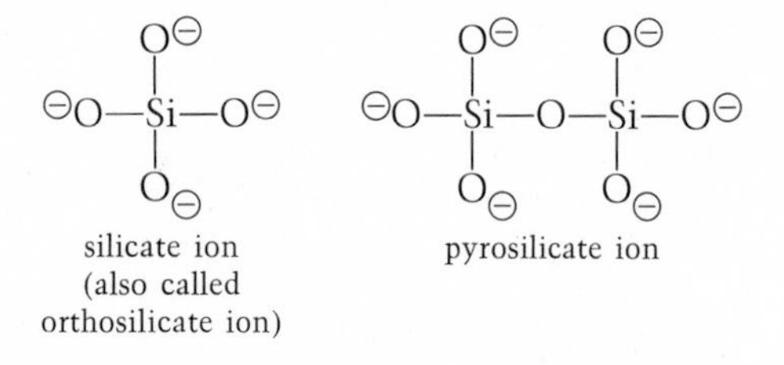

pyrosilicate ions, $Si_2O_7^{6-}$, but most silicate rocks contain two- or three-dimensional polymeric silicate ions which may contain hundreds of thousands or millions of atoms connected in chains such as that shown in Figure 18-6. Since these rocks are electrically neutral, they must also contain a sufficient number of cations to equalize all the negative charges carried by the oxygen atoms. Many cations fulfill this function, among them Na^+, K^+, Ca^{2+}, Mg^{2+}, Fe^{2+}, Mn^{2+}, Zn^{2+} and others. The combination of different forms of silicate ions, and different cations, produces many kinds of silicate rocks and minerals, and the situation is even more complicated by the fact that nature is seldom interested in preparing pure compounds, so that many individual rocks are mixtures.

Glass is a synthetic silicate, made by man. It is produced by melting SiO_2, usually in the form of sand, with carbonates (for example, Na_2CO_3, $CaCO_3$) and/or oxides. Different types of glasses require different proportions of carbonates or oxides.

Germanium oxide, GeO_2 (there is no GeO), is similar to SiO_2 in structure and properties.

OTHER COMPOUNDS OF SILICON AND GERMANIUM. Both silicon and germanium form simple hydrides and halides of the form EX_4. Silane (SiH_4) and germane (GeH_4) are gases like methane. The most important of the halides is silicon tetrachloride ($SiCl_4$), also a gas, which is used to make silicone polymers (Section

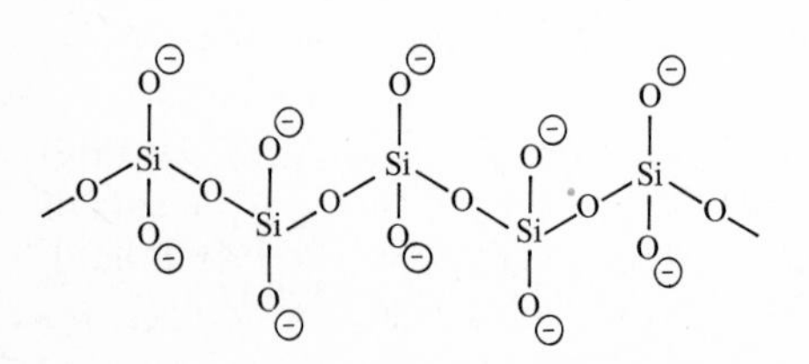

Figure 18-6 A linear (two-dimensional) form of a polymeric silicate ion.

22-5). Like carbon, silicon has a valence of four in all its stable compounds, but germanium can also form stable compounds in which its valence is two. An example is germanium difluoride (GeF_2), which is a white crystalline solid.

18-4 The Group VA Elements: N, P, As, Sb

The properties of the group IVA elements have already given us examples of how the elements become more metallic as we move down a column of the periodic table (Sections 2-10 and 13-13). We did not discuss the last two members, tin and lead, but they are obviously more metallic than germanium. In this section, we shall see additional examples of this tendency.

All the elements in this group form three covalent bonds per atom. In group IVA we saw that only carbon was able to form double and triple bonds. The situation is parallel here. Only nitrogen, but not phosphorus, arsenic, or antimony, can form double and triple bonds. However, if nitrogen is the only one of these elements that has this ability, it is also the only one that *lacks* another ability: the ability to form *five* covalent bonds. Phosphorus, arsenic, and antimony all form five covalent bonds in some compounds (10 electrons in the outer shell) and three in others (we shall see examples below), but nitrogen cannot expand its shell beyond eight, and has only a valence of three, not five.

Ionic bonds in this group are rare. There are a few ionic nitrides (for example, Li_3N, Ca_3N_2), which are made up of metal ions and nitride ions N^{3-}, and a few ionic phosphides (for example, Na_3P, Ca_3P_2), but the only member of this group to form positive ions is antimony, and even for this element, such ions are not common. There are only a few compounds containing Sb^{3+} ions, an example being $Sb_2(SO_4)_3$. No known compounds of this group contain +5 ions (even Bi^{5+} is unknown). The oxidation numbers in this group vary from +5 to −3.

THE ELEMENTS. We have already discussed N_2 (Section 17-3). Phosphorus, arsenic, and antimony all occur widely on Earth in various minerals, although phosphorus is much more abundant than the other two. Sb is sometimes found free, but As and P are not found in the free state. There are three allotropic forms of phosphorus, called *white, black,* and *red phosphorus;* all are solids, though white phosphorus melts at 44°C. White phosphorus is made up of P_4 molecules in the shape of tetrahedrons (note that each P atom has three covalent bonds).

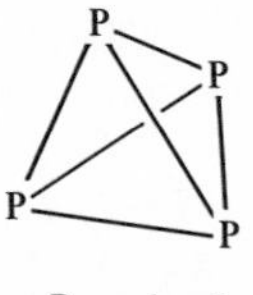

a P_4 molecule

The black and red forms have more complex structures. White phosphorus is the most reactive of the three and is normally stored under water, because in the dry state it reacts with the oxygen in the air and spontaneously ignites. It is very toxic. White phosphorus glows in the dark, a property that made it seem eerie when it was first discovered in 1669 by an alchemist named Brand. No one at that time had ever seen any substance give off light without heat, and for a time it was regarded as magical.

Arsenic and antimony also exist in several allotropic forms. The most common allotropic forms of these elements look like metals, but are not good conductors of electricity (if they were, we would consider them metals and not metalloids). Arsenic and its compounds are all poisonous, and have been used as poisons for centuries. One common arsenic poison is Paris green, a mixed salt of the formula $3Cu(AsO_2)_2 \cdot Cu(C_2H_3O_2)_2$, which is still used as an insecticide. Antimony and phosphorus are also poisonous.

Geometry of Group VA Compounds. The elements of this group have five electrons in their outer shells, and when they form three covalent bonds two electrons remain unbonded (the unshared pair). Consequently, the EPRT predicts that NX_3, PX_3, AsX_3 and SbX_3 molecules (where X is any atom) will have pyramidal shapes (Figure 18-7), in contrast to the planar shape for BX_3 molecules. This has been confirmed by many laboratory studies. Because of the unshared pair, all of those molecules are potential bases. When they combine with a proton or other Lewis acid, they form *positive* ions in which the central atom has *four* covalent bonds: for example,

$$NX_3 + H^+ \longrightarrow \overset{\oplus}{HNX_3}$$

EPRT tells us that the geometry of these positive ions is tetrahedral (Figure 18-7).

As mentioned above, all the elements except nitrogen can also form five covalent bonds: that is, they form compounds PX_5, AsX_5, and SbX_5. In such cases the molecule usually has the trigonal bipyramid structure (Section 14-13). These compounds are potential Lewis acids, and combine with bases to give *negative* ions in which the central atom has *six* covalent bonds: for example,

$$PX_5 + X^- \longrightarrow PX_6^-$$

These negative ions have an octahedral shape (Figure 18-8).

Hydrogen Compounds. Ammonia, NH_3, is an important base, though a weak one ($K_b = 1.8 \times 10^{-5}$). It is a colorless gas (bp $-33°C$), similar to water in its physical properties. In fact, NH_3 resembles water more than any other substance. The "ammonia" that we normally see in laboratories is actually a solution of NH_3 in water. NH_3 is quite soluble in water, which we would expect because of the hydrogen bonds that form between these two molecules. Because it is a weak base, the equilibrium

$$NH_3 + H_2O \rightleftharpoons NH_4^+ + OH^-$$

lies well to the left, and in a solution of NH_3 in water about 99% of the solute exists as NH_3 molecules and only about 1% as NH_4^+ ions (the exact percentages depend on the concentration—see the similar case of acids in water, Section 10-6). Liquid ammonia itself is an important solvent (see, for example, Section 10-14), and would undoubtedly be more important if it did not have to be cooled or pressurized in order to remain a liquid.

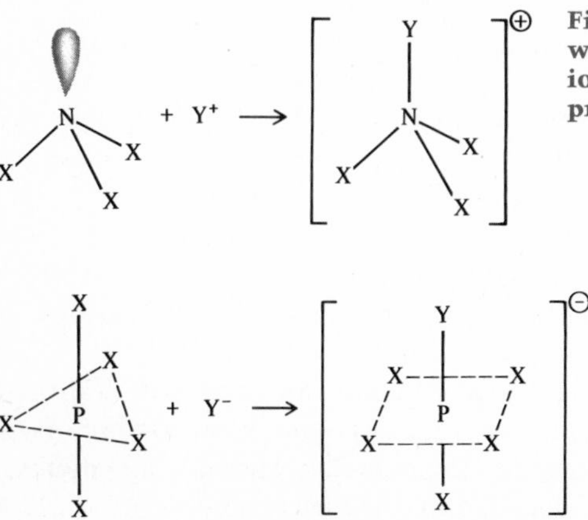

Figure 18-7 A pyramidal NX_3 molecule reacts with a positive ion to form a tetrahedral NX_3Y^+ ion. X is any atom or group; Y^+ is most often a proton.

Figure 18-8 A trigonal bipyramidal PX_5 molecule reacts with a negative ion to form an octahedral PX_5Y^- ion.

When ammonia reacts with an acid, the ammonium ion, NH_4^+, is produced. Many ammonium salts, $NH_4^+X^-$, can be crystallized, where X^- is a wide variety of anions. In much of its behavior the ammonium ion is analogous to the alkali metal ions Na^+ and K^+ (in Section 11-4, we saw that NH_4^+ is in the same group in the qualitative analysis scheme). In aqueous solution all three are very unreactive and generally remain in solution no matter what reagents are added, but NH_4^+ differs from the other two in that it reacts with OH^- ion to give NH_3.

Another compound of nitrogen and hydrogen is *hydrazine,* N_2H_4, which has the structure

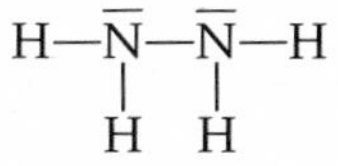

$$\begin{array}{c} \text{H}-\overline{\text{N}}-\overline{\text{N}}-\text{H} \\ \quad | \quad\;\; | \\ \quad \text{H} \quad \text{H} \end{array}$$

Although hydrazine is a simple molecule whose structure, like that of NH_3, is easily understandable on the basis of ordinary covalent bonds (valence of N = 3), the nitrogen in hydrazine has a different oxidation number. The oxidation number of nitrogen in NH_3 is -3; in N_2H_4 it is -2. This example emphasizes the fact that oxidation number is a purely *formal* concept, as compared to bonds, which are real. Hydrazine is a liquid, with a boiling point of 114°C. It is a base, but weaker than NH_3 (K_b of hydrazine $= 8.5 \times 10^{-7}$). As we would expect from its two unshared pairs, hydrazine is dibasic, and takes two protons. It is used as a rocket fuel because it burns in air with a high heat of combustion (in rocket fuels it is usually burned with pure oxygen rather than air).

Hydrides of the other group VA elements are much less important than NH_3. Phosphine (PH_3), arsine (AsH_3), and stibine (SbH_3) are all extremely poisonous. Arsine is easily decomposed by heat to arsenic, and this property is the basis of the Marsh test for the presence of arsenic. First the sample is reduced so that any arsenic compounds are converted to AsH_3. The AsH_3 is then heated until it decomposes, whereupon it deposits a film of arsenic on a test tube (an arsenic "mirror"). Stibine is even more easily decomposed than arsine.

HALIDES. All the elements of this group form compounds with all the halogens. For nitrogen, there is only one series of halides NX_3,* but for phosphorus, arsenic, and antimony, two series are known, EX_3 and EX_5. All 12 possible EX_3 compounds are known (E = P, As, or Sb; X = F, Cl, Br, or I), but not all the EX_5 compounds. The most important compounds in this group are phosphorus trichloride (PCl_3) and phosphorus pentachloride (PCl_5). All the EX_5 compounds are Lewis acids (see above), the most useful for this purpose being antimony pentachloride ($SbCl_5$). Most of the EX_3 compounds are prepared by direct treatment of the element with halogen. Excess halogen then gives the pentahalide: for example,

$$2As + 3Br_2 \longrightarrow 2AsBr_3$$

$$AsBr_3 + Br_2 \longrightarrow AsBr_5$$

OXIDES. The six known oxides of nitrogen are listed in Table 18-1. In nitrous oxide, N_2O, the oxidation number of nitrogen is $+1$. The molecules of N_2O are resonance hybrids:

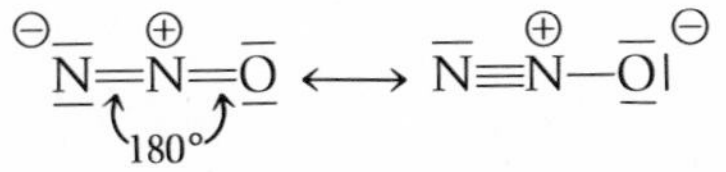

$$\overset{\ominus}{\underline{\overline{\text{N}}}}=\overset{\oplus}{\text{N}}=\underline{\overline{\text{O}}} \longleftrightarrow \overline{\text{N}}\equiv\overset{\oplus}{\text{N}}-\underline{\overline{\text{O}}}|^{\ominus}$$

$$(180°)$$

EPRT predicts linear geometry, and that is indeed the case. Like NO and NO_2, N_2O

*There are also a few nitrogen halides which contain two nitrogen atoms; similarly for phosphorus and arsenic.

Table 18-1 Oxides of Nitrogen

Formula	*Name*	*Oxidation number of nitrogen*
N_2O	Nitrous oxide	+1
NO	Nitric oxide	+2
N_2O_3	Dinitrogen trioxide	+3
NO_2	Nitrogen dioxide	+4
N_2O_4	Dinitrogen tetroxide	+4
N_2O_5	Dinitrogen pentoxide	+5

was one of the gases discovered by Joseph Priestley. In 1800, Humphry Davy discovered that inhalation of N_2O caused people to become light-headed and to laugh. It became known as laughing gas, and nitrous oxide parties were the social rage of London. Later it was discovered that it was an anaesthetic, and it is still used for that purpose today. It is also used in aerosol cans (see Section 17-5). Nitrous oxide can be prepared by heating molten ammonium nitrate, NH_4NO_3. It is relatively unreactive.

Nitric oxide (NO) is a colorless gas which can be prepared by heating a mixture of oxygen and nitrogen (Section 17-7—note that air is such a mixture). Once prepared, it reacts immediately with additional O_2 to give the brown gas nitrogen dioxide, NO_2. We have already seen the effects of these gases in air pollution. The oxidation number of nitrogen in NO and NO_2, respectively, is +2 and +4. Both of these compounds have an odd number of electrons. Consequently, they are both free radicals and both paramagnetic (Section 14-3). Their electronic structures, both resonance hybrids, may be represented as follows:

$$\dot{N}{=}O \longleftrightarrow {}^{\ominus}\dot{N}{=}O^{\oplus}$$

$$^{\ominus}O{-}\dot{N}^{\oplus}{=}O \longleftrightarrow O{=}\dot{N}^{\oplus}{-}O^{\ominus} \longleftrightarrow O{=}N{-}\dot{O} \longleftrightarrow \dot{O}{-}N{=}O$$

134°

NO_2 is a bent molecule (angle = 134°), as predicted by EPRT. Note that every canonical form of both compounds must possess one unpaired electron, although this electron can be on a nitrogen or an oxygen atom. Since the true structure is a hybrid of all the canonical forms, the single electron is actually present in an orbital which covers both atoms in NO and all three atoms in NO_2. For example, the bond order of the bond in NO is actually $2\frac{1}{2}$ (see Section 14-19). In the absence of O_2, NO is a stable compound which shows little tendency to dimerize, even though it is a free radical. NO_2 is less stable, and exists in equilibrium with its dimer, dinitrogen tetroxide (N_2O_4), which is colorless.

$$\underset{\text{brown}}{2NO_2} \rightleftharpoons \underset{\text{colorless}}{N_2O_4}$$

In the laboratory, NO and NO_2 can be made by the reaction between copper and nitric acid. Dilute HNO_3 gives mostly NO, while concentrated HNO_3 gives mostly NO_2, although neither compound is generally formed entirely free of the other.

with dilute HNO_3: $3Cu + 8H^+ + 2NO_3^- \longrightarrow 3Cu^{2+} + 2NO + 4H_2O$

with concentrated HNO_3: $Cu + 4H^+ + 2NO_3^- \longrightarrow Cu^{2+} + 2NO_2 + 2H_2O$

The other two oxides of nitrogen, dinitrogen trioxide (N_2O_3) and dinitrogen pentoxide (N_2O_5) (oxidation number of nitrogen = +3 and +5, respectively), are unstable at room temperature, and are chiefly of interest because they react with water to give nitrous and nitric acids, respectively.

There are several known oxides of phosphorus, arsenic, and antimony. The most important of these is phosphorus pentoxide, which despite its name actually has the formula P_4O_{10} (it is a dimer of the unknown compound P_2O_5). P_4O_{10}, which is made by burning white phosphorus in excess air, is a white solid which reacts with water to give phosphoric acid:

$$P_4O_{10} + 6H_2O \longrightarrow 4H_3PO_4$$

This reaction takes place so readily that P_4O_{10} is an important drying agent, which is used to remove water from gases and liquids. Its affinity for water is so great that it even removes the elements of water from many compounds; for example, it dehydrates pure HNO_3 to N_2O_5.

Oxygen Acids of Nitrogen. The two chief oxygen acids of nitrogen are nitrous acid (HNO_2) and nitric acid (HNO_3). These acids may be considered the hydrated forms of N_2O_3 and N_2O_5, respectively; as in these compounds, the oxidation numbers of nitrogen are +3 and +5, respectively.

$$N_2O_3 + H_2O \longrightarrow 2HNO_2$$
$$N_2O_5 + H_2O \longrightarrow 2HNO_3$$

Nitrous acid is not very stable at room temperature, either pure or dissolved in water, but many of its salts, called nitrites, are stable. Both nitrous acid and the nitrite ion NO_2^- have bent structures:

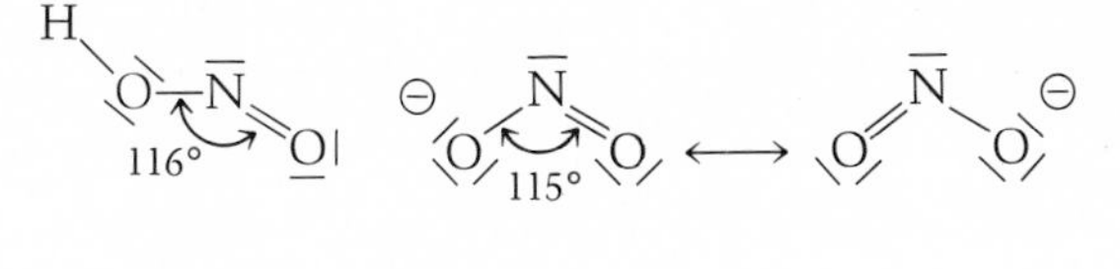

nitrous acid nitrite ion

The nitrite ion is a resonance hybrid, though nitrous acid is not. In contrast, both nitric acid and the nitrate ion are resonance hybrids.

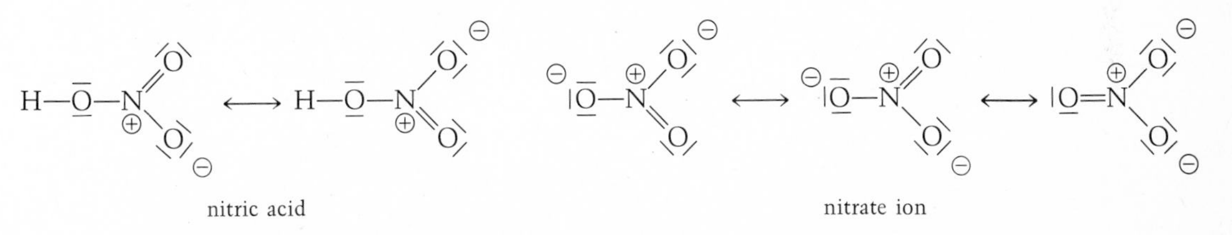

nitric acid nitrate ion

Nitrous acid is a weak acid ($pK_a = 4.6 \times 10^{-4}$), while nitric acid is one of our six strong acids. This is consistent with the generalization in Section 10-2 that the strength of acids of the form H_nAO_m increases with increasing m. We are now in a position to see why this is so in this case. Both nitrite and nitrate ions are resonance hybrids, but NO_3^- has three canonical forms, NO_2^- only two. In NO_3^- the negative charge is spread over three oxygen atoms, so that the actual charge on each atom is $-\frac{1}{3}$, while in NO_2^- it is $-\frac{1}{2}$. **Dispersed charges are more stable than concentrated charges,** so that NO_3^- is more stable than NO_2^-. Since NO_3^- is more stable, HNO_3 loses its proton (to give NO_3^-) more readily than HNO_2 (which gives NO_2^-), making it a stronger acid. Another closely related factor is that a charge of $-\frac{1}{2}$ attracts a proton better than a charge of $-\frac{1}{3}$. Similar reasoning can be applied in most cases of H_nAO_m acids. The principle that dispersed charges are more stable than concentrated charges is very important in chemistry and can be used to explain much chemical behavior.

Nitric acid is one of the most important inorganic compounds. It is a strong acid and a powerful oxidizing agent, and is widely used for both purposes. Many of the salts of nitric acid, called nitrates, are also important. Nitric acid is commercially prepared by the *Ostwald process*, which consists of a series of oxidation steps, the first of which is the oxidation of ammonia with a platinum catalyst to give NO. These processes are shown schematically (not balanced):

$$NH_3 + O_2 \xrightarrow{Pt} NO \xrightarrow{O_2} NO_2 \xrightarrow[H_2O]{O_2} HNO_3$$

The raw material, ammonia, is made by the Haber process (Section 17-3).

Oxygen Acids of Phosphorus. There are several oxygen acids of phosphorus in which the oxidation number of P is +5. The most common of these is phosphoric acid, H_3PO_4, which has the structure

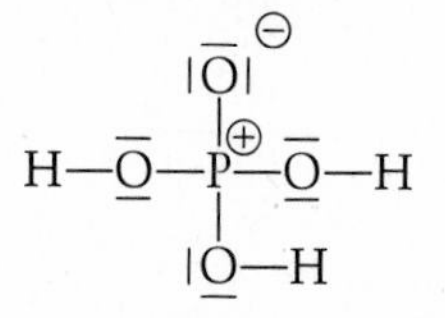

The P atom has a tetrahedral geometry. Phosphoric acid is a triprotic acid and there are three series of salts, phosphates, hydrogen phosphates, and dihydrogen phosphates: Na_3PO_4, Na_2HPO_4, and NaH_2PO_4, respectively. The three ions are also tetrahedral:

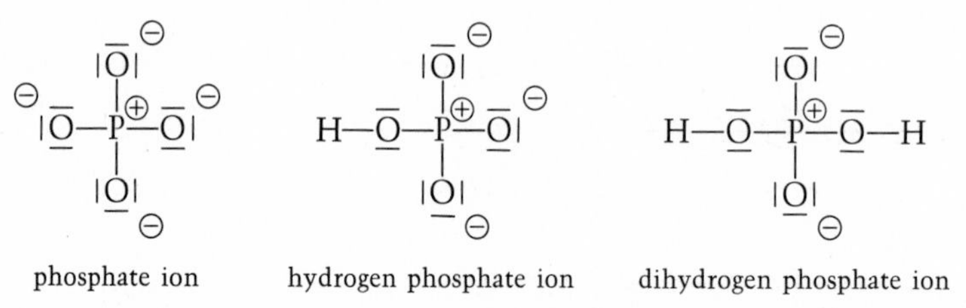

phosphate ion hydrogen phosphate ion dihydrogen phosphate ion

Phosphoric acid is a weak acid ($K_{a_1} = 7.5 \times 10^{-3}$) and has essentially no oxidizing ability, but is important because the phosphates (all three series) have many commercial uses, especially as fertilizers. H_3PO_4 is a solid (mp 42°C), but is usually encountered in laboratories and in industry as an 85% solution in water. It is made by treating phosphate rocks with sulfuric acid

$$2PO_4^{3-} + 3H_2SO_4 \longrightarrow 2H_3PO_4 + 3SO_4^{2-}$$

and by adding water to P_4O_{10}, which itself is made by the burning of white phosphorus.

Polyphosphoric acids are +5 oxygen acids of phosphorus which have more than one P atom per molecule. The first two members of the series are diphosphoric acid ($H_4P_2O_7$) and triphosphoric acid ($H_5P_3O_{10}$):

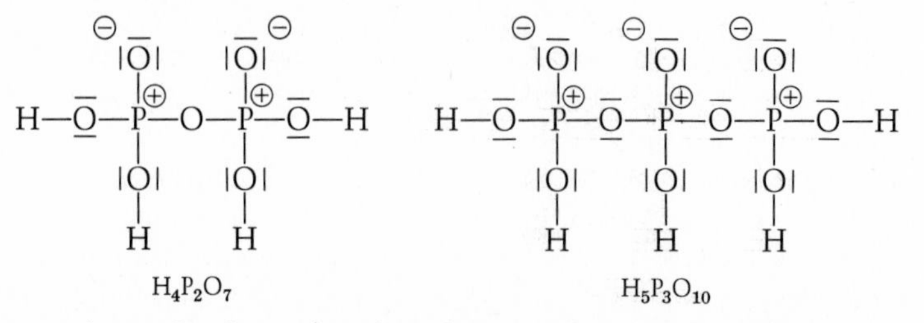

$H_4P_2O_7$ $H_5P_3O_{10}$

It can be seen from the formulas that these acids consist of two or more H_3PO_4 molecules which have been combined, with the removal of H from one molecule and OH from the next (the elements of water). They are also known as condensed phosphoric acids. They can be made by the direct dehydration of H_3PO_4 (by heating), or by the addition of small amounts of water to P_4O_{10}. We have already seen that salts of these acids (polyphosphates) are used as builders in detergents (Section 17-11). Phosphates and polyphosphates play vital roles in the functioning of living organisms (see Section 21-6).

The most important +3 oxygen acid of phosphorus is phosphorous acid H_3PO_3, which has a P—H bond.

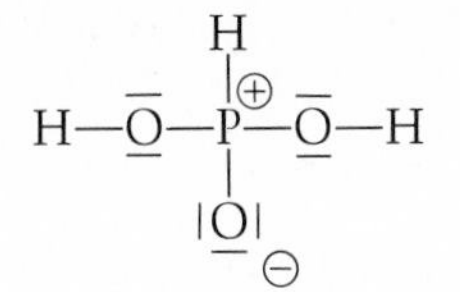

H_3PO_3 is a weak ($K_{a_1} = 1.0 \times 10^{-2}$) diprotic acid. The P—H bond is not acidic.

Other Compounds of Nitrogen. Nitrogen forms many other compounds, in all of which its valence is three (unless it has a formal charge or an unpaired electron). There are a large number of organic nitrogen compounds (we shall meet some of them in Chapters 20 and 21), and many of them are important constituents of living organisms. Nitrogen forms double and triple bonds, and its three valences can be distributed in three ways: 3 single bonds; 1 double and 1 single bond, or 1 triple bond. Some examples of each type are given:

CH_3—N(CH_3)—CH_3	H—N(H)—O—H	CH_3—C(=N—OH)—CH_3	H—C≡N
trimethylamine	hydroxylamine	acetoxime	hydrogen cyanide

18-5 The Group VIA Elements

The elements in this group are oxygen, sulfur, selenium, tellurium, and polonium. They all have six electrons in the outer shell, and all form compounds in which the oxidation number is −2. This occurs in three ways: (1) the central atom has two covalent bonds (and two unshared pairs) (for example, H_2O, H_2S); (2) the central atom gains two electrons to form ions with a charge of −2 (for example, O^{2-}, S^{2-}); and (3) the central atom forms one covalent bond and gains one electron, to form ions with a charge of −1 (for example, OH^-, SH^-). In this group both oxygen and sulfur can form double bonds, although the double bonds of sulfur are generally less stable. In those compounds in which the central atom has two single covalent bonds, the molecules are bent and not linear (these are $|\overline{A}B_2$ molecules): for example,

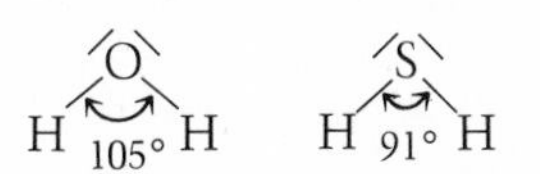

Except for the special case of OF_2 (where the oxidation number is +2), the highest possible oxidation number for oxygen is 0, so the oxidation number of oxygen can only be −2, −1, or 0. However, for the other elements of this group, the maximum oxidation number is +6.

The Elements. We have already discussed oxygen (Section 17-4) and its allotropic form ozone (Section 17-5). Sulfur occurs widely in nature—it is the fifteenth most abundant element on the earth's crust. It occurs as free sulfur, in many minerals and rocks as sulfides and sulfates, and in the atmosphere, as H_2S and SO_2 (Section 17-7). Two minerals that may be mentioned are *gypsum* ($CaSO_4$) and iron pyrite (FeS_2), also known as *fool's gold*, because it looks like gold and many people were deceived by it. Selenium and tellurium are much less abundant—they

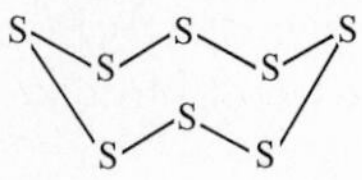

Figure 18-9 An S_8 molecule.

are often found in nature mixed with sulfur in its various forms, though in smaller amounts. Polonium is extremely rare, and not much is known about its chemistry. We shall not consider it further.

There are several allotropic forms of sulfur. The most stable of these is *orthorhombic sulfur,* a yellow solid made up of S_8 molecules, in which eight sulfur atoms, each with two covalent bonds, are arranged in a ring (Figure 18-9). When heated at 95.5°C, the orthorhombic form slowly changes into the monoclinic form, which also consists of S_8 rings, but arranged differently in the crystal lattice. Other allotropic forms of sulfur are known, in some of which there are S_6, S_{10}, or S_{12} rings; sulfur chain molecules are also known. To some extent, S_8 rings persist even when sulfur is melted, and even in the gaseous state. There are two allotropic forms of selenium but only one form of tellurium. As expected, the appearance of selenium is more metallic than that of sulfur; and that of tellurium is still more so.

OXIDES. We have previously mentioned that oxygen forms compounds with all elements except He, Ne, and Ar. Binary compounds of oxygen (when O is in the -2 oxidation state) are called **oxides.** We can distinguish three types of oxides, based on their bonding:

1. Ionic oxides, in which O_2^- is present in the crystal lattice along with positive ions. Only the most active metals form this type of oxide (examples are Na_2O, CaO, and MgO). The O^{2-} ion cannot exist in water solution because it reacts with water to give OH^- ions:

$$O^{2-} + H_2O \rightleftharpoons 2OH^- \qquad K > 10^{22}$$

Consequently, all ionic oxides that are soluble in water are converted to hydroxides in water solution.

2. Oxides that are small covalent molecules. We have already met several of these, including H_2O, NO, NO_2, CO, CO_2, etc.
3. Oxides that are giant covalent molecules. Examples are SiO_2 and B_2O_3. As we would expect, oxides in class 2 are generally liquids, gases, or low-melting solids, while those of classes 1 and 3 are high-melting solids.

Many oxides have acidic or basic properties when dissolved in water. In such cases, we can convert the formula of the oxide to that of a hydroxide by adding one or more water molecules: for example,

$$CaO + H_2O \longrightarrow CaO_2H_2 = Ca(OH)_2$$
$$SO_3 + H_2O \longrightarrow SO_4H_2 = (HO)_2SO_2 = H_2SO_4$$

Note that this is a purely formal process and does not necessarily correspond to the way the actual reaction takes place. If the result comes out with multiple coefficients, then we divide by the common factor: for example,

$$Na_2O + H_2O \longrightarrow Na_2O_2H_2 \longrightarrow NaOH$$
$$Cl_2O_7 + H_2O \longrightarrow Cl_2O_8H_2 = (HO)_2Cl_2O_6 \longrightarrow HClO_4$$
$$P_2O_5 + 3H_2O \longrightarrow H_6P_2O_8 = (HO)_6P_2O_2 \longrightarrow H_3PO_4$$

We saw in Section 10-2 that some of these hydroxides are basic, some acidic, and some amphoteric, and that basicity decreases and acidity increases in going from left to right across a row of the periodic table. Metallic character also decreases in this direction, and we can now state the following correlation:

Metal oxides give rise to basic hydroxides
Metalloid oxides give rise to amphoteric hydroxides
Nonmetal oxides give rise to acidic hydroxides

The borderlines between these groups are not hard and fast. For example, aluminum is usually classified as a metal, but it has an amphoteric hydroxide, which reacts with either acids or bases:

$$Al(OH)_3 + 3H_3O^+ \longrightarrow Al^{3+} + 6H_2O$$
$$H_3AlO_3 + 3OH^- \longrightarrow AlO_3^{3-} + 3H_2O$$

[Note that $Al(OH)_3$ and H_3AlO_3 are two ways of writing the same compound. We write it differently to emphasize the different types of reaction it undergoes.] This behavior shows that aluminum, although a metal in appearance, conductivity, and many other ways, does show some behavior expected of a metalloid, which is not surprising when we look at its position in the periodic table. Another type of exception to the above rule is found with some transition metals, whose higher oxides may not give basic hydroxides. An example is chromium, where CrO gives a basic hydroxide $Cr(OH)_2$, while CrO_3 gives the acid $H_2Cr_2O_7$.

Oxides are called *acidic* or *basic* oxides depending on the behavior of their corresponding hydroxides. Thus, SO_3, N_2O_5, and CrO_3 are acidic oxides, while MgO, K_2O, and CrO are basic oxides. Both kinds of oxides are often called **anhydrides.** An anhydride is the compound that results when water is removed (actually or formally) from the structure, by the reverse of the process described above. For example, SO_3 is the anhydride of H_2SO_4. Not all oxides are the anhydrides of acids or bases. Many, for example N_2O, CO, and MnO_2, show no such properties at all.

THE −2 OXIDATION STATE OF SULFUR, SELENIUM, AND TELLURIUM. Hydrogen sulfide, H_2S, is a poisonous gas, with an offensive odor which resembles that of rotten eggs. H_2S is even more poisonous than hydrogen cyanide, HCN. Hydrogen selenide, H_2Se, and hydrogen telluride, H_2Te, both of which are also gases with offensive odors, are even more poisonous than H_2S. All three are weak acids that dissolve in water. The acid strength of binary hydrides increases in going down the periodic table. H_2Te is the strongest of the three, though still a weak acid. Because they are weak acids, the three hydrides can be prepared by the action of strong acids on sulfides, selenides, and tellurides: for example,

$$S^{2-} + 2HCl \longrightarrow H_2S + 2Cl^-$$

Salts of H_2S are called **sulfides.** Many of them are known. Those of the alkali and alkaline earth metals are ionic sulfides, containing S^{2-} ions, but most of the others are nonionic and consist of three-dimensional covalent giant molecules (for example, FeS, CoS).

THE −1 OXIDATION STATE. Water is not the only hydride of oxygen. Another one is *hydrogen peroxide*, H_2O_2, in which the oxidation number of oxygen is −1. The electronic structure is

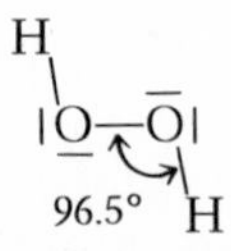

Note that both oxygens have a bent geometry, in accord with the EPRT. H_2O_2 is a liquid with a rather high boiling point of 152°C (because of extensive hydrogen bonding). However, it is usually encountered in water solution, most often as 30% H_2O_2 or as 3% H_2O_2. The 3% solution is used to bleach hair. H_2O_2 is a strong oxidizing agent (it is reduced to H_2O or OH^-) and is often used in laboratories for this purpose. It is also a reducing agent, being oxidized to O_2. There are many salts of H_2O_2; some of them (for example, sodium peroxide, Na_2O_2) contain the peroxide ion O_2^{2-}. The O—O bond is weaker than most other covalent bonds, but there are many known compounds which contain it. Compounds containing this bond are usually called **peroxides** or **peroxy compounds.** Most of them are quite reactive.

When sulfur is heated with an aqueous solution of sulfide ions, a mixture of **polysulfides** is formed. This mixture contains disulfides (S_2^{2-}) ions (analogous to peroxide ions, O_2^{2-}) as well as S_3^{2-}, and even larger ions.

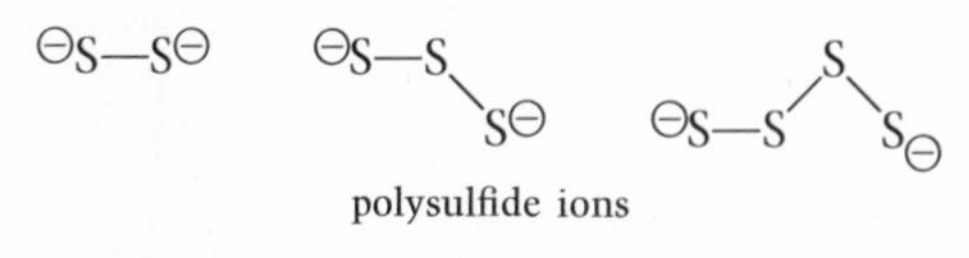

polysulfide ions

These ions may be crystallized as metallic polysulfides. Iron pyrite, mentioned earlier in this section, is a disulfide of iron.

Higher Oxidation States. Two oxides of sulfur were mentioned in Section 17-7. In sulfur dioxide (SO_2) the oxidation number of sulfur is +4; in sulfur trioxide (SO_3) it is +6. SO_2, which can be obtained by burning sulfur, is a gas with a powerful choking odor. [Note the difference between the odors of H_2S and SO_2. H_2S has a foul odor but does not choke, while the odor of SO_2 is not foul (it does not smell of rotten eggs) but it does choke. H_2S is more dangerous, since the choking quality of SO_2 is likely to drive people away before they can be harmed very much.] The molecule is bent, with the following electronic structure (note that sulfur has 10 electrons in its outer shell).

Sulfur trioxide is a low-boiling liquid that can be made by oxidizing SO_2 with O_2 in the presence of a catalyst:

$$2SO_2 + O_2 \xrightarrow{\text{Pt}} 2SO_3$$

It has a planar, triangular geometry:

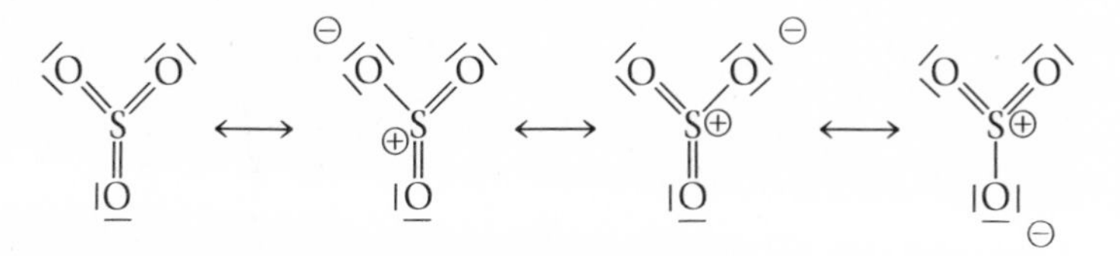

The corresponding oxides of Se and Te, SeO_2, SeO_3, TeO_2, and TeO_3 are all known. They are all solids.

SO_2 and SO_3 are both soluble in water. A solution of SO_2 in water is called *sulfurous acid.* We usually represent this as H_2SO_3 (considering SO_2 to be the anhydride of H_2SO_3), although studies have shown that it is mainly just SO_2 dissolved in water. However, it *behaves* as if H_2SO_3 was really present, losing one or two protons (it is a weak acid) to give two series of salts, called **bisulfites** (HSO_3^-) and **sulfites** (SO_3^{2-}). Sulfurous acid, sulfites, and bisulfites are all important reducing agents (the sulfur in them is oxidized to the +6 state).

SO_3 is the anhydride of *sulfuric acid* (H_2SO_4), which is a highly important laboratory and industrial compound (more tons of H_2SO_4 are produced each year than of any other chemical). It is produced industrially by two processes: the contact process and the lead chamber process. In both of these, SO_2 (obtained by the burning of sulfur) is oxidized to SO_3 by O_2 in the presence of catalysts (platinum in the former and oxides of nitrogen in the latter). The SO_3 is then added to water to give H_2SO_4. H_2SO_4 is a strong acid and a powerful dehydrating agent. To see its dehydrating ability, put a wooden splint into a beaker of concentrated H_2SO_4. The splint will turn black because the H_2SO_4 removes the elements of water from the wood (see the structure of cellulose in Figure 21-7), leaving black carbon as a residue. Since the same thing would happen if you put your finger in, it is important not to let any H_2SO_4 get on you or your clothing. The structure of H_2SO_4 is

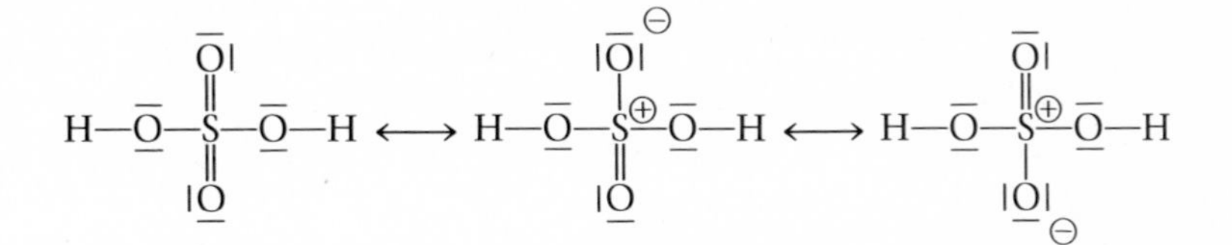

The central sulfur atom is tetrahedral. We can get pure H_2SO_4 by adding 1 mole of SO_3 to 1 mole of H_2O. However, this is not the limit of the solubility of SO_3 in H_2O. It is possible to add still more SO_3 to H_2SO_4. The liquid so obtained is called *fuming sulfuric acid* or *oleum*. Like H_2SO_3, H_2SO_4 forms two series of salts, in this case called *bisulfates* (for example, $NaHSO_4$) and *sulfates* (for example, K_2SO_4).

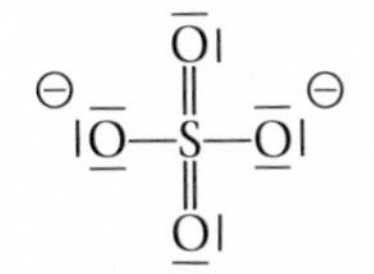

sulfate ion (one canonical form)

H_2SO_3 and H_2SO_4 are not the only oxygen acids of sulfur. A number of others are known, of which we will mention only two. H_2SO_5 is called peroxymonosulfuric acid, or Caro's acid. As its name implies, it has an O—O bond.

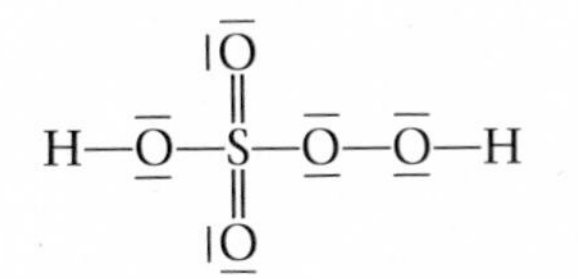

Caro's acid (one canonical form)

Thiosulfuric acid ($H_2S_2O_3$) is not important in itself (it is unstable at room temperature), but its sodium salt, sodium thiosulfate ($Na_2S_2O_3$), is an important reducing agent. When oxidized, it gives the tetrathionate ion ($S_4O_6^{2-}$):

$$2S_2O_3^{2-} \longrightarrow S_4O_6^{2-} + 2e^-$$

The structure of the thiosulfate ion is the same as that of SO_4^{2-}, except that a sulfur atom replaces an oxygen atom.

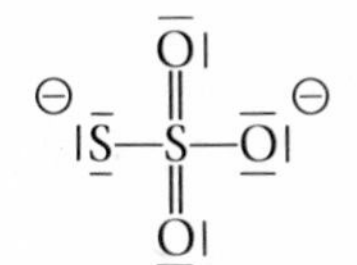

thiosulfate ion (one canonical form)

Sodium thiosulfate is also called *hypo*. It is used in photography, to fix the image to the negative during the developing process.

A number of oxygen acids of selenium are known, for example, selenous acid, H_2SeO_3, and selenic acid, H_2SeO_4.

18-6 The Halogens

The elements of group VIIA are called halogens. Four of them, fluorine, chlorine, bromine, and iodine, are very important. The fifth, astatine, has only been prepared artificially, in small quantities, and not much is known about it (though, of course, we can predict its properties from its position in the periodic table). We will not consider it further. As expected for elements with seven electrons in the outer shell, all the halogens form compounds with one covalent bond, as well as ions of the form E^-. They can all have oxidation numbers of 0 or -1. However, bromine, chlorine, and iodine also form compounds with higher oxidation numbers, up to a maximum of $+7$. In these, the halogen atom has more than one covalent bond. The highest number is found in the compound IF_7 where the iodine atom has seven covalent bonds, with 14 electrons in its outer shell. These higher-valent compounds are found only where the second element is oxygen or another halogen atom.

THE ELEMENTS. All the halogens exist as covalent nonpolar (why?) molecules of the formula E_2. Their properties change gradually as we go down column VIIA of the periodic table.

F_2 is a light greenish-yellow gas (bp $-188°C$)
Cl_2 is a green gas (bp $-35°C$)
Br_2 is a dark red-brown liquid
I_2 is a steel-gray solid, which sublimes to a violet vapor

Note that the color deepens, and the boiling point increases, as we go down the table. Iodine is purple in the vapor state and when dissolved in nonpolar liquids such as carbon tetrachloride, but is brown in many organic solvents, such as ethyl alcohol. The brown solutions of I_2 in ethyl alcohol (tincture of iodine) are sometimes used as disinfectants, but are generally regarded as too corrosive to the skin for this purpose. Fluorine is the most reactive of all the elements. It reacts, often violently, with every other element except N_2, Ar, He, and Ne, and with many compounds. This extreme reactivity is due to the very low F—F bond energy, 38 kcal/mole. The reactivity of the other halogens decreases in the order Cl_2, Br_2, I_2. All the halogens are very poisonous, especially F_2, because of its high reactivity.

None of the halogens is found free in nature. Fluorine is fairly abundant on the Earth's surface, being found in many minerals: for example, fluorapatite, $Ca_5F(PO_4)_3$, and fluorspar, CaF_2. Despite this abundance, the free element was first prepared in 1886 (by Henri Moissan, 1852–1907), sixty years after the first preparation of Br_2 (I_2 was first prepared in 1811; Cl_2 by Scheele, in 1774). The reason for this is obviously the extreme reactivity of F_2. The more reactive a substance is the harder it is to prepare it free from other substances. Many previous attempts to prepare F_2 had failed. F_2 today is usually prepared by essentially the same method used by Moissan: electrolysis of a solution of KF in liquid HF. Chlorine, bromine, and iodine are found on Earth chiefly in seawater, as Cl^-, Br^-, and I^- (Cl^- is by far the most abundant). Cl_2 is commercially obtained by electrolysis of seawater. Br_2 is also obtained from seawater, by the addition of Cl_2:

$$Cl_2 + 2Br^- \longrightarrow Br_2 + 2Cl^-$$

The Br_2 is then swept out by air. I_2 is prepared by oxidation of I^- by an oxidizing agent such as MnO_2.

All the halogens are oxidizing agents, whose oxidizing power decreases in the order F_2, Cl_2, Br_2, I_2 (they are reduced to F^-, Cl^-, Br^-, and I^-). For this reason, each halogen will oxidize any halide ion beneath it. The preparation of Br_2, given above, is an example.

BINARY HALIDES. All the elements except He, Ne, and Ar form halides, and many halides are common reagents in chemical laboratories. Most of the metallic halides are ionic compounds; most of the other halides are covalent. Halide ions are reducing agents, being oxidized to the free halogens. Obviously, the reducing power decreases in the order I^-, Br^-, Cl^-, F^-. (Why?) The F^- ion is an extremely poor reducing agent. An important reaction of many covalent halides is hydrolysis: they react with water, either at room temperature or above, to give compounds in which the halogens are replaced by OH groups. An example is

$$PCl_5 + 5H_2O \longrightarrow \underset{\text{not isolated}}{P(OH)_5} \xrightarrow[H_2O]{\text{loss of}} H_3PO_4$$
$$+ \; 5HCl$$

However, not all covalent halides can be hydrolyzed with water. Carbon tetrachloride, for example, does not react with water.

Binary compounds that contain two different halogens are called **interhalogen compounds.** Many of them are known, including five of the six possible AB compounds (the one that is not known is IF), as well as ICl_3, ClF_3, ClF_5, BrF_3, BrF_5, IF_3, IF_5, and IF_7. All the interhalogen compounds are fairly reactive oxidizing agents. The bonding in all of them is covalent.

OXYGEN ACIDS OF THE HALOGENS. A number of halogen oxides are known, in various oxidation states (for example, OF_2, Cl_2O_3, I_4O_9). We discussed the structure of one of these, ClO_2, in Section 14-3. However, some of the oxygen acids are more important. Although a number of oxygen acids of bromine and iodine are known (only one such acid is known for fluorine: HOF, which is very unstable), we will confine our discussion to those of chlorine. There are four such acids, listed here with their corresponding ions:

Oxidation number of chlorine				
+1	HOCl	hypochlorous acid	OCl^-	hypochlorite ion
+3	$HClO_2$	chlorous acid	ClO_2^-	chlorite ion
+5	$HClO_3$	chloric acid	ClO_3^-	chlorate ion
+7	$HClO_4$	perchloric acid	ClO_4^-	perchlorate ion

As we saw in Section 10-2, the acidity of these acids increases with increasing number of oxygen atoms: $HClO_4$ is the strongest, HOCl the weakest, of the four. HOCl is formed when Cl_2 is added to water:

$$Cl_2 + 2H_2O \rightleftharpoons HOCl + H_3O^+ + Cl^-$$

In this reaction Cl_2 is both oxidized and reduced (see if you can write the half-equations). If a base such as NaOH is added to the water, the HOCl gives up its proton to become the hypochlorite ion, OCl^-. Ordinary household bleach is a solution of sodium hypochlorite NaOCl in water, and is easily made by the addition to water of Cl_2 and NaOH.

HOCl is not stable in the free state, only in water solution. Although a weak acid ($K_a = 2.95 \times 10^{-8}$), it is a good oxidizing agent. Both chlorous acid ($HClO_2$) and chloric acid ($HClO_3$) are also unstable in the free state, and exist only in water solution. Both are oxidizing agents, and $HClO_3$ is also a fairly strong acid.

Perchloric acid ($HClO_4$) is one of the strongest acids known, as well as a powerful, hazardous, oxidizing agent when hot. Unlike the other oxygen acids of chlorine, $HClO_4$ does exist in the free state (though it is stable for only a few days). However, it is usually encountered in water solution.

18-7 The Noble Gases

The six elements of column 0 of the periodic table are called *noble gases*.

In 1868, Joseph Lockyer (1836–1920) and Pierre Janssen (1824–1907) examined the light coming from the sun by means of a spectroscope (Section 12-7). They noticed a line in the spectrum which did not correspond to that of any element then known and concluded that an element existed in the sun which had not yet been discovered on earth. Lockyer called this element *helium*, from the Greek *helios*, meaning sun.* The spectrum gave no information whatever about the element, except that it existed, and it was given the ending *ium*, which is usually reserved for metals because it was thought that since most elements are metals, this one probably was also.

When Mendeleev prepared his periodic table (Section 2-9) he did not include helium, because he did not know its atomic weight or other properties. Nor did he include any of the other noble gases, because none of these had yet been discovered. In 1892, Lord Rayleigh (1842–1919) was trying to prove or disprove Prout's hypothesis (Section 2-5) by measuring atomic weights very accurately to see if they really were exact integers. In doing so, he obtained what he thought was pure N_2 by removing all the O_2, CO_2, and H_2O from the air. He then compared the density (hence the molecular weight) of this N_2 with that of N_2 obtained by the decomposition of certain nitrogen-containing compounds and found that they differed by a small amount, no matter how carefully he carried out his experiments. He was unable to explain this, but William Ramsay (1852–1916) asked Lord Rayleigh for permission to work on this problem, and Ramsay and Rayleigh soon discovered that the "pure N_2" obtained from air was really not pure, and showed by means of a spectroscope that it contained a new element, which he called *argon*.†

Since argon did not fit into any space in the periodic table, this meant that a whole family of similar elements must exist, and by 1900, Ramsay had discovered helium, neon, krypton, and xenon (these were obtained mostly by distillation of liquified impure argon). It was then realized (how?) that the helium discovered in the sun by Lockyer and Janssen also existed here on earth. Radon was discovered in 1900 by Friedrick Dorn (1848–1916). Rayleigh and Ramsay both won Nobel prizes in 1904. Note that the discovery of all the noble gases depended on two previous discoveries: the spectroscope and the periodic table.

THE ELEMENTS. The six elements of group 0 are all monatomic gases, whose boiling points increase with increasing atomic weight (Table 18-2), as we expect for atoms with very small intermolecular forces (Section 6-6). All of them are found in the atmosphere; argon in comparatively large quantities, the others in much smaller amounts (Table 17-1). Helium is also found under the ground, mixed with certain deposits of methane (natural gas). Helium and radon are formed by radioactive decay (Chapter 23).

Helium is an element with several unique properties. It is the only known substance which has *no* solid–liquid–gas triple point (Section 15-6). That is, there is no temperature and pressure at which the solid, liquid, and gas can coexist. Helium is also the only known substance that cannot be solidified at atmospheric pressure.

*The discovery of an element on the sun that was unknown on earth led some people to speculate that the stars and planets might contain whole sets of elements different from those of earth, and maybe even from each other. However, today we know that this is not so. All stars and planets have the same elements.

†Henry Cavendish, in 1785, had removed all the oxygen from dry air by sending electric sparks through air containing excess O_2 ($N_2 + O_2 \rightarrow 2NO$), and was left wath a small bubble of gas that would no longer react with O_2. Cavendish, who of course had no spectroscope, gave no further explanation, but we now know that this bubble was mostly argon.

Table 18-2 Boiling Points of the Noble Gases

	Boiling point	
	°C	°K
He	−269	4
Ne	−246	27
Ar	−186	87
Kr	−152	121
Xe	−107	166
Rn	−65	208

At 1 atm pressure and absolute zero, helium is a liquid. All other substances are solids at 1 atm and 0°K. Solid helium is known, but only above 25 atm pressure. Liquid helium below 2°K has still more remarkable properties. It exhibits **superconductivity,** meaning that it has essentially *no* resistance to an electric current, and it has an extremely low viscosity, so that it flows out of any container holding it (Figure 18-10).

Helium is used in balloons and dirigibles, and, in the liquid form, as a low-temperature bath. Neon is used in neon signs. Argon and helium are used where inert atmospheres are required.

Compounds of the Noble Gases. All these elements have complete octets and are thus very unreactive. No compounds at all are known for helium, neon, and argon. Before 1962 no compounds were known for the other three elements either, but in that year Neil Bartlett (1932–) discovered that xenon could be made to form compounds. About 25 compounds of xenon are now known, in nearly all of which xenon is bonded to either fluorine, oxygen, or both. There are three xenon fluorides XeF_2, XeF_4, and XeF_6, all of which can be made by direct reaction of Xe with F_2, in different proportions. The two known oxides are xenon trioxide (XeO_3) and xenon tetroxide (XeO_4). Both are explosive. XeO_3 can be made by hydrolysis of XeF_6.

$$XeF_6 + 3H_2O \longrightarrow XeO_3 + 6HF$$

In all the xenon compounds, the central atom has more than eight electrons.

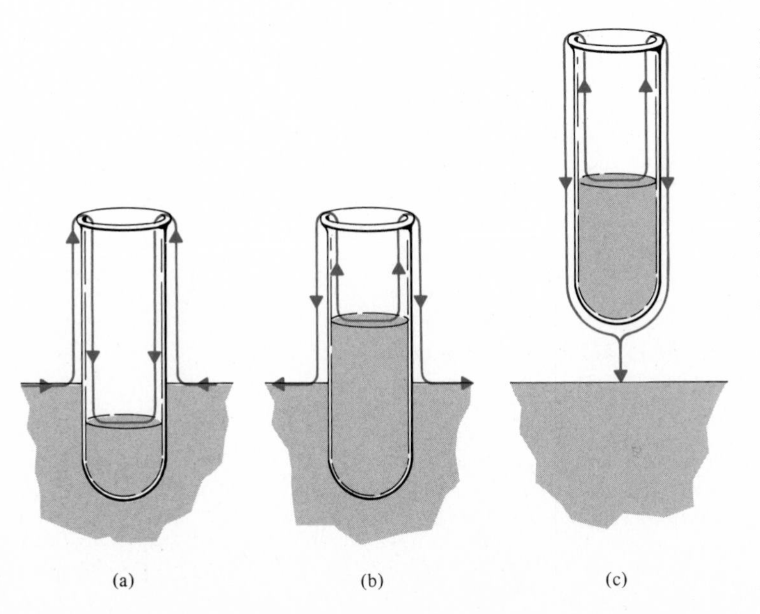

Figure 18-10 Helium flowing out of its container below 2°K. It will flow into (a) or out of (b) a flask lowered into a helium reservoir. (c) If the flask is taken completely out of the reservoir, the helium will flow over the sides and fall back into the reservoir. [Redrawn from "Superfluidity" by Eugene M. Lifshitz. Copyright © 1958 by Scientific American, Inc. All rights reserved.]

Electronic structures of some of them are

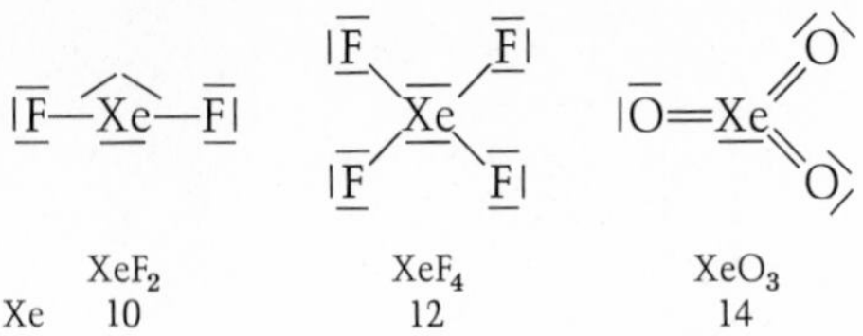

	XeF_2	XeF_4	XeO_3
Number of electrons in outer shell of Xe	10	12	14

The known oxidation states of Xe are 0, +2, +4, +6, and +8.

The chemistry of the noble gases is essentially that of xenon. A few krypton compounds are also known, the most important of which is KrF_2, which decomposes slowly at room temperature. From its position in the periodic table, we would expect that radon should form even more compounds than xenon, but the chemistry of radon has been studied very little, because it is radioactive and decays very rapidly (Chapter 23).

18-8 Oxidation States of the Nonmetals

At this point we may sum up the maximum and minimum possible oxidation states for the elements in groups IVA to 0 (Table 18-3). Note the regularity. Both the maximum and minimum oxidation numbers increase one step at a time as we move from left to right. Even though oxidation number is a purely formal concept, this regularity is related to the fact that the number of electrons in the outer shell of the atom also increases in the same way. In fact, the maximum oxidation number is the same as the number of valence electrons, while the minimum oxidation number is equal to the number of electrons needed to fill the octet. The regularity is not absolute. We recall that the maximum oxidation number for both oxygen and fluorine is 0, not +6 or +7, as is the case for the other elements of groups VIA and VIIA, respectively. Also, in group 0, it is only xenon that goes as high as +8. For three of the elements in this group the only oxidation number known is 0. Finally, we might expect the elements of group IIIA to have an oxidation range of +3 to −5, but the actual range here is +3 to −3.

Table 18-3 Maximum and Minimum Oxidation States

	Group				
	IVA	*VA*	*VIA*	*VIIA*	*0*
Maximum oxidation state	+4	+5	+6	+7	+8
Minimum oxidation state	−4	−3	−2	−1	0

Problems

18-1 Define each of the following terms. **(a)** hydride **(b)** electron-deficient compound **(c)** three center–two electron bond **(d)** bicarbonate **(e)** carbonate **(f)** oxide **(g)** anhydride **(h)** sulfide **(i)** peroxide **(j)** polysulfide **(k)** bisulfite **(l)** sulfite **(m)** noble gas **(n)** superconductivity

18-2 Why does hydrogen belong to a different class from all the other elements?

18-3 In which properties does hydrogen resemble the metals? The nonmetals?

18-4 Complete and balance (if no reaction, write NR):
(a) $H_2O + K \rightarrow$ **(b)** $H_3O^+ + Fe \rightarrow$ **(c)** $H_3O^+ + Cu \rightarrow$ **(d)** $H_2 + GeO_2 \rightarrow$
(e) $NaH + H_2O \rightarrow$

18-5 In B_4H_{10}, two of the boron atoms are connected to each other by an ordinary covalent bond; the other two are each connected to each of the first two by a three center-two electron bond. Draw the electronic structure for B_4H_{10}. How many ordinary covalent bonds and how many three center-two electron bonds are there in this molecule?

18-6 Which elements can form double bonds? Triple bonds?

18-7 In Section 18-3 it was mentioned that the CO molecule has a dipole moment. Which end is the positive end and which the negative end? To help you decide, draw the Lewis structure and see which end has the positive formal charge.

18-8 Two of the three canonical forms of graphite are shown in Figure 18-4. Draw the third canonical form.

18-9 Carbon forms two oxides CO and CO_2, but Si and Ge form only one each. Why is there no SiO or GeO?

18-10 Explain how graphite and diamond, both of which consist entirely of carbon, can have such vastly different properties (you cannot buy a graphite ring or a diamond pencil).

18-11 Draw Lewis structures and predict the shapes of the following species (some of them are resonance hybrids).
(a) ClO_2^- **(b)** ClO_3^- **(c)** SF_4
(d) SF_6 **(e)** BBr_3 **(f)** IF_5
(g) NO_2 **(h)** NO_2^- **(i)** NO_3^-
(j) SiF_6^{2-} **(k)** BBr_4^- **(l)** CCl_4
(m) CO_3^{2-} **(n)** NCl_3 **(o)** H_2Se

18-12 PBr_3 reacts with a proton to give the PBr_3H^+ ion. PBr_5 reacts with Br^- to give the PBr_6^- ion. Draw three-dimensional Lewis structures for all four of these species.

18-13 N_2O can be prepared by heating molten NH_4NO_3. What is the other product? Write the balanced equation.

18-14 When dilute nitric acid is added to a test tube containing copper metal, the space directly above the liquid is colorless, but above that a brown layer of gas is seen, and then a colorless space again. Explain.

18-15 Draw the Lewis structure of tetraphosphoric acid, $H_6P_4O_{13}$.

18-16 Write balanced equations for **(a)** the oxidation of H_2O_2 with PbO_2 in acid solution **(b)** the reduction of H_2O_2 with Cr^{3+} in acid solution.

18-17 For each of the following oxides, write the formula for the corresponding hydroxide (if any) and state whether you expect it to be acidic, basic, or amphoteric.
(a) SO_2 **(b)** K_2O **(c)** NO
(d) N_2O_3 **(e)** SiO_2 **(f)** As_2O_3
(g) BaO **(h)** MnO_2 **(i)** CO_2
(j) Cl_2O_3 **(k)** ClO_2

18-18 For each of the following acids or bases, write the anhydride, if any.
(a) H_2SO_4 **(b)** HNO_3 **(c)** LiOH
(d) H_3PO_4 **(e)** $Mg(OH)_2$ **(f)** HCl
(g) HOCl **(h)** H_3BO_3 **(i)** $Cr(OH)_3$
(j) $HMnO_4$ **(k)** $Sr(OH)_2$ **(l)** H_2SO_5
(m) H_2S **(n)** $Y(OH)_3$ **(o)** CH_3NH_2

18-19 Explain why H_2SO_4 is a stronger acid than H_2SO_3.

18-20 How many kilograms of H_2SO_4 can be obtained from 15,750 kg of S, by either the lead chamber or contact process? Is it necessary to have a balanced equation to perform this calculation?

18-21 Why can't the O^{2-} ion exist in water solution?

18-22 Draw a Lewis structure for the tetrathionate ion, $S_4O_6^{2-}$, given that it contains a chain of four sulfur atoms, and that each of the two sulfur atoms at the ends of the chain is connected to three oxygen atoms.

18-23 Why is Cl_2 more soluble in NaOH solutions than in pure water at the same temperature?

18-24 Write balanced equations to show how Cl_2 is reduced to Cl^- by each of the following.
(a) H_2SO_3 **(b)** SO_2 **(c)** SO_3^{2-}

18-25 One way to prepare Cl_2 is to oxidize Cl^- with PbO_2 in acid solution. **(a)** Write a balanced equation for this reaction (the PbO_2 goes to Pb^{2+}). **(b)** The same oxidation may also be carried out with H_2O_2, MnO_4^-, or $Cr_2O_7^{2-}$, all in acid solution. Br_2 and I_2 may be similarly prepared from Br^- and I^-, respectively, by any of these four oxidizing agents. However, F^- cannot be oxidized by any of them. Why not?

18-26 What color and physical state would you predict for astatine?

18-27 Write the half-equations for

$$Cl_2 + 2H_2O \rightleftharpoons HOCl + H_3O^+ + Cl^-$$

18-28 How was it proved that the helium discovered on the sun was the same element as the helium on Earth?

18-29 When an inert atmosphere is required, why are argon and helium usually used instead of neon or krypton?

18-30 Draw Lewis structures for XeF_6 and XeO_4.

18-31 Explain the fact that the only known binary compounds of the noble gases are of Kr, Xe, and Rn with O and F.

Additional Problems

18-32 Most of the rocks on the earth are silicates and carbonates. How do these two types differ in chemical structure?

18-33 Which element is the most abundant in the universe?

18-34 Explain why the boiling point of helium is less than that of H_2, although the molecular weight is higher.

18-35 In acid solution, hydrazine (N_2H_4) is oxidized to N_2 by IO_3^-, which is reduced to I^-. Write a balanced equation for this reaction, remembering that hydrazine is basic and exists in acid solution as $N_2H_5^+$.

18-36 Will H_2O_2 in basic solution reduce SO_4^{2-}? Will it reduce MnO_4^-? Write the balanced equation for the reaction (reactions) that takes place.

18-37 Where do we get the major supplies of the following pure elements: hydrogen, sulfur, bromine, neon, boron, nitrogen?

18-38 Draw a Lewis structure for the unstable compound formaldimine, CH_2NH.

18-39 Which of these molecules behaves as a Lewis acid and which as a base (Lewis or Brønsted)? Can any of them be amphoteric? **(a)** $SbCl_3$ **(b)** $AsBr_5$ **(c)** $(CH_3)_3N$ **(d)** PCl_5 **(e)** SbF_5 **(f)** N_2H_4

18-40 What is the percentage by weight of arsenic in the rat poison Paris green, $3Cu(AsO_2)_2 \cdot Cu(C_2H_3O_2)_2$?

18-41 A sample of impure Na_2O_2 weighing 0.375 g was allowed to react in acid solution with 2.500 g of I^-, which was more than enough to reduce the Na_2O_2 completely to H_2O. The excess I^- was oxidized by another oxidizing agent to I_2, which was titrated with an 0.1648 *N* sodium thiosulfate solution (which goes to the $S_4O_6^{2-}$ ion); 38.6 ml of thiosulfate solution was required. What is the purity of the Na_2O_2 sample, assuming that none of the impurities react with I^- or sodium thiosulfate?

18-42 A 500.0-ml solution of $HClO_4$ is reduced to Cl_2 by excess H_2O_2, which is itself oxidized to O_2. **(a)** Write the balanced equation. **(b)** If the total volume of dry gas ($Cl_2 + O_2$) obtained is 852 ml at STP, what was the initial concentration of $HClO_4$ in moles/liter?

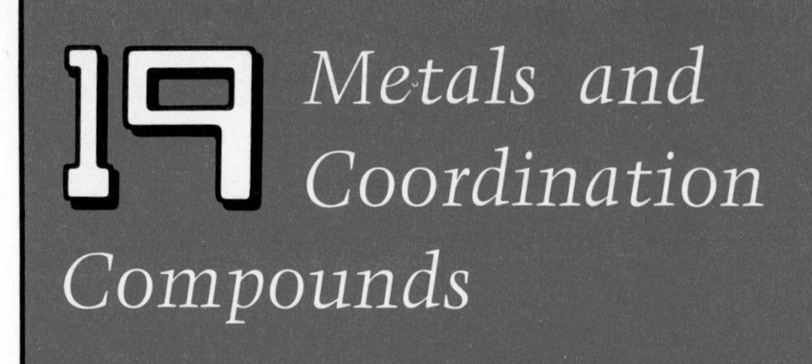

19 Metals and Coordination Compounds

Alfred Werner, 1866–1919

Most of the elements in the periodic table are metals. Metals have a shiny appearance—known as metallic *luster*. Most of them are silvery white in color, although other colors are known, for example, copper is red; gold is yellow. In finely divided form, most metals are black. Most metals are *malleable* (they can be hammered into thin sheets) and *ductile* (they can be drawn out into fine wires). Metals are good conductors of electricity and of heat. It is because they are good heat conductors (high thermal conductivity) that they feel cool to the touch, especially in cool weather (compare the touch of a piece of metal and a piece of wood outdoors on a cold day). The chief chemical property of metals is that they give up electrons, more or less easily, depending on the metal. Thus, most metals form positive ions. Some metals can also form covalent bonds. Metals have little or no tendency to form negative ions. In all these properties, metals differ considerably from the nonmetals and the noble gases. An **alloy** is a solid composed of two or more metals (for the method of preparing an alloy see Section 7-1). Alloys also have the properties listed above.

19-1 Metallurgy

A few metals are found pure in nature. As we might expect, these are the unreactive metals, the ones with very high standard reduction potentials. Examples are gold, platinum, and silver. With such metals, very little processing is required—nuggets or

IA	IIA	IIIB	IVB	VB	VIB	VIIB	VIIIB			IB	IIB	IIIA	IVA	VA			
3 Li	4 Be																
11 Na	12 Mg											13 Al					
19 K	20 Ca	21 Sc	22 Ti	23 V	24 Cr	25 Mn	26 Fe	27 Co	28 Ni	29 Cu	30 Zn	31 Ga					
37 Rb	38 Sr	39 Y	40 Zr	41 Nb	42 Mo	43 Tc	44 Ru	45 Rh	46 Pd	47 Ag	48 Cd	49 In	50 Sn				
55 Cs	56 Ba	57 La	72 Hf	73 Ta	74 W	75 Re	76 Os	77 Ir	78 Pt	79 Au	80 Hg	81 Tl	82 Pb	83 Bi			
87 Fr	88 Ra	89 Ac	104	105													

58 Ce	59 Pr	60 Nd	61 Pm	62 Sm	63 Eu	64 Gd	65 Tb	66 Dy	67 Ho	68 Er	69 Tm	70 Yb	71 Lu
90 Th	91 Pa	92 U	93 Np	94 Pu	95 Am	96 Cm	97 Bk	98 Cf	99 Es	100 Fm	101 Md	102 No	103 Lr

grains of the metal are physically separated from the accompanying rocks, and then washed free of dirt. Most metals, however, are not found free (or if they are, the amounts of free metal available are far from sufficient to supply all our needs) and must be converted to the free metal by chemical reactions. A mineral that contains a metallic compound is called an **ore,** and **metallurgy** is the process of converting an ore to the pure metal, or to an alloy of the metal, if that is what is desired. In most cases, the ores found are oxides of the metals, and the most common chemical process in metallurgy is reduction of metallic oxides. Many metals also occur as sulfides. The refining of these ores requires careful control of pollutant sulfur dioxide from the process.

The earliest metals known to man were those which occur free. Silver, gold, and copper were used in prehistoric times, mostly for small implements such as pins and fishhooks. About 3500 B.C. someone learned how to reduce copper oxides and carbonates to copper, and after this the supply of copper greatly increased. This era is known as the *bronze age*. Bronze is an alloy of copper and tin; and it is likely that copper and bronze were used indiscriminately at this time, with little distinction between them.*

Small amounts of iron, obtained from meteorites, were used in prehistoric times, but the true iron age did not begin until about 1200 B.C., when man learned how to get iron by reduction of iron oxides and carbonates. The metallurgy of iron was developed much later than that of copper, simply because the process is more difficult. Iron is more reactive than copper (Table 16-2) and therefore more reluctant to exist as a free element.

Many metals are, of course, still more reactive than iron, and their compounds are still more difficult to reduce. For example, aluminum is the third most abundant element on the surface of the earth, but it could not be used by ancient people in the way that they used copper and iron (which are less abundant), because they had no way of reducing aluminum ores. As we shall see, today it is done by electrolysis.†

We will discuss the metallurgy of three metallic elements: iron, aluminum, and copper. Industrially, these are three of our most important metals, and the metallurgy is quite different in each case.

IRON AND STEEL. The principal ores of iron are *hematite* (Fe_2O_3) and *magnetite* (Fe_3O_4) (magnetite is actually a mixture of Fe_2O_3 and FeO). These ores are placed in a *blast furnace* (Figure 19-1), where they are reduced to a crude form of iron called *pig iron*. Three other substances must also be present in the blast furnace. These are carbon (in the form of coke), limestone ($CaCO_3$), and oxygen (O_2). The coke and limestone are added through the top, and the oxygen is blown in through openings in the bottom. The oxygen oxidizes the carbon to carbon monoxide, and it is this compound which actually reduces the iron oxides:

$$2C + O_2 \xrightarrow{\Delta} 2CO$$

$$Fe_2O_3 + 3CO \longrightarrow 2Fe + 3CO_2$$

The limestone used in the blast furnace is called *flux*. The purpose of the flux is to remove impurities, chiefly SiO_2, which are present along with the iron ore. At the high temperatures present, the calcium carbonate is converted to calcium oxide, which combines with SiO_2 to give a mixture of silicates called *slag*.

$$CaCO_3 \xrightarrow{\Delta} CaO + CO_2$$

$$CaO + SiO_2 \longrightarrow CaSiO_3$$

*At first, the tin got into the bronze from tin oxides which were often present along with copper oxides in the copper ore. Later, when metallurgical skills increased, tin oxides were deliberately added to the copper ores.

†Aluminum was first prepared around 1825 independently by Hans Christian Oersted (1777–1851) and Friedrich Wöhler (1800–1882). They used potassium metal to reduce $AlCl_3$ to Al.

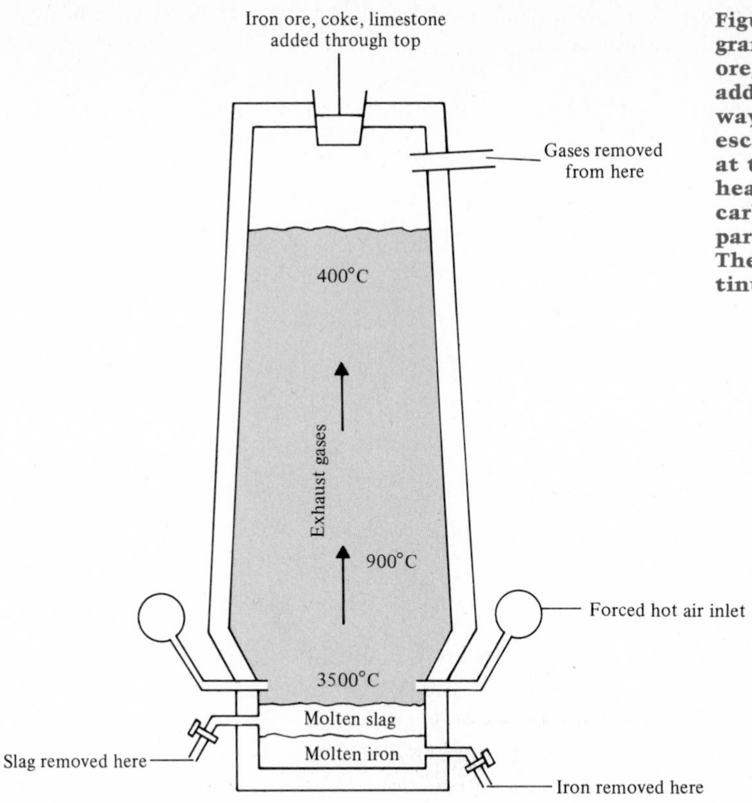

Figure 19-1 Schematic Diagram of a Blast Furnace. Iron ore, coke, and limestone are added at the top in such a way that the gases cannot escape. Hot air is introduced at the bottom. The air is heated by the burning of carbon monoxide, which is part of the exhaust gases. The furnace operates continuously, 24 hours a day.

The molten slag rests on top of the molten iron, and the layers are drawn off separately and allowed to solidify.

Pig iron is relatively soft and malleable, because it contains a high percentage of carbon (3 to 4%), as well as other impurities. In this form it is unsatisfactory for most uses, and almost all of it is converted to *steel*, which is tougher, stronger, and harder. Steel is iron containing selected impurities in selected concentrations. The most important impurity is carbon; it is this that gives steel its properties of hardness and strength. Of course, pig iron also contains carbon—the difference is in the concentration. Pig iron contains 3 to 4% carbon, while a typical steel contains about 0.5% carbon. Many other elements are also used as deliberately added impurities, among them Ni, Cr, W, and Mn.

There are many grades of steel, each of which contains a specific set of impurities in a specific concentration, and consequently has a specific set of properties, but they can be grouped into three main categories: (1) the largest category is *carbon steels*, which are general all-purpose steels; (2) *alloy steels* contain specific alloying elements needed to give certain properties: for example, W, Mo, and V increase hardness, Mo and Cr–Ni provide increased resistance to heat, and Si decreases electrical conductivity; and (3) *stainless steels* are resistant to rusting, the element which imparts this property being Cr.

It is obvious that to convert pig iron to steel, three things must be done: (1) the concentration of carbon must be reduced, (2) unwanted impurities must be removed, and (3) specific alloying elements must be added. Various processes have been used over the past century (e.g., the Bessemer converter, the open-hearth process), but the chief one used today is the *basic oxygen process*. The principle behind this process is very simple (Figure 19-2). A stream of pure O_2 is blown at high velocity onto the surface of the molten pig iron. The process uses high temperatures and is very rapid (less than half an hour). The oxygen oxidizes the excess carbon and other impurities, converting them to oxides. Some of these oxides are gases (chiefly CO_2) which are blown away, and others are solids (chiefly SiO_2, MnO_2). Fluxes are

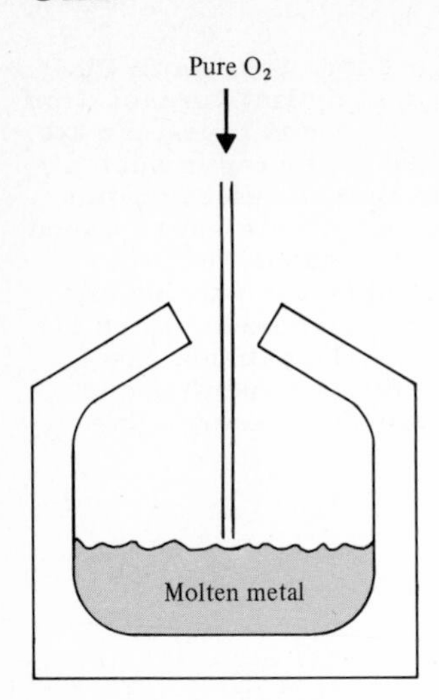

Figure 19-2 The basic oxygen process for the conversion of pig iron to steel.

added (for example, CaO, CaF_2) to combine with the solid oxides to form a slag. Desired impurities (for example, Cr, Mo, V) are added at certain points during the process, while the steel is molten.

ALUMINUM. Aluminum is not as strong as steel, but it has a much lower density and therefore has important uses where a lightweight metal is needed (for example, in airplanes, storm windows, and ladders). For these purposes, the strength of Al is increased by alloying it with small amounts of Cu, Mg, and other elements. Aluminum metal reacts rapidly with the oxygen in the air to give aluminum oxide (also called *alumina*):

$$4Al + 3O_2 \longrightarrow 2Al_2O_3$$

but this reaction is not a drawback, because Al_2O_3 forms a thin hard adherent film on the surface of the aluminum, which protects it from being oxidized further. The chief aluminum ore is *bauxite*, which is impure Al_2O_3. The Al_2O_3 is first purified and then reduced to Al. Because of the unreactivity of Al_2O_3, it is impractical to reduce it chemically, and it is done by an electrolytic method discovered independently in 1886 by Charles Hall (1863–1914) and Paul Héroult (1863–1914) when they were each 23 years old. Humphry Davy had previously (1807–1808) used electrolysis to isolate several active metal elements, including Na, K, Ba, Ca, Sr, and Mg, but Al proved more difficult. Davy and the chemists who came after him tried to obtain Al by electrolysis, but were not successful. The problem was that electrolysis requires the compound to be in a molten state, and the melting point of Al_2O_3 is too high (2045°C). Hall and Héroult solved the problem by finding that Al_2O_3 is soluble in a molten mineral called *cryolite* (Na_3AlF_6). Cryolite has a much lower melting point than Al_2O_3, and electrolysis is carried out on Al_2O_3 dissolved in cryolite at a temperature of about 960°C (Figure 19-3). Aluminum is deposited at the cathode, and O_2 comes off at the anode. This method is simple, but has a high cost because a considerable amount of electricity is required. One ton of purified Al requires 15,000 kilowatt hours of electricity.

COPPER. Copper is a good conductor of electricity (it is second only to silver in this respect) and is widely used to make electrical wires and other electrical equipment. Another important use is in coins. Copper is present in all current U.S. coins, as well as in many coins of other countries. Although native free copper exists on earth, there is not nearly enough for our needs, and so most of our copper must be obtained from ores. It is relatively easy to get copper from copper oxide ores, as the ancients did more than 5000 years ago, but again, the amounts of such ores are

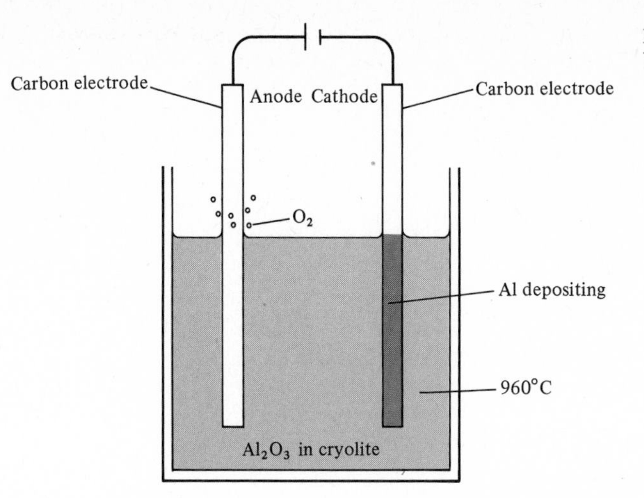

Figure 19-3 Production of aluminum by the Hall-Héroult process.

insufficient, and most copper today is obtained from copper sulfide ores, which usually contain some iron as well. The concentration of copper compounds in these ores is very low (about 1 to 2% copper) and the ores must first be concentrated by mechanical methods. The concentrated ores are then oxidized by heating in air, a process known as **roasting.** In this process the iron sulfides are converted to iron oxides, and the copper sulfides partially converted to copper oxides.

$$2FeS + 3O_2 \longrightarrow 2FeO + 2SO_2\uparrow$$
$$2Cu_2S + 3O_2 \longrightarrow 2Cu_2O + 2SO_2\uparrow$$

Some of the copper oxides react with some of the remaining copper sulfides to produce copper.

$$Cu_2S + 2Cu_2O \longrightarrow 6Cu + SO_2\uparrow$$

A flux of SiO_2 is added to combine with the FeO.

$$FeO + SiO_2 \longrightarrow \underset{\text{slag}}{FeSiO_3}$$

The molten slag is then removed. The remaining material, consisting chiefly of copper sulfides, copper oxides, and copper, is now strongly heated with oxygen, whereupon the oxides and sulfides are converted to copper by

$$Cu_2S + O_2 \longrightarrow 2Cu + SO_2\uparrow$$
$$Cu_2S + 2Cu_2O \longrightarrow 6Cu + SO_2\uparrow$$

At this point the copper, called "blister copper," is 99% pure, but still must be refined further, because some of the impurities lower the electrical conductivity of copper and so interfere with its major use. The final purification is done by electrolysis (Figure 19-4). In this process the impure copper is used as the anode, and a sheet of very pure copper is the cathode. The solution contains a low concentration of Cu^{2+} ions. At the anode, Cu goes into solution as Cu^{2+} ions.

$$Cu \longrightarrow Cu^{2+} + 2e^- \qquad \text{anode reaction}$$
$$Cu^{2+} + 2e^- \longrightarrow Cu \qquad \text{cathode reaction}$$

At the cathode, the Cu^{2+} ions are re-reduced to pure copper. The beauty of this method is that, when the voltage is adjusted properly, it removes virtually all metallic impurities from the copper, whether they are more active (lower $\mathcal{E}°$) or less active (higher $\mathcal{E}°$) than copper. Metals that are less active than copper—Ag, Au, Pt, etc.—cannot be oxidized at the anode as long as metallic copper is present. When atoms of these elements reach the surface of the anode, they drop off and fall to the

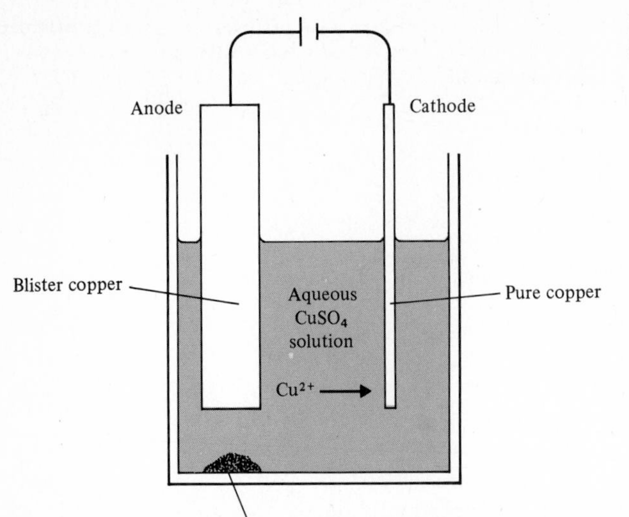

Figure 19-4 Final purification of copper by electrolysis.

bottom, forming a sludge. Because these are valuable metals, the sludge is later purified to obtain the gold, silver, etc. (in fact, the profits obtained from selling these metals are enough to pay for the entire cost of the electrolysis!). The less active metals go into solution at the anode but cannot plate out at the cathode as long as Cu^{2+} ions are present. Thus, both types of impurities are removed, and the copper produced in this way is 99.95% pure.

Iron, copper, and aluminum are important industrial metals, all are fairly abundant on earth, and each occurs in an ore that is relatively free of large concentrations of other metals. Things are quite different for many other metals. They are found in low abundance, often in low concentrations and mixed with relatively large amounts of other metals. As a result of this, many metals are produced as by-products in the refining of other metals. For example, cadmium is a by-product in the production of zinc; tantalum in the production of niobium, and, as we have seen, silver in the production of copper. Note that in all three of these cases, and in most other cases, the metals found together are in the same column of the periodic table.

19-2 The Metallic Bond

X-ray crystallography shows that the atoms in most solid metals are stacked in one of three patterns, all of which were described in Sections 15-2 and 15-3. These are hexagonal close packing (hcp), cubic close packing (ccp), and body-centered cubic (bcc). Figure 19-5 shows the most stable arrangement for many of the metals. In the

Li	Be												
bcc	hcp												
Na	Mg											Al	
bcc	hcp											ccp	
K	Ca	Sc	Ti	V	Cr		Fe	Co	Ni	Cu	Zn		
bcc	ccp	ccp	hcp	bcc	bcc		bcc	hcp	ccp	ccp	hcp		
Rb	Sr	Y	Zr	Nb	Mo	Tc	Ru	Rh	Pd	Ag	Cd		
bcc	ccp	hcp	hcp	bcc	bcc	hcp	hcp	ccp	ccp	ccp	hcp		
Cs	Ba	La	Hf	Ta	W	Re	Os	Ir	Pt	Au		Tl	Pb
bcc	bcc	hcp	hcp	bcc	bcc	hcp	hcp	ccp	ccp	ccp		hcp	ccp

Ce	Pr	Nd			Eu	Gd	Tb	Dy	Ho	Er	Tm	Yb	Lu
ccp	hcp	hcp			bcc	hcp	hcp	hcp	hcp	hcp	ccp	hcp	hcp

Figure 19-5 Crystal structures of the most stable solid forms of many of the metallic elements: hcp, hexagonal close packed; ccp, cubic close packed; bcc, body-centered cubic. Blank spaces indicate that the crystal structure is none of these three or that it is unknown.

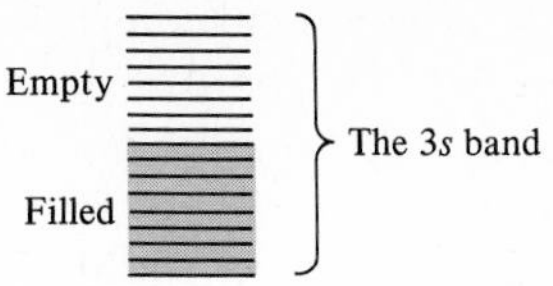

Figure 19-6 The energy levels of the 3s molecular orbitals in a piece of sodium are so close together that they form an essentially continuous band, of which the bottom half is filled with electrons.

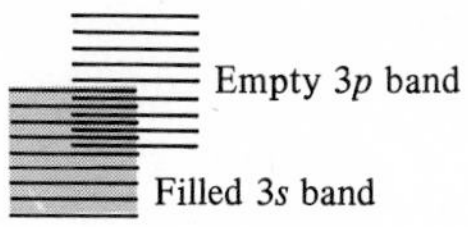

Figure 19-7 The valence bands of magnesium.

hcp and ccp arrangements, each atom is surrounded by 12 other atoms an equal distance away (nearest neighbors). In the bcc arrangement, each atom has eight nearest neighbors, though there are six more atoms not much farther away.

What type of bonding can it be that holds all of these atoms together in the crystal lattice? Let us examine some of the types of bonding we are familiar with. Ionic bonding is obviously impossible here: all the atoms would like to give up electrons, but none are willing to accept them. Ordinary covalent bonding will also not do: how could a beryllium atom, for example, with only two outer-shell electrons, form covalent bonds with 12 nearest neighbors? A third possibility is van der Waals forces, but a comparison with the noble gases will show us that this cannot be the explanation either. Take, for example, krypton. The atoms of this element *are* held together in the solid by van der Waals forces, and the melting point is -157°C. If the atoms of metals were also held together by van der Waals forces, we would expect that elements like Rb, Sr, Cu, and Zn, whose atomic weights are not much different from that of Kr, would also have such low melting points. In fact, their melting points are all very much higher than this.

Several theories have been proposed to explain the bonding in metals. We will qualitatively discuss one of these, called the **band theory.** Let us consider sodium, which has one electron in its outer shell, in a 3*s* orbital. In this case (bcc), each atom has eight nearest neighbors, and there is no reason why any one of them should be favored over the others. Therefore, the 3*s* orbital of each Na atom overlaps to an equal extent with the 3*s* orbitals of its eight nearest neighbors. But the 3*s* orbital of each of the eight nearest neighbors also overlaps with the orbitals of *its* seven other nearest neighbors, and these in turn overlap with *their* nearest neighbors. The result is that there is a continuous overlap of 3*s* orbitals throughout the entire piece of sodium.

Recall (Section 14-18) that overlap of any number of atomic orbitals gives rise to an equal number of molecular orbitals. In this case, the number of 3*s* orbitals overlapping is equal to the total number of Na atoms in the entire piece of sodium, so that an extremely large number of molecular orbitals are formed (6×10^{23} orbitals in a piece of Na weighing 23 g). Because there are so many molecular orbitals, their energies are very close together, and they form an essentially continuous **energy band,** which we can call the 3*s* band (Figure 19-6).

Once we have set up our molecular orbitals, the next step is to fill them. Each Na atom has one 3*s* electron, but each orbital holds two electrons. Since there are as many orbitals as there are Na atoms, this means that the bottom half of the band will be filled and the top half empty (Figure 19-6). Orbitals of the *inner* electrons of sodium (1*s*, 2*s*, 2*p*) are not near enough to overlap, and these electrons may be regarded as localized on each atom.

In a piece of sodium, then, there is a band of electrons covering the entire piece, which is half-filled and half empty. What happens in the case of a metal such as magnesium, where each atom has two electrons in the 3*s* orbital? In this case, the entire 3*s* band is filled, but there is another band (the 3*p* band) arising from overlap of the empty 3*p* orbitals, which partially overlaps the 3*s* band (Figure 19-7). That is, the lowest energy level of the $3p$ band has less energy than the highest energy level of the $3s$ band. All metals have an arrangement either like that of sodium or of magnesium. Either the highest-occupied band is half-filled, or, if it is filled, it overlaps with an empty band. We can now see what it is, according to the band theory, that holds the atoms of a solid metal together: it is *overlap of atomic orbitals to form molecular orbitals,* though to be sure, in quite a different way than is the case for ordinary covalent bonds.

The band theory not only explains the bonding but also the characteristic properties of metals which were mentioned at the beginning of this chapter.

1. *Electrical conductivity.* When a voltage is applied across the two ends of a piece of metal, an electrical current flows. This current is a stream of electrons. The electrons move easily, because filled bands (or parts of bands) are very close to empty bands (or parts of bands). The electrons move rapidly, from atom to atom, through the empty bands or the empty parts of bands (remember that these bands cover every atom in the piece of metal). For example, the electrons in Na move through the upper half of the $3s$ band, while those in Mg move through the $3p$ band (Figures 19-6 and 19-7). Of course, the metal does not lose any electrons while the current is flowing. Every time one electron moves away from an atom, its place is immediately taken by another electron right behind it. But it is the presence of empty bands (or parts of bands) lying very close in energy to filled bands (or parts of bands) which allows the current to flow.

2. *Thermal conductivity.* The relative freedom of electron motion also explains the high thermal conductivity of metals. When one end of any solid piece of matter is heated, the atoms or molecules vibrate faster, but the heat cannot spread rapidly throughout the piece unless a means is available for the transmission of energy (that is, motion of the atoms or molecules). In metals, the mobile electrons provide such a means. Atoms at the heated end of the piece vibrate faster, and thus collide more vigorously with neighboring electrons, giving them a higher energy (faster motion). The high-energy mobile electrons rapidly move away from the heated area, and impart their energy to atoms in their new locations, causing the latter to vibrate faster, which, of course, increases the temperature at that site. In a material with a low thermal conductivity (for example, glass), there is no means for the transmission of heat, and glass-blowers easily hold one end of a glass tube in their hands while five centimeters away (two inches), another part of the tube is immersed in the flame of an oxygen torch.

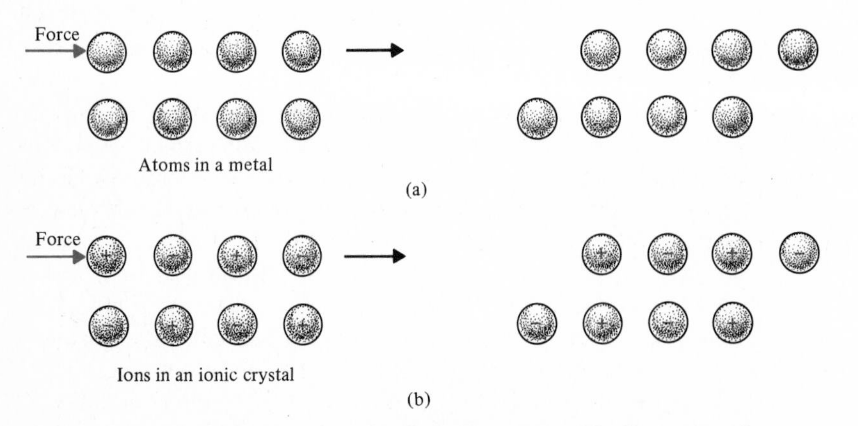

Figure 19-8 (a) When a stress is placed on a metal, the rows of atoms can slide past each other without affecting the attractive forces. (b) When the same thing is done to an ionic crystal, the attractive forces are completely disrupted. Ions of the same charge are now forced to be next to each other. If the stress is great enough, the crystal cleaves—it breaks apart, leaving smooth planes.

3. *Metallic luster.* Because the electron bands actually consist of many energy levels very close together, metals absorb light of a wide range of wavelengths. This light excites the electrons and causes them to move to higher levels. When they fall back to the ground state, the energy is given off as light, and it is this that causes the lustrous appearance of metals.

4. *Ductility and malleability.* It is relatively easy to deform a metal, that is, to change its shape (of course, metals differ in how easy this is to do). We can hammer a piece of gold into a thin leaf; we can bend a piece of sheet steel until it forms the shape of an automobile fender. This kind of deformation is impossible for ionic crystals. The reason is that nothing much is changed when the atoms of a metal slide past each other, but this is not true for an ionic solid. Figure 19-8 shows symbolically what happens when a stress is placed on a metal and on an ionic crystal. It is obvious that bonding in the crystal would be completely disrupted, while bonding in the metal is unchanged.

19-3 Semiconductors

Metals are good conductors of electricity because an empty band (or part of a band) is adjacent to or overlaps with a filled (or partly filled) valence band. In materials that do not conduct electricity (*insulators*) there is no empty band close to a filled one. The nearest empty band (or empty orbital) has an energy considerably higher than the filled one, and the electrons from the filled band are unable to reach the empty one (Figure 19-9 shows a schematic representation of this situation). **Semiconductors** are intermediate between these two situations. Here there is a relatively small energy between filled and empty bands. Typical intrinsic semiconductors are the metalloids selenium and boron. Semiconductors conduct electricity, but poorly.

A characteristic of semiconductors is that their electrical conductivity increases with increasing temperature, because the increasing temperature causes more electrons to jump to the empty bands or orbitals. This behavior is opposite to that of metals, where increasing temperature *decreases* electrical conductivity. This difference in behavior can be used to distinguish a material that is a conductor of electricity, although a poor one, from a true semiconductor.

Very pure germanium and silicon are extremely poor semiconductors, which is what we would expect from their diamond-like molecular structures (Section 18-3), in which all the bonding is covalent and free electrons are unavailable. The small degree of conductivity which they do display, especially at high temperatures, is caused by the breakdown of some of the covalent bonds, which frees some of the electrons. However, the conductivity of these elements can be greatly increased by adding very tiny controlled amounts of certain impurities. For example, the conductivity of very pure silicon at room temperature is increased by a factor of 10^5 by the addition of 1 ppm of boron. The addition of such impurities is known as **doping.** Silicon and germanium that have been doped in this manner are called

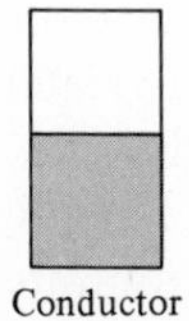
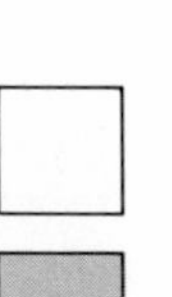
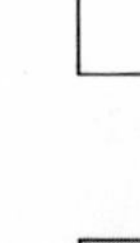
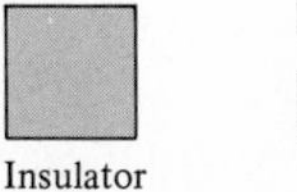

Figure 19-9 Schematic diagram of conductors, semiconductors, and insulators.

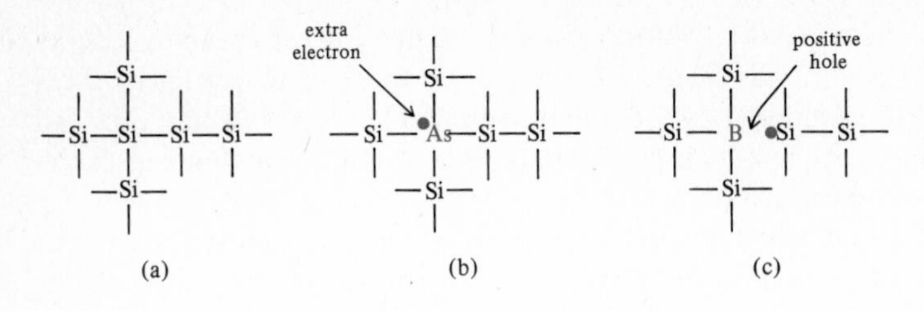

Figure 19-10 Schematic representations of pure silicon (a), silicon doped with arsenic (b), and silicon doped with boron (c).

impurity semiconductors. There are two types of these impurity semiconductors.

1. The addition of small amounts of a group VA element such as arsenic produces an **n-type semiconductor.** Each arsenic atom fits into the regular covalent diamond-like crystal structure of silicon or germanium, displacing a Si or Ge atom (the situation is shown schematically in Figure 19-10). The As atom has five outer-shell electrons, but only four are required to bond to the four silicons. The extra electron is left over and is free to travel through the crystal when a voltage is applied. The *n* in *n*-type semiconductor stands for negative, but this refers to the fact that electrons are free to move, not that there is an excess of electrons. In fact, there is no excess of electrons (As has an excess proton as well as an excess electron), and the entire crystal is neutral.

2. The addition of small amounts of a group IIIA element such as boron produces a ***p*-type semiconductor** (*p* for positive). In this case the boron atom has only three electrons and can form only three regular bonds (Figure 19-10c). The bond to the fourth Si atom contains only one electron (provided by the Si), and has room for another. This vacancy is called a **positive hole,** and the material can carry a current when a voltage is applied because a neighboring electron jumps into the hole, thus leaving behind another positive hole, which in turn is filled by another neighboring electron, etc. We can think of the current as being carried by the movement of positive holes, though, of course, only electrons are actually moving.

The usefulness of semiconductors is greatly increased when a *p*-type semiconductor is joined to an *n*-type semiconductor **(a *p-n* junction).** In this case, current will flow in only one direction: from the *n*-type side to the *p*-type side (by comparison, an ordinary piece of metal conducts equally well in any direction). This property allows *p-n* junctions to be used as rectifiers: devices that convert alternating to direct current. By far the most important application of semiconductors is their use in *transistors,* which perform the same functions as vacuum tubes and have the advantages of small size, lack of heat, low energy consumption, and long life. For these reasons they have replaced vacuum tubes in most applications, including radios, television sets, and computers. In addition, transistors have made possible a number of modern devices, including inexpensive calculators, digital watches, and hearing aids. Without the transistors contained in the electronic components involved, it would not have been possible for human beings to land on the moon in 1969.

Another use of semiconductors is also worthy of mention. We have pointed out that the electrical conductivity of semiconductors increases with increasing temperature. This increase is so great (about 4% per degree at room temperature for a typical semiconductor) that it provides a convenient way to measure temperature—the conductivity is measured and compared to a calibration curve. Semiconductors used to measure temperature are called **thermistors,** and they have the advantages over thermometers that they measure temperature with much greater precision (to 0.0001°C) and can be made very small so as to fit into small spaces. Semiconductors are also used as solar batteries, devices that convert light energy into electrial energy. Solar batteries are very useful in space satellites, but at this time are too expensive for routine use on earth.

19-4 Representative Metals

THE ALKALI METALS (GROUP IA). The alkali metals are soft (they can be cut with a knife), and have relatively low densities and melting points (Table 19-1). All of them are very reactive, and the reactivity increases in going down the periodic table, in accord with the electronegativities (Table 14-2), and first ionization energies (Table 13-13). All of these metals react with water, for example,

$$2Na + 2H_2O \longrightarrow H_2 + 2Na^+ + 2OH^-$$

The reactions with water become more violent in going down the periodic table. Potassium usually bursts into flame when added to water; rubidium and cesium usually explode. Since water in this reaction acts as an acid as well as an oxidizing agent, the alkali metals also react (violently) with any acid stronger than water. Because of their violence, students should not carry out these reactions without supervision.

All the alkali metals have one electron in their outer shells, and all of them bond almost exclusively by ionic bonds. The only ions formed are E^+ ions (of course, we predict from the second ionization energies in Table 13-13 that E^{2+} ions would be extremely unlikely). Therefore, the only possible oxidation states for group IA metals are 0 and +1. Since all the alkali metals readily give up their single valence electron, and loss of electrons is oxidation, the alkali metals are all very good reducing agents. The alkali metal ions E^+ are almost inert. They cannot be oxidized and are very hard to reduce. They are potential Lewis acids but are extremely weak in this respect. Two of the alkali metal ions, Na^+ and K^+, are very common, both in nature and in the chemical laboratory, and many salts of these ions are known, virtually all of which are soluble in water (Section 11-1). Lithium carbonate (Li_2CO_3) is an important drug that is used to treat various mental disorders, especially manic behavior.

THE ALKALINE EARTH METALS (GROUP IIA). These metals have properties similar to those of the alkali metals, but less extreme in each case. They are less reactive, less soft, and have higher densities, melting points (Table 19-1), and boiling points, because there are twice as many bonding electrons present. Beryllium forms mostly covalent bonds (valence = 2), but also Be^{2+} ions in some compounds. Magnesium also forms covalent bonds but has a greater tendency then Be to form E^{2+} ions. The other elements form only E^{2+} ions, and no covalent bonds. Oxidation numbers are 0 and +2. The most important element in this group is magnesium, which is used as a structural metal in applications that make use of its low density (lower than that of aluminum). However, metallic magnesium does have the disadvantage that it burns in air (with a bright light) to give magnesium oxide (MgO), and the ignition temperature is relatively low (this property is taken advantage of in magnesium flares). A suspension of magnesium hydroxide, $Mg(OH)_2$, in water is called *milk of magnesia,* and has been used for many years as

Table 19-1 Melting Points and Densities of the Group IA and IIA Elements

Melting points (°C)		*Densities (g/ml at 20°C)*	
IA	*IIA*	*IA*	*IIA*
Li 180	Be 1278	Li 0.534	Be 1.848
Na 98	Mg 651	Na 0.971	Mg 1.738
K 64	Ca 842	K 0.862	Ca 1.55
Rb 39	Sr 769	Rb 1.532	Sr 2.54
Cs 28	Ba 725	Cs 1.873	Ba 3.5
Fr —	Ra 700	Fr —	Ra ~5

a laxative. Magnesium is obtained from the sea, where it occurs as Mg^{2+} ions. Unlike many other materials, the potential supply of magnesium is virtually inexhaustible, as can be seen from the fact that each cubic mile of seawater contains about 6 million tons of Mg^{2+} ions, and there are about 330 million cubic miles of seawater.

The Group IIB Metals, Zn, Cd, and Hg. These elements are similar in some ways to the group IIA metals. They also have two electrons in their outer shells, though in this case there is a filled *d* subshell underneath rather than a complete shell as in the case of the IIA metals (Table 13-8). All these metals form E^{2+} ions, and all can form covalent bonds as well (valence = 2), so all have oxidation states of 0 and +2. Mercury, which is a liquid metal (an old name for mercury is *quicksilver*), also forms a series of compounds containing the Hg_2^{2+} ion (*mercurous ion*) [for example, Hg_2Cl_2, $Hg_2(NO_3)_2$]. In these compounds mercury is in the +1 oxidation state. The two Hg atoms are connected to each other by a covalent bond.

$$^{+}Hg—Hg^{+}$$

The Group IIIA Metals, Al, Ga, In, Tl. These metals behave as we would expect from the foregoing. All form E^{3+} ions, and all form covalent bonds (valence = 3) as well. In the covalent compounds (for example, $AlCl_3$, $GaBr_3$, InF_3), the central atom has only six electrons in the outer shell, and all of them are Lewis acids (Section 14-5). Thallium also forms stable +1 compounds (Tl^+ ions), the only member of this group to do so.

The Group IVA and VA Metals, Sn, Pb, Bi. Tin and lead, both known since antiquity, are important industrial metals. Tin is used to coat the steel in "tin" cans (this prevents corrosion of the steel), as well as in low-melting alloys such as solder. Lead is used in lead storage batteries (Section 16-9), in paints, in making lead tetraethyl (Section 17-9), and in other ways. In former times it was used to make pipes for carrying water (some of these pipes, dating from Roman times, are still in service), but is no longer used for this purpose because lead ions, some of which dissolve in the water, are poisonous. Lead and tin both have oxidation states of 0, +2, and +4. Both can form either two or four covalent bonds, as well as Sn^{2+} and Pb^{2+} ions. Pb^{4+} and Sn^{4+} ions are unknown. Bismuth, which is mostly used to make alloys with other metals, forms either three or five covalent bonds, as well as Bi^{3+} ions (not Bi^{5+} ions).

19-5 Transition Metals

In Section 13-8 we defined a transition element as an element with one or more incomplete inner shells. There are two kinds of transition elements. One kind contains atoms with only *one* incomplete inner shell; we can call these the *main transition elements*. Atoms of the other kind contain two incomplete inner shells; these are known as *inner transition elements* (next section). All transition elements of both kinds are metals. All of them are hard, strong, and have high melting and boiling points.

The periodic table shows that there are three series of main transition metals: (1) scandium to copper, (2) yttrium to silver, and (3) lanthanum to gold. Ground-state electronic configurations (from Table 13-8) of the outer shell and the incomplete inner shell of these 27 elements are shown in Table 19-2. Although the ground-state electronic configurations of the elements in a vertical column in most cases are not exactly parallel (for example, Fe and Os have d^6s^2, but Ru has d^7s^1), there are, nevertheless, many resemblances in properties within each column. In

Table 19-2 Ground-State Electronic Structures of the Two Outermost Shells of the 27 Main Transition Elements

Sc	$3d^14s^2$	Y	$4d^15s^2$	La	$5d^16s^2$
Ti	$3d^24s^2$	Zr	$4d^25s^2$	Hf	$5d^26s^2$
V	$3d^34s^2$	Nb	$4d^45s^1$	Ta	$5d^36s^2$
Cr	$3d^54s^1$	Mo	$4d^55s^1$	W	$5d^46s^2$
Mn	$3d^54s^2$	Tc	$4d^65s^1$	Re	$5d^56s^2$
Fe	$3d^64s^2$	Ru	$4d^75s^1$	Os	$5d^66s^2$
Co	$3d^74s^2$	Rh	$4d^85s^1$	Ir	$5d^76s^2$
Ni	$3d^84s^2$	Pd	$4d^{10}$	Pt	$5d^96s^1$
Cu	$3d^{10}4s^1$	Ag	$4d^{10}5s^1$	Au	$5d^{10}6s^1$

general, there is a greater resemblance in each column between the properties of the elements of the second and third series than there is of the first and second, or of the first and third. The electronic structures of the transition metals gives rise to several characteristic properties.

1. *Variable oxidation states.* All the main transition elements except Sc, Y, and La form compounds in two or more different oxidation states. Table 19-3 shows the oxidation states formed by these 27 elements. Table 19-4 shows some typical compounds in some of these states. The reason for the variable oxidation states is that both *d* and *s* electrons are available for bonding. For example, manganese can form the Mn^{2+} ion (oxidation number +2) by losing its two 4*s* electrons, but it can also form covalent bonds by using the two 4*s* electrons plus anywhere from one to all five of its 3*d* electrons (oxidation numbers +3 to +7). Sc, Y, and La easily form the corresponding +3 ions by loss of all three electrons, and no other oxidation state is known for these elements. It is customary to refer to the oxidation state of a metal by a Roman number either in parentheses or as a superscript. For example, manganese in the +2 oxidation state is referred to as Mn(II) or Mn^{II}.

2. *Constant oxidation state throughout a series.* Table 19-3 shows that all the elements from Ti to Cu have stable +2 oxidation states. The reason is that all but Cr and Cu have an outer shell consisting of $4s^2$. Loss or sharing of these two electrons results in a stable +2 state existing for all the metals in this series (Cr and Cu lose or share their single 4*s* electron and one of their *d* electrons). The same tendency is found with other oxidation numbers. Thus, Sc to Ni all have +3 oxidation states, as do La to Au; and Zr to Pd and Hf to Pt all have +4 oxidation states.

3. The *highest* oxidation state of the first five or six elements in each series is equal to the total number of *d* and *s* electrons in the two outer shells. Table 19-3 shows that this is true for Sc to Mn, Y to Ru, and La to Os. In each of these cases, the atom forms its maximum oxidation state by using all of the *d* and *s* electrons in the two outer shells (Table 19-2). It is largely because of these oxidation states that Mendeleev originally included these elements as the "B" groups of the periodic table. For example, the maximum oxidation state of S, Se, and Te is +6 (Section 18-5); just as it is for Cr, Mo and W. Because of this, Mendeleev put all of these elements into group VI of the periodic table. However, he also recognized that in most other respects, the properties of these two series of elements are totally different. Therefore, he called one set VIA and the other VIB.

Table 19-3 shows that for the last four or five elements in each series, the maximum oxidation number is *less* than the total number of *d* and *s* electrons in the two outer shells. This is because from this point on (except for Fe) each element has six or more *d* electrons (Table 19-2), which means that each element must have one or more filled *d* orbitals (that is, *d* orbitals containing two electrons). These *d* electron pairs are much less available for bonding than are *d* orbitals that contain only one electron.

4. Lower oxidation states are *less stable,* and higher oxidation states are *more stable,* in the second and third series than in the first. There are many examples of

Table 19.3 Oxidation Numbers of Transition Elements

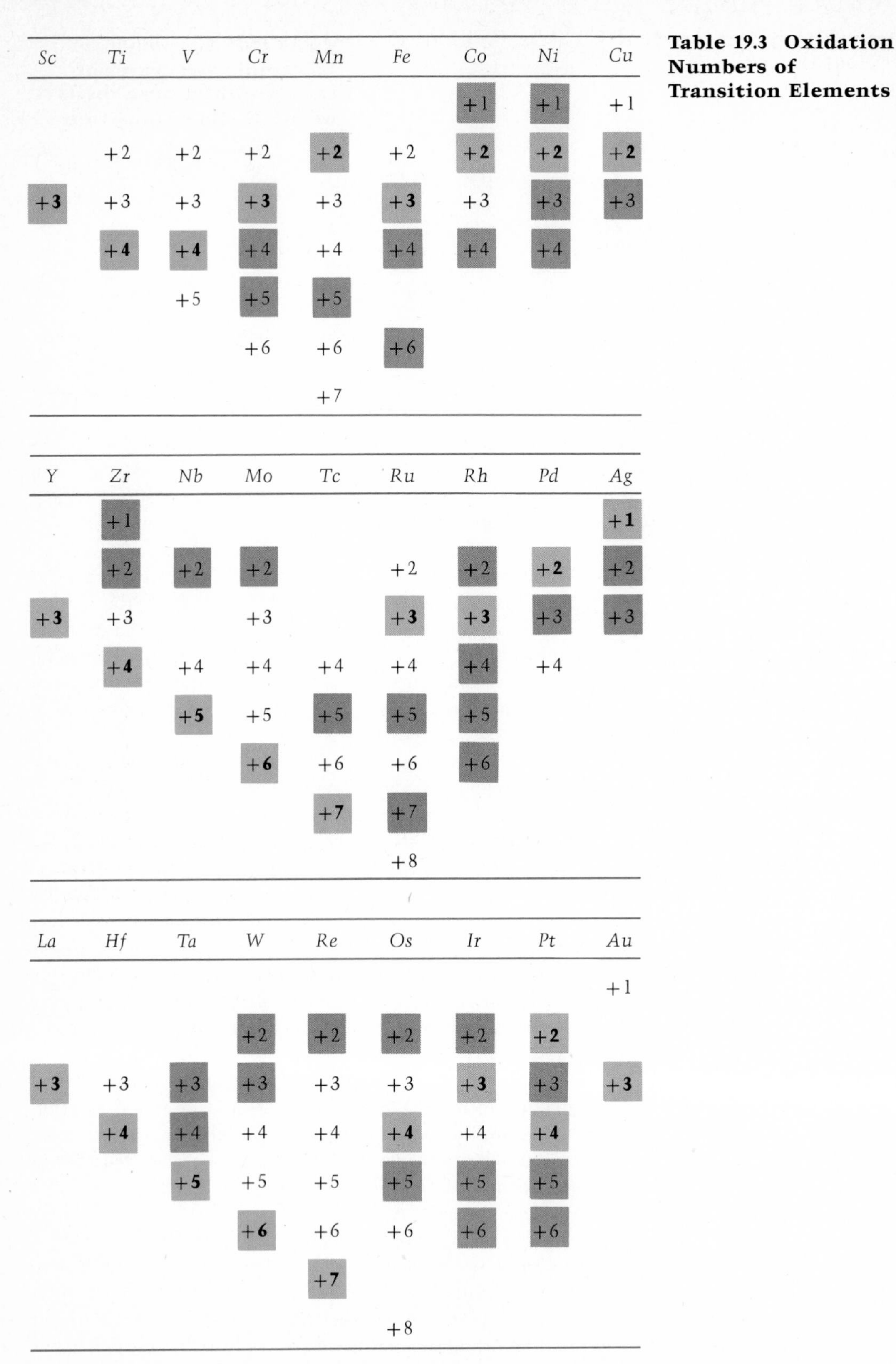

Sc	Ti	V	Cr	Mn	Fe	Co	Ni	Cu
						+1	+1	+1
	+2	+2	+2	**+2**	+2	**+2**	**+2**	**+2**
+3	+3	+3	**+3**	+3	**+3**	+3	+3	+3
	+4	**+4**	+4	+4	+4	+4	+4	
		+5	+5	+5				
			+6	+6	+6			
				+7				

Y	Zr	Nb	Mo	Tc	Ru	Rh	Pd	Ag
	+1							**+1**
	+2	+2	+2		+2	+2	**+2**	+2
+3	+3		+3		**+3**	**+3**	+3	+3
	+4	+4	+4	+4	+4	+4	+4	
		+5	+5	+5	+5	+5		
			+6	+6	+6	+6		
				+7	+7			
					+8			

La	Hf	Ta	W	Re	Os	Ir	Pt	Au
								+1
			+2	+2	+2	+2	**+2**	
+3	+3	+3	+3	+3	+3	**+3**	+3	**+3**
	+4	+4	+4	+4	**+4**	+4	**+4**	
		+5	+5	+5	+5	+5	+5	
			+6	+6	+6	+6	+6	
				+7				
					+8			

[a]Those on color boxes are the most stable states; those on grey boxes are uncommon or unstable.

Table 19-4 Some Compounds of the Main Transition Elements in Various Oxidation States

Oxidation state									
+1								NiCN	CuBr
+2		$TiCl_2$	VS	CrF_2	$MnCO_3$	FeO	CoS	$Ni(OH)_2$	$CuCl_2$
+3	Sc_2O_3	Ti_2O_3	V_2O_3	$CrCl_3$	MnF_3	$FeCl_3$	Co_2O_3		$KCuO_2$
+4		$TiCl_4$	VF_4	$CrBr_4$	MnO_2				
+5			VOF_3						
+6				CrO_3	K_2MnO_4				
+7					$KMnO_4$				
+1									AgCl
+2						$RuCl_2$	$RhCl_2$	PdO	AgF_2
+3	YBr_3	ZrF_3		MoF_3		$RuBr_3$	$RhCl_3$	Pd_2O_3	
+4		ZrO_2	NbS_2	MoO_2	TcS_2	RuF_4	RhF_4	PdO_2	
+5			Nb_2O_5	Mo_2S_5					
+6				MoO_3	TcF_6	K_2RuO_4			
+7					Tc_2O_7	$KRuF_6$			
+8						RuO_4			
+1									Au_2O
+2								$PtCl_2$	
+3	La_2S_3	$HfBr_3$			Re_2O_3	$Os(NO_3)_3$	$IrBr_3$	$PtBr_3$	Au_2O_3
+4		HfF_4	TaO_2	WO_2	$ReCl_4$	OsO_2	IrO_2	K_2PtCl_6	
+5			$TaCl_5$	WF_5	Re_2O_5	OsF_5	IrF_5	PtF_5	
+6				WCl_6	ReF_6	OsF_6	IrF_6	PtF_6	
+7					$KReO_4$				
+8						OsO_4			

this in Table 19-3. Thus, the +2 is an important state for all of the elements in the first series except Sc, but is only important for a few of the elements in the two lower series. Another example concerns the elements Cr, Mo, W, and Mn, Tc, Re. The +6 state is the most stable for Mo and W; the +7 for Tc and Re, but for Cr and Mn, these states are much less stable. Cr^{VI} exists only in a few oxygen compounds, such as chromium trioxide (CrO_3) and potassium chromate (K_2CrO_4), which are powerful oxidizing agents, being reduced to Cr^{3+}. The yellow chromate ion (CrO_4^{2-}) is converted in acid solution to the orange-red dichromate ion.

$$\underset{\substack{\text{chromate ion}\\ \text{yellow}}}{2CrO_4^{2-}} + 2H_3O^+ \rightleftharpoons \underset{\substack{\text{dichromate ion}\\ \text{orange-red}}}{Cr_2O_7^{2-}} + 3H_2O$$

The dichromate ion, in which the oxidation number of Cr is also +6, is also a strong oxidizing agent. Similarly, Mn^{VII} is also found in a few oxygen compounds, the most important of which is potassium permanganate $KMnO_4$, a deep purple compound, which, like K_2CrO_4, is a strong oxidizing agent, being reduced to the pink Mn^{2+} in acid solution and to the black insoluble MnO_2 in basic solution. The much greater stability of Mo^{VI}, W^{VI}, Tc^{VII}, and Re^{VII} is shown by the facts that (1) more compounds of these states exist (for example, WF_6, $MoOF_4$, Tc_2S_7, ReF_7, etc.) and (2) the corresponding oxygen compounds, such as K_2MoO_4, Na_2WO_4, $NaTcO_4$, and $KReO_4$, are much weaker oxidizing agents than K_2CrO_4 or $KMnO_4$.

5. In general, ionic bonding is important only for the first series, and only for the lower oxidation states within this series. All the elements from Ti to Cu form compounds containing E^{2+} ions. In water solution, these ions exist in hydrated form

Table 19-5 Hydrated Ions of the First Series of Transition Metals, and Their Colors in Aqueous Solution

$Ti(H_2O)_6{}^{3+}$	Violet
$V(H_2O)_6{}^{2+}$	Violet (unstable)
$V(H_2O)_6{}^{3+}$	Blue
$Cr(H_2O)_6{}^{2+}$	Blue
$Cr(H_2O)_6{}^{3+}$	Violet
$Mn(H_2O)_6{}^{2+}$	Pale pink
$Fe(H_2O)_6{}^{2+}$	Blue-green
$Fe(H_2O)_6{}^{3+}$	Yellow
$Co(H_2O)_6{}^{2+}$	Pink
$Ni(H_2O)_6{}^{2+}$	Green
$Cu(H_2O)_6{}^{2+}$	Blue

(Table 19-5). Ti^{+2} does not exist in water solution, because it is oxidized by the water to Ti^{3+}. V^{2+} is unstable in water solution. Some of the first series elements also form E^{3+} ions (especially Sc), but no E^{4+} or more highly charged ions are known, not even for Ti or V, where +4 is the most stable state. In higher oxidation states, the bonding is covalent. Only a few ions of any kind are known for the elements of the second and third series (for example, Ru^{2+}, Rh^{3+}, Pd^{2+}, Ag^+). In part, this is because the +2 and +3 oxidation states, which is where the ions are, are less important here. E^+ ions are found only for Cu^+ and Ag^+. However, Cu^+ ions, as well as other +1 copper compounds, are relatively unstable, and react with each other to give Cu^{2+} ions and Cu atoms:

$$2Cu^+ \longrightarrow Cu^{2+} + Cu$$

This type of reaction, in which one ion is oxidized by another identical ion (which is reduced), is called **disproportionation.** Au^I compounds also undergo disproportionation (to Au^0 and Au^{III}).

A few of the covalent compounds of the transition elements are simple covalent molecules (for example, TiF_4, $ZrCl_4$), but most are of the giant molecule type.

6. Paramagnetism and color. In filling in the electrons in a given series of transition elements (Table 19-2), electrons are added, one at a time, to the *d* subshell (ignoring irregularities). Because of Hund's rule (Section 13-6), all these elements (except Pd) contain one or more unpaired electrons. Therefore, transition elements are paramagnetic and so are most of their compounds. Another property which often results from the presence of unpaired electrons is color, and most compounds of transition elements are colored (Table 19-5 shows the colors of the hydrated aqueous ions of the first series; most other transition metal compounds are also colored). The element chromium provides an extreme example of this behavior, and in fact derives its name (from the Greek *chroma*, meaning color) from the many different colors of the compounds it forms. Chromic oxide (Cr_2O_3) is a green powder which is used to color U.S. dollar bills. In Section 19-12, we shall see an explanation for these colors.

7. Most transition metals readily form coordination compounds. These will be discussed in Sections 19-7 to 19-12.

8. One group of transition metals is worthy of special mention because of their low reactivity (they are often called **noble metals**). In this group are six of the nine elements of group VIII: Ru, Os, Rh, Ir, Pd, and Pt; and Au from group IB. The other 20 metals of the three transition series vary greatly in reactivity, but most are fairly reactive (though much less so than the group IA or IIA metals). However, these seven metals are relatively inert. They all have high values of $\mathcal{E}°$ (Table 16-2) and thus are hard to oxidize. All of them are resistant to attack by even very strong acids. The alchemists discovered that silver (which is also fairly inert, but less so) would dissolve in concentrated H_2SO_4 or HNO_3, but gold would not do so. They

Table 19-6 Ground-State Electronic Structures of the Three Outermost Shells of the Lanthanide Elements

Ce	$4f^2$	$6s^2$
Pr	$4f^3$	$6s^2$
Nd	$4f^4$	$6s^2$
Pm	$4f^5$	$6s^2$
Sm	$4f^6$	$6s^2$
Eu	$4f^7$	$6s^2$
Gd	$4f^7 5d^1$	$6s^2$
Tb	$4f^9$	$6s^2$
Dy	$4f^{10}$	$6s^2$
Ho	$4f^{11}$	$6s^2$
Er	$4f^{12}$	$6s^2$
Tm	$4f^{13}$	$6s^2$
Yb	$4f^{14}$	$6s^2$
Lu	$4f^{14} 5d^1$	$6s^2$

found that gold did dissolve in a mixture of concentrated HCl and HNO_3. They were so impressed with this fact that they called such a mixture (1 part HNO_3, 3 parts HCl) by the name of **aqua regia,** which means the water of kings. Today we know that aqua regia dissolves gold not because it is a strong acid or a strong oxidizing agent (although it is both), but because it converts gold to a soluble complex ion, $AuCl_4^-$. The HNO_3 is necessary for its oxidizing power; the HCl, to supply Cl^-. Platinum also dissolves in aqua regia, and palladium in concentrated HNO_3, but the alchemists would really have been impressed with ruthenium, osmium, rhodium, and iridium, had these metals been known at that time, because none of these four dissolve in any strong acid, including aqua regia.

19-6 Inner Transition Metals

There are two series of inner transition metals. The first series, Ce to Lu, is called the **lanthanide series,** because these elements follow lanthanum in the periodic table. The second series, Th to Lr, is called the **actinide series,** for a similar reason. Lanthanum has the ground-state configuration

$$[Xe]6s^2 5d^1$$

The next element, cerium, instead of continuing to fill the 5*d* subshell, begins to fill the 4*f* subshell, and the remaining elements continue to add to the 4*f* subshell until it is filled (Table 19-6). Since all of these elements have two 6*s* electrons, they have *three* unfilled shells: the 4, 5, and 6 levels. (Yb and Lu have filled 4*f* subshells. However, they are usually regarded as inner transition elements also, and behave in the same way). The electronic structures of the actinide elements are not as well known (most of the actinides are man-made elements; all are radioactive, and most have very short half-lives), but are probably fairly similar. The currently accepted electronic structures of the actinide elements are shown in Table 13-8.

The property of constant valence within a series, which we saw for the main transition elements, is even more extreme for the inner transition elements, because in this case for the most part, the two outer shells are identically filled, and the atoms only differ in the third-outermost shell. All the 14 lanthanides, like lanthanum itself, form +3 ions, by loss of the $6s^2$ electrons and one *d* or *f* electron. Some of them also form +2 ions, and a few form +4 ions, but these are always less stable than the +3 ions. Cerium is the only one with a stable +4 state; the most stable +2 is Eu^{2+}. The predominant chemistry is ionic, and few covalent compounds are known.

Because of these facts, the 14 lanthanides have very similar properties, and for many years, until modern techniques were developed, it was very difficult to separate them from each other (and from Y and La). For this reason, they were known as the **rare earth** elements, although they are not particularly rare. Thulium, the rarest of them (except for Pm, which does not occur on earth), is more abundant than As, Hg, or Cd, for example, none of which is normally considered to be a rare element. The lanthanides were rare only in pure form, because they were so difficult to separate from each other. Today, however, they are easily separated by ion exchange (Section 17-12), which can be done on a fairly large scale, and are no longer rare in any sense. The word "earth" is an old chemical name, meaning *oxide*. The lanthanides were called "earths" because most of them until recently were chiefly known in the form of their oxides.

The actinides have a lesser tendency to form ionic compounds then the lanthanides and a greater tendency to form covalent compounds. The most stable oxidation state for most actinides is +3 (this is so for all the elements from Am to No), but for others the most stable state is +4, +5, or +6. We see here the same tendency that we saw for the main transition elements: the second series has higher oxidation numbers than the first.

COORDINATION COMPOUNDS

19-7 Coordination Compounds: Introduction

In Section 14-5 we learned that metallic positive ions function as Lewis acids in the presence of suitable bases. Each base molecule carries an electron pair (or else it could not be a base), and uses this electron pair to form a coordinate covalent bond with the metallic ion. The compounds formed in this way are called **coordination compounds.** Let us look at a typical example, $Pt(NH_3)_2Cl_2$. We can consider that in this coordination compound there is a central Pt^{2+} ion which has formed four coordinate covalent bonds: two to NH_3 molecules and two to Cl^- ions:

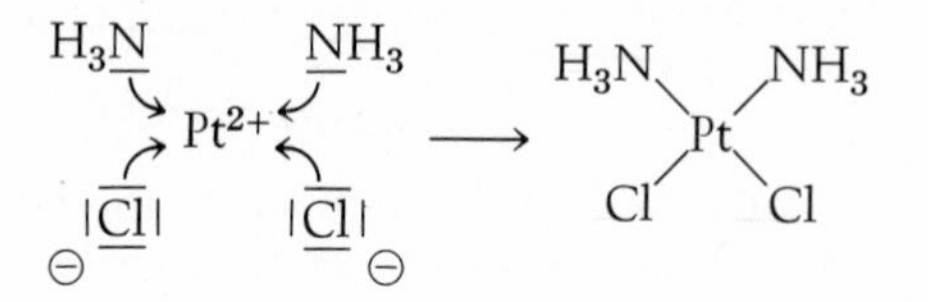

Once formed, the four covalent bonds are indistinguishable from any other covalent bonds. (Note that by our rule of formal charge, there should be a formal charge of +2 on the Pt and −1 on each nitrogen atom, but it is customary to omit these charges in drawing structures of coordination compounds.)

The bases that supply the electron pairs are called **ligands** (from the Latin *ligandus,* meaning to be bound). Each metallic ion can accept a certain number of electron pairs; this number is called the **coordination number** (do not confuse this with the oxidation number).* In the above example, the coordination number of Pt^{II} is 4. Each metallic ion has a typical coordination number (most often 4 or 6), but other numbers are usually possible also. For example, the coordination number in most complexes of Cr^{III} is 6, but a few Cr^{III} complexes are known with coordination numbers of 3, 4, and 5. In many cases, though obviously not all, the coordination number is twice the oxidation number.

In order to be a ligand, the only thing a molecule or ion needs is an unshared pair

*We have seen the term *coordination number* used before, to refer to the number of molecules surrounding a given liquid molecule (Section 6-7).

of electrons; consequently, there are hundreds of known ligands. They can be negative ions, such as Cl^-, CN^-, NO_3^-, or neutral molecules (for example, H_2O, NH_3, CO). This means that coordination compounds themselves can have varying charges; they can be positive ions, negative ions, or neutral molecules, depending on the oxidation state of the metal and on the nature of the ligands. *The total charge is the sum of the oxidation number of the metal and the charges on all the ligands.* For example, the compound mentioned above, $Pt(NH_3)_2Cl_2$ is neutral because it is made up of a +2 ion, two −1 ions, and two neutral molecules. The total charge is zero. The ligands around a central atom may be present in any combination. For example, Table 19-7 shows the seven Cr^{III} complexes in which H_2O and Cl^- are the ligands. Note that the charge on the total species increases by −1 each time a Cl^- ion replaces an H_2O molecule. When the coordinated species is an ion, it is known as a **complex ion,** but there is really no fundamental difference between a complex ion such as $Cr(H_2O)_5Cl^{2+}$ and the neutral molecule $Cr(H_2O)_3Cl_3$. Often coordination compounds in general are referred to as **complexes.** Of course, an ion such as $Cr(H_2O)_5Cl^{2+}$ cannot exist by itself, any more than any other ion can. Whether in water solution or in the solid state, there must be an equal number of negative charges present (these may be Cl^- ions or any other negative ions). Thus, the formula for a typical *compound* containing the complex ion $Cr(H_2O)_5Cl^{2+}$ might be $[Cr(H_2O)_5Cl]^{2+}SO_4^{2-}$ or $[Cr(H_2O)_5Cl]^{2+}2Cl^-$.

A number of these coordination compounds were known in the late nineteenth century, but their structures were completely misunderstood. Analysis of a compound such as $[Pt(NH_3)_3Cl_3]^+Cl^-$ showed that four atoms of chlorine were present for every atom of platinum, but when Ag^+ ion was added, only one of the Cl atoms precipitated as AgCl. Chemists of that time did not know what to make of such behavior, and were unable to come up with a satisfactory theory to explain the structures of these compounds. The man who finally explained them was Alfred Werner (1866–1919), who was born in Alsace, and lived most of his life in Switzerland. Werner showed that each central metallic atom had a given coordination number, that the ligands could be negative ions or neutral molecules, and that negative ions such as Cl^- could be present both as ligands and as ordinary negative ions associated with positive ions (Werner originated the terms "coordination compound" and "ligand"). Werner also deduced the three-dimensional structures of many of these compounds (Section 19-9), although final confirmation of these structures had to await the discovery of x-ray crystallography. For his work in this field, Werner was awarded the Nobel Prize in Chemistry in 1913.

Metals do not form coordination compounds equally well. In general, transition metals have the greatest tendency to form coordination compounds while the alkali and alkaline earth metals (group IA and IIA) have the least tendency to do so. All metallic ions in aqueous solution exist as coordination compounds in which the ligands are water molecules. Thus, Cr^{3+} in water is really $Cr(H_2O)_6^{3+}$ and Na^+ is really $Na(H_2O)_n^+$(the value of n in this case is not certain, though the most likely number is 4). However, these hydrated ions are not equally stable. The bonds in the

Table 19-7 Coordination Compounds of Cr^{III} in Which the Ligands Are Cl^- and H_2O

$Cr(H_2O)_6^{3+}$	complex ions (cations)
$Cr(H_2O)_5Cl^{2+}$	
$Cr(H_2O)_4Cl_2^+$	
$Cr(H_2O)_3Cl_3$	neutral molecule
$Cr(H_2O)_2Cl_4^-$	complex ions (anions)
$Cr(H_2O)Cl_5^{2-}$	
$CrCl_6^{3-}$	

[a]In each case the coordination number of Cr^{3+} is 6.

hydrated chromium ion are much stronger than those in the hydrated sodium ion. Thus, each Cr^{3+} remains bonded to the same six water molecules for long periods of time, but the water molecules in $Na(H_2O)_n^+$ do not remain attached to the Na^+, but constantly leave and are immediately replaced by other molecules*:

$$Na(H_2O)_n^+ + nH_2O \rightleftharpoons Na(H_2O)_n^+ + nH_2O$$

The alkali metals form very few coordination compounds other than the aqueous ones, but most of the transition elements form hundreds of coordination compounds, with many ligands. A metal does not have to exist as a free ion in order to form coordination compounds. In Sections 19-5 and 19-6 we saw that some transition metals form no ions at all, and that others do so only in lower oxidation states. Yet coordination compounds are known for many of the higher oxidation states as well as for the lower ones. For example, Pt^{4+} is not a known ion, either in aqueous solution or in any ionic compound, but platinum forms many stable coordination compounds in which its oxidation number is 4 [an example is $Pt(NH_3)_3Cl_3^+$, mentioned above]. It is obvious that such complexes must be made indirectly, and not by a direct reaction between the central ion and the ligands. In such cases, we often do not know the oxidation state of the central atom until we calculate it from the formula of the complex, by the reverse of the method mentioned above: by summing up the charges on the ligand and comparing this with the charge on the total species.

Example: Rhenium forms compounds in all oxidation states from +2 to +7 (Table 19-3). What is the oxidation state of rhenium in the $Re(CN)_8^{3-}$ ion?

Answer: The oxidation state is Re^{V}, since the addition of eight −1 ions gives a charge of −3.

The fact that free positive ions are not necessary for the formation of coordination compounds is emphasized by the fact that many coordination compounds of the transition metals are known in which the oxidation number of the central atom is 0, −1, and even lower! Some examples are given in Table 19-8.

We have defined a ligand as a species that supplies an electron pair to the central metallic ion. There are a large number of ligands that supply *two* electron pairs; such ligands are called **bidentate** ligands (bidentate = two teeth). Ligands that supply only one pair are **monodentate.** An important example of a bidentate ligand is ethylenediamine.

$$\bar{N}H_2—CH_2—CH_2—\bar{N}H_2$$

ethylenediamine

Note that ethylenediamine has two nitrogen atoms. Each nitrogen atom supplies one electron pair. In a bidentate ligand the two electron pairs are always attached to different atoms. Because of this, coordination compounds containing bidentate ligands always have rings of atoms. For example, when ethylenediamine is a ligand, a five-membered ring is always present, as in

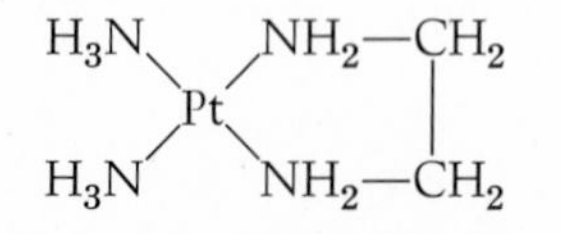

In this example, the coordination number of Pt^{II} is 4. Two of the four bonds are supplied by the ethylenediamine, the other two by two monodentate ligands. It is

*This kind of exchange can be detected by the use of heavy water: D_2O, also called deuterium oxide.

Table 19-8 Some Coordination Compounds in Which the Oxidation State of the Central Metal is Zero or Less[a]

Formula	Oxidation state
$Co(CN)_4^{4-}$	Co^0
$Ni(CO)_4$	Ni^0
$Ir_4(CO)_{12}$	Ir^0
$Cr_2(CO)_{10}^{2-}$	Cr^{-I}
$Mn(CO)_5^-$	Mn^{-I}
$Re(CO)_5^-$	Re^{-I}
$Fe(CO)_4^{2-}$	Fe^{-II}
$Mo(CO)_5^{2-}$	Mo^{-II}

[a]Note that CO is a neutral ligand.

also possible for bidentate ligands to supply all the pairs when the coordination number is 4 or 6, for example

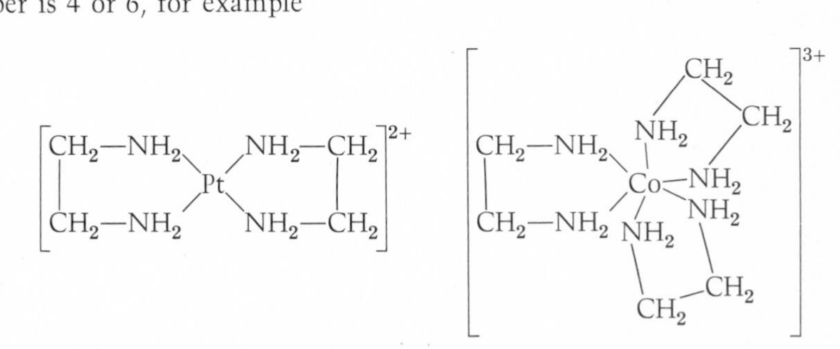

For use in formulas, ethylenediamine is abbreviated as en. Thus, the three compounds just shown can be written $Pten(NH_3)_2^{2+}$, $Pt(en)_2^{2+}$, and $Co(en)_3^{3+}$, respectively.

There are many other important bidentate ligands. Three that may be mentioned are acetylacetonate (acac), 2,2′-bipyridine (bipy), and the oxalate ion.

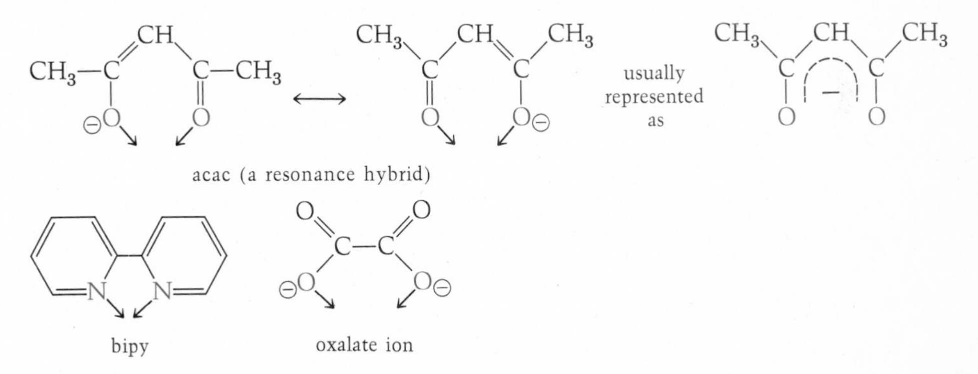

Many ligands are known which supply more than two electron pairs. These are called **tridentate, quadridentate,** etc. (the general name is **polydentate** or **multidentate**) ligands. Examples are diethylenetriamine (dien), a tridentate ligand, and ethylenediaminetetraacetate ($EDTA^{4-}$), a hexadentate ligand.

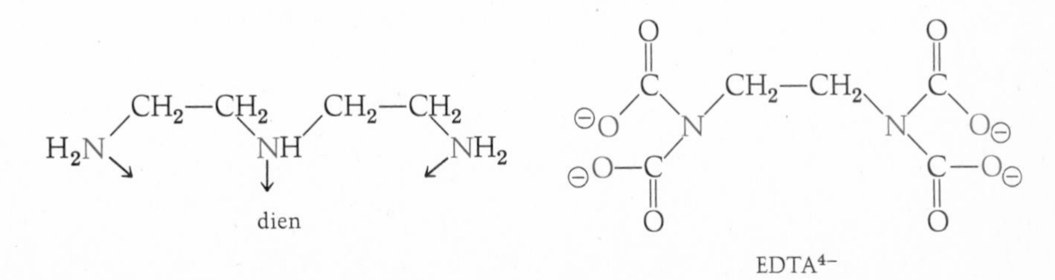

Bidentate and polydentate ligands are also called **chelating ligands** (from the Greek *chele,* meaning claw; a bidentate ligand is supposed to act like a crab, holding tightly to an object with its two claws), and coordination compounds of this type are often called **chelates.** In general, *chelates are more stable than coordination compounds that contain only monodentate ligands.* We have already seen an example of the use of chelates to sequester metallic ions (Section 17-11). The action of hemoglobin bonding to O_2 and CO (Section 17-7) is also a chelating action.

The increased stability of chelates is almost entirely an *entropy* effect, which can be explained as follows: Let us begin with $Cr(H_2O)_6{}^{3+}$ and convert it to $Cr(en)_3{}^{3+}$:

$$Cr(H_2O)_6{}^{3+} + 3\text{ en} \longrightarrow Cr(en)_3{}^{3+} + 6H_2O$$

Total number of molecules $1 + 3 = 4$ $\qquad 1 + 6 = 7$

The balanced equation shows us that only four molecules enter the reaction, but seven are produced. Since seven molecules can occupy more positions in space than four, the reaction results in increased entropy. This means, of course, that to drive the reaction backward requires a decrease in entropy, which requires energy. Compare the formation of $Cr(NH_3)_6{}^{3+}$, where only monodentate ligands are present.

$$Cr(H_2O)_6{}^{3+} + 6NH_3 \longrightarrow Cr(NH_3)_6{}^{3+} + 6H_2O$$

Total number of molecules $1 + 6 = 7$ $\qquad 1 + 6 = 7$

In this case, there is no change in the total number of molecules, so the entropy of the system is very little changed by reaction.

19-8 Nomenclature of Coordination Compounds

The formulas of coordination compounds can get very complex, and the names must also get very complex if the structure of the compound is to be derivable from the name. The International Union of Pure and Applied Chemistry (IUPAC) has devised a set of rules to name coordination compounds. We will give only a few of these rules which, however, will be sufficient to name most coordination compounds.

1. In the case of ionic compounds, the positive ion is named first, then the negative ion.
2. Ligands are named before the central metallic atom. Note that this is opposite to the practice in writing the formulas.
3. Anionic ligand names end in o: for example,

Cl^-	chloro
CN^-	cyano
OH^-	hydroxo
$NO_3{}^-$	nitro

4. Neutral ligands have their regular names:

CH_3NH_2	methylamine
$NH_2CH_2CH_2NH_2$	ethylenediamine
$(CH_3)_3P$	trimethylphosphine

with three important exceptions:

NH_3	ammine (note the double m)
H_2O	aqua
CO	carbonyl

5. Prefixes are used when identical ligands exist:

2	di or bis
3	tri or tris
4	tetra or tetrakis
5	penta or pentakis
6	hexa or hexakis

6. If two or more *different* ligands are present, they are listed in alphabetical order, regardless of the number. Thus, di**a**mmine is listed before **c**hloro.
7. The name of the metal is given in its usual form for neutral and cationic complexes. For anionic complexes, the metal is given the suffix "ate," in some cases in the Latin form (for example, ferrate, cuprate). The oxidation number of the metal is given in parentheses.

Examples:

$[Mn(NH_3)_4Cl_2]^+ \quad NO_3^-$	tetraamminedichloromanganese(III) nitrate
$[Cr(H_2O)_6]^{2+} \quad 2Cl^-$	hexaaquachromium(II) chloride (note that we do not say dichloride)
$Pt(NH_3)_2(CN)_2$	diamminedicyanoplatinum(II)
$K^+[Cr(NH_3)_2Br_4]^-$	potassium diamminetetrabromochromate(III)
$Na^+[Co(CO)_4]^-$	sodium tetracarbonylcobaltate(−I)
$[Cr(en)(NH_3)_3(OH)]^{2+} \quad 2Br^-$	diammineethylenediaminehydroxochromium(III) bromide
$[CuCl_3(CH_3NH_2)]^-$	trichloro(methylamine)cuprate(II) ion

An important group of coordination compounds consists only of transition metals coordinated with carbon monoxide. These compounds are usually named as metal carbonyls. Thus, the IUPAC name for $Ni(CO)_4$ is tetracarbonylnickel(0), but it is usually called nickel carbonyl. Nickel carbonyl is useful for inorganic and organic synthesis but is extremely toxic.

19-9 Geometry of Coordination Compounds

Like all covalent molecules, coordination compounds have definite shapes. These shapes depend on the coordination number. We will look at the shapes of compounds with the most common coordination numbers: 2, 4, 5, and 6.

Coordination Number 2. This coordination number is chiefly found in complexes of Cu^I, Ag^I, and Au^I. The geometry is *linear;* for example,

$$[H_3N{-}Ag{-}NH_3]^+ \qquad \text{diamminesilver(I) ion}$$

180°
linear geometry

Coordination Number 4. This coordination number is second in importance only to 6. In this case, there are two geometries found: *tetrahedral* and *square planar*. Square planar geometry is the most common for Pt^{II} and Pd^{II}. An example is $Pd(NH_3)_2Cl_2$, mentioned in the previous section; another is $PdCl_4^{2-}$. Tetrahedral geometry is more common than square planar, being found, for example, in many Fe^{II}, Cu^{II}, Co^{II}, Mn^{II} complexes, among others. An example is $CoBr_2[P(CH_3)_2]_2$. In some cases (for example, in some Ni^{II}, Co^{II}, and Cu^{II} com-

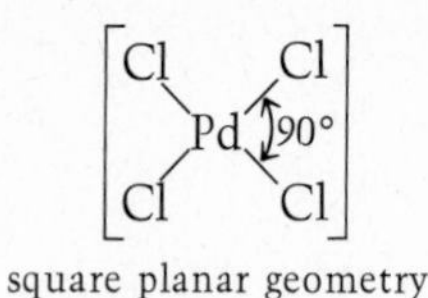

square planar geometry
tetrachloroplatinate(II) ion

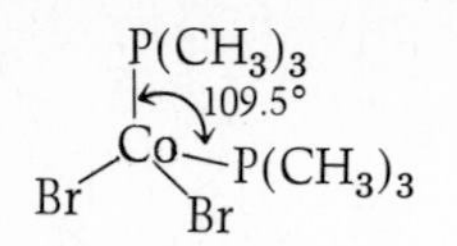

tetrahedral geometry
dibromobis(trimethylphosphine)cobalt(II)

plexes), the energy difference between the square planar and tetrahedral geometries is small, and which is preferred depends on the ligands.

Coordination Number 5. This coordination number is less common than 4, but many such cases are known, including some of Co^{II}, Cr^{II}, Cr^{III}, and Mo^{V}, among others. There are two principal geometries here, the *trigonal bipyramid* and the *square pyramid.* Examples of each are (the iron is Fe^0 and Fe^{III}, respectively)

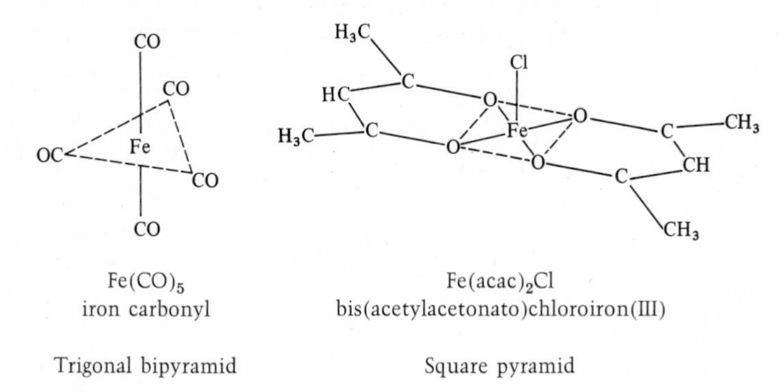

$Fe(CO)_5$
iron carbonyl

Trigonal bipyramid

$Fe(acac)_2Cl$
bis(acetylacetonato)chloroiron(III)

Square pyramid

The two geometries are similar in energy, and many complexes have a geometry that is intermediate between the two.

Coordination Number 6. This is by far the most common coordination number, and the geometry is practically always *octahedral.* The metal atom is at the center and the six ligand atoms occupy the corners of an octahedron: for example,

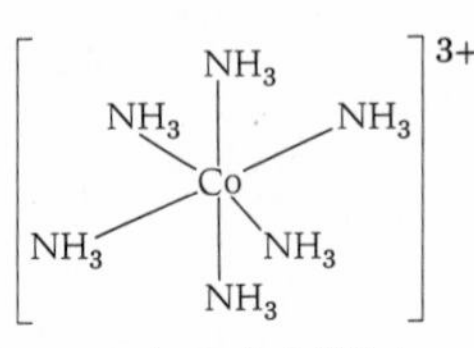

hexaamminecobalt(III) ion

Most people find it easier to visualize the geometry when they see the figure drawn this way:

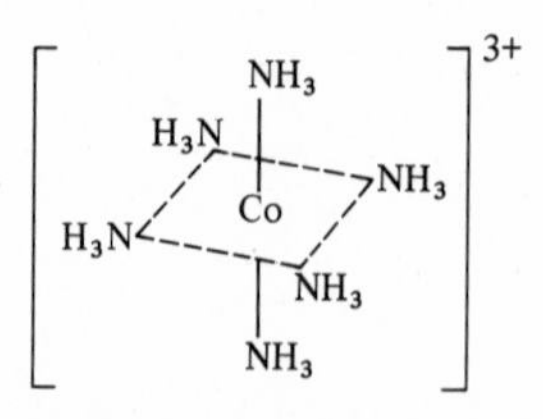

It must be remembered when looking at such a figure that there are *no* bonds between the NH_3 molecules. All six of the bonds are between Co and NH_3. The figure drawn in this way omits four of the bonds and shows four lines (connecting N atoms) which are not bonds. Another thing which is not immediately evident from the figure is that the top and bottom positions are no different from the other four. This can be seen if we allow a molecule of $Co(H_2O)_2(NH_3)_2Cl_2^+$ to tumble over:

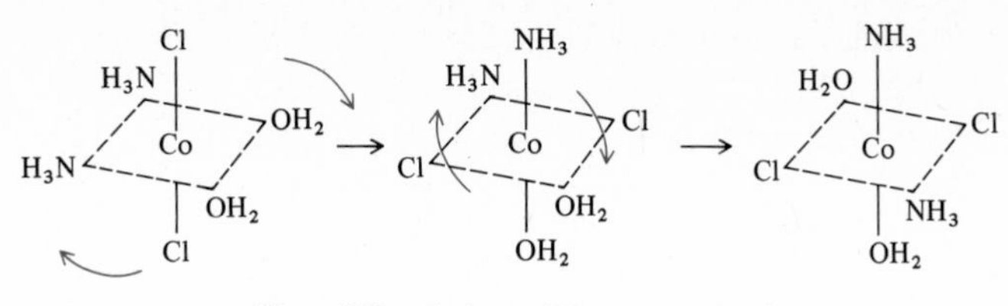

Three different views of the same molecule

19-10 Isomerism of Coordination Compounds

Two compounds which have the same molecular formulas but different structural formulas are called **isomers.** (By *molecular formula* we mean a formula that merely gives the number of atoms of each element present (for example, $PtCl_2N_2H_6$ or $MnCl_2O_8$). Coordination compounds show several types of isomerism, but we will discuss only three kinds here. Other types of isomerism will be encountered in Chapter 20.

Ionization Isomerism. We have seen that an anion like Cl^- can serve as a ligand or simply as a negative ion associated with a positive complex ion. This fact means that we can have one compound $[Co(H_2O)_4Cl_2]^+NO_2^-$ and another $[Co(H_2O)_4ClNO_2]^+Cl^-$. The two compounds are **ionization isomers.** The molecular formula for both is $CoH_8O_6Cl_2N$, but in the first both Cl^- ions are coordinated to the metal, while in the second, only one Cl^- is coordinated, and the other is a negative ion. Although the two isomers have the same molecular formulas, they have different properties: for example, the second will give a precipitate of AgCl when treated with Ag^+; the first will not.

Cis–Trans Isomerism of Square Planar Complexes. Square planar complexes of the form EA_2B_2 exhibit a type of isomerism called **cis–trans isomerism.** In each case two isomers can exist, one called *cis*, and the other *trans*. For example, these are the two isomers of $Pt(NH_3)_2Cl_2$:

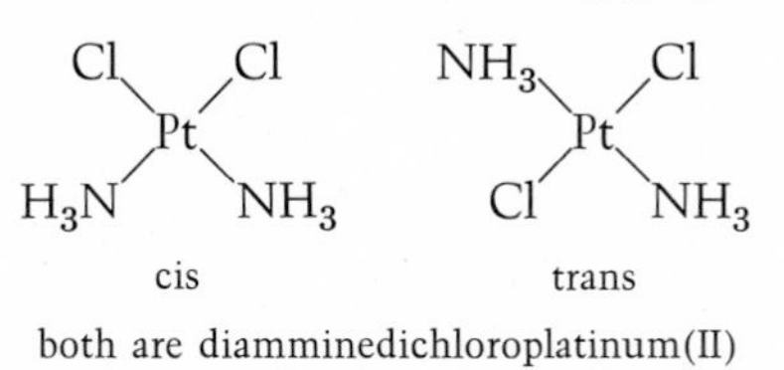

both are diamminedichloroplatinum(II)

The two compounds are different in structure and in properties. The isomer that has the two identical groups next to each other is called the cis isomer; the other is the trans. For isomers of this type, the cis isomer may (and usually does) have a dipole moment; the trans isomer never has. (Why?) Compounds of the form EA_4 and EA_3B do not show this type of isomerism (why not?); only one compound exists in each of these cases. The isomerism is a consequence of the square planar geometry. *Tetrahedral* compounds EA_2B_2 also do not show this isomerism.

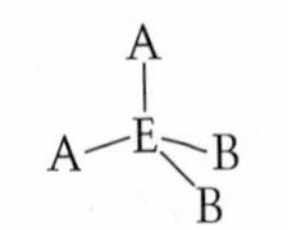

only one compound exists

Cis–Trans Isomerism of Octahedral Complexes. Certain octahedral complexes also exhibit cis–trans isomerism. As above, this does not depend on which ligands are present but on the pattern of the ligands. Thus, EA_6 and EA_5B

compounds do not show cis–trans isomerism (only one compound exists in each case), but EA_4B_2 and EA_3B_3 compounds do. In each case there are two isomers, and again, the cis isomer is the one where identical groups are nearest to each other:

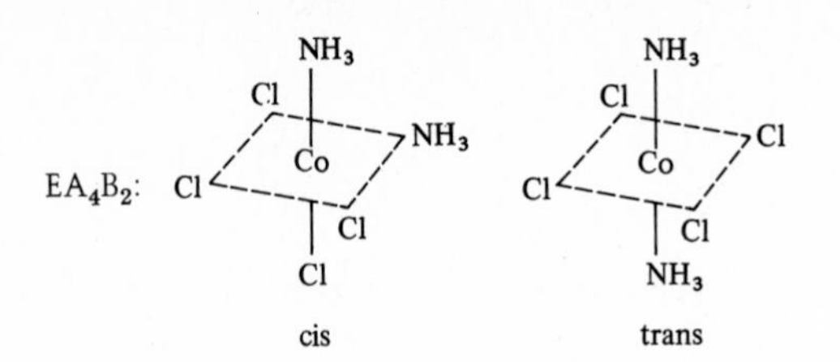

Note that

might appear to be a different isomer, but it is only the trans isomer "rolled over."

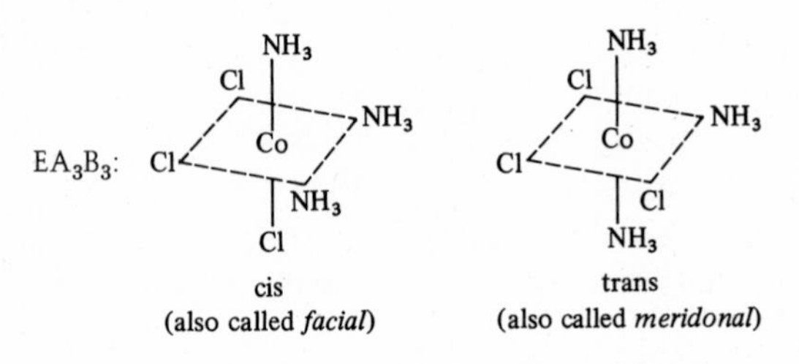

Most chelate complexes of the form EA_4B, where B is a bidentate ligand do not show cis–trans isomerism, because most bidentate ligands are simply not long enough to reach around; they must chelate cis:

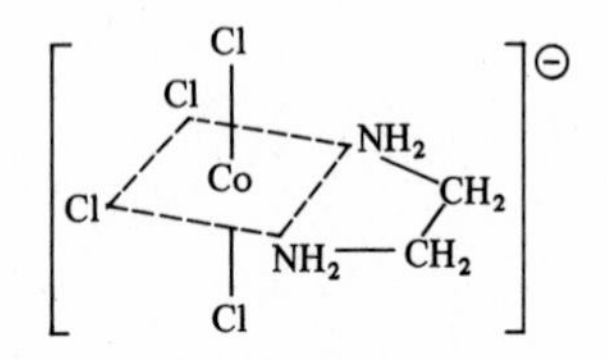

tetrachloro(ethylenediamine)cobaltate(III) ion

The only isomer

19-11 Instability Constants

We have mentioned (Section 19-7) that metals do not form coordination compounds equally well; the same can be said of ligands. Thus, each coordination compound has its own stability; some are very stable, some are quite unstable, and others have intermediate stabilities. These stabilities can be treated quantitatively, by the use of equilibrium constants, in a manner similar to that discussed in Chapter 11.

The complexes are formed one step at a time (stepwise), and we can write an equilibrium expression for each step. Each step has its own value of K, which in many cases (though not all) can be measured separately. Because they are equilibria, the reactions can be written in either direction, but it is customary to write them as dissociations rather than formations. For example, the following are the steps for the dissociation of $Cu(NH_3)_4^{2+}$ in water solution:

$$Cu(NH_3)_4^{2+} \rightleftharpoons Cu(NH_3)_3^{2+} + NH_3 \qquad K_1 = \frac{[Cu(NH_3)_3^{2+}][NH_3]}{[Cu(NH_3)_4^{2+}]} = 6 \times 10^{-3}$$

$$Cu(NH_3)_3^{2+} \rightleftharpoons Cu(NH_3)_2^{2+} + NH_3 \qquad K_2 = \frac{[Cu(NH_3)_2^{2+}][NH_3]}{[Cu(NH_3)_3^{2+}]} = 4 \times 10^{-4}$$

$$Cu(NH_3)_2^{2+} \rightleftharpoons Cu(NH_3)^{2+} + NH_3 \qquad K_3 = \frac{[Cu(NH_3)^{2+}][NH_3]}{[Cu(NH_3)_2^{2+}]} = 5 \times 10^{-4}$$

$$Cu(NH_3)^{2+} \rightleftharpoons Cu^{2+} + NH_3 \qquad K_4 = \frac{[Cu^{2+}][NH_3]}{[Cu(NH_3)^{2+}]} = 7 \times 10^{-5}$$

We have shown the final product of this sequence as Cu^{2+}. We know, of course, that Cu^{2+} does not exist as such in water solution; it is present as $Cu(H_2O)_4{}^{2+}$; however, for simplicity we generally ignore the water molecules. Note that all four of the equilibrium constants in this case are fairly low; thus, $Cu(NH_3)_4^{2+}$ has little tendency to dissociate.

The stepwise equilibrium expressions can be combined, and an overall constant obtained, by multiplying the expressions. For example:

$$\frac{[Cu(NH_3)_3^{2+}][NH_3]}{[Cu(NH_3)_4^{2+}]} \cdot \frac{[Cu(NH_3)_2^{2+}][NH_3]}{[Cu(NH_3)_3^{2+}]} \cdot \frac{[Cu(NH_3)^{2+}][NH_3]}{[Cu(NH_3)_2^{2+}]} \cdot \frac{[Cu^{2+}][NH_3]}{[Cu(NH_3)^{2+}]} = K_1K_2K_3K_4$$

$$\frac{[Cu^{2+}][NH_3]^4}{[Cu(NH_3)_4^{2+}]} = K_i = 8 \times 10^{-14}$$

This is the equilibrium expression for the overall reaction:

$$Cu(NH_3)_4^{2+} \rightleftharpoons Cu^{2+} = 4NH_3$$

The constant K_i is called the **instability constant,** or the **dissociation constant,** for the complex. The reciprocal of K_i is called a *stability constant* or a *formation constant*. A low value of K_i means a stable complex. Instability constants for a number of complex ions are given in Table 19-9.

If we know the value of the various *K*s, we can perform calculations similar to those we did in Chapter 11.

Table 19-9 Instability Constants of Some Complex Ions

Ion	*Instability constant*
$Co(NH_3)_6{}^{2+}$	1×10^{-5}
$Co(NH_3)_6{}^{3+}$	1×10^{-34}
$Cu(NH_3)_4{}^{2+}$	8×10^{-14}
$Fe(CN)_6{}^{4-}$	1×10^{-36}
$Fe(CN)_6{}^{3-}$	1×10^{-44}
$Ag(NH_3)_2{}^{+}$	5×10^{-8}
$Ag(CN)_2{}^{-}$	1×10^{-21}
$PtCl_4{}^{-}$	1×10^{-16}
$Ni(en)_3{}^{2+}$	1×10^{-18}
$Zn(OH)_4{}^{2-}$	4×10^{-16}

Example: To a 5.0×10^{-3} *M* solution of $AgNO_3$ is added enough gaseous NH_3 to make the concentration 1.0 *M* in NH_3. What are the concentrations of Ag^+, $Ag(NH_3)^+$, and $Ag(NH_3)_2{}^+$ in this solution?

$$K_1 = 4.2 \times 10^{-4} \text{ for } Ag(NH_3)_2{}^+ \rightleftharpoons Ag(NH_3)^+ + NH_3$$

$$K_2 = 1.2 \times 10^{-4} \text{ for } Ag(NH_3)^+ \rightleftharpoons Ag^+ + NH_3$$

Answer: Since the values of K_1 and K_2 are quite low and there is a large excess of NH_3, we first assume that almost all of the Ag^+ has been converted to $Ag(NH_3)_2^+$. Therefore, $[Ag(NH_3)_2^+] = 5.0 \times 10^{-3}\,M$.

The concentration of Ag^+ can be obtained from the overall equation.

$$K_i = 5.0 \times 10^{-8} = \frac{[Ag^+][NH_3]^2}{[Ag(NH_3)_2^+]}$$

We assume that very little NH_3 has been used up (actually $1.0 - 10^{-2}\,M$), so that $[NH_3]$ is still $1.0\,M$.

$$5.0 \times 10^{-8} = \frac{[Ag^+](1.0)^2}{5.0 \times 10^{-3}}$$

$$[Ag^+] = 2.5 \times 10^{-10}\,M$$

We may now get the concentration of $Ag(NH_3)^+$ from the equilibrium expression for either step.

(a)
$$K_1 = \frac{[Ag(NH_3)^+][NH_3]}{[Ag(NH_3)_2^+]}$$

$$4.2 \times 10^{-4} = \frac{[Ag(NH_3)^+](1.0)}{5.0 \times 10^{-3}}$$

$$[Ag(NH_3)^+] = 2.1 \times 10^{-6}\,M$$

(b)
$$K_2 = \frac{[Ag^+][NH_3]}{[Ag(NH_3)^+]}$$

$$1.2 \times 10^{-4} = \frac{(2.5 \times 10^{-10})(1)}{[Ag(NH_3)^+]}$$

$$[Ag(NH_3)^+] = 2.1 \times 10^{-6}\,M$$

The three concentrations are

$$[Ag^+] = 2.5 \times 10^{-10}\,M$$

$$[Ag(NH_3)^+] = 2.1 \times 10^{-6}\,M$$

$$[Ag(NH_3)_2^+] = 5.0 \times 10^{-3}\,M$$

In Section 11-3 we saw that some precipitates can be dissolved by the addition of compounds or ions that cause formation of a stable complex ion. A knowledge of the solubility product and of the instability constant allows us to calculate when this will occur.

Example: A flask contains one liter of water with 1.433 g (0.010 mole) of solid AgCl at the bottom. If 0.50 mole of NH_3 is added and the solution is well stirred, will this be enough to dissolve all the AgCl? (Assume that the total volume remains at one liter.)

$$K_{sp} \text{ for AgCl} = 1.6 \times 10^{-10}$$

Answer: If all the AgCl dissolves, what happens to the Cl^-? The Cl^- will be unchanged, so the solution will contain $0.010\,M$ of Cl^-. If we insert this value into the K_{sp} expression, we get

$$K_{sp} = [Ag^+][Cl^-]$$

$$1.6 \times 10^{-10} = [Ag^+](0.010)$$

$$[Ag^+] = 1.6 \times 10^{-8}\,M$$

This means that the free Ag^+ ion concentration cannot exceed 1.6×10^{-8} *M*, or else some of the AgCl would remain undissolved.

If all the AgCl dissolves, what happens to the Ag^+? This would be almost entirely converted to $Ag(NH_3)_2^+$, so we can assume that the concentration of $Ag(NH_3)_2^+$ will be 0.10 *M*. What concentration of NH_3 is required for the $[Ag(NH_3)_2^+]$ to be 0.010 *M* and the $[Ag^+]$ to be equal to 1.6×10^{-8} *M*? The equilibrium expression will tell us:

$$K_i = \frac{[Ag^+][NH_3]^2}{[Ag(NH_3)_2^+]}$$

$$5 \times 10^{-8} = \frac{(1.6 \times 10^{-8})[NH_3]^2}{0.010}$$

$$[NH_3]^2 = 3 \times 10^{-2}$$

$$[NH_3] = 2 \times 10^{-1} = 0.2\ M$$

Therefore, an NH_3 concentration of at least 0.2 *M* will be enough to dissolve all the AgCl. It is clear that 0.50 mole of NH_3 is sufficient to supply this concentration. Of the 0.50 mole of NH_3, $2 \times 0.010 = 0.020$ is used up in forming $Ag(NH_3)_2^+$. The rest (0.48 mole) remains dissolved in solution; far more than is required to maintain the $[Ag^+]$ below 1.6×10^{-8} *M*.

The instability constants we have been discussing are equilibrium constants and are of course valid only after equilibrium has been reached. In many of these cases, equilibrium is reached very rapidly; almost instantaneously, but this is not always so. Some of them take a very long time to become established. For example, the complex ion $Cu(NH_3)_4^{2+}$ immediately loses its NH_3 when nitric acid is added to a water solution of it.

$$Cu(NH_3)_4^{2+} \rightleftharpoons Cu^{2+} + 4NH_3$$

$$NH_3 + H_3O^+ \rightleftharpoons NH_4^+ + H_2O$$

The addition of nitric acid drives the equilibrium to the right by neutralizing (therefore removing) the NH_3. However, the complex ion $Co(NH_3)_6^{3+}$ reacts with nitric acid very slowly; a measurable reaction takes several days, even though the equilibrium constant for this reaction is highly favorable. Therefore, it cannot be safely assumed that all reactions in which complex ions are produced or dissociated will take place almost instantaneously. Recall (page 283) that we were able to make such an assumption for Brønsted acid–base reactions.

19-12 The Bonding in Coordination Compounds

A number of theories have been devised to explain the bonding in coordination compounds. One of the most fruitful is the **crystal field theory.** We will use it to describe the structure of octahedral complexes (coordination number = 6).

In the crystal field theory we assume that the central metal ion bears a positive charge and that the entire charge is concentrated at a single point (a **point charge**). The ligands are considered to be negative point charges. We thus picture an octahedral complex as being made up of a positive point charge at the center, and six negative point charges located on the *x*, *y*, and *z* axes (Figure 19-11). Ligands can be negative ions, or neutral molecules which donate an electron pair. It does not take a great stretch of the imagination to consider a negative ion to be a point change, but in the crystal field theory we consider neutral ligands to be point charges also. This is not unreasonable when we remember that neutral ligand molecules are dipoles (for

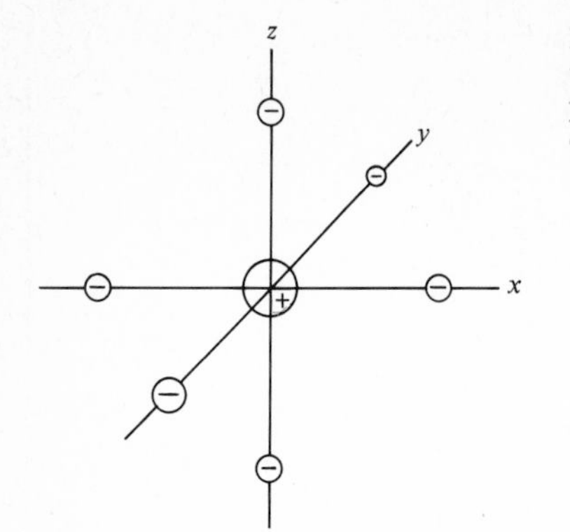

Figure 19-11 An octahedral complex as represented in the crystal field theory. The central metallic ion is a positive point charge. The six ligands are negative point charges.

example, H_2O, NH_3), and that it is the negative end of the dipole which approaches the central metallic ion. Thus, both neutral and negative ligands are treated as negative point charges.

Let us now look at the five *d* orbitals of the central metal, which are shown in Figure 19-12. Four of these have the same shape but project in different directions. The fifth (d_{z^2}) has a different shape. In the absence of any ligands, all five of these orbitals have the same energy (they are degenerate), and any *d* electrons that are present go into them one at a time, in accord with Hund's rule (Section 13-6). But in the presence of six ligands which are negative point charges, the five *d* orbitals do not have the same energy, because the electrons prefer those orbitals which are farthest away from the negative charges. An electron in an orbital close to a negative charge will have a higher energy then an electron farther away, because there is a greater repulsion between the electron and the point charge. Figure 19-13 shows that two of the five orbitals, the d_{z^2} and the $d_{x^2-y^2}$, have lobes that are directed along the axes; the other three do not. Therefore, the d_{xy}, d_{yz}, and d_{xz} orbitals get farther away from the negative point charges than do the other two, and so have a lower energy. Figure 19-13 also shows that the d_{xy}, d_{yz}, and d_{xz} orbitals are equivalent in energy to each other, since they all get equally far away from the three axes. It is not apparent from

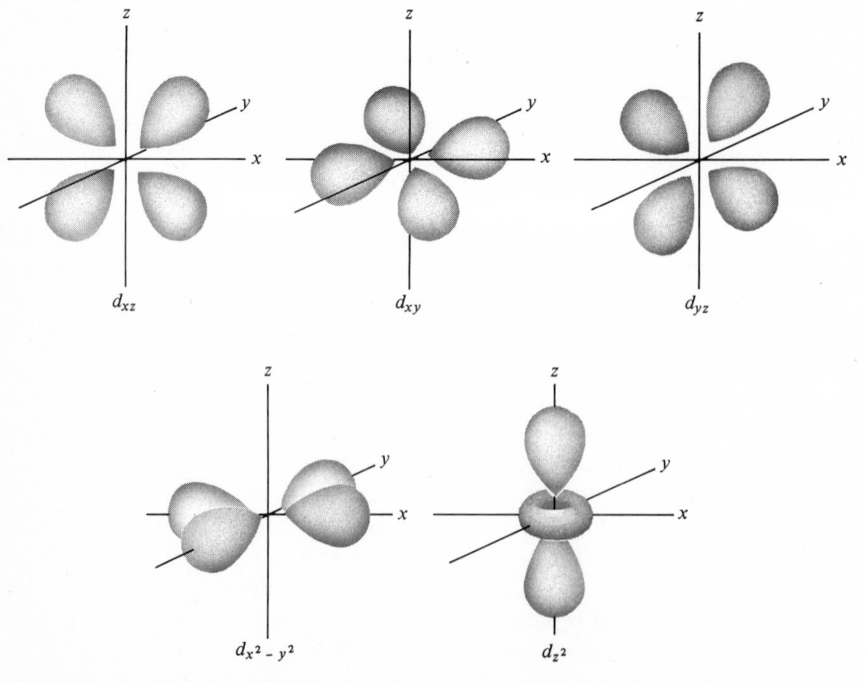

Figure 19-12 The five *d* orbitals.

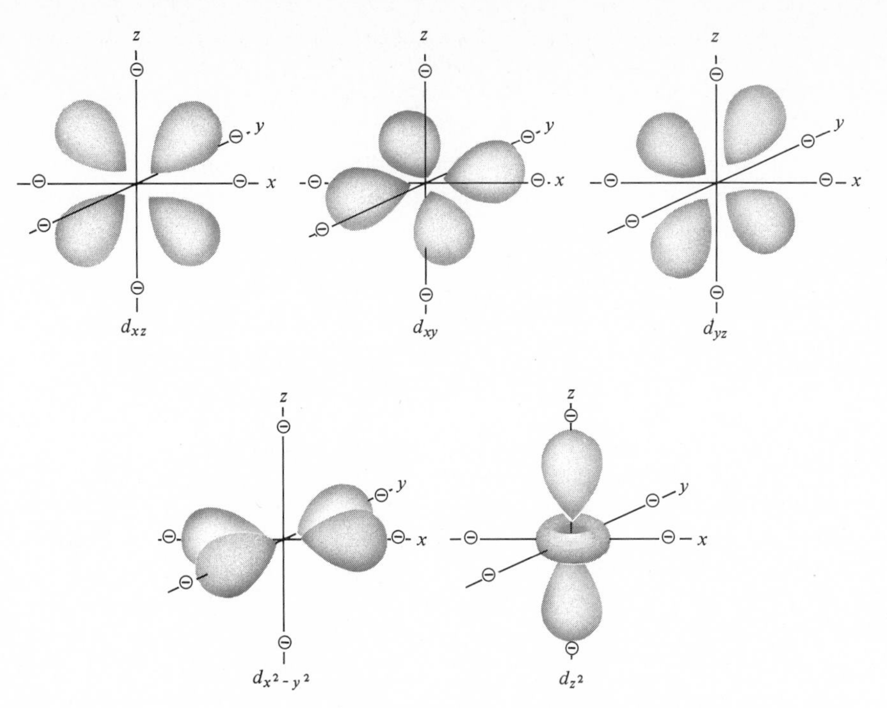

Figure 19-13 The five *d* orbitals and the six negative point charges.

Figure 19-13 that d_{z^2} and $d_{x^2-y^2}$ are equivalent in energy to each other, but this can be demonstrated.

The crystal field theory therefore predicts that *in an octahedral coordination compound, three of the d orbitals have a lower energy than the other two.* We say that the ligands have *split* the orbitals; the phenomenon is called **orbital splitting.** Figure 19-14 shows a representation of the two new levels. Because they are no longer equivalent, we give them new names. The three lower-energy orbitals are called t_{2g} and the two higher-energy ones e_g. The difference (Δ_0) in energy between the t_{2g} and the e_g levels depends on the ligands. Each ligand has a different capacity to cause splitting. Some of the common ligands can be ranked in this order.

$$CN^- > en > NH_3 > NCS^- > H_2O > OH^- > F^- > Cl^- > Br^- > I^-$$

Thus, a complex with six CN^- ions has a much higher Δ_0 than a complex with six I^- ions.

The crystal field theory is only a model. It obviously could not be completely accurate since the central metals and ligands are not point charges. Nevertheless, the splitting predicted by the crystal field theory is real and helps us to explain many of the properties of coordination compounds.

Let us now look at the bonding of octahedral complexes in terms of the molecular orbitals involved. If the central atom is a main transition element, there are nine orbitals it can use for bonding. These are the five *d*, three *p*, and one *s* orbitals of the outer two shells. For example, metals of the first series (Sc to Cu) can use these orbitals:

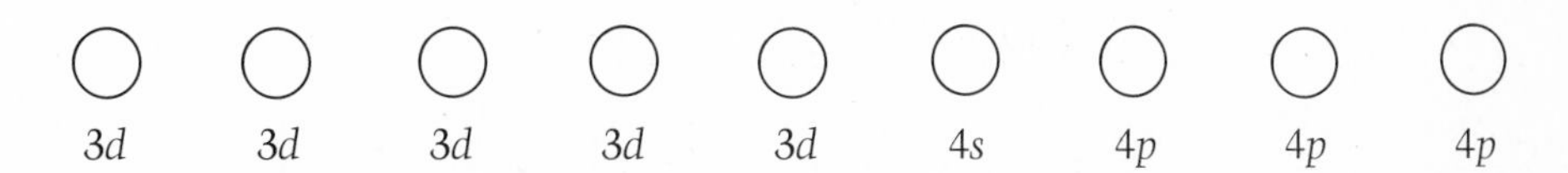

It is true that in the free atoms and their ions, only some of these orbitals are occupied, but remember that in a *coordinate* covalent bond the ligand supplies both

Figure 19-14 The two sets of *d* orbitals that have been split by an octahedral crystal field.

electrons, so all nine of these orbitals are potentially available for bonding. The six ligands overlap with the three 4*p* orbitals, the single 4*s* orbital, and *two* of the five 3*d* orbitals of the metal.

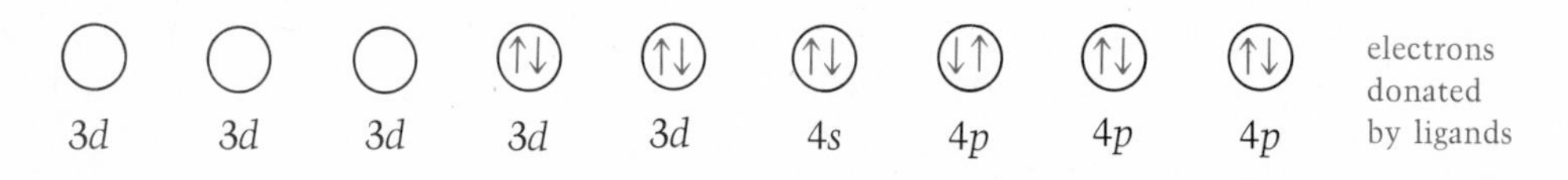

Figure 19-15 shows the molecular orbitals that result (remember that overlap of two orbitals gives two new orbitals, one bonding and one antibonding).

a. The three 4*p* orbitals overlap with three ligand orbitals to give six new orbitals: three bonding (t_{1u}) and three antibonding (t^*_{1u}).
b. The 4*s* orbital overlaps with one ligand orbital to give two new orbitals, one bonding (a_{1g}) and one antibonding (a^*_{1g}).
c. Two of the 3*d* orbitals overlap with the two remaining ligand orbitals to give four new orbitals, two bonding (e_g), and two antibonding (e^*_g).
d. The other three *d* orbitals do not overlap at all. They are still present, essentially unchanged. Hence, they are nonbonding orbitals. We call them t_{2g}.

Like the crystal field theory, the molecular orbital description shows that the *d* orbitals are split into two groups (we even give them the same labels). The three nonbonding orbitals are the same as the ones we called t_{2g} in the crystal field theory. However, the e_g orbitals are not quite the same. In the crystal field theory they are atomic orbitals; here they are molecular orbitals. In fact, it is the e^*_g orbitals here which are the equivalent of the e_g orbitals in the crystal field theory, since they are the ones which have a higher energy than the t_{2g} orbitals.

No matter what the metal, the six ligands supply a total of 12 electrons. These fill the t_{1u}, a_{1g}, and e_g orbitals. Since these are the bonding orbitals, this supplies most of the bonding. But in most cases there are also electrons which come from the metal, and these go into the t_{2g} and/or e^*_g orbitals. The pattern of filling these orbitals depends on how many electrons there are. If the metal ion has only one, two, or three *d* electrons (d^1, d^2, d^3), there is only one type of arrangement possible. The electrons go into the equivalent t_{2s} orbitals (Figure 19-16). This results in one, two, or three unpaired electrons, respectively. If the metal ion is d^8, d^9, or d^{10}, there is also only one arrangement possible in each case. As shown in Figure 19-17, this means that there are two, one, or no unpaired electrons, respectively.

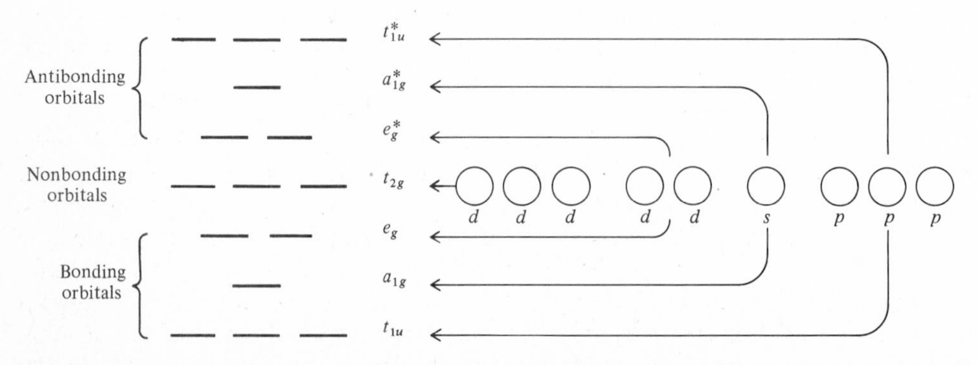

Figure 19-15 The molecular orbitals that arise when six ligands bond with a central metal. The three t_{2g} orbitals are nonbonding.

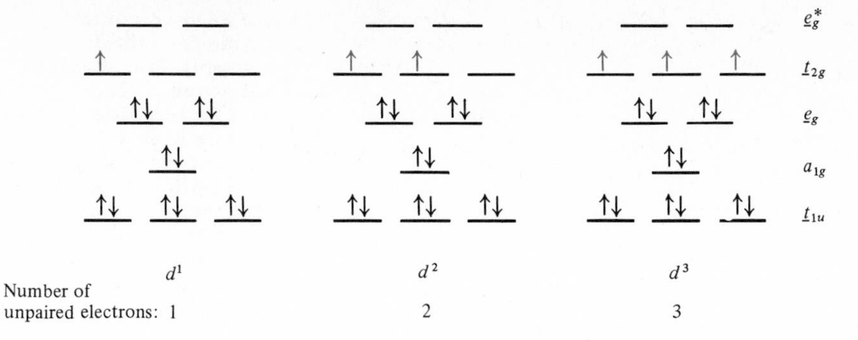

Figure 19-16 Electron diagram for metals that supply one, two, or three *d* electrons. The electrons in color are contributed by the metal, those in black by the ligands.

d^8 d^9 d^{10}

Number of unpaired electrons: 2 1 0

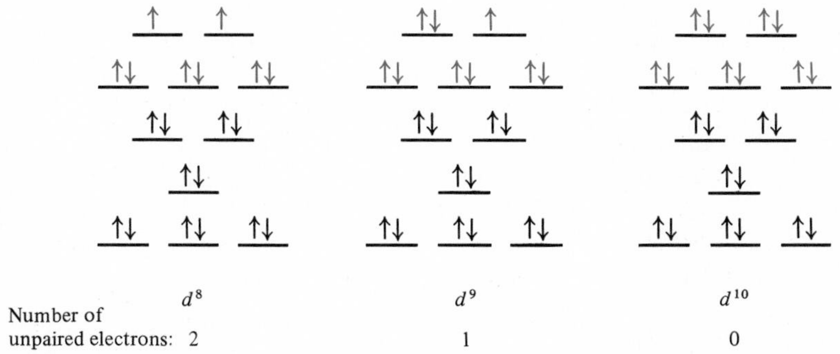

Figure 19-17 Electron diagrams for metals that supply eight, nine, or 10 *d* electrons. The electrons in color are contributed by the metal, those in black by the ligands.

However, in the cases of d^4, d^5, d^6, and d^7, there are two possible arrangements in each case (Figure 19-18), depending on how close together in energy the levels are (the value of Δ_0). The two possibilities are a consequence of Hund's rule, which says that in filling a set of equivalent orbitals, an electron finds it a lower-energy situation to go into an empty orbital than into an orbital that already contains one electron. In this case, the levels are not equivalent, but they may be close enough in energy so that for this purpose they act as if they were.

For example, let us look at d^4. The first three *d* electrons go into the t_{2g} orbitals. Where will the fourth go? If the e_g^* state is close in energy to the t_{2g} (low Δ_0), the fourth electron will go into one of the e_g^* orbitals (rather than pair up with one of the t_{2g} electrons), so that there will be one electron in each of four orbitals, and we will have four unpaired electrons. If Δ_0 is large, however, it would cost too much energy for an electron to go into the e_g^* level, and the fourth electron will pair with one of those in the t_{2g} level, resulting in only two unpaired electrons (Figure 19-18). These two possible arrangements are called **high-spin** and **low-spin,** and two similar arrangements are also possible for d^5, d^6, and d^7 (Figure 19-18).

From the molecular orbital theory alone, it is difficult to predict whether a given d^4, d^5, d^6, or d^7 compound will exist in a high-spin or a low-spin state. One of the advantages of the crystal field theory is that it allows us to make this prediction, based on the nature of the metal and of the ligands. Recall that certain ligands cause a large degree of splitting (high Δ_0), while others cause much less (low Δ_0). Thus, strong crystal fields favor low-spin states, while weak crystal fields favor high-spin states. The oxidation state of the metal also has an effect. Higher oxidation states (fewer electrons) favor low-spin states, perhaps because increasing the charge on the metal allows the ligands to get closer and exert more influence. Complexes of metals

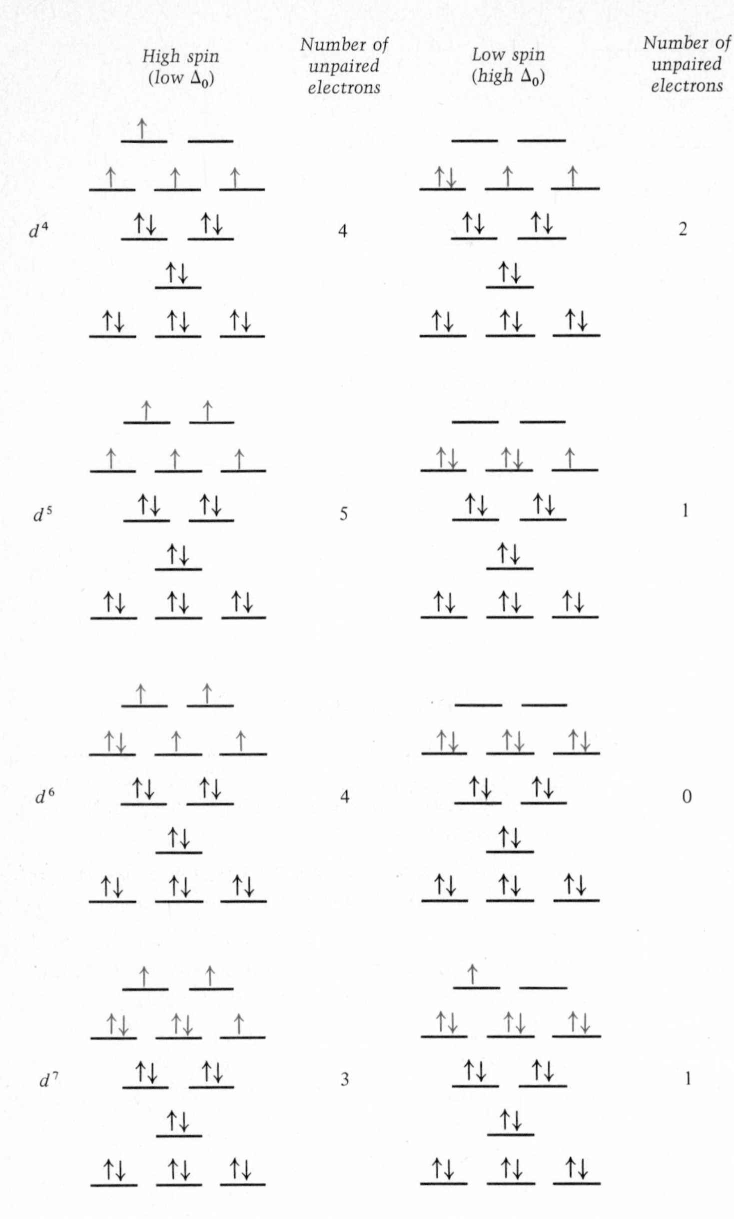

Figure 19-18 Electron diagrams for metals that supply four, five, six, or seven *d* electrons. The left side shows the high-spin states; the right side, the low-spin states. The electrons in color are contributed by the metal, those in black by the ligands.

in the second and third transition series have a greater tendency to exist in low-spin states than do those of the first series.

From these considerations, we can obtain the electronic structures of coordination compounds.

Example: Predict the electronic structure of the $Fe(CN)_6^{4-}$ ion, showing the number of unpaired electrons. Will it be paramagnetic?

Answer: From the number of ligands (six CN^- ions), and the total charge on the ion, we calculate that the charge on the iron is +2. An iron *atom* has $3d^64s^2$. An Fe^{2+} ion has lost two of these electrons (the $4s^2$), leaving $3d^6$. Thus, iron contributes six *d* electrons. The CN^- ion has a very strong crystal field, so that we would expect a low-spin state. Figure 19-18 shows that the low spin for d^6 is

↑↓ ↑↓ ↑↓ contributed by Fe^{2+}

↑↓ ↑↓
↑↓ } contributed by the CN^- ions
↑↓ ↑↓ ↑↓

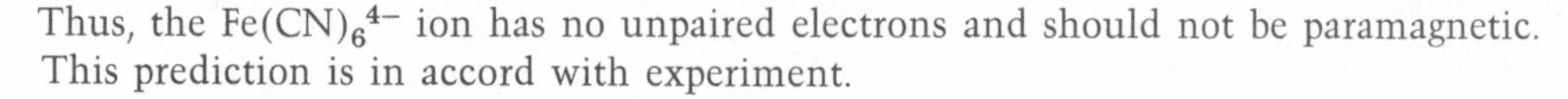

Thus, the $Fe(CN)_6^{4-}$ ion has no unpaired electrons and should not be paramagnetic. This prediction is in accord with experiment.

Most chemists who deal with coordination compounds use both the crystal field and molecular orbital approaches, as well as modified versions of these methods. The modified crystal field theory is often called **ligand field theory,** although this name is also applied to the entire set of theories which are used to explain coordination compound bonding, including the crystal field and molecular orbital theories. In any case, these theories explain many of the properties of coordination compounds. For example, we have already seen how the theory predicts the magnetism of coordination compounds, by giving the number of unpaired electrons. The theory also explains the colors of the complexes. For example, let us consider the simplest case, d^1. An example is the ion $Ti(H_2O)_6^{3+}$.

The titanium atom has $3d^24s^2$. The +3 ion has lost three electrons, and so is $3d^1$, with the electron pattern shown on the left in Figure 19-19. There is one electron in the t_{1g} level, but the e_g^* level is not far above it. Recall that an electron can jump from one level to a higher one, if exactly the right amount of light energy is supplied. In this case, the difference in energy Δ_0 between the t_{1g} and e_g^* levels is about 4×10^{-12} erg, which corresponds to light of wavelength of about 510 nm:

$$E = h\nu$$

$$\lambda = \frac{c}{\nu} = \frac{hc}{E} = \frac{(6.6 \times 10^{-27} \text{ erg·s})\ (3.0 \times 10^{10} \text{ cm/s})}{3.9 \times 10^{-12} \text{ erg}}$$

$$= 5.1 \times 10^{-5} \text{ cm} = 510 \text{ nm}$$

Light of this wavelength is green. When white light shines through a solution of $Ti(H_2O)_6^{3+}$, the green is absorbed (it is used to make the electrons jump to the excited state shown in Figure 19-19) so that the remaining light comes out without the green, and we see it as a mixture of red and violet, which appears mostly violet.

Similarly, most other transition metal complexes are colored (a few are shown in Table 19-5), because most of them have electrons in t_{2g} orbitals and vacancies in e_g^* orbitals. The colors differ because the value of Δ_0 differs (even for the same ion, with different ligands), and because most of the complexes have several d electrons, increasing the number of possible transitions. In fact, we usually obtain the values of Δ_0 from the spectra of the complexes (that is, the wavelengths of the light absorbed). Complexes of d^0 species, such as Sc^{3+} and Ti^{4+}, and d^{10} species, such as Zn^{2+} and

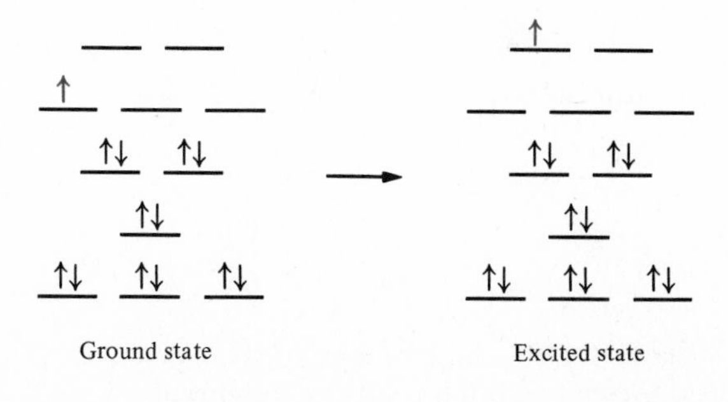

Figure 19-19 The ground state and first excited state of the Ti^{3+} ion.

Ag^+, are usually colorless, because the t_{2g} and e_g^* levels are either empty or completely filled.

We have shown how the crystal field and molecular orbital theories can be applied to octahedral complexes. These theories are also used for other types of complexes (for example, square planar, tetrahedral). For example, in a tetrahedral complex, the five *d* orbitals are divided into two sets, of which two orbitals have a lower energy than the other three.

Problems

19-1 Define each of the following terms. **(a)** alloy **(b)** metallurgy **(c)** ore **(d)** roasting **(e)** energy band **(f)** semiconductor **(g)** doping **(h)** *p-n* junction **(i)** disproportionation **(j)** lanthanide **(k)** earth **(l)** ligand **(m)** complex **(n)** bidentate **(o)** chelate **(p)** isomer **(q)** instability constant **(r)** point charge **(s)** orbital splitting **(t)** high spin

19-2 In what ways do metals differ from nonmetals?

19-3 Write the molecular formulas for each of the following. **(a)** cryolite **(b)** magnetite **(c)** bauxite **(d)** hematite **(e)** pyrite

19-4 If a modern aluminum plant uses a current of 100,000 amperes, how many kilograms of Al can it produce in an 8-hour period? How many kilograms of Al_2O_3 are required to produce that much Al?

19-5 Write equations for the reactions that take place in the blast furance.

19-6 If a 6-ft aluminum stepladder weighs 7.3 kg, estimate the weight of a steel ladder of the same size, assuming that it is all iron (density of aluminum = 2.70 g/ml; density of iron = 7.87 g/ml).

19-7 Is the cell in Figure 19-4 a voltaic cell or an electrolytic cell? Explain your answer.

19-8 Explain what holds the atoms together in a piece of potassium and in a piece of calcium. What differences are there, if any, in the bonding of these two metals?

19-9 Explain why graphite is a conductor of electricity and diamond a nonconductor.

19-10 Name two elements that germanium could be doped with to make it a *p*-type semiconductor, and two that would make it an *n*-type semiconductor.

19-11 From what we know about the properties of metals, is it surprising that the following metals are found in nature in these states? Explain. **(a)** Pb and Bi as sulfides **(b)** Ba and Sr as sulfates **(c)** Pt and Au as the free metal **(d)** Na^+ and Mg^{2+} in seawater **(e)** Cu and Hg both free and as the sulfides

19-12 Explain why each of the group IIA metals has a much higher melting point and density than the corresponding group IA metal (Table 19-1).

19-13 Predict which species in each of the following pairs of species has the higher ionization energy?

(a) Na, Mg **(b)** Na^+, Mg^+
(c) Na^+, Mg^{2+} **(d)** Rb, Cs
(e) Rb^+, K **(f)** Mg, Sr
(g) Cs, Ra^+

19-14 What reason can you suggest for the fact that Sn^{2+}, Pb^{2+}, and Bi^{3+} ions are all common, but Sn^{4+}, Pb^{4+}, and Bi^{5+} ions (each of which would have a complete outer shell) are unknown?

19-15 Why do Mo and W resemble each other chemically much more than either resembles Cr?

19-16 Why are there much greater differences in properties among the transition elements from Hf (72) to Au (79) than among the lanthanides Ce (58) to Lu (71)?

19-17 Explain with the help of Table 16-2 why Cu(I) and Au(I) are not found in aqueous solution.

19-18 Write the correct total charges for each of the following complexes, given the oxidation number of the metal.

(a) $PtCl_6$ (Pt^{IV})
(b) $Pt(NH_3)Cl_3$ (Pt^{II})
(c) $Cr(en)(NH_3)_2(OH)_2$ (Cr^{III})
(d) $CoBr_3(en)(H_2O)$ (Co^{III})
(e) $Ru(HSO_3)_4(NH_3)_2$ (Ru^{II})
(f) $CuCl_2(NH_3)(CH_3NH_2)$ (Cu^{II})
(g) $V(CO)_6$ (V^0)
(h) $Rh(CO)_2Cl_2$ (Rh^{III})
(i) $Zr_4(OH)_8(H_2O)_{16}$ (each Zr is Zr^{IV})

19-19 What is the oxidation number of the metal in each of the following complexes?
(a) $CuCl_3(NH_3)^-$ **(b)** $Co(en)_3^{2+}$
(c) CoF_6^{2-} **(d)** $Os(CO)_5$
(e) $Al(OH)_2(H_2O)_4^+$ **(f)** $PdCl_2(bipy)$
(g) $Co(CN)_4^{4-}$

19-20 For each case, state the number of outer-orbital *d* electrons contributed by the central metal.
(a) $Mn(H_2O)_6^{2+}$ **(b)** $PdCl_2(NH_3)_2$ **(c)** NbF_6^- **(d)** $[Co(en)_2Cl_2]^+$ **(e)** $NiBr_3[P(CH_3)_3]_2$
(f) VCl_5^- **(g)** $Au(CN)_2^-$

19-21 Draw structural formulas for each of the following complexes.
(a) $Ni(en)_3^{2+}$ **(b)** $Co(NH_3)_4(acac)^{2+}$
(c) $CrCl_2(bipy)_2^+$ **(d)** $Pd(C_2O_4)_2^{2-}$
(e) $Co(CN)_3(dien)$

19-22 Give suitable names, according to the IUPAC system, for each of the following.
(a) $2K^+ [NiCl_4]^{2-}$ **(b)** $Ru(CO)_5$
(c) $[Hf(OH)(NH_3)_3]^{3+}$ $3Br^-$ **(d)** $[Mn(NH_3)_3CN]^+$ Cl^-
(e) $[Co(en)_3]^{3+}$ $3Cl^-$ **(f)** $Na^+ [B(NO_3)_4]^-$
(g) $RuCl_2(CO)[P(CH_3)_3]_2$

19-23 Write the molecular formulas for each of the following compounds.
(a) hexaaquavanadium(III) ion **(b)** tetraaquadichlorocobalt(III) chloride
(c) bis(2,2′-bipyridine)chlororhodium(II) nitrate **(d)** sodium tetrafluoroargentate(III)
(e) *cis*-diiodobis(methylamine)platinum(II) **(f)** barium tetrafluorobromate(III)

19-24 Is the following reaction favored by an entropy effect? Explain.

$$NH_3\text{—}Pt(NH_3)_2\text{—}NH_2CH_2CH_2NH_2 \longrightarrow NH_3\text{—}Pt(NH_3)\text{—}(NH_2\text{—}CH_2\text{—}CH_2\text{—}NH_2\text{, chelate}) + NH_3$$

19-25 Draw the three-dimensional structure for each of the following complexes.
(a) $Os(dipy)_3^{3+}$ **(b)** $NiCl_2(PMe_3)_2$ (tetrahedral)
(c) $CuCl_2^-$ **(d)** $Zr(acac)_2Cl_2$
(e) $Pd(NH_3)_4^{2+}$ (square planar)

19-26 The complex ion $Ni(CN)_5^-$ can exist in either the square pyramid or the trigonal bipyramid configuration. Draw the two structures.

19-27 Of the following pairs, which are identical, which are isomers, and which are neither?

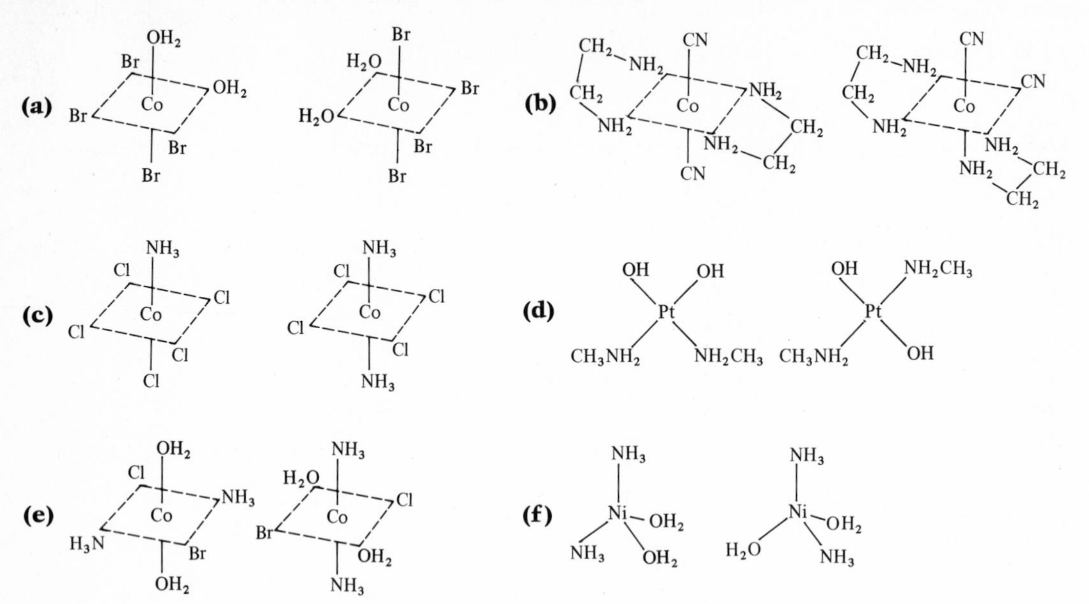

19-28 In each case, draw the other isomer or isomers, if any.

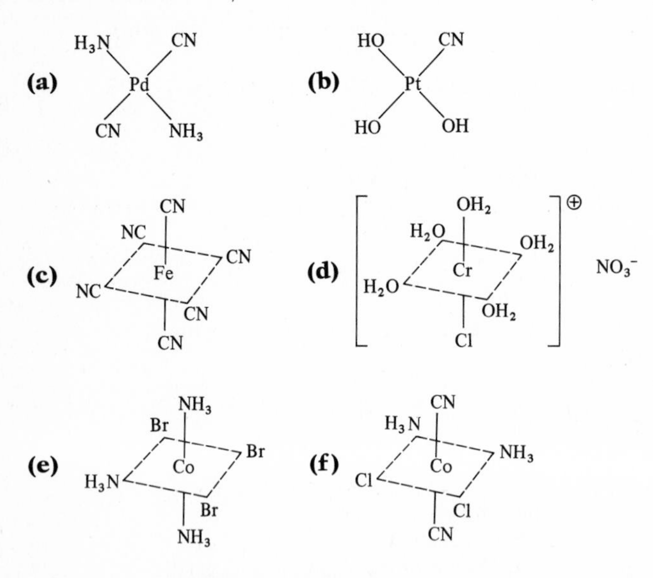

19-29 Write out the six separate steps for the dissociation of $Co(NH_3)_6^{3+}$ and the equilibrium expression for each step.

19-30 For the dissociation of $Fe(SCN)_3$ to Fe^{3+} and $3\,SCN^-$, the values of K_1, K_2, and K_3 are, respectively, 130, 16, and 1.0. Calculate the overall stability and instability constants.

19-31 3.00×10^{-3} mole of $Co(NH_3)_6^{3+}\ 3Cl^-$ is dissolved in 1.000 liter of water. What is the concentration of free Co^{3+} in this solution? (K_i for $Co(NH_3)_6^{3+} = 1 \times 10^{-34}$; neglect the intermediate dissociations.)

19-32 To a 1.00×10^{-3} *M* solution of $Cu(NO_3)_2$ is added enough gaseous NH_3 to make the concentration 0.50 *M* in NH_3. What is the concentration of free Cu^{2+} in the solution? (K_i for $Cu(NH_3)_4^+$ is 8×10^{-14}; neglect the intermediate dissociations.)

19-33 A solution is originally 0.10 *M* in Ag^+, and sufficient KCN is added to form $Ag(CN)_2^-$ and give an excess CN^- concentration of 0.020 *M*. **(a)** What is the concentration of Ag^+ in this new solution? **(b)** Will Ag_2CO_3 precipitate if the solution is made 1.10 *M* in CO_3^{2-}?

19-34 Draw electron diagrams, like those shown in Figures 19-16 to 19-18, for each of the following octahedral complexes. If there are two possibilities, show both.
(a) $Cr(NH_3)_6^{3+}$ **(b)** $AuBr_6^{3-}$
(c) VCl_6^{2-} **(d)** $Mn(CN)_6^{4-}$
(e) $Co(H_2O)_4Cl_2^+$

19-35 Predict whether each of the complexes in Problem 19-34 is paramagnetic or not.

19-36 Draw electron diagrams, like those in Figure 19-18 for the two possible spin states of each of the following octahedral complexes. In each case predict whether the high-spin or low-spin arrangement is favored.
(a) $Cr(H_2O)_6^{2+}$ **(b)** $[Co(en)_3]^{3+}$ **(c)** $[Os(CN)_6]^{3-}$ **(d)** $FeCl_6^{4-}$ **(e)** $IrCl_3(H_2O)_3^+$

19-37 Tell whether each of the complexes in Problem 19-36 is paramagnetic or not.

19-38 Given the number of *unpaired* electrons, decide whether each of these ions is high-spin or low-spin.
(a) $Mn(H_2O)_6^{2+}$ (5 unpaired)
(b) $Ru(bipy)_3^{2+}$ (no unpaired)
(c) NiF_6^{3-} (1 unpaired)

19-39 Show that when the crystal field theory is applied to linear complexes (coordination number = 2) it predicts that the five *d* orbitals are divided as follows:

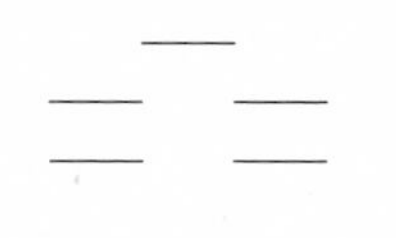

(Place the two ligands on the *z* axis.)

19-40 Show how the crystal field theory predicts that in a tetrahedral complex, the five *d* orbitals are split into two sets, of which two orbitals have a lower energy than the other three.

Additional Problems

19-41 Draw structural formulas for $Fe(en)_3^{3+}$ and $Fe(en)_3^{2+}$. Do they look different? What is the difference in the structure of these two ions?

19-42 The $Co(NH_3)_6^{3+}$ ion absorbs light whose wavelength is about 470 nm. What is the energy difference (Δ_0) between the t_{1_g} and e_g^* levels?

19-43 Explain why iron is a conductor of electricity, sulfur a nonconductor (insulator), and boron a semiconductor.

19-44 If OH^- is added to a solution containing Zn^{2+} ions, $Zn(OH)_2$ is precipitated when the ion product exceeds K_{sp}, but additional OH^- dissolves the precipitate because the $Zn(OH)_2$ is converted to $Zn(OH)_4^{2-}$. If the initial concentration of Zn^{2+} is 1.0×10^{-3} *M*, what concentration of OH^- is required to dissolve all the $Zn(OH)_2$ that was initially formed?

19-45 There are six possible octahedral isomers where Cr^{IV} is the central atom, and 4 NH_3, 2 Cl^-, and 2 Br^- groups are also present. Draw them all.

19-46 When Marie and Pierre Curie isolated radium from pitchblende, they carefully separated all the other elements until only radium was left. Which element do you think proved to be the most difficult to separate from the radium?

19-47 Why does the electrical conductivity of semiconductors increase with increasing temperature?

19-48 For a complex of coordination number 4 (whether square planar or tetrahedral), there are only four electron pairs contributed by the ligands. Draw a molecular orbital diagram, similar to Figure 19-15, for this case. Is it necessary for any such complex to use antibonding orbitals?

19-49 When the transistor was invented in 1948, the inventors had no idea that their invention would eventually make a moon landing possible. Is there a moral to this? Should scientific research only be carried out in order to solve a specific technical problem?

19-50 **(a)** If each cubic mile of seawater contains 6.00×10^6 tons of Mg^{2+} ions, what is the concentration of Mg^{2+} in the sea, in g/ml? **(b)** In 1974, the total world production of magnesium from the sea was 2.72×10^5 tons. If 100 times this amount were taken from the sea each year for the next 1 million years, what would be the concentration of Mg^{2+} in the sea at the end of that time? (Assume that the total volume of the sea is and will remain 3.30×10^8 cubic miles.)

20 Organic Chemistry

Jacobus van't Hoff, 1852–1911

As we have already pointed out, compounds that contain the element carbon are called **organic compounds,** while those which do not contain carbon are **inorganic compounds.*** Organic chemistry is the chemistry of carbon compounds. One might expect that there would be many more inorganic compounds known than organic ones, since all organic compounds must contain carbon, while inorganic compounds can contain any combination of the 104 other elements. However, such an expectation would not be borne out. There are more than 2 million known organic compounds, but only about 200,000 or 300,000 inorganic ones. The main reasons why so many organic compounds can exist are: (1) carbon atoms form stable bonds with other carbon atoms, so that short and long chains of carbon atoms are possible; and (2) the valence of carbon is four, which may be made up in different ways (see Figure 14-6) so that many combinations and arrangements of atoms are possible, as we shall shortly see.

We have used the expression "known compound." What does this mean? When a chemist recognizes that he has obtained a new compound, he reports this information, including how he obtained it, and some of its properties, by publishing a paper in a scientific journal. The compound is then considered to be known, even if the chemist destroys the compound (by using it to react with another compound) a day or two later. Thus, maybe only 10 or 20% of all known compounds actually exist at this moment. However, they are considered to be known because anyone who wishes can obtain them again by following the procedures in the published paper (indexing systems in chemistry are excellent, and such papers can usually be located in a comparatively short time). Known compounds are obtained from two sources:

1. Nature. Living things—animals and plants—make a great many organic compounds. Each species makes thousands. Compounds obtained from animals and plants are called **natural products.** Organic compounds are also obtained from petroleum, coal, and other fossil-fuel sources.
2. Synthesis. Most known compounds are not obtained from natural sources but are made in laboratories from other compounds. This is known as *organic synthesis.* We will discuss a few of the known synthetic methods in Section 20-8. One of the interesting things about organic chemistry as a science is that chemists have been able to synthesize so many fascinating and useful molecules that simply do not exist in nature. Of the hundreds of thousands of known synthetic organic compounds, a small percentage have proven so useful that they are manufactured commercially in large quantities. Some of them are used as drugs, others as dyes, explosives, insecticides, detergents, monomers for making polymers (see Chapter 22), and other uses. Some examples are aspirin,

*A few simple carbon compounds, CO, CO_2, Na_2CO_3, HCN, are often considered to be inorganic, even though they contain carbon.

librium, DDT, nitroglycerine, TNT, copper phthalocyanine (an important dye), soap, and ether. There are, of course, thousands more. As we have already seen, some of these compounds, useful as they are, are capable of causing environmental damage when added to the environment in large quantities (Sections 17-14 and 17-15).

Although there are more than 2 million known organic compounds, and more than 100,000 *new* synthetic organic compounds are made each year, we have not even begun to scratch the surface when it comes to making all possible organic compounds. Nobody knows how many of these there are, but the number is surely greater than many trillions (1 trillion = 10^{12}). Although we can predict the properties of a large number of these unknown compounds by analogy with the properties of known compounds, it is likely that some of them have surprises in store for us. Indeed, a number of compounds already synthesized have shown properties that were unexpected. It should be pointed out that these compounds remain unsynthesized not because we don't know how to make them (in fact, methods are known today for synthesizing almost any possible organic compound), but because the number is so vast that we simply have not gotten around to it yet! At this time, it seems unlikely that there will ever come a time when all possible organic compounds have been made.

The name "organic" is a holdover from history. In alchemical times, it was recognized that there were three sources of all matter: animal, vegetable, and mineral. When chemistry began as a science (around A.D. 1600), these categories continued to be used. After Scheele (1742–1786) discovered that lactic acid could be obtained both from animal and vegetable sources, these two categories were merged, and became "organic," while mineral became "inorganic." Around this time, the belief emerged that organic compounds (that is, those obtained from animal or vegetable sources) could not be obtained from inorganic compounds. This was called the Doctrine of Vital Force, since it was believed that a "vital force" is necessary to make organic compounds and that only living organisms possess this force. At that time, an "organic" compound was one produced by an animal or plant, or one that was synthesized from another organic compound.

In 1828, Friedrich Wöhler (1800–1882) accidentally prepared urea by treating ammonium chloride with silver cyanate.

$$NH_4Cl + AgNCO \xrightarrow{\Delta} \underset{\text{urea}}{NH_2CONH_2} + AgCl$$

Ammonium chloride and silver cyanate were undoubtedly inorganic, while urea was organic (it is found in human and animal urine), so that Wöhler had disproved the Doctrine of Vital Force. Although it took a few years for the entire chemical world to accept this conclusion, by 1845 it was accepted by everyone. The terms "organic" and "inorganic" were very useful, however, and the modern definitions were assigned by Kekulé around 1860. As we have seen, most organic compounds today do not come from living organisms. Today that branch of chemistry which deals with living organisms is called *biochemistry* (Chapter 21).

Only a few elements suffice to make up the vast majority of the more than 2 million known organic compounds. Besides carbon, the most important is hydrogen, which is present in almost all organic compounds. In fact, many thousands of known organic compounds contain *only* carbon and hydrogen (these are the **hydrocarbons,** discussed in Sections 20-1 to 20-5). Other important elements found in many organic compounds are oxygen, nitrogen, the four halogens, sulfur, and phosphorus. Although organic compounds have been made which contain iron, silicon, mercury, bismuth, and indeed almost every element in the periodic table

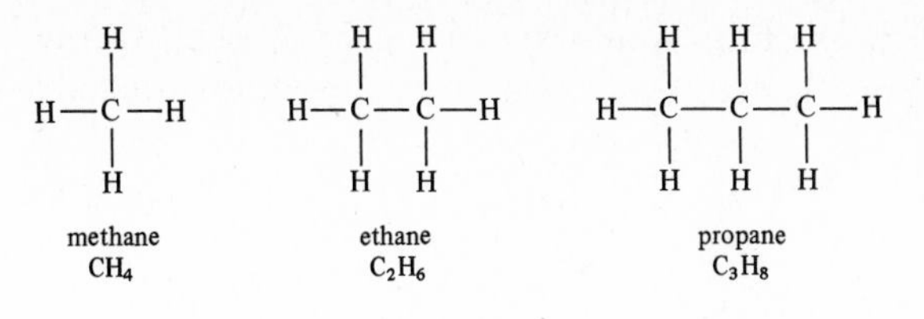

Figure 20-1 Structural and molecular formulas for the three simplest alkanes.

(other than the noble gases), more than 99% of all known organic compounds contain no elements other than the 10 listed above.

It would be difficult to study such an enormous number of compounds if they could not be divided into classes. Fortunately, this is easily done. The vast majority of organic compounds can be divided into a relatively small number of classes. What makes the subject amenable to study at all is that most or all of the compounds within a class have similar properties, or properties that change in a regular manner with changes in structure. We shall examine the most important of these classes in Sections 20-1 to 20-6.

20-1 Alkanes—Structure and Names

Alkanes are hydrocarbons that contain only single bonds. The three simplest alkanes are methane, ethane, and propane, which have one, two, and three carbon atoms, respectively. Figure 20-1 shows structural and molecular formulas for these compounds. Molecular formulas tell us how many atoms of each element are present in the molecule, but structural formulas not only tell us this, but show us the *connections* between the atoms. Organic chemists customarily use structural formulas rather than molecular formulas, for a reason we shall shortly encounter.

Although structural formulas are much better representations than molecular formulas, they still are a long way from showing the actual molecule, and indeed, they may lead us astray, if we are not careful. Structural formulas are two-dimensional, but molecules are three-dimensional. For this reason, organic chemists frequently use models, to represent the real molecule more closely than can be done on a flat sheet of paper or a blackboard. Figure 20-2 shows propane as represented by three kinds of models. Even here the models are not equally accurate. The space-filling type comes the closest to the real structure, though still not very close. Most often, organic chemists use shorthand structural formulas in which the C—H bonds are not actually shown but only implied: for example,

$$\text{H}-\underset{\text{H}}{\overset{\text{H}}{\underset{|}{\overset{|}{\text{C}}}}}-\underset{\text{H}}{\overset{\text{H}}{\underset{|}{\overset{|}{\text{C}}}}}-\underset{\text{H}}{\overset{\text{H}}{\underset{|}{\overset{|}{\text{C}}}}}-\text{H} \equiv CH_3-CH_2-CH_3$$

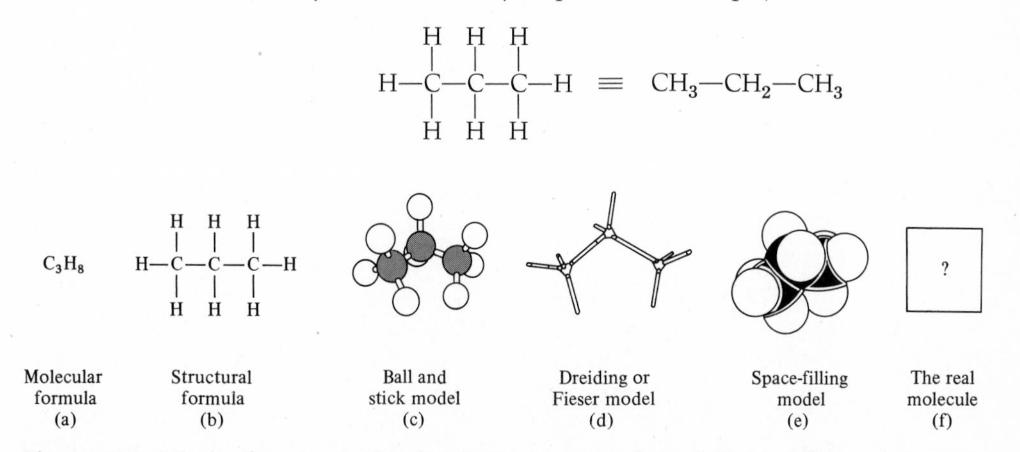

Figure 20-2 Ways of representing propane. As we go from left to right, we more and more closely portray the real molecule, which of course nobody has ever seen.

We learned in Section 15-13 that the four bonds of carbon are directed to the corners of a tetrahedron. The angles between any two bonds are approximately 109.5°.

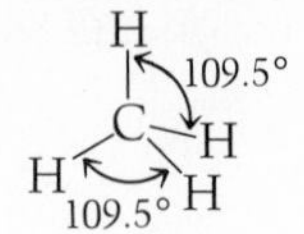

Note that the formula of propane as customarily written $CH_3—CH_2—CH_3$, therefore gives a wrong impression. The C—C—C angle is not 180°, but 109.5°. The models in Figure 20-2 clearly show this, but it is necessary to keep it in mind even when models are not immediately available.

After this brief digression about three-dimensional structures, let us return to the alkanes. There is only one alkane possible with one carbon, one with two carbons, and one with three carbons (methane, ethane, and propane, respectively), but when we come to four carbons there are *two* alkanes possible:

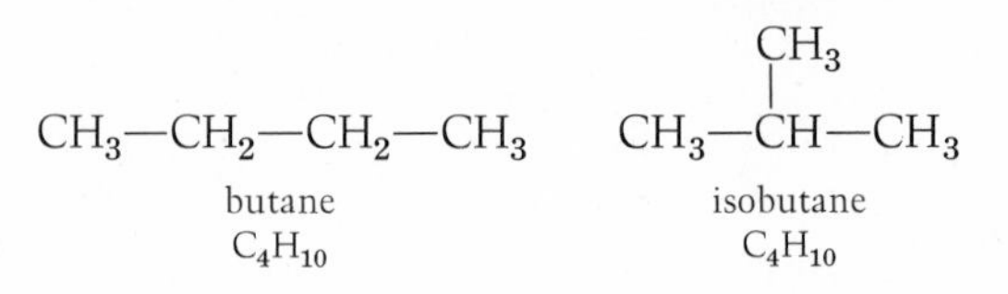

Both of these compounds have four carbons and 10 hydrogens, but their structural formulas (that is, the connections between atoms) are different. In particular, it may be noted that isobutane has a carbon atom connected to three other carbon atoms (we call this a **tertiary** carbon atom), while butane has no such carbon atom but does have two carbon atoms, each connected to two other carbon atoms (**secondary** carbon atoms), which isobutane does not have. Furthermore, isobutane has three **primary** carbon atoms (connected to one other carbon atom), while butane has two. The two compounds obviously have different structures, and laboratory investigation shows they have different properties as well. Their melting points, boiling points, reactivity, and all other properties are different. Compounds with the same molecular formulas but different structural formulas are called **isomers.** Isomers in organic chemistry are not rare; in fact, they are the rule rather than the exception, and this is the main reason chemists prefer to write structural formulas rather than molecular formulas.

When we came to alkanes with five carbons, we find that there are three isomers:

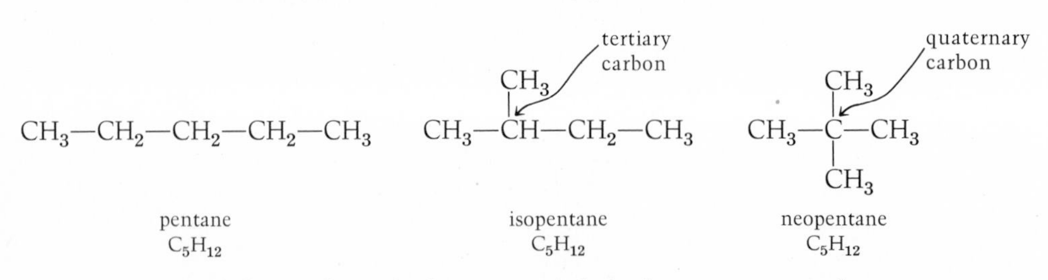

Pentane is called a *nonbranched* isomer, while both isopentane and neopentane are branched. In isopentane the branching is at the tertiary carbon atom, while in neopentane it is at the **quaternary** carbon atom (a carbon connected to four other carbons, which is, of course, the maximum possible). It does not matter how we write a structural formula, as long as the connections are accurate. Thus, all of these are correct ways to write isopentane:

$$\overset{\displaystyle CH_3}{\overset{|}{CH_3—CH—CH_2—CH_3}} \qquad \underset{\displaystyle CH_3}{\underset{|}{CH_3—CH—CH_2—CH_3}} \qquad \underset{\displaystyle CH_3}{\underset{|}{CH_3—CH_2—CH—CH_3}}$$

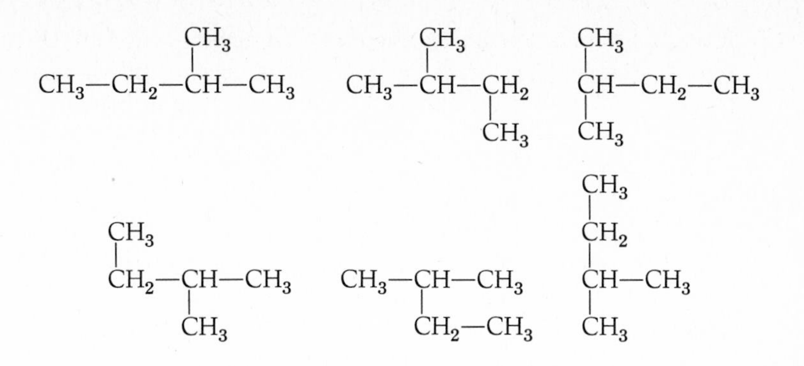

The three C_5H_{12} isomers, like the two C_4H_{10} isomers, are different compounds with different properties.

If we continued the process, we would find that there are five isomers with six carbons (all C_6H_{14}); nine with seven carbons (all C_7H_{16}), and the number of isomers increases rapidly with the number of carbons. For example, there are 75 $C_{10}H_{22}$ isomers and more than 300,000 compounds with the formula $C_{20}H_{42}$.

Because of the valences of carbon and hydrogen, the number of hydrogens that an alkane can have is fixed by the formula C_nH_{2n+2}. This means, for example, that all alkanes with 10 carbons must have $2(10) + 2 = 22$ hydrogens. There can be no alkanes with the formula $C_{10}H_{21}$, $C_{10}H_{23}$, $C_{10}H_{24}$, or any number of hydrogens other than 22. This becomes obvious when we realize that *all* alkanes can be built up (in our minds at least), by beginning with methane CH_4, and adding CH_2 units:

$$CH_3\text{—}H + CH_2 \longrightarrow CH_3\text{—}CH_2\text{—}H + CH_2 \longrightarrow CH_3\text{—}\underset{}{\overset{H}{\overset{|}{C}}}H\text{—}CH_3 + CH_2 \longrightarrow$$

$$CH_3\text{—}\overset{CH_3}{\overset{|}{C}}H\text{—}\underset{\underset{H}{|}}{C}H_2 + CH_2 \longrightarrow \text{etc.}$$

We cannot do this in practice, but we can in theory, and it shows us that *all* alkanes must have the formula C_nH_{2n+2}.

We called the three C_5H_{12} isomers pentane, isopentane, and neopentane. It is clear that we cannot go on like this. If we tried, we might find prefixes for the 5 C_6H_{14} compounds, and the 9 C_7H_{12} isomers (perhaps grouchohexane, chicohexane, harpohexane, etc.), but by the time we came to the 75 $C_{10}H_{22}$ compounds, we would run out of prefixes, and even if we did not, we could never remember them all. Fortunately, there is a better way. By use of a system called the IUPAC system (for the International Union of Pure and Applied Chemistry), we can give each alkane a name which corresponds to its structural formula, so that memory is kept to a minimum. The rules are fairly simple.

1. Unbranched alkanes (there can be only one for each carbon number) are given these names.

Number of carbons	*Name*	*Number of carbons*	*Name*
1	Methane	6	Hexane
2	Ethane	7	Heptane
3	Propane	8	Octane
4	Butane	9	Nonane
5	Pentane	10	Decane

It is important to memorize these names, because they are the basis of the entire IUPAC system. The system also includes names for unbranched alkanes with more than 10 carbons, but the names given above will be sufficient for our purposes. Note that the names of all alkanes end in **ane.** Unbranched alkanes of four carbons or more are also called *normal*-alkanes (or *n*-alkanes); for example, hexane is also called *n*-hexane.

2. When naming branched alkanes, select the longest continuous chain of carbons, and name it according to the list above. This is called the **parent** name.

Example:

$$\begin{array}{c} CH_3 \\ | \\ CH_3-CH_2-CH_2-CH_2-CH-CH_2-CH_3 \end{array}$$

In this example, the longest continuous chain (shown in color) has seven carbons. Therefore, the parent name is *heptane*.

3. Use one or more prefixes to name everything that is not part of the parent. In the above example the only thing that is not part of the parent is CH_3. CH_3 is called a *methyl* group, and we use the prefix *methyl*. The name would thus become methylheptane. However, this name is not sufficient, because there is more than one compound that would fit this name (in fact there are three; see below).

4. We use *numbers* to locate the position of the methyl group. In step 2 we found the longest continuous chain. A chain has two ends. *We must number the chain from whichever end gives the lowest number to the methyl group.*

$$\begin{array}{ccccccccccccc} & & & & & & & & CH_3 & & & & \\ & & & & & & & & | & & & & \\ CH_3 & - & CH_2 & - & CH_2 & - & CH_2 & - & CH & - & CH_2 & - & CH_3 \\ 1 & & 2 & & 3 & & 4 & & 5 & & 6 & & 7 \\ 7 & & 6 & & 5 & & 4 & & 3 & & 2 & & 1 \end{array}$$

If we number from left to right, the methyl group would be on the number 5 carbon, but numbering the other way puts the methyl group on the number 3 carbon. Since 3 is lower than 5, we have no choice, and the full name (and only correct name) of the compound is 3-methylheptane. Note that there are two other methylheptanes:

$$\begin{array}{c} CH_3 \\ | \\ CH_3-CH_2-CH_2-CH_2-CH_2-CH-CH_3 \end{array} \quad \text{2-methylheptane}$$

$$\begin{array}{c} CH_3 \\ | \\ CH_3-CH_2-CH_2-CH-CH_2-CH_2-CH_3 \end{array} \quad \text{4-methylheptane}$$

(Why isn't there a 5-methylheptane or a 1-methylheptane?)

5. If there is more than one group attached to the parent chain, we use prefixes: di, tri, tetra, penta, and so on. We must then supply numbers for each group, even if two are on the same carbon.

Examples:

$$\begin{array}{c} CH_3CH_3 \\ || \\ CH_3-CH-CH-CH_2-CH_3 \end{array} \quad \text{2,3-dimethylpentane}$$

$$\begin{array}{c} CH_3 \\ | \\ CH_3-CH_2-C-CH_3 \\ | \\ CH_3 \end{array} \quad \text{2,2-dimethylbutane}$$

$CH_3-C(CH_3)_2-CH_2-CH(CH_3)-CH(CH_3)-CH_3$ 2,2,4,5-tetramethylhexane

In the last example, 2,2,4,5 is lower than 2,3,5,5, which we would get if we counted from the other end of the chain. The rules say we must always use the lowest possible numbering.

Note that the prefix "di" must always be accompanied by two numbers, "tri" by three, etc., even if the same number must be written twice.

6. Unbranched groups with more than one carbon are given names derived from the corresponding alkanes by changing *ane* to *yl*.

CH_3-CH_2-	ethyl
$CH_3-CH_2-CH_2-$	propyl
$CH_3-CH_2-CH_2-CH_2-$	butyl etc.

The general name for all groups derivable from alkanes is **alkyl groups.**

7. When two different alkyl groups are present, both are given prefixes, and each group is separately numbered. Prefixes are listed in alphabetical order.

Examples:

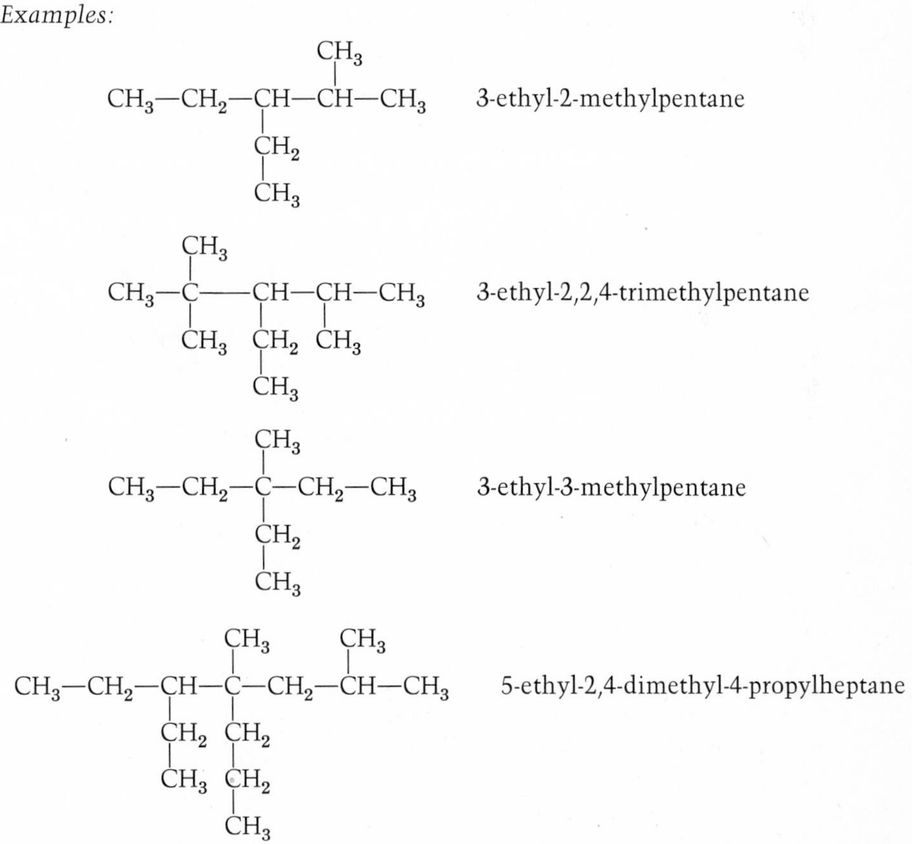

8. There are also names for a few branched alkyl groups which are seldom used in IUPAC names of alkanes, but are commonly used for other types of compounds. Two of the most common are

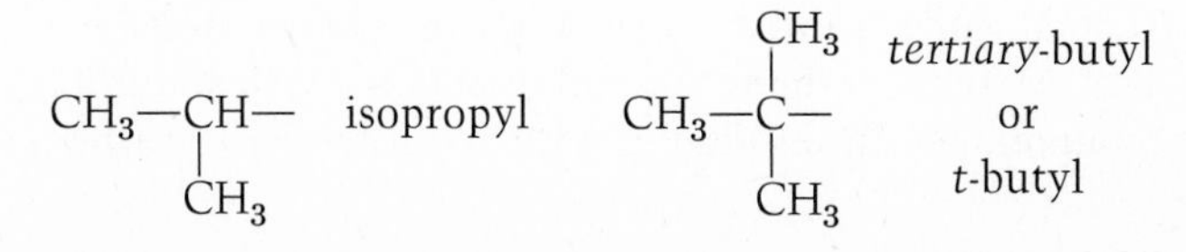

To avoid confusion, unbranched alkyl groups are often given the prefix *normal* or *n*; for example, $CH_3CH_2CH_2$— is called *n*-propyl as well as propyl.

It is apparent that we can name isobutane, isopentane, and neopentane by the IUPAC system:

	Common name	*IUPAC name*
$CH_3—CH(CH_3)—CH_3$	Isobutane	Methylpropane
$CH_3—CH(CH_3)—CH_2—CH_3$	Isopentane	Methylbutane
$CH_3—C(CH_3)_2—CH_3$	Neopentane	Dimethylpropane

(Why is it unnecessary to use numbers here?) These compounds each have *two* perfectly correct names—an IUPAC name and a common name. This is not an unusual situation. In fact, most organic compounds have two, three, or even more perfectly correct names, though this does not mean that any name thought up by any student is necessarily correct. For example, students sometimes choose the wrong longest continuous chain, resulting in an incorrect name.

Example:

$$CH_3—CH(CH_2CH_3)—CH_2—CH(CH_3)—CH_3$$

wrong chain: Five carbons
wrong name: 2-ethyl-4-methylpentane

$$CH_3—CH(CH_2CH_3)—CH_2—CH(CH_3)—CH_3$$

correct chain: Six carbons
correct name: 2,4-dimethylhexane

The longest continuous chain may be written in a straight line or it may be bent or twisted. The student must be careful to make sure that he or she has actually chosen the longest continuous chain in the molecule.

In alkanes there is free rotation around all carbon-carbon single bonds. These bonds are σ bonds (Section 14-16), and there is little to stop the molecules from rotating as shown in Figure 20-3. Except at very low temperatures (~50°K), alkane molecules have enough energy so that this kind of rotation is constantly going on. This is in contrast to double bonds, where, as we shall see in Section 20-3, free rotation cannot occur.

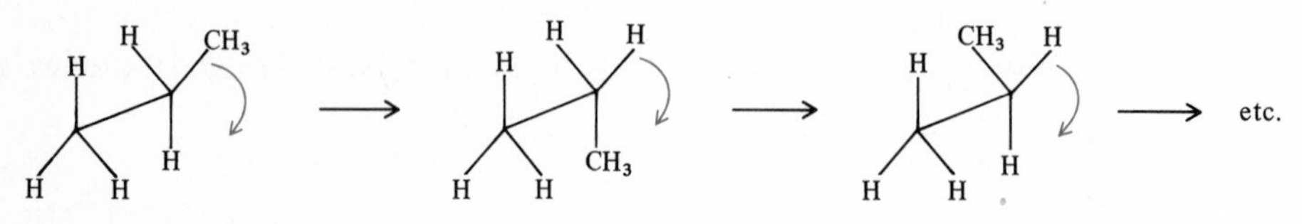

Figure 20-3 **Free rotation in the propane molecule.**

20-2 Alkanes—Properties. Petroleum

In general, boiling and melting points of alkanes increase with increasing molecular weight. Methane, ethane, and other alkanes with about five carbons or less are gases at room temperature. Alkanes with about five to 17 carbons are liquids, while those with about 17 or more carbons are waxy solids. Paraffin, which can be bought in many supermarkets, is a mixture of high-molecular-weight alkanes. All alkanes are insoluble in water and less dense than water (they float on top). Alkanes are soluble in each other.

Chemically, alkanes are inert to almost all reagents. They do not react with strong acids, bases, or oxidizing or reducing agents. However, they do undergo one kind of reaction that is of inestimable importance to our modern civilization. They burn. Petroleum and natural gas consist largely of mixtures of alkanes.

Petroleum is a thick black liquid which is found in large quantities underground in various parts of the earth. It is undoubtedly the result of decayed life forms, though exactly how it was formed is not certain. The composition of petroleum differs, depending on where it is found, but in all places, it is a mixture of hundreds, probably thousands, of different alkanes, ranging from methane up to alkanes with 40 or more carbons, with varying amounts of sulfur compounds and oxygen and nitrogen compounds also present. There are usually also some aromatic compounds, but virtually no alkenes or alkynes.

Because it is such a heterogeneous mixture, petroleum must be refined before it is used. The most important refinery process is distillation, in which the crude petroleum is separated into fractions, each with a different boiling range (hence a different average molecular weight). Figure 20-4 shows a typical refinery distillation column. The major fractions, each of which is still a mixture of many compounds, are as follows, in order of increasing boiling point:

1. Gases ($\sim C_1$ to C_4). These are not condensed into a liquid but are burned in order to supply heat for the distillation. The excess gases are simply burned in the atmosphere (anyone who has passed a petroleum refinery in operation has seen these fires burning at the top of the distillation towers).
2. Gasoline ($\sim C_4$ to C_{12}), boiling range 20 to 200°C. This of course is used to power cars and trucks, lawn mowers, snowmobiles, etc.
3. Kerosene ($\sim C_{10}$ to C_{14}), boiling range 200 to 275°C. At one time kerosene was the most important petroleum fraction; its principal use was in kerosene lamps. With the invention of the electric light, this use greatly diminished. Today it is much less important, but has a number of uses, one of which is jet fuel.
4. Fuel oil ($\sim C_{14}$ to C_{18}), boiling range 275 to 350°C. This fraction has two main uses: fuel oil for heating houses and other buildings; and fuel for diesel engines, in some cars, and in many trucks, trains, and ships.
5. Lubricating oil ($\sim C_{16}$ to C_{20}), boiling range >350°C, has a number of uses, including lubrication of automobile and other engines.
6. Residue ($> \sim C_{20}$). This fraction does not boil at all, but is recovered as a liquid that solidifies. A number of materials are made from it, including greases and asphalts.

In most cases the fractions cannot be used directly in the form in which they come from the distillation column, and they are subjected to several kinds of further treatment.

Gasoline is the most important of the six fractions, and it would be very convenient if a very large proportion of the petroleum alkanes had molecular weights which put them within this fraction. Unfortunately, nature has not made petroleum in this way; more than half of the petroleum consists of C_{13} and higher alkanes. Another disadvantage of natural petroleum is that a high percentage of molecules (more than half) are unbranched (normal) alkanes. This is a disadvantage because unbranched alkanes give rise to much more "knocking" than branched alkanes when used in automobile engines.* The tendency toward knocking can be measured quantita-

* **Knocking** takes place when the gasoline-air mixture in the cylinder (see Section 17-9) explodes before the spark goes off.

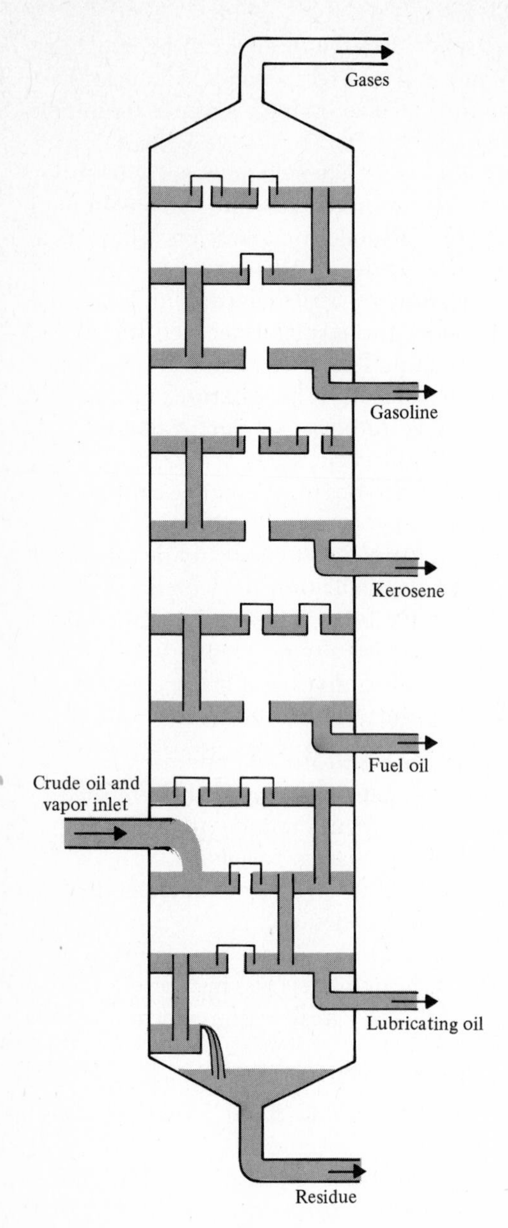

Figure 20-4 Distillation of petroleum. The tower may be about 100 feet tall.

tively; the number used for this purpose is called the **octane number.** Branched alkanes have much higher octane numbers than unbranched alkanes, and the octane number increases with increased branching.

Fortunately, there are several refinery operations that overcome both of these problems. **Cracking** is a process in which petroleum fractions (for example, crude gasoline, kerosene, etc.) are subjected to high heat and pressures in the presence of suitable catalysts. Under these conditions the molecules are broken into fragments, which, being very unstable, immediately reassemble into new alkanes. If the conditions have been chosen properly, the new alkanes will be smaller than the original ones (thus increasing the amount of gasoline), and more highly branched (thus increasing the octane number). The increased branching is a consequence of entropy, since there are many more branched alkanes possible than unbranched ones.

Another way to increase octane number is to add lead tetraethyl, but as we have seen (Section 17-9), this practice is being phased out for gasoline in the United States.

Natural gas is found underground, in large deposits, sometimes in the vicinity of petroleum, but more often not. It consists of more than 90% methane. The other components are largely hydrocarbon gases—ethane, propane, etc. Because of the low boiling point of methane (−162°C), natural gas makes an ideal fuel for transporting in pipes. It always flows, no matter how cold the weather. Natural gas, however, is not suitable for bottled gas; propane (obtained from the light fractions of petroleum) is used for this purpose. The reason is very simple. In order to ship a tank of bottled gas economically, the gas must be liquified (under pressure), because a tank can hold a much greater weight of a liquid than it can of a gas. But the critical temperature (Table 4-4) of methane, −82.8°C, is below room temperature, so that methane cannot be liquified at room temperature no matter how much pressure is applied. Propane, on the other hand, with a critical temperature of 96.7°C, is easily liquified at room temperature. Butane (bp 0°C), which is liquified even more easily (lower pressure required) can be used in the liquid form in cigarette lighters.

Although most petroleum is eventually used for fuel, not all of it is. About 2% of petroleum is used to synthesize various organic chemicals on a large scale. Even though only about 2% is used for this purpose, this represents a very large volume indeed, and more than 50% of all industrial synthetic organic compounds are made from this source (the other chief source is coal). These chemicals are eventually turned into dyes, explosives, drugs, plastics, artificial fabrics, detergents, and other materials. Petroleum is therefore not only important as a fuel, but also as a source of many of the products we use every day.

20-3 Alkenes and Alkynes

Hydrocarbons that contain double bonds are called **alkenes,** and all of their IUPAC names end in **ene.** Those which contain triple bonds are called **alkynes,** and their IUPAC names end in **yne.** In fact, we use the same rules we already had, with the addition of a few more. For double-bond compounds the *ane* ending is changed to *ene,* and the position of the double bond is given by a single number (the number of the lower-numbered carbon).

Examples:

$$\overset{1}{CH_2}{=}\overset{2}{CH}{-}\overset{3}{CH_2}{-}\overset{4}{CH_3} \qquad \text{1-butene}$$

$$CH_3{-}CH{=}CH{-}CH_3 \qquad \text{2-butene}$$

$$\overset{4}{CH_3}{-}\underset{\displaystyle CH_3}{\overset{3}{\underset{|}{CH}}}{-}\overset{2}{CH}{=}\overset{1}{CH_2} \qquad \text{3-methyl-1-butene}$$

$$CH_3{-}CH{=}CH_2 \qquad \text{propene (no number needed)}$$

Note that (1) in the case of 1-butene, the double bond goes from the 1-carbon to 2-carbon, but we only write the *lower* number. (2) We number the chain from the end that gives the *double bond* the lowest number even if that means that alkyl groups must get higher numbers (for example, 3-methyl-1-butene, and not 2-methyl-3-butene). The simplest alkene, ethene, is also called by a common name, *ethylene.*

Alkynes are named in a similar manner, with the ending *yne* instead of *ene.*

Examples:

$CH_3—C{\equiv}CH$	propyne
$CH_3—CH(CH_3)—CH(CH_3)—C{\equiv}C—CH_3$	4,5-dimethyl-2-hexyne

The simplest alkyne, $HC{\equiv}CH$ (ethyne), is more often called *acetylene*.

Many hydrocarbons are known which contain two or more double bonds. The suffix **diene** is used for two double bonds, **triene** for three, etc. In these cases, two, three, or more numbers are needed.

Examples:

$CH_2{=}CH—CH{=}CH_2$	1,3-butadiene
$CH_2{=}CH—C(CH_3){=}CH_2$	2-methyl-1,3-butadiene (common name: *isoprene*)
$CH_3—CH{=}CH—CH{=}CH—CH{=}CH_2$	1,3,5-heptatriene
$CH_2{=}CH—CH_2—CH_2—CH_2—CH_2—CH{=}CH_2$	1,7-octadiene

We can think of any alkene containing just one double bond as an alkane with two hydrogens missing. Therefore, if the general formula for an alkane is C_nH_{2n+2}, the general formula for an alkene with one double bond must be C_nH_{2n}. Similarly, the general formula for any alkyne with one triple bond, or alkene with two double bonds is C_nH_{2n-2}. The student may wish to verify these formulas by counting the hydrogens in the examples we have given.

Alkenes and alkynes are called **unsaturated** hydrocarbons, as opposed to alkanes, which are **saturated** hydrocarbons.

We have seen in Section 14-16 that ethylene is a planar molecule, with the sp^2 orbitals of the two carbon atoms resulting in 120° angles.

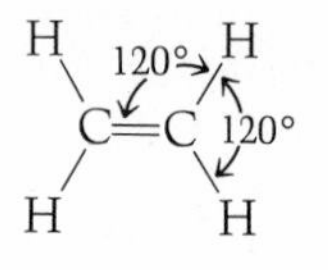

This is generally true for all alkenes: the two carbon atoms of each double bond, and the four atoms immediately attached to them, are all in a plane, and the six angles involved are all about 120°. Of course, the atoms not directly attached to the double-bond carbon atoms are generally not in this plane. For example, the three carbon atoms of propene $H_2C{=}CH—CH_3$ and three of the hydrogen atoms (those in color) are all in one plane, but the other three hydrogen atoms are not in this plane. Alkynes have the linear geometry we saw in Section 14-16 for acetylene. The two carbon atoms of the triple bond and the two atoms immediately connected to them are in a straight line: for example,

$$H_3C—C{\equiv}C—H \quad (180°)$$

There is no free rotation about double bonds, as there is about single bonds because a double bond consists of both a σ and a π orbital (Section 14-16). If rotation were to take place (Figure 20-5), the two p orbitals which originally formed the π orbital could no longer overlap, and the π bond would be destroyed. This means that the six atoms involved in the double bond are in a plane and do not change their relative positions. In some cases this results in cis and trans isomers (see Section

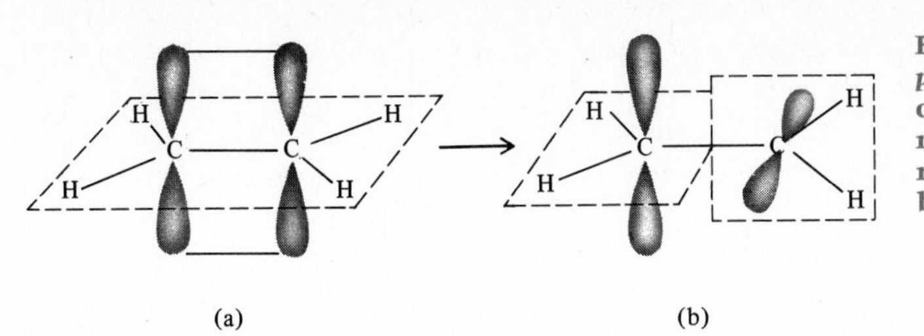

Figure 20-5 **(a) In ethylene, two *p* orbitals overlap to create a π orbital. (b) If one end of the molecule were to rotate with respect to the other, the *p* orbitals could no longer overlap.**

19-10): for example,

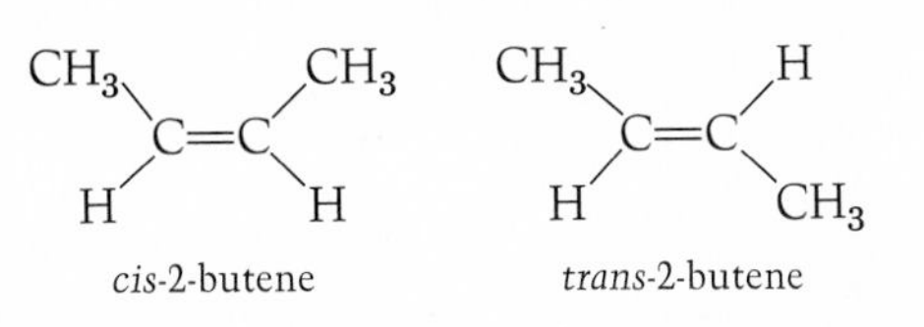

cis-2-butene *trans*-2-butene

Because there is no free rotation about the double bond, these isomers do not change into one another. Each has different melting points, boiling points, and all other properties. Not all double-bond compounds have cis and trans isomers. For example, there is only one 1-butene:

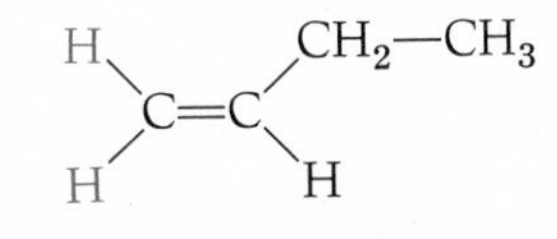

In this case, if we interchange the two hydrogens in color, we get the same compound back again. As is obvious from their geometry, alkynes do not have cis-trans isomerism either.

Cis and trans isomers have the same molecular formulas (as do all isomers), but they have the same *structural* formulas as well. Isomers of this kind are called **stereoisomers.**

The physical properties of alkenes and alkynes—boiling and melting points, density, and solubility—are essentially the same as those of the alkanes. Furthermore, these compounds also burn; in fact, with a higher temperature. The temperature of burning acetylene is high enough for it to be used in welding torches. However, the other chemical properties are quite different. Alkanes are essentially inert, but alkenes and alkynes undergo many reactions at their double and triple bonds. One reaction worth mentioning is the reaction with hydrogen and a catalyst: for example,

$$CH_3—CH=CH_2 + H_2 \xrightarrow{Pt} CH_3—CH_2—CH_3$$

By this reaction, called **catalytic hydrogenation,** the double or triple bond is converted to a single bond, and the alkene or alkyne to an alkane. The unsaturated hydrocarbon has become saturated; H_2 has been *added* to the double bond. This reaction is fairly easy to carry out in the laboratory or in industry. Many catalysts are known, including transition metals such as platinum, palladium, and nickel. Most of the other reactions of alkenes and alkynes are also addition reactions.

20-4 Cycloalkanes and Cycloalkenes

Many hydrocarbons are known which contain only single bonds (and are therefore alkanes), but the carbons are connected in a ring: for example,

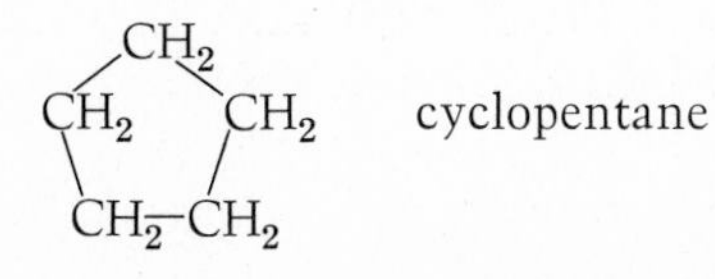

cyclopentane

Such compounds are called **cycloalkanes,** and are named by the IUPAC system in a manner similar to the noncyclic alkanes we discussed in Section 20-1. The parent part of the name indicates the number of carbon atoms in the ring, with the prefix **cyclo,** and numbers are used if necessary, to show the position of substituents.

Examples:

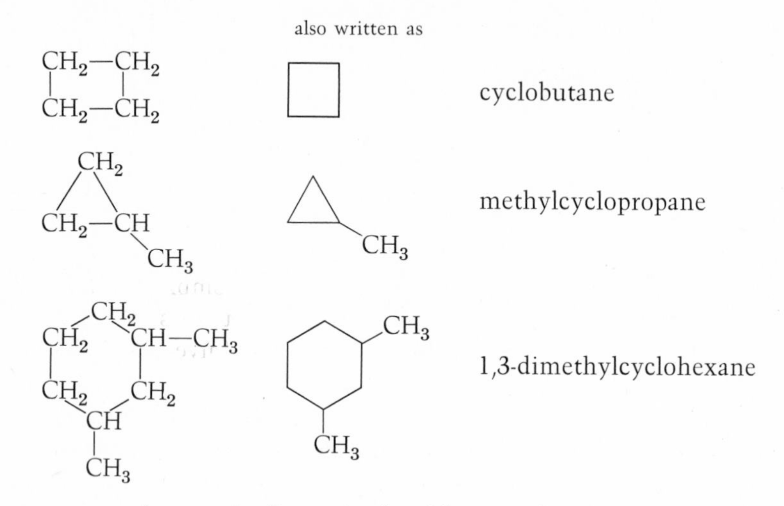

It is customary to draw cycloalkanes in the abbreviated manner shown on the right. In these compounds we have no chain with two ends, so a new rule is needed for numbering. The rule is: begin at any ring carbon, and proceed either clockwise or counterclockwise, but the numbers that result must be the lowest possible.

Examples:

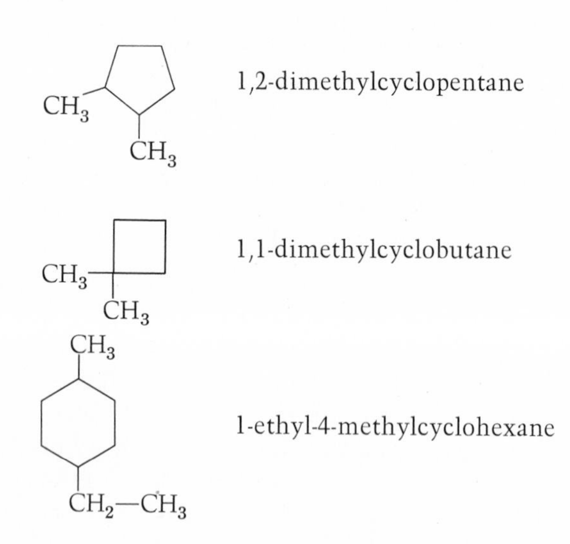

It does not matter how a given compound is drawn. Thus, all of these are equally correct ways of drawing 1,2-dimethylcyclohexane:

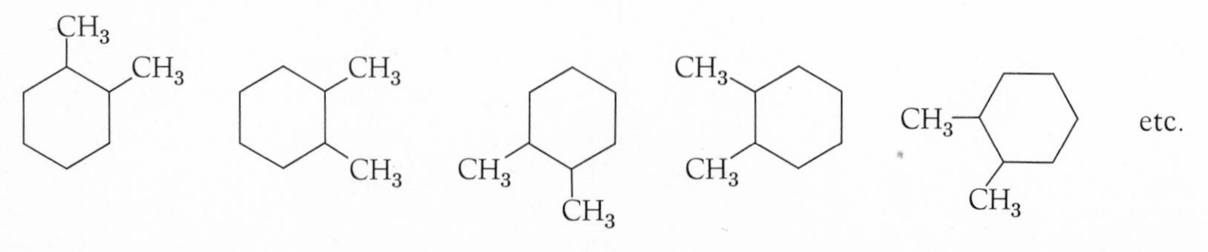

The general formula for a cycloalkane is C_nH_{2n} (why?), so, for example, all cycloalkanes containing a total of six carbons (not just ring carbons, but *all* carbons) have the molecular formula C_6H_{12}, and are therefore isomers of all alkenes that contain six carbons and just one double bond (remember that the general formula for monoalkenes is also C_nH_{2n}).

Cycloalkane rings are known in all sizes from cyclopropanes (three-membered) up to 50-membered and more. Cyclopropanes and cyclobutanes are **strained** rings, because their angles are smaller than the normal tetrahedral angle. For cyclopropanes, the angles must be ~60°, which means that the bonds are compressed from the normal 109.5° angles.

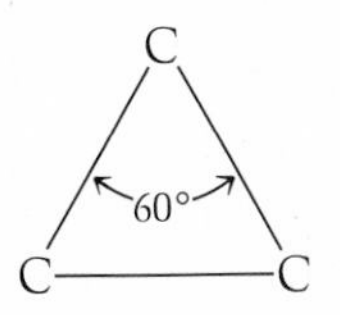

Cyclobutanes, with ~90° angles, are less compressed, but they, too, are strained. For this reason, cyclopropanes and cyclobutanes are more reactive than other alkanes. However, cycloalkanes with rings of five members and more do not show any of this type of strain.

Rings of five members and larger are not planar. The most important rings in chemistry are six-membered, and we will look at the structure of cyclohexane. The shape of this molecule is shown in Figure 20-6. This shape is known as the **chair form** (perhaps you can see the shape of a chair in this molecule). The diamond structure we saw in Figure 15-1a is made up of a large number of chair forms.

Double bonds can also exist in rings; these compounds are called **cycloalkenes.** The naming follows the rules we have already given.

Examples:

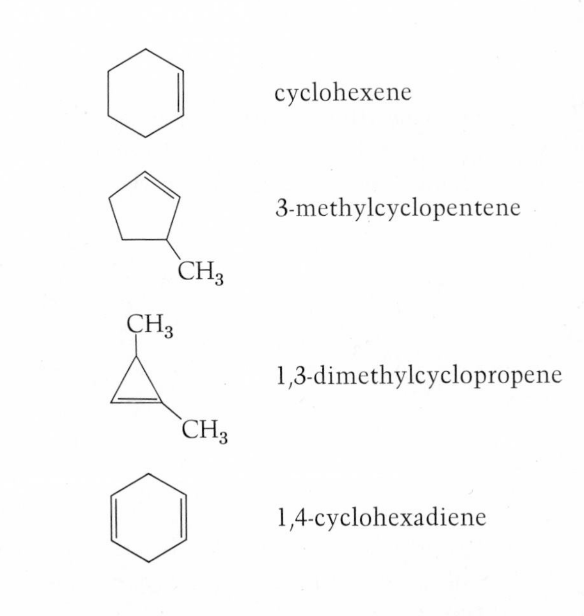

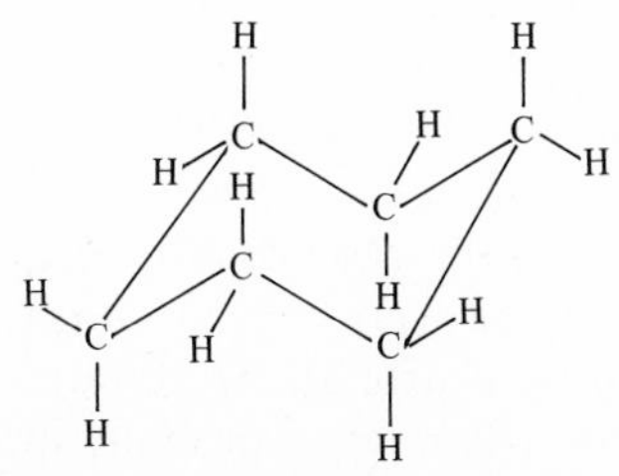

Figure 20-6 **The structure of cyclohexane (called the chair form).**

In most cases, it is unnecessary to write a number for the position of the double bond. It is understood that the double bond goes from 1 to 2.

Because of the geometries involved, triple bonds and 1,2-dienes (two double bonds on the same carbon) cannot exist in rings smaller than eight- or nine-membered.

Except for three- and four-membered rings, which are reactive because of their strain, the physical and chemical properties of cycloalkanes and cycloalkenes are essentially the same as those of the noncyclic alkanes and alkenes, respectively.

20-5 Aromatic Compounds

All of the hydrocarbons we have discussed so far are called **aliphatic** hydrocarbons. As we mentioned in Section 17-7, there is another class of hydrocarbons, called **aromatic.** The most important member of this class is *benzene,* which has the molecular formula C_6H_6, and almost all other aromatic compounds are closely related to benzene. The name "aromatic" was given to these compounds in the nineteenth century when only a few were known, and most of these had pleasant odors. Today we have a different definition for aromatic compounds (see below).

Laboratory experiments tell us that the six carbon atoms of benzene are connected in a ring, with each carbon atom connected to one hydrogen atom. If we draw a Lewis structure for benzene, we find that we need 12 pairs of electrons to account for these bonds

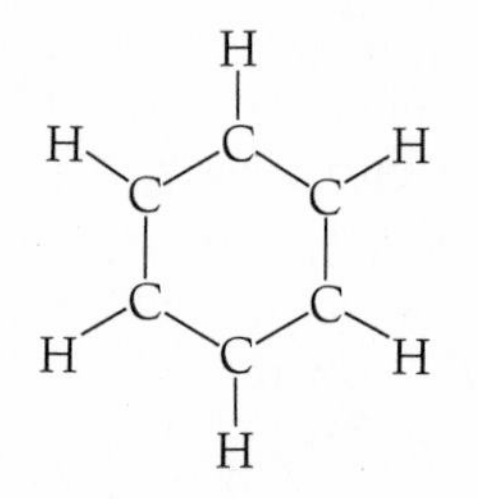

(out of the total of 30 electrons contributed by six carbons and six hydrogens), so that three pairs remain to be added. It is apparent that there are *two* ways to add these three pairs, both of which give every carbon an octet:

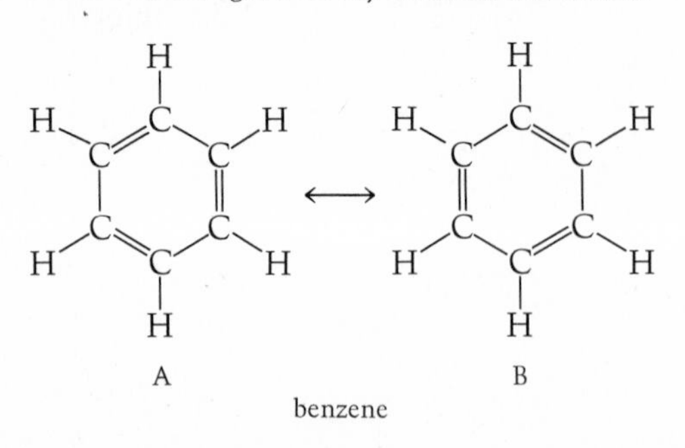

The structure of benzene can therefore not be satisfactorily represented by a single Lewis structure, and the real benzene is a resonance hybrid (Section 14-10) of the two canonical forms A and B.

We can also look at benzene in a different way, by examining its molecular orbitals. Each carbon is connected to three other atoms by sp^2 orbitals (Section 14-16). Since the three sp^2 orbitals of each atom are in a plane, this forces all 12 atoms of benzene to be in the same plane (Figure 20-7). As is the case for ethylene, each carbon atom has one *p* orbital left over and these overlap (Figure 20-7) to form three bonding π orbitals (there are also three antibonding π orbitals, which are unoccupied, and not shown in Figure 20-7). Each of the three bonding π orbitals, all

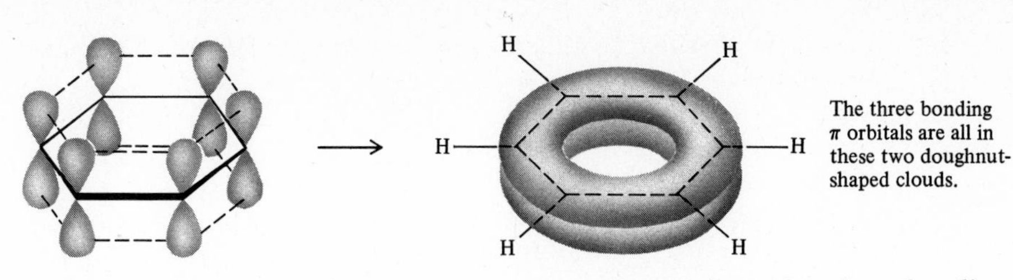

Figure 20-7 The six p orbitals of benzene overlap to form the three bonding π orbitals. The three antibonding π orbitals are not shown.

of which occupy approximately the same space at the same time (orbitals can do that because they are clouds), has the shape of two round donuts, one above the plane of the 12 atoms, and one below.

Thus, both the resonance and molecular orbital pictures show that benzene possesses a closed ring of six electrons, and this is what makes it different from aliphatic compounds. In fact, we define an **aromatic compound** as **any compound that possesses a closed ring of six electrons.*** Because neither of the two canonical forms shows the real molecule, benzene is often shown in this way:

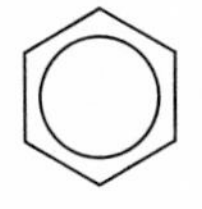

benzene

As mentioned in Section 14-10, resonance is a stabilizing phenomenon, and aromatic hydrocarbons are particularly stable. Although a single canonical form of benzene has three double bonds, benzene does not behave like an unsaturated molecule. Unlike alkenes, it does not easily undergo addition reactions. The typical reactions of aromatic compounds are substitution reactions, in which one of the hydrogens is replaced by another group: for example,

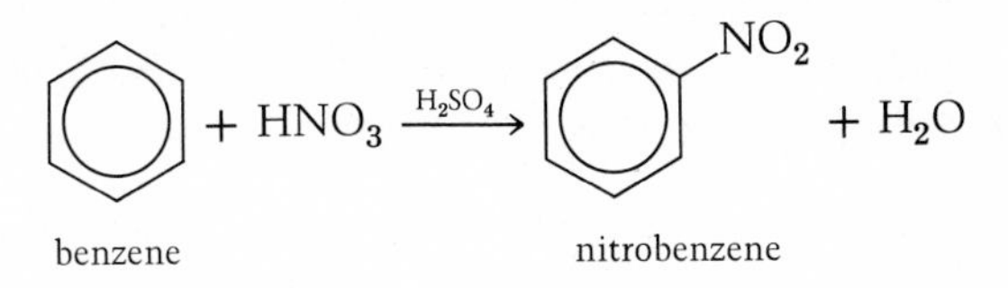

benzene nitrobenzene

Many aromatic compounds are named as derivatives of benzene, though some have common names.

Examples:

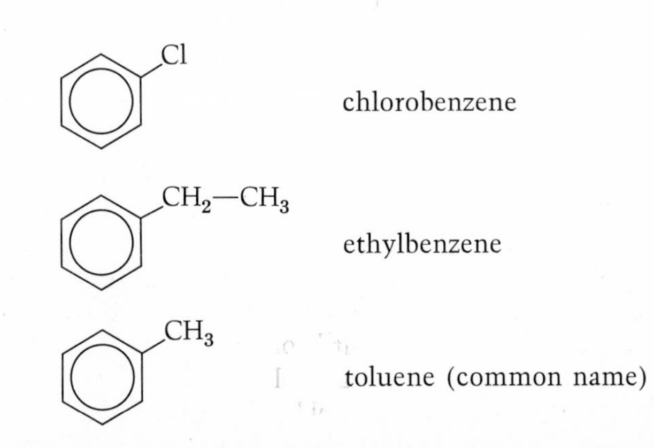

*A few aromatic compounds have been synthesized with closed rings of 10, 14, 18, or other numbers of electrons, but for the vast majority, this number is six.

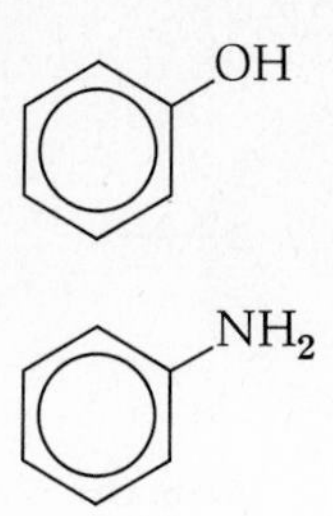

phenol (common name)

aniline (common name)

When a benzene ring has *two* hydrogens replaced by other groups, we may use numbers or, more commonly, the prefixes *ortho, meta,* and *para* for 1,2; 1,3; and 1,4, respectively.

Examples:

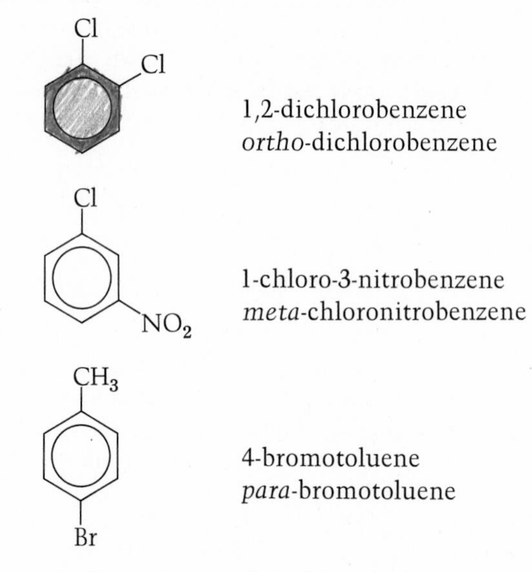

1,2-dichlorobenzene
ortho-dichlorobenzene

1-chloro-3-nitrobenzene
meta-chloronitrobenzene

4-bromotoluene
para-bromotoluene

When three or more hydrogens are replaced by other groups, we may not use *ortho, meta,* or *para*. We use numbers only.

Example:

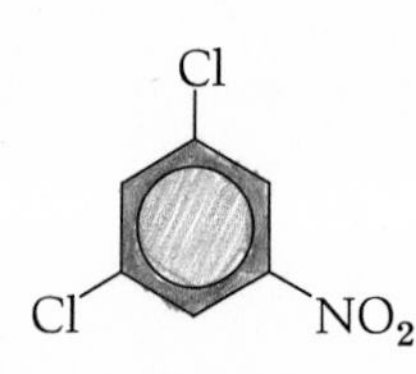

3,5-dichloronitrobenzene

When it is necessary to name the benzene ring as a group, it is called the *phenyl* group (*not* the benzyl group as you might suppose). The phenyl group is always C_6H_5.

Examples:

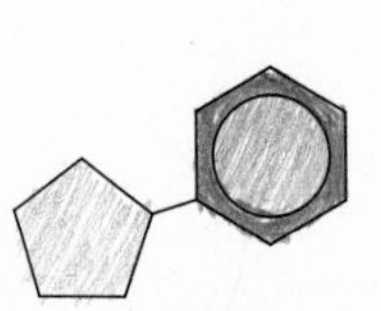

phenylcyclopentane

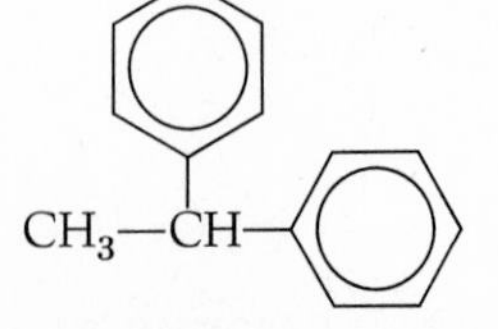

1,1-diphenylethane

Not all aromatic rings contain six carbon atoms. In some of them one or more of the carbons is replaced by another element (most often nitrogen). Such compounds are known as **heterocycles** or **heterocyclic rings.** Two examples are pyridine and pyrimidine, which contain one and two nitrogen atoms, respectively.

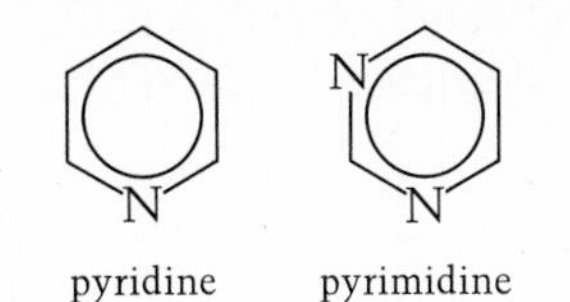

Because the valence of nitrogen is three, there is no hydrogen connected to the nitrogen in these compounds (that is, pyridine is C_5H_5N).

As we have already seen in Section 17-7, many aromatic compounds contain two or more fused benzene rings **(polynuclear aromatics)**: for example,

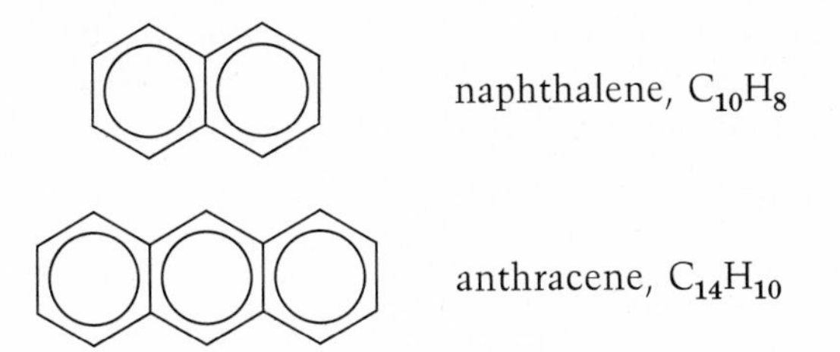

Aromatic hydrocarbons (for example, benzene, toluene, naphthalene) are called **arenes.**

20-6 Functional Groups

In Sections 20-1 to 20-5, we have discussed the hydrocarbons, and already it can be seen why there is such a vast number of possible organic compounds. In Section 20-1, we learned that there are hundreds of thousands of possible alkanes, even if we restrict ourselves to those containing no more than 20 carbons. In Section 20-4, we saw that cyclic alkanes were also possible, and this greatly increases the total number (for example, there are five alkanes C_6H_{14}, but 12 structural isomers of cycloalkanes with the formula C_6H_{12}). Of course, nearly all of these can have one or more double bonds, in many different positions, so the number of alkenes (including cis and trans isomers) is even greater. Add in the alkynes and aromatic hydrocarbons, and the total number of possible hydrocarbons becomes enormous. And yet this is only a tiny fraction of the total number of possible organic compounds, because every hydrocarbon can be substituted at virtually every position by a variety of functional groups. A **functional group** is that part of an organic molecule that is not alkane-like or arene-like. For example, two typical functional groups are hydroxy groups —OH and halogen atoms.

Examples:

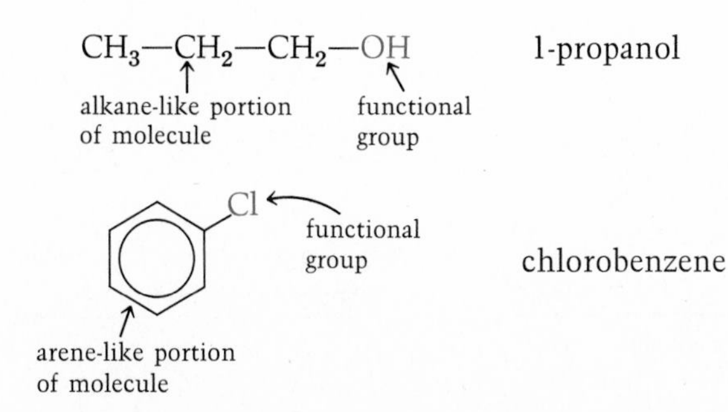

We can appreciate how the possibility of functional groups increases the total number of possible organic compounds by a simple example. There is only one alkane with one carbon atom (methane). But a simple counting exercise will show that there are 69 halogen derivatives of methane (for example, CH_3Br, CCl_4, CF_2Cl_2, CHI_3, etc.). The number of halogen derivatives increases still more with larger alkanes, where we also have the possibility of substitution at different positions: for example,

$CH_3—CH_2—CH_2—Cl$ 1-chloropropane

$CH_3—CH(Cl)—CH_3$ 2-chloropropane

Yet halogen is only one of several dozen known functional groups.

In the rest of this section we shall describe 10 of the most common functional groups (we have already dealt with two others: double and triple bonds). In writing organic formulas, it is customary to use the symbols R and Ar. **R** stands for any alkyl group—it may be methyl, ethyl, isopropyl, or any other alkyl group, even if it does not have a simple name: for example,

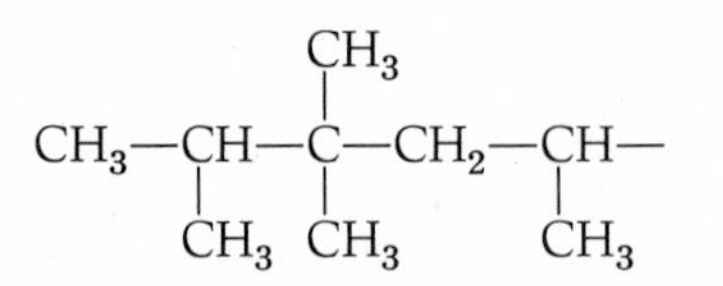

Similarly, **Ar** stands for any **aryl** group (that is, a group similarly derived from an arene). For example, the symbol ROH means not only 1-propanol (shown above), but *any* alkane that has been substituted in any position by an OH group; and ArCl means not only chlorobenzene, but any arene that has been substituted at a ring position by a chloro group. The groups to be discussed are shown in Table 20-1.

HALIDES. Aliphatic and aromatic compounds that contain any of the four halogens are called **halides:** alkyl halides or aryl halides, as the case may be. They are usually named by the IUPAC system, with chloro, bromo, etc. as the prefix (see the two chloropropanes shown above). The physical properties of halides are usually similar to those of the hydrocarbons, except that alkyl halides generally have high densities, often denser than water (so they sink to the bottom). Chemically, aryl halides are fairly inert (they undergo few reactions), but alkyl halides readily undergo substitution and elimination reactions (Section 20-8). Some important halides are chloroform, $CHCl_3$ (IUPAC name trichloromethane), which was once used as an anaesthetic; carbon tetrachloride, CCl_4, which was once extensively used as a dry cleaning agent, perchloroethylene, $Cl_2C{=}CCl_2$, which is the most important dry cleaning agent today; and Freon 12, CCl_2F_2, discussed in Section 17-5.

ALCOHOLS. Alcohols are called primary, secondary, or tertiary, depending on whether the OH group is connected to a primary, secondary, or tertiary carbon. Examples are

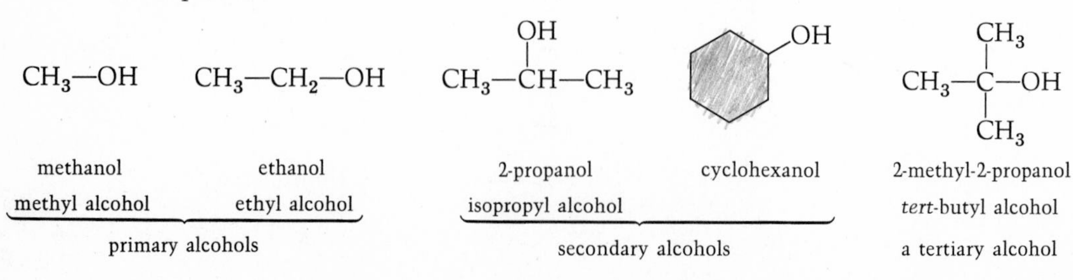

Table 20-1 The Most Important Functional Groups[a]

Functional group	Name of class of compounds possessing this group	Functional group	Name of class of compounds possessing this group
—C=C— (with single bonds down from each C)	Alkenes	R—C(=O)—OR′	Carboxylic esters
—C≡C—	Alkynes	R—C(=O)—NH_2	Amides
R—Cl (or R—Br, etc.)	Alkyl halides	R—C(=O)—NH—R′	Amides
Ar—Cl (or Ar—Br, etc.)	Aryl halides	R—C(=O)—NR'_2	Amides
R—OH	Alcohols	RNH_2	Amines
Ar—OH	Phenols	R_2NH	Amines
R—O—R	Ethers	R_3N	Amines
R—C(=O)—H	Aldehydes	RSH	Mercaptans
R—C(=O)—R′	Ketones		
R—C(=O)—OH	Carboxylic acids		

[a]In all cases, except ROH and RCl, R may also be Ar without changing the name of the class of compounds. R′ indicates that this group (alkyl or aryl) may be different from R.

As these examples show, alcohols may be named in two ways. In the IUPAC system, the OH group is indicated by the suffix **ol,** but otherwise the system is the same as we have previously seen. In the other system, the word *alcohol* is preceded by the name of the alkyl group. The physical properties of alcohols are quite different from those of the hydrocarbons and halides, in two ways: (1) their boiling points are much higher than hydrocarbons or halides of similar molecular weight (for example, boiling point of ethane −89°C; of methanol +65°C); and (2) the low-molecular-weight alcohols (up to three carbons) are completely soluble in water. Both of these differences are caused by hydrogen bonding. Alcohol molecules form hydrogen bonds with each other:

$$\mathrm{R{-}\underset{\displaystyle H}{\underset{|}{O}}{\cdots}H{-}\underset{\displaystyle R}{\underset{|}{O}}}$$

Since energy is required to break these bonds when the liquid is boiled, the boiling points are higher. Alcohols also form hydrogen bonds with water:

$$\mathrm{H{-}\underset{\displaystyle H}{\underset{|}{O}}{\cdots}H{-}\underset{\displaystyle R}{\underset{|}{O}}} \quad \text{or} \quad \mathrm{R{-}\underset{\displaystyle H}{\underset{|}{O}}{\cdots}H{-}\underset{\displaystyle H}{\underset{|}{O}}}$$

accounting for their solubility in water.

Alcohols undergo a number of reactions, one of which is a reaction with alkali metals, such as sodium, to give bubbles of hydrogen and **alkoxide ions.**

$$2ROH + 2Na \longrightarrow \underset{\text{alkoxide ions}}{2RO^-} + 2Na^+ + H_2\uparrow$$

This is the same reaction that water undergoes (Section 19-4), but alcohols react less vigorously, since they are weaker acids than water. Because they are weaker, their conjugate bases, alkoxide ions, are stronger bases than OH^- ions. The reaction with sodium to give hydrogen bubbles is used as a test for the presence of alcohols, but all traces of water must be absent. (Why?) Primary alcohols can be oxidized to carboxylic acids by a number of oxidizing agents, including $KMnO_4$ and $K_2Cr_2O_7$: for example,

$$\underset{\text{ethanol}}{3CH_3{-}CH_2{-}OH} + 2Cr_2O_7^{2-} + 16H^+ \longrightarrow \underset{\text{acetic acid}}{3CH_3{-}\overset{\displaystyle O}{\overset{\|}{C}}{-}OH} + 4Cr^{3+} + 11H_2O$$

The most important single alcohol is ethanol, CH_3CH_2OH, which is used industrially in large quantities, and is also the only alcohol present in alcoholic beverages. All alcoholic beverages, from beer and wine to whiskey, gin, and rum, with few exceptions, consist of 99% or more water and ethanol. The difference between one alcoholic beverage and another is: (1) the relative percentages of ethanol and water; and (2) the composition of the other 1%. Beer contains about 3 to 4% ethanol, table wine about 12%, and **spirits** (whiskey, rum, vodka, etc.) anywhere from 35 to 60%. In the United States, all spirits are labeled with a **proof number,** which is twice the percentage of ethanol by volume. Thus, an 86 proof whiskey contains 43% ethanol by volume. The use of alcoholic beverages dates back to prehistoric times when mankind first learned to ferment grain or fruits. The starch or sugar (see Sections 21-1 and 21-2) in the grain or fruit is converted to alcohol and CO_2 by enzymes present in yeast cells, which are used in the fermentation.

$$\underset{\text{glucose}}{C_6H_{12}O_6} \xrightarrow{\text{enzymes}} 2CH_3CH_2OH + 2CO_2$$

Although large quantities of ethanol are manufactured today by chemical processes, ethanol for drinking purposes is still made in the same old way: by fermentation of grain or fruits (in fact, U.S. law requires this). Because yeast cells die when the alcoholic content reaches about 13% by volume, the strongest alcoholic beverage known to ancient man was table wine. It was not until about A.D. 1200, when distillation was first applied to alcoholic beverages, that spirits were invented.

Other important alcohols are methanol (sometimes called wood alcohol), which has many uses, among them as a solvent for duplicating machines, and isopropyl alcohol, which is sometimes used as a rubbing alcohol. Many alcohols are known which have two or more OH groups (though not on the same carbon atom—such compounds are unstable). The most important of these are ethylene glycol and glycerol,

$$\underset{\text{ethylene glycol}}{\underset{OH\quad\;\; OH}{CH_2{-}CH_2}} \qquad \underset{\text{glycerol}}{\underset{OH\quad\;\; OH\quad\;\; OH}{CH_2{-}CH{-}CH_2}}$$

both of which are viscous liquids because of the extensive hydrogen bonding possible when two or more OH groups are present. Ethylene glycol is by far the most important antifreeze (Section 7-9), while glycerol (also called glycerine) is a highly important industrial compound which is used to make nitroglycerine, as a humectant (moistening agent), and for other purposes. Glycerol plays a major role in the chemistry of fats (Section 21-7).

PHENOLS. When an OH group is connected directly to an aromatic ring, the compound is not an alcohol but a **phenol.** The most common phenol is also called phenol.

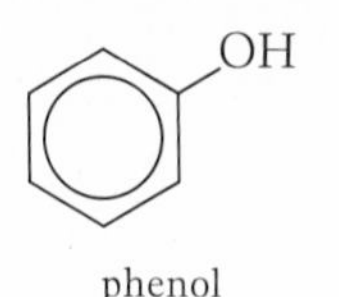

phenol

In this case, the same name is used for a single compound and for a class of compounds, a practice not uncommon in organic chemistry. Phenols are much stronger acids than alcohols; in fact, they are stronger acids than water (they turn litmus red). However, they are weaker acids than carboxylic acids (see below). A solution of phenol in water is known as *carbolic acid.* This solution was one of the first antiseptics to be used in medicine, but is no longer used for this purpose, because it burns the skin, and milder antiseptics have since been discovered.

ETHERS. These compounds are most often named by naming the alkyl or aryl groups and adding the word ether.

Examples:

$CH_3—O—CH_3$	$CH_3—O—CH_2—CH_3$	$C_6H_5—O—CH_2—CH_3$
dimethyl ether	ethyl methyl ether	ethyl phenyl ether

The physical properties of ethers are similar to those of alcohols except that their boiling points are lower than those of alcohols of the same molecular weight. (Why?) Note that ethers are isomers of alcohols (for example, compare ethanol and dimethyl ether). Chemically, ethers are quite unreactive, and undergo only a few reactions. By far the most important ether is diethyl ether, $CH_3CH_2—O—CH_2CH_3$, which is used as an anaesthetic and is also an important solvent, in the laboratory and industrially. Because of its low boiling point (35°C), diethyl ether is very flammable, and care must be taken to exclude flames whenever working with this compound.

ALDEHYDES AND KETONES. Aldehydes and ketones are often called **carbonyl compounds,** because they both possess the carbonyl group $—\underset{\underset{\displaystyle O}{\|}}{C}—$. In **ketones,** the carbonyl group is attached to two alkyl and/or aryl groups. In **aldehydes** at least one of these is hydrogen.

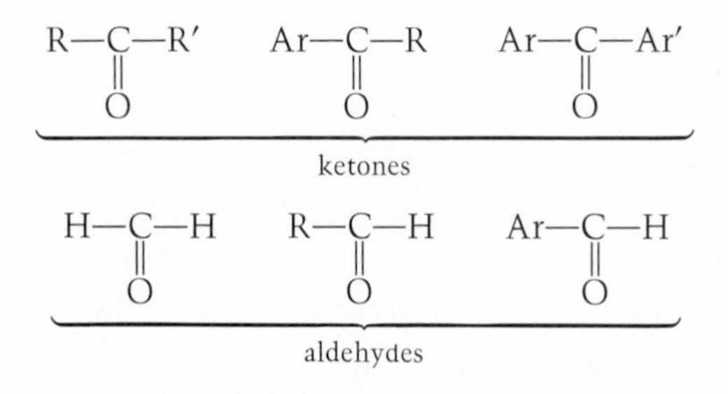

Ketones can be named by the IUPAC system: the suffix **one** is used to show the presence of this group, but otherwise the system is the same as before. Alternatively, ketones can be named by naming the alkyl or aryl groups and adding the word ketone (in a similar manner as with ethers). Ketones containing aryl groups are most easily named in the latter way.

Examples:

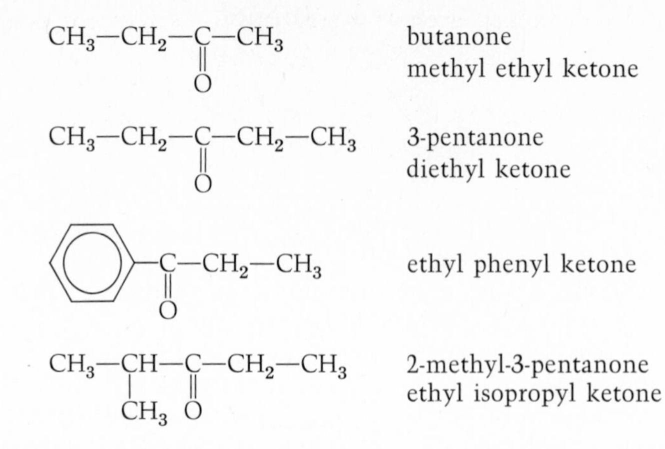

The most common ketone $CH_3{-}\underset{\Vert}{\overset{}{C}}{-}CH_3$ (with =O on the central C) is usually called acetone rather than propanone or dimethyl ketone.

Aldehydes may be named by the IUPAC system, with the suffix **al,** but are more often called by common names derived from the corresponding carboxylic acid (see below for the names of carboxylic acids).

Examples:

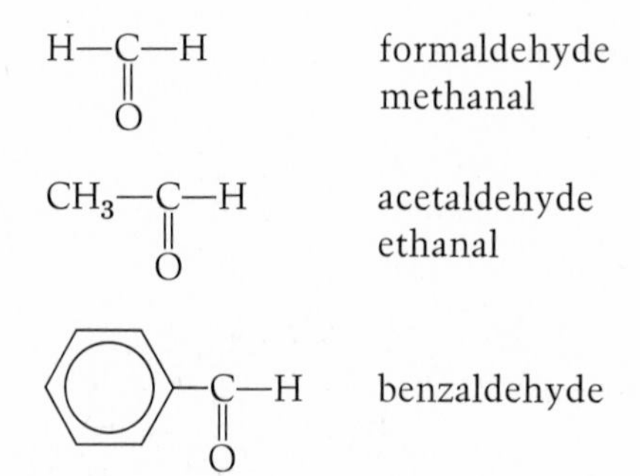

The physical properties of aldehydes and ketones are similar to those of the ethers. They are quite reactive chemically and undergo many reactions. Among the most important of these are oxidation–reduction reactions. In oxidation number, aldehydes stand midway between primary alcohols and carboxylic acids, and are easily oxidized to the acids or reduced to alcohols.

$$R{-}C({=}O){-}H \xrightarrow{Cr_2O_7^{2-} + H^+} R{-}C({=}O){-}OH$$

$$R{-}C({=}O){-}H \xrightarrow{H_2 + \text{catalyst}} R{-}CH_2OH$$

A common reducing agent for aldehydes is hydrogen and a catalyst (Section 20-3), although many other reducing agents could also be used. Ketones cannot normally be oxidized to give useful products, but can be reduced to secondary alcohols

$$R{-}C({=}O){-}R' + H_2 \xrightarrow{\text{catalyst}} R{-}CH(OH){-}R'$$

Formaldehyde, $H{-}C({=}O){-}H$, is a gas (bp $-21\,°C$), but is often used in water solution as a preservative in biological laboratories. A 40% solution of formaldehyde in water is called *formalin*. Acetone is an important solvent. Its sweet odor is familiar in glues, varnishes, paint removers, etc.

Carboxylic Acids. These compounds possess the **carboxyl group** $-\underset{\underset{\text{O}}{\|}}{\text{C}}-\text{OH}$. They can be named by the IUPAC system (the suffix is **oic acid**), but most often they are given common names.

Examples:

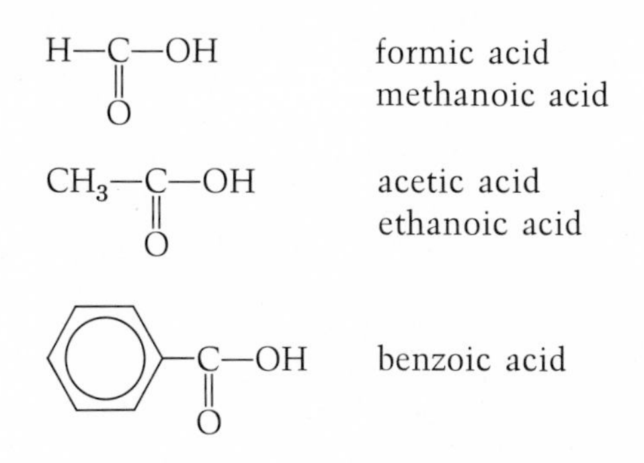

Note how the common names of the acids are related to those of the aldehydes.

The physical properties of the carboxylic acids are similar to those of the alcohols, except that the boiling points are higher due to double hydrogen bonding (for example, bp acetic acid 118°C; ethanol 78°C).

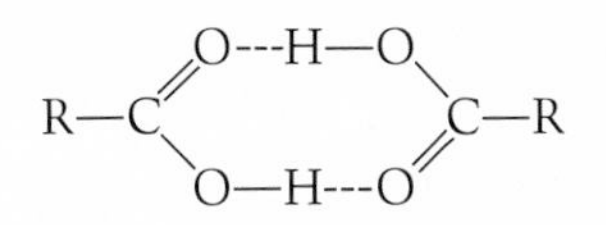

The outstanding chemical property of the carboxylic acids is implied by the name: they are acids. As we saw in Table 10-1, acetic, benzoic, and formic acids are intermediate acids, and not strong acids like HCl or H_2SO_4. Nevertheless, they are the strongest acids of all the compounds we are studying in this chapter (of the others, only phenols and mercaptans are stronger acids than water). The acidity of the various carboxylic acids RCOOH is fairly similar. With a few exceptions, all have pK_a values between about 4 and 5.

Because carboxylic acids are intermediate acids, they cannot exist as such in the presence of strong bases. Under these conditions they are converted to the corresponding ions (called carboxylate ions).

$$RCOOH + OH^- \longrightarrow RCOO^- + H_2O$$

This means that in water solution, carboxylic acids can only be present at low pH values. At pH values of about 9 or above, only the ion can exist, not the acid. Conversely, carboxylate ions cannot exist at low pH values; they are converted to the carboxylic acids.

$$RCOO^- + H_3O^+ \longrightarrow RCOOH + H_2O$$

Another important property of carboxylic acids is that they can be converted into compounds called **acid derivatives.** We will study two of these, carboxylic esters and amides, in this section. Carboxylic acids in which the R in RCOOH is an unbranched alkyl group are called **fatty acids,** because they can be obtained from fats (Section 21-7). A characteristic property of these acids is their foul odors. For example, the odor of stale perspiration is caused by the presence of butyric acid,

$CH_3CH_2CH_2COOH$. The odors of the higher unbranched carboxylic acids ($\sim C_7$ to C_{10}) are even more objectionable.

The most common carboxylic acid is acetic acid, which performs important functions in living organisms (for example, cholesterol is synthesized by the body from acetic acid) and has many industrial uses. It was discovered many centuries ago that if wine is allowed to stand in air, it turns to vinegar. Today we know that this is a result of the slow oxidation by air (autoxidation) of the ethanol in the wine to acetic acid. Vinegar is a solution of about 5% acetic acid in water.

A number of carboxylic acids possessing OH groups are found in various foods (nature put them there—they are not additives). A few of them are

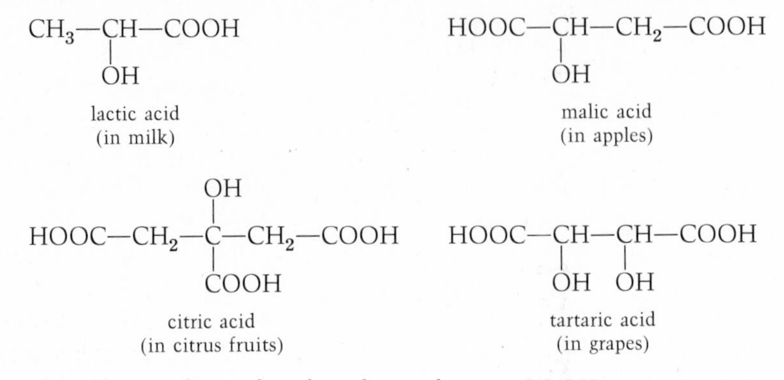

lactic acid (in milk) — malic acid (in apples) — citric acid (in citrus fruits) — tartaric acid (in grapes)

Malic and tartaric acids are dicarboxylic acids (two COOH groups); citric acid is a tricarboxylic acid.

Carboxylic Esters. A **carboxylic ester** can be thought of as being made up of a carboxylic acid and an alcohol (or phenol), with water being split out.

$$R-\underset{\underset{O}{\|}}{C}-\boxed{OH \quad H}O-R' \longrightarrow R-\underset{\underset{O}{\|}}{C}-O-R' + H_2O$$

Carboxylic esters can, in fact, be made in this way, if a strong acid such as HCl is present as a catalyst, but the reaction is an equilibrium,

$$RCOOH + R'OH \overset{H^+}{\rightleftharpoons} RCOOR' + H_2O$$

and some way must be found to drive it to completion. (One way to do this is to add a drying agent, which will absorb the water, thereby driving the equilibrium to the right).

To name carboxylic esters we first name the alkyl or aryl group that comes from the alcohol or phenol, then the carboxylic acid part. The *ic acid* ending is changed to **ate.**

Examples:

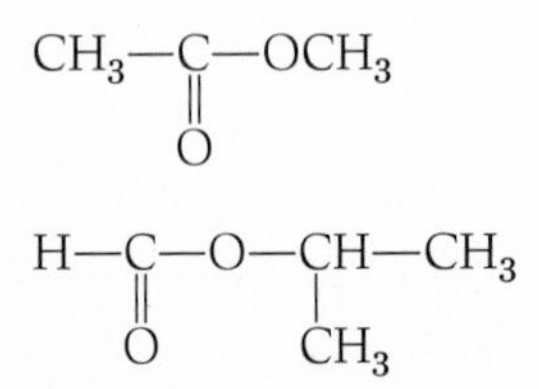

methyl acetate (from acetic acid)

isopropyl formate (from formic acid)

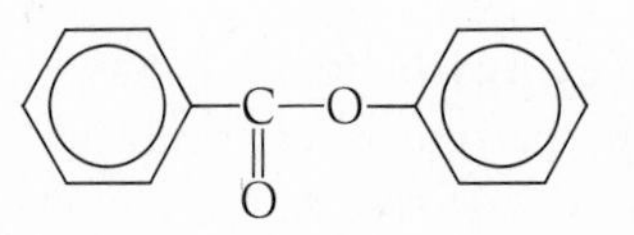

phenyl benzoate (from benzoic acid)

$$CH_3{-}CH_2O{-}\underset{\underset{O}{\|}}{C}{-}\underset{\underset{OH}{|}}{CH}{-}\underset{\underset{OH}{|}}{CH}{-}\underset{\underset{O}{\|}}{C}{-}OCH_2{-}CH_3$$ diethyl tartrate (from tartaric acid)

Because carboxylic ester molecules cannot hydrogen-bond to one another (why not?), the boiling points of carboxylic esters are usually lower than the carboxylic acids they are derived from (for example, bp ethyl acetate 77°C; acetic acid 118°C). However, their solubility in water is similar to that of the alcohols, ethers, aldehydes, ketones, and carboxylic acids. (Why?)

An important chemical property of carboxylic esters is that they can be hydrolyzed (**hydrolysis** = reaction with water) to give the carboxylic acids and alcohols from which they were derived. This can be done with either acidic or basic catalysts, but basic catalysts are more common, because acidic catalysts give the equilibrium shown above. The base-catalyzed hydrolysis of carboxylic esters is called **saponification.**

$$R{-}\underset{\underset{O}{\|}}{C}{-}OR' + H_2O \xrightarrow{OH^-} R{-}\underset{\underset{O}{\|}}{C}{-}O^- + R'OH$$

Because carboxylic acids cannot exist in the presence of a strong base, the product of saponification is not the acid, but the carboxylate ion. Saponification is used in making soap (Section 21-7).

In sharp contrast to the carboxylic acids, the carboxylic esters have odors that are sweet and pleasant. In fact, the sweet odor of many fruits, flowers, and other vegetation is due to the carboxylic esters therein. Examples are isopentyl acetate, which is found in bananas, methyl salicylate (oil of wintergreen), and menthyl acetate (in peppermint oil).

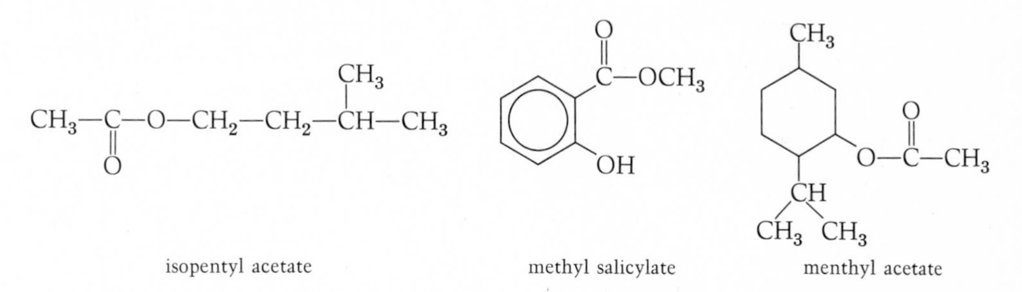

isopentyl acetate methyl salicylate menthyl acetate

Methyl salicylate is the methyl ester of salicylic acid, which is also a phenol.

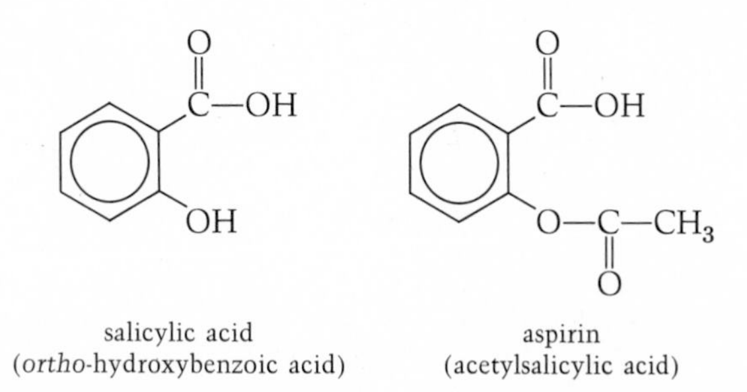

salicylic acid (*ortho*-hydroxybenzoic acid) aspirin (acetylsalicylic acid)

The common pain reliever aspirin is another ester of salicylic acid, but in this case it is the OH group of the salicylic acid which is esterified (with acetic acid), not the COOH group. Methyl salicylate is a naturally occurring compound, but aspirin is a man-made product and is not found in nature.

Because of their pleasant odors, carboxlic esters are used in perfumes, soaps, and cosmetics, as well as for food flavorings. Ethyl acetate, $CH_3{-}\underset{\underset{O}{\|}}{C}{-}OCH_2CH_3$ commercially the most important single ester, is used as a solvent for glues, paints, laquers, nail polish remover, etc.

AMINES. Alcohols, phenols, and ethers may be considered to be derivatives of water, with the hydrogens replaced by R or Ar (Table 20-2). Similarly, **amines** are derivatives of ammonia. As shown in Table 20-2, there are three kinds of amine—primary, secondary, and tertiary—depending on how many alkyl or aryl groups are directly connected to the nitrogen atom (note that this usage is not parallel to the primary-secondary-tertiary system for alcohols). Amines are named by naming the groups and adding the suffix **amine.**

Examples:

$CH_3—NH_2$	methylamine
$CH_3—CH_2—NH—CH_3$	ethylmethylamine
$CH_3—CH_2—NH—CH_2—CH_3$	diethylamine
$(CH_3)_3N$	trimethylamine
$CH_3—N(CH_3)—CH_2—CH_3$	ethyldimethylamine
$C_6H_5—NH_2$	aniline (common name)

The physical properties of amines are similar to those of the alcohols, except that boiling points are lower because hydrogen bonding is weaker (for example, CH_3NH_2 is a gas while CH_3OH is a liquid). Amines have odors that are weakly ammoniacal and at the same time fishlike. The most important chemical property of amines is that, like ammonia, they are basic—they are the only basic compounds among the compounds studied in this section; in fact, with a few exceptions, they are the only important organic bases at all. When they react with acids, they are converted to ions called substituted ammonium ions; for example,

$$\underset{\text{methylamine}}{CH_3—NH_2} + HCl \longrightarrow \underset{\text{methylammonium ion}}{CH_3—NH_3^+} + Cl^-$$

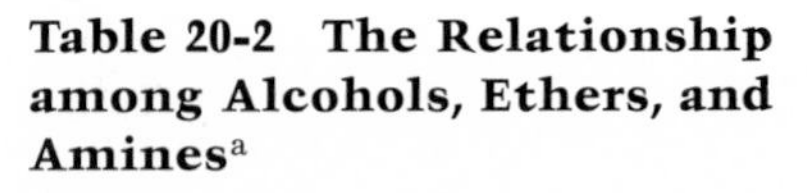
Table 20-2 The Relationship among Alcohols, Ethers, and Amines[a]

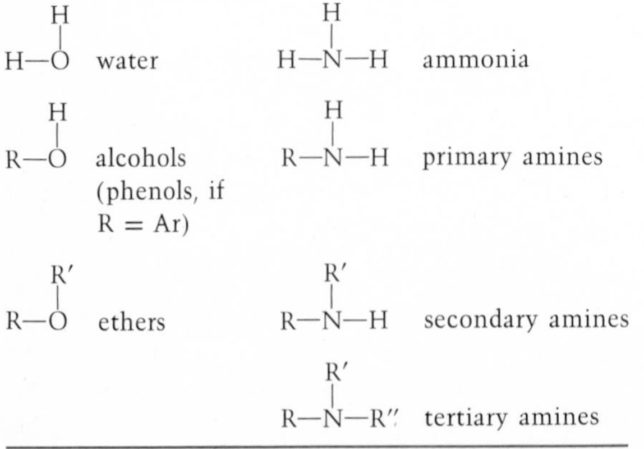

H—O—H	water	H—N(H)—H	ammonia
R—O—H	alcohols (phenols, if R = Ar)	R—N(H)—H	primary amines
R—O—R′	ethers	R—N(R′)—H	secondary amines
		R—N(R′)—R″	tertiary amines

[a]In all cases, R may be Ar.

With respect to pH, their properties are inverse to those of the carboxylic acids: at low pH values only the ion can exist, not the amine; while at high pH values only the amine can exist, not the ion.

AMIDES. Like carboxylic esters, amides are also acid derivatives. In this case, an **amide** may be thought of as being made up of a carboxylic acid and ammonia or a primary or secondary amine, with water split out.

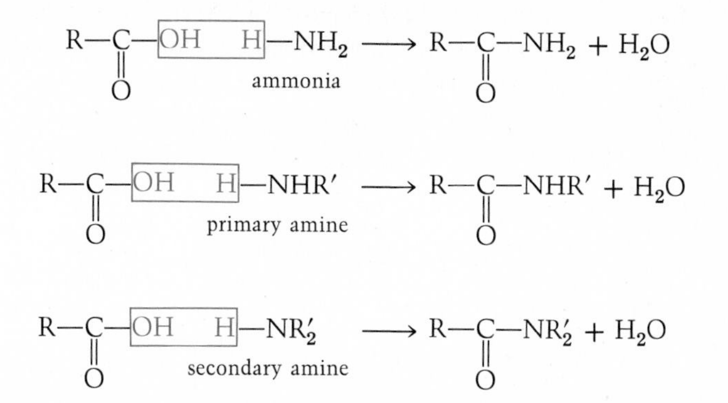

However, in this case it is not usually practical to make amides by a direct reaction, and other methods must be used. Since carboxylic acids are acidic and amines are basic, a direct reaction between them gives a salt, and not an amide; for example,

$$\mathrm{R{-}\underset{\underset{O}{\|}}{C}{-}OH + R'NH_2 \xrightarrow[\text{transfer}]{\text{proton}} R{-}\underset{\underset{O}{\|}}{C}{-}O^- \;\; R'NH_3^+}$$

Amides are named from the corresponding acid by dropping *ic acid* and adding **amide.** If alkyl or aryl groups are attached to the nitrogen, they are put into the prefix, with the letter *N* to signify attachment to nitrogen.

Examples:

$$\mathrm{CH_3{-}\underset{\underset{O}{\|}}{C}{-}NH_2} \qquad \text{acetamide}$$

$$\mathrm{C_6H_5{-}\underset{\underset{O}{\|}}{C}{-}NH{-}CH_3} \qquad N\text{-methylbenzamide}$$

$$\mathrm{H{-}\underset{\underset{O}{\|}}{C}{-}\underset{\underset{CH_3}{|}}{N}{-}CH_3} \qquad N,N\text{-dimethylformamide}$$

Amides have relatively high melting points. In fact, almost all of them are solids at room temperature (formamides are exceptions—they are liquids). Unlike amines, amides are neutral—they show neither acidic nor basic properties. Like carboxylic esters, they can be hydrolyzed, either with acidic or basic catalysts. The products are the carboxylic acid (or its ion) and ammonia or the amine (or its ion):

$$\mathrm{R{-}\underset{\underset{O}{\|}}{C}{-}NHR' + H_2O \xrightarrow{H^+} R{-}\underset{\underset{O}{\|}}{C}{-}OH + R'NH_3^+}$$

$$\mathrm{R{-}\underset{\underset{O}{\|}}{C}{-}NHR' + H_2O \xrightarrow{OH^-} R{-}\underset{\underset{O}{\|}}{C}{-}O^- + R'NH_2}$$

MERCAPTANS. **Mercaptans,** RSH, also called **thiols,** are the sulfur analogs of alcohols. They may be named simply by naming the alkyl or aryl group followed by the word mercaptan: for example, CH_3SH, methyl mercaptan. Mercaptans have two notable properties. They are weak acids (weaker than phenols but stronger than water), and they have foul odors. The major constituent of the liquid emitted by skunks is *n*-butyl mercaptan, $CH_3CH_2CH_2CH_2SH$.

A series of compounds all of which have the same functional group (or none at all), which differ in that each compound has one more CH_2 unit than the previous one is called a **homologous series.** For example, the unbranched alkanes CH_4, CH_3CH_3, $CH_3CH_2CH_3$, etc. form a homologous series, as do the unbranched primary alcohols CH_3OH, CH_3CH_2OH, $CH_3CH_2CH_2OH$, etc., the secondary alcohols,

$$CH_3—\underset{\displaystyle OH}{\underset{|}{C}H}—CH_3 \qquad CH_3—\underset{\displaystyle OH}{\underset{|}{C}H}—CH_2—CH_3 \qquad CH_3—\underset{\displaystyle OH}{\underset{|}{C}H}—CH_2—CH_2—CH_3 \qquad \text{etc.}$$

and the ketones,

$$CH_3—CH_2—\underset{\displaystyle O}{\underset{\|}{C}}—CH_2—CH_3 \qquad CH_3—CH_2—\underset{\displaystyle O}{\underset{\|}{C}}—CH_2—CH_2—CH_3$$

$$CH_3—CH_2—\underset{\displaystyle O}{\underset{\|}{C}}—CH_2—CH_2—CH_2—CH_3 \qquad \text{etc.}$$

In this section we have devoted most of our discussion to compounds containing only one functional group per molecule. Of course, molecules are by no means limited in this way. We have already seen compounds containing two or more functional groups of the same type (for example, glycerol), as well as those containing functional groups of different types (for example, aspirin, tartaric acid). There is really very little limit to this. Compounds exist in nature, or can be synthesized, which contain 5, 10, 15, or more functional groups, the same or different, in one molecule. When we realize that all these functional groups can be put into the vast number of possible hydrocarbons, in virtually any combinations or positions, the total number of possible organic compounds becomes astronomical.

20-7 Optical Activity and Chirality

In this section, we will look at another kind of isomerism, more subtle than any we have yet seen. In 1811, a French physicist Dominique Arago (1786–1853) discovered that a quartz crystal would rotate the plane of polarized light. Ordinary light vibrates in all planes, but polarized light in just one. Arago found that the plane of the light was rotated when it was sent through a piece of quartz (Figure 20-8). Later several other scientists, most notably Jean Baptiste Biot (1774–1862), showed that a number of other substances also rotated the plane of polarized light, though many other substances did not. A substance that has this property is called **optically active.** In 1848, Louis Pasteur (1822–1895), at that time a 26-year-old chemist and crystallographer, proposed an explanation for the phenomenon. He said: If molecules are nonsuperimposable on their mirror images, they will be optically active; otherwise, not. This was a bold prediction to make in 1848 when the structure of not one single molecule was known,* but it was proven correct in 1874 independently by Jacobus van't Hoff (1852–1911), who won the first Nobel Prize in Chemistry in 1901, and Joseph Le Bel (1847–1930).

*This prediction by Pasteur was not a guess. He had strong evidence for it, which is too involved to be discussed here.

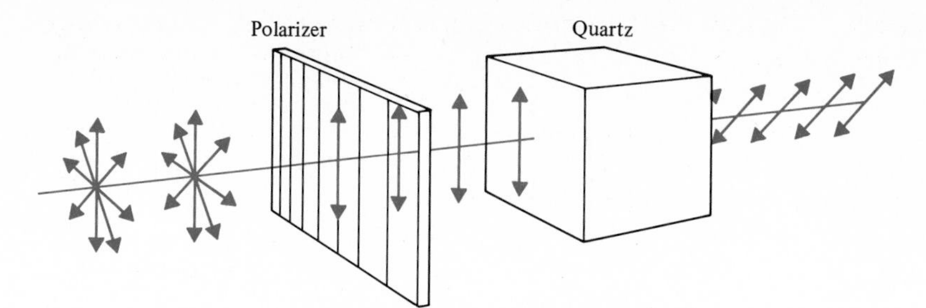

Figure 20-8 Ordinary light, which vibrates in all planes (indicated by arrows), becomes polarized when passed through a polarizer. When the polarized light passes through a transparent piece of quartz, the plane of the light is rotated.

If we look at the world around us, we will see that all objects can be divided into two classes: those that can be superimposed on their mirror images and those that cannot. The latter are called **chiral** (pronounced ky′ral). A good example of a chiral object is your left hand. If you hold it up to a mirror, you will see an image of your right hand (Figure 20-9), and you cannot superimpose the two. Some objects that are not chiral are golf balls, spoons, baseball bats, and wine glasses, all of which are superimposable on their mirror images (Figure 20-10).

Today we can state with certainty that Pasteur's proposal is correct: all optically active compounds are chiral; all optically inactive compounds are not chiral **(achiral).** Let us see what kind of molecules are chiral. We can easily prove to ourselves that a molecule which contains a carbon atom connected to four *different* groups must be chiral. Such a molecule is shown in Figure 20-11. Remember that carbon points its four bonds to the corners of a tetrahedron. As can be seen, the molecule on the left is the mirror image of the molecule on the right, just as a left-hand glove is the mirror image of a right-hand glove, but the molecules cannot be superimposed, any more than we can superimpose a left-hand glove on a right-hand glove (try putting a left-hand glove on your right hand). A carbon atom connected to four different groups is called an **asymmetric carbon atom** (also called a **chiral center**). Not only 2-butanol (in Figure 20-11), but any molecule containing a single asymmetric carbon atom is necessarily chiral (we shall see later in this section that molecules containing two or more asymmetric carbons are not always chiral).

Mirror

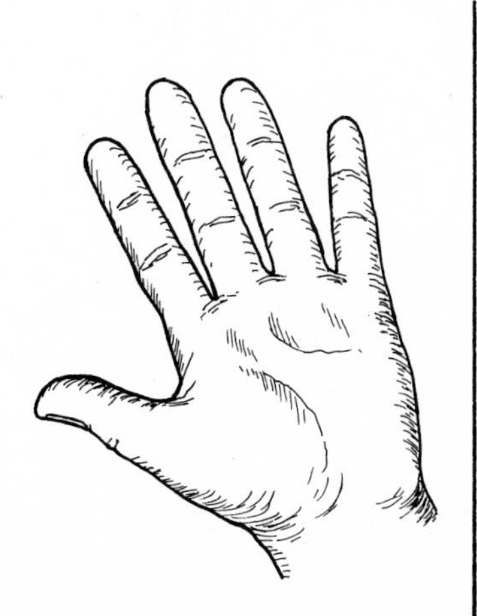

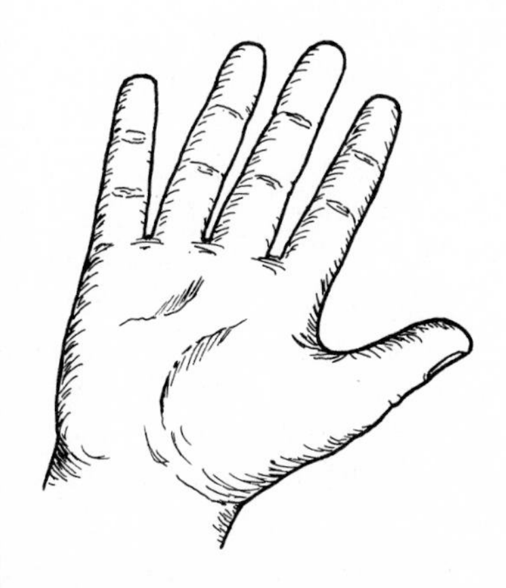

Figure 20-9 Left and right hands are nonsuperimposable mirror images (chiral).

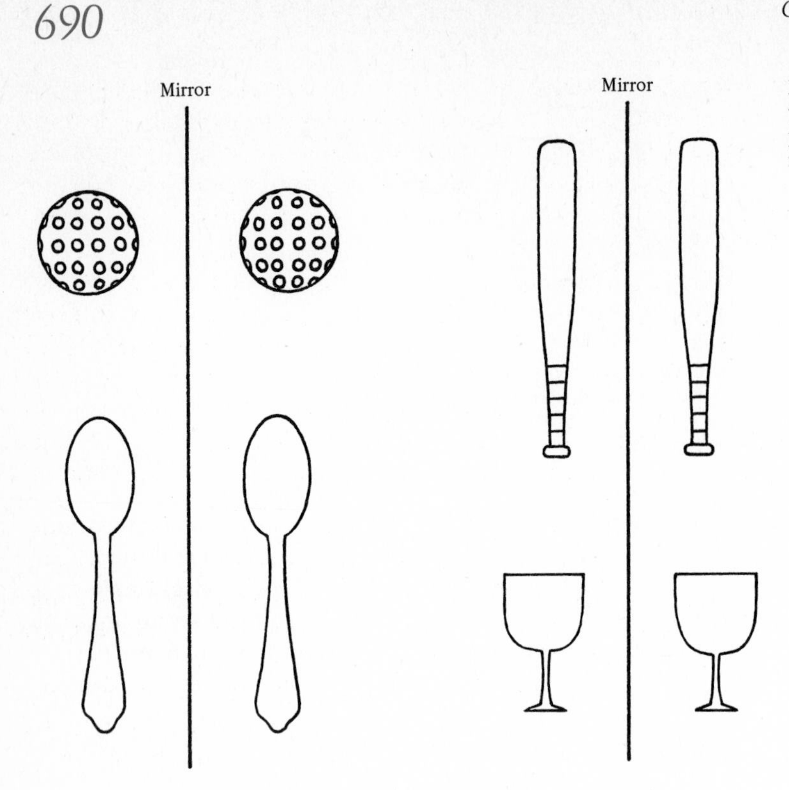

Figure 20-10 Some objects which are superimposable on their mirror images (achiral).

If the two molecules of 2-butanol shown in Figure 20-11 are not superimposable, they are not identical (superimposability is the same as identity). If they are not identical, they are different; so they must be stereoisomers (Section 20-3). However, they are the most subtle form of stereoisomers possible, since they are identical in all respects except direction: one is "left-handed" and the other "right-handed." Isomers that are nonsuperimposable mirror images of each other are called **enantiomers.** The similarity of shape shows itself in the similarity of properties. We have seen that structural isomers and cis–trans isomers differ in almost all properties. This is not the case for enantiomers. **All physical and chemical properties of enantioners are identical,** except for two. These are:

1. Enantiomers rotate the plane of polarized light in opposite directions. One enantiomer rotates the plane to the right (actually clockwise); we call this **dextrorotatory** and use the symbol (+). The other enantiomer must then rotate the plane of the light to the left (counterclockwise); we call this **levorotatory** and use the symbol (−). Although the enantiomers rotate the plane in opposite directions, the *amount* of rotation is the same, provided that the size of the sample is the same (the amount of rotation depends on the size of the sample). The amount and direction of rotation are measured on an instrument called a *polarimeter*.

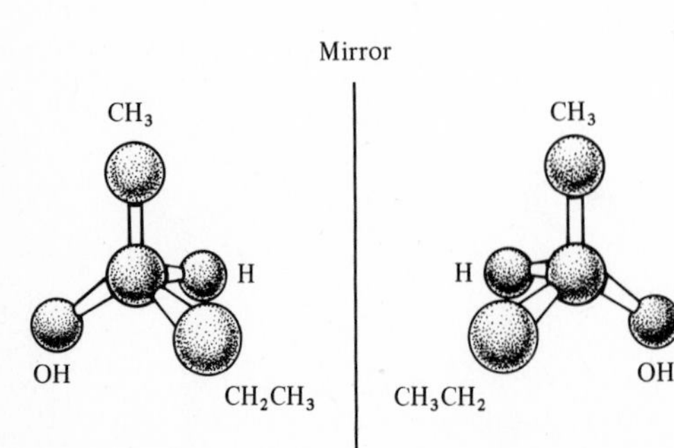

Figure 20-11 A molecule of 2-butanol and its mirror image. If we try to superimpose the two, we find that we cannot do so. The two molecules are a pair of enantiomers.

2. Two enantiomers always react at the same rate with any achiral compound, but *react at different rates with a chiral compound.* The most important examples here are biochemical. As we shall see in Chapter 21, many biological molecules are chiral, and living organisms usually possess only one of the two enantiomers. The food that we eat is digested by reacting with chiral molecules present in the body. Table sugar, sucrose, is chiral, and rotates the plane of polarized light to the right. We are able to digest this molecule, and gain energy from it, because the sucrose reacts with molecules in our body (enzymes) which also happen to be chiral, at a convenient rate. If we ate the (−) form of sucrose, we would not be able to digest it. Even though the two molecules are identical in all respects except handedness, the (+) sucrose and (−) sucrose react at very different rates with the chiral enzymes in our bodies, so different that the (+) isomer reacts readily, but the (−) isomer does not react at all. If these enzymes were achiral, they would react with both forms of sucrose at the same rate.

In all properties other than these two, enantiomers are identical. They have identical melting points, boiling points, densities, crystal shapes (though in a few cases the crystals too are nonsuperimposable mirror images), indexes of refraction, spectra, and all other properties. In fact, in most cases the only way to tell them apart is to put them into a polarimeter and see whether they rotate (+) or (−).

What about a carbon atom connected to only three different groups? Are molecules containing such atoms chiral? Figure 20-12 shows us that they are not. In order to be chiral, a molecule must have four different groups on one carbon atom; molecules that are more symmetrical than this are achiral.*

Although each enantiomer of a chiral substance rotates the plane of polarized light, a mixture of equal amounts of the two does not do so, because the rotations cancel. Such a mixture (called a **racemic mixture**) is always optically inactive. Racemic mixtures are fairly common in laboratories for the following reason. When a new asymmetric carbon atom is generated in a molecule that previously did not have any, equal amounts of the two enantiomers are always formed. For example, if we reduce butanone with H_2 and a catalyst,

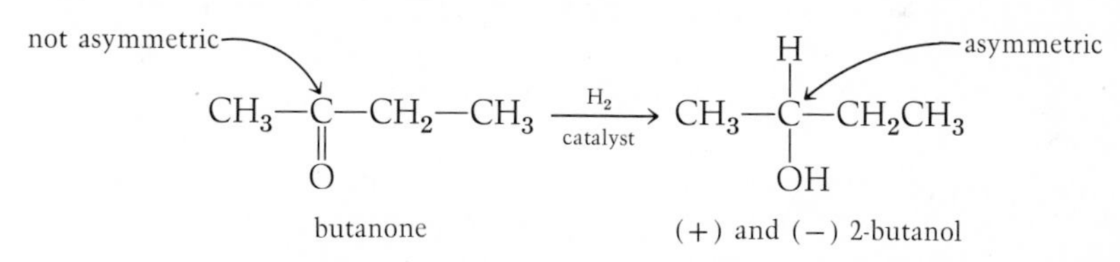

a carbon atom that is not asymmetric has been changed to one that is asymmetric. The 2-butanol formed in this reaction could be the (+) or the (−) enantiomer. In such cases, *racemic mixtures are always produced,* if all the reactants (butanone,

*Actually, there are compounds known whch are chiral even though they do not have an asymmetric carbon. These compounds are also optically active. We shall see an example on page 693.

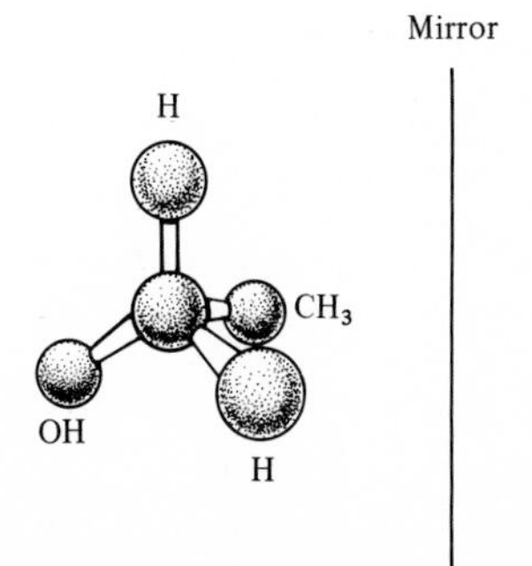

Figure 20-12 A molecule of ethanol is superimposable on its mirror image (just flip one of them over onto the other). The same is true for any carbon atom with at least two identical groups.

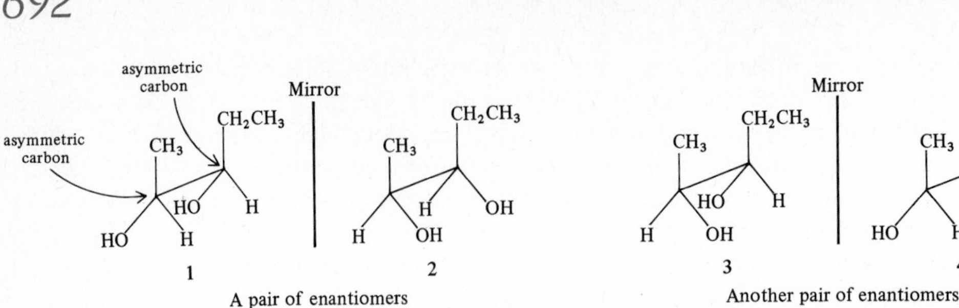

Figure 20-13 The compound 2,3-pentanediol has two asymmetric carbons. There are four isomers: two pairs of enantiomers. Note that isomer 1 is neither identical to 3 or 4 nor the mirror image of either of them. Stereoisomers that are not enantiomers are called *diastereomers*. In this example the two asymmetric carbons are connected to each other. However, the same number of isomers will be found even if they are not connected to each other.

H_2, and catalyst) are optically inactive. Optically active compounds or mixtures can only be produced from other optically active compounds or mixtures. Racemic mixtures, however, are comparatively uncommon in living organisms (although many examples are known). If a molecule in a living organism is chiral, then nature usually tends to make only one of the two enantiomers.*

Suppose that we have *two* asymmetric carbons in the same molecule. In most such cases *four* isomers are possible (Figure 20-13). Since a molecule can have only one mirror image; isomer 1, for example, can have only one enantiomer—in this case it is isomer 2. Similarly, isomers 3 and 4 are enantiomers of each other. However, neither isomer 1 nor isomer 2 are identical with or mirror images of isomers 3 or 4. Stereoisomers that are not enantiomers are called **diastereomers.** Isomer 1 is the enantiomer of isomer 2 and a diastereomer of isomers 3 and 4. Although enantiomers always have identical properties (except for the two previously noted), diastereomers do not, and melting points, boiling points, densities, and so on, of 1 are not the same as those of 3 or 4 (which are identical to each other).

In the more general case, a compound with 3 asymmetric carbon atoms has *eight* isomers possible; if 4 asymmetric carbons are present, there are *sixteen* isomers; etc. The formula for the total number of isomers is

$$\text{isomers} = 2^n$$

where n is the number of asymmetric carbons. The isomers occur in pairs, so that the 16 isomers in the case mentioned above exist as eight pairs of enantiomers.

However, there are cases known where fewer isomers than this are possible. Such a situation is shown in Figure 20-14. In this case the four groups on one asymmetric carbon are the same as the four groups on the other asymmetric carbon. When this situation holds, there are only three isomers: one pair of enantiomers and one achiral compound, called a **meso compound.** A meso compound is a compound that contains asymmetric carbons, but is nevertheless achiral. Meso compounds are of course optically inactive, *even though they have two asymmetric carbons.* Although meso compounds are relatively uncommon, it is well to be able to recognize one when you see it.

In a footnote on page 691 we mentioned that compounds are known which are chiral (and hence optically active) even though they have no asymmetric carbons. Among these are certain octahedral complexes (Section 19-7). For example, a molecule of *cis*-difluorobis(ethylenediamine)cobalt is nonsuperimposable on its mirror image, so that two enantiomers exist.

*This fact gives rise to an interesting question for which there is yet no certain answer. If many biological molecules are optically active, they must have been produced from other optically active molecules. In that case, how did the first living things on earth get their optically active molecules?

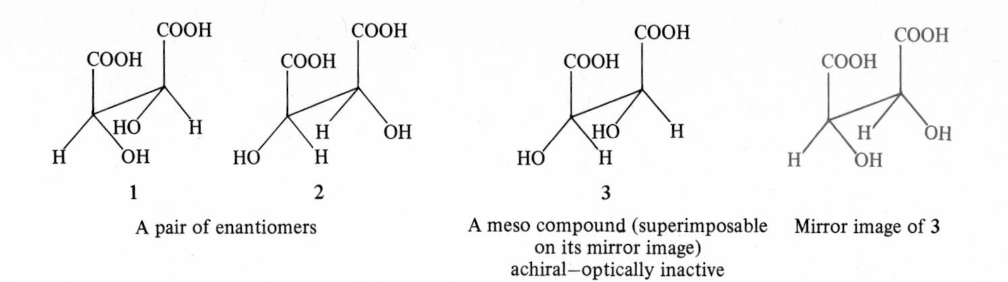

Figure 20-14 The three isomers of tartaric acid. Isomers 1 and 2 are a pair of enantiomers, with the properties previously discussed. Isomer 3 is a meso compound, achiral and optically inactive. The mirror image of 3 is shown in color. Note that this can be superimposed on 3 if we turn it around.

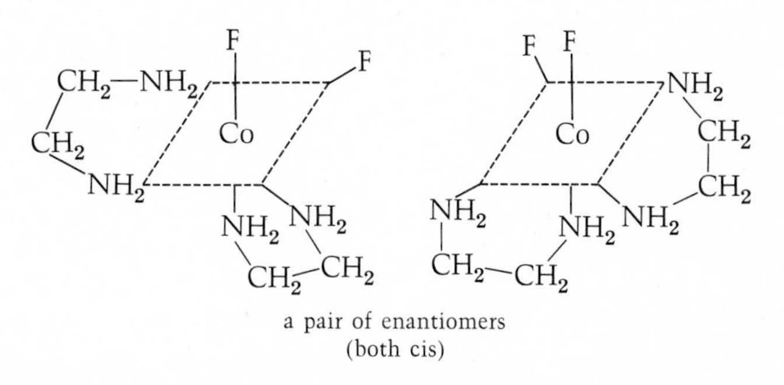

a pair of enantiomers
(both cis)

The properties of these enantiomers are identical to each other (with the two exceptions previously noted), just as are those of any two enantiomers. The *trans* isomer is superimposable on its mirror image, so that only one isomer exists, which is optically inactive.

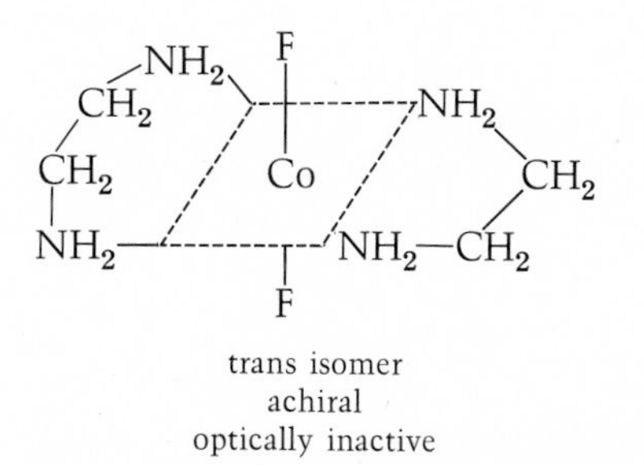

trans isomer
achiral
optically inactive

Thus, in the last page or two we have seen that asymmetric carbon atoms are neither a necessary nor a sufficient criterion for chirality. Compounds exist that are achiral even though they have asymmetric carbon atoms (although there must be at least two such atoms, suitably substituted), while other compounds exist that are chiral even though they have *no* asymmetric carbon atoms. Nevertheless, in most cases, chirality and asymmetric carbon atoms do go together. (In *all* cases, chirality and optical activity go together.)

20-8 Organic Reactions

In an organic reaction, one organic compound is converted to another, usually by the addition of one or more reagents or by the application of heat or light. We have already seen a number of examples (for example, the conversion of ketones to secondary alcohols; of secondary alcohols to ketones; of alkenes to alkanes). Organic reactions are carried out daily in chemical laboratories all over the world (in fact, running organic reactions is the chief activity of organic chemists, though they do other things also) and in industrial chemical plants. The difference between a reaction run in a laboratory and industrially is *scale*. A laboratory chemist may make 10 or 100 g of a compound, so that he works with 500-ml flasks and other equipment of similar size. This kind of apparatus would not do in industrial plants where compounds must be produced in amounts measured in tons. The apparatus is much larger, cannot be made of glass, and requires very different techniques to operate. It is usually designed and run by chemical engineers rather than chemists.

However, millions of years before there were human beings alive to carry out organic reactions of any kind, such reactions were taking place in biological organisms. Every living organism is a chemical factory, where scores, hundreds, or thousands of reactions, depending on the complexity of the organism, are taking place all the time, almost all of them catalyzed by proteins called enzymes (we shall discuss enzyme action in Section 21-4).

There are hundreds of known organic reactions, some simple, some complex, but almost all of them can be placed into one of five simple categories. Note that in any category, some reactions keep carbon skeletons intact, while in others the skeletons are altered (the carbon **skeleton** or **backbone** is the framework of carbon atoms in the molecule).

1. *Substitution*

$$\text{R—X} + \text{Y} \longrightarrow \text{R—Y} + \text{X}$$

In this type of reaction one group is replaced by another. Some examples are

$$\text{R—OH} + \text{SOCl}_2 \longrightarrow \text{R—Cl} + \text{SO}_2 + \text{HCl}$$

In this reaction, alcohols are converted to alkyl halides by treatment with thionyl chloride, $SOCl_2$. The carbon skeleton is unchanged.

$$\text{Ar—H} + \text{CH}_3\text{Cl} \xrightarrow{\text{AlCl}_3} \text{Ar—CH}_3 + \text{HCl}$$

Arenes react with chloromethane in the presence of aluminum chloride to give the corresponding methyl arenes. The carbon skeleton has been altered.

2. *Addition to a multiple bond*

$$\text{C=C} + \text{XY} \longrightarrow \begin{matrix}\text{X} & & \text{Y}\\ | & & |\\ \text{C} & \text{—} & \text{C}\end{matrix}$$

$$\text{C=O} + \text{XY} \longrightarrow \begin{matrix}\text{X} & & \text{Y}\\ | & & |\\ \text{C} & \text{—} & \text{O}\end{matrix} \qquad \text{etc.}$$

In this type of reaction, two groups (X and Y) are added to a double or triple bond. The catalytic hydrogenation of alkenes (Section 20-3) and the reduction of aldehydes and ketones (Section 20-6) belong to this category. Other examples are

$$\text{R—}\underset{\underset{\text{O}}{\|}}{\text{C}}\text{—H} + 2\text{R'OH} \xrightarrow{\text{H}_3\text{O}^+} \text{R—}\underset{\underset{\text{OR'}}{|}}{\overset{\overset{\text{OR'}}{|}}{\text{C}}}\text{—H} + \text{H}_2\text{O}$$

an acetal

Aldehydes react with alcohols in the presence of a strong acid to give **acetals** (compounds with two OR groups on the same carbon).

$$\text{—}\underset{|}{\text{C}}\text{=}\underset{|}{\text{C}}\text{—} + \text{CO} + \text{H}_2 \xrightarrow[\text{catalyst}]{\text{pressure}} \text{—}\underset{|}{\overset{\overset{\text{H}}{|}}{\text{C}}}\text{—}\underset{|}{\overset{\overset{\text{CHO}}{|}}{\text{C}}}\text{—}$$

Alkenes may be converted to aldehydes by treatment with carbon monoxide and hydrogen under pressure, in the presence of a catalyst. This is called the oxo process and is an important industrial reaction. Note that the length of the carbon skeleton of the alkene has been increased by one carbon.

3. *Eliminations*

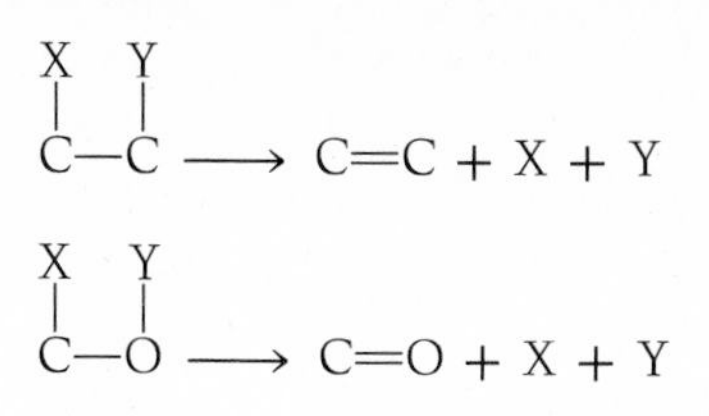

This is the reverse of the previous type. A new double or triple bond is created. Examples are

$$-\underset{\text{H}}{\overset{|}{\underset{|}{\text{C}}}}-\underset{\text{Cl}}{\overset{|}{\underset{|}{\text{C}}}}- + OH^- \xrightarrow{\text{ethanol}} -\overset{|}{\text{C}}=\overset{|}{\text{C}}- + H_2O + Cl^-$$

Alkyl halides can be converted to alkenes by treatment with KOH dissolved in ethanol, if they have a hydrogen atom in the proper position.

$$R-\underset{\text{O}}{\underset{\|}{\text{C}}}-NH_2 \xrightarrow{P_4O_{10}} \underset{\text{a nitrile}}{R-C\equiv N} + H_2O$$

Amides which have no *N*-alkyl or *N*-aryl groups (see Table 20-1), can be dehydrated by treatment with phosphorous pentoxide, to give **nitriles:** compounds with a CN group.

4. *Rearrangements.* In these reactions one part of a molecule appears to become detached and then reconnected to another part of the same molecule. A large variety of such reactions is known. A few examples are

$$CH_3-\underset{\text{OH}}{\overset{CH_3}{\overset{|}{\underset{|}{\text{C}}}}}-\underset{\text{OH}}{\overset{CH_3}{\overset{|}{\underset{|}{\text{C}}}}}-CH_3 \xrightarrow{\text{HCl}} CH_3-\underset{\text{O}}{\underset{\|}{\text{C}}}-\underset{CH_3}{\overset{CH_3}{\overset{|}{\underset{|}{\text{C}}}}}-CH_3 + H_2O \qquad \text{"the pinacol rearrangement"}$$

Here a diol is treated with HCl, causing a methyl group to move from one carbon atom to another. The carbon skeleton has of course been altered.

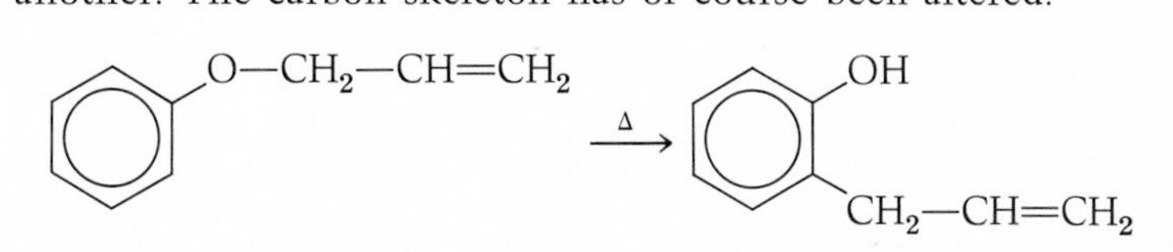

In this reaction a certain type of ether (with a double bond in the position shown) is heated. The product is a phenol, with the group formerly attached to the oxygen now on the ring, in a position ortho to the OH group.

5. *Oxidation–reduction reactions.* We have already seen that primary alcohols can be oxidized to carboxylic acids, and that aldehydes can be oxidized or reduced. Many other oxidation–reduction reactions are also known.

We have stated that the principal activity of organic chemists is the running of organic reactions. Another important activity is the determination of the mechanism of these reactions, because a knowledge of mechanisms would enable us to predict which reactions will succeed and which will not. Organic mechanisms have been studied intensively for more than 40 years in hundreds of laboratories around the world, and a great deal of knowledge has been gained, although there is much that is still unknown. Mechanisms are not easy to study because we cannot actually see the molecules react, and because each individual molecule reacts extremely rapidly (though in a given flask, the entire reaction may take place slowly, if only a few molecules are reacting in each second) so that the molecules don't hold still for investigation, as they do, for example, in x-ray diffraction. Consequently, everything we know has been determined by inference. A number of methods have been developed for the investigation of reaction mechanisms, the most important of which is measurement of reaction rates (kinetics). In Section 9-8, we discussed how chemists use reaction rates to investigate mechanisms.

Problems

20-1 Define each of the following terms. **(a)** hydrocarbon **(b)** tertiary carbon **(c)** primary carbon **(d)** isomers **(e)** branched alkane **(f)** *normal*-alkane **(g)** alkyl group **(h)** knocking **(i)** octane number **(j)** cracking **(k)** unsaturated **(l)** saturated **(m)** catalytic hydrogenation **(n)** cis and trans **(o)** stereoisomer **(p)** strain **(q)** the chair form **(r)** aromatic compound **(s)** arene **(t)** functional group **(u)** R **(v)** Ar **(w)** proof number **(x)** spirits **(y)** carbonyl compounds **(z)** fatty acid **(aa)** hydrolysis **(bb)** saponification **(cc)** homologous series **(dd)** optical activity **(ee)** chiral **(ff)** asymmetric carbon atom **(gg)** enantiomers **(hh)** dextrorotatory **(ii)** racemic mixture **(jj)** diastereomer **(kk)** meso compound **(ll)** carbon skeleton

20-2 Draw structural formulas for each of the following alkanes. **(a)** 3-methyloctane **(b)** dimethylpropane **(c)** 2,3,4-trimethylheptane **(d)** 3,3-diethylhexane **(e)** 3-ethyl-2,7-dimethylnonane **(f)** 5-butyldecane **(g)** 3,5,7-triethyl-2,4,6,8-tetramethylnonane **(h)** 5,5-diisopropyldecane

20-3 Name each of the following alkanes by the IUPAC system.

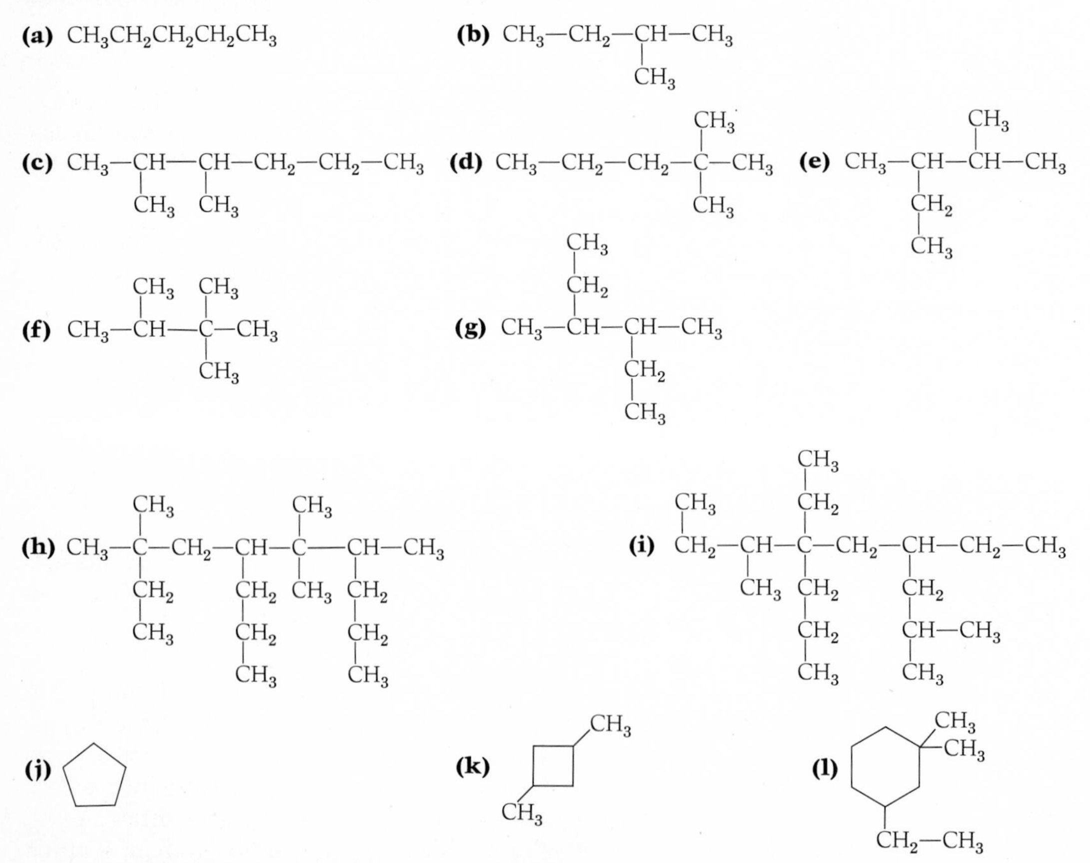

20-4 Draw structural formulas for each of the following.
(a) all five isomeric alkanes with six carbons (C_6H_{14}) **(b)** all nine isomeric alkanes with seven carbons (C_7H_{16})

20-5 Give IUPAC names to the alkanes whose structural formulas you drew in Problem 20-4.

20-6 What are the principal petroleum fractions?

20-7 Draw structural formulas for each of the following alkenes and alkynes. **(a)** 2-butyne **(b)** 2,3-dimethyl-2-butene **(c)** 3,3-diethylcyclopentene **(d)** *trans*-2-pentene **(e)** 2,4-dimethyl-1,5-heptadiene

20-8 Give IUPAC names for each of the following alkenes and alkynes.

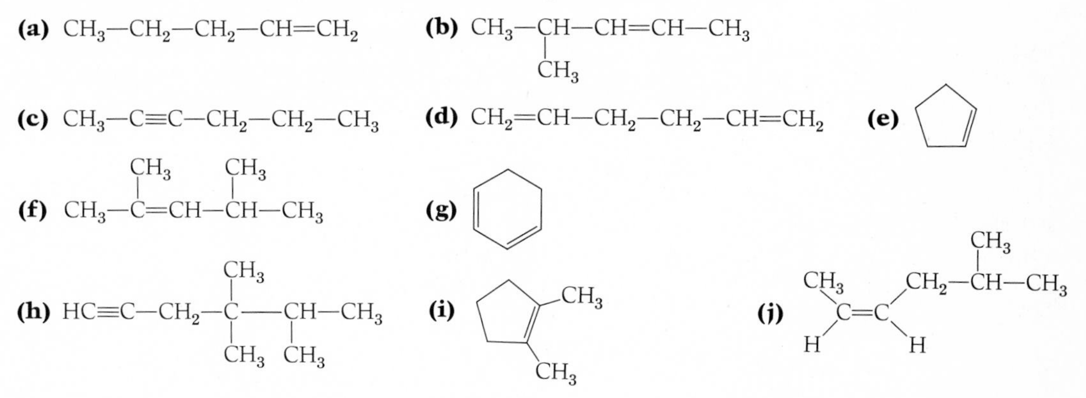

20-9 Draw structural formulas for all compounds with the formula C_5H_{10}.

20-10 Draw the 12 structural isomers of *cycloalkanes* with the formula C_6H_{12}.

20-11 For which of the following are cis and trans isomers possible?

(a) $CH_3CH{=}CHCH_2CH_3$ **(b)** $CH_2{=}CHCH_2CH_3$ **(c)** $CH_3C(CH_3){=}CHCH_3$ **(d)** $CH_3CH_2C(CH_3){=}CHCH_3$ **(e)** $CH_3C(Cl){=}CHCH_3$ **(f)** 2-Pentene **(g)** 2-Methyl-2-pentene **(h)** 2,3-Dimethyl-2-pentene **(i)** 3-Hexene **(j)** 1,1-Dichloro-1-butene **(k)** 2-Pentyne

20-12 What can you say about the number of double bonds, triple bonds, and rings in each of the following formulas?
(a) C_7H_{14} **(b)** $C_{11}H_{24}$ **(c)** C_9H_{16} **(d)** $C_{17}H_{30}$ **(e)** $C_{13}H_{26}$ **(f)** $C_{19}H_{40}$

20-13 Draw structural formulas for each of the following aromatic compounds.
(a) aniline **(b)** *meta*-iodophenol **(c)** *ortho*-nitrotoluene **(d)** 2,3-dichlorophenol **(e)** 2,5-dimethylbromobenzene **(f)** 1,3-diphenylbutane

20-14 Name each of the following aromatic compounds.

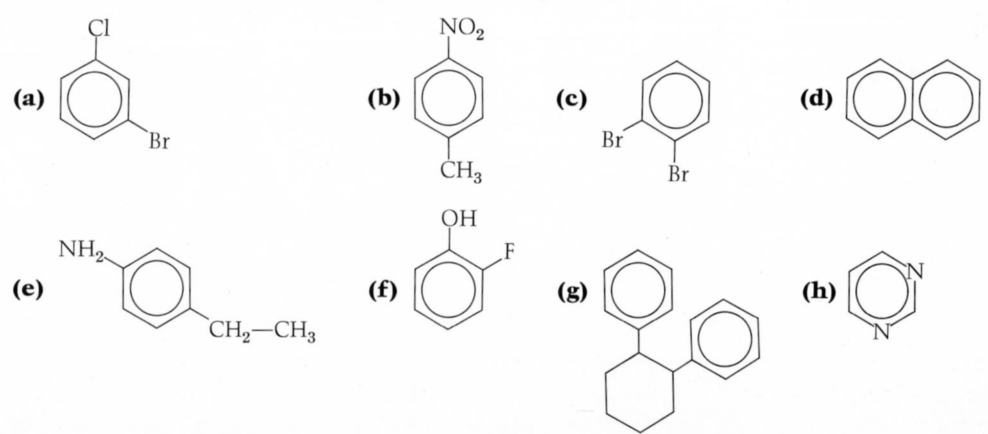

20-15 State which class each of the following compounds belongs to.

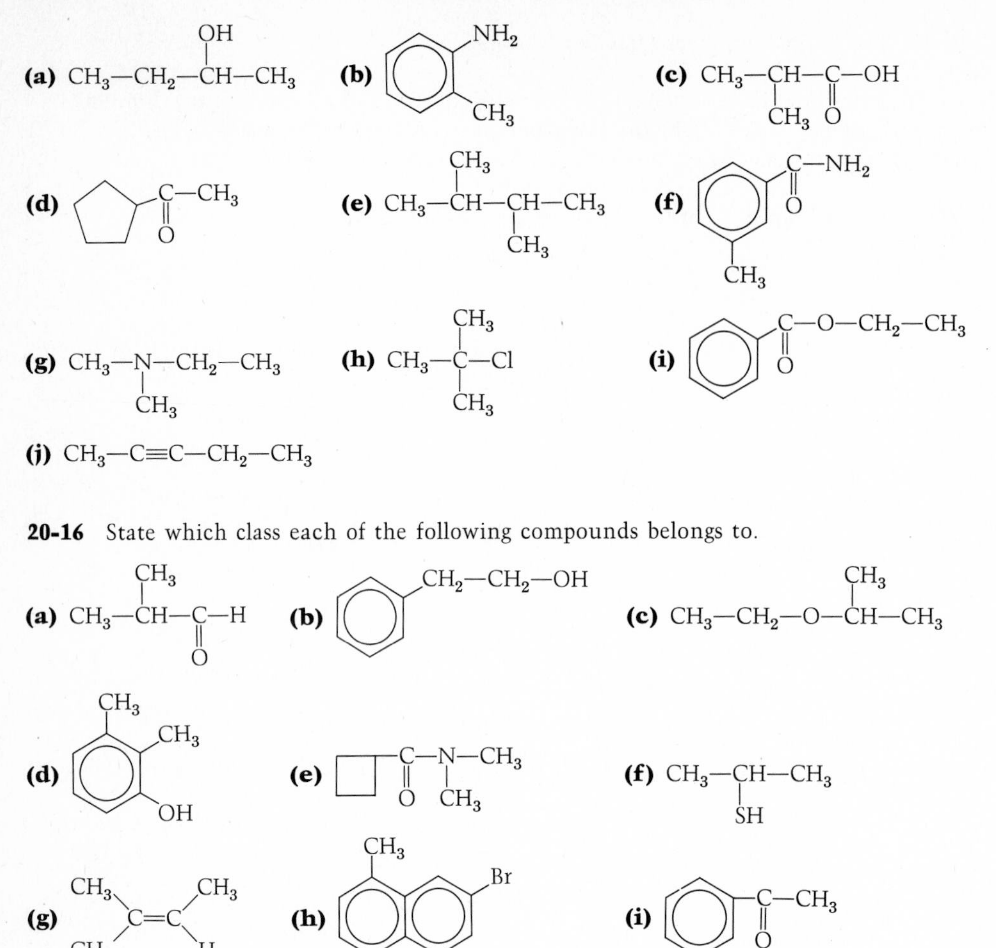

20-16 State which class each of the following compounds belongs to.

20-17 Each of the following compounds has two or three functional groups. Put each compound into as many classes as appropriate.

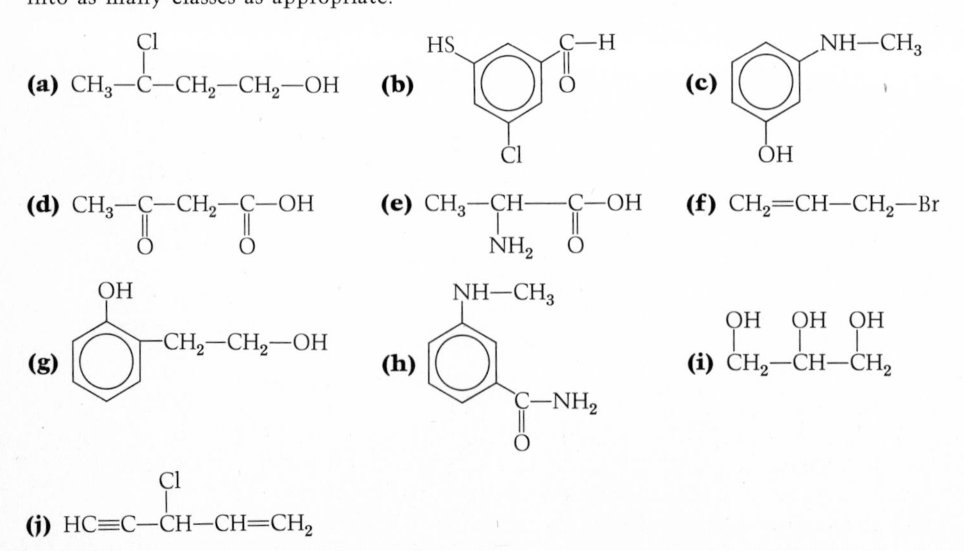

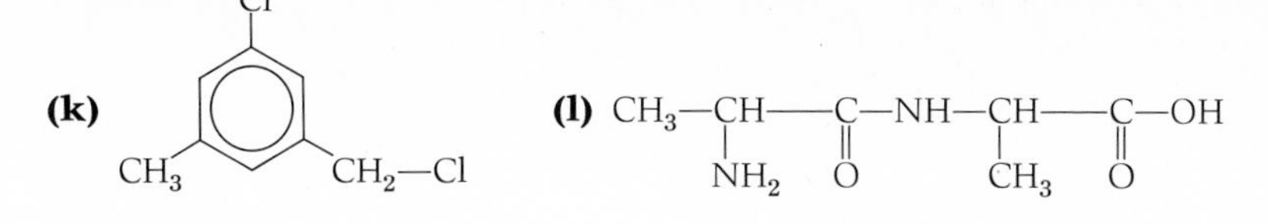

20-18 Name each of the following compounds. Give two names where possible.

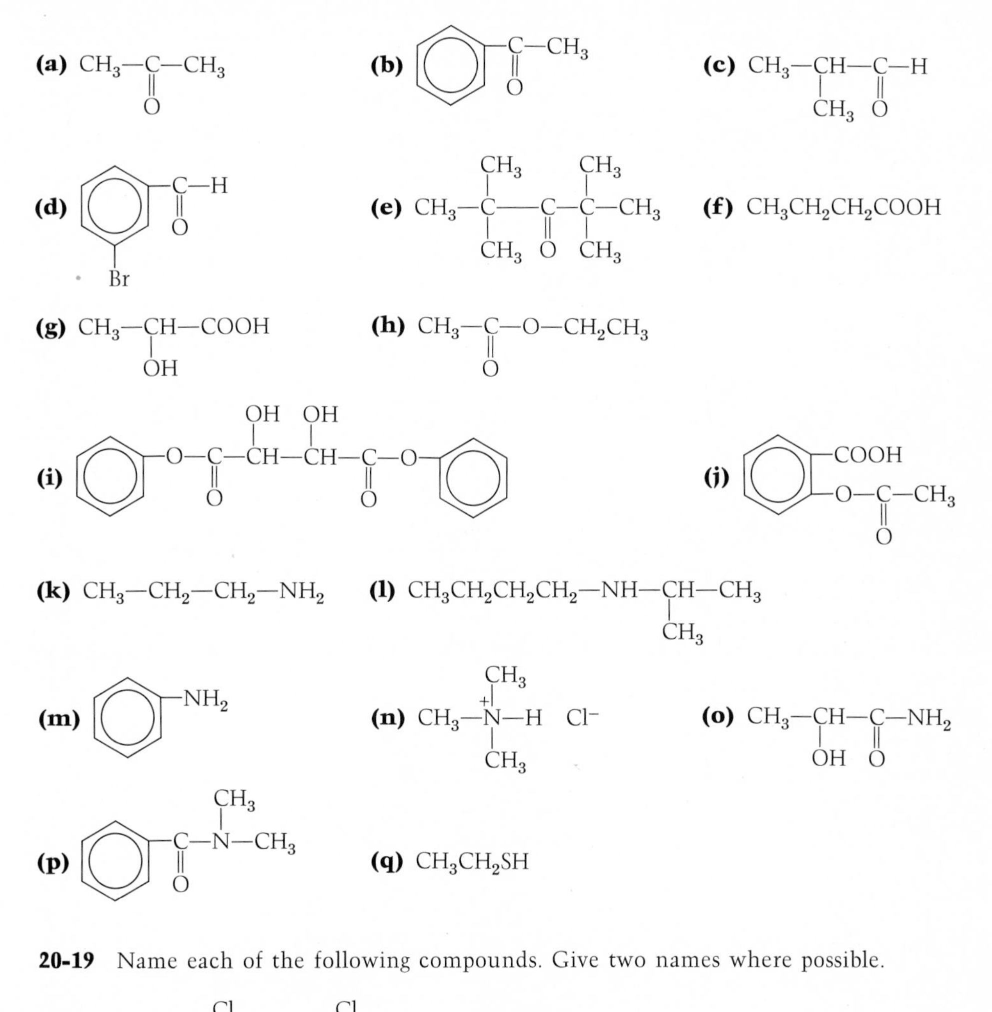

20-19 Name each of the following compounds. Give two names where possible.

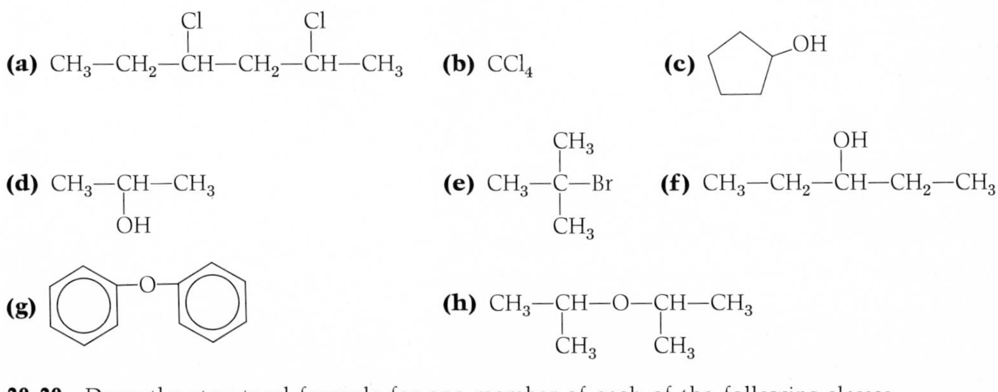

20-20 Draw the structural formula for one member of each of the following classes.
(a) alkene **(b)** ether **(c)** phenol **(d)** carboxylate ion **(e)** aldehyde
(f) carboxylic ester **(g)** diene **(h)** arene **(i)** mercaptan **(j)** secondary alcohol
(k) tertiary amine **(l)** alkoxide ion **(m)** fatty acid **(n)** dicarboxylic acid
(o) amide **(p)** thiol **(q)** acetal **(r)** nitrile

20-21 Which of the classes of compounds we have met in this chapter are **(a)** fairly inert (undergo few reactions) **(b)** acidic **(c)** basic.

20-22 Draw structural formulas for each of the following groups. **(a)** *n*-butyl **(b)** isopropyl **(c)** *n*-pentyl **(d)** *t*-butyl **(e)** phenyl **(f)** hydroxy **(g)** bromo **(h)** carbonyl **(i)** carboxyl

20-23 Classify each of the following as a primary, secondary, or tertiary alcohol or amine, or not an alcohol or amine at all.

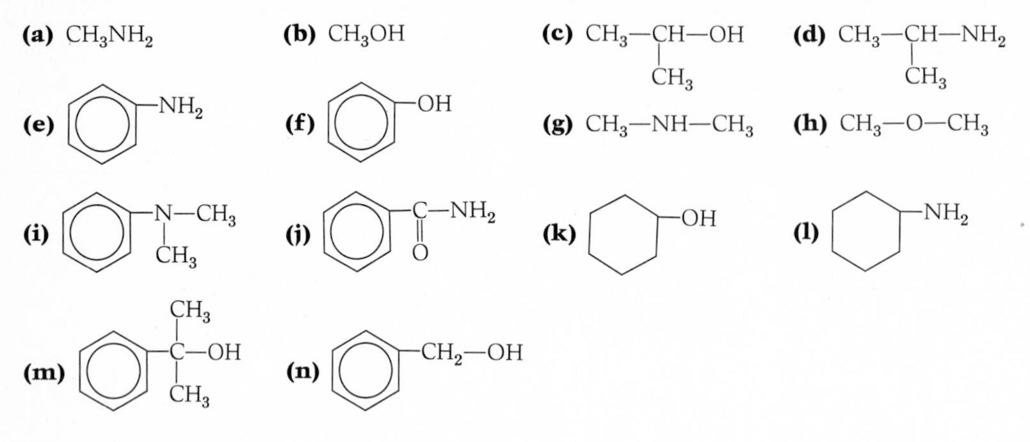

20-24 Without looking them up in tables, arrange the following compounds in order of increasing boiling point.
(a) CH_3OH **(b)** $CH_3—CH_3$ **(c)** CH_3COOH **(d)** CH_3CH_2OH **(e)** $CH_3CH_2CH_2CH_2CH_3$

20-25 Write the structural formulas for the first four members of the homologous series beginning with formic acid H—COOH.

20-26 Which of these compounds are chiral and which are not chiral?

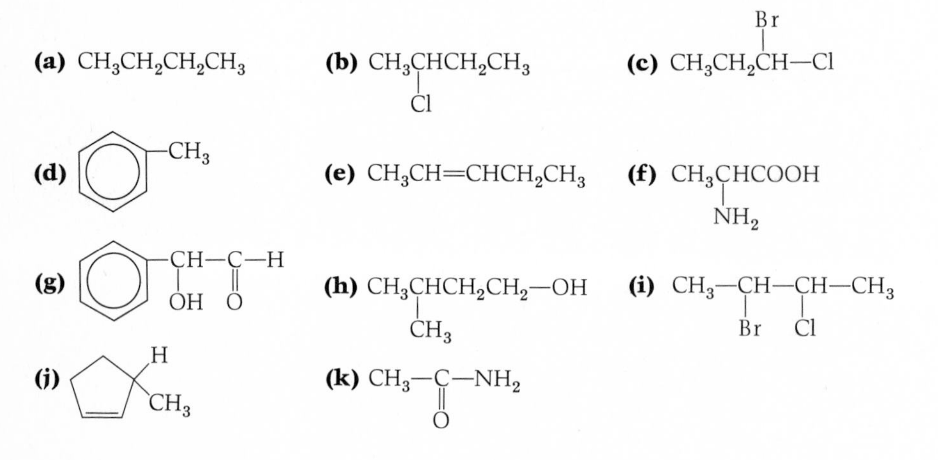

20-27 Draw all the stereoisomers possible in each case.

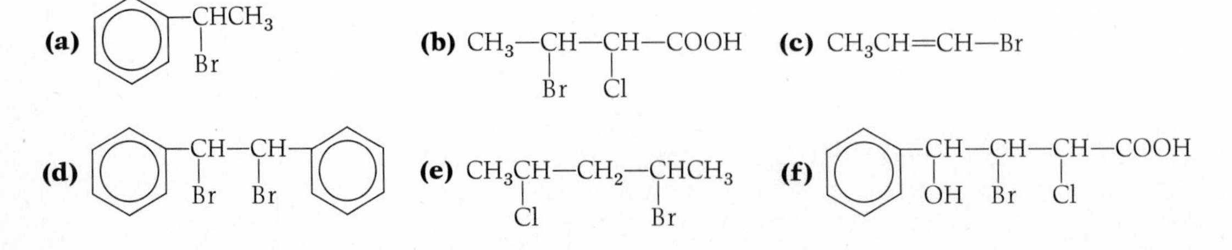

20-28 In each case, predict if the product will rotate the plane of polarized light.

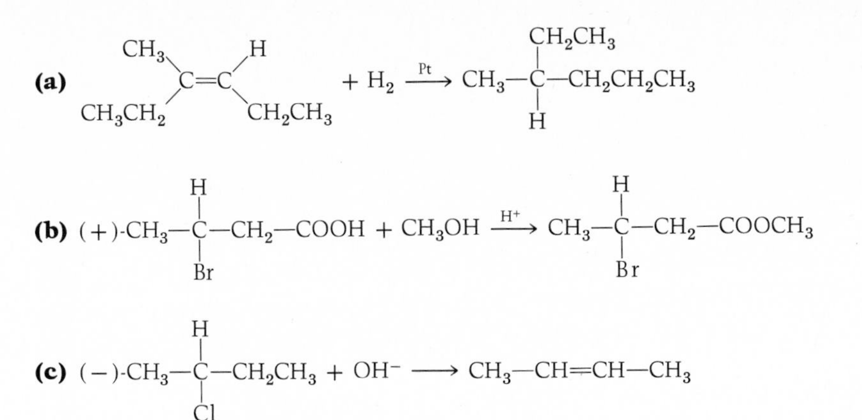

20-29 Complete and balance each of the following. If there is no reaction, write NR.

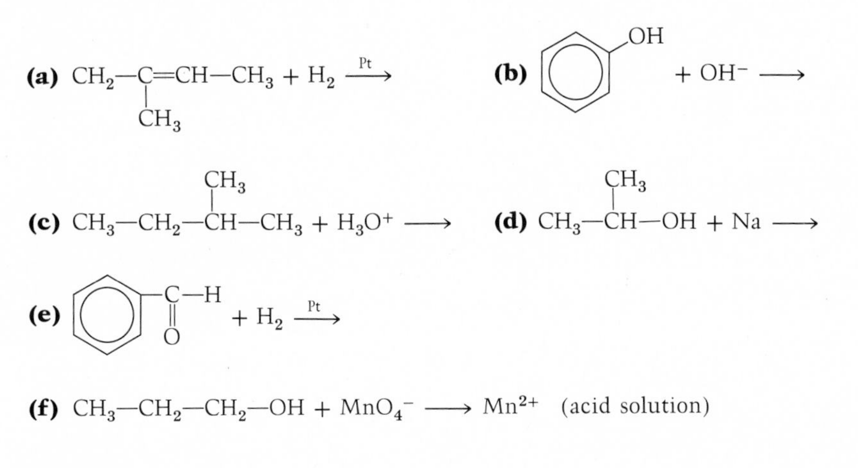

(g) $C_6H_{12}O_6 \xrightarrow{\text{enzymes}}$
glucose

20-30 Complete and balance each of the following. If there is no reaction, write NR.

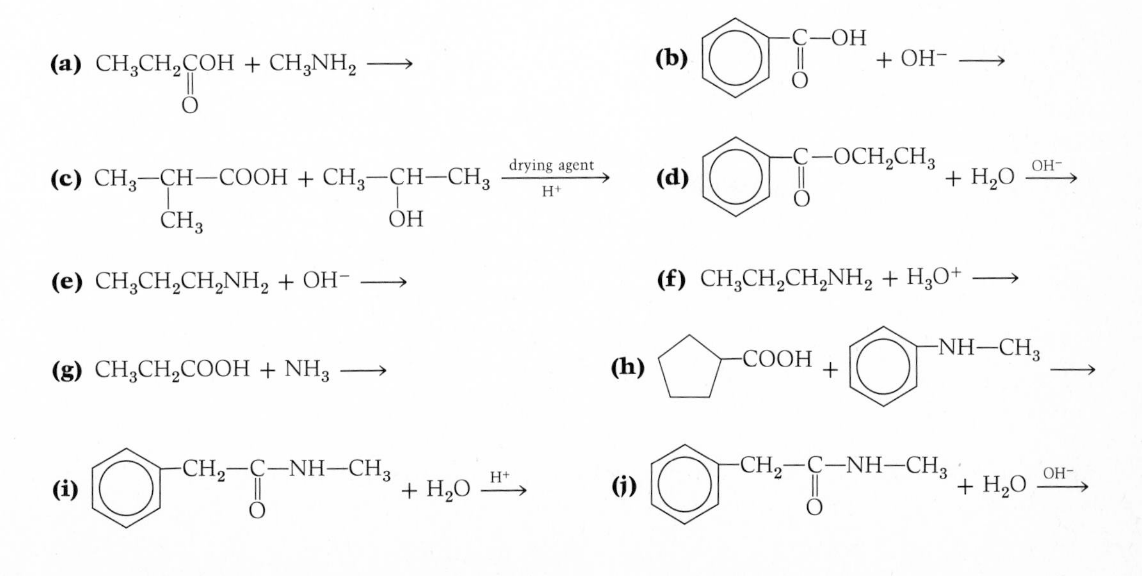

Additional Problems

20-31 The following names are all incorrect. State what is wrong with each. Do not try to give correct names.

(a) 2-dimethylbutane **(b)** 4-methylpentane
(c) 2-ethyl-4,4-dimethylpentane **(d)** *trans*-1-pentene
(e) *trans*-2-pentyne **(f)** 3-methyl-3-butene
(g) 2,3-methylhexane **(h)** 3-methyl-3-hexyne
(i) *cis*-2-methyl-2-pentene **(j)** 2-methylcyclohexane
(k) 2-methylcyclohexene **(l)** 4-bromo-*meta*-dichlorobenzene
(m) *para*-phenol

20-32 Draw structural formulas for each of the following compounds. **(a)** isobutane **(b)** ethylene **(c)** acetylene **(d)** benzene **(e)** pyridine **(f)** anthracene **(g)** chloroform **(h)** Freon-12 **(i)** ethylene glycol **(j)** glycerol **(k)** formaldehyde **(l)** acetone **(m)** formic acid **(n)** acetic acid **(o)** citric acid **(p)** aspirin

20-33 Compounds with the formula C_nH_{2n-2} must have two double bonds, or two rings, or one double bond and one ring, or one triple bond. Draw all compounds with the formula C_4H_6. There are nine of them.

20-34 The following compounds are dissolved in water, and the pH adjusted to 2 by the addition of sulfuric acid. Write formulas for the actual species present in each case.

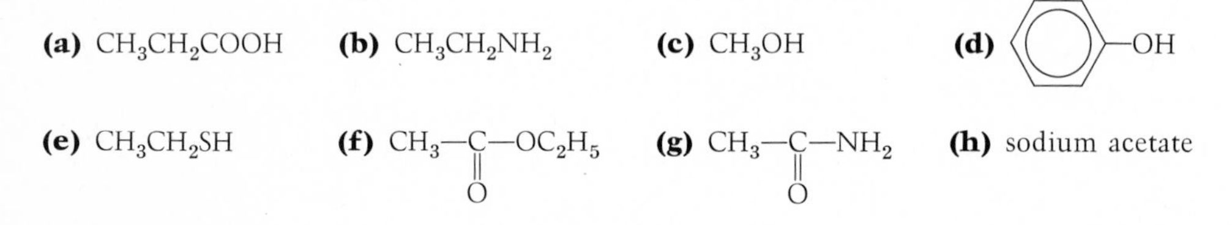

(a) CH_3CH_2COOH **(b)** $CH_3CH_2NH_2$ **(c)** CH_3OH **(d)** C_6H_5—OH
(e) CH_3CH_2SH **(f)** CH_3—C(=O)—OC_2H_5 **(g)** CH_3—C(=O)—NH_2 **(h)** sodium acetate

20-35 The aqueous solutions in Problem 20-34 are now adjusted to pH 13, by the addition of NaOH. Write formulas for the actual species present now.

20-36 Carboxylic esters resemble carboxylic acids of similar molecular weight in solubility in water, but not in boiling point. Explain why.

20-37 Petroleum can be used as a fuel or as a raw material for making many products (Section 20-2). Do you think we should continue our current practice of burning as much of it as we can, or should we try to conserve it for use as a raw material?

20-38 Why is the boiling point of ethyl methyl ether, $CH_3OCH_2CH_3$, lower than that of its isomer 1-propanol, $CH_3CH_2CH_2OH$?

20-39 State which compounds (and reaction conditions if necessary) will allow each of the following reactions to take place.

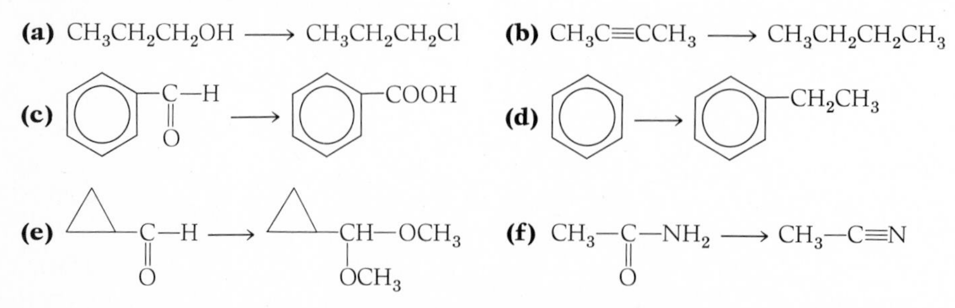

(a) $CH_3CH_2CH_2OH \longrightarrow CH_3CH_2CH_2Cl$ **(b)** $CH_3C{\equiv}CCH_3 \longrightarrow CH_3CH_2CH_2CH_3$
(c) C_6H_5—C(=O)—H $\longrightarrow$ C_6H_5—COOH **(d)** $C_6H_6 \longrightarrow C_6H_5$—$CH_2CH_3$
(e) C_3H_5—C(=O)—H $\longrightarrow$ C_3H_5—CH(OCH_3)—OCH_3 **(f)** CH_3—C(=O)—$NH_2 \longrightarrow CH_3$—C≡N

20-40 See if you can write formulas for the 69 halogen derivatives of methane.

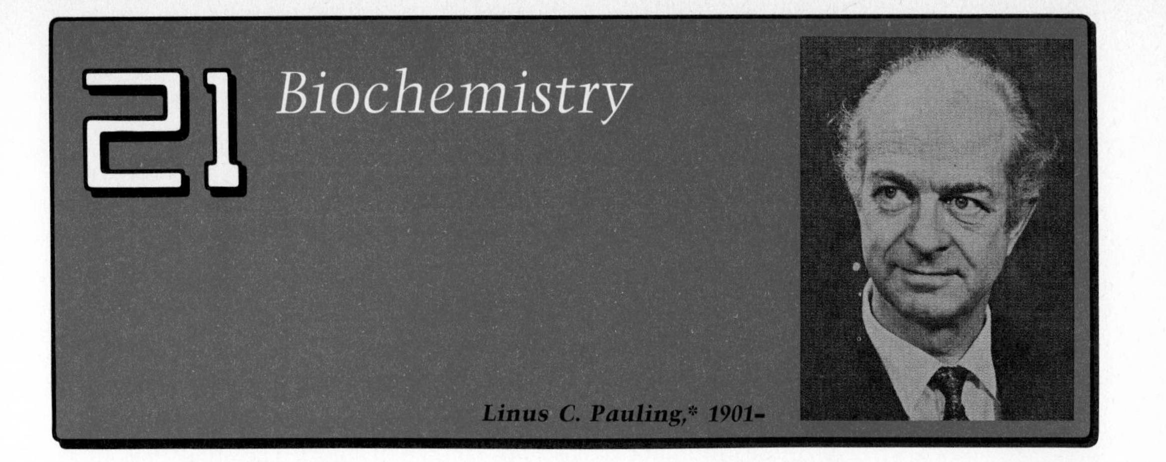

Biochemistry is the chemistry of living organisms. It is an area that is difficult to investigate because there are so many different compounds present and so many reactions taking place, all in the space of very small cells. For this reason our basic understanding of what is going on has come much later than it has in the other major branches of chemistry, and is still far from perfect. Nevertheless, a great deal has been learned in recent years, including one of the most important discoveries of all: the molecular basis of genetics—the actual molecular structure of genes, and how it is that genes are able to carry out their functions. This is discussed in Section 21-5. With all the knowledge that has been gained, there is still a great deal that is unknown, and biochemists have a long way to go before they can say that they have a fairly complete knowledge of the chemistry of any living organism above the level of a virus. In this chapter we will be able to discuss only a few aspects of the chemistry of living things. We will confine ourselves mostly to the three major components of our foods (other than water), which are also major components of our bodies: carbohydrates, proteins, and fats, but we will also discuss nucleic acids and other biological compounds.

21-1 Monosaccharides

In the early nineteenth century, it was found that foods derived from plants contained a type of compound which was given the name **carbohydrate** because analysis showed that only carbon, hydrogen, and oxygen were present, and the ratio of hydrogen atoms to oxygen atoms was always 2:1. Thus, they could be looked upon as being made up of carbon and water (hence: carbohydrate). As we shall see, we know today that water is not actually present in carbohydrates. The oxygen is in the form of OH and ether groups.

The simplest kind of carbohydrates are called **monosaccharides** (we shall define this word later). The most important monosaccharide is D-*glucose*, which, as shown in Figure 21-1, is made up of a six-membered ring containing one oxygen atom, four OH groups directly on the ring (at the carbons numbered 1, 2, 3, and 4), and a CH_2OH group at carbon 5. Like cyclohexane (Figure 20-6) the six-membered ring is in the chair form. This type of six-membered ring is called a **pyranose** ring. An examination of its structure will show that D-glucose has five asymmetric carbons (all except carbon number 6). The total number of stereoisomers is therefore $2^5 = 32$, or 16 pairs of enantiomers. These pairs differ from each other in whether the OH and CH_2OH groups are pointed up or down. For example, Figure 21-2 shows three of the 16 pairs. The enantiomer of D-glucose is called L-glucose.

*From *Structural Chemistry and Molecular Biology* edited by Alexander Rich and Norman Davidson. W. H. Freeman and Company. Copyright © 1975.

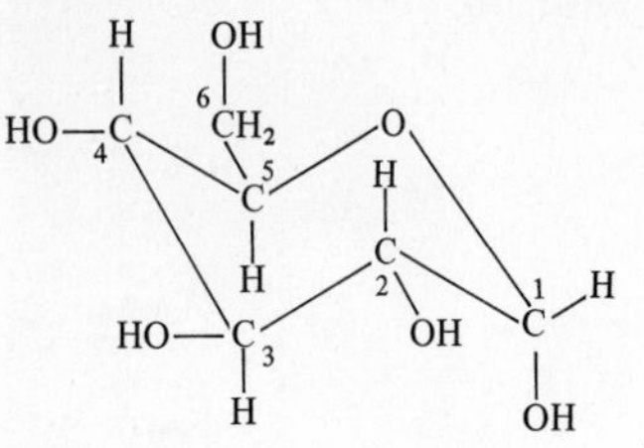

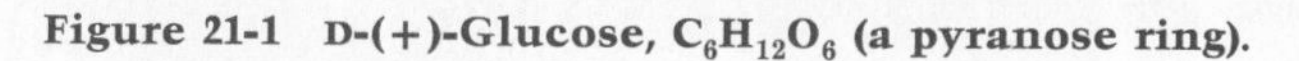

Figure 21-1 D-(+)-Glucose, $C_6H_{12}O_6$ (a pyranose ring).

Obviously, in this molecule, the OH groups (and the CH_2OH group) must be in the same positions as in D-glucose, or it would not be the mirror image. But D-galactose differs from D-glucose in that the OH group at carbon 4 is pointed up, rather than down, while D-gulose differs at two positions (shown in color). The other 26 isomers differ at all other possible combinations. Note that the names of almost all carbohydrates end in **ose.**

The letter D is used for those compounds which have the same configuration as D-glucose *at carbon* 5. L is used for their mirror images. D-Glucose is called D-(+)-glucose because it rotates the plane of polarized light to the right (clockwise). L-Glucose must therefore rotate the plane to the left. Note that there is no correlation between the D designation, which indicates a structural relationship to D-glucose, and the direction of rotation of the plane of polarized light. As indicated in Figure 21-2, D-gulose rotates the plane to the left.

D-Glucose is a solid that is very soluble in water. When it dissolves in water, the pyranose ring opens up at carbon 1, to give a noncyclic compound, which has an aldehyde functional group and five OH groups (Figure 21-3). As shown in the figure, this ring opening takes place when the hydrogen of the C-1 OH group (shown in

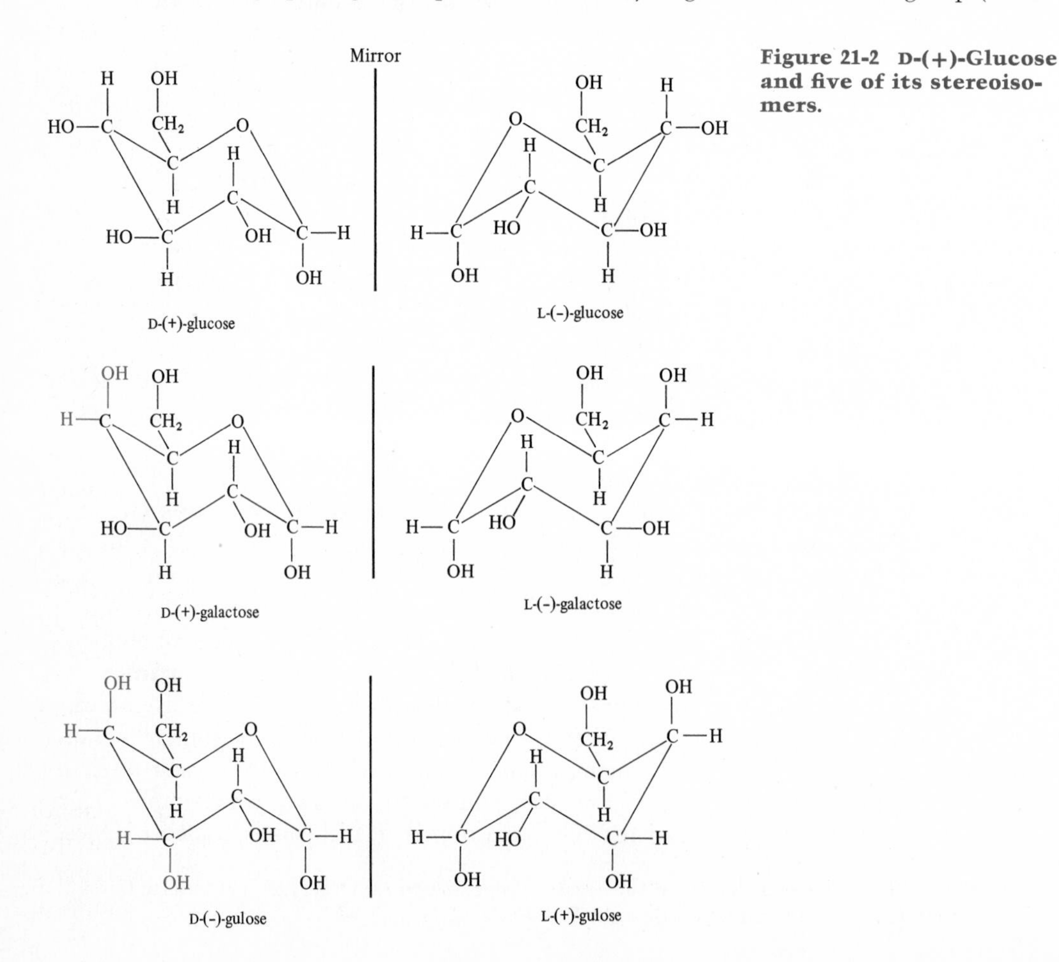

Figure 21-2 D-(+)-Glucose and five of its stereoisomers.

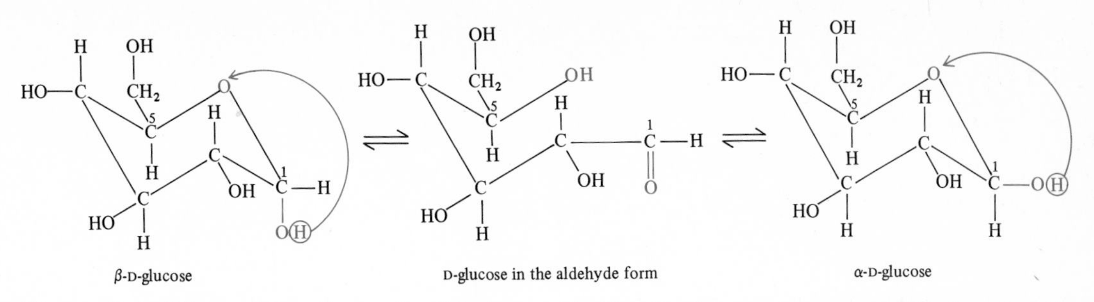

Figure 21-3 When either form of solid glucose is dissolved in water, an equilibrium is set up among the three forms shown. Note that the hydrogen attached to the C-1 OH group (shown in color) of either the α or the β form has migrated to the C-5 OH group in the aldehyde form.

color) migrates to the ring oxygen, which now becomes the C-5 OH group in the aldehyde form. The aldehyde form has only *four* asymmetric carbons: carbon 1 is no longer asymmetric.

another way to write the aldehyde form of D-glucose

H
|
^{1}C=O
|
H—^{2}C—OH
|
HO—^{3}C—H
|
H—^{4}C—OH
|
H—^{5}C—OH
|
$^{6}CH_2OH$

The aldehyde form is not very stable and immediately closes up again to restore the six-membered pyranose ring. But when it does so, the newly formed OH group at C-1 can either go down (which restores the original molecule), or up, which produces a new stereoisomer, shown on the right in Figure 21-3. Although these two stereoisomers are different, they are both called D-glucose. The one with the OH down is given the prefix β; the other is called α. The new one is of course another one of our 32 stereoisomers. Note that this ring opening and closing does not affect the stereochemistry at the other carbons—only at carbon 1. Thus, the configurations at carbons 2, 3, 4 and 5 are the same in all three compounds shown in Figure 21-3. Both α-D-glucose and β-D-glucose can be obtained pure in the solid form. When dissolved in water, either one produces an equilibrium mixture of all three compounds. The aldehyde form cannot be obtained as a solid, and even in solution is present only in very low concentration.

Each of the 32 stereoisomers behaves the same way. They are all solids, and when dissolved in water each forms an equilibrium with an aldehyde form and with another one of the original 32 which differs from the first only in the configuration at C-1. Since all 32 of these compounds have six carbons, they are called **hexoses.** Since they all have an open chain form that contains an aldehyde group, they are called **aldoses.** A word that expresses both ideas is **aldohexose,** and these 32 compounds are aldohexoses.

As we have indicated, D-glucose (including both the α and the β form, since both are present in water solution) is the most important monosaccharide. It is synthesized by almost all living organisms, both animals and plants. As we shall see, it is also a major constituent of disaccharides and polysaccharides. An older name for D-glucose is *dextrose*. This name was given when it was first discovered that D-glucose rotates the plane of polarized light to the right. It is obvious that this name does not apply to L-glucose. Of the other 30 aldohexoses, most are not found in

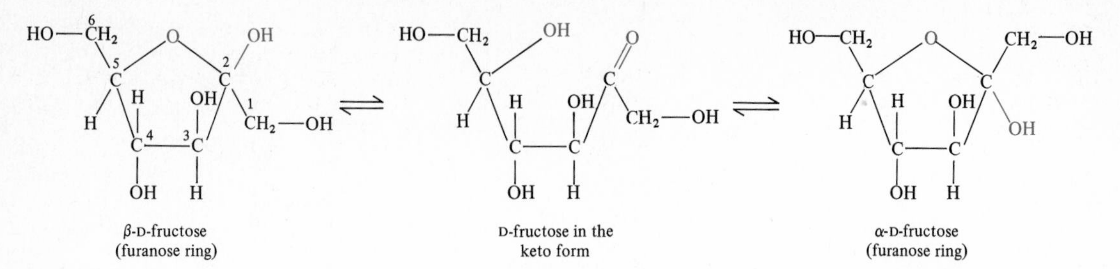

Figure 21-4 When either form of solid fructose is dissolved in water, an equilibrium is set up among the three forms shown.

nature but have been synthesized in laboratories. Despite the widespread occurrence of D-glucose, its mirror image, L-glucose, is not found in nature. This example emphasizes that when nature makes a chiral compound, it usually makes only one of the two enantiomers.

Most monosaccharides are aldoses, but there is another type, called **ketoses.** When these compounds dissolve in water, they open up to give ketones rather than aldehydes. The most important is D-(−) fructose (Figure 21-4), also called *levulose.* In this case, it is C-2 which loses its configuration when the ring is opened. Since it has six carbons, D-fructose is a **ketohexose.** (How many ketohexoses are there?) Instead of a pyranose ring, D-fructose has a five-membered ring, called a **furanose** ring.

In addition to the hexoses, there are several important monosaccharides with only five carbons **(pentoses).** One of these, shown in Figure 21-5, is called D-*ribose.* D-Ribose is an aldopentose. Like D-fructose, D-ribose has a furanose ring. We shall meet D-ribose again in Section 21-5.

We are now ready to define monosaccharide. A **monosaccharide** is a compound containing an unbranched chain of carbons, each of which has an OH group, except that there is a C=O group at either C-1 (aldoses) or C-2 (ketoses); or is a cyclic derivative of such a compound. Monosaccharides are also called **sugars,** and most of them are sweet, though the degree of sweetness varies; some are sweeter than others. Two of them, D-(+)-glucose and D-(−)-fructose, are used to sweeten various foods, including cakes and candies. However, ordinary table sugar is not a monosaccharide at all, but a disaccharide, as we shall see in the next section.

21-2 Disaccharides

Disaccharides are made up of two monosaccharide units joined together in an ether linkage. That is, two OH groups, one from each monosaccharide, combine to form water and an ether group:

Figure 21-6 shows the structures of four important disaccharides. Let us first look at maltose. Maltose is made of two molecules of D-glucose. The ether linkage has been

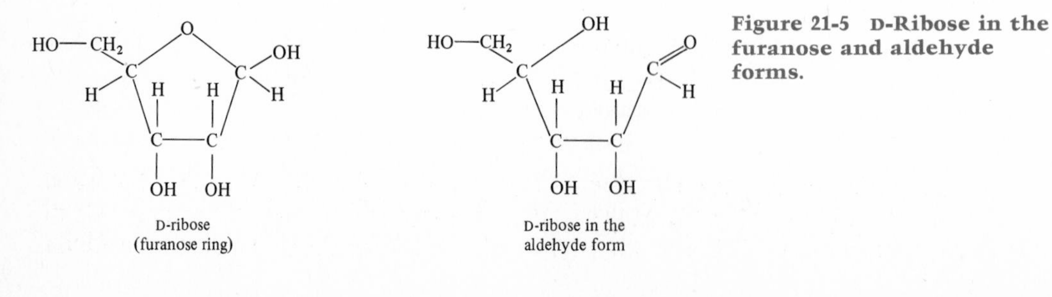

Figure 21-5 D-Ribose in the furanose and aldehyde forms.

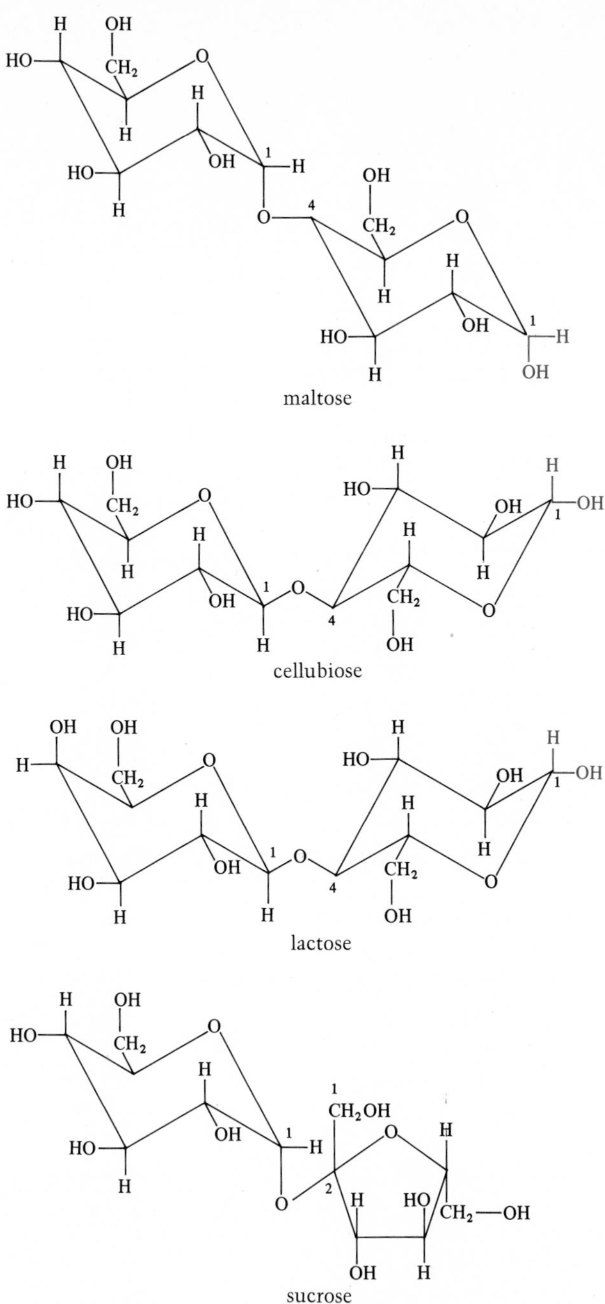

Figure 21-6 Some important disaccharides. The groups shown in color will be interchanged when the compounds are dissolved in water.

formed between the C-1 OH group of the ring on the left and the C-4 OH group of the ring on the right. When a monosaccharide aldose uses its C-1 OH group to form an ether linkage, as in maltose, it "locks" the ring. The original OH group has been converted to an OR group (where R is the entire second ring):

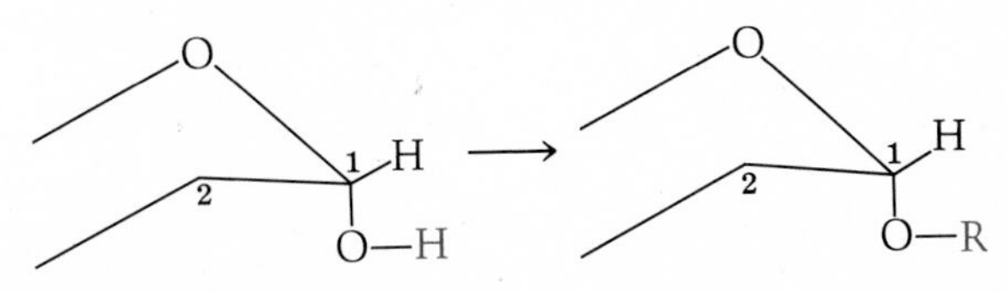

Note that the C-1 carbon is now connected to two OR groups, so that an acetal function (Section 20-8) is present. This change from an OH group to an acetal produces a major change in the properties of this ring. Since the H that formerly migrated to the ring oxygen has been replaced by R, the ring can no longer open up when dissolved in water. It is now "locked," so the OR is frozen into the α position. In maltose, the *second* ring (the one on the right) is not locked. This ring has used its C-4 OH group to form the ether, so the C-1 still has its OH group, and both α and β forms will be present in water solution. In all disaccharides at least one ring is locked.

Thus, for any disaccharide, the questions are: (1) what two monosaccharide molecules are present; (2) which OH groups are connected; (3) are both rings locked, or just one; and (4) is the locked ring (or rings) locked in the α or β position? Table 21-1 supplies the answer to these questions for the four disaccharides shown in Figure 21-6.

Maltose, which is a hydrolysis product of starch, is added to many foods, including baby formulas, and is also used in the manufacture of beer. As we have noted, it is made up of two units of D-glucose, connected 1–4; and the locked ring has the C-1 frozen in the α position (OR pointing down). *Cellubiose,* which is a hydrolysis product of cellulose, is also made up of two units of D-glucose, also connected 1–4, but in this case C-1 of the locked ring is frozen in the β position (H pointing down). This difference is by no means trivial. It makes a considerable difference in the properties of these two disaccharides. As can be seen in Figure 21-6, the fact that the OR points up in cellubiose means that the second ring is upside down with respect to the first rather than rightside up, as in maltose.

Lactose, also called *milk sugar,* is found in cow's milk and in human milk. The locked ring here is D-galactose, the unlocked ring D-glucose, the connection is 1–4, and the configuration at C-1 of the locked ring is β.

The fourth disaccharide in Figure 21-6 and Table 21-1 is *sucrose*. This is the sugar we are all familiar with. It is often called *cane sugar,* though it is obtained from sugar beets as well as from sugar cane. Sucrose differs from the three other disaccharides we have mentioned in that both rings are locked, and that one monosaccharide is a ketose. The C-1 of a D-glucose ring (in the α position) is connected to the C-2 of a D-fructose ring (in the β position). It was discovered more than a century ago that sucrose can be hydrolyzed by certain enzymes or by dilute acids to give D-glucose and D-fructose. This process is called *inversion of sucrose,* because sucrose rotates the plane of polarized light to the right, while an equimolar mixture of D-glucose and D-fructose rotates the plane to the left (D-glucose rotates +, but D-fructose rotates −, and by a greater amount). Because of this, an equimolar mixture of D-glucose and D-fructose is called *invert sugar.*

The four disaccharides we have discussed all have the molecular formula $C_{12}H_{22}O_{11}$, which in fact must be the case for any disaccharide made up of two hexose units:

$$[C_6H_{12}O_6 + C_6H_{12}O_6] - H_2O = C_{12}H_{22}O_{11}$$

Table 21-1 Description of the Four Disaccharides Shown in Figure 21-6

	Locked ring	*Configuration at C-1*	*Rings connected at:*	*Unlocked ring*
Maltose	D-Glucose	α	1–4	D-Glucose
Cellubiose	D-Glucose	β	1–4	D-Glucose
Lactose	D-Galactose	β	1–4	D-Glucose
Sucrose	D-Glucose	α	1–2	D-Fructose (locked in the β position)

Sucrose is not the only disaccharide that can be hydrolyzed. The acetal linkage of all disaccharides can be hydrolyzed either with dilute acid or by various enzymes. When disaccharides are digested in the body, the first step consists in this hydrolysis, which converts one molecule of a disaccharide to two molecules of monosaccharide.

21-3 Polysaccharides

In the disaccharide maltose, one ring is locked, the other unlocked. We can conceive that the unlocked end could be connected to a third monosaccharide ring, giving a *trisaccharide;* the addition of a fourth ring would give a *tetrasaccharide;* and so on. A long chain of rings connected in this way is called a **polysaccharide.** A polysaccharide is an example of a **polymer,** which is a large molecule made up of many repeating units (we will discuss other polymers later in this chapter and in Chapter 22).

A number of polysaccharides are found in nature, some containing D-galactose, some containing D-fructose, some containing other monosaccharides, but by far the most important are *starch* and *cellulose,* both of which contain only D-glucose. Starch is synthesized by both animals and plants (animal starch is called *glycogen*) and is used by both for food storage. Many of the foods that we eat are rich in starch, including potatoes, rice, spaghetti, bread, and corn. Starch makes up a large fraction of the total food consumed by human beings and by animals. It is easily digested in the body, and we gain most of our food energy from this source. Cellulose is even more widespread in nature than starch. It is synthesized by practically all plants (not animals), and is used by them as a structural component: it is the substance that provides rigidity to a tree trunk, a flower stem, or a blade of grass. The purest natural form of cellulose is cotton. Wood is mostly cellulose (along with a noncarbohydrate polymer called *lignin,* which imparts increased rigidity). We take advantage of the structural properties of cellulose by using wood as a building material, and paper, which is made from wood, in books, cardboard boxes, and so on. Unlike starch, cellulose cannot be digested by human beings or indeed by any animal above the level of a microorganism. (Cows eat grass and termites eat wood, but these animals do not themselves digest the cellulose. It is microorganisms that live inside them which accomplish this job.)

The properties of starch and cellulose are thus entirely different. Neither could fulfill the other's function. Neither the plants nor we could use starch as a structural material. We cannot digest cellulose. Yet the *structures* of both are remarkably similar. As we have already noted, both starch and cellulose are entirely composed of D-glucose units. In both of them, the C-1 position of each locked ring is usually connected to the C-4 position of the next ring. What, then, is the difference? The difference is that each unit in starch is connected to the next by an α linkage, while each unit in cellulose is connected by a β linkage. Remember that maltose and cellubiose, obtained from starch and cellulose, respectively, also differ in this way, and even though only two units are present in each molecule, their properties are quite different. In starch and cellulose, the difference between the α and β connections causes a major difference in the entire shapes of the two chains. As shown in Figure 21-7, starch is a steplike molecule with all the units parallel to each other, while cellulose has a more linear pattern, and every other ring is upside down.

There is also a difference in the average size of starch and cellulose molecules. In both starch and cellulose not all molecules are the same size. This is typical of most polymers (though not of most proteins or nucleic acids; see Sections 21-4 and 21-6). A given molecule might consist of 827 units while the one next to it might have 953 units, and the next one 785 units. When we measure their sizes, the best we can get

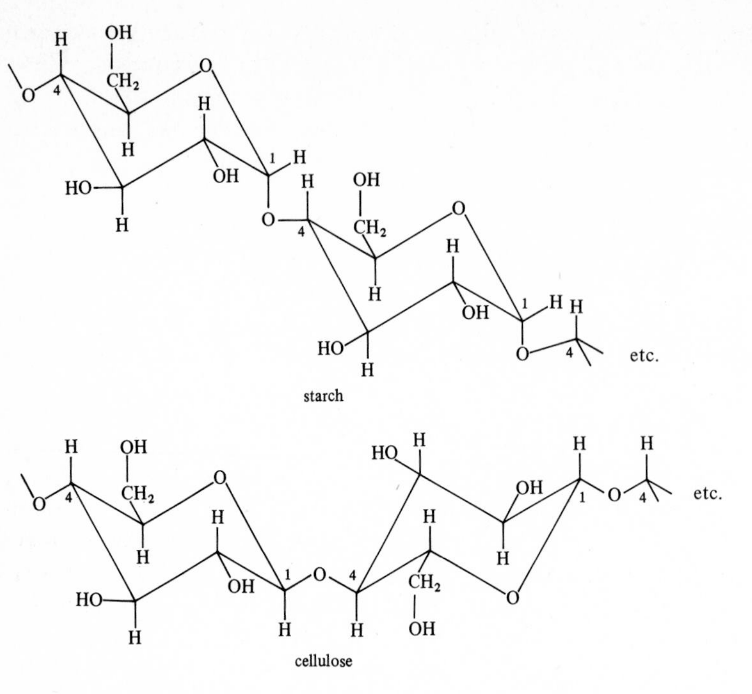

Figure 21-7 Portions of starch and cellulose chains.

is an average. The average size of cellulose molecules depends on which plant we get it from, but a typical average might be 5000 glucose units per molecule. Starch molecules are smaller, with perhaps 1000 to 4000 units in each molecule. Both starch and cellulose molecules also contain *branching*. At various points in the chain another glucose unit forms an ether linkage with the OH group at C-6, to form a whole side chain (Figure 21-8).

Human beings do not possess the enzymes necessary to digest cellulose, but like the disaccharides, cellulose can be completely hydrolyzed to D-glucose by heating with dilute acids, and we can, of course, digest D-glucose. Why, then, don't we solve the world's food problems (at least partially; we would still need proteins) by hydrolyzing wood, grass, cornhusks, and so on? The answer very simply is cost. The heat necessary for the hydrolysis is expensive, and so is the cost of separating the D-glucose from the lignin and other components of the vegetation. At this time it is much cheaper to grow wheat, corn, and other forms of starch than it is to hydrolyze cellulose. However, this may not always be true in the future, and it is well to remember that there is a great deal of cellulose growing all over the world.

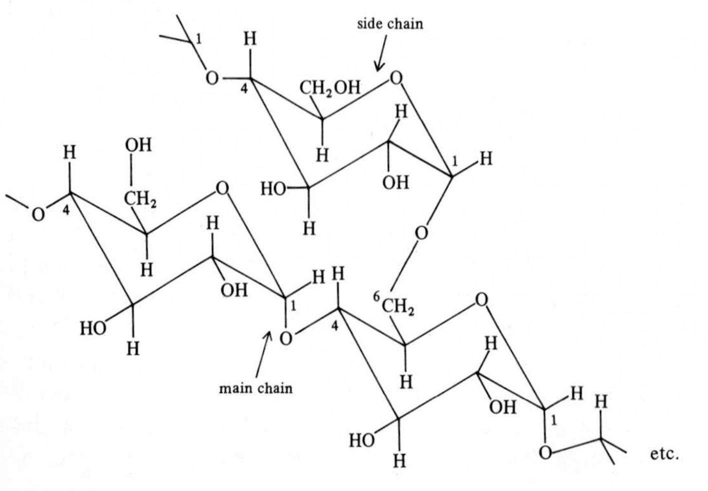

Figure 21-8 A portion of a starch molecule with a side chain.

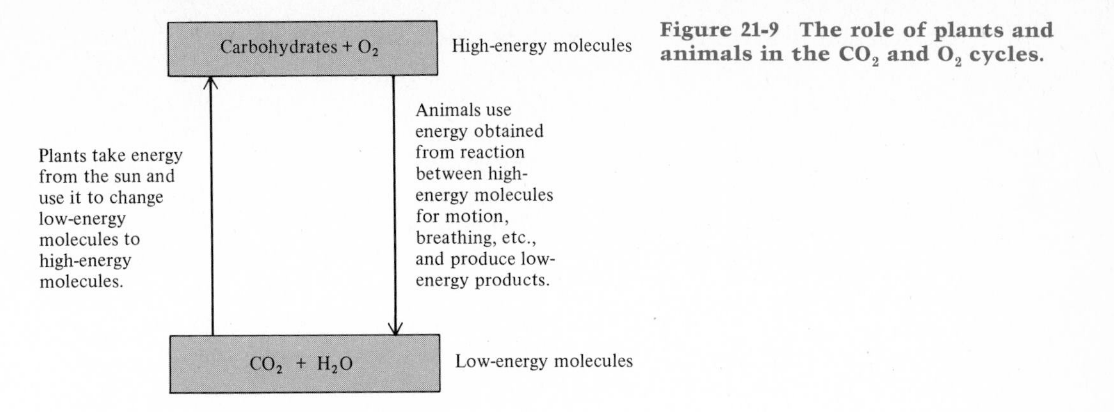

Figure 21-9 The role of plants and animals in the CO_2 and O_2 cycles.

The basic source of carbohydrates in nature is the process called **photosynthesis.** Plants synthesize monosaccharides, chiefly glucose, and then combine the monosaccharide molecules into disaccharides and polysaccharides. The photosynthesis of glucose is very complicated, consisting of many individual reaction steps. Because of this it has not been easy to unravel, and not all of the process is completely understood. The overall reaction may be written

$$6CO_2 + 6H_2O \xrightarrow[\text{chlorophyll}]{\text{light}} C_6H_{12}O_6 + 6O_2$$

Chlorophyll, which is a mixture of green pigments found chiefly in the leaves of plants, is absolutely necessary for the process, as is sunlight. The chlorophyll absorbs energy from the light, raising it to an excited state and transfers this energy to certain molecules present in the leaf cell: these molecules are then capable of combining with CO_2 and H_2O to produce other molecules, which are subsequently converted to D-glucose. Molecules of O_2 are given off in the process (for a further discussion, see Section 23-10). As we have already mentioned, animals also synthesize glucose, but this is an entirely different route, usually beginning with lactic acid. Neither chlorophyll nor ultraviolet light is required.

The photosynthesis of carbohydrates by plants is the major way in which energy enters the world of living things. Energy comes from the sun, is absorbed by the plants, and is stored in the form of carbohydrates. Animals obtain their energy by eating the plants, and metabolizing the carbohydrates, or by eating animals that have eaten the plants. These processes are accompanied by cycling of CO_2 and O_2. Plants emit oxygen, which is formed in the photosynthesis reaction; animals breathe it in and use the oxygen to metabolize the carbohydrates they have eaten, deriving energy from this reaction. Animals breathe out CO_2, which is then reabsorbed by the plants and turned into carbohydrates (Figure 21-9).

21-4 Amino Acids and Proteins

Proteins are biological polymers which are found in all living organisms, from viruses to human beings, and which perform many functions. All **enzymes** are proteins, and every living cell contains hundreds or thousands of different enzymes. These enzymes catalyze virtually every chemical reaction that takes place within the organism. All activities of the organism, both conscious (for example, moving an arm, opening an eye, reading a book) and unconscious (for example, digestion of food, manufacture of new blood cells, maintenance of body temperature), are conducted with the aid of enzymes. Enzymes are remarkably *specific* catalysts. Many different enzymes are needed because each enzyme generally catalyzes only one reaction, and a different enzyme is needed for each process. Some enzymes act on only one specific compound, others on one specific class of compounds. A compound acted on by an enzyme is called a **substrate.** Enzymes are very efficient catalysts. Chemists in laboratories can almost never carry out a given reaction as fast as an enzyme can catalyze that same reaction within an organism. Enzyme-catalyzed reactions are often 10^9 to 10^{20} times faster than reactions run in laboratories. Furthermore, enzymes function at body temperatures, while chemists often must heat reactants to relatively high temperatures in order to get them to proceed at a convenient rate.

It is not too much to say that we are what our enzymes allow us to be. If we have brown eyes, it is because we have the enzymes necessary to synthesize the brown pigment. If the microorganisms found in a termite can digest cellulose, it is because they have the enzymes that can catalyze the reactions necessary for this purpose. Except for identical twins, no two people have exactly the same enzymes, though most individual human enzymes are present in all human beings. If we look at other species, differences in the enzymes present become greater as the species get further away from each other on the evolutionary scale.

Not all proteins are enzymes. Proteins also carry out other functions, one of which is structural. Such things as animal hair, fur, fingernails, horns, etc. are made of proteins. These proteins are called **fibrous** proteins. Wool and silk are two important fabrics which consist of fibrous proteins (note the difference from cotton, which is cellulose). Enzymes and other proteins found in cells are called **globular** proteins.

All proteins are polymers made up of chains of amino acids connected to each other by amide linkages. An **amino acid** is a compound containing a carboxylic acid functional group $-\underset{\underset{\displaystyle O}{\|}}{C}-OH$ and an amino group $-NH_2$ or $-NHR$. Although chemists have synthesized hundreds of amino acids, nature uses only about 20 of these to make proteins. All of these 20 (shown in Table 21-2) are α-amino acids, meaning that the amino group is attached to the same carbon as the COOH group. Therefore, all 20 of the amino acids found in proteins have the formula

$$R-\underset{\underset{\displaystyle NH_2}{|}}{\overset{\overset{\displaystyle H}{|}}{C}}-\overset{\overset{\displaystyle O}{\|}}{C}-OH$$

except proline, and even proline is not very different: the same formula can represent proline if we connect R and NH_2. Each of the 20 amino acids has a COOH group and an NH_2 or NHR group. We saw on page 687 that a COOH compound and an NH_2 or NHR compound could, in theory, combine, with water splitting out, to give an amide. If the two molecules that combine in this fashion are amino acids, the resulting amide (called a **dipeptide**) has a COOH and an NH_2 or NHR group still remaining.

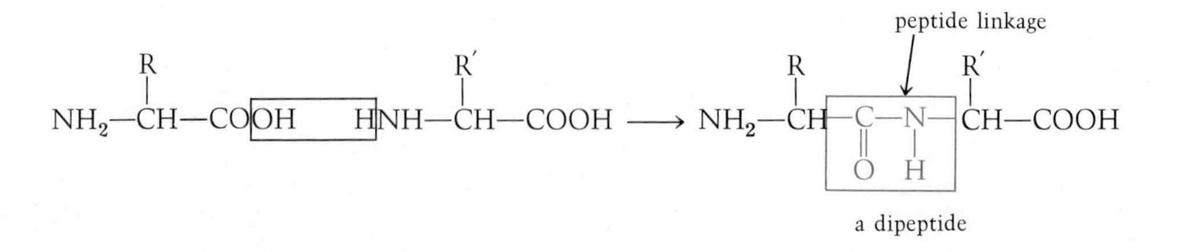

Table 21-2 The Twenty Amino Acids Commonly Found in Proteins[a]

Name	Abbreviation	Structure
Glycine	Gly	$H—CH—COOH$ NH_2
Alanine	Ala	$CH_3—CH—COOH$ NH_2
*Valine	Val	$CH_3—CH—CH—COOH$ CH_3 NH_2
*Leucine	Leu	$CH_3—CH—CH_2—CH—COOH$ CH_3 NH_2
*Isoleucine	Ile	$CH_3—CH_2—CH—CH—COOH$ CH_3 NH_2
*Phenylalanine	Phe	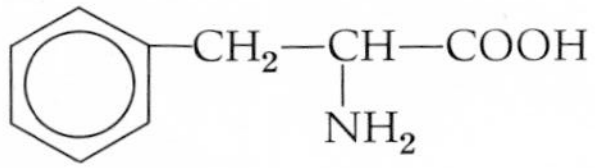 $—CH_2—CH—COOH$ NH_2
Tyrosine	Tyr	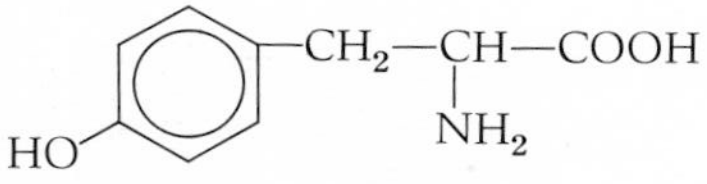 $—CH_2—CH—COOH$ HO NH_2
Serine	Ser	$HO—CH_2—CH—COOH$ NH_2
*Threonine	Thr	$CH_3—CH—CH—COOH$ OH NH_2
Cysteine	Cys	$HS—CH_2—CH—COOH$ NH_2
*Methionine	Met	$CH_3—S—CH_2CH_2—CH—COOH$ NH_2
*Tryptophan	Trp	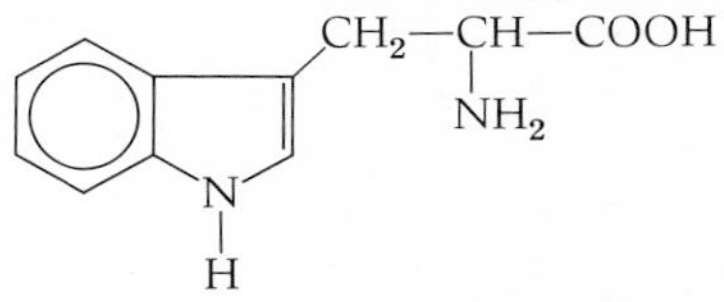 $—CH_2—CH—COOH$ NH_2 N—H
Proline	Pro	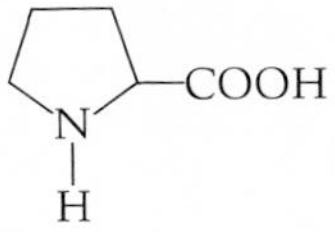 —COOH N—H
Asparagine	Asn	$NH_2—C—CH_2—CH—COOH$ O NH_2
Aspartic acid	Asp	$HO—C—CH_2—CH—COOH$ O NH_2
Glutamine	Gln	$NH_2—C—CH_2CH_2—CH—COOH$ O NH_2
Glutamic acid	Glu	$HO—C—CH_2CH_2—CH—COOH$ O NH_2
*Lysine	Lys	$NH_2—CH_2CH_2CH_2CH_2—CH—COOH$ NH_2
*Arginine	Arg	$NH_2—C—NH—CH_2CH_2CH_2—CH—COOH$ NH NH_2
*Histidine	His	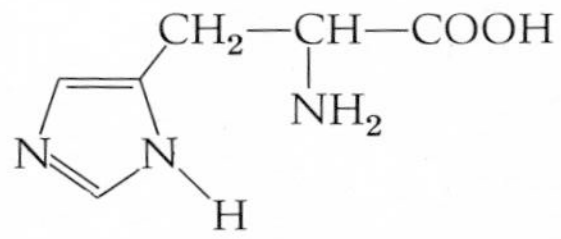 $CH_2—CH—COOH$ NH_2 N N—H

[a] Those marked with an asterisk are *essential amino acids,* meaning that the human organism cannot synthesize them in sufficient quantities, so they must be present in the diet. The others are also essential to life, but the human organism can make its own supply of these.

Consequently, a third amino acid can in theory be added, to give a tripeptide.

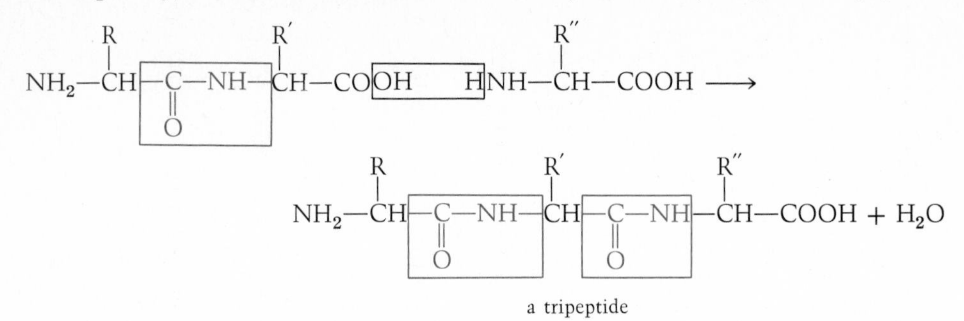

The CONH group (shown in color) is called a **peptide linkage.** The R groups are called *side chains*. In a similar manner, we can have tetrapeptides, pentapeptides, and so on, for very long chains. It is these molecules that are proteins. Note that it is customary to write peptide chains with the free NH_2 group on the left and the free COOH group on the right.

We will come back to proteins shortly, but first there are a few more things to be said about the approximately 20 amino acids that nature uses to make proteins. As can be seen in Table 21-2, the 20 R groups are extremely varied and seemingly random. Some are simple, nonpolar groups such as H, CH_3, and $(CH_3)_2C$ (but note the absence of CH_3CH_2); others are much larger and more complicated. Some have additional functional groups containing oxygen, nitrogen, or sulfur. The last three amino acids on the list (arginine, lysine, and histidine) possess an additional basic group (besides the NH_2 or NHR group which is present in all of them); these are called **basic amino acids.** Two others (glutamic acid, aspartic acid) possess an extra COOH group; these are **acidic amino acids.** The monosodium salt of glutamic acid (monosodium glutamate; MSG) is used as a spice to improve the flavor of meats and other foods. There is no good explanation why nature uses these particular 20 amino acids and not others. Ten of the amino acids (starred in Table 21-2) are called **essential amino acids.** Actually, all 20 are essential for life, but these 10 must be present in the human diet, because the human body cannot synthesize enough of them to meet its needs.

Since all 20 amino acids have the formula

$$R\text{—}\underset{\displaystyle NH_2}{\overset{\displaystyle H}{C}}\text{—}COOH$$

(arrow to C: asymmetric carbon (except when R = H))

then all of them except glycine have an asymmetric carbon, and each can exist in two nonsuperimposable forms. As we might predict, nature uses only one of these, and it is the same one for all the amino acids. This form is called L and is shown in Figure 21-10. Virtually all amino acids found in proteins will superimpose on the L model shown in Figure 21-10. D-Amino acids are almost never found. Some of the amino acids also have another asymmetric carbon (which ones?).

Although it is customary to write free amino acids (that is, not in proteins) as $R\text{—}\underset{\displaystyle NH_2}{CH}\text{—}COOH$, this is not their actual formula. From our discussions on pages 683 and 686, we can see that COOH as acids are strong enough to react with NH_2 as bases. Therefore, the real formula of amino acids is

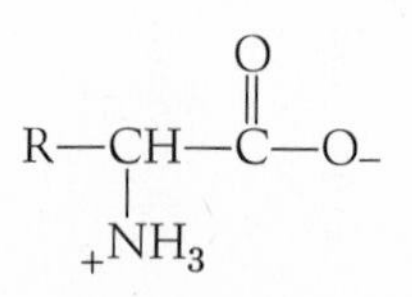

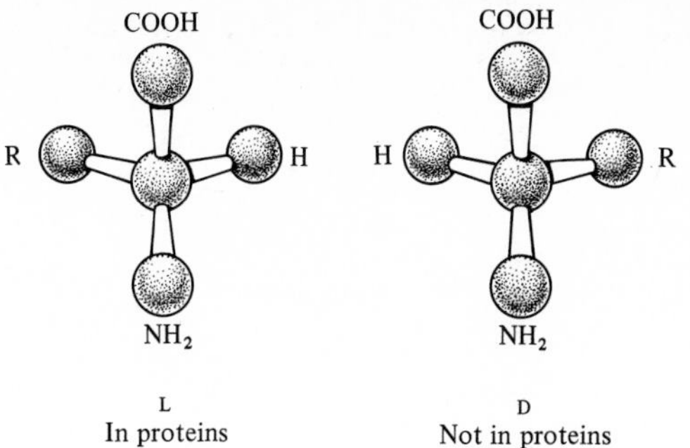

Figure 21-10 Virtually all naturally occurring amino acids have the L configuration.

That is, a proton has been transferred from the COOH to the NH_2 group. Ions that have + and − charges on different atoms are called **zwitterions.**

Some students may be wondering why we said that there are *approximately* 20 amino acids found in proteins. The answer is that it is hard to know just how to count them. For example, asparagine is the amide of aspartic acid (note that NH_2 has replaced OH in a COOH group). Some people would say these count as two; others count them as one. Another example is cysteine. This molecule is usually found in proteins in a dimeric form, called *cystine.*

$$\underset{\text{cysteine}}{2\text{HOOC—}\underset{\text{NH}_2}{\underset{|}{\text{CH}}}\text{—CH}_2\text{—SH}} \longrightarrow \underset{\text{cystine}}{\text{HOOC—}\underset{\text{NH}_2}{\underset{|}{\text{CH}}}\text{—CH}_2\text{—S—S—CH}_2\text{—}\underset{\text{NH}_2}{\underset{|}{\text{CH}}}\text{—COOH}}$$

Do these count as one or two? Your decision is as good as anyone else's. Cystine is very important in the structure of proteins, because it can tie two chains together:

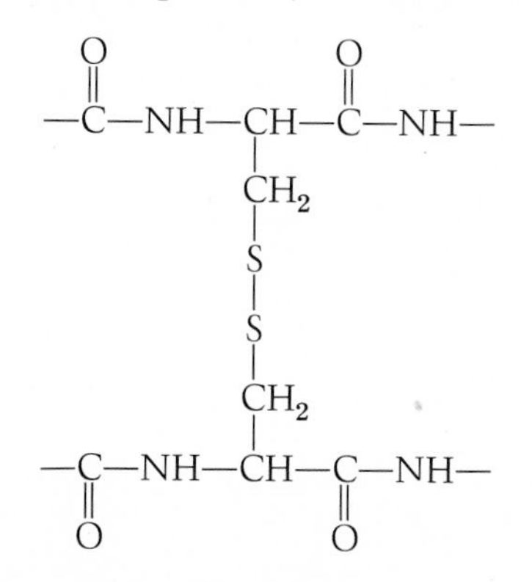

These linkages are called **disulfide bridges.**

We have seen that each individual has thousands of enzymes, and that each enzyme is highly specific in its action, so that an organism can only carry out those reactions for which it has enzymes. All enzymes are proteins, and all proteins are chains of the same 20 amino acids (though not every protein has all 20 amino acids) connected by peptide linkages. What then is the difference between one enzyme molecule and another? The answer is **the sequence in which the amino acids are arranged on the chains.** This fact was unknown 30 years ago, because no protein structures were known at that time. But in 1955, Frederick Sanger (1918–) after more than 10 years of work, was able to establish the structure of insulin (Figure 21-11), and by use of his methods, and of further improvements, the structures of many other proteins have since been determined. In fact, it is now possible to synthesize many proteins, including insulin, in the laboratory (not an easy task, since it will be recalled—page 687—that amides cannot be formed by direct treatment of a —COOH compound with an $—NH_2$ compound).

^{1}Phe		^{1}Gly	
^{2}Val		^{2}Ile	
^{3}Asn		^{3}Val	
^{4}Gln		^{4}Glu	
^{5}His		^{5}Gln	
^{6}Leu		^{6}Cys	—
^{7}Cys	—S—S—	^{7}Cys	S
^{8}Gly		^{8}Ala	|
^{9}Ser		^{9}Ser	S
^{10}His		^{10}Val	|
^{11}Leu		^{11}Cys	—
^{12}Val		^{12}Ser	
^{13}Glu		^{13}Leu	
^{14}Ala		^{14}Tyr	
^{15}Leu		^{15}Gln	
^{16}Tyr		^{16}Leu	
^{17}Leu		^{17}Glu	
^{18}Val		^{18}Asn	
^{19}Cys	—S	^{19}Tyr	
^{20}Gly	S—	^{20}Cys	
^{21}Glu		^{21}Asn	
^{22}Arg			
^{23}Gly			
^{24}Phe			
^{25}Phe			
^{26}Tyr			
^{27}Thr			
^{28}Pro			
^{29}Lys			
^{30}Ala			

Figure 21-11 Schematic diagram of the insulin molecule. See Table 21-2 for the three-letter abbreviations. The amino acids numbered 1 have the free COOH groups; those at the other end have the free NH_2 groups.

Note that the insulin molecule consists of two chains, one of 21 units, the other of 30 (if we count a cystine molecule as being made up of two cysteine units), held together by two disulfide bridges. A third disulfide bridge connects two parts of the shorter chain. Insulin contains 17 of the 20 amino acids in Table 21-2, so that the 51 units have many duplications (for example, there are 6 leucines, 3 phenylalanines, etc.). In these respects, insulin is a typical protein molecule. In general, enzymes and other protein molecules (except fibrous proteins) have *specific* chain lengths. Most polymers are like starch and cellulose in that each molecule in a given sample has a different number of units. Proteins (except fibrous proteins) are exceptions. Every molecule of any individual protein has the same number of amino acid units, and the amino acids are in the same order. In most proteins, as in insulin, many amino acids are duplicated, and parts of the chain (or chains) are connected by disulfide bridges.

We have said that the action of enzymes is highly specific and that each has a different sequence of amino acids. We may conclude that it is the particular amino acid sequence an enzyme has which allows it to catalyze the one specific reaction that it does catalyze. In doing so, the amino acid sequence performs two functions:

1. It enables the chain to coil up in one and only one specific three-dimensional

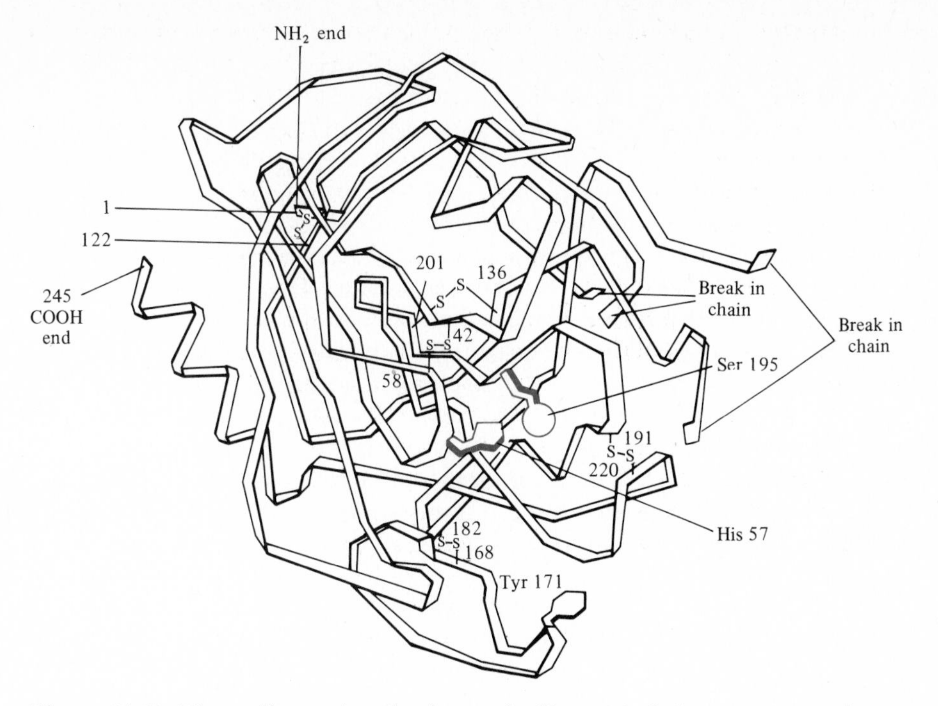

Figure 21-12 Three-dimensional schematic diagram of the enzyme α-chymotrypsin. The single chain of 245 units must be cleaved in two places, with loss of four amino acids, in order for the enzyme to be active. There are five disulfide bridges. The active site is made up of the serine at 195 and the histidine at 57.

shape. For example, Figure 21-12 shows a three-dimensional schematic diagram of the enzyme α-chymotrypsin (this enzyme catalyzes the hydrolysis of other proteins). The α-chymotrypsin molecule consists of a single chain of 245 amino acids, but the enzyme cannot act unless it is coiled exactly in the manner shown in Figure 21-12. In this particular case, action of the enzyme also requires that the chain be broken in two places, with the loss of four amino acids, but this does not affect the coiled shape.* The peptide chain is held in the coiled shape by the disulfide bridges, by hydrogen bonding between various OH, NH groups, etc., on the amino acid side chains, and by van der Waals forces and other electrostatic attractions and repulsions. In this connection it should be mentioned that enzymes act in water solution, which has an important effect on the structure, since polar side chains, as in serine, lysine, glutamic acid, etc., are attracted to the water, and hence turn toward the outer surface of the coil, while nonpolar side chains, as in alanine, leucine, and phenylalanine, turn away from the water and try to bury themselves within the coil. If an enzyme molecule is uncoiled to any significant extent, its catalytic action is completely lost, even though the amino acid sequence remains intact. This process, called **denaturation,** can be brought about by heat, by strong acids or bases, or by a variety of other reagents. A common example of protein denaturation is the boiling of eggs. Some denaturated proteins can spontaneously reverse the denaturation process, and coil up again; others cannot.

2. Although an enzyme chain may contain two or three hundred amino acids,

*The cleavage of the chain in two places raises a very interesting point. The molecule is synthesized in one continuous chain, and then coils up, but is still inactive in this form. It is transported to a different location, where it is activated by the cleavage. It then goes to work, hydrolyzing proteins taken in as food. If it were active in the original cell in which it was synthesized, it would hydrolyze the proteins of that cell, thus destroying it! Thus, it seems that nature has required that the chain be broken before activation in order to protect the organism from self-destruction.

only a few of these are directly involved in the catalytic action of the enzyme. This part of the enzyme is called the **active site.** In α-chymotrypsin (Figure 21-12) the active site consists primarily of two amino acid units—a histidine at 57 and a serine at 195. The other 243 amino acids are necessary to coil the molecule into exactly that shape that brings the 57 His and 195 Ser near each other in exactly that position into which the substrate fits. The substrate, or a portion of it, fits into a cavity at the active site of the enzyme in the same way that a key fits into a lock (this is called the lock-and-key theory of enzyme action). Other nearby portions of the enzyme also attract the substrate, by specific electrostatic attractions. The substrate molecule is able to get into the cavity, undergo reaction, and get out again, very quickly, accounting for the very fast rates of reaction. Substrate molecules with slightly different shapes cannot fit, and hence do not react, accounting for the specificity of enzyme action.

The lock-and-key theory also explains why enzymes in general react with only one enantiomer of a chiral compound and not the other. The cavity in the enzyme is always chiral, and only one enantiomer will fit (Figure 21-13). In order to use the other enantiomer, nature would have to make a complete enzyme which is the enantiomer of the one in nature. Nature does not do this (for one thing, it would require that only D-amino acids be used, and these are not generally found in nature), and this explains the fact that we have previously mentioned: that when making a chiral molecule, nature in general makes only one enantiomer.

The case of α-chymotrypsin illustrates another feature of enzyme specificity. This enzyme has its maximum activity at pH 8; above or below this its activity rapidly decreases. The reason is that below pH 8, the 57 histidine becomes protonated (the free base is necessary for activity), while above pH 8, the NH_2 group at the end of the chain (at position 1) becomes deprotonated. For reasons that are not completely understood, it is necessary that this NH_2 group be in its ionic form, NH_3^+.

The mechanism of enzyme action has been intensively investigated for several years, and only recently have the main outline become clear. There is a great deal still to be learned.

Many enzymes and other proteins carry out their functions without assistance, but others need the help of certain other molecules, called **coenzymes.** Most *vitamins* are either coenzymes or are converted by the body to coenzymes (vitamins are defined in Section 21-8).

Unlike enzymes and other globular proteins that are present in aqueous solutions within cells, fibrous proteins have molecules that are not equal in size. These

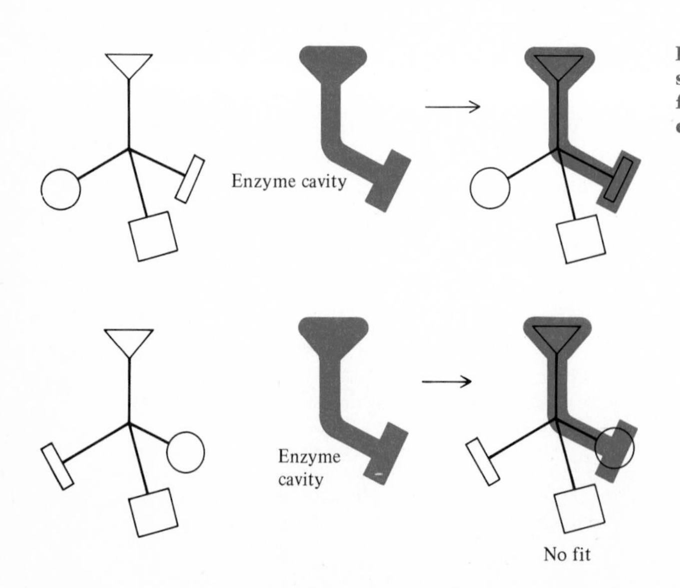

Figure 21-13 Schematic diagram showing how one chiral molecule fits into an enzyme cavity, but its enantiomer does not.

proteins also differ from those in solution in that their three-dimensional structures are not globular coils, but more regular extended geometrical shapes, as in cellulose, which is not surprising since the functions performed by these proteins (they lend structural rigidity) are similar to those of cellulose. The most common structures of fibrous proteins are the **alpha-helix** (Figure 21-14) and to a lesser extent, the **pleated sheet** (Figure 21-15). Both of these structures were discovered by the x-ray crystallography work of Linus Pauling (1901–) and Robert B. Corey (1897–). Among the fibrous proteins that adopt the alpha-helix configuration is *keratin*, which is the chief compound of hair. As shown in Figure 21-14, the alpha-helix is held together by hydrogen bonds, which connect the C═O of each peptide unit to the N—H of the peptide unit four amino acids farther down. Note that the amino acid proline cannot be present in an alpha-helix. If a proline is present in the chain, it disrupts the helix. It may also be noted that an alpha-helix can only be constructed if all the amino acids are L (or all D). It cannot be made from a mixture of D and L amino acids. One fibrous protein that has the pleated sheet structure is silk. Note (Figure 21-15) that hydrogen bonding is present here, too, but in this case between chains.

The sequence of amino acids in a protein is called the **primary** structure of the protein. The way the chain is folded or coiled (for example, into a helix or pleated sheet) is the **secondary** structure of the protein. In most enzymes and other proteins found in cells, portions of the chains adopt the alpha-helix structure, perhaps for ten or twenty units. The further folding or twisting of the already coiled chain is called the **tertiary** structure.

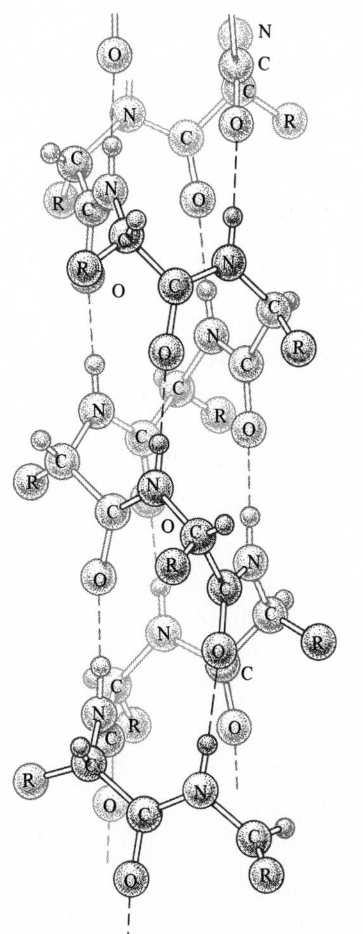

Figure 21-14 The alpha-helix configuration of hair. [From *Chemistry* by Linus Pauling and Peter Pauling. W. H. Freeman and Company. Copyright © 1975.]

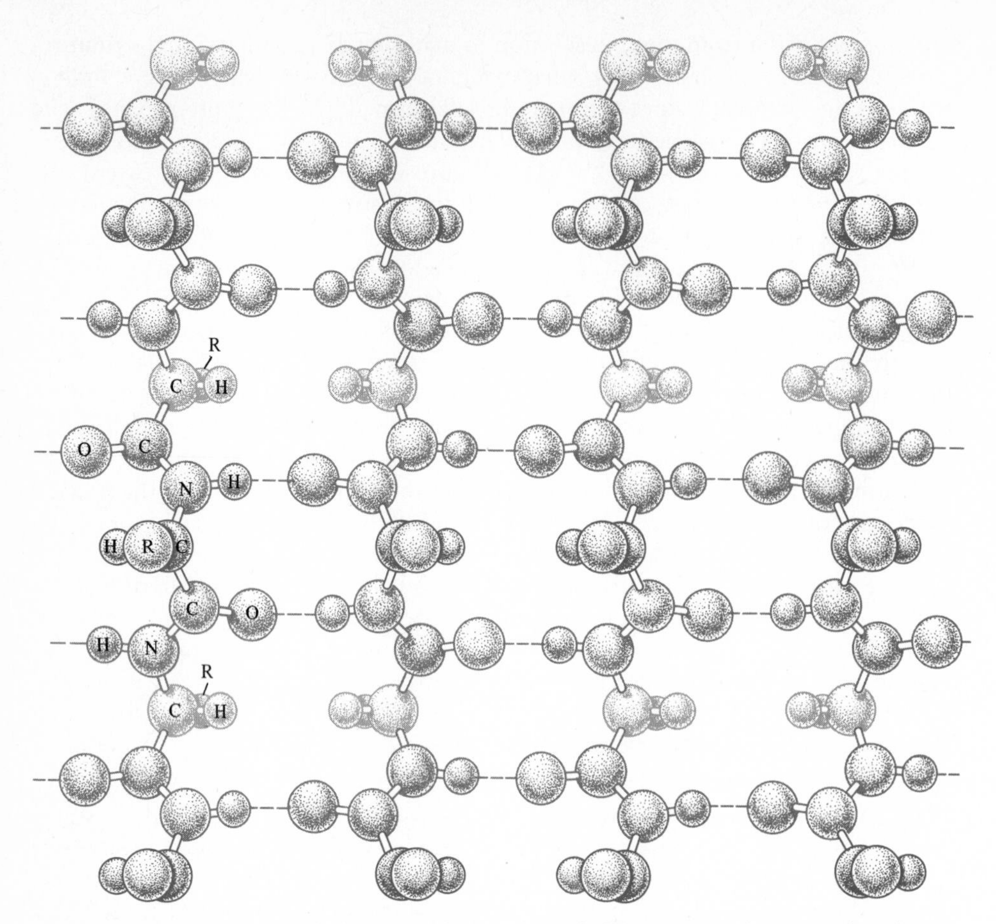

Figure 21-15 The pleated-sheet configuration of silk. [From *Chemistry* by Linus Pauling and Peter Pauling. W. H. Freeman and Company. Copyright © 1975.]

21-5 Nucleic Acids

Nucleic acids are polymers found in the nucleus of every living cell. There are basically two kinds: **ribonucleic acid (RNA)** and **deoxyribonucleic acid (DNA).** The polymer chain in RNA is made up of alternating units of the monosaccharide D-ribose (Figure 21-5) and phosphoric acid. As shown in Figure 21-16, each phosphoric acid is connected, by loss of water, to the C-3 OH group of one ribose molecule and to the C-5 OH group of another. Each ribose unit is connected to two phosphate units, one at C-3 and the other at C-5. DNA has the same basic structure as RNA, except that the sugar involved is deoxyribose (Figure 21-17), which is the same as ribose, except that the C-2 OH group has been replaced by H. The OH group at the C-1 in ribose and deoxyribose is not found in RNA or DNA. It has been replaced by certain heterocyclic molecules, called *bases* (see below). Thus, RNA has only one free OH group per ring, while DNA has none at all. The molecules are acidic because one of the three OH groups of each phosphoric acid unit is still there (although two have been replaced). This is why they are called nucleic *acids*.

At each C-1 position of DNA and RNA are found certain bases. Just as nature uses 20 amino acids to make proteins, it uses four bases to make RNA. These, shown in Figure 21-18, are two *purines*, called *adenine* and *guanine*, and two pyrimidines (Section 20-5), called *cytosine* and *uracil*. Three of these four are also found in DNA, but DNA does not contain uracil. The fourth base in DNA is a methyl

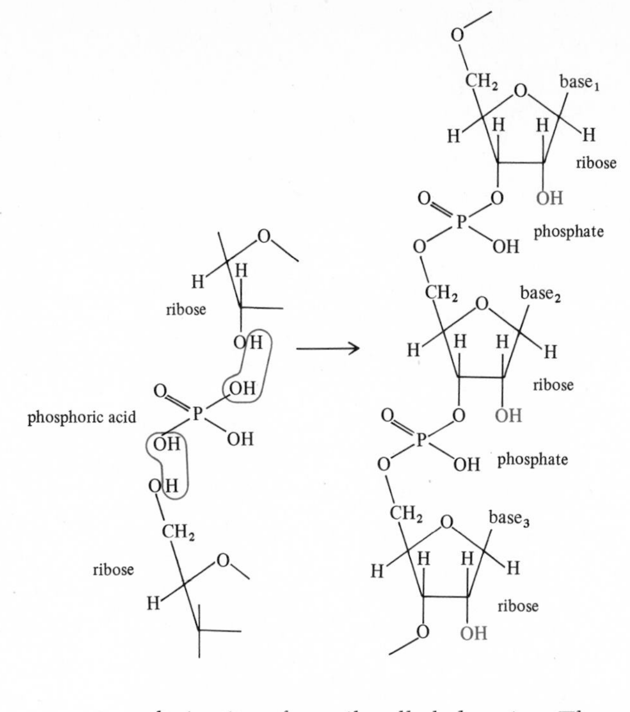

Figure 21-16 A portion of an RNA molecule. A DNA molecule has the same structure, except that the OH group at the 2 position in each ribose (shown in color) is replaced by H.

derivative of uracil, called *thymine*. These bases are attached to the C-1 carbons of DNA and RNA at the positions shown in color in Figure 21-18. The purines are attached by one of the nitrogens on the five-membered ring, while the pyrimidines use one of the two nitrogens of the six-membered ring. Figure 21-19 shows a portion of a DNA molecule with bases attached. All five of these bases can exist in two different forms in the free state (Figure 21-18), and the actual compounds are equilibrium mixtures of the two forms. In one form, the six-membered ring is aromatic; in the other, it is not. When the pyrimidines are part of the DNA or RNA molecules only the nonaromatic form can exist (why?), and cytosine, uracil, and thymine are always in this form when part of the nucleic acid chain. The purines can exist in both forms even when they are part of the chain. (Why?) The phosphate and ribose units make up the *backbone* of the chain, from which the bases project.

In Section 21-4, we saw that the difference between one protein molecule and another was the different sequence of amino acid units. The situation is the same with nucleic acids. The difference between one DNA (or RNA) molecule and another is the different sequence of bases. In Section 21-6, we shall see that there is a fundamental relationship between the amino acid sequence in proteins and the base sequence in DNA.

In 1953, James Watson (1928-   ) and Francis Crick (1916-   ) used x-ray crystallography data obtained by Maurice Wilkins (1916-   ) and other methods to show that DNA exists as a **double helix.** Each DNA molecule consists of *two* chains wrapped around each other, as shown in Figure 21-20. Each of the two chains has the structure shown in Figure 21-19. The backbone of each chain is a helix, and the chains are held together by hydrogen bonds between the bases. The most important

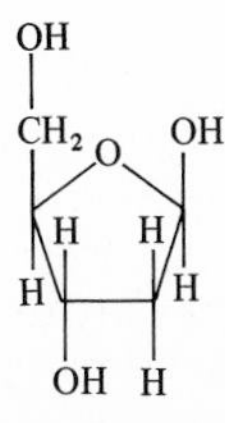

Figure 21-17 Deoxyribose.

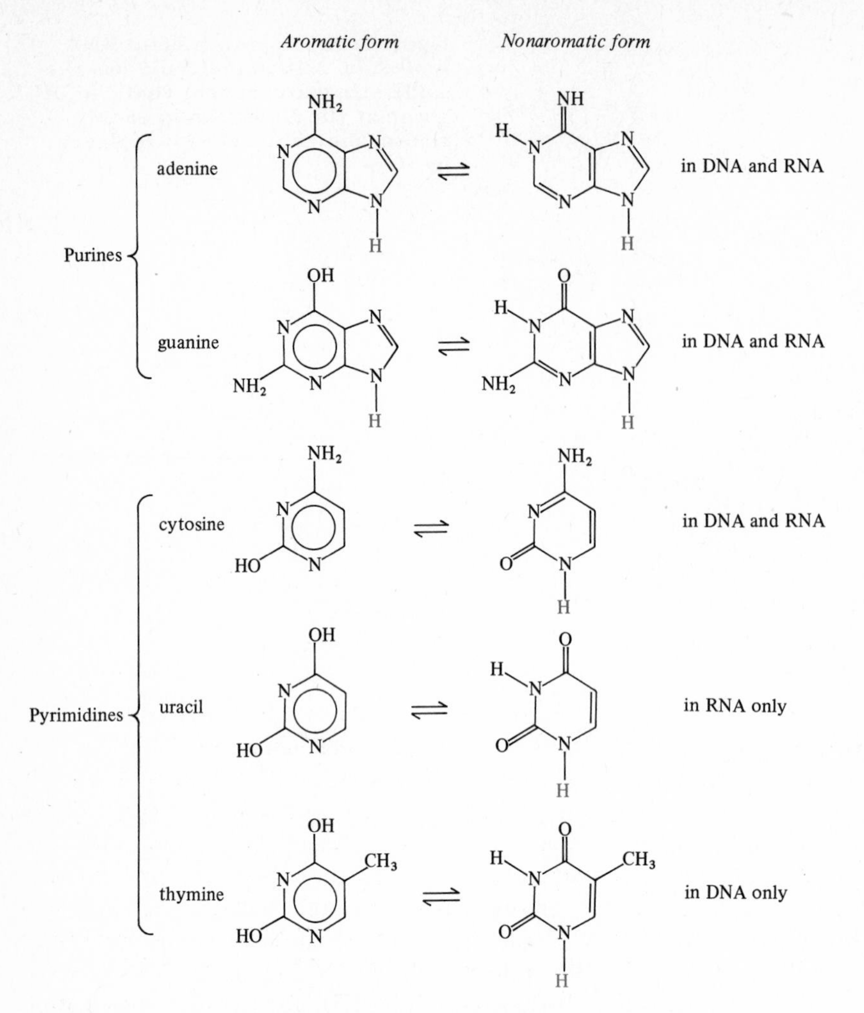

Figure 21-18 The five bases that are found in RNA and DNA. When free, each can exist in the two forms shown. The colored hydrogen is replaced when the bases are present in RNA or DNA.

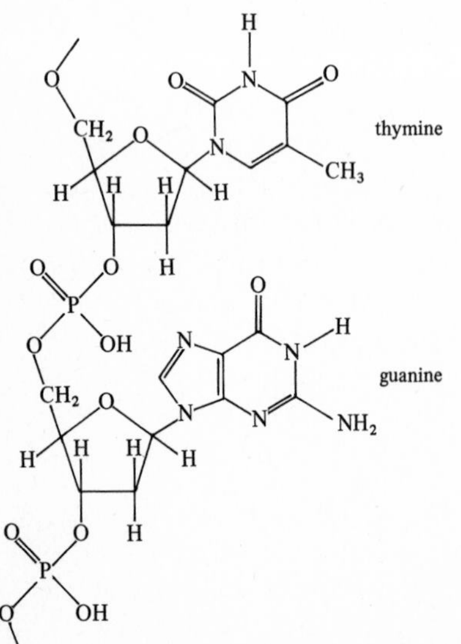

Figure 21-19 A portion of a DNA molecule showing the attachment of two bases.

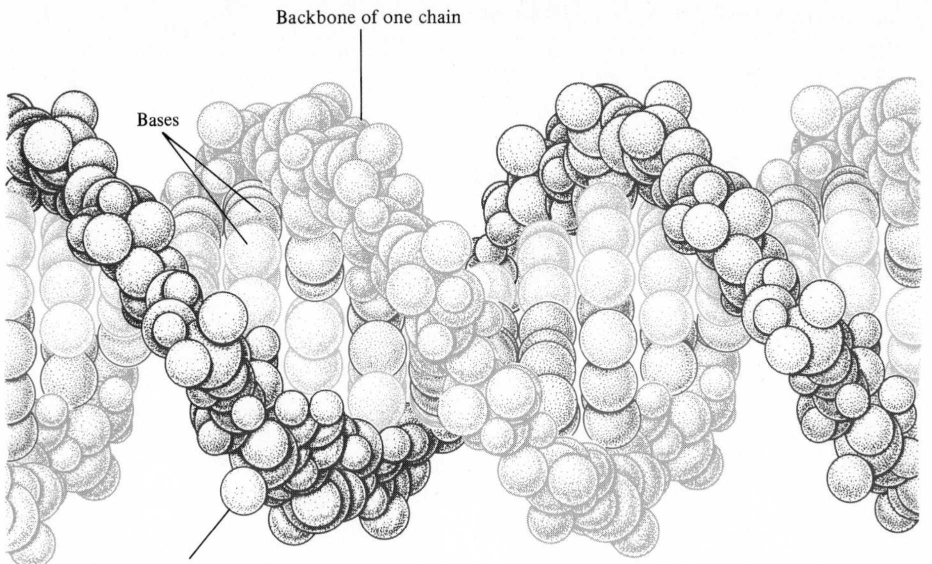

Figure 21-20 The double-helix structure of DNA. The bases are shown in color. [From "The Synthesis of DNA" by Arthur Kornberg. Copyright © 1968 by Scientific American, Inc. All rights reserved.]

part of the discovery of Watson and Crick was that when a given base is present in one chain, only *one* of the four bases could fit into the opposite position on the other chain! That is, a guanine must always be opposite a cytosine, and a thymine always opposite an adenine. Figure 21-21 shows a schematic representation of a portion of a DNA double helix. Note that there are two hydrogen bonds between adenine and

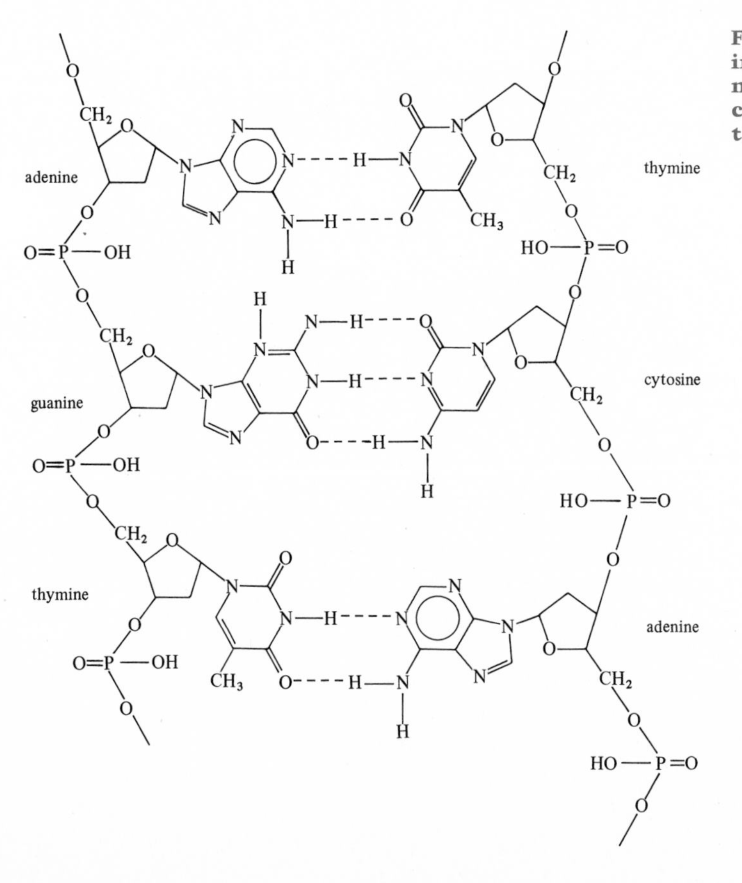

Figure 21-21 Schematic drawing of a portion of a DNA molecule, showing two chains, with the bases held together by hydrogen bonds.

thymine, and three between guanine and cytosine. The realization that each base could only have one partner was the key that unlocked the secret of heredity, as we shall see in the next section.

21-6 Heredity and the Genetic Code

It has been known for centuries that offspring tend to resemble their parents in many ways. The work of Gregor Mendel (1822-1884) put heredity on a quantitative basis and eventually led to the discovery of genes. Genes, which are present in the chromosomes of every living cell, determine the characteristics of the organism and are transmitted from parents to offspring. The genes possessed by any individual determine whether that individual will be tall, short, black, white, male, female, have blue eyes or brown, good or poor coordination and musical ability, be color-blind, have blood type A, B, AB, or O, or have any of a number of inherited diseases, some of which are fatal at birth. Furthermore, genes determine the very species to which an individual belongs. It is the genes that decide whether an organism will be a human being, a fish, a bird, a toad, a fig tree, a bacterium, or a fern.

It is one of the great triumphs of the scientific method that we now understand at least the main outlines of what genes are and how they actually carry out their functions. We now know that **genes are molecules of DNA or portions of molecules of DNA.** DNA has two functions:

1. It duplicates itself. A large organism such as a mammal or tree contains millions of cells. The nucleus of every one of these cells contains the *same* DNA molecules (except the sperm and egg cells, which contain only half the number of DNA molecules found in all other cells and mammalian red blood cells, which contain no DNA). At conception only one cell was present, and that one usually very small. That original cell divided and redivided many thousands of times to produce all the cells of the adult organism. Obviously, there must be a mechanism whereby the individual DNA molecules present in the original cell can duplicate themselves to produce enough of them for every cell in the adult organism. Since the difference between one DNA molecule and another is the sequence of four bases, this mechanism must ensure that when duplication takes place, the same sequence is obtained in the duplicate that was present in the original.

2. It controls the synthesis of proteins. In Section 21-4, we stated that each individual organism has many different protein molecules, including the enzymes that catalyze the reactions which take place in that organism. It is the DNA molecules that determine which protein molecules can be present. Each DNA molecule or portion of a DNA molecule contains the information necessary to produce one enzyme or other protein molecule. If an individual organism has that particular DNA molecule, it will be able to synthesize that particular protein. If it does not have that DNA molecule, it will not be able to synthesize the protein, which means, of course, that whatever function that protein molecule carries out cannot take place in that organism. Individuals get their particular DNA molecules through inheritance from their parents, and from occasional mutations (a mutation is a change in a DNA molecule induced by outside effects; for example, a cytosine might be replaced by a thymine, which, of course, forces a change in the opposite base). There is no other source of DNA.

Let us now see how DNA duplication (also called **replication**) actually takes place. As we have seen, DNA is a double helix. Two strands are intertwined, the bases connected by hydrogen bonds. When the DNA is about to duplicate, the strands separate. Each strand now constructs a new partner, using molecules of deoxyribose, phosphoric acid, and the four bases, all of which are present within the nucleus of the cell. Because of the fact that only guanine fits opposite cytosine and only thymine opposite adenine, **the new partners must have the same order of bases as the old partners** (Figure 21-22). Thus, two new double-helix molecules are now present, each one identical to the original molecule. The construction of the new DNA molecules requires that reactions take place between the deoxyribose, phosphates, and the four bases. These reactions are, of course, catalyzed by enzymes.

The second function of DNA is to control the synthesis of proteins. This is accomplished by a complex mechanism not yet completely understood, but whose main outlines are as follows. The DNA directs the synthesis of a type of RNA called **messenger-RNA.** Messenger-RNA is a single-stranded molecule; the order of bases on this molecule is identical to the order of bases on one of the strands of the DNA (except that uracil replaces thymine), and, of course, complementary to the other. Messenger-RNA molecules are not very stable and are manufactured by the cell as needed. The messenger-RNA combines with a part of the cell called a **ribosome.** Ribosomes are composed of proteins and a second type of RNA, called ribosomal-RNA. Within the cell are all 20 of the amino acids that will be assembled into the protein. A third type of RNA exists, called **transfer-RNA.** There are at least 20 different transfer-RNA molecules, one or more for each amino acid. Each transfer-RNA molecule is specific for only one amino acid. Protein synthesis takes place at the place where the messenger-RNA is attached to the ribosome. Each transfer-RNA molecule picks up its specific amino acid molecule somewhere in the cell, and carries it to the ribosome–messenger-RNA junction, where it is attached to the growing peptide chain. This process continues, each transfer-RNA molecule bringing a specific amino acid to the growing chain, until the protein is complete.

How do the different transfer-RNA molecules know which one is supposed to go next? The answer to this question is supplied by the **genetic code,** in which *the sequence of bases in DNA and RNA corresponds to the sequence of amino acids in the protein*. Let us consider the possibilities of this code. Is it possible that each base could correspond to one amino acid? No, because there are only four bases and 20 amino acids. How about two bases for each amino acid? This is nearly enough, but not quite. There are only 16 combinations of four bases taken two at a time. It has been shown that the genetic code uses a sequence of three adjacent bases to code each amino acid (such a sequence is called a **codon**). There are 64 combinations of four bases taken three at a time, more than enough for 20 amino acids, and there is considerable redundacy in the code: that is, most of the amino acids correspond to

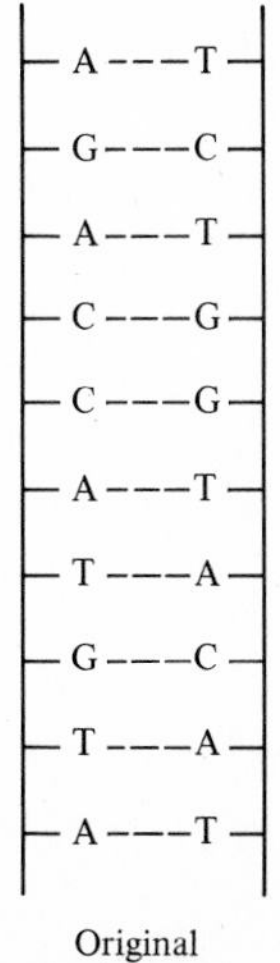

Original double helix

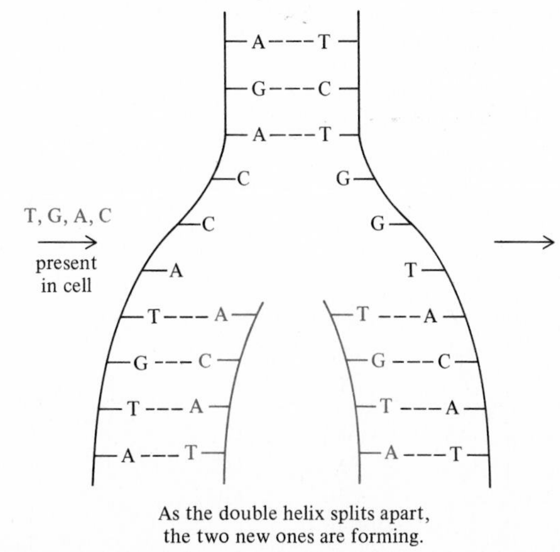

As the double helix splits apart, the two new ones are forming.

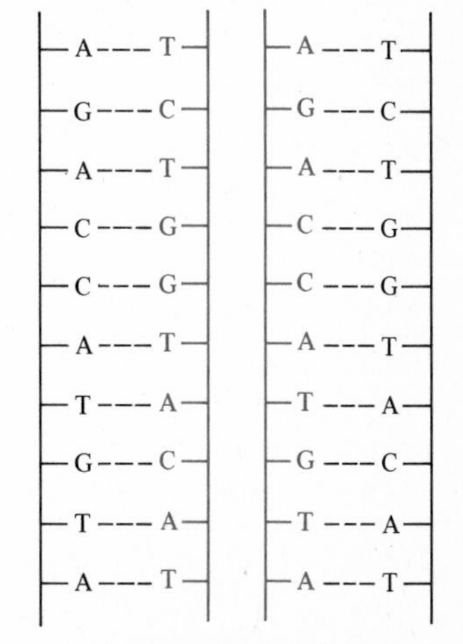

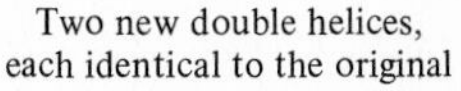
Two new double helices, each identical to the original

Figure 21-22 Replication of DNA.

two or more codons. Because a sequence of three bases is required, the genetic code is called a **triplet code.**

The genetic code is shown in Table 21-3. Here is how the genetic code would operate in a given situation. Consider the portion of a DNA molecule shown in Figure 21-23. The strands forming the double helix would be separated, and one of the strands (the one shown on the right) would serve as a template for the synthesis of the messenger-RNA with the sequence shown. As can be seen in Table 21-3, the genetic code specifies that GCC is a codon for the amino acid alanine. Each molecule of transfer-RNA is about 70 to 80 units long, but one specific section of the transfer-RNA molecule which picks up alanine has the sequence CGG. This sequence is, of course, complementary to GCC, so that only this molecule of transfer-RNA (carrying alanine) and no other can fit into the chain at this point. It is likely that 20 different transfer-RNA molecules, each carrying a different amino acid, come along to try to add their particular amino acid to the chain, but only one has the correct sequence for the next open place on the messenger RNA. After the alanine is added, the sequence GAA is next to be exposed, and only the transfer RNA carrying glutamic acid has the complementary sequence (CUU) which is necessary to fit here. Thus, the sequence of the bases in both messenger-RNA and transfer-RNA is important in protein synthesis. As far as we know, base sequence in ribosomal-RNA plays no part in the process. All these reactions are catalyzed by enzymes, and the synthesis is usually remarkably rapid. It has been shown that a bacterium can synthesize a protein of 150 amino acid units in about 20 seconds! As far as we know, the genetic code shown in Table 21-3 is universal; that is, the same code applies to all living things on this planet.

In summary, it is DNA, the actual molecules that genes are made of, which carry the information necessary for the synthesis of proteins. This information is in the form of specific sequences of four bases. The DNA transfers this information to messenger-RNA, which actually directs the synthesis of proteins, using ribosomes and transfer-RNA (as well as enzymes) to assist in the process. The information in the DNA and messenger-RNA is converted to specific sequences of amino acids by means of the genetic code. Each codon (sequence of three bases) codes a specific amino acid. The amino acid sequence in the protein enables it to carry out one specific function. If it is an enzyme, it can catalyze one specific reaction. Besides

Table 21-3 The Genetic Code[a]

UUU	Phe	UCU	Ser	UAU	Tyr	UGU	Cys
UUC	Phe	UCC	Ser	UAC	Tyr	UGC	Cys
UUA	Leu	UCA	Ser	UAA	END[a]	UGA	END[a]
UUG	Leu	UCG	Ser	UAG	END[a]	UGG	Trp
CUU	Leu	CCU	Pro	CAU	His	CGU	Arg
CUC	Leu	CCC	Pro	CAC	His	CGC	Arg
CUA	Leu	CCA	Pro	CAA	Glu	CGA	Arg
CUG	Leu	CCG	Pro	CAG	Gln	CGG	Arg
AUU	Ile	ACU	Thr	AAU	Asn	AGU	Ser
AUC	Ile	ACC	Thr	AAC	Asn	AGC	Ser
AUA	Ile	ACA	Thr	AAA	Lys	AGA	Arg
AUG	Met	ACG	Thr	AAG	Lys	AGG	Arg
GUU	Val	GCU	Ala	GAU	Asp	GGU	Gly
GUC	Val	GCC	Ala	GAC	Asp	GGC	Gly
GUA	Val	GCA	Ala	GAA	Glu	GGA	Gly
GUG	Val	GCG	Ala	GAG	Glu	GGG	Gly

[a]END refers to signals indicating chain endings.

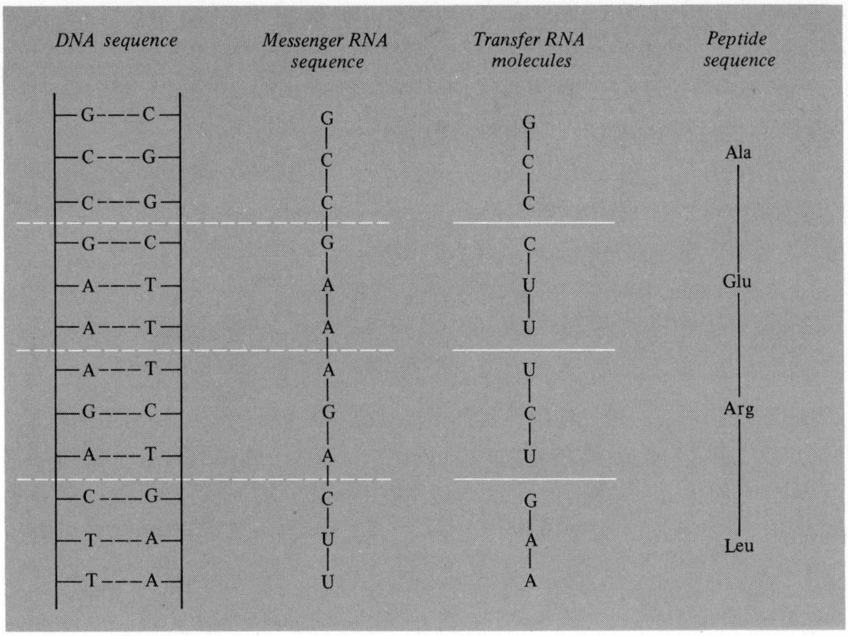

DNA sequence	Messenger RNA sequence	Transfer RNA molecules	Peptide sequence
G---C, C---G, C---G	G–C–C	G–C–C	Ala
G---C, A---T, A---T	G–A–A	C–U–U	Glu
A---T, G---C, A---T	A–G–A	U–C–U	Arg
C---G, T---A, T---A	C–U–U	G–A–A	Leu

Figure 21-23 Operation of the genetic code.

carrying the information necessary for protein synthesis, the DNA also duplicates itself. There in a nutshell are the main outlines of the chemical processes responsible for genetic inheritance, and for the great diversity of living species and of individuals within each species. Virtually all of this information was unknown 30 years ago. It is a tribute to the many skilled chemists and biologists who have worked in this field that so much could be learned in such a short time.

21-7 Fats

Compounds obtained from animals or plants that are insoluble in water but soluble in typical organic nonpolar solvents such as ether or benzene are called **lipids.** In this section, we will discuss one of the major types of lipids, called **fats.**

Fats are carboxylic esters. In Section 20-6, we learned that a carboxylic ester is made up of a carboxylic acid and an alcohol. In fats the alcohol is always glycerol.

$$\begin{array}{ccc} CH_2 - & CH - & CH_2 \\ | & | & | \\ OH & OH & OH \end{array}$$

glycerol

Since glycerol has three OH groups, three carboxylic acid units can be attached, so that fats are also called **triglycerides.** The carboxylic acids that are found in fats are typically those with long unbranched chains. They may or may not contain double bonds. Some carboxylic acids found in fats are

$$CH_3CH_2CH_2CH_2CH_2CH_2CH_2CH_2CH_2CH_2CH_2CH_2CH_2CH_2CH_2COOH$$

palmitic acid—16 carbons

$$CH_3CH_2CH_2CH_2CH_2CH_2CH_2CH_2CH_2CH_2CH_2CH_2CH_2CH_2CH_2CH_2CH_2COOH$$

stearic acid—18 carbons

$$CH_3CH_2CH_2CH_2CH_2CH_2CH_2CH_2CH{=}CHCH_2CH_2CH_2CH_2CH_2CH_2CH_2COOH$$

oleic acid—18 carbons, 1 double bond

$$CH_3CH_2CH{=}CHCH_2CH{=}CHCH_2CH{=}CHCH_2CH_2CH_2CH_2CH_2CH_2CH_2COOH$$

linolenic acid—18 carbons, 3 double bonds

The three carboxylic acid units found in any individual fat molecule may be the same or different. For example, a typical fat molecule might be

$$\begin{array}{l} CH_2{-}O{-}\overset{\overset{\displaystyle O}{\|}}{C}{-}CH_2CH_2CH_2CH_2CH_2CH_2CH_2CH_2CH_2CH_2CH_3 \quad \text{12 carbons, lauric acid} \\ | \\ CH{-}O{-}\overset{\overset{\displaystyle O}{\|}}{C}{-}CH_2CH_2CH_2CH_2CH_2CH_2CH_2CH_2CH_2CH_2CH_2CH_2CH_2CH_2CH_2CH_2CH_3 \quad \text{18 carbons, stearic acid} \\ | \\ CH_2{-}O{-}\overset{\overset{\displaystyle O}{\|}}{C}{-}CH_2CH_2CH_2CH_2CH_2CH_2CH_2CH{=}CHCH_2CH{=}CHCH_2CH_2CH_2CH_2CH_3 \quad \text{18 carbons, 2 double bonds, linoleic acid} \end{array}$$

Any sample of a fat is a mixture of many different molecules. All of them are esters of glycerol, but the carboxylic acids vary from molecule to molecule.

Note that all the carboxylic acids we have shown here contain an even number of carbon atoms. Carboxylic acids with an odd number of carbon atoms are virtually never found in fats. The reason for this is that living organisms synthesize these acids entirely from building blocks of acetate ions, so that the carbon atoms are put on in units of two.

Nearly all fats contain both saturated and unsaturated carboxylic acids. Those that contain only a small degree of unsaturation are called **saturated fats.** Those with more double bonds are **unsaturated fats.** The melting point of a fat decreases with increasing unsaturation, and saturated fats are usually solids, while unsaturated fats are usually liquids. Liquid fats are also called **oils.** Some typical saturated fats are butter, lard, tallow, and the fat in beefsteak. Typical unsaturated fats are olive oil, peanut oil, linseed oil, and corn oil. Coconut oil is an exception. Although it is a liquid, it is saturated. It is a liquid because the carboxylic acids making it up are mostly short-chain. Solid fats usually come from animals, liquid fats from plants.

If the only difference between solid and liquid fats is usually the degree of unsaturation, it should be possible to convert liquid fats to solid fats by adding hydrogen and a catalyst (Section 20-3). In fact, this is done, and the product is sold in supermarkets under brand names such as Crisco, Spry, and Fluffo. The unsaturated fats are not completely hydrogenated, since this would make them brittle. Some double bonds are allowed to remain.

We saw in Section 20-6 that a carboxylic ester can be hydrolyzed with a basic catalyst to produce the alcohol and carboxylate ion. This process, called **saponification,** was first applied to fats more than 4000 years ago in ancient Mesopotamia, and is thus one of the oldest organic reactions known to mankind. When fats are saponified with sodium hydroxide (the ancients used wood ashes, which contain KOH), the products are the alcohol glycerol and a mixture of the sodium salts of the acids present.

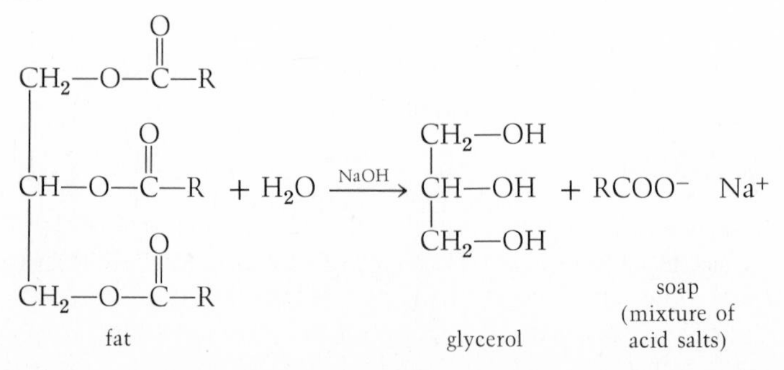

This mixture of the salts of fatty acids is called *soap*. We have already seen (Section 17-11) how soap carries out its washing action.

21-8 Vitamins, Steroids, and Alkaloids

There are many important biological compounds, in addition to those we have already discussed. In this section, we will mention three additional types of compounds: vitamins, steroids, and alkaloids.

VITAMINS. In 1906 Frederick Hopkins (1861–1947), following up on earlier work of Christiaan Eijkman (1858–1930), discovered that certain organic compounds (other than proteins, fats, and carbohydrates) had to be present in the diet, or else illness and even death could result. Casimir Funk (1884–1967) gave these compounds the name **vitamins.** This was an important concept because the germ theory of disease (itself not very old at that time) held that all illness was caused by microorganisms. Unlike carboxylic acids, proteins, fats, etc., vitamins do not have any structural features in common. Vitamins are defined by their biological functions and not by their chemical structures. Because the structures of vitamins can be rather complicated, and because many of their structures were unknown at the time that vitamins were discovered, the custom arose of calling many of them by letters: vitamin A, vitamin C, and so on. Some of the most common vitamins are shown in Figure 21-24. Lack of vitamin A causes eye diseases and night blindness; lack of the B vitamins causes beri-beri, pellegra, and pernicious anemia; lack of vitamin C causes scurvy; and lack of vitamin D causes rickets.

STEROIDS. Steroids are very important compounds, which are synthesized by living organisms and which carry out many functions in the body. All of them have this basic four-ring skeleton:

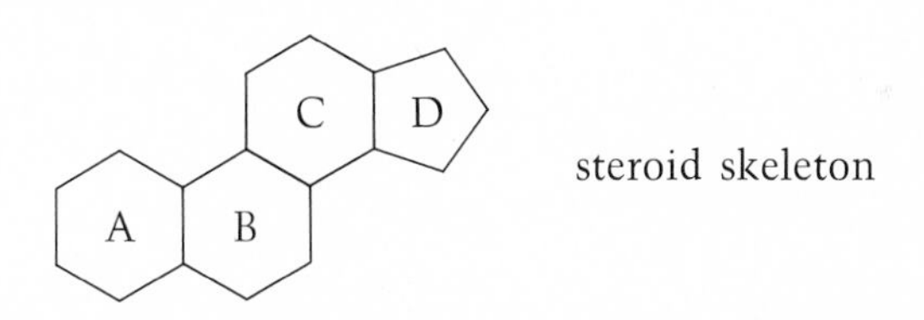

though in some of them, the skeleton is modified slightly [for example, vitamin D_2 (Figure 21-24) is a steroid, with ring B opened up]. The most important steroid is cholesterol,

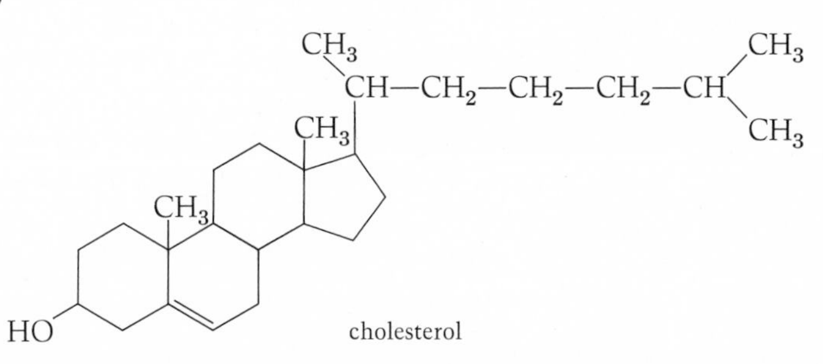

because the body uses it as a starting compound with which to make all the other steroids. Cholesterol is an essential compound; without it we would die. Fortunately, there is no chance of that, because even if we had no cholesterol in our diet, our bodies would synthesize all that we needed. However, most of us do get cholesterol in our diets, and some people get excessive amounts. Although it is not yet completely proven, there is strong evidence that excess cholesterol in the blood can lead to cholesterol deposits in the arteries, which in turn can lead to heart disease and

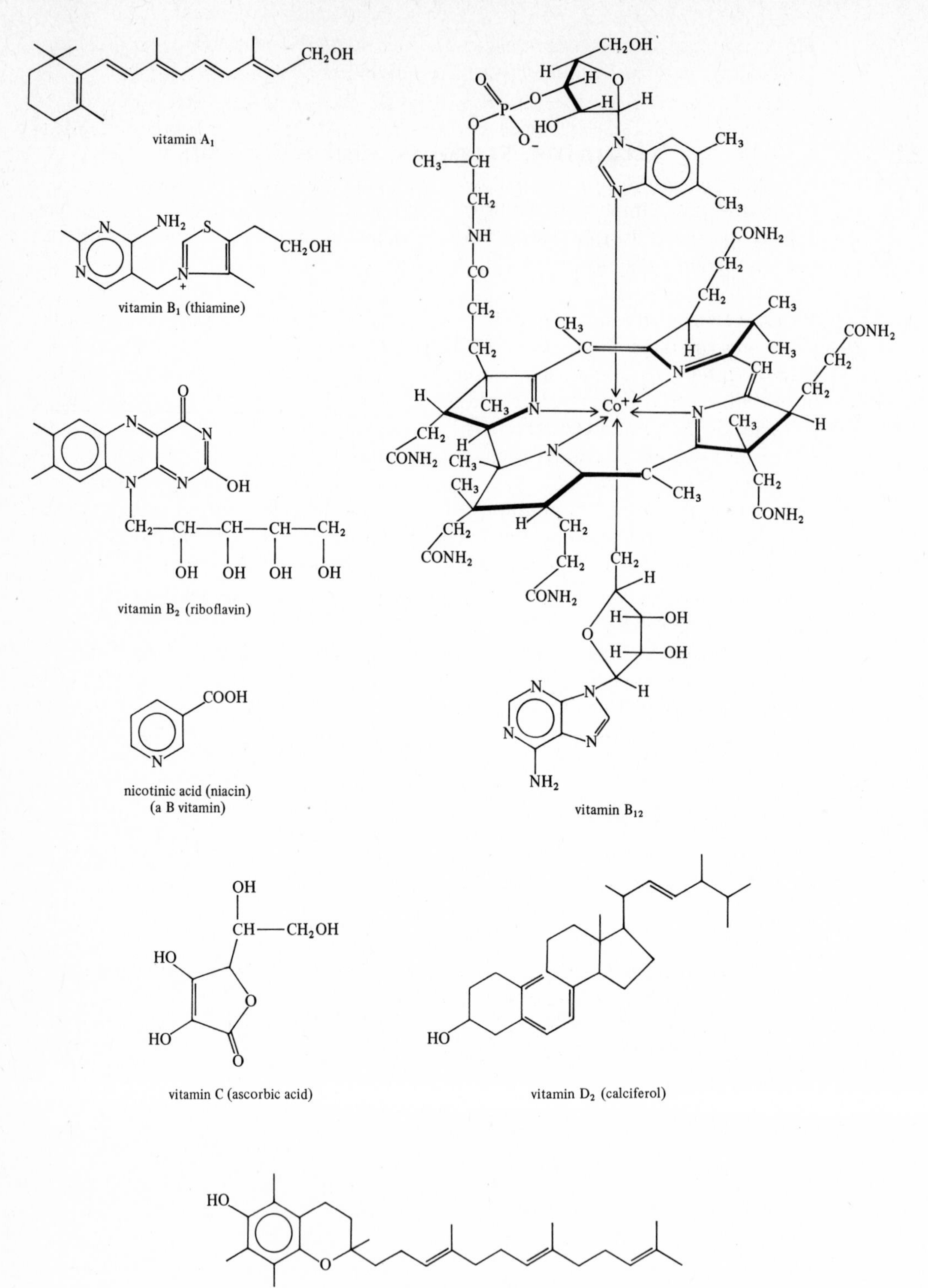

vitamin E (tocopherol)

Figure 21-24 Some important vitamins.

early death. Cholesterol has eight asymmetric carbon atoms (can you find them all?), so there are $2^8 = 256$ steroisomers. As we might expect, only one of these is the cholesterol found in living organisms. The only organic starting compound used by the body to synthesize cholesterol is acetic acid. The 27 carbon atoms come from 14 molecules of acetic acid (the extra carbon atom is used for other purposes).

Among other important steroids are cholic acid, cortisone,

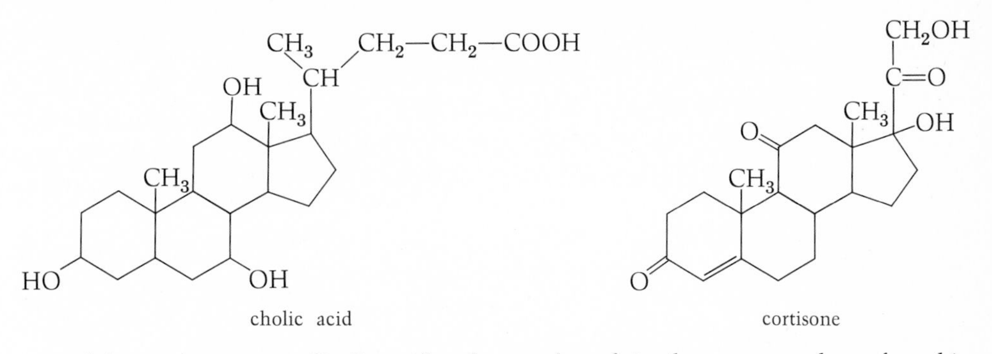

cholic acid cortisone

and the sex hormones. Cholic acid and a number of similar compounds are found in bile, where they aid in the digestion of lipids. Cortisone is one of the adrenal cortical hormones that aids in regulating carbohydrate metabolism. Lack of sufficient cortisone causes rheumatoid arthritis, and cortisone is sometimes used in treatment of this disease. There are three main sex hormones. Testosterone is the principal male sex hormone and estradiol the principal female sex hormone. (There is no single compound called estrogen. Estradiol is the most important of a group of hormones called estrogens.)

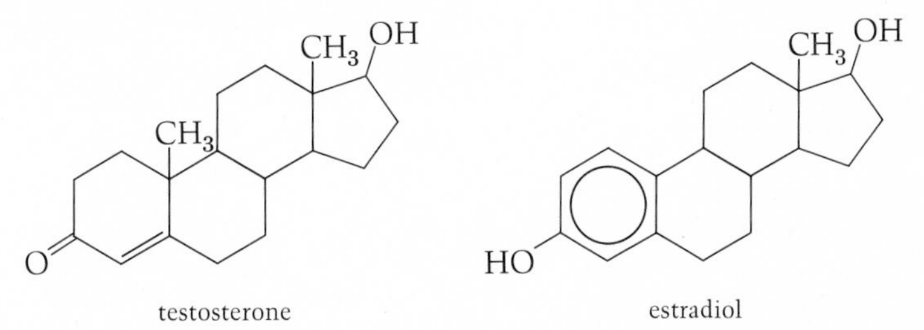

testosterone estradiol

These hormones regulate the development of the sexual organs, and such secondary sexual characteristics as facial hair and breast development. Note the similarity of the two structures. They differ only in ring A, which is aromatic in estradiol, but not in testosterone. The third major sex hormone is progesterone, a female sex hormone that controls the reproductive cycle and pregnancy.

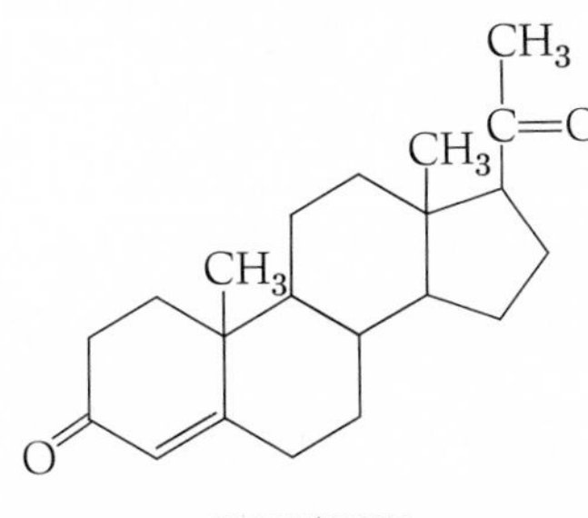

progesterone

All normal human beings manufacture all three of these hormones, but males make mostly testosterone, while females make mostly estradiol and (in their reproductive years) progesterone. Some individuals have too much or too little of one or more of these, resulting in unwanted physical characteristics. Such conditions must be treated by a physician.

Birth-control pills are synthetic steroids (that is, they have the steroid skeleton but are not made by any living organism) that resemble progesterone. An example is norethindrone.

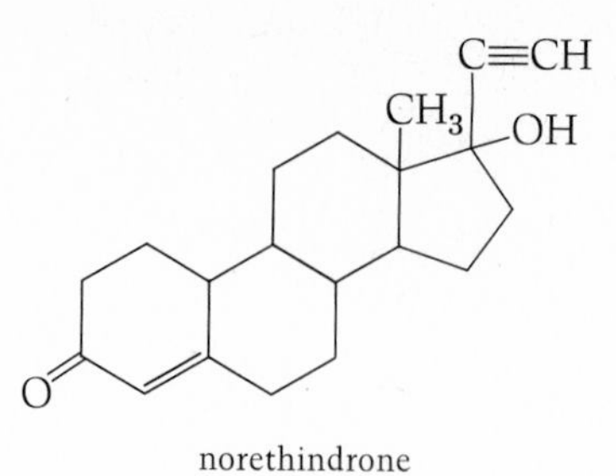

norethindrone

ALKALOIDS. Alkaloids are defined as organic bases that can be isolated from plants. In practically all of them, the basic group is an amino group. Thousands of different alkaloids have been isolated from plants, with structures ranging from very simple to very complex. A number of important alkaloids are shown in Figure 21-25: Note that many drugs, both in the medical and nonmedical senses, are alkaloids, including nicotine and caffeine, which are two of the three most important drugs taken nonmedicinally in the United States (the other of course is ethanol). Nicotine

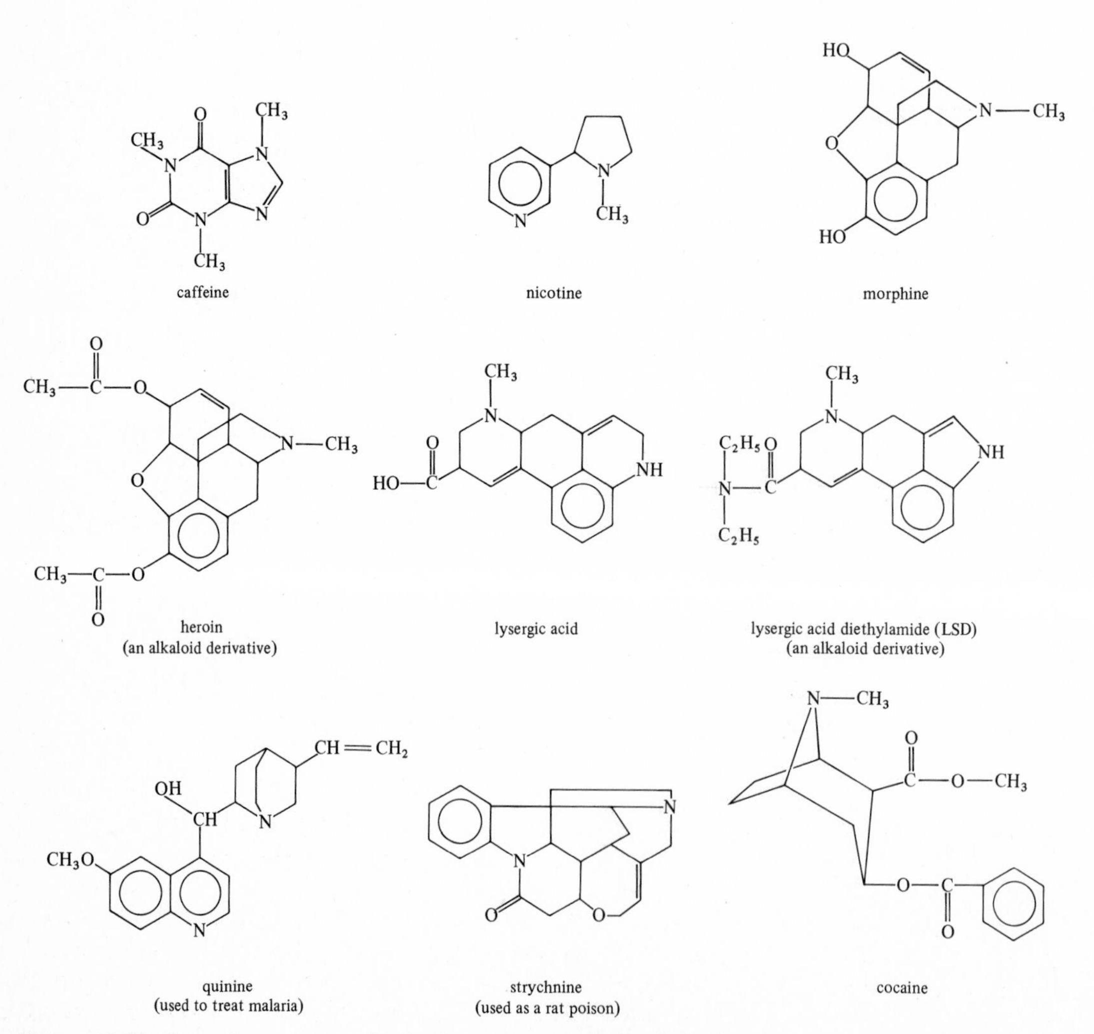

Figure 21-25 Some important alkaloids, and two alkaloid derivatives.

is present in cigarette smoke, and caffeine in coffee, tea, cocoa, and many nonprescription headache remedies, cold preparations, and stimulants. Caffeine is also a purine (Section 21-5). Other alkaloidal drugs are morphine, cocaine, and quinine. Heroin and LSD are not alkaloids, because they are not found in plants, but are very simple synthetic derivatives of alkaloids.

Problems

21-1 Define each of the following terms. **(a)** carbohydrate **(b)** monosaccharide **(c)** disaccharide **(d)** polysaccharide **(e)** pyranose **(f)** furanose **(g)** aldohexose **(h)** ketohexose **(i)** dextrose **(j)** pentose **(k)** invert sugar **(l)** polymer **(m)** glycogen **(n)** photosynthesis **(o)** enzyme **(p)** substrate **(q)** protein **(r)** amino acid **(s)** dipeptide **(t)** MSG **(u)** essential amino acid **(v)** zwitterion **(w)** disulfide bridge **(x)** denaturation **(y)** active site **(z)** alpha-helix **(aa)** pleated sheet **(bb)** DNA **(cc)** double helix **(dd)** mutation **(ee)** codon **(ff)** lipid **(gg)** fat **(hh)** triglyceride **(ii)** vitamin **(jj)** steroid **(kk)** alkaloid

21-2 Of the following pairs, which are enantiomers, which are diastereomers, which are structural isomers, and which are none of these? **(a)** D-gulose and L-galactose **(b)** D-gulose and D-galactose **(c)** D-glucose and D-fructose **(d)** D-galactose and L-galactose **(e)** α-D-glucose and β-D-glucose **(f)** D-glucose and sucrose **(g)** sucrose and lactose **(h)** D-glucose and D-ribose

21-3 What is the difference in structure between α-D-glucose and β-D-glucose?

21-4 Draw structural formulas for L-glucose in the pyranose form and in the open-chain form, and show how they interconvert.

21-5 The aldohexose D-allose rotates the plane of polarized light to the right. **(a)** Can you predict which direction the plane will be rotated by L-allose? **(b)** Can you predict which direction the plane will be rotated by D-idose, another aldohexose?

21-6 Classify each of the following as aldohexoses, aldopentoses, ketohexoses, ketopentoses, or none of these. **(a)** D-glucose **(b)** D-ribose **(c)** L-galactose **(d)** L-fructose **(e)** dextrose **(f)** levulose **(g)** maltose **(h)** D-galactose **(i)** L-glucose

21-7 **(a)** How many ketohexoses are there? **(b)** How many aldopentoses? **(c)** How many ketopentoses?

21-8 Draw structural formulas for each of the following in the furanose forms. **(a)** D-ribose **(b)** L-fructose

21-9 Draw structural formulas for each of the following. **(a)** sucrose **(b)** lactose **(c)** starch **(d)** cellulose

21-10 Which monosaccharide is the one chiefly synthesized in the green leaves of plants? What catalyst is necessary for the process?

21-11 **(a)** What is the molecular formula for any disaccharide made up of two pentose units? **(b)** Of one pentose and one hexose?

21-12 All the monosaccharides are very soluble in water but insoluble in benzene or diethyl ether. Can you suggest why?

21-13 Draw structural formulas for each of the following peptides. **(a)** Gly–Ala **(b)** Ser–Val **(c)** Asn–Pro **(d)** Leu–Phe–Arg

21-14 Draw structural formulas for **(a)** the acidic amino acids **(b)** the basic amino acids

21-15 **(a)** Which amino acids have more than one asymmetric carbon atom? Label the asymmetric carbons. **(b)** Which amino acid has no asymmetric carbon?

21-16 The real structure of glycine is $H_3\overset{+}{N}—CH_2—COO^-$. **(a)** Show what happens to glycine when it is added to water and excess HCl is added. **(b)** Show what happens to the aqueous solution when excess aqueous NaOH is added. **(c)** Is there any pH at which glycine is not an ionic compound?

21-17 What is the difference in structure between cotton and wool?

21-18 Show how the lock-and-key theory explains the specificity of enzyme action and the extremely rapid rates of enzyme-catalyzed reactions.

21-19 Show how the bases adenine and cytosine are attached to the RNA chain.

21-20 Why is it that the pyrimidines can be attached to the DNA or RNA chains only in the nonaromatic form, but the purines can be attached in either form?

21-21 What are the two functions of DNA and how are they carried out?

21-22 What are the functions of RNA?

21-23 Draw the complete structure of a single-strand DNA chain that codes for the amino acid methionine.

21-24 How many different fat molecules can be made if one molecule of glycerol is combined with one molecule each of stearic acid, oleic acid, and palmitic acid?

21-25 Draw the structure of a fat molecule made up of glycerol, one molecule of lauric acid in the middle, and one each of oleic and linolenic acids on the ends.

21-26 Show how the fat molecule in Problem 21-25 can be converted to a saturated fat molecule.

21-27 Why do the carboxylic acid molecules found in fats almost always have an even number of carbon atoms?

21-28 Which vitamin(s) in Figure 21-24 **(a)** has a pyrimidine ring **(b)** can be made from nicotine **(c)** prevents night-blindness **(d)** is an alcohol **(e)** is a carboxylic acid?

21-29 Find the eight asymmetric carbons in cholesterol.

21-30 Certain organisms have the ability to synthesize one of the vitamins in Figure 21-24 from a steroid. Which vitamin do you think this is?

21-31 Marijuana is obtained from a plant. One of the main physiologically active constituents of marijuana is tetrahydrocannabinol, $C_{21}H_{30}O_2$. Why is this not an alkaloid?

Additional Problems

21-32 Why will putting sugar in an automobile gasoline tank cause trouble?

21-33 What is the genetic code and how does it operate?

21-34 Vitamin C (Figure 21-24) rotates the plane of polarized light to the right. Its enantiomer is also a known compound. Would you expect that the enantiomer can also prevent scurvy?

21-35 What is the difference between fibrous and globular proteins?

21-36 If pure α-D-glucose is dissolved in water and the optical rotation measured immediately it is found that a solution of standard concentration rotates the plane of polarized light 112° to the right (+112°). However, if the solution is allowed to stand at constant temperature and the rotation measured again, the value slowly decreases, until after a time it reaches +52.5°, after which it no longer changes. Pure β-D-glucose behaves the same way, except that the rotation starts at +19° and then slowly increases to +52.5°, after which it no longer changes. Explain these results.

21-37 Draw the structural formula for a typical soap molecule.

21-38 What is the basic difference in structure between maltose and cellubiose? Between starch and cellulose?

21-39 **(a)** How many asymmetric carbons are there in cortisone? **(b)** How many asymmetric carbons are there in estradiol? **(c)** How many stereoisomers are possible in each case? **(d)** How many of these do you suppose have physiological activity?

21-40 Suppose that we set out to synthesize all possible proteins made up of the 20 amino acids in Table 21-2. (This would be very useful for research purposes, because we could then add any conceivable protein to any organism we wished.) At first we limit ourselves to a small part of this task, and make only those that are 10 amino acid units long (decapeptides). **(a)** How many different decapeptides are there, using only the 20 amino acids in Table 21-2? **(b)** If we learned how to make them very rapidly, so that we could make one decapeptide a day, how long would it take to make all of them? **(c)** Do the same calculations for proteins that are 20 units long (eicosapeptides), but assume that we can make 1 million of them each day. **(d)** What is the likelihood that we will ever synthesize all possible proteins of typical lengths—say, between 100 and 300 units?

22 Polymers

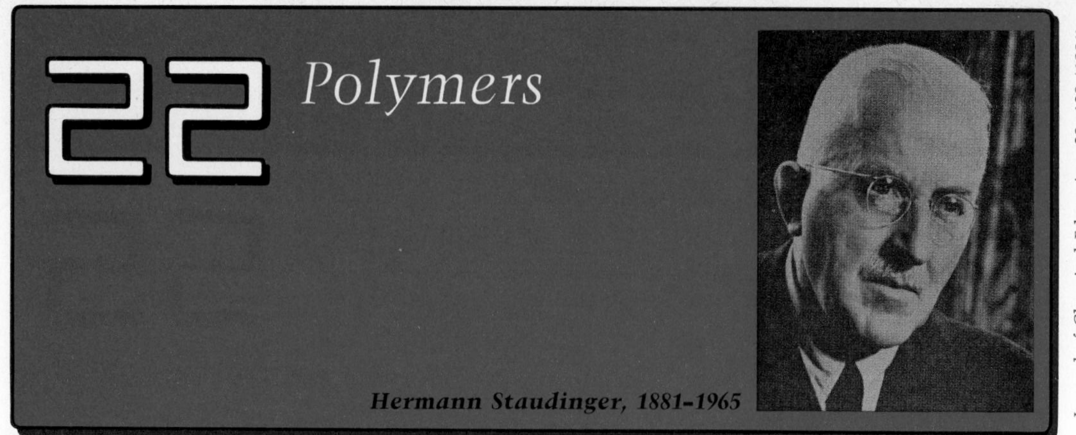

Hermann Staudinger, 1881–1965

Journal of Chemical Education, **28**, 120 (1951).

We have defined a **polymer** as a large molecule made up of many repeating units. Another word for them is **macromolecules.** Most of the matter that we see in the world around us (except for water and metals) consists of polymers. The earth and rocks under our feet, the bricks, concrete, and wood that we use as building materials, the fabrics and leather with which we clothe ourselves, trees and other plant life, and our very bodies themselves are largely or entirely composed of polymers. We have already discussed a number of naturally occurring polymers: proteins, nucleic acids, starch and cellulose; and diamond, graphite, sand, and other inorganic giant molecules. In this chapter, we shall largely deal with man-made polymers. Because of their large molecular weights and of their often complex structures, chemists came rather late to the study of polymers. They preferred at first to work with smaller molecules, which were easier to study. Chemical industry also avoided polymers for similar reasons. This began to change in 1909 with the invention of the first commercial synthetic polymer, Bakelite. Much progress was made in the 1920s and 1930s, and this period was capped by the invention of nylon, which was first marketed in 1938. After World War II, the chemical industry intensively began to produce polymers, and today polymers make up more than half of the output of chemical industry.

22-1 Types of Polymers

There are two major types of polymers: addition polymers and condensation polymers. In an **addition polymer,** molecules containing double bonds are connected to each other in the following manner:

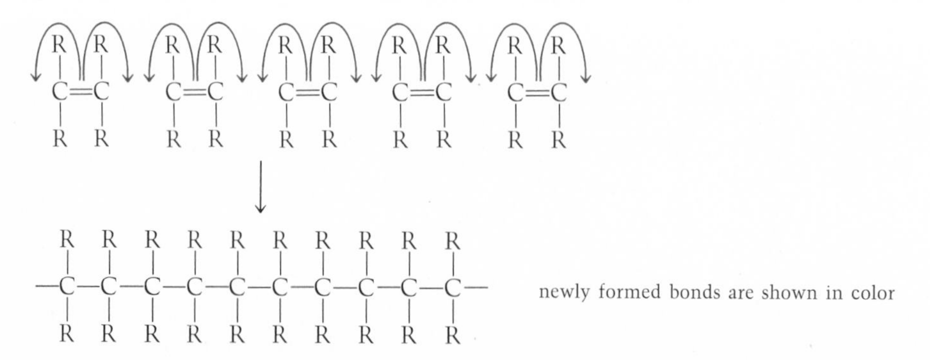

newly formed bonds are shown in color

The double-bond compound $R_2C{=}CR_2$ is called the **monomer,** and in the process is **polymerized** to give the polymer. It is obvious that any compound containing a double bond can, in theory at least, produce this type of polymer. Addition polymers will be discussed in Sections 22-2 and 22-3. In a **condensation polymer,** the monomers must have at least two functional groups of the type that are capable of forming connections with other functional groups. Polymerization is accomplished when each functional group connects with another in this fashion:

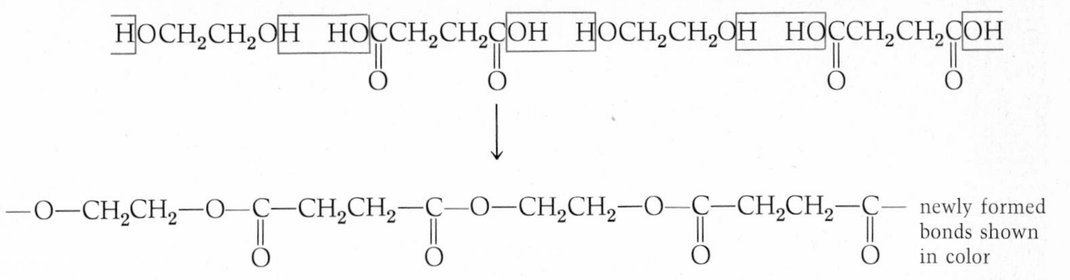

In most cases, small molecules (usually water) are lost when the functional groups combine with each other. In the example shown, the polymer is composed of two different monomers, but condensation polymers can also be made from a single monomer, if it has the right functional groups: for example,

$$\text{HO—CH}_2\text{CH}_2\text{—}\underset{\underset{\text{O}}{\|}}{\text{C}}\text{—OH} \quad \text{HO—CH}_2\text{CH}_2\text{—}\underset{\underset{\text{O}}{\|}}{\text{C}}\text{—OH} \quad \text{HO—CH}_2\text{CH}_2\text{—}\underset{\underset{\text{O}}{\|}}{\text{C}}\text{—OH}$$

$$\downarrow$$

$$\text{—O—CH}_2\text{CH}_2\text{—}\underset{\underset{\text{O}}{\|}}{\text{C}}\text{—O—CH}_2\text{CH}_2\text{—}\underset{\underset{\text{O}}{\|}}{\text{C}}\text{—O—CH}_2\text{CH}_2\text{—}\underset{\underset{\text{O}}{\|}}{\text{C}}\text{—}$$

Starch, cellulose, proteins, and nucleic acids are all condensation polymers. Other condensation polymers will be discussed in Sections 22-4 and 22-5.

Another way to classify polymers is by the overall shapes of the molecules. Polymer structure can be linear or three-dimensional. **Linear polymers** have one long continuous backbone. The examples shown above, as well as proteins and nucleic acids, are linear, or one-dimensional polymers. Three-dimensional shapes can be brought about by **cross-linking** of linear polymers or by an overall three-dimensional structure. In a cross-linked polymer, linear chains are connected by bridges (cross-linkages), which may be bonds or may consist of one or more atoms (Figure 22-1). We have already seen examples of crossed-link polymers in proteins with disulfide bridges, and in starch and cellulose. Diamond (Figure 15-1a) is an example of a polymer with an overall three-dimensional structure.

Polymers that are largely or entirely linear have an important difference in properties from those which are mostly three-dimensional. When heated, the former get soft and pliable, while the latter remain hard. The first type is called **thermoplastic;** the other **thermosetting.** Thermoplastic polymers can be molded and cast, but thermosetting polymers must be formed into the desired shape as the monomers become polymerized. Once formed, the shapes cannot be changed, except by cutting.

The monomer units that go to make up a given polymer need not be the same, whether the polymer is of the addition or condensation type. For example, an addition polymer can be made by polymerizing a mixture of $R_2C{=}CR_2$ and $R_2C{=}CH_2$, in any proportions. The resulting polymer will then have some

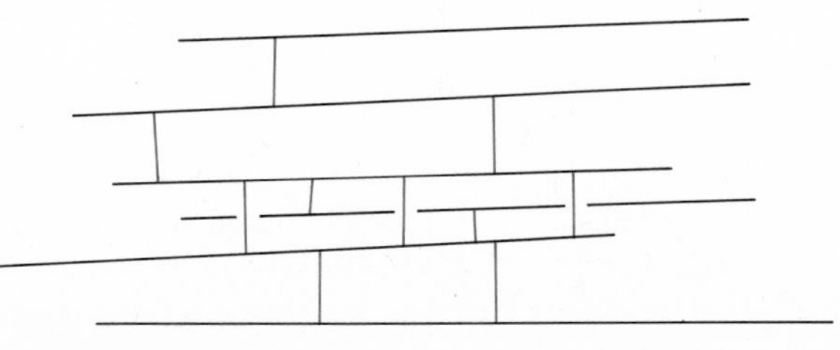

Figure 22-1 Schematic diagram of a cross-linked polymer. Linear chains are shown in black, cross-linkages in color.

$—R_2C—CR_2—$ units and some $—R_2C—CH_2—$ units. A polymer made up of two or more different monomers is called a **copolymer.**

As indicated above, man-made polymers are very important in our modern lives. Despite the fact that we have been using them only since 1909, we would find it hard to get along without them. As we shall see in the rest of this chapter, man-made polymers are used in plastics (both rigid and in thin sheets), artificial fibers, synthetic rubbers and leathers, adhesives, coatings (for example, the glazed surface of refrigerators), and other uses. Sometimes the same polymer can be used in two or more ways, depending on how it is fabricated. For example, nylon is an artificial fiber, but it can also be formulated as a plastic. Incidentally, the word **plastic** originally applied only to thermoplastic polymers, but it is often used today to refer to the thermosetting type as well.

The properties, and hence the usefulness, of a given polymer depend on a great many variables. One of these is chain length. Some polymers contain average chain lengths of two or three hundred units, others of two or three thousand, or two or three million units. (Note that we say *average* chain lengths. In the typical polymer the chain length of each molecule is different.) Polymers made up of the same monomer, but with different average chain lengths, may have greatly different properties. Even polymers with the same average molecular weights may have different properties if their molecular-weight distributions are different. Another factor that arises in many (though not all) cases is head-to-tail vs. head-to-head orientation. For example, polymerization of a monomer $R_2C=CH_2$ can take place in two basic ways:

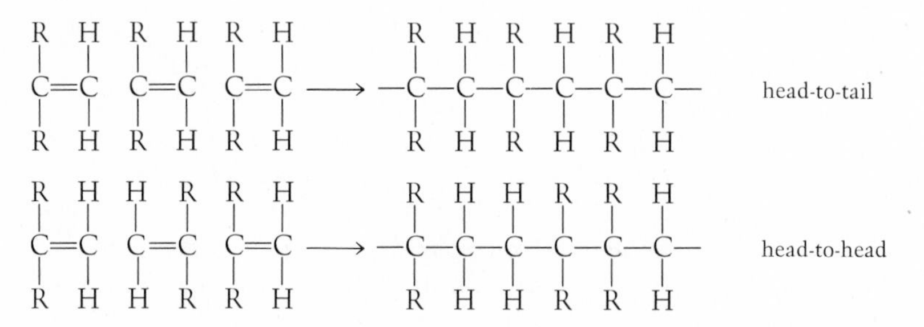

It is also possible that the polymerization will be random, with some portions of the chain in the head-to-head and others in the head-to-tail orientation. The properties of all of these polymers may differ considerably. Among other variables are the amount of cross-linking and the presence and proportions of copolymers.

The effects of all these variables on the properties of the polymers can often be determined only by trial and error. Therefore, companies must spend considerable time and effort to develop a polymer that is commercially useful.

The impact of polymers on our modern lives has been largely favorable, because we now have materials that perform functions which naturally occurring materials cannot do at all, or can do only at greater cost. However, there is a negative side as well. Because man-made polymers did not evolve through natural processes, living organisms do not possess the enzymes necessary to decompose them; that is, they are not biodegradeable. Because of this, discarded polymers represent a form of pollution of our environment, and are a problem with which society will increasingly find it necessary to deal.

22-2 Addition Polymers

Simple addition polymers are widely used to make many products. Some of the most important are shown in Table 22-1. The simplest in terms of chemical structure, as well as the largest in terms of tons annually produced, is polyethylene. This polymer

Table 22-1 Some Important Addition Polymers

	Monomer	*Polymer*	
Ethylene	$CH_2{=}CH_2$	$-CH_2-CH_2-CH_2-CH_2-$	Polyethylene
Propylene	$CH_2{=}CH(CH_3)$	$-CH_2-CH(CH_3)-CH_2-CH(CH_3)-$	Polypropylene
Styrene	$CH_2{=}CH(C_6H_5)$	$-CH_2-CH(C_6H_5)-CH_2-CH(C_6H_5)-$	Polystyrene
Vinyl chloride (chloroethene)	$CH_2{=}CH(Cl)$	$-CH_2-CH(Cl)-CH_2-CH(Cl)-$	Polyvinyl chloride (PVC)
Acrylonitrile (cyanoethene)	$CH_2{=}CH(CN)$	$-CH_2-CH(CN)-CH_2-CH(CN)-$	Polyacrylonitrile
Methyl methacrylate	$CH_2{=}C(CH_3)(COOCH_3)$	$-CH_2-C(CH_3)(COOCH_3)-CH_2-C(CH_3)(COOCH_3)-$	Acrylic plastics
Tetrafluoroethene	$CF_2{=}CF_2$	$-CF_2-CF_2-CF_2-CF_2-$	Teflon
Vinylidine chloride + Vinyl chloride (copolymer)	$CH_2{=}C(Cl)(Cl)$ + $CH_2{=}CH(Cl)$	$-CH_2-CCl_2-CH_2-CH(Cl)-$	Saran

is essentially a very long chain unbranched alkane (though branched polyethylenes are also available). Because of its chemical inertness (like the alkanes) and light weight (low density), it is used to make bottles, drums and other containers, as well as pipes, toys, coatings, and many miscellaneous articles. In the form of thin sheets it is used for such things as plastic bags to cover garments that have just been dry-cleaned. However, polyethylene has a relatively low softening temperature (though this depends on how it is made), and in general cannot be used where temperatures as high as 100°C are required. For example, hospital items that must be sterilized with boiling water cannot be made from polyethylene. Polypropylene has properties similar to those of polyethylene and is used for many of the same purposes, but has a higher softening point, making it suitable for hot-water pipes and other higher-temperature applications. It is also used to make plastic hinges.

Polystyrene is used to make various items such as toys, refrigerator dishes, wall tiles, electrical parts, and lenses. Many students are familiar with polystyrene models. Polyvinyl chloride (PVC) has many uses. Virtually all phonograph records are now made of this material, as well as pipes, bottles, floor tiles, valve parts, toothbrush handles, ice-cube trays, and many other objects. In thin sheets and rigid films it is also used to wrap foods. Most rigid plastic packages of luncheon meats are made of PVC. A few years ago it was discovered that some workers in factories where

PVC is made developed a rare form of liver cancer caused by the monomer vinyl chloride. This discovery raised the possibility that people who ate foods in PVC bottles or wrapped with PVC sheets or film might have ingested very tiny amounts of unpolymerized vinyl chloride which are trapped in the PVC and then absorbed by the food. Should we stop packaging foods in PVC? The question is very complex and not easily answered. We don't want to expose people to the danger of liver cancer, but are the very small amounts of vinyl chloride that are present in the PVC enough to do this? These amounts (~3 to 5 ppm) are so small that a few years ago we would not have been able to detect them with the analytical methods then known. The factory workers were exposed to amounts of vinyl chloride vapor thousands of times greater than this for periods of many years, before the hazards of this compound were known. Questions of this kind pose a real dilemma for modern society, which would like to remain safe from poisoning and other technological hazards but still wants to enjoy the fruits of technical progress. One thing that could be done and was done in this case was to reduce the amount of vinyl chloride trapped in the PVC, but it is at present not possible to make PVC that is entirely free of the monomer.

Polyacrylonitrile is manufactured as a fiber under such brand names as Orlon and Acrilan. Pure polyacrylonitrile is difficult to dye, so acrylonitrile is usually polymerized with small amounts of a copolymer. Orlon and Acrilan are used for carpets, blankets, pillow stuffing, and for sweaters and other clothing. Acrylic plastics, of which polymethyl methacrylate (shown in Table 22-1) is one of the most important, are transparent plastics that have rigidity and excellent resistance to weathering. They are used for outdoor signs, canopies, windows, and automobile parts, as well as such products as combs, brushes, dentures, and faucet handles.

Teflon is extremely inert, even more so than polyethylene. It also has a much higher softening point. It is therefore used to coat cooking vessels such as frying pans, and to make valves, electrical insulators, wire coverings, and so on. Saran is a copolymer of vinylidene chloride $CH_2{=}CCl_2$ (IUPAC name 1,1-dichloroethene) and vinyl chloride, usually containing 75% or more of vinylidine chloride. Saran is used to make pipes, to line steel pipes, and for automobile seat covers. In thin sheets it is also used to wrap foods. Approximately one-third of the flexible plastic used to package foods in grocery stores is Saran (PVC and polyethylene are also widely used).

22-3 Rubber and Other Polydienes

In Section 20-3, we learned that a hydrocarbon with two double bonds is called a diene. If the two double bonds are separated by only one single bond, the compound is called a **conjugated diene.** A conjugated diene can be polymerized in the following way:

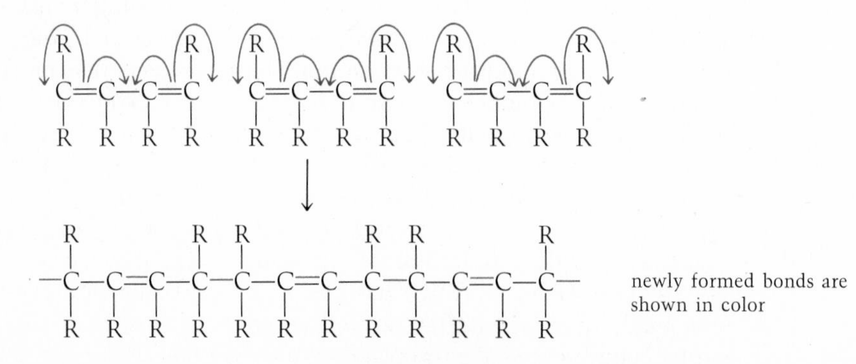
newly formed bonds are shown in color

The resulting polymer is an addition polymer, but every fourth bond in the chain is a double bond. Note that the double bonds in the polymer are not in the same

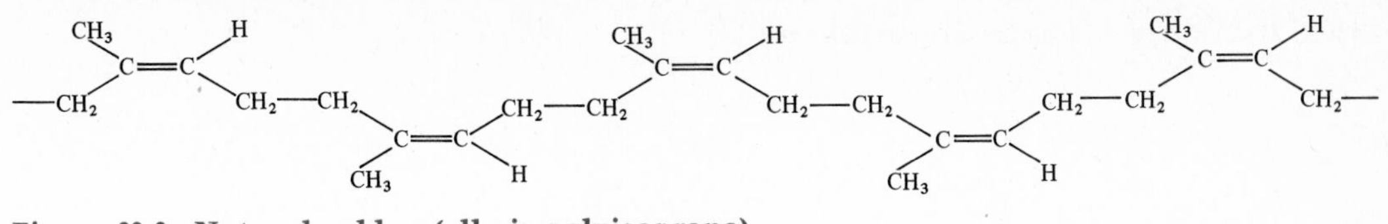

Figure 22-2 Natural rubber (all-cis polyisoprene).

positions as they were in the monomer. The most important polymer of this type is natural rubber, which is obtained upon coagulation of the sap (called *latex*) of the rubber tree (also called the *Hevea* tree). In natural rubber the monomer is 2-methyl-1,3-butadiene (isoprene), and the polymer is called polyisoprene.

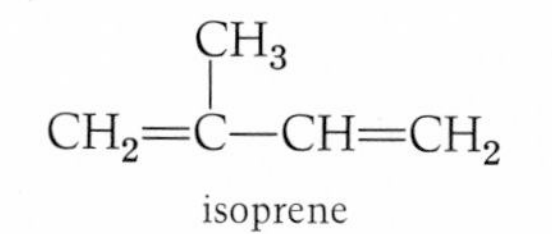

isoprene

Remember that suitably substituted double bonds can be cis or trans. In natural rubber the isoprene molecules are coupled in a head-to-head manner, and all the double bonds have the cis geometry (Figure 22-2). This geometry posed problems to chemists during World War II. When rubber supplies fell short, the U.S. government began a program to try to synthesize rubber from isoprene, which was abundantly available (from petroleum). Unfortunately, all their polymerizations of isoprene produced only polyisoprenes with mostly trans double bonds, since trans double bonds are generally more stable than the cis type. Polyisoprenes with trans double bonds do not have the same elastic properties as natural rubber and are generally much less useful (*gutta-percha,* which is used to make golf balls and as insulation for cables is a natural all-*trans*-polyisoprene). Despite the large amounts of money the U.S. government put into this program (it was a "crash" program), a synthetic all-*cis*-polyisoprene was not synthesized during the war, and this goal was only reached several years later (1955). Scientific "crash" programs do not always succeed (though sometimes there may be an end run around them; see below).

The most important property of rubber is its elasticity; it can be extended a considerable amount, and then, when released, it snaps back to its original dimensions. However, raw rubber is too soft and tacky to be of much use and it is fortunate that Charles Goodyear (1800–1860) discovered in 1839 that rubber could be hardened by heating it with sulfur. This process, called **vulcanization,** produces sulfur cross-links that connect the chains to each other, and convert a linear polymer to a cross-linked polymer (Figure 22-3). Two to three percent sulfur gives soft rubber which is used for rubber bands, erasers (which is where rubber got its name from: it was used to rub out pencil marks), shoe soles, and so on. Somewhat more sulfur (5 to 8%) gives rubber that can be used for tires. The fact that rubber still has double bonds leaves it open to attack by air. The oxygen molecules cross-link the chains even more, making the rubber get brittle with age. Various antioxidants are added to prevent this. (The antioxidants work by being more oxidizable than the rubber. Therefore, the oxygen attacks them and not the rubber.)

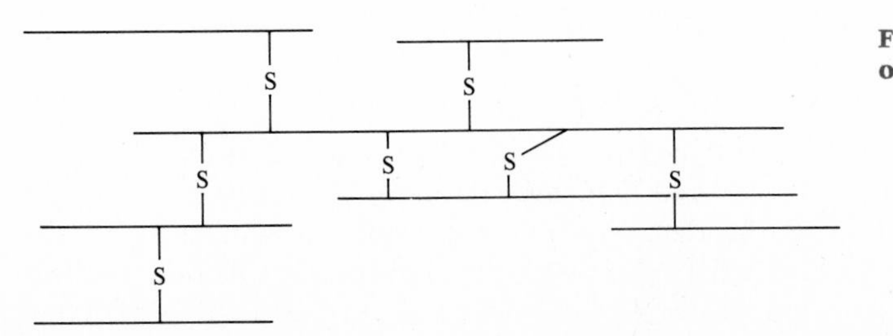

Figure 22-3 Schematic diagram of vulcanized rubber.

Table 22-2 Some Synthetic Rubbers

	Monomer	Polymer	
2-Chlorobutadiene (chloroprene)	$CH_2{=}CH{-}C(Cl){=}CH_2$	$-CH_2{-}CH{=}C(Cl){-}CH_2{-}CH_2{-}CH{=}C(Cl){-}CH_2-$	Neoprene
Isoprene + Isobutene (copolymer)	$CH_2{=}CH{-}C(CH_3){=}CH_2$ + $CH_2{=}C(CH_3){-}CH_3$	$-CH_2{-}C(CH_3)_2{-}CH_2{-}CH{=}C(CH_3){-}CH_2{-}CH_2{-}C(CH_3)_2-$	Butyl rubber
1,3-Butadiene + Styrene (copolymer)	$CH_2{=}CH{-}CH{=}CH_2$ + $CH_2{=}CH{-}C_6H_5$ 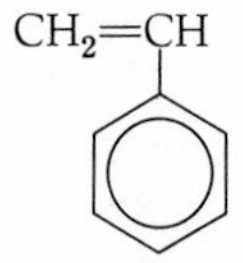	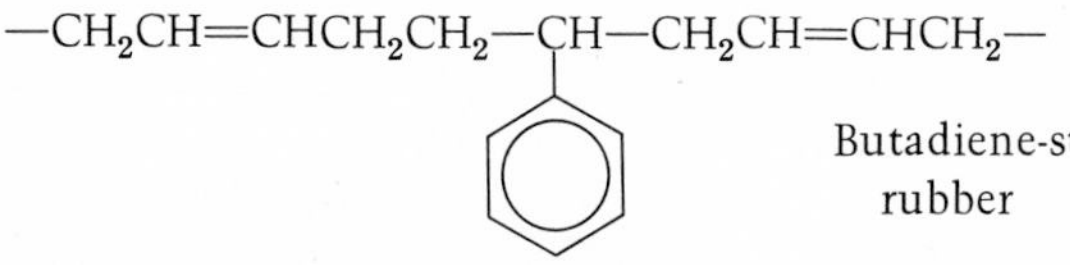 $-CH_2CH{=}CHCH_2CH_2{-}CH(C_6H_5){-}CH_2CH{=}CHCH_2-$	Butadiene-styrene rubber

Although natural rubber is a highly useful material, it nevertheless is not the best material for all applications where an elastic substance is needed. Elastic polymers made from other dienes are called **synthetic rubbers.** When it was realized during World War II that the effort to duplicate natural rubber would not succeed, it was decided to tackle the problem by producing various synthetic rubbers (some synthetic rubbers were already in use before the war). This constituted an "end run" on the problem. Several of the synthetic rubbers made in this way have proven to be superior to natural rubber for certain purposes (it would be surprising if the rubber obtained from the sap of the *Hevea* tree turned out to be best for all applications), and today less than 50% of the rubber used in the world is natural rubber. Three important synthetic rubbers are shown in Table 22-2. Butyl rubber and butadiene-styrene rubber are copolymers of a diene and a compound with one double bond. In butyl rubber the major ingredient is isobutene, which means that the polymer has only a few double bonds. When it is vulcanized, virtually all the double bonds are used up, so the rubber is very resistant to air. Butyl rubber is used to make inner tubes. On the other hand, butadiene-styrene rubber contains much more butadiene than styrene. Neoprene is resistant to heat and flames because of its high chlorine content. It is used in many applications, including gasoline hose.

Though all of these rubbers are based on dienes, there are other synthetic rubbers with different chemical structures, though they are all polymers (for example, see silicone rubber, Section 22-5).

22-4 Linear Condensation Polymers

We will discuss three important types of linear condensation polymers: polyamides, polyesters, and polyurethanes.

POLYAMIDES. Wool and silk have been used for centuries to make clothing and other fabrics. When it was discovered that these naturally occurring materials are polyamides, the possibility arose that synthetic polyamides might be made that would also be useful as fabrics, and might be cheaper than wool or silk, or have other advantages. This possibility was realized in 1938 when *nylon,* invented in 1935 by a research team headed by Wallace Carothers (1896–1937; he was also a coinventor of neoprene), was first put on the market. At first, nylon was used only to make women's stockings, but later was used in all kinds of clothing, as well as in carpets, tire cord, rope, parachutes, and paintbrushes, as well as for nonfabric uses such as electrical parts, valves, gears, rifle stocks, clips, and fasteners, etc. Nylon is very tough and strong, nontoxic, nonflammable, and resistant to chemicals. There are several kinds of nylon, but all are polyamides. The first, and still the most important type, is a polymer of hexamethylenediamine and adipic acid:

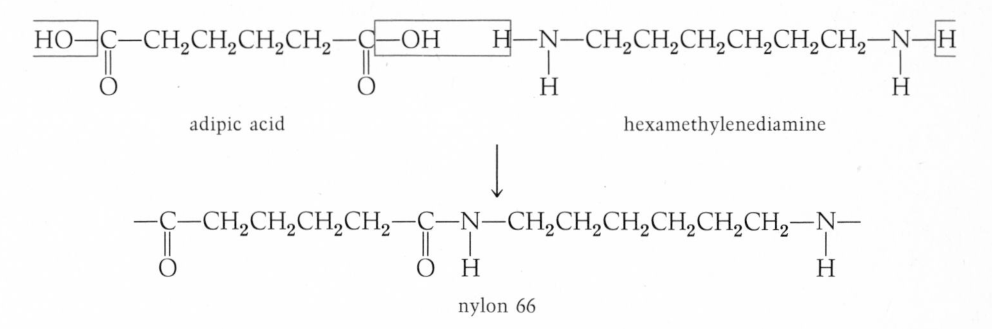

This nylon is called nylon 66 because the carboxylic acid and the diamine each have six carbons. Other nylons are polymers of single monomer, containing an amino and carboxyl group. The most important of these is nylon 6, also called polycaprolactam:

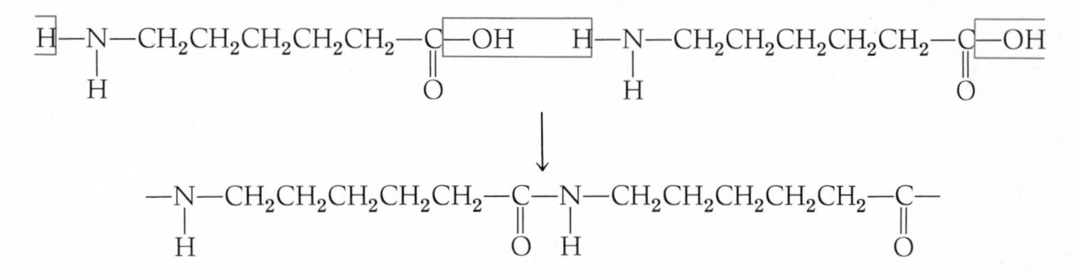

POLYESTERS. These polymers are generally made from a dicarboxylic acid and a dihydroxy compound. One of the most important is Dacron, made from ethylene glycol and terephthalic acid:

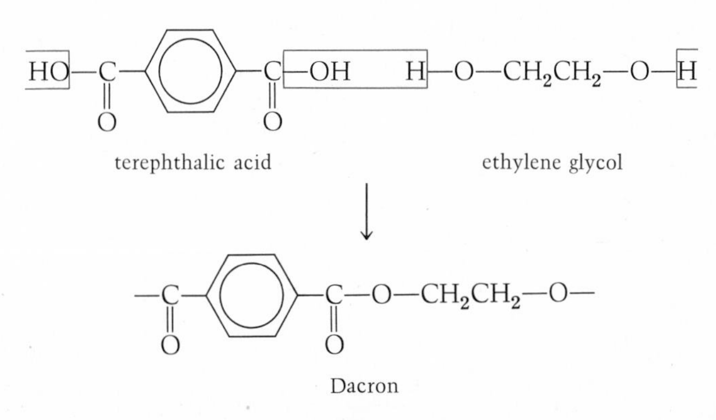

This polyester, and similar ones, are widely used in clothing of all kinds, as well as in tire cords, carpets, etc. Dacron can also be produced in thin sheets, in which case it is called Mylar. Polyesters can be made with small amounts of unsaturated monomers, so that the polymer chains contain some double bonds. These are cross-linked with certain catalysts, to produce solid polyester polymers that are used to make furniture and furniture parts (they can be made to look like wood), as well as boats, truck cabs, motor homes, and other uses.

Polyurethanes. When an alcohol is added to an **isocyanate,** a urethane is produced:

$$\underset{\text{an isocyanate}}{R{-}N{=}C{=}O} + R'OH \longrightarrow \underset{\text{a urethane}}{R{-}\underset{H}{\underset{|}{N}}{-}\underset{OR'}{\underset{|}{C}}{=}O}$$

Polyurethanes are made by adding a compound that has two OH groups to a diisocyanate (generally an aromatic one). The compound, which has two OH groups, is itself a polymer, usually with a molecular weight of 1000 to 2000. A typical polyurethane would be made as follows:

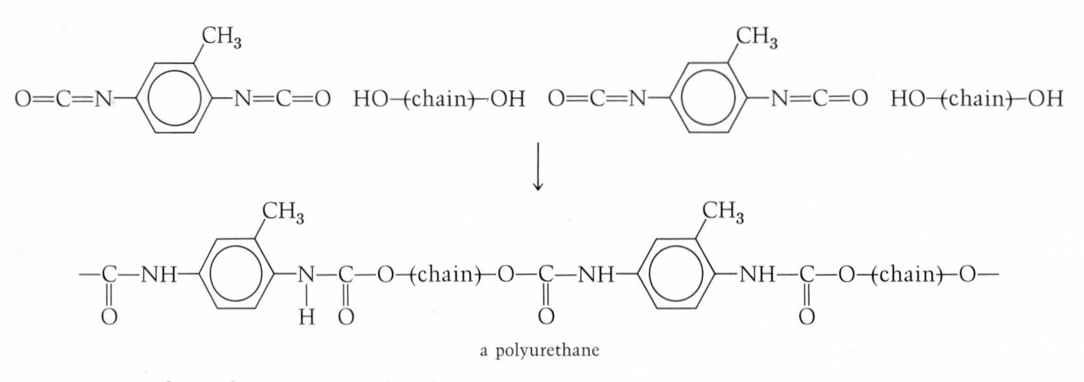

Polyurethanes are used as foam rubbers as well as in coatings, adhesives, and other applications.

22-5 Cross-linked Condensation Polymers

The first commercial man-made polymer was invented by Leo Baekeland (1863–1944) in 1909.* This polymer is made from phenol and formaldehyde, which react as follows, with water being lost:

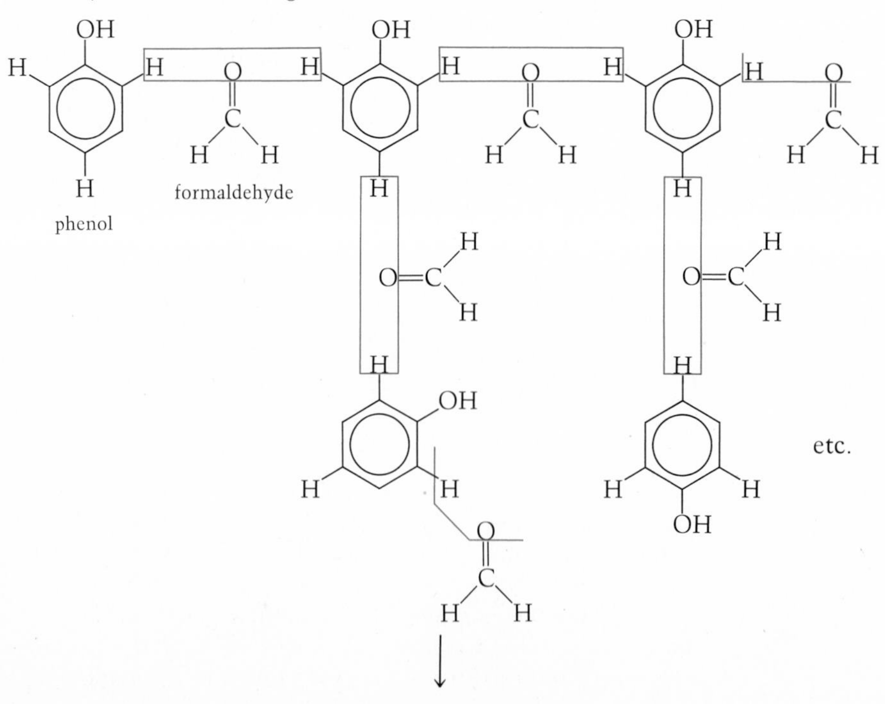

*Celluloid was invented earlier, but this is not strictly a man-made polymer. It is made by nitrating cotton (cellulose).

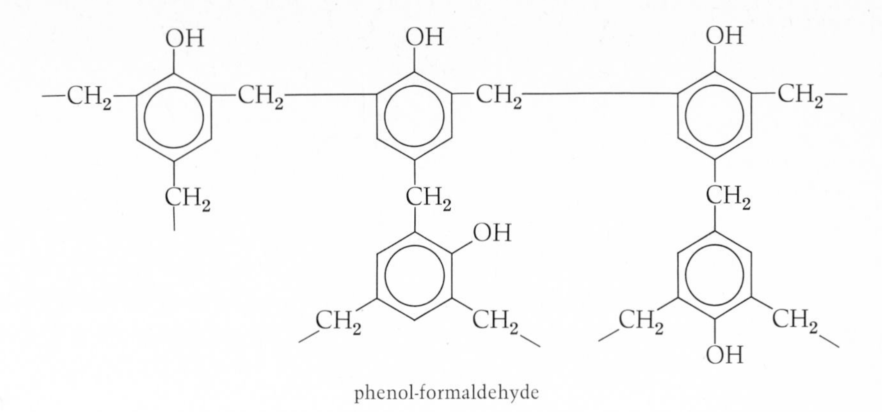

phenol-formaldehyde

Since each phenol molecule can react at three positions (two ortho and one para), the result is a three-dimensional array of benzene rings connected to each other by CH_2 groups. These polymers are called phenol-formaldehyde resins (Baekeland gave his product the trade name Bakelite, a name that is still used). Other phenols may also be used (for example, *para*-cresol, which is *para*-methylphenol), and the polymer can be formulated in various ways so that a wide variety of types is possible. Most of these are thermosetting, and phenol-formaldehyde resins are used to make hard plastic objects that can withstand heat. Among these are knobs and handles for cooking utensils, washing machine parts, distributor caps, telephones, radio cabinets, camera cases, etc. Phenol-formaldehyde resins are also used as an adhesive, primarily to bond thin strips of wood together to form plywood.

Some other important three-dimensional condensation polymers are:

Glyptal and *alkyds* are three-dimensional polyesters. In the formation of glyptal, phthalic acid (with two carboxyl groups) reacts with glycerol (with three OH groups) to give a polyester. Dacron (Section 22-4) is a linear polyester, because the carboxylic acid and alcohol each have two functional groups, but here there is an extra OH group, so a three-dimensional network can form

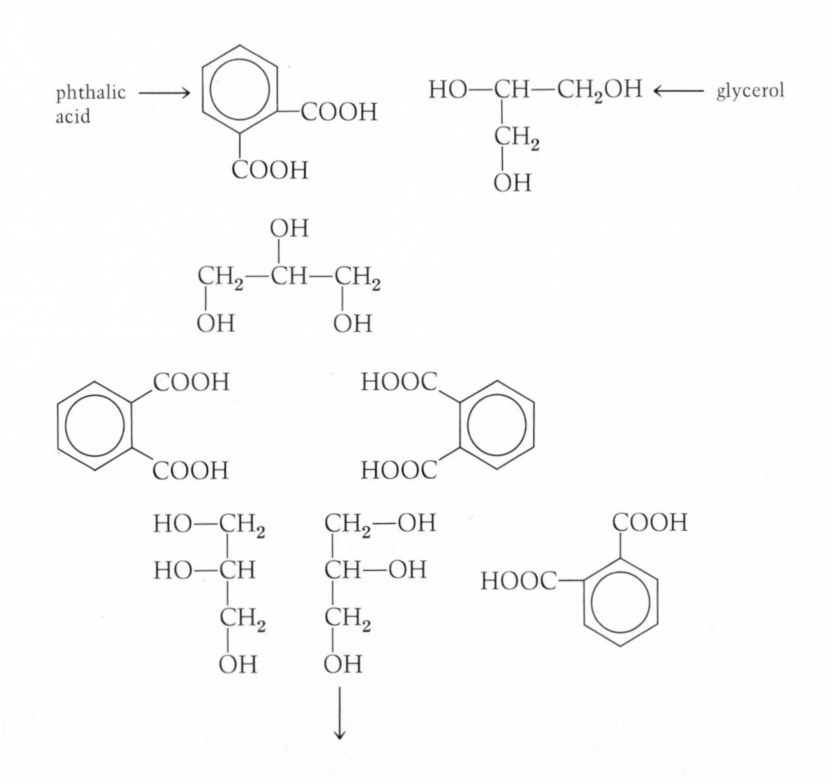

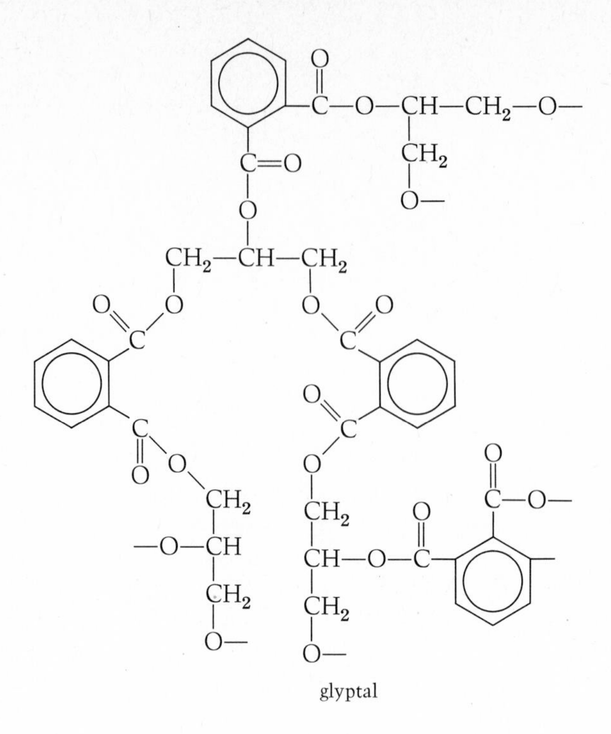
glyptal

Similar polymers can be prepared from other dicarboxylic acids and other compounds with three or more OH groups. The general name for these polymers is **alkyd resins.** They are widely used as coatings for such things as refrigerators and washing machines. When the polymer is made, only some of the ester links are formed. After it has been applied to the surface, it is baked to complete the esterification process, giving a hard, shiny surface, resistant to heat and chemicals. These polymers can also be reinforced with fiberglass or other materials. Such reinforced polyesters are used to make boats, camping trailers, etc.

Epoxy resins are first made by allowing a diphenol (usually bisphenol A) to react with epichlorohydrin in the presence of a base, to give a linear polymer.

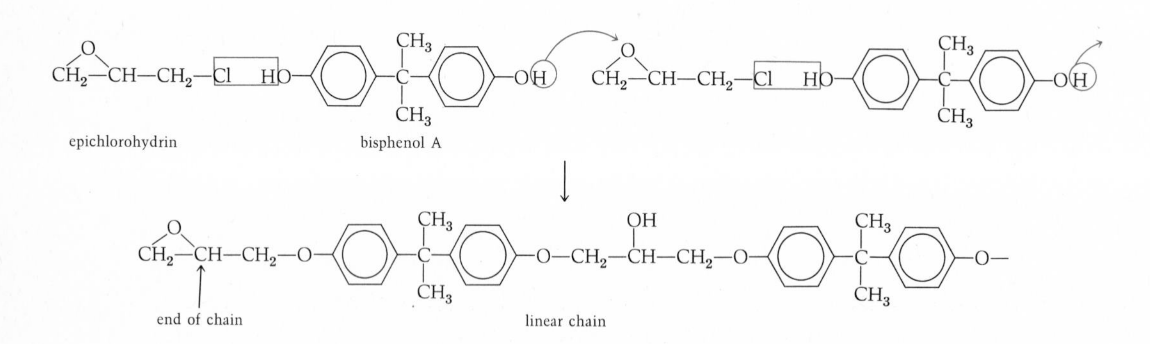

These chains are relatively short and have epoxy rings (three-membered rings containing one oxygen and two carbons) at each end. The chains can be cross-linked by reaction with a compound containing three amino groups. The amino groups combine with the epoxy rings to cross-link the chains, as shown in Figure 22-4. Epoxy resins are used as adhesives (epoxy glues) and are reinforced with fiberglass to give plastics that are used for the same purposes as the reinforced polyesters.

Epoxy glues are sold in two packages, one containing the chains (this is called the cement) and the other the amino compound (called the hardener). The two packages are mixed to set the polymer.

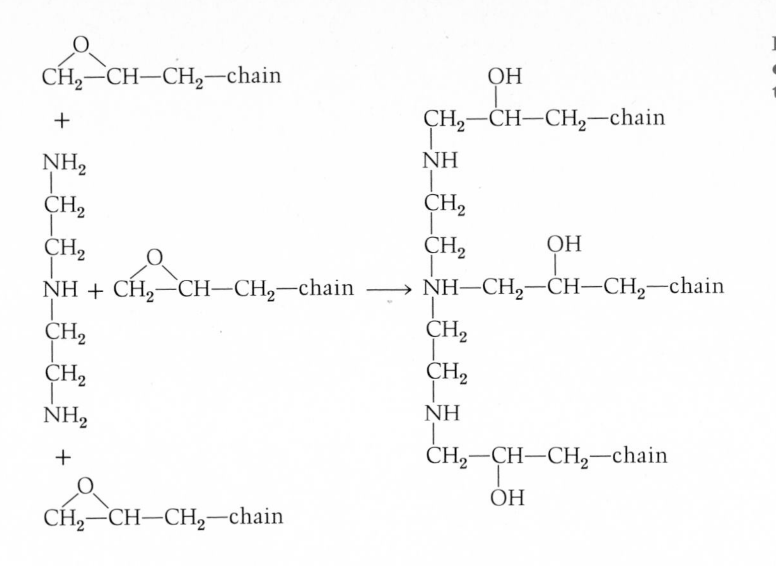

Figure 22-4 Amino groups cross-link with epoxy rings to form epoxy resins.

Silicones. Silicon, being just under carbon in the periodic table, has a valence of four, and many compounds are known in which silicon is bonded to two halogen (usually chlorine) atoms and two alkyl or aryl groups, or to three halogen atoms and one alkyl or aryl group:

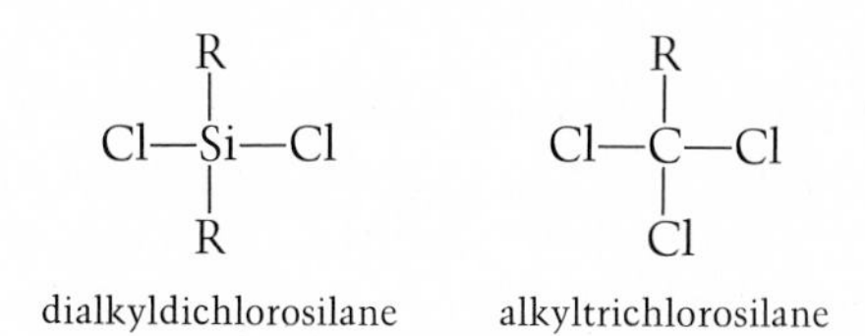

These chlorinated silanes readily react with water to give Si—O—Si compounds. If we start with a dialkyldichlorosilane, we get a linear polymer, called a **siloxane.**

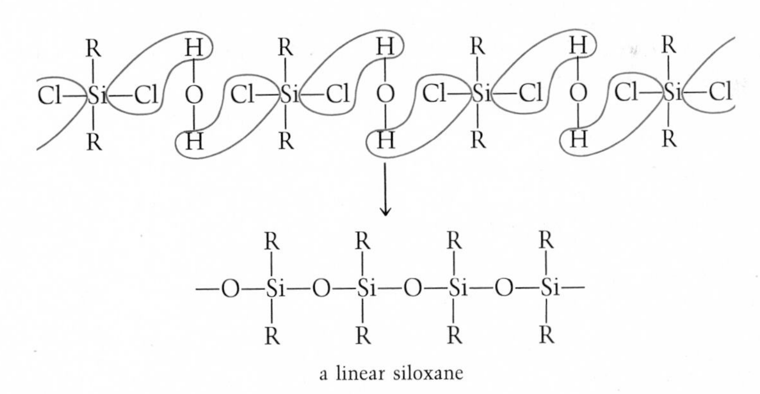

If we start with an alkyltrichlorosilane ($RSiCl_3$) or with a mixture of R_2SiCl_2 and $RSiCl_3$, we get a three-dimensional siloxane.

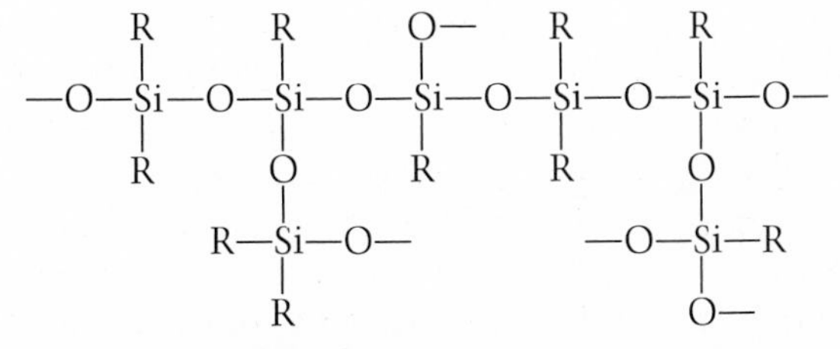

a three-dimensional siloxane made from a mixture of R_2SiCl_2 and $RSiCl_3$

A large number of different polymers can be made by varying the proportions of R_2SiCl_2 and $RSiCl_3$ and by using various combinations of alkyl and/or aryl groups. Although siloxane is the proper chemical name for these polymers, they are usually called **silicone polymers.** The three-dimensional structure of silicone polymers resembles that of silicon dioxide (Figure 18-5), and these polymers have some of the same properties. They are inert to heat and to cold, to weather, and to the action of many chemicals. They also have very high resistance to electricity and their surface properties keep other materials from sticking. Mostly or entirely linear dimethyl siloxane is a liquid which is used as a mold-releasing agent, where its antisticking properties allow a plastic or rubber molded object to be easily removed from the mold. The high heat stability and electrical resistance allow solid silicones to be used for structural and electrical parts on airplanes and missiles. Silicone polymers were originally developed for the coating of electrical wires, and they are still extensively used for this purpose. Silicones are also used to coat fabrics and glass to make them water-repellent. Solid silicones can be produced in the form of rubber. These silicone rubbers are used in applications requiring high or low temperatures, where ordinary rubber melts or becomes brittle. "Silly putty" is a silicone rubber.

Problems

22-1 Define each of the following terms. **(a)** macromolecules **(b)** monomer **(c)** polymerization **(d)** cross-linking **(e)** thermoplastic **(f)** thermosetting **(g)** copolymer **(h)** plastic **(i)** PVC **(j)** conjugated diene **(k)** vulcanization **(l)** urethane **(m)** isocyanate **(n)** alkyd **(o)** silane **(p)** silicone

22-2 What is the difference between addition and condensation polymers?

22-3 Draw structural formulas for addition polymers made from the following monomers. **(a)** $CH_2{=}CH{-}Br$ **(b)** $ClCH{=}CHCl$ **(c)** $CH_3CH{=}CHCl$

22-4 Draw a structural formula for a 50:50 copolymer of $F_2C{=}CF_2$ and $Cl_2C{=}CCl_2$. Compare its structure with that of a polymer of $Cl_2C{=}CF_2$.

22-5 The compound vinyl alcohol $CH_2{=}CH{-}OH$ does not exist because it is unstable. Yet polyvinyl alcohol is an important industrial polymer. Suggest how polyvinyl alcohol might be made.

22-6 What single monomer could give rise to the following polymer?

$$-O-CH_2-O-CH_2-O-CH_2-$$

22-7 $Cl_2C{=}CBr_2$ can be polymerized in two main ways **(a)** head-to-head and **(b)** head-to-tail. Draw the structure of the polymer that results in each case.

22-8 Which of the vitamins in Figure 21-24 contains conjugated double bonds?

22-9 Draw structural formulas for polymers of each of the following monomers. **(a)** butadiene $CH_2{=}CH{-}CH{=}CH_2$ **(b)** 2,3-diphenylbutadiene $CH_2{=}\underset{C_6H_5}{\underset{|}{C}}{-}\underset{C_6H_5}{\underset{|}{C}}{=}CH_2$ **(c)** a 50:50 copolymer of butadiene and 2,3-diphenylbutadiene **(d)** a 50:50 copolymer of 2,3-diphenylbutadiene and vinyl chloride $CH_2{=}CHCl$

22-10 Draw structural formulas for condensation polymers made from the following monomers or pairs of monomers.
(a) $H_2NCH_2CH_2NH_2$ and $HOOCCH_2CH_2COOH$ **(b)** $H_2NCH_2CH_2CH_2COOH$

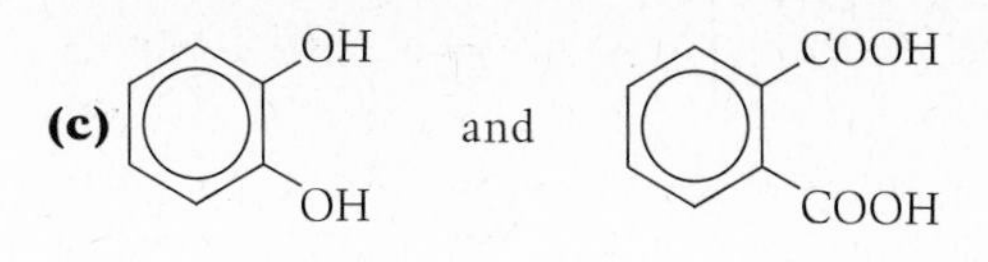

(d) $HOCH_2CH_2CH_2OH$

22-11 Calculate the percentage by weight of C, H, O, and N in nylon 66. Is it necessary to know the molecular weight in order to do this?

22-12 Draw the structure of a polymer made from 1,3-propanediol $HO—CH_2CH_2CH_2—OH$ and

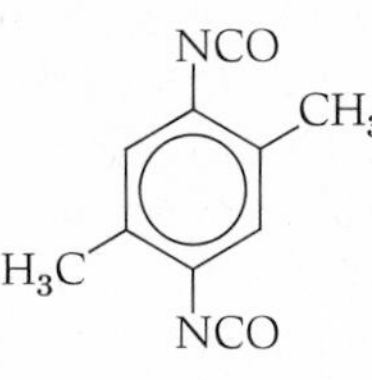

22-13 Draw the structural formula for each of the following. **(a)** nylon 55 **(b)** nylon 7

22-14 Draw the structure of a polymer that you might expect to form if succindialdehyde $H—C(=O)—CH_2CH_2—C(=O)—H$ were mixed with 1,3-propanediol $HO—CH_2CH_2CH_2—OH$, in the presence of H_3O^+. (*Hint:* See Section 20-8.)

22-15 What is the small molecule, if any, which is eliminated when the following condensation polymers are made? **(a)** nylon **(b)** Dacron **(c)** a polyurethane **(d)** a phenol-formaldehyde resin **(e)** an epoxy resin **(f)** a silicone

22-16 Draw the structure of a phenol-formaldehyde type polymer made from *para*-methylphenol and formaldehyde, and show why there is less cross-linking in this polymer than in phenol-formaldehyde itself.

22-17 Draw the structural formula for an alkyd resin made from ethylene glycol, $HOCH_2CH_2OH$, and trimesic acid,

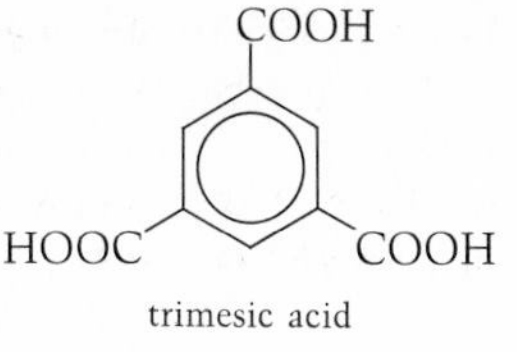

trimesic acid

22-18 Draw the structure of a polymer made from each of the following. **(a)** $(CH_3)_2SiCl_2$ and water **(b)** a mixture of $(CH_3)_2SiCl_2$ and $CH_3CH_2SiCl_3$ and water

Additional Problems

22-19 Why can't a phenol-formaldehyde type polymer be made using 2,4,6-trimethylphenol instead of phenol?

22-20 For each of the following, tell whether it would be a suitable monomer for (1) an addition polymer, (2) a condensation polymer or (3) neither. **(a)** $CH_3CH{=}CHCH_2CH_3$

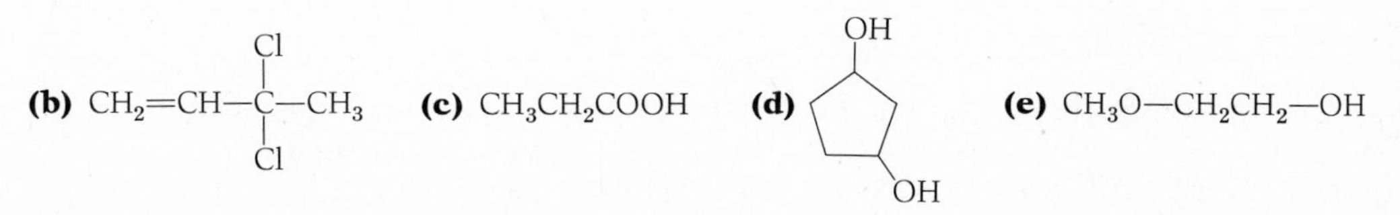

(f) 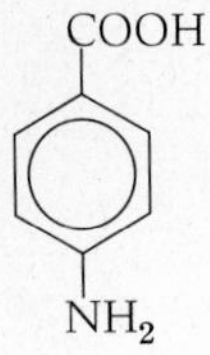 p-H_2N–C_6H_4–COOH (COOH at top of benzene ring, NH_2 at bottom)

22-21 The following are structures of polymers. Draw structures for the monomers (there may be one or two in each case).

(a) $-O-CH_2CH_2-O-CH_2CH_2-O-$

(b) $-CH_2-\underset{Br}{\underset{|}{CH}}-CH_2-\underset{Br}{\underset{|}{CH}}-CH_2-\underset{Br}{\underset{|}{CH}}-$

(c) $-\underset{O}{\underset{\|}{C}}-CH_2CH_2CH_2-\underset{O}{\underset{\|}{C}}-NH-CH_2CH_2-NH-\underset{O}{\underset{\|}{C}}-CH_2CH_2CH_2-$

(d) $-CH_2-CH{=}CH-\underset{OCH_3}{\underset{|}{CH}}-CH_2-CH{=}CH-\underset{OCH_3}{\underset{|}{CH}}-CH_2-CH{=}CH-\underset{OCH_3}{\underset{|}{CH}}-$

22-22 Assuming that no cheap substitute can be found, do you think society should stop making PVC? (See pages 739–740.)

22-23 Draw structural formulas for each of the following. **(a)** phenyltrichlorosilane **(b)** diethyldichlorosilane **(c)** trimethylchlorosilane

22-24 The following polymer could theoretically be made in two different ways. What are they?

$$-CH_2-CH_2-\underset{Cl}{\underset{|}{CH}}-\underset{Cl}{\underset{|}{CH}}-CH_2-CH_2-\underset{Cl}{\underset{|}{CH}}-\underset{Cl}{\underset{|}{CH}}-CH_2-CH_2-$$

22-25 Show why the molecular weight of a large polymer (mol wt $\approx$ 100,000) cannot be determined by the freezing-point-lowering method discussed in Section 7-9. (*Hint:* Calculate by how many degrees the freezing point of 1000 g of water would be lowered by 100 g of such a polymer, assuming that 100 g would dissolve in the water.) Why can't the molecular weight be determined by the vapor-density method (Section 4-6)?

23 Nuclear Chemistry

Marie Sklodowska Curie, 1867–1934

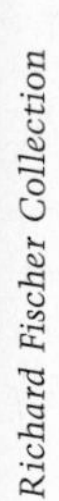

In the other chapters of this book we discuss the chemical reactions of inorganic and organic compounds. As we have seen, it is only electrons that are involved in these reactions. Chemical bonds are made up of electrons; and all chemical reactions involve bond breaking and/or bond formation; so all chemical reactions consist only of changes in the electronic structures of atoms and/or molecules. The nuclei of the atoms are not affected by chemical reactions. It makes little or no difference to a nucleus of chlorine whether it is in a chlorine atom, with 17 electrons; a Cl_2 or CCl_4 molecule, in which one of the 17 electrons is in a sigma bond; or in a chloride ion Cl^-, in which there are 18 electrons around the nucleus. The behavior of the chlorine *nucleus* is the same in all cases. What does this behavior consist of? In most cases, other than the attraction for the electron clouds, the answer is—nothing. Most of the nuclei that are present on the earth are inert—they never change. If a nucleus of chlorine consists of a bundle of 17 protons and 18 neutrons, then that bundle of 35 nucleons will remain intact and unchanged forever, despite the changes that may be taking place in the electron clouds surrounding it. However, there are exceptions. Some nuclei undergo spontaneous reactions, and others can be made to undergo reactions. It is these *nuclear reactions* that are the subject of this chapter.

23-1 The Discovery of Radioactivity

In 1896, Henri Becquerel (1852–1908) was doing experiments involving a uranium compound [potassium uranyl sulfate, $K_2UO_2(SO_4)_2$] and photographic plates. In the course of this work he put a sample of the solid uranium salt atop a sealed photographic plate in a drawer and let it remain there for several days. Since the plate was completely wrapped up, no light could get to it and it should have remained unexposed. However, when Becquerel developed the plate he found, much to his surprise, that it was fully exposed. He reasoned that some kind of radiation coming from the uranium salt had the power to penetrate through the wrapping and expose the plate. Further investigation showed that the uranium was the source of the radiation, not the potassium or sulfate. Uranium metal itself, and all other uranium compounds, also behaved in the same way.

This phenomenon was named **radioactivity** by Marie Sklodowska Curie (1867–1934) who, along with her husband, Pierre Curie (1859–1906), began to study it. In 1898, the Curies found that thorium and its compounds were also radioactive. In the same year, they discovered two new radioactive elements: polonium and radium, which they isolated from uranium ore.

The fact that atoms like uranium, thorium, polonium, and radium give off radioactivity proved that these atoms could not be the hard indivisible particles that Dalton had thought them to be. The radiation had to come from somewhere, and it could only be from the atom itself. It was the discovery of radioactivity, along with the work of J. J. Thomson (Section 12-3) that convinced the scientific world that atoms were indeed composed of particles smaller than themselves. We know today that radioactivity comes from the *nucleus* of atoms and not from the electron clouds.

Becquerel and the two Curies shared the 1903 Nobel Prize in Physics for the discovery of radioactivity. In 1911, Madame Curie was awarded a second Nobel Prize, this time in chemistry, for the discovery of radium and polonium.

23-2 Kinds of Radioactivity

Becquerel, the Curies, and Ernest Rutherford all showed that there were three different kinds of natural radioactivity. This can be demonstrated in the following way. Drill a hole in a lead block and put a sample of radioactive material in the hole. The effect of this is to produce a stream of radioactivity in one direction (Figure 23-1), since radiation in all other directions is absorbed by the lead. Then allow the stream of radioactivity to pass through an electric field. The three kinds of radioactivity behave differently in the presence of this field. One kind is attracted to the negative side, one to the positive side, and the third is not attracted to either side. These are called alpha, beta, and gamma, respectively. Not all radioactive substances give off all three kinds of radioactivity.

1. *Alpha particles*. These are the most massive of the three types. They are positively charged and hence attracted to the negative side of the field. They consist of helium nuclei He^{2+}; that is, each alpha particle is a bundle consisting of two protons and two neutrons (Figure 23-2). Because of the large mass, alpha particles have low energies and low penetrating power. They are absorbed by or bounce off of the outer layer of skin and are not particularly harmful, unless ingested.
2. *Beta particles*. These are electrons. A "beta ray" is a stream of electrons, e^-. Their negative charge attracts them to the positive side of an electrical field. They have much higher energies than alpha particles, penetrate farther, and are more harmful.
3. *Gamma rays*. This type of radioactivity does not consist of matter at all. It is simply electromagnetic radiation (Section 12-1) of a very high frequency. Being neutral, it is not attracted by an electric field. The energy of gamma radiation is the highest of the three kinds; it has the most penetrating power and is the most harmful to plants, animals, and human beings.

When alpha, beta, or gamma radiation strikes matter, it can cause various effects. For example, if it passes through a gas, the impact on the gas molecules can be great

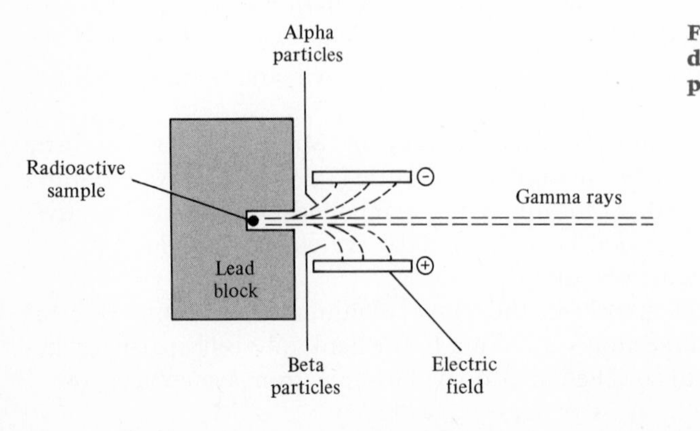

Figure 23-1 The three kinds of radiation behave differently in the presence of an electric field.

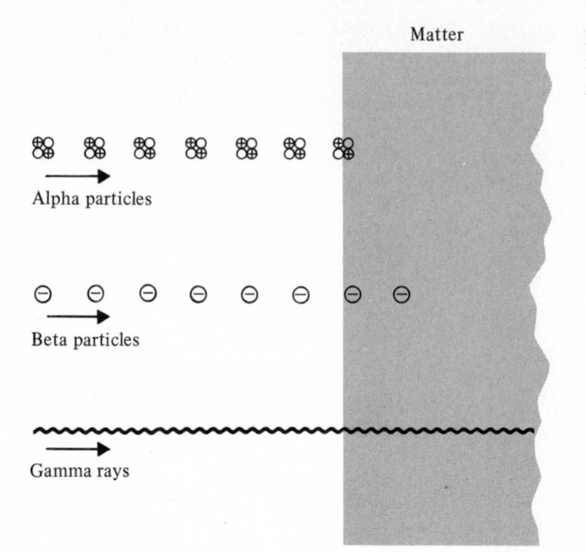

Figure 23-2 The three kinds of radioactivity. Alpha particles penetrate the least, gamma rays the most.

enough to knock out one or more electrons, converting the atoms of the gas to positive ions. Several devices for the detection and quantitative measurement of radiation depend on this property. One of these is the Geiger–Müller counter, whose operation is shown in Figure 23-3. Radioactivity enters the tube through a thin window and ionizes the argon atoms in its path. The presence of the Ar^+ ions causes current to leap from the cathode to the anode, producing a pulse of current, which is amplified and counted. The Wilson cloud chamber is another device that depends on this property (Figure 23-4). Here the chamber contains water vapor. When radiation creates a path of ions, each ion becomes a nucleus around which the water vapor condenses. The track of water droplets can be seen with the naked eye.

With some materials, the impact of radioactivity causes quite a different effect: instead of knocking electrons out completely, it promotes them to an upper energy level. They do not remain in the upper level for long but immediately drop down again, which means of course that they must give up the same amount of energy that they absorbed in moving to the upper level. Depending on the material, this energy may be in the form of heat or light. In a *scintillation counter,* the compound present is ZnS, which gives off light. The light is converted to electricity, and the radiation is counted in this manner. It is not only scintillation counters that make use of this effect. The same effect is used in luminous watches, whose hands are painted with an extremely small amount of a radium salt along with zinc sulfide or another compound that behaves similarly. The radium constantly gives off radiation, which is absorbed by the ZnS, which then gives off light.

As everybody knows, radioactivity is harmful to living things, and the extent of the harm increases with the extent of the exposure. Both Becquerel and Pierre Curie found that radium in contact with the skin would cause burns. Radioactivity is

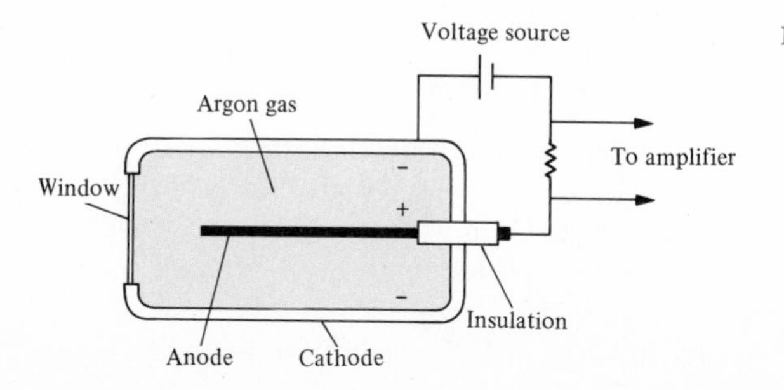

Figure 23-3 A Geiger-Müller counter.

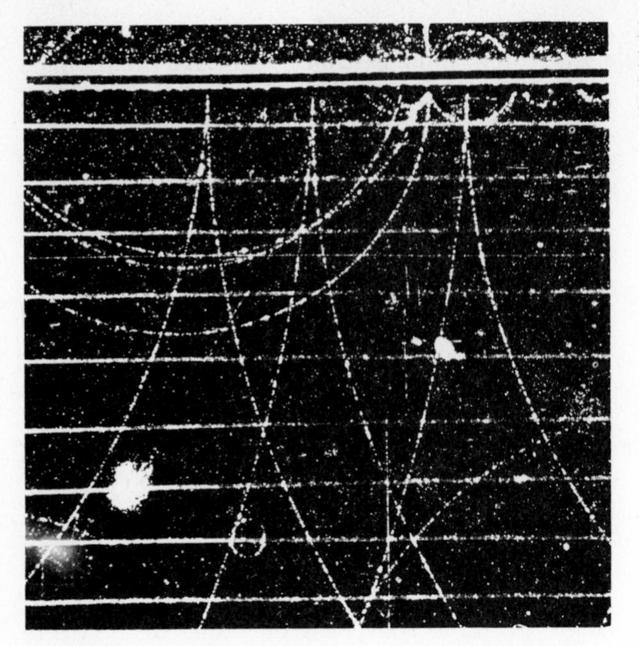

Figure 23-4 Tracks of electron-positron pairs in a Wilson cloud chamber. [Courtesy Lawrence Radiation Laboratory.]

invisible and silent. We cannot hear it, see it, feel it, or smell it. It is quite possible to receive enough radioactivity to kill you within a few days and yet not to know that you have been exposed to it. People who work in areas where radioactivity is found must wear dosimeters or radiation badges, which show how much radioactivity they are exposed to over a period of time. It is clear that a unit is required that will let us know the total amount of exposure to radioactivity. Such a unit is the **roentgen,** which is the dose of radiation that will produce 1.6×10^{12} ion pairs in 1 g of air. This amount of radiation corresponds to 83 ergs of ionization energy. A similar unit is the **rad,** which is based on 100 ergs instead of 83. Because the different types of radiation have different effects on humans, a more common unit is the **rem** (*r*adiation *e*quivalent, *m*an). Rems are related to roentgens as follows:

Gamma rays	1 rem = 1 roentgen
X rays	1 rem = 1 roentgen
Beta particles	1 rem = 0.1 roentgen
Alpha particles that get into the body	1 rem = 0.01 roentgen

The harm caused to human beings by radiation is cumulative over a lifetime. Therefore, what counts is the total received. It is not possible to live on earth and be totally free of radiation. The presence of naturally radioactive rocks and minerals, as well as cosmic rays, which come to us from outer space, make sure that everybody gets some exposure to radiation. This is called **background radiation** and amounts to a total of about 6.4 rem for a 70-year lifetime spent at sea level (the amount is higher at higher altitudes because of greater exposure to cosmic rays). As far as we can tell, the background radiation has little or no effect on the health of human beings. However, higher doses do have effects, as follows:

1. A single dose of about 400 or 500 rems results in death within a day or two. This happened to many people who escaped the immediate effects of the blasts and fire at Hiroshima and Nagasaki in 1945, as well as to a few scientists and workers in nuclear industries who have accidentally been exposed to such radiation.
2. A single dose of about 100 to 200 rems causes radiation sickness, whose

symptoms include nausea, high fever, vomiting, and loss of hair. Recovery is usually very slow. People who work with radioactive materials are usually limited to an exposure of no more than 5 rems/yr.

3. Small amounts of radiation over a long period of time can cause cancer, which may not show up until years later. From about 1915 to 1925 thousands of women were employed to paint radium solutions on watch dials, so that they would glow in the dark. Many of the women licked the brushes to put a fine point on them, not realizing that radium was harmful. A substantial number of these women died early deaths from cancer, generally cancer of the mouth and lips.
4. Small doses of radioactivity can have quite a different effect: they can cause mutations of the genes (Section 21-6) and thus affect the heredity of the individual and the species. There is no lower limit to this effect. The background radiation of the earth certainly produces a number of mutations in the various species each year, and this is probably an important source of the mutations that have been happening on earth since life began. However, most mutations are either lethal or otherwise deleterious to the organism, and any increase in radiation by man-made sources can only be harmful in the long run. The Federal Research Council has set a limit of 0.17 rem/yr (12 rems per 70-year lifetime) above background radiation as the maximum radiation allowable for the general population.

Another unit used in measuring radioactivity is the **curie.** One curie is defined as 3.70×10^{10} radioactive disintegrations per second (this is the rate of radioactivity from 1.00 g of radium). A curie is a large amount of radioactivity, and more common units are millicuries and microcuries.

23-3 Radioactive and Stable Isotopes

In Section 2-6, we defined isotopes as atoms of the same element which contain different numbers of neutrons. There are more than 300 different isotopes found on earth. Some of these isotopes are stable; others radioactive. Besides these, hundreds of other isotopes have been artificially synthesized by man, in ways that are discussed in Section 23-6. All of the artificial isotopes are radioactive.

Stable isotopes give off no radioactivity, and exist forever, as far as we can tell. At least one stable isotope is known for all elements up to bismuth (Z = atomic number 83), except for technetium ($Z = 43$) and promethium ($Z = 61$). No stable isotope is known for any element with an atomic number greater than 83, that is, for polonium and higher elements. Altogether, there are 264 stable isotopes of 81 elements. These isotopes exist on earth in various abundances. For example, in Table 2-4 we saw that 80.22% of naturally occurring boron is $^{11}_{5}B$, while the rest is $^{10}_{5}B$.

There is no satisfactory theory to explain why some isotopes are stable and others are not, but we do know certain things. One thing is that neutrons are necessary to keep protons together. The only stable nucleus without neutrons is $^{1}_{1}H$, which has only a single proton. Since protons are positively charged, it might be expected that two or more protons close together in a nucleus would repel each other and fly apart. Of course, they do not do so. Some force much more powerful than the repulsive force of Coulomb's law holds them together. We do not completely understand the nature of this force, which physicists call the *strong interaction*, but the neutrons have something to do with it.

If we examine the 264 stable isotopes, we find some regularities.

1. At the low end (up to about $Z = 20$), the most stable nuclei have about equal numbers of protons and neutrons. Beyond this, the neutron/proton ratio

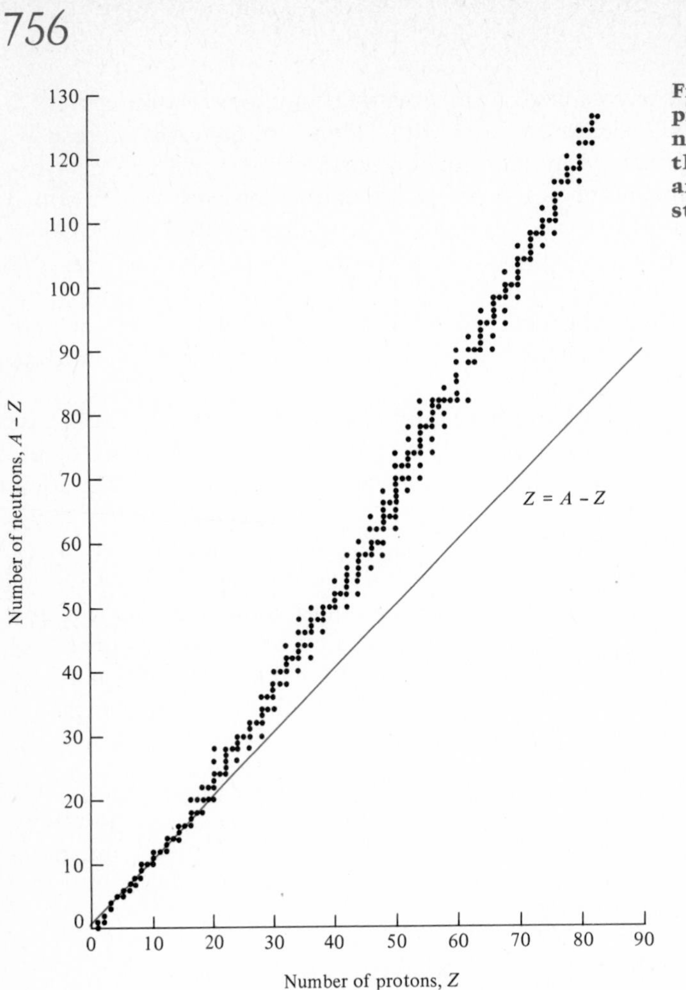

Figure 23-5 Neutron/proton ratio of the stable nuclei. Each dot shows the number of protons and neutrons in a known stable nucleus.

increases, until at bismuth it reaches about 1.5. Figure 23-5 is a plot of the stable nuclei, with number of protons plotted against number of neutrons. From this graph we can see that for the light nuclei, the number of neutrons required to keep the nucleus intact is about equal to the number of protons, but beyond about $Z = 20$, this is insufficient, and more neutrons are required.

2. Most of the stable nuclei have even numbers of protons and/or neutrons. As shown in Table 23-1, of the 264 stable nuclei 157 have both an even number of protons *and* an even number of neutrons. Only 5 ($^{2}_{1}H$, $^{6}_{3}Li$, $^{10}_{5}B$, $^{14}_{7}N$, $^{180}_{73}Ta$) have both an odd number of neutrons and an odd number of protons.
3. Certain numbers seem to be particularly stable. These are 2, 8, 20, 28, 50, 82, and 126. In general, atoms with these numbers of protons have a larger number of stable isotopes than others, and atoms whose numbers of protons and neutrons both belong to this list show an especially high stability (for example, $^{4}_{2}He$, $^{16}_{8}O$, $^{40}_{20}Ca$, $^{208}_{82}Pb$). This fact reminds us of the stable electronic configurations (2, 8, 18, 32, . . .) we saw in Chapter 13, and has led to the hypothesis that

Table 23-1 Number of Stable Isotopes

Protons	*Neutrons*	*Number of stable isotopes*
Even	Even	157
Even	Odd	52
Odd	Even	50
Odd	Odd	5

perhaps there are shells of nucleons in the nucleus similar to the electron shells outside the nucleus.

4. Above $Z = 83$, no nucleus is stable, no matter how many neutrons are present. This is the upper limit for stable nuclei.

At this time, our knowledge of the nucleus is less than our knowledge of electrons, and we do not completely understand why even numbers should be so stable, why a proton/neutron ratio of 1 is stable for the lower elements but more neutrons are needed for the higher elements, or any of the other facts listed above.

23-4 Half-life

Radioactive nuclei are not stable. As we shall see in Section 23-5, when a nucleus emits an alpha or beta particle, it changes and becomes the nucleus of a different element. For this reason, radioactivity is also called **radioactive decay** or **radioactive disintegration.**

Radioactive decay is always a first-order reaction. As we saw in Section 9-4, the integrated rate expression for a first order reaction is

$$\log \frac{[A]_0}{[A]} = \frac{kt}{2.303}$$

where $[A]_0$ = initial concentration
$[A]$ = concentration at time t
k = rate constant for a given radioactive isotope

However, the rate of radioactive decay is more often given in terms of half-life (Section 9-5):

$$\text{half-life} = \tau = \frac{0.693}{k}$$

As with all first-order reactions, the half-life of radioactive decay is independent of the initial concentration. For example, assume that we have a sample of 10 g of $^{131}_{53}I$, which has a half-life of 8 days. At the end of 8 days only 5.0 g will remain. Eight days later, there will be 2.5 g left, and 8 days after that 1.25 g. The process continues, with half of each quantity remaining after every period of 8 days, until, finally, the number of $^{131}_{53}I$ atoms becomes so small that the rate law is no longer followed. For practical purposes we can say that any sample of a radioactive isotope will completely decay by the end of about 8 or 10 half-lives.

Not only is the half-life independent of initial concentration, but also of temperature and pressure. The fact that the rate of radioactive decay does not change with changing temperature means that the activation energy (Section 9-6) for the reaction is zero. The rate and half-life of radioactive decay are also independent of the chemical environment of the nucleus. It does not matter if $^{131}_{53}I$ nuclei are present in I_2, CH_3I, BI_3, or SI_2 molecules; in I^-, I^+, or I_3^- ions; or in any other form of iodine. The half-life is 8 days in all cases. In fact, rates and half-lives of radioactivity are constants under all conditions. Despite a great deal of effort, no one has been able to discover any way to change them. One-half of any sample of $^{131}_{53}I$ will decay in 8 days. There is no way known to make this happen faster or slower than that.

The half-lives of radioactive isotopes, being constants, can be found in tables in handbooks such as the *Handbook of Chemistry and Physics* (CRC Press, Cleveland, Ohio). From these values we can easily calculate how long it will take for any portion of a radioactive isotope to decay and how much will be left at the end of any given time.

Example: The half-life of $^{90}_{38}Sr$ is 28.1 yr. How long will it take a 10.0-g sample of $^{90}_{38}Sr$ to be reduced to 0.100 g?

Answer: To solve this problem, first find k from the equation relating k and half-life:

$$\tau = \frac{0.693}{k}$$

$$k = \frac{0.693}{28.1 \text{ years}} = 0.0247/\text{yr}$$

Now that we have k, we can use the first-order rate equation given above:

$$\log \frac{[A]_0}{[A]} = \frac{kt}{2.303}$$

$$\log \frac{10.0}{0.100} = \frac{(0.0247/\text{yr})t}{2.303}$$

$$t = \frac{(2.303)(\log 1.00 \times 10^2)}{0.0247/\text{yr}} = 186 \text{ yr}$$

Thus, it takes 186 years for 10.0 g of $^{90}_{38}Sr$ to be reduced to 0.100 g or for any amount of this isotope to be 99% decayed. This problem has particular revelance because $^{90}_{38}Sr$ is present in the fall-out from nuclear bomb explosions, and some of it found its way into grass, the bodies of cows, and the milk produced by these cows.

Example: The half-life of $^{137}_{58}Ce$ is 9.0 h. How much of a sample of 10.0 g will remain after 24 h?

Answer:

$$\tau = \frac{0.693}{k}$$

$$k = \frac{0.693}{9.0 \text{ h}} = 0.077/\text{h}$$

$$\log \frac{[A]_0}{[A]} = \frac{kt}{2.303}$$

$$\log \frac{10.0}{[A]} = \frac{(0.077/\text{h})\ (24 \text{ h})}{2.303}$$

$$\log \frac{10.0}{[A]} = 0.80$$

$$\frac{10.0}{[A]} = \text{antilog } 0.80 = 6.3$$

$$[A] = \frac{10.0}{6.3} = 1.6 \text{ g of } ^{137}_{58}\text{Ce will remain}$$

As far as we can tell, radioactive decay seems to be a *random* process. Assume that we have a sample of 1 million atoms of $^{147}_{64}Gd$, which has a half-life of 35 h. We know that by the end of 35 h approximately 500,000 of the million atoms will have decayed; the other 500,000 will not (though about 250,000 of these will decay in the next 35 h). But what is it that makes one particular atom decide to decay now, while its neighbor waits for later? We do not know. To the extent of our knowledge, all the

Table 23-2 Half-lives of Some Radioactive Isotopes

$^{13}_{8}O$	8.7×10^{-3} s
$^{115}_{47}Ag$	20 s
$^{183}_{78}Pt$	7 min
$^{226}_{89}Ac$	29 h
$^{95}_{41}Nb$	35 days
$^{63}_{28}Ni$	92 yr
$^{14}_{6}C$	5730 yr
$^{50}_{23}V$	6×10^{15} yr

million atoms appear to be identical, and indeed, eventually they all will decay, but some decay sooner, others later. If we have a sample of an isotope with a really long half-life (for example, $^{238}_{92}U$, half-life = 4.51 billion years), then the atoms we hold in our hands have been in existence for billions of years, yet now, while we are holding the sample, some atoms suddenly decay, while others will wait yet another billion years before doing so. Why this should be is a mystery that we have not yet solved. As long as large numbers of atoms are present, the laws of statistics tell us that half-lives will be precisely obeyed. Just as we know, for example, the percentage of 65-year-old males who will die in the United States in any given year, we know how long it will take for half of any sample of a radioactive isotope to decay. But when the sample becomes very small—a few hundred or a few thousand atoms—statistics are no longer obeyed, and after that the number of atoms which decay in a given period can no longer be accurately predicted.

The thousand or more radioactive isotopes (most of which are man-made and not found in nature) vary enormously in their half-lives. Some have half-lives as short as that of $^{8}_{4}Be$: 2×10^{-16} s. Any sample of this isotope is completely gone in much less than a second. At the other extreme are atoms such as $^{238}_{92}U$, mentioned above, with a half-life of 4.51×10^{9} years. Some half-lives are expressed in minutes, some in hours, some in years. Table 23-2 shows examples. Since we do not know the reason why any particular nucleus is stable or unstable, we obviously do not know the reason for the wide variation in half-life.

An isotope with a short half-life decays very quickly. Therefore, the radioactive isotopes found in nature must belong to one of two categories.

1. Those with very long half-lives. The earth was formed perhaps 4 billion years ago. Any radioactive isotope that began its existence on the earth at that time will long since have completely decayed, unless its half-life is more than about half a billion years. A number of isotopes with long half-lives have been on the earth since the beginning. Among them are $^{238}_{92}U$ and $^{232}_{90}Th$ (half-life 1.41×10^{10} yr).
2. Those which are formed by some process that occurs in nature. Most of the naturally occurring radioactive isotopes belong to this category. They are formed by decay of other radioactive isotopes (Section 23-5), or by interaction of certain kinds of atoms in the upper atmosphere with radiation coming from the sun or outer space. For example, $^{14}_{6}C$ decays with a half-life of 5730 yr, but the supply is constantly being replenished, as we shall see in Section 23-11.

23-5 The Products of Radioactive Decay. Natural Transmutation

So far we have focused our attention on the radioactivity coming out of the nuclei of radioactive atoms. We will now turn to the residue—what is left behind when radioactivity is emitted. Around 1900, Ernest Rutherford and Frederick Soddy realized that when an atom gives off an alpha or beta particle it turns into an atom of a different element. For example, when a uranium atom loses an alpha particle, it becomes a thorium atom. The new atom is called the **daughter** of the original one. The transmutation of one element to another was the essential goal of alchemy (in this case it was "base metals" that alchemists hoped to change into gold). By 1900, transmutation had long been regarded as impossible by all scientists. Yet Rutherford and Soddy had proof that nature was spontaneously changing uranium into thorium! When publishing their work they deliberately avoided using the word "transmutation," but that is exactly what they had discovered. In this section, we will deal with natural transmutation; in Sections 23-6 to 23-9, we will see that man-made transmutation is also possible.

ALPHA PARTICLES. When a nucleus loses an alpha particle, it has lost two protons and two neutrons. In losing two protons its atomic number has decreased by two, which of course means that it is now the nucleus of a different element, two places to the left in the periodic table. Because it has lost two neutrons also, the new mass number is always four units less than the old one. For example, $^{225}_{89}Ac$ is an alpha emitter. The daughter nucleus must have a mass number of 221 and an atomic number of 87. The atomic number of 87 makes it a nucleus of francium. We can write the equation for the nuclear reaction as

$$^{225}_{89}\text{Ac} \longrightarrow {}^{221}_{87}\text{Fr} + {}^{4}_{2}\text{He} \qquad \text{(half-life 10.0 days)}$$

In a nuclear equation, the lower numbers represent the charge on the nucleus (Z) and the upper numbers the total number of protons plus neutrons (the mass number, A). In all nuclear equations both Z and A must balance. (Why?) Electrons are ignored in nuclear equations. In general, alpha particles are emitted mostly by nuclei with high atomic numbers—greater than $Z = 82$. We have already seen that such atoms are unstable, and the loss of alpha particles is nature's method of reducing their atomic numbers.

BETA PARTICLES. There are no electrons in the nucleus. Yet beta particles, which are electrons, are emitted from the nucleus by many radioactive isotopes (beta emitters). The emitted electron comes from a neutron, which upon losing the electron is turned into a proton, in a process we do not fully understand. Since the nucleus now has one more proton, its atomic number has been *increased* by one, and it has now moved one place to the *right* in the periodic table; however, the total number of neutrons and protons has not changed, and A remains the same.

Example: $^{78}_{32}Ge$ is a beta emitter. Write the nuclear equation for the process.

Answer:

$$^{78}_{32}\text{Ge} \longrightarrow {}^{78}_{33}\text{As} + {}^{0}_{-1}e \qquad \text{(half-life 1.47 h)}$$

In order to make the lower numbers balance, we must include the charge of -1 on the electron.

Many nuclei are beta emitters, especially those which lie *above* the band of dots in Figure 23-5. Nuclei in this position have too high a neutron : proton ratio, and loss of a beta particle makes this ratio more favorable.

GAMMA RAYS. Gamma rays are usually emitted along with alpha or beta particles, though some isotopes emit gamma radiation without an accompanying particle. Unlike loss of alpha or beta particles, loss of gamma radiation causes no change in atomic number or mass number. It results from a realignment of the nucleons in the nucleus. Loss of an alpha or beta particle often leaves the new nucleus in an excited state (analogous to the excited electron states discussed in Chapter 13). When it drops down to a lower state, the extra energy is given off as gamma rays. An example is

$$\underset{\text{excited state}}{[^{77}_{34}\text{Se}]^*} \longrightarrow \underset{\text{ground state}}{^{77}_{34}\text{Se}} + \gamma \qquad \text{(half-life 17.5 s)}$$

The three types of radioactivity mentioned above are the only kinds known for isotopes that occur in nature. However, it has been found that two other types of radioactive process are possible, which occur in certain man-made isotopes. In both of these processes the neutron/proton ratio increases, so they are generally found in isotopes which lie *below* the band of dots in Figure 23-5. The two processes are electron capture and positron emission.

Electron Capture. This process (also called **K capture**) is the opposite of beta emission. The nucleus captures an electron from the *K* or *L* shell of the electron cloud and uses the electron to convert a proton to a neutron. Thus, the mass number is unchanged, but the atomic number has decreased by one.

Example: $^{71}_{32}$Ge undergoes electron capture. Write the nuclear equation.

Answer:

$$^{71}_{32}\text{Ge} + {}^{0}_{-1}e \longrightarrow {}^{71}_{31}\text{Ga} \qquad \text{(half-life 11.4 days)}$$

The daughter isotope $^{71}_{31}$Ga must then demote one of its outer electrons to the *K* or *L* shell to make up for the loss.

Positron Emission. A **positron** is a particle that has the same mass as an electron but a charge of +1 rather than −1. When a positron is emitted from the nucleus, a proton is turned into a neutron, so that as in the case of electron capture, the mass number is unchanged while the atomic number decreases by one.

Example:

$$^{122}_{53}\text{I} \longrightarrow {}^{122}_{52}\text{Te} + \underset{\text{positron}}{^{0}_{1}e}$$

Antimatter. A positron is an antiparticle. After positrons were first discovered in 1932 by Carl Anderson (1905–), it seemed likely that antiprotons and antineutrons might also be possible. These two particles were indeed later discovered. When a positron collides with an electron, both are completely annihilated, and their total mass is converted to energy in the form of gamma rays. The same thing happens when any antiparticle meets a normal particle. We can conceive of a hypothetical planet made up entirely of antimatter: positrons, antiprotons, and antineutrons. Such a planet would function in exactly the same way as a planet made up of ordinary matter as long as there is no contact. Indeed, it is conceivable that some stars and planets in our universe are completely composed of antimatter. However, any contact between antimatter and ordinary matter results in complete annihilation. We know, therefore, that there is no antimatter normally present on our planet. Any such particles that are produced in laboratories can survive only until they touch a piece of ordinary matter.

Radioactive Decay Series. In many cases, the daughter nucleus of a radioactive isotope, either natural or man-made, is itself radioactive. Its daughter, in turn, may also be radioactive, and in this way a long series of isotopes may be formed, starting with just one. Three such series are found in nature, beginning with $^{238}_{92}U$, $^{235}_{92}U$, and $^{232}_{90}Th$. The $^{238}_{92}U$ series is shown in Figure 23-6. If one of the members of the series is a gas (for example, radon) it may escape, but otherwise all the members of series will be found together in one place. For example, analysis of a rock containing $^{238}_{92}U$ will show the presence of $^{234}_{90}Th$, $^{234}_{91}Pa$, $^{230}_{90}Th$, etc. All the members of such a series continue to decay, each with its own half-life, until a stable isotope is reached. In the case of the three natural series, this is an isotope of either lead or bismuth. This explains how the Curies found radium and polonium in a uranium ore.

A knowledge of the half-lives in these series allows us to estimate the age of rocks. Many of the rocks on the earth contain at least some uranium. From an analysis of the relative proportions of $^{238}_{92}U$ and $^{206}_{82}Pb$ present in a rock, it is possible to estimate how long ago the rock was formed. Even though the quantities involved may be very tiny, our modern methods of analysis are good enough to measure them. For rocks that do not contain uranium, other elements have been used, most notably $^{40}_{19}K$, which decays to $^{40}_{18}Ar$ with a half-life of 1.3×10^9 yr. The oldest rocks on earth, as determined by these methods, are about 4 billion years old, meaning that the earth is at least as old as this.

From Figure 23-6 it can be seen that alternative pathways may exist in some series. For example, most nuclei of $^{218}_{84}Po$ decay by alpha emission, but a small percentage emit beta particles instead.

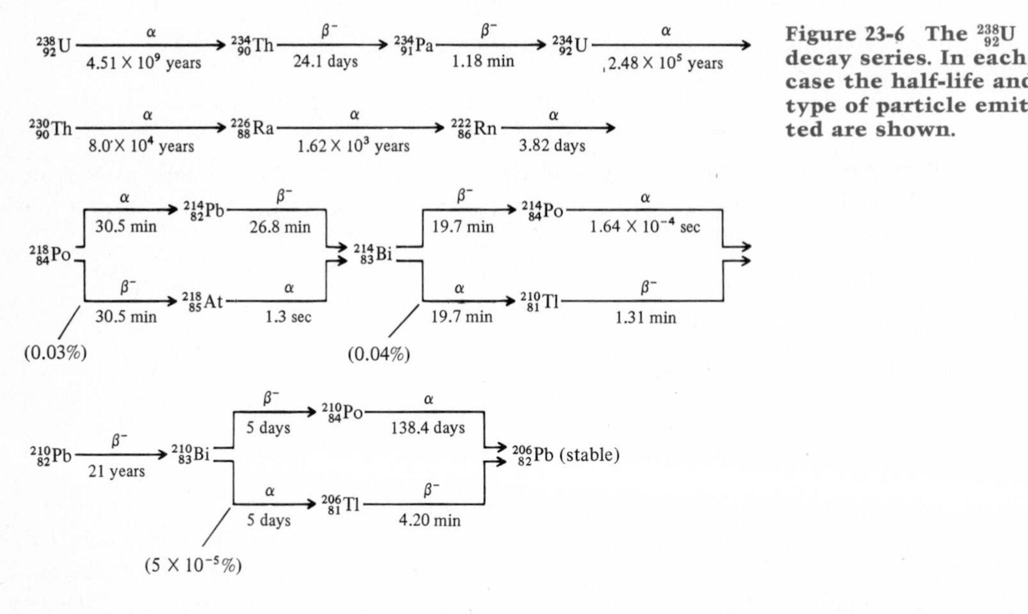

Figure 23-6 The $^{238}_{92}U$ decay series. In each case the half-life and type of particle emitted are shown.

23-6 Artificial Transmutation

After Rutherford and Soddy demonstrated that transmutation was not impossible, but in fact had been occurring in nature for billions of years, the idea naturally arose that perhaps man-made transmutations might also be possible. In 1919, Rutherford reported the first successful artificial transmutation. He changed nitrogen-14 into oxygen-17, by sending a stream of alpha particles into a container of nitrogen gas. The nuclear equation is

$$^{14}_{7}N + ^{4}_{2}He \longrightarrow ^{17}_{8}O + ^{1}_{1}H$$

The other product is hydrogen nuclei ^1_1H. Rutherford thus became the first person to realize the dream of the alchemists: to turn one element into another, though not in the way they had hoped for (he didn't produce gold).

Since 1919, many other artificial nuclear reactions have been run. In general, the method is to allow a stream of small particles, which may be alpha particles, protons, deuterium nuclei (called **deuterons**), or other small positive ions, to collide with larger target molecules. Some examples are:

$$^7_3Li + ^1_1H \longrightarrow ^7_4Be + ^1_0n \qquad (p,n)$$
$$^{31}_{15}P + ^2_1H \longrightarrow ^{32}_{15}P + ^1_1H \qquad (d,p)$$
$$^{14}_7N + ^1_1H \longrightarrow ^{15}_8O + \gamma \qquad (p,\gamma)$$

It is customary to classify such reactions by indicating the particle which bombards, and that which is given off. Thus, the first equation shown is a (p,n) reaction, because protons collide with lithium atoms, and neutrons are given off. The symbol d is used for deuterons and α for alpha particles. The original Rutherford reaction is an (α,p) reaction. The third reaction listed shows that a particle is not always given off. In this case, the $^{14}_7N$ nucleus captures the proton and keeps it, turning into a $^{15}_8O$ nucleus. Only gamma rays are given off.

Alpha particles given off by a radioactive nucleus make inefficient projectiles because they do not travel very fast, and are repelled by the target nuclei (because like charges repel), especially by nuclei of the heavier elements. This problem arises with any positively charged projectile. There are two ways to overcome it.

1. Accelerate the particle to very high speeds. The higher the speed, the greater the force with which the particles strike the target nucleus, and the easier it is to overcome the repulsion. Positive particles can be accelerated in a number of ways, but the most important is by use of the **cyclotron,** invented by Ernest O. Lawrence (1901–1958) in 1930. In a cyclotron (Figure 23-7), the particle is forced to move in a spiral path by the influence of electrical and magnetic fields until it emerges at a high velocity. Linear accelerators can also be used.
2. Use a neutral particle, the obvious choice being the neutron. Neutrons, being neutral, are not repelled by a positive nucleus. Many nuclear reactions have been run with neutrons, examples being

$$^{45}_{21}Sc + ^1_0n \longrightarrow ^{45}_{20}Ca + ^1_1H \qquad (n,p)$$
$$^{27}_{13}Al + ^1_0n \longrightarrow ^{24}_{11}Na + ^4_2He \qquad (n,\alpha)$$
$$^{34}_{16}S + ^1_0n \longrightarrow ^{35}_{16}S + \gamma \qquad (n,\gamma)$$

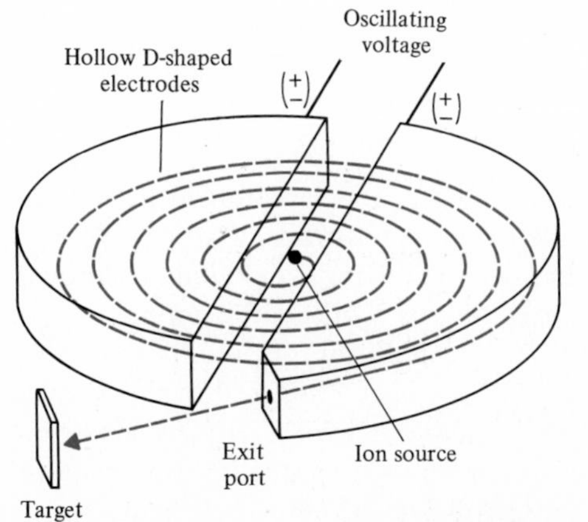

Figure 23-7 Schematic diagram of a cyclotron.

Where do we get the neutrons from? One way to get neutrons is from another nuclear reaction which gives off neutrons. A good example is a mixture of beryllium oxide with a small amount of radium sulfate. The $^{226}_{88}Ra$ in the mixture decays with a half-life of 1620 yr, emitting alpha particles.

$$^{226}_{88}Ra \longrightarrow ^{222}_{86}Rn + ^{4}_{2}He$$

The alpha particles collide with beryllium nuclei, giving neutrons.

$$^{4}_{2}He + ^{9}_{4}Be \longrightarrow ^{12}_{6}C + ^{1}_{0}n \qquad (\alpha,n)$$

The neutrons come out of the mixture, and will continue to come out as long as the radium lasts. They may then be used to collide with any suitable target. Another important source of neutrons is nuclear reactors, which we will discuss in Section 23-8.

What actually happens in a nuclear reaction of the type we have been discussing is not completely understood, but it is fairly certain that in such cases the bombarding particle is captured by a target nucleus, which thus becomes bigger (and is now called a **compound nucleus**). As we have seen, sometimes the reaction ends here, with only energy given off, but in most cases, the compound nucleus breaks down by giving off a small particle. The important thing is that the compound nucleus lives for a comparatively long time before it breaks down. The lifetime is only about 10^{-12} s, but this is a very long time on the nuclear scale, and it means we can consider the formation and breakdown of the compound nucleus as two entirely separate events. Evidence is that the reaction between $^{27}_{13}Al$ and neutrons gives four different pairs of products, and that the same pairs of products, in the same proportions, are obtained by bombardment of $^{27}_{13}Al$ with deuterons and of $^{31}_{15}P$ with neutrons.

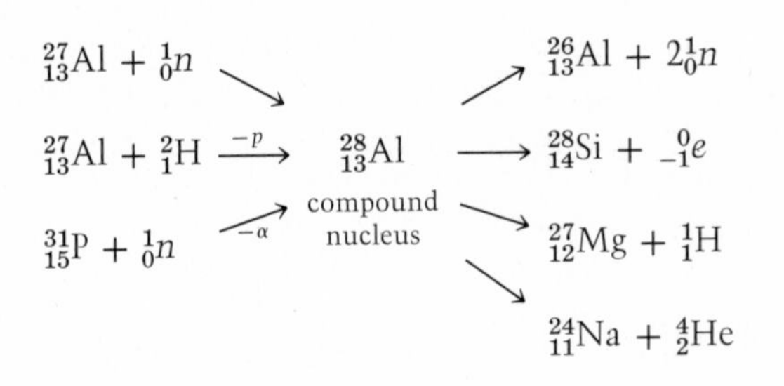

These results are explained by the formation of a compound nucleus $^{28}_{13}Al$ which can decay four different ways, depending on which particle is ejected.

The isotope produced by Rutherford in 1919, $^{17}_{8}O$, is a naturally occurring isotope. The first artificial isotope was made in 1934 by Frédéric (1900–1958) and Irène (1897–1956) Joliot-Curie (the latter was the daughter of Pierre and Marie Curie), who bombarded $^{27}_{13}Al$ with alpha particles to produce $^{30}_{15}P$.

$$^{27}_{13}Al + ^{4}_{2}He \longrightarrow ^{30}_{15}P + ^{1}_{0}n \qquad (\alpha,n)$$

$^{30}_{15}P$ was the first of more than a thousand artificial isotopes produced by nuclear reactions. Like all the others, it is radioactive (it emits positrons, with a half-life of 2.5 min). At least one radioactive isotope has been made for every element known.

However, perhaps the most exciting aspect of this work is the production of isotopes of elements which are not found in nature at all (or found only in traces). Elements 43 (Tc), 61 (Pm), 85 (At), and 87 (Fr) have been produced in this way, as have all the elements above uranium (92). The elements with $Z = 93$ or above are called **transuranium elements.** The first of these, neptunium, was made in 1940 at the University of California, by bombardment of $^{238}_{92}U$ with neutrons.

$$^{238}_{92}U + ^{1}_{0}n \longrightarrow ^{239}_{92}U + \gamma \qquad (n,\gamma)$$
$$^{239}_{92}U \longrightarrow ^{239}_{93}Np + ^{0}_{-1}e \qquad \beta \text{ emission}$$

This work was done by Edwin McMillan (1907–) and Philip Abelson (1913–). Within a few years several other transuranium elements were made by Glenn T. Seaborg (1912–) and coworkers, also at the University of California. Some of the nuclear reactions involved were

$$^{239}_{94}Pu + ^{2}_{1}H \longrightarrow ^{240}_{95}Am + ^{1}_{0}n \qquad (d,n)$$
$$^{238}_{92}U + ^{12}_{6}C \longrightarrow ^{244}_{98}Cf + 6\,^{1}_{0}n$$
$$^{252}_{98}Cf + ^{10}_{5}B \longrightarrow ^{257}_{103}Lr + 5\,^{1}_{0}n$$

McMillan and Seaborg were awarded the Nobel Prize in Chemistry in 1951.

Most of the transuranium isotopes have short half-lives, and some are very unstable indeed. For example, the longest-lived isotope of nobelium ($Z = 102$), $^{255}_{102}No$, has a half-life of 3 min. Beyond element 100, the discoveries of transuranium elements have involved experiments in which the isotopes were produced and identified one atom at a time! For example, no actual sample of lawrencium (element 103) has ever been made, either of the element or any of its compounds. Only a few atoms were made (by the equation shown above), and they were indentified by counting the radiation given off and measuring its energy.

23-7 Nuclear Energy

In 1905, Albert Einstein published his famous equation

$$E = mc^2$$

where c is the velocity of light. The equation states that matter can be converted into energy (or vice versa), and E is the amount of energy given off when a mass m disappears. This equation caused a major revolution in the thinking of scientists because it had previously been believed that matter and energy were two completely independent things. Einstein showed that they were not; that matter could be regarded as a kind of potential energy, since it could, in theory, be converted into energy. The equation merged the older laws of conservation of matter and conservation of energy into a new law, which states that **matter-energy can neither be created nor destroyed.**

Calculations involving the Einstein equation are simple, but we must be careful about the units. If the velocity of light is taken to be cm/s, then E will come out in ergs; otherwise, in some other energy units.

Example: If 1.00 g of matter is completely converted to energy, how much water at 20°C could be boiled by this amount of energy? The velocity of light is 3.00×10^{10} cm/s.

Answer:

$$E = mc^2$$
$$= 1.0\text{ g}(3.00 \times 10^{10}\text{ cm/s})^2$$
$$= 9.0 \times 10^{20}\text{ g}\cdot\text{cm}^2/\text{s}^2 = 9.0 \times 10^{20}\text{ ergs}$$

We now convert this into calories:

$$9.0 \times 10^{20}\text{ ergs} \times \frac{1\text{ cal}}{4.19 \times 10^{7}\text{ ergs}} = 2.1 \times 10^{13}\text{ cal}$$

From Sections 5-2 and 5-3 we know that 620 cal are required to convert 1 g of water at 20°C to 1 g of steam at 100°C.

$$\frac{2.1 \times 10^{13}\ \text{cal}}{620\ \text{cal/g water}} = 3.4 \times 10^{10}\ \text{g water}$$

This amount of water occupies 3.4×10^{10} ml or 3.4×10^{7} liters. That is, if a mass of 1.0 g was completely converted to energy, the amount of energy would be sufficient to boil 34,000,000 liters of water at 20°C. This is an enormous quantity of energy to be obtained from so small an amount of matter as 1.0 g. Note that the equation does not specify the kind of matter, only the quantity: 1.0 g of any kind of matter would give 9.0×10^{20} ergs.

Einstein's 1905 equation answered what had previously been a troublesome question: Where does the energy of radioactivity come from? The answer is that part of the mass of the nucleus is being converted to energy. All nuclear reactions, both natural and artificial, involve a change in energy. Because the masses of many nuclear particles are accurately known, we can calculate the energy change in a nuclear reaction from just the balanced equation and Einstein's mass-energy formula (note that we cannot do this for ordinary chemical reactions). Table 23-3 lists some atomic masses.

Example: How much energy was released (or absorbed) in the reaction in which Rutherford converted $^{14}_{7}N$ to $^{17}_{8}O$?

$$^{14}_{7}N + ^{4}_{2}He \longrightarrow ^{17}_{8}O + ^{1}_{1}H$$

Answer: The $^{14}_{7}N$ and $^{17}_{8}O$ are atoms. The $^{4}_{2}He$ and $^{1}_{1}H$ are nuclei (no electrons). The mass of a helium nucleus is the mass of the atom (4.00260 g/mole) minus the mass of the two electrons ($2 \times 0.0005486 = 0.0010972$ g/mole) = 4.00150 g/mole. For $^{1}_{1}H$ we use the mass of the proton, 1.007277 g/mole.

Total mass on left:

$^{14}_{7}N$	13.999231 g/mole
$^{4}_{2}He$	4.00150 g/mole
	18.00073 g/mole

Total mass on right:

$^{17}_{8}O$	16.99913 g/mole
$^{1}_{1}H$	1.007277 g/mole
	18.00641 g/mole

Mass difference:

$$\begin{array}{r} 18.00073\ \text{g/mole} \\ -18.00641\ \text{g/mole} \\ \hline -\ 0.00568 = -5.68 \times 10^{-3}\ \text{g/mole} \end{array}$$

For 1 mole (6.02×10^{23}) of $^{14}_{7}N$ atoms, 5.68×10^{-3} g of mass was *gained* in the reaction. To find the corresponding amount of energy, we use Einstein's equation:

$$\begin{aligned} E &= mc^2 \\ &= (5.68 \times 10^{-3}\ \text{g/mole})(3.0 \times 10^{10}\ \text{cm/sec})^2 \\ &= 5.1 \times 10^{18}\ \frac{\text{g}}{\text{mole}}\frac{\text{cm}^2}{\text{s}^2} = 5.1 \times 10^{18}\ \text{ergs/mole} \end{aligned}$$

Thus, 5.1×10^{18} ergs or 1.2×10^{11} cal are *absorbed* when 1 mole of $^{14}_{7}N$ is converted to $^{17}_{8}O$. For a single $^{14}_{7}N$ molecule,

$$\frac{5.1 \times 10^{18} \text{ ergs/mole}}{6.02 \times 10^{23} \text{ molecules/mole}} = 8.5 \times 10^{-6} \text{ erg/molecule}$$

Is the mass of a nucleus equal to the sum of the masses of the nucleons that make up the neucleus? Let us choose an example from Table 23-3, $^{11}_{5}B$, and see if this is so. A nucleus of $^{11}_{5}B$ contains five protons and six neutrons. To get the mass of the nucleus, we subtract the mass of five electrons from the mass of the atom: $11.00931 - (5 \times 0.0005486) = 11.00657$ amu.

five protons	$5 \times 1.007277 =$	5.036385 amu
six neutrons	$6 \times 1.008665 =$	6.051990 amu
total mass of individual particles		11.088375 amu
mass of $^{11}_{5}B$ nucleus		11.00657 amu
difference (mass defect)		0.08181 amu

A nucleus of $^{11}_{5}B$ has a *smaller* mass than the particles that make it up! This difference, 0.08181 amu in the case of $^{11}_{5}B$, is called the **mass defect.** If we made the same calculation for any other isotope, we would find a mass defect in every case. For all atoms the mass of the nucleus is less than the sum of the masses of the nucleons in the free state. Why is this so? To understand the reason, let us compare the energy of the individual particles with that of the total nucleus. Which has the higher energy? We know that in nature it is always the lowest energy situation that is favored. Any nucleus must have a lower energy than the particles from which it is made. Otherwise, it would split up and become free particles with a lower energy.

Table 23-3 Masses of the Three Elementary Particles and of Some Atoms[a]

Particle	*Mass (g/mole or amu)*
Electron	0.0005486
Proton	1.007277
Neutron	1.008665
$^{1}_{1}H$	1.007825
$^{2}_{1}H$	2.01410
$^{4}_{2}He$	4.00260
$^{7}_{3}Li$	7.01600
$^{7}_{4}Be$	7.0169
$^{9}_{4}Be$	9.01218
$^{10}_{4}Be$	10.01353
$^{11}_{5}B$	11.00931
$^{12}_{6}C$	12 (a defined number)
$^{14}_{6}C$	14.00324
$^{14}_{7}N$	13.999231
$^{16}_{8}O$	15.99491
$^{17}_{8}O$	16.99913
$^{19}_{9}F$	18.99840
$^{32}_{16}S$	31.97207
$^{33}_{16}S$	32.97146
$^{34}_{16}S$	33.96786
$^{35}_{17}Cl$	34.96885
$^{37}_{17}Cl$	37.04983

[a]The masses of the atoms are total masses, including the nuclei and the electrons. The units are amu or g/mole.

When a group of protons and neutrons come together to form a nucleus, energy must be given up (Figure 23-8). For the particles to separate again, the same amount of energy would have to be added. The difference in energy between the nucleus and the particles that make it up is called the **binding energy** of the nucleus. We do not know the nature of the forces holding the protons and neutrons together in the nucleus. All we can say is that the binding energy serves as the "nuclear glue." As in the similar case of the covalent bond (Section 14-4), the nucleons must remain together unless they can obtain an amount of energy equal to the binding energy.

We can now understand the reason for the mass defect. If energy is equivalent to mass, then the energy that is lost when protons and neutrons come together to form a nucleus shows up as a decrease in mass. In the case of $^{11}_{5}B$, the 11 particles have a total mass of 11.088375 amu. When they form $^{11}_{5}B$, they lose 0.08181 amu of this mass (the mass defect). **The energy equivalent of the mass defect is the binding energy,** which we can calculate by the Einstein equation.

Example: What is the binding energy in the $^{11}_{5}B$ nucleus?

Answer:

$$\begin{aligned} E &= mc^2 \\ &= (0.08181 \text{ g/mole})(3.0 \times 10^{10} \text{ cm/s})^2 \\ &= 7.4 \times 10^{19} \text{ ergs/mole} \end{aligned}$$

or

$$E = \frac{7.4 \times 10^{19} \text{ ergs/mole}}{6.02 \times 10^{23} \text{ nuclei/mole}} = 1.2 \times 10^{-4} \text{ erg/nucleus}$$

The 11 nucleons in an $^{11}_{5}B$ nucleus must remain together unless 1.2×10^{-4} erg of energy is added, in which case they could convert this energy into 0.08181 amu of mass, allowing the five protons and six neutrons to fly apart. Of course, less energy would be needed to remove only one or two protons or neutrons, since in this case some of the binding energy would remain. In fact, when a radioactive nucleus loses an alpha or beta particle, it *decreases* in energy: the daughter nucleus is more stable

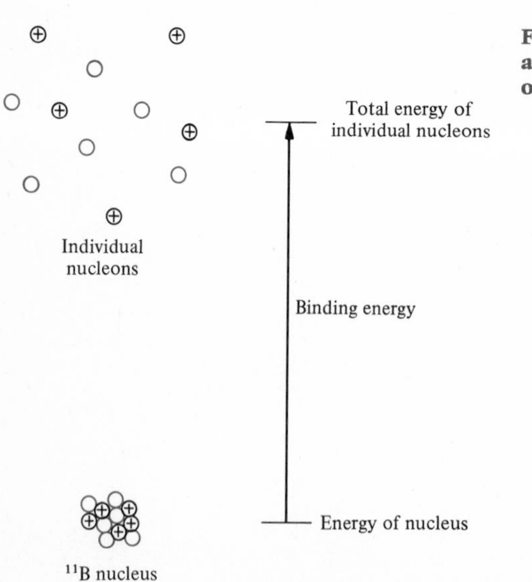

Figure 23-8 Any nucleus has a lower energy and a smaller mass than the individual nucleons that make it up.

than the parent. Nevertheless, even these nuclei have a binding energy, which is the amount of energy necessary to separate completely all the neutrons and protons.

Earlier in this section we calculated that 1.0 g of mass would yield 9.0×10^{20} ergs if completely converted to energy. At present, there is no way known to do this. We cannot as yet convert any amount of matter *completely* to energy. However, it is not necessary to convert completely in order to get energy from nuclear reactions. Because the amount of energy released is so great when even a small quantity of matter is converted, we can get useful amounts of energy even if we are able to convert only a small fraction of any sample of matter.

To see what kind of nuclear reactions might be feasible for this purpose, it is useful to draw a curve showing the binding energy per nucleon for the known isotopes. For example, the binding energy of $^{11}_{5}B$ is 1.2×10^{-4} erg. Since there are 11 nucleons, the binding energy per nucleon is 1.1×10^{-5} erg/nucleon. Such a curve is shown in Figure 23-9. Note that the highest part of the curve is at mass number ≈ 40 to 80, and that the curve drops off at both ends, especially at the low end. From this curve we can see which nuclear reactions are feasible for the purpose of releasing energy. For example, suppose that we took an isotope of mass number 240 and split it into two approximately equal pieces, of mass number about 120 each. The graph shows that each of the two pieces has a higher binding energy per nucleon than the original isotope. Since the total number of nucleons is the same in the two pieces as in the original, the amount of binding energy has increased, which means that the mass defect has increased: the total mass of the two pieces is *less* than the mass of the original nucleus. The difference is given off as energy. The same is theoretically true for cleavage of any nucleus of high mass number. At the other end of the curve we have the opposite situation. If we could force two or more very small nuclei to come together to make a larger nucleus, the new nucleus would have more binding energy per nucleon than the original ones. The mass of the new nucleus would be *less* than that of the original pieces, and energy would be released. This means that in theory at least we can make the following statements:

1. We can get energy by splitting large nuclei. This is called **nuclear fission.**
2. We can get energy by combining very small nuclei into large ones. This is called **nuclear fusion.**
3. We can *never* get nuclear energy by either fission or fusion of the intermediate nuclei (those with mass numbers of about 40 to 100).

Although, in theory, we can get nuclear energy from fission of any large nucleus or fusion of any very small ones, in practice, we know how to do it for only a few nuclei. We will discuss fission in Section 23-8, fusion in Section 23-9.

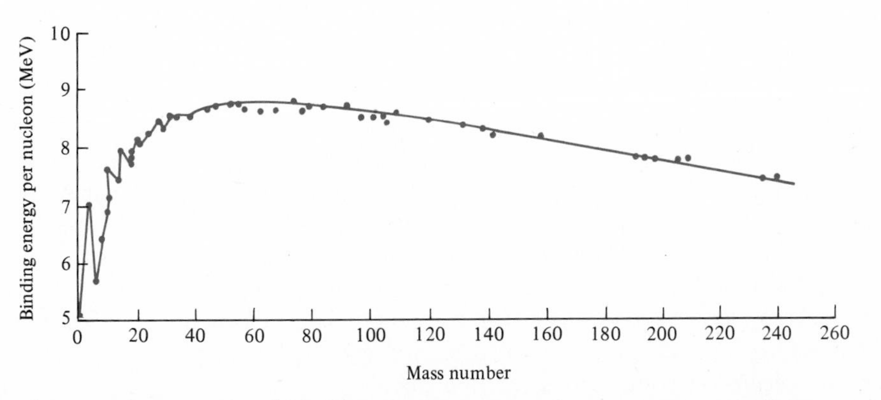

Figure 23-9 A curve of binding energy per nucleon vs. mass number.

23-8 Nuclear Fission

In the late 1930s, Otto Hahn (1879–1968) and his coworkers Lise Meitner (1878–1968) and Fritz Strassman (1902–) were attempting to prepare transuranium elements by bombarding uranium with neutrons. Although this approach would later be successfully used by McMillan and Abelson in 1940 (Section 23-6), Hahn and his coworkers were unable to isolate any transuranium elements. Instead, when they analyzed the samples, they found that *lighter* elements were present rather than the heavier elements they had hoped to synthesize. In 1939, Hahn and Strassman reported the presence of these elements, especially barium, but did not give an explanation of how they might have been formed. Meitner, who had been forced to leave Germany in 1938, published a paper from Sweden, in which she showed that barium and other light elements were formed by cleavage of uranium atoms (after being hit by neutrons) into smaller pieces, and named the process *nuclear fission*. Hahn was awarded the 1944 Nobel Prize in Chemistry for his discovery.

Further experiments showed that only one isotope of natural uranium undergoes fission. This is ^{235}U, which constitutes only 0.720% of natural uranium. The other natural isotope, ^{238}U, does not undergo fission. When a ^{235}U nucleus captures a neutron, the resulting compound nucleus is unstable, and cleaves in a variety of ways, two typical ones being

$$^{235}_{92}U + ^{1}_{0}n \longrightarrow ^{82}_{35}Br + ^{151}_{61}Pm + 3\,^{1}_{0}n + 4\,^{0}_{-1}e + \gamma$$
$$^{235}_{92}U + ^{1}_{0}n \longrightarrow ^{139}_{56}Ba + ^{94}_{36}Kr + 3\,^{1}_{0}n$$

The second reaction accounts for the barium found by Hahn. Altogether, more than 35 fission products have been observed, including elements from germanium to gadolinium. However, the fission products are less important than the neutrons which are given off. Each fission requires one neutron but gives off two or three. Each neutron given off can then collide with another ^{235}U nucleus, causing another cleavage, which generates two or three more neutrons, and the process can continue as long as ^{235}U nuclei are present. This is called a **chain reaction** (Figure 23-10). Why does each fission give off two or three neutrons? The answer to this question may be seen by looking at Figure 23-5, where we see that ^{235}U, being a heavy nucleus, has a much higher neutron/proton ratio than the lighter fission products. When fission takes place, the excess neutrons are released. Also released is a great deal of energy, as we saw in Section 23-7. The following is a list of the products resulting from fission of 1 g of ^{235}U:

989.0 mg fission products
10.2 mg neutrons
0.1 mg radiation
0.7 mg converted to kinetic energy } 0.08% of mass converted to energy

The energy released shows up as kinetic energy of the fragments, that is, as heat. Although 0.8 mg out of each gram is a very small amount of mass, it converts into a very large amount of energy (Section 23-7), and the scientists who read the papers by Hahn and Meitner in 1939 saw the possibility that fission of ^{235}U might be used in a bomb. A letter from Albert Einstein alerted President Roosevelt to the possibility, and large amounts of money were poured into the project (called the Manhattan Project). As we all know, it was successful, and atomic bombs were first exploded in 1945. The energy released is often called atomic energy, but nuclear energy is a better term, because it comes from the nucleus.

The first problem the scientists on the Manhattan Project had to solve was to separate the small amount of ^{235}U from the natural uranium, which is mostly ^{238}U.

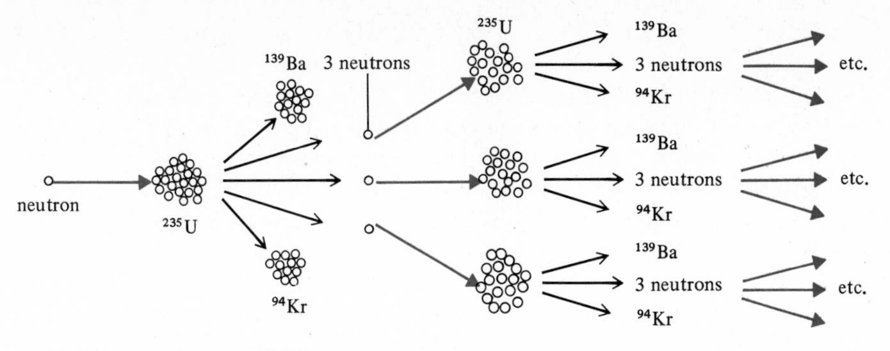

Figure 23-10 Fission of $^{235}_{92}U$.

^{238}U does not undergo fission when struck with neutrons, but it does absorb neutrons, which means that a chain reaction is hard to sustain, since most of the neutrons coming from fission of ^{235}U nuclei are absorbed by ^{238}U nuclei (which constitute more than 99% of naturally occurring uranium) rather than by ^{235}U nuclei. Working on a crash program, several methods of separating the two isotopes were developed, the most important of which was gas diffusion (see Section 4-10), and a bomb was constructed. All that is necessary for an atomic bomb is a piece of ^{235}U in a compact mass of a certain size (called the **critical size**). If the size is too small, or the uranium too thinly spread out, too many neutrons escape from the surface and a chain reaction cannot be sustained. In the actual bomb, two or more pieces of ^{235}U, each below the critical size, are kept separate, and the bomb is triggered by bringing them together.

The methods of separating the uranium isotopes proved long, tedious, and costly, and scientists looked for an end run around this problem. What else undergoes fission? Later it was discovered that very heavy nuclei (mass numbers greater than 240) undergo *spontaneous* fission, that is, without being struck by anything. Examples are $^{242}_{95}Am$, $^{242}_{97}Bk$, and $^{244}_{96}Cm$. These reactions have very short half-lives. For the longest one known ($^{242}_{95}Am$), the half-life is 1.4×10^{-3} s, and some others have half-lives as short as 10^{-10} s. However, these nuclei are too scarce to be useful. Many more-abundant nuclei, especially with mass numbers above 200, will cleave if struck by very fast neutrons. However, this is too inefficient to be useful, because the neutrons are moving so fast that only a tiny fraction of them can be captured by a nucleus. What was needed was another isotope which, like ^{235}U, would sustain a chain reaction when struck by slow neutrons. This was found in $^{239}_{94}Pu$. Workers on the Manhattan Project were already familiar with this isotope. We mentioned that ^{238}U absorbs neutrons, though it does not undergo fission. Absorbtion of a neutron gives ^{239}U, which quickly decays by beta radiation as follows:

$$^{238}_{92}U + {}^{1}_{0}n \longrightarrow {}^{239}_{92}U$$

$$^{239}_{92}U \longrightarrow {}^{239}_{93}Np + {}^{0}_{-1}e \qquad \text{(half-life 23.5 min)}$$

$$^{239}_{93}Np \longrightarrow {}^{239}_{94}Pu + {}^{0}_{-1}e \qquad \text{(half-life 2.35 days)}$$

$^{239}_{94}Pu$ has a half-life of 24,400 yr. Later, one other nucleus was discovered which undergoes fission when struck with slow neutrons. This is ^{233}U. At this time, these three isotopes, ^{235}U, ^{239}Pu, and ^{233}U, are the only nuclei known from which it is feasible to get nuclear energy by fission.

NUCLEAR REACTORS. When World War II ended in August 1945 (and the rapid ending of the war was the immediate result of the first use of atomic bombs), attention turned to the possibility of using nuclear fission to generate electricity. The problem here is a little different because what we want is a nuclear chain reaction that sustains itself but does not explode. The first such **nuclear reactor** was constructed in 1942 in a squash court of the University of Chicago by a team headed by Enrico Fermi (1901–1954). Since then, nuclear reactors of many different types have been made, and research to design new types is still going on (Figure 23-11 shows one type of nuclear reactor). Unlike nuclear bombs, nuclear reactors can use natural uranium, even though less than 1% of it is ^{235}U. The problem of the absorption of neutrons by ^{238}U can be solved by slowing down the neutrons that are produced by fission. These neutrons are fast when emitted, but if the reaction is run in the presence of a **moderator,** the neutrons are slowed by collision with the molecules of the moderator. The two most important moderators for natural uranium reactors are heavy water (D_2O) and graphite. Ordinary water is not useful because it captures neutrons too readily. The fuel and moderator must be kept in separate compartments.

For greatest efficiency a nuclear reactor should have a neutron emission/capture ratio of 1:1. That is, each fissioning nucleus throws out two or three neutrons, but we want only one of these to be used to cleave another nucleus. If more than one does so, the reactor will get out of control (automatic safety features stop the reaction if this happens). The problem is solved by the use of control rods, which are filled with material that captures neutrons. The most common materials used are cadmium and boron, both of which capture neutrons more efficiently than ^{235}U. The control rods can be pushed in or pulled out (Figure 23-11), as is necessary to control the process. When the control rods are pushed all the way in, the reaction stops.

A second type of reactor uses **enriched fuel,** which is natural uranium to which some ^{235}U or ^{239}Pu has been added. Reactors using enriched fuel can be made smaller and are easier to build and less expensive (although still very costly). They can use ordinary water as moderator, and they permit the moderator to be mixed with the fuel.

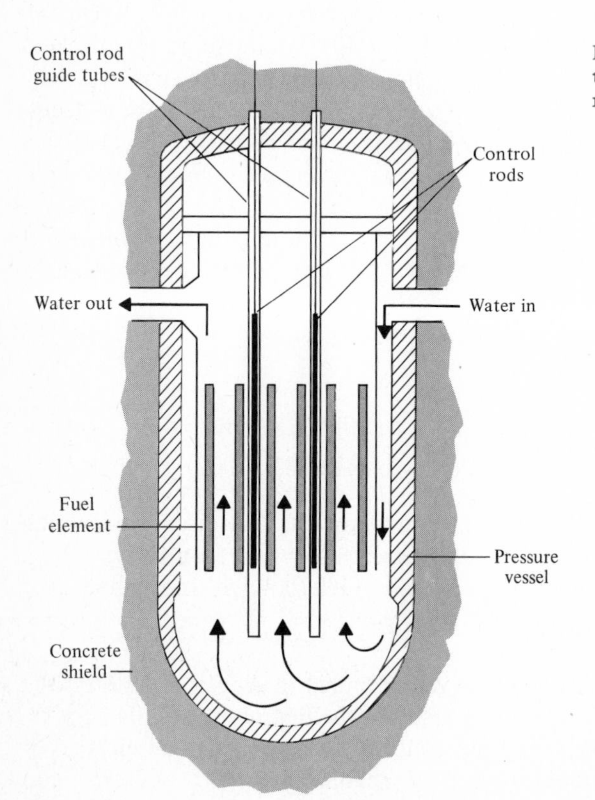

Figure 23-11 A nuclear reactor. This particular reactor uses enriched uranium, not the natural material.

The nuclear reaction in a reactor acts as a source of heat. Therefore, it can be used in exactly the same way as a coal-fired or oil-fired electricity generating plant. Only the heat source is different. The heat is used to boil water; the steam drives a turbine that generates electricity. Figure 23-12 shows such a plant. Note that the steam never comes in contact with radioactive material. The first commercial nuclear power plant came into use in California in 1957. By 1976, there were 60 such plants in operation in the United States, generating about 9% of all our electricity. This percentage is expected to grow. In some European countries the figure is even higher. Nuclear generating plants have two major advantages: (1) they do not use up oil or coal, and (2) they produce no air pollution. However, they do present problems.

1. Although a nuclear explosion is impossible, accidents or sabotage might result in the release of radioactive material.
2. The problem of radioactive wastes. After a reactor has been running for some time, the fission products build up. They must be periodically removed because they absorb neutrons. When the reaction is stopped, the fuel is removed and the remaining uranium and plutonium separated. The rest is "high-level" radioactive waste. This waste presents a serious problem because it is highly radioactive, and some of the isotopes have long half-lives. A material with a half-life of say 20,000 years will require 160,000 to 200,000 years before it is harmless. We can bury the waste (in fact, we do), but who can be sure that it will remain undisturbed 50,000 years from now when there will undoubtedly be a totally different kind of civilization on earth, if there is any life at all. This is an entirely new problem to human beings. Ordinary pollution, being chemical in nature, is dealt with by natural processes in a period of perhaps a few hundred years. Even as serious an event as that at Seveso (Section 17-15) will cause the area to be uninhabitable for only a few score years at most. But radioactive pollution is different because it will remain for so long a time, and because we do not know any way to speed up the process. All we can do is wait for it to decay at its own rate. Material that has been in contact with the reactor must also be dealt with (for example, the clothing of workers, solvents, tools, trash, etc.). This is called low-level waste and represents a less serious problem.
3. It is possible that someone could steal plutonium and use it to make a bomb, or a government might do this with a reactor that is supposed to be used for peaceful purposes. Plutonium is not only radioactive but also highly toxic. This problem is especially important with breeder reactors (see below).
4. The costs of building nuclear reactors have proven to be much higher than anyone expected.

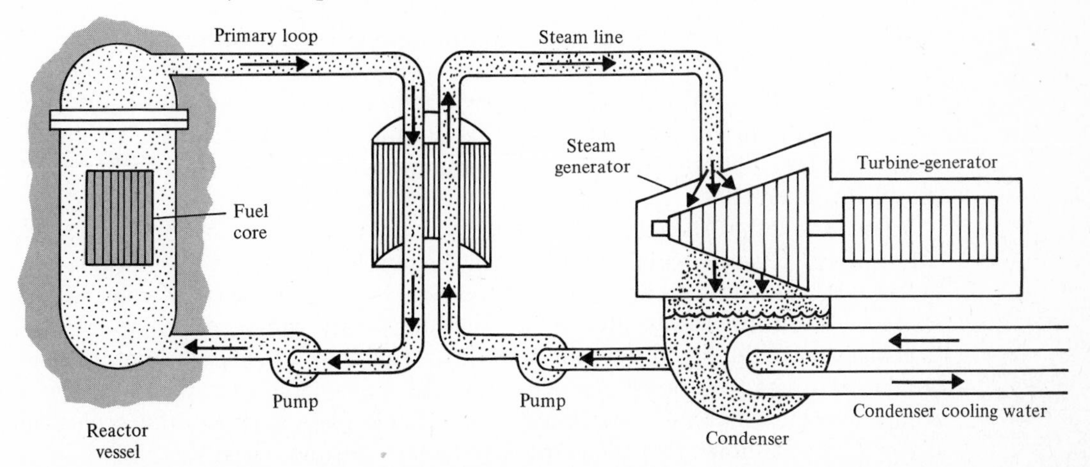

Figure 23-12 A nuclear fission power plant. [Modified from *Energy*, copyright 1976 by American Chemical Society, California Section.]

5. The only naturally occurring fissionable nucleus is ^{235}U. The supply of high-grade uranium ore is not that great, and some day must run out. How much is left is uncertain. Some estimates say that there is enough only for 25 years or so; other estimates are considerably greater.

Because of all these problems, the number of new reactors being built in the United States has declined in recent years, and there are many people who feel that no more should be built.

There is a solution available for at least one of the five problems listed above, the one of availability of uranium fuel. This problem can be solved by the **breeder reactor.** In a breeder reactor, the reaction is surrounded by a shell of natural uranium or, more economically, of natural uranium from which some of the ^{235}U has been removed in a previous reactor. Some of the neutrons escaping from the reaction are captured by the ^{238}U, which is thereby converted to ^{239}Pu by the reaction given on page 771.

$$^{238}_{92}U + ^{1}_{0}n \longrightarrow ^{239}_{92}U \xrightarrow{\beta} ^{239}_{93}Np \xrightarrow{\beta} ^{239}_{94}Pu$$

The plutonium can then be collected and used as fuel in another reactor. In a breeder reactor, more fuel is made than is used up! (Remember that the reactor has been generating electricity all the time that it has been making ^{239}Pu.) Some of ^{238}U present in ordinary reactors is also converted to ^{239}Pu, but the amount is much less. Instead of a shell of ^{238}U, a shell of ^{232}Th can also be used. This absorbs neutrons to give $^{233}_{92}U$, the third fissionable nucleus.

$$^{232}_{90}Th + ^{1}_{0}n \longrightarrow ^{233}_{90}Th \xrightarrow{\beta} ^{233}_{91}Pa \xrightarrow{\beta} ^{233}_{92}U$$

By the use of breeder reactors it is possible to more than double the amount of nuclear fuel naturally present on earth. In fact, because they would enable us to use economically uranium ore with a very low uranium content, breeder reactors would multiply by 50 the energy available from nuclear fission. However, development of breeder reactors has run into many technical problems, and none are yet in use. Perhaps none ever will be.

Small nuclear reactors have proven useful in submarines and aircraft carriers. The advantage of using nuclear reactors for these applications is that the vessels do not have to return to port for refueling but can remain at sea for several years.

23-9 Nuclear Fusion

Except for nuclear energy, all of the energy on the earth comes ultimately from the sun. Every day some 1.5×10^{18} kcal of energy comes to the surface of the earth from this source, and of course this is only a tiny fraction of the total amount emitted by the sun. Furthermore, this vast output has continued for billions of years and is likely to go on for billions of years longer. From where does the sun (and other stars) get all this energy?

Scientists in the nineteenth century, most notably Hermann von Helmholtz (1821–1894), began to wonder about this question. Before that, it had always been believed that the sun was made of a combustible material like coal or wood, and it was simply burning away. But by the middle of the nineteenth century, this explanation would no longer do, because it could be calculated from heat of combustion data that no matter what fuel the sun might be made of, it would completely burn out in a few hundred thousand years at most, and geological discoveries had shown that the earth (and hence the sun) was at least 100 million years old. Julius von Mayer (1814–1878) and Helmholtz proposed that the sun got its energy from gravitation. If it had begun as a larger body and then contracted, energy would be released. However, this too was insufficient, since it gave a lifetime to the sun of only 25 million years.

The answer to the problem was finally supplied by Hans Bethe (1906–), who was awarded the 1967 Nobel Prize in Physics. The sun is composed almost entirely of hydrogen and helium, and the energy comes from the nuclear fusion of four nuclei of ${}^{1}_{1}H$ into one of ${}^{4}_{2}He$. The process occurs in steps (it is very unlikely that four protons could all come together at one time), and more than one pathway is possible. One such pathway is

$$\begin{aligned} {}^{1}_{1}H + {}^{1}_{1}H &\longrightarrow {}^{2}_{1}H + {}^{\ 0}_{+1}e \\ {}^{2}_{1}H + {}^{1}_{1}H &\longrightarrow {}^{3}_{2}He \\ {}^{3}_{2}He + {}^{1}_{1}H &\longrightarrow {}^{4}_{2}He + {}^{\ 0}_{+1}e \\ \hline \text{overall:} \quad 4\,{}^{1}_{1}H &\longrightarrow {}^{4}_{2}He + 2\,{}^{\ 0}_{+1}e \quad \text{(positrons)} \end{aligned}$$

Every second the sun loses 4.3×10^{9} kg (or 4,200,000 tons) of mass by this reaction, but its total mass is so great that the loss of mass is imperceptable even over millions of years. We can expect the sun to keep pouring out energy for billions of years, since it now contains about 73% H and 26% He by mass (the rest is other elements).

We saw in Section 23-7 that nuclear fusion produces energy when the nuclei to be fused are on the low end of the scale. The sun derives its energy by fusion of the lightest of all nuclei, the proton. The idea of using nuclear fusion to generate energy here on earth is a very attractive one, for reasons that we shall soon see, and scientists have been working in this field for more than 20 years. Various light nuclei could be fused. Among the possibilities are

$$\begin{aligned} {}^{2}_{1}H + {}^{2}_{1}H &\longrightarrow {}^{3}_{1}H + {}^{1}_{1}p + 6.4 \times 10^{-6} \text{ erg} \\ {}^{2}_{1}H + {}^{2}_{1}H &\longrightarrow {}^{3}_{2}He + {}^{1}_{0}n + 5.1 \times 10^{-6} \text{ erg} \\ {}^{2}_{1}H + {}^{3}_{1}H &\longrightarrow {}^{4}_{2}He + {}^{1}_{0}n + 2.8 \times 10^{-5} \text{ erg} \end{aligned}$$

The latter process, fusion of deuterium and tritium, is a particularly efficient one. The energy emitted, 2.8×10^{-5} erg per deuterium nucleus (or 4×10^{8} kcal/mole), corresponds to about 0.4% of the mass of the two particles. This is much higher than the 0.08% we saw coming from nuclear fission of ${}^{235}U$.

Nuclear fusion has several important advantages over nuclear fission as a means of generating electricity:

1. The supply of deuterium is essentially unlimited, for all practical purposes. Although only 0.015% of the naturally occurring hydrogen atoms are ${}^{2}_{1}H$, there is such an enormous amount of water on the earth that it seems unlikely that we could ever run out of enough deuterium to supply our energy needs.
2. As pointed out earlier, the efficiency is greater for fusion than for fission; a larger fraction of the mass is converted to energy.
3. The products of fusion are, for the most part, not radioactive. Therefore, the problem of radioactive wastes is removed entirely, or at least greatly minimized. Considering the fusion reactions shown above, ${}^{3}_{2}He$ and ${}^{4}_{2}He$ are nonradioactive, and in fact, completely harmless in all other ways. Tritium (${}^{3}_{1}H$) is radioactive (half-life 12.26 yr), but because it is valuable it will be recovered and cannot be considered as waste. However, neutrons are emitted by fusion reactions, and these are captured by the walls and other components of the reactor, as well as by cooling fluids, etc., making these materials radioactive and weakening them. They will have to be changed periodically.

Because of these advantages, it is likely that some day fusion reactors will completely replace fission reactors. However, that day is not here yet, because so far no one has perfected a self-sustaining fusion device that will produce more energy than is put into it. The difficulty is that the nuclei must be heated to very high temperatures so that they will have enough kinetic energy to overcome the nuclear repulsion. After all, we must force two positively charged nuclei to fuse together. For the reactions between two deuterium nuclei, the temperature needed is about 200 million °C! (By comparison, the temperature at the surface of the sun is 10,000°C). The mixture of deuterium and tritium requires a lower temperature: only 30 million °C, but this is still very high. Because of the high temperatures required, fusion reactions are called **thermonuclear reactions.** At these temperatures electrons are separated from nuclei, and they all float together in a sea called a **plasma.**

Actually, the problem is not so much how to heat the deuterium to such high temperatures, but what to keep it in. In order to get nuclear fusion, the mass has to stay together, but if it touches any part of a solid container, it will cool. The main approach taken so far to meet this problem is the "magnetic bottle." The plasma is confined in a small space by a magnetic field so that it does not have to touch any solid matter. So far, temperatures of up to 20 million °C have been achieved in this way for small fractions of a second, but there is still a long way to go.

A more recent approach involves the use of lasers. A laser is a device that emits light energy in a highly coherent form. In this approach, small pellets (~1 mm in diameter) containing frozen deuterium and tritium are exposed to light from lasers. The idea is that the high laser energy will rapidly raise the temperature while the fuel is in concentrated form.

So far, no practical fusion reactor has yet been developed. Mankind can only hope that one will be developed before too long, because if it is, it has the potential of solving all of man's energy needs forever (and matter needs—see Section E-5).

Although we are still a long way from a practical fusion reaction, fusion has been used in a different way—in the hydrogen bomb. In this bomb, also called the thermonuclear bomb, a fusion reaction is made to go by using the energy of a fission bomb to supply the high temperature. A hydrogen bomb consists of a fission bomb together with a conainer of fusionable hydrogen isotopes. The fission is the "trigger" that causes the fusion reaction to take place. Hydrogen bombs can be made much more powerful than ordinary fission bombs. They give off much less radioactive fallout in proportion to their destructive force and are sometimes considered to be "clean" bombs. However, they cannot be entirely "clean" because the fission trigger will always produce some fallout. Fortunately, hydrogen bombs have only been tested and have not been used in actual warfare. If they are ever used, it may mean the end of our civilization on the earth.

23-10 Uses of Radioactive Isotopes

Radioactive isotopes have found many uses in chemistry, medicine, and technology. In this section we will look at a few of them.

1. In cancer treatment. When physicians first learned that radioactivity can damage living cells (Section 23-2), it occurred to them that this property could be put to use against cancer cells. At first they used radium for this purpose, but today this has been largely replaced with $^{60}_{27}Co$, which gives off higher-energy radiation and is cheaper. Since the radioactivity kills normal cells as well as cancer cells, it must be applied in such a way as to be directed as much as possible to the cancer cells only. In most cases it is not possible completely to avoid some damage to normal cells, and many patients suffer some symptoms of radiation sickness. This problem can be minimized for certain types of cancer, where circumstances permit. For example, the body requires small amounts of iodine in the diet, which mostly goes to the thyroid gland (this is why many people eat iodized salt). This fact makes available a convenient treatment for patients with thyroid cancer. They are given doses of

radioactive iodine isotopes ($^{128}_{53}I$ or $^{131}_{53}I$). The body carries these to the thyroid, where they can irradiate the cancer while causing minimal harm to other areas of the body.

2. Radioactive tracing in chemistry. All isotopes of a particular element have virtually identical chemical properties. If we add a radioactive isotope of any element to an ordinary sample of the same element, the added isotope will mix with sample **(radioactive labeling),** and from that point on will do whatever the sample does. This fact is taken advantage of in the use of radioactive tracing to study the mechanisms of organic, inorganic, and biochemical reactions (see Sections 9-8 and 20-8). The method is best shown by examples. The salts of carboxylic acids react with cyanogen bromide to give alkyl cyanides and carbon dioxide.

$$R{-}\underset{\underset{O}{\|}}{C}{-}O^- + \underset{\text{cyanogen bromide}}{BrCN} \xrightarrow{\Delta} R{-}C{\equiv}N + CO_2 + Br^-$$

In studying the mechanism of this reaction, the first question to be asked is: Does the carbon in the RC≡N come from the acid salt or from BrCN? It is not easy to answer this question by any other method, but radioactive labeling gives a quick and simple answer. If the reaction is carried out with an acid salt in which some of the carbon in the acid salt consists of radioactive ^{14}C, then it is only necessary to separate the RCN and CO_2 at the end of the reaction and to see which is radioactive. When this experiment was carried out, it was found that the alkyl cyanide was radioactive, and not the CO_2.

$$R{-}\underset{\underset{O}{\|}}{C}{-}O^- + BrCN \xrightarrow{\Delta} R{-}C{\equiv}N + CO_2 + Br^- \qquad \text{the carbon in color is } ^{14}C$$

This experiment proved unequivocally that the carbon in the BrCN went to the CO_2, and not to the alkyl cyanide, which was a surprise to the investigators, who expected the opposite result. Other evidence was now needed to learn further details of the mechanism, but the most important fact was now established.

Radioactive tracing has frequently been used in biochemistry. One important example was the use of CO_2 containing ^{14}C to study the mechanism by which plants manufacture glucose in photosynthesis (Section 21-3). Melvin Calvin (1911–) put plants into an atmosphere containing labeled CO_2. After a period of time, the plants were removed from the atmosphere and the leaves analyzed for whichever compounds they contained. Those compounds that were radioactive must have been formed from CO_2 during the period of time that the plant was exposed to the labeled CO_2. By constantly decreasing the exposure time, Calvin was able to discover which compounds were formed first, which were then formed from these, etc., until the basic pathway of the whole complex process was unraveled. Among the facts learned was that many of the reaction steps have already taken place within a few seconds of exposure to CO_2. Calvin was awarded the Nobel Prize in Chemistry in 1961.

3. Radioactive tracing in medicine. Sometimes a physician can find out exactly what is wrong with a patient by using radioactive tracing. For example, if a patient is suspected of having poor blood circulation in a hand, the doctor can inject the hand with a solution of NaCl containing some radioactive Na^+ and thereby follow the path that the NaCl takes in traveling through the blood vessels of the hand. Any abnormal circulation is easily detected. This kind of diagnosis is not easily made in other ways.

4. Isotopic dilution. This technique is useful when it is difficult to separate completely two components of a mixture but we would like to know the concentrations. We add to the mixture a small amount of a radioactive isotope of one of the components, mix thoroughly, and then remove a portion of that component and

measure its radioactivity. This technique is useful in cases where we cannot easily separate two compounds completely, but can remove a pure sample of one of them.

For example, suppose that we have a mixture of $Zn(NO_3)_2$ and $Ca(NO_3)_2$. We can dissolve the mixture in water and add a small known quantity of radioactive $^{62}Zn^{2+}$ to the mixture. After thorough mixing, we precipitate some of the Zn^{2+} as ZnS, weigh it, and measure its radioactivity. This will tell us what percentage of the ^{62}Zn has been recovered, which must be the same percentage of the total zinc that has been recovered.

Example: To 10.0 g of a mixture of $Zn(NO_3)_2$ and $Ca(NO_3)_2$ dissolved in water was added 0.050 microcurie of $^{62}Zn^{2+}$. A small amount of H_2S was added in the presence of NH_3, enough to precipitate 0.75 g of pure ZnS. The radioactivity of the precipitate was 0.027 microcurie. What percentage of the original sample was $Zn(NO_3)_2$?

Answer: First we determine how many g of zinc are in 0.75 g of ZnS.

$$\text{at wt of Zn} = 65.38 \text{ g/mole}$$

$$\text{mol wt of ZnS} = 97.44 \text{ g/mole}$$

$$\text{wt of Zn} = \frac{65.38}{97.44} \times 0.75 = 0.50 \text{ g Zn}$$

Thus, 0.50 g of Zn contained 0.027 microcurie. Since the total radioactivity of the Zn is 0.50 microcurie, the 0.50 g of Zn must make up 0.027/0.050 of the Zn in the whole sample. The total weight of zinc in the sample is found by solving

$$\frac{0.027}{0.050} x = 0.50 \text{ g}$$

$$x = 0.93 \text{ g Zn}$$

The original sample contained 0.93 g of Zn, and hence

$$\frac{\text{mol wt of } Zn(NO_3)_2 = 189.39}{\text{at wt of Zn} = 65.38} \times 0.93 = 2.7 \text{ g of } Zn(NO_3)_2 \text{ in original sample or } 27\%$$

5. Neutron activation analysis. This technique is a useful analytical tool for analyzing trace impurities. The sample is irradiated with neutrons, which convert some of the nuclei present to radioactive isotopes (Section 23-6). The radioactivity of the sample is then measured. From the type and intensity of the radiation, it can be determined which elements are present and in what concentrations. The technique is extremely sensitive. Only milligrams of sample are required and impurities can often be detected in concentrations of 1 ppm or lower. Neutron activation analysis also has the advantage that it is nondestructive to the sample. Only a very small number of nuclei are affected, and the sample is unharmed. This can be very useful, for example, if we wish to determine the authenticity of a 400-year-old painting by analyzing paint to see if its pattern of impurities corresponds to that found in authentic paintings of the same period, or if it is a later forgery. Neutron activation has also been used in medicine. For example, an abnormally high sodium or copper content in fingernails may be a sign of cystic fibrosis.

23-11 Radioactive Dating

We have already discussed the determination of the age of rocks by analysis of the ratio of mother and daughter nuclei present (Section 23-5). Another type of **radioactive dating** is based on the isotope ^{14}C, which has a half-life of 5730 yr. The method was discovered by Willard Libby (1908–), who was awarded the Nobel Prize in Chemistry for this discovery. With a half-life of 5730 yr, all ^{14}C would have vanished from the earth long ago if new supplies were not constantly being created. This happens in the upper atmosphere, where neutrons produced by cosmic rays collide with ordinary ^{14}N.

$$^{14}_{7}N + ^{1}_{0}n \longrightarrow ^{14}_{6}C + ^{1}_{1}H$$

Once formed in this way, the ^{14}C mixes with the ^{12}C normally present in the CO_2 in the atmosphere. The rates of ^{14}C formation and ^{14}C decay are such that the ratio of ^{14}C in atmospheric CO_2 is about 1 part in 10^{12} parts of ^{12}C. All plants absorb atmospheric CO_2 and use it to construct all of their organic molecules. Animals eat the plants and get their carbon in this manner. As long as a plant or animal remains alive, it continues to acquire new carbon compounds, either by absorbing CO_2 from the air or by eating plants that have done so. This means that the ratio of $^{14}C/^{12}C$ in all parts of the plant or animal must remain at 1 part in 10^{12} as long as the organism lives. But when it dies, the situation changes. The ^{14}C decays, with a half-life of 5730 yr, but the nonradioactive ^{12}C remains unchanged. No new ^{14}C can enter the dead organism.

Libby saw that if we measure the $^{14}C/^{12}C$ ratio of any object that had once been part of a living thing, we would know how long it has been since it died. The ^{14}C concentration is very low but can be measured by counting its radioactivity. This discovery has proven invaluable to historians and archeologists, since they now have a means of assigning an approximate date to ancient objects made of wood, bone, fabric, skin, or anything that was once part of a living organism. It is reasonable to assume, for example, that a wooden statue was carved from a tree that had not been dead very long. The radioactivity counting method can be used for objects no older than about 60,000 yr, or about 10 half-lives of ^{14}C. In objects older than this, the ^{14}C content is so low that its radioactivity can no longer be distinguished from background radiation. However, a new method, which does not measure radioactivity but uses a mass spectrometer (Section 2-7) to measure the $^{14}C/^{12}C$ ratio, is more sensitive and may be used to date objects up to 100,000 yr old. By either method, the precision of the result decreases with increasing age. At the other end of the scale, the method cannot be applied to objects less than two or three hundred years old, because the loss in ^{14}C content is too small to measure.

Example: A bone taken from a garbage pile found buried deep under a Turkish hillside had a $^{14}C/^{12}C$ ratio 0.477 times the ratio in a living plant or animal. What was the date of the prehistoric village in which the bone was discovered?

Answer: We use essentially the same method as in Section 23-4. First we calculate k and then use the first-order rate equation.

$$k = \frac{0.693}{5730 \text{ yr}} = 1.21 \times 10^{-4}/\text{yr}$$

$$\log \frac{[A]_0}{[A]} = \frac{kt}{2.303}$$

$$\log \frac{1.000}{0.477} = \frac{1.21 \times 10^{-4}/\text{yr}}{2.303} t$$

$$t = \frac{(2.303)(\log 2.09)}{1.21 \times 10^{-4}/\text{yr}} = 6.1 \times 10^{3} = 6100 \text{ yr}$$

The bone was tossed away (more precisely, the animal whose bone it was died) about 6100 years ago, or about 4100 B.C. We can thus be sure that a village was in existence at that place at that time.

The method as outlined above depends on the $^{14}C/^{12}C$ ratio in the atmospheric CO_2 remaining constant for thousands of years, and Libby at first thought that this was indeed the case. However, it has been discovered that this is not correct: the ratio has fluctuated because the amount of ^{14}C produced is affected by sunspots and changes in the earth's magnetic field. For example, if the magnetic field becomes weaker, its ability to repel cosmic rays decreases, and more neutrons are produced, leading to an increase in atmospheric ^{14}C production. Fortunately, a method has been found to get around this difficulty. Certain trees have been alive for thousands of years. Each year a tree forms another ring, which now becomes a part of its trunk. That ring is essentially dead—it takes in no more ^{14}C. If we then take a very old tree which has recently died and analyze each ring for ^{14}C, we will be able to calculate what the $^{14}C/^{12}C$ ratio in the atmosphere was in each year. This is known as **dendrochronology,** and the tree that has proven to be most satisfactory is the bristlecone pine, which grows in the White Mountains of California and lives to be more than 4000 yr old. In essence, dendrochronology is used to calibrate the ^{14}C radioactive dating method.

The technique of radioactive dating is a spectacular example of the help that chemists can supply to historians. A book* on the history of archaeology states: "It is no exaggeration to say that the discovery of radiocarbon dating is the most important development in archaeology since the discovery of the antiquity of man and the acceptance of a system of three technological ages." Radioactive dating has been used to date cave paintings, the timbers in Stonehenge, the Dead Sea scrolls, Egyptian mummies, and thousands of other relics of the past.

Problems

23-1 Define each of the following terms. **(a)** radioactivity **(b)** alpha particle **(c)** beta particle **(d)** gamma ray **(e)** roentgen **(f)** rad **(g)** rem **(h)** curie **(i)** background radiation **(j)** radioactive decay **(k)** daughter isotope **(l)** electron capture **(m)** K capture **(n)** positron **(o)** antimatter **(p)** deuteron **(q)** cyclotron **(r)** compound nucleus **(s)** transuranium element **(t)** mass defect **(u)** binding energy **(v)** nuclear fission **(w)** nuclear fusion **(x)** chain reaction **(y)** moderator **(z)** nuclear reactor **(aa)** enriched nuclear fuel **(bb)** breeder reactor **(cc)** thermonuclear reaction **(dd)** radioactive labeling **(ee)** isotopic dilution **(ff)** neutron activation analysis **(gg)** radioactive dating **(hh)** dendrochronology

23-2 List the following nuclei in order of what you predict should be increasing stability. **(a)** $^{106}_{46}Pd$ **(b)** $^{152}_{76}Os$ **(c)** $^{16}_{8}O$ **(d)** $^{58}_{25}Mn$ **(e)** $^{9}_{2}He$ **(f)** $^{85}_{37}Rb$

23-3 The half-life of $^{139}_{60}Nd$ is 5.2 h. How long will it take a 15.0-g sample of it to be reduced to 0.25 g?

23-4 The half-life of $^{85}_{36}Kr$ is 10.76 yr. How long will it take for 98% of any sample of $^{85}_{36}Kr$ to decay?

23-5 The half-life of $^{47}_{23}V$ is 33.0 min. How much of a sample of 37.5 g will remain after 2 h 30 min?

23-6 A scientist prepared a 73.0-g sample of pure $^{200}_{81}Tl$. At the end of 101 h only 5.00 g was left. Calculate the half-life of $^{200}_{81}Tl$.

23-7 A sample of a uranium ore was found to contain 0.0365% U and 0.00856% Pb. Assuming that all the lead came from disintegration of ^{238}U, how old is the ore? (The half-life of $^{238}U = 4.51 \times 10^9$ yr; ignore other isotopes and other decay products.)

* *The Origins and Growth of Archaeology,* by Glyn Daniel, Thomas Y. Crowell Company, New York, 1968.

23-8 Write nuclear equations to show how the following act as alpha emitters. **(a)** $^{210}_{83}Bi$ **(b)** $^{238}_{94}Pu$ **(c)** $^{248}_{100}Fm$ **(d)** $^{233}_{92}U$ **(e)** $^{174}_{72}Hf$

23-9 Write nuclear equations to show how the following act as beta emitters. **(a)** $^{159}_{63}Eu$ **(b)** $^{141}_{56}Ba$ **(c)** $^{242}_{95}Am$ **(d)** $^{45}_{19}K$ **(e)** $^{98}_{43}Tc$

23-10 Write nuclear equations to show how the following act as positron emitters. **(a)** $^{116}_{51}Sb$ **(b)** $^{74}_{36}Kr$ **(c)** $^{188}_{79}Au$ **(d)** $^{8}_{5}B$

23-11 Write nuclear equations to show how the following undergo electron capture. **(a)** $^{84}_{39}Y$ **(b)** $^{113}_{50}Sn$ **(c)** $^{246}_{99}Es$ **(d)** $^{41}_{20}Ca$

23-12 Which of the following would be most likely to act by (1) alpha emission, (2) beta emission, (3) electron capture or positron emission? **(a)** $^{116}_{47}Ag$ **(b)** $^{103}_{48}Cd$ **(c)** $^{146}_{64}Gd$ **(d)** $^{245}_{96}Cm$ **(e)** $^{24}_{10}Ne$

23-13 Complete and balance each of the following equations.
(a) $^{137}_{58}Ce + ? \rightarrow ^{137}_{57}La + ^{1}_{1}H$
(b) $^{109}_{47}Ag + ^{1}_{0}n \rightarrow \gamma + ?$
(c) $^{227}_{91}Pa \rightarrow ^{4}_{2}He + ?$
(d) $? + ^{2}_{1}H \rightarrow ^{36}_{18}Ar + ^{1}_{0}n$
(e) $^{11}_{5}B + ? \rightarrow ^{13}_{7}N + 2^{1}_{0}n$
(f) $^{255}_{102}No + ^{16}_{8}O \rightarrow ? + 5^{1}_{0}n$

23-14 Write nuclear equations, given the starting isotope and reaction notation for each of the following. **(a)** $^{12}_{6}C$ (d,n) **(b)** $^{40}_{18}Ar(\alpha,p)$ **(c)** $^{99}_{44}Ru(p,n)$ **(d)** $^{80}_{34}Se$ (d,p) **(e)** $^{55}_{25}Mn$ (n,γ) **(f)** $^{130}_{52}Te$ $(d,2n)$

23-15 Calculate how much energy in cal/mole is gained or lost in each of the following nuclear reactions. (Assume that $^{1}_{1}H$ and $^{4}_{2}He$ are nuclei and the other species are atoms.)
(a) $^{7}_{3}Li + ^{1}_{1}H \rightarrow ^{7}_{4}Be + ^{1}_{0}n$
(b) $^{4}_{2}He + ^{9}_{4}Be \rightarrow ^{12}_{6}C + ^{1}_{0}n$
(c) $^{14}_{6}C \rightarrow ^{14}_{7}N + ^{0}_{-1}e$ (beta emission)
(d) $^{32}_{16}S + ^{4}_{2}He \rightarrow ^{35}_{17}Cl + ^{1}_{1}H$

23-16 Using the data in Table 23-3, calculate the mass defect and binding energies (erg/nucleus) for each of the following nuclei. **(a)** $^{14}_{6}C$ **(b)** $^{32}_{16}S$ **(c)** $^{4}_{2}He$ **(d)** $^{35}_{17}Cl$

23-17 Given the following binding energies, calculate the mass of 1.000 mole of each of the following atoms. **(a)** $^{153}_{63}Eu$, 2.904×10^{10} kcal/mole **(b)** $^{27}_{13}Al$, 4.973×10^{9} kcal/mole **(c)** $^{238}_{92}U$, 4.155×10^{10} kcal/mole

23-18 For the combustion of ethane, C_2H_6,

$$C_2H_6(g) + 3\tfrac{1}{2}O_2(g) \longrightarrow 2CO_2(g) + 3H_2O(l) \qquad \Delta H_{298} = -368.4 \text{ kcal}$$

how much mass is lost or gained when 1.000 mole of C_2H_6 is completely burnt with $3\frac{1}{2}$ moles of O_2? Can this amount of mass be measured on modern balances, which can weigh to four decimal places?

23-19 **(a)** The mass of an electron is 9.110×10^{-28} g. A positron has the same mass. Calculate the amount of energy released when a positron encounters an electron. **(b)** Make the same calculation for a proton–antiproton collision (mass of proton and of antiproton $= 1.673 \times 10^{-24}$ g).

23-20 Calculate binding energies and binding energies per nucleon for each of the following nuclei, given the mass of the atoms **(a)** $^{23}_{11}Na$, 22.9898 g/mole **(b)** $^{28}_{14}Si$, 27.97693 g/mole **(c)** $^{86}_{38}Sr$, 85.9094 g/mole **(d)** $^{169}_{69}Tm$, 168.9344g/mole **(e)** $^{240}_{96}Cm$, 240.0555 g/mole **(f)** What happens to the binding energy per nucleon as the size of the nucleus increases?

23-21 The reaction

$$^{27}_{13}Al + ^{2}_{1}H \longrightarrow ^{28}_{14}Si + ^{0}_{-1}e + ^{1}_{1}H$$

is endothermic, requiring 2.218×10^{11} cal/mole of $^{27}_{13}Al$ in order to take place. The mass of $^{27}_{13}Al$ atoms is 26.98153 g/mole. Calculate the mass of a mole of $^{28}_{14}Si$ atoms. (Assume that $^{2}_{1}H$ and $^{1}_{1}H$ are nuclei; use 2.998×10^{10} cm/s for the velocity of light.)

23-22 What are the advantages and disadvantages of obtaining energy from nuclear reactors?

23-23 What are the advantages and disadvantages of breeder reactors as compared to other nuclear reactors?

23-24 What are the advantages and disadvantages of nuclear fusion as a source of energy compared to other sources?

23-25 Calculate how much energy, in kcal, is lost or gained in the conversion of a mole of ^{6}Li to tritium, by the reaction

$$^{6}_{3}Li + ^{1}_{0}n \longrightarrow ^{3}_{1}H + ^{4}_{2}He$$

(Masses: $^{6}_{3}Li$ atom = 6.01512 g/mole; $^{3}_{1}H$ atom = 3.01605 g/mole; assume that $^{4}_{2}He$ is a nucleus.)

23-26 One possible reaction for nuclear fusion is

$$^{6}_{3}Li + ^{2}_{1}H \longrightarrow 2\,^{4}_{2}He$$

How much energy is emitted per lithium atom? What percent of the mass of the two atoms is converted to energy? (Assume that a $^{6}_{3}Li$ atom reacts with a $^{2}_{1}H$ atom to produce 2 He atoms; the mass of $^{6}_{3}Li$ = 6.01512 g/mole.)

23-27 In studying the mechanism of the reaction

$$CH_3-\underset{\displaystyle\overset{\|}{O}}{C}-O-CH_3 + H_2O \xrightarrow{H^+} CH_3-\underset{\displaystyle\overset{\|}{O}}{C}-O-H + CH_3-O-H$$

we need to know whether the O found in the CH_3OH comes from the CH_3COOCH_3 or from the water. Suggest a method for finding this out.

23-28 Potassium ions are found inside the red cells of the blood as well as outside, in the plasma. Suggest a method for determining whether or not any K^+ ions move from outside to inside or vice versa, and if so, at what rate.

23-29 It was desired to learn the percentage of ^{60}Co in a certain soil, but the concentration was too low for ordinary analytical methods. The following technique was used: A sample of soil weighing 150.0 g was thoroughly mixed with 1.00 g of ^{59}Co (a nonradioactive isotope), and 0.275 g of cobalt was then extracted from the soil, purified, and found to have an activity of 0.250 microcurie. **(a)** What was the total activity of the soil sample? **(b)** If 0.010 g of ^{60}Co has an activity of 1.0 millicurie, what was the concentration of ^{60}Co in the soil, in g ^{60}CO/g soil? (Assume that the soil contains no other radioactive isotope.)

23-30 If a tree dies and the trunk remains undisturbed for 13,750 yr, what percentage of the original ^{14}C content is still present? (The half-life of ^{14}C = 5730 yr.)

23-31 An animal found frozen in the arctic ice had a $^{14}C/^{12}C$ ratio 0.00450 times the ratio in a living animal. How long has the animal been dead?

23-32 An archaeologist wishes to determine if some hundred-year-old pottery excavated on the east coast of the United States was made in Devon, England, where a famous pottery-making firm was situated, or was merely styled after the original pattern but made in the United States. Suggest a method from this chapter by means of which this may be determined. (Assume that an authentic sample of hundred-year-old Devon pottery is available.)

23-33 Could ^{14}C dating be used to determine the age of petroleum deposits on the earth? Explain

Additional Problems

23-34 Calculate how much energy is gained or lost when a single $^{234}_{92}U$ atom undergoes alpha decay.

$$^{234}_{92}U \longrightarrow {}^{4}_{2}He + {}^{230}_{90}Th$$

(Masses: ^{234}U atom = 234.0409 g/mole; ^{230}Th atom = 230.0331 g/mole.)

23-35 Most radioactive isotopes are either alpha or beta emitters, but not both. However, a typical sample of radioactive rock, when placed in a lead block such as that in Figure 23-1, will emit particles that are deflected to both the positive and negative sides of an electric field. Explain.

23-36 A parchment found in an Egyptian tomb had a $^{14}C/^{12}C$ ratio 0.635 times the ratio in a living plant or animal. How old is the parchment?

23-37 The supplies of uranium and thorium on this planet are limited. Iron and silicon are in much greater supply. Why don't we begin research on how to get energy from fission or fusion of iron or silicon?

23-38 The half-life of ^{230}Th is 8.0×10^4 yr. If a scientist in 375 B.C. had been able to obtain a 2.75-g sample of pure ^{230}Th, how much of it would remain in the year A.D. 2000?

23-39 Complete and balance each of the following.

(a) $^{59}_{27}Co + {}^{1}_{0}n \rightarrow {}^{56}_{25}Mn + ?$
(b) $^{78}_{33}As \rightarrow {}^{0}_{-1}e + ?$
(c) $^{26}_{12}Mg + ? \rightarrow {}^{27}_{12}Mg + {}^{1}_{1}H$
(d) $^{43}_{20}Ca + {}^{4}_{2}He \rightarrow {}^{1}_{1}H + ?$
(e) $^{245}_{95}Am + {}^{12}_{6}C \rightarrow 2\,{}^{1}_{0}n + ?$
(f) $^{235}_{92}U + {}^{1}_{0}n \rightarrow {}^{139}_{54}Xe + {}^{94}_{38}Sr + ?$

23-40 For the reaction

$$4\,{}^{1}_{1}H \longrightarrow {}^{4}_{2}He + 2\,{}^{0}_{+1}e$$

the energy given off is 6×10^8 kcal/mole of $^{1}_{1}H$. How much mass is lost when 1.0 mole of $^{1}_{1}H$ is fused to give $\frac{1}{4}$ mole of $^{4}_{2}He$ and $\frac{1}{2}$ mole of positrons?

23-41 The half-life of ^{107}Pd is 7×10^6 yr. How long will it take for a 25.0-g sample of it to be reduced to 1.75 g?

23-42 Assume that you wish to know the number of fish in a rather large lake. Suggest a method, similar to the isotopic dilution method, for finding out.

23-43 The percentage of ^{235}U in naturally occurring uranium has been found to be $0.7202 \pm 0.0006\%$ in most places on earth. However, analysis of the uranium taken from a mine near Oklo in Gabon, Africa, showed a ^{235}U percentage of only $0.7171 \pm 0.001\%$, and the analysis was not in error. Suggest a possible reason why the $^{235}U/^{238}U$ ratio was abnormally low [for a discussion, see an article by R. West, in the *Journal of Chemical Education,* **53,** 336 (1976)].

23-44 A 5.000-g sample of biological material is to be analyzed for vitamin B_{12}. This vitamin contains cobalt, and 0.20 microcurie of radioactive cobalt was added to the sample. After the radioactive cobalt thoroughly mixed with the cobalt present in the vitamin, 0.038 g of the vitamin was isolated from the mixture and found to have a radioactivity of 0.0057 microcurie. What percentage of the original sample was vitamin B_{12}?

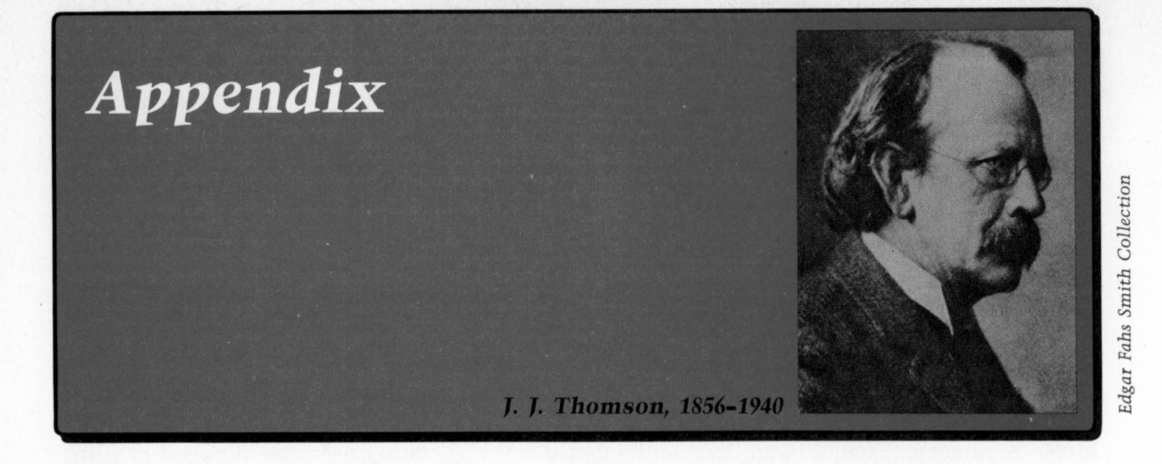

J. J. Thomson, 1856–1940

A Nomenclature of Elements and Simple Inorganic Compounds

THE ELEMENTS. The names and symbols of the elements are given on the inside back cover. It has been customary for the names of elements to be assigned by their discoverers, and many of them have been named for countries, groups of countries, or localities within countries. Examples are gallium (for France), germanium, americium, californium, scandium, and europium. Most of the element names are the same in all languages, but some differ, especially the most common ones. Nitrogen is *azote* in French and *stickstoff* in German. Oxygen is *sauerstoff* in German and кислород (kislorod) in Russian. Aluminum is spelled aluminium in Britain and Canada.

The **symbols** of the elements are the same in all languages, even those which have other alphabets, such as Russian, Arabic, and Chinese. Because the names may differ in the various languages, it is impossible for the symbol to come directly from the name in every language. For English, the symbols do correspond to the name for all but 11 elements. These 11 are:

Element	*Symbol*	*Latin or German name*
Antimony	Sb	Stibium
Copper	Cu	Cuprum
Gold	Au	Aurum
Iron	Fe	Ferrum
Lead	Pb	Plumbum
Mercury	Hg	Hydroargyrum
Silver	Ag	Argentum
Tin	Sn	Stannum
Potassium	K	Kalium (German)
Sodium	Na	Natrium (German)
Tungsten	W	Wolfram (German)

The Latin names were used as the basis for the symbols of eight of these precisely because the names are different in different languages. It was Jöns Jacob Berzelius (1779–1848), inventor of the one- and two-letter symbols, who suggested Latin, so as to give no living language any advantage. However, the suggestion was not followed for elements such as K and Na, where the symbols correspond to the German, and N and O, where the symbols correspond to the English.

SIMPLE INORGANIC COMPOUNDS. Many inorganic compounds can be regarded as consisting of two parts. The rule is that the more electropositive part is given first, and then the more electronegative part, in both the name and the formula. This means that metals come before nonmetals or negative ions. Examples are:

$NaCl$	sodium chloride
$Al_2(SO_4)_3$	aluminum sulfate
$(NH_4)_3PO_4$	ammonium phosphate
$Ba(OH)_2$	barium hydroxide

The names of some common polyatomic ions are given in Table 2-5.

As is evident in the above examples, the *proportions* of the constituents are often not indicated but simply understood. Thus, chemists know that the valence of aluminum is 3 and that of sulfate is 2, so aluminum sulfate must be $Al_2(SO_4)_3$. However, where elements exhibit variable valence, it is often not possible to know the proportions without being told. Thus, "nitrogen oxide" might be NO, N_2O, NO_2, N_2O_4, etc., all known compounds. The earliest way to handle this type of problem was to use different suffixes for different valences. When two valences are possible for a metal, the suffix *ic* stands for the higher valence and *ous* for the lower valence, whatever these valences are. Examples are:

$FeCl_2$	ferrous chloride
$FeCl_3$	ferric chloride
SnO	stannous oxide
SnO_2	stannic oxide
$CuNO_3$	cuprous nitrate
$Cu(NO_3)_2$	cupric nitrate

Although this system is still widely used, it is imperfect, because it does not apply to cases where three or more valences are possible. The IUPAC (see Appendix B) discourages use of *ic* and *ous*. Instead, the IUPAC provides two systems.

1. Use the prefixes *di, tri, tetra, penta, hexa* to indicate 2, 3, 4, 5, or 6 identical units. This system is especially good for binary compounds of the nonmetals. Examples are:

N_2O	dinitrogen oxide (common name: nitrous oxide)
NO	nitrogen oxide (common name: nitric oxide)
NO_2	nitrogen dioxide
S_2Cl_2	disulfur dichloride
Fe_3O_4	triiron tetraoxide
SF_6	sulfur hexafluoride

2. Indicate the oxidation number of the electropositive component by a roman number. This is called the **Stock system.** Examples are:

$FeCl_2$	iron(II) chloride
$FeCl_3$	iron(III) chloride
BaO_2	barium(II) peroxide
As_2O_3	arsenic(III) oxide (can also be called diarsenic trioxide)

Binary acids (these contain hydrogen and one other element) may be named by the preceding system, but it is also common practice to use the prefix *hydro* and the suffix *acid*. The corresponding anions end in the suffix *ide*. Examples are:

HCl	hydrogen chloride	hydrochloric acid	Cl^- chloride ion
HI	hydrogen iodide	hydroiodic acid	I^- iodide ion
H_2S	hydrogen sulfide	hydrosulfuric acid	S^{2-} sulfide ion

Almost all ternary acids contain oxygen as one of the elements, so they consist of hydrogen, oxygen, and one other element. Where the other element forms ternary acids in two oxidation states, the suffix *ic acid* is used for the higher and *ous acid* for the lower oxidation state. For the corresponding anions, the suffixes are *ate* and *ite*, respectively. Examples are:

H_2SO_4	sulfuric acid	SO_4^{2-} sulfate ion
H_2SO_3	sulfurous acid	SO_3^{2-} sulfite ion
HNO_3	nitric acid	NO_3^- nitrate ion
HNO_2	nitrous acid	NO_2^- nitrite ion

Where there is only one oxidation state, the suffixes are *ic acid* and *ate*. Examples are:

H_2CO_3	carbonic acid	CO_3^{2-} carbonate ion
H_3BO_3	boric acid	BO_3^{3-} borate ion

Some elements form acids in three or more oxidation states. For these the prefix *hypo* and the suffix *ous acid* are used for the lowest oxidation state. For elements in groups VIIA or VIIB of the periodic table, the prefix *per* is used for the highest oxidation state. Examples are:

$HOCl$	hypochlorous acid	ClO^-	hypochlorite ion
$HClO_2$	chlorous acid	ClO_2^-	chlorite ion
$HClO_3$	chloric acid	ClO_3^-	chlorate ion
$HClO_4$	perchloric acid	ClO_4^-	perchlorate ion
H_2MnO_4	manganic acid	MnO_4^{2-}	manganate ion
$HMnO_4$	permanganic acid	MnO_4^-	permanganate ion

Where an acid has two protons to give, it can form ions by giving one or both. The IUPAC uses the prefix *hydrogen* for the ion formed when only one proton is given. A much older system, not sanctioned by the IUPAC, but still widely found, uses the prefix *bi*. Examples are:

H_2CO_3	carbonic acid
HCO_3^-	hydrogen carbonate ion bicarbonate ion
$NaHCO_3$	sodium hydrogen carbonate sodium bicarbonate
CO_3^{2-}	carbonate ion
Na_2CO_3	sodium carbonate

For triprotic acids, the prefix *bi* is never used; only the prefix *hydrogen*, with another prefix to indicate the number. Examples are:

H_3PO_4	phosphoric acid
$H_2PO_4^-$	dihydrogen phosphate ion
NaH_2PO_4	sodium dihydrogen phosphate
HPO_4^{2-}	hydrogen phosphate ion
Na_2HPO_4	disodium hydrogen phosphate
PO_4^{3-}	phosphate ion
Na_3PO_4	trisodium phosphate or sodium phosphate

For the naming of coordination compounds, see Section 19-8. For the naming of organic compounds, see Chapter 20.

B The SI System of Units

The International Union of Pure and Applied Chemistry (IUPAC) is a member of the International Organization for Standardization (ISO), along with other scientific organizations. Because the number of units used in the various areas of science was growing, the ISO in the 1960s adopted the International System of Units (called the SI system) and called upon all scientific organizations to adopt this system. Many have done so. In the SI system there are seven basic units and two supplementary units.

	Physical quantity	*Unit*	*Symbol*	
	Length	meter	m	
	Mass	kilogram	kg	
	Time	second	s	
	Electric current	ampere	A	
	Temperature	kelvin	K	
	Amount of substance	mole	mol	
	Luminous intensity	candela	cd	do not concern us in this book
Supplementary units	Plane angle	radian	rad	do not concern us in this book
Supplementary units	Solid angle	steradian	sr	do not concern us in this book

These units may be used with the following prefixes to indicate multiples or fractions.

Prefix	*Symbol*	*Factor*	*Prefix*	*Symbol*	*Factor*
tera	T	10^{12}	deci	d	10^{-1}
giga	G	10^{9}	centi	c	10^{-2}
mega	M	10^{6}	milli	m	10^{-3}
kilo	k	10^{3}	micro	μ	10^{-6}
hecto	h	10^{2}	nano	n	10^{-9}
deca	da	10^{1}	pico	p	10^{-12}
			femto	f	10^{-15}
			atto	a	10^{-18}

In addition to the basic units, shown above, there are a number of derived units, some of which are as follows:

Physical quantity	*Unit*	*Symbol*	*Definition of unit*
Area	square meter	m^2	
Volume	cubic meter	m^3	
Density	kilogram/cubic meter	kg/m^3	
Concentration	mole/cubic meter	$mole/m^3$	
Force	newton	N	$kg \cdot m/s^2$
Pressure	pascal	Pa	N/m^2
Energy	joule	J	$kg \cdot m^2/s^2$
Power	watt	W	J/s
Electric charge	coulomb	C	$A \cdot s$
Electric potential difference	volt	V	$J/A \cdot s$
Frequency	hertz	Hz	cycles/second

Note the following:

1. The liter is not actually a part of the SI system. It is considered as just a special name for $(dm)^3$ or $10^{-3}\ m^3$.
2. In addition to liter, the following units are acceptable for use with the SI system: for time, minute, hour, day; and for angles, degree, minute, and second.
3. Temperature is measured in kelvins, *not* degrees Kelvin. That is, 273 K; not 273°K. However, it is also correct to use degrees Celsius; 50°C. 50 C is not correct.
4. In the case of mass, although kg is the basic unit, the prefixes apply to grams. Thus, milligrams means $\frac{1}{1000}$ g, not $\frac{1}{1000}$ kg.
5. Unit symbols are always used in the singular form and without periods, e.g. 35 g; not 35 g. or 35 gs.

Many of the units that chemists have used for many years, and which we have used in this book, are not part of the SI system and are to be gradually phased out. Among these are

atm	erg	faraday
torr	dyne	roentgen
Å	mole/liter	curie
cal	amu	rad

When you think of how often we have used some of these units in this book, you will see why American chemists have been so slow to adopt the SI system.

In Appendix C are given some physical constants in SI units as well as some conversion factors.

C Tables

1. Conversion factors

Length	
1 Å	= 10^{-8} cm (exact)
1 in	= 2.54 cm (exact)
1 m	= 39.37 in
1 mile	= 5280 ft (exact)
1 mile	= 1.609 km

Volume	
1 qt	= 0.9463 liter
1 gal	= 3.785 liters
1 cubic ft	= 28.316 liters
1 liter	= 1000 cm^3 (exact)
1 m^3	= 100 liters (exact)

Mass	
1 amu	= 1.6605×10^{-24} g
1 ounce (avdp)	= 28.3495 g
1 lb (avdp)	= 453.59 g
1 ton	= 907.180 kg

Pressure	
1 mm Hg	= 1 torr (exact)
1 atm	= 760 torr (exact)
1 atm	= 1.01325×10^6 dynes/cm^2 (exact)
1 atm	= 101,325 pascals (exact)
1 atm	= 14.696 lb/sq. in.

Energy	
1 erg	= 1 g · cm^2/s^2 (exact)
1 cal	= 4.184 J (exact)
1 cal	= 4.184×10^7 ergs (exact)
1 J	= 10^7 ergs (exact)
1 electron volt	= 1.6021×10^{-19} J
1 liter · atm	= 24.2179 cal
1 kWh	= 3.6×10^6 J
1 Btu (British thermal unit)	= 1055.056 J (exact)

2. Some Basic Physical Constants

			SI units
Speed of light in vacuum	c	2.997925×10^{10} cm/s	2.997925×10^{8} m/s
Charge on electron	e	1.60219×10^{-19} coulombs	1.60219×10^{-19} coulombs
Avogadro's number	N	6.02204×10^{23} molecules/mole	6.02204×10^{23} molecules/mole
Atomic mass unit	amu	1.66057×10^{-24} g	1.66057×10^{-27} kg
Mass of electron	m_e	9.10953×10^{-28} g	9.10953×10^{-31} kg
Faraday's number	$\mathcal{F}$	9.64846×10^{4} coulombs/equiv.	9.64846×10^{4} coulombs/mole
Planck's constant	h	6.6262×10^{-27} erg · s	6.6262×10^{-34} joule · s
Gas constant	R	0.08206 liter · atm/mole · °K 1.987 cal/mole · °K	8.3144 joule/kelvin · mole
Rydberg constant (for hydrogen)	R_H	1.09677576×10^{5} cm^{-1}	1.09677576×10^{7} m^{-1}

3. Vapor Pressure of Water

Temperature (°C)	Pressure (torr)	Temperature (°C)	Pressure (torr)	Temperature (°C)	Pressure (torr)
0	4.58	20.5	18.1	32	35.7
1	4.93	21	18.7	33	37.7
2	5.29	21.5	19.2	34	39.9
3	5.69	22	19.8	35	42.2
4	6.10	22.5	20.4	40	55.3
5	6.54	23	21.1	45	71.9
6	7.01	23.5	21.7	50	92.5
7	7.51	24	22.4	55	118.0
8	8.05	24.5	23.1	60	149.4
9	8.61	25	23.8	65	187.5
10	9.21	25.5	24.5	70	233.7
11	9.84	26	25.2	75	289.1
12	10.5	26.5	26.0	80	355.1
13	11.2	27	26.7	85	433.6
14	12.0	27.5	27.5	90	525.8
15	12.8	28	28.3	95	633.9
16	13.6	28.5	29.2	100	760.0
17	14.5	29	30.0	105	906.1
18	15.5	29.5	30.8	110	1074.6
19	16.5	30	31.8	120	1489.1
20	17.5	31	33.7		

4. Common Logarithms

N	0	1	2	3	4	5	6	7	8	9
10	0000	0043	0086	0128	0170	0212	0253	0294	0334	0374
11	0414	0453	0492	0531	0569	0607	0645	0682	0719	0755
12	0792	0828	0864	0899	0934	0969	1004	1038	1072	1106
13	1139	1173	1206	1239	1271	1303	1335	1367	1399	1430
14	1461	1492	1523	1553	1584	1614	1644	1673	1703	1732
15	1761	1790	1818	1847	1875	1903	1931	1959	1987	2014
16	2041	2068	2095	2122	2148	2175	2201	2227	2253	2279
17	2304	2330	2355	2380	2405	2430	2455	2480	2504	2529
18	2553	2577	2601	2625	2648	2672	2695	2718	2742	2765
19	2788	2810	2833	2856	2878	2900	2923	2945	2967	2989
20	3010	3032	3054	3075	3096	3118	3139	3160	3181	3201
21	3222	3243	3263	3284	3304	3324	3345	3365	3385	3404
22	3424	3444	3464	3483	3502	3522	3541	3560	3579	3598
23	3617	3636	3655	3674	3692	3711	3729	3747	3766	3784
24	3802	3820	3838	3856	3874	3892	3909	3927	3945	3962
25	3979	3997	4014	4031	4048	4065	4082	4099	4116	4133
26	4150	4166	4183	4200	4216	4232	4249	4265	4281	4298
27	4314	4330	4346	4362	4378	4393	4409	4425	4440	4456
28	4472	4487	4502	4518	4533	4548	4564	4579	4594	4609
29	4624	4639	4654	4669	4683	4698	4713	4728	4742	4757
30	4771	4786	4800	4814	4829	4843	4857	4871	4886	4900
31	4914	4928	4942	4955	4969	4983	4997	5011	5024	5038
32	5051	5065	5079	5092	5105	5119	5132	5145	5159	5172
33	5185	5198	5211	5224	5237	5250	5263	5276	5289	5302
34	5315	5328	5340	5353	5366	5378	5391	5403	5416	5428
35	5441	5453	5465	5478	5490	5502	5514	5527	5539	5551
36	5563	5575	5587	5599	5611	5623	5635	5647	5658	5670
37	5682	5694	5705	5717	5729	5740	5752	5763	5775	5786
38	5798	5809	5821	5832	5843	5855	5866	5877	5888	5899
39	5911	5922	5933	5944	5955	5966	5977	5988	5999	6010
40	6021	6031	6042	6053	6064	6075	6085	6096	6107	6117
41	6128	6138	6149	6160	6170	6180	6191	6201	6212	6222
42	6232	6243	6253	6263	6274	6284	6294	6304	6314	6325
43	6335	6345	6355	6365	6375	6385	6395	6405	6415	6425
44	6435	6444	6454	6464	6474	6484	6493	6503	6513	6522
45	6532	6542	6551	6561	6571	6580	6590	6599	6609	6618
46	6628	6637	6646	6656	6665	6675	6684	6693	6702	6712
47	6721	6730	6739	6749	6758	6767	6776	6785	6794	6803
48	6812	6821	6830	6839	6848	6857	6866	6875	6884	6893
49	6902	6911	6920	6928	6937	6946	6955	6964	6972	6981
50	6990	6998	7007	7016	7024	7033	7042	7050	7059	7067
51	7076	7084	7093	7101	7110	7118	7126	7135	7143	7152
52	7160	7168	7177	7185	7193	7202	7210	7218	7226	7235
53	7243	7251	7259	7267	7275	7284	7292	7300	7308	7316
54	7324	7332	7340	7348	7356	7364	7372	7380	7388	7396
N	0	1	2	3	4	5	6	7	8	9

N	0	1	2	3	4	5	6	7	8	9
55	7404	7412	7419	7427	7435	7443	7451	7459	7466	7474
56	7482	7490	7497	7505	7513	7520	7528	7536	7543	7551
57	7559	7566	7574	7582	7589	7597	7604	7612	7619	7627
58	7634	7642	7649	7657	7664	7672	7679	7686	7694	7701
59	7709	7716	7723	7731	7738	7745	7752	7760	7767	7774
60	7782	7789	7796	7803	7810	7818	7825	7832	7839	7846
61	7853	7860	7868	7875	7882	7889	7896	7903	7910	7917
62	7924	7931	7938	7945	7952	7959	7966	7973	7980	7987
63	7993	8000	8007	8014	8021	8028	8035	8041	8048	8055
64	8062	8069	8075	8082	8089	8096	8102	8109	8116	8122
65	8129	8136	8142	8149	8156	8162	8169	8176	8182	8189
66	8195	8202	8209	8215	8222	8228	8235	8241	8248	8254
67	8261	8267	8274	8280	8287	8293	8299	8306	8312	8319
68	8325	8331	8338	8344	8351	8357	8363	8370	8376	8382
69	8388	8395	8401	8407	8414	8420	8426	8432	8439	8445
70	8451	8457	8463	8470	8476	8482	8488	8494	8500	8506
71	8513	8519	8525	8531	8537	8543	8549	8555	8561	8567
72	8573	8579	8585	8591	8597	8603	8609	8615	8621	8627
73	8633	8639	8645	8651	8657	8663	8669	8675	8681	8686
74	8692	8698	8704	8710	8716	8722	8727	8733	8739	8745
75	8751	8756	8762	8768	8774	8779	8785	8791	8797	8802
76	8808	8814	8820	8825	8831	8837	8842	8848	8854	8859
77	8865	8871	8876	8882	8887	8893	8899	8904	8910	8915
78	8921	8927	8932	8938	8943	8949	8954	8960	8965	8971
79	8976	8982	8987	8993	8998	9004	9009	9015	9020	9025
80	9031	9036	9042	9047	9053	9058	9063	9069	9074	9079
81	9085	9090	9096	9101	9106	9112	9117	9122	9128	9133
82	9138	9143	9149	9154	9159	9165	9170	9175	9180	9186
83	9191	9196	9201	9206	9212	9217	9222	9227	9232	9238
84	9243	9248	9253	9258	9263	9269	9274	9279	9284	9289
85	9294	9299	9304	9309	9315	9320	9325	9330	9335	9340
86	9345	9350	9355	9360	9365	9370	9375	9380	9385	9390
87	9395	9400	9405	9410	9415	9420	9425	9430	9435	9440
88	9445	9450	9455	9460	9465	9469	9474	9479	9484	9489
89	9494	9499	9504	9509	9513	9518	9523	9528	9533	9538
90	9542	9547	9552	9557	9562	9566	9571	9576	9581	9586
91	9590	9595	9600	9605	9609	9614	9619	9624	9628	9633
92	9638	9643	9647	9652	9657	9661	9666	9671	9675	9680
93	9685	9689	9694	9699	9703	9708	9713	9717	9722	9727
94	9731	9736	9741	9745	9750	9754	9759	9763	9768	9773
95	9777	9782	9786	9791	9795	9800	9805	9809	9814	9818
96	9823	9827	9832	9836	9841	9845	9850	9854	9859	9863
97	9868	9872	9877	9881	9886	9890	9894	9899	9903	9908
98	9912	9917	9921	9926	9930	9934	9939	9943	9948	9952
99	9956	9961	9965	9969	9974	9978	9983	9987	9991	9996
N	0	1	2	3	4	5	6	7	8	9

D Balancing Oxidation-Reduction Equations by the Oxidation-Number Method

In Section 16-3 we described the half-equation technique for balancing oxidation–reduction equations. In this appendix, we will describe another technique, the **oxidation-number method,** with which some students may be more comfortable.

Consider the reaction of permanganate ion (MnO_4^-) with oxalate ion ($C_2O_4^{2-}$) to give manganous ion (Mn^{2+}) and carbon dioxide (CO_2) in acid solution:

$$MnO_4^- + C_2O_4^{2-} \longrightarrow Mn^{2+} + CO_2$$

The first step in this procedure is to determine the oxidation number of each element on both sides of the equation so that we can learn which element has been oxidized and which reduced. Using the rules we previously adopted for determining oxidation numbers (Section 16-1), we find that

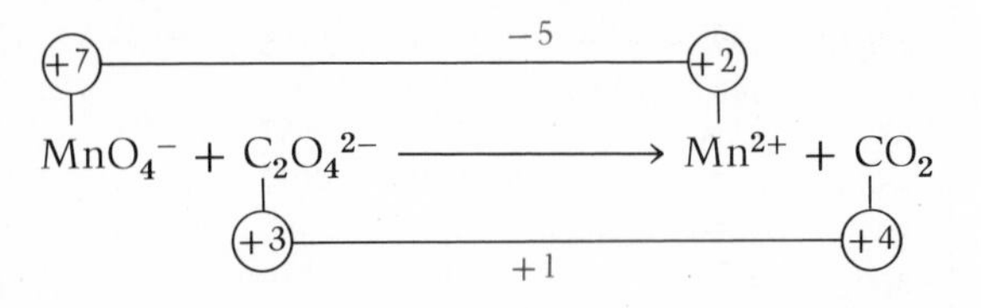

The oxidation number of Mn in MnO_4^- is $+7$, while in Mn^{2+} it is $+2$. It has decreased by 5 and MnO_4^- has been reduced. For carbon, the oxidation number has gone from $+3$ to $+4$, so the change is $+1$, and the carbon has been oxidized.

The next step is to adjust the coefficients of the oxidized and reduced species so that the total increase in oxidation number is equal to the total decrease in oxidation number. Note that the $C_2O_4^{2-}$ ion has two carbon atoms, while CO_2 has only one. We place the coefficient 2 in front of the CO_2 so that the carbon atoms are balanced (this will avoid fractional coefficients). The equation is now

$$MnO_4^- + C_2O_4^{2-} \longrightarrow Mn^{2+} + 2CO_2$$

(Mn: $+7 \rightarrow +2$, -5; C: $+3 \rightarrow +4$, $2 \times (+1) = +2$)

The oxidation number of manganese has changed by -5 while the carbon oxidation number changed by $+1$. There are two carbon atoms in each oxalate ion, and each changes by $+1$, so we have a total change of $+2$ in the carbon oxidation number. To balance we multiply MnO_4^- and Mn^{2+} by 2 and $C_2O_4^{2-}$ and ($2CO_2$) by 5.

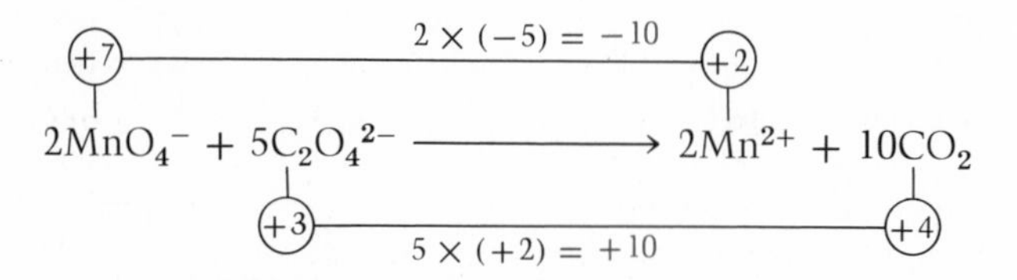

Note that the total oxidation number change is balanced: the oxidation-number change ($+10$) of the species that is oxidized equals the oxidation number change (-10) of the species reduced.

The equation is almost balanced. We have balanced everything but oxygen, hydrogen, and the charges. In acid solution we do this as follows. First we count the total charges on both sides and determine the difference.

$$2MnO_4^- + 5C_2O_4^{2-} \longrightarrow 2Mn^{2+} + 10CO_2$$

In this case the total charge on the left is -12; on the right, $+4$. There has been a change of $+16$. We balance this by adding as many H^+ ions to whichever side needs them to balance the charges. In this case, we add $16H^+$ ions to the left side.

$$16H^+ + 2MnO_4^- + 5C_2O_4^{2-} \longrightarrow 2Mn^{2+} + 10CO_2$$

The charges are now balanced: $+4$ on each side. As the final step, we add enough water molecules to complete the balancing, in this case, $8H_2O$ to the right side.

$$16H^+ + 2MnO_4^- + 5C_2O_4^{2-} \longrightarrow 2Mn^{2+} + 10CO_2 + 8H_2O$$

At this point we check to make sure that the final equation is balanced, with respect to both atoms and charge. We will, of course, have the same equation as if we had used the half-equation method.

The procedure in basic solution is the same, except for the last two steps. For example, we will balance the equation

$$CrO_4^{2-} + I^- \longrightarrow Cr^{3+} + IO_3^-$$

in basic solution. First we tag those elements whose oxidation numbers have changed.

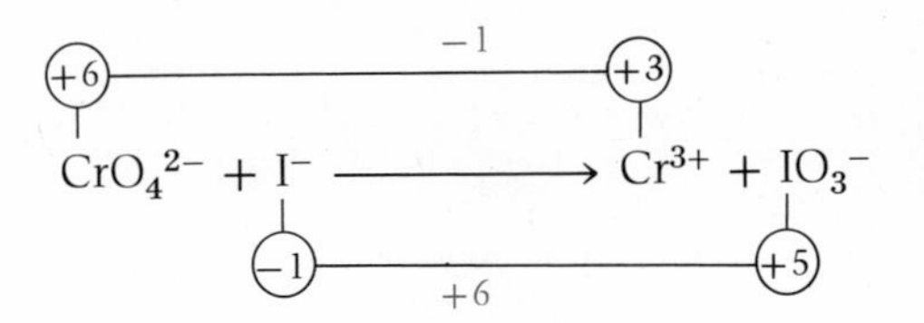

The I^- ion is oxidized while the CrO_4^{2-} is reduced. Adjusting the coefficients so that the total oxidation number change is equal gives us

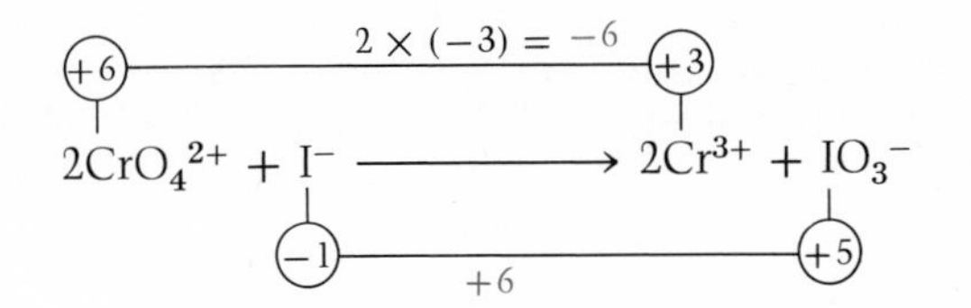

In basic solution we balance the charge difference by adding OH^- ions to the side with the excess positive charge.

$$2CrO_4^{2-} + I^- \longrightarrow 2Cr^{3+} + IO_3^-$$

Here we have a total charge of -5 on the left and $+5$ on the right, so we need $10OH^-$ ions on the right in order to balance the charge.

$$2CrO_4^{2-} + I^- \longrightarrow 2Cr^{3+} + IO_3^- + 10OH^-$$

Finally, we balance the oxygens by adding $5H_2O$ to the left side.

$$5H_2O + 2CrO_4^{2-} + I^- \longrightarrow 2Cr^{3+} + IO_3^- + 10OH^-$$

As usual, the last step is to check that the equation is balanced, both atomically and electrically. As with the half-equation method, the oxidation-number method gives us balanced ionic equations. We can convert to molecular equations by the method described at the end of Section 16-3.

The Fifth Solvay Physics Conference, Brussels, Belgium, October 23-29, 1927. One of the most remarkable gatherings of great scientists in one place at one time. Seventeen of these scientists won Nobel prizes in Physics or Chemistry, some of them before and some after this picture was taken.

***Front row seated, left to right:* I. Langmuir, M. Planck, M. Curie, H. A. Lorentz, A. Einstein, P. Langevin, C. E. Guye, C. T. R. Wilson, O. W. Richardson.**

***Second row seated, left to right:* P. Debye, M. Knudsen, W. L. Bragg, H. A. Kramers, P. A. M. Dirac, A. H. Compton, L. de Broglie, M. Born, N. Bohr.**

***Standing in back, left to right:* A. Piccard, E. Henriot, P. Ehrenfest, E. Herzen, T. De Donder, E. Schrödinger, E. Verschaffelt, W. Pauli, W. Heisenberg, R. H. Fowler, L. Brillouin.**

E Nobel Prize Winners in Chemistry

1901 **Jacobus H. van't Hoff** (1852–1911), Dutch, for his discovery of the laws of osmotic pressure (Section 7-10) and other discoveries involving chemical equilibrium. [The Nobel committee really wanted to honor van't Hoff for his 1874 discovery of the tetrahedral nature of carbon (Sections 14-13 and 20-7), but under the terms of Nobel's will, the prize was only supposed to go for recent work.]

1902 **Emil Fischer** (1852–1919), German, for his synthesis of sugars and purines. [Fischer, one of the greatest organic chemists that ever lived, made many fundamental discoveries in the fields of carbohydrates, amino acids and proteins, as well as purines.] (Chapter 21).

1903 **Svante Arrhenius** (1859–1927), Swedish, for the discovery that many substances form ions when dissolved in water (Section 3-2).

1904 **William Ramsay** (1852–1916), British, for the discovery of the noble gases and of their place in the periodic system (Section 18-7). [Rayleigh was given the Nobel Prize in Physics the same year.]

1905 **Adolf von Baeyer** (1835–1917), German, for his investigations on organic dyes, his advancement of the dye industry, and his studies of strain in cyclic organic compounds (Section 20-4).

1906 **Henri Moissan** (1852–1907), French, for his discovery of the element fluorine, and for inventing an electric-arc furnace, permitting chemists to reach much higher temperatures than had previously been attainable.

1907 **Eduard Buchner** (1860–1917), German, for the discovery of cell-free fermentation.

1908 **Ernest Rutherford** (1871–1937), British, for his investigations of radioactivity and of the chemistry of radioactive substances (Chapter 23). [Much of Rutherford's greatest work came *after* he won the Nobel prize.]

1909 **Wilhelm Ostwald** (1853–1932), German, for his work on equilibrium, kinetics, and catalysis, and for the understanding that enzymes are catalysts and can therefore be studied by the techniques of physical chemistry.

1910 **Otto Wallach** (1847–1931), German, for his pioneering work in the field of cyclic nonaromatic organic compounds, especially those members of a class of naturally occurring compounds called terpenes.

1911 **Marie Sklodowska Curie** (1867–1934), French (born in Poland), for the discovery of radium and polonium, as well as the study of the properties of these elements and their compounds. [Mme Curie had previously shared the 1903 Nobel Prize in Physics with her husband, Pierre Curie, and with Henri Becquerel, for the discovery of radioactivity (Section 23-1).]

1912 **Victor Grignard** (1871–1935), French, for his discovery of magnesium-containing organic compounds (called Grignard reagents) and of the important organic reactions in which they can be used, and **Paul Sabatier** (1854–1941), French, for his discovery of the catalytic hydrogenation of unsaturated compounds (Section 20-3).

1913 **Alfred Werner** (1866–1919), Swiss, for his discovery of the nature of coordination compounds (Section 19-7).

1914 **Theodore W. Richards** (1868–1928), American, for his accurate determination of many atomic weights.

1915 **Richard M. Willstätter** (1872–1942), German, for his research on the structure of chlorophyll and other plant pigments.

1916–1917 No award.

1918 **Fritz Haber** (1868–1934), German, for the discovery of the Haber process for ammonia synthesis (Section 17-3).

1919 No award.

1920 **Walther H. Nernst** (1864–1941), German, for his work in thermochemistry, particularly for the discovery of the third law of thermodynamics (Section 8-11).

1921 **Frederick Soddy** (1877–1956), British, for the discovery of isotopes (Section 2-6).

1922 **Francis W. Aston** (1877–1945), British, for the invention of the mass spectroscope (Section 2-7), its use to discover isotopes in a large number of nonradioactive elements and for his enunciation of the rule that the mass numbers of isotopes are integers.

1923 **Fritz Pregl** (1869–1930), Austrian, for his invention of new methods for the analysis of micro quantities (3 to 5 milligrams) of organic compounds.

1924 No award.

1925 **Richard A. Zsigmondy** (1865–1929), German, for his invention of the ultramicroscope and his demonstration by means of it that colloidal solutions are heterogeneous.

1926 **The Svedberg** (1884–1971), Swedish, for his invention of the ultracentrifuge, and its use to study the nature of colloidal solutions.

1927 **Heinrich Otto Wieland** (1877–1957), German, for his investigations on the structure of steroids (Section 21-8), in particular a group of steroids called bile acids.

1928 **Adolf O. R. Windaus** (1876–1959), German, for his investigations on the structure of steroids, especially on cholesterol and vitamin D (Section 21-8).
[*Note:* The 1927 and 1928 prizes were awarded at the same time. The Nobel Committee was especially struck by the fact that a compound which Wieland had produced by the reduction of a bile acid (cholanic acid) was also produced by Windaus from cholesterol. This was the first indication that bile acids and cholesterol were chemically related.]

1929 **Arthur Harden** (1865–1940), British, and **Hans von Euler-Chelpin** (1873–1964), Swedish, for their independent investigations on the mechanism of sugar fermentation.

1930 **Hans Fischer** (1881–1945), German, for the discovery of the structure of and the synthesis of hemin, a part of the hemoglobin molecule, and his research into the structure of chlorophyll.

1931 **Karl Bosch** (1874–1940), German, for improvements in the Haber process (Section 17-3), and **Friedrich Bergius** (1884–1949), German, for his discovery of methods for converting coal by hydrogenation into a liquid fuel (Section E-3). These awards were combined because both involved the use of high pressures.

1932 **Irving Langmuir** (1881–1957), American, for his discoveries in the area of surface chemistry. He studied thin films of molecules adsorbed onto the surface of metals.

1933 No award.

1934 **Harold C. Urey** (1893–), American, for his discovery of deuterium (Section 2-6).

1935 **Frédéric Joliot-Curie** (1900–1958), French, and his wife, **Irène Joliot-Curie** (1897–1956), French, for the first synthesis of artificial isotopes (Section 23-6).

1936 **Peter J. W. Debye** (1884–1966), American, for contributions to the knowledge of molecular structure through his investigations on dipole moments (Section 14-7) and on x-ray and electron diffraction.

1937 **Walter N. Haworth** (1883–1950), British, for investigations on carbohydrates (Section 21-1) and vitamin C (Section 21-8), and **Paul Karrer** (1889–1971), Swiss, for investigations on vitamins A and B_2 and related compounds (Section 21-8).

1938 **Richard Kuhn** (1900–1967), German, for his work on vitamins A and B and related compounds (Section 21-8).

1939 **Adolph F. J. Butenandt** (1903–), German, for his work on determining the structures of the sex hormones (Section 21-8), and **Leopold Ružička** (1887–1976), Swiss, for his studies on naturally occurring organic compounds called terpenes.

1940–1942 No award.

1943 **George von Hevesy** (1885–1966), Hungarian, for his idea that radioactive isotopes could be used as tracers (Section 23-10).

1944 **Otto Hahn** (1879–1968), German, for the discovery of nuclear fission (Section 23-8).

1945 **Artturi I. Virtanen** (1895–1973), Finnish, for the discovery that green fodder could be preserved by acid treatment.

1946 **James B. Sumner** (1887–1955), American, for his discovery that enzymes can be crystallized, and **John H. Northrop** (1891–) and **Wendell M. Stanley** (1904–1971), Americans, for their preparation of enzymes and virus proteins in pure form.

1947 **Robert Robinson** (1886–1975), British, for his investigations of the structures and reactions of alkaloids (Section 21-8). Robinson determined the structures of morphine and strychnine, among other compounds.

1948 **Arne W. K. Tiselius** (1902–1971), Swedish, for his improvements in a method for the separation of polymers, called electrophoresis, and for the use of this method to separate protein mixtures.

1949 **William F. Giauque** (1895–), American, for his contributions in the field of chemical thermodynamics, particularly concerning the behavior of substances at extremely low temperatures. He devised methods for reaching temperatures much lower than 1°K (Section 8-11) and for measuring them.

1950 **Otto P. H. Diels** (1876–1954), and **Kurt Alder** (1902–1958), both German, for their discovery of a new organic reaction known as the diene synthesis or the Diels–Alder reaction. The reaction proved very useful for the synthesis of chlorinated hydrocarbon insecticides (see dieldrin, aldrin, Section 17-15).

1951 **Edwin M. McMillan** (1907–), and **Glenn T. Seaborg** (1912–), both American, for their discovery of several transuranium elements (Section 23-6).

1952 **Archer J. P. Martin** (1910–), British, and **Richard L. M. Synge** (1914–), British, for their invention of an important separation technique called paper chromatography. [Martin later invented a still more important technique called gas chromatography.]

1953 **Hermann Staudinger** (1881–1965), German, for his discoveries in the field of polymers.

1954 **Linus Pauling** (1901–), American, for his studies of the nature of the chemical bond (for example, resonance, electronegativity, hybrid orbitals; Sections 14-10, 14-6, and 14-15) and the elucidation of the structures of proteins (Section 21-4). [Pauling won the Nobel Peace Prize in 1962.]

1955 **Vincent du Vigneaud** (1901–1978), American, for the synthesis of a small protein molecule (oxytocin, an octapeptide).

1956 **Cyril N. Hinshelwood** (1897–1967), British, and **Nikolai N. Semenov** (1896–), Russian, for their independant research in the mechanism of a certain type of reaction, in which free radicals are intermediates. Both of them used the techniques of kinetics to study reaction mechanisms (Section 9-8).

1957 **Alexander R. Todd** (1907–), British, for the synthesis of components of DNA and RNA (base–sugar–phosphate combinations; see Section 21-5), and for his elucidating how the components fitted together to make up the chains.

1958 **Frederick Sanger** (1918–), British, for his determination of the structure of insulin (Section 21-4).

1959 **Jaroslav Heyrovský** (1890–1967), Czech, for his discovery of polarography, a method of qualitative and quantitative analysis for ions that uses voltaic cells.

1960 **Willard F. Libby** (1908–), American, for the invention of the technique of radiocarbon dating (Section 23-11).

1961 **Melvin Calvin** (1911–), American, for his work on the mechanism of photosynthesis (Section 21-3).

1962 **Max F. Perutz** (1914–), British, and **John C. Kendrew** (1917–), British, for the discovery of the secondary and tertiary structures of the proteins hemoglobin and myoglobin (Section 21-4), by the use of x-ray crystallography.

1963 **Karl Ziegler** (1898–1973), German, for the discovery of a low-pressure method for the polymerization of ethylene to polyethylene (Section 22-2), and **Giulio Natta** (1903–), Italian, for his studies on the three-dimensional stereochemistry of polymers.

1964 **Dorothy Crowfoot Hodgkin** (1910–), British, for determination of the structures of vitamin B_{12} (Figure 21-24) and penicillin by x-ray crystallography.

1965 **Robert Burns Woodward** (1917–), American, for the total synthesis of many naturally occurring organic compounds, including cholesterol, strychnine, chlorophyll, lysergic acid, and quinine. [Since winning the prize, he also synthesized vitamin B_{12}.]

1966 **Robert S. Mulliken** (1896–), American, for his discovery of molecular orbitals (Section 14-18).

1967 **Manfred Eigen** (1927–), German, **Ronald G. W. Norrish** (1897–1978), British, and **George Porter** (1920–), British, for their discovery of methods of measuring rates of extremely fast reactions (including some that take less than a billionth of a second). Norrish and Porter worked together, Eigen independently.

1968 **Lars Onsager** (1903–1976), American, for his work on the thermodynamics of irreversible processes.

1969 **Derek H. R. Barton** (1918–), British, and **Odd Hassel** (1897–), Norwegian, for their independent contributions to the development of the study of the conformations of organic molecules (see for example, cyclohexane; Section 20-4).

1970 **Luis F. Leloir** (1906–), Argentinian, for his discovery of the sugar nucleotides and their role in the biosynthesis of carbohydrates.

1971 **Gerhard Herzberg** (1904–), Canadian, for his work on the structure of free radicals.

1972 **Christian B. Anfinsen** (1916–), **Stanford Moore** (1913–), and **William H. Stein** (1911–), all Americans, for the determination of the amino acid sequence and three-dimensional structure in the enzyme ribonuclease.

1973 **Geoffrey Wilkinson** (1921–), British, and **Ernst O. Fischer,** (1918–), German, for studies on the structures of certain organometallic coordination compounds (ferrocenes; also called sandwich compounds).

1974 **Paul Flory** (1910–), American, for his studies on the physical chemistry of man-made polymers (Chapter 22).

1975 **John W. Cornforth** (1917–), British, and **Vladimir Prelog** (1906–), Swiss, for their independent contributions to the study of the stereochemistry of organic compounds (Section 20-7).

1976 **William Lipscomb** (1919–), American, for his work in determining the structures and properties of the boranes (Section 18-2), and in the discovery of the two electron–three center bond.

1977 **Ilya Prigogine** (1917–), Belgian, for his contributions to the thermodynamics of nonequilibrium systems.

1978 **Peter Mitchell** (1920–), British, for his contribution to the understanding of biological energy transfer.

Answers to Selected Problems

Chapter 1

1-5(a) 2.15 dm, 2.15×10^8 nm **(b)** 6.44×10^{-9} megagram, 6.44×10^3 micrograms **(c)** 1.35×10^{-5} kiloliter, 1.35 centiliters **(d)** 23 m, 2.3×10^{11} Å **(e)** 4.1×10^{-2} g, 41 mg **(f)** 8.2×10^4 ml, 82 liters **(g)** 2.54×10^{-1} nm, 2.54×10^{-7} mm, 2.54×10^{-8} cm **1-7** 25.8 ml **1-8(a)** 111 g **(b)** 207 ml **1-10** 6.12×10^4 g **1-12** 1.25 g/ml **1-14(a)** 41.9°F **(b)** −459°F **(c)** 81°F **1-15(d)** −73.3°C **(e)** −240°C **(f)** 538°C **(g)** 0°C **1-16** −40°C = −40°F **1-18(a)** 2 **(b)** 4 **(c)** 1 **(d)** 4 **(e)** 4 **(f)** 3 **1-19(a)** 26 **(b)** 3.3 **(c)** 0.037 **(d)** 47,000 **(e)** 0.79 **(f)** 3.8 **1-23(a)** 0.52 **(b)** 1.50×10^4 **(c)** 0.64 **(d)** 2×10^3 or 1.7×10^3 **1-24(a)** 2991 **(b)** 4.8 **(c)** 0.000162 **(d)** 0.04 **(e)** 43000 or 4.3×10^4 **1-25(a)** 100,062.1 **(b)** 204 **(c)** 404 **1-27(e)** 117 m **(f)** 4.5×10^2 liters, 4.8×10^2 quarts **(g)** 6.6×10^{-3} gal, 8.8×10^{-4} ft^3 **(h)** 86 cal, 3.6×10^2 J **(i)** 1.79×10^4 torr, 2.38×10^6 pascals **1-30** 0.882 cent/g **1-31** 0.470 dollar/liter, 1.78 dollars/gal **1-32** 1.98×10^6 dyn/cm^2 **1-34(a)** 1.00×10^2 **(b)** 1.34×10^{-4} **(c)** 7.430×10^3 **(d)** 1.10×10^2 **(e)** 3.49×10^{-8} **(f)** 6.395×10^2 **(g)** 4.732×10^0 **(h)** 7×10^0 **(i)** 4×10^{14} (assuming that only one figure is significant. If two, then it would be 4.0×10^{14}, etc.) **(j)** 2.2222×10^3 **1-35(a)** 3.755×10^9 **(b)** 5.62×10^4 **(c)** 3.4×10^{-2} **(d)** 8.36×10^{-3} **(e)** 7.7213×10^1 **1-37(a)** 1.74×10^5 **(b)** −86 **(c)** 71.3 **(d)** -7.57×10^4 **(e)** 1.15×10^{29} **1-38(d)** −7.9547 **(e)** 1.0360 **(f)** 8 **(g)** 4 **(j)** −11.4264 **(k)** 14.2561 **(l)** 14.0963 **1-39(a)** 1×10^{14} **(b)** 1×10^{100} **(c)** 10 **(d)** $10^{-3.0}$ **(e)** 1 **(f)** 2.09×10^5 **(g)** 5.89×10^{23} **1-40(a)** 1.309 **(b)** 7.785 **(c)** −11.86 **(d)** undefined **(e)** 0 **1-41(a)** −0.08, −5.68 **(b)** 2.3, −0.51 **(c)** 7, −0.5 **1-43** 4.9×10^7 dyne **1-44** 9.9×10^4 J, 9.9×10^{11} ergs **1-45** 2.30×10^{39} **1-46** 1.73×10^3 cm **1-48** 7.557×10^{20} eV/s **1-51** 78.5% gold **1-52** 8×10^{14} s^{-1} **1-54** 197 **1-55** 63 erg/cm^2

Chapter 2

2-2(a) 11.89 g Fe, 5.110 g O **(b)** 2.327 g Fe/g O **2-3** 88.89% O, 11.11% H **2-6(a)** 47.9982 **(c)** 63.0128 **(d)** 36.461 **(e)** 261.34 **(f)** 242.09 **(i)** 288.43 **(j)** 893.51 **(l)** 544.1 **2-7** elements: (b), (g), (h), (k); compounds: (c), (e), (j), (m); mixtures: (a), (d), (f), (i), (l) **2-8** chemical changes: (a), (d), (e), (g), (h) **2-10(a)** $^{74}_{32}Ge$ **(b)** $^{55}_{25}Mn^{2+}$ **(c)** $^{2}_{1}H^{+}$ **(d)** $^{243}_{96}Cm$ **(e)** $^{31}_{15}P^{3-}$ **(f)** $^{159}_{65}Tb^{3+}$ **2-12(a)** 31p, 31e, 32n **(b)** 17p, 18e, 18n **(c)** 13p, 10e, 14n **(d)** 92p, 92e, 143n **(e)** 84p, 86e, 118n **(f)** 10p, 10e, 7n **(g)** 48p, 50e, 48n **2-15** 80.22% $^{11}_{5}B$, 19.78% $^{10}_{5}B$ **2-16** 28.12 **2-21(a)** 3.35 moles NH_3 **(d)** 0.258 mole $C_{20}H_{24}N_2O_2$ **2-22(b)** 1.1 moles Si atoms **(c)** 2.9×10^5 moles O atoms **(d)** 23.28 moles Na^+ ions **(g)** 0.153 mole I^- ions **2-23(c)** 1.55×10^{25} molecules C_6H_6 **(d)** 4.18×10^{22} molecules Al_2Cl_6 **(e)** 4.3×10^{19} molecules N_2O_4 **2-24(a)** 46 atoms Fe **(b)** 3.47×10^{25} atoms O **(c)** 8.7×10^{18} atoms Fe **(g)** 3.11×10^{23} Cl^- ions **2-26(a)** $SbCl_3$ **(b)** Br_2 **(c)** C_3H_7I **2-29** Ar, K; Co, Ni; Te, I **2-31(a)** KBr **(b)** MgI_2 **(d)** H_2Se **(e)** Li_3N **(j)** Rb_2O **(k)** GeI_4 **(l)** PI_3 **(n)** AlF_3 **2-34** 1.00015 **2-35(a)** $^{65}_{29}Cu$ **(b)** $^{33}_{16}S$ **(d)** $^{233}_{90}Th$ **(e)** $^{232}_{90}Th$ **(g)** $^{212}_{83}Bi$ **2-36(a)** 5.551 moles, 3.34×10^{24} molecules, 1.00×10^{25} atoms **(c)** 1.611 moles, 9.70×10^{23} molecules, 9.70×10^{24} atoms **(e)** 0.7568 mole, 4.56×10^{23} molecules, 6.84×10^{24} atoms **2-38** 3.246 g I for each g K **2-39** 2.99×10^{-22} g/molecule **2-41(b)** 159.69 **(d)** 342.14 **(e)** 227.087

Chapter 3

3-3(a) $2KClO_3 \rightarrow 2KCl + 3O_2$ **(c)** $MnCl_2 + 2KOH \rightarrow Mn(OH)_2 + 2KCl$ **(f)** $3Ca(OH)_2 + 2H_3PO_4 \rightarrow Ca_3(PO_4)_2 + 6H_2O$ **(j)** $2C_{24}H_{50} + 73O_2 \rightarrow 48CO_2 + 50H_2O$ **3-4(b)** $2C_2H_2 + 5O_2 \rightarrow 4CO_2 + 2H_2O$ **(d)** $P_4 + 5O_2 + 6H_2O \rightarrow 4H_3PO_4$ **(f)** $MnO_2 + 4HCl \rightarrow MnCl_2 + Cl_2 + 2H_2O$ **(i)** $4FeS + 9O_2 + 4H_2O \rightarrow 2Fe_2O_3 + 4H_2SO_4$ **(j)** $Fe_2O_3 + 3CO \rightarrow 2Fe + 3CO_2$ **3-5(a)** $NH_4^+ + OH^- \rightarrow NH_3(g) + H_2O$ **(c)** $OH^- + H^+ \rightarrow H_2O$ or $OH^- + H_3O^+ \rightarrow 2H_2O$ **(d)** $Ag^+ + Cl^- \rightarrow AgCl\downarrow$

(f) $SO_3^{2-} + 2H^+ \rightarrow SO_2\uparrow + H_2O$ or $SO_3^{2-} + 2H_3O^+ \rightarrow SO_2\uparrow + 3H_2O$ **3-6(a)** $C_7H_{10}N_2O$ **(b)** CH_2O **(f)** N_2O_5 **(g)** $Na_2S_2O_3$ **3-7(a)** H_2O_2: 34.0146 g/mole **(b)** Hg_2Cl_2: 472.09 g/mole **(e)** $C_{24}H_{40}O_4$: 392.578 g/mole **3-8(a)** 91.44120% Rb, 8.55880% O **(d)** 13.035% Li, 46.9016% As, 40.0631% O **(g)** 26.729% Cu, 40.417% C, 5.93526% H, 26.9189% O **(h)** 12.3684% K, 18.6430% Co, 26.5854% N, 1.9130% H, 40.49016% O **3-9(a)** 3.97×10^{23} O atoms **(c)** 136.4 g **3-10(b)** $LaCl_3$ **(c)** $PbCr_2O_7$ **(e)** $Na_2S_2O_3$ **3-11(a)** $MgSO_4$ **(b)** C_4H_5N **(c)** $C_7H_8N_2O_3$ **3-12(a)** 285.342 g/mole **(c)** 3.6×10^{-10} g **3-13(b)** 5.57×10^{23} molecules **3-16** 30.05% **3-17** $BaWO_4$ **3-20** 45% **3-22** 8.1×10^3 kg **3-23** 1.17 g **3-26(a)** 28.63 g **(b)** 0.2555 mole **3-27** 47.4% **3-29(a)** 5.0 g Fe_3O_4 will remain **(b)** 16.2 g **3-30** 236 g **3-31(a)** 31.0 g **(b)** 0.375 mole **3-34** 7878 tons **3-35(a)** 362 g **3-37** 65.6% **3-38** $C_{13}H_{18}N_2O_4$ **3-40** 10.6 h **3-41** 652.6 g **3-42(b)** 755.8 g **3-43** 0.566 mole **3-45** 2.97×10^4 g **3-46(a)** 1.19×10^{23} molecules **3-48** 139 g **3-49(a)** 637 g **(b)** 677 g **(c)** 2.23×10^3 g

Chapter 4

4-3(a) 230 torr **(b)** 3.0×10^5 dyn/cm^2 **(c)** 4.4 lb/in^2 **4-5(a)** 1030 torr, 1.355 atm **(c)** 154 torr, 0.203 atm **4-7** 4.94 liter · atm **4-10(a)** 23.0 ml **(c)** 0.77 liter **(d)** 97.7 cm^3 **(f)** 9×10^{-3} liter **4-12** 59°C **4-13** 94 liters **4-14(a)** 2.0 g **(c)** 3.10×10^{-3} g **(d)** 3.7×10^{-9} g **4-18** 0.224 liter **4-19(b)** 2.20 moles, 49.3 liters **(d)** 1.88 moles, 42.1 liters **4-20(a)** g/mole **(b)** g/mole **(c)** molecules/mole **(d)** liters/mole (at STP) **4-23(a)** 30.0 g/mole **(b)** 32.0 g/mole **(d)** 83 g/mole **(g)** 16.0 g/mole **4-24(b)** 0.89 g/liter **(c)** 2.95 g/liter **4-25** 1.64 g/liter **4-26** 395°C **4-28** 411 liters **4-30** 191 liters **4-32** 2.37 kg **4-33** 15.63 liters **4-36** 0.0334 mole, 0.0673 g **4-38(a)** 580 liters **(b)** 16.7 atm **4-39(a)** O_2: 0.25, N_2: 0.675, CO_2: 0.075 **(d)** 1.97×10^3 atm **4-41(a)** 3.1639 **(c)** 0.79778 **(e)** 0.106632 **4-44(a)** O_2 at 300°K: 4.84×10^4 cm/s, 1080 miles/h; CCl_4 at 300°K: 2.20×10^4 cm/s, 492 miles/h **(b)** O_2 at 300°K: 3.74×10^{10} ergs/mole; CCl_4 at 300°K: 3.74×10^{10} ergs/mole **4-45** H_2: 8090°K, O_2: 128,000°K **4-47** 1.20 atm, 1.21 atm **4-48(b)** 0.195 liter **4-49** 3.725×10^5 g **4-50** C_3H_8O **4-53** 0.556 atm **4-55(a)** 4.4×10^4 g **4-56** 14 g **4-57** 16.9 liters **4-58** 15.1 atm total pressure **4-59(a)** 3.38×10^{10} ergs/g, 1.84×10^5 cm/s, 4120 miles/h **4-62** EF_6 **4-63** 52.0 g/mole **4-64(a)** 15.25 ml **4-66** 611°C **4-67(b)** 3.06 atm **4-68(b)** 6.34×10^{21} molecules **4-69(a)** 1.94×10^{22} molecules **4-70(a)** 10,100°K **(b)** 160,000°K **(c)** H_2: 470°K, O_2: 7460°K **4-71(b)** 8.67 atm

Chapter 5

5-4 $q = \Delta E + w$ **5-5(b)** 0 cal **(d)** 0 cal **(e)** −100.0 cal **5-6** 5.26 cal **5-8(a)** 609 torr **(b)** −73 cal **5-9** 44°C **5-11** 63°C **5-13** 29.5 cal/mole · °C **5-14** 12.96 kcal **5-15** 4.420×10^4 cal **5-17(a)** 5.2×10^4 cal **(b)** 1.08×10^3 cal **(e)** 7.00×10^5 cal **5-18** 62°C **5-19(b)** −188.028 kcal **5-21** −133 kcal **5-22(a)** −6408 cal **(b)** −26,806 cal **(c)** +22,080 cal **(g)** +54,127 cal **5-23(d)** −372,824 cal **(e)** 71 cal **(g)** +54,190 cal **(h)** −22,130 cal **5-24(a)** −196,447 cal **(b)** −17,467 cal **5-25(a)** −24,836 cal **(c)** −326,714 cal **5-27(a)** −1348 kcal **5-28(c)** +24,930 cal **5-29** −129.20 kcal **5-30(c)** 0.0236 oz **5-31(a)** $w = 0$, $q = \Delta E = -212.8$ kcal **5-33** −530.5 kcal **5-34** 8.5°C **5-35(a)** 14 kcal/mole **5-36** 89°C **5-37(c)** 13.8 liters CO_2 **5-40** −651 kcal **5-42(a)** 488 torr **(b)** +3.7 cal **5-43(a)** −13,509 cal **(c)** −41,690 cal **(e)** +94,521 cal **(g)** −780,992 cal **5-44** zero

Chapter 6

6-2 23.8 torr **6-3(a)** 737 torr **6-8(a)** H_2O **(b)** Kr **6-10(a)** 7.8×10^4 cal **(b)** 8.3×10^4 cal **6-11(b)** 2.73×10^3 cal **(c)** 3.05×10^3 cal **6-13** 1.04×10^4 cal **6-14** 479 torr **6-16(a)** 122°C **(b)** 108°C **6-17** 4.6×10^3 cal/mole **6-18(a)** Xe **(b)** ethanol **(e)** SiH_4 **6-19(a)** increased **(b)** decreased **(c)** unchanged **(d)** increased **(e)** increased **6-22(a)** hydrogen bonding, dipole interactions **(e)** London forces **(g)** London forces **(h)** dipole interactions **6-24** yes: (b), (d), (g), (h), (i), (j) **6-26(a)** 5.0×10^3 cal/mole **(b)** 9.6×10^{23} **6-29(a)** 3.33×10^4 cm/s **(b)** 2.37×10^4 cm/s **6-30** 320°K **6-33** 27°C **6-35** 126°C **6-37** 86 cal/g

Chapter 7

7-3(a) vinegar **(b)** honey **(c)** household ammonia **7-5** 872 g **7-6(a)** 60% **(b)** 47.0 **7-7** $\chi_{glycine} = 0.65$, $\chi_{alanine} = 0.032$, $\chi_{phenylalanine} = 0.33$ **7-9(a)** 15 g **(c)** 31 g **(d)** 0.073 g **7-10(b)** 2.20 millimoles, 139 mg **(d)** 4.5×10^{-4} millimole, 3.5×10^{-2} mg **7-11(a)** 8.14 milliequiv **(c)** 1.28×10^4 milliequiv **7-12(a)** 0.0727 mole/liter **(c)** 9.39 mole/liter **7-13(a)** 0.218 *N* **(c)** 18.8 *N* **7-14** 49.9 ml **7-15(b)** 45.018 g/equiv **(d)** 64.042 g/equiv **(e)** 44.931 g/equiv **7-16(c)** 0.0139 equiv **(d)** 1.00 equiv **(f)** 220 equiv **7-17(a)** 93.0 g **(d)** 3.06×10^3 g **7-18(a)** 0.311 *N* **(b)** 0.856 *N* **(c)** 2.2 *N* **7-19(a)** 0.11 mole, 8.1 g, 0.22 equiv **(b)** 0.088 equiv, 0.088 mole, 5.5 g **7-22** 117 ml **7-23** 0.689 *N* **7-24(a)** 4.18% **(d)** 0.127 *m* **7-25(a)** 0.554 *m* **(c)** 0.0632 *m* **(d)** 0.684 *m* **(f)** 0.0524 *m* **7-26(d)** 18.0 *M* **(e)** 2.7×10^2 *m* **7-29** more soluble in benzene: (b), (c), (d), (f), (h) **7-30** 74.9 torr **7-31(c)** 248 torr **7-32** 149 g/mole **7-33** 1.8×10^2 g/mole **7-35** 301 ml, 10.8 *m* **7-36** 1.23°C **7-38** 1.37 *m* **7-39** 4 ions **7-42(a)** A **(b)** A **(c)** neither **(d)** B **(e)** A **7-43** 6.4×10^3 g/mole **7-44(a)** −0.0150°C **(b)** 100.00413°C **(c)** 150 torr, 204 cm H_2O **7-47(a)** 0.306 g/ml **(c)** 3.13 *M* **(d)** 0.48 liter **7-49** methanol **7-51(a)** 129°C **(b)** 2.1°C **(c)** 80.3 atm **7-52(a)** 0.677 *M* **(c)** 0.1587 *M* **(e)** 0.584 *M* **7-53** 3.0×10^2 g/mole **7-54(c)** 1.1 *N* **(d)** 6.30 *N* **(e)** 9.0×10^{-3} *N* **7-56** 67 torr, 650 torr **7-59(a)** g/mole **(b)** equiv/mole **(c)** dimensionless **(d)** g/equiv **7-61** 157 torr **7-62(a)** 64.2 g acetic acid, 42.8 g water **(d)** 10.7 *M* **(e)** 25.0 *m* **7-63** 70.0% **7-64(a)** 0.074 g/ml **(c)** 1.0 *M* **(d)** 75 ml

Chapter 8

8-3(a) $K = \dfrac{[NH_3]^2}{[N_2][H_2]^3}$ **(c)** $K = \dfrac{[CO_2][H_2]}{[CO][H_2O]}$ **(e)** $K = \dfrac{[NO_2]}{[N_2O_4]^{1/2}}$ **(g)** $K = \dfrac{[H]^2}{[H_2]}$ **(h)** $K = \dfrac{[CO_2]^4[H_2O]^6}{[C_2H_6]^2[O_2]^7}$ **(j)** $K = \dfrac{[S]^8}{[S_8]}$ **8-5(a)** $\dfrac{[HBr]^2}{[H_2][Br_2]} = 3.50 \times 10^4$ **(b)** $\dfrac{[H_2]^{1/2}[Br_2]^{1/2}}{[HBr]} = 5.35 \times 10^{-3}$ **(d)** $\dfrac{[HBr]^4}{[H_2]^2[Br_2]^2} = 1.23 \times 10^9$ **8-8** For a 700-ml flask: Cl_2O_5 = 0.076 mole/liter, ClO_2 = 0.064 mole/liter, O_2 = 0.54 mole/liter **8-9(a)** $K = \dfrac{[NO_2]^2}{[N_2O_4]}$

(b)

Equilibrium concentration (moles/liter)		*Percent dissociation of N_2O_4*
N_2O_4	*NO_2*	
1.05	0.0784	4%
0.0559	0.018	−×−
0.249	0.038	−×−

8-11(a) [CO] = 2.63 moles/liter, $[H_2O]$ = 3.95 moles/liter, $[CO_2]$ = 1.27 moles/liter, $[H_2]$ = 32.8 moles/liter

8-12

	Initial composition (moles/liter)				*Equilibrium composition (moles/liter)*			
	CO_2	*H_2*	*H_2O*	*CO*	*CO_2*	*H_2*	*H_2O*	*CO*
(a)	1	9	0	0	0.07	8.07	0.93	0.93
(b)	0	0	1	9	0.86	0.86	0.14	8.14
(c)	1	1	1	0	0.59	0.59	1.41	0.41
(d)	0	1	9	9	3.7	4.7	5.3	5.3

8-13 1.69 **8-14** 4.2 moles **8-16** $[CO_2]$ = 0.0115 mole/liter, $[H_2]$ = 0.0015 mole/liter, [CO] = 8.5×10^{-3} mole/liter, $[H_2O]$ = 8.5×10^{-3} mole/liter, *K* = 4.2
8-17 $[CH_3COOH]$ = 0.62 *M*, $[CH_3OH]$ = 0.12 *M*, $[CH_3COOCH_3]$ = 0.38 *M*, $[H_2O]$ = 0.38 *M*

8-19 For the concentration of CO_2: **(a)** decrease **(b)** decrease **(c)** increase **(d)** no change **(e)** decrease **(f)** no change **(g)** no change **(h)** increase. For the equilibrium constant K: The only one of these things that can affect the equilibrium constant is (a) an increase in temperature. K will increase. **8-21(a)** $[NO] = 0.0089$ mole/liter, $[N_2] = 0.0372$ mole/liter, $[O_2] = 0.0372$ mole/liter **(b)** $[NO] = 0.0184$ mole/liter, $[N_2] = 0.0741$ mole/liter, $[O_2] = 0.0741$ mole/liter **8-22** Lowering the temperature will drive the following equilibria to the right: (b), (c), (d) **8-23(a)** 0.72 mole **(b)** greater **8-24** 0.0111 **8-25** 0.0711 **8-26** 0.639 **8-27** 0.639 **8-28** increase: (a), (b), (d), (i), decrease: (c); no change: (e); cannot tell: (f), (g), (h) **8-30** true **8-31(b)** -27.69 cal/deg **(d)** 5.77 cal/deg **8-32(a)** At 25°C: 1200 cal **(c)** 54°C **8-33** 611.500 kcal **8-35(a)** $-54{,}635$ cal **(d)** $-61{,}452$ cal **(e)** 20,719 cal **(f)** $-16{,}658$ cal **8-37(b)** -4762 cal **(c)** $-191{,}390$ cal **(d)** $-52{,}448$ cal **(e)** 1340 cal **8-38(b)** 3.0×10^3 **(c)** 10^{140} **(d)** 2×10^{38} **(e)** 0.105 **8-41(a)** 7940 cal **8-43(a)** $K_p = 1.07 \times 10^{-3}$ **(b)** $K_p = 0.786$ **8-44(b)** $-19{,}400$ cal (0°C), $-18{,}300$ cal (35°C), $-17{,}800$ cal (50°C), $+3600$ cal (1000°K) **8-45** H_2: 1.05 moles/liter, I_2: 0.050 mole/liter, HI: 1.90 moles/liter, $K = 69$ **8-49** 158 **8-51(a)** 44 **(b)** The "K" we calculated does not equal 55.6 because the system has not yet reached equilibrium. **8-53(a)** 29,700 cal/mole **(b)** at 1300°K, $K = 2.5 \times 10^{-3}$ **8-55(b)** left **8-56** 6.9 **8-58(a)** -5.3 cal/mole · °K **(b)** 23.3 cal/mole · °K **(c)** 0.24 cal/mole · °K

Chapter 9

9-2(a) overall third order **(c)** overall first order **(e)** zero order **9-3(b)** liter/mole · s **(c)** mole/liter · s **(f)** 1/s **9-5** $k_1 = 2k_2 = 2k_3$ **9-7** 2.0×10^{-5} mole/liter · s

9-8(a) $\frac{-\Delta[NO]}{\Delta t} = k_1[H_2][NO]^2$, $\frac{-\Delta[H_2]}{\Delta t} = k_2[H_2][NO]^2$, $\frac{\Delta[N_2]}{\Delta t} = k_3[H_2][NO]^2$, $\frac{\Delta[H_2O]}{\Delta t} = k_4[H_2][NO]^2$ **(b)** $k_1 = k_2 = 2k_3 = k_4$ **9-9** $n = 1$, $m = 2$, $k = 1.00$ liter²/mole² · s **9-10** $m = 1$, $n = 2$, $k = 6.60 \times 10^{-3}$ liter²/mole² · s **9-12** 6.93×10^7 1/s **9-14** zero order **9-15(a)** 25.0 s **(b)** 69.3 s **(c)** 200 s **9-16** 13.28 g **9-18(a)** $k_2/k_1 = 2.2$, approximately double **9-20** at 298°K, $K = 2.0 \times 10^3$ moles/liter **9-21** No. It is usually possible to find several mechanisms that are consistent with any given rate law. **9-23(a)** rate $= k[H_2]$, not consistent **(b)** rate $= k[H_2][I_2]$, consistent **9-24** 9-22: I, IH_2; 9-23(a): H, I; 9-23(b): H_2I_2; 9-23(c): H, I. **9-27(a)** 4.6×10^5 s **(b)** 5% **9-29** 9800 cal **9-31** $k = 5.90 \times 10^{-5}$ s^{-1}, $\tau = 1.18 \times 10^4$ s **9-33(a)** zero **(b)** fourth **(c)** first **d)** zero **(e)** minus first **(f)** second **9-36(a)** second order **(b)** $k = 8.71 \times 10^{-4}$ **9-38** $E_a = 3.37 \times 10^4$ cal, $A = 8 \times 10^{26}$

Chapter 10

10-2(a) NH_4^+ **(b)** NH_3 **(c)** HOAc **(d)** H_2O **(e)** H_3O^+ **(f)** HF **(g)** H_2F^+ **(h)** HPO_3^{2-} **(i)** H_3PO_4 **(j)** $H_2PO_4^-$ **(k)** HCl **(l)** HCN **(m)** HSO_4^- **(n)** H_2SO_4 **10-3(a)** NO_3^- **(b)** $H_2PO_4^-$ **(c)** HPO_4^{2-} **(d)** PO_4^{3-} **(e)** NH_3 **(f)** NH_2^- **(g)** F^- **(h)** HS^- **(i)** S^{2-} **(j)** $H_2BO_3^-$ **(k)** OH^- **(l)** H_2O **(m)** OAc^- **(n)** HF **(o)** HCO_3^- **(p)** CO_3^{2-}

10-4(e) $\underset{base_1}{H_2O} + \underset{acid_2}{H_2O} \rightleftharpoons \underset{acid_1}{H_3O^+} + \underset{base_2}{OH^-}$ **(k)** $\underset{acid_1}{NH_4^+} + \underset{base_2}{NH_2^-} \rightleftharpoons \underset{base_1}{NH_3} + \underset{acid_2}{NH_3}$

10-5 to the right: (c), (d), (f); to the left: (a), (b), (e), (g) **10-6** reactions that will take place: (a), (b), (d), (f), (j), (k); reactions that will not take place: (c), (e), (g), (h), (i) **10-8** $Br^- < HSe^- < H_2As^- < GeH_3^-$ **10-10** $H_2O < H_2S < H_2Se < H_2Te$ **10-12(a)** H_2Te **(b)** HBr **(c)** HBr **(d)** HCl **(e)** HI **10-13(f)** $HClO_3$ **(g)** H_2SeO_4 **(h)** H_3PO_4 **(i)** HIO_4 **(j)** HF **(k)** H_2S **(l)** NH_4^+ **10-14(h)** KOH **(i)** CsOH **(j)** HS^- **(k)** ClO_2^- **(l)** HS^- **(m)** NO_2^- **(n)** CO_3^{2-} **10-15(a)** 10^{-3} **(b)** 10^{-7} **(c)** 10 **(d)** 2.9×10^{-6} **10-17(a)** $[H_3O^+] = 1.8 \times 10^{-5}\,M$, $[OH^-] = 5.6 \times 10^{-10}\,M$ **(b)** $[OH^-] = 1.8 \times 10^{-5}\,M$, $[H_3O^+] = 5.6 \times 10^{-10}\,M$ **(c)** $[H_3O^+] = 0.250\,M$, $[OH^-] = 4.00 \times 10^{-14}\,M$ **(d)** $[H_3O^+] = 7.9 \times 10^{-1}\,M$, $[OH^-] = 1.3 \times 10^{-14}\,M$ **10-19** 6.81 **10-21(a)** 0.699 **(b)** 2.72 **(c)** 5.00 **(d)** 4.11 **10-23** $K_a = 0.0494$, pH $= 0.70$ **10-24** pH $= 10.70$, degree of dissociation $= 0.33\%$ **10-26** $K_a = 2.9 \times 10^{-12}$, p$K_a = 11.54$ **10-27** $K_b = 6.3 \times 10^{-9}$, p$K_b = 8.20$ **10-28** $1.5 \times 10^{-2}\,M$ **10-29(a)** $C_7H_7O^-$, 1.6×10^{-4} **10-30(a)** $HCH_4N_2O^+$, 7.94×10^{-1} **10-34(a)** pH $= 1.92$, $[SO_3^{2-}] = 1.0 \times 10^{-7}\,M$ **(b)** $[SO_3^{2-}] = 1.5 \times 10^{-8}\,M$ **10-35(a)** neutral **(b)** basic **(c)** acidic **(d)** acidic

(e) basic **(f)** acidic **(g)** neutral **(h)** acidic **(i)** acidic **(j)** acidic **(k)** basic **(l)** neutral **10-36(a)** $[H_3O^+] = 8.59 \times 10^{-5}\ M$, $[OH^-] = 1.16 \times 10^{-10}\ M$, pH = 4.07, pOH = 9.93, percent dissociation = 0.0344% **(d)** $[H_3O^+] = 3.8 \times 10^{-5}\ M$, $[OH^-] = 2.6 \times 10^{-10}\ M$, pH = 4.42, pOH = 9.58, percent hydrolysis = 0.0152% **10-37(a)** 7 **(b)** 8.52 **(c)** 7.88 **(d)** 10.80 **(e)** 5.48 **(f)** 12.60 **10-38(a)** $[H_3O^+] = 1.2 \times 10^{-5}\ M$, $[OH^-] = 8.3 \times 10^{-10}\ M$, pH = 4.92, pOH = 9.08, percent hydrolysis = 0.0048% **(b)** no hydrolysis. pH = pOH = 7, $[H_3O^+] = [OH^-] = 10^{-7}\ M$ **(c)** $[H_3O^+] = 3.3 \times 10^{-9}\ M$, $[OH^-] = 3.0 \times 10^{-6}\ M$, pH = 8.48, pOH = 5.52, percent hydrolysis = 0.00075% **(e)** $[H_3O^+] = 2.0 \times 10^{-3}\ M$, $[OH^-] = 5.0 \times 10^{-12}\ M$, pH = 2.70, pOH = 11.30, percent hydrolysis = 0.27% **10-39** 0.2218 N **10-40** 64.3 g/equiv, no **10-41** 104 g/mole **10-44(a)** 13 **(b)** 12.91 **(c)** 12.82 **(d)** 12.52 **(e)** 11.72 **(f)** 11.00 **(g)** 10.00 **(h)** 9.00 **(i)** 5.00 **(j)** 3.00 **(k)** 1.70 **10-46** 4.05 **10-47(a)** 1.1 **(b)** 0.0056 **(c)** 0.44 **10-49** 39 g **10-50(a)** 7.67 **(b)** 7.37 **(c)** 5.37 **10-52(a)** 1.7 **(b)** 4.4 **(c)** 9.5 **10-54** $H_2SO_4 + HNO_3 \rightarrow HSO_4^- + H_2NO_3^+$ **10-56** 0.689 N, 0.689 M **10-57** $K_b = 1.8 \times 10^{-5}$, $pK_b = 4.74$ **10-59(a)** 2.52 **(b)** 4.92 **(c)** 5.35 **(d)** 4.57 **10-61** 162 g/equiv **10-62(a)** All Arrhenius acids are Brønsted acids. **(b)** CH_3OH; all acids weaker than water fit into this category. **(c)** All Arrhenius bases are Brønsted bases. **(d)** Cl^-; all bases weaker than water fit into this category. **10-63** $4.6 \times 10^{-17}\ M$ **10-67(a)** pH = 2.65, percent dissociation = 1.51% **(b)** pH = 4.36, percent dissociation = 44% **10-68** pH = 4.3, pOH = 9.7 **10-69** $K_b = 4.4 \times 10^{-10}$, $pK_b = 9.36$ **10-70** 3.4×10^{-2} mole HCl **10-71** 9.5×10^{-7} mole H_3O^+

Chapter 11

11-2 insoluble in water: AgS, AgCl, $BaSO_4$, $Fe_2(SO_4)_3$, MoS_2; all others are soluble **11-3(a)** $Ba^{2+} + SO_4^{2-} \rightarrow BaSO_4\downarrow$ **(c)** $Pb^{2+} + CrO_4^{2-} \rightarrow PbCrO_4\downarrow$ **(d)** $CO_3^{2-} + 2H_3O^+ \rightarrow CO_2\uparrow + 3H_2O$ **(e)** $SO_3^{2-} + 2H_3O^+ \rightarrow SO_2\uparrow + 3H_2O$ **(g)** $Ag^+ + Cl^- \rightarrow AgCl\downarrow$ **(i)** $Ag^+ + 2NH_3 \rightarrow Ag(NH_3)_2^+$ **(k)** $HOAc + OH^- \rightarrow H_2O + OAc^-$ **(l)** $NH_4^+ + CO_3^{2-} \rightarrow NH_3 + HCO_3^-$ **(o)** $S^{2-} + 2H_3O^+ \rightarrow H_2S + 2H_2O$ no reaction: (b), (f), (h), (j), (m), (n) **11-5(a)** $K_{sp} = [Ag^+]^2[S^{2-}]$, $Q = [Ag^+]^2[S^{2-}]$ **(c)** $K_{sp} = [Fe^{3+}][OH^-]^3$, $Q = [Fe^{3+}][OH^-]^3$ **11-6** 1.1×10^{-8} **11-7** 3.26×10^{-70} **11-9** 7.5×10^{-3} mole/liter, 0.13 g/100 ml **11-11** yes: (a), (c), (d); no (b) **11-12** $[Ba^{2+}]$, $5.0 \times 10^{-2}\ M$, $[SO_4^{2-}] = 2.2 \times 10^{-9}\ M$ **11-13** $[Ag^+] = 2.9 \times 10^{-24}\ M$, $[S^{2-}] = 1.85 \times 10^{-2}\ M$ **11-14(a)** 3.4×10^{-13} g/ml **(b)** 4.7×10^{-23} g/ml **11-15** 1.6×10^{-5} **11-17** AgBr will precipitate first. **11-19(a)** $9.1 \times 10^{-9}\ M$ **(b)** $8.9 \times 10^{-16}\ M$ **11-20** 1.9 **11-21** $1.7 \times 10^{-2}\ M$ **11-23** (b), (c), (f) **11-24** 1.54×10^{-16} **11-25** 8.3×10^{-2} mole **11-27** 8.9 **11-31** $[Ag^+] = 4.4 \times 10^{-2}\ M$, $[H_3O^+] = 1.6 \times 10^{-2}\ M$, $[NO_3^-] = 6.0 \times 10^{-2}\ M$, $[Cl^-] = 3.5 \times 10^{-9}\ M$ **11-33** 4.5×10^{-10} mole/liter, 4.8×10^{-8} mg/ml **11-34** 1.9×10^{-5} g/100 ml **11-35** $2.5 \times 10^{-4}\ M$ **11-36** yes: (a), (d), (e), (f) **11-37** 3×10^{-8} g **11-38** 9.54

Chapter 12

12-3(a) 4.20×10^{-5} cm **(b)** 1.92×10^4 cm **(c)** 1.00×10^{-12} cm **12-4(a)** blue visible light **(b)** radio waves **(c)** cosmic rays **(d)** ultraviolet light **(e)** infrared light **12-5(a)** $2.38 \times 10^4\ cm^{-1}$ **(b)** $5.21 \times 10^{-5}\ cm^{-1}$ **(c)** $1.00 \times 10^{12}\ cm^{-1}$ **12-6(a)** 4.73×10^{-12} erg, 1.13×10^{-19} cal **(b)** 1.03×10^{-20} erg, 2.46×10^{-28} cal **(c)** 1.99×10^{-4} erg, 4.76×10^{-12} cal **12-7(a)** $4.41 \times 10^{14}\ s^{-1}$ **(b)** $1 \times 10^{20}\ s^{-1}$ **(c)** $1.04 \times 10^8\ s^{-1}$ **12-8(a)** red visible light **(b)** gamma rays **(c)** FM-television waves **(d)** microwaves **(e)** x rays **12-9(a)** $1.471 \times 10^4\ cm^{-1}$ **(b)** $3 \times 10^9\ cm^{-1}$ **(c)** $3.47 \times 10^{-3}\ cm^{-1}$ **12-10(a)** 2.92×10^{-2} erg, 6.98×10^{-20} cal **(b)** 7×10^{-7} erg, 2×10^{-14} cal **(c)** 6.89×10^{-19} erg, 1.65×10^{-26} cal **12-13** 1.0×10^3 Khz **12-15(a)** -5.15×10^3 coulombs **(b)** -2.32×10^3 coulombs **(c)** -5.26×10^{-1} coulomb **12-17** 2.1×10^{-6} erg **12-18** at 2000 Å, K.E. = 2.94×10^{-12} erg, λ = 2830 Å; at 3500 Å, no electron will be emitted **12-20(a)** 82,258.1820 cm^{-1} **(b)** 97,491.1787 cm^{-1} **(c)** 107,963.864 cm^{-1} **12-24** 2.179×10^{-11} erg/molecule, 8.716×10^{-13} erg/molecule **12-25(a)** -1.634×10^{-11} erg/molecule **(b)** -2.043×10^{-11} erg/molecule **12-26(a)** 1.24×10^{-4} cm **(b)** 1.23×10^{-7} cm **(c)** 2.85×10^{-9} cm **12-28** energy in ergs: 8.01×10^{-8} erg, velocity: 3.10×10^8 cm/s, momentum: 5.18×10^{-16} g · cm/s, wavelength: 1.28×10^{-3} Å **12-29** wavelength: 1.5×10^{-28} Å, frequency: $2.0 \times 10^{46}\ s^{-1}$ **12-31(a)** 5×10^{-31} cm **(b)** 1×10^{-25} cm **(c)** 1×10^{-6} cm **12-33(a)** no electron will be emitted **(b)** K.E. = 2×10^{-9} erg, wavelength = 0.3 Å **12-34** 2.8×10^{-5} cm, 3.6×10^{-5} cm **12-36** 7.5×10^{14} 1/s, 3.8×10^{14} 1/s **12-38** 8.0×10^{-12} erg, 4.0×10^{-12} erg **12-39** 5331.54883 cm^{-1}, 7799.29429 cm^{-1} **12-40** 8.8 min, 42.0 min **12-41** 2.0×10^{9} cm

Chapter 13

13-3(a) 0, 1, 2, 3, or 4 **(b)** 10 electrons **(c)** 10 electrons **(d)** no electrons **(e)** 14 electrons **(f)** 11 values **(g)** 1 value **13-5(a)** 7 orbitals **13-6(a)** 3*s* **(b)** not possible; *n* must be greater than ℓ **(c)** 4*p* **(d)** 5*g* **(e)** not possible; *n* must be greater than ℓ **(f)** 6*f* **(g)** 5*f* **13-7(a)** $n = 2, \ell = 0$ **(b)** $n = 3, \ell = 2$ **(c)** there is no 2*d* orbital **(d)** $n = 7, \ell = 1$ **(e)** $n = 6, \ell = 4$ **(f)** there is no 3*f* orbital **(g)** $n = 5$, $\ell = 3$ **(h)** there is no 1*p* orbital **13-9(a)** 5*s*: 2, 5*p*: 6, 5*d*: 10, 5*f*: 14 **(b)** 5*f*: 7, 6*f*: 7 (there are 7 orbitals in any *f* subshell) **13-13(a)** Sc^{2+} (19 electrons): $1s^22s^22p^63s^23p^63d^1$ **(b)** Zn^{2+} (28 electrons): $1s^22s^22p^63s^23p^63d^{10}$ **(c)** Cr^{3+} (21 electrons): $1s^22s^22p^63s^23p^63d^3$ **(d)** K^{2+} (17 electrons): $1s^22s^22p^63s^23p^5$ **(e)** S^{2-} (18 electrons): $1s^22s^22p^63s^23p^6$ **(f)** I^- (54 electrons): $1s^22s^22p^63s^23p^63d^{10}4s^24p^64d^{10}5s^25p^6$ **13-15(a)** The 2*s* subshell can hold only 2 electrons, not 3. **(b)** There is no 1*p* subshell, so no electrons should be put there. **(c)** There is no 2*d* subshell, so no electrons should be put there. **(d)** Only *s* orbitals have been filled. This is all right for 1*s* and 2*s*, but 2*p* has a lower energy than 3*s* and will fill first. Similarly, 3*p* will fill before 4*s*. **(e)** The 3*p* orbitals have lower energies than 3*d*, and will fill first. **13-18(b)** Na **(d)** S^- **(e)** Ti (f) Fe **(h)** Zn^+ **(j)** Fe^{2+} **(k)** Fe^{3+} **(l)** Pb **(n)** Te^{2+} **3-20(a)** a neutral atom in the ground state **(b)** a negative ion H^- in the ground state **(c)** a positive ion Ne^+ in the ground state **(d)** a positive ion Ne^+ in an impossible state (The 2*s* subshell can hold a maximum of two electrons.) **(e)** a neutral atom in the ground state **(f)** a positive ion Sc^+ in an excited state (One electron has been promoted from a 3*d* to a 4*s* orbital.) **13-23(a)** S **(b)** Te **(c)** Na **(d)** Al **(e)** S^{2-} **(f)** Cl^- **(g)** K^- **(h)** Na **(i)** O^{2-} **(j)** Kr **(k)** I^- **(l)** Ga **(m)** Li^+ **(n)** Na **(o)** Pb **(p)** Se^{2-} **(q)** Si **13-24(a)** 0.73 Å **(b)** 1.14 Å **13-27(a)** Cl **(b)** Se **(c)** Na^+ **(d)** P **(e)** S **(f)** Ar **(g)** Ge **(h)** Be **(i)** Si **(j)** N **(k)** S^+ **(l)** C^{3+} **(m)** Be^{2+} **(n)** B^{3+} **13-28(a)** K, Ca^+, Sc^{2+} **(b)** Ge, Si, C **13-31** The following are paramagnetic, with the number of unpaired electrons given: **(a)** 2 **(b)** 4 **(c)** 5 **(e)** 1 **13-32(a)** 50 electrons **(b)** 7 4*f* orbitals **(c)** 7 6*f* orbitals **13-33** 1.51 Å **13-35(a)** praseodymium **(b)** chlorine **(c)** chromium **(e)** polonium **(f)** fluorine **13-37(a)** $n = 5$ **(b)** 9 *g* orbitals **(c)** 18 electrons **(d)** 4 nodes **13-38(a)** S, S^-, S^{2-} **(b)** S, Si, Na **13-39(a)** 15 **(b)** 7 **(c)** 2 **(d)** 8 **(e)** 2 **(f)** none (This is impossible.) **(g)** 1 **13-40** P^{3-}, S^{2-}, Cl^-, K^+, Ca^{2+}, Sc^{3+}, Ti^{4+}

Chapter 14

14-2(a) -1 **(c)** -2 **(d)** $+2$ **(g)** -1 **(h)** $+1$ **14-4** (b), (d), (e), (g), (h), (i), (j), (k) **14-6** (a), (c), (g), (h), (i) **14-9** ionic bonds: (d), (e); polar covalent bonds: (a), (b), (c), (f), (g), (j); "pure" covalent bond: (k); both ionic and polar covalent bonds: (h), (i) **14-10(a)** 41% **(b)** 4.9% **(c)** 1.8% **(d)** 82% **14-11** no dipole moments: (b), (f), (g), (i), (j) **14-14** in order of increasing bond energy: N—I < N—Br < N—Cl < N—F. **14-15(a)** 8 **(b)** 20 **(d)** 17 **(f)** 32 **(g)** 14

14-16 (a) |I̅—I̅| **(d)** ⊖C≡N| **(h)** |F̅—B(—|F̅|)—F̅| **(i)** |F̅—B⊖(—|F̅|)(—|F̅|)—F̅| **(l)** H—C·(—H)(—H)

14-17 (c) |C̅l—P⊕(—|O̅|⊖)(—|C̅l|)—C̅l| **(d)** ⊖|O̅—Si⊕(—|O̅|⊖)—O̅|⊖ **(g)** H—N⊕≡C⊖ **(k)** H—O̅—B(—|O̅|⊖)—O̅—H

14-18 nothing wrong: (e), (i) **14-19 (a)** H—C(=O̅|)—O̅|⊖ ⟷ H—C(—|O̅|⊖)=O̅|

(c) |O̅=O⊕—O̅|⊖ ⟷ ⊖|O̅—O⊕=O̅|

(g) 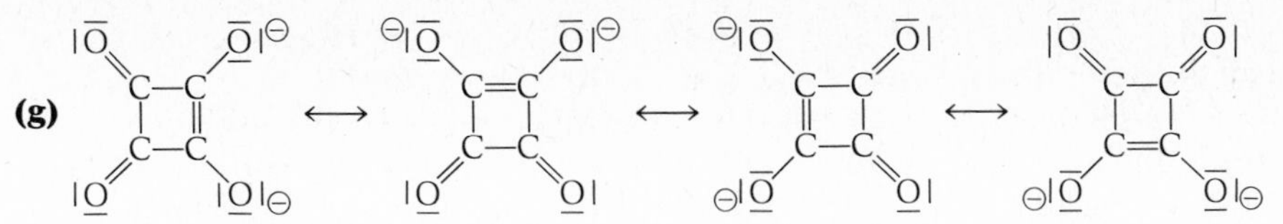

14-21 correct forms: (b), (e); valid but unreasonable: (a), (f); incorrect: (c), (d)

14-23 (a) H—CH₂—C(=O)—O⁻ ⟷ H—CH₂—C(O⁻)=O **(b)** neither **14-24(c)** NCl_5 **(d)** CCl_2

(k) OF_4 **14-26** soluble in water: (a), (h), (l); in benzene: (b), (f), (j), (m); in both: (d), (g), (k); in neither: (c), (e), (i) **14-27** strong: (a), (c), (f), (h); weak (b), (e), (g), (j), (k)
14-28(a) triangular (AB_3) **(b)** trigonal bipyramid (AB_5) **(d)** octahedral (AB_6)
(e) tetrahedral (AB_4) **(i)** trigonal bipyramid ($|\bar{A}B_2$) **(m)** octahedral ($|\bar{A}B_4$)
14-29(a) tetrahedral ($\bar{A}B_3$) **(d)** angular ($\bar{A}B_2$) **(f)** angular ($\bar{A}B_2$) **(i)** tetrahedral (AB_4)
(k) triangular (AB_2) **14-31(a)** 109°28′ **(b)** 120° **(c)** 180° **14-32(a)** sp^2 **(b)** sp^3
(c) sp **(d)** sp **14-40(a)** $\sigma_{1s}^2\sigma_{1s}^{*2}\sigma_{2s}^2\sigma_{2s}^{*2}\pi_{2p_y}^1$, bond order = $\frac{1}{2}$, paramagnetic
(c) $\sigma_{1s}^2\sigma_{1s}^{*2}\sigma_{2s}^2\sigma_{2s}^{*2}\pi_{2p_y}^2\pi_{2p_z}^2$, bond order = 2, not paramagnetic **(f)** $\sigma_{1s}^2\sigma_{1s}^{*2}\sigma_{2s}^2\sigma_{2s}^{*2}\pi_{2p_y}^2\pi_{2p_z}^2\sigma_{2p_x}^1$, bond order = 2.5, paramagnetic **14-41** ionic bonds: (b), (e), (i), (m); polar covalent bonds: (a), (d), (f), (g), (h), (j), (l); "pure" covalent bonds: (c) (k) **14-43(b)** sp^3 **(c)** sp^2 **(e)** sp^3
14-45(a) NO_2^-: |O̅=N̅—O̅|⁻ ⟷ ⁻|O̅—N̅=O̅|,
NO_2: |O̅=Ṅ⁺—O̅|⁻ ⟷ ⁻|O̅—Ṅ⁺=O̅| ⟷ |O̅=N̅—O̅| ⟷ |O̅—N̅=O̅| **14-48(a)** $\sigma_{1s}^2\sigma_{1s}^{*1}$, bond order = $\frac{1}{2}$, paramagnetic **(c)** $\sigma_{1s}^2\sigma_{1s}^{*2}$, bond order = 0 **(g)** $\sigma_{1s}^2\sigma_{1s}^{*2}\sigma_{2s}^2\sigma_{2s}^{*2}\pi_{2p_y}^1\pi_{2p_z}^1$, bond order = $1\frac{1}{2}$, paramagnetic **14-49** NaI **14-50** yes **14-51(b)** tetrahedral ($\bar{A}B_3$) **(d)** tetrahedral ($|\bar{A}B_2$) **(g)** trigonal bipyramid ($|\bar{A}B_2$)

14-52 (c) ⁻|O̅—S⁺(—O̅|⁻)—O̅|⁻ ⟷ |O̅=S̅(—O̅|⁻)—O̅|⁻ ⟷ ⁻|O̅—S̅(—O̅|⁻)=O̅| ⟷ ⁻|O̅—S̅(=O̅|)—O̅|⁻

(e) ⁻C̅≡N⁺—O̅|⁻ **(f)** |O̅=C=N̅|⁻ ⟷ ⁻|O̅—C≡N|

(h) H—O̅—S(=O̅|)(=O̅|)—O̅—H ⟷ H—O̅—S⁺(—O̅|⁻)(=O̅|)—O̅—H ⟷ H—O̅—S⁺(=O̅|)(—O̅|⁻)—O̅—H ⟷ H—O̅—S²⁺(—O̅|⁻)(—O̅|⁻)—O̅—H

14-53 no dipole moment: (d), (f) **14-54** (c), (d), (e), (i)

Chapter 15

15-2 molecular crystals: (a), (e); ionic solids: (d), (f); giant molecules: (b), (e) **15-3(a)** 47.64% empty space **(b)** 31.94% empty space **(c)** 25.95% empty space **15-5(b)** face-centered cubic **15-6** (a), (b), (c), (e) **15-8** 68.06% **15-9** 2.16 Å **15-11** tungsten
15-12(a) 3.93×10^{-8} cm **(d)** 21.4 g/cm³ **15-13(b)** 2.72 g/cm³ **15-16** cubic
15-17 3.17 Å **15-20(a)** 6 **(b)** 3 **15-25** 3.62 g/cm³ **15-28** 4 Ca^{2+} ions, 4 O^{2-} ions, face-centered cubic **15-30(a)** 6 **(b)** 8.68 g/cm³ **15-33(a)** 8 nearest neighbors **(b)** 12 nearest neighbors **(c)** 12 nearest neighbors

Chapter 16

16-2(a) −2 **(b)** +6 **(c)** +4 **(d)** 0 **(e)** +2.5 **(f)** +6 **16-3(a)** I +5, F −1
(d) B +3, H −1 **(e)** O +2, F −1 **(f)** Bi +3 **(k)** P +5, Cl −1 **(l)** C $-\frac{8}{3}$, H +1
(p) Cr +6, O −2 **(v)** H +1, Se −2 **(w)** C −2, H +1, O −2 **(x)** Mn +4, O −2
16-7(a) $2Cr + 3Cu^{2+} \rightarrow 2Cr^{3+} + 3Cu$ **(c)** $H_2O + PbO_2 + OH^- + Cl^- \rightarrow Pb(OH)_3^- + ClO^-$
(d) $5CO + I_2O_5 \rightarrow 5CO_2 + I_2$ **(f)** $2S_2O_3^{2-} + Br_2 \rightarrow S_4O_6^{2-} + 2Br^-$
(g) $H_2O_2 + 2HI \rightarrow 2H_2O + I_2$ **(k)** $2ClO_2 + O_2^{2-} \rightarrow 2ClO_2^- + O_2$
16-8(a) $5H_2O + 2CrO_4^{2-} + I^- \rightarrow 2Cr^{3+} + 10OH^- + IO_3^-$
(c) $4H_2O + 2MnO_4^- + 3S^{2-} \rightarrow 2MnO_2 + 8OH^- + 3S$
(d) $8Al + 18H_2O + 3KNO_3 \rightarrow 8Al(OH)_3 + NH_3 + 3KOH$
(e) $3I_2 + 6OH^- \rightarrow 5I^- + IO_3^- + 3H_2O$
(g) $2NaOH + 2CoSO_4 + 2H_2O + Na_2O_2 \rightarrow 2Co(OH)_3 + 2Na_2SO_4$
(h) $4HCl + 3WO_3 + SnCl_2 \rightarrow W_3O_8 + H_2O + H_2SnCl_6$
(j) $24H_2SO_4 + 6NaTcO_4 + 10Ti \rightarrow 6TcSO_4 + 24H_2O + 5Ti_2(SO_4)_3 + 3Na_2SO_4$

16-9(b) 7.99970 g/equiv **(d)** 107.868 g/equiv **16-10(a)** 170.467 g **16-11(a)** 1.45 g **(b)** 1.74 g **16-12** 59.4 ml **16-14(a)** 63.0128 g/equiv **(b)** 21.0043 g/equiv **16-16** 67.0 g/equiv **16-18(a)** $Mg|Mg^{2+}||Cl^-|Cl_2|Pt$ **16-19(a)** -1.14 V **(b)** $+2.20$ V **(d)** $+1.01$ V **(g)** $+0.46$ V **(h)** $+0.09$ V **(j)** $+0.13$ V **(l)** $+0.47$ V **(o)** -0.21 V **16-20(b)** $Na|Na^+||S^{2-}|S$ **(j)** $Pb|PbSO_4|SO_4{}^{2-}||Ni^{2+}|Ni$ **(l)** $Ag|Ag(CN)_2{}^-||Cu^{2+}, Cu^+|Pt$ **16-21** (b), (d), (g) **16-23(a)** F_2 **(b)** $MnO_4{}^-$ **(c)** PbO_2 **(d)** Ag^+ **(e)** Fe^{3+} **(f)** Cu^{2+} **(g)** Cu^+ **(h)** MnO_2 **(i)** Li^+ **(j)** F_2 **(k)** H_2O **(l)** PbO_2 **16-25(a)** (1) -0.19 V **16-26(a)** $\mathcal{E}_{cell} = +0.14 - \frac{0.05916}{2}\log\frac{[Sn^{2+}][H_2]}{[Sn][H^+]^2}$

(c) $\mathcal{E}_{cell} = +1.06 - \frac{0.05916}{2}\log\frac{[Br^-]^2[H^+]^2}{[Br_2][H_2]}$ **(f)** $\mathcal{E}_{cell} = 0.00 - \frac{0.05916}{2}\log\frac{[H_2][H^+]^2}{[H_2][H^+]^2}$

16-27(a) $+0.14$ V **(c)** $+1.27$ V **(f)** -0.059 V **16-28(a)** $\Delta G = -6.5$ kcal/mole, $\Delta G° = -6.5$ kcal/mole, $K = 5 \times 10^4$ **(c)** $\Delta G = -58.6$ kcal/mole, $\Delta G° = -48.9$ kcal/mole, $K = 6.3 \times 10^{35}$ **(f)** $\Delta G = +2.7$ kcal/mole, $\Delta G° = 0$, $K = 1.00$ **16-29** 1.7 M **16-30** 1.1×10^{-5} M **16-32** $\Delta G° = +46.95$ kcal/mole, $K = 3.7 \times 10^{-35}$ **16-35** 6.73×10^{-4} coulomb **16-36(a)** 3.46 g Mg **(b)** 387 g Pb **16-37(a)** 3 **(b)** 5 **(c)** 2 **16-39(b)** 0.334 faraday **(d)** 0.00788 faraday **(e)** 1.8 faradays **16-40(a)** 0.018799 g H_2 **(d)** 2.0120 g Ag **16-41** 14.2 g **16-43** 222 ml **16-44** F^-, Au^+, Fe^{2+}, OH^-, Cu, Pb, Mg **16-46** 185 ml **16-49** 185.8 ml **16-50(a)** -1 **(b)** $+7$ **(c)** $+1$ **(d)** 0 **(e)** -1 **(f)** $+3$ **(g)** $+5$ **(h)** $+7$ **(i)** -1 **(j)** -1 **(k)** $+3$ **(l)** -1 **16-52** 0.18 M **16-54** Cu^{2+} **16-55** $H_2O + IO_3{}^- + Cl_2 \rightarrow IO_4{}^- + 2H^+ + 2Cl^-$
$2OH^- + IO_3{}^- + Cl_2 \rightarrow IO_4{}^- + H_2O + 2Cl^-$
16-57 Al: 1.663×10^4 s, Ga: 6437 s. The unknown salt contained Ga^{3+}.

Chapter 17

17-3 at 750 torr: N_2 577 torr, O_2 155 torr, H_2O 11.3 torr; at 300 torr: N_2 231 torr, O_2 61.8 torr, H_2O 4.50 torr **17-4(a)** the thermosphere **(b)** the troposphere
17-6 $3Mg + N_2 \rightarrow Mg_3N_2$
$2Al + N_2 \rightarrow 2AlN$
$3Zn + N_2 \rightarrow Zn_3N_2$
$2B + N_2 \rightarrow 2BN$
17-7 $K_c = 0.901$, $K_p = 2.95 \times 10^{-4}$; K_c will not change if the pressure is increased, but will change if the temperature is increased. **17-8** 2.4×10^4 g **17-9** 1.9×10^5 cal/mole **17-12** 8.3×10^{14} s^{-1} **17-14** $N_2 + O_2 \rightarrow 2NO$ **17-15** the burning of coal and fuel oil and certain industrial processes **17-16** 78 kg limestone ($CaCO_3$); 110 kg $CaSO_4$ will be produced. **17-23** 15 g $Ca(OH)_2$ **17-25** 109 g $Ca(HCO_3)_2$, 98.8 g $Mg(HCO_3)_2$ **17-28** physical processes: sedimentation, coagulation, filtration, removal of tastes and odors, fluoridation; chemical processes: removal of hardness, chlorination **17-32** BOD = 5.7 ppm **17-35** 1.4×10^{-4} g Hg **17-36** 6.3×10^{20} moles NaCl **17-37** 651 cal more than the distilled water **17-39** -2.31°C **17-42** 64.4 g soap **17-44** $N_2{}^+$: $\sigma_{1s}^2\sigma_{1s}^{*2}\sigma_{2s}^2\sigma_{2s}^{*2}\pi_{2p_y}^2\pi_{2p_z}^2\sigma_{2p_x}^1$ **17-46** 6.0×10^7 moles NaCl **17-47** It can't. This example points up the difficulty in controlling pollution in our modern industrial society.

Chapter 18

18-4(a) $2H_2O + 2K \rightarrow H_2 + 2K^+ + 2OH^-$ **(b)** $2H_3O^+ + Fe \rightarrow H_2 + Fe^{2+} + 2H_2O$ **(c)** $H_3O^+ + Cu \rightarrow$ NR **(d)** $GeO_2 + 2H_2 \rightarrow Ge + 2H_2O$ **(e)** $NaH + H_2O \rightarrow H_2 + OH^- + Na^+$ **18-6** double bonds: C, N, O, S; triple bonds: C, N, O (oxygen only if it carries a charge) **18-7** $^{\oplus}\bar{O}{\equiv}\bar{C}^{\ominus}$ **18-13** other product is water: $NH_4NO_3 \rightarrow N_2O + 2H_2O$ **18-15**

```
           ⊖|Ō|   ⊖|Ō|   ⊖|Ō|   ⊖|Ō|
             |      |      |      |
H—Ō—P⊕—Ō—P⊕—Ō—P⊕—Ō—P⊕—Ō—H
             |      |      |      |
            |O|    |O|    |O|    |O|
             |      |      |      |
             H      H      H      H
```

18-16(a) $H_2O_2 + PbO_2 + 2H^+ \rightarrow O_2 + Pb^{2+} + 2H_2O$
(b) $3H_2O_2 + 2Cr^{3+} + H_2O \rightarrow Cr_2O_7{}^{2-} + 8H^+$ **18-17(a)** H_2SO_3 acidic **(b)** KOH basic **(c)** none **(d)** HNO_2 acidic **(e)** H_2SiO_3 acidic **(f)** $HAsO_2$ amphoteric **(g)** $Ba(OH)_2$ basic **(h)** none **(i)** H_2CO_3 acidic **(j)** $HClO_2$ acidic **(k)** none **18-18(a)** SO_3 **(b)** N_2O_5 **(c)** Li_2O **(d)** P_2O_5 **(e)** MgO **(f)** none **(g)** Cl_2O **(h)** B_2O_3 **(i)** Cr_2O_3 **(j)** Mn_2O_7 **(k)** SrO **(l)** none; SO_4 is not a known compound **(m)** none

(n) Y_2O_3 **(o)** none **18-20** 48,180 kg H_2SO_4 **18-21** It reacts with water to give OH^-.
18-24(a) $H_2O + Cl_2 + H_2SO_3 \rightarrow 2Cl^- + H_2SO_4 + 2H^+$
(b) $H_2O + Cl_2 + SO_2 \rightarrow 2Cl^- + SO_3 + 2H^+$ **(c)** $H_2O + Cl_2 + SO_3^{2-} \rightarrow 2Cl^- + SO_4^{2-} + 2H^+$
18-27 $2H_2O + Cl_2 \rightarrow 2HOCl + 2e^- + 2H^+$ $Cl_2 + 2e^- \rightarrow 2Cl^-$ **18-28** Their line spectra were identical. **18-29** They are more abundant and consequently cheaper. **18-33** hydrogen, by a wide margin **18-35** $3N_2H_5^+ + 2IO_3^- \rightarrow 3N_2 + 3H^+ + 2I^- + 6H_2O$
18-36 $3HO_2^- + 2MnO_4^- + H_2O \rightarrow 3O_2 + 2MnO_2 + 5OH^-$ **18-38** H—C—H (C=N) N—H
18-40 44.34127% **18-41** 72% **18-42(a)** $2H^+ + 2ClO_4^- + 7H_2O_2 \rightarrow Cl_2 + 8H_2O + 7O_2$
(b) 0.0190 *M* $HClO_4$

Chapter 19

19-3(a) Na_3AlF_6 **(b)** Fe_3O_4 **(c)** Al_2O_3 **(d)** Fe_2O_3 **(e)** FeS_2 **19-4** 2.7×10^5 g Al, 510 kg Al_2O_3 **19-6** 21 kg **19-10** for *p*-type: B, Al, Sc; for *n*-type: As, Sb **19-13(a)** Mg **(b)** Na^+ **(c)** Mg^{2+} **(d)** Rb **(e)** Rb^+ **(f)** Mg **(g)** Ra^+ **19-18(a)** 2− **(b)** 1− **(c)** 1+ **(d)** no charge **(e)** 2− **(f)** no charge **(g)** no charge **(h)** 1+ **(i)** 8+ **19-19(a)** II **(b)** II **(c)** IV **(d)** 0 **(e)** III **(f)** II **(g)** 0 **19-20(a)** 5 **(b)** 8 **(c)** 0 **(d)** 6 **(e)** 7 **(f)** 1 **(g)** 10 **19-22(a)** potassium tetrachloronickelate(II)
(b) pentacarbonylruthenium(0) or ruthenium pentacarbonyl
(c) triamminehydroxohafnium(IV) bromide **(d)** triamminecyanomanganese(II) chloride
(e) tris(ethylenediamine)cobalt(III) chloride **(f)** sodium tetranitroborate(III)
(g) carbonyldichlorobis(trimethylphosphine)ruthenium(II) **19-23(a)** $V(H_2O)_6^{3+}$
(b) $[Co(H_2O)_4Cl_2]^+ Cl^-$ **(c)** $[Rh(bipy)_2Cl]^+ NO_3^-$ **(d)** $Na^+ [AgF_4]^-$ **19-27** identical: (a), (e), (f); isomers: (b), (d); neither: (c) **19-28** one other: (a), (d), (e); no other: (b), (c); four others: (f) **19-30** instability constant: 2.1×10^3, stability constant: 4.8×10^{-4}
19-31 1×10^{-6} *M* **19-32** 2×10^{-15} *M* **19-33(a)** 3×10^{-19} *M* **(b)** no precipitate will form **19-38(a)** high spin **(b)** low spin **(c)** low spin **19-42** 2.6×10^{12} ergs/mole
19-44 0.2 *M* **19-46** barium **19-50(a)** 1.31×10^{-3} g/ml **(b)** 1.29×10^{-3} g/ml

Chapter 20

20-3(a) pentane **(b)** methylbutane (no number needed because there is only one methylbutane) **(c)** 2,3-dimethylhexane **(d)** 2,2-dimethylpentane
(e) 2,3-dimethylpentane **(f)** trimethylbutane **(g)** 3,4-dimethylhexane
(h) 3,3,6,6,7-pentamethyl-5-propyldecane **(i)** 4,6-diethyl-2,7-dimethyl-6-propylnonane
(j) cyclopentane **(k)** 1,3-dimethylcyclobutane **(l)** 1-ethyl-3,3-dimethylcyclohexane
20-8(a) 1-pentene **(b)** 4-methyl-2-pentene **(c)** 2-hexyne **(d)** 1,5-hexadiene
(e) cyclopentene **(f)** 2,4-dimethyl-2-pentene **(g)** 1,3-cyclohexadiene
(h) 4,4,5-trimethyl-1-hexyne **(i)** 1,2-dimethylcyclopentene **(j)** *cis*-5-methyl-2-hexene
20-11 Cis and trans isomers are possible for (a), (d), (e), (f), and (i). **20-12(a)** A noncyclic alkane with 7 carbons must have 16 hydrogens. This formula has only 14. Therefore, any compound with the formula C_7H_{14} must have either one ring or one double bond, but not both. **(b)** No rings, no double or triple bonds. Any compound with the formula $C_{11}H_{24}$ is a noncyclic alkane. **(c)** There must be either two rings, two double bonds, one ring *and* one double bond, or one triple bond. **20-15(a)** secondary alcohol **(b)** aromatic amine **(c)** carboxylic acid **(d)** ketone **(e)** alkane **(f)** amide **(g)** tertiary amine **(h)** alkyl chloride **(i)** carboxylic ester **(j)** alkyne **20-17(a)** primary alcohol, alkyl chloride
(b) aldehyde, aryl chloride, mercaptan (or thiol) **(c)** secondary amine, phenol
(d) carboxylic acid, ketone **(e)** carboxylic acid, primary amine **(f)** alkyl bromide; also contains a double bond, so it is usually called an unsaturated alkyl bromide **(g)** primary alcohol, phenol **(h)** secondary amine, amide **(i)** alcohol (called a triol, because there are 3 OH groups) **20-18(a)** acetone, propanone, dimethyl ketone (only the first of these is commonly used) **(b)** methyl phenyl ketone (this compound is also called acetophenone)
(c) 2-methylpropanal **(d)** *meta*-bromobenzaldehyde, 3-bromobenzaldehyde
(e) 2,2,4,4-tetramethyl-3-pentanone, di-*t*-butyl ketone **(f)** butanoic acid, butyric acid
(g) lactic acid, 2-hydroxypropanoic acid **20-21(a)** fairly inert: alkanes, cycloalkanes, aryl halides, ethers **(b)** acidic: carboxylic acids, mercaptans, phenols **(c)** basic: amines (primary, secondary, and tertiary) **20-23** primary alcohols: (b), (n); secondary alcohols: (c), (k); tertiary alcohol: (m); primary amines: (a), (d), (e), (l); secondary amine: (g); tertiary amine: (i); others are neither amines nor alcohols **20-26** chiral: (b), (c), (f), (g), (i), (j); others not chiral

20-28(a) no **(b)** yes **(c)** no **20-31(h)** If the formula corresponding to this name is drawn, there will be 5 bonds on a carbon atom. Therefore such a compound could not exist. **(i)** 2-Methyl-2-pentene does not exist in cis and trans isomers. **(j)** If there is only one substituent on cyclohexane, it gets no number (the number 1 is understood). It could not be 2 in any case. **(k)** There is no such compound. The only methylcyclohexenes are the 1-methyl, 3-methyl, and 4-methyl isomers. Draw them and see. **(l)** The prefixes ortho, meta, and para must not be used if three or more groups are on a benzene ring. **(m)** The name does not tell what group is para to the OH group. **20-34(a)** CH_3CH_2COOH **(b)** $CH_3CH_2NH_3^+$ **(c)** CH_3OH **20-35(a)** $CH_3CH_2COO^-$ **(b)** $CH_3CH_2NH_2$ **(c)** CH_3OH **20-39(a)** $SOCl_2$ **(b)** H_2 and a catalyst such as Pt **(c)** $Cr_2O_7^{2-} + H^+$ **(d)** $CH_3CH_2Cl + AlCl_3$ **(e)** $CH_3OH + H_3O^+$ **(f)** P_2O_5

Chapter 21

21-2 enantiomers: (d); diasteriomers: (a), (b), (e); structural isomers: (c), (g); neither: (f), (h) **21-5(a)** L-allose rotates it to the left **(b)** you cannot make this prediction **21-6** aldohexoses: (a), (c), (e), (h), (i); ketohexoses: (d), (f); aldopentose: (b); none of these: (g) **21-7(a)** 8 **(b)** 8 **(c)** 4 (In this problem we are not considering furanose and pyranose rings.) **21-10** D-glucose; chlorophyll is the necessary catalyst, and sunlight is also necessary **21-11(a)** $C_{10}H_{18}O_9$ **(b)** $C_{11}H_{20}O_{10}$ **21-16(a)** $H_3\overset{+}{N}-CH_2-COO^- + H_3O^+ \rightarrow H_3\overset{+}{N}-CH_2-COOH + H_2O$ **(b)** $H_3\overset{+}{N}-CH_2-COO^- + OH^- \rightarrow H_2N-CH_2-COO^- + H_2O$ **(c)** no **21-24** three different fat molecules **21-28(a)** vitamins B_1, B_2, and B_{12} **(b)** nicotinic acid **(c)** vitamin A **(d)** vitamins A_1, B_1, B_2, B_{12}, C, and D_2 **(e)** nicotinic acid. **21-30** vitamin D_2 **21-39(a)** 6 **(b)** 5 **(c)** cortisone: 64 isomers; estradiol: 32 isomers **(d)** It is very likely that only one stereoisomer has physiological activity in each case. **21-40(a)** 1.02×10^{13} decapeptides **(b)** 27.9 billion yr **(c)** 1.05×10^{26} eicosapeptides, 2.88×10^{17} yr **(d)** It is so unlikely as to be flatly impossible.

Chapter 22

22-3 (a) $-CH_2-\underset{\underset{Br}{|}}{CH}-CH_2-\underset{\underset{Br}{|}}{CH}-$ **(b)** $-\underset{\underset{Cl}{|}}{CH}-\underset{\underset{Cl}{|}}{CH}-\underset{\underset{Cl}{|}}{CH}-\underset{\underset{Cl}{|}}{CH}-$ **(c)** $-\underset{\underset{CH_3}{|}}{CH}-\underset{\underset{Cl}{|}}{CH}-\underset{\underset{CH_3}{|}}{CH}-\underset{\underset{Cl}{|}}{CH}-$ **22-6** formaldehyde, $H_2C{=}O$ **22-7(a)** $-CCl_2-CBr_2-CBr_2-CCl_2-CCl_2-CBr_2-CBr_2-CCl_2-$ **(b)** $-CCl_2-CBr_2-CCl_2-CBr_2-CCl_2-CBr_2-CCl_2-CBr_2-$ **22-8** Vitamins A_1 and D_2 have systems of conjugated C=C double bonds. There are conjugated double bonds with C=O or C=N in vitamins B_1, B_2, B_{12}, and C. **22-11** 63.6856% C, 9.79763% H, 14.1389% O, 12.3779% N; no **22-15(a)** H_2O **(b)** H_2O **(c)** none **(d)** H_2O **(e)** HCl **(f)** HCl **22-20(a)** addition polymer **(b)** addition polymer **(c)** neither **(d)** condensation polymer **(e)** neither **(f)** condensation polymer **22-21(a)** $HOCH_2CH_2OH$ **(b)** $CH_2{=}CHBr$ **(c)** $NH_2CH_2CH_2NH_2 + HOOC-CH_2CH_2CH_2-COOH$ **(d)** $CH_3O-CH{=}CH-CH{=}CH_2$ **22-24** (1) a head-to-head polymer of vinyl chloride $CH_2{=}CHCl$ (2) a 50:50 copolymer of ethylene $CH_2{=}CH_2$ and 1,2-dichloroethylene $ClHC{=}CHCl$.

Chapter 23

23-2 (b) $\approx$ (e) $<$ (d) $<$ (f) $<$ (a) $<$ (c) **23-3** 32 h **23-5** 1.6 g **23-7** 1.37×10^9 yr **23-8(a)** $^{210}_{83}Bi \rightarrow {}^{206}_{81}Tl + {}^{4}_{2}He$ **(b)** $^{238}_{94}Pu \rightarrow {}^{234}_{92}U + {}^{4}_{2}He$ **23-9(a)** $^{152}_{63}Eu \rightarrow {}^{152}_{64}Gd + {}^{0}_{-1}e$ **(b)** $^{141}_{56}Ba \rightarrow {}^{141}_{57}La + {}^{0}_{-1}e$ **23-10(a)** $^{116}_{51}Sb \rightarrow {}^{116}_{50}Sn + {}^{0}_{1}e$ **(b)** $^{74}_{36}Kr \rightarrow {}^{74}_{35}Br + {}^{0}_{1}e$ **23-11(a)** $^{84}_{39}Y + {}^{0}_{-1}e \rightarrow {}^{84}_{38}Sr$ **(b)** $^{113}_{50}Sn + {}^{0}_{-1}e \rightarrow {}^{113}_{49}In$ **23-12** 1: (d); 2: (a), (e); 3: (b), (c) **23-13(a)** $^{1}_{0}n$ **(c)** $^{223}_{89}Ac$ **(e)** $^{4}_{2}He$ **23-15(a)** 5.0×10^{10} cal/mole gained **(b)** 1.08×10^{11} cal/mole lost **(c)** 7.43×10^{10} cal/mole lost **(d)** 5.50×10^{10} cal/mole gained **23-16(a)** mass defect = 0.11303 amu, $E = 1.69 \times 10^{-4}$ erg/nucleus **(c)** mass defect = 0.03147 amu, $E = 4.70 \times 10^{-5}$ erg/nucleus **23-17(a)** 152.921 amu **(c)** 238.051 amu **23-19(a)** 1.6375×10^{-6} erg **(b)** 3.007×10^{-3} erg **23-20(a)** binding energy = 2.99×10^{-4} erg/nucleus, 1.30×10^{-5} erg/nucleon **(c)** binding energy = 1.20×10^{-3} erg/nucleus, 1.40×10^{-5} erg/nucleon **(e)** binding energy = 2.90×10^{-3} erg/nucleus, 1.21×10^{-5} erg/nucleon **23-21** 27.99758 g/mole **23-25** 1.36×10^8 kcal lost **23-29(a)** 0.909 microcurie **(b)** 6.06×10^{-8} g ^{60}Co/g soil **23-30** 19% **23-31** 4.47×10^4 yr **23-34** 9.5×10^{-6} erg lost/atom **23-38** 2.69 g **23-39(b)** $^{78}_{34}Se$ **(d)** $^{46}_{21}Sc$ **(f)** $3{}^{1}_{0}n$ **23-40** 0.03 g/mole **23-41** 3×10^7 yr **23-44** 26.7%

Index

B

C

D

E

F

G

H

I

J

K

M

N

Q

R

S

X

Y

Z

Conversion Factors

Length
1 Å = 10^{-8} cm (exact)
1 in = 2.54 cm (exact)
1 m = 39.37 in
1 mile = 5280 ft (exact)
1 mile = 1.609 km

Pressure
1 mm Hg = 1 torr (exact)
1 atm = 760 torr (exact)
1 atm = 1.01325×10^{6} dynes/cm^{2} (exact)
1 atm = 101,325 pascals (exact)
1 atm = 14.696 lbs/sq. in.

Volume
1 qt = 0.9463 liter
1 gallon = 3.785 liters
1 cubic ft = 28.316 liters
1 liter = 1000 cm^{3} (exact)
1 m^{3} = 1000 liters (exact)

Energy
1 erg = 1 g · cm^{2}/s^{2} (exact)
1 cal = 4.184 J (exact)
1 cal = 4.184×10^{7} ergs (exact)
1 J = 10^{7} ergs (exact)
1 electron volt = 1.6021×10^{-19} J
1 liter · atm = 24.2179 cal
1 kWh = 3.6×10^{6} J
1 Btu = 1055.056 J (exact)

Mass
1 amu = 1.6605×10^{-24} g
1 ounce (avdp) = 28.3495 g
1 lb (avdp) = 453.59 g
1 ton = 907.180 kg

Some Basic Physical Constants

			SI units
Speed of light in vacuum	c	2.997925×10^{10} cm/sec	2.997925×10^{8} m/s
Charge on electron	e	1.60219×10^{-19} coulomb	1.60219×10^{-19} coulomb
Avogadro's number	N	6.02204×10^{23} molecules/mole	6.02204×10^{23} molecules/mole
Atomic mass unit	amu	1.66057×10^{-24} g	1.66057×10^{-27} kg
Mass of electron	m_e	9.10953×10^{-28} g	9.10953×10^{-31} kg
Faraday's number	$\mathfrak{F}$	9.64846×10^{4} coulombs/equiv.	9.64846×10^{4} coulombs/mole
Planck's constant	h	6.6262×10^{-27} erg · s	6.6262×10^{-34} J · s
Gas constant	R	0.08206 liter · atm/mole · °K 1.987 cal/mole · °K	8.3144 J/kelvin · mole
Rydberg constant (for hydrogen)	R_H	1.09677576×10^{5} cm^{-1}	1.09677576×10^{7} m^{-1}